GLACIERS&
GLACIATION

SECOND EDITION

Douglas I. Benn
Department of Geology, University Centre
in Svalbard and School of Geography and
Geoscience, University of St Andrews

David J.A. Evans
Department of Geography, Durham University

Routledge
Taylor & Francis Group

LONDON AND NEW YORK

First published 1998
This edition published 2010
By Hodder Education

Published 2013 by Routledge
2 Park Square, Milton Park, Abingdon, Oxon OX14 4RN
711 Third Avenue, New York, NY, 10017, USA

Routledge is an imprint of the Taylor & Francis Group, an informa business

British Library Cataloguing in Publication Data
A catalogue record for this title is available from the British Library

ISBN 13: 978-0-340-90579-1 (pbk)

Cover photo: Englacial cave at an altitude of 5,000 metres in the Khumbu Glacier,
on the south side of Mt Everest © Doug Benn.
Back cover photo: Descending into a moulin, Hansbreen, Svalbard © Jason Gulley.

Typeset in 9.5/12.5 pt Sabon by MPS Limited, A Macmillan Company, Chennai, India

CONTENTS

PREFACE TO THE FIRST EDITION

For most people living in temperate countries, ice is usually encountered only in small quantities on cold days or as small blocks in drinks. In this everyday form, ice appears to be a rigid, brittle and usually slippery material, but in glaciers and ice sheets ice can exhibit a wide variety of surprising and fascinating behaviour. It can flow plastically, like toothpaste, mould itself around hills and valleys, and creep and slide from high snowfields down towards lower ground. It can curve out huge troughs and alter entire landscapes, scouring away soil and other surficial material or blanketing large tracts with glacial deposits. Glaciers and ice sheets can alter the Earth's climate, chilling the atmosphere and the oceans, and profoundly affecting the global hydrological cycle. They also preserve valuable records of past climate change locked up in ice crystals and air bubbles, and the oscillations of glacier margins provide direct indicators of recent climatic shifts. Furthermore, glaciers can store and suddenly release immense quantities of meltwater, producing catastrophic floods such as those which occurred in Iceland in late 1996 which subglacial volcanic activity melted the base of a portion of Vatnajökull. In heavily populated regions, huge volumes of glacigenic material are continuously excavated by people to use in buildings and roads, and many of the holes left behind are being filled by the ever-increasing mountains of domestic and industrial waste. Thus, either directly or indirectly, glaciers and glaciation impact upon a large proportion of the planet's population.

As students in the 1970s and 1980s we were influenced by the glacial textbooks by Clifford Embleton and Cuchlaine King (*Glacial Geomorphology,* 1975) and David Sugden and Brian John (*Glaciers and Landscape,* 1976), and as researchers and teachers of the 1990s we became increasingly frustrated by the lack of up-to-date versions of these fine volumes. As subscribers to the axiom 'if you want something done, do it yourself', we embarked upon *Glaciers and Glaciation* in 1993. Our aim was to produce a modern synthesis, covering all important aspects of glaciers and their effects. While the focus of the book is on glacial geomorphology and sedimentology, we have also tried to encompass important work in related fields such as glaciology and Quaternary studies. Glaciology as a discipline has expanded enormously in the past few decades, and its importance to glacial geologists and geomorphologists is increasingly clear. It has also become a very rigorous, mathematical subject which can be rather impenetrable to those without a background in physics, and students attempting to explore the research literature can rapidly find themselves lost in a forest of frightening equations. In response, we have tried to present the more important results of modern glaciology in an accessible way, with a minimum of mathematics. Some equations have been included, however, as a means of expressing fundamental relationships in a concise way and as a gentle introduction to the more technical aspects of the literature. Contemplation of these relations should be rewarded by increased insight, although those who feel numerically paralysed should be able to skim them and still grasp the essentials from the text.

We hope that students will find *Glaciers and Glaciation* as interesting and stimulating to read as it was for us to write. It is intended for this book to provide substance for all levels of undergraduate training and to be utilized by postgraduate researchers and university teachers. Given the recent prolific production rate of glacier-related literature, we found it difficult to be comprehensive in our approach, and it is inevitable that some recent research papers will not figure in the text by the time it researches the bookstores. We may also have missed something or given insufficient space to subjects that you regard as fundamental to the field. If so, please let us know and we will endeavour to put it right in following editions.

PREFACE TO THE SECOND EDITION

Glacier science has undergone enormous changes in the decade since the publication of the first edition of *Glaciers and Glaciation*. Improved remote-sensing and geophysical techniques have vastly increased our knowledge of glacier dynamics and mass balance, and have allowed the great ice sheets to be viewed as complete, integrated systems for the first time in history. Stunning images of former glacier beds – both on land and under the sea – have become available, while previously hidden processes at the ice–bed interface are coming into focus as the result of down-borehole instrumentation, sophisticated geophysical data processing and carefully designed experiments. These advances in observational techniques have occurred in parallel with breakthroughs in glaciological theory and numerical modelling, which now form major subdisciplines in their own right. This burst of activity has been fuelled by increased concerns about the impacts of anthropogenic climate change, particularly how a reduction in ice volume might affect sea levels, water resources and natural hazards. Glacier science at the beginning of the twenty-first century, therefore, is a vigorous, thriving discipline, with clear societal relevance.

Updating *Glaciers and Glaciation* to reflect these changes in the subject has been challenging, to say the least. Much of the book has been completely rewritten or restructured, with a greater emphasis on glaciological processes and their impacts. We have added a new chapter on the Greenland and Antarctic Ice Sheets, and major new sections on glacier dynamics and volume changes, and their implications for sea-level change. To incorporate this wealth of new material, measures had to be taken to ensure our book did not inflate to a size that would necessitate a free wheelbarrow with every sale. This problem is not new, as the following quote from J.K. Charlesworth's great book *The Quaternary Era* (1957) illustrates: 'J. Geikie, in preparing the first edition of his *Great Ice Age* (1874), had to abandon his plan of preparing a bibliography since it would have required a volume in itself. Today after the lapse of a further 80 years and with the vastly accelerated output of Quaternary literature, such a task is well nigh impossible.' So, following in Charlesworth's footsteps, we had to make the difficult decision to remove large numbers of older examples and references from the text to make way for the new. Most of the papers cited in this new edition post-date the mid 1990s, and very few pre-date the 1980s. This is not to imply that the older literature is no longer relevant. On the contrary, many 'new' ideas have long pedigrees, and valuable insights into current issues can often be found in the writings of former generations. Our view is that interested readers could (and should) seek out the older literature, using the new reference list as a gateway to the entire canon. We hope that the authors of 'deleted' papers will understand our reasons for doing so; and will rest assured that the first edition of this book, with its extensive references to the older literature, will still be found on library shelves.

This edition also contains rather more equations than the first, and more in-depth discussion of the physics underlying important aspects of glacier behaviour, such as mass and force balance, hydrology and dynamics. This shift was motivated by the conviction that certain fundamental concepts are essential to understanding modern issues in glaciology and glacial geology. Simplified versions of these ideas may be appealing to teachers and students alike, but they can only carry one so far before much glacier behaviour becomes incomprehensible. Progress can only be made when the fundamentals are upgraded, and this will inevitably entail 'unlearning' much of what was previously learned. It is far better, we believe, to avoid such false trails by investing more time in the fundamentals at the outset. We have therefore extended the philosophy of the first edition, with step-by-step discussions of a greater range of key equations and concepts. While readers can still choose to skim this more technical material, we strongly recommend otherwise. Although many of the equations in Part I of this book may appear very daunting at first, time spent working through them will not be wasted and contemplation of their mysteries will eventually be rewarded by deeper insights into the ways of the glacier.

ACKNOWLEDGEMENTS

Since the publication of the first edition of *Glaciers and Glaciation*, I have continued to be fortunate enough in my research and teaching to visit many interesting and often enchanting localities. Always I am in the company of dedicated and friendly scientists or students, sharing not just a passion for a fascinating subject of study, but also the trials, tribulations and triumphs of field-based research. Over the last ten years I have spent long hours scrutinizing glacial sediments and landforms with Dave Twigg, Colm Ó Cofaigh, John Hiemstra, Brice Rea and Dave Roberts, living out of a tent and a Land Rover and usually sampling far more ale than is healthy for us, but always enjoying the process of research as much as the product. John England also continues to generously allow me to savour what it's like to use a helicopter rather than walk – and what would we both do if we no longer had the opportunity to throw Frisbees and chip golf balls at each other's tents or recite *Fawlty Towers* at each other ad nauseam? I make no apologies for enjoying what I do for a living – I had it in my sights when I was doing A level geography at school – all those beautiful images of glaciers and mountains, conveying a pioneering atmosphere that I still feel to be irresistibly evocative, even though some of the answers, I appreciate, probably lie in the laboratory!

I remain indebted to the funding bodies that make it all possible, especially the Royal Society in the UK and the Polar Continental Shelf Project in Canada. Since moving to Durham in 2004 I have been part of a large, productive and friendly Quaternary Environmental Change research team, in a professional atmosphere that cannot fail to be conducive to creative thinking – an entertaining bunch too, especially at the impromptu meetings of the 'grumpy old men' at tea breaks! A large number of people at Arnold have been involved with this edition and they have been incredibly patient with us, especially Liz Wilson, who must have wondered exactly what sort of monster we were writing at times. Many colleagues around the world, too numerous to mention, have generously donated photographs, figures and advice.

Finally, the most important acknowledgements, as always, must go to my family. My mother, father and sister are always happy to receive me when I, all too rarely, pass through their neighbourhood, even though they often get bullied into having a quick look at the relict pingos down the road or checking out the glacial sediments in the cliffs at their nearby coast. My wife Tessa and my daughters Tara and Lotte continue to support me unfailingly, often joining my research expeditions, living in their tents, socializing with my students and colleagues and often making the flapjacks. Their companionship has been priceless and I regret not one minute of hanging around camp some days to indulge in storytelling, going for a swim or taking a picnic. They are my permanent reminder that life is far too short to take academia too seriously!

Dave Evans

It is now 27 years since a penniless climbing bum got talking to Danny McCarroll at Froggat Edge in Derbyshire, and realized that a living could be made studying glaciers and glaciation. For that life-changing insight (and many fine rock climbs over the years), thank you, Danny. A career in glaciology has not only allowed me to spend a large proportion of my time in wild and beautiful landscapes – as was the original intention – but also to become part of a thriving academic community working on one of the most fascinating, challenging and important of the natural sciences. I wish to thank my teachers and mentors, John Rice, Colin Ballantyne and Chalmers Clapperton, who helped me focus my raw enthusiasm and develop a more measured approach to research. Colin remains one of my most valued friends and collaborators, and my contribution to this book owes much to his influence. My thinking has been deeply influenced by interactions with numerous colleagues and students. There is not space to acknowledge them all, but some may recognize certain passages and diagrams in this book, and recall their scribbled origins on various coffee napkins and beer mats. Particular thanks are due to Nick Hulton and Garry Clarke for patiently helping me make the transition from dirt to ice. Friendships that survive the rigours and perils of fieldwork at high latitudes and altitudes are worth more than gold, and I raise my rather battered hat to all those who have walked yak trails, threaded through crevasse fields or shared ice-cave bivouacs with me. *Skål* to: Artur Adamek, Alison Banwell, Annelie Bergstrom, Nicole Davis, Guglielmina Diolaiuti, Jason Gulley, Kat Hands, Anne Hormes, Endre Gjermundsen Nick Hulton, Martin Kirkbride, Lene Kristensen, Adrian Luckman, Sven Lukas, Martin Machiedo, Ruth Mottram, Tavi Murray, Lindsey Nicholson, Faezeh Nick, Lewis Owen, Tony Prave, Ruth Robinson, Claudio Smiraglia, Monica Sund, Sarah Thompson, Maria Temminghof, Charles Warren, Seonaid Wiseman and Katleen Van Hoof. Drafts of parts of this book were read by Regine Hock and Alastair Dawson, and their helpful comments are gratefully acknowledged.

Perhaps the most significant development in my professional life in the last decade has been joining the staff of the University Centre in Svalbard (UNIS). This unique institution provides all the facilities of a modern university in the midst of the high Arctic, allowing one to conduct fieldwork on calving and surging glaciers within a short snowmobile or boat ride of a fast internet connection, a world-class library and a sauna. For financial and logistical support, and for all the fun times indoors and out, *tusen takk* to all my friends, colleagues and students at UNIS. Particular thanks are due to Hanne Christiansen and Ole Humlum for encouraging me to come here, and to Piotr Glowacki and Jacek Jania for generously supporting our collaboration at the Polish Research Station at Hornsund. Throughout my time at UNIS, I have remained on the faculty of the University of St Andrews, and I owe a big thank you to my colleagues there for supporting my applications for leave and putting up with my lengthy absences in the Far North. And of course, none of it would have been possible without the sacrifices made by Sue and Andrew, who will always be a part of me.

Doug Benn

PART ONE
GLACIERS

The calving front of Wahlenbergbreen, Svalbard (Doug Benn)

CHAPTER 1

INTRODUCTION

Ice arch on the Ngozumpa Glacier, Nepal, looking towards Cho Oyu (8,201 m) (Doug Benn)

1.1 GLACIER SYSTEMS

The life and death of glaciers and ice sheets have wide-reaching impacts, and can profoundly affect natural ecosystems and human communities. Together with sea ice, lake ice, snow cover and ground ice, glaciers and ice sheets comprise the *cryosphere*, from the Greek word *kryo* meaning 'cold'. Mass and energy are constantly exchanged between the cryosphere and the other major components of the Earth system, the hydrosphere, atmosphere, biosphere and lithosphere. Glaciers are sensitive barometers of climate change, constantly growing and shrinking in response to changes in temperature, snowfall and other factors. In recent decades, a reduction of global ice volume has raised concerns about vanishing water supplies, increased hazards from outburst floods and avalanches, and sea-level change. In earlier times, however, inhabitants of mountain regions suffered loss of land and life as glaciers expanded during the Little Ice Age. To those who lost their farms to advancing ice, glaciers were like threatening monsters (fig. 1.1). Now, glaciers are commonly viewed as endangered species, victims of human-induced climate change (fig. 1.2; Carey, 2007).

Glaciers have shaped the landscapes of huge areas of the Earth's surface, scouring out rock and sediment and depositing thick accumulations of glacial debris. Moraines, laid down like tidemarks at former glacier limits, provide valuable sources of information on past glacier activity and climate change, while ice entombed in Greenland, Antarctica and smaller glaciers and ice caps contains rich archives of former environmental conditions.

Figure 1.2 Images such as this portrayal of the retreat of Gangotri Glacier, India, are widely employed to convey messages about the impact of anthropogenic climate change (NASA, http://earthobservatory.nasa.gov).

Figure 1.1 In this 1892 engraving by H.G. Willinck, the Mer de Glace in the French Alps is portrayed as an icy dragon creeping down from the high mountains to threaten the valley below (Collection Payot, Conseil Général de la Haut Savoie).

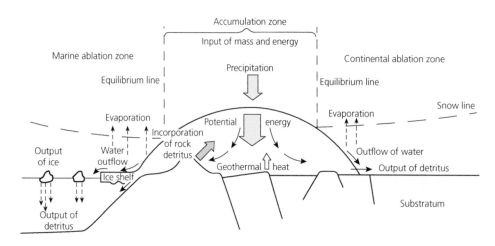

Figure 1.3 Mass and energy fluxes in an idealized glacier system (Brodzikowski and van Loon, 1991).

Glaciers are also endlessly fascinating, beautiful, powerful and wild.

To understand how glaciers behave, it is helpful to view them as *systems*, with inputs and outputs, and interactions with other systems, such as the atmosphere, the oceans, rivers and the landscape (fig. 1.3). Mass enters the system in the form of snowfall and rock debris. Because it occupies an elevated position in the Earth's gravitational field, this mass has potential energy, which is expended as the glacier flows downslope. The energy expended is used to warm or melt ice, and must then be dissipated from the system in the form of heat or water. All the while, potential energy is turned into work, inexorably transferring ice and rock from the continents towards the sea.

1.1.1 MASS BALANCE

The gain and loss of ice in glacier systems is known as the *mass balance*. Snow and ice from direct snowfall, blown snow and avalanching from slopes above the glacier surface are collectively termed *accumulation*. This snow and ice are then transferred downvalley by glacier movement until they reach areas where they are lost to the system, either by melting, evaporation or the breakaway of ice blocks or icebergs, collectively known as *ablation* (fig. 1.3). Glaciers grow where climatic and topographic conditions allow the inputs to exceed the losses (i.e. where accumulation exceeds ablation), and glaciers recede where the outputs are greater than the inputs. Energy exchanges between the glacier and the atmosphere above and the solid Earth below also modify the temperature of the ice, so that the total energy store in the glacier can change through time even if the mass remains constant.

Most glaciers can be subdivided into two zones, an inner or upper zone, where annual accumulation exceeds losses by ablation, and an outer or lower zone, where ablation exceeds accumulation. These two zones are known as the *accumulation zone* and the *ablation zone* respectively and are separated by the *equilibrium line*, where annual accumulation and ablation are equal. The position of the equilibrium line (the *equilibrium line altitude* or ELA) is dictated by local and regional climate and topography. Glacier mass and energy balance form the subject of chapter 2, while iceberg calving is discussed in detail in chapter 5.

1.1.2 MELTWATER

Meltwater is an extremely important component of glacier systems. Outputs of water from melting glaciers exert a very strong influence on the hydrology of proglacial areas, and feed water into other parts of the global hydrological system, including the oceans and the atmosphere. As it flows from glacier margins towards the sea, meltwater shapes the land, carving out gorges and depositing broad spreads of gravel, sand and silt. Within and beneath glacier and ice sheets, liquid water profoundly affects glaciers behaviour, controlling rates of glacier flow and influencing the processes and rates of erosion and deposition. The behaviour and work of meltwater is described in chapter 3.

1.1.3 GLACIER MOTION

Snow and ice are transferred from areas of accumulation to areas of ablation by glacier flow. Flow takes place by a variety of processes, which can be grouped together as *sliding*, *deformation of the ice* and *deformation of the glacier bed*. By one or more of these processes, glaciers move through the landscape, delivering snow and ice to areas where ablation exceeds accumulation, and meltwater and icebergs can leave the system. Rates and patterns of glacier motion depend on the balance between the driving forces (the downslope component of gravitational acceleration) and resisting forces (drag at the bed and margins of the glacier). When the whole system is in balance, rates of glacier flow match the rates that snow and ice is added and lost in the accumulation and ablation areas, so that the inputs, throughputs and outputs exist in a state of dynamic equilibrium. However, various factors can throw the system out of balance, causing glaciers and ice sheets to exhibit all kinds of surprising behaviour, such as rapid advances or retreats that are unrelated to climate. The processes and patterns of glacier motion are discussed in chapter 4. The dynamic behaviour of glaciers and ice caps

forms the topic of chapter 5, while that of the Greenland and Antarctic Ice Sheets is examined in chapter 6.

1.1.4 GLACIERS AND SEA-LEVEL CHANGE

Water, in liquid, solid or gaseous form, is constantly moving between the oceans, the atmosphere, the cryosphere and the hydrosphere. These fluxes rarely balance, resulting in changes of *storage*. Increased storage in glaciers and ice sheets means less in the oceans, and vice versa, so glacier fluctuations can directly affect sea level. Changes in ice volume also influence patterns of loading on the Earth's crust and the distribution of mass around the planet, further influencing local and global sea levels. The influence of glacier fluctuations on past, present and future sea-level change forms the subject of chapter 7.

1.1.5 EROSION AND DEBRIS TRANSPORT

Glaciers are among the most effective agents of erosion on Earth, excavating impressive troughs and fjord basins, and scouring broad areas clear of soil and debris. They are also very efficient transporters of debris, carrying vast amounts of silt, sand, gravel and boulders up to several hundreds of kilometres from their source areas. This debris is then deposited in many types of environment, ranging from the glacier sole to the ocean floor. Processes and forms of glacial erosion form the subject of chapter 8, and the throughput of debris by ice and water is described in chapter 9.

1.1.6 GLACIAL SEDIMENTS, LANDFORMS AND LANDSCAPES

In the mid and high latitudes, the most obvious legacy of glaciers and ice sheets is the range of sediments and landforms left behind after deglaciation, including corries, troughs, drumlins and moraines. These sediments and landforms can be used to reconstruct the extent and behaviour of former ice masses, and provide important clues to the past and present workings of the global climate system. Glacial sediments and landforms are also important from an engineering point of view, because so much human activity takes place in terrain affected by glacier erosion or deposition. Chapter 10 discusses depositional processes and sediments; chapter 11 examines depositional landforms; and chapter 12 zooms out to the widest scale to describe the overall impact of glaciation on entire landscapes.

1.2 GLACIER MORPHOLOGY

The form glaciers take is a function of climate and topography, and the morphology of any one glacier is unique to its location on the Earth's surface. Consequently, glacier morphologies form a broad continuum, from the smallest niche glacier to the largest ice sheet. For the purposes of description and study, however, it is convenient to subdivide this continuum into different glacier types, based on size, morphology and relationship to topography (Sugden and John, 1976).

1.2.1 ICE SHEETS AND ICE CAPS

Ice sheets and ice caps submerge the landscape, at least in their central portions, and major patterns of ice flow are largely independent of undulations in the bed. However, in the outer regions of ice sheets and ice caps, faster-moving ice streams and outlet glaciers are commonly located in troughs. A size of 50,000 km^2 is adopted as the threshold between an ice cap and an ice sheet. Thus the ice masses presently covering most of Antarctica and Greenland are designated as ice sheets, whereas the largest ice masses in Nordaustlandet (Svalbard), Ellesmere Island, Baffin Island and Iceland are referred to as ice caps (fig. 1.4).

Ice sheets and ice caps can be subdivided into ice domes, which are high areas of relatively slowly moving ice, and ice streams and outlet glaciers, which are more rapidly flowing zones through which most of the ice is discharged towards the periphery.

Ice domes

An ice dome is a broad, upstanding area of an ice sheet or ice cap. The underlying land surface may be either a topographic high (e.g. East Antarctica) or a vast topographic low (e.g. the centres of the former Laurentide and Scandinavian Ice Sheets). Ice thicknesses can often exceed 3000 m in ice-sheet domes, but thicknesses are only several hundreds of metres beneath ice-cap domes.

Ice divides and dispersal centres

Mass is evacuated radially outwards from a central ice dome or complex of coalescent domes, which thus act as *dispersal centres*. On either side of *ice divides*, flow is directed in two opposing directions. At the divide, the horizontal component of velocity is zero, although vertical flow occurs in response to ice compaction and deformation. Ice sheets are rarely simple domes, and usually contain several divides. Ice divides will change position as an ice sheet develops and decays. Consequently, patterns of ice flow will change considerably, and ice-directional indicators like drumlins and flutings will overprint or even destroy each other. These mega-scale imprints of glaciation are discussed in detail in chapter 12.

Outlet glaciers and ice streams

The movement of ice within ice sheets and ice caps can be subdivided into *sheet flow*, which occurs within the ice dome areas, and *stream flow*, which occurs in outlet glaciers and ice streams. Outlet glaciers and ice streams

First-order classification	Second-order classification
Ice sheet and ice cap (unconstrained by topography)	Ice dome Ice stream Outlet glacier
Glaciers constrained by topography	Ice field Valley glacier Transection glacier Cirque glacier Piedmont lobe Niche glacier Glacieret Ice apron Ice fringe
Ice shelves	Confined ice shelf Unconfined ice shelf Ice rise

Table 1.1 Classification scheme for glacier morphology

Figure 1.4 Landsat false-colour image of Vatnajökull, Iceland. The snowline is delimited by the boundary between bare ice (mid to dark blue) and snow (light blue). Note the small outlet glaciers of the subsidiary dome of Oraefi to the south, which lies between the outlet glaciers Breiðamerkurjökull (right) and Skeiðarárjökull (left) (NASA, https://zulu.ssc.nasa.gov).

are rapidly moving, channelled ice radiating out from the interiors of ice sheets and ice caps (fig. 1.5). Outlet glaciers occupy troughs or valleys, whereas ice streams are flanked by slowly moving ice (Bentley, 1987; Truffer and Echelmeyer, 2003). Some fast-moving parts of ice sheets exhibit characteristics of both ice streams and outlet glaciers, such as the Rutford Ice Stream in Antarctica, which is bounded by the Ellsworth Mountains on one side and a slow-moving part of the ice sheet on the other. The world's longest glacier, the 700 km Lambert Glacier, Antarctica, and one of the world's fastest glaciers, Jakobshavn Isbrae, Greenland, are both ice streams along some parts of their length, and outlet glaciers along others. Ice streams can be differentiated from the surrounding ice due to the presence of heavily crevassed

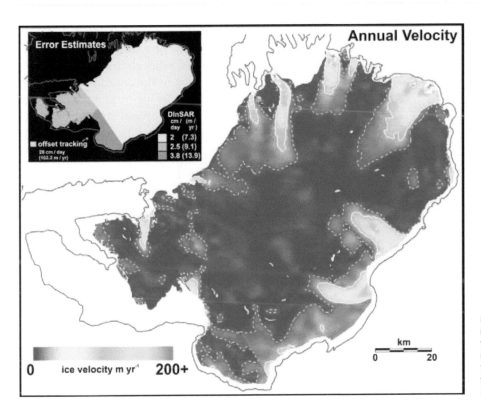

Figure 1.5 Ice surface velocities on the ice cap Austfonna, Svalbard, obtained from synthetic aperture radar (SAR) interferometry. Most of the ice flux from the ice cap is focused in narrow zones of streaming flow (Dowdeswell et al., 2008a.).

zones along their margins, where rapidly streaming ice shears past sluggish surrounding ice.

Outlet glaciers and ice streams commonly occupy depressions in the glacier bed, which can form deep troughs. The heads of ice streams (at least in some cases) coincide with a step in the subglacial topography separating thin, slowly moving ice from thicker, rapidly streaming ice. The gradients of outlet glaciers tend to be less steep than those of ice domes (Bentley, 1987).

Rapid flow through outlet glaciers and ice streams accounts for most of the discharge from ice sheets, and they are responsible for the majority of icebergs and ice-rafted debris reaching the world's oceans. Their importance in draining ice sheets, taken together with their dynamic nature, means that ice streams and outlet glaciers exert a considerable influence on the behaviour and stability of ice sheets, and has become the focus of a major research effort in recent years (chapter 6).

1.2.2 GLACIERS CONSTRAINED BY TOPOGRAPHY

Ice fields

Ice fields differ from ice caps in that they do not possess a simple, dome-like surface, and their flow is influenced by the underlying topography (fig. 1.6). Ice fields will develop in any area with generally gentle but locally fretted topography and at an altitude sufficient for ice accumulation. Examples of such ice masses are the Columbia Ice Fields in the Canadian Rocky Mountains, the ice fields

Figure 1.6 Salyut 6 photograph of the Southern Patagonian Ice Field of Chile and Argentina, 10 March 1978 (Williams and Ferrigno, 1999).

of the St Elias Mountains in the Canadian Yukon Territory/Alaska, the Tien Shan/Kunlun Shan ice fields in China, and the Patagonian Ice Fields, all of which are drained by large valley glaciers.

Valley glaciers

Wherever ice is discharged from an ice field or cirque into a deep bedrock valley it forms a valley glacier (fig. 1.7).

Figure 1.7 Brenva Glacier on the south side of Mont Blanc, Italian Alps. Avalanches transfer ice from the upper glacier onto the lower tongue (Doug Benn).

Such glaciers may possess a simple, single-branched planform, or form dendritic networks similar to those of fluvial systems. Like rivers, valley glaciers can be arranged according to their position in the drainage basin hierarchy. The form of valley glacier networks is often strongly influenced by bedrock lithology and structure. Bedrock slopes beneath valley glaciers are often steep and the altitudinal ranges of the glaciers can be very large. The most distinctive characteristic of valley glaciers is the presence of ice-free slopes overlooking the glacier surface. Such slopes are important sources of snow and ice accumulation in the form of avalanches, and of rock debris falling on to the glacier surface (sections 2.5.4 and 9.3.1).

Transection glaciers

Where glacierized mountain landscapes are deeply dissected, ice cover tends to form transection glaciers, or interconnected systems of valley glaciers. In such situations, glaciers flow down from several directions into a system of radiating valleys, but the ice overspills the pre-existing drainage divides (fig. 1.8). Transection complexes form web-like patterns, with ice *diffluences*, where ice flow splits and sends branches down two or more

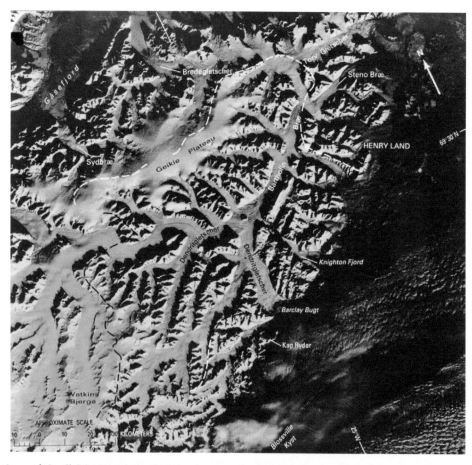

Figure 1.8 Landsat image of King Christian IX Land, Greenland, showing transection glaciers crossing the regional drainage divide (dashed line) (Canada Centre for Remote Sensing, provided by Richard Williams, USGS).

Figure 1.9 Cirque glaciers, Lahul Himalaya, India (David Evans).

Figure 1.11 Piedmont lobes fed by a plateau ice cap near Dobbin Bay, Ellesmere Island, Arctic Canada (David Evans).

Figure 1.10 Slettmarkbreen, a small cirque glacier in central Norway (Doug Benn).

Figure 1.12 A niche glacier on the northern flank of Botnafjellet, near Sandane, southern Norway. Note the Little Ice Age moraine and the rock glacier at the base of the slope (David Evans).

channels, and *confluences*, where ice flow from two or more tributaries converges to form a single unit.

Cirque glaciers

The form of a cirque glacier is dictated by the armchair-shaped bedrock hollow which acts as an accumulation basin, especially for wind-driven snow. The various sizes and forms of cirque glaciers constitute a continuum, from glaciers that are entirely confined by their hosting bedrock hollows (figs 1.9 and 1.10), to glaciers that form the heads of larger valley glaciers in more heavily glacierized terrain.

Piedmont glaciers

Wherever valley glaciers debouch onto lowland areas after travelling through bedrock troughs they will form piedmont glaciers or lobes. Piedmont glaciers such as the Malaspina Glacier, Alaska and Skeiðarárjökull, Iceland, have large areas below the equilibrium line altitude. The Malaspina Glacier lies entirely in the ablation zone of the Lower Seward Glacier, is 600 m thick and occupies a vast depression up to 250 m below sea level. Numerous smaller examples exist in the Canadian High Arctic,

where subpolar glaciers debouch from plateau ice fields onto the lowlands of U-shaped valleys (fig. 1.11). Clearly, the discharge of mass from the accumulation zones of the feeder ice fields is sufficient to maintain relatively large glacier surface areas below the equilibrium line.

Niche glaciers, glacierets, ice aprons and ice fringes

The smallest glacier ice masses, *ice aprons*, are thin snow and ice accumulations adhering to mountain sides (fig. 1.12). Similar accumulations occupying small depressions along coasts are referred to as *ice fringes*. Thin ice patches occupying depressions on less precipitous terrain are often referred to as *glacierets* and are produced by snow drifting and avalanching. Glacierets that exist due to ice avalanching from ice falls at steep plateau edges have been called *fall glaciers*. Where the location of an ice body is controlled by a niche or rock bench in a mountain or valley side it is termed a *niche glacier*. Niche glaciers and ice aprons differ from large snowpatches because they undergo significant movement as the result of internal deformation or basal sliding. Glacier motion is initiated when the forces imposed by the weight and surface gradient of the ice overcomes the internal resistance. Theoretical considerations suggest that this threshold will be reached when a snowpatch extends between 30 m and 70 m from the base of its backwall (Ballantyne and Benn, 1994a).

Icefalls

Icefalls are steep zones on a glacier where ice flow is extremely rapid, sometimes exceeding ten times that of the glacier elsewhere along its course. The acceleration of a glacier as it enters an icefall creates a zone of extreme extending flow, where the ice thins and stretches under large tensile stresses, opening up large numbers of crevasses and breaking the ice surface into unstable ice blocks known as seracs. The deceleration of ice flow at the base of an icefall creates a zone of compressive flow, where crevasses are closed up and the ice thickens.

Ice cover is not necessarily continuous through icefalls, and there are sometimes expanses of bare rock slabs or cliffs between the top of the icefall and the rest of the glacier below (fig. 1.7). Ice flow through the icefall takes the form of intermittent avalanches triggered by the collapse of seracs, piling up cones of broken ice blocks at the base. Such discontinuous icefalls are very common in high mountain environments, where accumulation areas can be separated from the lower glacier by rock cliffs

many hundreds of metres high. Clearly, icefalls are extremely hazardous places to be, and some, like the notorious Khumbu Icefall on Mount Everest, present serious obstacles to mountaineering expeditions and cost many lives.

1.2.3 ICE SHELVES

Ice shelves are low-gradient, floating glacier tongues, produced where glaciers flow offshore or where sea ice thickens by surface accumulation and bottom accretion. Ice shelves are confined to high polar settings, and are most extensive around the coast of continental Antarctica. Smaller ice shelves fringe the Antarctic Peninsula, northern Greenland and Ellesmere Island (fig. 1.13). The characteristics of ice shelves, and the controls on their stability, are discussed in section 5.6.

Wherever floating ice shelves thicken to the point where they can ground on offshore shoals, the margins of islands or coastal shelves, they produce *ice rises*. These can form either by ice flow over the obstacle, local thickening due to surface accumulation, or some combination of the two. Where an ice rise forms by local accumulation, it can have its own radial flow pattern, independent of the general flow direction of the ice shelf that surrounds it (fig. 1.14). Wherever glacier ice shelves override an offshore shoal they become heavily crevassed, forming *ice rumples* (fig. 1.15).

Figure 1.13 Ice shelves and ice islands of the Canadian Arctic. (a) The Ward Hunt Ice Shelf off the north coast of Ellesmere Island, as it appeared in 1950. Large calving events have subsequently removed most of the shelf. Note the ridge and trough pattern (rolls) picked out by surface meltwater in the troughs. (Energy, Mines and Resources, Canada) (b) An ice island floating on the Arctic Ocean, surrounded by sea ice. This early spring view was taken before meltwater had collected in the troughs. The north coast of Ellesmere Island is visible in the distance. (Polar Continental Shelf Project, EMR Canada) (c) The Cape Alfred Ernest Ice Shelf on the Wooton Peninsula, as it appeared in 1950. This composite ice shelf has components of sea ice (largely bottom left) and glacier ice. The size of the ice shelf has since been reduced considerably (Energy, Mines and Resources, Canada).

Figure 1.14 The Gipps Ice Rise in the former Larsen Ice Shelf, Antarctic Peninsula, 1 January 1972. At that time the ice rise measured 9 × 18 km and stood 300 m above the surrounding shelf (Charles Swithinbank).

Figure 1.15 Ice rises and ice rumples in the Wordie Ice Shelf, Antarctic Peninsula, 28 November 1966 (US Navy; photo provided by Richard Williams, USGS).

1.3 PRESENT DISTRIBUTION OF GLACIERS

Glacier ice presently covers almost 16 million km² of the Earth's surface, most of which is contained within the Antarctic (~13.5 million km²) and Greenland (~1.74 million km²) Ice Sheets. The remaining 3 per cent or 500,000 km² exists as ice caps, ice fields and glaciers located at high latitudes or in mountainous regions around the world. These are located predominantly in the northern hemisphere, specifically around the Arctic Ocean basin, in mountainous maritime localities like Norway and Alaska, and in continental high mountain terrains like the Alps and the Himalaya. In the southern hemisphere, significant glacier ice cover exists in the Andes, New Zealand and sub-Antarctic islands such as South Georgia. The distribution of glaciers reflects the interplay between effective precipitation, temperature and topography. These factors vary systematically around the globe, most notably with latitude, altitude and distance from a moisture source.

1.3.1 INFLUENCE OF LATITUDE AND ALTITUDE

Other factors being equal, glaciers will be more extensive closer to the poles because the low solar angle at high latitudes means that less energy is available to melt snow and ice. Similarly, for any given latitude, the likelihood of glacier survival increases with altitude because mean temperatures decrease with air density. The interaction of latitudinal and altitudinal factors creates a broad global pattern of glaciation in which glaciers only exist in equatorial regions at high altitudes, and occupy progressively lower and lower altitudes towards the poles. For example, glaciers only exist above 4500–5000 m in the tropical mountains of Irian Jaya, central Africa and Ecuador, but can form at sea level in high-latitude regions such as the Canadian and Greenland High Arctic. This pattern clearly emerges in figure 1.16, which shows present and past glacier equilibrium line altitudes (ELAs) in a north–south transect through the North and South American Cordilleras. Glacier ELAs rise from both poles towards the equator, reaching a maximum in the subtropical Mexican Highlands and Bolivian Andes. The simple rise towards the equator is complicated by variations in precipitation, so that glacier ELAs are lower than expected in areas with high snowfall. This is the case for the heavily glacierized southern slope of the Alaska range and the Patagonian Ice Fields, where snowfall can exceed 7 m per year (Carrasco et al., 2002). Similarly, ELAs are relatively lower in Ecuador, within the Intertropical Convergence Zone, than in arid areas such as Bolivia and Mexico.

1.3.2 INFLUENCE OF ASPECT, RELIEF AND DISTANCE FROM A MOISTURE SOURCE

The survival of glaciers is also influenced by aspect, relief and distance from moisture sources. Mountain landscapes such as the Himalaya, the European Alps, the North American Rockies and Coast Ranges, the New Zealand Southern Alps and the South American Andes are at sufficient altitude and are large enough to host considerable glacier ice cover, but at local scales, mountain shape, aspect and continentality, as well as size, need to be taken into account. Slope aspect plays a significant role in the location of glaciers at local scales, especially in marginal settings where the regional snowline is located just below mountain summits. Snowline altitudes may vary by several hundreds of metres in high-relief terrain, due to

(a)

(b)

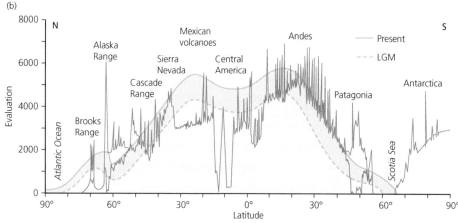

Figure 1.16 Transect from pole to pole along the American Cordilleras and Antarctica, showing the general altitude of glacier equilibrium lines for the present (solid line) and LGM (dashed line) (Broecker and Denton, 1989).

the differential receipt of both solar radiation and precipitation. The lowest snowlines in many northern hemisphere ranges are in north-east-facing basins, due to the fact that north-facing slopes receive the least solar radiation and are in the lee of prevailing south-westerly winds, thereby acting as snow traps. Even in areas where the regional snowline lies above the highest summits, glaciers can be produced by snow blow.

Even in the most glaciated regions, precipitous peaks often remain free of snow despite the fact that adjacent highland plateaux at lower elevations host large ice fields or ice caps (fig. 1.17). This is a function of relief, which involves summit area and shape as well as altitude. Generally, higher mountains will be able to accumulate enough snow to form a glacier, but narrow summits with steep sides will not have sufficient space for snow accumulation. Where mountain slopes are too steep to hold snow, avalanches can transfer snow to lower elevations, sometimes nourishing glaciers well below the regional ELA (section 2.4).

Figure 1.17 The Dent du Géant, Mont Blanc massif, European Alps. Steep mountain tops such as this remain ice-free even though they stand above the glacier equilibrium line (Doug Benn).

Even though a particular environment may be cold enough for glacier survival, it also may lie at a considerable distance from the nearest moisture source. Although many continental interiors experience extremely low temperatures during the winter, they receive very small amounts of precipitation. In such areas, precipitation totals may be insufficient to maintain even mountain ice fields, so the proximity to open water may be critical to glacierization. A good example of this is the distribution of glacier ice in the Canadian Arctic, where large ice fields on south-east Ellesmere Island and Devon Island are nourished by the open 'north water' of Baffin Bay, but ice thicknesses and glacier cover diminish in a westerly direction due to moisture starvation (Koerner, 1977). Similarly, the continental interior of Antarctica is extremely arid, and most ice sheet accumulation occurs close to the coast (section 6.3.2).

1.4 PAST DISTRIBUTION OF GLACIERS

The mass of glacier ice on Earth is constantly changing, on timescales from less than one day to millions of years. At the present time, glaciers and ice sheets cover approximately 10 per cent of the Earth's surface, but this figure has undergone repeated dramatic changes throughout Earth's history. At the peak of the great Quaternary glaciations, for example, ice cover was over three times as extensive as now, whereas during long intervals it was considerably less. Great concern has been expressed about the possible consequences of glacier retreat in recent decades, but in geological terms the recent glacier recession is just the latest episode in a long history of incessant change. To provide a sense of perspective on modern changes in glacier extent, it is essential to have

an appreciation of the magnitude of past glacier fluctuations. It is beyond the scope of this book to review the vast literature on the past distribution of glaciers, and the following brief account is intended to highlight the main controls on glacier extent on different timescales, and to direct the reader towards some of the recent literature.

1.4.1 'ICEHOUSE' AND 'GREENHOUSE' WORLDS

On timescales of hundreds of millions of years, the Earth has undergone several swings between periods of high average glacier ice cover ('icehouse' conditions) and periods of little or no ice cover ('greenhouse' conditions) (Frakes et al., 1992; Deynoux et al., 2004). Shorter-term fluctuations are superimposed on these long-term cycles, as glaciers constantly adjust their volume, extent and rates of flow to shifting environmental conditions. Geologically speaking, the Earth is currently in an icehouse state, with large ice sheets in the polar regions and widespread mountain glaciers. The current icehouse began around 34 million years ago, at the beginning of the Late Cenozoic, since which time ice sheets and glaciers have waxed and waned, but never totally disappeared (Zachos et al., 2001; Barker et al., 2007). Prior to the Late Cenozoic, the Earth experienced a long 'greenhouse' period, when glacier ice was either very restricted or absent. Before that, major glaciations occurred during the Permo-Carboniferous (326–267 million years ago (Ma)), the Late Devonian to early Carboniferous (361–349 Ma) and the late Ordovician (445.6–443.7 Ma). All three of these glaciations occurred when parts of the supercontinent of Gondwana occupied the south polar regions (Crowell, 1999).

The two most severe icehouse events occurred much earlier, during the Proterozoic (Hoffman and Schrag, 2002; Kopp et al., 2005). During the Makganyene glaciation in the Palaeoproterozoic (2.3–2.2 billion years ago) and the Cryogenian period in the Neoproterozoic (740–630 Ma), glacier ice extended into equatorial latitudes, average global temperatures were well below freezing, and the global hydrological cycle was much diminished. Both of these icehouse events have attracted a great deal of attention, not only because of their unusual severity, but because they were associated with significant changes in the evolution of life and atmospheric oxygen levels (Maruyama and Santosh, 2008).

Many factors are important in controlling the average surface temperature of the Earth. On billion-year timescales, there has been a gradual increase in solar luminosity, and hence the amount of shortwave radiation reaching the Earth. Four billion years ago, solar output was only 75 per cent of its present value (Gough, 1981; Beer et al., 2000), explaining why the Earth was more susceptible to major icehouse episodes early in its history.

It is a remarkable fact that, despite this large, long-term change in solar output, the Earth's climate system has remained within the limits required to support life, an enigma that Sagan and Chyba (1997) termed the 'early faint sun paradox'. It is now widely accepted that global climate is modulated by the concentration of greenhouse gases in the atmosphere, which strongly influence the near-surface energy balance (IPCC, 2007). On timescales of tens to hundreds of millions of years, carbon dioxide (CO_2) concentrations in the atmosphere are strongly influenced by tectonic processes. Carbon is removed from the atmosphere by weathering of silicates and subsequent burial in oceanic sediments (section 3.9). Sedimentary rocks can then be subducted at convergent plate margins, and after a long period of storage within the lithosphere, CO_2 is eventually re-released into the atmosphere by volcanic outgassing. Carbon dioxide concentrations in the atmosphere thus reflect the balance between removal by weathering and release from volcanoes. Superimposed on this geologic carbon cycle are variations in carbon uptake and release driven by biological (including anthropogenic) activity and other processes.

It has been argued that low atmospheric CO_2 levels and icehouse conditions are favoured by the existence of a supercontinent and low volcanic activity, whereas high CO_2 'greenhouse' conditions are favoured by continental rifting and multiple dispersed continents (Eyles and Young, 1994). The Permo-Carboniferous and Late Cenozoic glaciations were associated with consistently low levels of atmospheric CO_2 (<500 ppm), whereas the prolonged Mesozoic to Early Cenozoic warm interval coincided with CO_2 levels generally greater than 1000 ppm (fig. 1.18; Royer et al., 2004; Royer, 2006; Fletcher et al., 2008). The short-lived and regionally restricted Ordovician glaciation, however, apparently occurred when atmospheric CO_2 levels were more than ten times higher than now.

Tectonic processes have also been invoked to explain the large icehouse–greenhouse climate transitions in the Proterozoic. It has been argued that the Cryogenian icehouse was initiated in response to CO_2 drawdown following the break-up of the supercontinent of Rodinia (Hoffman and Schrag, 2002; Donnadieu et al., 2004). At its peak, glaciation may have been extensive enough to lock the world in a 'Snowball Earth' state, in which subfreezing temperatures were maintained globally by large losses of shortwave radiation from highly reflective ice surfaces (an *ice-albedo feedback*; cf. section 2.2). Escape from this extreme icehouse might have required build-up of atmospheric CO_2 by volcanic outgassing over millions of years, until ice melting and reduction of shortwave losses initiated a warming feedback and the establishment of extreme greenhouse conditions (Hoffman and Schrag, 2002). This model has inspired vigorous debate, and

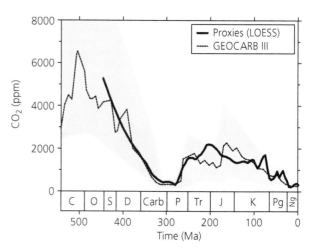

Figure 1.18 Reconstructed atmospheric CO_2 levels through the Phanerozoic. C: Cambrian, O: Ordovician, S: Silurian, D: Devonian, Carb: Carboniferous, P: Permian: Tr: Triassic. J: Jurassic, K: Cretaceous, P: Palaeogene, Ng: Neogene. The Palaeozoic comprises the Cambrian through Permian, the Mesozoic comprises Triassic through Cretaceous, and the Cenozoic consists of the Palaeogene and Neogene (Royer, 2006).

several variations and alternatives have been proposed (e.g. Young, 2004; Maruyama and Santosh, 2008; van Loon, 2008). Finding a solution to the enigma of the great Proterozoic glaciations, which is both comprehensive and consistent with all of the geologic evidence, is liable to keep geochemists and Earth system modellers off the streets for some time to come.

Climate is also strongly influenced by the distribution of continents, mountain chains and sea-ways, which constrain patterns of oceanic and atmospheric circulation, and the ways in which energy is redistributed around the Earth. For example, it has been argued that the onset of glaciation in the Mid Cenozoic was driven by uplift of the Tibetan Plateau, the Himalaya and the western Cordillera of North and South America, which increased CO_2 drawdown by weathering (Ruddiman, 1997; Pearson and Palmer, 2000). A further factor in the cooling of Antarctica was the deepening of the Drake Passage between the Antarctic Peninsula and South America around 34–30 Ma, encouraging development of the Antarctic Circumpolar Current and reduced heat flux from lower latitudes (Kennett, 1977; Livermore et al., 2005).

Superimposed on icehouse periods are shorter-term periodic oscillations between colder conditions when glaciers and ice sheets are more extensive, and warmer conditions when ice cover is much reduced. The timing of these oscillations is much better known for the Late Cenozoic than for earlier icehouse periods for which dating control is considerably poorer. For the last 2 million years or so, the sequence of *glacial* and *interglacial* periods is strongly linked to cyclical changes in the Earth's orbit around the sun, although the relationship is not simple and many interacting factors modulate ice sheet

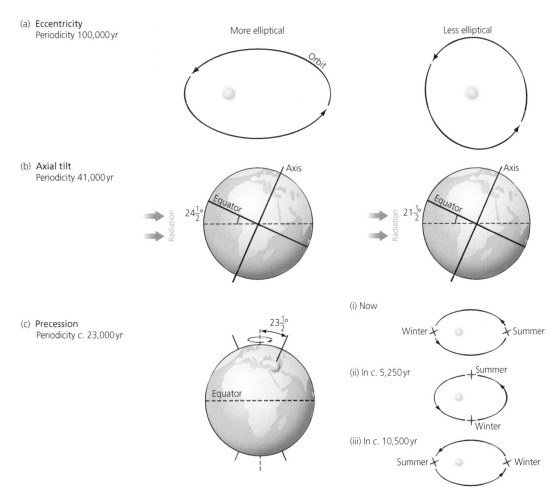

(a) Eccentricity
Periodicity 100,000 yr

More elliptical

Less elliptical

Orbit

(b) Axial tilt
Periodicity 41,000 yr

Axis

Equator

Radiation

$24\frac{1}{2}°$

Axis

Equator

Radiation

$21\frac{1}{2}°$

(c) Precession
Periodicity c. 23,000 yr

$23\frac{1}{2}°$

Equator

(i) Now

Winter   Summer

(ii) In c. 5,250 yr

Summer

Winter

(iii) In c. 10,500 yr

Summer   Winter

Figure 1.19 Milankovitch cycles in the Earth's orbit, which modulate the amount of solar radiation reaching the surface. 'Summer' and 'winter' in (i)–(iii) refer to the northern hemisphere seasons (Modified from Lowe and Walker, 1997).

growth and decay (Imbrie et al., 1992, 1993; Raymo and Nisancioglu, 2003; Jouzel et al., 2007; Bintanja and van de Wal, 2008). Three orbital cycles have been identified (fig. 1.19). First, the shape or *eccentricity* of the Earth's orbit fluctuates, becoming more or less elliptical on a cycle of about 100,000 years; second, the tilt or *obliquity* of the Earth's axis relative to the orbital plane fluctuates over a 41,000-year cycle; and third, the direction of tilt of the Earth's axis relative to the distant stars undergoes a 23,000-year cycle (*precession*) which alters the timing and variability of the seasons. Taken together, these cycles cause variations in the amount of solar radiation received throughout the year on different parts of the Earth's surface, thus altering the most fundamental input to the Earth's climate system. Additionally, equator-to-pole insolation gradients influence atmospheric circulation patterns and moisture fluxes towards high latitudes (Raymo and Nisancioglu, 2003).

Changes in radiation receipts in response to orbital variations bring about responses and chain reactions in the Earth's internal elements (e.g. the atmosphere, oceans, hydrological cycle, vegetation cover, and glaciers and ice

sheets) which act as a linked system (Imbrie et al., 1992, 1993; Bradley, 1999; Clark et al., 1999). Changes in one part of the Earth system in response to an external change can bring about responses in all the other elements because they are coupled or connected. This can set up feedback loops which can amplify or dampen the original signal, causing a non-linear response of the global climate system. Ice sheets play an important role in global climate change, serving to amplify or dampen climatic variations. Therefore, glacier growth may be both a consequence and a cause of climatic change, complicating any attempt to understand the controls on glacial cycles (Bintanja and van de Wal, 2008).

1.4.2 CENOZOIC GLACIATION

Evidence that extensive glaciers and ice sheets existed in the recent geological past has been recognized since the mid nineteenth century (Chorley, 1973; Cunningham, 1990). Since that time, glacial chronologies have been developed for many parts of the world, although it is only comparatively recently that these have been placed on a

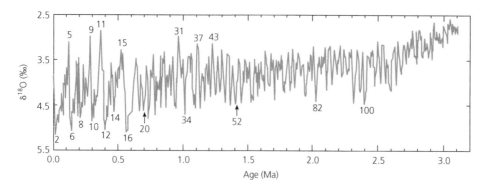

Figure 1.20 Deep-sea oxygen isotope record from core ODP-607 from the equatorial Atlantic. Selected marine isotope stages are labelled. Note the increased importance of the ~100,000-year cycle in the later part of the record (Bradley, 1999, fig. 6.16).

secure quantitative footing. Since the 1960s, radiocarbon dating has been used to constrain the timing of recent glacial events (<40,000 years), although this technique depends on locating suitable organic material. Quaternary glacial chronology has been revolutionized in the first decade of the twenty-first century by the advent of cosmogenic radionuclide (CRN) and optically stimulated luminescence (OSL) dating (Aitken, 1998b; Gosse and Philips, 2001). These techniques date the duration of surface exposure and the age of sediment burial, respectively, allowing glacial chronologies to be placed on an absolute timescale and correlated with long marine and ice core records (see sections 2.7 and 7.3).

Global ice volumes can be calculated from oxygen isotope ratios in marine sediments and other methods (sections 7.2 and 7.3). The marine oxygen isotope record also provides a benchmark chronology for the Quaternary period (the last 2 million years or so, up to and including the present time; Pillans, 2007). Marine isotope stages (MIS) are numbered back in time from the present, with cold periods (*glacials* or *stades*) assigned even numbers and warm periods (*interglacials* or *interstades*) assigned odd numbers. The current warm period (the Holocene) is identified with MIS-1 and the last period of full glacial conditions with MIS-2 (fig. 1.20). During all periods characterized by glaciation, climate varies in a quasi-periodic way regardless of the location or extent of ice sheets. The dominant wavelength since the onset of ice sheet glaciation in the early Oligocene corresponds to the obliquity cycle (~40,000 years). Around 800 to 900 ka, the dominant wavelength of climate cycles switched to approximately 100,000 years, close to the length of the eccentricity cycle. This appears paradoxical, since of all the Milankovitch cycles, eccentricity exerts the weakest influence on solar radiation receipts at the Earth's surface. Many mechanisms have been proposed to account for this switch in the periodicity of global climate cycles, most of which focus on how weak eccentricity forcing might be amplified by non-linear Earth system processes (Maslin et al., 2001). Recently, it has been argued that the ~100,000-year climate cycle actually reflects the 23,000-year precession cycle, with major glacial cycles related to four to five precessional cycles

(Maslin and Ridgwell, 2005). Under this interpretation, eccentricity serves to pace the system by modulating precession forcings, rather than being the primary driver.

As noted above, the Late Cenozoic glaciation began earliest in Antarctica (fig. 1.21). There is evidence that mountain glaciers existed in Antarctica throughout most of the Cenozoic, and for ice sheet growth after the Eocene–Oligocene boundary at ~34 Ma (e.g. Barrett, 1996; Hambrey and McKelvey, 2000; Anderson and Shipp, 2001; Anderson et al., 2002; Escutia et al., 2005). The Antarctic Ice Sheet expanded and retreated several times until the Middle Miocene 'climatic optimum' at 14–10 Ma, after which it became a persistent feature, though still subject to oscillations. At the Last Global Glacial Maximum (LGM) (~20,000 years ago), the ice sheet extended close to the edge of the continental shelf, and had an area of over $15 \times 10^6 \, km^2$.

Glaciation of the northern hemisphere was initiated around 3.2 Ma (fig. 1.21; Zachos et al., 2001). The largest of the northern hemisphere ice sheets was the Laurentide Ice Sheet, which, at its maximum extent, covered large parts of northern North America (fig. 1.22; Dyke et al., 2002; Mickelson and Colgan, 2004). At glacial maxima, the Laurentide Ice Sheet was confluent with the Cordilleran Ice Sheet in the west and the Innuitian Ice Sheet in the north, although at times of lesser ice extent the latter two behaved as independent entities (Booth et al., 2004; England et al., 2006). During the LGM, the Cordilleran Ice Sheet extended from the Puget Lowlands in Washington State to the western tip of the Aleutian Islands (Briner and Kaufman, 2008). At this time, the Innuitian Ice Sheet covered the Canadian Arctic Archipelago, and was composed of an alpine sector to the east and a lowland sector to the west. Several palaeo-ice streams drained the ice sheets (section 12.4.2). The Greenland Ice Sheet also expanded over the surrounding continental shelf during glacial periods (e.g. Evans et al., 2009), but unlike the Laurentide Ice Sheet was able to persist through some, but perhaps not all, interglacial periods. There is evidence that the Greenland Ice Sheet was smaller in the last (Eemian) interglacial (~125,000 years ago) than at present (section 7.3.1; Otto-Bliesner et al., 2006).

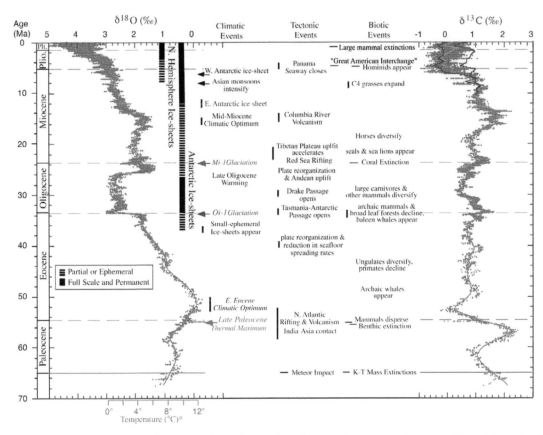

Figure 1.21 Global deep-sea oxygen and carbon isotope records, showing long-term climate trends through the Cenozoic (Zachos et al., 2001, fig. 2).

Figure 1.22 Reconstruction of the Laurentide Ice Sheet for the LGM, when it was confluent with the Cordilleran Ice Sheet in the west and the Innuitian Ice Sheet in the Canadian Arctic Archipelago (Dyke et al., 2002).

A large ice sheet also developed over Eurasia on several occasions during the Quaternary. At its maximum extent, the Eurasian Ice Sheet extended from Ireland in the west to Novaya Zemlya in the east and Svalbard in the north, and its southern margin lay through England, the Netherlands, Denmark, northern Germany, Poland and Russia (Svendsen et al., 2004; fig. 1.23). Much of the northern part of this ice sheet was grounded below sea level in the Barents and Kara Sea areas, and may have been similar to the present-day West Antarctic Ice Sheet (Mangerud et al., 2002; Svendsen et al., 2004; Andreassen et al., 2008). Evidence for numerous ice streams has been identified on the continental shelf along the northern and western parts of the former ice sheet (e.g. Andreassen et al., 2004, 2008; Ottesen et al., 2005a, b, 2007; Golledge and Stoker, 2006; Bradwell et al., 2007).

A smaller ice sheet, with a maximum area of around 300,000 km² formed in Iceland during glacial periods, in addition to ice caps and expanded local glaciers in the north-west of the island (Ingolfsson et al., 2010). At glacial maxima, glaciers in the European Alps coalesced into a large ice field/transection glacier complex that extended into the surrounding lowlands in some areas (fig. 1.24; Ivy-Ochs et al., 2006, 2008). Elsewhere in Europe and

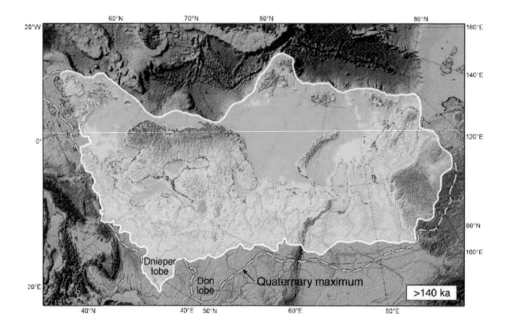

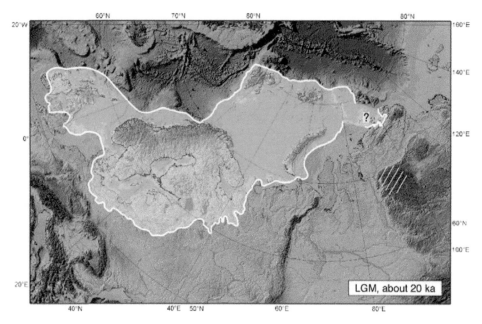

Figure 1.23 Reconstructions of the Eurasian Ice Sheet for (a) the Late Saalian (MIS-6; c.160–140 kyr) and (b) Late Weichselian (LGM) (MIS-2; c.25 kyr) (Svendsen et al., 2004, figs 13 and 16).

Asia, mountain glaciers expanded from numerous local centres (e.g. Ehlers and Gibbard, 2004a, c; Hughes and Woodward, 2008). It has been claimed that an ice sheet developed over Tibet and the Himalaya, with major impacts on regional and global climate (e.g. Kuhle, 2007). However, a large number of CRN and OSL dates for moraines throughout the region show that Quaternary glaciations were characterized by the expansion of local glaciers in separate mountain massifs, in many cases extending only a few kilometres from present-day glacier limits (fig. 1.25; Owen et al., 2008, 2009). This evidence demonstrates that no Tibetan Ice Sheet can have existed during at least the last few full glacial cycles.

In the Americas, several mountain ranges supported ice fields and/or valley and cirque glaciers during glacial periods. These include the Ahklun Mountains and Brooks Range of Alaska (Briner and Kaufman, 2008), the Rocky Mountains and Sierra Nevada (Kaufman et al., 2004; Pierce, 2004), the highest mountains in Mexico, Guatemala, Costa Rica and Venezuela (Lachniet and Vazquez-Selem, 2005), and the tropical and subtropical Andes (Smith et al., 2008). In Patagonia, an ice sheet developed along the Andean Mountains from 30° S to the southern tip of the continent (Rabassa, 2008; Glasser et al., 2008). The most extensive Quaternary glaciation in southern South America (the Greatest Patagonian Glaciation) occurred 1.1 Ma, much earlier than the global Quaternary temperature minimum and ice-volume maximum (Hulton et al., 2002; Kaplan et al., 2009), possibly due to greater elevation of accumulation areas prior to deep glacial erosion.

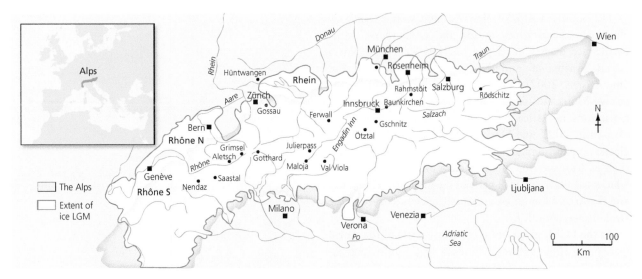

Figure 1.24 Extent of glacier ice in the European Alps at the LGM (Ivy-Ochs et al., 2008).

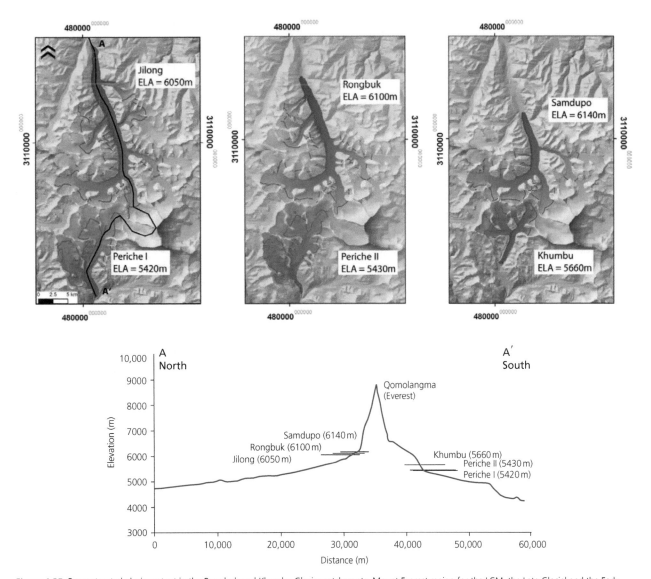

Figure 1.25 Reconstructed glacier extent in the Rongbuk and Khumbu Glacier catchments, Mount Everest region for the LGM, the Late Glacial and the Early Holocene (Owen et al., 2009).

The Southern Alps of New Zealand were extensively glaciated several times during the Quaternary, while smaller ice masses developed on New Zealand's North Island, Tasmania and New Guinea (Colhoun and Shulmeister, 2007). Small glaciers also formed in the Snowy Mountains of Australia. Evidence for Quaternary glaciations in Africa has been reviewed by Kaser and Osmaston (2002) and Mark and Osmaston (2008). Outlet glaciers expanded radially from ice caps on high mountains such as Kilimanjaro, Mount Kenya and the Ruwenzori. Glaciers also developed in parts of the Ethiopian Highlands and the Atlas Mountains, and small glaciers formed in the Drakensberg, South Africa.

At the LGM, glacier ELAs in many parts of the world were up to 1000 m lower than today (e.g. Kaufman et al., 2004; Mark et al., 2005). In some areas, such as the Mount Everest region in the east-central Himalaya, ELA lowering was only 200–300 m, probably because the effects of lower temperatures were partially offset by reduced precipitation (Owen et al., 2009).

For the period since the LGM, there is abundant evidence for glacier oscillations on timescales of hundreds to thousands of years. One of the most intensively studied periods of glacier expansion is the Younger Dryas Stade (12,900–11,600 years ago), named after the mountain avens *Dryas octopetala* which was a prominent component of the flora of southern Scandinavia at that time. This interval saw a large cooling over Greenland, widespread growth of ice caps and glaciers in the British Isles, a major ice sheet re-advance in Scandinavia, and smaller glacier advances in other parts of the world (fig. 1.26; Peteet, 1995; Lowe and Walker, 1997; Golledge et al., 2008b). The timing of this event is inconsistent with orbital forcing, so some other mechanism must be responsible. One possibility is that Younger Dryas cooling was triggered when drainage of the huge proglacial Lake Agassiz disrupted the thermohaline circulation in the North Atlantic, thus reducing poleward heat transport in the region (e.g. Teller et al., 2002). Mismatches between the onset of cooling and the chronology of lake drainage cast some doubt on this appealing model (e.g. Lowell et al., 2005), and other plausible mechanisms have been proposed (e.g. Renssen et al., 2000).

The largest glacier oscillation in historical times occurred during the Little Ice Age (Grove, 1988, 2001; Matthews and Briffa, 2005; Solomina et al., 2008). Following the warm climates of the medieval period, glaciers began to advance in the thirteenth century in many mountain regions of the world. In some areas, glaciers reached approximately the same limits on more than one occasion (section 5.4.2). One of the coldest parts of the Little Ice Age coincides with a prolonged minimum in sunspot activity and solar output known as the Maunder minimum (fig. 1.27; Beer et al., 2000), and several lines of evidence indicate that variations in solar activity were a

Figure 1.26 (a) Reconstructed Younger Dryas glacier limits in Scotland. (b) Oblique view of reconstructed glaciers in the Cuillin Hills, Isle of Skye (Ballantyne, 1989).

primary driver of the Little Ice Age and other Holocene cool periods (Shindell et al., 2003; Wiles et al., 2004; Damon and Peristykh, 2005; Holzhauser et al., 2005). Additional cooling appears to have been driven by stratospheric loading by volcanic aerosols (Free and Robock, 1999). Solar and volcanic forcings can account for part of the post-Little Ice Age warming, but modelling results indicate that most of the observed recent warming trend reflects anthropogenic greenhouse gas emissions (IPCC, 2007). The large impact of natural climate forcing processes during the pre-industrial era strongly suggests that solar variations and volcanic eruptions will continue to play a significant but unknown role in future climate changes.

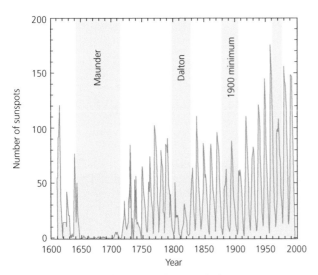

Figure 1.27 Time series of sunspot numbers over the last 400 years. The 11-year Schwabe cycle is superimposed on longer-term variations (Beer et al., 2000).

Porter (1989) has argued convincingly that for most of the Quaternary, glacier extent was intermediate between full glacial maxima and the interglacial minima experienced during the Holocene, and that such intermediate conditions are the most relevant to long-term analyses of environmental change and landscape evolution. Using long palaeoclimatic records from ocean cores, Porter argued that in Britain and Iceland average ice cover was similar to that during the Younger Dryas, whereas in Scandinavia ice cover consisted of large mountain ice fields covering almost all of Norway and the highlands of Sweden. The Laurentide Ice Sheet probably consisted of three coalescent domes centred over Keewatin, Labrador and Baffin Island, covering approximately the same area as at 9000 years BP. The Greenland and Antarctic ice sheets were slightly larger than at present, whereas glaciers in major mid- to high-latitude mountain ranges, such as the southern Alps of New Zealand, the Himalaya, the southern Andes, the European Alps, and the Cascade, Brooks and Alaska Ranges, probably extended a few kilometres to a few tens of kilometres beyond their present margins. According to Porter, this average ice cover represents the principal boundary conditions for much of Quaternary landscape evolution, including the erosion of classic glacial mountain landscapes of cirques and fjords, the development of erosional shore platforms, tropical atolls (due to the correlation between ice cover and global sea level), and long-term sediment yields.

CHAPTER 2

SNOW, ICE AND CLIMATE

A perfect day on Breiðamerkurjökull, with Öræfajökull (2110 m), Iceland's highest peak in the background (Doug Benn)

2.1 INTRODUCTION

Glaciers and ice sheets are large, dynamic stores of water, constantly exchanging mass and energy with the atmosphere, hydrosphere and other parts of the earth system. Glaciers gain mass by the input of snow and other forms of ice accumulating on their surfaces, and undergo mass loss (ablation) by melting, iceberg calving and other processes. The total glacier mass evolves through time depending on the balance between accumulation and ablation, which in turn depend on climate and local topographic factors. Similarly, glacier temperature evolves in response to the balance between inputs and outputs of energy at the surface and the bed, and heat generated within the glacier by ice flow. Changes in mass and ice temperature are strongly coupled glacier

responses to climate. The amount of snow cover and melting and refreezing processes strongly influence ice temperature evolution; and, conversely, near-surface ice temperature determines whether surplus energy will result in ablation or warming of the glacier. Additionally, frictional heat sources within and beneath the ice depend on ice velocity, which in turn is partly controlled by surface mass balance (section 5.2.1). Thus, ice temperatures and mass balance need to be considered together if glacier response to climate change is to be fully understood.

Glacier mass balance and thermal structure impact on many other parts of glacier systems and the extraglacial environment. At the catchment scale, changes in snow and ice storage influence river discharge and thus water resources and flood risks, while at the global scale, fluctuations in glacier volume impact directly on sea-level change (Bamber and Payne, 2004). Ice temperature has a profound effect on many glacial processes, including surface ablation, englacial and subglacial water flow, rates of glacier motion, and patterns of glacial erosion and deposition. Understanding the links between glacier mass balance, thermal regime and climate also allow earth scientists to reconstruct long-term environmental changes from evidence of past glacier fluctuations.

This chapter begins by considering the energy sources at glacier surfaces; section 2.3 examines the controls on ice temperature and the thermal structure of glaciers; section 2.4 deals with processes of ice accumulation and melting; and section 2.5 considers how spatial and temporal variations in these processes determine glacier mass balance.

2.2 SURFACE ENERGY BALANCE

2.2.1 CHANGES OF STATE AND TEMPERATURE

Energy exchanges between the atmosphere and a snow or ice surface can result in either *changes in the state of water* (melting, freezing, etc.) or *temperature changes*. If the temperature of the snow or ice is below freezing, a net surplus of energy at the glacier surface will raise its temperature, and if its temperature reaches 0 °C, any further energy surplus will result in melting. Conversely, a net deficit in energy at the surface can chill the ice or lead to ice accumulation through condensation of vapour or freezing of water. Mass loss can also occur at sub-zero temperatures by sublimation if the atmospheric vapour content is sufficiently low.

For a given input of energy, the temperature change experienced by a substance depends on its *specific heat capacity*, denoted by the letter c. For ice, c is 1.751 J g^{-1} K^{-1} at −50 °C, rising to 2.000 J g^{-1} K^{-1} at −10 °C and 2.050 J g^{-1} K^{-1} at 0 °C. That is, for ice at −10 °C (263 K), 2 Joules of energy are required to raise the temperature of one gram of ice by one degree. Water at 0 °C has a specific heat capacity of 4.217 J g^{-1} K^{-1}. Melting and evaporation involve breaking hydrogen bonds between water molecules, which requires an input of energy. Conversely, freezing and condensation release an equivalent amount of energy. Because the energy released during freezing or condensation can be thought of as having been latent within the water or vapour, it is known as *latent heat*. This is actually a misnomer, because 'heat' is a manifestation of energy flow, not a material property, and 'latent heat' is more properly termed *enthalpy of fusion (or condensation)*. However, because 'latent heat' is in such widespread use, it is retained in this book. Melting requires 334 Joules per gram, whereas freezing releases an equivalent amount of energy. Thus, the latent heat of fusion/melting (L_f = 334 J g^{-1}. Over eight times as much energy is required by evaporation, or released during condensation, as the latent heat of condensation/vaporization L_v = 2500 J g^{-1} (fig. 2.1).

Rates of energy flow are measured in Watts (W), where 1 Watt = 1 Joule per second (J s^{-1}). For many purposes, including glacier mass balance studies, it is useful to consider energy exchanges across a given surface area, so energy fluxes are measured in Watts per square metre (W m^{-2}). The sum of all energy fluxes over any given time interval is known as the *energy balance*, and can be used to calculate the evolution of near-surface temperatures and glacier ablation rates (Röthlisberger and Lang, 1987; Paterson, 1994; Oerlemans, 2001; Greuell and Genthon, 2004). The most important components of the energy balance at glacier surfaces are: (1) solar radiation (shortwave); (2) terrestrial and atmospheric radiation (longwave); (3) sensible heat exchange with the atmosphere; (4) latent heat transferred during condensation, evaporation and sublimation; (5) heat supplied by rain; (6) heat used to change the temperature of the ice; and (7) latent heat consumed or released during melting and freezing.

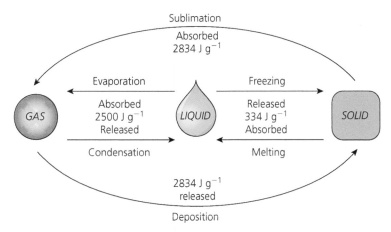

Figure 2.1 Phase changes between ice, water and vapour, showing the amount of energy (latent heat) consumed or released during transformation.

Since energy is neither created nor destroyed, all of the energy balance components must sum to zero:

$$SW + LW + QH + QE + QR - QT - M = 0$$

$$(2.1)$$

where SW is net shortwave radiation flux, LW is net longwave radiation flux, QH is sensible heat transfer, QE is latent heat transfer, QR is energy from rain, M is energy used to melt ice or freeze water, and QT is energy used for temperature change in the ice. The first five components represent exchanges of energy between the atmosphere and the glacier, and are examined in turn in the following sections. The last two components represent energy changes within the glacier-meltwater system, and are discussed in sections 2.3 and 2.4.

2.2.2 SHORTWAVE RADIATION

All materials radiate electromagnetic energy, which can be visualized as either particles (photons) or waves. Cool materials radiate low-energy photons (longer wavelengths) such as radio waves and infra-red radiation, while hotter materials radiate higher energy photons (shorter wavelengths), including visible light, ultra-violet radiation, x-rays and gamma rays. Radiation from the sun is predominantly in the wavelengths between 0.2 and 4.0 micrometres (μm), peaking in the visible part of the spectrum ($0.4-0.7$ μm), but also including significant amounts of ultra-violet and near infrared radiation. Solar radiation is termed *shortwave radiation* by meteorologists, and provides the primary source of energy for the world's climate system. Radiation emitted from terrestrial materials, such as rocks, atmospheric gases and ice, is predominantly in the wavelengths between 4 and 120 μm, and is known as *longwave* or *far infrared radiation*. Taken together, solar and terrestrial radiation are the most important components of the energy balance on most glaciers (Hock, 2005).

At the top of the Earth's atmosphere, the intensity of solar radiation is ~1368 W m^{-2} on a surface at right angles to the solar beam. Closer to the Earth's surface, increasing proportions are scattered by air molecules, cloud droplets and other particles, so that shortwave radiation reaches the surface as direct (but filtered) sunshine and diffuse (scattered) radiation. The decrease (attenuation) of radiation intensity depends on the *optical depth* of the atmosphere, which is a function of distance, humidity, cloudiness and the concentration of other particles. Below thick clouds, shortwave receipts may be only a few per cent of the value at the top of the atmosphere, but in the thin, clear air on high mountains, shortwave receipts can be as high as 1200 W m^{-2} on suitably oriented surfaces.

The amount of direct shortwave radiation received by a surface depends on its orientation relative to the

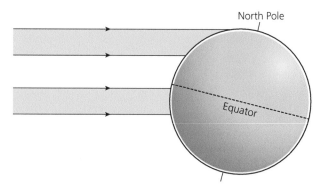

Figure 2.2 Relationship between latitude and receipt of solar radiation. Near the equator, where the midday sun is high in the sky, solar energy is distributed over a smaller area than near the poles. As a result, more energy is available for heating the surface and atmosphere per unit area at low latitudes. Attenuation of the solar beam by scattering in the atmosphere further reduces the available energy at high latitudes.

position of the sun, and will be highest when the sun's rays make an angle of 90° with the surface, and zero when the solar beam is parallel to the surface. This means that shortwave receipts will depend on (1) latitude; (2) slope gradient and aspect; and (3) time of day (fig. 2.2). If all of these factors are known, the clear-sky solar radiation receipts can be calculated from standard equations (Iqbal, 1983). The low solar angle in the high latitudes, and in the mid latitudes during the winter, means that midday radiation receipts are much lower than in the tropics. At a local scale, this pattern is modified by slope gradient and aspect, and topographic shading by mountains and other barriers, which can result in very different amounts of incoming radiation on different parts of a glacier.

At the Earth's surface, only part of the incoming solar radiation is absorbed, the remainder being reflected or transmitted through the material. The proportion of shortwave radiation reflected from a surface is given by the *albedo*:

$$\alpha = SW_{out}/SW_{in} \qquad (2.2)$$

where SW_{in} and SW_{out} are the incoming and outgoing solar radiation, respectively. The net shortwave component of the surface energy balance is thus:

$$SW = SW_{in}.(1 - \alpha) \qquad (2.3)$$

Albedo values are high for fresh snow surfaces and low for bare and dirty ice surfaces (Röthlisberger and Lang, 1987; Paterson, 1994; table 2.1). The very high albedo of snow means that much less energy is available for ablation on snow-covered glacier surfaces than on bare or debris-covered ice. Thus, for given radiation inputs, the amount of ablation will increase during the ablation season on the lower parts of glaciers, as winter snow cover gives way to bare ice surfaces. It is therefore very important to consider spatial and temporal variations in albedo

	Range
Dry snow	0.80–0.97
Melting snow	0.66–0.88
Firn	0.43–0.69
Clean ice	0.34–0.51
Slightly dirty ice	0.26–0.33
Dirty ice	0.15–0.25
Debris-covered ice	0.10–0.15

Table 2.1 Albedos for snow and ice (Paterson, 1994)

when calculating radiation receipts on glaciers (Brock et al., 2000; Brock, 2004).

Shortwave radiation is the most important energy balance component in continental environments where clear-sky conditions are common (Braithwaite and Olesen, 1990; Braithwaite et al., 1998). Shortwave receipts are particularly high on low-latitude, high-altitude glaciers, such as in the northern and central Andes and the Himalaya, where the midday solar angle is high and the filtering and scattering effect of the atmosphere is at a minimum (Kayastha et al., 1999; Benn et al., 2001). In maritime climates, shortwave radiation contributes less to the overall energy balance, because of the high frequency of cloudy conditions (Hock, 2005).

2.2.3 LONGWAVE RADIATION

Glacier surfaces receive *longwave radiation* emitted from the atmosphere and surrounding terrain. The glacier also emits longwave radiation, so the net longwave radiation is the balance of the incoming and outgoing components:

$$LW = LW_{in} - LW_{out} \qquad (2.4)$$

The radiation flux depends on temperature, according to the Stephan-Bolzmann equation:

$$I = \varepsilon \sigma T^4 \qquad (2.5)$$

where T is the temperature in Kelvins, σ is the Stephan-Bolzmann constant (5.67×10^{-8}) and ε is the emissivity, which for snow and ice has a value ranging from 0.980 to 0.995. Longwave radiation losses from ice at 0 °C (273 K) are therefore ~309 to 311 W m^{-2}, surprisingly large values which greatly exceed the intensity of incoming shortwave radiation on an overcast day. Debris-covered glacier surfaces can become very warm during daytime, with correspondingly high longwave losses. For example, for a debris surface temperature of 20 °C (293 K), outgoing longwave flux equals 413 W m^{-2}.

Incoming longwave radiation depends primarily on the near-surface air temperature (Oke, 1987; Ohmura, 2001), and can be higher or lower than the outgoing

Figure 2.3 Warm rock surfaces emit longwave radiation, adding to the energy available for melting adjacent snow and ice (D.I. Benn).

longwave flux, depending on the temperature of the glacier surface, the atmosphere and adjacent terrain. Net longwave radiation at the glacier surface can therefore be either positive or negative, but it is usually negative when ice is melting (and hence relatively warm). Longwave radiation receipts tend to be high at the margins of valley glaciers overlooked by rockwalls, where radiation emitted from warm rock surfaces can accelerate ablation (fig. 2.3).

Water vapour is a particularly efficient greenhouse gas, absorbing and re-radiating longwave radiation and thus increasing net energy fluxes from the atmosphere to the surface. Therefore, longwave radiation can be a very important component of the energy budget when the air is humid, such as during hazy or overcast weather. Clear, dry air has a much lower capacity to absorb and re-radiate longwave radiation, resulting in lower longwave receipts at the surface during the day and very rapid reductions in air temperature at sunset, when incoming shortwave radiation is cut off.

2.2.4 SENSIBLE AND LATENT HEAT: TURBULENT FLUXES

Thermal energy directly transferred between materials is known as *sensible heat* because it can be felt with the senses. In the context of glacier energy balance, sensible heat flux refers to the transfer of molecular momentum from the atmosphere to the ice surface, or vice versa, depending on the temperature gradient between the atmosphere and the glacier. Air is a very good insulator, so efficient sensible heat transfer depends on the bulk movement of near-surface air (the *boundary layer*) by turbulence or convection. A familiar example of this effect is wind chill, in which warm air is stripped away from exposed skin surfaces by cold winds, so that heat losses from the body are much larger than they would be in still air. Sensible heat transfer to glaciers is most efficient when the atmosphere is much warmer than the snow or ice, and strong winds and a rough glacier surface encourage turbulence (Paterson, 1994). Warm winds can be associated with local air circulation, such as *valley winds* that blow up mountain valleys during the day or *Föhn*-type winds on the lee side of mountain ranges (Barry, 1992; Meier et al., 1994), or with large-scale low-pressure systems.

In glacier energy balance studies, 'latent heat flux' refers to energy exchanges associated with evaporation/condensation and sublimation/deposition (fig. 2.1). Latent heat exchanges *within* the glacier-meltwater system (i.e. associated with melting/freezing) are considered to be separate energy balance terms (section 2.4.3). The processes of evaporation, condensation, sublimation and deposition are primarily controlled by the humidity of the air and windspeed. Wind is important for the same reasons as for sensible heat transfer: turbulence strips away the near-surface boundary layer (which above ice will be saturated with water vapour) and encourages efficient mixing with the air mass above.

Sensible and latent heat fluxes are collectively termed *turbulent heat fluxes*. They are difficult to determine accurately because of the complex role played by surface roughness and atmospheric turbulence, and are generally estimated using near-surface gradients in temperature and humidity, together with windspeed and empirically determined *bulk transfer coefficients*. Several alternative turbulent heat flux equations have been proposed, which yield reasonable values for most glaciological purposes (e.g. Paterson, 1994; Braithwaite et al., 1998; Kayastha et al., 1999; Oerlemans, 2001). Examples of bulk transfer functions are:

$$QH = \rho_A c_P A_B u_W (T_A - T_S) \qquad (2.6)$$

$$QE = -0.622 L_V A_B u_W (e_A - e_S) \rho_0 / P_0 \qquad (2.7)$$

in which QH and QE are sensible and latent heat flux, respectively, ρ_A is air density, c_P is the specific heat capacity of air (1.0035 J g^{-1} K^{-1} for dry air at 0 °C), A_B is a bulk transfer coefficient, u_W is wind speed, T_A is air temperature, T_S is surface temperature (273 K for a melting ice surface), L_V is the latent heat of vaporization (2500 J g^{-1}), e_A and e_S are vapour pressure of the air and at the surface, and ρ_0 and P_0 are reference values of air density and pressure. Wind speed, air temperature and air vapour pressure are all measured at some specified elevation above the surface ('screen height'), the value of which influences the bulk transfer coefficient A_B.

Turbulent heat fluxes are highest on glaciers close to the western seaboard of North America, southern Iceland, western Norway and Chile, which are strongly influenced by humid, maritime air streams (Laumann and Reeh, 1993). Latent heat fluxes are significant in cold, dry conditions, such as those found on high-altitude glaciers during the dry season, where sublimation losses are high (e.g. Mölg and Hardy, 2004; Cullen et al., 2007).

2.2.5 ENERGY SUPPLIED BY RAIN

Relatively warm rainwater falling on a glacier surface will be chilled when it contacts cold snow or ice, adding energy to the glacier in the process. If the snow or ice is already at the melting point, then the energy will be used for melting. The energy flux resulting from lowering the temperature of the rainwater is given by:

$$QR = \rho_w \dot{r} c_w (T_r - T_i) \qquad (2.8)$$

where ρ_w is water density, $\dot{r}$ is rainfall rate (in metres per second), c_w is the specific heat capacity of water (4.2 J g^{-1} K^{-1}), and T_r and T_i are the temperature of the rain and ice, respectively. If the snow or ice is below the melting point, the rainwater will be cooled then frozen, and latent heat transfer will warm the ice. In general, rain contributes only a very minor component of the energy balance of glaciers, although it can be significant in the short term during the passage of warm fronts in maritime environments.

2.2.6 WHY IS GLACIER ICE BLUE?

Visitors to glaciers are often struck by the intense blue colour of clean, freshly fractured ice. This is sometimes wrongly attributed to Rayleigh scattering, which makes the sky appear blue. In the atmosphere, Rayleigh scattering occurs because air molecules preferentially scatter short-wavelength electromagnetic radiation (blue light), while longer wavelengths (red and yellow light) are transmitted through the air. This is not the case with ice, which appears blue because the longer visible wavelengths are preferentially *absorbed*. The more energetic (blue) wavelengths are not absorbed, but are scattered back, just like light from many other types of surface (Bohren, 1983). This effect is greatest for bubble-free ice with large crystals, and when the ice surface is smooth. Consequently, the blue colour tends to be most intense in the walls of fresh fractures or calved icebergs. Preferential absorption does not occur on rough, weathered ice and snow surfaces. Because their albedo is not strongly wavelength-dependent, they appear white to the viewer.

2.3 ICE TEMPERATURE

Glaciers are not uniformly cold. In high polar continental environments, ice temperatures can be as low as −40 °C, but large parts of many glaciers are at or close to the melting point, even in permafrost environments. There are also large variations in temperature within individual glaciers, so that some parts may be well below freezing while others are at the melting point. Glaciologists recognize an important distinction between ice that is at the melting point (*temperate* or *'warm' ice*) and ice at lower temperatures (*'cold' ice*). (A glaciologist's idea of 'warm' is probably as skewed as a geologist's idea of 'recent' or an astronomer's idea of 'near'.) In the following sections, we consider how and why temperature varies in glacier ice, and outline a thermal classification of the world's glaciers.

2.3.1 THE MELTING POINT OF ICE

The temperature at which ice melts is not constant at $0\,°C$, but decreases as the ice is placed under increasing pressure at a rate of $0.072\,°C$ per million Pascals (MPa). For example, the pressure at the base of a glacier $2000\,m$ thick is approximately $17.6\,MPa$, enough to lower the melting point to $-1.27\,°C$. Consequently, the melting point of ice is referred to as the *pressure-melting point*. This is actually a slight misnomer, as the melting point of ice is also influenced by the presence of impurities such as solutes. The freezing point of seawater depends on its salinity, and is typically around $-2\,°C$ at atmospheric pressure. Because the melting point of ice is such an important physical threshold, it is usual to consider ice temperature relative to the local pressure-melting point. The importance of pressure melting and refreezing in regulating glacier motion is examined in section 4.4.2, and its role in subglacial sediment entrainment is discussed in section 9.3.2.

2.3.2 CONTROLS ON ICE TEMPERATURE

The temperature distribution within a glacier depends on energy exchanges at the upper and lower boundaries, internal sources of heat within the glacier and its flow history. The primary controls on ice temperature are:

(1) energy exchange between the glacier surface and the atmosphere;
(2) the geothermal heat flux at the bed;
(3) frictional heat generated englacially by ice creep, and at the bed by sliding and sediment deformation.

Energy exchange with the atmosphere, by any of the energy fluxes described in section 2.2, can raise or reduce the temperature of near-surface snow or ice. In addition, refreezing of meltwater within cold snow releases latent heat, providing an important source of energy (section 2.2.1). Because freezing of one gram of water releases $334\,J$ of energy (fig. 2.1), this is a particularly important process for raising the temperature of snow from the cold values typical of winter up to or close to the pressure-melting point. Refreezing of meltwater can create extensive areas

of 'warm' ice, even in areas where the mean annual air temperature is below freezing. In contrast, where melting is unimportant (such as the interior of Antarctica) the temperature of ice $\sim\!10\,m$ below the surface is close to the mean annual air temperature (e.g. Liston et al., 1999).

Thermal energy is conducted from warmer to colder areas within a glacier, following the steepest temperature gradient. The efficiency of heat conduction through a material is measured by the *thermal conductivity*, measured in Watts per metre per degree of temperature difference ($W\,m^{-1}\,K^{-1}$) and denoted by the letter k. For ice, $k = 2.22\,W\,m^{-1}\,K^{-1}$ at $0\,°C$, rising to $2.76\,W\,m^{-1}\,K^{-1}$ at $-50\,°C$. The thermal conductivity of snow is much smaller because trapped air is such a poor conductor of heat, so k is strongly dependent on snow density. Snow with a density of $600\,kg\,m^{-3}$ has a k value of $\sim\!0.55\,W\,m^{-1}\,K^{-1}$, whereas for a density of $100\,kg\,m^{-3}$, k may be as low as $0.2\,W\,m^{-1}\,K^{-1}$ (Sturm et al., 1997). This means that temperature fluctuations at the surface can be transmitted by conduction through several metres of ice over the course of a season, but to a lesser depth through snow, especially fresh, low-density snow. The penetration depth of a *cold wave* in winter thus depends on the amount of snow cover, and will be least where snow accumulation is high. Conduction will also transmit *warm waves*, but this process may be less efficient than conductive cooling because the ice surface cannot rise above $0\,°C$, limiting temperature gradients compared to those possible in winter.

Climatic change can alter energy exchanges between the atmosphere and glacier surfaces, resulting in multi-annual shifts in glacier temperature. For example, Rabus and Echelmeyer (2002) found that mean annual ice temperatures at a depth of $10\,m$ in the ablation area of McCall Glacier, Alaska, rose by over $1\,°C$ between 1972 and 1995. Near-surface ice temperature profiles can also be influenced by glacier thinning, because ablation brings formerly deeper ice closer to the surface. Even when this factor is taken into account, however, it is clear that the observed temperature increase at McCall Glacier was the result of changes in climate and the associated energy fluxes. In deep ice cores, ice deposited during the last glacial cycle is colder than that formed during the present interglacial period (Dahl-Jenssen and Johnsen, 1986; Johnsen et al., 2001; Vimeux et al., 2002; Blunier et al., 2004).

Geothermal heat flux is the upwards transfer of heat resulting from the decay of radioactive isotopes within the Earth. Values of the heat flux are dependent on the tectonic setting; rates lower than $0.04\,W\,m^{-2}$ are typical for continental shield areas which have old, thick crust, and rates in excess of $0.09\,W\,m^{-2}$ may be recorded in recent volcanic terrains. The geothermal heat flux raises the temperature of basal ice wherever heat is produced faster than it can be conducted towards the surface. Once the ice has been raised to the pressure-melting point, excess heat is used to melt ice. The global average geothermal heat flux of $0.06\,\mathrm{W\,m^{-2}}$ is enough to melt a 6 mm thickness of ice at its pressure-melting point each year (Paterson, 1994).

Frictional heat can be generated by internal deformation within the glacier, sliding at the base, or the deformation of subglacial sediments, when gravitational potential energy is converted into thermal energy as ice moves downslope. For ice deforming by simple shear (section 4.2.2), the rate of temperature change can be calculated from:

$$\frac{dT}{dt} = \frac{\tau\dot{\varepsilon}}{\rho_i c_i} \tag{2.9}$$

where T is ice temperature, t is time, τ is the shear stress, $\dot{\varepsilon}$ is the shear strain rate, ρ_i is the ice density and c_i is the specific heat capacity of ice. Since the strain rate is dependent on the shear stress, heat generation by internal shear is greatest where τ is large, that is, at depth within the ice (Ashwanden and Blatter, 2005; section 4.3.1). Once the ice has been raised to the pressure-melting point, further internal shearing will result in melting, increasing the intergranular water content (Fowler, 1984). Hubbard et al. (2003) reported water contents of ~10 per cent in ice near the bed of Glacier de Tsanfleuron, compared with only ~1 per cent at higher levels. The melt rate $\dot{M}$ can be calculated from:

$$\dot{M} = \frac{\tau\dot{\varepsilon}}{\rho_i L_f} \tag{2.10}$$

Note the close similarity of this expression to equation 2.9: the only difference is that the energy dissipated by shearing is used to provide the latent heat of melting (L_f) instead of temperature change.

Heat generated during sliding and subglacial sediment deformation can be conducted up through the ice, although melting from this source will only occur at the glacier bed (Fowler, 1984). For basal motion, the melt rate is calculated from:

$$\dot{M} = \frac{\tau U_b}{\rho_i L_f} \tag{2.11}$$

where U_b is the rate of ice motion over the bed. Sliding rates of $20\,\mathrm{m\,yr^{-1}}$ can generate as much heat as the average geothermal heat flux. Rates of basal motion are not directly proportional to shear stress, however, and fast sliding can occur at low shear stresses if the basal traction is low (sections 4.4 and 4.5). If τ is small, rates of meltwater production will be low even if U_b is large.

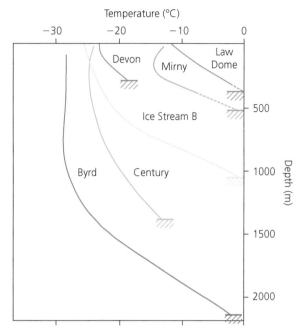

Figure 2.4 Selected temperature profiles from the Greenland and Antarctic Ice Sheets and the Devon Ice Cap (Modified from Paterson, 1994).

Temperature distributions are also influenced by glacier flow, which transports ice from high to low elevations or from continental interiors towards marginal zones. Ice at progressively greater depths in a glacier will have originated as snow further and further upglacier. In the Greenland and Antarctic Ice Sheets, the deepest, farthest-travelled ice is likely to have had the lowest initial temperatures, since it formed at high elevations in the continental interior. As a result, a decrease in temperature with depth is commonly observed in the upper parts of boreholes located far from ice divides (Paterson, 1994; van der Veen, 1999a; fig. 2.4). The deeper parts of boreholes in Antarctica and Greenland show a temperature increase with depth, because strain and geothermal heating have overprinted the inherited temperature characteristics.

The temperature of ice at any given location therefore depends on its initial temperature at time of formation, its subsequent thermal evolution by latent heat fluxes and strain heating as it is buried and advected through the system, and the dissipation of heat by conduction towards colder areas (van der Veen, 1999a).

2.3.3 THERMAL STRUCTURE OF GLACIERS AND ICE SHEETS

Glaciologists use a threefold classification for glaciers, based on whether the ice is at or below the pressure-melting point: (1) *temperate glaciers*, which are everywhere at the melting point except for a surface layer a few metres thick which is subject to seasonal temperature cycles; (2) *cold glaciers*, which are everywhere below the melting point and are frozen to their beds; and

Figure 2.5 ASTER image of cold glaciers in the Dry Valleys, East Antarctica (NASA: http://earthobservatory.nasa.gov).

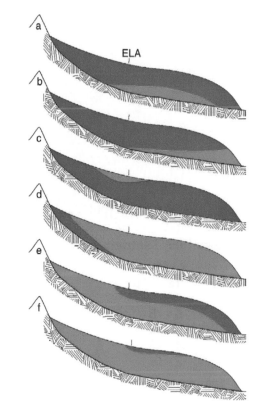

Figure 2.6 Polythermal glacier types, with warm ice shown in red and cold ice in blue (Modified from Pettersson, 2004).

(3) *polythermal glaciers*, which are composed of both cold and warm ice.

Temperate glaciers only occur where the winter cold wave is everywhere completely eliminated the following summer. This is most likely to be the case where both winter snowfall and summer melt rates are high, because: (a) thick snow cover will insulate the underlying ice from low winter air temperatures; (b) latent heat release associated with refreezing of meltwater will effectively raise snow temperatures in the accumulation zone; and (c) chilled near-surface ice in the ablation zone is effectively removed by ablation as the melt season progresses. Temperate glaciers thus tend to occur in temperate maritime areas with high precipitation and summer melting, such as the western coastal ranges of North America, western Norway, southern Iceland, New Zealand and Chilean Patagonia.

Cold glaciers occur where surface, englacial and subglacial heat sources are too small to raise the ice to the pressure-melting point, and thus are found exclusively in cold, arid environments where snow accumulation is small. Well-known examples include the Suess, Canada and Commonwealth Glaciers in the McMurdo Dry Valleys, Antarctica (Fountain et al., 2006; fig. 2.5). This region is a cold desert, with a mean annual temperature in the valley floors of $\sim -20\,°C$ and mean precipitation totals of only $\sim 60\,mm\,yr^{-1}$. Positive temperatures

do occur in summer, allowing limited surface melting, although sublimation and dry calving are important mass loss processes.

Polythermal glaciers are the most geographically widespread of the three glacier types, and exhibit a wide range of thermal structures depending on the balance of surface and subsurface warming processes (Blatter and Hutter, 1991; Pettersson, 2004). Figure 2.6 shows a range of polythermal glacier structures. Types (a) and (b) are found in cold climates, where surface melt rates are small. All of the ice formed is cold, but can be raised to the pressure-melting point at depth by strain heating. Polythermal glaciers of this type include Trapridge Glacier, Yukon (Clarke et al., 1984), and John Evans Glacier, Ellesmere Island (Copland and Sharp, 2001). In type (c), the snowpack is warmed in the lower part of the accumulation zone by refreezing of meltwater. Both cold and warm ice are produced in the near-surface zone, and both can occur in the ablation area. Type (d) forms in a similar way, but cold ice is restricted to cold, high-altitude parts of the glacier. Predominantly warm glaciers of this type occur in the western European Alps (e.g. Haeberli, 1976; Suter and Hoelzle, 2002; fig. 2.7). In type (e), warm ice is created in the accumulation zone in spring by refreezing of meltwater, but winter chilling creates a near-surface layer of cold ice in the ablation area. Polythermal glaciers of this type are widespread in the polar maritime climate

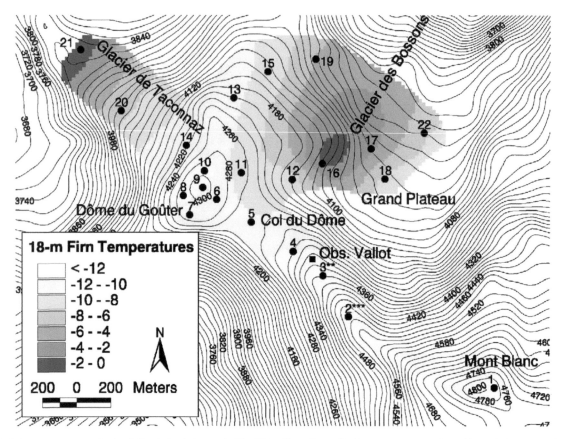

Figure 2.7 Ice temperatures at depths of 18 m in boreholes, north side of Mont Blanc, France. The lower, temperate reaches of these glaciers are not shown (Suter and Hoelze, 2002).

of Svalbard (Dowdeswell et al., 1984) and in the eastern side of the Scandinavian mountains. A very well studied example of this type is Storglaciären in Sweden (Pettersson et al., 2003). If ablation rates are high, the cold layer can be stripped off the lower part of the ablation zone in summer (type f). Polythermal glaciers can be conveniently subdivided into *predominantly cold* (types a–c) and *predominantly warm glaciers* (types d–f).

The Greenland and Antarctic Ice Sheets are also thermally complex. Throughout most of continental Antarctica, air temperatures are low at all times of year and surface melting is rare (Das and Alley, 2005). Consequently, only cold ice is formed at the ice sheet surface. In some coastal regions, melting and refreezing is sufficient to raise a shallow surface layer to the pressure-melting point in summer, but this is chilled the following winter (Liston et al., 1999). Below the West and East Antarctic ice sheets, basal ice can be raised to the pressure-melting point by geothermal heat flux (below thick ice) and strain heating (in areas of focused ice discharge such as ice streams). Using satellite magnetic data, Maule et al. (2005) inferred anomalously high geothermal heat fluxes below parts of West Antarctica, which likely play an important role in initiating and maintaining some ice streams. Spatially complex basal thermal regimes in the Siple Coast may record transient flow instabilities triggered by episodes of subglacial volcanism (Engelhardt, 2004).

The relatively warm temperatures near the ELA of the Greenland Ice Sheet mean that in some parts of the ice sheet, melting and refreezing processes are capable of raising surface snow to the pressure-melting point, while only cold ice is formed in higher and colder locations (Greuell and Konzelmann, 1994). As is the case in Antarctica, geothermal and strain heating raise the basal ice to the pressure-melting point below thick or rapidly flowing ice (e.g. Funk et al., 1994). Basal melt rates indicate particularly high geothermal heat fluxes in the initiation zone of the north-east Greenland ice stream (Fahnestock et al., 2001).

2.4 PROCESSES OF ACCUMULATION AND ABLATION

2.4.1 SNOW AND ICE ACCUMULATION

The most important primary source of ice accumulation on most glaciers is snowfall, the amount of which varies a great deal from place to place and throughout the year. Whether precipitation falls as snow or rain clearly depends on the near-surface air temperature, but there is

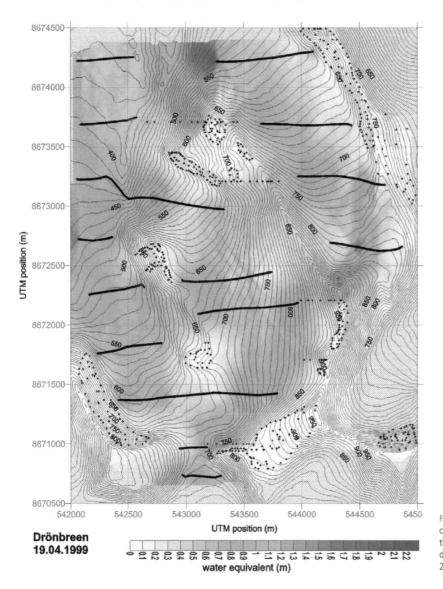

Figure 2.8 Measured snow accumulation patterns on Drønbreen, Svalbard. The heavy black lines show the location of radar profiles from which snow depths were interpolated (Jaedicke and Sandvik, 2002).

not a straightforward transition at 0 °C. For example, on a small Himalayan glacier, the probability that precipitation falls as snow was found to vary from 1 to 0 across an air temperature range of over 4 °C:

$$p_s = p(0.8 - 0.23\ T_a) \quad (-0.8° \leqslant T_a \leqslant 3.4°) \tag{2.12}$$

where p is the precipitation, p_s is snowfall and T_a is the air temperature (Kadota et al., 1997). This is because the temperature of the air is not the same as the temperature of the precipitation, which falls rapidly from higher, cooler elevations.

The highest accumulation rates occur in mountainous maritime regions in the mid latitudes, such the Coast Range of Alaska, southern Iceland, western Patagonia, South Georgia and the west coast of New Zealand's South Island, where frequent cyclonic weather systems and orographic uplift combine to produce very heavy snowfall. Accumulation rates as high as 14 m yr^{-1} (water equivalent)

have been reported for parts of the Patagonian Ice Fields (Warren and Sugden, 1993; Schwikowski et al., 2006). Conversely, snowfall is lowest far from any oceanic moisture source, such as the interior of Antarctica, which is one of the most arid regions on Earth. On a more local scale, accumulation rates are strongly influenced by redistribution processes such as wind-blowing or avalanching. Snow is entrained by high winds and redeposited where windspeed is low, profoundly influencing patterns of accumulation (Jaedicke and Sandvik, 2002; fig. 2.8). In steep, mountainous areas, avalanches can transfer large quantities of snow onto glacier surfaces. Indeed, on many high mountain glaciers, almost all accumulation is from avalanches, and almost none from direct snowfall (Benn and Lehmkuhl, 2000; fig. 2.9).

Ice crystals can also form on glacier surfaces by the freezing of supercooled water droplets carried by the wind. The most important type of ice formed by this process is *rime ice*, which can assume strange plant-like shapes built up by the eddying moisture-laden air. Rime ice

Figure 2.9 The west face of Nuptse, Khumbu Himal, Nepal. The debris-covered glacier at the foot of the face is mainly fed by snow and ice avalanches, including dry calving from hanging glaciers. At ~2.5 km high, this wall was unsurprisingly described by Reinhold Messner as a 'death zone', and was first climbed by Tomaž Humar and Janez Jeglič in 1997 (D.I. Benn).

Figure 2.10 Snow accumulation layers on Zongo Glacier, Bolivia. Successive ablation surfaces are picked out by dark dust layers. Note the unconformity near the top of the left-hand ice cliff, where ablation has removed some layers. Refreezing of surface meltwater is indicated by the icicles on the right (D.I. Benn).

accumulation is most rapid in cool, humid conditions on surfaces which are most exposed to the wind (Sugden and John, 1976). Some direct accumulation may occur by the freezing of rainwater or groundwater that comes into contact with a glacier, although this tends to be small relative to other components.

2.4.2 TRANSFORMATION OF SNOW TO ICE

If the yearly total of snow or ice accumulating on a glacier surface exceeds local losses by ablation, net accumulation occurs. Year after year, successive accumulation layers are built up, and the deeper layers eventually turn to glacier ice (fig. 2.10). The transformation of snow to ice occurs as the volume of air-filled pores is reduced and the bulk density increases. Freshly fallen snow has a density of 50–200 kilograms per cubic metre (kg m^{-3}), compared with 830–910 kg m^{-3} for glacier ice, and 1000 kg m^{-3} for pure liquid water at 0 °C. Pure ice has a density of 917 kg m^{-3}. Snow that has survived one melt season and has begun this transformation is known as *firn* and has a density of 400–830 kg m^{-3} (Paterson, 1994). The transition between firn and ice occurs when interconnected air passages become sealed off, isolating air in separate bubbles. Additional increases in density beyond this point are achieved by compression of the bubbles, placing the enclosed air under pressure.

The metamorphic processes by which snow is transformed into ice, and the time taken for the transformation to occur, depend on climate. Where melting rarely or never occurs, the principal mechanisms leading to an increase in density are (a) restructuring by wind; (b) the movement of crystals relative to one another; (c) changes in crystal size and shape; and (d) internal deformation of crystals. Wind has the effect of breaking up snowflakes into smaller ice crystals, and redepositing them in drifts which have much higher density than snow deposited in

still air. Once deposited, snow increases in density due to the pressure exerted by overlying snow, which causes crystals to move relative to one another and adopt stronger, more stable packings. As a result, compaction increases with increasing depth of burial. This process is encouraged by progressive changes in the size and shape of crystals in the snowpack, because denser packings are possible for compact particles than for irregular shapes. Fresh snow crystals have elaborate, complex shapes with many branches and re-entrants, but the crystals gradually assume more equant forms because molecules migrate from one part of the crystal to another, so as to reduce energy gradients (Paterson, 1994; Sturm and Benson, 1997). Changes in shape are most important in the early stages of transformation, when overburden pressures are low and void spaces are common. Further volume change can result from the internal deformation, or *creep*, of ice crystals when the pressure of the overlying snow is sufficiently high (section 4.3.1). Models of the ice densification process are important for the estimation of mass balance from ice sheet elevation changes (section 2.5.2), and have been developed by Arthern and Wingham (1998) and Zwally and Li (2002).

The transformation of snow to ice is greatly accelerated by melting and refreezing processes. Where melting occurs at the surface, meltwater percolates downwards and fills pore spaces within the snowpack, and the displaced air escapes upwards. If the surrounding snow is below 0 °C, the meltwater will refreeze, producing compact ice bodies (Wakahama et al., 1976; Bøggild et al., 2005). Clearly, this process will produce high-density ice much more rapidly than the dry processes described above. The contrasts in time taken for snow to transform into glacier ice in the presence and absence of melting are well illustrated by the variations in density with depth for the Upper Seward Glacier, Yukon, Canada, and part of the Greenland Ice Sheet (fig. 2.11). The Seward

Glacier profile shows transformation from firn to ice ($830 \, kg \, m^{-3}$) at only 13 m below the surface, or within 3–5 years of burial. In contrast, in the Greenland profile, from a cold continental setting, this change does not occur until a depth of 66 m, equivalent to more than 100 years since burial (Paterson, 1994).

The crystal structure of glacier ice is also modified by *dynamic recrystallization* processes during ice flow. For example, Tison and Hubbard (2000) found systematic variations in ice crystallography in an Alpine glacier, reflecting near-surface metamorphic processes and dynamic recrystallization. Four crystallographic units were identified.

(1) Within 20 m of the surface in the accumulation zone, homogeneous, fine-grained ice is formed by ice recrystallization in the absence of significant stresses.

(2) At greater depths, coarser crystals begin to develop due to dynamic recrystallization at low strain rates.

(3) At depths greater than 33 m throughout the glacier, there is an abrupt increase in crystal size, marking the onset of widespread dynamic recrystallization.

(4) Within 10 m of the bed, the ice consists of large, interlocking grains reflecting intense, continuous deformation in the basal zone.

Near the glacier surface, varying processes of snow metamorphism produce a range of *snow and ice facies* (or *zones*), which migrate sequentially upglacier in the course of the ablation season (Müller 1962; Nolin and Payne, 2007; fig. 2.12). The *dry snow facies* forms where air temperature is always below 0 °C and melting does not occur. This facies is found only in very cold regions, such as the interior of the Greenland and Antarctic Ice Sheets and at very high altitudes in Alaska, Yukon and central Asia. The *percolation facies* develops where there is some surface melting, and meltwater percolates down through the snow where it refreezes in horizontal *ice lenses* or vertical *ice glands* (fig. 2.13). With decreasing elevation, the depth of meltwater percolation increases until it penetrates the full thickness of the snowpack, creating the *wet snow facies*. Widespread refreezing at the base of the snowpack can form a continuous mass of *superimposed ice*, the outcrop of which at the surface defines the *superimposed ice zone* (fig. 2.14). Latent heat is released during refreezing (section 2.2.1), raising the temperature of the surrounding snow. When the entire snowpack has been raised to 0 °C, refreezing will cease and additional meltwater will either accumulate within the snow or run off. Where glacier surface gradients are small, meltwater can accumulate in large areas of slush and surface ponds (Lüthje et al., 2006; Sneed and Hamilton, 2007). Where the previous winter's accumulation is entirely removed, old glacier ice is exposed in the *bare ice facies*. The most complete development of ice facies is found on polythermal glaciers, which can have bare ice, superimposed ice, wet snow, percolation and, in some regions, dry snow facies.

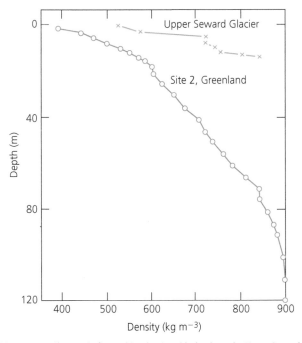

Figure 2.11 Changes in firn and ice density with depth on the Upper Seward Glacier, St Elias Mountains, Canada, where melting and refreezing occurs within the snowpack, and a site on the Greenland Ice Sheet where there is little or no melting (Paterson, 1994).

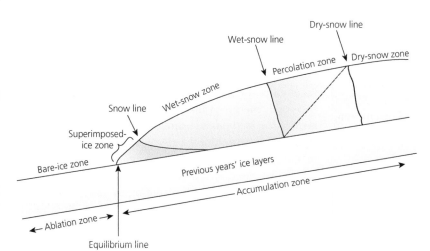

Figure 2.12 Near-surface ice and snow facies (Nolin and Payne, 2007).

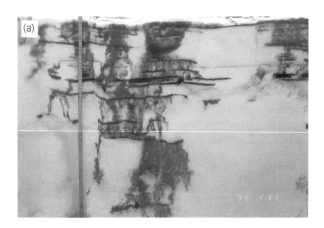

Figure 2.13 The geometry of ice lenses and glands was highlighted by introducing blue dye at the snow surface and then excavating a pit. (a) The horizontal bands were more permeable layers in the snow, and dye percolating along these layers refroze to form ice lenses. (b) This excavated ice gland formed in a zone of concentrated vertical dye flow (C.E. Bøggild).

Superimposed ice that survives the ablation season is known as *internal accumulation* (Reijmer and Hock, 2008), and in polar maritime environments can account for a significant proportion of the total accumulation. Koerner (1970) found that about 90 per cent of the ice accumulating on the Devon Ice Cap, Canada, formed by the transformation of snow to superimposed ice. In an elegant study of superimposed ice formation, Wadham and Nuttall (2002) concluded that it accounts for 16–25 per cent of the total net accumulation on Midre Lowénbreen, Svalbard.

2.4.3 MELTING OF SNOW AND ICE

Ablation refers to all of the processes by which snow and ice are lost from a glacier, and includes melting, evaporation, sublimation, scouring by wind, calving of icebergs

Figure 2.14 Ice facies on Fuglebreen, Svalbard, showing snow (left), superimposed ice (centre), and foliated glacier ice (right) (Doug Benn).

into water and avalanching of ice blocks from terrestrial ice cliffs (dry calving). Melting is the dominant process of ablation on most land-terminating glaciers. Sublimation is only important in cold, dry environments, and is discussed in section 2.4.4. Calving results from a complex set of dynamic processes and is considered in detail in section 5.5.

Ice melt rates can be calculated if the energy fluxes from all sources are known (Oerlemans, 2001). Rearranging equation 2.1, the energy flux available for melting is:

$$SW + LW + QH + QE + QR - QT = M \qquad (2.13)$$

where all energy balance components are in W m^{-2} (i.e. J s^{-1} m^{-2}). The melt rate $\dot{M}$ is this energy flux divided by the latent heat of melting ($L_f = 3.34 \times 10^5$ J kg^{-1}):

$$\dot{M} = M/L_f \, (M > 0) \qquad (2.14)$$

It is useful to express melt rates in terms of surface lowering: $dz/dt = M/L_f \, \rho$ where ρ is the snow or ice density. This formula yields surface lowering rates dz/dt in metres per second, so to convert this to more convenient units such as millimetres per day, appropriate scaling factors must be used. Penetration of radiation below the surface can cause significant *internal melt* within snow or below *weathering crusts* in ice (Munro, 1990; Cutler and Munro, 1996). Weathering crusts are produced by differential melting along grain boundaries, and consist of porous ice with loosely interlocking crystals.

Where detailed meteorological data are not available (as is often the case) it will not be possible to calculate the total energy available for melting. In such cases, *melt-index methods* are often used, which are based on statistical relationships between melt rates and some readily available measure of energy inputs (Hock, 2003). One widely used melt index is the sum of *positive degree days*, defined as the sum of the mean daily temperature $\bar{T}_d$ (°C) for all days in which $\bar{T}_d > 0$ °C (e.g. Braithwaite and Olesen, 1989; Braithwaite, 1995; Arendt and Sharp,

1999; Braithwaite and Zhang, 2000). The constant of proportionality in such relationships, or *degree-day factor*, varies widely, and measured values range from 2.7 to 11.6 mm of melt (water equivalent) per positive degree day (mm day^{-1}°C^{-1}) for snow, and from 5.4 to 20.0 mm day^{-1}°C^{-1} for ice (Hock, 2003; Zhang et al., 2006). The lower degree-day factors for snow surfaces reflect their higher albedo, which means that less melting will occur at any given temperature. Other variations in degree-day factors result from differences in the relative importance of individual energy components. High sensible heat fluxes are associated with low degree-day factors, and vice versa (Ambach, 1988). Because of their dependence on albedo and the energy balance components, degree-day factors for specific locations will vary with time, so melt-index methods need to be tuned to evolving local conditions.

Melt-index models can be criticized on the grounds that major components of the energy flux at a glacier surface, such as net shortwave radiation, are independent of, or correlated only indirectly with, air temperature. In an examination of the physical basis of melt-index models, Ohmura (2001) has argued that their success arises from the importance of incoming longwave radiation and sensible heat flux (which mainly depend on air temperature) in the overall energy balance. It is clear, however, that the applicability of melt-index models depends on local meteorological conditions, and they may not perform well in areas with high shortwave radiation fluxes (such as high-altitude environments), where melt totals are only weakly correlated with temperature (e.g. Kayastha et al., 2000a).

To account for other contributions to melt, multiple regression techniques have been used to predict melt rates from two or more meteorological variables (Hock, 2003). Such *combination methods* form a gradational series from simple degree-day approaches to full energy balance expressions. One widely used approach is to add a radiation term to temperature index models (e.g. Martinec, 1989; Brubaker et al., 1996; Aizen et al., 1997; Hock, 1999). For example, Konya et al. (2004) proposed a melt function of the form:

$$\dot{M} = aT + b(1 - \alpha)SW_{in} + c \quad (T > T_{crit})$$

$$\dot{M} = b(1 - \alpha)SW_{in} + c \quad (T \leq T_{crit}) \qquad (2.15)$$

where $(1 - \alpha)SW_{in}$ is net shortwave radiation, T_{crit} is a critical air temperature at which temperature-driven melting begins, and a, b and c are empirically determined coefficients. T_{crit} can be set to 0 °C, or some higher value. An advantage of this approach over degree-day models is that melting can occur in response to solar radiation receipts even if air temperatures are low. Combination methods are capable of predicting hourly melt with a high degree of accuracy, provided the model coefficients are tuned to local conditions (Hock, 1999, 2003; fig. 2.15).

2.4.4 SUBLIMATION AND EVAPORATION

Sublimation and evaporation will only occur if the air humidity at a glacier surface is greater than that in the free atmosphere above. If this is the case, water vapour will be driven along a *humidity gradient* away from the surface, a process that is much more efficient if there is significant turbulent mixing (section 2.2.5). The processes of sublimation and evaporation require much more energy than melting (fig. 2.1), and this can be supplied from radiation receipts, sensible heat transfer from the air or chilling of the ice. Thus, mass losses by sublimation and evaporation are determined by atmospheric humidity and windspeed, and entail a net subtraction of energy from the total budget (QE in equation 2.1).

It follows that sublimation is a significant process of ablation only in cold, dry environments, such as the polar ice sheets and high-altitude glaciers. For example, Lewis et al. (1998) found that sublimation accounted for 40–80 per cent of the observed summer ablation from a glacier surface in the Dry Valleys of Antarctica. Sublimation can also be a very important process during snow-blow, when moisture is transferred to the atmosphere from ice particles in suspension (Pomeroy and Jones, 1996). Bintanja (1998) found that sublimation of suspended snow accounts for up to 17 cm of ablation per year (water equivalent) in windy coastal regions of Antarctica.

Sublimation on high-altitude, low-latitude glaciers can produce distinctive pinnacles of ice or snow called *penitentes* (fig. 2.16). These range from a few centimetres to around 5 m in height and are consistently tilted towards the midday sun; their name derives from their perceived resemblance to the robed, hooded forms of religious penitentes. The first scientific observations of penitentes were made by Charles Darwin (1809–82) during the voyage of the *Beagle*, and key processes in their formation were identified in the 1950s by the great French glaciologist Louis Lliboutry. The formation of penitentes requires a combination of low relative humidity, high shortwave radiation receipts and a high solar angle, conditions common in parts of the Himalaya and Andes (Naruse and Leiva, 1997; Corripio and Purves, 2005). When the sun is high in the sky, net shortwave receipts are highest in hollows in snow and ice surfaces due to multiple reflections off the walls. Differential sublimation losses therefore amplify topographic irregularities in a powerful positive feedback. When hollows become very deep, reduction of wind turbulence allows pockets of saturated air to form. This encourages melting which further deepens the hollows, while the peaks continue to ablate more slowly by sublimation alone.

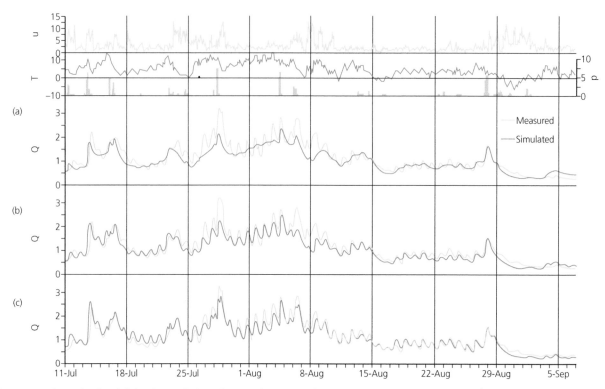

(a)

(b)

(c)

Figure 2.15 Observed and modelled meltwater discharge from Storglaciären, Sweden. Top panel: hourly wind speed (u: m s^{-1}), temperature (T: °C) and precipitation (P: mm hr^{-1}). Melt calculations are based on (a) degree-day model; (b) temperature index model, including potential direct solar radiation; (c) energy balance model (Hock, 2003).

Figure 2.16 Snow penitentes at an elevation of 4700 m, Paso de Agua Negra, Chile, December 2007 (Lindsey Nicholson).

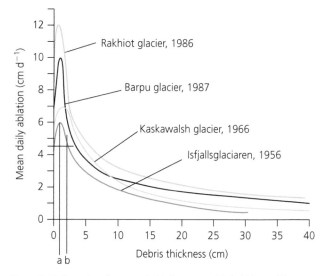

Figure 2.17 Examples of measured ablation rate vs debris thickness (Østrem curves). For the Isfjallsglaciären curve, (a) shows the debris thickness under which maximum melt occurs, and (b) indicates the thickness at which melt equals that for bare ice (Nicholson and Benn, 2006).

2.4.5 THE INFLUENCE OF DEBRIS COVER

The presence of superficial debris on snow or glacier ice profoundly influences surface energy balance and ablation rates. Numerous field experiments have demonstrated that there is a consistent, non-linear relationship between debris thickness and melt rates (e.g. Østrem, 1959; Mattson et al., 1993; Kayastha et al., 2000b; fig. 2.17). Under very thin debris, ablation rates rise with increasing debris thickness to a peak value when debris thickness is ~2 cm. Under thicker debris, ablation rates

decline exponentially. This pattern reflects two opposing effects of debris cover. First, rock surfaces generally have a lower albedo than snow or bare ice, so they will absorb more shortwave radiation and increase the energy available for melting. Second, debris acts as a thermal barrier between ice and the atmosphere, reducing the energy flux to the ice surface. The albedo effect dominates for thin or

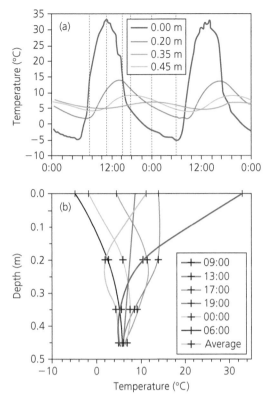

Figure 2.18 (a) Diurnal temperature oscillations at different depths in supraglacial debris on the Ngozumpa Glacier, Nepal. Temperature waves from the surface are damped with depth, with an increasing time lag. (b) Instantaneous vertical temperature profiles are non-linear, although the daily mean (heavy purple line) is almost linear. The vertical lines in (a) mark the times of the profiles shown in (b) (Nicholson and Benn, 2006).

patchy debris cover, whereas the insulation effect is most important where debris is thick. The increase of melt rates with debris thickness for very thin debris layers most likely reflects the transition from discontinuous debris cover to a continuous layer, and an associated shift in mean surface albedo.

A quantitative understanding of these processes has developed only recently. Nakawo and Young (1981, 1982) modelled a debris layer as a simple steady-state system, with a linear temperature gradient between the top and bottom of the layer. The conductive heat flux Q_c through the layer is given by:

$$Q_c = k\,(T_s - T_i)\,/\,h_d \qquad (2.16)$$

where T_s and T_i are the temperatures at the top and base of the debris layer, respectively, h_d is the debris layer thickness, and k is the thermal conductivity. However, there are practical problems with using this simple model to calculate melt rates in real situations. Most importantly, in nature, instantaneous temperature profiles are non-linear and constantly changing in response to fluctuating energy receipts at the surface (fig. 2.18). During the day in summer, when there is a net gain of energy at the top of the debris layer, a warming wave migrates

downwards. Conversely, at night, when there is an energy deficit at the surface, the near-surface temperature gradient reverses and stored heat is conducted upwards. Deeper in the debris layer, stored heat may continue to be conducted downwards, and ice melting may continue even if the temperature at the surface drops below 0 °C.

Nicholson and Benn (2006) have shown that although instantaneous temperature gradients are non-linear, daily mean temperature gradients are usually linear, except during times of rapidly changing weather conditions. This indicates that, if daily mean surface temperatures are used as input, the Nakawo and Young model gives reasonable values of daily average heat flux through a debris layer, from which melt rates can be calculated using equation 2.14. Obtaining representative field measurements of debris surface temperature is clearly impractical, however, especially in remote areas. Thermal data from satellites have been used, although these yield instantaneous temperatures rather than daily means (Nakawo and Rana, 1999). The most versatile approach is to solve the energy balance equation for mean daily surface temperature using meteorological data as inputs, and to use the result to calculate melt rates (Nicholson and Benn, 2006). Successful application of this approach requires estimates of the thermal conductivity of the debris. Methods of doing this are discussed by Conway and Rasmussen (2000) and Nicholson and Benn (2006).

In permafrost environments, near-surface sediments thaw each summer, whereas deeper material remains permanently frozen. The zone of seasonal freezing and thawing, which is typically a few tens of centimetres to a metre thick, is known as the *active layer* (Ballantyne and Harris, 1994). If supraglacial debris is thicker than the active layer thickness, therefore, melting of the underlying ice will cease altogether, until either the debris cover is removed or the climate warms.

2.5 MASS BALANCE

2.5.1 DEFINITIONS

Mass balance refers to change in the mass of all or part of a glacier over some specified time period. Most mass balance studies consider changes over one year, although two definitions of 'year' are in use. First, the *balance year* is defined as the interval between two successive annual minima in the mass of the glacier, and provides the best overall measure of annual changes in mass storage (Paterson, 1994). In the mid and high latitudes, annual minima occur in the autumn after the end of the ablation season but before the first significant snowfalls of winter. This is not the case in the tropics and subtropics, but where there is a pronounced dry season, the end of this period provides a convenient end point for the balance year. In glacier

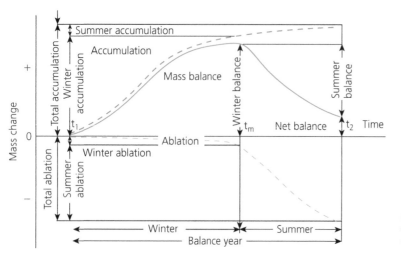

Figure 2.19 Annual cycles of accumulation and ablation on mid- and high-latitude glaciers, showing definitions of mass balance terms (Paterson, 1994).

accumulation areas, annual minima correspond to boundaries between annual accumulation layers in the snow stratigraphy, so defining the balance year in terms of annual minima is known as the *stratigraphic method*. The second definition of 'year' is the *fixed date* or *calendar year*, defined in the mid and high latitudes as the period between 1 October and 30 September. In practice, it is not always possible to visit field sites on pre-defined dates, so many programmes adopt a compromise between stratigraphic and fixed-date methods, known as the *combined method*.

The net annual mass balance b_n is the sum of accumulation and ablation over the calendar or balance year. For a particular location on the glacier, this is known as the *specific mass balance*. In the mid and high latitudes, it is useful to define the *winter balance* for the period that accumulation exceeds ablation, and the *summer balance* for the period of net mass loss (fig. 2.19). The net, winter and summer mass balance for the entire glacier can be calculated from the specific mass balances measured at a sample of points on the surface. The *mean specific mass balance* (also known as the *glacier-wide* or *area-averaged mass balance*) is the total change in mass divided by the area of the glacier.

2.5.2 MEASUREMENT OF MASS BALANCE

Direct measurements

Direct or glaciological methods for determining surface mass balance are based on field measurement of ablation and accumulation totals (Østrem and Brugman, 1991; Kaser et al., 2003a; Hubbard and Glasser, 2005). *Accumulation* can be measured in pits excavated into the snowpack or from cores. Annual increments of snow are identified from changes in density or crystal size, or by a dirt layer representing a summer ablation or non-accumulation surface (fig. 2.12). The net annual accumulation b_a, expressed in terms of water equivalent, is

Figure 2.20 Ablation stake on Austre Broggerbreen, Svalbard. This stake is part of the network monitored on the glacier by Norsk Polarinstittutt (Doug Benn).

determined from the thickness h and average density ρ_i of the annual layer:

$$b_a = h(\rho_i/\rho_w) \qquad (2.17)$$

where ρ_w is the density of water. Winter accumulation totals can be obtained using the same method, except that measurements are made at the end of the accumulation season instead of the end of the balance year. Direct measurements of *ablation* can be made using stakes drilled into the ice, which are used to measure the drop in the ice surface over the measurement period (fig. 2.20). Continuous measurements of ablation can be made using an electronic rangefinder mounted on a cross-arm between two poles, or a pressure transducer placed in a drill hole and linked to a surface fluid reservoir (Bøggild et al., 2004).

Direct measurement methods obtain the mass balance at points on the glacier surface, so values must be interpolated before the total and specific net balance of the whole glacier can be assessed (fig. 2.21). This procedure may introduce large errors if sampling points are too few

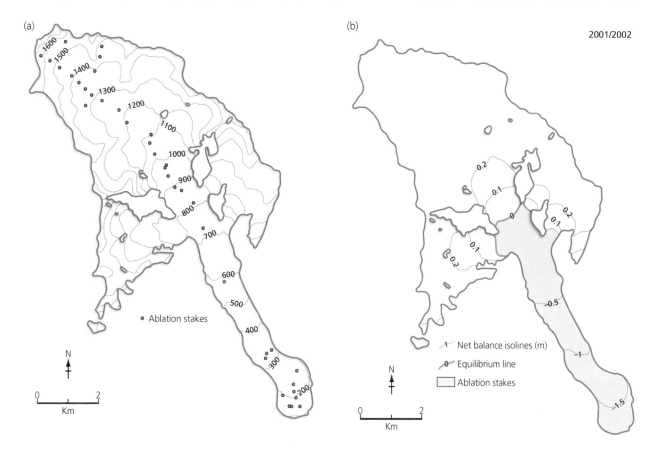

2001/2002

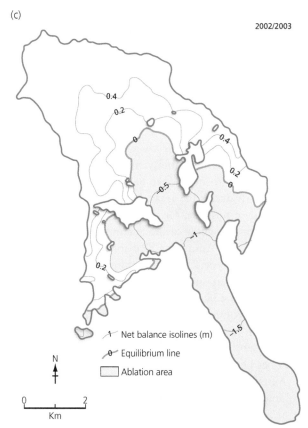

2002/2003

Figure 2.21 Mass balance measurements on White Glacier, Canada. (a) Stake network; (b, c) interpolated pattern of mass balance for two measurement years (World Glacier Monitoring Service, 2005a).

or widely spaced (Cogley, 1999; Fountain and Vecchia, 1999).

Direct measurements of mass balance have been made for around 280 of the world's glaciers, but only 86 records extend over 10 years or more (Braithwaite, 2002; Dyurgerov, 2002). Two of the most complete mass balance records available are those for Storglaciären in Sweden and Storbreen in Norway, which were started in 1946 and 1949, respectively (Østrem and Haakensen, 1993). In North America, the longest mass balance studies have been undertaken on the South Cascade and Blue Glaciers in Washington State since 1956, and on Peyto, Place and Sentinel Glaciers in western Canada since 1965 (Letreguilly and Reynaud, 1989). In the European Arctic, mass balance measurements have been collected for Bröggerbreen, Midre Lövenbreen and Voringbreen on Svalbard since 1966/67 (Lefauconnier and Hagen, 1990). The World Glacier Monitoring Service (WGMS) collects and publishes standardized information on glacier mass balance (Haeberli et al., 1998, 2005; World Glacier Monitoring Service, 2005a).

Hydrological methods

The annual mass balance of a glacier represents a change in water storage in the catchment (Collins, 1984; Singh and Singh, 2001; Jansson et al., 2003). This means that if other components of the annual water balance in a catchment can be quantified, the net balance of a glacier

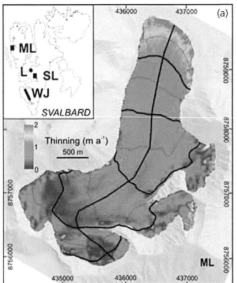
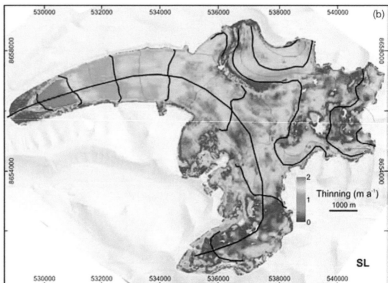

Figure 2.22   Changes in glacier elevation derived from digital elevation models based on aerial photographs and airborne laser altimetry data. (a) Midre Lowénbreen, 2003–2005; (b) Slakbreen, 1990–2003. Both of these Svalbard glaciers have lost mass over their entire surfaces; thinning rates are shown in metres per year (Kohler et al., 2007).

can be determined:

$$p_a - R_a - E_a + \Delta S_g + \Delta S_o = 0 \qquad (2.18)$$

where p_a, R_a and E_a are annual totals of precipitation, runoff and evaporation, respectively, ΔS_g is the change in glacial storage and ΔS_o is the change in other water stores (e.g. non-glacial snow cover, supraglacial, englacial and subglacial water, and subsurface aquifers). To calculate net glacier mass balance by this method, accurate estimates of all other components must be obtained. Runoff can be determined comparatively easily from a gauging station on the main meltstream, but other components present considerable challenges. Evaporation is particularly difficult to measure, and in mountain areas precipitation commonly exhibits very complex spatial variations and representative measurements are difficult to obtain, especially for high altitudes (Barry, 1992). Nevertheless, hydrological methods have been used to estimate glacier mass balance in many parts of the world (e.g. Collins, 1984; Ribstein et al., 1995; Kaser et al., 2003b), although mostly on glaciers smaller than 10 km² in area due to sampling problems in larger catchments. Because of the difficulty of distinguishing glacier mass balance from other forms of storage change, hydrological methods are usually employed in conjunction with ablation measurement programmes and/or modelling studies (e.g. Arnold et al., 1998; Kuhn, 2003). Such integrated studies can yield important insights into the interactions between different components of glacierized basins, and provide important inputs for water resource prediction and management programmes. Glacier hydrology is discussed in detail in chapter 3.

Geodetic methods

Increasingly, glacier mass balance is determined by using aerial photographs or satellite data to calculate changes in glacier volume, which are then converted into changes in mass using estimates of the average densities of snow, firn and ice on different parts of the glacier (Krimmel, 1999; Bamber and Kwok, 2004; Bamber and Rivera, 2007; fig. 2.22). Such *geodetic* methods are in many ways superior to direct glaciological measurements, because of better areal coverage, especially in complex terrain (e.g. Krimmel, 1999; Østrem and Haakensen, 1999; Kohler et al., 2007). Equally important, geodetic methods allow the total mass balance of a glacier system to be determined, including dynamic changes and calving losses. In recent years, the availability of stereo satellite data has allowed geodetic methods to be applied over large, previously poorly accessible areas. In particular, such methods have made it possible, for the first time, to estimate mass balance for the whole of the continent-scale ice sheets of Greenland and Antarctica (Rignot and Thomas, 2002; Bamber and Payne, 2004). An important limitation of this approach is that snow and ice density are often poorly known, and are likely to change through time. As a result, changes in volume may have uncertain relationships with changes in mass. For example, surface melting followed by refreezing within snow will result in surface lowering, but does not constitute net ablation.

Aerial photographs have been used to determine ice volume changes in many parts of the world (Haeberli et al., 1998). Because of the expense of commissioning flights, however, aerial photographs are not commonly available at suitable annual intervals, and consequently mean annual mass balance is generally calculated from volume changes over longer time intervals. Some workers have extended the temporal range of geodetic methods by comparing aerial photographs or satellite images with older topographic maps or ground surveys (e.g. Paterson and Reeh, 2001). Although useful, such exercises are inevitably associated with large and often unquantifiable errors because of inaccuracies in the older data.

Satellite-borne radar altimeter data (e.g. Seasat, Geosat and European Remote Sensing Satellites ERS-1 and ERS-2) have been used to measure ice elevation changes on the polar ice sheets since the late 1970s (e.g. Zwally et al., 1989; Wingham et al., 1998; Li et al., 2003). Greater precision can be achieved with laser altimeter systems, as has been demonstrated in aircraft-borne surveys, which are capable of resolving even small-scale topographic detail (e.g. Arendt et al., 2002; Krabill et al., 2002; Bamber et al., 2005; Kohler et al., 2007). Satellite-borne laser altimeters (e.g. the geosciences laser altimeter system (GLAS) aboard NASA's Ice, Cloud and Land Elevation Satellite (ICESat) can now deliver more accurate and representative data on ice sheet mass balance than ever before (Zwally et al., 2002b; Thomas et al., 2005).

Gravimetric methods

A very important new approach to assessing mass balance changes over large regions employs the Gravity Recovery and Climate Experiment (GRACE), to make direct measurements of changes in the Earth's mass distribution. Launched in 2002, the GRACE mission uses variations in the distance between two identical polar-orbiting spacecraft to detect small changes in the Earth's gravitational field (Chen et al., 2005).

Major advantages of this approach are that it determines changes in mass rather than volume, and that it integrates mass balance over very large areas. Care must be taken to separate the ice sheet mass balance signal from other mass distribution changes, especially other parts of the hydrological system and isostatic rebound, and some disagreement exists regarding the optimum analytical methods (Luthcke et al., 2006, 2008; Wahr et al., 2006). GRACE measurements are so sensitive that they can detect annual variations in glacier mass resulting from winter accumulation and summer ablation (Luthcke et al., 2006, 2008; Velicogna and Wahr, 2006; fig. 2.23).

2.5.3 ANNUAL MASS BALANCE CYCLES

The amount of snow and ice stored in glaciers undergoes systematic changes throughout the year, following cycles of gain and loss depending on the seasonal distribution of accumulation and ablation. Several types of cycle are possible, depending on the climatic regime, especially the timing of warm and cold seasons, maximum precipitation and variations in the proportion of precipitation falling as snow (Ageta and Higuchi, 1984; Kaser and Osmaston, 2002). Three basic types of mass balance cycle are discussed here:

(1) *winter accumulation type*, with a well-defined winter accumulation season and summer ablation season;
(2) *summer accumulation type*, with maxima in accumulation and ablation occurring simultaneously during the summer months;
(3) *year-round ablation type*, with one or two accumulation maxima coinciding with wet seasons.

Winter accumulation type

This is the most familiar type of glacier mass balance cycle, and is characteristic of mid- and high-latitude glaciers, such as those in the European Alps, Scandinavia, North America, Patagonia, the western Himalaya and New Zealand. In these areas, there are pronounced seasonal variations in temperature, resulting in distinct winter accumulation and summer ablation seasons (fig. 2.24a). Accumulation occurs mainly during the winter, when most precipitation falls as snow and ablation is at a minimum. As a result, the mass of the glacier as a whole increases during the winter, and is at a maximum in the spring, just prior to the onset of the ablation season. During the summer months (June to September in the northern hemisphere, November to March in the southern), melting predominates over accumulation, particularly on the lower part of the glacier. Much of the precipitation may fall as rain, although significant snowfalls may still occur at higher altitudes. The mass of the glacier is at a minimum at the end of the ablation season, just prior to the first snows of winter. It should be noted however, that on some low-altitude glacier tongues, such as those in southern Iceland, ablation may also occur in the winter during the passage of warm fronts.

Summer accumulation type

In this type of mass balance cycle, maxima in accumulation and ablation occur more or less simultaneously during the summer (fig. 2.24b). Glaciers of this type occur in high-altitude areas in the outer tropics, with a pronounced summer precipitation maximum and a cold, dry winter, such as the monsoon-dominated parts of the Himalayan chain in India and Nepal, and the high Andes of Peru and Bolivia (Ageta and Higuchi, 1984; Francou et al., 1995; Wagnon et al., 1999; Kaser and Osmaston, 2002). In these areas, there is little precipitation in winter, and summer precipitation falls as snow on the upper parts of the glaciers. Ablation is also at a maximum during the summer, when temperatures are highest. As a result, when annual ablation and accumulation totals are comparable, the total mass of the glacier undergoes little change throughout the year, unlike the large cyclic changes typical of winter accumulation type glaciers.

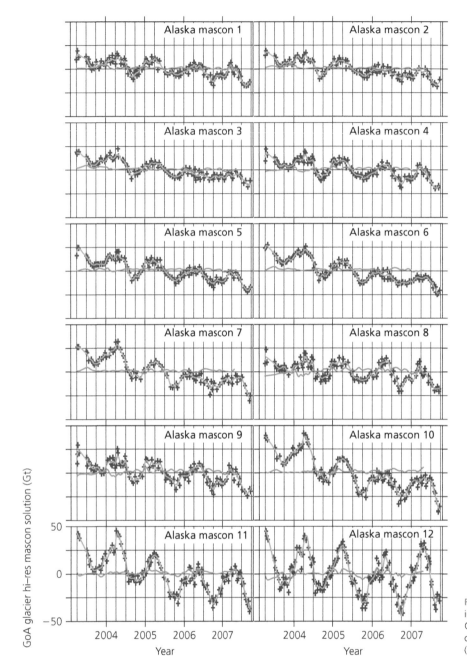

Figure 2.23 Temporal trends in glacier mass in 12 subregions of Alaska, derived from GRACE gravimetric data. Note the annual cycles superimposed on longer-term trends (Luthcke et al., 2008).

Summer accumulation type glaciers, however, are sensitively dependent on wet-season precipitation totals, and the annual net balance may vary widely from one year to the next (Francou et al., 1995).

Year-round ablation type

This type of cycle is characterized by ablation throughout the year on the lower parts of glaciers, and is typical of the inner tropics where seasonal variations in temperature are much less than diurnal variations. Cyclic variation in the amount of ablation can occur, however, due to seasonal changes in cloudiness, affecting radiation balance (Mölg et al., 2008). Accumulation can occur throughout the year on the upper parts of this type of glacier, although

distinct peaks may occur coincident with one or two wet seasons (fig. 2.24c). Double wet season peaks affect the high-altitude glaciers in the Ruwenzori Mountains and Mount Kenya, east-central Africa, where snowfall is highest during the northward and southward passage of the Inter-Tropical Convergence Zone in March to June and September to December, respectively (Hastenrath, 1984; Kaser and Osmaston, 2002). One precipitation peak may be larger than the other, and inter-annual variability can be high. Single accumulation peaks occur on the glaciers of Mount Jaya in New Guinea, Kilimanjaro in East Africa, and on Ecuadorian volcanic peaks (Hope et al., 1976; Hastenrath, 1984). Above the mean 0 °C isotherm, almost all precipitation falls as snow, while below it lies a

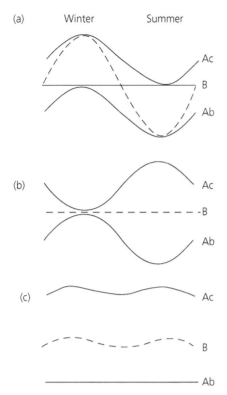

Figure 2.24 Annual mass balance cycles for glaciers of (a) winter accumulation type, (b) summer accumulation type, and (c) year-round ablation type. Ac: accumulation; Ab: ablation; B: balance (Modified from Ageta and Higuchi, 1984).

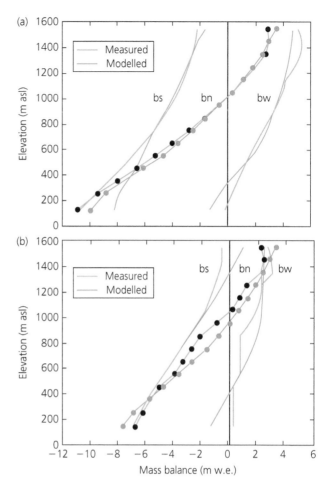

Figure 2.25 Measured and modelled mass balance gradients for Engabreen for (a) 1997 and (b) 1982. Note that the net mass balance is the sum of ablation and accumulation (Schuler et al., 2005).

zone where snow or rain may fall, depending on transient weather conditions (Kaser and Osmaston, 2002). Near the terminus, almost all precipitation falls as rain.

2.5.4 MASS BALANCE GRADIENTS

On many glaciers, the amounts of annual ablation and accumulation vary systematically with altitude, although this simple pattern is often complicated by local influences. The rates at which annual ablation and accumulation change with altitude are termed the *ablation gradient* and the *accumulation gradient*, respectively. Taken together, they define the *mass balance gradient* (fig. 2.25). Mass balance gradients for several North American glaciers are illustrated in Figure 2.26.

Ablation gradients

The total thermal energy in a mass of air is proportional to its density, so in the troposphere air temperature tends to decrease with altitude. The average *environmental lapse rate* is about 6.5 °C km^{-1}, varying between 4° and 9 °C km^{-1} depending on air-mass characteristics (Barry and Chorley, 2003). Decreases in temperature cause corresponding decreases in important components of the energy balance, particularly incoming longwave radiation and sensible heat flux (Ohmura, 2001). Although shortwave receipts may increase with altitude, on valley glaciers temperature effects are usually more important, and

ablation generally decreases approximately linearly with increasing elevation. Ablation gradients are steepest where the air temperature is frequently above 0 °C on the lower glacier, falling to lower values higher up. This is often the case on mid-latitude glaciers in the summer, but the very steepest ablation gradients are found in the tropics, where ablation occurs throughout the year (Kaser and Osmaston, 2002; fig. 2.27). Shallow ablation gradients are typical of cold regions where the main control on ablation is shortwave radiation, which is only weakly dependent on altitude (Kuhn, 1984).

Ablation gradients can be determined from direct observations or derived from energy balance models. Theoretical gradients compare very well with observations for many glaciers, increasing confidence that they can be used to explore the behaviour of vanished glaciers, or to predict the response of existing glaciers to future climate change (e.g. Oerlemans, 2001; Kaser and Osmaston, 2002; Schuler et al., 2005; fig 2.25).

Non-linear ablation gradients can result from altitudinal variations in cloudiness and humidity (which influence radiation receipts, latent and sensible heat transfer, and rates of evaporation and sublimation), proximity to

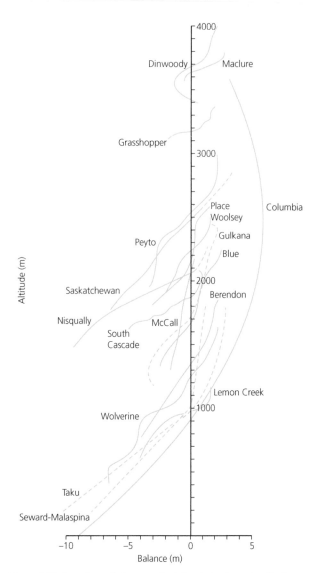

Figure 2.26 Annual mass balance gradients for glaciers in western North America (Mayo, 1984).

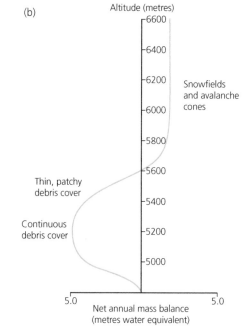

Figure 2.28 (a) The debris-covered ablation zone of Khumbu Glacier, Nepal; (b) Schematic mass balance gradient for the Khumbu Glacier (Benn and Lehmkuhl, 2000)

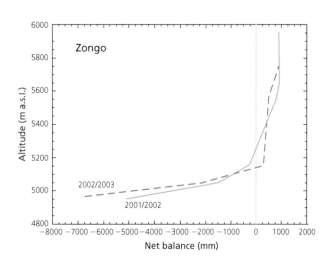

Figure 2.27 Mass balance gradients for Zongo Glacier, Bolivia, for two measurement years (World Glacier Monitoring Service, 2005a).

rockwalls (which influence amounts of available long-wave radiation and sensible heat transfer), the amount of shading and glacier aspect (Oerlemans, 2001). The commonest cause of non-linear gradients, however, is the presence of debris on the lower parts of glaciers. On debris-covered glacier tongues, increases in debris thickness towards the terminus offset the effects of increasing air temperature, causing a reversal of the ablation gradient. Indeed, ablation rates may be negligible near the terminus, where debris is thickest, and highest in the upper part of the ablation area, where debris cover is thin or patchy. Few direct measurements have been made of the specific annual balance of debris-covered glaciers, and published ablation gradients tend to be based on very few data points (fig. 2.28; Benn and Lehmkuhl, 2000). Recent advances in modelling ablation beneath supraglacial debris, however, allow theoretical gradients to be constructed (Nicholson, 2004).

Accumulation gradients

On many cirque and valley glaciers, the amount of net annual accumulation generally increases with altitude, rising from zero at the equilibrium line (fig. 2.25). This is because the passage of moisture-bearing winds over mountain barriers causes uplift, cooling the air and increasing precipitation, and because the overall proportion of precipitation that falls as snow increases at cooler, higher altitudes. On some mountain ranges, however, the amount of snowfall does not increase with height, and may even decrease (see Columbia Glacier, fig. 2.26). This tends to be the case in cold, high-latitude or high-altitude environments where summits rise above the main moisture-bearing air masses.

The shape of accumulation gradients can also be strongly influenced by local topography. In particular, the redistribution of snowfall by avalanching from steep slopes and wind scouring from exposed areas can result in accumulation patterns that differ markedly from original climatically controlled snowfall distributions. In high mountain environments such as the Himalaya, many glaciers gain most of their mass in huge avalanche cones located at the base of steep headwalls (Benn and Lehmkuhl, 2000; fig. 2.29).

On large ice caps and ice sheets, horizontal distance from moisture sources is a much more important control on accumulation totals than altitude. For example, on the Antarctic Ice Sheet many low-lying coastal areas have higher precipitation than the high continental interior (section 6.3.2). However, accumulation is not a simple function of distance from the coast, as moisture fluxes depend on atmospheric circulation patterns.

Balance ratios

As they are controlled by different climatic variables, the accumulation and ablation gradients on any one glacier generally have different values, with the ablation gradient somewhat steeper than the accumulation gradient. (Here, 'steeper' refers to the rate of change of mass balance with altitude. Confusion can arise because altitude is commonly plotted on the vertical axis of mass balance diagrams, so that 'steeper' mass balance gradients are represented by *less steep* curves.) The difference between accumulation and ablation gradients means that most mass balance curves show an inflection at the equilibrium line altitude (fig. 2.26). The ratio between the two gradients is known as the balance ratio, defined as:

$$BR = \frac{db_{nb}}{dz} / \frac{db_{nc}}{dz} \qquad (2.19)$$

where b_{nb} and b_{nc} are mass balance in the ablation area and accumulation area, respectively (Furbish and Andrews, 1984).

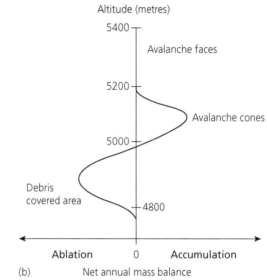

Figure 2.29 (a) The Ama Dablam Glacier, Nepal. Most of the accumulation is supplied by avalanches from the backwall of the glacier. (b) Schematic mass balance gradient (D.I. Benn; (b) Benn and Lehmkuhl, 2000).

The balance ratio ignores any non-linearity that may exist in the respective mass balance gradients, but it remains a useful generalization that summarizes the overall mass balance curve of a glacier. For a sample of 22 glaciers in Alaska, Furbish and Andrews (1984) found that balance ratios average about 1.8, indicating that the vertical change in mass balance is 1.8 times as large in the glacier ablation areas than in their accumulation areas. In contrast, the balance ratios of tropical glaciers are much higher because of year-round intense melting in their ablation zones and very poor dependence of accumulation with altitude. A very useful review of balance ratio data from around the world has been provided by Rea (2009).

Glaciers with high balance ratios have small ablation areas compared with the area of the glacier, because only a small area of ablating ice is required to balance inputs from snowfall higher up on the glacier. In contrast, glaciers with lower balance ratios require larger ablation areas to balance inputs, although the ablation area will still generally be less than half of the whole area of the

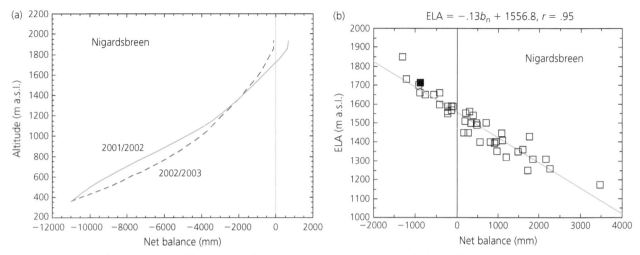

Figure 2.30 (a) Mass balance data for Nigardsbreen, Norway, for two successive measurement years. In both years the mass balance was negative. In 2001/2002, the glacier lost an average of 890 mm (water equivalent) over its entire area, and the ELA was at an elevation of 1715 m. In 2002/2003, the mass balance was even more negative (1160 mm water equivalent) and the ELA lay above the glacier summit. (b) Equilibrium line altitude vs net balance for all measurement years. The fitted curve shows that a net balance of zero corresponds to an ELA of ~1550 m. This is the elevation of the steady-state ELA for the contemporary geometry of the glacier (World Glacier Monitoring Service, 2005a).

glacier. Hence, tropical glaciers tend to have smaller ablation areas than mid-latitude glaciers with similar geometry, and their equilibrium line altitudes will be correspondingly closer to the terminus.

2.5.5 THE EQUILIBRIUM LINE

Definitions

The *equilibrium line* joins points on a glacier where annual accumulation exactly balances annual ablation, that is, where the net annual mass balance $b_n = 0$. Despite the apparent simplicity of this definition, the idea of the equilibrium line is complicated in practice. First, the altitude of the equilibrium line is rarely constant across a glacier, but varies with patterns of snow accumulation, shading and other factors. It rarely runs smoothly from one side of a glacier to the other and may even form closed loops in places (Oerlemans, 2001). The commonly used term *equilibrium line altitude* or *ELA*, therefore, is always at least partially a theoretical construct that involves simplification of reality. Second, equilibrium line altitudes for any given year reflect transient climatic conditions, and the glacier as a whole may have gained or lost mass over the preceding year (fig. 2.30). The *steady-state ELA* is the average altitude at which $b_n = 0$ for a glacier with zero net balance as a whole. The steady-state ELA most closely corresponds to the theoretical ELAs calculated for former glaciers (Benn et al., 2005). It is important to note, however, that during periods of sustained positive or negative mass balance, a steady-state ELA may never actually occur on a particular glacier. If glacier geometry is constantly changing, the idea of a steady-state ELA is therefore of questionable value. In such cases, the

mean ELA, defined for some specified time period, is a more meaningful indicator of the average relationship between a glacier and its climatic environment.

Determining equilibrium line altitudes

Glacier equilibrium line altitudes can be determined directly from mass balance observations. When data are available for several years, steady-state ELAs can be identified by plotting annual ELA against total mass balance and finding the value of the ELA that corresponds to $b_n = 0$ (fig. 2.30).

The ELA for any given year can also be identified by observing the distribution of snow and ice on the glacier surface at the time of year when glacier mass is at a minimum. This information can be obtained from field surveys, aerial photographs or satellite images. On temperate glaciers, the ELA coincides with the *snowline*, defined as the lower limit of the previous winter's snow on the glacier surface at the end of the ablation season. This is identifiable as the boundary between white snow cover and off-white glacier ice or old firn. The *firn line*, or boundary between firn and glacier ice, may coincide with the snowline if the ELA has maintained a constant position for a number of years. For glaciers with negative mass balance, the snowline lies above the firn line. Where superimposed ice forms at the base of the snowpack, the ELA will lie at the lower limit of the superimposed ice zone at the end of the ablation season, which may be some distance lower down the glacier than the snowline (section 2.4.2).

Spectral analysis of satellite data greatly facilitates the identification of snow and bare ice facies, allowing monitoring of glacier ELAs over large areas (e.g. Williams et al., 1991; Klein and Isacks, 1999; Winther et al., 1999; Bindschadler et al., 2001). Radar imagery can also be used to identify the boundary between snow and ice surfaces (e.g. Bindschadler et al., 1987; Adam et al., 1997). Locating the equilibrium line on glaciers in polar regions can be complicated by the presence of superimposed ice. On the ground, superimposed ice can be easily identified because it lacks the foliation and other structures typical of old glacier ice (Fig. 2.14), but it is very difficult to identify using satellite data (Engeset et al., 2002). One promising method for identifying superimposed ice from satellite data is the use of concurrent measurements of near-infrared albedo and surface roughness (Nolin and Payne, 2007).

Precipitation–temperature relationships at the ELA

Because the equilibrium line is located where annual accumulation totals exactly balance ablation totals, the ELA is very closely linked to local climate, particularly solid precipitation totals and the energy available for melting. The ELA is sensitive to perturbations in these variables, and rises in response to decreasing snowfall and/or increasing energy flux to the ice surface, and vice versa. Because winter precipitation is correlated with accumulation, and ablation season temperature is correlated with melting (section 2.4.3), for any population of glaciers there tends to be a positive correlation between winter precipitation and summer temperature at the ELA. The precise form of this relationship varies from region to region, reflecting differences in the mix of energy balance components that drive ablation.

For a large global dataset, Ohmura et al. (1992) found that best-fit relationship between mean summer air temperature (T_{sum} in °C) and total annual precipitation (p_a in mm yr^{-1}) at the ELA was:

$$p_a = 645 + 296\, T_{sum} + 9\, T_{sum}^2 \qquad (2.20)$$

(fig. 2.31). The standard error of this relationship is quite large (±200 mm), but the predictive power was further improved by including the amount of incoming shortwave and net longwave radiation at the ELA. Glaciers with higher radiation receipts (i.e. those at lower latitudes or facing towards the equator) will experience greater amounts of ablation relative to air temperature than those with low receipts, and therefore require larger precipitation totals to balance the annual mass budget. Braithwaite (2008) has proposed an alternative, powerful approach to analysing precipitation–temperature relationships using degree-day factors for snow (section 2.4.3).

Precipitation–temperature relationships explain regional rises in glacier ELAs with distance from moisture sources. In areas of lower snowfall, less energy is required to melt

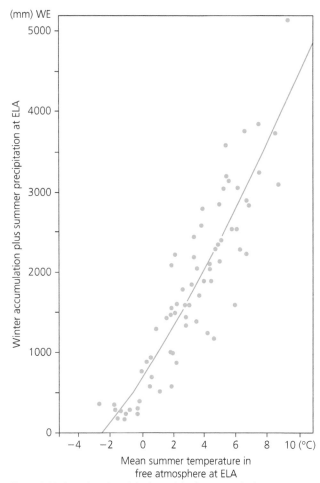

Figure 2.31 Annual total precipitation and the free-atmospheric temperature observed at the equilibrium line altitudes of 70 glaciers (Redrawn from Ohmura et al., 1992).

the annual accumulation at the ELA than where snowfall is high. Thus, in areas of low precipitation, ELAs will tend to be at higher altitudes, where temperatures are lower. Such relationships are very useful for palaeoclimatic reconstructions, allowing precipitation at the ELA to be estimated if mean summer air temperatures can be reconstructed from proxy data (Nesje and Dahl, 2003; Benn and Ballantyne, 2005).

2.5.6 GLACIATION LEVELS OR GLACIATION THRESHOLDS

The *glaciation level* or *threshold* is a theoretical surface in a glacierized terrain which separates ice-free and ice-covered summits. Glaciation levels are usually derived by calculating the average difference between the highest unglacierized and lowest glacierized summits (Østrem 1966; fig. 1.17). Because regional variations exist, glaciation levels are calculated for localized areas or at the largest scale possible, and then contoured (*iso-glacihypses*) at a regional or smaller scale. The mapping of glaciation levels using the summit method is relatively quick and simple and can provide an overview of local and regional trends, which in turn reflect the interaction of all climatic and topographic variables. In an assessment of the glaciation level of the southern Coast Mountains of British Columbia, Canada, Evans (1990) defined the *all-sided glaciation level* which lies up to 300 m higher than local glaciation levels. The all-sided glaciation level lies at the altitude at which glaciers will be formed on all aspects rather than just the northerly slopes of mountains.

2.5.7 GLACIER SENSITIVITY TO CLIMATE CHANGE

Changes in the amount of precipitation or any of the energy balance components discussed in section 2.2 will impact on the mass balance of a glacier. For a glacier initially in equilibrium, increases in snowfall and/or reductions in the energy available for melting will lead to expansion of the accumulation zone, lowering of the ELA and positive net mass balance. Conversely, a decrease in snowfall and/or an increase in the energy available for melting will result in a reduction in the size of the accumulation area, a rise of the ELA and negative net balance. The mean specific mass balance of a glacier depends on its mass balance gradient and its hypsometry, or area distribution with altitude, so different glaciers within the same region may have contrasting responses to the same climate signal (Furbish and Andrews, 1984). In extreme cases, climate changes may cause the ELA to rise above the highest point of a glacier, so that net mass loss occurs over the entire surface (fig. 2.30). Clearly, if this situation persists for many successive years, rapid glacier recession or complete disappearance will result (Hastenrath and Kruss, 1992; Schneeberger et al., 2003). Occasionally, glacier mass balance can also be modified by non-climatically driven changes in surface characteristics. Slope failures can deliver large amounts of debris onto a glacier surface, reducing ablation until the debris is evacuated by glacier flow (Gardner and Hewitt, 1990; D'Agata and Zanutta, 2007). Mass losses by calving are influenced by a complex web of dynamic factors as well as climate, and are discussed in detail in section 5.5.

The change in specific mass balance of a glacier in response to a given change in climate is known as its *climate sensitivity* (Oerlemans and Fortuin, 1992; Oerlemans, 2001). Climate sensitivities can be assessed using either full energy balance or degree-day approaches to model the impact of specified climate shifts (section 2.4.3). The climate sensitivity to a temperature change of ±1 °C can be defined as:

$$C_T = \frac{B(+1°C) - B(-1°C)}{2} \qquad (2.21)$$

that is, half of the difference between the total mass balance (*B*) of the glacier under a one degree rise and fall in

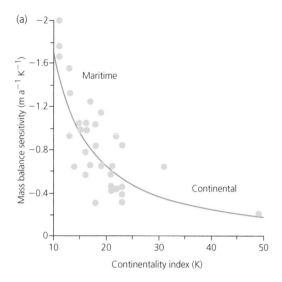

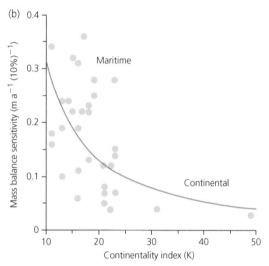

Figure 2.32 Mass balance sensitivity vs a continentality index (based on the amplitude of the annual temperature cycle) for 32 glaciers, to (a) a temperature increase of 1 °C and (b) a precipitation increase of 10 per cent (De Woul and Hock, 2005).

temperature. Similar indices can be defined for glacier sensitivity to changes in precipitation, summer temperature or other climatic variables. For example, climate sensitivity to a 10 per cent variation in precipitation is:

$$C_P = \frac{B(+10\%) - B(-10\%)}{2} \qquad (2.22)$$

Climate sensitivity to both temperature and precipitation change is greatest in maritime areas with high precipitation (Oerlemans, 2001; De Woul and Hock, 2005; fig. 2.32). For given increases in temperature, maritime glaciers will experience relatively large reductions in mass balance because:

(a) a change in the partitioning of precipitation between snow and rain has a larger effect where precipitation is higher;

(b) increases in the length of the ablation season have the greatest impact on glaciers that extend down into temperate climatic zones;

(c) feedbacks between melting and albedo amplify the initial climate forcing.

For a reduction in precipitation by a given percentage, maritime glaciers will experience the greatest reduction in mass balance because the absolute reduction in accumulation is greater, and feedbacks between reduced snow cover and albedo amplify the initial signal. In an analysis of 42 arctic glaciers and ice caps, De Woul and Hock (2005) found that a 1 °C increase in summer temperature corresponds to a change in mean specific mass balance of −0.75 to −1.75 m yr^{-1} for glaciers in maritime Iceland, but only −0.08 to −0.20 m yr^{-1} for glaciers in the much more arid Canadian arctic archipelago. The contrasting sensitivity of maritime and continental glaciers to changes in precipitation is very well illustrated by recent glacier mass balance trends in Norway (Chinn et al., 2005). Since the mid 1980s, strengthening of winter westerly airflow over Norway has resulted in increased winter precipitation and higher winter balances on many glaciers. In the west, where precipitation is highest, glaciers entered a phase of strong positive mass balances (e.g. Ålfotbreen and Nigardsbreen, fig. 2.33), but in the drier east there was only a short-lived and minor increase in mass balance in the late 1980s, which had little impact on the longer-term trend of negative balances (e.g. Hellstugubreen and Storbreen, fig. 2.33). Recent mass balance records for many glacierized regions show an increase in both summer and winter balances in response to warming and higher precipitation. This intensification of the hydrologic cycle has steepened glacier mass balance gradients, with corresponding increases in mass throughput and climatic sensitivity (Dyurgerov, 2003).

In recent years, several studies have demonstrated relationships between glacier mass balance and inter-annual variability in large-scale atmospheric circulation patterns. McCabe and Fountain (1995) showed that negative cumulative mass balance of the South Cascade Glacier, Washington State, reflects reduced winter snowfall associated with shifts in atmospheric circulation over the North Pacific Ocean and northern North America. Since the 1970s, greater persistence of the Aleutian Low over the north-west Pacific and the associated occurrence of a high-pressure system over western Canada has led to frequent dry north or north-easterly airflow over the glacier, leading to anomalously low snowfall. This circulation pattern also reduces the winter balance of other monitored glaciers in the region, such as the Sentinel and Peyto Glaciers in western Canada, but is associated with higher than normal winter balances in southern Alaska due to an increase in the onshore advection of warm, moist air (Bitz and Battisti, 1999; Watson et al., 2006). In Peru and Bolivia, years of low mass balance coincide with higher sea-surface

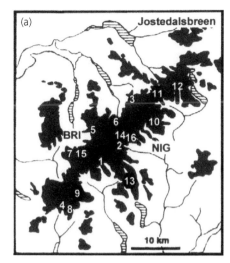

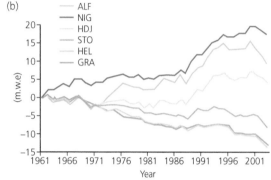

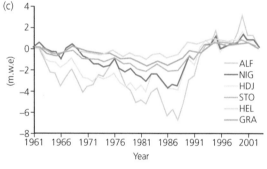

Figure 2.33 (a) Map of western south Norway, showing the location of glaciers for which mass balance data are available. (b) Cumulative net balances of six glaciers located on a west–east profile across western south Norway. (c) Cumulative deviations from the average winter balance for the six glaciers (After Chinn et al., 2005).

temperatures in the eastern Pacific associated with El Niño events (Thompson et al., 1984; Francou et al., 1995).

2.6 GLACIER–CLIMATE INTERACTIONS

2.6.1 EFFECTS OF GLACIERS AND ICE SHEETS ON THE ATMOSPHERE

So far in this chapter we have considered the ways that the state of the atmosphere affects snow and ice. The presence of glaciers and ice sheets can also have significant effects on the atmosphere, modifying weather and climate from the

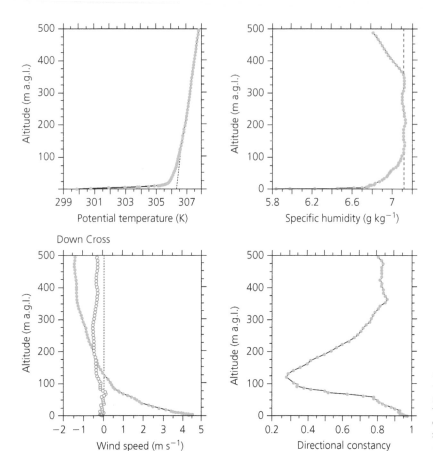

Figure 2.34 Mean climatic conditions over Pasterze Glacier, Austria, in summer 1994. The data show a persistent boundary layer with low potential temperature (i.e. measured temperature adjusted to constant pressure conditions), high humidity and a strong downslope wind component (After van den Broeke, 1997).

local to the global scale. Snow and ice surfaces influence the energy balance of the lower atmosphere in two main ways. First, because the temperature of clean snow and ice surfaces cannot rise above 0 °C, the maximum upwards longwave radiation flux is ~310 W m^{-2} (section 2.2.3). Consequently, warming of near-surface air by longwave radiation tends to be small, and is often less than that above adjacent ice-free surfaces. Second, turbulent exchanges of sensible heat and water vapour over cold, moist ice surfaces commonly chill near-surface air (section 2.2.4). Both of these processes encourage the development of a relatively cool, stable atmospheric boundary layer above a glacier, and an inversion of the near-surface air temperature gradient, particularly when regional winds are light.

Air density increases with decreasing temperature (Barry and Chorley, 2003), so cold air will sink below adjacent warm air masses. Thus, where a chilled near-surface layer forms over a glacier surface, it will tend to flow downslope as a *gravity current* typically up to around 100 m thick (van den Broeke, 1997; fig. 2.34). This effect is well known to inhabitants of glacierized regions as the *glacier wind*, and is known by the technical term *katabatic wind*, from the Greek *katabatikos* meaning 'going downhill'. Katabatic winds are strongest on sunny days, when the contrast between air temperatures above ice-covered and ice-free surfaces is at its greatest. There is typically a pronounced diurnal variation in wind

speed, with a peak in the afternoon when wind speeds commonly reach around 5–10 m s^{-1} (Obleitner, 1994; Strasser et al., 2004).

Continental-scale ice sheets have a correspondingly larger impact on air circulation. The high albedo and high elevation of ice sheet centres encourage very low temperature values, especially during the polar night; winter values plunge to −70 °C over Antarctica and to −40 °C over Greenland. The sheer size of these ice bodies ensure that they exert a large influence over global- as well as local-scale climate patterns. Patterns of near-surface air circulation over Antarctica are shown in figure 2.35. The mean winds spread outwards from the high, cold continental interior, flowing downslope towards the coast. Topographic control of wind vectors produces marked areas of confluence, where air from large areas of the interior is channelled into relatively narrow zones, increasing wind intensity and persistence (Liu and Bromwich, 1997; van Lipzig et al., 2004). The persistent surface airflow over Antarctica is commonly referred to as a katabatic wind. However, the Antarctic winds are not simply generated by differential buoyancy, as is the case for true katabatic winds (Parish and Cassano, 2003; Parish and Bromwich, 2007). In fact, the pattern of near-surface airflow forms the equatorward component of a large-scale circulation pattern in the southern high latitudes known as the *Polar Cell* (Parish and Bromwich, 2007). Airflow

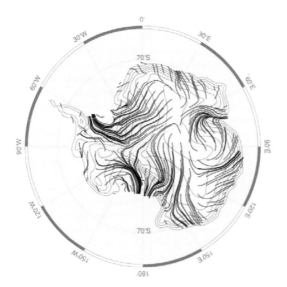

Figure 2.35 Mean wind streamlines over Antarctica 1980–93 (van Lipzig et al., 2004).

out from the Antarctic continent is sustained by subsidence of chilled air over the interior. In turn, this is fed by poleward upper-level transport of air uplifted over the mid-latitude circumpolar trough. Air circulation in the Polar Cell is largely driven by large-scale pressure patterns, locally modified by ice-sheet topography, and downslope surface winds therefore reflect a much more complex suite of processes than the locally driven katabatic winds on valley glaciers.

The Greenland Ice Sheet is much smaller, and occupies a lower-latitude position than the Antarctic Ice Sheet, and it has a smaller impact on hemispheric air circulation. It is more affected by travelling depressions embedded in the circumpolar westerlies, although the cold, high ice sheet interior encourages a semi-permanent high-pressure system, particularly in winter, discouraging inflow of warm maritime air (Steffen and Box, 2001). This large-scale pressure pattern means that average wind fields radiate outwards from the ice sheet interior, locally enhanced by buoyancy-driven katabatic winds.

Ice caps and ice sheets also modify precipitation patterns, although the magnitude and sign of change vary with ice-mass size. For ice domes less than 50 km across, precipitation tends to be enhanced all the way to the summit due to orographic factors. In contrast, where an ice sheet expands to form a dome of continental proportions, the increase in precipitation with altitude only occurs at the outer margins of the ice sheet, and nearer to the summit the precipitation drops off considerably. For example, the accumulation rate for the Antarctic ice sheet is ten times greater at most parts of the coast than at the ice sheet centre (section 6.3.2).

The influence of former ice sheets on global climate cannot be observed directly, but can be studied using palaeoclimatic data and computer simulations of global atmospheric circulation (e.g. Broccoli and Manabe, 1987; Clark et al., 1999; Bromwich et al., 2004b, 2005; Marshall et al., 2004). Such models have shown that, at glacial maxima, the great northern hemisphere ice sheets – particularly the Laurentide Ice Sheet – profoundly influenced global energy balance and atmospheric circulation patterns. The western portion of the Laurentide Ice Sheet anchored a permanent blocking anticyclone, causing the winter jet stream to split into a northern branch over the Canadian Arctic and a southern branch impacting southern North America. In spring, this split flow transitioned into a single branch that migrated north over the ice sheet during summer. Reorganization of storm tracks profoundly affected precipitation patterns and resulted in substantial cooling downwind, causing an expansion of sea ice in the North Atlantic. Enhanced cooling at high latitudes associated with albedo and other feedbacks steepened the equator–pole temperature gradient and increased the strength of east–west (zonal) and north–south (meridional) atmospheric circulation.

2.7 ICE CORES

Layers of snow and ice laid down each year in glacier accumulation zones preserve a stratigraphic record of past environmental conditions. The isotopic composition of the water molecules making up the ice provides a means of reconstructing former air temperatures, while trapped air bubbles represent samples of the atmosphere at the time of ice formation. Additionally, glacier ice contains impurities such as volcanic tephra, windblown dust or radioactive fallout, providing high-resolution archives of changing snow composition.

For ice cores to provide useful records of environmental change, it is necessary to date each level by some means. In the upper layers of cores, annual accumulation layers commonly stand out clearly as alternate bands of clear and bubbly ice, which can be dated by counting down from the surface, like tree rings. Annual layers may also be identified from chemical or isotopic signatures, reflecting seasonal variations in solute loading or air temperature. For deeper, older parts of cores, individual layers may not be apparent, and dating is achieved by other means. These include: identifying horizons of known age, such as volcanic tephra; matching features of the ice core record to other dated palaeo-environmental archives; and calculations based on ice flow models (Paterson, 1994; Bradley, 1999).

2.7.1 ICE CORING PROGRAMMES

Major deep-drilling programmes in the Greenland and Antarctic Ice Sheets were begun in the 1970s, since which time long ice cores have yielded immensely valuable records of environmental change in both polar regions

Figure 2.36 Deep ice core sites in (a) the Antarctic and (b) Greenland (National Snow and Ice Data Center, http://nsidc.org).

	Latitude	Longitude	Altitude (m)
Aggasiz, Ellesmere Island	80.70 N	73.1 W	1730
Devon Island	75.42 N	82.50 W	1800
Penny Ice Cap, Baffin Island	67.25 N	65.75 W	1900
Austfonna, Svalbard	79.85 N	24.14 E	750
Vetreniy, Franz Josef Land	80.78 N	63.55 E	505
Academii Nauk, Severnaya Zemlya	80.5 N	94.8 E	810
Guliya, Tibet	35.28 N	81.48 E	6710
Dunde, Tibet	38.1 N	96.40 E	5325
Sajama, Bolivia	18.1 S	68.88 W	6542
Huascaran, Peru	9.1 S	77.60 W	6048
Quelccaya, Peru	13.9 S	70.80 W	5670

Table 2.2 Selected ice core sites in Arctic and high-altitude ice caps

(fig. 2.36). Between 1979 and 1981, the Greenland Ice Sheet Project (GISP) drilled the Dye 3 ice core, which totalled 2037 m in length. GISP2 focused on a glaciologically better location on the summit of the ice sheet, and produced an ice core 3053.44 m long, the deepest ice core recovered in the world at the time. The Greenland Ice Core Project (GRIP) was a multinational European research project, organized through the European Science Foundation. Between 1989 and 1992, GRIP successfully drilled a 3028 m ice core to the bed of the Greenland Ice Sheet at 72° 35′ N, 37° 38′ W. Subsequently, the North Greenland Ice Core Project (NGRIP) set up a drill site at 75.1° N, 42.32° W. Bedrock was reached in 2003, below an ice thickness of 3085 m. The NGRIP core provided a continuous environmental record of the last 123,000 years, including part of the last (Eemian) interglacial period (NGRIP members, 2004).

Ice cores from Antarctica can extend back much farther in time, reflecting lower accumulation rates and the great stability of the ice sheet. The European Project for Ice Coring in Antarctica (EPICA) core from Dome C, Antarctica, goes back an astounding 740,000 years, spanning the last eight full glacial cycles (EPICA community members, 2004). Shorter, but no less valuable records have been obtained from a number of other sites in Antarctica, including the Japanese Dome Fuji Station (Dome F) and the West Antarctic Ice Sheet Divide (WAIS Divide) Ice Core Drilling Project.

Ice core records have also been obtained from ice caps on Arctic islands (e.g. Kekonen et al., 2002; Fisher and Koerner, 2003; Kotlyakov et al., 2004; Virkkunen et al., 2007), and at high-altitude sites in Asia, South America and Africa (e.g. Thompson et al., 1984, 1986, 2000, 2002; Thompson, 2000; table 2.2). Some of the drilling sites are at altitudes of over 6000 m, an extreme working environment by anyone's standards.

2.7.2 STABLE ISOTOPES

The water molecule H_2O is made up of two hydrogen atoms and one oxygen atom. In common with other elements, atoms of oxygen and hydrogen always contain the same number of protons, but the number of neutrons can vary, producing different *isotopes* or forms of the element. Since the combined number of protons and neutrons in a nucleus defines the *atomic mass*, different isotopes have different masses. Oxygen atoms always have eight protons, but can have eight, nine or ten neutrons, yielding three isotopes with atomic masses of 16, 17 and 18 (^{16}O, ^{17}O and ^{18}O). Hydrogen atoms always have one proton, but in nature can have no or one neutron, resulting in two isotopes, ^1H and ^2H. (^2H is also known as deuterium, D. A third hydrogen isotope ^3H or tritium was produced in the atmosphere during the atmospheric bomb-testing era of the 1950s and early 1960s.) Water molecules may therefore consist of any one of nine possible combinations of the five naturally occurring isotopes, ranging from molecules with three light isotopes ($^1H_2{}^{16}O$) to those with three heavy isotopes ($D_2{}^{18}O$). However, only three of these combinations are common: $^1H_2{}^{16}O$, $^1HD^{16}O$ and $^1HD^{18}O$ (Hubbard and Sharp, 1989). In nature, the global relative abundance of oxygen isotopes is 99.76 per cent (^{18}O), 0.04 per cent (^{17}O) and 0.2 per cent (^{16}O); for hydrogen the figures are 99.984 per cent (^1H) and 0.016 per cent (D). The isotopic composition of glacier ice varies systematically with several factors, the most important of which are: (1) the composition of the precipitation from which the ice was formed, and (2) the history of melting and refreezing within the ice. Isotope analyses can therefore yield important insights into past climatic conditions and glacial processes.

The isotopic composition of ice from successive levels in a core is determined by melting the ice in the laboratory and analysing it with a mass spectrometer. Variations in the proportions of ^{16}O and ^{18}O are expressed as $\delta^{18}O$ (delta^{18}O) values, which measure the difference between the observed $^{18}O/^{16}O$ ratio and that in a standard water sample, known as Standard Mean Ocean Water (SMOW). Similarly, different proportions of ^1H and ^2H (D) are expressed as δD values, compared with the same standard.

The isotopic composition of the precipitation falling on a glacier depends on its history of evaporation and condensation as part of the hydrologic cycle. During evaporation, water molecules composed of light isotopes turn to vapour more readily than those composed of heavy isotopes, a process known as *fractionation*. The resulting vapour is relatively depleted in deuterium and ^{18}O compared with the initial water. Conversely,

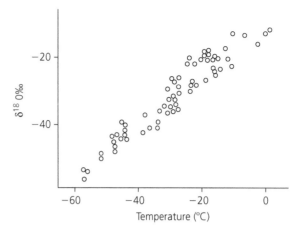

Figure 2.37 Present-day relationship between $\delta^{18}O$ and mean annual air temperature, based on values in firn 10–12 m below the surface of the Greenland and Antarctic Ice Sheets (Data from Dansgaard et al., 1973).

when water vapour condenses, molecules containing heavy isotopes tend to pass from the vapour to the liquid state more readily than 'light' molecules, resulting in precipitation that is enriched in heavy isotopes compared with the remaining vapour. As condensation proceeds, more of the remaining heavy isotopes will be removed, and the vapour will become increasingly depleted of ^{18}O and deuterium. Therefore, progressive cooling of water vapour (e.g. as moist air passes from a warm ocean over a cold land mass) will result in precipitation with increasingly lighter isotopic composition, as the parent vapour becomes more and more depleted of heavy isotopes. Thus the isotopic composition of precipitation reflects both air mass history and the temperature at which condensation occurs (Dansgaard, 1964; Jouzel et al., 2000).

Within many regions, the isotopic composition of snow can be expressed as a simple linear function of surface temperature, T_A:

$$\delta^{18}O\ _{SNOW} = a\,T_A + b \qquad (2.23)$$

where a and b are empirically determined constants (fig. 2.37). The slope, a, of the $\delta^{18}O$ / T_A relationship varies globally, and is highest in high-latitude areas. For Greenland, $a = 0.67\text{‰}\,°C^{-1}$ (Dansgaard et al., 1973; Jouzel et al., 2000), and at Byrd Station, Antarctica, $a = 0.99\text{‰}\,°C^{-1}$ (Robin, 1983). In tropical regions, where most precipitation is associated with deep convective storms, oxygen isotope ratios are mostly controlled by the amount of precipitation and air mass history, and may be only weakly correlated with surface temperature (Grootes et al., 1989; Yao et al., 1996; Rozanski et al., 1997; Thompson, 2001).

Equation 2.23 has been widely used to reconstruct continuous palaeotemperature records from $\delta^{18}O$ variations in ice cores, especially in the polar regions. It has become clear, however, that the present-day gradient of the $\delta^{18}O$ / T relationship (the 'spatial slope') did not necessarily apply in the past. Variations in the slope of the $\delta^{18}O$ / T relationship can result from changing sea-surface temperatures in precipitation source regions, shifting atmospheric circulation patterns, and the seasonal distribution of precipitation (Jouzel et al., 1997, 2000; Werner et al., 2000; Huber et al., 2006). In the case of the Greenland Ice Sheet, oxygen isotope variations in the Late Glacial and Early Holocene may also reflect the abundance of isotopically light meltwater from the Laurentide Ice Sheet in precipitation source regions (Fisher and Koerner, 2003). Various approaches have been developed to calibrate 'stable isotope thermometers', including isotope fractionation modelling and direct measurement of ice temperatures in boreholes (e.g. Johnsen et al., 2001; Vimeux et al., 2002; Blunier et al., 2004). The latter approach works well for ice younger than the LGM, but thermal diffusion processes blur the record beyond recognition for older ice. With careful analysis, isotopic fluctuations can be used to determine temperature variations in the oceanic source regions, as well as at the core site itself (Vimeux et al., 2002).

Figure 2.38 shows reconstructed temperature changes at Summit, Greenland, based on $\delta^{18}O$ data from the GRIP ice core, calibrated using borehole ice temperatures (Johnsen et al., 2001). The core spans the last full glacial–interglacial cycle, and shows that during the last (Eemian) interglacial, temperatures were ~5° higher than today. The onset of full glacial conditions occurred stepwise, with alternating cooling and warming episodes. The period between ~70,000 and 12,000 years BP is characterized by highly dynamic climatic conditions, with repeated temperature fluctuations of ~15°C (Dansgaard-Oeschger events; Dansgaard et al., 1993). The rapid warming at ~12,000 years BP marks the termination of glacial conditions, and the beginning of the Holocene. Three high-altitude ice cores from Tibet and two from South America extend back to LGM, allowing comparison with the polar ice sheet records (Thompson, 2000; fig. 2.39). Some events, such as the Younger Dryas cooling around 12,000 years BP, can be seen in several of the records, but many isotopic trends may reflect regional factors such as variations in precipitation (Pierrehumbert, 1999).

2.7.3 ANCIENT ATMOSPHERES: THE GAS CONTENT OF GLACIER ICE

During the transformation of snow into ice below the surface of a glacier, air bubbles become isolated and trapped. These bubbles represent 'fossil air', preserving samples of the atmosphere from the time they were sealed off. Analysis of the composition of such bubbles in ice cores thus provides a means of tracing changes in atmospheric chemistry over many thousands of years (Lorius et al., 1990). Isolation of bubbles (*gas occlusion*) in polar ice typically occurs at depths of 50–150 m, so the trapped air may be a few centuries to several millennia younger than the adjacent ice, depending on ice temperature and accumulation rate (Blunier et al., 2004). Uncertainties about the age difference mean that caution must be exercised when inferring causal links between atmospheric chemistry and palaeoclimatic proxies in ice cores.

Particular attention has been paid in recent years to the record of variations in 'greenhouse gases' such as carbon dioxide (CO_2) and methane (CH_4), due to their regulating effects on global temperature. Variations in CO_2 and CH_4 and palaeotemperatures derived from $\delta^{18}O$ values have been determined for the last eight glacial–interglacial cycles from a deep core drilled at Vostok, Antarctica (Petit et al., 1999; EPICA project members, 2004; Loulergue et al., 2008; Lüthi et al., 2008; fig. 2.40). Oscillations in the greenhouse gas concentrations mirror the climatic fluctuations to a remarkable degree, peak concentrations occurring during warm periods, and low concentrations marking cold episodes. The parallelism between climate change and greenhouse gas concentrations reflects feedbacks between glacial, oceanic, atmospheric and biologic systems, whereby small climate changes initiate changes in the production and sequestering of trace gases by oceanic and terrestrial plants, which then forces further climatic changes (Köhler and Fischer,

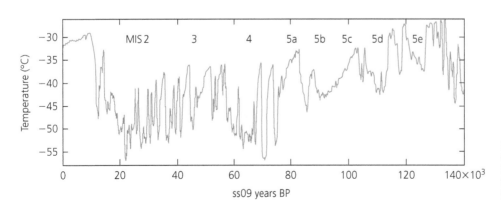

Figure 2.38 Past temperature changes for Summit, Greenland, based on GRIP $\delta^{18}O$ and borehole thermometry calibration (Johnsen et al., 2001).

2006). The Vostok record also shows that current concentrations of CO_2 and CH_4 (~360 parts per million and ~1,700 parts per billion, respectively) are unprecedented for at least the last 700,000 years. The ice core record thus provides an extremely important perspective on the current debate about anthropogenic greenhouse gas emissions and the patterns and mechanisms of climatic change.

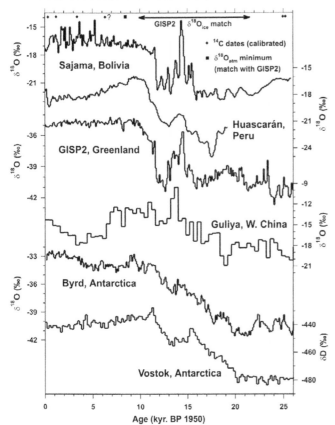

Figure 2.39 Comparison of oxygen isotope records from South America, Greenland, China and Antarctica (Thompson, 2000).

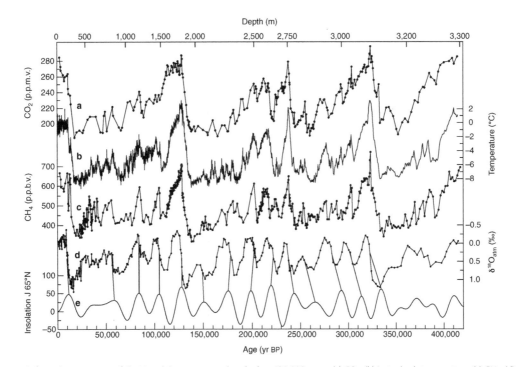

Figure 2.40 Records from the upper part of the Vostok ice core, spanning the last 420,000 years. (a) CO_2. (b) Isotopic air temperature. (c) CH_4. (d) $\delta^{18}O$. (e) Mid June insolation at 65° N (Petit et al., 1999).

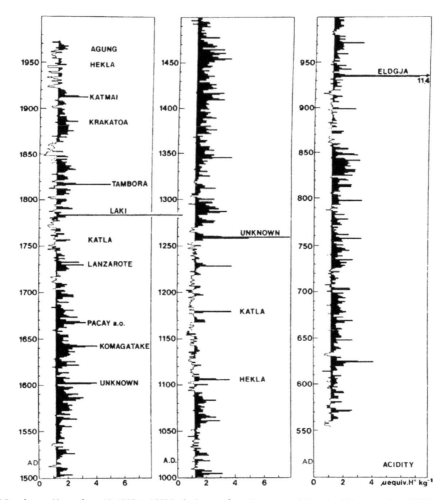

Figure 2.41 Mean acidity of annual layers from AD 1500 to 1972 in the ice core from Crete, central Greenland (Hammer et al., 1980).

2.7.4 SOLUTES AND PARTICULATES

Impurities contained within glacier ice also provide records of former environmental conditions. Windblown dust and soluble ions can yield data on the strength and direction of atmospheric circulation (e.g. Biscaye et al., 1997; Delmonte et al., 2002), and tephra horizons and peaks in acidity record the deposition of aerosols injected into the atmosphere by volcanic eruptions (e.g. Hammer et al., 1980; Zielinski et al., 1994; Zielinski, 1995; Cole-Dai et al., 2000). Figure 2.41 shows the acidity of annual layers from AD 1500 to AD 1972 in the ice core from Crete, central Greenland. Acidity peaks reflect the production of sulphur dioxide by major eruptions, and its transformation to sulphuric acid aerosol in the atmosphere. Most of the acidity peaks can be matched with historical eruptions, including Tambora and Krakatao

in the East Indies. Particularly prominent is the peak associated with the 1783 Laki Fissure Eruption, Iceland, which caused 10,000 deaths in Iceland and elevated mortality across Europe (Grattan et al., 2003; Stone, 2004).

Ice cores worldwide contain radioactive fallout from atmospheric nuclear bomb tests in the 1950s and early 1960s and the Chernobyl reactor accident in 1986 (e.g. Koerner and Taniguchi, 1976; Warneke et al., 2002). Apart from acting as a reminder of the folly of detonating nuclear weapons, such fallout provides useful stratigraphic marker horizons. For example, the mass balance of the glacier complex Kongsvegen/Sveabreen in Svalbard has been reconstructed by Lefauconnier et al. (1994) using radioactive fallout from Russian nuclear tests in 1961/62 and the Chernobyl reactor accident in 1986.

CHAPTER 3

GLACIER HYDROLOGY

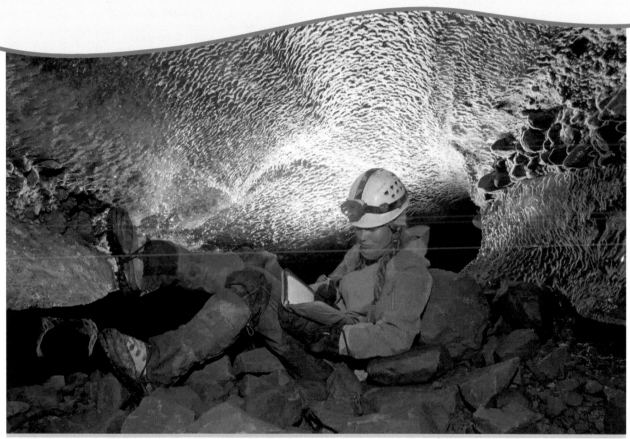

Taking notes during a subglacial conduit survey, Rieperbreen, Svalbard (Jason Gulley)

3.1 INTRODUCTION

The production, storage and transport of water exerts a profound influence on glacier behaviour. The rate of motion of glaciers and ice sheets is sensitively dependent on the presence of water at the bed or at crystal boundaries within the glacier ice. Water contributes substantially to glacial erosion, debris transport and deposition, both as a direct agent and through its impact on the action of ice. Furthermore, water released from glaciers presents both benefits and hazards for human populations. In glacierized regions with low summer rainfall, such as the high mountains of central Asia, glacier melt provides an important source of water during the growing season, allowing cultivation of valleys that would otherwise be too arid for agriculture. Conversely,

the sudden release of stored water in catastrophic floods constitutes a serious and recurrent threat in regions such as the high Andes, Himalaya and Iceland. Glacial meltwater has also been harnessed as a source of hydroelectric power, particularly in Scandinavia and the European Alps, and the successful building and running of such schemes demands a clear understanding of the principles of water flow and sediment transport.

This chapter begins with a review of the basic principles controlling the flow of water through glaciers. Then in sections 3.3 to 3.5, we examine the range of drainage systems known or inferred to occur on, in and below glaciers. Section 3.6 discusses runoff from glaciers, and the ways in which it is influenced by climate change. Glacial lakes and outburst floods are explored in section 3.7, while section 3.8 looks at the rapidly evolving disciplines of glacio-biology and ecology. The chapter concludes with a review of glacial hydrochemistry and chemical weathering. The influence of water on glacier motion is discussed in chapter 4, and the sediments and landforms produced by flowing water are described in chapters 9, 10 and 11.

3.2 BASIC CONCEPTS

3.2.1 WATER SOURCES AND ROUTING

Water may enter glacial drainage systems by melting of snow or ice at the glacier surface, at the bed or within the ice. Additionally, water can directly enter glaciers from extraglacial sources, such as rain, subaerial runoff or groundwater (fig. 3.1). Spatial and temporal patterns of surface meltwater production vary enormously, depending on local climate and other factors (section 2.4.3). In the ablation zones of tropical glaciers, melting can occur on every day of the year, whereas melt days rarely, if ever, occur in continental Antarctica. Between these extremes, most glaciers experience annual melting cycles, in addition to shorter-term variations associated with changing weather conditions. Melting within or beneath glaciers in response to frictional heat generated by ice deformation or sliding is much more constant through time, and contributes little to seasonal variations in water production. Large variations in internal and basal melting, however, may occur on longer timescales.

Water can flow over glacier surfaces (supraglacially), through the ice (englacially), along or within the bed (subglacially), or follow various combinations of these flow paths. Although transit through glacier systems can be very efficient, with little time delay between meltwater production and runoff, water can also be retained on, in or beneath the glacier for varying amounts of time. Storage may be either in liquid form or as ice following refreezing (Jansson et al., 2003). In consequence, runoff from glaciers is highly variable in space and time, depending on variations in meltwater production, the character and efficiency of transport paths, and changing amounts of storage. The precise route that water will take over, through or below a glacier depends on the balance between (1) the gradients in hydraulic potential that drive water flow, and (2) resistance to flow. These two factors are discussed in the following sections.

3.2.2 HYDRAULIC POTENTIAL

Water flows from one place to another following gradients in *hydraulic potential*. The idea of hydraulic potential is central to understanding water flow through glaciers, so it is worth examining in some detail. For surface streams, including those flowing over a glacier,

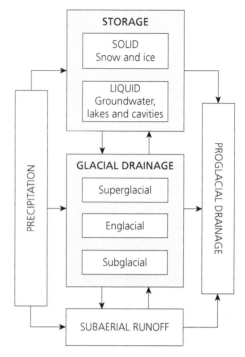

Figure 3.1 Water sources and routing in glacierized catchments

hydraulic potential depends only on the water's mass and elevation. The situation is more complex for the case of water in enclosed passages within or beneath a glacier, because water can be subject to variations in *pressure* as well as elevation. The elevation and pressure components of hydraulic potential can be expressed as follows:

$$\phi = \rho_w g z + P_w \tag{3.1}$$

where ϕ is hydraulic potential, ρ_w is water density, g is gravitational acceleration, z is elevation and P_w is water pressure. Hydraulic potential is also known as *head*. In this terminology, the first term on the right-hand side is called *elevation head*, and second term is *pressure head*. The pressure term depends on the load that must be supported by the water, and is taken to be zero at a free surface. In a standing water body such as a lake, water pressure at any elevation z is given by:

$$P_w = \rho_w g(h_w - z) \tag{3.2}$$

where h_w is the elevation of the water surface.

If we insert this expression into equation 3.1, we can see that the sum of the elevation and pressure terms is the same for all points in the water column, provided the water density is constant. In other words, as one goes deeper below the surface of a lake, the pressure head term increases at the same rate that the elevation term goes down. This means that there are no gradients in hydraulic potential, and water will not flow from one part of the column to another. There is *hydrostatic equilibrium* (fig. 3.2a). Differences in water density due to variations

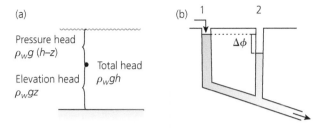

Figure 3.2 Hydraulic potential: (a) in a standing body of water of constant density, hydraulic potential is equal everywhere, because the sum of the pressure head and elevation head is constant. (b) When water flows along a pipe, friction between the flowing water and the walls results in a head loss, illustrated here by the lower water level in tube 2 than tube 1.

in temperature, salinity or suspended sediment concentration may upset the hydrostatic equilibrium and induce vertical motion in a standing water body.

For the case of water flowing along a pipe or crack, friction between the water and the walls supports some of the weight of the overlying water, resulting in a pressure drop (or *head loss*) $\Delta\phi_f$:

$$P_w = \rho_w g(h_w - z) - \Delta\phi_f \tag{3.3}$$

(Röthlisberger and Lang, 1987; fig. 3.2b). The rate of head loss (or *head gradient*) depends on the resistance to flow resulting from drag on the walls. Drag or resisting forces are considered in the next section.

3.2.3 RESISTANCE TO FLOW

Resistance to fluid flow through any system depends on the properties of both the fluid and the flow path. Fluids resist flow because energy is required to overcome bonds between molecules, a property known as *viscosity* and denoted by the Greek letter μ (mu). Water has a viscosity of approximately 1.00×10^{-3} Pascal seconds (kg m^{-1} s^{-1}) at 20°C, and flows more easily than olive oil, with a viscosity of 8.1×10^{-2} Pa.s at the same temperature. The viscosity of water increases with decreasing temperature, and is 1.79×10^{-3} Pa.s at 0°C.

The resistance offered by the flow path depends on the dimensions and roughness of the channel or passage and, in porous media, on the size, tortuosity and connectedness of pore spaces. Resistance to flow in porous media is measured by *hydraulic conductivity*, K, or *permeability*, κ (Greek letter kappa). Note that both of these variables decrease with increasing flow resistance. Permeability depends only on the material properties of the flow path, and has units of area (m^2), whereas hydraulic conductivity also includes fluid properties such as density and viscosity, and has units of velocity (m s^{-1}). The two properties are related to each other by:

$$K = \kappa \frac{\rho g}{\mu} \tag{3.4}$$

where ρ is fluid density, g is gravitational acceleration and μ is viscosity. Materials with large, well-connected pores – such as well-sorted gravels – have high permeabilities (low resistance), whereas those with small, poorly connected pores – such as clay-rich till or ice – have low permeabilities (high resistance). The influence of permeability on water flow through porous media is expressed in *Darcy's Law*:

$$q = \frac{-\kappa}{\mu} \frac{d\phi}{dL} \tag{3.5}$$

where q is the flux (discharge per unit area: m sec^{-1}) and $d\phi/dL$ is the gradient in hydraulic potential. This shows that flow rates increase with the hydraulic gradient and permeability (Freeze and Cherry, 1979). Analogous expressions can be used for flow through other types of system, such as fractures or pipes. For open-channel flow, the *Manning formula* relates flow velocity to the hydraulic gradient and flow resistance, as follows:

$$\bar{V} = \frac{1}{\tilde{n}} R^{*\,2/3} S^{1/2} \tag{3.6}$$

where $\bar{V}$ is the cross-sectional mean velocity, $\tilde{n}$ is Manning's roughness coefficient, R^* is the hydraulic radius of the channel and S is the channel slope. Because the channel slope is the same as the gradient in elevation head, S is identical to the hydraulic gradient $d\phi/dL$ for open-channel flow. The *hydraulic radius* is defined as:

$$R^* = A_r \big/ p^* \tag{3.7}$$

where A_r is the cross-sectional area of the channel and p^* is the 'wetted perimeter'.

Flow of water over or through a glacier reflects its permeability at two scales: *primary permeability*, or the permeability of intact ice or snow, and *secondary permeability*, associated with fractures and other passageways. For water at the bed, resistance to flow is governed by the permeability of subglacial materials (unconsolidated sediment or rock) and the size and distribution of gaps at the ice–bed interface. In this sense, glacial drainage is directly analogous to water flow through rocks, which can take place through pore spaces within intact rock (primary permeability), or along joints and other fissures (secondary permeability).

Snow and firn have relatively large, well-connected spaces between ice crystals, and have permeabilities in the range 10^{-9} to 10^{-10} m^2, about the same as sorted sand (Albert and Perron, 2000). This means that surface meltwater can drain easily into and through *snow* or *firn aquifers* (section 3.3.1). Interconnected spaces also exist

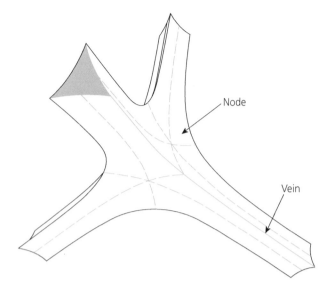

Figure 3.3 Geometry of veins.

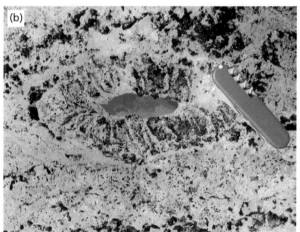

Figure 3.4 (a) Englacial passage being squeezed shut by ice overburden pressure, Hansbreen, Svalbard. The passage walls are coated with hoar frost. Note the line of suture top left. (b) Crystal quirk on the surface of Breiðamerkurjökull, Iceland. The crystal structure is picked out by thin layers of volcanic ash (J. Gulley and D. Benn).

between crystals in temperate glacier ice, in the form of water-filled *veins* at triple junctions between crystals, *lenses* between two crystals, and *nodes* where veins meet (fig. 3.3; Nye, 1989; Mader, 1992). Some studies have concluded that integrated networks of veins and lenses in temperate ice can transport significant volumes of water, but others have argued that this is unlikely to be the case (see Röthlisberger and Lang, 1987; Lliboutry, 1996; Nye, 1997). Recent work shows that the primary permeability of glacier ice is extremely low, in the order of 10^{-18} m^2, or about the same as granite (Jordan and Stark, 2001). This means that rates of water flow through vein networks must be very small indeed, except where hydraulic gradients are exceptionally large (equation 3.5). Thus, intact ice has such low permeability that, for practical purposes, it can be regarded as impermeable. This means that on snow-free glacier surfaces, meltwater will flow over the ice unless secondary permeability – resulting from fractures, unfrozen debris bands and other weaknesses – allows it to be routed to the subsurface. The role of secondary permeability in the evolution of englacial drainage systems is considered in section 3.3.2.

Many glacier beds are composed of unconsolidated sediments or permeable rocks, and are thus potential aquifers. In mountain regions, such aquifers tend to be thin, but in lowland areas they may be tens of metres in thickness (Boulton et al., 1995, 2007a; Piotrowski, 2006). The permeability of subglacial aquifers is highly variable, ranging from 10^{-8} m^2 for sorted gravels to 10^{-18} m^2 for compact, clay-rich tills. Where glacier beds are at the pressure-melting point, the ice–bed interface is an important discontinuity along which water can flow. Flow resistance depends to a large degree on the configuration of the basal drainage system, and is discussed in section 3.4.

3.2.4 CHANNEL WALL PROCESSES: MELTING, FREEZING AND ICE DEFORMATION

Because ice can melt, refreeze and deform, the character and dimensions of glacial drainage pathways can adjust rapidly in response to changing conditions. Depending on the balance between melting, freezing and ice deformation, surface channels or conduits in or under the ice can either enlarge or contract through time (fig. 3.4a). Conduits that have been closed or constricted by adfreezing of water onto their walls are sometimes exposed on glacier surfaces, and have distinctive radial ice formations known as *crystal quirks* (Stenborg, 1969; fig. 3.4b).

Melting of ice (or freezing of water) occurs in response to energy exchanges between water and the surrounding ice. *Frictional heat* is generated when water flows along a channel or passage, due to dissipation of the energy required to overcome fluid viscosity. Some of this energy is used to raise the temperature of the water, but this will be offset by losses of energy used to melt ice. Isenko et al.

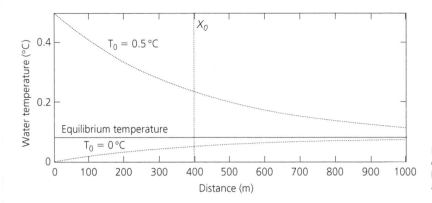

Figure 3.5 Changes in water temperature along a circular ice-walled conduit for two initial values. In both cases, the water temperature asymptotically approaches the equilibrium value (Isenko et al., 2005).

(2005) have shown that for any given flow conditions, there is an equilibrium water temperature for which viscous heat generation is exactly balanced by losses due to wall melting. Water with a higher (or lower) starting temperature will cool (or warm) asymptotically towards this equilibrium value (fig. 3.5). Field observations show that where channels are floored by sediments, water temperature is higher than predicted because the sediment layer inhibits heat exchange between the water and the ice.

Modelling the spatial and temporal evolution of water temperature along a glacial stream is a complex problem, and has been examined in detail by Clarke (2003). If the water is already at equilibrium with its surroundings and all the energy produced by viscous heating is used to melt ice, the melt rate $\dot{M}$ along a reach of length ΔL can be calculated from:

$$\dot{M} = \left(\frac{\rho_w g}{\rho_i L_f}\right) Q \left(-\frac{d\phi}{dL}\right) \Delta L \qquad (3.8)$$

where ρ_i is the ice density ($\sim 900\,\mathrm{kg\,m^{-3}}$), L_f is the latent heat of melting ($3.34 \times 10^5\,\mathrm{J\,kg^{-1}}$; section 2.2.1) and Q is the discharge. Because the terms inside the first pair of brackets can be taken to be constants, equation 3.8 shows that the primary controls on frictional melt rate are the discharge and head loss. The minus sign means that as head decreases downflow the melt rate is positive.

Melting or freezing can also occur at the walls of channels or passages if there is a temperature difference between the water and the surrounding ice. During glacier lake outburst floods, initial lake water temperatures can exceed 0 °C and advect significant amounts of heat into glacier drainage systems (section 3.7). Alternatively, surface meltwater at or close to 0 °C can come into contact with colder ice, encouraging freeze-on. A less intuitively obvious process is that melting of conduit walls or adfreezing of water can occur in response to changes in *pressure*. As explained section 2.3.1, the melting point of ice decreases with increasing pressure at a rate of 0.072 K per MPa. If, for example, ice and meltwater are both at a

pressure of 3 MPa at some point within a glacier (equivalent to an ice thickness of 340 m or a water column of 304 m), they can coexist at the pressure-melting point of −0.216 °C. However, if the water should experience a sudden reduction in pressure to 1 MPa (either through advection to a point with lower pressure head or by some other mechanism), its pressure-melting point will rise to −0.072 °C. Since the water is still at −0.216°, it is *supercooled* relative to its new situation and will begin to freeze (Alley et al., 1998; Lawson et al., 1998). Freezing releases latent heat, warming the refrozen ice (and remaining water) towards the new pressure-melting point until a new equilibrium is reached. Conversely, increases in pressure can lower the pressure-melting point of water, encouraging passage enlargement by melting. The rate of melting (or freezing) $\dot{M}_p$ as water adjusts to its new pressure-melting point along a flow path of length ΔL can be calculated from:

$$\dot{M}_p = c_w \cdot \dot{T}_m \left(\frac{\rho_w^{\,2} g}{\rho_i L_f}\right) Q \left(-\frac{dP_w}{dL}\right) \Delta L \qquad (3.9)$$

where c_w is the specific heat capacity of water ($4.217\,\mathrm{J\,g^{-1}\,K^{-1}}$), and $\dot{T}_m$ is the change in pressure-melting point with pressure (Röthlisberger and Lang, 1987).

The dimensions of englacial and subglacial passages can also adjust by ice deformation. As explained in sections 4.3 and 4.4, ice deformation will occur in response to spatial variations in pressure. Thus, a passage inside or below a glacier will contract (or enlarge) if there is a difference in pressure between the passage and the surrounding ice. This pressure difference is known as the *effective pressure*, N:

$$N = P_i - P_w \qquad (3.10)$$

To a good approximation, the ice pressure P_i is a simple function of the weight of the overlying ice:

$$P_i = \rho_i g(h_i - z) \qquad (3.11)$$

where h_i is the elevation of the ice surface and z is the elevation of the point in question. (In reality, horizontal pushes and pulls within the glacier may cause the ice pressure to differ somewhat from this value, although in most places the difference will be small; see section 4.3.1.)

If passages are not water-filled, P_w is zero and $N = P_i$. In water-filled passages, P_w can be less than, equal to or in excess of ice pressure. When $P_w = P_i$, $N = 0$, and when $P_w > P_i$, N is negative. The rate of passage closure (or expansion) can be calculated by inserting the effective pressure into Glen's Flow Law (section 4.4.1), adapted for the specific passage geometry (e.g. Röthlisberger and Lang, 1987; Hooke, 2005). For a circular passage, the rate of change of passage radius can be approximated by:

$$\frac{dr^*}{dt} = r^* \cdot A \cdot n^{-n} N^n \qquad (3.12)$$

where r^* is the passage radius, and A and n are the flow law parameters. Since n is approximately equal to 3, equation 3.12 shows that passage closure rate increases with the cube of the effective pressure. Closure rates will therefore be greatest for low-pressure passages below thick ice.

Passage adjustment by ice deformation will continue until $N = 0$. For this reason, it is assumed in some steady-state models that water pressure is equal to the ice overburden pressure (Röthlisberger, 1972; Shreve, 1972; Arnold et al., 1998; Pálsson et al., 2001). Thus, the hydraulic potential is defined as:

$$\phi = \rho_w g z + \rho_i g (h_i - z) \qquad (3.13)$$

where the first term on the right-hand side is the elevation head and the second term is the pressure head.

Where glacial drainage systems are fed by surface meltwater, however, rapid fluctuations in recharge rates mean that steady-state conditions will hardly ever occur (Röthlisberger and Lang, 1987). At times of rising recharge, glacial drainage systems may be too restricted to evacuate the accumulating water fast enough, and water will back up the system, causing water pressure to rise. Conversely, at times of falling recharge, passages are likely to be larger than they need to be. In this case, storage will decrease, effective pressures will rise in response to falling water pressure, and passages will contract by ice deformation. During the melt season, the timescale of recharge fluctuations is shorter than that required for passage adjustment, so steady state is never achieved. At times and places where surface meltwater does not reach the bed, drainage systems are more likely to be in equilibrium. Recent evidence from Antarctica, however, suggests that even below polar ice sheets drainage systems are not in steady state (Fricker et al., 2007). Equation 3.13 therefore represents a highly idealized situation which is

usually a poor match for real conditions. Because it is easy to apply, equation 3.13 has been very widely used in glacial-hydrological theory, although readers should be aware that such simple 'steady-state' formulations are likely to produce misleading results.

3.3 SUPRAGLACIAL AND ENGLACIAL DRAINAGE

3.3.1 SUPRAGLACIAL WATER STORAGE AND DRAINAGE

The contrasting permeabilities of snow and ice (section 3.2.3) mean that supraglacial drainage systems on snow-covered surfaces and bare ice are very different in character. When melting begins on snow surfaces, water readily percolates downwards through interconnected pore spaces. If deeper parts of the snowpack are below the freezing point, percolating water will refreeze and form *superimposed ice* (section 2.4.2). This process releases latent heat which gradually warms the snowpack, and when the entire snowpack is raised to the melting point refreezing will cease. Meltwater then accumulates within the snow. *Firn aquifers* can be several metres thick and can store a large amount of water, significantly delaying runoff from the glacier (Schneider, 2000; Jansson et al., 2003). Where hydraulic gradients are small (i.e. in areas of low surface slope), extensive *slush swamps* can form, making glacier travel difficult and unpleasant. Integrated drainage systems can be slow to develop in firn aquifers, and runoff can be delayed by several days or weeks. Initially, water tends to drain laterally through interconnected pores in the snow, then rills and surface channels form as water content increases. Water flowing in snow-banked channels can entrain rafts of snow or masses of slush which melt in transit, adding to discharge. Drainage becomes more efficient through time, releasing stored water and leading to a rapid increase in runoff from the glacier (Jansson et al., 2003).

Because intact glacier ice is essentially impermeable (section 3.2.3), meltwater does not percolate downwards, but runs off over the surface. Runoff can form sheet flows or follow incised channels. The rate of downcutting by a stream flowing over glacier ice can be calculated using an expression similar to equation 3.8, modified for open-channel conditions (Fountain and Walder, 1998):

$$\dot{a} = \frac{1}{2} \left(\frac{\pi}{2\tilde{n}} \right)^{3/8} \left(\frac{\rho_w g}{\rho_i L_f} \right) S^{19/16} Q^{5/8} \qquad (3.14)$$

This shows that incision rates are approximately linearly proportional to channel slope S (i.e. the rate of head loss), and less strongly dependent on discharge Q and channel roughness $\tilde{n}$. Observations of supraglacial streams

Figure 3.6 Supraglacial channel on Longyearbreen, Svalbard. This channel is a perennial feature and is reoccupied every melt season (D.I. Benn).

confirm that incision rates are highest at steep reaches, and channels tend to evolve rapidly by the upstream migration of nickpoints (Gulley et al., 2009a). Because ice-walled channels have low roughness, flow velocities tend to be large relative to sediment- or rock-floored channels. Falling into supraglacial streams is not recommended.

For supraglacial channels to persist, channel downcutting rates must exceed the rate of ablation of the adjacent ice surfaces (Gulley et al., 2009a). Where subaerial melt rates are low, such as in cool environments or on debris-covered glaciers, deep, perennial channels can form (fig. 3.6). Such channels commonly meander, and can exhibit great regularity of wavelength and amplitude (Knighton, 1972). In sediment-floored streams, meandering develops in association with erosion of the outer part of bends and deposition in the inner part. There is no equivalent depositional process in ice-walled supraglacial channels, and meanders evolve by a combination of erosion and downcutting. Meander amplitude tends to increase through time, and the abandoned upper parts of channels are progressively removed by surface melting. Surface drainage patterns can also be strongly guided by structures such as crevasses and foliation, sometimes forming channel networks with strong linear elements.

3.3.2 ENGLACIAL DRAINAGE

It has long been recognized that englacial drainage systems can convey surface meltwater to the beds of temperate glaciers, where increased water storage can trigger accelerated basal motion (Iken and Bindschadler, 1986; Willis, 1995; sections 4.4 and 4.5). Until recently, it was believed that this could not occur where ice is below the pressure-melting point, since 'cold' ice was thought to act as an effective barrier to water flow (Hodgkins, 1997). It has now become clear that High Arctic glaciers and many

parts of the Greenland Ice Sheet can exhibit dynamic responses to surface melting on short timescales, indicating efficient routing of surface meltwater to the bed via englacial conduits (section 6.2.3; Zwally et al., 2002a; Boon and Sharp, 2003; Copland et al., 2003a; van de Wal et al., 2008). The realization that climatic signals can influence ice sheet dynamics has led to renewed interest in englacial drainage systems, and their role in coupling the surface and basal boundary conditions of glaciers and ice sheets.

For many years, glaciological ideas about the character and evolution of englacial drainage systems were deeply influenced by the theoretical model developed by Shreve (1972). This model is based on three main assumptions:

(1) the englacial drainage system is in steady state;
(2) englacial water will flow along the steepest hydraulic gradient within the glacier;
(3) pressure head is equal to the pressure of the surrounding ice minus a small component due to melting of the walls (equation 3.13; section 3.2.2).

The assumption of steady state means that discharge is unchanging and that the rate of tunnel enlargement by melt must be everywhere exactly equal to the rate of tunnel closure by ice creep.

Importantly, Shreve (1972) noted that this conception of englacial hydraulic potential strictly only applied to water-filled passages, but added that 'with suitable caution it is permissible and more convenient for most purposes to treat it as if it were defined throughout the ice' (p. 207). Proceeding on this basis, Shreve reached the following conclusions. First, equipotential surfaces within the ice (i.e. planes along which ϕ is constant) will dip upglacier at approximately 11 times the surface gradient. (Röthlisberger and Lang (1987, p. 251) showed that this is an approximation which only applies where the ice surface slope is small; the factor is not always 11, but decreases with increasing surface slope.) Second, large conduits will be at lower pressure than small ones, because the wall melting rate increases with discharge, which in turn increases with conduit radius. Third, water will flow through the glacier at right angles to the equipotential surfaces, which are locally distorted by the presence of large, low-pressure conduits. From the above, englacial drainage systems were predicted to form downglacier-dipping, branching conduits fed by networks of veins (fig. 3.7).

The Shreve model has been widely adopted as a fundamental component of englacial drainage theory (e.g. Röthlisberger and Lang, 1987; Fountain and Walder, 1998; Hooke, 2005). There is no evidence, however, that the model provides a realistic picture of actual drainage systems within glaciers. Observed englacial drainages show no discernible tendency to follow theoretical

(a)

(b)

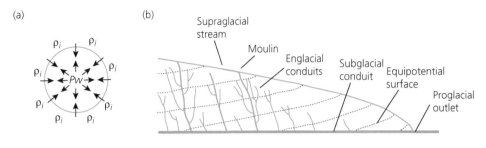

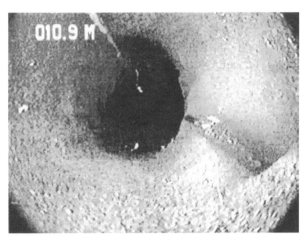

Figure 3.8 Englacial conduit imaged by a down-borehole video camera, Haut Glacier d'Arolla. The dark hole near the centre of the image is the borehole (looking down), and the conduit enters from the right-hand side. The conduit had a 'keyhole' shape and was 15 cm high and 4 cm wide (Copland et al., 1997).

Figure 3.7 The Shreve conception of englacial drainage. (a) The conduit will contract if water pressure is less than ice pressure, and enlarge due to wall melting. (b) Arborescent system of conduits aligned normal to equipotential surfaces.

Figure 3.9 A moulin on Midre Lovénbreen, Svalbard. Dye is being poured into the moulin by Tris Irvine-Fynn as part of the tracing experiment

potential gradients (e.g. Vatne, 2001; Stuart et al., 2003; Gulley and Benn, 2007; Gulley et al., 2009) and generally do not behave as predicted by the model. This is a reflection of the highly idealized nature of its assumptions.

Crucially, the model depends on the assumption that water can flow freely through glacier ice along the steepest potential gradient, whereas under most circumstances the extremely low primary permeability of glacier ice provides a formidable barrier to water flow. Lliboutry (1996) has argued that infiltration of water through intact glacier ice will be negligible, except where internal water production and hydraulic gradients are very large – conditions that are met near the beds of glaciers sliding fast over a rough bed. In such cases, water is thought to migrate through the vein network from high-pressure areas above the upstream side of bumps, where melting occurs, to low-pressure areas above the lee side, where it refreezes (Lliboutry, 1993; section 4.5.2). Elsewhere on glaciers, however, water flow through intact glacier ice is probably negligibly small.

Important insights into the character of englacial drainage systems have been provided by examination of ice cores and down-borehole photography and video imaging (e.g. Raymond and Harrison, 1985; Pohjola, 1994; Harper and Humphrey, 1995; Copland et al.,

1997; Fountain et al., 2005). Voids with a variety of forms have been observed, ranging from millimetre-scale tubes to planar fracture-like features up to 20 cm across (fig. 3.8). In an extensive drilling programme in Storglaciären, Sweden, Fountain et al. (2005) found that 80 per cent of observed voids had fracture-like geometries, leading them to argue that the englacial hydrological system of the glacier consists of a distributed network of fractures. Boreholes and cores, however, sample only a tiny part of the total volume of a glacier, and provide an inevitably incomplete picture of the three-dimensional form of englacial drainages.

The role of fractures in routing surface meltwater into glaciers has long been recognized. On glacier surfaces, sinkholes or *moulins* (from the French word meaning 'mill') are often associated with crevasses or crevasse traces (fig. 3.9). Detailed field studies by Stenborg (1969, 1973) demonstrated that moulins tend to form at the upglacier end of crevasse fields, when fractures capture drainage from supraglacial streams. The formation of new moulins leads to the abandonment of older moulins downstream when their water supply is cut off. The strong influence exerted by crevasse distribution on the location of moulins means that englacial drainage can be routed in directions very different from those predicted by Shreve's (1972) theory.

Figure 3.10 Artur Adamek prepares to descend a vertical shaft in Hansbreen, Svalbard (J. Gulley).

Moulins and other hydrologically significant passages on glaciers are commonly large enough to permit human entry, opening up the possibility of direct exploration by ice caving or *glacio-speleology* (fig. 3.10). This important source of information on the formation and characteristics of englacial conduits has, until recently, received very little attention in the English-language glaciological literature (e.g. Holmlund and Hooke, 1983; Holmlund, 1988), although it has received much more recognition in other traditions (e.g. Jania, 1996). Indeed, glacio-speleology was dismissed in the first edition of this book as a 'questionable practice'. One of the authors, however, has now recanted and is an enthusiastic proponent of ice caving (although the other author believes he has taken leave of his senses). Glacio-speleology actually has a very long history and an extensive literature (e.g. Pulina and Rehak, 1991; Anderson et al., 1994; Badino, 2002; Badino et al., 2007), and has much to teach glaciology about the rich diversity of drainage systems hidden below the surface. It must be emphasized, however, that glacier caving is potentially dangerous, and should only be attempted by those with experience of both ice climbing and speleological rope work.

Speleological observations indicate that englacial passages can form in three main ways:

(1) incision of surface streams followed by roof closure;
(2) hydrologically enhanced ice fracturing;
(3) the exploitation of pre-existing permeable structures within the ice (Gulley et al., 2009).

The idea that englacial conduits can form by incision and closure was proposed by Fountain and Walder (1998), who envisaged that streams flowing along the bottom of crevasses would progressively cut down towards the bed and become isolated from the surface when abandoned levels of the passage were squeezed shut by ice creep. There is now abundant evidence that cut and closure is an effective mechanism of conduit formation, although the presence of crevasses is neither a necessary nor a sufficient

condition for channel initiation (Gulley et al., 2009a). The crucial factor is that incision rates must exceed subaerial melt rates on the surrounding ice, and provided this condition is met, perennial surface streams will become more deeply entrenched through time. While crevasses may locally guide drainage patterns, all of the examples studied to date were initiated as supraglacial melt streams on uncrevassed parts of glaciers.

Cut-and-closure channel formation is favoured where surface ablation rates are low (i.e. cool environments or debris-covered glacier surfaces) and incision rates are high (i.e. in areas of high stream discharge and surface slope; equation 3.8). Since supraglacial stream discharges are a consequence of surface melting, this implies that relatively large catchments are necessary to provide sufficient water. As predicted by Fountain and Walder (1998), conduit roofs can squeeze shut by ice creep, but the upper parts of passages can also become blocked by drifted snow, refrozen meltwater, and masses of rafted ice blocks (fig. 3.11). The processes of channel incision and roof closure result in distinctive meandering vadose canyons with prominent sutures (lines of closure) running along the roof. Tubular morphologies can develop by the re-enlargement of restricted reaches during pipe-full (phreatic) flow conditions. Cut-and-closure conduits typically have long low-gradient sections interrupted by flights of steep steps marking nickpoints in the stream. Incision can proceed until channels reach the glacier bed, or to the base level of the glacial hydraulic system (such as the elevation of a riegel or the top of an impermeable sediment barrier). Such conduits, however, are prone to blockage during the winter months by refreezing of meltwater and ice creep, so that the flow is re-routed to a higher exit point in the following summer, bypassing blocked reaches (fig. 3.12). Downward incision typically proceeds at rates of a few metres per year, but blockages can cause the position of the active conduit to switch upwards by tens of metres in a single season. Thus, cut and closure is a relatively inefficient mechanism for routing surface water to the bed, except where glaciers are very thin (Gulley et al., 2009a).

A much more rapid mechanism for routing surface water to a glacier bed involves the propagation of water-filled crevasses (Boon and Sharp, 2003; Gulley et al., 2009b). It has long been theorized that water-filled crevasses can penetrate all the way to the bed of a glacier, and the possible role of this process in englacial drainage evolution was discussed at length by Röthlisberger and Lang (1987). In recent years, interest in the process has been stimulated by the observation that accelerated basal motion follows the onset of surface melting on the Greenland Ice Sheet (Zwally et al., 2002a; van de Wal et al., 2008). The role of water in the propagation of crevasses has been modelled by van der Veen (1998a, 2007)

Figure 3.11 Cut-and-closure englacial conduits in Svalbard. (a) Meandering plug of snow on the surface of Longyearbreen, marking the trace of a closed canyon. (b) The upper levels of this conduit in Longyearbreen have been sutured by ice creep, while refrozen meltwater has accumulated on the floor. Closure rates were determined using the pairs of stakes installed in the walls. (c) A spectacular nickpoint inside Scott Turnerbreen. (d) Tubular passage formed by the re-enlargement of a sutured canyon, Longyearbreen (F. Strozzi, N. Hulton, J. Gulley).

and Alley et al. (2005b), and is discussed in detail in section 4.8.1. In essence, crevasses will grow if the forces tending to open them (tensile stresses in the ice) are greater than the resisting forces (ice strength and compressive stresses associated with overburden pressure). Ice overburden pressure increases rapidly downwards, and the resulting compression increasingly offsets any tensile stresses tending to pull the crevasse open, thus limiting the depth of penetration of surface crevasses. The presence of water in a crevasse, however, critically alters the force balance by pressing outwards on the walls, cancelling out the overburden pressure effect. Thus, where sufficient water is available, crevasses can keep on extending downwards until they meet the bed. Theoretical analyses indicate that this can happen on a timescale of hours to days.

Ice cave exploration has provided spectacular evidence in support of this model. Benn et al. (2009a) described examples of englacial conduits formed by hydrologically assisted fracturing (or 'hydrofracturing') in widely differing glacial environments. The common factor at all sites is that the key conditions required for deep fracturing are met – an abundant water supply and ice subject to tensile stresses. The bed was reached at one of the sites, Crystal Cave in Hansbreen, Svalbard, which has been explored several times since its discovery in the 1980s (Schroeder, 1995). Crystal Cave lies at the confluence of two glaciers, where water from lateral surface streams fills a large hollow early in each melt season. The resulting lake breaks into the snow-blocked cave entrance as the melt season progresses, and every few years creates a new vertical shaft by hydrofracturing. Older abandoned shafts are slowly carried downglacier by ice flow, and gradually close by a combination of ice infilling and creep. In October 2006, one shaft was followed to the glacier bed 65 m below the surface, where it led into a till-floored subglacial conduit (fig. 3.13). Within 10 m of the bed, the ice was at the pressure-melting point. Water oozing from the walls soaked gloves and clothing, which then refroze

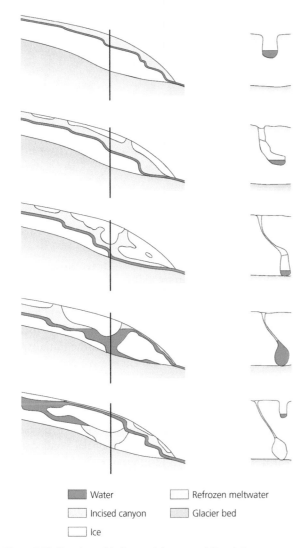

Figure 3.12 Generic model of cut-and-closure conduit evolution (Gulley et al., 2009a).

Water

Incised canyon

Ice

Refrozen meltwater

Glacier bed

on reascent to higher levels of the glacier. When revisited in April 2007, the cave system was flooded to within 15 m of the surface, indicating englacial storage of at least $1.3 \times 10^6 \, \text{m}^3$ of water.

Exploration of other sites has shown that fracture-type englacial conduits are very commonly associated with ephemeral lakes at glacier confluences. Water draining from such lakes can also directly enter subglacial transport by penetrating along the ice–bed interface, a process that also appears to operate at places such as Gornersee in Switzerland (Huss et al., 2007). Although the conditions for surface-to-bed fracturing are often met at glacier confluences, the essential ingredients of sufficient water supply and stressed ice can be brought together in a wide range of situations, including shear zones at glacier margins, the upglacier ends of crevasse fields, and below supraglacial lakes in areas of extending or compressive flow (Stenborg, 1973; Holmlund and Hooke, 1983; Holmlund, 1988; Boon and Sharp, 2003; Box and Ski,

Figure 3.13 Surface-to-bed drainage via hydrologically assisted fracturing, Hansbreen, Svalbard. (a) Moulin developed along a fresh fracture. (b) Older vertical shaft, greatly modified by flowing water. (c) Sheared, temperate ice near the glacier bed. (d) Subglacial continuation of the drainage system (J. Gulley).

Figure 3.14 Englacial conduit developed along permeable debris structures, Ngozumpa Glacier, Nepal. (a) Three linked conduit entrances formed along a debris band. (b) Inner part of passage, with phreatic upper level and incised floor (D.I. Benn and J. Gulley).

2007; Catania et al., 2008; Benn et al., 2009a). There is good reason to suspect that hydrofracturing is a very widespread process, allowing surface meltwater rapid access to the beds of both temperate and polythermal glaciers. The role of this process in modulating flow of the Greenland Ice Sheet is discussed in section 6.2.3.

Hydrofracturing can also form englacial passages upwards from the glacier bed, when pressurized water forces new escape routes to the surface. Spectacular evidence for this process is the sudden appearance of fountains of water rising above glacier surfaces (e.g. Roberts et al., 2000; Pälli et al., 2003). Situations where 'bottom-up' fracturing appears to be important are at the boundary between warm and cold basal ice in polythermal glaciers, where downglacier movement of subglacial water is impeded by a permafrost barrier at the margin (Murray et al., 2000; Wadham et al., 2001a), and during outbursts of subglacial lakes (Roberts et al., 2000; Roberts, 2005). Fractures (either through the proglacial sediment or the glacier tongue) essentially introduce a secondary permeability, increasing the capacity of the system to permit higher discharges. Low, wide englacial passages formed along probable 'bottom-up' fractures on the Aldegonda Glacier, Svalbard, have been described by Mavlyudov (2005).

The final mechanism of englacial conduit formation for which there is speleological evidence is the exploitation of pre-existing secondary permeability. Gulley and Benn (2007) described several examples of englacial conduits that extend between supraglacial lake basins on glaciers in the Everest region of Nepal. The conduits follow relict debris-filled structures, which provide permeable pathways through otherwise impermeable ice. Where such structures bridge between a lake floor and a lower point on the glacier surface, water is driven out of the lake along the gradient in hydraulic potential. As water leaks along the debris band, water flow enlarges a passage, and a positive feedback between water discharge and passage size eventually creates a conduit. Initially, such conduits

tend to have tubular (phreatic) cross sections, but then vadose incision towards base level creates a distinctive 'keyhole' morphology (fig. 3.14). This process is directly analogous to speleogenesis in soluble rocks, where fissures such as joints and bedding planes allow the initial water flow that leads to the development of cave systems. Open fractures in clean ice can also be exploited by meltwater, and nucleate englacial drainage systems (Gulley, 2009).

An important development in the study of englacial drainage systems combines speleological exploration with surface geophysical surveys. Englacial voids can be detected by careful high-resolution radar surveys, and their characteristics inferred by comparing observations with model predictions based on a range of alternative geometries and water levels (Stuart et al., 2003). Ice cave exploration can greatly facilitate this type of work by providing ground truth in accessible areas, whereas geophysical surveys can be used to map inaccessible parts of an englacial drainage system. Pioneering work of this kind was conducted by Stuart et al. (2003) on Austre Brøggerbreen, Svalbard, where Vatne (2001) had mapped part of the englacial drainage system. The characteristics of the system indicate that it is of cut-and-closure type.

3.4 SUBGLACIAL DRAINAGE

Subglacial hydrology is one of the most important branches of glaciology, because of the profound influence it exerts on ice dynamics (chapter 5). Great advances in understanding subglacial drainage systems have been made in recent years, mainly as a result of carefully designed field experiments in conjunction with theoretical analyses, but the inaccessibility of most glacier beds and potential ambiguities in the available data mean that important problems still remain. A complete theory of subglacial drainage evolution remains an important but elusive goal.

Subglacial drainage systems can be subdivided into two main categories: *channelized systems*, in which water

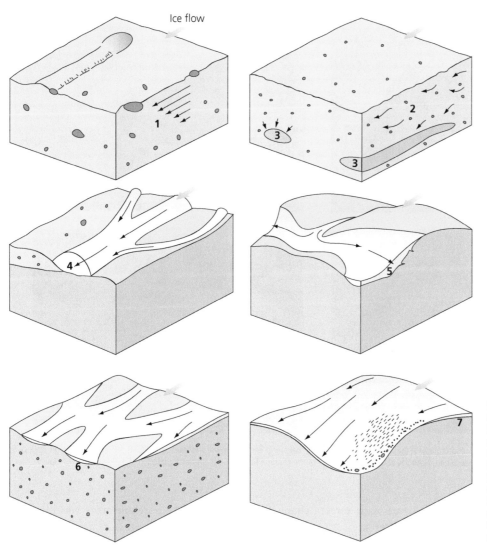

Figure 3.15 Possible configurations of subglacial drainage systems. (1) Bulk movement of water with deforming till. (2) Darcian pore-water flow. (3) Pipe flow. (4) Dendritic channel network. (5) Linked cavity system. (6) Braided canal network. (7) Thin film at the ice–bed interface. Types 1, 2, 5, 6 and 7 are distributed systems, and types 3 and 4 are channelized systems.

is confined to relatively narrow conduits, and *distributed systems* which extend over larger proportions of the bed. Within these categories, a number of drainage configurations have been identified or inferred from theoretical considerations (fig. 3.15):

1. Channelized systems:
    (a) Röthlisberger (R-) or Hooke (H-) channels – incised up into ice
    (b) Nye channels (N-channels) – incised down into rock or sediment
    (c) Tunnel valleys – large channels incised down into rock or sediment.
2. Distributed systems:
    (a) water film – between ice and rock or sediment
    (b) linked cavity network – between ice and rock
    (c) braided canal network – between ice and sediment
    (d) groundwater flow – within subglacial sediment or rock.

Channelized systems are generally relatively efficient pathways, allowing rapid flow through well-connected

networks. In contrast, distributed systems are generally less efficient, and meltwater follows more tortuous routes through or over the bed. The type of drainage system has important implications for glacier dynamics, by controlling the distribution of stored water and frictional resistance at the bed (sections 4.5 and 4.6).

3.4.1 SUBGLACIAL CHANNELS

Subglacial channels may be incised upwards into the ice (*Röthlisberger* and *Hooke channels*), cut down into the glacier bed (*Nye channels* and *Tunnel valleys*), or a combination of the two. Examples of both upward- and downward-incised channel reaches have been observed at glacier beds, sometimes within a single system (fig. 3.16).

The characteristics of subglacial channels were first investigated theoretically by Röthlisberger (1972). By assuming steady-state conditions in which channel enlargement by wall melting is exactly balanced by creep closure (section 3.1.4), he was able to model relationships between discharge and water pressure along idealized

Figure 3.16 Subglacial conduits below Svalbard glaciers. (a) Upward-incised conduit, Bakaninbreen. (b) Upward-incised conduit, Fuglebreen. Note large scallops in the roof. (c) Nye channel incised in till, Rieperbreen. Note the very high roughness of the channel floor (d). Nye channel incised in rock, Paulabreen (J. Gulley).

conduits. Röthlisberger argued that because melt rates increase with water flux (equation 3.8), water pressure will decrease with discharge. Thus, water pressure in large channels will be lower than in smaller ones, leading to the development of branching (arborescent) subglacial drainage systems, in which large channels draw in water from their surroundings. There is good evidence that branching channel networks do occur below glaciers, because tracer studies demonstrate that water from multiple surface input points can be transported rapidly to a single outlet at the glacier terminus (e.g. Fountain, 1993; Nienow et al., 1998).

While the assumption of steady state allowed important insights to be gained in the early years of modern glacier hydrology, it is unlikely to be representative of natural conditions in most subglacial conduits. Below mountain glaciers, subglacial conduit systems develop to allow the efficient evacuation of surface-derived meltwater, routed to the bed via moulins (section 3.3.2). Surface water inputs to the bed fluctuate dramatically in response to diurnal and weather-related variations in surface melting. Hubbard et al. (1995) elegantly demonstrated pressure variations in the vicinity of a subglacial conduit by monitoring water levels in an array of boreholes in the Glacier d'Arolla, Switzerland. A subglacial

channel running below the borehole array was identified from the presence of a *variable pressure axis* (VPA) characterized by low minimum water pressures and high diurnal water pressure variations (fig. 3.17). Part of the record from five of the boreholes is shown in figure 3.18, and reveals a fascinating picture of water-pressure variations in the conduit and its environs. Holes 29 and 35, located 10 m and 5 m, respectively, from the centre of the VPA, show the largest diurnal variation, with water-level fluctuations equivalent to a pressure range of ~1 MPa. In contrast, holes 42 and 43, located 50 m and 75 m from the VPA, show only small diurnal variations, but a higher mean. Hole 40, 25 m from the VPA, has an intermediate pattern of pressure variation. Importantly, the records show episodic reversal of the pressure gradient between holes 29, 35 and 40, on one hand, and 42 and 43, on the other. At times of high discharge during the afternoon, water pressure in the channel is higher than in the surrounding bed, so water is driven out of the channel. Conversely, at times of low discharge at night and during the morning, water pressure in the channel is lower than in the surrounding bed, so water at the bed flows towards the channel. Hubbard et al. (1995) argued that water exchange between the channel and the surrounding bed occurred through a permeable till layer, although it is

(a)

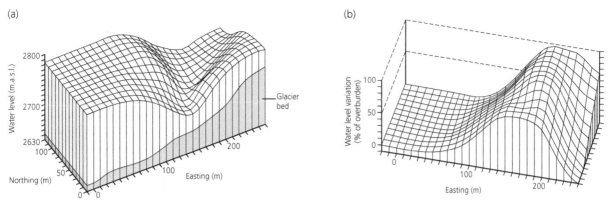

(b)

Figure 3.17 The variable pressure axis at Glacier d'Arolla. (a) Minimum borehole water levels (metres above sea level). (b) Amplitude of water level variation (Hubbard et al., 1995).

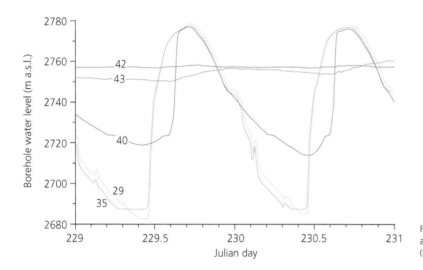

Figure 3.18 Diurnal variation in borehole water levels at different distances from the variable pressure axis (Hubbard et al., 1995).

possible that at least some of the flow was carried in a water film or similar system at the ice–till interface. It should be noted that the daily mean pressure in the channel was lower than in the surrounding bed, indicating that on diurnal timescales the channel was a net sink for water produced at the bed.

Röthlisberger (1972) envisaged that subglacial conduits would be semicircular in cross section (although his initial model dealt with the computationally simpler case of cylindrical channels). On the basis of dye-tracer returns and borehole investigations, Hooke et al. (1990) and Hock and Hooke (1993) argued that subglacial conduits are more likely to have low, broad cross sections. They reasoned that such cross profiles will be favoured because: (a) when conduits are not full, melting will be focused at the margins rather than the ceiling; and (b) closure by creep will be less efficient at conduit margins than the ceiling because of ice–bed friction. Direct observations indicate that conduit cross profiles are quite variable. Near their lower entrances (*portals*), many conduits have semicircular cross sections, although this mainly reflects roof collapse rather than wall melting by water. (*This also underlines the very high risks involved in entering portals,*

especially during the melt season.) At greater depths inside glaciers, subglacial conduits tend to be low and broad, as predicted by Hooke (fig. 3.19a). In some cases, however, conduits are not simple, single-branched tunnels, but display anastomosing side branches that appear to accommodate excess water at times of peak flow (fig. 3.19b). Such observations are important, as they show that channelized drainage systems may have more complex forms than is commonly assumed in modelling studies.

Further insights into the character of subglacial drainage can be gained from the study of Nye channels incised into former glacier beds. Nye channels can occur as single, isolated features or in braided networks covering larger areas (Walder and Hallet, 1979; Sharp et al., 1989b). The presence of Nye channels implies that water flow is consistently focused along the same route. This is most likely to occur where bedrock topography exerts a strong control on the hydraulic gradient at the bed, such as steep-sided valleys or rough glacier beds. Indeed, incised subglacial meltwater channels do commonly occur along the axes of valleys, wind between bedrock bumps or cut across cols on ridges.

Figure 3.19 (a) Low, wide subglacial conduit, Bakaninbreen. (b) Nye channel cut in till, with the entrance of an upward-incised side branch to the left of the figure, Rieperbreen (J. Gulley).

The fact that hydraulic potential at glacier beds is determined by water pressure as well as elevation (equation 3.1) means that subglacial channels need not follow the bed slope, but can travel across the slope or even uphill. This is an important difference between subglacial channels and surface streams, the flow of which is determined only by differences in elevation.

3.4.2 WATER FILMS

There is abundant evidence that, where temperate ice is in contact with a hard bed, ice and rock are typically separated by a thin film of water (e.g. Hallet, 1979a; Lappegard et al., 2006). Such films have limited ability to transport water and are believed to carry only water produced by local basal melting. Meltwater will be produced at glacier beds if the ice is at the pressure-melting point, and more energy is generated by geothermal and frictional sources than is conducted away through the ice (section 2.3.2). Melt rates are locally enhanced on the upglacier sides of bumps, where ice–bed contact pressures are higher than average, and the pressure-melting point is lower. Conversely, on the downglacier side of bumps, contact pressures are low and the pressure-melting point is elevated. This encourages refreezing, a process known as *regelation* (section 4.5.2; Hubbard and Sharp, 1993). The principal role of the subglacial water film below many glaciers appears to be to transfer water from areas of pressure-melting to areas of regelation; for this reason it is sometimes known as the *regelation water film* (Hallet, 1979a). Amounts of melting and freezing may not exactly balance, however. Isotopic evidence indicates that, where there is an excess of melting over freezing, the excess probably enters channels or other drainage networks and is lost to the water film (Sharp et al., 1990; Hubbard and Sharp, 1993). Conversely, water films may gain water from adjacent channels at times of high discharge (section 3.4.1).

Although subglacial water films have been observed below many glaciers, their characteristics are very difficult to measure directly. Consequently, much of what we know about subglacial water films has been determined by indirect methods. For example, calcite ($CaCO_3$) deposits on recently deglaciated limestone surfaces have been widely used to reconstruct the thickness, extent and chemistry of former subglacial water films (e.g. Hallet, 1979a; Sharp et al., 1990; Hubbard and Hubbard, 1998; Ng and Hallet, 2002; Carter et al., 2003). On glacier beds composed of limestone, pressure-melting on the upglacier sides of bumps produces undersaturated, chemically aggressive water which dissolves calcite. Freezing on the lee side of bumps elevates the solute concentration in the remaining water, which results in precipitation. The resulting calcite deposits incorporate fine rock particles, which are inferred to have been transported to the site of precipitation in the water film. The size distribution of the particles thus provides a limiting value on film thickness. Included particles are generally <20 μm in diameter, with only a few per cent exceeding 50 μm. Rarely, sand-sized particles are found (Hallet, 1979a; Carter et al., 2003; fig. 3.20). This implies that regelation water films are usually only around 20 μm thick, with localized thickening to several tens or even hundreds of micrometres. The distribution of the coarsest particles in lenses suggests that flow may be concentrated in irregular microchannels separated by protuberances of the bed.

According to detailed modelling by Ng and Hallet (2002), spatial variations in film thickness also result from interactions between bed roughness, refreezing efficiency, calcite precipitation and film water pressure. On carbonate rockbeds these variations are amplified through time as a result of preferential precipitation. Clearly this feedback does not operate on non-soluble rockbeds, although the Ng and Hallet model also explains how variations in film thickness and pressure will arise in response to small bed irregularities.

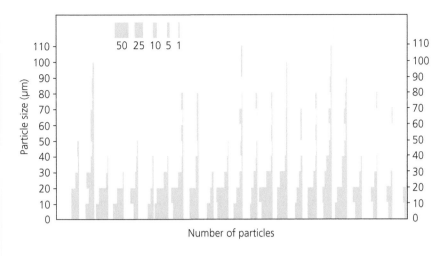

Figure 3.20 Size distribution of rock fragments in subglacial carbonate deposits, indicating typical thickness of the subglacial water film (After Hallet, 1979a).

3.4.3 LINKED CAVITY SYSTEMS

Water-filled cavities will open up at the interface between a glacier and its bed where basal water pressure exceeds the local ice pressure. This situation is favoured on the downglacier side of bumps, where ice pressure normal (at right angles) to the bed is at a minimum (fig. 3.21). Two basic cavity geometries can be defined: step cavities, where the downglacier side of a bump has an abrupt step, and wave cavities, where the bed varies smoothly (Kamb, 1987; fig. 3.22). The mechanisms of cavity formation are essentially the same in both cases, although the threshold pressure conditions for cavity formation and growth differ. Mechanisms of cavity formation were discussed in detail by Lliboutry (1979) and Fowler (1987).

The idea that subglacial water can flow through systems of interconnected (linked) cavities was first advanced by Lliboutry (e.g. 1979), and was developed further by Walder (1986) and Kamb (1987). Linked cavity systems are envisaged to consist of numerous lee-side cavities connected by narrow connections or *orifices* (fig. 3.23). Water transit velocities will be low because of throttling of flow in orifices, the tortuous nature of the network, and temporary storage in poorly connected cavities. Indeed, not all water-filled cavities need to be integrated into a single network at any one time. The extent of the active network will vary with water storage, as cavities increase or decrease in size and the number of open orifices changes (Kamb, 1987; Sharp et al., 1989b). Striking evidence for linked cavity networks can be found on recently deglaciated limestone surfaces, where former water-filled cavities are readily identifiable from solutional features (section 8.4.4; Walder and Hallet, 1979; Sharp et al., 1989b).

Theoretical analyses indicate that discharge-pressure relationships in linked cavity networks differ greatly from those in channelized systems (Fowler, 1987; Kamb, 1987). For small discharges, ice melting tends to be an unimportant mechanism of cavity enlargement, because

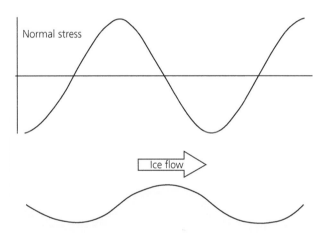

Figure 3.21 Variation of normal stress over a wavy bed. Normal stress fluctuates around the average value, and is greatest on the upstream sides of bumps and lowest on the downstream sides.

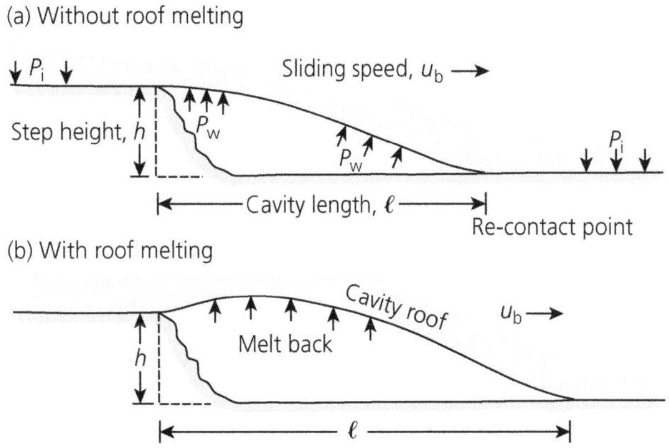

(a) Without roof melting

(b) With roof melting

Figure 3.22 Geometry of an ideal cavity in the lee of a step on the glacier bed ('step cavity') (Hooke, 2005, after Kamb, 1987).

of the large surface area of the drainage system compared with its volume. Cavity dimensions, therefore, are mainly determined by the mechanics of ice–bed separation and are defined by the condition that $P_w > P_i$. Therefore, increases in water pressure will increase the volume and

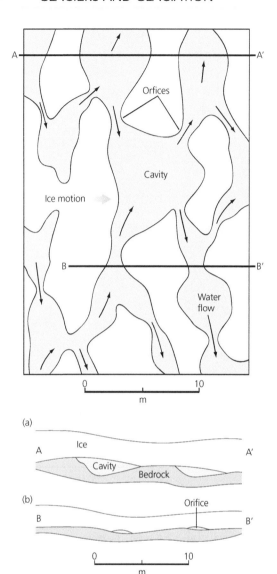

Figure 3.23 Schematic diagrams of a linked cavity network. (a) Plan view. (b) Cross sections through cavities (A-A') and orifices (B-B') (Hooke, 2005, after Kamb, 1987).

the carrying capacity of the system. This positive relationship between pressure and discharge is precisely the opposite of the steady-state behaviour of channels, where increases in discharge are associated with decreases in water pressure (section 3.4.1). Whereas large, low-pressure channels tend to capture drainage from smaller, high-pressure ones, no such tendency exists for water-filled cavities at low discharges. Up to a point, the pressure in larger cavities will be greater than in smaller ones, and there will be no tendency for such cavities to capture drainage or undergo growth at the expense of others. Within limits, therefore, linked cavity systems are expected to be stable features, with no tendency to evolve into branching channel systems.

This relationship begins to break down at high discharges, when ice melting becomes more important than water pressure in determining the size of cavities. As is

the case for channels, wall melting offsets cavity closure resulting from ice creep, so where melting rates are high P_w can fall below P_i. Ice melt rates increase with discharge, so increasing water discharges through a cavity will result in a larger pressure drop. Therefore, large, low-pressure cavities will capture drainage from smaller, high-pressure cavities, and grow. With continued cavity growth, drainage will be increasingly carried by large, well-connected cavities, which may evolve into a channel. The instability of linked cavity systems at high discharges provides a mechanism whereby subglacial drainage can 'switch' to a more efficient mode, allowing large amounts of water to be evacuated from the bed.

These theoretical results imply that linked cavity drainage systems are most likely to occur where (1) water inputs to the system are small, or (2) fast basal motion over rough glacier beds encourages extensive cavity formation. The former condition is typical of winter, when little or no water is routed from the surface and subglacial drainage consists of basally produced meltwater and water stored at the bed during the preceding melt season.

Experiments conducted at the Svartisen Subglacial Laboratory (SSL) below Engabreen, Norway, have provided good evidence for the existence of a linked cavity network during the winter months (Lappegard and Kohler, 2005; Lappegard et al., 2006). Built as part of a subglacial hydroelectric scheme, SSL is one of the few localities where routine observations can be made of a glacier bed below thick ice (fig. 3.24). Artificial cavities can be melted in the basal ice by hot water injected from a tunnel below, allowing instruments to be installed in the rockbed before the cavity closes again by ice creep. Lappegard and Kohler (2005) conducted a series of pump tests, during which pressurized water was introduced to the ice–bed interface and the resulting pressure fluctuation measured by an array of load cells fixed into the bed. Propagation of pressure waves during winter pump tests was consistent with a high-pressure drainage system with a high degree of spatial connectivity. Records of natural pressure variations over longer time periods also indicated the existence of a low-capacity, distributed drainage system during the winter months (Lappegard et al., 2006). In contrast, during the summer months, pressure fluctuations at the bed were consistent with the close proximity of a high-capacity, low-pressure channel.

3.4.4 GROUNDWATER FLOW

Groundwater flow through hard rockbeds is generally insignificant owing to the low permeability of most rocks. Important exceptions occur where glaciers are underlain by limestones, and subglacial water can be routed into cave systems such as Castleguard Cave below the Columbia Ice Field, Alberta (Ford et al., 2000). Unconsolidated sediments occur below parts of many

Figure 3.24 Svartisen Subglacial Laboratory. (a) Tunnel in bedrock beneath the glacier. Access to the glacier bed can be gained via the vertical shaft in the back of the picture. (b) Miriam Jackson taking a sample of basal ice from the bed of Engabreen (Miriam Jackson and John Telling).

glaciers and ice sheets, forming potential aquifers through which water can flow. Two basic mechanisms have been proposed for groundwater flow through such aquifers. First, if the sediment undergoes shear as the result of glacially imposed stresses, bulk movement of water will occur as it is carried along with the mineral grains (Clarke, 1987a). Second, pore water can flow relative to the mineral skeleton under a hydraulic gradient (section 3.2.3). This process is known as *Darcian flow* after the French civil engineer Henry Darcy (1803–58) who conducted pioneering work on water flow through soils. Subglacial deforming layers are usually thin (section 4.6), so the second mechanism is generally much more important than the first.

Darcy's Law (equation 3.5) shows that pore-water discharge is proportional to the hydraulic potential gradient and the permeability and thickness of the aquifer. The permeability of subglacial aquifers is highly variable, depending on grain size distribution and porosity. For tills, measured permeabilities span a huge range, from $\sim 10^{-8}$ to $10^{-18}\,\mathrm{m^2}$; Piotrowski, 2006). In situ tests (e.g. Kulessa et al., 2005), generally yield lower values than lab testing, reflecting the fact that the bulk properties of geological materials differ from their small-scale properties. For deforming tills, permeability can increase through time in response to dilation, or decrease as a result of compaction (section 4.2.2; Hubbard and Maltman, 2000;

Tulaczyk et al., 2000a). Below mountain glaciers, tills are generally thin ($\sim 1\,\mathrm{m}$), so will not be capable of transmitting large volumes of water (Hubbard and Nienow, 1997). Where glaciers overlie thicker valley fills or other sediment accumulations, however, groundwater flow may account for substantial discharges, especially where permeable gravels and sands are present.

One well-studied example of a subglacial aquifer is at Breiðamerkurjökull, Iceland, which consists of a thick succession of gravels and sand capped by a thin till layer. To obtain data on subglacial groundwater flow, Boulton et al. (2001a) installed arrays of pressure sensors in the till and the subtill aquifer in anticipation of a glacier advance over the site. As the glacier began to advance over the arrays, pore-water pressures were higher in the subtill aquifer than in the till, but as the advance progressed, the pressure gradient reversed, with generally higher pressures in the till than in the underlying sands and gravels (fig. 3.25). From this, Boulton deduced that meltwater produced at the bed of the glacier flows downwards through the till, then moves downglacier through the higher permeability aquifer, until it wells upwards close to the margin. This model of groundwater flow was used to interpret patterns of deformation observed in the till at Breiðamerkurjökull, as described in section 4.6.3.

The great Pleistocene ice sheets extended over large areas of unconsolidated sediments, which could have

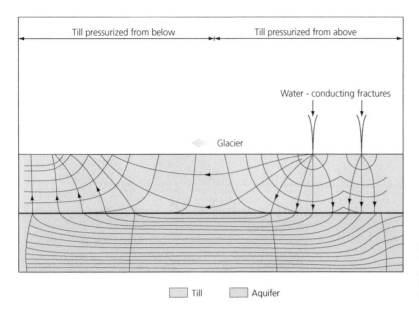

Figure 3.25 Hypothetical patterns of porewater flow through subglacial till (with low permeability) and a sub-till aquifer (with higher permeability). The system is fed via fractures extending from the surface (Boulton et al., 2001a).

accommodated significant amounts of groundwater flow (e.g. Boulton et al., 1995; Piotrowski, 2006). Old glacial groundwater has been identified in areas occupied by Laurentide and Fennoscandian Ice Sheets from its low ^{18}O and ^{2}H values and high dissolved oxygen content (e.g. Remenda et al., 1994; Boulton and Caban, 1995; Piotrowski, 2006). During glacial periods, the presence of ice sheets meant that these aquifers were subject to high hydraulic gradients, pushing water into and through the system. On deglaciation, this glacial water can remain trapped in sediments. A full understanding of long-term groundwater behaviour, therefore, must take account of reorganization of drainage patterns between interglacial and glacial conditions (van Weert et al., 1997).

3.4.5 WATER AT THE ICE–SEDIMENT INTERFACE

If water supply exceeds the capacity of a subglacial aquifer, excess water may flow along the ice–bed interface. Conditions at the interface between ice and sediment are more complex than those at hard glacier beds (sections 3.4.2 and 3.4.3) because of the effect of ice regelation at the scale of individual sediment grains. At low water pressures, melting and refreezing (regelation) in response to small-scale pressure variations around particles allows ice to infiltrate into subglacial sediments, creating a zone of ice-cemented sediment at the bed (Iverson et al., 1995, 2007; section 9.3.2). Ice infiltration is suppressed when basal water pressure is close to the ice pressure, or a thick layer of ice-cemented sediment has already formed. Iverson et al. (2007) presented evidence that a zone of very high porosity will exist between the ice-cemented sediment and the underlying unfrozen sediment, which can thicken to form a film at high water pressures. Film thickness appears to be sensitively dependent on water pressure, with typical values lying in the range 0.1–0.2 mm (100–200 μm), rising to ~1.00 mm at very high water pressures. The existence of a water film at the ice–bed interface leads to *decoupling*, where the local basal shear stress falls to zero. This loss of frictional drag has important implications for basal motion, as discussed in sections 4.5 and 4.6.

On the basis of modelling studies, Walder and Fowler (1994) argued that water films at the ice–sediment interface will become unstable once they exceed a critical thickness, and will break up into anastomosing networks of broad, shallow *canals*. These networks are analogous to linked cavity systems, and are believed to exhibit a similar relationship between water pressure and discharge – that is, higher discharges will be associated with higher water pressures, such that there will be no tendency for large canals to grow at the expense of small ones. Thus, in canal systems, drainage will tend to remain distributed over wide areas. More recent modelling by Ng (2000) supports these conclusions, while emphasizing the important role of sediment transport in determining drainage conditions.

3.5 GLACIAL HYDROLOGICAL SYSTEMS

The supraglacial, englacial and subglacial drainage elements described in sections 3.3 and 3.4 combine in various ways to evacuate water from glaciers. Glacial hydrological systems are highly variable in space and time, depending on climate, thermal regime, glacier geometry, and other factors. In the following sections we briefly describe the hydrological systems believed to be typical of temperate and polythermal glaciers. The hydrology of the Greenland and Antarctic ice sheets is discussed in chapter 6.

3.5.1 TEMPERATE GLACIERS

Numerous studies have been conducted on the hydrology of glaciers that are entirely or predominantly temperate (e.g. Willis et al., 1990; Fountain, 1993; Nienow et al., 1998; Schuler et al., 2002; Kulessa et al., 2005). Most of these studies have focused on the character of englacial and subglacial drainage systems, and their relationship with changing inputs of surface meltwater. Most of our current knowledge relates to small glaciers with terrestrial termini (which simplifies field logistics), in which surface meltwater is routed towards the bed via moulins or crevasses. Detailed tracer studies have demonstrated the existence of distinct catchments within some glaciers, consisting of regions of the surface which drain via groups of moulins to different outlets at the terminus (e.g. Fountain, 1993; fig. 3.26). There can be large changes in relative discharges from different portals at glacier termini from one year to the next, indicating that catchments within glaciers can adjust rapidly in response to changes in glacier geometry and other factors.

The character of the englacial and subglacial drainage systems that link inputs (moulins) and outputs (portals) has largely been deduced from dye tracing experiments, supplemented by borehole water-pressure records and geophysical studies. In dye tracing studies, a solution of fluorescent dye (such as rhodamine) is introduced into a moulin or other input point, then detected below the stream portal using a detection device such as a fluorimeter. Transit times and dye concentration and dispersion are then used to make inferences about the intervening drainage system. This approach has revealed that drainage systems in warm-bedded glaciers typically exhibit distinctive seasonal cycles (Hubbard and Nienow, 1997). In winter, when inputs of surface water are infrequent or absent, recharge of subglacial drainage systems is mainly by basal melting and the relocation of stored water. Drainage is inefficient, and basal water pressures are generally high. Although it is reasonable to assume some kind of distributed drainage system at the bed, in no case is its precise configuration known, and it is probably highly variable in time and space. Analysis of borehole water levels at Haut Glacier d'Arolla by Gordon et al. (1998) suggests that, prior to significant water inputs from the surface, the bed consists of hydraulically impermeable patches interspersed with poorly connected storage spaces.

At the onset of the melt season, surface-derived meltwater begins to reach the glacier bed via moulins beneath the supraglacial snowpack. Initially, this can result in an increase in stored water at the bed, as indicated by vertical uplift of glacier surfaces (e.g. Iken et al., 1983; Sugiyama and Gudmundsson, 2004). Tracers poured down moulins early in the melt season typically emerge at the glacier terminus in diffuse waves, indicating slow, relatively inefficient transport (fig. 3.27). As surface meltwater

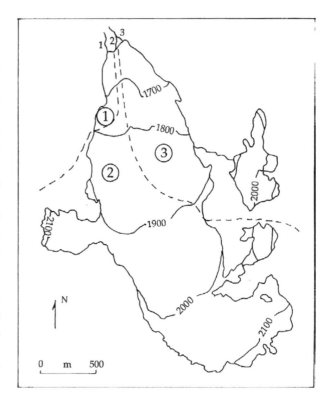

Figure 3.26 The catchment areas of three meltstreams emerging from the South Cascade Glacier, Washington State, USA (Redrawn from Fountain, 1993).

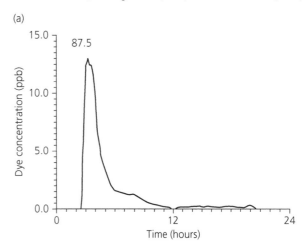

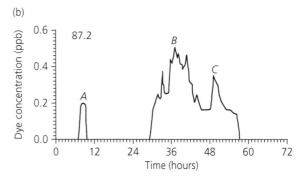

Figure 3.27 Results of dye-tracing experiments on Midtdalsbreen, Norway, highlighting the difference between efficient and inefficient subglacial drainage systems. (a) A single pulse of dye emerging from the glacier after a short lag, indicating transport through an efficient conduit. (b) Several dye peaks with lags of up to 48 hours, indicating a more complex pattern of storage and release. Note the different timescales (Adapted from Willis et al., 1990).

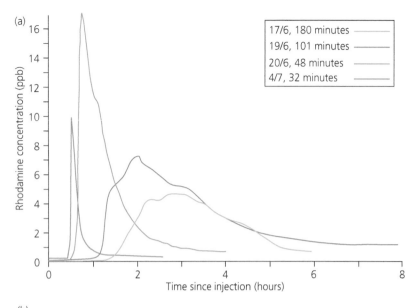

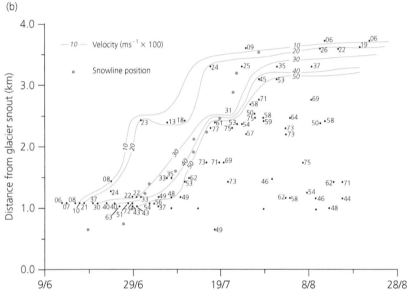

Figure 3.28 Results of dye-tracing experiments on Glacier d'Arolla, Switzerland. (a) Sequence of four dye return curves for injections from a single moulin between 17 June and 4 July 1990. (b) Contoured plot of water flow velocity as a function of location and date of injection. Note how the snowline position parallels the velocity isolines, indicating that efficiency of water flow reflects water inputs to the system (Nienow et al., 1998).

inputs increase, the englacial–subglacial drainage system undergoes a marked transition, after which tracers emerge at the glacier terminus in a single, sharp-peaked slug. Transit speeds are typically around $0.5\,\mathrm{m\,s^{-1}}$, an order of magnitude greater than pre-transition flow rates. These results indicate an increase in the efficiency of the drainage system as the melt season progresses, with a transition from a 'slow', inefficient system to a 'fast' system capable of accommodating larger meltwater fluxes. The 'fast' system consists of high-capacity conduits, which experience large fluctuations in discharge and water pressure on diurnal cycles (Hubbard et al., 1998; section 3.3.1).

The exact timing of the transition from the 'slow' to 'fast' system varies from year to year, and may coincide with several consecutive days of high melt rates or a heavy rainstorm. Nienow et al. (1998) showed that on Haut Glacier d'Arolla, the boundary between the two systems migrates upglacier as the melt season progresses, closely

following the position of the transient snowline (fig. 3.28). When snow cover disappears, near-surface water storage is dramatically reduced (section 3.3.1) and melt rates increase (section 2.2.2), resulting in an increase in peak water fluxes into moulins. The picture that emerges from Haut Glacier d'Arolla is of progressive integration of the drainage system, involving the growth of a branching conduit system in response to higher recharge rates. New conduit branches are added sequentially, linking newly reactivated moulins with the conduit system already established further downglacier. These subglacial conduits, therefore, represent the continuation at the bed of supraglacial and englacial flow paths (Fountain and Walder, 1998). Between major conduits, basal drainage is apparently slow and inefficient, with narrow corridors adjacent to channels exhibiting a damped response to pressure fluctuations in the 'fast' system (Hubbard et al., 1998).

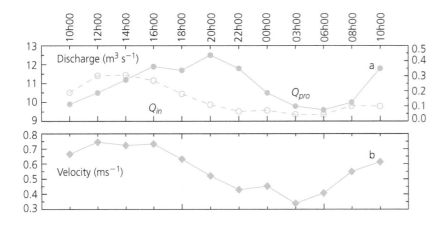

Figure 3.29 Diurnal variations in water flow velocity through Unteraargletscher. (a) Discharge into a moulin Q_{in} at the time of tracer injection (dashed line, right scale) and bulk discharge in the proglacial stream Q_{pro} at the time of tracer detection (solid line, left scale). (b) Transit velocity (Schuler et al., 2004).

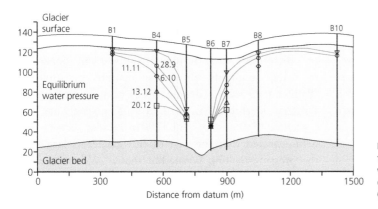

Figure 3.30 Transverse profile of Breiðamerkurjökull, Iceland, in the vicinity of a subglacial conduit. The red lines indicate borehole water levels in the autumn and early winter (dates shown). Note the decrease in hydraulic gradient towards the channel through time (Boulton et al., 2007a).

Significant diurnal variations in the efficiency of a 'fast' conduit system below Unteraargletscher, Switzerland, were detected by Schuler et al. (2004). By repeating tracer injections at intervals of a few hours, they found that water transit speeds varied by almost an order of magnitude in a single diurnal cycle (fig. 3.29). Two possible explanations were proposed for this behaviour.

(1) Conduit size could vary significantly over the daily cycle by the competing processes of creep closure and melt enlargement. Ice thickness at the tracer input site was 350 m, so significant closure could occur on a timescale of hours.

(2) Shifting pressure gradients between the conduit and the feeder moulin could alter the degree of coupling between the two systems ('inflow modulation').

This work highlights the complexity of subglacial drainage systems, and the difficulty of inferring drainage system evolution from a few tracer experiments conducted at long time intervals.

As surface meltwater inputs cease at the end of the ablation season, subglacial discharges diminish. The efficient drainage network developed during the previous melt season will therefore tend to close down in response to rising effective pressure and more rapid ice deformation. There is a lack of data on spatial patterns of drainage

shutdown in the autumn, although it seems reasonable to suppose that conduit closure will be most rapid where the ice is thickest, usually near the centre of the glacier. Spatially variable closure of cavities will tend to trap water at the bed, providing possible nuclei for the next year's drainage system.

The structure of the drainage system below Breiðamerkurjökull was investigated in field and modelling studies by Boulton et al. (2007a, b). As noted in section 3.4.5, the glacier is underlain by a thick succession of gravels and sand which act as a major subglacial aquifer. Figure 3.30 shows water-pressure records from boreholes drilled to the glacier bed in the vicinity of a large subglacial channel. Water pressure rises with distance from the channel, and during the summer it is close to overburden (i.e. effective pressure is close to zero) in the inter-channel zone. During autumn, water pressures progressively fall, and the pressure gradient towards the channel decreases. Boulton et al. (2007a, b) argued that most of the basally produced meltwater passes through the subtill aquifer, and that channels develop to accommodate discharges in excess of groundwater system capacity. In combination, rates of basal meltwater production and aquifer permeability determine a characteristic channel spacing, so that the combined aquifer–channel system has sufficient capacity to prevent pore-water pressures from exceeding the overburden pressure.

At Breiðamerkurjökull, high geothermal heat flux and large subglacial catchments maintain significant base flow during the winter months, maintaining perennial channels which are then enlarged each summer to accommodate surface-derived meltwater.

3.5.2 POLYTHERMAL GLACIERS

Because polythermal glaciers have a wide range of possible thermal structures (section 2.3.3), their drainage systems are much more diverse than those of temperate glaciers. Predominantly warm polythermal glaciers with only thin, cold surface layers in their ablation zones, such as Storglaciären, Sweden, have drainage systems that are closely similar to those of temperate glaciers (e.g. Seaberg et al., 1988; Fountain et al., 2005). At the other extreme, predominantly cold polythermal glaciers with only limited near-surface areas of warm ice, have drainage systems more akin to those of polar glaciers. Between these extremes, a range of drainage types is possible. However, the hydrology of only a few polythermal glaciers is known in any detail, making it difficult to present a general model at this stage.

Where ablation rates are relatively low, perennial supraglacial channels tend to form on uncrevassed regions of polythermal glaciers, especially along topographic lows close to the lateral margins. These can evolve into cut-and-closure-type englacial conduits following blockage of their upper parts by snow or ice, or closure by ice creep (section 3.3.2; Gulley et al., 2009a). Although such conduits can incise to the glacier bed, hydrologically enhanced fracturing provides a much more efficient means of linking supraglacial and subglacial drainage systems (Boon and Sharp, 2003; Gulley et al., 2009b). Moulins formed by fracturing are particularly common at glacier confluences, but can form wherever a plentiful water supply coincides with sufficiently stressed ice. Once water reaches the bed, its subsequent routing is strongly influenced by the thermal regime of the bed. If warm basal ice extends to (or close to) the margin, subglacial water can be discharged throughout the year. In permafrost areas, wintertime release of stored water results in the formation of *proglacial icings*, also known as *naled* or *aufeis* (fig. 3.31). In some cases, these have been shown to originate by the slow release of subglacial water (Wadham et al., 2000), although proglacial icings can also be fed by water released from englacial storage (Hodgkins et al., 1998, 2004). Water can persist over winter even in cold regions of polythermal glaciers if the melting point is depressed by high solute concentrations (Wadham et al., 2000).

Where a barrier of cold ice forms at glacier margins over winter, this can present a significant barrier to subglacial outflow in the following summer. Breakthrough

Figure 3.31 Proglacial icing, Brooks Range, Alaska (Ole Humlum).

of pressurized water results in the sudden appearance of water fountains on glacier margins or forefields (Pälli et al., 2003). A detailed study of outburst events at Finsterwalderbreen, Svalbard, in 1995 and 1999 has been presented by Wadham et al. (2001a). Radar surveys have demonstrated that the bed of this glacier is largely at the pressure-melting point, but that a cold zone exists below the terminus (fig. 3.32). The warm-bedded region forms a *confined aquifer* that becomes increasingly pressurized during the melt season as water is added from the surface. A subglacial water pressure record from 1995 is shown in figure 3.33. Pressure rose steadily from Day 139 (19 May), and was close to overburden by Day 221 (9 August). A sudden drop occurred on 20 August, coincident with a dramatic increase in stream discharge from the western side of the glacier, apparently the result of breakthrough of stored water past the thermal barrier.

On John Evans Glacier, a predominantly cold polythermal glacier on Ellesmere Island, Bingham et al. (2005) have documented seasonal evolution of the drainage system, similar to that known from warm-bedded glaciers (section 3.5.1). However, the formation and persistence of efficient drainage paths at the bed is sensitively dependent on surface melt rates. In the warm summer of 2000, outflow from the glacier commenced on 22 June, 1–2 days after the sudden drainage of ponded surface water down moulins following surface-to-bed fracturing (Boon and Sharp, 2003; section 3.3.2). As is the case at Finsterwalderbreen, breakthrough at the terminus appears to have been accomplished by pressurized water forcing a path through cold ice or frozen sediments (Skidmore and Sharp, 1999). For the two weeks following breakthrough, water throughflow times from moulin to terminus became progressively lower, suggesting the development of an efficient channel system. In the cooler summer of 2001, an efficient drainage system took

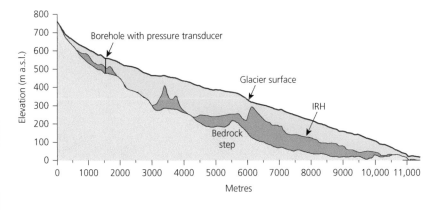

Figure 3.32 Thermal structure of Finsterwalderbreen, determined from ground-penetrating radar survey by Tore Tonning. Cold ice is indicated in blue, and temperate ice in red (Wadham et al., 2001a; Wadham published this figure with permission).

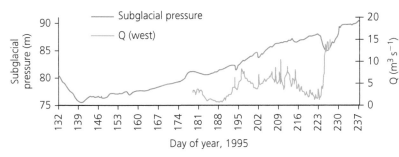

Figure 3.33 Variation in subglacial water pressure (borehole water level) and discharge from the western meltstream of Finsterwalderbreen, summer 1995 (Wadham et al., 2001a).

longer to develop following breakthrough, apparently in response to lower recharge rates.

3.5.3 MODELLING GLACIAL HYDROLOGICAL SYSTEMS

Because glacial hydrological systems play such an important role in regulating subglacial pressure conditions and proglacial runoff, much effort has been invested in developing predictive numerical models. Currently available models are considerably more sophisticated than the early analytical studies by Röthlisberger (1972) and Röthlisberger and Lang (1987), although major simplifying assumptions must still be made in order to represent some important physical processes. In particular, determining where surface water should reach the bed, and finding realistic ways of evolving the subglacial drainage system through time present major challenges. Contrasting approaches to addressing these problems can be illustrated using the work of Arnold et al. (1998) and Flowers and Clarke (2002a).

Arnold et al. (1998) developed a hydrological model for Haut Glacier d'Arolla, calibrated using the extensive observational database for the glacier (Richards et al., 1996). The model calculates meltwater production from the surface energy balance, then computes the subsequent transport of water over and through the glacier to the outlet stream. Known moulin positions were used to specify points where surface meltwater reaches the bed and to infer the structure of the basal drainage network (fig. 3.34). The character of the network was evolved

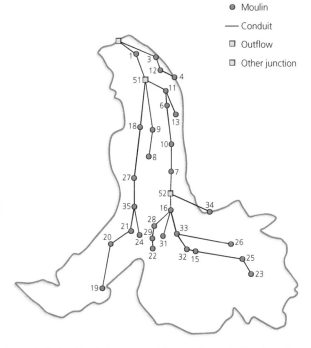

Figure 3.34 Specified drainage network for modelling water flow through Glacier d'Arolla (Arnold et al., 1998).

through time by solving equations for melt enlargement and creep closure, thus allowing the capacity of the system to respond to changing inputs from the surface. The model captured the essential features of observed proglacial hydrographs, including diurnal, seasonal and intermediate-scale fluctuations, and subglacial water

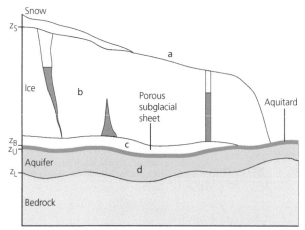

Figure 3.35 Definition sketch showing components of the Flowers and Clarke dynamic, multicomponent drainage model (Flowers and Clarke, 2002a).

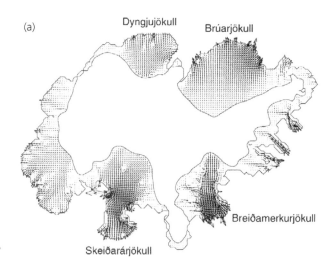

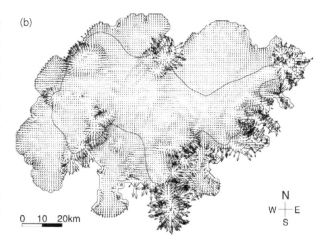

Figure 3.36 Modelled drainage structure of Vatnajökull, Iceland. The interior contour marks the glacier equilibrium line. (a) Dynamic multicomponent model. (b) Static, hydraulic potential model (Flowers et al., 2003).

pressure records. The best matches were for the middle of the melt season, when the conduit network was well established, but the model performed less well for the beginning and end of summer (i.e. when the drainage system is enlarging or contracting).

A different approach was taken by Flowers and Clarke (2002a), who built on the lumped element model developed by Clarke (1996). Rather than specifying a drainage network, they parameterized basal drainage as a two-layer distributed system, analogous to a porous subglacial sheet perched above a subtill aquifer (fig. 3.35). The subglacial sheet is allowed to expand vertically to accommodate flow, and its hydraulic conductivity increases with water sheet thickness. Although the physical picture of a porous sheet may not provide a particularly strong match to reality, the important point is that the system of equations behaves in appropriate ways to external forcings. The subglacial sheet and the underlying aquifer are coupled to supraglacial and englacial drainage systems, so that water can flow through or be stored within the system, depending on recharge rates and antecedent conditions.

The model has been applied to Trapridge Glacier, Yukon (Flowers and Clarke, 2002b) and, with some modifications, to Vatnajökull in Iceland (Flowers et al., 2003). The modelled basal drainage structure for Vatnajökull is shown in figure 3.36a, compared with results of the static *hydraulic-potential model*. The essential difference between these two models is that the latter assumes that hydraulic potential is a simple function of bed elevation and ice overburden pressure (equation 3.13), whereas in the former it depends on the balance between recharge rates and the ability of the basal drainage system to evacuate stored water. In the Flowers et al. (2003) model, supraglacial meltwater below the ELA is transferred to the bed, while meltwater above the ELA is routed over the surface to the ELA and then to the bed. Water is also

produced subglacially in response to the geothermal heat flux, although basal melting in response to frictional heating is neglected. Despite its simplifying assumptions, the Flowers model yields a much more realistic picture of the ice-cap drainage system than the static hydraulic-potential method. Areas of high water flux occur below the major outlet glaciers, and predicted regions of low subglacial effective pressure (equation 3.10) coincide with areas prone to episodic flow acceleration, or surges (section 5.7).

In the last decade or so, considerable advances have been made in our understanding of glacial hydrological systems, and our ability to represent them in glacier models has improved immensely. However, the statement made by Arnold et al. (1998), that 'much work remains to be done in terms of field observations, modelling and development of theory of glacier hydrological systems', is no less true today than when it was written.

3.6 PROGLACIAL RUNOFF

Because glaciers can store and release water on such a wide range of timescales, the presence of glaciers in a catchment profoundly influences the amount and timing of runoff in the basin. In the following sections, we examine the role of glaciers in modulating runoff on two timescales: (1) intra-annual cycles, and (2) longer-term variations driven by climate change.

3.6.1 SEASONAL AND SHORTER-TERM CYCLES

The discharge of proglacial rivers responds to a wide range of factors, including rainfall, snow and ice melt, the release of water stored in the glacier and non-glacial groundwater flow. Variations in the amount of water delivered from sources result in discharge fluctuations on daily and annual cycles, and on an irregular basis associated with the passage of weather systems.

The basic daily pattern of discharge from glaciers into proglacial rivers is driven by the diurnal temperature cycle. During the melt season, a cycle of rising and falling discharge is superimposed on base flow, or daily minimum discharge (fig. 3.37). Base flow comes from a variety of sources, including (1) subglacial meltwater; (2) water released from storage in snow, firn or subglacial and englacial cavities; and (3) groundwater (Röthlisberger and Lang, 1987). Generally, these components vary little on a day-to-day basis, but can exhibit large variations on longer timescales. Superimposed on base flow are the components of the day's meltwater that drain rapidly through the glacier system, such as meltwater from bare ice surfaces routed via efficient englacial and subglacial conduits. Daily discharge peaks lag behind the time of maximum melting by a few hours, the length of delay depending on the distance the water has to travel and the configuration of the drainage pathways. Lag times are long for snow-covered glacier surfaces and poorly developed subglacial drainage systems, and short for ice surfaces and well-developed subglacial conduits. Thus, as the ablation season progresses, there is a general trend towards shorter lag times (Röthlisberger and Lang, 1987; Jansson et al., 2003).

Seasonal runoff cycles depend to a large degree on the climatic regime, particularly the relative timing of precipitation and temperature maxima. In regions with a winter precipitation maximum and a summer temperature maximum, such as the European Alps and western North America, melting of snow and ice stored in glaciers provides the main component of river runoff in summertime, thus shifting the discharge maximum relative to the precipitation peak (fig. 3.38). This shift occurs to an even greater degree in low-latitude mountain regions with

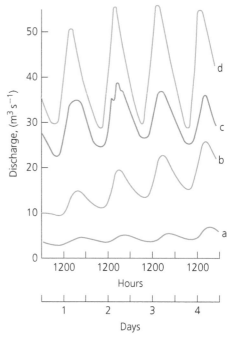

Figure 3.37 Diurnal variations in discharge from the proglacial stream of Gornergletscher, Switzerland, showing progressive increase in base flow and diurnal cycle amplitude during the ablation season of 1959. (a) 17–20 May. (b) 14–17 June. (c) 23–26 June. (d) 19–22 July (Patterson, 1994).

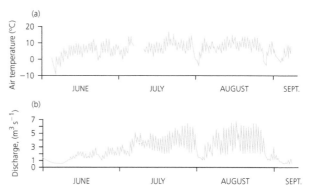

Figure 3.38 Variations in (a) air temperature and (b) proglacial discharge for Haut Glacier d'Arolla for the ablation season of 1989. Note the low discharges during a cold weather spell in early August (Gurnell et al., 1992).

pronounced wet and dry seasons, such as the Cordillera Blanca in Peru, where glacial runoff provides an essential source of water for almost half of the year (Juen et al., 2007; fig. 3.39). On the other hand, in areas with a summer precipitation maximum and a winter minimum, such as monsoon-dominated parts of the Himalaya, the glacial influence on the seasonal runoff cycle is not so marked, and discharge closely follows the seasonal pattern of precipitation (e.g. Hasnain, 1999; fig. 3.40).

The onset of discharge from a glacier early in the melt season often comes in the form of a sudden flood, known as the *spring event* (Röthlisberger and Lang, 1987; Flowers and Clarke, 2000; Mair et al., 2003). Spring events tend to be triggered by high surface melt rates or heavy rainfall, which are routed to the bed and flood the poorly developed winter drainage system of the glacier. Following breakthrough to the margin, an integrated supraglacial, englacial and subglacial drainage system develops, and marked diurnal cycles in discharge are established for the remainder of the melt season (fig. 3.41). Discharges decline in late summer, when supplies of stored englacial and subglacial water and low-altitude snowbeds have dwindled. Runoff falls rapidly to low winter levels with the onset of cold weather in the autumn.

During the melt season, flood events can be associated with weather systems, changes in the glacial drainage system and outbursts from lakes. Periods of rapid ablation occur in response to high air temperatures, high solar radiation or enhanced turbulent heat flux in warm, windy weather (e.g. Boon et al., 2003). During sustained periods of enhanced melting, proglacial hydrographs exhibit a steady rise in base flow, reflecting the gradual release of water from the snow and firn aquifers. Superimposed on this trend are larger than average diurnal fluctuations, representing quickflow from areas of bare ice. The highest weather-related discharges, however, tend to be associated with high rainfall during summer and autumn storms. Rainfall adds directly to runoff in the basin, and also contributes to snow and ice melting (section 2.4.3). Glacier lake outburst floods are discussed in detail in section 3.7.

3.6.2 RUNOFF AND CLIMATE CHANGE

Historical records show that runoff from glacierized basins can undergo significant changes in response to climate forcing (e.g. Collins, 2008; Cassassa et al., 2009; fig. 3.42), and it is clear that the same was true in prehistoric times. Given the important role of glaciers in modulating the seasonal patterns of runoff, concerns have been expressed that, in a warming world, glacier shrinkage will have serious implications for water resources in many parts of the world (e.g. Barnett et al., 2005). These concerns were highlighted in a very influential report published by the Worldwide Fund for Nature (WWF, 2005), which paints a grim picture of vanishing glaciers and water shortages on a massive scale, affecting hundreds of millions of people. Although widely reported by the world's media, the scientific basis of such dire prognoses is rather slender. In fact, comparatively few studies have addressed the complex problem of predicting runoff response to climate change at the regional scale. Although changes in the amount and seasonal distribution of runoff are hard to predict in practice, they are relatively easy to understand in principle. Broadly, glacier runoff response to climate change reflects three factors:

(1) energy-driven changes in the amount of melting;
(2) glacier volume changes, which affect the area over which melting can occur;
(3) changes in the volume and stability of glacier lakes.

Factors (1) and (2) are considered in this section, and (3) is discussed in section 3.7.

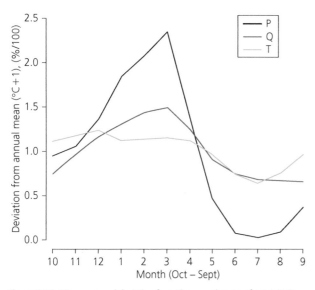

Figure 3.39 Mean seasonal deviation from the annual mean of precipitation (P), glacier runoff (Q) and air temperature (T), in the Cordillera Blanca, Peru. The discharge curve has a lower amplitude than the precipitation due to net glacial storage in the wet season and release in the dry season (Juen et al., 2007).

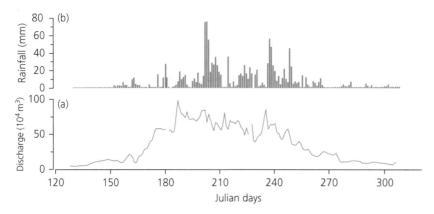

Figure 3.40 Precipitation and proglacial runoff in the Dokriani catchment, monsoonal Garhwal Himalaya, India (Hasnain, 1999).

In a warming climate, increased glacier melt results in a component of runoff in excess of contemporary precipitation, termed the *deglaciation discharge dividend* (Collins, 2008). As deglaciation proceeds, the ice reservoir decreases in area and the amount of the dividend declines. The dividend vanishes when deglaciation is complete, and thereafter the magnitude of runoff (and its seasonal distribution) reflects patterns of precipitation and temporary storage in seasonal snow cover. Conversely, in a cooling climate, a net increase in the amount of water stored in glacier ice will expand the glacial reservoir at the expense of runoff. Total runoff will be less than the precipitation, and its seasonal distribution will largely reflect seasonal energy receipts.

Predicting future changes in glacial runoff, however, is fraught with uncertainty. First, choices have to be made about appropriate climate scenarios. Current global circulation models disagree about the magnitude (and even

sign) of precipitation changes that can be expected in the coming decades (IPCC, 2007). The models are in closer agreement regarding likely temperature changes, and it is usual to adopt one or more of the scenarios defined by the Intergovernmental Panel on Climate Change (IPCC) as the basis for hydrological modelling studies. Second, runoff models are highly variable in their sophistication and scope. Many runoff predictions are based on calculated glacier melt totals using some specified temperature increase, and assume that the glacier area remains unchanged (e.g. Singh et al., 2006). More sophisticated models evolve the area of ice cover through time, or take a series of 'snapshots' based on prescribed changes in glacier extent (e.g. Horton et al., 2006).

Hagg et al. (2007) modelled possible future changes in runoff in the semi-arid Tien Shan of central Asia by assuming a doubling of CO_2 relative to pre-industrial levels and specifying changes in glacier area. The model predicts

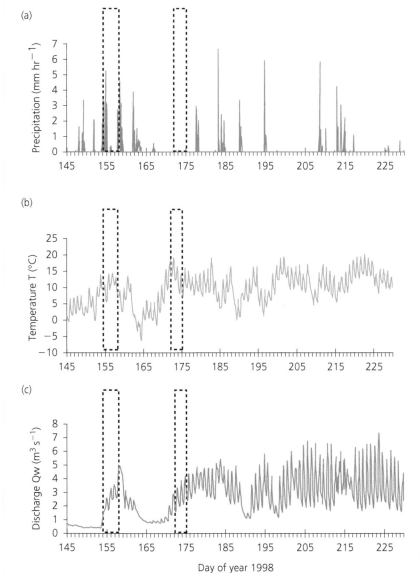

Figure 3.41 Spring events at Haut Glacier d'Arolla, 1998. (a) Precipitation. (b) Air temperature. (c) Proglacial stream discharge. The boxes highlight the 'spring event' time periods (Mair et al., 2003).

Figure 3.42 Disused field systems in the Hunza Valley, Pakistan. The presence of glaciers is crucial to human activity in areas with dry summers, by providing water for irrigation during the growing season. Agriculture was abandoned in this area when a small glacier in the catchment disappeared (L.A. Owen).

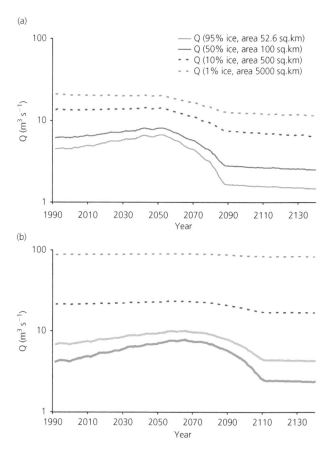

Figure 3.43 Modelled change in annual mean flow under a climate warming of $0.06°\ yr^{-1}$ for idealized catchments in (a) the western Himalaya and (b) the eastern Himalaya (Rees and Collins, 2006).

enhanced snowmelt during spring and higher flood risk in summer, as melting increases in response to higher temperatures. Complete loss of ice cover is predicted to result in increased spring runoff (snowmelt), but lower summer runoff. Juen et al. (2007) modelled runoff in the Cordillera Blanca, Peru, for four different IPCC climate change scenarios, and found that mean annual runoff remains virtually unchanged, but with a considerable increase in seasonality. Runoff increases during the wet season, but dramatically decreases during the dry season because the 'smoothing' effect of glaciers is reduced or removed.

The ways in which changes in glacier runoff might affect larger regions are explored in a very useful modelling study by Rees and Collins (2006), who simulated runoff from nested groups of catchments for climate regimes equivalent to parts of the Karakoram and Nepal Himalaya. These regions have similar temperature cycles, but greatly differing precipitation patterns. In the Karakoram, precipitation totals are low, with a maximum in winter–spring, while in Nepal totals are high, with a pronounced maximum during the summer monsoon. Under a warming scenario, both regions experience initial increases in discharge, followed by a decline as the volume of glacier stores diminishes (fig. 3.43). The magnitude of the reduction is greatest in the Karakoram, where

glacier meltwater contributes most to summer discharges. However, in both regions the relative reduction in runoff is much greater in small, heavily glacierized catchments, and diminishes as catchment size increases and the proportion of the catchment occupied by glaciers goes down. In Nepal, the impacts of warming are barely noticeable in larger catchments due to the increasing influence of rainfall-fed, non-glacial runoff, but in the Karakoram the runoff reduction extends much farther downstream. This study indicates that a reduction in glacier area will result in reduced summer stream discharges close to the glacier source areas, but the impact will be felt further downstream (where larger population centres are located), principally in the semi-arid western parts of the Himalayan range. This detailed study, therefore, provides little support for the idea that reduction of glacier volume will result in regional-scale water shortages in monsoonal Asia.

3.7 GLACIAL LAKES AND OUTBURST FLOODS

3.7.1 INTRODUCTION

Bodies of water can be stored in subglacial, supraglacial, englacial or proglacial environments, wherever free drainage is prevented by some form of ice, sediment or rock barrier. Such water bodies vary greatly in size, from small pockets to lakes hundreds of thousands of square kilometres in area (e.g. Teller et al., 2002; Mangerud et al., 2004). Some lakes expand and contract in response to glacier fluctuations on timescales of tens of years to millennia, whereas others fill and drain over periods of days to years. Lakes in overdeepened rock basins are stable and long-lived features of glaciated environments, whereas water bodies that owe their existence to ice or sediment dams are necessarily ephemeral.

Glacial lakes can be classified according to their topographic position and the nature of the dam (Tweed and Russell, 1999; Clague and Evans, 2000; Roberts et al., 2005). *Supraglacial lakes* occupy closed hollows on a glacier surface. *Englacial water bodies* are usually small, typically occupying blocked conduits or moulins (e.g. Gulley et al., 2009b). *Subglacial lakes* can form at temperate glacier beds, where an area of relatively low hydraulic potential is surrounded by regions with higher potential (e.g. Siegert, 2000; Björnsson, 2002).

Ice-dammed lakes form wherever a glacier creates a barrier to local or regional drainage. In mountain regions, this can happen in the following situations:

(1) where ice-free side valleys are blocked by a glacier in the trunk valley;
(2) where trunk valleys are blocked by glaciers spilling out from side valleys;
(3) at the junction between two valley glaciers.

In lowland areas, ice-dammed lakes can form where glaciers advance into areas that slope down towards the ice margin, blocking preglacial drainage outlets. Lake level can be controlled by a spillway on the watershed or by the ice itself, depending on the relationship between the glacier and the surrounding terrain. The lifetime of ice-dammed lakes is clearly controlled by the persistence and integrity of the ice dam, and drainage will inevitably occur when the glacier retreats. Longer-lived dams can be created if the glacier deposits moraines, outwash fans or other sediment accumulations along its margins. In mountain areas, *moraine-dammed lakes* form in much the same range of topographic positions as ice-dammed lakes (fig. 3.44; Clague and Evans, 2000).

Catastrophic drainage of glacial lakes can produce river flows orders of magnitude larger than normal peak discharges in a catchment, with major consequences for landscape change and human life and infrastructure. Outburst floods present serious hazards in many glacierized regions (e.g. Clague and Evans, 2000; Björnsson, 2002; Quincey et al., 2007), and there is abundant geomorphological and sedimentological evidence that this was also the case in the past. Given the potentially great destructiveness of outburst floods, they are well known to the inhabitants of glacierized regions and have names

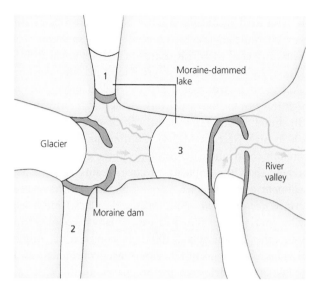

Figure 3.44 Sketch map showing principal locations of moraine-dammed lakes. (1) Between glacier and its own moraine. (2) In a side valley, dammed by a moraine deposited by glacier in the trunk valley. (3) In the trunk valley, dammed by a moraine deposited by tributary-valley glacier (Clague and Evans, 2000).

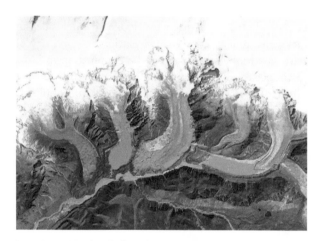

Figure 3.45 ASTER image of moraine-dammed lakes in Bhutan. The large lake to the right (Luggye Tsho) partially drained in 1994, carving the prominent flood track and causing loss of life and infrastructure downstream.

in many languages. For example, outburst floods are known as *jökulhlaup*s in Iceland, *débâcles* in the European Alps, *alluviones* in the Andes, and *tschoscrup* in Tibetan-speaking parts of the Himalaya. Although the Icelandic *jökulhlaup* (literally, 'glacier-leap') is widely used in the English-language literature to refer to all outburst floods from glaciers, here we adopt the more internationally recognized term *glacier lake outburst flood*, or *GLOF*.

3.7.2 MORAINE-DAMMED LAKES

In recent decades, retreat of mountain glaciers from their Little Ice Age maxima has resulted in the formation of lakes in basins between the glacier margin and frontal and lateral moraines (Yamada, 1998; Clague and Evans, 2000; Iturrizaga, 2005; fig. 3.45). GLOFs from such lakes have resulted in considerable loss of life and property, and they continue to present serious hazards in some areas (Lliboutry et al., 1977; Richardson and Reynolds, 2000a; Carey, 2008). The largest proglacial moraine-dammed lakes form in front of debris-covered glaciers, because high rates of sediment delivery at their margins can rapidly build up moraine dams, which can be well over 100 m high (section 11.3.4). Continuous moraine dams can only form where debris supply at the glacier margin is greater than the capacity of meltstreams to transport sediment away. Glacier margins of this type have been termed *decoupled margins*, because the glacial and fluvial sediment transport systems are essentially decoupled from each other (Benn et al., 2003). In contrast, at *coupled margins*, the fluvial system can transport sediment away as fast as it is delivered by the glacier. In this case, the meltstream can maintain an open breach through the terminal moraine, allowing water to be continuously evacuated from the system. Large proglacial moraine-dammed lakes will therefore tend to develop where debris supply is large compared with fluvial discharge – for example, in front of glaciers overlooked by high side- and backwalls, and relatively small glaciers in cool, dry climatic environments. Lakes can be dammed up in basins on the ice-distal side of lateral moraines irrespective of whether the frontal margins are coupled or decoupled.

As well as providing sediment for the moraine dam, debris cover also exerts a profound influence on the early stages of the evolution of proglacial moraine-dammed lakes, through its effects on glacier mass balance and hydrology (Benn et al., 2001; Hambrey et al., 2008). On glaciers with extensive covers of supraglacial debris, the mass balance gradient is commonly reversed in the lower part of the ablation zone due to the insulating effect of thick debris (section 2.4.5). As a result, during periods of negative mass balance, a greater amount of thinning occurs in the upper part of the ablation zone, where debris is relatively thin, compared with the terminal zone where debris cover is thicker. This pattern of ice-surface lowering results in a reduction in the overall ice-surface gradient, encouraging ice stagnation and retention of meltwater (fig. 3.46; Benn et al., 2003). Differential melting and local subsidence causes the widespread development of water-filled hollows on the stagnating glacier surface (Kirkbride, 1993). Two types of supraglacial lakes have been identified on debris-covered glaciers, which have contrasting relationships with the hydrological system (Benn et al., 2001). First, *base-level lakes* form at the elevation of water outflow from the glacier. In high mountain environments, large terminal moraine loops constitute effectively impermeable barriers to water flow, so the minimum elevation of their crests determines the base level for water draining from the glacier. The evolution of base-level lakes is therefore tied to that of the moraine dam, and lakes will partially or completely drain only when the level of the dam is lowered. Second, *perched lakes* form in closed supraglacial hollows above base level. These will persist if they are floored by intact (effectively impermeable) ice, but will drain if a hydraulically efficient connection is made between the lake and an area with lower hydraulic potential (fig. 3.47). This can occur either when permeable debris-filled crevasse traces

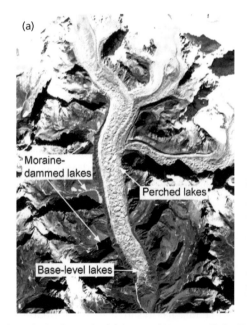

Figure 3.46 Lakes on the debris-covered Ngozumpa Glacier, Nepal. (a) ASTER image of the glacier in November 2005, showing different types of glacial lakes. Lakes appear brown in this false-colour image. (Image processing: A. Luckman) (b) View of the glacier from Gokyo Ri, 2008. The ribbon-shaped water body near the glacier terminus (left) is a base-level lake dammed behind the terminal moraine; numerous perched lakes occupy supraglacial hollows farther upglacier (left). The large blue lake at bottom right is in a side valley, dammed by the lateral moraine of the glacier (Elke Morgner).

are exploited and enlarged by water flow (Gulley and Benn, 2007), or by hydrologically driven fracture propagation where the ice is under sufficient stress (Gulley et al., 2009a). Observations on Himalayan glaciers indicate that perched lakes typically persist for a few years, and drain before attaining large volumes ($\sim 10^4$ to $10^5 \, \mathrm{m}^3$). Thus, drainage of a single perched lake is unlikely to make a significant impact on proglacial discharge. Perched lakes do, however, have a considerable impact on ablation of the glacier, since ablation rates around lake margins are typically many times greater than those beneath continuous debris cover (Sakai et al., 2000; Benn et al., 2001). Thus, the formation of perched lakes accelerates glacier wastage, progressively lowering the ice surface towards base level.

Potentially hazardous base-level lakes, therefore, tend to develop on low-gradient, stagnating glacier tongues where continuous moraine dams provide a high base level for the glacier hydrological system (Richardson and Reynolds, 2000a; Benn et al., 2003; Hambrey et al., 2008). These characteristics can be readily identified using remote-sensing techniques, allowing at-risk glaciers to be identified over large, remote areas where extensive fieldwork would be impractical (Quincey et al., 2007; Bolch et al., 2008a). Once formed, base-level lakes can grow rapidly by melting and calving, both around the lake margins and below water level (Kirkbride, 1993; Benn et al., 2001; Warren and Kirkbride, 2003; Röhl, 2006, 2008). A base-level lake in front of Imja Glacier, Nepal, is shown in figure 3.48. In the 1960s, a few small ponds existed in the terminal zone of the glacier, which by the mid 1970s had coalesced to form a lake that extended across the full width of the glacier. Subsequently, the lake increased in area, principally by calving of the active front of the Imja-Lhotse Shar Glacier, and by 2007 the lake was 2 km long with an area of $1.03 \, \mathrm{km}^2$. There has been less change of the shoreline position at the downglacier end, where the lake is dammed by a mass of thickly debris-mantled ice and moraine some 500–700 m across (Hambrey et al., 2008). The great extent of the ice dam relative to its height ($\sim 50 \, \mathrm{m}$) means that the lake is unlikely to pose an imminent GLOF threat (cf. Watanabe et al., 1995).

The likelihood of a GLOF from a moraine-dammed lake depends on both the integrity of the dam and the nature of possible triggers. In general, broad dams (e.g. former ice-contact outwash fans) are less susceptible to failure than narrow ones (e.g. sharp-crested moraine

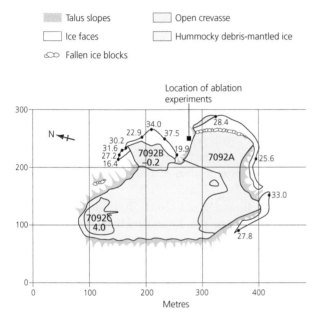

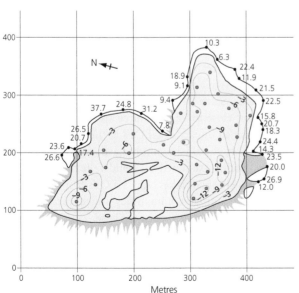

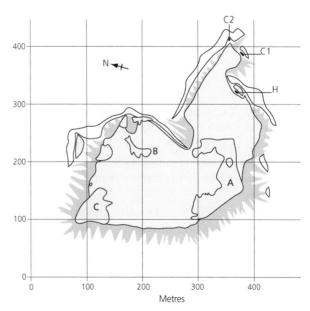

Figure 3.47 Evolution of a perched lake on Ngozumpa Glacier. Top: 1998; bottom left: 1999; bottom right: 2000. In the space of two years this lake filled, expanded, then drained through an englacial conduit (Benn et al., 2001).

Figure 3.48 Imja Lake, Khumbu Himal, Nepal. The downwasting Imja-Lhotse Shar Glacier can be seen behind the lake, many tens of metres below its old lateral moraines. The peak on the left is Makalu (8463 m) (D.I. Benn).

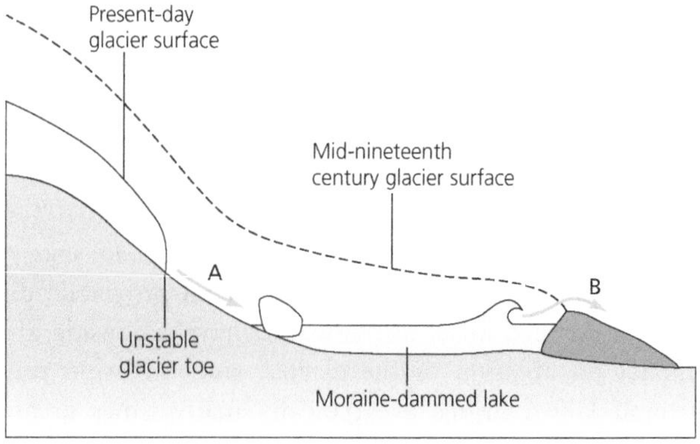

Figure 3.49 Ice avalanches from hanging glaciers are common sources of seiche waves, which overtop and erode moraine dams and trigger GLOFs (Clague and Evans, 2000).

ridges). Melting of ice cores may lead to progressive degradation of dams through time. Moraine dam failure is commonly triggered by *seiche waves* generated by rock and ice avalanches or seismic shocks (Lliboutry et al., 1977; Clague and Evans, 2000; Carey, 2008; fig. 3.49). If a wave overtops the sediment barrier and initiates erosion, lake water escaping through the newly formed gap can lead to a catastrophic positive feedback process, in which water discharge and dam erosion are mutually reinforcing. GLOFs from moraine-dammed lakes release highly destructive mixtures of water, rock, mud, glacial ice and other debris, which may be augmented by bank erosion along the flood path. Lake drainage may continue until all water is released from storage, or cease while some remains. In the latter case, a second flood may be released from the same lake at some later date.

One of the most destructive GLOFs in recent times occurred in the Cordillera Blanca, Peru, in 1941. The flood was initiated when an ice avalanche crashed into Lake Palcacocha, triggering giant waves that overtopped and eroded the moraine dam. Flood volume increased when it flowed into a second moraine-dammed lake further downvalley, causing it to burst as well. By the time the flood reached the town of Huaraz, 23 km downstream from Lake Palcacocha, it contained 8 million m³ of water and debris. The rapidly flowing slurry destroyed one-third of the town, killing an estimated 5000 people. During the twentieth century, GLOFs in the Cordillera Blanca have resulted in considerable loss of life, due to the presence of large population centres in glacierized catchments.

Numerous GLOFs have also occurred in the Himalaya, but generally these have not been as destructive as the worst Peruvian events because they have occurred in less highly populated areas. Local impacts, however, can still be considerable. For example, in August 1985 an outburst from Dig Tsho, in front of Langmoche Glacier in the Khumbu region of Nepal, was triggered when a large mass of ice avalanched from the

glacier (Vuichard and Zimmerman, 1987). Local witnesses reported multiple surges of dirty water, carrying along boulders and trees. Significant damage was done along 80 km of river course, including erosion of cultivated land, collapse of buildings into the flood, and the destruction of 14 bridges. A nearly completed hydroelectric plant was completely destroyed, with an estimated cost of at least US$2 million. The loss of human life was remarkably small, with five reported deaths. This was attributed to the fact that the flood occurred during the 3-day Sherpa festival of Phagnhi, when the affected villages were largely unoccupied.

Cenderelli and Wohl (2001, 2003) have shown that recent GLOFs in the Khumbu Himal have peak discharges which are much greater than seasonal high-flow floods (SHFFs) near their source, but become decreasingly so with distance downstream. This is because the hydrographs of GLOFs attenuate downstream from the source due to frictional diffusion of the flood wave, whereas meteorologically triggered floods tend to increase downstream as the contributing catchment area increases. Near the lake source area, GLOF discharges are ~50 times higher than SHFFs, but only ~5 times higher 27 km downstream. Presumably, there is a cross-over at some point, below which GLOF peak flows are less than seasonal floods. Since there is little human activity in the valley bottoms close to GLOF source areas, this explains the relatively small impact of GLOFs in this region compared with Peru. It should be emphasized, however, that the outburst floods examined by Cenderelli and Wohl were from relatively small lakes. Any GLOFs from larger lakes will impact human communities much farther downstream than has hitherto been the case.

Measures for alleviating GLOF risks have been introduced in a number of places. In Peru, the Control Commission of Cordillera Blanca Lakes (CCLCB) was established in 1951, and conducted pioneering research into GLOF mechanisms, in addition to implementing hazard mitigation work. Since its formation, CCLCB has

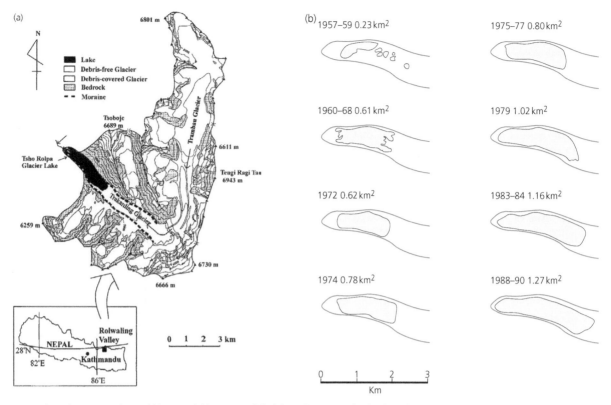

(a)

Figure 3.50 Tsho Rolpa moraine-dammed lake, Nepal. (a) Location of the lake at the margin of Trakarding Glacier. (b) Change in surface area of Tsho Rolpa 1957–1990.

contained 35 moraine-dammed lakes by constructing erosion-resistant dams at lake outlets and/or lowering water levels. Remediation works, however, may be limited by financial and logistical considerations. For example, remedial measures have reduced the risk of a GLOF from Tsho Rolpa, in the Rolwaling district of Nepal (Reynolds, 1999; Rana et al., 2000; Kattelmann, 2003). Within the space of less than 50 years, this lake had grown from a few small ponds on the lower tongue of the Trakarding Glacier to the largest moraine-dammed lake in Nepal (fig. 3.50). By 1999, it was 3.5 km long, 0.5 km wide and up to 135 m deep, and $\sim 80 \times 10^6 \, m^3$ of water was dammed by an ice-cored moraine complex of doubtful stability. In the late 1990s, a sluice-gate system was installed in the moraine, and lake level lowered by 3 m at a cost of US$2.7 million. This reduced the volume of the lake by one-third, but has not altogether removed the risk of future floods. As an additional precaution, an early warning system has been installed to alert villagers to sudden drops in water level.

3.7.3 ICE-DAMMED LAKES

Although superficially similar, there are significant differences between lakes dammed by moraines and those dammed by glacier ice. First, ice-dammed lakes commonly form when drainage becomes blocked by advancing ice, so GLOF risks can be high during periods

Figure 3.51 Ice-dammed lake at the eastern margin of Breiðamerkurjökull, Iceland. The lake has partially drained and stranded icebergs litter its surroundings (R.H. Mottram).

of glacier growth. This contrasts with GLOF risks from proglacial moraine-dammed lakes, which tend to be greatest during periods of glacier retreat. In the European Alps, for example, there was a high frequency of *débâcles* from ice-dammed lakes during the Little Ice Age (Vivian, 2001), with great loss of life and property.

Second, drainage of an ice-dammed lake does not necessarily involve destruction of the ice dam (fig. 3.51). Many ice-dammed lakes drain via subglacial passages, which can rapidly close by ice creep when discharges fall.

Some ice-dammed lakes drain by the erosion of supraglacial spillways, cutting down through the ice dam in an analogous way to the incision of a moraine dam (Raymond and Nolan, 2000). Such spillways, however, can also be closed by ice flow much more readily than a moraine dam can be reformed. This means that ice-dammed lakes can undergo recurrent filling and drainage cycles, on annual or longer timescales, so that repeated GLOFs can originate from the same site. Well-known examples of lakes that episodically drain below their ice dam include: Hazard Lake, Yukon, Canada (Clarke, 1982); Hidden Creek Lake, dammed by Kennicott Glacier, Alaska, USA (Anderson et al., 2003a); Graenalón, dammed by Skeiðarárjökull, Iceland (Roberts et al., 2005); and Gornersee, dammed at the confluence between Gornergletscher and Grenzgletscher, Switzerland (Huss et al., 2007).

An influential theory of ice-dammed lake drainage was developed by Nye (1976), building on insights by Mathews (1973) and Björnsson (1974). The geometry of the lake and ice dam used in the model is shown in figure 3.52. The lake will persist if there is a region (known as the *seal*) below the glacier where the basal ice pressure exceeds the pressure exerted by the lake water. Subglacial water pressure below the seal is assumed to be:

$$P_w = \rho_w g(Z_w - Z_s) \qquad (3.15)$$

where Z_w and Z_s are the elevation of the lake surface and the bed at the seal, respectively (this is equivalent to the *pressure head* defined in equation 3.2, and assumes no head loss due to friction between the lake and the seal). Increasing water storage in the lake will increase Z_w, until at the flotation level the subglacial water pressure at the seal is equal to the cryostatic pressure. At this point water is able to break through the seal and escape beneath the glacier. Frictional heat dissipated by the escaping water enlarges the subglacial conduit, allowing continuing drainage even though Z_w, and hence the driving water pressure, fall below the critical threshold. Eventually, water pressure falls to the extent that closure of the conduit by ice creep exceeds wall melting rates, and the

seal is reformed. This may happen following complete lake drainage or while some water still remains in the lake.

In Nye's original model, numerous simplifying assumptions were made. For example, the effects of warm lake water were ignored, and water was assumed to escape beneath the glacier via a pre-existing cylindrical conduit, driven by the potential gradient averaged along the whole flow path. Over the years, numerous improvements have been made to this model, while retaining its core ideas. Clarke (1982) modified the Nye theory to take account of the effects of lake temperature and reservoir geometry. A key assumption of this early work is that discharge during outburst floods is controlled by restriction (or bottleneck) at a single point in the flow path, which must be located sufficiently close to the reservoir that the water temperature is essentially that of the lake.

Rather than adopt simplifying assumptions about the existence of flow bottlenecks, Spring and Hutter (1981, 1982) developed a complete mathematical description of the evolving water discharge, temperature and conduit cross-sectional area along the entire length of the conduit, from inlet to outlet. The complexity of their model, however, meant that efforts to obtain numerical solutions to their full system of equations met with limited success. The Spring–Hutter formulation was further developed by Clarke (2003), who derived a rigorous but tractable model of ice-dammed lake drainage that allows important aspects of flood evolution to be examined in detail. Clarke applied his model to GLOFs from three recurrently draining ice-dammed lakes for which reservoir and ice geometry and flood discharge hydrographs are well known. Contrary to the assumptions of bottleneck models, in no case did the dominant flow constriction remain fixed at the seal. Rather, the model runs showed that the controlling constriction tends to be located at the outlet, and in some cases migrates up-flow as the flood progresses. Thus, although the seal may play a role in controlling the flood onset, it ceases to function as the controlling flow constriction once the flood is in progress.

Clarke (2003) observed that many important processes remain poorly parameterized, including the thermal evolution of the flood waters and channel roughness. In addition, some observed characteristics of GLOFs from ice-dammed lakes are at variance with fundamental model assumptions. For example, many lakes are known to drain before water level is high enough to float the ice dam (Clarke, 1982; Anderson et al., 2003a; Roberts, 2005; Roberts et al., 2005). For example, Roberts et al. (2005) found that GLOFs from Graenalón, Iceland, have occurred by three distinct drainage regimes, and that shifts from one regime to another have involved major changes in flood magnitude and frequency. In recent years, lake water has been able to escape beneath the ice barrier well before theoretical flotation conditions are

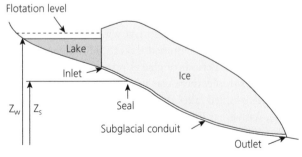

Figure 3.52 Geometry of an ice-dammed lake and the glacier seal used in outburst flood models.

met, and drainage system geometry apparently differs considerably from the classical view of a pressure-coupled lake and conduit system. These issues are examined further in the following section, which focuses on some of the most spectacular GLOFs of modern times.

3.7.4 ICELANDIC SUBGLACIAL LAKES

As in many mountainous glacierized regions, ice-dammed lakes are common around glacier margins in Iceland (e.g. Roberts et al., 2005). However, Iceland also sits astride the tectonic spreading zone between the North American and Eurasian Plates, and has numerous active volcanic centres. Several of these are overlain by ice caps, and the juxtaposition of volcanic activity and glacier ice creates an exceptional capacity for glacier lake outburst floods. Five volcanic systems exist under the largest ice cap, Vatnajőkull, each containing central volcanoes and fissure swarms (fig. 3.53; Björnsson, 2002). In addition, numerous volcanic centres exist under the second largest ice cap, Myrdalsjőkull, including the Katla volcanic system.

Locally high geothermal heat fluxes in these volcanic centres continuously melt basal ice, and subsidence of the overlying glacier creates permanent depressions in the ice cap surface. Basal meltwater from surrounding areas tends to build up below the depressions, because relatively low ice overburden pressures create a pressure sink at the glacier bed. The interface between the subglacial lake and the overlying ice is commonly dome-shaped, and lakes even cap mountains in some cases (Björnsson, 2002). Particularly rapid melting occurs during volcanic eruptions, due to efficient heat flux from molten lava to the ice (Höskuldsson and Sparks, 1997;

Wilson and Head, 2002). This means that subglacial reservoirs rapidly increase in volume during eruptions, with correspondingly high GLOF potential.

Many destructive events have occurred in historical times, such as the 1362 eruption of ice-capped stratovolcano Öraefajőkull, when debris-laden floods escaped below outlet glaciers and several farmsteads were washed away. Contemporary accounts (quoted by Björnsson, 2002) are remarkably detailed, and recount how 'the glacier itself slid forwards ... just like melted metal poured out of a crucible'.

One of Iceland's best-known volcanigenic subglacial lakes is Grimsvötn, situated below the western part of Vatnajőkull (fig. 3.54). In keeping with its name ('Grim's Lake', after the vengeful Norse god Grim), Grimsvötn is one of the most hellish places on Earth, with meltwater supplied by high geothermal heat and intermittent eruptions, about 30 of which have been reported in the last 400 years. During the refilling phase, the lake lies below a caldera-like depression on the ice surface, 10 km wide and 300 m deep, and is capped by a floating ice shelf (Björnsson, 2002). The level of the lake water rises until the lake drains subglacially at intervals of 1–10 years. Astonishingly, this extreme environment harbours life (Gaidos et al., 2004).

Typically, floods are triggered when lake level is 60–70 m lower than the level required to float the ice dam (fig. 3.55). The flood waters escape southwards below the glacier Skeiðarárjőkull and emerge via portals onto the outwash surface to the south, where their arrival is announced by a sulphurous smell in the glacial river. As the flood increases in volume, vast quantities of muddy water and tumbling ice blocks surge across the outwash plain towards the sea. The flood track crosses the only

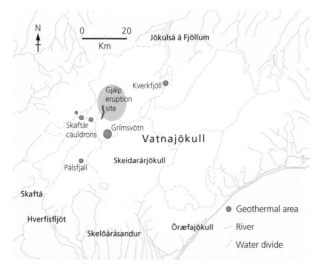

Figure 3.53 Map of Vatnajőkull, Iceland, showing glacial drainage basins and the location of subglacial volcanic centres (Björnsson, 2002).

Figure 3.54 Ice cauldron beginning to form on the surface of Vatnajőkull, Iceland, as the result of geothermal heating in the Grimsvötn caldera in August 1972 (Aerial photo supplied by Landmaelingar Islands).

road along the southern Icelandic coast, and GLOFs commonly disrupt transport between south-east Iceland and the capital.

GLOFs from Grimsvötn form two distinct populations, depending on flood magnitude. Relatively small floods (with peak discharges $<3 \times 10^3 \, \text{m}^3 \, \text{s}^{-1}$) reach their peak in 2 to 3 weeks, and terminate around 1 week later. Typically, their discharge hydrographs exhibit a non-linear, approximately exponential rising limb, and a more rapid falling stage (fig. 3.56). The general form of the rising limb of these hydrographs can be explained in terms of expansion of a single tunnel by frictional melting (Nye, 1976; Spring and Hutter, 1981; Clarke, 2003), although

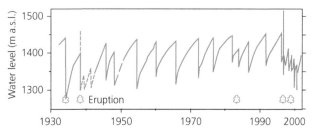

Figure 3.55 The lake level of Grimsvötn, 1930–2000 and the timing of subglacial volcanic eruptions. The lake level rises until rapid drainage occurs. Note the unusually high lake level in 1996 (Björnsson, 2002).

the precise details depend crucially on choice of model parameters.

The larger events commonly do not follow this pattern, and some form a distinct class of *linearly rising jökulhlaups*. These have a rapid, approximately linear rising limb, and a more gradual, non-linear falling limb, with peak discharges up to $4 \times 10^4 \, \text{m}^3 \, \text{s}^{-1}$. A well-studied example of a large, linearly rising *jökulhlaup* is the Grimsvötn flood of November 1996. Between late September and mid October 1996, an eruption from the Gjalp fissure in the Grimsvötn drainage basin led to ~2.7 km^3 of meltwater being added to the ~0.5 km^3 of water already in the lake. On the basis of previous events, it was expected that drainage would occur when the lake level reached 1450 m above sea level. However, the lake continued to fill for a further 60 m, until hydraulic potential below the lake seal equalled the ice overburden pressure. At this point, a large portion of the ice dam floated, and flood waters rushed out towards the south. This was the first time in observational history when the dam was floated off the bed.

Breakthrough of water to the glacier margin occurred on the morning of 5 November. Sediment-laden water burst from multiple ice-roofed vents and fractures along the entire 23 km ice margin (fig. 3.57; Roberts et al.,

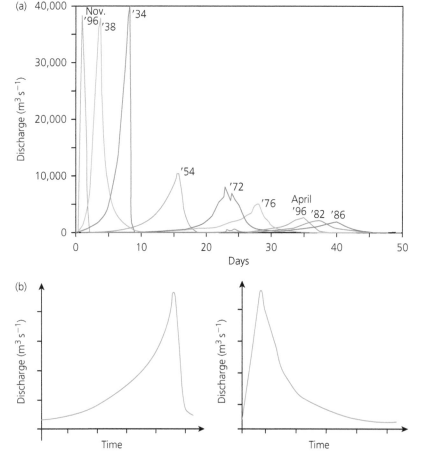

Figure 3.56 (a) Typical hydrographs of GLOFs from Grimsvötn (1934, 1938, 1954, 1972, 1976, 1982 and 1996). (b) Left: non-linearly rising hydrograph; right: linearly rising hydrograph (Björnsson, 2002).

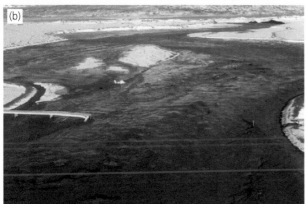

Figure 3.57 Grimsvötn jökulhlaup, 5th November 1996. (a) Flood waters inundating the outwash plain in front of Skeiðarárjökull (top left). The road linking south-east Iceland to the west of the country is in the middle distance. (b) Road bridge breached by flood waters. Helicopter for scale. (c) Large ice blocks rafted in the turbulent, sediment-laden flood water (Photographs kindly supplied by M.T. Gudmundsson, University of Iceland).

2000). Angular blocks of ice in excess of 30 m in diameter were dislodged from the glacier adjacent to the fractures, producing extensive embayments in the glacier surface. The flood reached a peak discharge of 45–53 m^3 sec^{-1} within 14 hours, much sooner than any previously monitored GLOF from Grimsvötn; in all, 3.2 km^3 of water drained from the lake within a period of 40 hours.

In part, the rapid onset of the 1996 flood reflects the high temperature of the lake water. During lake drainage,

a depression 6 km long, 1 km wide and 100 m deep was created across the ice dam by collapse of the flood flow path, and Björnsson (2002) calculated that the energy required to melt the 'missing' ice implies an average lake water temperature of 8 °C. Lake drainage was so rapid, however, that melting alone could not have enlarged the outlet routes fast enough (Björnsson, 2002). Instead, the flood wave appears to have formed a sheet-like flow, creating accommodation space by lifting the ice off large areas of the bed (Roberts, 2005). In addition, flow paths were created and enlarged by hydrofracturing, and large blocks of ice were prised from the glacier and swept out across the foreland (Russell et al., 2006).

These observations highlight the limitations of existing glacial hydrological theory, which does not incorporate many of the processes known to be important in linearly rising jökulhlaups. It should also be remembered that few detailed data are available for GLOFs of any kind. Of the floods that have been observed, many key variables – such as flow path geometry, channel roughness and evolution of water temperature – are impossible to observe directly and remain poorly understood. Taken together with the logistical difficulties and dangers of conducting fieldwork around draining glacial lakes, these problems continue to limit our ability to understand and predict the most destructive of glacial phenomena.

3.7.5 ESTIMATING GLOF MAGNITUDES

Because GLOFs can pose such severe threats to life and property, it is important to be able to estimate the magnitude of floods that might be released from lakes of known volume. Several methods have been devised to predict peak flood discharges, the most widely used of which is the *Clague–Mathews relation* (Clague and Mathews, 1973). This empirical *scaling law* describes the volume–discharge relationship for ice-dammed lakes that drain subglacially:

$$Q_{max} = KV_t^b \qquad (3.16)$$

where Q_{max} is the peak discharge and V_t is the total water volume drained during the flood. In their original study based on ten ice-dammed lakes, Clague and Mathews obtained values of $K = 75$ and $b = 0.67$. Subsequent studies have obtained different values for the prefactor K, but for datasets covering several different lakes, the exponent b is remarkably consistent at around 2/3 (fig. 3.58; Walder and Costa, 1996; Ng and Björnsson, 2003; Huss et al., 2007). A physical basis for the Clague–Mathews relation has been proposed by Ng and Björnsson (2003), based on the model of Nye (1976). The change in lake volume V_t during a flood corresponds with a change in effective pressure at the seal from a low value N_1 at the

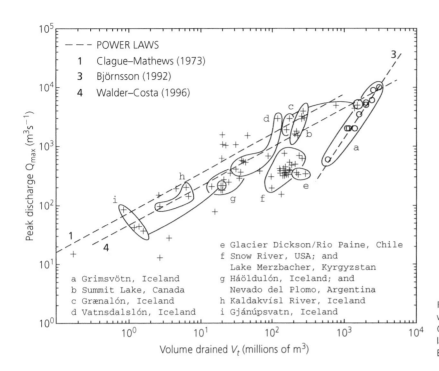

Figure 3.58 Peak flood discharge plotted against volume drained for subaerial lakes (crosses) and Grimsvötn (circles), and alternative empirical power laws derived from subsets of the data (Ng and Björnsson, 2003).

time of flood initiation (when lake depth and water pressure are high) to a higher value N_2 at the time of flood termination (when lake depth and water pressure are lower). For any given lake, therefore, larger floods will be associated with a greater change in effective pressure $N_2 - N_1$. Since the evolution of discharge during the flood is intimately related to the changing effective pressure (equation 3.10), this results in a predictable relationship between lake volume change and peak discharge. For a population of lakes, complications are introduced by differing lake geometries. Ng and Björnsson (2003) showed that scaling effects alone indicate a value of $b \approx 1$, and conjecture that the lower empirically determined value mainly reflects the effects of different flood initiation mechanisms.

In a detailed study of a 40-year flood record from Merzbacher Lake in the Tien Shan, Ng et al. (2007) showed that flood magnitude is sensitively dependent on weather conditions, particularly air temperature, which modulates both the rate of recharge to the lake as it drains and lake water temperature. They showed that regional climatic warming is the primary cause of a rising trend of peak discharges from the lake, and argued that climatic warming is likely to promote higher-impact GLOFs in many parts of the world.

3.8 LIFE IN GLACIERS

Wherever water exists in liquid form on, in or under glaciers, some kind of life can usually be found. The adage

'where there is water, there is life', therefore, appears to apply as much to glacial environments as to any other extreme habitat on Earth (Tranter et al., 2005). An astonishing diversity of organisms make their living in the world's glaciers and ice sheets, and it is now known that glacial ecosystems play a key role in rock weathering and biogeochemical cycles. An excellent introduction to the rapidly expanding literature on glacial ecosystems has been provided by Hodson et al. (2008).

A fundamental distinction can be made between *supraglacial ecosystems* and *subglacial ecosystems*. The former are primarily fuelled by solar energy, in common with the majority of Earth surface ecosystems, whereas in permanently dark subglacial environments, organisms must either recycle organic carbon imported from the surface or utilize chemical or geothermal sources of energy. Microbial life can also exist in englacial environments, but limited nutrient and energy availability mean that although microorganisms may remain viable for millennia, they appear to play only a minor role in the overall biogeochemical cycles of glaciers.

3.8.1 SUPRAGLACIAL ECOSYSTEMS

Representatives of all five kingdoms of the living world (Animals, Plants, Fungi, Protista and Bacteria) are to be found on glacier surfaces. The diversity and composition of supraglacial ecosystems is very variable, depending on whether the glacier surface consists of snow, bare ice or a debris mantle. On snow or ice surfaces, the vast majority of the biota are microorganisms, but on debris-covered

Figure 3.59 Svalbard poppies (*Papaver dahlianum*) on stagnant debris-covered ice, Larsbreen (D.I. Benn).

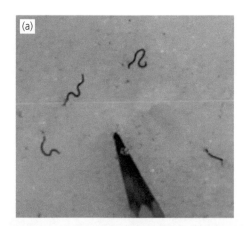

Figure 3.60 Glacier ice worms. (a) On the surface of Columbia Glacier. (b) In close-up (Mauri Pelto).

Figure 3.61 Cryoconite holes (Andy Hodson).

surfaces more complex ecosystems can include mosses, flowering plants, insects and even trees (fig. 3.59). The key factors are the availability of sunlight, nutrients and the stability of the substrate (Stibal et al., 2006).

Like snow in non-glacial environments, glacial snow contains wind-blown nutrients that can support psychrophilic (cold-loving) microorganisms (Jones, 1999). One of the most well-known signs of life on snow surfaces is 'red snow', or blooms of the algae *Chlamydomonas nivalis*. The red colour is from a carotenoid pigment that this photosynthesizing organism contains as well as chlorophyll. Many other species of algae flourish on glacier surfaces, and Takeuchi et al. (1998) have shown that the species composition of supraglacial algal communities varies systematically with surface mass balance. In the accumulation area of their study glacier in the Himalaya, *Trochiscia* species are dominant, but they are gradually replaced by *Cylindrocystis* in the higher biomass ablation area.

Among the more curious denizens of glaciers are *glacier ice worms* (fig. 3.60; Shain et al., 2001). These are annelids of the species *Mesenchytraeus solifugus*, which populate low-elevation temperate glaciers in the Pacific Northwest of North America. Growing to ~15 mm in length, at times they occur in high densities and can sometimes be found in tight, twisting bundles of up to 100 individuals. They feed on algae and can be observed 'fishing' while anchored to the ice banks of ponds and streams. Hartzell et al. (2005) studied over 100 populations of glacier ice worms, and found that there are two geographically distinct lineages in the Pacific Northwest, with no evidence of gene flow between the two.

Conditions for life on glaciers are especially favourable where surface debris provides a ready source of nutrients. Thin accumulations of fine-grained debris (*cryoconite* or 'cold dust') on ablating glacier surfaces tend to gather in clots, which then melt holes into the ice due to their low albedo (Takeuchi et al., 2001; Fountain et al., 2004). This results in distinctive water-filled *cryoconite holes*, which commonly occur in large groups (fig. 3.61). Depending on climate, cryoconite holes can remain open during the ablation season (open holes) or freeze over at the surface

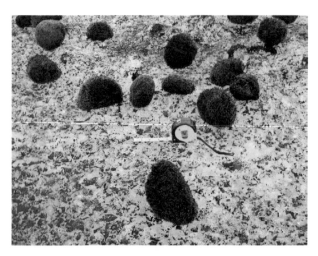

Figure 3.62 Moss polsters or 'glacier mice' on Breiðamerkurjökull, Iceland (D.I. Benn).

Figure 3.63 Mature spruce and hemlock trees on the stagnant debris-covered surface of Fairweather Glacier, Alaska (B. Molnia, USGS).

(closed holes). Both types can harbour a rich diversity of life, including cyanobacteria, green algae and fungi, while diatoms, tardigrades, rotifers and nematodes have also been observed. Rates of photosynthesis are significant in open holes, but photosynthetic organisms can also exploit sunlight penetrating through ice lids. In closed holes, biological activity is generally limited by nutrient availability, especially when they have been sealed off from the atmosphere for many years (Fountain et al., 2004). Extensive mats of cryoconite debris can form on glacier ablation zones, and can be of sufficient thickness for warm, anaerobic conditions to develop (Hodson et al., 2008). The conditions in these mats differ markedly from those in well-ventilated open cryoconite holes, and biological productivity can be surprisingly high. Bacterial respiration in cryoconite releases carbon into the atmosphere, and Hodson et al. (2007) calculated carbon fluxes of 12–14 kg of carbon $km^{-2} yr^{-1}$ from glacier surfaces in Svalbard.

Debris covers thicker than a few centimetres retard melting of the underlying ice (section 2.4.5), encouraging a degree of substrate stability and providing a platform for plant growth. On medial moraines and similar debris-covered areas, lichens and mosses can thrive. If moss mats are broken up during glacier ablation, free-rolling balls of moss, or *polsters*, can form. These are especially common in association with tephra outcrops on Icelandic glaciers, and were given the evocative name of *jökla-mys* (glacier mice) by Eythorsson (1951). Each 'mouse' is ovoid in shape, a few centimetres across, and contains a core of tephra (fig. 3.62). They appear to have everything they need: a nutrient supply, abundant moisture and all-over exposure to sunlight as ablation of the underlying ice periodically causes them to roll over.

Insects are also quite common on debris-covered glacier surfaces. For example, on Forni Glacier, Italian Alps, Gobbi et al. (2006) found pioneer species such as the wolf-spider (*Pardosa saturatior*) and ground beetle

(*Oreonebria castanea*). Himalayan debris-covered glaciers provide a wide diversity of habitats, ranging from stable areas of fine-grained sediment to supraglacial lakes. Some flowering plants can thrive on the poor, mineral-rich soils, including the daisy *Anaphilis triplinervis* (Gaur et al., 2003). Relatively long-lived supraglacial lakes with stable margins can contain diverse algae and diatoms, as well as animals such as copepods, fly and midge larvae, and daphnids (Takeuchi and Kohshima, 2000).

In permafrost regions, glacier melting ceases once debris cover exceeds the active layer thickness (i.e. the depth of annual thawing), and a range of flowering plants can flourish on the resulting stable surfaces. Even in temperate environments, ablation rates below thick debris cover are so low that surfaces can be stable enough for complete vegetation cover to develop (Krüger and Kjær, 2000). Supraglacial debris can be sufficiently stable to allow tree growth. In Alaska, for example, stands of old-growth spruce can occur on stagnant, debris-covered ice (fig. 3.63). Stagnant ice is not a requirement for tree growth, however. In Chile, Veblen et al. (1989) described forest ecosystems developed on ice that was flowing at rates of up to 22 m yr^{-1}. Three tree and six shrub species were common, as were numerous other vascular plants, including ferns and lianas. The stands are dominated by the southern beech (*Nothofagus dombeyi*), with trees up to 40 cm in diameter and around 70 years in age.

3.8.2 SUBGLACIAL ECOSYSTEMS

Viable microbes have been found in frozen soils beneath the cold-based glacier Longyearbreen, Svalbard, where they have lain dormant since being overridden over 1000 years ago (e.g. Humlum et al., 2005). For these organisms, glaciation is a temporary inconvenience and they simply remain inert while encased in ice and suspend biological activity until the return of more favourable conditions. Other organisms, however, are specially adapted to living at glacier beds, and carry out their entire life cycle in the cold and dark. The existence of such subglacial microbial communities was first demonstrated by Sharp et al. (1999), who recovered living bacteria from boreholes to the bed of two Swiss glaciers. Subsequently, microbial life has been found below glaciers in many parts of the world, including valley glaciers in New Zealand (Foght et al., 2004), Icelandic subglacial lakes (Gaidos et al., 2004) and the Antarctic Ice Sheet (Mikucki et al., 2004; Lanoil et al., 2009).

The permanent absence of solar radiation means that subglacial biological activity must rely on organic carbon derived from overridden soils or introduced from the surface, or a non-solar energy source such as geothermal heat or redox (reduction–oxidation) reactions. *Chemolithotrophic* organisms – which derive their energy from rock-weathering processes – appear to be very widespread below glaciers, and exert a strong influence on weathering rates and the chemistry of subglacial waters (e.g. Tranter et al., 2002; Wadham et al., 2004; Skidmore et al., 2005; Wynn et al., 2006; Hodson et al., 2008). The kinds of chemical reactions that can be sustained depend sensitively on the availability of oxidants (e.g. O_2, NO_3) and other factors (section 3.9.4), suggesting that the composition of subglacial microbial communities varies systematically with the type of drainage system (Tranter et al., 2005). Where surface water reaches the bed, the efficient delivery of dissolved oxygen will maintain aerated environments, and microorganisms living in sediments flanking surface-fed channels will be able to exploit aerobic weathering processes such as oxidation of sulphides. In contrast, in inefficient, distributed drainage systems away from major channels, anoxic conditions are more widespread and subglacial microorganisms must utilize anaerobic reactions such as reduction of manganese, iron and sulphates, and – where organic carbon is available – methanogenesis. A fascinating issue is the possible role of subglacial microbial communities in cycling organic carbon reservoirs overrun by the great Pleistocene ice sheets. It is estimated that ~10^{17} grams of carbon are currently stored in soils in the areas that were ice covered at the LGM. If a similar amount of carbon was stored in these areas during the last interglacial, this means that a vast amount of carbon must have been cycled through the glacial system (Sharp et al., 1999). The fate of this carbon is still not well understood. If organic carbon was oxidized microbially beneath the ice sheets, it could have been quite rapidly re-released to the atmosphere through meltwaters. On the other hand, if it was stored subglacially, there could have been a considerable lag before release.

In recent years, much attention has been paid to the possibility of life in Antarctic subglacial lakes, including Lake Vostok (section 6.3.5; Siegert et al., 2003; Studinger et al., 2003). So far, such lakes have not been sampled directly, owing to the risk of contamination by microbes introduced from the surface. However, at Lake Vostok, freezing at the lake–ice interface has resulted in ~210 m of ice accretion onto the ice sheet base. Coring of this ice has allowed former lake water to be sampled, without the need to penetrate to, and potentially contaminate, the lake itself. The small numbers of microbes discovered so far have DNA profiles closely similar to those of contemporary surface microbes. It is therefore possible that these microbes have simply been recycled through the lake, having melted out from the ice sheet base (Siegert et al., 2003). It remains possible, however, that unique communities exist within deeper lake water and associated sediments.

3.9 GLACIER HYDROCHEMISTRY

3.9.1 OVERVIEW

Rivers in glacierized catchments commonly contain high concentrations of dissolved ions (negatively or positively charged atoms or molecules), mostly derived from chemical weathering of bedrock and unlithified sediments. Meltstreams also contain ions (or *solutes*) derived from the atmosphere, originally deposited on glacier surfaces within snow or rain, then released into glacifluvial systems. Common dissolved chemical species include positively charged *cations* such as calcium (Ca^{2+}), sodium (Na^+), potassium (K^+) and magnesium (Mg^{2+}); negatively charged *anions* such as bicarbonate (HCO_3), sulphate (SO_4^{2-}), nitrate (NO_3^-) and chlorine (Cl^-); aqueous protons (hydrogen ions: H^+); and dissolved gases such as oxygen (O_2), nitrogen (N_2) and carbon dioxide (CO_2). Solutes can be used as natural tracers that can help identify water sources and flow routing, and the solute content of glacial runoff is often measured by scientists trying to make sense of glacial hydrological systems. As noted in section 3.8.2, there is increasing recognition of the role of microorganisms in glacial environments, and analysis of the composition of glacier meltwater provides important information about subglacial ecosystems (Tranter et al., 2002; Hodson et al., 2008). Good reviews of the hydrochemistry of glaciers have been provided by Brown (2002) and Tranter (2003).

The solute content of waters emerging from glaciers varies greatly in time and space. It has long been recognized that the solute concentration of alpine glacier runoff has an inverse correlation with discharge. Figure 3.64 shows variations in solute content (as indicated by electrical conductivity) and discharge for the meltwater stream emerging from Gornergletscher, Switzerland, for a 6-day period in 1975. Solute content is high at times of low discharge (base flow) and low during peak discharge in the afternoons. In simple terms, these variations reflect the mixing of two meltwater components:

(1) relatively dilute meltwater routed rapidly from the glacier surface via surface streams and efficient conduits (*quickflow*);

(2) solute-rich water that has been transported through less efficient components of the englacial and subglacial drainage system (*delayed flow*).

The dilute quickflow component dominates proglacial runoff in the afternoons when surface meltwater is flushed through the system, whereas the concentrated delayed flow component dominates at times of diminished meltwater input. Similar patterns occur on annual cycles (fig. 3.65). Solute concentrations tend to be high in winter when little or no surface melting occurs, total discharges are low and water residence times are long.

Sharp et al. (1995b) reported that solute concentrations in the stream draining Haut Glacier d'Arolla are three to seven times higher in winter than in summer. However, because the stream discharge is very small in winter, the total solute flux is low, despite the high concentrations. Metcalf (1986) estimated that for Gornergletscher, less than 10 per cent of the annual total solute load leaves the glacier between September and May.

Early models of glacial solute fluxes were based on the idea that distinct 'englacial' and 'subglacial' components can be identified in proglacial streams, and that variations in solute content can be used to quantify their relative contributions to runoff (e.g. Collins, 1979; Gurnell and Fenn, 1985). If the solute content of the two components is known, their relative proportions can be determined using the following relation:

$$Q_t C_t = Q_{sub} C_{sub} + Q_{en} C_{en} \qquad (3.17)$$

where Q is discharge, C is solute concentration, and the subscripts t, sub and en denote total, subglacial and englacial components, respectively. Although two-component mixing models yield a useful general picture of glacial drainage, their assumptions are too restrictive to adequately capture the complexity of glacier hydrochemistry (Sharp et al., 1995b). Both quickflow and delayed flow components of glacial runoff typically contain water from multiple sources, and evolve through time as they pass through the glacial hydrological system (Tranter and Raiswell, 1991). Since the 1990s, detailed analyses of the solute load of runoff from both temperate and polythermal glaciers have been used to develop new models of solute acquisition and transport through glaciers (e.g. Tranter et al., 1996; Hodgkins et al., 1998; Wadham

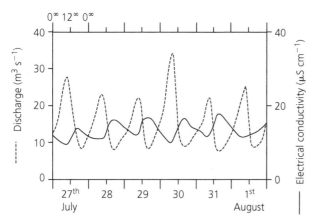

Figure 3.64 Temporal variations in electrical conductivity (which is directly proportional to solute concentration) and discharge for the proglacial stream of Gornergletscher, Switzerland (Redrawn from Collins, 1979).

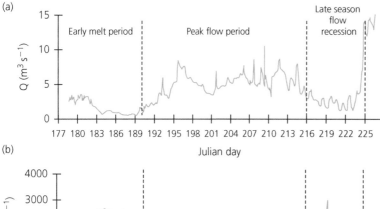
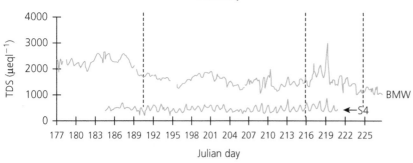

Figure 3.65 Temporal variation in (a) bulk discharge, and (b) total dissolved solids (TDS) in bulk meltwater and supraglacial meltwater at Finsterwalderbreen, Svalbard, during the ablation season of 1995. Note how the bulk meltwater (BMW) has a high concentration of total dissolved solids in the early melt period when runoff is low. Supraglacial runoff (S4) shows no long-term trend, but varies inversely with discharge on a diurnal cycle. The sudden increase in discharge on JD 225 is due to an outburst event (Wadham et al., 1998).

et al., 1998; Tranter, 2003; Mitchell et al., 2006). Such work is labour-intensive, and the database is still relatively small, but has revealed a rich, complex and often puzzling picture of chemical weathering and solute transport in glacierized catchments.

The chemical composition of glacial runoff reflects processes operating in all of the environments through which snow, ice and water pass on their journey through the glacier system, from the atmosphere to the proglacial stream. In the following sections we examine the chemistry of snow, then consider chemical weathering in subglacial and proglacial environments, and conclude with a discussion of rates of chemical erosion in glacierized catchments.

3.9.2 SNOW CHEMISTRY

Snow does not consist of pure water but also contains a variety of solutes, depending on air mass history and location of deposition relative to source regions (Fountain, 1996; Kuhn, 2001; Tranter and Jones, 2001). For example, snow in maritime regions commonly contains high concentrations of sodium and chlorine derived from sea salt, whereas proximity to anthropogenic pollution sources increases the concentration of compounds such as sulphate and nitrate. Once snow is deposited on a glacier surface, its initial chemical composition is modified by melting and refreezing processes. Solutes tend to be preferentially flushed out of melting snow, contributing relatively enriched waters to glacier systems early in the melt season. Field and laboratory studies have shown that up to 80 per cent of the snowpack solute load may be released in the initial ~25 per cent of meltwater runoff (Bales et al., 1989). Sulphate and nitrate tend to be preferentially leached, leaving the snowpack temporarily enriched in chlorine and sodium. Concentrations of Cl^- also decrease exponentially in the early melt season, as elution flushes the remaining ions from the snowpack. At the end of the melt season, any remaining firn tends to be depleted of solutes relative to the original snow, and the remaining solutes are present in different proportions (Eichler et al., 2001).

Snowpack-derived meltwater therefore tends to be relatively solute-rich at the beginning of the melt season, and becomes progressively dilute as melting proceeds. Rainfall or snowfall events during the ablation season, however, can introduce a fresh input of solutes to the glacier surface.

3.9.3 CHEMICAL WEATHERING PROCESSES

Chemical weathering rates, and the types of reactions that can occur, depend on a number of factors, including:

(1) existing concentrations of ions in the water;
(2) the area of water–rock contact;
(3) the length of time water is in contact with reactive rock surfaces;
(4) the availability of dissolved atmospheric gases and other reactants.

The maximum possible concentration of dissolved ions in a solution is known as the *saturation point*, which depends on pressure and temperature. Ions can be dissolved if the water is undersaturated with respect to those ions. Conversely, ions will be precipitated out of the solution if solute concentration rises above saturation, which can happen if water is lost through evaporation or freezing, if the temperature rises or if pressure falls. Importantly for glacial processes, solutes are rejected during freezing, which can cause the remaining waters to become supersaturated.

Chemical weathering is promoted by large rock–water contact areas, which provide potential *reaction surfaces*. Fine-grained sediments such as glacial rock flour have very large surface areas relative to their volume, and weather much more rapidly than coarse sediments or intact bedrock surfaces, if other factors are equal. The influence of reaction-surface area and rock–water contact times has been systematically explored in experimental work by Brown et al. (1996). Figure 3.66 shows how solute concentrations increase most rapidly when undersaturated water is in contact with fine-grained sediments. Rates of solute uptake are initially high (when water is far from saturation), then increase more slowly as saturation is approached.

Many important chemical weathering processes involve oxygen (O_2) and carbon dioxide (CO_2). These gases are present in dissolved form in well-aerated glacial meltwater, but tend to become depleted where water throughflow rates are small (section 3.9.4). Hydrogen cations (H^+) are required for a number of important weathering reactions. Hydrogen ion concentration is measured on the pH (potential hydrogen) scale. Distilled water has a pH of 7, acidic water (with a higher concentration of hydrogen ions) has lower pH values, and alkaline water (with a lower hydrogen concentration) has higher pH values. The relationship between mineral solubility and pH is not simple, however, and is commonly non-linear. Silica is slightly soluble at all commonly occurring pH values, whereas alumina (Al_2O_3) is readily soluble only in water below pH 4 and above pH 9.

From the above, efficient dissolution and uptake of ions will occur where undersaturated waters come into contact with extensive reaction surfaces, water–rock contact times are high and suitable reactants are available. Taken together, these factors place major constraints on where and when certain reactions can occur, and their rates of action. Some of the more important reactions involved in glacial chemical weathering are summarized below (Tranter et al., 2002).

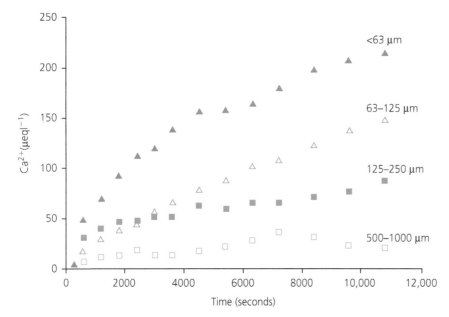

Figure 3.66 Concentrations of Ca^{2+} as a function of time in laboratory dissolution experiments, in which rock particles were stirred continuously in de-ionized water in free contact with the atmosphere. The particle size fractions used in each experiment are indicated (Brown et al., 1996).

Carbonation refers to reactions in which dissolved carbon dioxide reacts with carbonate (e.g. calcite) or silicate (e.g. feldspar and mica) minerals, leading to the release of ions. Examples of carbonation reactions are as follows:

$$CaCO_3 + H_2O + CO_2 \Leftrightarrow Ca^{2+} + 2HCO_3$$
(calcite) (calcium ion + 2 bicarbonate ions)

$$CaAl_2Si_2O_8 + H_2O + 2CO_2 \Leftrightarrow Ca^{2+} + 2H_2O + H_2Al_2Si_2O_8$$
(calcium feldspar) (calcium ion) (weathered feldspar)

The availability of carbon, either in the form of CO_2 in air bubbles or dissolved in meltwater, or from organic sources, is an important constraint on carbonation rates.

Reduction–oxidation or *redox* reactions involve exchanges of electrons between atoms. Oxidation reactions remove electrons from an atom, whereas reduction reactions add electrons to an atom. Despite its name, oxidation need not involve oxygen, and many other ions are effective oxidizing agents. A familiar example of oxidation involving oxygen is rusting of iron to form iron oxide:

$$4Fe + 3O_2 \Rightarrow 2FeO_3$$

In the absence of free oxygen, oxidation can be accomplished by other ions. For example, iron sulphide can be oxidized by ferric ions:

$$FeS_2 + 14Fe^{3+} + 8H_2O + 16CaCO_3$$
$$\Leftrightarrow 15Fe^{2+} + 16Ca + 2SO_4 + 16HCO_3$$

Reactions such as this can be mediated by anaerobic microbes, allowing efficient weathering to occur in anoxic environments (Tranter et al., 2002).

Coupled sulphide oxidation and carbonate/silicate dissolution (SO-CD or SO-SD) refers to linked reactions in which sulphide minerals (such as pyrite or 'fool's gold') and carbonate or silicate minerals are both weathered in the presence of dissolved oxygen. SO-CD reactions can be summarized as follows:

$$4FeS_2 + 16CaCO_3 + 15O_2 + 14H_2O$$
(pyrite) (calcite)
$$\Leftrightarrow 16Ca^{2+} + 16HCO_3 + 8SO_4 + 4Fe(OH)_3$$
(ferric oxyhydrofides)

while the following is an example of an SO-SD reaction involving potassium feldspar:

$$4FeS_2 + KAlS_3O_8 + 15O_2 + 86H_2O$$
(pyrite) (K-feldspar)
$$\Leftrightarrow 16K + 4Al_4Si_4(OH)_8 + 32H_4SiO_4 + 4Fe(OH)_3$$
(kaolinite) (ferric oxyhydroxides)

The amount of SO-SD weathering beneath glaciers is believed to be small relative to SO-CD weathering (Tranter et al., 2002).

Many other types of weathering reaction are possible. One important group of reactions is known collectively as *hydrolysis*, which involves the release of a cation from a mineral in exchange for a hydrogen atom from water. The following are examples of hydrolysis of potassium feldspar:

$$KAlSi_3O_8 + H_2O \Leftrightarrow K^+ + HAlSi_3O_8 + OH$$
(K-feldspar) (weathered feldspar)

and calcite:

$$CaCO_3 + H_2O \Leftrightarrow Ca^{2+} + HCO_3 + OH$$

Chemical reactions in subglacial environments cannot be observed directly, so weathering processes must be

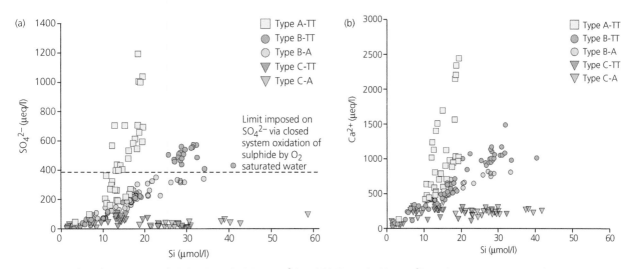

Figure 3.67 Co-plots of concentration of (a) silica (Si) and sulphate (SO_4^{2-}); and (b) silica and calcium (Ca^{2+}) in subglacial water recovered from boreholes in Haut Glacier d'Arolla. The ratios between these major ions highlight different populations of water. TT denotes the 'true type' of water, whereas A denotes artefacts caused by contamination of the water from other sources during sampling (Tranter et al., 2002).

inferred from water chemistry. This is not always straightforward, because the presence of particular ions in meltwater can potentially be explained by several alternative processes. One widely used method is to calculate the ratios between concentrations of pairs of major ions, and to determine which processes are best able to explain the observed combinations (e.g. Tranter et al., 2002). Even then, there may be no unique solution, because glacial meltwaters may have complex histories involving various combinations of chemical processes at different times.

3.9.4 SUBGLACIAL CHEMICAL WEATHERING

Chemical weathering can take place wherever undersaturated waters come into contact with reactive rock surfaces, and numerous studies have demonstrated that it can be very efficient in subglacial environments (e.g. Tranter et al., 1996; Anderson et al., 1997; Wadham et al., 1998, 2000; Hodson et al., 2002). It should be noted, though, that significant chemical weathering can also take place on glacier surfaces. In the Indian Himalaya, for example, monsoon rainfall during July and August enhances chemical weathering rates within supraglacial debris cover, introducing significant amounts of sulphate into meltstreams (Hasnain and Thayyen, 1999a). Supraglacial chemical weathering also occurs in moraine areas on otherwise clean glaciers, although bulk meltwaters tend to be relatively dilute (Wadham et al., 1998).

Glacier beds can be regarded as complex, dynamic mosaics of chemical weathering environments, controlled by bedrock and till lithology, the availability of reactants such as O_2 and CO_2, and other factors. It has become clear that the availability of free oxygen and carbon dioxide are particularly important controls on subglacial

chemical weathering processes (Tranter, 2003). At glacier beds, O_2 and CO_2 can be supplied from several possible sources. First, dissolved atmospheric gases can be transferred from the surface in turbulent, aerated meltwater. High partial pressures of CO_2 are also encouraged by low water temperatures. Waters with high concentrations of dissolved oxygen and carbon dioxide, therefore, will come into contact with glacier beds in the vicinity of major surface-fed channels, which are likely to be major sites of carbonation and oxygen-driven oxidation reactions. Elsewhere on the bed, O_2 and CO_2 can be released from air bubbles in the ice during regelation (section 4.5), although the supply from this source is limited, and in the absence of surface meltwater glacier beds tend to become anoxic (Tranter et al., 2005). Additionally, organic carbon can be washed down from the glacier surface, or released from overridden vegetation or soils. The latter may be particularly important early in glacial cycles when ice sheets override boreal forests (Sharp et al., 1999; Skidmore et al., 2000).

From the foregoing, it can be seen that contrasting subglacial chemical weathering environments can be expected in areas accessed by surface meltwater and in more hydraulically isolated areas. This has been elegantly demonstrated by Tranter et al. (2002), who sampled subglacial water in a borehole array around a subglacial channel at Haut Glacier d'Arolla (see section 3.4.1 for details of the sampling site). Three types of subglacial water were found, reflecting different chemical weathering environments (fig. 3.67). First, *Type A water* was found in very isolated areas of the bed, where there was apparently little or no contact with surface-derived water. Type A water had high solute concentrations, with high SO_4, Ca and HCO_3 relative to Si. This type of water has long residence times and little free oxygen. Under these anoxic conditions, sulphide oxidation appears to be driven by microbial activity, employing dissolved iron as an oxidizing agent. Second, *Type B water* is similar to Type A, but with a greater availability of free oxygen. Its chemistry indicates weathering mainly by carbonate and silicate hydrolysis, followed by coupled SO-CD and SO-SD reactions, apparently in relatively efficient parts of a distributed drainage system. There is evidence, however, that parts of this system may become anoxic. Finally, *Type C water* was the most dilute, and was found in the channel margin zone which experienced regular contact with oxygen-rich, surface-derived waters. Lower concentrations of the major ions and higher Si concentrations suggest a weathering environment in which the glacier bed is depleted of reactive sulphides, and solute is acquired mainly by the slow dissolution of carbonate and silicates.

The strong influence of drainage system configuration on chemical weathering processes means that subglacial weathering environments can be expected to evolve on annual and other timescales (Anderson et al., 2003b). The growth of the subglacial drainage system below Haut Glacier d'Arolla during the ablation season (section 3.5.1) leads to systematic seasonal variation in the chemistry of bulk meltwaters (Tranter et al., 1996). Concentrations of both SO_4 and HCO_3 are highest during the early melt season, and decrease thereafter in response to the development of an efficient, surface-fed drainage system.

Evidence for annual cycles in chemical weathering below Finsterwalderbreen, a polythermal valley glacier in Svalbard, has been presented by Wadham et al. (1998, 2000). Water emerges from the terminus of this predominantly warm-based glacier throughout the winter, where it builds up in a large proglacial icing. Samples taken from the icing and associated early spring upwelling water indicate that meltwater emerging from the glacier bed is relatively dilute during the winter months, apparently due to limited sources of oxygen and carbon. Once surface meltwater begins to access the bed in late spring, the potential for subglacial solute acquisition increases due to a renewed supply of dissolved atmospheric gases for oxidation and carbonation reactions. The solute content of water emerging from the bed is high in the early summer, but progressively declines thereafter in response to increasing dilution by surface meltwater and reduced transit times (fig. 3.68).

Due to the great variety of basal thermal regimes (section 2.3.3), polythermal glaciers exhibit a very broad range of hydrochemical characteristics. On one hand, the hydrochemistry of predominantly warm-based polythermal glaciers closely resembles that of temperate Alpine glaciers (Tranter et al., 1996; Wadham et al., 1998); on the other hand, the crustally derived solute load of small, cold-based glaciers may mainly reflect weathering in ice-marginal environments (Hodgkins et al., 1998).

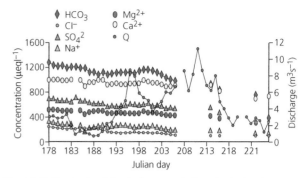

Figure 3.68 Temporal variation in the concentration of major ions in subglacial water upwelling at the terminus of Finsterwalderbreen during summer 2005. Data have been adjusted to remove snowpack-derived solutes (Wadham et al., 2000).

It should be noted, however, that channelized subglacial drainage is possible below cold-based glaciers, because cut-and-closure type conduits can reach the glacier bed (section 3.3.2; Yde et al., 2008; Gulley et al., 2009a). Water flowing in vadose cut-and-closure channels will tend to be well aerated, encouraging efficient, if highly localized, subglacial chemical weathering.

3.9.5 PROGLACIAL ENVIRONMENTS

Proglacial areas are also important weathering environments (Anderson et al., 2000; Tranter, 2003). Recently deposited sediments on glacier forelands are typically water-saturated for part of the year, so reaction-surface areas and rock–water contact times tend to be high. In high latitudes, seasonal saturation of the ground is also encouraged by the presence of permafrost. In permafrost areas, the upper few tens of centimetres of the ground freeze and thaw on a seasonal basis (the active layer), whereas deeper levels are permafrozen. Summer thawing of the active layer produces meltwater, which is prevented from draining by underlying permafrost. Consequently, the active layer is often waterlogged early in the melt season, and some areas may remain saturated throughout the summer months. Seasonal freezing of the active layer serves to concentrate solute in residual waters, which can be mobilized during the next spring thaw. In areas with a dry summer season, solute enrichment can also occur by evaporation. As the ground surface dries out, water is drawn to the surface by capillary action. Evaporation progressively concentrates solutes in near-surface water, and eventually ions may precipitate out of the solution. White crusts of evaporitic salts are widespread on the forelands of Svalbard glaciers, which experience little summer rainfall.

Exchange of water between stream channels and proglacial sediment occurs on various timescales. When the proglacial zone is inundated by rising glacifluvial discharge, solutes can be transferred from groundwater or evaporite deposits and enter fluvial transport (Anderson

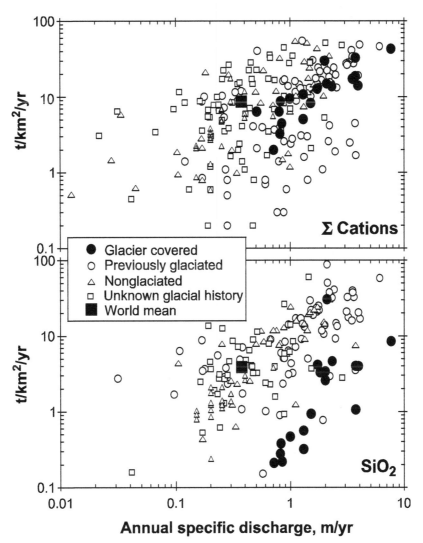

Figure 3.69 Chemical denudation rates from active glaciers compared with a global data set of non-glacierized catchments (Anderson, 2005).

et al., 2000). In consequence, the solute load of glacier meltstreams can increase significantly with distance across glacier forelands (e.g. Wadham et al., 2001b). This has important implications for the location of sampling sites: unless water is sampled close to the glacier terminus, its hydrochemistry will have a proglacial signature superimposed on the glacial signal.

3.9.6 RATES OF CHEMICAL EROSION

Despite generally low temperatures, chemical erosion in glacierized environments proceeds at rates comparable to those in temperate catchments (Tranter, 2003). Solute fluxes from glaciers therefore constitute a major component of landscape denudation (Anderson et al., 1997; Anderson, 2005). In terms of catchment area, solute fluxes from glacierized basins tend to be higher than the global average, although when glacierized catchments are compared with non-glacierized basins with similar water discharges, a more complex picture emerges. Data compiled by Anderson (2005) shows that, for given specific annual discharges, total cation fluxes from glacierized basins are near the global mean, while silica fluxes are at the lower end of the range (fig. 3.69). Both total cations and silica fluxes increase with discharge, with maximum observed values of 30–40 tonnes per square kilometre per year.

Anderson (2005) has argued that these patterns reflect competing controls on chemical erosion below glaciers. As discussed in section 3.9.3, chemical weathering rates are directly proportional to the area of reactive rock surfaces, so high weathering rates are encouraged in glacial environments where undersaturated meltwater comes into contact with abundant fresh, fine-grained debris. Anderson (2005) has estimated that for mechanical erosion rates and grain-size distributions typical of temperate glaciers, in the order of $10^4 km^2$ of new mineral surface area is produced per square kilometre per year.

This represents a vast rock-surface area over which weathering reactions can take place, releasing ions that can then be evacuated by meltstreams.

Second, many chemical weathering processes are less efficient at low temperatures, tending to reduce solute fluxes from glacierized catchments relative to warmer environments. For silicate minerals, chemical weathering rates are nearly an order of magnitude lower at 0 °C than at 20 °C (White et al., 1999; Anderson, 2005). This is not the case for carbonate weathering, however, due to the high solubility of CO_2 in cold water. Consequently, a substantial portion of the total cation flux from the world's glaciers is derived by dissolution of carbonates (e.g. Jacobson et al., 2002; Tranter et al., 2002). For example, at Bench Glacier, Alaska, carbonate minerals constitute only ~1 per cent of the bedrock area, but carbonate dissolution contributes 70 per cent of the total cation flux (Anderson et al., 2000). This explains why total cation fluxes from glacierized basins are similar to those from non-glacierized catchments with similar discharges, while silica fluxes are substantially less.

The correlation between chemical weathering rates and specific annual discharge appears to reflect an increase in reactive mineral surface area with increasing water flux. Anderson (2005) argued that the key process in this relationship is the rate of glacial mechanical erosion, which controls rock flour production. Subglacial mechanical erosion requires basal motion (section 8.2), so subglacial erosion is likely to be more efficient where substantial amounts of meltwater reach the bed. Available data support the idea that chemical weathering rates are directly proportional to mechanical erosion rates, as indicated by suspended sediment concentrations in proglacial rivers. However, the data set is small and further field studies are required.

Carbonation of both silicate and carbonate minerals extracts CO_2 from glacial meltwater (section 3.9.3), so these weathering processes act as a sink for carbon dioxide. In the case of carbonates, however, the products of carbonation are bicarbonate (HCO_3) and calcium (Ca^{2+}) ions, which recombine in the oceans to form new carbonate minerals, re-releasing CO_2. Weathering of carbonates, therefore, has no net effect on atmospheric CO_2 levels, and only the carbonation of silicates results in long-term drawdown of atmospheric CO_2. It has been suggested that silicate weathering in glacierized catchments could perturb atmospheric CO_2 concentrations over glacial timescales (Gibbs and Kump, 1994; Sharp et al., 1995a). Modelling work, however, indicates that chemical weathering below ice sheets probably had a very small impact on the global CO_2 budget, and that glacigenic CO_2 drawdown is unlikely to have contributed to global cooling during Quaternary cold periods (Ludwig et al., 1999; Jones et al., 2002).

Very few studies have been conducted on the CO_2 budgets of glacierized basins. Data presented by Hodson et al. (2000) indicates that rates of contemporary CO_2 drawdown in glacierized basins are highly variable, depending on catchment lithology and discharge, but that the global mean is similar to that estimated for the world's largest river basins. Estimates of CO_2 drawdown are subject to major uncertainties, however, and the impact of glacier fluctuations on atmospheric carbon dioxide levels remains poorly understood.

CHAPTER 4

PROCESSES OF GLACIER MOTION

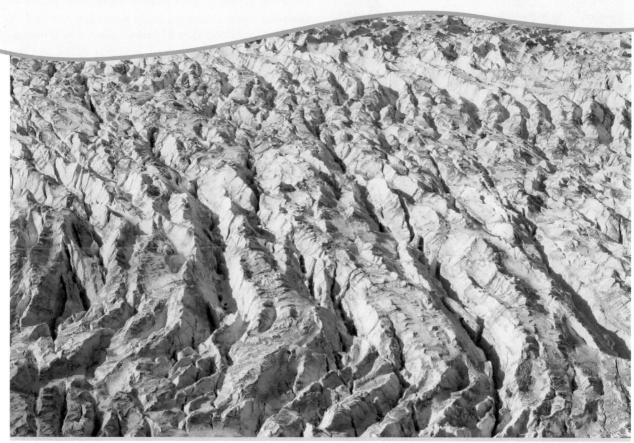

Crevassed ice on Kronebreen, Svalbard (Doug Benn)

4.1 INTRODUCTION

One of the most fundamental characteristics of glaciers and ice sheets is their ability to move (fig. 4.1). Glacier flow transfers ice from high elevation accumulation areas and continental interiors to areas where ice is lost by melting and calving, and hence plays a major role in the hydrological cycle. The response of the cryosphere to climate change – and future sea-level rise – therefore depends to a large degree on the mechanics of glacier flow. In addition, glaciers are effective agents of landscape change because of their ability to flow, transporting rock debris into depositional basins from areas of net erosion. Because of its central role in glacier behaviour, the physics of glacier motion has exercised the minds of scientists since the dawn of glaciology as a subject, and

it remains an important research field to this day (Clarke 1987b, 2005). Recent work has yielded a huge increase in knowledge about the processes of glacier motion, and our present understanding of glacier movement has resulted from a creative interplay between measurements made beneath modern glaciers, geophysical studies, experimentation with samples of ice or sediment in the laboratory, numerical modelling, and interpretation of sediments and landforms in formerly glaciated areas. Our understanding is still incomplete, and many questions remain about the detailed processes that control flow rates. This is partly because the forces controlling flow depend on several interacting factors – such as temperature, debris content of the ice, bed roughness and water pressure – but mainly because data on actual conditions within and below glaciers are extremely difficult to obtain.

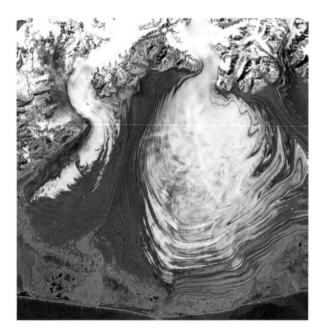

Figure 4.1 This spectacular false-colour Landsat image of the Malaspina Glacier, Alaska, clearly reveals patterns of ice flow (NASA/USGS).

In this chapter, we examine processes of glacier flow, beginning in section 4.2 with the physical controls on ice velocities, particularly the stresses driving and resisting flow and their implications for rates of strain. Ideas introduced in this section are fundamental to understanding ice flow and many other aspects of glacial action, including erosion, subglacial sediment transport and glacitectonics. In sections 4.3, 4.4 and 4.5, we examine the mechanisms of glacier flow – ice creep, sliding over rigid beds, and glacier motion over deformable beds. Section 4.6 examines ways in which glaciologists calculate rates of basal motion, and discusses their strengths and weaknesses. Section 4.7 concludes the chapter with a discussion of structural glaciology, focusing on crevasses, foliation and related structures, and what they can reveal about the flow history of a glacier. Larger-scale flow dynamics of glaciers and ice sheets are covered in chapters 5 and 6.

4.2 STRESS AND STRAIN

The words 'stress' and 'strain' are often used interchangeably in everyday language, but in scientific usage their meanings are quite distinct. *Stress* is a measure of how hard a material is being compressed, stretched or twisted as the result of applied forces, whereas *strain* measures the amount of deformation that occurs as the result of stress. A familiar example of this distinction is what happens when a tube of toothpaste is squeezed. The toothpaste comes out due to deformation (strain) resulting from exerting a higher pressure on the surface of the tube than that at the nozzle (the stress). Similarly, glacier

motion (strain) occurs in response to stresses arising from the weight of the ice. Readable accounts of relationships between stress and strain in a wide variety of familiar materials and structures can be found in two books by Gordon (1976, 1978).

A complete analysis of the stress and strain in a glacier (or toothpaste tube) requires consideration of the forces acting in all directions on the material, and the whole range of resulting deformations. Excellent guides to analysing stress and strain in ice have been provided by van der Veen (1999a) and Hooke (2005). The following discussion of stress and strain is largely based on these texts, but focuses on explaining the key concepts rather than the details of analytical techniques. The more important equations are examined at some length, to help develop insights into how glaciers work and to provide a solid basis for more advanced studies.

4.2.1 STRESS

To understand stress, it is useful to begin with the concept of force. *Force* can be thought of as the physical influences which change, or tend to change, the state of motion of a mass, and is defined as *mass times an acceleration*. For example, the downwards force at the base of a block of ice is the product of the mass of the block and the gravitational attraction (an acceleration) between the block and the Earth. The SI unit of mass is the kilogram (kg) and that of acceleration is metres per second per second ($m\,s^{-2}$), so the unit of force is $kg\,m\,s^{-2}$, or Newton (N). When a force acts on a surface, the intensity of its action depends on the area over which it is distributed. This distributed force is a *stress*, which is defined as *force per unit area*. A given force acting on a small area results in a larger stress than the same force acting over a large area. For example, a 10,000 kg mammoth supported by a table leg $0.25\,m^2$ in area would exert a stress 100 times larger than the same mammoth resting on a block with a surface area of $25\,m^2$ (fig. 4.2). The same effect explains why girls in high heels cause more damage to floors than heavy men in flat-soled shoes. The unit of stress is the Pascal, or Newton per square metre ($1\,Pa = 1\,N\,m^{-2}$), although for the high stresses encountered at the base of glaciers, it is more useful to employ the kiloPascal ($1\,kPa = 1000\,Pa$). Another unit of stress which is often seen in the older glaciological literature is the bar ($1\,bar = 100\,kPa$).

Stresses on a surface can be thought of in terms of two components: the stress acting at right angles to the surface (the *normal stress*) and that acting parallel to the surface (*shear stress*; fig. 4.3). The stress on a surface consists of two equal and opposite *tractions*. In the case of normal stresses, the two opposing tractions either press together across the surface (*compressive stress*) or pull away from it (*tensile stress*). For shear stresses, the tractions are parallel, but act in opposite directions. In this book, we use the Greek letter σ (sigma) as the symbol for stress, with subscripts indicating its direction of action and the plane on which it acts (fig. 4.4). Following common convention, we also use the symbol τ (tau) for specified shear stresses. Readers should be aware, however, that many other conventions are also in use, and that different symbols may be used for the same variables in other texts.

At the base of a glacier, the normal stress acting on the bed is principally due to the weight of the overlying ice. If

the bed is horizontal, the normal stress (or *pressure*) exerted by an ice column with basal area a_b is:

$$\sigma_{zz} = \rho_i\, g(H \times a_b)/a_b, \text{ or:}$$
$$\sigma_{zz} = \rho_i\, gH \tag{4.1}$$

where σ_{zz} is the normal stress acting in the vertical (z) direction, ρ_i is the density of the ice, g is gravitational acceleration, and H is the ice thickness.

In glaciers, normal stresses also act in all other directions, due to the tendency of ice to spread under its own weight. In some cases, the normal stresses can be the same in all directions, but if the ice is subject to other pushes or pulls they can differ significantly from the average value. The mean normal stress is sometimes known as the *cryostatic pressure*, by analogy with hydrostatic and lithostatic pressure in water and rock respectively. The difference between any normal stress and the cryostatic pressure is termed the *deviatoric stress*, because it represents a deviation from the mean. For example, the deviatoric stress in the x direction, σ'_{xx} (sigma prime xx), is defined as:

$$\sigma'_{xx} = \sigma_{xx} - 1/3(\sigma_{xx} + \sigma_{yy} + \sigma_{zz}) \tag{4.2}$$

Figure 4.2 Monty the mammoth is safely supported by a block with cross-sectional area A_1, but breaks a table leg with the much lower cross-sectional area A_2. Because Monty's weight must be supported by a smaller area, he exerts a greater stress and thus falls to the floor (With acknowledgements to Twiss and Moores, 1992).

Diagrams		Definitions
(a) Force	Force components	A push or a pull
(b) Traction	Traction components	Force per unit area on a surface of a specified orientation (a measure of force intensity)
(c) Surface stress	Surface stress components	A pair of equal and opposite tractions acting across a surface of specified orientation
(d) Shear stress		A pair of tractions acting parallel to a surface
(e) Normal stress		A pair of tractions acting at right angles to a surface
Compressive	Tensile	

Figure 4.3 Definition of stress components (Modified from Twiss and Moores, 1992).

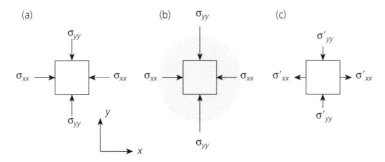

Figure 4.4 Normal stresses on a square element. (a) Normal stresses in the *x* and *y* directions. The first subscript denotes the direction in which the stress acts, and the second subscript denotes the direction normal (at right angles) to the surface on which the stress acts. (b) Unequal normal stresses. The shaded circle indicates the mean (cryostatic) pressure. (c) Deviatoric stresses resulting from the unequal normal stresses shown in (b).

The deviatoric stress in a given direction is compressive if the normal stress in that direction is greater than the mean, and tensile if it is less than the mean. Thus, tensile deviatoric stresses can occur even when all the absolute normal stresses are compressive, provided the normal stresses are smaller in some directions than others (fig. 4.4). A useful analogy is that of four children standing in a circle pushing against a beach ball in the centre. If one pair of children (pushing against opposite sides of the ball) exerts a greater push than the other, the weaker pair will be forced outwards, even though they are still pushing against the ball.

Normal and shear stresses are not independent, but vary together in systematic ways. Figure 4.5a shows that ice subject to a shear stress is pulled apart in some directions and pushed together in others. These pulling (tensile) and pushing (compressive) stresses are greatest at an angle of 45° from the shear plane, and zero across the shear plane itself. Conversely, ice subject to unequal normal stresses will also experience shear stresses in directions oblique to the normal stress axes (fig. 4.5b). This shear stress is zero parallel to the normal stress maxima, and is at a maximum at an angle of 45° to them. Thus, *a shear stress in one direction results in deviatoric stresses in others, and vice versa.* Ice is usually subject to some combination of shear and normal stresses, which combine to produce a unique set of shear and deviatoric stresses. The maximum and minimum normal stresses in any given situation are known as the *principal stresses*.

A useful means of visualizing the relationship between normal and shear stresses is *Mohr's circle* (fig. 4.6), introduced by Otto Mohr in 1892. To represent stresses in two dimensions, the normal stresses σ_{xx} and σ_{yy} (measured in any arbitrary pair of orthogonal directions) are plotted on the horizontal axis, and the corresponding shear stress σ_{xy} plotted on the vertical axis. The mean normal stress plots midway between σ_{xx} and σ_{yy} and is labelled *B*. A line connecting *B* with point *A* (σ_{xx}, σ_{xy}) defines the radius of a circle, the circumference of which denotes the combination of normal and shear stresses acting in all possible directions in the material. Where the absolute value of the shear stress is at a maximum (at the top and bottom of the circle), both normal stresses are equal to the mean and the deviatoric stresses are zero. Conversely, where the

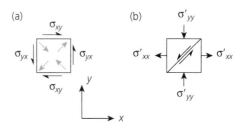

Figure 4.5 Shear stresses on a square element. (a) Shear stress definitions. In the same way as normal stresses, the first subscript denotes the direction in which the stress acts, and the second subscript denotes the direction normal to the surface on which the stress acts. The gray arrows indicate the deviatoric stresses resulting from the applied shear stresses. (b) Deviatoric stresses result in shear stresses, which are at a maximum on the diagonals of the square element shown. Only one diagonal is shown for clarity.

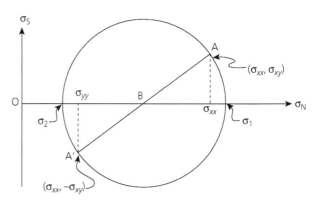

Figure 4.6 Mohr's circle (Hooke, 2005).

shear stress is zero (at the right and left sides of the circle), both normal stresses are furthest from the mean and the deviatoric stresses are at a maximum. These points define the *principal stresses*, labelled σ_1 and σ_2. Figure 4.6 shows the stresses in only two dimensions, but of course, in reality, materials are also subject to stresses in the third dimension. In this case, there is a third principal stress, acting at right angles to the other two. In three dimensions, the maximum principal stress is denoted by σ_1, the minimum principal stress by σ_3 and the intermediate one by σ_2.

Angles on Mohr's circle (measured from point B relative to the horizontal axis) represent directions in the material (measured relative to the principal stress axes), with the convention that angles on the diagram are *twice* the corresponding angles in the material. It can therefore be seen that the direction where the shear stress is at a

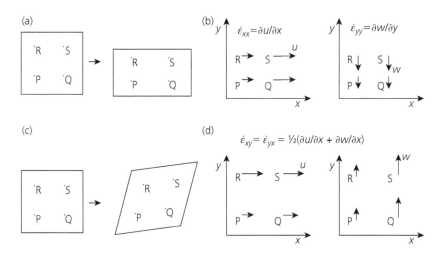

Figure 4.7 Strain rate definitions. (a) Pure shear: the square is stretched and flattened, but not skewed. (b) The strain rates are defined in terms of velocity gradients (i.e. how the velocity component u varies in the x direction, and w varies in the y direction). (c) Simple shear: the square is skewed, but not stretched or flattened. (d) The strain rate is defined in terms of the gradient of the velocity component u in the y direction, and w in the x direction. In real-world situations, strain is commonly a mixture of pure and simple shear, but can still be resolved into these components.

maximum (90° on Mohr's circle) is at 45° to the principal stress axes. The use of Mohr's circle (and the equivalent set of equations) in the analysis of stresses is discussed in detail by Hooke (2005).

4.2.2 STRAIN

Strain is the change in shape and dimensions of a material due to stress. Different materials – such as ice, water, sediments and rocks – exhibit a wide variety of strain responses to a given stress. Some will deform quickly, some will deform slowly, some will break and yet others may appear to be unaffected. In fact, all materials undergo strain of one type or another when placed under stress, no matter how small. We can divide strains into two basic types: recoverable or *elastic strains*, and irrecoverable or *permanent strains*. Elastic strain refers to a temporary change in shape of a material, which lasts only as long as stress is applied. Once the stress is removed, the material goes back to its original shape. This type of strain is not confined to familiar 'springy' materials such as rubber or the spring in a shock absorber, but is also a characteristic of apparently rigid materials, such as a concrete pavement or bedrock at the base of a glacier. Although the temporary change in shape may be very small, stresses applied to elastic materials always cause compression or stretching of molecular bonds, resulting in the storage of *strain energy*. This stored energy provides the resisting force to balance the applied force, which is why you don't sink into the pavement when you walk along the street (Gordon, 1976). At some critical level of stress, which varies widely between different materials, the stored strain energy will be released, resulting in *permanent deformation* or *failure*. The stress that marks the onset of permanent deformation defines the yield strength, discussed in detail in section 4.3.3. Permanent deformation can take the form of *brittle failure*, where the material breaks along a fracture, or *ductile deformation*, where the material undergoes flow or creep.

Deformation of a material may or may not involve changes in its volume. Some materials can undergo either contraction or expansion (*dilation*) in response to deformation, whereas others can deform without volume change (*constant-volume deformation*). As noted in section 2.4.2, the transformation of snow to ice in the accumulation area of a glacier results in a decrease in volume and an increase in density of the snow. Where melting and refreezing are negligible, this contraction (or consolidation) occurs in response to the normal stresses imposed by the weight of the overlying snow. Once formed, glacier ice is essentially incompressible, and most deformation is constant-volume strain. An exception to this is an increase in the volume of the surface layers of a glacier when crevasses open in response to tensile stresses (section 4.7).

The amount of strain experienced by a material is measured by comparing its shape and size before and after deformation. The patterns of deformation associated with glacier flow may be highly irregular, but are often relatively simple, particularly if small time intervals or volumes of material are considered. The two fundamental types of strain are *pure shear* and *simple shear* (fig. 4.7). Pure shear involves flattening or stretching of a material under compressive and tensile deviatoric stresses. In this type of strain, the orientation of lines drawn parallel and normal to the applied stress remains constant as the material is flattened or stretched. In contrast, deformation by simple shear resembles the shearing of a pack of cards. This type of deformation involves the rotation of all lines drawn through a material except those parallel to the shear plane. The strain occurring near the beds or lateral margins of glaciers often closely resembles simple shear.

It is important to distinguish two ways of measuring strain. First, *strain rate* measures the amount of strain that occurs per unit of time. Second, *cumulative strain* refers to the net amount of strain that takes place in a given time interval, and depends on both the rate and duration of strain. High cumulative strains can result

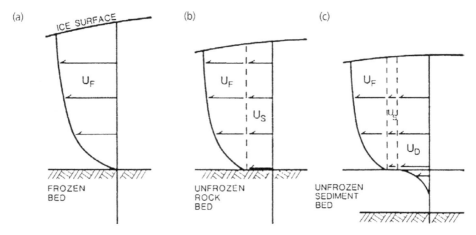

Figure 4.8 Schematic diagram showing the vertical distribution of velocity for different types of glacier motion. (a) Ice deformation only. (b) Ice deformation and basal sliding. (c) Ice deformation, basal sliding and deformation of subglacial sediments. Note that the vertical scale in the sediment bed is greatly exaggerated relative to that of the overlying ice (Boulton, 1996a).

from low strain rates sustained over a long period, or high strain rates lasting only a short time.

Strain rates are defined in terms of *velocity gradients*, or the relative motion of different parts of a material. In pure shear, consisting of stretching or compression of a material, the velocity gradient is measured along the direction of motion (fig. 4.7b), so that the strain rate in the x direction is:

$$\dot{\varepsilon}_{xx} = \frac{\partial u}{\partial x} \qquad (4.3)$$

In contrast, for simple shear, the velocity gradients are measured at right angles to the direction of motion (fig. 4.7d):

$$\dot{\varepsilon}_{xy} = \frac{1}{2}\left(\frac{\partial v}{\partial x} + \frac{\partial u}{\partial y}\right) \qquad (4.4)$$

If the velocities u and v (motion in the x and y directions, respectively) are measured in metres per year, the strain rate $\dot{\varepsilon}$ is in units of metres per year per metre, or simply yr^{-1}. The dot above the Greek letter epsilon ε indicates that it is a rate, using the notation introduced by Isaac Newton (1643–1727) to signify changing quantities. The *differential notation* used on the right-hand side of equations 4.3 and 4.4 is another way of denoting changing quantities (in this case, velocity) and was developed by Gottfried Leibniz (1646–1716), co-inventor of the calculus. For brevity, doubled subscripts are commonly replaced with a single subscript, so that $\dot{\varepsilon}_{xx}$ becomes $\dot{\varepsilon}_x$.

4.2.3 RHEOLOGY: STRESS–STRAIN RELATIONSHIPS

Glacier motion occurs by strain within the ice or the bed, or by sliding at the interface between the two (fig. 4.8). The relationships between stress and strain in glaciers and their beds are therefore vital to understanding glacier dynamics and glacial geologic processes. The way in which strain rate varies with applied stress for a given material is known as its *rheology*, a term coined by Eugene C. Bingham in 1920, inspired by Heraclitus' expression *panta rei*, 'everything flows' (which is certainly true in glacial environments). Material rheology can be understood in terms of two basic properties: the yield strength and the stress–strain relationships at higher stresses.

Yield strength

The yield strength of a material is defined as the value of the applied stress at the onset of permanent deformation, and is measured in Pascals. The yield strength can be zero, meaning that permanent deformation will occur at any stress, no matter how small. This appears to be the case for ice, although it is sometimes convenient to assume a yield strength in simple models (see section 5.3.2). Glaciers are underlain by a great variety of materials, ranging from weak clay to igneous rocks, the strength of which strongly influences processes of glacier motion, as well as rates of erosion and sediment transport.

The yield strength of subglacial sediments or rocks can be understood as the sum of two properties: cohesion and friction. *Cohesion* refers to the forces binding a material together, including chemical bonds and electrostatic forces between particles. For unlithified materials such as subglacial till, chemical bonding between grains is usually unimportant and cohesion often is due entirely to electrostatic forces between clay particles. Electrostatic attraction is negligibly small for particles larger than about 1 micron (μm), and therefore cohesion only contributes towards yield strength when materials are lithified or significant amounts of clay are present. The cohesive strengths of selected sediments and rocks are shown in table 4.1.

Frictional strength is the macroscopic outcome of interactions between rough material surfaces at microscopic

Material	Cohesion (kPa)	Friction angle (°)
Dense sand (well-sorted)	0	32–40
Dense sand (poorly sorted)	0	38–46
Gravel (well sorted)	0	34–37
Gravel (poorly sorted)	0	45–48
Bentonite clay	10–20	7–13
Soft glacial clay	30–70	27–32
Stiff glacial clay	70–150	30–32
Till (mixed grain size)	150–250	32–35
Soft sedimentary rock	1000–20,000	25–35
Igneous rock	35,000–55,000	35–45

Table 4.1 Typical cohesion values and friction angles for some geologic materials (Selby, 1993)

scales. Frictional strength is directly proportional to the forces pressing surfaces together – that is, the normal stresses. An everyday example of this effect is the friction between sandpaper and the surface of a piece of wood – the harder the sandpaper is pressed down (normal stress), the harder it is to slide along because of the increase in frictional resistance. For each material, the specific frictional strength is measured by the ratio between the normal stress and the applied shear stress at the onset of permanent deformation. This ratio is known as the *coefficient of friction* and is written as tan Φ, where Φ is the *angle of internal friction*. It has a unique value for each material and is influenced by the roughness and resistance to crushing of the interacting surfaces. The angle of internal friction for some common geological materials is shown in table 4.1.

Study of the yield strength of materials was pioneered by Charles Augustin de Coulomb (1736–1806), and further developed by Otto Mohr (1835–1918). The contribution of cohesion and frictional strength to yield strength is summarized in what is now known as the Mohr–Coulomb equation:

$$\tau_{yield} = c + N \tan \Phi \qquad (4.5)$$

This important equation states that the yield strength of a material (σ_{yield}) is given by the cohesion (c) plus the product of the effective pressure, N (also known as effective normal stress) and the coefficient of friction (tan Φ) (fig. 4.9). The *effective pressure* is equal to the total normal stress acting across a surface, minus the water pressure (equation 3.10). At low water contents, surface tension at wet contacts serves to pull surfaces together, increasing the effective pressure and causing a rise in frictional strength. This effect, known as *pore-water tension* or negative pore-water pressure, explains, among other things, why damp sand is better than dry for building sandcastles. For saturated materials, surface tension forces disappear, and if the water cannot drain away freely, positive pore-water pressures develop and support part of the weight of the overlying material. The effective

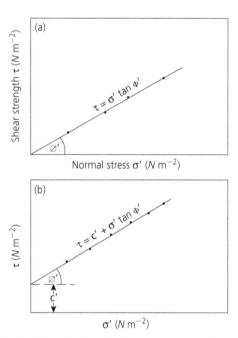

Figure 4.9 Graphs showing the relationship between normal stress and shear strength for (a) cohesionless and (b) cohesive materials. The angle of internal friction is the slope of the line (Selby, 1983).

pressure is therefore reduced. This causes a reduction in strength because confined water can support compressive stresses (and hence reduce N), but cannot support a shear stress. This is why saturated sand is weaker than either damp or dry sand and is useless as sandcastle-building material.

In the Mohr–Coulomb equation, cohesion is regarded as a constant, independent of N. This is not strictly true in all cases: where cohesion is due to electrostatic forces (such as for clay-rich tills), cohesion decreases with increasing distance between particles. This is a function of the sediment *porosity*, n^*:

$$n^* = V_v/V \qquad (4.6)$$

or the volume of pores (i.e. air- or water-filled voids in the sediment) divided by the total volume of the material.

An alternative measure of the pore volume of a sediment is the *voids ratio*:

$$e^* = V_v/V_s \qquad (4.7)$$

or the volume of pores divided by the volume of solid particles. Porosity increases as effective pressure decreases (Clarke, 1987a; Tulaczyk et al., 2000a), so one should expect the cohesive strength of even clay-rich tills to be small where the effective pressure approaches zero.

Deformation above the yield strength

Once the yield strength is exceeded, materials can respond to increasing stresses in a variety of ways (fig. 4.10). In *viscous materials*, strain rate increases linearly with the applied stress. The constant of proportionality between stress and strain rate is the *viscosity*, signified by the Greek letter μ (mu):

$$\mu = \frac{(\sigma - \tau_{yield})}{\dot{\varepsilon}} \quad (\sigma > \tau_{yield}) \qquad (4.8)$$

If the yield stress is zero, fluids of this type are referred to as *Newtonian*; otherwise they are known as *Bingham* (or viscoplastic) materials. Where strain rates increase non-linearly with applied stress, the viscosity is not constant, but is a function of the strain rate. In this case, the stress–strain relationship takes the form:

$$\dot{\varepsilon} = a\sigma^b \qquad (4.9)$$

Materials of this type (which include ice) are variably known as non-linear viscous or power law fluids.

Perfectly plastic materials only begin to undergo irrecoverable strain at the yield stress, at whatever rate is necessary to keep the applied stress from exceeding the yield stress. There is therefore no unique value of the strain rate when the applied shear stress equals the yield stress, and strain cannot be predicted from simple flow laws such as equation 4.9. In *Coulomb-plastic* materials, the yield stress is a linear function of the effective normal pressure (equation 4.5). As discussed in section 4.4.1, ice exhibits non-linear viscous behaviour, although in simple models it is sometimes assumed to have perfectly plastic behaviour (section 5.3.2).

4.2.4 FORCE BALANCE IN GLACIERS

Glacier flow is driven by gradients in the force exerted by the ice (the *driving stress*) and resisted by drag at the glacier boundaries and ice viscosity (*resistive stresses*). In almost all glaciological situations, the driving and resistive stresses are close to being in balance, and accelerations can be ignored. This forms the foundation of the *force balance* approach, which provides a very powerful framework for studying glacier dynamics (van der Veen and Whillans, 1989a; van der Veen, 1999a). A lucid and

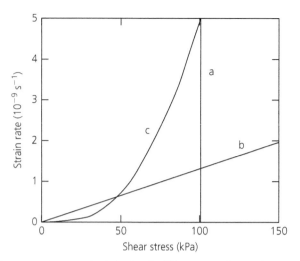

Figure 4.10 Stress–strain relationships for different types of material. (a) Perfectly plastic material, which remains rigid until the shear stress reaches the yield stress (100 kPa in this case). The material then deforms at whatever rate is required to prevent the shear stress from exceeding the yield stress. (b) Newtonian, linear-viscous material, for which strain rate is linearly proportional to the shear stress. (c) Non-linearly viscous material, such as ice (Modified from Paterson, 1994).

readable account of force balance in glaciers, and its application in modelling studies has been provided by van der Veen and Payne (2004).

The driving stress is defined as:

$$\tau_d = \rho_i g H \tan \alpha \qquad (4.10)$$

where α is the surface slope of the ice. In some publications, this formula is given with the sine of the surface slope in place of the tangent. This is because early formulations of the driving stress assumed a parallel-sided slab of ice resting on a planar slope; the older form of the driving stress equation is correct for this geometry, but not for others. Equation 4.10 is correct where H is measured vertically, for a glacier with surface slope α and any bed configuration (van der Veen and Payne, 2004; Hooke, 2005).

Resistance to flow arises from drag at the glacier bed (*basal drag*), the sides of the glacier or ice stream (*lateral drag*), and from spatial variations in pushing or pulling forces (*longitudinal stress gradients*). In the simplest case, the driving stress (or, more correctly, the driving traction) is exactly balanced by an equal and opposite traction at the glacier bed. In this case, equation 4.10 gives the value of the *basal shear stress*, τ_b, as well as the driving stress. Basal drag is very difficult to measure directly, and in force balance calculations it is generally inferred from the difference between other, more easily calculated terms (e.g. Whillans and van der Veen, 1997; Rolstad et al., 2000; O'Neel et al., 2005).

Lateral drag arises where bed resistance varies in a cross-flow direction; this can occur at the margins of a valley or trough, or where there are lateral changes in bed roughness, substrate type, basal temperature or water pressure (Raymond, 1996; Raymond et al., 2001). Beneath a floating ice shelf, bed resistance is zero and almost all of the driving stress can be supported by lateral drag at the margins. On most glaciers, both the bed and the margins contribute to resistance, and the importance of lateral drag varies with trough width, shape and other factors (Nye, 1957).

The *longitudinal stress gradient* is the result of variations in along-flow directed deviatoric stresses. This can be understood by returning to our analogy of children pushing on a beach ball. If two children push equally hard on opposite sides of the ball, neither child is pushed back by the other, although the ball will bulge out at the sides, top and bottom. In technical language, there is a compressive deviatoric stress in the direction of push, and a tensile deviatoric stress in the other two directions (side to side, and up and down). If one child pushes harder than the other, the deviatoric stresses are unequal in the direction of push. Therefore, there is a longitudinal stress gradient and one child is pushed back. The same principle applies to glaciers. Where along-flow compressive deviatoric stresses decrease in a downglacier direction, the glacier is pushed from behind. On the other hand, if along-flow compressive deviatoric stresses increase downglacier, the ice is held back from in front. Longitudinal stress gradients are usually regarded as a resistive stress, but they can equally validly be considered as forces driving flow.

In combination, the forces driving and resisting ice flow define the force balance (van der Veen and Whillans, 1989a; fig. 4.11). In the x direction (measured downflow), this can be stated as:

$$\rho_i g H \tan \alpha = \tau_b - \frac{\partial (H \bar{\tau}_{xy})}{\partial y} - \frac{\partial (H \bar{\sigma}'_x)}{\partial x} \qquad (4.11)$$

The driving stress (equation 4.10) is on the left-hand side, and is balanced by the combined terms on the right. The first of these is basal drag (basal shear stress), the second is lateral drag (the gradient in the product of ice thickness and resistance in the cross-flow (y) direction), and the third term is the resistance arising from longitudinal stress gradients (the gradient in the product of ice thickness and along-flow deviatoric stress in the direction of flow). $\bar{\tau}_{xy}$ is the shear stress acting parallel to the sides of the glacier, and $\bar{\sigma}'_x$ is the longitudinal deviatoric stress acting in the direction of ice flow. Both $\bar{\tau}_{xy}$ and $\bar{\sigma}'_x$ are vertical averages, as indicated by the overbars. Note that for consistency with other parts of this book, we use a different notation for the stresses than that used by van der Veen and Whillans (1989a, b) and van der Veen (1999a).

The force balance equation provides a rigorous method of analysing the controls on glacier dynamics (e.g. van der Veen and Whillans, 1989b; Rolstad et al., 2000; van der Veen and Payne, 2004; O'Neel et al., 2005). Many glaciological studies adopt simpler approaches, however, in which one or more terms on the right-hand side of equation 4.11 are either ignored or parameterized

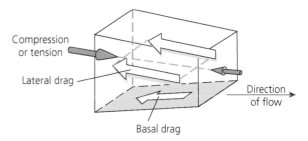

Figure 4.11 Resistive stresses opposing the driving stress. Flow resistance is associated with friction at the sides of the glacier (lateral drag), drag at the bed (basal drag) and gradients in longitudinal compression or tension (van der Veen, 1999a).

in some way. For example, the longitudinal stress term is often assumed to be negligible if glacier thickness (H) and slope (α) are averaged over distances several times the glacier thickness (e.g. Bindschadler et al., 1977). For glaciers in bedrock troughs, the effects of lateral drag can be approximated using a shape factor f, which varies with channel geometry (Nye, 1957). If both longitudinal and lateral stress terms are dealt with in this way, the force balance equation reduces to:

$$f \rho_i g \bar{H} \tan \bar{\alpha} = \tau_b \qquad (4.12)$$

where $\bar{H}$ and $\bar{\alpha}$ denote spatial averages. The shape factor f is defined as:

$$f = \frac{A_r}{H p^*} \qquad (4.13)$$

where A_r is the area of the glacier cross section and p^* is the 'glacierized perimeter' of the cross section. For ice that is far from lateral boundaries (such as much of the interior of Antarctica), the shape factor is equal to 1. Equation 4.12 has the advantage of ease of use, but at the expense of physical realism and conceptual clarity, and modern glacier models include all terms in the force balance equation (section 5.3.3).

4.3 DEFORMATION OF ICE

The deformation of ice in response to stress can occur by either *creep* or *fracture*. Of the two, creep is the much more important process of glacier motion, but fracture underpins many significant glaciological processes, such as iceberg calving, glacitectonic deformation, and weakening of ice stream and ice shelf margins. In this section, we examine creep processes and how they influence patterns of glacier flow. Fracture and associated structures are discussed in section 4.7.

4.3.1 GLEN'S FLOW LAW

Ice creep is defined as deformation resulting from movement within or between individual ice crystals.

It resembles the deformation of metals at temperatures close to their melting point, and involves the interaction of several complex processes (Weertman, 1983; Alley, 1992a). Movement within crystals may occur by gliding along *cleavage planes* (planes of weakness related to the molecular structure of the crystal) or movement along *crystal defects*, whereas movement between crystals involves *recrystallization* at grain boundaries. Deformation patterns resulting from ice creep are strikingly illustrated by fold structures exposed on glacier surfaces, which record substantial changes in the shape of layered ice during flow (fig. 4.12; Hambrey and Lawson, 2000).

In laboratory tests, samples of ice initially exhibit changing stress–strain relationships, but eventually settle down to steady behaviour. Relationships describing the long-term, steady creep of ice in response to stress are known as *flow laws*, the most widely used being *Glen's Flow Law* (Glen, 1955). This was first adapted for glaciers by Nye (1957), and is more fully referred to as *Nye's Generalization of Glen's Flow Law*. In its full form, Glen's law relates the strain rate to all of the stresses acting on the ice (i.e. all of the deviatoric and shear stress components; Hooke, 2005). For example, simple shear at the bed of a glacier is a function of both the basal shear stress τ_b (which can also be written as σ_{xz}) and an *effective stress* σ_e, which incorporates all of the stress components:

$$\dot{\varepsilon}_{xz} = A\sigma_e{}^{n-1}\tau_b \qquad (4.14)$$

Similar expressions can be written for strain rates in other directions. In situations where the basal shear stress is considered to balance the driving stress, Glen's law can be framed in a simpler, approximate form:

$$\dot{\varepsilon} = A\tau^n \qquad (4.15)$$

A and n are parameters that define the relationship between stress and strain. The rate parameter A increases exponentially with ice temperature (Hooke, 1981; van der Veen and Payne, 2004; fig. 4.13), so that ice deforms much more readily as it warms towards its pressure-melting point. This is mainly because creep processes are most effective when melting occurs at grain boundaries. A can be calculated from Hooke's (1981) modification of the Arrhenius relation:

$$A = A_0 \exp\left[-\frac{Q^*}{R_g T_i} + \frac{0.49836}{(T_0 - T_i)^k}\right] \qquad (4.16)$$

where $A_0 = 9.302 \times 10^{-2}$ Pa^{-3} yr^{-1}, Q^* is the activation energy for creep (7.88×10^4 J mol^{-1}), R_g is the gas constant (8.314 J mol^{-1} K^{-1}), T_i is the ice temperature in Kelvins (K), $T_0 = 273.39$ K and $k = 1.17$. Note that with the exception of time (years), SI units are used for all the values quoted here. Different units may be employed in other publications.

The flow law exponent n also varies, but is usually close to 3. This means that ice is a non-linear viscous material, because strain rates increase non-linearly with the applied stresses (fig. 4.10). Note that cryostatic pressure does not figure in Glen's law. This reflects the fact that, unlike

Figure 4.12 Patterns of ice deformation on Skeiðerárjökull, Iceland, revealed by surface outcrops of volcanic ash layers in the ice. The ash layers were originally deposited as extensive blankets on the upper glacier, where they were buried by snow, and are now melting out of the ice (NASA: http://earthobservatory.nasa.gov).

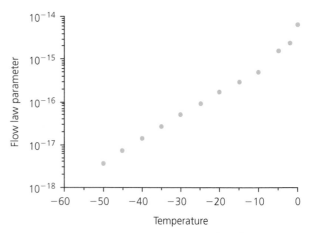

Figure 4.13 Variation in the flow law parameter A (σ^{-1} kPa^{-3}) with ice temperature. Note the logarithmic scale on the y axis, and how the log-linear relationship breaks down near the melting point (Data from Paterson, 1994).

many materials, the strain rate of ice is independent of the mean normal stress and depends simply on the deviatoric and shear stress components, and the properties of the ice.

4.3.2 CRYSTAL FABRIC, IMPURITIES AND WATER CONTENT

In practice, ice creep rates will differ somewhat from values calculated using Glen's Flow Law, because factors other than temperature can influence the creep rate of glacier ice. These include the size and orientation of crystals, the presence of impurities, and water content. Ice crystals deform much more easily in one direction than any other, due to the presence of cleavage planes controlled by the molecular structure (fig. 4.14). In fact, ice crystals will deform along such cleavage planes 100 to 1000 times easier than any other direction (Hooke, 2005). Therefore, in polycrystalline ice, strain rates are sensitively dependent on the aggregate orientation of ice crystals, a property known as the *ice crystal fabric*. A sample of polycrystalline ice in which large numbers of crystals have cleavage planes aligned parallel to the shear stress (i.e. with a strong fabric) will deform more easily than ice with randomly oriented crystals (a weak fabric; Lile, 1978; Duval, 1981). Favourable crystal fabrics may develop in ice under steady applied stresses as the result of crystal rotation and recrystallization (Alley, 1992a; Paterson, 1994). In addition, crystal defects may propagate parallel to the direction of shear, adding to the progressive weakening of ice or *work softening*. Because different parts of a glacier will have been deformed at different rates and at different times, crystal fabrics and strain responses often vary markedly from place to place. Some studies have noted that strong crystal fabrics are characteristic of ice at depth, where stresses are highest (e.g. Blankenship and Bentley, 1987), although a general model of fabric development and ice deformation remains elusive (Alley, 1992a).

Impurities exert a strong influence on the strain response of glacier ice, although precise flow laws are very difficult to establish. Three main kinds of impurities occur in natural glacier ice: dissolved ions (solutes), gas bubbles and solids (rock particles). The influence of solutes is variable, some tending to harden the ice, some to soften it, and others having no effect (Nakamura and Jones, 1973). Concentrations of gas bubbles usually soften ice, by creating lines of weakness. The influence of rock particles on the flow characteristics of ice is of great importance, due to the widespread occurrence of debris in the basal layers of glaciers (section 9.3.2). However, different studies have reached conflicting conclusions, some indicating that strain rates increase with debris content, others suggesting that they decrease (Hubbard and Sharp, 1989). Laboratory studies by Nickling and Bennett (1984) concluded that the strength of debris-ice mixtures increases as debris contents increase from 0 to 75 per cent, then falls sharply at higher debris contents, probably as the result of the varying influence of friction between rock particles and the cohesive effect of the ice. These results are broadly consistent with data on natural ice-debris mixtures obtained by Echelmeyer and Zhongxiang (1987) and Fitzsimons et al. (2001), although the influence of other factors, such as solute content and ice temperature, complicate the picture.

Intergranular water content has a large effect on the rheology of temperate ice (Lliboutry and Duval, 1985; Hooke, 2005). B. Hubbard et al. (2003) found that basal ice can have a water content an order of magnitude higher, and be an order of magnitude softer, than ice at higher levels in the glacier. Because stresses are highest near the glacier bed, this has a major impact on ice flow rate and glacier dynamics. Detailed observations of the deformation of basal ice in the Svartisen Subglacial Laboratory below Engabreen, Norway, by Cohen (2000) indicated that pronounced rheological softening occurs due to the presence of unbound water at the interfaces between ice and sediment layers. Given the wide range of deformational mechanisms and influences on ice rheology, it is unrealistic to think in terms of a single flow law for ice (Alley, 1992a). For practical reasons, however, these complications are ignored and a simple approach is taken in most glacier modelling studies (e.g. Marshall et al., 2004).

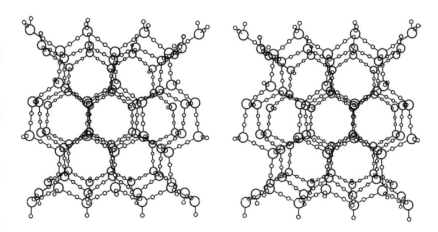

Figure 4.14 Stereographic view of the structure of an ice crystal. Oxygen sites are shown by large circles, and possible hydrogen sites (only half of which are occupied) by small circles. Cleavage planes are parallel to the page (Hooke, 2005, after Hamilton and Ibers, 1968).

4.3.3 ICE CREEP VELOCITIES

At any level within a glacier, the velocity resulting from ice creep is the cumulative effect of the strain rates (i.e. velocity gradients) in the underlying ice column. This means that, where creep results from shearing parallel to the bed, velocities are zero at the bed and increase towards the surface. Glen's Flow Law indicates that strain rates will be largest near the base of the glacier, where the shear stress is at a maximum. Thus, the increase in velocity with height is most rapid near the bed, while velocity increases more slowly with height closer to the surface, because strain rates are small in shallower layers of the ice (fig. 4.8). The velocity at the surface of a glacier U_i can be calculated by inserting the formula for driving stress (equation 4.10) into Glen's Flow Law (equation 4.15) and integrating with respect to height. If A and n are assumed to be constant, this yields:

$$U_i = \frac{2A}{n+1}(\rho_i g \tan\alpha)^n H^{n+1} \qquad (4.17)$$
$$\text{(Paterson, 1994)}$$

The mean velocity due to ice creep is found by integrating this expression in turn and dividing by H:

$$\bar{U}_i = \frac{2A}{n+2}(\rho_i g \tan\alpha)^n H^{n+1} \qquad (4.18)$$

Actual velocities may differ from these calculated values because A and n can vary with depth in the glacier, due to differences in ice temperature or water content. Additionally, these expressions are based on the assumption that the only significant stress deforming the ice is the bed parallel shear stress (taken to be equal and opposite to the driving stress). In some situations, such as the non-streaming parts of ice sheets, this approximation is reasonable, but in valley glaciers and ice streams other stresses are usually important. Where the assumptions of equations 4.17 and 4.18 are inappropriate, numerical methods must be used to solve the full flow law equation (e.g. Blatter, 1995; Hubbard, 2000).

4.4 SLIDING

Sliding refers to slip between a glacier and its bed. Everyday experience tells us that wet ice is extremely slippery, and that the friction between wet ice and a smooth surface (such as the sole of a shoe) is very low indeed. Similarly, if glacier beds consisted of wet ice resting on perfectly smooth surfaces, basal resistance to the driving stresses would be effectively zero. If there was no traction to oppose the driving stress, glaciers would accelerate without limit and would soon cease to exist. Since glaciers do not do this, we can conclude that some other factor must be limiting the sliding speed of wet-based glaciers. The force balance approach (section 4.3.4) teaches us that the most important question is not what drives glacier sliding, but how sliding is resisted. In this section, we look at how local processes acting at the glacier bed can control sliding. Sliding is not simply controlled by local processes, however, and non-local influences on sliding rates are considered in section 4.6.2.

4.4.1 FROZEN BEDS

The adhesive strength of frozen surfaces is very high, even where they are smooth. This is why it is so difficult to scrape the ice from a car windscreen on a cold morning. For glacier beds below the pressure-melting point, therefore, the bed can often support larger shear stresses than the ice itself, and movement more readily occurs by ice creep above the bed. Until recently, it was thought that sliding cannot occur where glacier beds are frozen, but theoretical work and careful field observations indicate that limited sliding is in fact possible (e.g. Fowler, 1986a; Weber, 2000). Below Urumqi Glacier 1, western China, Echelmeyer and Zhongxiang (1987) measured sliding rates of around $0.01\,\text{mm}\,\text{d}^{-1}$ (less than $4\,\text{mm}\,\text{yr}^{-1}$) at $-5\,°\text{C}$. Similar sliding rates were measured by Cuffey et al. (1999) below Meserve Glacier, Antarctica, where the temperature of the ice–bed interface was $-17\,°\text{C}$. Sliding is made possible by the fact that, even at such low temperatures, thin water films can be maintained due to high solute contents. Although sliding rates are very small, they can add up to considerable displacements over long time periods. Sliding at cold glacier beds is therefore insignificant in terms of short-term glacier dynamics, but can be important on geological timescales (Waller, 2001).

4.4.2 SLIDING OF WET-BASED ICE

Where the bed of a glacier is at the pressure-melting point, resistance to sliding arises from a variety of sources. First, bumps on rockbeds provide obstacles around which ice must be transferred by some mechanism. Resistance arising from obstacles is known as *form drag*. Second, debris entrained in basal ice generates friction between a glacier and the bed, creating *frictional drag*.

Glacier ice flows past obstacles on the bed by two main mechanisms: *regelation* and *enhanced creep*. The word *regelation* means refreezing, and in the present context it refers to processes that allow glacier ice to slide over rough beds by melting on the upglacier side of obstacles and refreezing on the downglacier side. Regelation sliding occurs because most resistance to glacier movement is provided by the upstream sides of obstacles, resulting in locally high pressures and a consequent lowering of the pressure-melting point (section 2.3.1). This encourages the melting of ice immediately upglacier of obstacles.

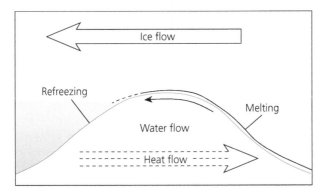

Figure 4.15 Regelation sliding mechanism, showing how the energy and mass fluxes associated with melting and freezing are balanced by water and heat flow.

The resulting meltwater migrates to the low-pressure zone downglacier of the obstacle, where it refreezes due to the higher pressure-melting point there (fig. 4.15). The ice therefore bypasses obstacles by temporarily turning to water and back again. This process can be demonstrated by looping a weighted wire over an ice block. The small surface area of the wire means that the normal stress beneath it is very high, lowering the pressure-melting point and allowing the wire to melt its way through the ice. Above the wire, the normal stress is low, raising the melting point and causing refreezing. By this means, the wire can pass right through the block without breaking it.

The importance of regelation in glacier sliding has been recognized for many years on theoretical grounds (Clarke, 1987b), and has been confirmed many times by direct observation in subglacial cavities (e.g. Kamb and LaChapelle, 1964; Hallet et al., 1978; Hubbard and Sharp, 1993). The role of regelation in forming basal debris-rich ice and the characteristics of regelation ice are described in section 9.3.2.

The classical theory of subglacial regelation was first developed by Weertman (1964), and later modified by Lliboutry (1968, 1987) and Kamb (1970). Consideration of the heat balance shows that regelation will be most effective when latent heat released by freezing on the downstream side can be conducted through the obstacle to assist in melting on the upstream side. This implies that regelation should be most effective where obstacles are small, due to the inefficiency of heat conduction through larger obstacles. The flow of meltwater associated with regelation sliding is usually thought to occur in a thin film between the ice and its bed (Weertman, 1964; Hallet, 1979a; section 3.4.2), although Lliboutry (1993) has argued that, in temperate ice, water can also flow through a vein network within a basal ice layer some 20 cm thick. According to Lliboutry, pressure-melting at the upstream sides of obstacles occurs at grain boundaries *within* the ice as well as at the ice–bed interface. The associated refreezing is thought to occur in a thin layer (~3.5 cm thick) of the ice above the low-pressure downstream sides

of obstacles. This view implies that regelation sliding is closely related to ice creep processes. Further complications of the regelation process arise as the result of chemical interactions between the glacier and its bed. Basal ice can entrain and transport ions dissolved from the bed (section 3.9.2; Hallet, 1976a, b; Souchez and Lorrain, 1978), thus lowering the melting temperature and influencing the regelation process.

The second mechanism that allows ice to flow past bedrock obstructions is *enhanced creep*. It will be remembered that the strain rate of ice varies non-linearly with the applied stress (equations 4.14 and 4.15). Therefore, stress concentrations around the upstream sides of obstacles result in locally high strain rates, causing the ice to efficiently deform around and (to a lesser extent) over the bump. The basal ice thus continually modifies its shape to allow continued slip between the ice and the bed. As noted above, theoretical work by Lliboutry (1993) suggests that, in temperate ice, the mechanism of enhanced creep involves an element of regelation, indicating that ice flow mechanisms may be strongly interrelated. A form of enhanced creep is also important at the base of cold-based glaciers, allowing the ice to circumvent basal obstacles without sliding. It should be remembered, however, that creep rates are much lower for cold ice due to the temperature dependence of the flow law (section 4.4.1).

Enhanced creep is most effective around large obstacles, because larger volumes of the basal ice experience enhanced stresses and higher strain rates. Because regelation is most effective around small obstacles, it follows that enhanced creep and regelation will vary in relative importance according to the size of obstacles on the bed. For small obstacles, regelation will be the dominant process, whereas enhanced creep will dominate for larger ones. Thus there should be some intermediate *controlling obstacle size* for which neither mechanism is efficient, and which presents maximum resistance to sliding. Theoretical analyses and observations suggest that this critical obstacle size is in the region of 0.5 m (Kamb and LaChapelle, 1964; Boulton, 1975).

4.4.3 GLACIER–BED FRICTION

Basal ice commonly contains rock debris, and if this debris is in contact with the bed, the resulting *frictional drag* will add to the overall resistance to flow offered by the bed. Quantitative theories of frictional drag beneath glaciers were first developed in the 1970s by Geoffrey Boulton and Bernard Hallet. The theory proposed by Boulton (1974, 1975, 1979) is based on the Mohr–Coulomb equation (section 4.3.3), and assumes that basal drag τ_b is a simple function of the effective pressure N and the fraction A_F of the bed covered with debris:

$$\tau_b = \tan \Phi \, A_F N \qquad (4.19)$$

where tan Φ is the friction coefficient. A modified form of this model (the 'sandpaper friction' model) was developed by Schweizer and Iken (1992).

Coulomb-type models are strictly only applicable for friction between rigid bodies. Basal ice is deformable, however, so the friction between particles enclosed in ice and underlying bedrock will also depend on the way that ice flows over and around particles. This idea was developed by Hallet (1979a, 1981), who argued that because ice can deform completely around subglacial particles, contact forces will be independent of the effective pressure. Hallet reasoned that the contact force between a particle and the bed will be the sum of two components: (1) the buoyant weight of the particle, and (2) drag force resulting from ice flow towards the bed. The buoyant weight is the weight of the rock particle minus the weight of the same volume of ice, and is therefore proportional to the mass of the particle and the difference between the densities of ice and rock. Ice flow towards the bed produces an additional contact force between the bed and particles embedded in the basal ice. The faster the flow towards the bed, the higher the contact force, and the higher the resulting drag. Ice flow towards the bed can result from a combination of melting due to geothermal and frictional heat, and vertical straining of the ice. Therefore, the Hallet model predicts that high friction between particles and the bed will occur below large, heavy particles and/or where basal melting rates are high and ice is straining rapidly towards the bed.

Recently, pioneering experiments conducted in the Svartisen Subglacial Laboratory, in combination with improved theoretical analysis, have resulted in significant advances in understanding how rock debris influences subglacial drag (Iverson et al., 2003; Cohen et al., 2005; Iverson et al., 2007). Carefully designed experiments show that measured values of frictional drag (shear traction) are strongly positively correlated with effective normal pressure, increasing at times of low subglacial water pressure and vice versa (fig. 4.16). The form of this relationship is consistent with Coulomb-type models, but the measured values of frictional drag are an order of magnitude higher than those predicted by equation 4.19. A new theory of subglacial friction developed by Cohen et al. (2005) accounts for this discrepancy by incorporating the effects of both the effective pressure and ice flow towards the bed. An important feature of the model is the incorporation of water-filled cavities on the lee side of particles. Such cavities have been observed in laboratory experiments by Iverson (1993), and play a very important role in modulating the contact forces between particles and the bed, and thus in controlling basal drag at larger scales.

4.4.4 THE ROLE OF WATER

In the preceding sections, we have seen that water plays an important role in the regulation process and in modulating

frictional drag. Subglacial water has an additional, very wide-reaching effect on basal sliding, by determining the areal extent of cavities between the ice and the bed. Cavity formation, or *cavitation*, governs the degree of coupling at the ice–bed interface at a wide range of scales, and thus exerts a fundamental control on bed resistance and sliding velocities.

Water-filled cavities will form when the subglacial water pressure (P_w) exceeds the local normal pressure exerted by the ice (P_i). This is most likely to occur on the lee side of obstacles, where normal stresses are lower than average (fig. 3.21). Cavity formation thus depends on bed roughness, which governs the magnitude of the normal pressure fluctuation across bumps. The threshold water pressure for cavity formation is known as the *separation pressure*, and is given by:

$$P_s = P_i - \frac{\lambda \tau}{a\pi} \qquad (4.20)$$

where P_s is separation pressure, P_i is normal pressure exerted by the ice, λ is bump wavelength, τ is basal shear stress, a is bump amplitude and π is 3.1427... This relation shows that the separation pressure is low for low-wavelength, high-amplitude bumps, and high for long-wavelength, low-amplitude bumps (Schweizer and Iken, 1992). Pressure fluctuations in subglacial cavities, and the extent of cavity formation, generally reflect variations in the supply of meltwater from the surface (Willis, 1995; section 5.3.1).

The presence of water-filled cavities affects sliding behaviour in several ways (fig. 4.17). First, at small scales, cavity formation reduces the effective bed roughness, by submerging bumps on the bed. Second, pressurized water in cavities can exert a downglacier traction on upglacier-facing cavity walls, increasing the downglacier driving forces (Röthlisberger and Iken, 1981). This is known as the 'hydraulic jack' mechanism. Third, and probably most importantly, by reducing the area of ice–bed coupling, cavity formation affects the distribution of stress at the glacier bed. Water cannot support a shear stress, so the basal shear stress is locally zero above water-filled cavities. This means that the driving stress must be supported by those parts of the bed that remain in contact with the ice (i.e. the crests and stoss sides of obstacles). The consequent stress concentrations at these points increase the efficiency of regulation by reducing the pressure-melting point, and enhances ice deformation, both of which will increase the sliding rate (Willis, 1995). With increasing water pressure, cavities extend over larger areas of the bed, reducing the area of ice–bed contact. Furthermore, cavities in the lee of one bump can drown bumps located downglacier, so that traction is lost over increasingly large areas of the bed (Fowler, 1986a, 1987; Schoof, 2005). Numerical models of this

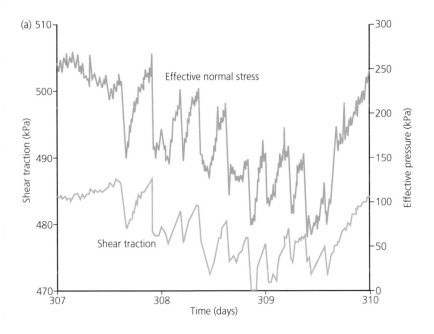

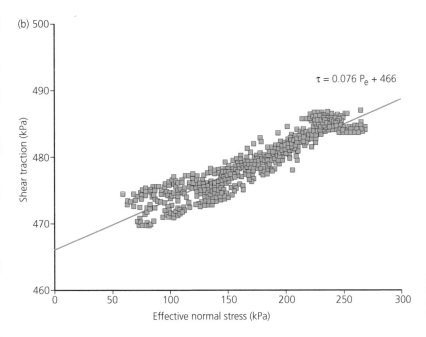

Figure 4.16 Relationships between shear traction and effective normal stress determined from experiments in the Svartisen Subglacial Laboratory. (a) Time series; (b) data plot (Cohen et al., 2005).

process, and the associated relationships between effective pressure, velocity and basal traction, are discussed in section 4.6.2.

4.5 DEFORMABLE BEDS

The sliding mechanisms discussed in section 4.4 apply where glaciers sit directly on solid rock. Many glaciers, however, are underlain by unlithified sediments or poorly consolidated sedimentary rocks, which can deform in response to stresses imposed by the overlying ice. Until the 1980s, little consideration had been given to the role

that soft bed materials might play in ice motion, but following work by G.S. Boulton and co-workers in Iceland (Boulton, 1979; Boulton and Jones, 1979; Boulton and Hindmarsh, 1987) and the discovery of saturated, weak sediment beneath Ice Stream B (Whillans Ice Stream), West Antarctica (Alley et al., 1986, 1987; Blankenship et al., 1987), it was widely realized that soft glacier beds could deform at relatively low stresses, and might therefore play an important role in controlling ice flow rates. Following these discoveries, deformation of subglacial sediments was used to explain a wide range of phenomena, and the implications of this new perspective appeared to be so wide-reaching that it was hailed as a 'paradigm

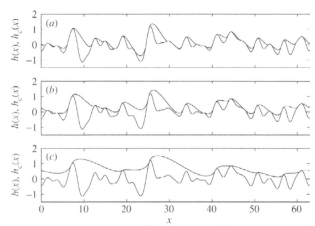

Figure 4.17 Modelled geometry of basal cavities for different values of basal velocity and effective normal pressure. Note that as cavities become more extensive, the bed roughness is reduced and the driving stress must be supported by smaller upglacier-facing areas of the bed. Ice flow is from left to right (Schoof, 2005).

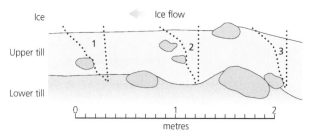

Figure 4.18 The displacement of segmented rods placed into the till below Breiðamerkurjökull, Iceland, showing bed deformation (Adapted from Boulton and Hindmarsh, 1987).

shift' in glaciology (Boulton, 1986a). In the initial wave of enthusiasm, the deforming bed concept was sometimes employed rather uncritically, and a balanced understanding of the key physical processes has emerged only gradually, as the result of a small number of well-designed field campaigns, in conjunction with laboratory experiments and analyses of material recovered from boreholes. The set of observations is still very small, however, and many important issues remain the subject of ongoing research (e.g. Murray, 1997; Boulton et al., 2001a; Fischer and Clarke, 2001; Fowler, 2003; Boulton and Zatsepin, 2006; Kavanaugh and Clarke, 2006).

4.5.1 THE BOULTON–HINDMARSH MODEL

The first direct evidence for deformation of subglacial sediment was obtained below Breiðamerkurjökull, an outlet of the ice cap Vatnajökull in Iceland (Boulton, 1979; Boulton and Hindmarsh, 1987). From tunnels excavated into the ice margin, segmented rods and water-pressure sensors were inserted into the unfrozen till beneath the glacier, and ice motion was measured using markers in the tunnel and on the glacier surface. Additional measurements of basal motion were made by using a wire spool in the tunnel, which was attached to an anchor emplaced in the stiff deeper layers of the subglacial till. After the conclusion of some experiments, the till beneath the tunnel was drained and excavated. Rod segments were found to have been displaced downglacier by varying amounts, showing that deformation of a layer of till approximately 0.5 m thick accounted for 80–90 per cent of the movement of the glacier (fig. 4.18). Concurrent pore-water pressure and velocity data displayed a strong positive correlation, but with velocity peaks lagging pore-water peaks by a few hours. The methods used to measure basal motion could not determine the relative contributions of till deformation and

sliding at the ice–till interface – which may have varied through time – but it was assumed in the subsequent analysis that velocity peaks equate to peak strain rates in the deforming till.

The experimental results from Breiðamerkurjökull were used to derive the first quantitative 'flow laws' for subglacial till. Boulton and Hindmarsh (1987) fitted the data to a function of the form:

$$\varepsilon = K(\tau - \tau_{yield})^a N^{-b} \quad (\tau > \tau_{yield}) \qquad (4.21)$$

where K, a and b are empirically fitted constants. This relationship states that the strain rate $\dot{\varepsilon}$ rises as the shear stress τ becomes increasingly greater than the yield strength τ_{yield}, and decreases with the effective pressure N. Two alternative versions of this function were fitted to the data, one with $\tau_{yield} = 0$, and the other with τ_{yield} calculated from equation 4.5. In both cases, the exponent a was close to unity, indicating a till viscosity almost independent of strain rate. That is, the till appeared to exhibit either a slightly non-linear viscous or Bingham-type viscous rheology (see section 4.2.3).

This rheological model was widely accepted as a description of the behaviour of deforming till, and was employed in numerous interpretations of the glacial geologic record (e.g. Boulton, 1987; Hart et al., 1990; Hicock and Dreimanis, 1992; Hart, 1995a; Benn and Evans, 1996; Hart and Rose, 2001) and theoretical studies (e.g. Alley et al., 1987; Hindmarsh, 1998a; Fowler, 2000). Recent investigations of the properties of subglacial tills, however, have indicated that there are difficulties with the Boulton–Hindmarsh model as a general flow law. First, the values of basal shear stress used in the initial model formulation are open to question because the original experiments were conducted very close to an ice margin, where basal shear stress is not a simple function of local ice thickness and gradient (Hooke et al., 1997; Murray, 1997; van der Veen, 1999a). This means that one of the fundamental input parameters for the flow law, τ, is subject to unknown and possibly large errors. Second, laboratory testing of till samples and data from instruments emplaced in glacier beds via boreholes have gradually revealed a different and more complex picture of till rheology, as described in the following sections.

4.5.2 LABORATORY TESTING OF SUBGLACIAL TILLS

Important data have been obtained from laboratory experiments designed to quantify the response of till samples to a range of applied stresses. Kamb (1991) investigated samples of till recovered from the bed of Ice Stream B (Whillans Ice Stream), West Antarctica (Engelhardt et al., 1990). The till samples were subjected to rate-controlled and stress-controlled shear box tests, which showed that the till behaved in a nearly Coulomb-plastic manner. That is, the till underwent sudden failure at a critical value of the applied shear stress (yield or shear strength), which increased linearly with the effective normal pressure (see section 4.3.3). This behaviour was so unlike the viscous rheology proposed by Boulton and Hindmarsh (1987) that doubts were expressed whether shear box tests provide a representative picture of in situ till behaviour. Subsequent work employing a *ring-shear apparatus*, however, has confirmed the initial results. This device consists of two rings which can rotate relative to each other, between which a sample of sediment is placed (Iverson et al., 1998; Hooyer and Iverson, 2000; fig. 4.19). Unlike traditional shear boxes, this apparatus allows samples to be deformed to high cumulative strains, allowing sediment properties to be investigated with greater rigour. Iverson et al. (1998) investigated a basal till collected from below the margin of Storglaciären, Sweden, and samples of the Wisconsin-age Two Rivers till, Michigan. Although their grain-size distributions are very different, both tills exhibited almost Coulomb-plastic behaviour, with cohesive strengths and friction coefficients of 5 kPa and 0.49 (Storglaciären) and 14 kPa and 0.32 (Two Rivers). The yield strength of the tills is almost independent of strain rate, although a slight weakening occurred at high strain rates, possibly resulting from increased till voids ratios.

Ring-shear and triaxial tests of undisturbed till cores and remoulded samples from below Whillans Ice Stream and other West Antarctic ice streams show convincingly that these tills also exhibit Coulomb-plastic or nearly Coulomb-plastic rheology (Tulaczyk et al., 2000a; Kamb, 2001). Although the Ice Stream B tills are quite clay-rich (~35 per cent by weight), their cohesive strength c is very small (<1 kPa), reflecting high voids ratios (~0.6) and hence relatively large inter-particle spacing. Their coefficient of friction (tan ϕ) is ~0.45. At the low effective pressures beneath Antarctic ice streams, the yield strength of the till is around 2 kPa, but varies by about an order of magnitude on either side of this figure.

Applying concepts from soil mechanics, Tulaczyk et al. (2000a) concluded that the behaviour of the Ice Steam B tills (and other tills) can be described using just three variables: shear strength (σ_{yield}), effective pressure (N), and voids ratio (e). The relationship between effective pressure

and shear strength is given by the Mohr-Coulomb equation (equation 4.5), and that between effective pressure and voids ratio is given by:

$$e = e_0 - C \log(N/N_0) \qquad (4.22)$$

The reference voids ratio e_0 is the value of e at some specified value of the effective pressure (N_0). For the Ice Stream B tills, e_0 was found to be ~0.75 at 1 kPa. Equation 4.22 shows that the voids ratio increases logarithmically as effective pressure decreases, with the precise form of the relationship dependent on the coefficient of compressibility C. This has different values depending on whether the till is normally consolidated or overconsolidated.

When the effective pressure is higher than any the sediment has experienced in the past, the sediment is in the *normally consolidated* or 'virgin' state. When such a sediment is subjected to progressively higher effective pressures, the voids ratio will steadily decrease, following the *normal consolidation line* (NCL) (fig. 4.20). For the Ice Stream B tills, the slope of this line (C) equals 0.12 to 0.15. Volume reduction during the consolidation process mainly involves permanent (irrecoverable) rearrangements of particles, so if effective pressure begins to fall, the sediment will remain in an *overconsolidated* state. Unloading (and reloading) of overconsolidated sediments mostly results in elastic (recoverable) volume changes, following the *unloading–reloading line* (URL) in figure 4.20. For the Ice Stream B tills, the slope of this line is 0.02. The patterns of volume change described by the NCL and URL are further modified by shearing of the till. For a Coulomb-plastic material, failure occurs when the applied shear stress equals the yield strength, which increases with effective pressure (equation 4.5). This critical state is plotted in figure 4.20 as the *critical state line* (CSL), which lies parallel to and below the NCL. A sediment subjected to increasing shear stresses will move towards the critical state. This means that a normally consolidated till (plotting above the CSL) will undergo additional consolidation when subjected to a shear stress, as grains adopt closer packings. On the other hand, an overconsolidated till (plotting below the CSL) will increase its voids ratio during shear – behaviour known as *dilation* (or dilatation).

4.5.3 DIRECT OBSERVATIONS OF DEFORMABLE GLACIER BEDS

Over the last two decades, a range of instruments has been developed to allow in situ measurement of strain, pore-water pressure, sliding rates and other properties of soft glacier beds, to determine how tills actually behave in the subglacial environment (Fischer and Hubbard, 2006). The design of instruments and successful data

(a)

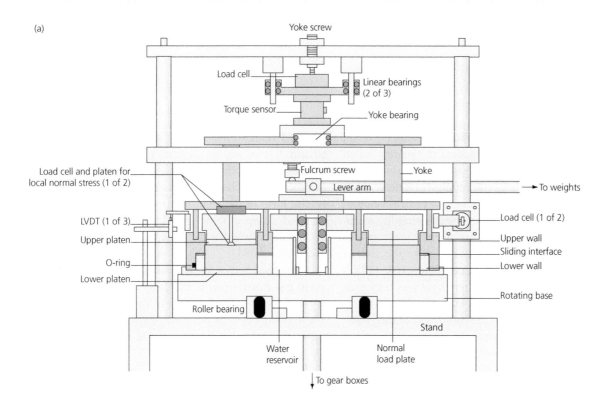

(b)

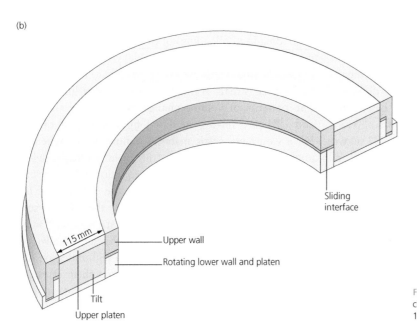

Figure 4.19 Line drawing of a ring-shear device and close-up of part of the sample chamber (Iverson et al., 1998).

capture present considerable challenges, and the current state of the art is the outcome of a long, recursive process of experiment and redesign (Clarke, 1987b; Fischer and Clarke, 2001; Kavanaugh and Clarke, 2006). The five main types of instrument are: (1) tiltmeters; (2) drag spools; (3) ploughmeters; (4) dragometers; and (5) autonomous subglacial probes. *Tiltmeters* measure the tilt of cells relative to the vertical (fig. 4.21). Changes in tilt angle through time are used to determine the relative displacement of the sediment at the top and bottom of the cell,

and hence strain rates in the till (equation 4.4). Tilt cells are commonly deployed in sets of two or more to obtain vertical strain rate profiles in deforming till. *Drag spools* are designed to measure motion of the ice relative to an anchor in the underlying sediment (fig. 4.22). Drag spool anchors are usually emplaced at shallow depths within the subglacial sediment, with the aim of measuring sliding at the ice–bed interface. If anchors are placed at deeper levels, a component of the measured displacement may be due to deformation within the sediment itself. In this case,

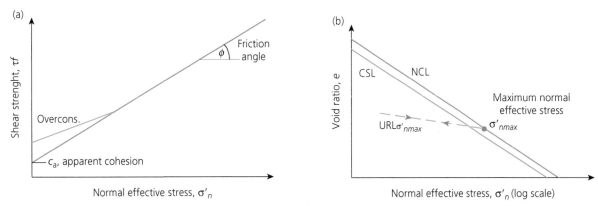

Figure 4.20 Schematic diagrams showing relationship between till properties. (a) Relationship between normal effective stress and shear strength in a Coulomb-plastic material, showing the effect of overconsolidation (cf. fig. 4.9). (b) Relationship between normal effective stress and void ratio (Tulaczyk et al., 2000).

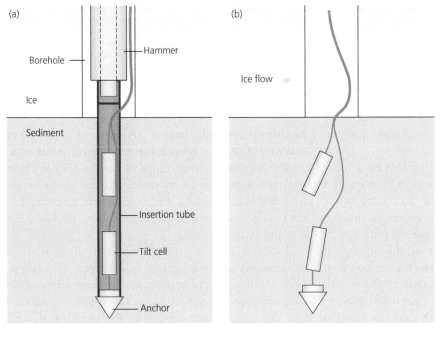

Figure 4.21 Tiltmeters. (a) Insertion into subglacial till. (b) Rotation of tilt cells in deforming till after withdrawal of the insertion tube (Modified from Fischer and Hubbard, 2006).

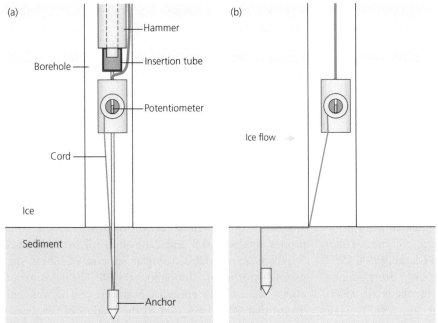

Figure 4.22 Drag spool. (a) Insertion of instrument into still subglacial till. (b) Measurement of sliding following removal of insertion tool (Modified from Fischer and Hubbard, 2006).

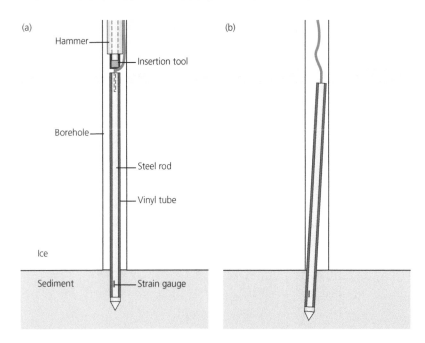

Figure 4.23 Ploughmeter. (a) Insertion of ploughmeter into subglacial sediment. (b) Force is exerted on the ploughmeter as it is dragged through the bed, recorded by a strain gauge near the tip (Fischer and Hubbard, 2006).

however, interpretation of the records is complicated by the possibility that the cable may intermittently slice into the till like a cheese-wire.

The design of *ploughmeters* was inspired by a lucky accident reported by Humphrey et al. (1993), in which a drill stem became jammed in basal sediments below Columbia Glacier, Alaska. The stem was dragged through the glacier bed for 5 days before it was recovered, and its bent and striated state indicated that the uppermost 0.65 m of the bed consisted of deformable sediment. Subsequent analysis allowed an estimate of the in situ strength of the till to be calculated. Intentional plough-meters consist of steel rods fitted with strain gauges that record elastic bending as the rod is dragged through the till (Fischer and Clarke, 1994; fig. 4.23), allowing till strength to be quantified. *Dragometers* work on a similar principle, and consist of a small cylinder ('fish') dragged behind a pipe inserted into the till. The forces experienced by the fish (and hence till strength) are measured by a load cell located in the borehole at the top of the pipe. *Autonomous subglacial probes* can be fitted with a range of instruments, including pressure sensors and tiltmeters, and can transmit data to a receiver on the surface or in the borehole (Hart et al., 2006; Truffer and Harrison, 2006). They have the major advantage of being free of cables, potentially allowing them to behave like clasts in the till.

Down-borehole data have been obtained from below several glaciers, including Tapridge Glacier, Yukon (Blake et al., 1992, 1994; Fischer and Clarke, 1994, 1997, 2001; Kavanaugh and Clarke, 2006), Storglaciären, Sweden (Iverson et al., 1994, 1995; Hooke et al., 1997), Bakaninbreen, Svalbard (Porter et al., 1997), Haut Glacier d'Arolla, Switzerland (Mair et al., 2003), Black

Rapids Glacier, Alaska (Truffer and Harrison, 2006), Breiðamerkurjökull, Iceland (Boulton et al., 2001a; Boulton, 2006) and Briksdalsbreen, Norway (Hart et al., 2006). Although a wide range of behaviours have been observed, a number of recurrent patterns have emerged from these observations, shedding important new light on the relationships between till deformation, ice–bed coupling and glacier motion.

Tiltmeter records provide evidence of patterns of deformation within subglacial tills. Deformation is typically confined to the uppermost few centimetres of the bed, although some studies have found that deformation can also occur at greater depths. For example, tilt cells emplaced below Haut Glacier d'Arolla recorded intermittent shear strain at shallow depths (<~7 cm) but little or no shear deeper in the till (Mair et al., 2003; fig. 4.24). More complex patterns of deformation were detected beneath Breiðamerkurjökull by Boulton et al. (2001a) and Boulton (2006), using an ingenious development of the drag spool concept. An array of spool anchors was pre-packed in a core of till, which was then emplaced beneath the glacier via a borehole, potentially allowing the simultaneous measurement of strain rate profiles within the till and sliding at the ice–till interface. The records indicate that most displacement occurred at or close to the ice–till interface, with intermittent deformation events at deeper levels (fig. 4.25).

Truffer and Harrison (2006) found that autonomous probes emplaced 0.8 and 2.1 m into till beneath Black Rapids Glacier exhibited similar patterns of tilting, suggesting deformation throughout the till. The deformation events were highly episodic, however, and it was estimated that 50–70 per cent of the motion of the glacier occurred by sliding or deformation of the uppermost

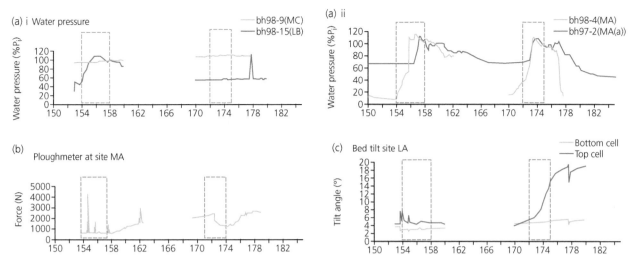

Figure 4.24 Borehole instrument data from Haut Glacier d'Arolla for Julian days 150–184 (May 30–July 6), 1998. The dashed boxes indicate two 'spring events' with locally high basal water pressures and enhanced basal motion. (a) Borehole water pressures. (b) Ploughmeter records showing transient episodes of high drag during the first event, and a longer-lived fall in bed strength during the second. (c) Tiltmeter records, showing little bed deformation during the first event, and substantial deformation of the upper part of the bed during the second. The bottom cell was emplaced at a depth of approximately 10 cm into the till, with the top cell 6.5 cm above (Mair et al., 2003).

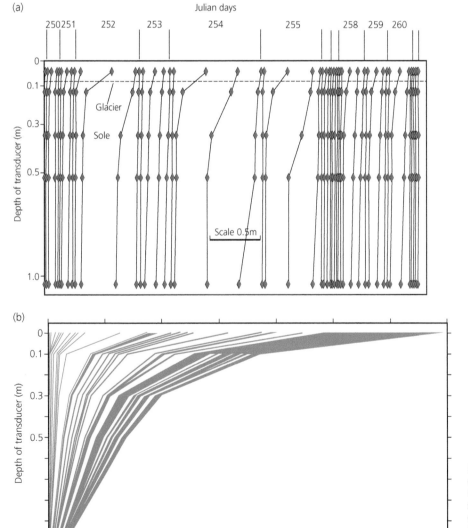

Figure 4.25 (a) Six-hourly patterns of displacement in subglacial till below Breiðamerkurjökull, inferred from drag spool records. Note displacement concentrated near the top of the till on days 252 and early in days 254 and 255, and displacement at deeper levels late in days 254 and 255. (b) Progressive cumulative strains over the time period shown in (a) (After Boulton, 2006).

20 cm of the till. Earlier work at Black Rapids Glacier had seemed to show that most basal motion occurred at depths greater than 2 m beneath the ice–till interface, either by sliding of till over bedrock or along shear zones in the till (Truffer et al., 1999). This conclusion, however, was based on the assumption that the cables between the surface and sensors in the till would have broken if significant slip had occurred at the ice–till interface. Subsequent tests indicated that cable stretch was much greater than the manufacturer's specifications, casting doubt on the idea that failure occurred deep within the till. This illustrates the fact that Murphy's Law (If anything can go wrong, it will) is particularly applicable to down-borehole instrumentation (Hooke et al., 1997).

In keeping with the above results, Engelhardt and Kamb (1998) found that most basal motion of Whillans Ice Stream (Ice Stream B) occurs by basal sliding or deformation in the uppermost 5 cm of the till. This result is highly significant, as it was previously believed that fast motion of this ice stream was accomplished by deformation distributed through several metres of till (e.g. Alley et al., 1987). Drag spool records from below Kamb Ice Stream (Ice Stream D), in contrast, have been interpreted as evidence for significant deformation at depths > 60 cm within the till (Kamb, 2001), although this conclusion is based on the assumption that the spool anchor remained firmly lodged throughout the experiment.

Several studies have shown that although till strength is reduced at times of high water pressure, till strain rates do *not* increase at such times. Instead, concurrent tiltmeter and water-pressure measurements indicate that till strain rates tend to *decrease* when water pressure is high (Iverson et al., 1995; Hooke et al., 1997; Fischer and Clarke, 2001; Kavanaugh and Clarke, 2006).

These results are at odds the patterns reported from Breiðamerkurjökull by Boulton and Hindmarsh (1987), and were not expected by most of the research community. Iverson et al. (1995) and Hooke et al. (1997) showed that reduced deformation at times of high water pressure can be explained in terms of the degree of *coupling* between the glacier and the till. When water pressure in the till is low, the glacier is likely to be strongly coupled to the bed by a process of ice infiltration into pore spaces (section 9.3.2), and shear stresses can be supported by the underlying till (Iverson, 1993; Iverson and Semmens, 1995). In contrast, when water pressure is high, ice infiltration is suppressed, decoupling the glacier from its bed and reducing the traction across the ice–bed interface. Thus, even though the till is weaker at times of high water pressure, till deformation is reduced. At the same time, decoupling encourages rapid sliding along the ice–till interface. Drag spool records from Trapridge Glacier, for example, show that peak sliding rates ($\sim$12 cm day^{-1}) occur at times of rapidly rising water pressure (Fischer and Clarke, 2001; Kavanaugh and Clarke, 2006).

Another recurrent feature of tiltmeter records is repeated, short-lived intervals of *negative* strain rates at times of falling effective pressure and high sliding speed (fig. 4.26; Blake et al., 1992; Iverson et al., 1995, 2007; Fischer and Clarke, 1997, 2001; Hooke et al., 1997) – that is, tilt cells rotate *upglacier*, indicating strain in the opposite direction to that expected for downglacier till deformation. This unexpected pattern can be explained in terms of Coulomb-plastic till behaviour outlined in section 4.5.2 (Tulaczyk et al., 2000a). If effective pressure is reduced, the till will be overconsolidated with respect to the new pressure conditions, and if the shear stress falls as the result of decoupling, the till will respond

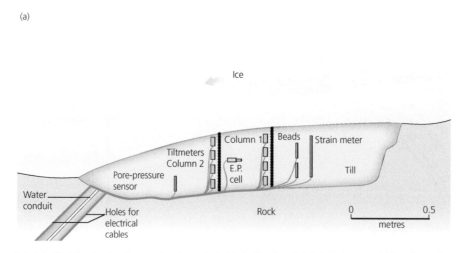

Figure 4.26 Patterns of strain in till, determined during experiments conducted in the Svartisen Subglacial Laboratory. (a) Experimental setup. A prism of till was installed beneath the glacier, containing arrays of tiltmeters and other instruments. (b) Upper panel: water pressure in the till was elevated in four 'pump tests', by injecting water from the tunnel below. Lower panels: cumulative strain records derived from the two upper tiltmeters in columns 1 and 2. The distances are the height of each tiltmeter above the base of the till prism. The reverse slopes of the curves during pump tests 3 and 4 indicate decreasing cumulative strain (i.e. upglacier rotation of the cells) (After Iverson et al., 2007).

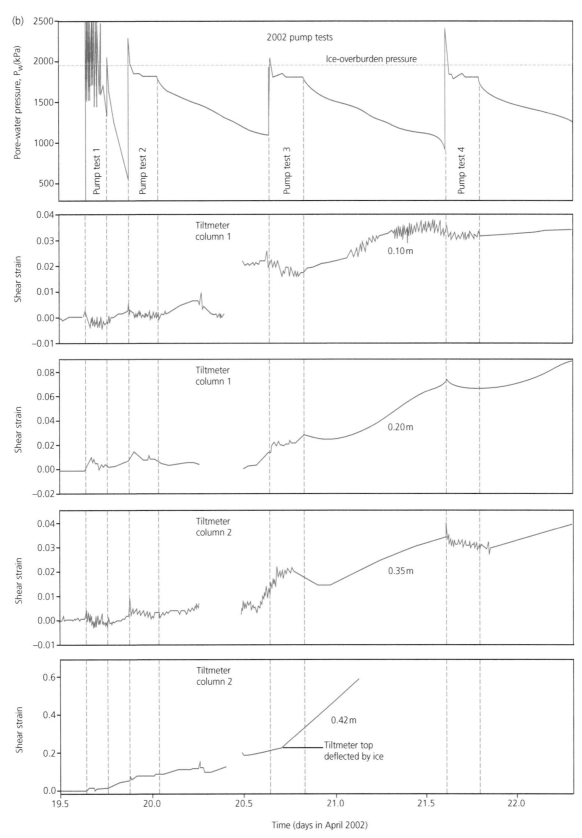

Figure 4.26 (*Continued*)

to the reduced effective pressure by dilating, following the URL (fig. 4.20). This increase in till volume is expected to result in vertical extension of the till, so that tilt cells will rotate closer to the vertical. However, Iverson et al. (2007) have shown that periods of negative strain rates can also occur when effective pressure is either steady or increasing, and have argued that they reflect elastic (recoverable) straining.

4.5.4 RHEOLOGY OF SUBGLACIAL TILL

The evidence from laboratory experiments and instrumentation of glacier beds is more consistent with a Coulomb-plastic model of till behaviour than the viscous rheology proposed by Boulton and Hindmarsh (1987). This has been most conclusively demonstrated by Kavanaugh and Clarke (2006), who compared the predictions of a range of rheological models with concurrent records of pore-water pressure, basal sliding (drag spool), till deformation (tiltmeters) and till strength (ploughmeter) below Trapridge Glacier. The Coulomb-plastic model best represents many key elements of the observational record, particularly the reduction of till strength and increased sliding speed at times of high water pressure (fig. 4.27). Records from a tiltmeter installed at a depth of 0.15 m within the till indicated deformation throughout the diurnal cycle, with a weak tendency towards lower strain rates at times of high water pressure. The tiltmeter

records qualitatively match the results of a linear-viscous till deformation model, although the observed strain rates are an order of magnitude lower than modelled values.

Evidence for deformation at depth within tills has been regarded as a problem for the Coulomb-plastic model, which seemed to imply that deformation should be confined to a very thin layer just below the ice–till interface. This is because effective pressures – and hence sediment strength – are expected to be smallest at the top of the till, so that till deformation should not occur at depth where sediment yield strength will be higher than the shear stress. Two solutions have been proposed to explain this discrepancy. First, if a pulse of meltwater passes through a till layer as a wave, then peak water pressures (and minimum effective stresses) can occur at different depths within the till during the diurnal cycle (Tulaczyk et al., 2000a). This means that the weakest part of the till layer will not always be at the ice–bed interface, but will vary in position through time. Kavanaugh and Clarke (2006) incorporated this effect in their model, with water pulses originating at the base of the ice and passing downwards through the till. Where till is underlain by an aquifer (as is the case at Breiðamerkurjökull, where the till overlies a thick sequence of gravels and sands), water pressure waves can also originate from below, encouraging deformation at depth within the till (Boulton et al., 2001a; Boulton, 2006). A second explanation for distributed strain in Coulomb-plastic tills has been proposed by

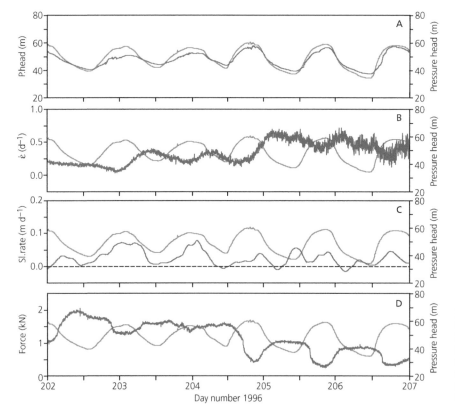

Figure 4.27 Down-borehole instrument records from Trapridge Glacier during 20–25 July (days 202–207), 1996. (a) Water pressure. (b) Strain rate in till derived from tiltmeter records. (c) Sliding rate from drag spool. (d) Till strength derived from forces measured by ploughmeter (Kavanaugh and Clarke, 2006).

Iverson and Iverson (2001), focusing on the behaviour of the till at a granular level. During shearing, intergranular forces are supported by chains of particles extending down into the till. Significant stress fluctuations occur in response to the formation and failure of particle chains at different depths within the till. A numerical model of this process predicted a distributed pattern of strain that closely matches measured till displacement curves (cf. Boulton and Dobbie, 1998). The evidence for deformation at depth within tills has led Hindmarsh (1997) to argue that their bulk response to stress can be described by a viscous rheological model, even though they exhibit Coulomb-plastic behaviour at small scales.

The viscous deformation and Coulomb-plastic models have very different implications for modelling rates of ice flow. In the viscous model, till deformation directly controls the velocity of the overlying ice because the strain rate is assumed to be uniquely determined by the applied stress and the effective pressure (equation 4.21). In contrast, the Coulomb-plastic model implies that once till strength has been exceeded, the till will fail and the ice will slide over the till. The rate of sliding is not determined locally by properties of the till, but is controlled by non-local sources of resistance, such as 'sticky spots' on the bed or the glacier margins. In this case, the till is a passive part of the system, and is simply overridden by the active ice above. To distinguish sliding over soft beds from sliding over hard substrates, the term *ploughing* has been introduced, in recognition of the fact that soft beds may be swept by clasts or other protuberances in the bed of the ice, disturbing and displacing the till surface (Brown et al., 1987; Tulaczyk et al., 2001; Clark et al., 2003).

4.6 RATES OF BASAL MOTION

Calculating rates of basal motion (including sliding, ploughing and/or deformation of subglacial sediments) is much more difficult than calculating ice creep rates, owing to the complex mix of processes involved and the wide range of possible controlling variables. The importance of basal motion, however, especially below fast-flowing glaciers, has prompted a search for realistic, workable 'sliding laws' for use in predictive numerical models. In practice, 'realistic' and 'workable' appear to be mutually exclusive requirements, and success in one is usually at the expense of the other. In the following sections we discuss the pros and cons of simple 'sliding laws' widely used to parameterize basal motion, and then consider 'friction laws' derived from modern theories of basal motion (e.g. Fowler, 1987; Schweizer and Iken, 1992; Schoof, 2005; Gagliardini et al., 2007).

4.6.1 'SLIDING LAWS'

Empirical relationships between basal velocity, driving stress and subglacial effective pressure (sliding laws) have been derived by fitting observational data to simple functions of the form:

$$U_b = k_s \tau_d^p N^{-q} \qquad (4.23)$$

where U_b is basal velocity, τ_d is the driving stress, N is effective pressure (equation 3.10), and k_s, p and q are empirically determined constants. Equation 4.23 reflects the facts that sliding rate increases with the driving stress and is inversely proportional to the effective pressure (i.e. sliding increases with P_w) (fig. 4.28). On the basis of laboratory experiments, Budd et al. (1979) proposed that values of $p = 3$ and $q = 1$ yield the best results. Very similar values were obtained by Bindschadler (1982), using data from Variegated Glacier for summer 1973, although the best-fit parameters for winter velocity data from this glacier over the period 1973–81 are $p = 5$ and $q = 1$ (Raymond and Harrison, 1988). It has become increasingly clear, however, that the sliding law parameters vary very widely, depending on which data set is used (e.g. Nick et al., 2007a). Moreover, Iken and Truffer (1997) found that the exponents in the sliding law differed even during a single advance–retreat cycle of Findelengletscher, Switzerland (section 5.4.1).

A fundamental shortcoming of equation 4.23 is that it yields infinite velocities when $N = 0$ and $\tau_d > 0$. This problem arises because basal drag is equated with the driving stress, with the consequence that basal shear stress is allowed to be finite even if effective pressure (and bed strength) is zero. In reality, part of the driving stress is also supported by lateral drag and/or longitudinal stress gradients (section 4.3.4), so unlimited glacier acceleration will not occur if basal drag vanishes. In practical terms, this means that sliding laws such as equation 4.23 are only useful in situations where basal drag provides most of the resistance to flow (and when $N > 0$), but will yield unrealistic results when other sources of resistance are significant.

A further problem with using pressure-dependent sliding laws in ice sheet models is that they require subglacial water pressures as inputs. Although recent models of subglacial hydraulic systems can yield reasonable values of subglacial water pressures (e.g. Flowers and Clarke, 2002a, b; Flowers et al., 2003), such models have not yet been incorporated into time-evolving models of ice flow. Instead, ice sheet modellers have tended to use simpler 'sliding laws' that either neglect subglacial effective pressures (e.g. Marshall et al., 2002) or adopt simple parameterizations of basal water pressure where the glacier bed is below sea level (e.g. Huybrechts and de Wolde, 1999).

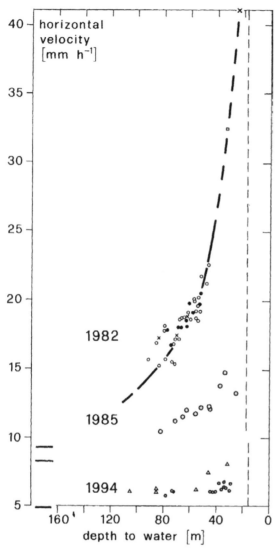

Figure 4.28 Velocity of Findelengletscher as a function of water pressure in different years. Water pressure is given as depth to water from the top of a borehole (lesser depths are equivalent to higher pressures) (Iken and Truffer, 1997).

4.6.2 LOCAL AND NON-LOCAL CONTROLS ON ICE VELOCITY

An important point to bear in mind about the sliding laws discussed in section 4.6.1 is that they are not derived from a rigorous analysis of the physics of glacier motion, but are simply convenient functions that mimic some (but not all) of the observed relationships between basal motion and its glaciological controls. In addition to the shortcomings discussed above, such sliding laws are based on the assumption that rates of basal motion at any point are determined by *local* driving stress and basal water pressure. In reality, rates of basal motion are controlled by both local and non-local factors. That is, the rate at which ice slides or ploughs over its bed reflects not only the local basal traction, but also the effect of more distant sources of resistance, and the ways in which spatial variations in

stress are accommodated by the ice. This can be illustrated by considering basal motion over a small water-filled cavity. Above the cavity, basal drag is zero and there is no local resistance to flow; here, the ice is *decoupled* from the bed. However, the ice above the cavity does not accelerate without limit, as implied by sliding laws such as equation 4.23, because it is restrained by the surrounding ice, which, in turn, is held back by drag elsewhere on the bed. Thus, basal velocities in areas of zero drag are controlled by more or less distant higher-strength regions. Areas of relatively high basal drag are known as *sticky spots*, which may take the form of bedrock bumps or lodged boulders, frozen patches within mainly thawed beds, well-drained areas or regions of the bed that are not connected to significant water sources (Alley, 1993; Stokes et al., 2007). At a larger scale, the lateral margins of a glacier or ice stream are regions of high drag, which exert a powerful non-local influence on rates of basal motion across the whole system.

The restraining influence of regions of high drag is extended into regions of lower drag (including decoupled regions of the bed) via longitudinal and lateral stresses in the ice. These stresses (collectively termed *membrane stresses* by Hindmarsh, 2006) redistribute resistance to flow through the ice mass, smoothing out variations in basal motion and, by extension, ice surface speed. A full understanding of patterns of glacier motion, therefore, requires knowledge of spatial patterns of drag at the bed and ice margins, together with full three-dimensional stress–strain relationships within the ice body.

The role of lateral drag in controlling basal velocities across a glacier or ice stream can be illustrated using a simple model. We make the initial assumption that basal drag and longitudinal stress gradients are negligible, and that drag at the lateral margins provides the only significant source of resistance to flow. This approximates the stress conditions on floating ice shelves confined within embayments, and some ice streams (Whillans and van der Veen, 1997; van der Veen, 1999a; Raymond et al., 2001). In this case, the centreline velocity U_C can be approximated from:

$$U_C = \frac{2A}{n+1}\left(\frac{\tau_d}{H}\right)^n W^{n+1} \qquad (4.24a)$$

and the transverse velocity profile from:

$$U(y) = U_C\left(1 - \left(\frac{y}{W}\right)^{n+1}\right) \qquad (4.24b)$$

where A and n are the parameters in Glen's Flow Law (equation 4.15), τ_d is the driving stress, H is ice thickness, W is the half-width of the flow unit and y is the horizontal coordinate with the origin on the centreline (van der Veen, 1999a). U_C is the same at the ice surface and the bed, since basal traction is zero and no vertical shearing occurs within the ice. Equation 4.24a is closely similar to that used for calculating glacier surface velocities arising from ice deformation (equation 4.17). This is because it is derived in exactly the same way, except that it is obtained by integrating the flow law *horizontally*, from the glacier margin to the centreline, rather than *vertically* from the bed to the surface. This reflects the fact that the rate of ice flow is controlled by lateral shear strain in the ice, in response to opposing tractions arising from the driving stress and drag at the flow-unit margins. Note that the centreline velocity is proportional to the fourth power of the half-width of the glacier, so that ice will flow much more rapidly where the channel margins are far apart than in narrow channels. Transverse velocity profiles predicted by equation 4.24b are shown in figure 4.29 (black curve). The important point to note is that basal velocities are controlled by a non-local property: drag at more or less distant lateral margins. Points close to the margins have low sliding speeds, whereas points further from the margins have greater sliding speeds because they are more remote from the source of resistance to flow.

Where basal traction is non-zero, bed properties also influence sliding speeds by introducing an additional source of drag, which further retards the ice. Equation 4.24 can be modified to derive an approximate expression for basal velocity where resistance to flow is from a combination of lateral and basal drag (Raymond, 1996; Raymond et al., 2001; Benn et al., 2007a):

$$U_B = \frac{2A}{n+1}\left(\frac{\tau_d - \tau_b}{H}\right)W^{n+1} \qquad (4.25)$$

To illustrate the behaviour of this interesting function, Benn et al. (2007a, b) defined basal drag τ_b as a simple function of basal effective pressure. When $N = 0$, basal drag is set to zero, and equation 4.25 reduces to equation 4.24a. When N equals the ice overburden pressure (i.e. $P_w = 0$), basal drag is assumed to equal the driving stress, in which case no basal motion occurs. The relationship between effective pressure and basal drag at intermediate values of N was defined using a tunable parameter, allowing flow to be retarded by basal drag to greater or lesser degrees (fig. 4.29).

Non-local controls on rates of basal motion are also introduced by along-flow gradients in basal and/or lateral drag. For example, where basal drag diminishes downflow, ice in the low-drag region will be retarded by the ice located upglacier, whereas ice in the high-drag region will be 'pulled' by the ice located downglacier. In this case, longitudinal membrane stresses tend to smooth variations in basal sliding rates across the bed.

From the above, it is clear that a realistic representation of basal sliding in glacier flow models requires tractable, physically based 'friction laws' to replace the 'sliding laws' discussed in section 4.6.1. The simplest 'friction laws' are

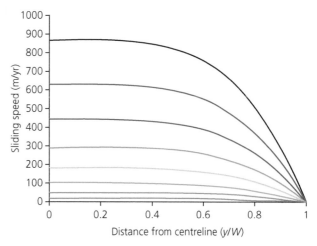

Figure 4.29 Output of simple model to illustrate the influence of local and non-local factors on the basal velocities along a transverse profile. The black curve shows the case for zero basal drag, while the lower curves show the effect of increasing basal drag.

based on a Coulomb-plastic yield criterion, and specify the maximum traction that can be supported on a sediment bed by intergranular friction and cohesion (section 4.2.3). Friction laws of this type may be appropriate for low-roughness till beds, and form the basis of the undrained plastic bed model of ice stream flow (Tulaczyk et al., 2000b; Bougamont et al., 2003a; section 6.3.6). Modelling basal drag for rough rockbeds is a much harder problem. As discussed in section 4.4.2, a basal shear traction or *form drag* arises as a consequence of normal stresses on upglacier-facing parts of bedrock bumps, and is modulated by the extent of water-filled cavities on bump lee sides. Relationships between basal drag, effective pressure and sliding velocity have been modelled by Fowler (1986b, 1987) and Schoof (2005), and explored further by Gagliardini et al. (2007). An important conclusion of this work is that basal drag is not independent of velocity. This is because the sliding rate influences the extent of lee-side cavities, and therefore the area of the bed that is able to support stresses. Thus, the Fowler–Schoof 'friction law' determines combinations of U_b and τ_b that can occur under specified normal stress and bed geometry. The friction law can be written as:

$$\tau_b = N.C\left(\frac{\Lambda}{\Lambda + \Lambda_0}\right)^{\frac{1}{n}} \qquad (4.26)$$

where τ_b is basal drag, N is the spatially averaged effective pressure at the bed, C is a constant related to bed geometry, n is the exponent in Glen's Flow Law (equation 4.15), Λ (Greek capital letter lambda) is:

$$\Lambda = \frac{U_b}{N^n} \qquad (4.27)$$

and Λ_0 is a geometrically controlled maximum value of Λ (Schoof, 2005). An important feature of this friction law is that it is *multi-valued*. That is, a given basal velocity may be associated with more than one value of basal drag (fig. 4.30). At low velocities and high effective pressures, cavities do not form on the bed and τ_b is independent of N. At higher velocities and/or effective pressures, cavities form on the lee side of bumps and τ_b increases as the upstream side of bumps support more and more of the total stress at the bed (the rising limb of the curve in fig. 4.30). Beyond a critical point, τ_b *decreases* with further increases in velocity and/or N as more and more bumps on the bed are flooded by cavities extending from larger bumps upstream (the falling limb of the curve in fig. 4.30; cf. fig. 4.17).

The incorporation of realistic friction laws into three-dimensional, time-evolving models of glacier flow, and coupling such models with realistic treatments of sub-glacial hydrology, remain important but elusive goals.

4.7 CREVASSES AND OTHER STRUCTURES: STRAIN MADE VISIBLE

Deformation of glacier ice can form crevasses and other distinctive structures, which can convey information about past and present patterns of strain that are difficult or impossible to obtain by other means. Indeed, crevasse patterns have been described as 'the writings in a glacier's history book' (Hertzfeld et al., 2004; fig. 4.31). Fracturing also plays a key role in a very wide range of important glaciological processes, including the formation of deep moulins (section 3.3.2), calving (section 5.5) and shearing of ice stream margins (section 6.3.4). In this section we examine the factors responsible for crevasse formation, and the ways in which strain histories can be reconstructed from fractures and other structures.

4.7.1 CREVASSES

Crevasses are among the most dramatic features of glaciers and ice sheets, and their cold blue depths have fascinated and appalled mountain travellers through the ages. Despite being very widespread, crevasses have attracted relatively little attention from glaciologists, and our current understanding is largely based on a few key studies, summarized in an excellent review by van der Veen (1999b). Crevasses will open up wherever the forces pulling ice apart exceed the strength of the ice, triggering brittle failure. They can develop in three basic ways, or *fracture modes* (fig. 4.32):

(1) Mode I fracture (opening mode) occurs as a result of *tensile stresses* pulling the walls of the crack apart.

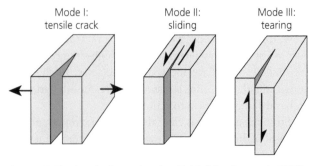

Figure 4.31 Heavily crevassed ice on Breiðamerkurjökull, Iceland: (a) from the air and (b) on the surface. Note the person for scale (R. Mottram and D.I. Benn).

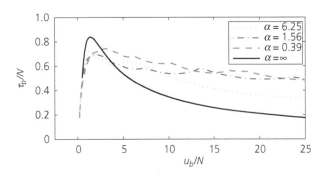

Figure 4.30 Modelled relationships between basal velocity, effective pressure and basal shear stress. The parameter α is a measure of bed morphology. $\alpha = \infty$ corresponds to a regular, sinusoidal bed, and smaller values represent increasingly irregular beds (Schoof, 2005).

Figure 4.32 The three fundamental modes of brittle failure (Benn et al., 2007b).

(2) Mode II fracture (sliding mode) occurs in response to *shear stresses* applied parallel to the crack plane, and the crack grows in the same direction as the direction of shear.

(3) Mode III fracture (tearing mode) also occurs in response to shear stresses, but with crack growth at right angles to the direction of shear.

It is clear that crevasses on glaciers can develop in two or more of these ways simultaneously (*mixed-mode fracture*), although most analyses are based on the assumption that fracturing is entirely in Mode I.

The tensile strength of ice is highly variable, depending on factors such as temperature, water content, density and structure, but generally lies in the range 90–320 kPa (Vaughan, 1993; van der Veen, 1998a, 1999b). Although applied tensile stresses may exceed these values through the full thickness of a glacier, crevasses are usually confined to a shallow surface layer because at depth the *deviatoric* tensile stress is reduced by other forces. Below the surface of a glacier, any longitudinal tensile stresses tending to pull the ice apart are at least partly offset by compressive stresses arising from the weight of the overlying ice (equation 4.1). These compressive stresses increase steeply with depth, so the deeper one goes, the greater the chance that they will be high enough to offset any tensile stress. Crevasse depth can thus be defined as the point where the tensile stress tending to open the crevasse is exactly balanced by the compressive stress arising from the weight of the ice (fig. 4.33). Nye (1957) used this insight to derive a formula for crevasse depth d for any given surface strain rate:

$$d = \frac{2}{\rho_i g} \left(\frac{\dot{\varepsilon}_{xx}}{A} \right)^{\frac{1}{n}} \tag{4.28}$$

where ρ_i is ice density, g is gravitational acceleration, $\dot{\varepsilon}_{xx}$ is the longitudinal strain rate and A and n are the flow law parameters. Because ice overburden pressures increase rapidly downwards, crevasses rarely exceed 25–30 m on temperate glaciers where ice is relatively soft, but can be much deeper in polar regions where ice is colder and stiffer (Paterson, 1994). In a field study on Breiðamerkurjökull, Iceland, Mottram and Benn (2009) found that measured crevasse depths were generally close to those using equation 4.28.

Where crevasses contain water, ice overburden pressures are also opposed by water pressure (van der Veen, 1998a; fig. 4.33). A modified form of the Nye crevasse depth model, incorporating the effect of water pressure, was proposed by Benn et al. (2007a and Otero *et al.* (2010):

$$d = \frac{1}{\rho_I g} \left[2 \left(\frac{\dot{\varepsilon}_{xx}}{A} \right)^{\frac{1}{n}} + (\rho_w g d_w) \right] \tag{4.29}$$

where d_w is the depth of water in the crevasse. Adding water to a crevasse will cause it to penetrate deeper, because water pressure acts in the same direction as the longitudinal tensile stress tending to pull the crevasse apart. If more water is added, d_w will increase further, allowing more crevasse growth. If sufficient water is available, the crevasse may be able to propagate through the full thickness of a glacier or ice shelf, with important implications for penetration of meltwater through glaciers (section 3.3.2) and iceberg calving processes (section 5.4.2; Benn et al., 2007b).

Equations 4.28 and 4.29 ignore the effect of *stress concentrations*, or locally increased tensile stresses near the tip of crevasses. The presence of a crack in a stressed material causes pronounced stress concentrations around the crack tip, magnifying the externally applied force (fig. 4.34). The magnitude of this effect is measured by the *stress intensity factor*, which depends on crack length and other factors. A detailed analysis of the role of stress concentrations on crevasse depths, employing the methods of Linear Elastic Fracture Mechanics, has been developed by van der Veen (1998a, b). For single crevasses, stress

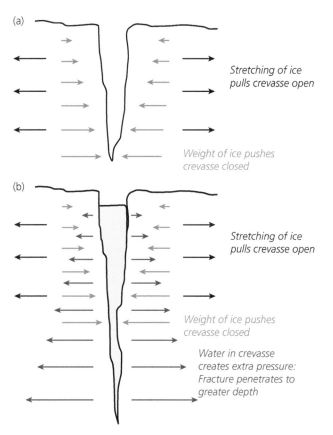

(a)

Stretching of ice pulls crevasse open

Weight of ice pushes crevasse closed

(b)

Stretching of ice pulls crevasse open

Weight of ice pushes crevasse closed

Water in crevasse creates extra pressure: Fracture penetrates to greater depth

Figure 4.33 Factors controlling the depth of penetration of a surface crevasse. (a) Extending flow pulls the crevasse open, whereas the compressive stress arising from the weight of the ice pushes it shut. The weight of the ice increases downwards, placing a limit on crevasse depth. Stress concentration effects at the crack tip are not shown. (b) Water in a crevasse provides an additional outward-acting force, allowing the crevasse to penetrate to greater depths.

(a)

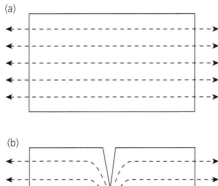

(b)

Figure 4.34 Idealized stress patterns in a rectangular block under tensile stress: (a) without and (b) with a crack. The presence of the crack focuses stress trajectories (i.e. the paths along which stress is passed from molecule to molecule) around the crack tip, increasing the likelihood of failure there (Modified from Gordon, 1976).

Figure 4.35 Crevasses on Mer de Glace, Mont Blanc, France. Bergschrunds are clearly visible, curving across the steep ice slopes below the rock peaks (D.I. Benn).

concentrations may significantly increase the depth of penetration, but for fields of crevasses, stress concentration effects are comparatively small (van der Veen, 1998a).

A type of surface crevasse that is characteristic of cirque and valley glaciers is the *randkluft*, a fissure separating glacier ice from the rockwall at the head of the valley. Randklufts form due to movement of the ice away from the rockwall, and may be enlarged by preferential summer ablation adjacent to warm rock surfaces (Mair and Kuhn, 1994). *Bergschrunds* are related features, consisting of deep, transverse crevasses near the heads of valley glaciers (fig. 4.35). One particularly large example almost swallowed the mountaineer Joe Simpson, as related in his classic book *Touching the Void*. It is often assumed that bergschrunds separate immobile, cold-based ice at the head of the glacier from active, sliding ice below, although Mair and Kuhn (1994) have shown that on Daunferner, a glacier in the Stubai Alps, Austria, ice was

sliding both above and below a large bergschrund, which formed in a region where ice thickness and velocity were rapidly increasing. Comparative studies in other areas are few, but it seems likely that similar situations also occur on other glaciers.

Basal crevasses are fractures that extend upwards from the base of glaciers or ice shelves. A theoretical analysis by van der Veen (1998b) indicates that they require both high tensile stresses and very high basal water pressures to overcome the large compressive stresses at depth, and so only tend to form where ice is both extending rapidly and floating, or almost so. Few data are available on basal crevasses below extant glaciers, but the frequent occurrence of basal crevasse-squeeze ridges on the beds of surging glaciers is consistent with the theoretical results (section 11.2.9).

Faults are fractures that develop parallel to shear planes, where the fracture walls move relative to each other but remain in contact. Thrust faults form where the compressive stresses in the flow direction exceed the ice overburden pressure (i.e. where the longitudinal deviatoric stress is compressive, and the vertical deviatoric stress is tensile (fig. 4.36). Normal faults and strike slip faults also occur on glaciers, depending on the relative values of the applied stresses.

4.7.2 CREVASSE PATTERNS

According to classical theory, crevasses form at right angles to the principal tensile stress direction (Nye, 1952; Paterson, 1994). As shown in section 4.2.1, tensile stresses can arise from applied normal stresses, shear stresses or a combination of the two. Figure 4.37 shows how common crevasse patterns observed on glaciers reflect different stress configurations. *Chevron crevasses* are linear fractures aligned obliquely upvalley from the margins of a valley glacier towards the centreline, and form in response to simple shear associated with drag at the valley walls. Similar crevasse patterns can be found near the margins of ice streams, where there are strong lateral gradients in ice velocity. *Transverse crevasses* form in regions of extending flow (pure shear). Near the centre of an extending valley glacier, the principal tensile stress is parallel to glacier flow, so that transverse crevasses open up at right angles to the centreline. Towards the margins, where the stress pattern is influenced by the drag of the valley walls, the principal tensile stress progressively rotates, producing crevasses that are concave downflow. Where lateral drag is unimportant, transverse crevasses may extend straight across the ice for great distances. *Splaying crevasses* form where valley glaciers are subject to compressive flow. In this case, the crevasses are roughly parallel to the flow direction towards the centre and bend outwards to meet the margins at angles of less than 45°. Piedmont lobes can have striking *radial* crevasse patterns related to lateral spreading of the ice. In heavily crevassed areas such as ice

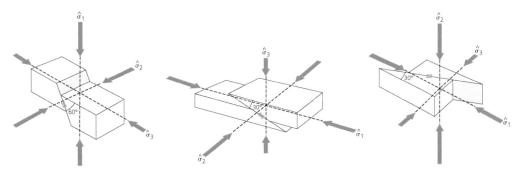

Figure 4.36 Relationship between fault type and principal stress orientation. (a) Normal fault, with the maximum compressive stress vertical. (b) Thrust fault, with the minimum compressive stress vertical. (c) Strike-slip fault, with intermediate compressive stress vertical (Twiss and Moores, 1992).

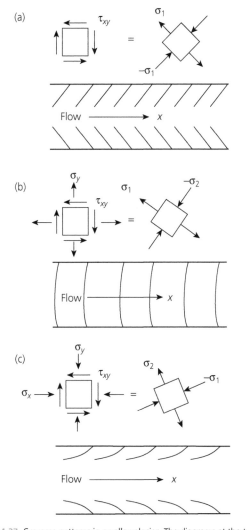

Figure 4.37 Crevasse patterns in a valley glacier. The diagrams at the top of each panel show shear stresses and normal stresses at the glacier surface near the upper margin (left) and the associated principal stresses (right). (a) Chevron crevasses resulting from lateral shear stresses at the margins. (b) Curved transverse crevasses resulting from a combination of lateral shear stresses and longitudinal tensile stress (extending flow). (c) Splaying crevasses due to a combination of lateral shear stresses and longitudinal compressive stress (compressive flow) (Modified from Nye, 1952).

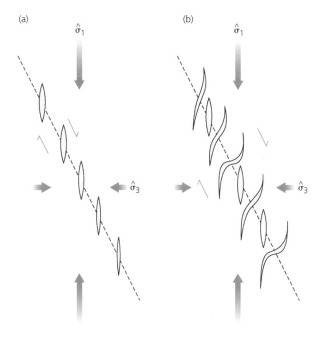

Figure 4.38 Model for the formation of *en echelon* gash fractures. (a) An array of crevasses forms along a shear zone, with each fracture perpendicular to the minimum principal stress. (b) Ductile deformation along the shear zone rotates the central portion of the fractures, deforming the originally linear fracture into a sigmoidal shape. Different generations of fracture show differing amounts of rotation (Twiss and Moores, 1992).

A detailed analysis of the relationship between stress fields and crevasse patterns has been presented by Hertzfeld et al. (2004).

Once formed, crevasses will be carried by glacier flow into areas where stress conditions differ from those which formed them. Crevasses at the margins of valley glaciers and ice streams will tend to rotate because the part nearest the centreline will have the highest velocity. Under conditions of simple shear, offset chains known as *en echelon crevasses* typically form in response to the combined effects of simple shear and rotation (fig. 4.38). However, rotated crevasses will not necessarily remain open as stress conditions change, and may close up or be replaced by others with more suitable orientations. Crevasse patterns can change dramatically at times of changing stress regime, such as when flow is accelerating or decelerating

falls, blocks of ice between crevasses may form spectacular broken pinnacles known as seracs, which often collapse under the influence of glacier movement and ablation, contributing to the downglacier transfer of ice.

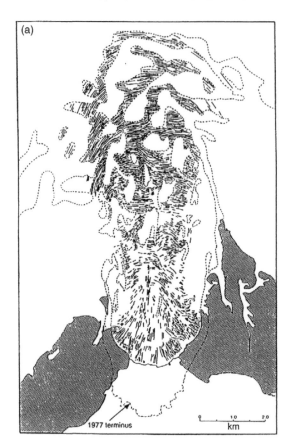
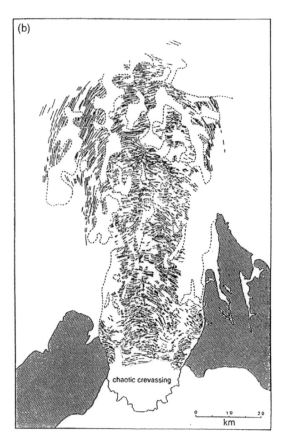

Figure 4.39 Crevasse pattern maps of Bodleybreen: (a) before and (b) after the arrival of a surge at the terminus. In the first map, the crevasse pattern indicates extending flow in the upper glacier and compressive flow in the lower part, whereas in the second map, extensional flow on the lower glacier has overprinted the older splaying crevasses with transverse crevasses (Hodgkins and Dowdeswell, 1994).

through time (Hambrey and Lawson, 2000). Figure 4.39 shows crevasse patterns on Bodleybreen, Svalbard, before and during a surge. In the pre-surge pattern, the terminal area displays radial crevasses indicative of lateral spreading, whereas during the surge it is traversed by numerous transverse crevasses reflecting large longitudinal tensile stresses (Hodgkins and Dowdeswell, 1994).

On closing, crevasses leave linear scars known as *crevasse traces* (Hambrey, 1975, 1994; Hambrey and Lawson, 2000). These commonly take the form of prominent layers of blue ice formed by the freezing of meltwater in the crevasse prior to closure. Alternatively, thin layers of white, bubbly ice can mark the former position of snow infills. In both cases, rock debris may be present within the crevasse trace, composed of material that fell into the original open crevasse. A more subtle type of trace forms during brittle fracture of the ice, but without separation of the walls. If the rate of extension is not too great, an open crevasse does not form; instead, the stretching is taken up by the recrystallization of ice parallel to the fracture creating a *tensional vein* analogous to those in deformed rocks (Hambrey and Müller, 1978). If a crevasse fills with debris and then closes, a *debris-filled crevasse trace* is formed (Gulley and Benn, 2007). These are common on debris-covered glaciers in high mountain

environments, where they provide permeable pathways for englacial water flow (section 3.3.2). Crevasse traces and tensional veins are usually initially close to vertical, but are rotated into other orientations by glacier flow.

4.7.3 LAYERING, FOLIATION AND RELATED STRUCTURES

Glacier ice commonly exhibits various types of layering, imprinted at different stages of its history. In accumulation areas, the most common form of layering is *sedimentary stratification*, which reflects annual cycles of snow accumulation and melting. The layers are generally parallel to the glacier surface and are often beautifully exposed in the walls of crevasses (fig. 2.10). Usually, the strata consist of relatively thick layers of coarse-grained, white, bubbly ice, separated by thinner layers of clear, bubble-free ice. The bubbly ice represents compacted winter snow, whereas the intervening clear layers originate from snow in which the original pore spaces have been filled by refrozen meltwater (superimposed ice; section 2.4.2). Dirt layers, formed when wind-blown dust or other debris is concentrated during summer melting, are often a prominent feature of sedimentary layering. Successive layers are generally broadly parallel, although unconformities

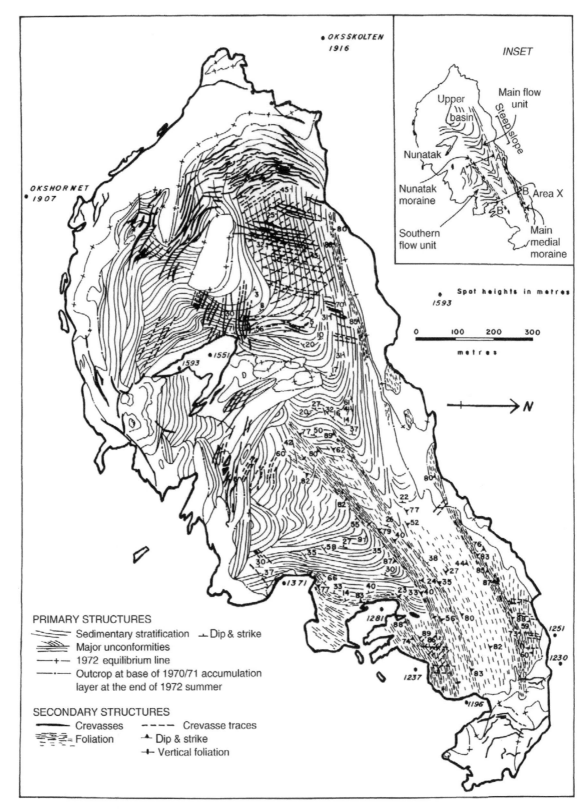

Figure 4.40 Surface foliation and crevasse patterns on Charles Rabotsbreen, Norway (Hambrey, 1976).

Figure 4.41 Ice breccia formed from a frontal apron of ice blocks, Ellesmere Island (D.J.A. Evans).

can occur, reflecting periods of irregular melting or wind scouring on the glacier surface prior to renewed accumulation. Annual layers can be a few millimetres to several metres thick, depending on the specific mass balance at that point on the glacier.

Foliation refers to pervasive planar structures found in almost all glaciers (Hambrey and Lawson, 2000). Individual layers consist of fine-grained white ice, coarse-grained clear (blue) ice or coarse-grained bubbly ice, with crystals in the fine layers usually less than 5 mm and those in the coarse layers around 1–15 cm (Hambrey, 1994; Paterson, 1994). Foliation can develop from sedimentary stratification or crevasse traces which have undergone high strains and/or have been rotated by glacier flow (Hambrey, 1975, 1994; Hooke and Hudleston, 1978; Hambrey et al., 2005). However, some foliation is thought to be an entirely metamorphic structure, produced by alteration of the ice under high strains (Pfeffer, 1992; Hambrey and Lawson, 2000). In detail, foliation can display structural variations related to the strain history of the ice. Isolated, tight fold hinges may be the only remnants of large-scale folds that otherwise have been stretched and deformed beyond recognition. Stretching of stiff ice layers embedded within relatively soft ice can break up the stiffer ice, producing sausage-like structures known as *boudins*.

The orientation of foliation is determined by the original disposition of the primary or other structures from which the foliation is derived, and/or the history of ice deformation. It is commonly possible to recognize two main orientations, transverse and longitudinal, which may coexist with a cross-cutting relationship (fig. 4.40; Hambrey, 1976, 1994; Hambrey et al., 2005). *Transverse foliation* commonly consists of rotated primary stratification or crevasse traces located downglacier from zones of transverse crevasses and icefalls. In valley glaciers, ice flow causes the progressive deformation of transverse foliation into a series of arcs. In three dimensions, the

foliation dips downward towards the centre of the arc, the whole resembling a set of nested spoons. Glaciers that pass through a number of icefalls may have several intersecting sets of arcuate foliation (Paterson, 1994). *Longitudinal foliation* is aligned parallel to glacier flow and originates by the rotation and folding of sedimentary layering, crevasse traces or other structures during ice flow, particularly where ice converges from wide basins into narrow tongues (Hambrey and Müller, 1978; Hambrey and Lawson, 2000). Patterns of folding in foliation can allow ice deformation history to be reconstructed, sometimes in considerable detail (e.g. Hambrey et al., 1999, 2005).

Ice breccia is a distinctive type of ice formed below intensively crevassed areas or ice cliffs, where fractured ice reconsolidates (fig. 4.41). Voids between blocks can be infilled with refrozen meltwater or snow to create an aggregate seamed by randomly oriented veins.

Ogives are repetitive arcuate bands or waves formed below icefalls on some glaciers. They are convex downflow, with the amount of curvature increasing in each successive band as the result of the greater velocity of the central part of the glacier compared to the margins (fig. 4.42). Two basic types of ogive have been recognized (Waddington, 1986; Paterson, 1994; Goodsell et al., 2002). *Band ogives* are alternating bands of dark and light ice, and are the surface expression of three-dimensional spoon-shaped structures dipping upstream within the glacier. The dark bands consist of highly foliated, rather dirty ice, whereas the light bands tend to be composed of more uniform bubbly ice. *Wave* or *swell-and-swale ogives* are similar in planform, but consist of alternating ridges and troughs.

A widely accepted mechanism for the formation of ogives was proposed by John Nye, who argued that they reflect seasonal variations in the passage of ice through icefalls. Because ice moves faster through an icefall than elsewhere, it is stretched and thinned as it accelerates into

Figure 4.42 Ogives on Saskatchewan Glacier, Canada (D.J.A. Evans).

result, ice flowing through an icefall in summer will undergo relatively high ablation, and will tend to collect more windblown dust and other superficial debris. At the base of the icefall, ice discharged during the summer will form troughs or dark bands as the ice is compressed once more. In contrast, ice passing through the icefall in winter will collect excess snow and become wave crests or light bands. The amplitude of wave ogives tends to decrease downglacier from a maximum of about 5 m near the base of the icefall. This is probably due to differential melting, the troughs being protected from ablation by infills of drifted snow or debris. According to this model, ogives are formed annually, each light-dark or ridge-trough pair representing one year of ice movement. This implies that they can be used to estimate ice velocities, as first suggested by the Scottish glaciologist James Forbes in the mid nineteenth century (ogives are sometimes known as *Forbes bands*).

An alternative model for the formation of band ogives has been proposed by Goodsell et al. (2002), based on studies on the Bas Glacier D'Arolla, Switzerland. Structural analysis and ground-penetrating radar surveys showed that dark bands on the surface correspond to prominent upglacier-dipping structures within the glacier, interpreted as layers of intense foliation. It was argued that these represent zones of enhanced shear, formed annually at the base of an icefall.

the upper part of the icefall. This thinning, plus the high concentration of crevasses in icefalls, means that ice passing through an icefall has a larger surface area than equivalent volumes of ice elsewhere on the glacier. As a

CHAPTER 5

GLACIER DYNAMICS

Paulabreen, Svalbard, in spring 2006, three months after surge termination. The ridge in the foreground is an ice-cored esker (Doug Benn)

5.1 INTRODUCTION

Understanding glacier dynamics – how the motion of glaciers varies in time and space – involves knowledge of all of the topics covered in chapters 2–4 of this book. Surface melting and accumulation cause changes in ice thickness, altering the stress distribution within the glacier. Resisting forces at the glacier bed are strongly influenced by glacier hydrology, particularly the volume and configuration of water at the bed. Taken together, surface mass balance and glacier hydrology constantly shift the balance of forces within and beneath the ice, causing glaciers to speed up or slow down on shorter or longer timescales. The calving of icebergs from water-terminating glaciers and the collapse of seracs from mountainsides involve an intricate web of dynamic processes coupled with fracture mechanics.

In this chapter we will draw extensively on concepts developed in the foregoing chapters, to develop an understanding of the dynamic behaviour of the world's glaciers and ice caps. The dynamics of the Greenland and Antarctic Ice Sheets are covered separately in chapter 6. This chapter begins with a discussion of the main principles underlying glacier dynamics. Section 5.3 then discusses how these principles are used to construct mathematical models, which are among the most important tools used by glaciologists to understand how glaciers respond to environmental change. Section 5.4 examines the dynamics of valley glaciers, focusing on short-term velocity fluctuations and longer-term advance and retreat cycles. In section 5.5, we turn to the curious case of calving glaciers, and in section 5.6 we consider the fragile lives of floating ice shelves. Finally, section 5.7 addresses one of the most spectacular and enigmatic of dynamic phenomena – glacier surges.

5.2 UNDERSTANDING GLACIER DYNAMICS

5.2.1 BALANCE VELOCITIES

Averaged over long time periods, ice discharges are governed by glacier mass balance. This idea can be simply illustrated using the 'wedge' concept, which represents mass gained and lost as two wedges (Sugden and John, 1976). In the highly idealized glaciers shown in figure 5.1, net ablation increases from zero at the equilibrium line to a maximum at the terminus, producing a wedge of ablation. Similarly, net accumulation increases from zero at the equilibrium line towards higher elevations, producing a wedge of accumulation. The rate at which mass is added to or subtracted from each wedge is controlled by the mass balance gradient (section 2.5.3). To maintain a steady state, the glacier must transfer mass gained in the accumulation wedge to replace that lost in the ablation wedge, and ice discharge through any cross section on the glacier must equal the rate of net mass gain upglacier (total accumulation minus ablation). This can be expressed quantitatively as:

$$Q_x = \sum_{i=1}^{x} \dot{b}_i w_i \qquad (5.1)$$

where Q_x is the discharge through a cross section at a distance x from the highest point on the glacier, Σ represents 'the sum of', and $\dot{b}_i$ and w_i are the net balance rate and width, respectively, of successive portions of the surface from the head of the glacier to x. Note that mass losses by ablation upglacier of x are assumed to pass through the cross section as meltwater, and hence are not included in the ice discharge.

The average velocity, $\bar{u}_x$, through the cross section is equal to:

$$\bar{u}_x = Q_x / A_x \qquad (5.2)$$

where A_x is the cross-sectional area. Because of their dependence on mass balance, velocities calculated in this way are known as *balance velocities* (Clarke, 1987c). Two important implications stem from this simple model:

(1) Ice discharges increase from the head of the glacier to the equilibrium line, then decrease towards the terminus. For a glacier of constant cross-sectional area, average velocities increase and decrease in a similar way. These simple patterns, however, will be modified by changes in glacier cross section area. Constrictions in a valley will be associated with higher balance velocities, so that the discharge of ice can be maintained through a smaller cross-sectional area. Similarly, higher velocities are also often associated

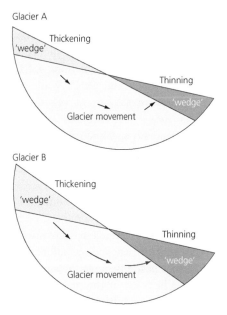

Figure 5.1 The idea of balance velocity, illustrated using the wedge concept. Glacier B has a steeper mass balance gradient than Glacier A, so requires higher velocities to balance the mass gained and lost in the two wedges (Modified from Sugden and John, 1976).

with glacier steepenings, such as icefalls, where ice tends to be thinner.
(2) Balance velocities will be highest on glaciers with steep mass balance gradients, or where the accumulation from wide basins is focused into narrow channels. The influence of the mass balance gradient can be understood by considering two glaciers of identical shape but differing mass balance gradients (fig. 5.1). Where the mass balance gradient is steeper, the mass of the ablation and accumulation wedges will be greater, thus requiring greater rates of throughflow to maintain equilibrium.

Regional trends in mass balance gradients mean that glaciers in humid, maritime areas should flow faster than glaciers in arid, cold climates, if other factors are held constant. Steep mass balance gradients reflect heavy snowfall in the upper accumulation area and high ablation rates near the terminus, and correspond with high rates of ice throughput (Andrews, 1972; Kuhn, 1984). Such high-activity glaciers occur in moist, mid-latitude areas such as southern Alaska, western Norway, New Zealand and Patagonia. In contrast, shallow mass balance gradients indicate small differences in specific mass balance across the altitudinal range of a glacier, and are associated with slow-moving, low-activity glaciers. Examples of this type are the Devon Ice Cap and White Glacier in Arctic Canada, and the glaciers around the Dry Valleys, Antarctica. The strong influence of glacier geometry on balance velocities, however, means that there is no simple relationship between mass balance gradients and ice velocity.

Cold-based glaciers, which are everywhere frozen to their beds, can only flow by internal deformation of the ice and the upper part of the bed, because sliding is very small at sub-freezing temperatures (section 4.4.1; fig. 5.2; Echelmeyer and Zhongxiang, 1987; Waller, 2001). For example, the Meserve Glacier in the Dry Valleys of Antarctica has maximum velocities of only $2\,\mathrm{m\,yr^{-1}}$ (Chinn, 1988). Generally speaking, glaciers in cold regions have low mass balance turnover, so that the low velocities associated with the creep of cold ice are sufficient to discharge the annual ice accumulation. However, where mass turnover is higher, such as where ice from large accumulation areas is focused into narrow outlets, or below thick ice, basal ice may be warmed to the pressure-melting point, thus allowing efficient sliding and increasing velocities to levels sufficient to discharge the annual balance.

Where glacier beds are at the pressure-melting point, efficient basal sliding or subsole deformation can occur, leading to very high glacier velocities in some cases. The world's fastest glaciers are large tidewater outlets of the Greenland Ice Sheet, such as Jacobshavns Isbrae and Kangerdlugsuaq, where flow speeds in excess of 12 km per year have been measured near the calving front (section 6.2.4; Howat et al., 2005, 2007; Rignot and Kanagaratnam, 2005). Velocities in excess of $7000\,\mathrm{m\,yr^{-1}}$ have also been measured near the calving terminus of San Rafael Glacier, an outlet of the North Patagonian ice field, Chile (fig. 5.3; Warren, 1993). Both Jacobshavns Isbrae and Kangerdlugsuaq have high balance velocities (Bamber et al., 2000), largely due to the funnelling of ice from very large catchments (>92 million and 51 million $\mathrm{km^2}$, respectively; Rignot and Kanagaratnam, 2005). For dynamic reasons, the high velocities observed in recent years are in excess of the balance velocities, and are clearly not sustainable in the long term. For San Rafael Glacier, high velocities result from a combination of topographic focusing and very high mass turnover rates (Warren, 1993).

Glaciers in most parts of the world have surface velocities intermediate between the very slow values typical of thin, cold-based glaciers and the very high values associated with ice streams and some calving glaciers. For wet-based valley glaciers, surface velocities are typically a few tens of metres per year, attributable to a combination of basal sliding and ice creep (e.g. Iken and Truffer, 1997; MacGregor et al., 2005). The wide range of glacier velocities has led to a classification of glaciers into *normal glaciers*, with velocities typically in the order of 10^1–$10^2\,\mathrm{m\,yr^{-1}}$, and *fast glaciers*, with velocities in the order of 10^2–$10^3\,\mathrm{m\,yr^{-1}}$ (Clarke, 1987b). For normal glaciers, driving stresses are relatively high (40–120 kPa; Paterson, 1994), associated with relatively high drag at the bed, and velocity increases with the shear stress. In contrast, for fast outlet glaciers and ice streams, driving stresses are generally very low (~20 kPa), associated with low basal drag. The low bed strength of fast glaciers appears always to be due to the presence of pressurized water at the bed in some form of distributed drainage system (Kamb, 2001; Alley et al., 2004; Bamber et al., 2007; section 6.3.4). This distinction reflects two fundamentally different types of flow: one where basal drag is high and contributes a large proportion of the resistance to flow; and the other where large portions of the bed have very low strength, and resistance to flow is mainly from lateral drag, in addition to isolated 'sticky spots' on the bed. Spatial switching from one mode to another occurs at ice stream onset zones (section 6.3.4), whereas temporal switching occurs during glacier surge cycles (section 5.7).

This tendency to fast flow is due to long-term interactions between climate, glacier activity and landscape evolution. Deep lake basins and fiords typically form in deeply dissected montane environments in maritime climatic settings, such as southern Alaska and Patagonia, where high snowfall and ablation drives the ice flux required for rapid erosion (Hallet et al., 1996). Glaciers in such environments, therefore, have the mass throughput necessary

Figure 5.2 Cold-based valley glacier, NW Ellesmere Island, Arctic Canada (D.J.A. Evans).

Figure 5.3 The intensely crevassed terminus of San Rafael Glacier, Chile. Upwelling meltwater has cleared floating ice from a large area in front of the glacier (Charles Warren).

to sustain fast glacier flow, and occupy overdeepened erosional basins. The presence of deep water at glacier margins provides a high base-level for subglacial and englacial drainage, encouraging high basal water pressures and rapid sliding. The presence of fast, water-terminating glaciers in maritime mountain ranges is thus part of a self-reinforcing process of landscape change.

5.2.2 DEVIATIONS FROM THE BALANCE VELOCITY

Measured velocities on glaciers usually differ from calculated balance velocities. The differences may be small, representing temporary deviations from the average on a timescale of hours, days, weeks or months (Willis, 1995), or large, representing major departures from equilibrium conditions over years, decades or even centuries (Murray et al., 2003; O'Neel et al., 2005). Deviations from the balance velocity occur because actual glacier flow rates depend on the magnitude of the driving stress and the nature of the forces resisting flow (section 4.3.4). Over long periods of time, the driving and resisting forces are kept in a natural and dynamic balance by glacier flow, because glaciers tend to adopt configurations that generate high enough driving stresses to discharge the ice accumulation. Both driving and resisting forces, however, may change on a variety of timescales, causing the velocity to differ from that predicted from mass balance alone.

Because the driving stress is a function of the thickness and surface gradient of the ice (equation 4.10), it will increase or decrease in response to changes in the geometry of the glacier. For example, higher than average snowfall in the accumulation area will thicken the glacier. If the glacier surface slope also increases or remains constant, thickening will lead to higher shear stresses and increased ice velocities, which serve to discharge the excess mass towards the ablation area. Conversely, excessive ablation will thin the glacier, potentially reducing shear stresses and velocities.

As described in section 4.2.4, resisting stresses result from drag between the glacier and its bed, drag at the glacier margins, and the ways in which variations in these drag terms are distributed through the ice by longitudinal and lateral stress gradients. The single most important cause of variations in resisting stresses on short timescales is the distribution and pressure of water at the bed (Willis, 1995). Elevated water pressures reduce drag by increasing the spatial extent of water-filled cavities and reducing the strength of subglacial sediments (sections 4.4.4 and 4.5.3; Willis, 1995; Boulton, 2006; Kavanaugh and Clarke, 2006). Water at the glacier bed does not directly affect ice creep rates, which are mainly controlled by shear stress, internal water content and ice temperature. Creep rates may, however, be indirectly influenced by subglacial water pressures, because water-pressure controlled variations in

basal drag on one part of a glacier will alter the longitudinal stresses elsewhere (Hutter, 1983; van der Veen and Whillans, 1989a; Kavanaugh and Clarke, 2001). Resisting stresses can also change due to other factors. For example, the removal of floating ice shelves may suddenly reduce upglacier-directed forces, or *backstress*, and trigger ice acceleration. The role of backstress in the dynamics of calving glaciers is discussed in section 5.5.1.

5.2.3 CHANGES IN ICE THICKNESS: CONTINUITY

Changes in ice thickness at a particular location can occur as a result of ice accumulation and ablation, or dynamic processes. The factors influencing glacier thickness change can be understood most clearly with reference to the *continuity equation*, which sums all possible mass inputs and outputs to a given part of a glacier. For ease of understanding, we present this important equation in a simplified form, based on two assumptions:

(1) ice flow is in the x direction only, with no transverse flow convergence or divergence
(2) the ice is incompressible, so that changes in ice mass are equivalent to changes in volume.

More complete forms of the continuity equation are described by van der Veen (1999a).

Figure 5.4 shows a column of glacier ice. Ice flows in from the left and out on the right, and mass can accumulate or ablate at the upper and lower surfaces. The change in ice thickness through time ($\partial H/\partial t$) is therefore the sum of these inputs and outputs:

$$\frac{\partial H}{\partial t} = \frac{\partial Q}{\partial x} + \dot{b}_s + \dot{b}_B \qquad (5.3a)$$

where Q is ice discharge, and $\dot{b}_s$ and $\dot{b}_B$ are the balance rates at the upper surface and the bed, respectively. The

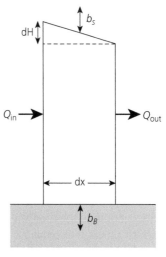

Figure 5.4 Definition sketch illustrating terms in the continuity equation.

mass balance terms are positive for the case of mass gain (accumulation) and negative for mass loss (ablation). For grounded glacier ice, basal melting and freeze-on are generally negligible as mass balance terms and are usually ignored, but below floating ice shelves $\dot{b}_B$ can be large, with a significant influence on shelf dynamics (section 5.6.1).

The term $\partial Q/\partial x$ is the gradient in ice discharge in the downflow direction, and represents the dynamic component of thickness change. Since discharge equals the vertically averaged velocity times ice thickness, the continuity equation can be written:

$$\frac{\partial H}{\partial t} = -\frac{\partial(\bar{U}H)}{\partial x} + \dot{b}_s + \dot{b}_B \qquad (5.3b)$$

A very useful form of the continuity equation can now be obtained by expanding the dynamic term, using the product rule of differentiation:

$$\frac{\partial H}{\partial t} = -\left[H\frac{\partial\bar{U}}{\partial x} + \bar{U}\frac{\partial H}{\partial x}\right] + \dot{b}_s + \dot{b}_B \qquad (5.3c)$$

We have now split the dynamic component of thickness change into two, both of which have clear physical meaning and are worth considering carefully. The first term inside the square brackets is the change in column thickness associated with downstream variations in ice velocity. The velocity gradient ($\partial\bar{U}/\partial x$) is the longitudinal strain rate (section 4.2.2), which measures the rate of downglacier stretching or compression of the ice column. Thus, if the velocity at the downstream side is greater than that on the upstream side, the ice in the column is stretched in the direction of flow. Since the ice is assumed to be incompressible, stretching in this direction must result in thinning in the vertical direction. The opposite is true if ice flow speed decreases downglacier. The thinning or thickening rate resulting from this process is simply the velocity gradient multiplied by the ice thickness. The second term on the right-hand side is the change in column thickness due to the advection of (usually) thicker ice from upglacier, and is equal to the thickness gradient ($\partial H/\partial x$) multiplied by the velocity. So, equation 5.3 states that thickness changes through time are equal to the combined effects of (1) stretching or shortening, (2) the transport of ice from upglacier, and (3) addition or subtraction of mass at the surface and the base of the ice. In more complete forms of the continuity equation, more terms are added to account for transverse flow divergence or convergence, and density changes.

For equilibrium conditions, all the terms on the right-hand side of the continuity equation must sum to zero. For example, ice thickness will remain constant if negative mass balance in the ablation area is offset by dynamic

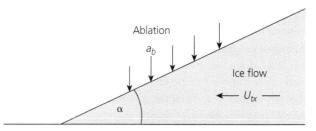

Figure 5.5 Sketch showing the opposing effects of velocity and ablation on the position of an ice margin. The thickness gradient $\partial H/\partial x$ in equation 5.3 is equal to the tangent of the surface slope α.

thickening, or positive mass balance in the accumulation area is offset by dynamic thinning. Similarly, the terminus of a glacier will maintain a constant position if all terms in the continuity equation sum to zero. This will be the case if the thickening rate due to the delivery of ice from upglacier exactly balances the ablation rate, and the stretching rate is zero (fig. 5.5). Clearly, the mass balance term varies on a seasonal basis, and many glaciers consequently undergo cycles of winter advance and summer retreat. If the sum of the mass balance and dynamic terms at the terminus is positive or negative on annual or longer timescales, the glacier will undergo net advance or retreat, respectively.

5.2.4 THERMODYNAMICS

It is instructive to consider glacial dynamics in terms of energy exchanges within the glacier; and between the glacier and the atmosphere above and the solid earth below (section 2.3.2). Convection and other dynamic processes in the atmosphere deliver snow to high elevations on glacier surfaces. By virtue of its elevation, glacier ice possesses gravitational potential energy, which is transformed into thermal energy during ice flow. This thermal energy can raise the temperature of ice or be used for melting, and is ultimately dissipated to the surroundings by conduction to the surface (if the air temperature is less than the deep ice temperature) or runoff of meltwater.

Rates of glacier motion are intimately linked to energy fluxes, especially near the bed. As explained in section 2.3.2, changes in temperature or phase (melting and freezing) near the bed of a glacier depend on the balance between the production and dissipation of energy. The rate of temperature change at the bed of a glacier can be written in terms of the energy balance:

$$\frac{\partial T}{\partial t} = \frac{\tau\dot{\varepsilon} + G - k_i\Theta}{c_i\rho_i} \qquad (5.4)$$

and similarly, for ice at the pressure-melting point, the energy balance determines rates of melting or freezing:

$$\dot{M} = \frac{\tau\dot{\varepsilon} + G - k_i\Theta}{L_f\rho_i} \qquad (5.5)$$

The numerator in both of these expressions contains three terms: two heat sources and a heat sink. The first term, the product of shear stress and strain rate (or, in the case of a temperate glacier bed, basal sliding speed U_b) is strain heating; the second term G is the geothermal heat flux; and the third term, the product of thermal conductivity and temperature gradient in the ice, is heat conducted towards the surface. (For simplicity, advection of colder or warmer ice is omitted here.) The denominators represent the energy requirements of warming (equation 5.4) or melting (equation 5.5), and convert the energy budget to changes of temperature or state (see also section 2.3.2.).

The relationships between glacier flow and sources or sinks of heat give rise to some interesting feedbacks which have important consequences for glacier dynamics. First, frictional heating resulting from ice flow can warm or melt basal ice, both of which will encourage more rapid motion. In turn, this can generate more frictional heat, further increasing ice flow speeds. Thus, an initial perturbation can become amplified by a positive feedback loop, leading to unstable glacier acceleration. This process will not continue without limit, however, because of the role played by basal traction in frictional heating. Equations 5.4 and 5.5 show that frictional heating at the bed of a glacier is directly proportional to the product of the shear stress and strain rate (or basal motion). Thus, even if basal velocities are very large, frictional heating will not occur if there is no traction between the glacier and its bed. This means that frictional heating shuts down when ice decouples from the bed, preventing runaway basal melting.

The second feedback involves interactions between flow speeds and ice thickness. Acceleration of the lower reaches of a glacier can lead to longitudinal extension and dynamic thinning of the ice (equation 5.3), reducing the distance between the bed and the surface. In a cold-climate environment, this will steepen the temperature gradient and increase heat diffusion away from the bed, thus tending to decrease basal ice temperature. This relationship between strain rate and ice thickness is therefore a negative feedback, which has a stabilizing influence on glacier flow. In contrast, acceleration of the upper reaches of a glacier relative to the lower part will lead to dynamic thickening of the intermediate zone, increasing the bed–surface distance and inhibiting heat diffusion. This process can promote faster flow, encouraging the velocity increase to propagate downglacier. A third feedback involves the relationship between ice flow and heat advection. Increased ice flow velocities can lead to more efficient advection of colder ice from upglacier, which tend to reduce velocities and damp the initial velocity signal.

The strong coupling between ice dynamics and heat fluxes underpins a wide range of glacier behaviour, including surge cycles and ice stream acceleration and shutdown, which are discussed in detail in sections 5.7 and 6.3.6, respectively.

5.3 GLACIER MODELS

5.3.1 OVERVIEW

Because the dynamic behaviour of glaciers is governed by complex webs of processes, it is difficult to predict how a glacier or ice sheet will respond to a particular change in climate or other environmental variable. Similarly, it can be far from simple to extract precise environmental information from physical evidence of past glacier behaviour, such as a sequence of moraines. Since the 1970s, glaciologists increasingly have used computer simulations to predict the future behaviour of glaciers, reconstruct their past, and explore the fundamental processes governing their dynamics. The power and sophistication of such simulations has increased greatly, and computer modelling is now well established as one of the most important tools of modern glaciology. Despite this, the potential and limitations of glacier models are not widely appreciated beyond the confines of the modelling community. In this section, we discuss the main types of models used in glaciology and the ways that they are used to represent aspects of reality. Our aim is not to present a comprehensive review, but to enable readers to better understand what models can and cannot do, and to help them interpret model results found in the technical literature. Some excellent overviews of glacier models are available (e.g. van der Veen, 1999a; Marshall et al., 2004; van der Veen and Payne, 2004), and the reader is referred to these for more detailed information.

In essence, glacier models represent physical quantities and/or processes by sets of equations, which are then solved to determine how the system will behave under specified conditions. For example, the set of equations may describe the surface energy balance (equation 2.1), snow accumulation (equation 2.12), basal shear stress (equation 4.10) and relationships between stress and ice flow rates (equation 4.14). The governing equations of a model can be solved using either analytical or numerical methods. *Analytical methods* are used to solve equations that describe smooth, continuous variations in quantities. An example is given in section 5.3.2, in which the equilibrium surface profile of an ice cap is specified in a single, compact equation. However, analytical methods are impractical for the majority of glaciological problems, either because the governing equations are too complex, or the required inputs do not vary in predictable ways. *Numerical methods* overcome these difficulties by calculating quantities only at distinct points in space or time, then proceeding towards the desired

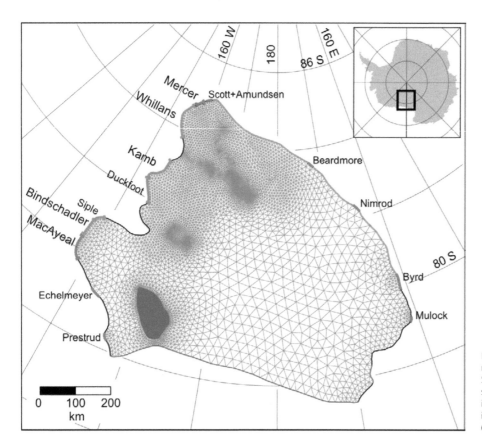

Figure 5.6 Finite-element mesh used to model flow variations across the Ross Ice Shelf. Note the variable mesh resolution, so areas of complex flow can be resolved in greater detail. The model 'feeds' ice into the shelf through the coloured gates (Image courtesy of Christina Hulbe).

solution in a series of steps. In the case of the ice surface profile in section 5.3.2, this involves calculating the ice surface elevation at fixed intervals along the flow line, using each calculation as the springboard for the next. In glaciology, numerical methods are mainly used to calculate the evolution of a glacier through a series of steps in time (section 5.3.4).

Ice sheet models typically have a horizontal grid spacing of many kilometres, so they can only resolve processes that operate on scales larger than this. If small-scale processes are of interest, a smaller grid spacing is adopted, although due to demands on computer time this will tend to limit the overall size of the model domain. In some models, the grid spacing can be varied across the model domain, to allow greater resolution in areas of interest (such as ice streams) and more rapid running of the model in more homogeneous regions (such as the interior of an ice sheet) (fig. 5.6). The stepwise character of numerical methods means that the accuracy of each calculation depends on that of the last, and there is a possibility that errors will accumulate through time. Consequently, modellers devote a great deal of attention to ensuring that their numerical methods and computer codes do not generate spurious results or *artefacts*. Model testing and validation is discussed in section 5.3.4.

Glacier models can be subdivided into two main types, according to their intended purpose. First, *diagnostic models* focus on particular processes, and are used to explore the workings of certain aspects of glacier systems. Second, *prognostic models* are used to simulate the evolution of glaciers through time, and to study how they might respond to specified environmental inputs. One characteristic shared by all models is that they must adopt simplifying assumptions to make problems soluble. In reality, the flow of a glacier involves the deformation of trillions of ice crystals, each subject to a shifting array of forces. Resistance to flow varies with every bump and wrinkle on the glacier bed, and alters moment by moment as water meanders and pulses through the system in ever-changing patterns. It is impossible to fully describe every detail of such a system, or keep track of every grain of ice as it moves and deforms. The best that can be achieved is an approximate description of the glacier and the aggregate behaviour of its constituent parts. The key issue is how well these approximations represent the system, and how much of the essential system behaviour is retained and how much is lost. Glaciologists endeavour to incorporate all relevant factors into their models, although in some cases decisions are constrained by what is actually achievable. For example, it is agreed that water flow at the glacier bed exerts a fundamental control on basal motion, but to date no method has been found to accurately represent the role of basal water in time-evolving models of glacier flow. Consequently, modellers generally choose to adopt simple parameterizations of basal sliding, with a consequent loss of model fidelity. It is very important to bear this in mind when assessing model predictions of glacier response to climate change, because key dynamic processes may be omitted or poorly represented (van der Veen, 2002b). Models do not faithfully replicate reality, but with care and skill they can be used to gain valuable insights into how glaciers function and how they might evolve through time.

5.3.2 EQUILIBRIUM GLACIER PROFILES

Glacial geologists commonly wish to define the surface form of former ice masses as input for palaeoclimatic reconstructions (e.g. Rea and Evans, 2007). A good approximation of the equilibrium profile of a glacier or ice cap can be obtained from a simple steady-state model, based on the assumption that the ice has perfectly plastic rheology (section 4.2.3). In this case, ice deformation will only occur if the basal shear stress is equal to the yield stress. Where the basal shear stress is below this value, the ice will not move, but will thicken or steepen until the yield stress is reached. Thereafter, the ice will flow at whatever rate is required to prevent the basal shear stress from exceeding the yield stress. Thus, the ice surface profile will constantly adjust so that the basal shear stress is everywhere equal to the yield stress. In the simplest form of the model, it is further assumed that all of the driving stress is opposed by basal drag. The basal shear stress is therefore defined as:

$$\tau_b = \tau_{yield} = \rho_i\, g\, H\, \frac{\partial h}{\partial x} \qquad (5.6)$$

(Note that the ice surface gradient $\partial h/\partial x$ is equal to $\tan \alpha$ in equation 4.10.)

To find the equilibrium surface profile, we aim to find the values of ice thickness H all along a flow line (along the x direction, positive upglacier) which satisfy the model assumptions. If the bed is horizontal this can be achieved using an analytical approach, integrating equation 5.6 to yield:

$$H = \left[\frac{2\tau_b}{\rho_i g} x\right]^{\frac{1}{2}} \qquad (5.7)$$

By inserting an appropriate value for the basal shear stress (yield stress) into this expression, we obtain a unique value of H for every value of x. Equation 5.8 thus fully describes the form of the whole ice surface, which in this case is a parabola (fig. 5.7a). For land-terminating glaciers and ice caps, appropriate values of τ_b might lie in the range 50–100 kPa.

This kind of analytical approach cannot be taken if the bed elevation or yield stress varies along the flow line. Instead, numerical methods must be used, which define

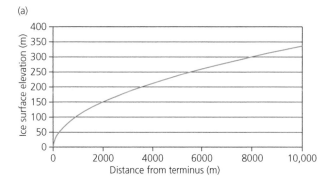

(a)

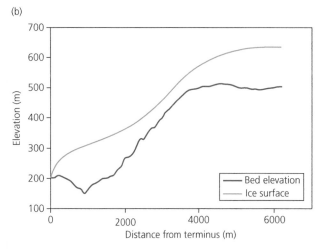

(b)

Figure 5.7 Steady-state glacier profiles, based on perfectly plastic rheology. (a) Parabolic surface profile of a glacier resting on a horizontal bed, derived using equation 5.7. (b) Surface profile of a glacier with an irregular bed and varying basal shear stress, derived using equation 5.8.

the ice surface elevation step by step, beginning at the ice front and proceeding upglacier. Rather than defining a continuous curve, this approach calculates the elevation of discrete points on the ice surface, in a series of steps numbered 1, 2, 3, ..., i, $i+1$, $i+2$, ... etc. As with any numerical approach, it is important to use the correct function, otherwise there will be a tendency for calculated points to either overshoot or undershoot the desired value. Small errors might then accumulate with each iteration, introducing unwelcome artefacts into the reconstructed ice surface. This is the case with most published methods for defining ice surface profiles (e.g. Schilling and Hollin, 1981; Rea and Evans, 2007).

An exact method for calculating equilibrium profiles was derived by van der Veen (1999a, pp. 148–50). His function for defining the ice surface elevation at point $i+1$ is:

$$h_{i+1}^2 - h_{i+1}(Z_{Bi} + Z_{Bi+1})$$
$$+ h_i(Z_{Bi+1} - H_i) - \frac{2\Delta x \bar{\tau}_{yield}}{\rho_I g} = 0 \qquad (5.8)$$

This is a quadratic equation, which can be solved using standard methods, allowing the ice surface profile to be determined step by step if the bed elevation Z_B is known. Benn and Hulton (2009) developed a user-friendly spreadsheet program for calculating former glacier profiles using this method (fig. 5.7b). Balance velocities can be estimated for such reconstructed glaciers using equations 5.1 and 5.2 and assumed mass balance gradients as inputs (e.g. Carr and Coleman, 2007; Rea and Evans, 2007).

5.3.3 TIME-EVOLVING GLACIER MODELS

To explore issues in glacier dynamics, glaciologists use prognostic models of greater or lesser complexity, which simulate the evolution of an ice mass through a series of time steps. Since the landmark work of Huybrechts (1990), time-evolving glacier models have been developed by several research groups (e.g. Marshall and Clarke, 1997; Payne, 1999; Pattyn, 2003; Parizek et al., 2005). Although varying in detail, all of these models link some form of the continuity equation (section 5.2.2) with functions for calculating mass balance and ice fluxes, to determine how ice thickness and other properties change in response to specified climatic inputs. The key components of time-evolving glacier models are as follows.

Mass balance

Snow accumulation is usually parameterized using functions describing how precipitation and air temperature vary with altitude or distance from the ice sheet margin. In practice, however, realistic treatments of snow accumulation are very difficult to achieve, due to large uncertainties concerning past, present or future precipitation, and the role of wind in redistributing snow (Marshall et al., 2004). Ablation rates are simpler to model, and can be calculated either from the energy balance equation (section 2.2.1) or some type of melt-index function (section 2.4.3) (Greuell and Genthon, 2004). The latter approach is often preferred in prognostic models because less information is required for model inputs. Although full energy balance approaches represent the relationship between the atmosphere and ice melt in a much more realistic way than melt-index methods, there are inevitably great uncertainties regarding the state of the atmosphere in the past and the future. This means that uncertainties over the model inputs can easily outweigh the advantages of greater realism and precision. Therefore, most models parameterize melt rates using degree-day functions. Rates of basal melting or freezing are very important for the mass balance of floating ice shelves, but are difficult to model due to the complexity of the interactions between ice shelf and ocean. Examples of contrasting approaches to this major problem are presented by Grosfield and Sandhäger (2004) and Walker and Holland (2007), and are discussed further in section 5.6.1. Calving is often very poorly represented in glacier models, despite its importance in the mass balance of the cryosphere. Until

very recently, calving losses have been represented by empirical relationships between calving rate and depth, or other simple parameterizations (van der Veen, 2002; Zweck and Huybrechts, 2003; Benn et al., 2007b). Recent progress in this important area is reviewed in section 5.5.3.

Thermodynamics

The simplest ice flow models assume that the ice is everywhere at the pressure-melting point (isothermal models). More realistically, *thermo-mechanically coupled models* evolve the ice temperature through time, depending on sources and sinks of energy (e.g. van der Veen, 1999a; sections 2.3.2 and 5.2.4). Because ice temperature influences rates of ice flow and vice versa, such models can exhibit ice velocity–temperature feedbacks and mimic certain kinds of dynamic behaviour, such as oscillatory flow regimes and ice stream initiation (e.g. MacAyeal, 1993; Hulton and Mineter, 2000; Boulton and Hagdorn, 2006; fig. 5.8).

Stresses

A complete description of the stresses acting on an ice mass must include nine stress components (three normal and six shear stresses) (section 4.2.1). Determining the full stress field and its evolution through time involves solving the *Stokes* (or *Navier–Stokes*) *equations* developed by George Gabriel Stokes (1819–1903), the Irish mathematician and physicist. These differential equations are the cornerstone of fluid dynamics, and have very many applications in science and engineering. Finding ways of solving the full Stokes equations for glacier systems, and performing the calculations repeatedly and efficiently for each time step of a model run, have been among the biggest challenges that glacier modellers have had to meet. Indeed, the necessary techniques and computing power have become available only in the last few years (e.g. Hindmarsh, 2004; Pattyn et al., 2008), and the majority of glacier models solve simplified (or reduced) forms of the Stokes equations to approximate the glacier stress field.

Until very recently, most glacier models adopted the simplifying assumption that all resistance to flow is at the basal boundary of the ice (van der Veen and Payne, 2004). In this case, the only stresses considered are shear stresses acting on horizontal planes, τ_{xz} and τ_{yz}, greatly simplifying the Stokes equations. This is known as the *shallow ice approximation* or SIA (Hutter, 1983), and yields reasonable results for non-streaming regions of ice sheets. It has also been shown to perform well at simulating the fluctuations of valley glaciers on timescales of several years or more, and with considerably less demand for computer resources than full-stress models (Leysinger Vieli and Gudmundsson, 2004). However, the accuracy

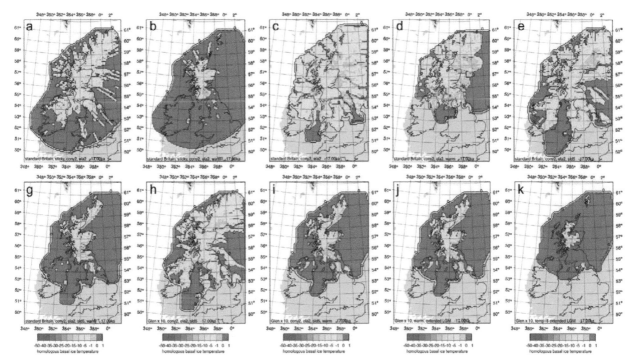

Figure 5.8 Modelled basal ice temperatures for the last British-Irish Ice Sheet, using a variety of input parameters (Boulton and Hagdorn, 2006).

of the SIA decreases as the contribution of basal slip to glacier motion increases (Gudmundsson, 2003), and it is inappropriate for many important situations, such as ice divides, ice streams and the transition zones between grounded and floating ice. Glacier models that incorporate additional stress terms, especially longitudinal stresses, are termed *higher-order models* (Hindmarsh, 2004; Pattyn et al., 2008). These adopt a variety of strategies for solving the stress field and its evolution through time (e.g. Blatter, 1995; Hubbard, 2000; Pattyn, 2003). The modelling community is currently investing considerable effort in the development of these models, and applying them to pressing glaciological problems.

Ice flow

Ice fluxes resulting from internal deformation can be calculated from Glen's Flow Law (section 4.3.1) in combination with the Stokes equations for conservation of mass and momentum. Thermo-mechanically coupled models allow the rate factor A to vary with ice temperature, although other possible controls on ice rheology (e.g. crystal fabric and water content) are usually ignored. Basal motion is still poorly represented in ice sheet models, which generally either adopt pressure-dependent 'sliding laws' or other simple parameterizations (section 4.6.1; fig. 5.9). These can be tuned to yield good approximations of observed surface velocities, but general, simple 'sliding laws' remain elusive (Iken and Truffer, 1997). As discussed in section 4.6.3, rates of basal motion depend not only on the local driving stress and basal drag, but also on sources of resistance elsewhere on the glacier

bed and/or margins, and the ways that these non-local stresses are accommodated by strain of the ice. Realistic treatment of basal motion in glacier models therefore requires knowledge of the spatial variation in basal and lateral drag, as well as consideration of the full stresses in the ice. Significant progress has been made in this direction (e.g. Pattyn, 2003; Nienow et al., 2005), although more work is required before such basal sliding schemes can be routinely incorporated in time-evolving glacier models. Compounding the already formidable difficulties in developing realistic models of basal sliding is the problem of modelling basal drainage and its influence on drag at the glacier bed. Integrated, glacier-scale hydrological models have been developed in recent years (e.g. Flowers and Clarke, 2002a, b; Flowers et al., 2003), but modelling the influence of basal hydrology on ice flow is still at an early stage (e.g. Parizek and Alley, 2004a).

Isostatic adjustment

When simulating the evolution of ice sheets over millennia, it is important to consider the effect of ice sheet loading and unloading on the Earth (see section 7.2.3.). Depression of the Earth's crust below an ice sheet has two major effects on the ice mass. First, the elevation of the ice sheet surface is reduced, with consequent impacts on surface mass balance. Second, isostatic depression can increase the area of an ice sheet grounded below sea level, making it more susceptible to collapse at times of rapid sea-level rise or increasing heat flux from the oceans. Isostatic loading and unloading can be simulated by coupling models of an ice sheet and the underlying lithosphere and

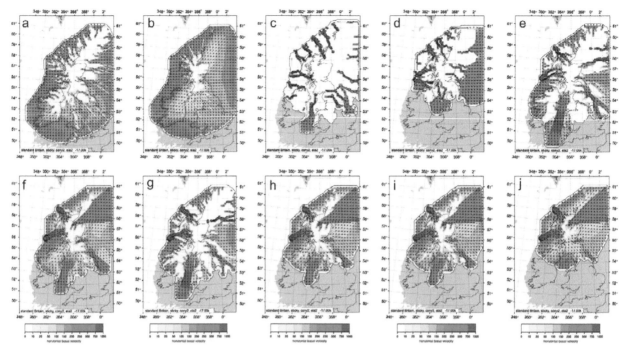

Figure 5.9 Modelled ice velocities for the last British-Irish Ice Sheet, using the same set of input parameters illustrated in figure 5.8 (Boulton and Hagdorn, 2006).

asthenosphere. The choice of Earth model is important, because lithospheric rigidity can have a large influence on ice sheet evolution. Different methods of modelling isostatic adjustment have been compared by Le Meur and Huybrechts (1996). The most realistic models are computationally demanding spherical self-gravitating viscoelastic (SGVE) or 'full Earth' models, although Le Meur and Huybrechts found that simpler elastic lithosphere/relaxed asthenosphere (ELRA) models can also give acceptable results (see also Parizek and Alley, 2004b). Recent experiments, however, suggest that ELRA models cannot successfully model ice thickness and volume changes and lithospheric response over a full glacial cycle (van den Berg et al., 2008).

Bed topography

All glacier and ice sheet models require the initial topography of the bed to be specified. For reconstructions of Quaternary ice sheets, digital elevation models of the present-day topography can be used, but for models of extant glaciers and ice sheets, the elevation of the bed must be established by seismic profiling or (more usually) radio echo sounding (e.g. Welch et al., 1998). Data coverage is increasing rapidly (e.g. Vaughan et al., 2006), but the lack of topographic data remains one of the major impediments to modelling the behaviour of many present-day glaciers. The topography of the glacier bed will change over a glacial cycle in response to subglacial erosion or sedimentation, and some interesting studies have explored how these processes create characteristic glacial stratigraphy and landforms (e.g. Boulton 1996a, b; Alley

et al., 2003b) or influence glacier dynamics (Nick et al., 2007a).

Coupling ice sheet and climate models

Small ice masses have relatively little influence on climate, so models of valley glaciers generally assume that the glacier passively responds to climate forcing. However, large ice caps and ice sheets considerably modify their climatic environment by altering the surface topography, albedo and other factors. During a glacial cycle, therefore, ice sheets and climate evolve together, each influencing the other. Modelling this interactive system demands considerable computer resources, because it involves coupling General Circulation Models (GCMs) of the atmosphere with ice sheet models. A variety of approaches have been taken to this daunting problem (Marshall et al., 2004). One of the most widely used is the *snapshot method*, in which a GCM is run to simulate climate for a particular set of boundary conditions (orbital parameters, sea-surface temperature, ice sheet configuration, etc.), which then provides the mass balance forcing for several 'years' of ice sheet simulation, after which an updated GCM simulation is used and the cycle repeated (e.g. Marshall and Clarke, 1999). Intermediate climate states between 'snapshots' can be simulated in various ways, to obtain a continuously varying mass balance forcing (e.g. Huybrechts et al., 2004). Alternatively, *matrix methods* use a library of generic GCM simulations for various combinations of boundary conditions, and the most appropriate simulation is chosen as the modelled ice sheet evolves. Other studies have adopted a combination of snapshot and matrix methods, in attempts to find the best compromise between realism and practicability (DeConto and Pollard, 2003; Pollard and DeConto, 2005). The best solution involves direct coupling between ice sheet models and GCMs, so that the ice sheet and its climatic environment co-evolve in 'real time' (e.g. Ridley et al., 2005) although this makes considerable demands on computer resources.

In assessing the validity of glacier models, it is important to distinguish between those aspects of the output that reflect the model physics and those which are artefacts of the numerical methods employed to implement the model. To this end, a series of benchmark studies have been conducted, in which different models are applied to a set of common tasks. The EISMINT project (European Ice Sheet Modelling INiTiative) was initiated in the 1990s to test the relative performance of ice sheet models (Huybrechts et al., 1996; Payne et al., 2000), and was extended to models of ice shelves (MacAyeal et al., 1996). Recently, a similar initiative was launched for higher-order models (Pattyn et al., 2008).

The highly technical nature of the subject means that those outside the small modelling community rarely have the opportunity to use time-dependent ice sheet models as research and teaching tools. Recently, some efforts have been made to remedy this situation. Pattyn (2006) has developed a thermo-mechanical flow-line model using the widely available Microsoft Excel® spreadsheet program. This model, named GRANTISM (GReenland and ANTarctic Ice Sheet Model), allows users to see the effect of air temperature changes on ice sheet dimensions and exhibits many of the features of more complex models, such as asymmetric growth and decay cycles. It is therefore a very effective and user-friendly student teaching tool. Source code for some three-dimensional flow models is now freely available. GLIMMER is a thermo-mechanical flow model using the shallow ice approximation to represent the stress field, and has been widely used in ice sheet reconstructions (Payne and Dongelmans, 1997; Boulton and Hagdorn, 2006; Rutt et al., 2009). Code and documentation for GLIMMER can be downloaded from: http://forge.nesc.ac.uk/projects/glimmer/. SICOPOLIS (SImulation COde for POLythermal Ice Sheets) is based on a similar set of governing equations (Greve, 1997), and can be accessed at: http://sicopolis.greveweb.net/.

5.4 DYNAMICS OF VALLEY GLACIERS

The scientific study of valley glacier dynamics has a long history, dating back to the work of pioneer glaciologists Louis Agassiz (1807–73) and James Forbes (1809–68). Because of their relatively small size and ease of access, valley glaciers in Europe and North America have long

been forcing grounds for the study of glacier dynamics, allowing many important principles to be established. Even so, until recently, data on glacier motion could only be obtained by very labour-intensive survey methods or expensive air-photo time series, so knowledge about glacier dynamics consisted of hard-won 'snapshots' of limited spatial and temporal extent. With the advent of global positioning systems and remote-sensing techniques, however, data are now much more easily and cheaply obtainable, even for very remote glaciers (e.g. Reeh et al., 1999; Kääb, 2005; Vieli et al., 2006). In combination with hot-water drilling and geophysical techniques, this new technology has revolutionized our understanding of the dynamics of valley glaciers. At the same time, glaciologists have also been presented with many more puzzling questions, and valley glaciers have lost none of their capacity to surprise and intrigue those who study them.

5.4.1 INTRA-ANNUAL VELOCITY VARIATIONS

On timescales of less than one year, most fluctuations in the velocity of valley glaciers reflect variations in basal drag modulated by changes in subglacial hydrology. Cold-based glaciers, which are everywhere frozen to their beds, flow at constant low speeds throughout the year, whereas both temperate and warm-based polythermal glaciers commonly exhibit distinct intra-annual variations in basal motion following changes in the volume and areal extent of subglacial water (Willis, 1995). Basal melt rates do not vary significantly on short timescales on most glaciers, so fluctuations in the amount of water stored at the bed are mainly due to variations in inputs of surface meltwater and reorganizations of the basal drainage system. Variations in water delivered from the surface follow daily or seasonal ablation cycles, or result from transient weather events such as prolonged warm weather or periods of rainfall. Müller and Iken (1973) conducted pioneering work on White Glacier, Arctic Canada, which demonstrated close correlations between surface ablation rates and ice velocities (fig. 5.10). Velocities increase during the short summer period, with short-lived peaks occurring at times of high melting, indicating that surface meltwater is rapidly routed to the bed via moulins (section 3.3.2).

Since the 1970s, transient velocity increases in response to changes in surface meltwater supply have been observed on many glaciers (e.g. Iken and Bindschadler, 1986; Naruse et al., 1992; Willis, 1995; Iken and Truffer, 1997; Rabus and Echelmeyer, 1997; Rippin et al., 2005; Jobard and Dzikowski, 2006). Increasing water supply from the surface can add to the amount of water stored at the bed in distributed basal drainage systems (section 3.4), increasing the area of ice–bed decoupling and altering the magnitude and spatial patterning of basal drag (Schweizer and Iken, 1992; Kavanaugh and Clarke, 2001;

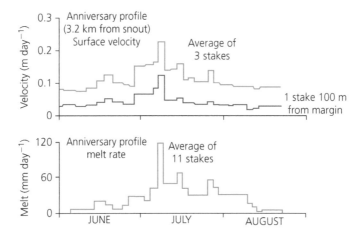

Figure 5.10 Velocity fluctuations of White Glacier, Axel Heiberg Island. The close association between velocity and melt rate shows that surface meltwater reaching the bed exerts a strong control on sliding speed (Willis, 1995, after Müller and Iken, 1973). Ground photograph of the snout of White Glacier (M.J. Hambrey).

sections 4.4.4 and 4.5.3). At times of falling water supply from the surface, water may be able to drain from the system faster than it is recharged, so that the volume of stored water falls and sliding velocities are reduced. Evidence that stored water plays a key role in episodes of accelerated sliding is provided by short-lived periods of uplift of glacier surfaces accompanying velocity increases (Iken et al., 1983; Iken and Bindschadler, 1986; Willis, 1995; Anderson et al., 2004; Sugiyama and Gudmundsson, 2004; MacGregor et al., 2005). Transient uplift events can occur by two main mechanisms:

(1) pressurized water in lee-side cavities and other voids can lift the glacier off the bed,

(2) vertical thickening of the ice can occur in response to longitudinal shortening, as the glacier adjusts to changing resisting forces at the bed (Sugiyama and Gudmundsson, 2004).

Surface velocity data can be used to calculate the strain thickening effect, and to determine the amount of uplift due to water storage (fig. 5.11; Anderson et al., 2004; MacGregor et al., 2005).

Although correlations exist between ice velocity and basal water pressure (section 4.6.1), it is clear that water pressure is not the only controlling variable and that a key role is played by the configuration of the basal drainage system. When an efficient channelized subglacial drainage system exists, surface-derived water is confined to a small percentage of the bed and transit times are small. Most of the bed is unaffected by short-term discharge fluctuations, so that glacier sliding rates can remain low even when there are high water pressures in the channels. Therefore, 'switches' from distributed to channelized networks during the course of a melt season are commonly mirrored in changes in sliding velocity (Willis, 1995; Anderson et al., 2004).

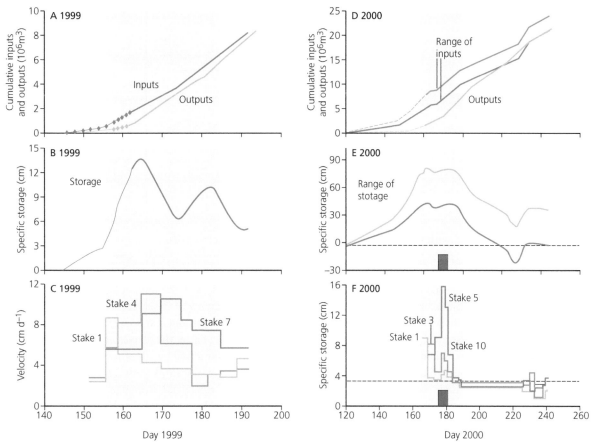

Figure 5.11 Evidence for the role of water storage on basal velocities, Bench Glacier, Alaska, during the ablation seasons of 1999 and 2000. (a, d) Cumulative water inputs and outputs. (b, e) Calculated subglacial water storage. (c, f) Measured ice surface velocities (MacGregor et al., 2005).

The influence of drainage configuration on glacier velocity is clearly highlighted by *spring events*, or high inputs of water from melting or rainstorms early in the ablation season, when basal drainage systems are poorly developed. For example, MacGregor et al. (2005) documented short-lived speed-ups on Bench Glacier, Alaska, in June and early July of 1999 and 2000, with peak velocities two to four times mean annual rates (fig. 5.11). The speed-ups were initiated when the glacier was still snow-covered and large amounts of early-season meltwater were still retained within the glacier system. The velocity pulses first appeared close to the terminus, and propagated upglacier at 200–250 m day^{-1}, with velocities returning to pre speed-up values within 5–25 days, depending on location. The speed-ups were accompanied by uplift of the glacier surface, and were clearly associated with an increase in water storage over large areas of the glacier bed. At the end of the winter, the subglacial drainage system would have had very low capacity due to creep closure, so the renewed influx of surface-derived meltwater in the spring could not be evacuated quickly enough to balance recharge rates. In consequence, water was forced along large areas of the ice–bed interface, reducing drag at the bed. Enlargement of the basal

drainage system by roof melting led to an increase in efficiency, evacuating water from the bed, increasing drag and terminating the high-velocity event.

A very comprehensive study of the linkages between intra-annual velocity variations and changing hydrological conditions was conducted by Copland et al. (2003a) on John Evans Glacier, a predominantly cold polythermal glacier on Ellesmere Island. The thermal structure of this glacier is Type D in the Blatter and Hutter (1991) classification (section 2.2.3), consisting of a thick mantle of cold ice overlying a limited area of temperate ice. Measured surface velocities average 3.5 cm day^{-1} in the winter months, rising to an average of 5.3 cm day^{-1} during summer. This seasonal increase, however, mainly reflects concentrated velocity events lasting 2–4 days, with speeds of up to 6.7 cm day^{-1}. In 1998 and 1999, velocity events occurred around a week after the onset of intense melting at the surface, and immediately following establishment of a drainage connection between the surface and the bed. Boon and Sharp (2003) showed that this connection was made by hydrologically driven fracturing below the surface of an ephemeral supraglacial lake (section 3.3.2). In the early part of the melt season, meltwater was trapped at the bed behind the frozen terminus, but

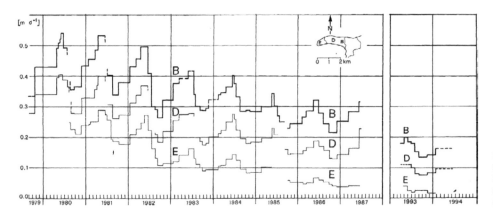

Figure 5.12 Velocity of three poles along the centreline of Findelengletscher during advance (1979–1986) and retreat (after 1987) (Iken and Truffer, 1997).

breakthrough was achieved within 24 hours of surface waters reaching the bed, presumably because the additional water pressurized the subglacial reservoir, increasing the stress on the thermal dam. Following establishment of an outflow at the terminus, an efficient subglacial drainage pathway was established and the velocity events were terminated. In 1999, a second velocity event occurred later in the ablation season, when melting resumed following a cold period. It appears likely that the established subglacial conduit system had contracted by ice creep during the cold period when recharge from the surface was low (section 3.2.4), so that renewed meltwater input to the bed once more exceeded the capacity of the system.

The complexity of the relationships between basal drainage and velocity fluctuations is well illustrated by the behaviour of Findelengletscher, Switzerland, during a recent advance–retreat cycle (Iken and Truffer, 1997). Figure 5.12 shows velocity records for three markers on the glacier surface. Annual cycles are clearly apparent, with maxima in the early part of the melt season, but both summer and winter velocities show a steady decline throughout the observation period. During the early 1980s, when the glacier was advancing most rapidly, velocities are much higher for equivalent driving stresses and basal water pressures than at the beginning of the 1990s, after the advance had terminated (fig. 5.12). Iken and Truffer (1997) attributed this behaviour to the degree of connectedness of the basal drainage system, with periods of more rapid motion relating to more extensive interconnected cavities (early melt season) and a greater density of conduits (late summer). The reasons for this change are not clear, although these results clearly show that measurements of 'water pressure' fail to capture the complexity of the influence of water on basal motion.

Recently, a number of detailed studies have demonstrated spatially heterogeneous velocity fields during speed-up events, reflecting the changing distribution of areas of low basal drag (slippery spots) and areas of high drag (sticky spots). For example, Kavanaugh and Clarke (2001) documented reorganizations of the flow field of Trapridge Glacier, Yukon, during spring speed-up events.

Episodes of accelerated basal motion were initiated by the gradual erosion of a sticky spot when the area became hydraulically connected to the basal drainage system. The loss of basal drag in the uncoupled region was accommodated by a transfer of basal stress to unconnected areas of the bed, via longitudinal and lateral stress gradients in the overlying ice. Hence, the glacier adjusted to spatial variations in basal traction by adopting a new stable stress configuration. Relationships between surface velocities and an evolving subglacial drainage pathway below Haut Glacier d'Arolla, Switzerland, have been examined by Mair et al. (2003). A network of velocity markers was established on the glacier surface in the vicinity of the *variable pressure axis* (VPA) described in section 3.4.1. The VPA is a narrow longitudinal zone of the bed where basal water pressures have low diurnal minima and high diurnal variability, and marks the position of a subglacial conduit which develops seasonally in response to surface meltwater inputs via moulins. Surface velocities during two spring speed-up events in 1998 and one in 1999 exhibited varying spatial patterns, apparently reflecting contrasting bed responses to increasing water influxes (fig. 5.13). In the first 1998 event and the 1999 event, high water pressures in a broad zone around the VPA were associated with high surface velocities in that area, apparently reflecting widespread decoupling of the ice–bed interface. In contrast, in the second 1998 event, high water pressures were confined to a narrower zone centred on the VPA. Surface velocities increased throughout the study area, with the highest values occurring on the glacier centreline. Instruments inserted into the bed show that much of the velocity increase was accomplished by subglacial sediment deformation, indicative of glacier–bed coupling. These fascinating observations clearly show how the non-linear response of subglacial sediments to water-pressure fluctuations (section 4.5.3) are translated into complex large-scale flow patterns in space and time.

Transient speed-ups can also occur in response to subglacial routing of glacier lake outbursts, as has been elegantly demonstrated by Magnusson et al. (2007), who

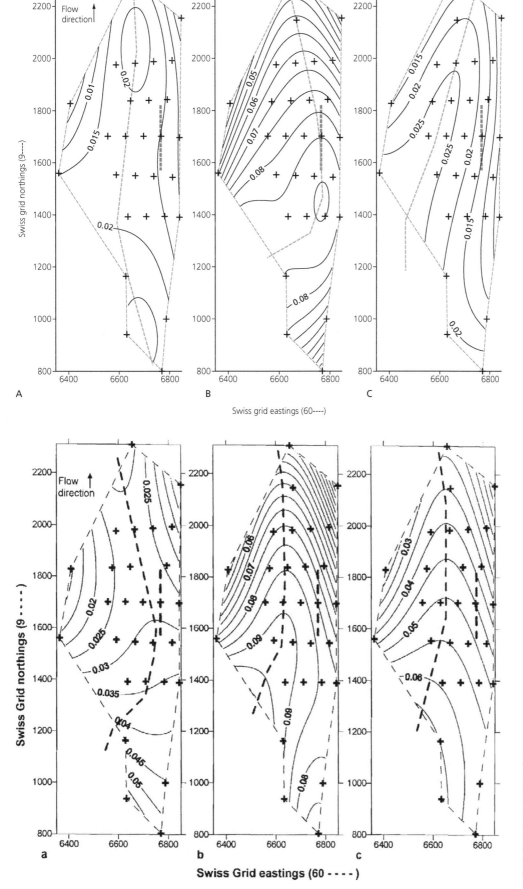

Figure 5.13 Plots of ice velocities (m day⁻¹) in the region of the 'variable pressure axis' on Haut Glacier d'Arolla, before, during and after the spring event. Top: 1998 event 1, Bottom: 1998 event 2, Overleaf 1999. See also figures 3.17 and 3.18 (Mair et al., 2003).

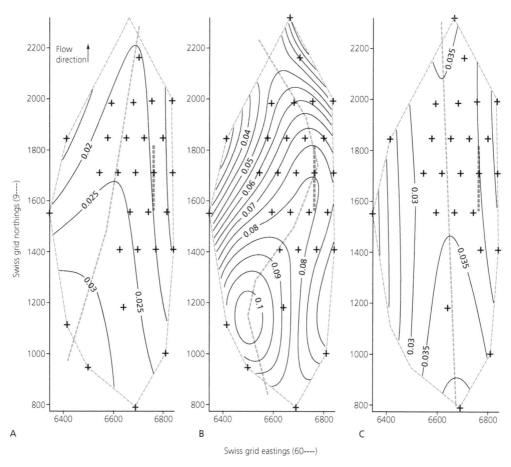

Figure 5.13 (Continued)

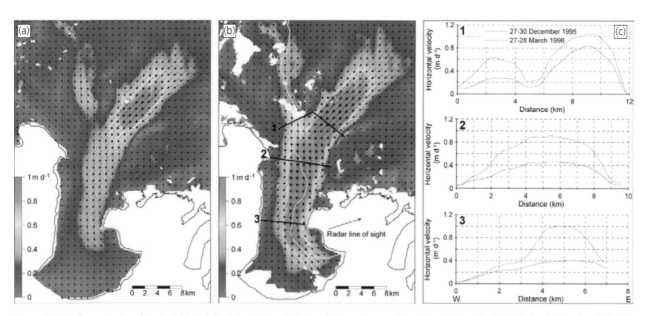

Figure 5.14 Surface velocities of Skeiðarárjökull, Iceland, (a) at the end of December 1995 and (b) the end of March 1996, during the beginning of a subglacial outburst flood. The light blue curve shows the calculated flood track. (c) Three cross profiles of December and March glacier velocities, for locations shown in (b) (Magnusson et al., 2007).

studied impacts of jökulhlaups on the surface velocity fields of Skeiðarárjökull and Tungnaárjökull, two adjacent outlets of Vatnajökull, Iceland. These glaciers are periodically affected by outburst floods from Grimsvötn and the Skaftá cauldrons, respectively, when stored geothermally melted water escapes along their beds (cf. section 3.7). During a jökulhlaup in March 1996, the surface velocity of Skeiðarárjökull increased up to threefold in an area up to 8 km wide along the flood path (fig. 5.14), indicating widespread decoupling at the bed. An earlier outburst flood during October 1995 triggered a fourfold increase in the velocity of Tungnaárjökull over an area 9 km wide. On this glacier, the maximum in ice velocity occurred before the flood waters reached the terminus, and following breakthrough the velocity fell to pre-flood values. These observations indicate reorganization of the basal drainage system from sheet flow during the early stages of the flood to channelized flow following breakthrough. Both floods happened outside of the main melt season, and peak discharges were less than or equal to normal summer peak flows, dramatically demonstrating that absolute magnitudes of water discharge are less important than the discharge relative to the capacity of the subglacial drainage system. Because the floods occurred at times when the basal drainage systems were poorly developed, system capacity was rapidly exceeded during the rising limb of the flood hydrograph, increasing water storage over broad swathes of the glacier beds. Once the subglacial drainage systems had adjusted to high discharges by forming efficient channels, water storage was reduced and basal drag increased once more.

Warm-based glaciers do not necessarily exhibit significant seasonal velocity variations. For example, the sliding velocity of Grinell Glacier, Montana, USA, remains remarkably constant, at rates of 33 mm day^{-1} during the summer months (Anderson et al., 1982). The lack of significant variations in the basal motion suggests that surface meltwater is routed efficiently through the glacier, and much of the bed is unaffected by water pressure fluctuations.

5.4.2 MULTI-ANNUAL VARIATIONS

On timescales of several years, non-surging glaciers can undergo large dynamical changes in response to climatic forcing. Most obviously, increases or decreases in surface mass balance can trigger terminus advance or retreat, and a great deal of research effort has focused on determining the relationships between forcing and the magnitude and timing of response.

The timescale over which a glacier responds to a climatic change is known as the *response time*. Unfortunately, this term has been used in several different ways in the literature, generating a degree of confusion. Bahr et al. (1998, p. 9777) defined response time as 'the time required to move from one steady state – in which the downslope mass flux is in equilibrium with the glacier's mass balance – to another steady state following a change in the mass balance environment'. A similar but less precise definition was adopted by Paterson (1994, p. 319). In contrast, Jóhannesson et al. (1989a) and Oerlemans (2001) defined response time (or *memory time*) as the time constant of an exponential function

describing the evolution of the system. This definition is based on the idea that a glacier will approach a new equilibrium state asymptotically – that is, the response is initially rapid and becomes progressively damped through time as the glacier gets closer to the new equilibrium. The characteristic timescale for changes of this type is known as the *e-folding time*, usually defined as the time taken to achieve a proportion ($1/e$) of the total change, where e is the base of the natural logarithms. Another useful concept is the *reaction time*, which refers to the time taken for the terminus to react to a change in climate – that is, the time lag between a climate signal and the first signs of change in glacier length. Readers should beware that 'response time' has sometimes been used as a synonym for 'reaction time'.

Several attempts have been made to calculate glacier response times using analytical methods, and useful summaries have been provided by Paterson (1994), Bahr et al. (1998) and Oerlemans (2001). Early work (e.g. Nye, 1963) used *kinematic wave theory* to analyse the propagation of mass perturbations on glaciers (cf. van de Wal and Oerlemans, 1995; van der Veen, 1999a). In its general form, this theory has many applications and can be used to describe, among other things, traffic flow on busy roads. The key concept is that in a flowing medium, any perturbation in flow will propagate through the medium at a rate determined by the principle of conservation of mass. In the case of an increase in accumulation on a glacier, a zone of increased ice discharge will propagate downglacier faster than the ice velocity. Individual packets of ice will first accelerate, then decelerate as the wave passes. The effects of the original perturbation will therefore arrive at the terminus much faster than the rate of ice flow. Kinematic waves are difficult to distinguish on real glaciers, because the surface profile is constantly adjusting to changes in mass balance and basal conditions on several timescales. However, bulges of increased thickness, travelling downglacier faster than the ice itself, have been observed on several glaciers, including the Mer de Glace, Glacier des Bossons and Miage Glacier in the Mont Blanc massif, European Alps, and Nisqually Glacier, USA (Paterson, 1994; Thompson et al., 2000; fig. 5.15). Van de Wal and Oerlemans (1995) have argued that many such discharge perturbations may not be true kinematic waves, but reflect variations in basal sliding unrelated to mass balance forcings. Given the importance of subglacial hydrology in glacier dynamics, kinematic wave theory is likely to give only an approximate indication of the propagation of discharge perturbations through glaciers.

Results obtained using kinematic wave theory predict that the response times of even small glaciers are of the order of hundreds to thousands of years (e.g. Nye, 1963), suggesting a degree of stability in the cryosphere which is not supported by records of climate and glacier fluctuations (van der Veen, 1999a; Oerlemans, 2001).

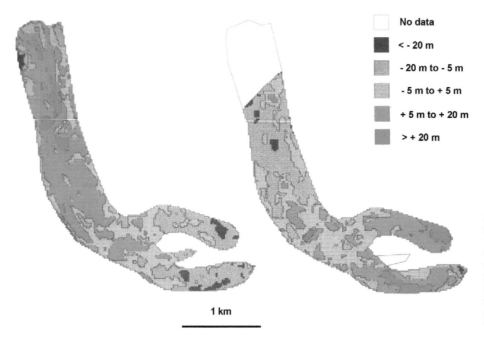

No data

< - 20 m

- 20 m to - 5 m

- 5 m to + 5 m

+ 5 m to + 20 m

> + 20 m

1 km

Figure 5.15 Surface elevation changes on Miage Glacier, Italy, (left) 1967–75 and (right) 1975–99. In the first period, the upper ablation area was thickening, with thinning on the lower tongues. In the second period, this pattern had reversed, indicating the movement of a wave of mass downglacier (Modified from Thompson et al., 2000).

Jóhannesson et al. (1989a) showed that this unrealistically long timescale is a consequence of the assumed dynamics, especially in the lower regions of the glacier. They took an alternative approach, based on the time necessary to accumulate or ablate the volume of ice required to bring the glacier to its new equilibrium (Jóhannesson et al., 1989b), and proposed that the response time t_R can be estimated using:

$$t_R \propto [h] / (-[\dot{b}]) \qquad (5.9)$$

where $[h]$ is the characteristic glacier thickness and $\dot{b}$ is the balance rate at the terminus. Bahr et al. (1998) showed that equation 5.9 leads to the counter-intuitive result that response time decreases as glacier size increases. This is because larger glaciers push further into the ablation zone, and ablation at the terminus increases more rapidly than the characteristic ice thickness. Thus, $[h] / (-[\dot{b}])$ becomes smaller as the glacier increases in size. However, Bahr et al. (1998) argued that glacier response time also depends on the *mass balance index*, $\partial \dot{b}/\partial x$, or the rate of change of mass balance with horizontal distance along the glacier surface. The mass balance index reflects the way a particular glacier 'samples' the climatic environment, and is typically high for small, steep glaciers, and small for large, low-gradient glaciers. The response time is inversely proportional to the mass balance index, so small glaciers will tend to have shorter response times than longer ones. Thus, there are two competing effects of glacier size: as glacier length increases, $[h] / (-[\dot{b}])$ becomes smaller, decreasing the response time; however, $\partial \dot{b}/\partial x$ also becomes smaller, increasing the response time. Bahr et al. (1998) showed that the latter effect dominates, so that small glaciers will tend to have the shortest response times.

Such theoretical approaches are valuable inasmuch as they provide insights into the factors governing glacier response to climate forcing, but their simplifying assumptions limit their usefulness as quantitative predictive tools. Crucially, real meteorological conditions undergo continuous change, and the response of a glacier to a change in climate is unlikely to be complete before some other significant shift occurs. Furthermore, glacier responses can also vary widely depending on the nature of the climate signal, where on the glacier the initial impact is felt, and how that impact is transferred to the terminus by dynamic processes. Even for a single glacier, some scenarios may provoke an immediate response at the glacier terminus, whereas others involve considerable time lags. The idea that a glacier has a single, well-defined response time for all climate perturbations is a misconception.

While analytical approaches ('minimal models'; Oerlemans, 2008) can offer insights into the controls on glacier behaviour, most recent work on glacier response to climate change has focused on time-dependent numerical models. By using climate data as inputs and the known history of a valley glacier as the target output, the model can be used to explore the intermediate dynamic processes (e.g. van de Wal and Oerlemans, 1995; Pfeffer et al., 1998; van der Veen, 1999a; Oerlemans, 2001, 2005; van der Veen and Payne, 2004; Sugiyama et al., 2007). It should be remembered, however, that achieving a good fit between model output and observed behaviour does not mean that the model faithfully represents reality, merely that it is possible to tune the model parameters to arrive at the desired result. As noted in section 5.3.3, subglacial hydrology and basal motion are still very poorly represented in numerical models, meaning that dynamic effects such as the long-term slowdown of

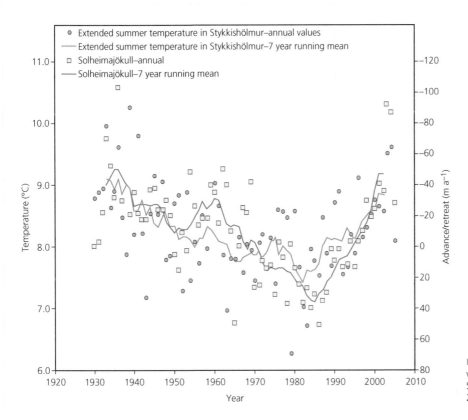

Figure 5.16 Time series of air temperature variations and ice-front position of Solheimajökull, Iceland (Sigurdsson et al., 2007).

Findelengletscher (Iken and Truffer, 1997) will not feature in model simulations. Well-designed model experiments take such limitations into account, and systematically explore the impact that different variables have on the model output, thus helping to generate new hypotheses and research strategies (e.g. Leysinger Vieli and Gudmundsson, 2004).

A simpler empirical approach to studying glacier response to climate is based on statistical correlations between glacier length changes and climatic variables. For example, Sigurdsson et al. (2007) correlated glacier length changes with mean summer temperature (May–September) for a sample of Icelandic glaciers. Data for Solheimajökull are shown in figure 5.16, where a very strong relationship is apparent ($r^2 = 0.66$). Interestingly, there is no time lag built into this relationship, indicating that this glacier has a very short reaction time (although not necessarily a short response time; section 5.2.5). Nesje (2005) correlated ice front oscillations of Briksdalsbreen, Norway with a 'climate index' incorporating both summer temperature and winter precipitation, and found that periods of retreat were mainly forced by summer temperature, whereas advances reflected periods of heavy winter snowfall.

Recently, *neural network models* have been employed to optimize the fit between glacier length changes and climatic variables, using different combinations of weightings and lags (Steiner et al., 2005, 2008; Nussbaumer et al., 2007). High-resolution series of glacier length variations were established using historical photographs,

Figure 5.17 The Mer de Glace in 1823, shortly after its nineteenth-century maximum extension. Accurately observed paintings such as this provide valuable records of past glacier fluctuations. (Samuel Birmann, *au village des Prats, août 1823*. Reproduced by permission of Kunstmuseum Basel, Kupferstichkabinett. Photo Credit: Kunstmuseum Basel, Martin P. Bühler.

maps and paintings spanning the Little Ice Age and subsequent warming period (fig. 5.17). Reconstructed fluctuations of the Mer de Glace, France, show that the glacier front reached similar advanced positions on seven occasions between AD 1600 and 1850, since when a general pattern of retreat has been interrupted by short-lived readvances (fig. 5.18). Nussbaumer et al. (2007) showed that temperature and precipitation were roughly equal in importance as drivers of fluctuations of the Mer de Glace throughout the study period. In contrast, for Lower

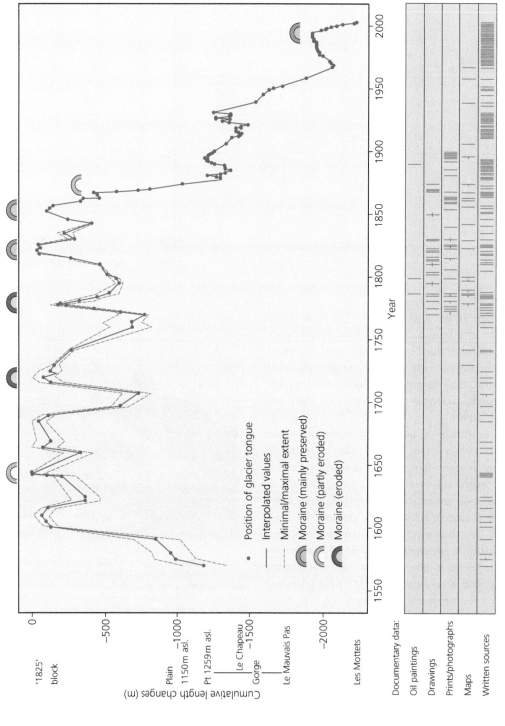

Figure 5.18 Fluctuations of Mer de Glace, France, during and following the Little Ice Age, reconstructed from a variety of sources (Nussbaumer et al., 2007).

Grindelwald Glacier, Switzerland, advances or retreats appear to have been forced by various combinations of seasonal temperatures and precipitation, while for Nigardsbreen, Norway, winter and spring precipitation is the dominant forcing variable.

5.5 CALVING GLACIERS

Globally, calving is a very important ablation process, accounting for most of the mass loss from the Antarctic Ice Sheet, over 50 per cent of the losses from the Greenland Ice Sheet, and substantial amounts from smaller glaciers in the Arctic, Antarctic, Alaska, Patagonia and other regions. Many calving glaciers are fast-flowing and undergo sudden changes in speed and terminus position, and therefore can have a disproportionate influence on the rate at which ice is transferred into the oceans (Bamber et al., 2007). Tidewater glaciers in many parts of the world have recently accelerated and retreated, including glaciers in the Antarctic Peninsula (Cook et al., 2005; Pritchard and Vaughan, 2007), Alaska (Meier et al., 1994; Arendt et al., 2002) and Patagonia (Rignot et al., 2003). The recent collapse of the Larsen Ice Shelf in the Antarctic Peninsula is perhaps the most widely known and dramatic example of how rapidly calving glaciers can respond to external triggers (Scambos et al., 2000; MacAyeal et al., 2003; Glasser and Scambos, 2008). Ice shelf collapse and calving glacier dynamics may play a key role in amplifying ice sheet response to climate triggers, introducing major uncertainties over likely future sea-level change (e.g. Vaughan and Spouge, 2002; IPCC, 2007; Vaughan, 2008; sections 6.2.4 and 6.3.7).

Despite their importance, a comprehensive understanding of calving glaciers has been slow to develop, for several reasons. First, calving glaciers are very diverse. Their margins may be fully buoyant, forming floating ramps or ice shelves, or remain grounded on the bed. They may terminate in the sea (*tidewater glaciers*) or proglacial lakes (*freshwater glaciers*). Some are slow-flowing and calve rarely, whereas others are fast-flowing and release icebergs almost continuously from their intensely fractured termini. Some, such as Columbia Glacier, Alaska, and San Rafael Glacier, Chile, are among the fastest and most active glaciers on Earth (Meier and Post, 1987; Warren and Sugden, 1993; Warren et al., 1995; figs. 5.19 and 5.20). Second, calving is not a single process, but a family of related processes, complicating attempts to develop general models. Although numerous calving models have been proposed (e.g. Reeh, 1968; Iken, 1977; Hughes, 2002; Hanson and Hooke, 2003), they tend to apply to specific glacier geometries, and relatively few studies have attempted to identify general, unifying principles (Benn et al., 2007b). Third, calving processes

Figure 5.19 A calving event from San Rafael Glacier, Chile. The calving front rises to ~45 m above the waterline and the water depth is in excess of 250 m (C.R. Warren).

interact with glacier dynamics in complex ways, so understanding calving necessarily involves addressing other long-standing problems in glaciology, such as subglacial hydrology, basal motion, ice fracture, and energy exchanges between ice sheets and the oceans. Finally, and not least, there are considerable challenges associated with making detailed observations in remote, life-threatening environments, so field data are seldom as comprehensive as glaciologists would like. By far the most comprehensive data set is for Columbia Glacier, Alaska, which is the focus of a major research programme undertaken by the United States Geological Survey (USGS)

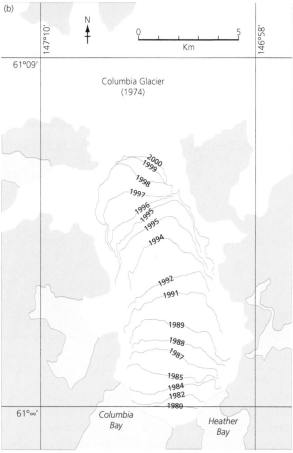

Figure 5.20 Columbia Glacier, Alaska. (a) Aerial view of the glacier in 1980 at the beginning of its recent retreat. (Austin Post, USGS, http://earthobservatory. nasa.gov/images/imagerecords/5000/5668/glacier_columbia_c1980.jpg) (b) Glacier limits in 1974 and terminus positions between 1980 and 2000 (Redrawn from Krimmel, 2001).

(fig. 5.20; see Pfeffer, 2007a). Important data have also been collected for calving glaciers in many parts of the world, although there are still large gaps in the observational record.

Meier (1997, p. 112) observed that: 'calving is a problem in fracture mechanics coupled to ice dynamics'. In this section, we emphasize this dual nature of the 'calving problem' by considering calving processes in their dynamic context. We begin by examining how the presence of lake water or seawater at a glacier terminus affects the force balance and velocity structure of calving glaciers, and then go on to consider how varying stress patterns can trigger ice fracturing and calving events. Controls on the advance and retreat of calving glaciers are examined in section 5.5.3. The mass balance of ice shelves and the factors that influence ice shelf stability are examined in section 5.6, and the behaviour of calving glaciers in Greenland and ice streams and ice shelves in continental Antarctica are considered in chapter 6.

5.5.1 FLOW OF CALVING GLACIERS

The wide range of behaviour of calving glaciers, and their differences with land-terminating glaciers, can be understood with reference to the forces that drive and resist flow, especially in the crucial zone where ice meets water. In this section, we focus on how basal drag, lateral drag and longitudinal stress gradients vary in calving glaciers, and how their relative magnitudes control ice velocity and strain rates.

First, *basal drag* can be profoundly affected by the presence of water at the margin, especially if it is deep enough to fully or partially float the ice. Ice will float if it is less than a critical *flotation thickness*, H_F:

$$H_F = \frac{\rho_W}{\rho_I} D_W \qquad (5.10)$$

where D_W is water depth, ρ_i is ice density ($\approx 900\,\mathrm{kg\,m^{-3}}$ and ρ_W is water density. The flotation thickness is less in fresh water ($\rho_w \approx 1000\,\mathrm{kg\,m^{-3}}$) than in seawater ($\rho_w \approx 1030\,\mathrm{kg\,m^{-3}}$):

$$H_F \approx 1.11\,D_W \text{ (fresh water)}$$
$$\approx 1.14\,D_W \text{ (seawater)}$$

Where a calving margin is greater than the flotation thickness the terminus will be grounded, but part of its weight will be supported by water pressure. At the terminus, the basal water pressure is:

$$P_W = \rho_W g D_W \qquad (5.11)$$

This affects the subglacial hydraulic potential further upglacier, because standing water at the terminus determines the base level for the whole subglacial drainage system of the glacier. That is, if basal or surface-derived meltwater is to be evacuated from the glacier bed, the

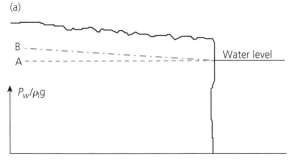

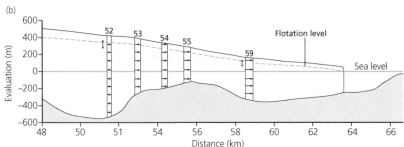

Figure 5.21 Hydraulic potential in tidewater glaciers. (a) Sketch illustrating hypothetical piezometric surfaces (i.e. elevation that water would reach in a borehole connected to glacier bed). (A) Minimum surface at sea level. This is the base level for the whole subglacial drainage system, and represents that commonly used in modelling studies. (B) Sloping piezometric surface associated with water flow through the system. (b) Longitudinal profile of Columbia Glacier in 1987, showing the ice surface, bedrock elevation (stippled), sea level and the flotation level. Vertical exaggeration is 4:1. Ice velocities at five locations are indicated by horizontal arrows (distance travelled in a 53-day measurement interval). The range of water levels observed in boreholes are indicated by bars at km 52 and 59, providing an indication of hydraulic potential at these locations. See figure 5.24 for the locations of these points (Meier et al., 1994).

hydraulic potential in all parts of the system must be greater than the value at the terminus (fig. 5.21). Water pressure beneath a water-terminating glacier can usefully be defined in terms of a bed elevation component and an additional component associated with the storage and flux of meltwater:

$$P_W = \rho_W g Z_B + \phi \quad (Z_B \leq 0) \qquad (5.12)$$

where Z_B is the bed elevation (measured downwards from proglacial water level), and ϕ represents any additional water pressure associated with water storage. This expression does not take account of any density differences between the glacier meltwater and the proglacial water body. Where significant differences occur, an expanded form of equation 5.12 should be used (Meier and Post, 1987; Benn et al., 2007b). The additional water pressure term ϕ depends on the hydraulic gradient in the subglacial drainage system, which in turn is a function of water recharge rates and the hydraulic conductivity of the system (section 3.2.3). Understanding of subglacial hydrology is not yet sufficiently advanced to allow calculation of the hydraulic gradients below real glaciers from first principles, although it is clear that hydraulic gradients can be quite large (Kamb et al., 1994). Thus, the hydraulic effects of lake water or seawater at the terminus can reach far into the interior of wet-based glaciers.

Basal drag is inversely proportional to effective pressure on both hard- and soft-bedded glaciers (e.g. Tulaczyk et al., 2000a; Cohen et al., 2005; Schoof, 2005), so if ice thickness decreases relative to water depth, traction at the bed will be reduced. This effect explains why the speed of calving glaciers tends to increase as ice thins towards the terminus (fig. 5.22; Kamb et al., 1994; Vieli et al., 2004). On several glaciers, however, the highest velocity gradients occur some distance upglacier of the terminus, and diminish close to the calving front. For example, on LeConte Glacier, Alaska, O'Neel et al. (2001) noted that strain rates increase to within about 200 m of the terminus, where an abrupt decrease occurs. Similar patterns have been observed on Columbia Glacier (Krimmel, 2001) and Breiðamerkurjökull, Iceland (Björnsson et al., 2001). This reduction in strain rate probably reflects very low basal drag where the glacier is close to buoyancy, and a correspondingly large role played by longitudinal stresses in the force balance of the terminal zone.

Because basal drag tends to be small where glaciers are grounded in deep water, lateral drag and longitudinal stress gradients play a correspondingly large role in the force balance. Figure 5.23 shows calculated force balance components for Columbia Glacier for summer 1985 (O'Neel et al., 2005). A region of high driving stress occurs near km 54 (panel a), which is largely supported by basal and lateral drag (panels c and d: note that negative values of lateral drag and longitudinal stress gradient oppose the driving stress). This high stress region coincides with a relatively narrow and shallow part of the trough (fig. 5.24). Both basal and lateral drag decrease downglacier of the constriction, and basal drag has particularly low values in the terminal zone where lateral drag supports a greater proportion of the driving stress. Longitudinal stress gradients fluctuate in sign along the glacier, indicating alternating compressive and extensional stress gradients in response to variations in basal and lateral drag (panel b).

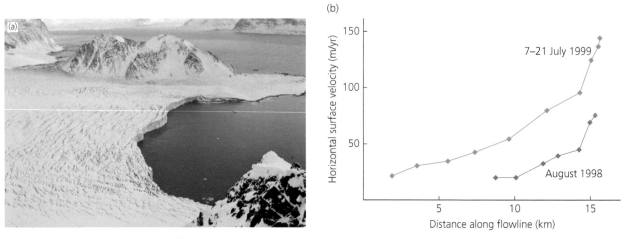

Figure 5.22 Hansbreen, Svalbard. (a) View of the glacier in October 2005 (D.I. Benn). (b) Along-flow variations in surface velocity. Note the increase in velocities towards the terminus (on right) and the large difference between years (Benn et al., 2007b, after Vieli et al., 2004).

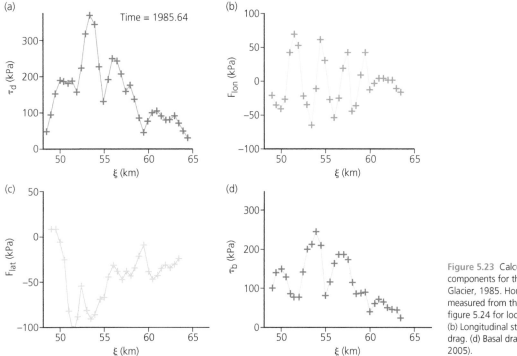

Figure 5.23 Calculated force balance components for the centreline of Columbia Glacier, 1985. Horizontal scale is in km, measured from the head of the glacier (see figure 5.24 for locations). (a) Driving stress. (b) Longitudinal stress gradient. (c) Lateral drag. (d) Basal drag (After O'Neel et al., 2005).

Figure 5.23 neatly illustrates the correspondence between trough constrictions and regions of high basal and lateral drag. From equation 5.12, it can be seen that basal water pressure will be relatively low where Z_B is small. If thick ice occurs at such locations, effective pressures (and basal drag) will be large. Similarly, the restraining influence of lateral drag is greatest in narrow reaches of a trough (section 4.6.2). The effect of trough constrictions can be transmitted upglacier by longitudinal stress gradients, allowing ice above the constriction to be thicker than it would otherwise be. Narrow and shallow reaches of a trough are known as *pinning points*, in

recognition of the major role they play in determining the stability of calving glaciers. The importance of pinning points in the advance and retreat of calving glaciers is discussed in section 5.5.3.

As in the case of land-terminating glaciers, calving glaciers commonly exhibit short-term velocity variations in response to fluctuations in subglacial water storage and drainage system configuration. Figure 5.25 shows time series of meteorological data, glacier velocities and borehole water levels for two points on Columbia Glacier in summer 1987 (Kamb et al., 1994; Meier et al., 1994). Borehole water levels at both sites indicate that

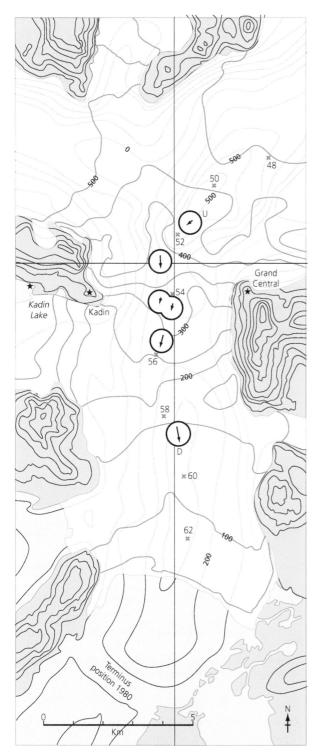

Figure 5.24 Bed (lighter lines) and surface (darker lines) topography of Columbia Glacier in 1987. The numbered points mark the km markers used as horizontal scales in figures 5.21 and 5.23 (Meier et al., 1994).

after the noontime peak in ablation rates. Borehole water levels also display strong diurnal cyclicity, strongly suggesting that velocities are controlled by variations in storage of water at the glacier bed. Timing of the water-level peaks, however, shows that the relationship between local water pressures and velocities is not straightforward. At km 52, the water-level peaks occur approximately 1 hour before the velocity peaks, around 7 hours after ablation-rate peaks. In contrast, at km 59 the water-level peaks occur about 6 hours ahead of the velocity peaks, and only about 2–3 hours after the time of maximum ablation. These data indicate that the effect of changing surface-water inputs on basal hydrology is time-transgressive, with the timing of water-pressure pulses depending on factors such as drainage configuration and location relative to efficient drainage pathways. The synchroneity of the velocity peaks, however, indicates that the glacier was responding to some distributed property of the drainage system, rather than local conditions. Analysis by Kamb et al. (1994) suggests that this is the total volume of water stored beneath the glacier, which reached a maximum around 2–3 hours before the velocity peaks.

Further insight into the relationship between basal hydrology and glacier dynamics is provided by longer-term velocity fluctuations. During the measurement period, there were four events during which velocities were higher than average (fig. 5.25). Events 2 and 3 coincided with periods of high rainfall, whereas events 1 and 4 occurred at times of high windspeed, when warm föhn winds increased ablation by turbulent heat transfer (section 2.2.4). All four velocity peaks, therefore, closely coincide with increased inputs of water from the surface. Interestingly, ice velocities following events 1 and 2 were lower that their pre speed-up values, even though surface ablation rates were similar to before. Meier et al. (1994) termed these *extra slowdowns*, and attributed them to reorganization of the basal drainage system in response to higher discharges. This appears to be similar to the switches in drainage efficiency identified below alpine glaciers (e.g. Nienow et al., 1998; section 3.5.1). The effects of the enhanced water input events or drainage reorganizations on Columbia Glacier, however, are not apparent in the borehole water-level records. The reasons for this are unclear, although Kamb et al. (1994) argued that local effects appear to dominate the data on timescales greater than one day.

The complexity of the relationship between the Columbia Glacier borehole water level and ice velocity records reflects the fact that the boreholes unavoidably sampled essentially random points on what must be a very heterogeneous drainage system. A complementary picture emerges from a study by Vieli et al. (2004) on Hansbreen, a tidewater glacier in Svalbard.

basal effective pressures were consistently low, episodically becoming zero or negative. At such times, local basal traction would have been zero, and the driving stress must have been supported non-locally. Ice velocities at both sites exhibit pronounced diurnal cycles, with the highest values occurring almost simultaneously about 8 hours

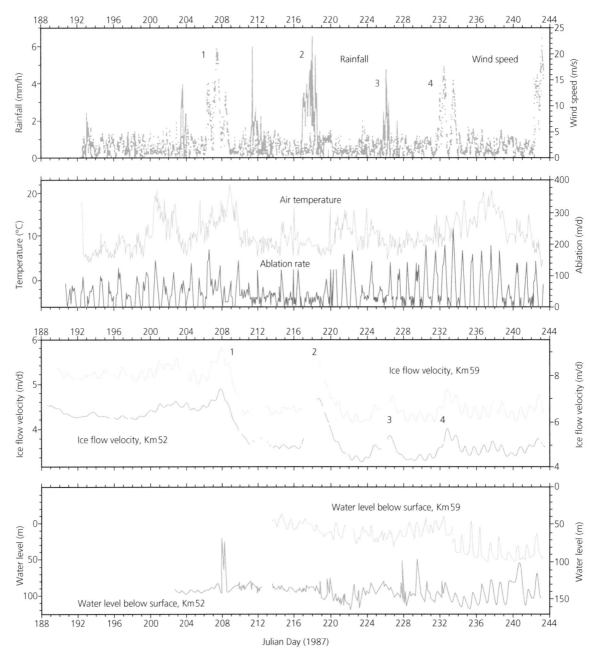

Figure 5.25 Water inputs and velocity fluctuations, Columbia Glacier. The panels show (from top to bottom): rainfall and windspeed, air temperature and ablation rate, surface velocities measured at km 52 and km 59, borehole water levels at the same two sites (After Meier et al., 1994).

Figure 5.26 shows horizontal and vertical velocity records for three stations near the centreline of the glacier, together with water levels measured in a moulin situated 500 m upglacier of the highest station (D). The pronounced velocity peak on 16 July was triggered by a föhn wind event, similar to events 1 and 4 in the Columbia Glacier record. High water levels in the moulin indicate that inputs of surface meltwater exceeded the capacity of the subglacial drainage system, increasing water storage and locally causing effective pressures to fall to zero. Rising water levels in the moulin coincided with uplift of the glacier surface at station D, indicating the growth of

subglacial cavities and ice–bed decoupling. Lesser amounts of uplift also occurred at stations B and A, located closer to the glacier terminus. The velocity peak occurred earliest at station D and propagated quickly downglacier, with all velocity peaks occurring just before moulin water levels began to fall rapidly. The relatively simple correlation between moulin water levels and ice surface uplift and velocity probably reflect the fact that the moulin was directly connected to hydrologically significant drainage pathways at the glacier bed, and therefore provided a more representative indicator of subglacial effective pressure than the Columbia Glacier boreholes.

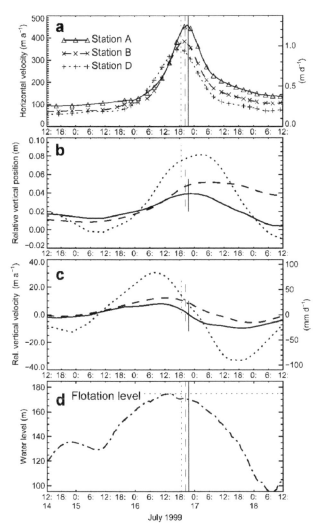

Figure 5.26 A speed-up event on Hansbreen, Svalbard, in July 1999. (a) Horizontal velocities at three stations. (b) Vertical position of markers on glacier surface. (c) Vertical velocities. (d) Moulin water level (Vieli et al., 2004).

Longitudinal stress gradients can also undergo rapid changes following variations in basal drag or the boundary condition at the ice front, forcing short-term velocity fluctuations on tidewater glaciers. On Columbia Glacier and LeConte Glacier, velocities near the glacier termini fluctuate in harmony with the tidal cycle, with the highest velocities occurring at low tide (fig. 5.27; Meier and Post, 1987; Walters and Dunlap, 1987; O'Neel et al., 2001). The effect is quite large. At Columbia Glacier, a 1 per cent variation in water depth at the terminus is associated with a 4 per cent fluctuation in speed, whereas at LeConte Glacier, a 1.5 per cent variation in water depth results in a 5.5 per cent fluctuation in speed. This pattern reflects the varying longitudinal force balance at the glacier terminus – at high tide, the back-pressure exerted by the seawater on the ice front is at a maximum, providing greater opposition to the driving stresses. The tidal influence on velocity declines upglacier, with a damping length (i.e. decay to 1/e of the initial value) of 2 km at Columbia

Glacier and 0.5 km at LeConte Glacier, the contrast reflecting the differing longitudinal stress gradients at the two glaciers (O'Neel et al., 2001). A mathematical analysis of the effects of tides on glacier velocities has been presented by Thomas (2007).

5.5.2 CALVING PROCESSES

All calving events occur when tensile stresses close to a glacier margin are large enough to propagate fractures right through the ice, isolating blocks from the main body of the glacier. The factors that control the timing and location of individual calving events, however, are clearly very complex, and attempts to correlate calving frequency with environmental variables have generally failed to identify any clear relationships (Warren et al., 1995; O'Neel et al., 2003). To a large extent, this must reflect the multivariate nature of calving processes, and the fact that the proximal cause of iceberg calving may simply be the final stage in a long chain of events (O'Neel et al., 2007). However, a useful overview of calving mechanisms and their controls can be obtained by considering scenarios that can, singly or in combination, generate the stresses required for fracturing (Benn et al., 2007b). These include:

- longitudinal stretching associated with large-scale velocity gradients
- steep stress gradients at ice cliffs or floating ice fronts
- ice cliff undercutting by melting at or below the waterline
- bending forces at buoyant glacier margins.

Longitudinal stretching associated with velocity gradients

On glaciers with grounded calving margins, ice velocities typically increase towards the terminus because faster sliding is encouraged by diminishing basal drag as the glacier approaches flotation (section 5.5.1). This downstream increase in velocity stretches the ice, potentially opening surface crevasses transverse to flow, and if net tensile stresses are large enough, crevasses will propagate through the ice and trigger calving. The velocity structure of the terminus, and the associated crevasse patterns, should therefore act as an important first-order control on the location of the calving margin (Benn et al., 2007b). Where longitudinal extension occurs in conjunction with simple shear near the glacier margins, distinctive arcuate crevasse patterns are created, determining the form of concave *calving bays* (fig. 5.22a). Crevasses formed upglacier from the terminus can play an important role in calving processes by acting as pre-existing lines of weakness along which the glacier will break when subjected to additional forces (Dowdeswell, 1989; Warren et al., 1995; O'Neel et al., 2007). Indeed, crevasses formed in

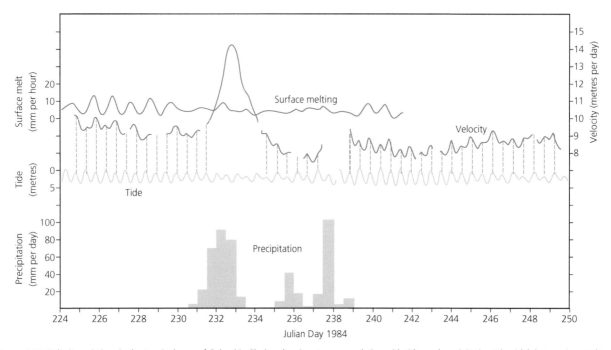

Figure 5.27 Velocity variations in the terminal zone of Columbia Glacier, showing strong correlation with tides and precipitation. The tidal data are inverted to highlight the coincidence of velocity peaks with low tide (Krimmel and Vaughn, 1987).

Figure 5.28 The terminus of Kronebreen, a fast-flowing calving glacier in north-western Svalbard. Calving events exploit crevasses formed near the terminus and old fractures advected from upglacier (D.I. Benn).

areas of accelerating flow many kilometres upglacier can be advected towards the terminus by ice flow, where they can be reactivated (fig. 5.28). On ice shelves, full-depth crevasses or *rifts* form in areas of high-velocity gradients, such as shear margins or where longitudinal or lateral spreading rates are high (fig. 5.29; Lazzara et al., 1999; Joughin and MacAyeal, 2005). Once rifts have formed, a variety of factors can cause them to propagate across the shelf until they intersect other rifts or the shelf edge, eventually isolating tabular bergs, sometimes many hundreds of kilometres across (section 5.6.2).

The role of velocity gradients in determining the location of calving margins can explain why some calving glaciers have grounded termini, whereas others form floating tongues or ice shelves. We have seen that basal drag decreases with basal effective pressure, and that resistance from lateral drag is inversely proportional to trough width (chapter 4 and section 5.5.1). Longitudinal velocity

gradients – and crevasse propagation – will therefore be influenced by downglacier gradients in basal effective pressure or trough width. Where ice is grounded and the ice thins rapidly in the downstream direction, this will decrease the basal effective pressure and encourage faster flow. Conversely, if lateral drag contributes significantly to resistance, and downglacier gradients in basal effective pressure are small, loss of basal drag as the glacier becomes buoyant will not have such a large effect on the overall force balance. In this case, velocity gradients in the vicinity of the grounding line may be insufficient to cause deep crevassing, and ice can become fully buoyant without calving. In this case, the glacier will grade into a floating tongue or ice shelf, and the calving margin will be some distance downstream of the grounding line (Benn et al., 2007a, b). On the other hand, a reduction of lateral drag where troughs increase in width downglacier will lead to accelerating flow, explaining why calving margins

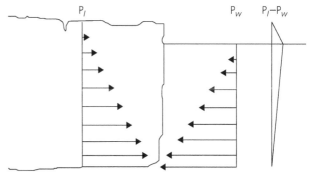

Figure 5.31 Schematic view of outward-directed cryostatic forces and backward-directed hydrostatic forces at the terminal cliff of a floating ice shelf. The triangle on the right represents the net outward-directed force (Benn et al., 2007b).

Steep stress gradients at ice cliffs or floating ice fronts

Individual calving events reflect local stress patterns superimposed on the overall velocity structure of the glacier. Large stress gradients typically occur at calving margins due to an imbalance between outward- and inward-directed forces at the frontal ice cliff (fig. 5.31). Cryostatic pressure increases downward at almost 9 kPa per metre, and at a subaerial ice cliff the outward-directed component of this pressure is practically unopposed by atmospheric pressure. The resulting tensile deviatoric stress can be very large, especially near the base of tall subaerial ice cliffs (Hanson and Hooke, 2003; fig. 5.32). The resulting ice velocity gradients encourage widening and propagation of suitably oriented fractures (fig. 5.33). For a subaqueous ice cliff, the outward-directed pressure is partially opposed by water pressure, although this will always be less than the ice pressure where the ice is grounded.

One of the earliest models of calving was based on an analysis of stress imbalances at the margin of a floating ice shelf (Reeh, 1968). Recently, Larour et al. (2004) provided striking evidence that this model can explain calving along the margins of rifts in ice shelves. Although of much smaller magnitude than the release of tabular bergs by rifting, calving by this mechanism progressively eats away at the ice margins, filling rifts with a melange of ice blocks and sea ice. The presence of this melange can influence rift propagation, and therefore the frequency of large-scale calving events.

Hughes (e.g. 2002) argued that calving from subaerial portions of ice cliffs occurs by failure along concave, overhanging shear bands, in a manner likened to slippage between the pages of a book bent around its binding. This forward-bending theory invokes a highly specific ice margin geometry, and the stress patterns modelled by Hughes differ from those obtained by Hanson and Hooke (2000, 2003). Moreover, the Hughes model appears to have limited empirical support, as the geometry of blocks calved from subaerial ice cliffs rarely conform to predictions (e.g. Benn et al., 2007b).

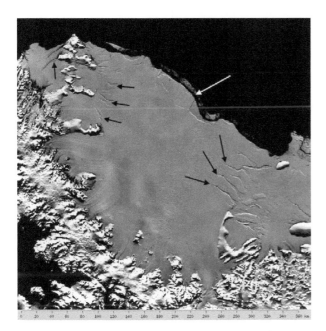

Figure 5.29 MODIS image of Larsen C ice shelf, Antarctic Peninsula. Prominent rifts are indicated by black arrows. The white arrow indicates where rifting has almost isolated a large tabular block (Haran et al., 2005, http://nsidc.org).

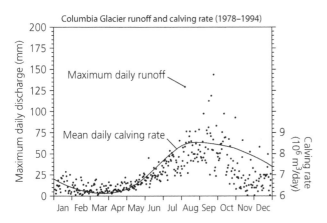

Figure 5.30 Relationship between calving rate (solid line) and meltwater runoff, Columbia Glacier (Tangborn, 1997).

are commonly located at fjord widenings and similar locations.

Meltwater stored in surface crevasses will cause them to penetrate deeper than dry crevasses, providing an additional, transient control on the position of the calving margin. The most spectacular example of meltwater-triggered calving is the catastrophic break-up of the Larsen Ice Shelf following meltpond formation (Scambos et al., 2000; section 5.6.3). The association between surface melt and calving is less clear on short timescales (e.g. O'Neel et al., 2003), although it appears likely that meltwater inputs to crevasses are at least partly responsible for annual calving cycles on some glaciers (e.g. van der Veen, 1997a; fig. 5.30).

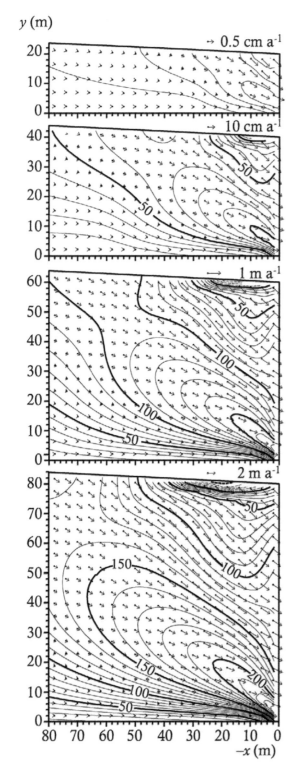

Figure 5.32 Variation of the longitudinal deviatoric stress σ'_{xx} (contours, 10 kPa interval) and velocity (vectors) for subaerial ice cliffs, varying in height from 20 m (top) to 80 m (bottom) in 20 m increments (Hanson and Hooke, 2003).

Ice cliff undercutting by melting at or below the waterline

Melt rates at or below the waterline commonly exceed subaerial melt rates, due to efficient transfer of energy from warm, circulating water. Melting is commonly focused

Figure 5.33 Flake of ice spalling from an ice cliff at the margin of Miage Glacier, Italy. A frozen-over lake lies beneath the snow (D.I. Benn).

close to the waterline, due to the presence of a buoyant layer of warmer water, although in some situations melt-water erosion can occur at greater depths. Subaqueous melting progressively undercuts terminal ice cliffs, increasing force imbalances at the margin (fig. 5.34). Calving can then occur by outward toppling of undercut pillars, collapses of melt-notch roofs along arcuate fractures, and other processes (Iken, 1977; Benn et al., 2001; Warren and Kirkbride, 2003; Haresign and Warren, 2005). Calving events are commonly associated with the reactivation of pre-existing fractures. Kirkbride and Warren (1997) observed that calving can follow a cyclic pattern, beginning with the development of a waterline notch, followed by upwards growth of an overhang by spalling of progressively larger lamellae, then renewed notch development.

Thermo-erosional undercutting of both freshwater and tidewater margins typically follows seasonal cycles, and effectively ceases during winter, especially if lake or sea ice is present (e.g. Vieli et al., 2002). Detailed measurements of melt rates in waterline notches have been made by Röhl (2006) at the terminus of Tasman Glacier, New Zealand. Notch formation rates varied between 10 and 30 cm per day in summer, corresponding to a mean melt rate of 34 m yr^{-1}. During summer, however, melt rates are poorly correlated with water temperature, because energy exchanges between the water body and the ice are strongly dependent on water circulation, which evacuates cooled water from the ice face and replaces it with relatively warm water. *Free convection*, driven only by vertical temperature variations in the water column, is relatively slow and this has led to the common perception that subaqueous ice melt rates are likely to be small

Figure 5.34 Wintertime view of a waterline notch at the margin of an ice-contact lake, Miage Glacier, Italy (M.P. Kirkbride).

(e.g. Russell-Head, 1980). However, at many calving fronts – especially in temperate tidewater settings – vigorous upwelling of meltwater from the bed drives *forced convection*, resulting in much higher energy fluxes and melt rates than is possible by free convection. Recent evidence from Patagonia and Alaska indicates that subaqueous melt rates can be very high, even reaching rates comparable with total calving losses (Motyka et al., 2003; Haresign and Warren, 2005).

When calving is driven by melting at or below the waterline, the long-term calving rate will equal the rate at which the subaerial part of the cliff is undermined by subaqueous melting. Calving rate will therefore be a function of water body properties such as temperature, density structure and circulation patterns. As shown by Röhl (2006), the rate of overhang development is not a simple function of total ice melt, but also depends on how that melt is focused at particular levels. A given melt amount distributed over a broad elevational range will be less effective in undercutting the ice margin than the same amount focused over a narrow range. The distribution of melt will depend on water-level fluctuations, water stratification, circulation patterns and other factors.

Bending forces at buoyant glacier margins

Glacier termini may make the transition from grounded to buoyant if surface melting thins the ice below the flotation thickness. Removal of ice from the surface means that the glacier margin can become *super-buoyant*, no longer in hydrostatic equilibrium and subject to net upward buoyant forces. As surface melting proceeds, the glacier terminus can become increasingly out of equilibrium, producing large bending forces near the junction with grounded ice. The buoyant margin can slowly move back into equilibrium by ice creep, but can also fracture catastrophically, producing tabular icebergs which can be hundreds of metres across (e.g. Warren et al., 2001). This

process has been documented in impressive detail by Boyce et al. (2007) for the lake-calving Mendenhall Glacier, Alaska. In the late twentieth century, this glacier had strong negative mass balance, with melt rates of around $10\,\mathrm{m\,yr^{-1}}$ in the terminal zone. By the spring of 2004, large parts of the glacier tongue were super-buoyant, and the terminus underwent upwarping, raising melt notches above lake level (fig. 5.35). The rate of uplift, however, was not enough to restore hydrostatic equilibrium, allowing considerable bending stresses to build up. Between June and August 2004, a series of four large calving events occurred, causing the terminus to retreat by more than 200 m. The calving events appear to have been triggered by sudden increases in stress, following rises in lake level and/or surface melting, and may have been further encouraged by loss of contact between the glacier and the eastern valley side, which formerly provided some additional lateral drag. Boyce et al. (2007) argued that the crucial factor triggering calving is the rate at which buoyant forces acting on the tongue are increased. Slow perturbations can be accommodated by ice creep, whereas rapid perturbations are more likely to trigger calving events. This type of buoyant calving has been observed on a number of freshwater glaciers, but never on tidewater glaciers, which may reflect the greater magnitude and frequency of perturbing events (such as tides) in saltwater settings.

Buoyant forces can also trigger calving of a projecting *ice foot* below the water surface, formed in response to melting near the waterline or preferential calving of the subaerial parts of an ice face. For obvious reasons, there have been few direct observations of ice feet. Hunter and Powell (1998) used a submersible remotely operated vehicle (ROV) to explore the submerged part of the front of Muir Glacier, Alaska, and found that an ice foot projected 52 m beyond the base of the subaerial ice cliff. The density difference between ice and water renders ice feet potentially unstable, and blocks may shoot up many tens of metres in front of the subaerial terminus (Warren et al., 1995). Detailed observations of subaqueous calving have been made by Motyka (1997) at the deep-water margin of LeConte Glacier, Alaska. Subaqueous calving events were commonly preceded by the collapse of subaerial portions of ice cliff, whereupon large blocks suddenly rose violently parallel to the ice front. Some blocks emerged as much as 200–300 m from the subaerial calving front, and isolated 'shooters' up to 500 m from the ice front have been reported. O'Neel et al. (2007) have suggested that the close association between subaerial and subaqueous calving events reflects the sudden reduction of ice overburden pressure on the submerged portion, and the consequent increase in net upward forces. Long-term rates of buoyancy-driven calving will be controlled by the rate of ice loss above the waterline, either by

(a)

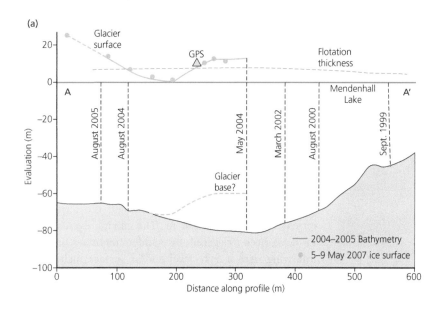

(b)

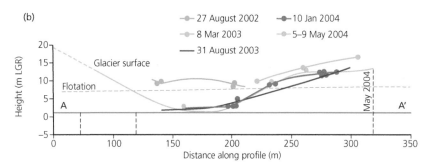

Figure 5.35 The terminal zone of Mendenhall Glacier, showing evidence of buoyant disequilibrium and uplift of the glacier front (After Boyce et al., 2007).

calving of the subaerial part of the ice cliff or by surface melting.

Dry calving

Calving is also an important process of mass loss on some land-terminating glaciers. Small dry calving events are common along the fronts of cold glaciers in parts of the High Arctic and Antarctica, where steep ice cliffs are maintained by ice deformation and low melting rates (Diolaiuti et al., 2004). The largest dry calving events, however, are from hanging glaciers in high mountain regions, where they can constitute major hazards. For example, failure of part of a hanging glacier on Huascaran in the Cordillera Blanca, Peru, in 1962, triggered a catastrophic mass flow that buried the town of Ranrahirca and killed 4000 people. Another avalanche from the same glacier in 1970 was the most deadly in world history, killing an estimated 15,000 people (Carey, 2008). Relatively small ice masses frequently calve from hanging glaciers, and constitute a major cause of death among high-altitude mountaineers.

The mechanics of dry calving from hanging glaciers in the European Alps have been studied by Pralong and Funk (2006) and Faillettaz et al. (2008). Calving events are preceded by the opening of arcuate crevasses on the glacier surface, which isolates a portion of glacier front

(wedge fracture) or defines the upper limit of a larger mass sliding along a basal surface (slab fracture). As fractures propagate, the ice mass undergoes a phase of unstable acceleration before final failure. In parts of the European Alps, the evolving geometry of potentially dangerous hanging glaciers can be monitored in detail, allowing calving events to be predicted. In the absence of such data, however, those venturing below hanging glaciers are essentially playing a form of Russian roulette.

5.5.3 'CALVING LAWS'

The importance of calving as an ablation process has prompted the search for 'calving laws' that can be used in time-evolving glacier models. Calving rate can be defined as:

$$U_C = \bar{U}_T - \frac{\delta L}{\delta t} \qquad (5.13)$$

where U_c is the calving rate, $\bar{U}_T$ is the vertically averaged glacier velocity, L is glacier length and t is time. Calving rate is therefore defined as the difference between ice velocity at the glacier terminus and glacier length change over time, showing that calving losses are intimately linked to the rate at which ice is delivered to the glacier terminus. This simple equation can be viewed in two distinct but complementary ways (van der Veen, 1996; fig. 5.36). First, one can consider the controls on ice velocity and calving rate, and regard an advance or retreat of the terminus as a consequence of an imbalance between these variables. Alternatively, one can focus on the controls on ice velocity and terminus position, and regard the calving rate as a passive outcome of glacier dynamic factors. Alternative 'calving laws' that have been proposed reflect these different viewpoints.

Brown et al. (1982) found a strong correlation between calving rates and water depth for a sample of 12 Alaskan tidewater glaciers. This correlation, and subsequent work with larger data sets, formed the basis of *water-depth calving laws* of the form:

$$U_C = a + b\, D_W \qquad (5.14)$$

where D_W is the width-averaged water depth and a and b are empirically determined coefficients. Both a and b are consistently larger for tidewater glaciers than for freshwater glaciers (e.g. Warren et al., 1995; Warren and Kirkbride, 2003), although there is also large variation within these populations (Benn et al., 2007b; fig. 5.37). A number of attempts have been made to explain the correlation between water depth and calving rate, and the differing relationships for tidewater and freshwater glaciers. Van der Veen (1996) convincingly argued that equation 5.14 actually reflects the correlation between water depth and glacier velocity, rather than indicating

any direct causal relationship between water depth and calving rate. Because the rate at which glacier termini change position is usually small compared with ice velocity, a strong correlation exists between velocity and calving rate. For a constant ice thickness, increasing water depths will be associated with decreasing basal effective pressure. Therefore, in general, glaciers grounded in deep water will tend to flow faster than glaciers in shallow water. Ice will be delivered to the calving front at a greater rate, so the calving rate is greater (equation 5.13). The greater density of seawater than fresh water means that tidewater glaciers will flow more rapidly for the same depth of water at the terminus (Benn et al., 2007a, b).

As an alternative to water-depth 'calving laws', van der Veen (1996) focused on the factors that control the *position* of a calving terminus rather than those that might control the calving *rate*. By examining the detailed data series collected by USGS on Columbia Glacier, van der Veen noted that the calving front tends to be located where the ice is approximately 50 m greater than the flotation thickness, H_f (equation 5.10). If the glacier thinned (due to either melting or longitudinal extension), the calving front tended to retreat to a position where this *height-above-buoyancy calving criterion* was again satisfied. A more general form of the buoyancy criterion was introduced by Vieli et al. (2000), replacing the fixed value of 50 m with a fraction of the flotation thickness. Time-evolving glacier models incorporating the height-above-buoyancy criterion successfully reproduce much of the behaviour of calving glaciers (Vieli et al., 2001, 2002; Nick et al., 2007a, b). An important shortcoming of this calving criterion, however, is that it cannot account for the existence of ice shelves (where the height above buoyancy is zero). As a result, the model cannot be used to predict the behaviour of glaciers or ice streams that flow into ice shelves, or to explore the evolution of marine ice sheets from floating ice.

Another approach to modelling calving behaviour was proposed by Benn et al. (2007a, b), in which calving-front position is predicted from the velocity structure of the glacier. As explained in section 4.7.1, the depths of

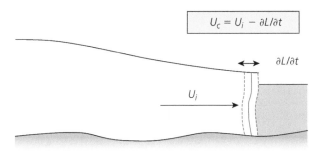

Figure 5.36 Definition of calving rate.

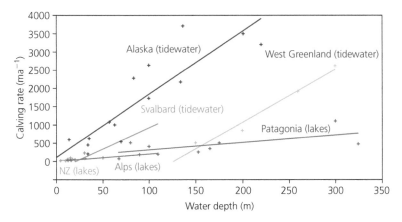

Figure 5.37 Relationships between calving rates and water depth for samples of glaciers from different regions (Haresign, 2004 (unpublished PhD thesis); previously in Benn et al., 2007b).

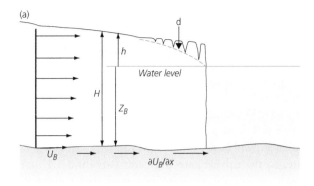

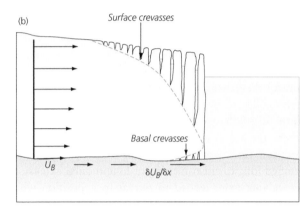

Figure 5.38 Definition sketch illustrating variables used in the crevasse-depth calving model. (a) Waterline crevasse-depth criterion; (b) full depth crevasse criterion as suggested by C.J van der Veen (Benn et al., 2007a).

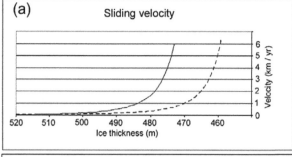

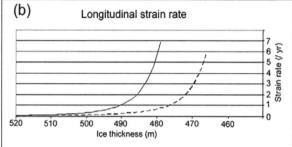

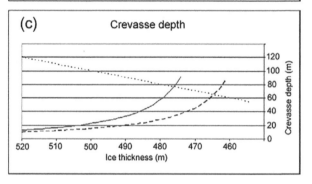

Figure 5.39 Modelled effects of sliding velocity on calving margin position and calving rates for tidewater (solid lines) and freshwater (dashed lines) calving glaciers. Water depth = 400 m, glacier surface slope = 1.15°. (a) Sliding velocities. (b) Longitudinal strain rates. (c) Depths of surface crevasses. Calving is assumed to occur when the crevasse depth curves cross the ice height above the waterline (dashed line). Due to the greater buoyancy of ice in salt water than fresh water, tidewater glaciers will flow faster and have higher longitudinal strain rates than freshwater glaciers in any given water depth. For two otherwise identical glaciers, the tidewater glacier will have a higher calving front and a greater calving rate (Modified from Benn et al., 2007a).

surface crevasses increase with the longitudinal strain rate (velocity gradient) at the glacier surface. If the longitudinal strain rate is sufficiently high, therefore, crevasses may propagate far enough to trigger calving. The question is, how deep do crevasses need to be for calving to occur? Observations on many calving glaciers show that surface crevasses near the terminus penetrate close to the waterline. Calving commonly occurs by collapse of the subaerial part of the calving front, followed after some delay by buoyant calving of the subaqueous part (e.g. O'Neel et al., 2007). Alternatively, if crevasses extend below the waterline and water is able to flood into them from the surface, sea or lake, then full-depth crevassing could occur by hydrofracturing (equation 4.29). This is probably reasonable for fjord glaciers where crevasses near the margin commonly intersect the ice front. So, as a first approximation, the calving margin can be defined as the point at which surface crevasses penetrate as far as the waterline, at which point calving is assumed to occur (fig. 5.38). Different crevasse-depth thresholds may be appropriate in other circumstances. Obviously, the *crevasse-depth calving criterion* is built on a highly idealized picture of the calving process, in which the calving margin position is determined solely by crevasse depth, which in turn is a simple function of the large-scale velocity structure of the glacier. It is not intended to be a realistic representation of individual calving events, but to

provide a means of predicting the likely position of the calving margin from the first-order controls on crevasse propagation.

By relating calving to the longitudinal strain rate, the crevasse-depth model predicts that the calving margin position will be determined by the velocity structure of the glacier, which in turn depends on spatial patterns of basal and lateral drag. For example, if a rapid downglacier decrease in ice thickness brings the glacier close to flotation, the resulting increase in velocity will encourage deep crevassing and calving (fig. 5.39). This provides a physical basis for the height-above-buoyancy calving criterion. However, if most of the resistance to flow is provided by lateral drag, a moderate thinning of the ice may have only a small effect on speed. As a result, it is possible for ice to become fully buoyant without longitudinal strain rates ever becoming high enough to trigger calving. Instead, the position of the calving front will be strongly influenced by the width of the trough occupied by the glacier, because the centreline velocity is proportional to the fourth power of channel half-width (equation 4.25). A downglacier increase in trough width will therefore result in longitudinal extension of the glacier and encourage calving, whereas a downglacier decrease in width will encourage longitudinal compression, suppressing calving and encouraging ice front stability or advance. Variations in trough width also influence patterns of flow divergence and convergence, which also affect glacier stability. If trough width increases downglacier, the ice must diverge laterally. This thins the ice and encourages crevassing. Conversely, a downglacier decrease in trough width will lead to lateral flow convergence, causing dynamic thickening. The crevasse-depth calving model thus provides additional insight into the functioning of topographic pinning points.

5.5.4 ADVANCE AND RETREAT OF CALVING GLACIERS

Calving glaciers commonly respond to environmental change in very different ways to nearby land-terminating glaciers, in terms of the timing, magnitude and even the sign of terminus position changes. For example, the majority of Alaskan tidewater glaciers have retreated in response to twentieth-century climatic warming, but the onset of retreat occurred at different times (Meier and Post, 1987; Post and Motyka, 1995; Wiles et al., 1995; Calkin et al., 2001). Moreover, several large calving glaciers have defied the trend and gained mass, and advanced instead of retreating (Arendt et al., 2002; Trabant et al., 2002). Taku Glacier advanced 7.3 km between 1890 and 1988, and Hubbard Glacier, the largest glacier in North America, advanced at a rate of 16 m yr⁻¹ between 1895 and 1948, increasing to 32 m yr⁻¹ after 1948 (Mayo, 1988; Motyka and Begét, 1996; Trabant et al., 2003).

Advance–retreat cycles of calving glaciers also tend to be asymmetric, with protracted periods of slow advance alternating with short-lived periods of rapid retreat (Meier and Post, 1987). This pattern can be illustrated by the well-documented Columbia Glacier (Pfeffer, 2007a). Between 1880 and 1979, the glacier terminus abutted an island in Columbia Bay, maintaining a remarkably stable position (fig. 5.20). Between 1950 and 1970, however, negative surface mass balance led to thinning of the glacier, and by the early 1970s deep embayments had formed in the terminus. These changes prompted Austin Post to predict that retreat of the glacier was imminent (see Tangborn, 1997; Pfeffer, 2007a). The retreat began in 1979, when the glacier separated from the island, and the terminus receded by 14 km in the next 25 years. At the

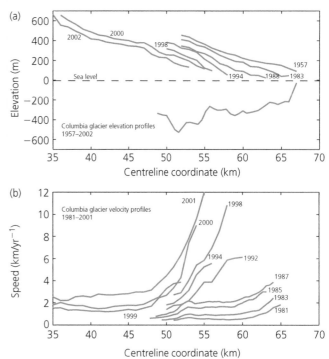

Figure 5.40 Changes in the ice surface profile and velocity of Columbia Glacier during its recent retreat. Note the very large increase in velocity as the glacier terminus retreated into the overdeepening between km 50 and 55 (Pfeffer, 2007a).

same time, the glacier thinned by up to 20 m yr⁻¹ at low elevations, and underwent dramatic acceleration. Speeds near the terminus were around 2 km yr⁻¹ in 1981, but had increased to approximately 12 km yr⁻¹ in 2001 (fig. 5.40). In combination, ice acceleration and calving retreat increased the discharge of icebergs into the fjord, reaching values in excess of 6.5 km³ yr⁻¹ (O'Neel et al., 2005). Retreat of the glacier slowed in 2000 when the terminus reached a narrowing of the trough, although there are clear indications that renewed rapid retreat is imminent, and is likely to affect a further 15 km of the glacier.

This rapid retreat contrasts with the slow rate of advance of Columbia Glacier during the last millennium. From subfossil trees contained within till exposed by the retreat, it has been possible to establish a detailed chronology for the period prior to historical times (Kennedy, 2003; Nick et al., 2007a). Tree-ring ages from in situ logs indicate an average advance rate of 36 m yr⁻¹ between AD 1060 and 1808, with a significant stillstand or minor retreat around AD 1450. The recent retreat of the glacier, therefore, was an order of magnitude faster than its advance.

The advance and retreat of calving glaciers involves both the dynamic processes discussed in section 5.5.1 and the calving processes described in section 5.5.2. As we have seen, ice dynamics and calving are not independent, because calving is influenced by the velocity structure of the glacier, and ice velocity is sensitive to changes in terminus position and force balance (Meier and Post, 1987;

Benn et al., 2007a, b). Changes to this intricately coupled system can be thought of in terms of two end-member scenarios. First, changes in ice dynamics can lead changes in terminus stability; and second, changes in margin stability can initiate dynamic changes that then propagate upglacier. In the first case, the dynamics can be regarded as the 'master' and calving the 'slave'; whereas in the second case, calving is the 'master' and dynamics the 'slave'. Of course, these scenarios are not mutually exclusive, and feedbacks between dynamics and calving mean that many real-world situations contain elements of both.

An analytical model of the dynamic response of calving glaciers has been developed by Pfeffer (2007b). The model builds on earlier kinematic wave theory (e.g. Bahr et al., 1998; Pfeffer et al., 1998; section 5.2.5), and incorporates a pressure-dependent sliding law in which glacier motion is a function of driving stress and basal effective pressure (equation 4.23). Pfeffer argued that because driving stress and effective pressure both depend on ice thickness (equations 3.10 and 4.10), a calving glacier can respond to an initial, climatically forced thinning in one of two ways. When water depth is small relative to ice thickness, a reduction in ice thickness will reduce the driving stress more than effective pressure, thereby decreasing ice velocity. On the other hand, when ice is close to flotation, any reduction in ice thickness will cause the effective pressure to fall faster than the driving stress, resulting in an increase in sliding speed. Ice acceleration is accompanied by an increase in longitudinal strain rate, which stretches and thins the ice, leading to a further reduction in effective pressure. Velocity increase and thinning therefore form a positive feedback loop, in which an initial surface lowering and speed-up are amplified and propagated upglacier by dynamic processes. Benn et al. (2007a) also recognized that pressure-dependent sliding implies that glacier margins will be dynamically unstable when close to flotation. However, they pointed out that pressure-dependent sliding laws implicitly assume that basal drag is the only source of resistance, which is rarely the case for real calving margins. If a substantial proportion of the driving stress is supported by lateral drag, an externally imposed thinning is less likely to trigger flow acceleration because basal effective pressure has a smaller influence on the force balance. If a glacier margin advances or retreats into a wider section of a trough, however, the change in lateral restraint will result in flow acceleration. Furthermore, changing spatial patterns of basal and lateral drag can alter the magnitude and sign of longitudinal stress gradients, which may amplify or damp the initial dynamic response (O'Neel et al., 2005).

Benn et al. (2007a, b) showed how longitudinal stretching of a glacier tongue will encourage calving retreat (fig. 5.41). Higher extensional strain rates will lead to deeper crevasses (equation 4.28), and dynamic

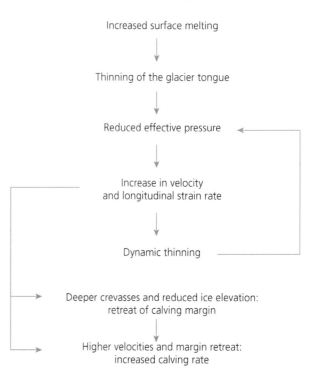

Figure 5.41 Relationships between thinning, acceleration and calving retreat, resulting from effective pressure-dependent basal motion and the influence of longitudinal strain rate on dynamic thinning and calving (Benn et al., 2007b).

thinning will reduce the ice height above the waterline, thus making it more likely that crevasses will extend down to water level and trigger calving. Increased stretching will therefore encourage retreat of the calving margin to a location where longitudinal strain rates are lower. An increase in surface melting will amplify this effect, making it easier for crevasses to reach the threshold depth. Rapid terminus retreat will occur if terminus thinning and stretching cause a glacier to pull back from a pinning point into deeper water or a wider reach of a trough, and will continue until the margin restabilizes at another pinning point.

The advance and retreat of calving glaciers, therefore, will be strongly influenced by along-flow variations in trough depth and width, and glaciers will tend to retreat more rapidly or advance more slowly through deeper and/or wider reaches. Where the bed slopes downward away from the ice margin, advances and retreats will tend to be rather restricted in magnitude. This is because any advance of the glacier into deeper water will lead to flow acceleration, longitudinal extension and crevassing, limiting the advance, whereas any retreat of the terminus will reduce calving losses, limiting the retreat. Conversely, where the bed slopes downwards upglacier, a small increase in ice thickness will initiate unstable advance into shallowing water, and a decrease in ice thickness will cause accelerating retreat into deeper water. Because of the influence of topography on ice margin stability, a population of calving glaciers can exhibit a range of

(a)

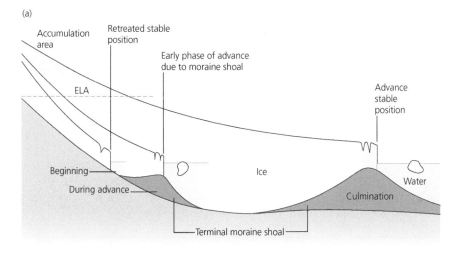

(b)

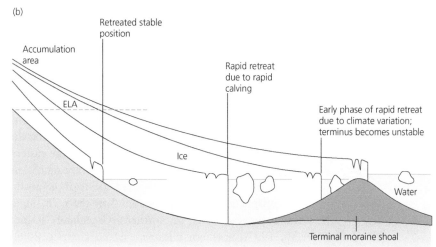

Figure 5.42 Patterns of glacier advance and retreat in association with grounding-line shoals. (a) Slow advance behind a mobile shoal which reduces calving losses at the margin. (b) Rapid retreat into deep water, leaving the shoal abandoned at the limit of the advance (Redrawn from Warren, 1992).

non-linear reactions to climatic forcing, whereby a given climatic signal will trigger large or small glacier response (Warren, 1992; Wiles et al., 1995).

The advance and retreat of calving glaciers is also influenced by sediment deposition and erosion, which change the distribution of pinning points and overdeepenings. The accumulation of grounding-line fans, morainal banks or deltas at a glacier margin can reduce calving losses and encourage ice margin stability (Warren, 1992; Hunter et al., 1996). Sediment shoals can advance, conveyor-belt fashion, in front of the glacier by a combination of subglacial erosion on the upglacier side and deposition on the downglacier side (fig. 5.42). This mobile pinning point can allow the glacier to advance slowly through deep water behind the advancing shoal. Subglacial erosion of sediment behind the shoal can rapidly create an overdeepening beneath the glacier (Wiles et al., 1995), creating a situation where a small recession of the margin can initiate catastrophic calving retreat into deepening water. Thus, glacier retreat from grounding line shoals tends to be rapid, in contrast with advance by shoal migration, which is limited by sediment

throughput rates. Rapid subglacial sediment erosion exerted a strong influence on the late twentieth-century retreat of Breiðamerkurjökull, Iceland (Björnsson, 1996; Nick et al., 2007b). During the Little Ice Age, the glacier advanced over a thick package of fluvial sediments and terminated close to the sea. Sediment erosion beneath the glacier created an overdeepened basin, which has encouraged rapid calving retreat of part of the margin.

The effects of sedimentation, in combination with climatic and glaciological factors, can explain the asynchronous behaviour of Alaskan tidewater glaciers noted above (Trabant et al., 2002). The advancing glaciers tend to have unusually small ablation areas, and prior to their twentieth-century advance experienced a major retreat phase. This earlier retreat removed most of the glaciers' ablation areas, so that once the glaciers stabilized in shallow water they had positive mass balances. Sediment shoals building up in front of the glacier margins then allowed the glaciers to readvance by reducing calving losses. This process can be reversed as the result of sediment dynamics in the shoal. Rapid retreat of Muir Glacier, for example, appears to have been triggered by

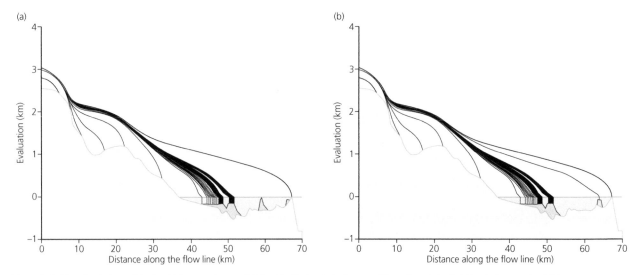

Figure 5.43 Two simulations of the late Holocene advance of Columbia Glacier, including a mobile sediment shoal at the glacier margin. Calving rates are parameterized using (a) water depth and (b) height above buoyancy functions. The ice surface profiles are shown for 20-year time intervals, so the spacing between profiles provides an indication of advance rate (After Nick et al., 2007a).

collapse of its morainal bank during the 1899 earthquake (Hunter et al., 1996). The effects of ice-marginal sedimentation on the dynamics of tidewater glaciers have been modelled by Nick et al. (2007a), who simulated the late Holocene advance of Columbia Glacier using a pressure-dependent sliding law and both water-depth and height-above-buoyancy calving rules (fig. 5.43).

5.6 ICE SHELVES

Ice shelves have attracted a great deal of scientific and media attention in recent years, partly because spectacular ice shelf break-up events in the Antarctic Peninsula have been interpreted by some as ominous symptoms of climate change. There are concerns that ice shelf removal may impact on the stability of inland ice, increasing the flux of land ice to the oceans. It is important, therefore, to understand the mass balance and dynamics of ice shelves and their relationship with feeder glaciers.

Three broad types of ice shelves can be recognized:

- *glacier ice shelves*, consisting of the floating margins of outlet or valley glaciers;
- *sea ice shelves*, or areas of locally grounded fast ice (i.e. sea ice attached to a landmass), fed by surface snowfall and basal freezing;
- *composite ice shelves*, which contain significant amounts of both glacier ice and sea ice (Jeffries, 2002; Jezek and Liu, 2005).

Ice shelves are confined to high polar settings where glacier ice is below the pressure-melting point. As noted in section 5.5.2, floating ramps can extend up to a few hundred metres in front of temperate glaciers, although these are ephemeral features which are not classed as ice

shelves. The largest ice shelves are the Ross and Filchner-Ronne Ice Shelves, which occupy huge embayments in the coast of continental Antarctica (section 6.3.1). Many smaller ice shelves exist around Antarctica, including the Antarctic Peninsula as far as around 66° S on the eastern side and 70° S on the western. Ice shelves are much less extensive along the warmer glaciated coasts of the northern hemisphere, and are only found in the cold, high-latitude environments of northern Greenland (Rignot et al., 1997), northern Ellesmere Island (Jeffries, 2002) and Franz Josef Land (Dowdeswell et al., 1994a).

The boundary between grounded glacier ice and a floating ice shelf is known as the *grounding line*. Where the ice surface gradient is small, this boundary may consist of a transition zone of lightly grounded and barely floating ice, rather than a well-defined line, and in such cases the term *grounding zone* is more appropriate. By definition, ice thickness at the grounding line is equal to the flotation thickness (equation 5.10), although it should be noted that the mean density of high-latitude ice shelves can be much less than $900\,\mathrm{kg\,m^{-3}}$, because their near-surface layers are composed of compacted snow, not solid glacier ice. The grounding line of a glacier can be identified from surface elevation data by a pronounced downglacier decrease in surface slope, reflecting the transition to zero basal drag below the floating portion (Dowdeswell et al., 1994; Rignot et al., 1997).

It is often asserted that melting of ice shelves does not contribute to sea-level rise because ice shelves are afloat and already displace water. This is not actually true, because ice shelf melting produces fresh water, which has a lower density than seawater. This means that the volume of seawater displaced by an ice shelf is less than the volume occupied by the same mass of fresh water.

The volume difference is around 2.6 per cent, and it has been calculated that if all extant sea ice and ice shelves melted, average global sea level would rise by ~4 cm (Noerdlinger and Brower, 2007).

5.6.1 MASS BALANCE OF ICE SHELVES

Ice shelves can gain mass by glacier inflow, surface snowfall and basal freezing, and lose mass by iceberg calving, surface melt and basal melting (fig. 5.44). Ice shelves are particularly sensitive to energy inputs from the atmosphere and the ocean, which determine the extent of surface and basal melting, respectively, and they can respond rapidly to changes in either weather or oceanic circulation patterns (Scambos et al., 2000; Rignot and Jacobs, 2002).

Snow accumulation on ice shelves is highest where low temperatures coincide with proximity to oceanic moisture sources. Significant surface melting occurs during summer on some ice shelves, particularly those at relatively low latitudes, such as around the Antarctic Peninsula. Because of the low gradients of ice shelves, prolonged surface melting can result in the extensive development of meltponds (Scambos et al., 2000). The accumulation of surface water can trigger a positive feedback effect, because the relatively low albedo of water increases the absorption of shortwave radiation and accelerates melting of the surrounding snow. Widespread melting and pond formation preceded the recent collapse of some Antarctic Peninsula ice shelves (van den Broeke, 2005), probably due to the role of water in increasing the penetration depth of surface crevasses (Scambos et al., 2000; MacAyeal et al., 2003; section 5.6.4). This sensitivity to surface melting means that ice shelves may be vulnerable even if the surface mass balance is not strongly negative. In the Antarctic Peninsula, the viability limit for ice shelves coincides with the −9 °C mean annual isotherm (Morris and Vaughan, 2003).

Basal melting and freezing processes contribute substantially to the mass balance of ice shelves. The freezing point of water decreases with pressure and salinity, so basal melting and accumulation are controlled by the characteristics of the water in the sub-ice shelf cavity. Basal freezing will be favoured by factors that reduce water temperature, salinity or pressure, while melting will be encouraged by increases in these variables. For example, buoyant subglacial meltwater emerging at the grounding line will rise along the ice shelf base, reducing its pressure and leading to freeze-on. Large-scale water circulation patterns also have a major effect on sub-ice shelf mass balance (Holland, 2002; Walker and Holland, 2007). Sea ice formation beyond an ice shelf produces high-salinity water, due to brine rejection during freezing. This relatively dense water sinks and enters the ice shelf cavity at depth. Although the water is near the surface

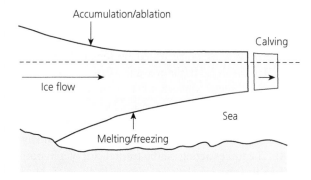

Figure 5.44 Sketch of an ice shelf, showing inputs and outputs.

freezing point, the higher pressure at depth means that the water is warm enough to melt ice at the ice shelf base. The addition of meltwater forms cold, fresh and buoyant *ice shelf water*. This water forms frazil ice as it ascends, which accretes onto the ice shelf base. Freezing at the base of an ice shelf produces a layer of *marine ice*, which can be readily identified in ice cores from its characteristic crystallographic and isotopic signatures (Souchez and Lorrain, 1991; Oerter et al., 1992). Marine ice thickness can be determined remotely using radio-echo sounding, because of the dielectric contrast between meteoric and marine ice (Thyssen et al., 1993). For example, using a combination of satellite radar altimetry and airborne radio echo sounding, Fricker et al. (2001) found that marine ice accounts for approximately 9 per cent of the volume of the Amery Ice Shelf. Marine ice is concentrated in the north-west of the shelf, reflecting clockwise water circulation. There is strong evidence that melting below ice shelves has recently increased in response to incursion of warmer ocean waters (Rignot and Jacobs, 2002; Shepherd et al., 2004). This is an important process because ice shelf thinning influences the dynamics of feeder glaciers and ice streams, so that oceanic warming signals can be transmitted deep into ice sheets (sections 5.6.4 and 6.3.6).

As discussed in section 5.5.2, two main processes are involved in calving from ice shelves. First, bending associated with the force imbalance at the ice front leads to frequent low-magnitude events that eat into the ice shelf edge. Second, rare high-magnitude events occur when rifts (full-depth crevasses) meet the shelf edge or other rifts and isolate large tabular bergs (Lazzara et al., 1999; Joughin and MacAyeal, 2005; fig. 5.29). Propagation of rifts is generally slow and episodic, and occurs mainly in response to internal glaciological stresses rather than external forces imposed by storms and tides (Larour et al., 2004; Joughin and MacAyeal, 2005; Bassis et al., 2008). Calving during ice shelf break-up can occur by an extreme form of fracture propagation, in which large sections of the shelf break into innumerable small fragments. This process is discussed in detail in section 5.6.4.

5.6.2 FLOW OF ICE SHELVES

Where ice is fully afloat, basal drag is zero and all resistance to flow must be provided by longitudinal stresses and/or lateral drag at the margins. Patterns of ice shelf flow reflect these sources of resistance. Longitudinal stresses arise from gravitational and buoyant forces acting on the floating ice, and are accommodated by *creep spreading* (van der Veen, 1999a; van der Veen and Payne, 2004). The forces acting on the frontal cliff of a hypothetical free-floating ice shelf are shown in figure 5.31. The vertically averaged outward-directed stress resulting from the weight of the ice is:

$$\bar{P}_I = \frac{1}{2}\rho_I g H \qquad (5.15)$$

and the vertically averaged backward-directed stress due to water pressure is:

$$\bar{P}_W = \frac{1}{2}\rho_W g(H-h) \qquad (5.16)$$

where H is the ice thickness and h is the elevation of the ice surface above water level. For free-floating ice, $h = (1 - \rho_i/\rho_W)H$, so the net outward-directed stress (the mean longitudinal deviatoric stress) is:

$$\bar{\sigma}'_x = \frac{1}{2}\rho_I g\left(1 - \frac{\rho_I}{\rho_W}\right)H \qquad (5.17)$$

(van der Veen and Payne, 2004). Thus, even though the whole system is in hydrostatic equilibrium, there is a net outward-directed deviatoric stress at the ice front. This is accommodated by outward creep of the ice, and by inserting equation 5.17 into Glen's Flow Law (equation 4.15) the longitudinal strain rate is found to be:

$$\frac{\partial U}{\partial x} = A\left[\frac{1}{4}\rho_I g\left(1 - \frac{\rho_I}{\rho_W}\right)\right]^n H^n \qquad (5.18)$$

This shows that the creep spreading rate depends only on the ice thickness H and the relative densities of ice and water.

Most, if not all, ice shelves are grounded at the sides by fjord margins, islands or promontories. These grounded margins exert drag on the ice, increasing the resistance of the system to flow in much the same way as the margins of a valley glacier (section 4.6.2). Ice shelves with grounded margins are referred to in the literature as *confined ice shelves*, in distinction to (hypothetical) *unconfined ice shelves*. Lateral resistance typically dominates the force balance of confined ice shelves, so that transverse velocity profiles resemble those predicted by equation 4.24b and

shown in figure 4.29. The theoretical velocity profiles, however, are based on the assumption that the flow law parameters A and n are constant across the shelf, which is unlikely to be the case in reality. Indeed, inverse modelling studies of ice shelf flow (e.g. Vieli et al., 2006, 2007) have shown that ice is much 'softer' at flow-unit boundaries due to strain heating, crevassing and other factors, increasing velocity gradients near the shelf margins and along longitudinal bands between major flow units (*suture zones*; fig. 5.45).

Lateral drag at ice shelf margins allows the shelf to be thicker than it would be if it were simply free-floating. Surface slopes will be steepest for narrow confined ice shelves, and gentlest in wide ice shelves where drag at the margins is relatively unimportant, a prediction confirmed by field data (fig. 5.46). A related effect is that drag around ice shelf margins helps to retard the flow of ice located upglacier. This upglacier-directed force is known as *backstress* (van der Veen, 1997b), defined as: *resistance to flow at the lateral margins (and locally grounded areas) of an ice shelf, which is transferred upglacier by longitudinal stresses.* A free-floating, unconfined ice shelf has no drag at the lateral margins or the bed, so it exerts no backstress on the grounded ice upglacier. In this case, the average upglacier-oriented force at the grounding line is exactly the same as it would be if no ice shelf were present, so removal of the shelf will not affect the force balance of the grounded part of the glacier. In the (real-world) case of confined ice shelves, removal of the shelf would result in a loss of backstress, and cause the grounded part of the glacier to accelerate. Numerical modelling of glacier response to ice shelf removal (DuPont and Alley, 2005; Hindmarsh, 2006) has shown that the acceleration is a transitory effect, continuing only until the glacier finds a new equilibrium. Essentially, the loss of backstress causes the glacier to stretch and thin until the driving stress matches the reduced resistance.

Glacier response to the break-up of ice shelves around the Antarctic Peninsula is discussed in the following section. The implications of this process for the stability of the West Antarctic Ice Sheet are discussed in section 6.3.6.

5.6.3 ICE SHELF BREAK-UP

In recent years, ice shelves in both the northern and southern hemispheres have undergone rapid retreat. In most cases, ice shelf retreat followed two distinct phases, with an initial phase of progressive, climatically driven frontal retreat lasting for several years (and sometimes decades), followed by a second phase of catastrophic collapse. Ice shelf extent on the northern coast of Ellesmere Island, Canada, contracted by 90 per cent during the period 1906–82, by calving at the seaward edge (Vincent et al., 2001). The ice shelf remnants then stabilized until 2000–2, when the largest remnant – the ~3000-year-old

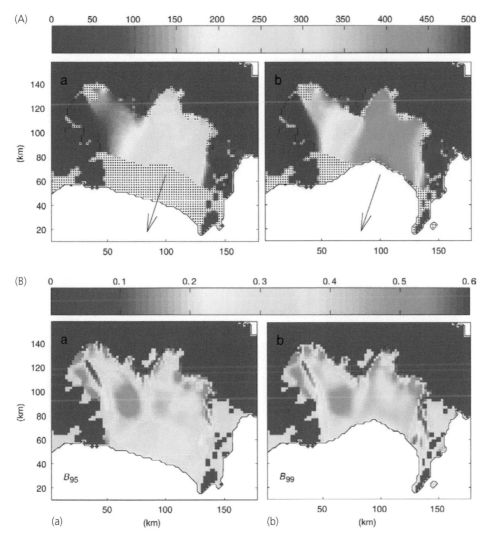

Figure 5.45 Flow of Larsen B ice shelf prior to its collapse. (A) Surface flow velocities (m yr^{-1}) in (a) 1995 and (b) 1999. Black dotted areas indicate regions without velocity data. (B) Best estimates of the flow law parameter B (Pa yr$^{0.33}$ × 10^6) derived from inverse modelling. The parameter B is related to the prefactor A in Glen's Flow Law and increases with ice stiffness (Vieli et al., 2007).

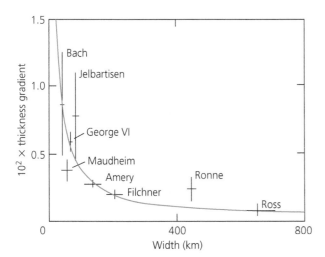

Figure 5.46 Relationship between the ice shelf width and longitudinal gradient in ice thickness. Narrower ice shelves are steeper due to the restraining effects of lateral drag (Paterson, 1994).

Ward Hunt Ice Shelf – broke in two, followed by additional fissuring and calving (Mueller et al., 2003). Ice shelf break-up led to the drainage of an ice-dammed lake (Disraeli Fjord), a rare ecosystem type.

Much more extensive ice shelf loss has occurred around the Antarctic Peninsula, largely in response to climatic warming in the region. Summer temperatures measured at Faraday/Vernadski station have increased by 0.56 °C per decade over the last 50 years (Turner et al., 2005a). On the western side of the peninsula, the Wordie and Müller Ice Shelves began to retreat around 1960, culminating in the rapid disintegration of the former in the late 1980s (Vaughan and Doake, 1996). Before final disintegration, the Wordie Ice Shelf was split by numerous longitudinal rifts, essentially changing one confined ice shelf into several unconfined shelves which then rapidly broke up. Wilkins Ice Shelf began to experience large losses in the early 1990s, with a major break-up event occurring in February and March 2008. George VI Ice Shelf, pinned between the mainland and Alexander Island, was reduced in area by almost 1000 km^2 between 1974 and 1995 (Lucchitta and Rosanova, 1998). The northern margin of this ice shelf marks the southernmost boundary of recent ice shelf retreat on the Antarctic Peninsula, and it is believed to be particularly vulnerable to future atmospheric and oceanic warming (Smith et al., 2007).

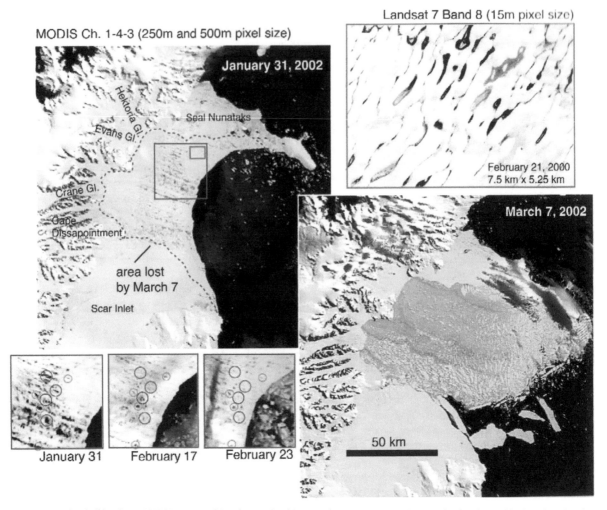

Figure 5.47 Larsen ice shelf break-up, 2002 (Courtesy Ted Scambos, National Snow and Ice Data Center, University of Colorado, Boulder, based on data from MODIS).

The most spectacular and widely known break-up events affected the Larsen Ice Shelf on the eastern side of the peninsula. Formerly the largest of the Antarctic Peninsula shelves, the Larsen Ice Shelf consisted of four sectors, labelled A, B, C and D, from north to south. Of these, only the south-west corner of Larsen B and the relatively small Larsen C and D remain at the time of writing (2009). Between 1975 and 1989, the edge of the northernmost portion, Larsen A, retreated at ~1 km yr^{-1}. Retreat accelerated in the early 1990s, culminating in rapid disintegration of most of the remaining ice shelf in January 1995 (Rott et al., 1996, 2002; Skvarca et al., 1999; MacAyeal et al., 2003). In only a few days, 1200 km^2 of shelf fragmented into a mixture of kilometre-scale tabular icebergs and innumerable smaller pieces. This pattern of disintegration was repeated on an even larger scale when Larsen B broke up in 2002. Most of the 3370 km^2 of ice shelf break-up occurred between 23 February and 7 March 2002 (Scambos et al., 2003; Rack and Rott, 2004; Glasser and Scambos, 2008; fig. 5.47).

The break-up of the Larsen Ice Shelf was preceded by a decade of surface lowering, during which surface elevation was reduced by up to $0.27 \pm 0.11\,\mathrm{m\,yr^{-1}}$ (Shepherd et al., 2003). Ice shelf thinning was probably the result of both surface ablation and increased melting at the ice shelf base. The final break-up of both Larsen A and B followed unusually warm summers during which large numbers of meltponds formed on the ice shelf surfaces. Scambos et al. (2000) argued that this surface water allowed crevasses to penetrate the full thickness of the shelf. As noted in section 4.7.1, water-filled crevasses can propagate downwards essentially without limit, because water pressures at crack tips are higher than cryostatic pressures, allowing tensile stresses to pull the ice apart (van der Veen, 1998a). In the absence of surface water, longitudinal strain rates over much of the Larsen Ice Shelf were not sufficiently high to promote rift formation, but when surface crevasses were filled with water, the altered force balance led to full-depth rifting nearly simultaneously over large portions of the ice shelves. The force exerted by large blocks as they turned over during calving also appears to be a significant factor in the near-simultaneity of ice shelf collapse, in a kind of 'domino effect' (MacAyeal et al., 2003; fig. 5.48). Vieli et al. (2007)

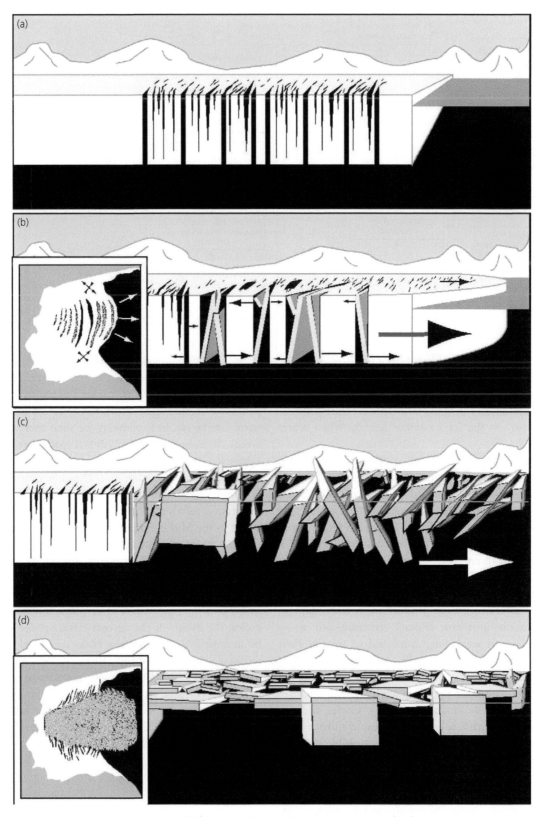

Figure 5.48 Hypothetical mechanism of catastrophic ice shelf break-up. (a) Longitudinal extension and inputs of surface meltwater promote widespread full-depth crevassing. (b) Narrow, free-floating fragments begin to capsize, exerting additional forces on the remaining shelf. (c) Climax of the break-up event, in which a chaotic mass of icebergs spreads outwards. (d) Terminal condition, in which a significant fraction of the icebergs have capsized (MacAyeal et al., 2003).

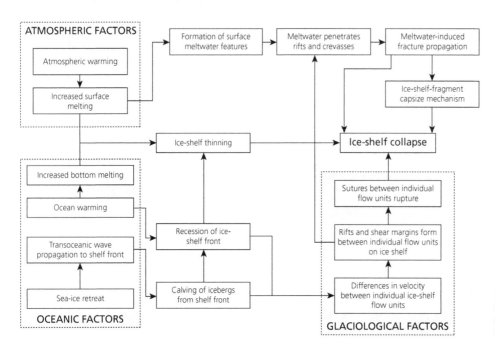

Figure 5.49 Flow chart showing the network of factors influencing ice shelf collapse (Redrawn from Glasser and Scambos, 2008).

showed that Larsen B accelerated prior to its collapse, due to softening of shear zones along margins. The cause of the softening appears to be the penetration of water into marginal crevasses, which causes deeper crevasse penetration and warms the ice by latent heating when water refreezes.

Glasser and Scambos (2008) conducted a structural glaciological analysis of Larsen B, and showed that pre-existing lines of weakness played a crucial role in the 2002 break-up. Analysis of satellite imagery demonstrated that the former shelf consisted of four flow units fed by separate glacier systems, between which lay relatively weak *suture zones* of thinner ice. Within the last 20 years, new rift systems developed within two of the suture zones, further weakening the ice. Final collapse occurred when increased surface meltwater production, bottom melting, changes in the upstream velocity structure and widespread structural weakening combined to tip the shelf over a critical threshold (fig. 5.49). Jezek and Liu (2005) have pointed out that many of the ice shelves along the south-eastern Antarctic Peninsula are composite features made up of heavily fractured floating glacier tongues bound together with fast ice. These areas are likely to be mechanically weak, and particularly susceptible to rapid break-up.

Break-up of Larsen A and B has also affected the dynamics of outlet glaciers that formerly fed into the shelf (Rott et al., 2002; De Angelis and Skvarca, 2003; Scambos et al., 2004). Following the 2002 collapse of Larsen B, several glaciers underwent dramatic acceleration and thinning. The greatest speed-up occurred on Hektoria and Green Glaciers, where a large acceleration began within months of ice shelf collapse. Glacier speed-up was greatest near the terminus, leading to high longitudinal strain rates and crevassing. The acceleration and ice surface lowering propagated upstream over the ensuing months, and continued for at least 2 years (fig. 5.50). This behaviour is almost certainly the result of loss of back-stress on the glaciers following ice shelf removal (section 5.6.2). It is noteworthy in this respect that Flask and Leppard Glaciers, which flow into the remaining shelf areas, showed no speed-up over the same time period.

Marine sediments in areas of recently vanished Antarctic ice shelves show that open-water conditions prevailed in the Mid Holocene, prior to ice shelf regrowth in the last few millennia (Hjort et al., 2001; Pudsey and Evans, 2001). Larsen B, however, appears to have existed continuously since before the LGM until its recent disintegration (Domack et al., 2005). Thus, while ice shelf retreat and regrowth clearly occur in response to natural climate cycles, the extent of the current retreat phase appears to be unprecedented within the Holocene. In global terms, recent calving losses from Antarctic Peninsula glaciers have had only a small effect on mean sea level. In some parts of continental Antarctica, ice shelf thinning has been linked to increased ice discharge from the interior, with much bigger implications for future sea-level rise. This important topic is addressed in section 6.3.7.

5.7   GLACIER SURGES

5.7.1   OVERVIEW

Some glaciers switch between phases of rapid and slow flow on timescales of a few years to several decades. Such *surging glaciers* have attracted a great deal of attention because, like many instances of unusual or pathological behaviour, they offer a very instructive perspective on

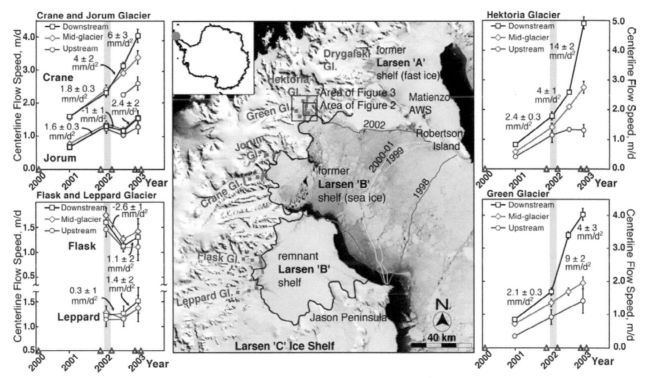

Figure 5.50 Glacier response to removal of Larsen B ice shelf. Centre panel: MODIS image showing ice shelf extents and location of tributary glaciers. Surrounding panels: Time series of centreline velocities of six glaciers in the Larsen B catchment (Scambos et al., 2004).

'normality'. As originally defined, glacier surges are cyclic phenomena which are not directly triggered by external events, but instead result from internally driven oscillations in conditions at the bed of the glacier (Meier and Post, 1969; Sharp, 1988a). In some of the literature, however, the term 'surge' is used more loosely to refer to any dramatic glacier speed-up (e.g. the 'surge' of tidewater glaciers following the collapse of Larsen A in the Antarctic Peninsula; De Angelis and Skvarca, 2003; section 5.6.3). Such externally forced events, however, fall outside the normally accepted definition of surges, and in this book we maintain a distinction between internally triggered surges and externally triggered *speed-ups*. The distinction between internal and external triggers, however, may be difficult to apply in some cases because internal dynamic processes are clearly not independent of external conditions. In addition, glaciers exhibit a wide range of surging behaviours, and the distinction between surge-type and non surge-type glaciers may not be clear cut (Meier and Post, 1969; Murray et al., 2003; Nolan, 2003; Frappé and Clarke, 2007).

Surging glaciers undergo systematic changes in morphology and behaviour during a surge cycle (fig. 5.51; Meier and Post, 1969; Raymond, 1987; Murray et al., 2003). The period of slow flow between surges is known as the *quiescent phase*. Ice velocities are less than the balance velocity, and ice builds up in a *reservoir area* in the upper part of the glacier. The increase of mass in the upper part of the glacier and mass loss near the terminus result in an increase in the ice surface gradient, which continues until the *surge* or *active phase* is initiated. During the surge, ice is rapidly transferred from the reservoir area to the *receiving area* in the lower glacier. Removal of ice from the reservoir area results in rapid surface lowering, and remnants of pre-surge ice are commonly left stranded on valley sides, high above the new glacier surface (fig. 5.52). At the same time, transfer of ice to the receiving area results in thickening and, in some cases, dramatic advance of the glacier front (fig. 5.53). However, not all surges lead to ice front advances, and many come to an end before reaching the glacier terminus. Maximum velocities during the active phase are typically one or two orders of magnitude greater than the velocity during the quiescent phase.

On land-terminating glaciers, surges often begin in the upper reaches of the glacier, then propagate downglacier. The boundary between fast-flowing ice and the slower ice downglacier is a zone of intense compression, and usually takes the form of a steep bulge or *surge front*. The front behaves like a kinematic wave, travelling downglacier faster than the ice velocity, and as it passes downglacier individual packets of ice are first accelerated then decelerated. This pattern of surge propagation has been documented on numerous glaciers, including Variegated Glacier, Alaska (Kamb et al., 1985), Bakaninbreen, Svalbard (Murray et al., 1998) and Trapridge Glacier, Yukon (Frappé and Clarke, 2007). Surges can also propagate upglacier from the point of initiation. In Iceland, for example, surges commonly begin in the upper ablation zone, then propagate up- and downglacier from there (Björnsson et al., 2003). On some

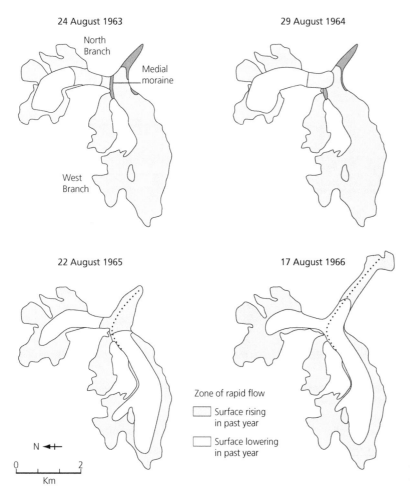

24 August 1963

North
Branch

Medial
moraine

West
Branch

29 August 1964

22 August 1965

17 August 1966

Zone of rapid flow

☐ Surface rising
   in past year

☐ Surface lowering
   in past year

N ◄━┼━

0   2
     Km

Figure 5.51 Maps of Tyeen Glacier, Alaska, showing shifting zones of ice thickening and thinning associated with surges of two major branches of the glacier (After Meier and Post, 1969).

Figure 5.52 A depleted reservoir area following a surge; Skobreen, Svalbard in June 2006. The dark band above the glacier on the left side of the picture is pre-surge ice, indicating drawdown of up to 50 m. This dynamic thinning and the extensive development transverse crevasses are consequences of longitudinal stretching. See figure 5.60 for location of Skobreen (D.I. Benn).

tidewater glaciers, surges have started in broad areas of the lower tongue, then propagated up-ice from there. This pattern has been documented on several tidewater glaciers in Svalbard, including Osbornebreen (Rolstad et al., 1997), Monacobreen (Luckman et al., 2002), Fridtjofbreen (Murray et al., 2003) and Perseibreen (Dowdeswell and Benham, 2003).

Pritchard et al. (2003, 2005) have presented a detailed analysis of velocity and elevation changes on Sortebrae, a tidewater glacier in east Greenland, during its 1992–95 surge. The surge began between November 1992 and January 1993 in the central part of the glacier, and then propagated rapidly upglacier. The active phase lasted for 28–32 months, with flow rates of up to 24 m day^{-1}.

Figure 5.53 The front of Paulabreen, the receiving area of the surge of Skobreen. The picture was taken in September 2005 before surge termination, and shows heavily crevassed ice calving into the fjord. See figure 5.60 for location of Paulabreen (L. Kristensen).

In total, the surge transferred $24.3 \pm 9.5 \, km^3$ of ice from the reservoir zone, with an average surface lowering of 180 m on the upper glacier. In contrast, there was relatively little thickening on the lower glacier, but the terminus advanced by 10 km and an estimated $12.1 \, km^3$ of ice was lost by calving.

For each surging glacier, the active and quiescent phases tend to be of relatively constant length, resulting in a quasi-periodic cycle, although there are large variations in cycle length between glaciers and between regions (Dowdeswell et al., 1991; Björnsson et al., 2003). In Svalbard, for example, the active phase of surging glaciers typically lasts for 4–10 years, compared with only 1–3 years for surging glaciers in north-west Alaska, Iceland and the Pamirs. However, although the active phase is longer in Svalbard, maximum ice velocities are comparatively low, ranging between 1.3 and 16 m day^{-1}, compared with velocities of 50 m day^{-1} measured on Variegated Glacier in Alaska. Thus, during a surge in Svalbard, mass is transferred over a longer period but at a lower rate than in Alaska. The quiescent phase is also relatively long for Svalbard glaciers (50–500 years) compared with other areas (20–40 years). For some glaciers, the length of the surge cycle has been shown to reflect the time required for snow accumulation to refill the reservoir zone (e.g. Eisen et al., 2001), so the surge cycle will be shorter where snowfall rates are high (such as coastal regions of Alaska) compared with more arid regions (such as Svalbard). On some Icelandic glaciers, however, surges occur at highly irregular intervals, indicating that climate is not the only control.

It has been argued that glaciers may pass from surging to non-surging type in response to climate change if, for example, there is insufficient mass accumulation to recharge the reservoir zone. Dowdeswell et al. (1995) showed that 18 Svalbard glaciers were surging in 1936–38, compared with only 5 in 1990, and argued that this trend reflects sustained negative glacier mass balance during the late twentieth century. The statistical significance of this result remains to be tested, but it is clear that many small former surge-type glaciers are now cold-based with strongly negative mass balances, and are unlikely to surge again under present conditions. Similarly, Vernagtferner in the Austrian Alps experienced four strong, surge-like advances during the Little Ice Age, but is not currently of surge-type (Hoinkes, 1969). On the other hand, climate change may be increasing surge frequency in the Karakoram Mountains of Pakistan (Hewitt, 2007). These observations show that it is logically possible for a glacier to undergo a single surge during a short-lived climate cycle. Although the surge itself may be triggered by internal dynamic processes, these are embedded within a wider set of controls on glacier behaviour, thus blurring the distinction between internal and external controls.

Surges leave distinctive structural imprints on glaciers. One of the most obvious is the presence of *looped medial moraines*, which are striking teardrop shaped loops of debris on the surface of many surging glaciers (Meier and Post, 1969; Croot, 1988a). Looped moraines record cyclic differences in velocity between a trunk glacier and its tributaries (fig. 5.54). If tributary glaciers surge when the main trunk glacier is quiescent, they advance across the trunk glacier forming moraine loops. These are then carried downglacier by the next surge of the trunk. The intense shear characteristic of the margins of surging glaciers can form longitudinal foliation in the ice (Pfeffer, 1992), whereas compression at the surge front causes folding, thrust faulting, crevassing and thickening of basal debris sequences (fig. 5.55; Sharp et al., 1988; Lawson et al., 1994; section 4.7.3).

Lawson (1996) and Lawson et al. (2000) used aerial photographs and surface velocity data to study the structural evolution of Variegated Glacier, Alaska, over its recent surge cycles. It was found that surging creates distinctive crevasse patterns on the glacier surface, reflecting patterns of cumulative strain on the glacier. In the upper parts of the glacier, which were unaffected by surges, transverse crevasses record predominantly extending flow. In contrast, the middle zone of the glacier has a complex crevasse pattern, including longitudinal crevasses formed by compressive flow in advance of the surge velocity peak, and transverse crevasses formed by extending flow behind the velocity peak. Finally, the lower zone, which experienced only compressive flow, is

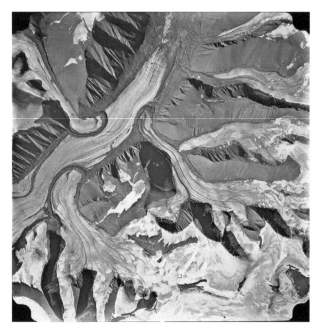
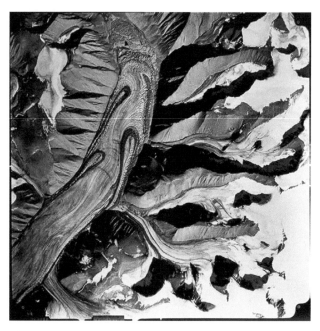

Figure 5.54 Abrahamsbreen, Svalbard: (a) in 1969 before a surge, and (b) in 1990 after a surge. The transport of tributary ice lobes by the trunk glacier has formed a series of looped moraines (Aerial photographs provided by Norsk Polarinstitutt).

Figure 5.55 Paulabreen during its surge in September 2005. The heavily crevassed surge front is advancing into old ice, which is thickening and cracking in response to high compressive stresses (L. Kristensen).

characterized by longitudinal crevasses. Structural relationships on the glacier in general do not reflect the highly complex deformation histories of the ice, and evidence of substantial cumulative strain can be overprinted by later events.

5.7.2 DISTRIBUTION OF SURGING GLACIERS

Surging glaciers are not evenly distributed around the world's glaciated regions, but tend to cluster in particular areas. Concentrations of surging glaciers are found in Alaska, Yukon and British Columbia in north-western North America (Meier and Post, 1969; Clarke, 1991; Frappé and Clarke, 2007); Ellesmere Island and Axel Heiberg Island in Arctic Canada (Copland et al., 2003b);

Svalbard, East Greenland and Iceland in the north Atlantic region (Jiskoot et al., 2000, 2003; Björnsson et al., 2003); and the Pamir, Caucasus, Tien-shan and Karakoram mountains in Asia (Dolgushin and Osipova, 1975; Hewitt, 2007). The Arctic Canada and Svalbard clusters consist entirely of polythermal glaciers; the Icelandic surging glaciers are all warm based; whereas the north-western North America cluster contains both temperate and polythermal glaciers.

A number of statistical analyses have been conducted to identify possible environmental or glaciological controls on surging. While several statistically significant correlations have been found *within* regions, the results tend to be rather inconsistent *between* regions. For example, Clarke (1991) found that, for glaciers in the Yukon, surge-type glaciers tend to be longer, wider and less steep than normal glaciers, and that glacier length shows the strongest individual correlation with surge tendency. Glacier length is also significantly correlated with surge behaviour for Svalbard glaciers (Hamilton and Dowdeswell, 1996; Jiskoot et al., 2000), but in East Greenland there is no direct correlation between glacier length and surge behaviour (Jiskoot et al., 2003). In Iceland, where almost all of the major outlet glaciers draining the ice caps are surge-type, the average surface slopes of Icelandic surging glaciers lie in the range 1.6–4°, compared with 2.9–25.7° for non-surging glaciers (Björnsson et al., 2003). In contrast, Jiskoot et al. (2000) found that gently sloping glaciers in Svalbard are *less* likely to surge than steeper glaciers. There is a tendency for surging glaciers to cluster on certain rock types, but correlations are much stronger in some regions than others. In Svalbard, surging glaciers are most likely to occur on sedimentary rocks (Hamilton and Dowdeswell,

1996; Jiskoot et al., 2000), and in Alaska many surging glaciers are found within the fractured rocks of the Denali Fault system, but are absent from the hard, granitic rocks of the Coast Mountains (Post, 1969; Truffer et al., 1999). On the other hand, there is no apparent association between surging glaciers and substrate type in East Greenland or Iceland (Björnsson et al., 2003; Jiskoot et al., 2003). Some of the statistical relationships can be explained with reference to existing models of surge processes, but as yet no model is capable of explaining all of the observed patterns. One drawback of statistical studies is that glaciers may be misclassified as non-surge type because they have never been observed to surge and display no overt signs of having done so. This is a classic example of 'absence of evidence is not evidence of absence', and may be particularly problematic in areas where surge cycles are very long and observational records short.

Detailed studies have been conducted on only a very small sample of the world's surging glaciers. It is clear, however, that temperate and polythermal surging glaciers are strikingly different in terms of their dynamics. Examples of each are presented in the following sections.

5.7.3 TEMPERATE GLACIER SURGES

The classic example of a temperate surging glacier is Variegated Glacier, on the coastal side of the St Elias Mountains in Alaska (fig. 5.56). The focus of a pioneering and still unsurpassed field study during its 1982–83 surge, Variegated Glacier is the benchmark against which all other temperate surging glaciers are compared. Seven surges are known to have occurred in the twentieth century, with the interval between surge initiation ranging from 13 to 18 years (Eisen et al., 2005). The length of the surge cycle is modulated by surface mass balance in the accumulation zone, variations in which determine the time required to recharge the reservoir zone to a critical ice thickness (Eisen et al., 2001).

Evolution of the geometry, velocity and drainage of Variegated Glacier were monitored in detail between 1973 and 1986, spanning a complete surge cycle (Kamb et al., 1985; Raymond, 1987). During the quiescent phase prior to the 1982–83 surge, the reservoir area thickened and the receiving area thinned, resulting in a steepening of the glacier. At this time, surface velocities progressively increased, with maximum summer values rising from around 0.2 m day^{-1} in 1973 to around 1 m day^{-1} in 1981. These increases were largely due to rising ice creep rates in response to increasing driving stress, but a small contribution was also made by increased sliding rates. Throughout this period, measured velocities were much less than calculated balance velocities (Bindschadler et al., 1977; Raymond and Harrison, 1988). Each June and July between 1978 and 1981, the thickening reservoir area underwent four to six waves of accelerated motion similar

Figure 5.56 Variegated Glacier, shortly after the surge of 1982–83. Martin Sharp for scale (M.J. Hambrey).

to surges, but on a smaller scale (Kamb and Engelhardt, 1987). These *minisurges* propagated downglacier as waves of elevated ice velocity, with the wave peaks passing downglacier at 250–400 m hr^{-1}, or approximately 1000 times the ice velocity. As each wave passed through a point on the glacier, ice velocities increased abruptly (over a few hours), then decayed more slowly, over a day or so. The velocity waves were accompanied by waves of increased basal water pressure, with similar abrupt rises and slow falls. The minisurges were similar to short-term episodes of accelerated motion observed on non-surging glaciers (section 5.4.1), and were attributed to temporary increases in sliding velocity associated with a pulse of meltwater under high pressure. There was no apparent meteorological cause for increased meltwater, and Kamb and Engelhardt (1987) concluded that the water had been released from storage in reservoirs (probably subglacial). Hydrological aspects of the meltwater pulses have been discussed in detail by Humphrey et al. (1986).

The glacier surged in two phases, the first beginning early in 1982 and terminating in July of that year, and the second beginning in the winter of 1982–83 and lasting until early summer (Kamb et al., 1985). The second phase was the more extensive, and affected almost all of the glacier. Each phase took the form of a wave of enhanced velocities propagating downglacier, the leading edge of which was a dramatic ice bulge or *surge front*. Ice at the advancing surge front was subjected to intense longitudinal compression and vertical thickening, causing the ice surface to rise by up to 7 m day^{-1} (Raymond et al., 1987; fig. 5.57). Peak velocities occurred a short distance upglacier from the surge front, and farther upglacier the ice was subject to large tensile stresses and was extensively crevassed. With the passage of the surge waves, the ice surface in the reservoir area fell, leaving blocks of ice stranded on the valley walls. During the surge phases, the velocity fluctuated on hourly, daily and multi-day timescales (fig. 5.58). On several occasions, velocity peaks were followed by abrupt slowdowns, the last of which

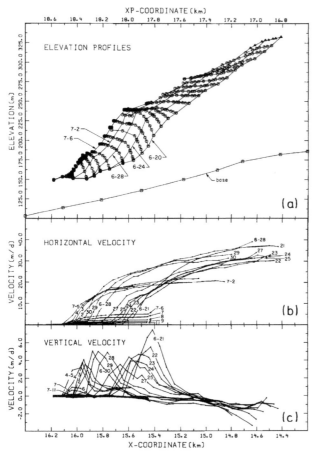

Figure 5.57 Propagation of the surge front, Variegated Glacier, 1983.
(a) Longitudinal profiles and ice trajectories (circles). (b) Horizontal velocities.
(c) Vertical velocities. Note the wave of strong uplift moving downglacier in (c)
(Raymond et al., 1987).

marked the termination of the surge. The slowdown events coincided with drops in subglacial water pressure, indicating that they reflect hydraulic controls on the sliding rate. Dye injected into boreholes during the surge phase indicate low meltwater transit velocities (approximately $0.02\,\mathrm{m\,s^{-1}}$) and temporary water storage at multiple sites below the glacier, indicative of an inefficient, distributed subglacial drainage system (section 3.4; Kamb et al., 1985; Kamb, 1987). Surge termination in early July was accompanied by the rapid release of large amounts of stored water and a marked increase in water transit velocities to around $0.7\,\mathrm{m\,s^{-1}}$, indicating the development of an efficient conduit system beneath the glacier.

The subsequent surge of Variegated Glacier followed a similar pattern to the 1982–83 event, although it was not monitored in the same level of detail (Eisen et al., 2005). The surge began in the winter of 1994–95 and terminated in June 1995, following two days of record high temperature. The surge appears to have consisted of this single phase, although it is possible that there was a much weaker second phase the following season.

The pattern of cold season initiation and summer termination also applies to most other Alaskan surging glaciers for which observations are available. For example, the single-phase 1987–88 surge of West Fork Glacier began in late August, after the main Alaskan melt season, and terminated the following July (Harrison et al., 1994), while the two phases of the 1993–95 surge of Bering Glacier began in winter or early spring and terminated in midsummer (Fatland and Lingle, 2002; Roush et al.,

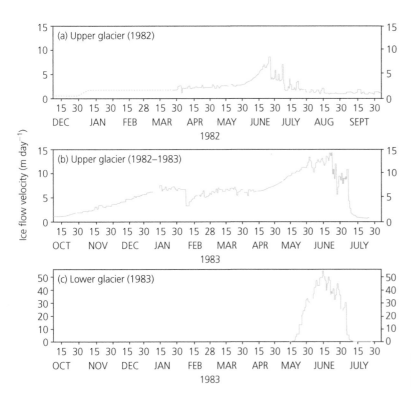

Figure 5.58 Time series of surface velocity on the upper and lower Variegated Glacier during the 1982/83 surge. Note different scales (Kamb et al., 1985).

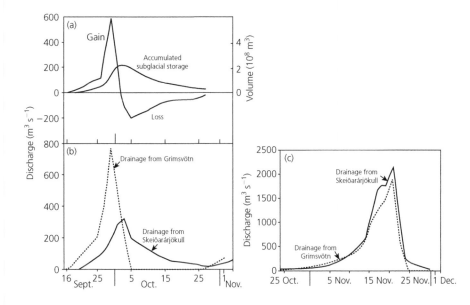

Figure 5.59 Contrasting hydrology of Skeiðarárjökull during and following its surge in 1991. (a) Change in water storage during the surge. (b) Drainage from Grimsvötn and discharge from the glacier terminus after the surge. Note the different scale (Björnsson, 1998).

2003). This pattern does not appear to be universal, however, as the 1986–87 surge of Peters Glacier probably terminated in winter. The hydrology of other Alaskan surging glaciers also appears to evolve in similar ways to that observed at Variegated Glacier. For example, there is evidence for extensive stored water at the bed of Bering Glacier during its 1993–95 surge (Fatland and Lingle, 2002), and termination of the first phase of the surge coincided with a massive flood, sufficiently great to float and transport ice blocks 25–30 m in diameter. The outrushing waters eroded an ice canyon (250 m wide and 70 m deep) headward into the glacier front, at a rate of 150–200 m per day for 3–4 days (Fleisher et al., 1998).

Further insight into the subglacial hydrology of temperate surging glaciers was provided by Björnsson (1998), using observations of the 1991 surge of Skeiðarárjökull, an outlet of the ice cap Vatnajökull in Iceland (fig. 5.59). The surge began in late March, about 10 km from the terminus, and propagated both down- and upglacier. In early May, the downward-propagating surge front reached the eastern side of the glacier terminus, which advanced until mid July, then stopped. In late September, the western side of the terminus began to advance, coincident with the input of water to the glacier bed during partial drainage of the Grimsvötn subglacial lake (see section 3.7.4). This drainage event was unlike other known floods from Grimsvötn, as there was a lag of 4 days between the initiation of drainage and the appearance of flood water at the glacier terminus, and for over a month considerable amounts of water were retained below the glacier. The advance of the western terminus terminated in early November, when there was only a small remaining volume of stored water. At around this time, a second flood from Grimsvötn occurred. This time, the flood followed the usual pattern, with a strongly peaked discharge hydrograph indicative of drainage through an enlarging

conduit. Björnsson concluded that the contrasting flood hydrographs reflect different configurations of the subglacial drainage system during surge and non-surge conditions. During the surge, water escaping from the lake encountered a distributed drainage system, and was able to spread out into multiple temporary storage sites. A conduit system did not develop, despite considerable input of water from the lake. After surge termination, on the other hand, a lake outburst was rapidly followed by conduit development and efficient evacuation of water from beneath the glacier.

5.7.4 POLYTHERMAL SURGING GLACIERS

Polythermal glaciers are very diverse in terms of their thermal structure (section 2.3.3), and this is also true of their surging behaviour. Surging glaciers of warm polythermal type, with only limited areas of cold ice in their upper reaches (Types d and e, fig. 2.9), appear to be similar to temperate surging glaciers in all important respects (e.g. West Fork Glacier; Harrison et al., 1994). In this section, we focus on polythermal surging glaciers that are at least partly underlain by cold ice during the quiescent period. Detailed studies have been conducted on two polythermal surging glaciers: Bakaninbreen in Svalbard and Trapridge Glacier in the northern St Elias Mountains, Yukon Territory, Canada. These are discussed in turn below.

Bakaninbreen is a 17 km long valley glacier in central Spitsbergen, Svalbard, that last surged between 1985 and 1995. About 7 km from its terminus, Bakaninbreen is confluent with Paulabreen, and the lower ablation zones of the two glaciers are separated by a prominent medial moraine (fig. 5.60). In 1985, a steep ramp developed at the downglacier boundary of faster flowing ice, around 8 km from the terminus of Bakaninbreen (Dowdeswell et al., 1991). The ramp was 25 m high during the early stages of the surge, but grew to 60 m as the surge propagated downglacier during the next few years. Between 1985 and 1989, the rate of surge front propagation varied between ~1 and ~1.8 km yr^{-1}, but after 1989, the rate progressively declined, and by 1994–95 was only 1.8–3.0 m yr^{-1} (Murray et al., 1998).

The mechanisms of surge propagation at Bakaninbreen have been studied in detail by Murray et al. (1997, 1998, 2000), Murray and Porter (2001) and Barrett et al. (2008). Ground-penetrating radar soundings in 1986 and 1996 and temperature measurements in 1994 and 1995 showed that the ice ahead of the surge front was frozen to the bed, whereas the rapidly flowing ice behind the

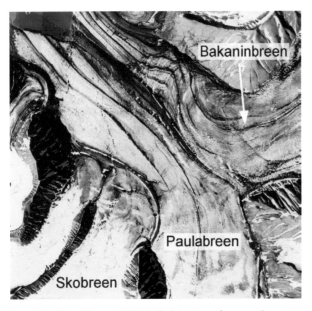

Figure 5.60 Bakaninbreen in 1990, in the late stages of a surge. The contrast between contorted and undeformed medial moraines clearly delimits the surge front. The locations of Skobreen and Paulabreen, which surged around 2001–5 are also shown (Aerial photograph from Norsk Polarinstitutt).

front was warm-based, with the lowermost ~40 m of the glacier at or very close to the pressure-melting point (fig. 5.61). The thermal boundary appears to have advanced with the surge front throughout the event. The glacier bed consisted of a mixture of glacimarine muds and glacial till, at least 0.2 m thick. Murray et al. (2000) argued that these sediments did not completely thaw during the surge, and that basal water was trapped between the glacier sole and permafrost. The resulting elevated water pressures reduced the frictional strength of the bed, encouraging basal motion. Downglacier propagation of the surge bulge was also associated with longitudinal shortening and vertical thickening of the ice, at least part of which was accommodated by brittle failure. Murray et al. (1997, 2000) documented a series of coherent englacial structures dipping upglacier at around 21–35°, both ahead of and behind the surge front, which were interpreted as thrust faults. Most of the structures terminated englacially, but a few cropped out on the glacier surface, where they contained basal sediment and intermittently discharged pressurized water. Further evidence for brittle failure at the surge front was provided by analysis of seismic emission recordings made in 1987 (Stuart et al., 2005). Three distinct types of emissions were identified, interpreted as the result of shallow faulting upglacier of the surge front, and brittle failure of cold ice and resonance in water-filled cracks ahead of the surge front. Propagation of the surge was therefore associated with thawing of the basal ice and the uppermost part of the bed, and the development of high basal water pressures. Water intermittently escaped from the bed to the surface along thrust faults, encouraging lower basal water pressures and heat loss from the bed. As a result, this 'leaky surge front' appears to have limited the rate of surge propagation and contributed to surge termination (Murray et al., 2000). Escape of water through gaps in the

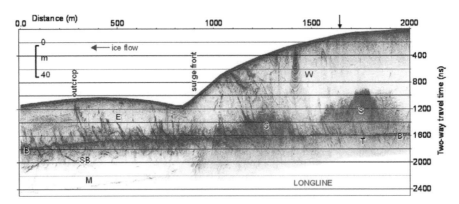

Figure 5.61 Ground Penetrating Radar profile from Bakaninbreen, showing the surge front and forebulge (on the left). B = basal reflection, W = wire to instrument at glacier bed, S = scattering zone (indicating warm ice), T = trough in basal reflection where scattering zone increases in thickness, E = shallowly dipping englacial features (probable thrust faults), SB = subglacial reflections which indicate that the ice downglacier of the surge front is underlain by permafrost, M = first multiple of basal reflection. The depth scale bar shown assumes a wave propagation velocity of 0.167 m ns^{-1} (Previously unpublished image kindly provided by Tavi Murray).

subglacial permafrost also appears to have played a role in surge termination (Smith et al., 2002).

The adjacent glacier, Paulabreen, surged in 2003–5 (figs. 5.52 and 5.53). The surge began in the tributary glacier Skobreen, then propagated down Paulabreen at around $3 \, km \, yr^{-1}$, approximately double the maximum rate of surge propagation on Bakaninbreen. Average ice velocities during the surge were around $700 \, m \, yr^{-1}$. The surge bulge reached the glacier terminus early in 2005, after which the glacier advanced 1.9 km into a fjord until termination in winter 2005–6. The surge behaviour of Paulabreen was therefore rather different from that of its neighbour. Another interesting aspect of the surges of Bakaninbreen and Paulabreen is that neither surge propagated across the medial moraine to affect the other glacier. Benn et al. (2009). showed that a persistent subglacial conduit exists below the medial moraine, and argued that this allows subglacial water to escape, thus forming a hydraulic barrier between the two glaciers.

Yet another variant of polythermal glacier surge is provided by the most recent surge of Trapridge Glacier, which began around 1980 and terminated in 2002 (Frappé and Clarke, 2007). Although the glacier was monitored in great detail for much of this period (e.g. Clarke et al., 1984; Clarke and Blake, 1991), the surge was only recognized as such with hindsight. This is because the peak velocity was much lower than expected, and was believed at the time to represent a transient speed-up during the quiescent period. It is now recognized that Trapridge Glacier underwent a 'slow surge', adding to the growing menagerie of surge phenomena.

Evolution of the glacier during surge onset is not known in detail, but by 1980 an impressive wave-like bulge had formed on the glacier surface, the front of which steepened as it advanced downglacier (figs. 5.62 and 5.63). The bulge first developed at the boundary between cold-based ice on the lower part of the glacier ('the apron') and the warm-based ice upglacier. The thermal boundary also migrated downglacier during the surge, but at a slower rate than the bulge. Downglacier propagation of the bulge appears to have occurred mainly in response to longitudinal stress coupling and strain thickening, and thawing of the bed appears to have been a slower, less important process than at Bakaninbreen. Between 1980 and 1988, the bulge advanced downglacier at mean rate of $30 \, m \, yr^{-1}$, then reached the limit of the former ice apron and became the terminus. The bulging glacier front then advanced 450 m, at an average rate of $14 \, m \, yr^{-1}$, terminating in 2002.

Before onset of the surge, creep and basal motion contributed approximately equally to ice motion in the warm-based central part of the glacier (Zones A and B,

Figure 5.62 Trapridge Glacier in 2006, after the termination of its 'slow surge' (M.J. Hambrey).

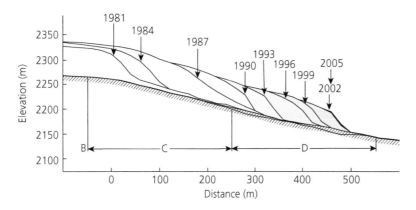

Figure 5.63 Propagation of the surge front of Trapridge Glacier during its 'slow surge' (Frappé and Clarke, 2007).

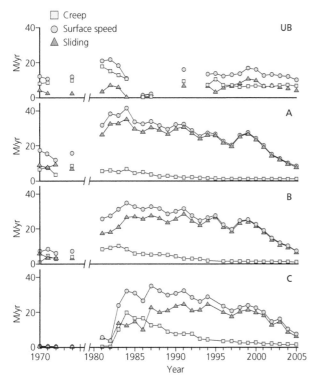

Figure 5.64 Time series of velocities for four zones on Trapridge Glacier. The locations of Zones B and C are shown in figure 5.63. Zones UB and A are located farther upglacier (Frappé and Clarke, 2007).

fig. 5.64), but in the cold-based apron (Zone C) 100 per cent of flow was due to ice deformation. After surge onset, >80 per cent of the ice velocity in Zones A and B was due to basal motion. After 1986, when most of the apron ice was incorporated into the surge bulge, flow in Zone C also switched to dominantly basal motion. The average velocity in Zone A increased from $16\,\mathrm{m\,yr^{-1}}$ in 1974 to $39\,\mathrm{m\,yr^{-1}}$ in 1980. The peak velocity ($42\,\mathrm{m\,yr^{-1}}$) was reached in 1984, after which the glacier underwent a series of velocity oscillations superimposed on a gradual deceleration. After 1999, the glacier gradually slowed to pre-surge velocities, and was less than $9\,\mathrm{m\,yr^{-1}}$ in 2005. Basal motion of the glacier was by a combination of sliding and shallow deformation of subglacial till (Fischer and Clarke, 2001; Kavanaugh and Clarke, 2006).

Measurements of subglacial water pressure showed that the basal drainage system was highly variable, both in space and time. This resulted in large variations in bed strength and ice–bed coupling, with efficient coupling in response to low water pressures at some times and places, and decoupling in response to rising water pressures at others. Diurnal variations in water pressure indicate that surface meltwater was able to access the bed, and that water could drain from the glacier through porous subglacial materials. The dissipation of subglacial water thus prevented high water pressures from developing over large areas of the bed, and is the likely reason why high velocities were not attained during the surge.

5.7.5 SURGE MECHANISMS

The strong dynamical contrasts between temperate and polythermal surging glaciers has motivated the development of two distinct models of surging (see Murray et al., 2003). The *hydrologic switch model* attributes temperate glacier surges to changes in basal hydrology, whereas the *thermal switch model* explains surges of polythermal glaciers in terms of changes in basal thermal regime. Both of these ideas have a long pedigree, but here we focus on recent formulations of these models, developed with the advantage of data sets from Variegated Glacier, Trapridge Glacier, Bakaninbreen and elsewhere. Despite the differences between temperate and polythermal glacier surges, there are also many similarities, and both temperate and polythermal surging glaciers can occur within the same geographical cluster. Frappé and Clarke (2007, p. 14) suggested that this points to an underlying dynamical unity, and concluded that models of surge mechanisms 'should be able to explain the entire spectrum of surge speed and structure observed within a geographical cluster, irrespective of the thermal regime of the glaciers'. A unified theory of glacier surges remains elusive, however.

The *hydrologic switch model* was developed to explain the concomitant changes in sliding speed and basal hydrology observed during the 1982–83 surge of

Variegated Glacier, and has been developed further following observations of other temperate surges (Kamb et al., 1985; Raymond, 1987; Eisen et al., 2005). According to this model, surge initiation and termination occur in response to a switch in basal drainage configuration from an efficient conduit system to an inefficient linked cavity system, and back again (section 3.4.3). The dramatic fluctuations in ice velocity during the course of the surge cycle are therefore explained by processes which modulate the storage, distribution and pressure of water at the bed. Kamb (1987) conducted a detailed mathematical analysis of the narrow connecting passages (orifices) in a linked cavity system, to identify possible threshold behaviour. According to his analysis, a linked cavity system is stable when sliding velocities are high and the hydraulic gradient is low. When the hydraulic gradient is high, however, orifices are susceptible to unstable growth because roof melting by viscous heat dissipation outstrips closure by ice creep. Kamb concluded that this instability means that a linked cavity system will collapse into a series of conduits when water inputs are large, draining water from the bed. This model, therefore, can explain why a sudden release of stored water – and surge termination – should occur in summer following the onset of surface melting. It is also consistent with the highly dispersed dye returns during the surge of Variegated Glacier, and the unusual pattern of water storage below the surging Skeiðarárjökull following the September 1991 jökulhlaup.

While Kamb's theory provides a mechanism for destabilizing a distributed drainage system at high hydraulic gradients, thus explaining surge termination, it does not demonstrate why the opposite switch (conduit to linked cavity network) should occur, and how a surge might be initiated. The observation that temperate glacier surges usually begin in winter led Raymond (1987) to argue that rapid sliding was triggered by rising water pressures at the bed, resulting from the closure of the conduit system that drained the glacier during summer. One possible way this switch might work has been discussed by Eisen et al. (2005), who argued that water will leak from conduits into low-pressure lee-side cavities if low discharges coincide with high basal shear stress (and a correspondingly large pressure fluctuation over bedrock bumps). This, they suggested, could trigger a hydraulic switch when the basal shear stress reaches a threshold value (i.e. following a period of ice build-up during the quiescent period) and discharges are low (i.e. during winter).

A distributed drainage system, however, is not a sufficient condition for fast glacier flow, because widespread ice–bed decoupling also requires a large *volume* of pressurized water. It has been argued that water released from englacial storage could provide the necessary water source in winter (e.g. Fatland and Lingle, 2002). While it is certainly true that large amounts of water can be stored in englacial voids, there is no convincing evidence that drainage of englacial water to glacier beds is implicated in surging, nor any explanation why this process should be important on some glaciers but not others. Another unresolved issue with the present form of the hydrologic switch model is that the linked cavity system considered by Kamb (1987) may not be the most appropriate picture of the subglacial plumbing of a surging glacier. The few surging glaciers to have been investigated in detail are underlain by till (Harrison and Post, 2003; Truffer and Harrison, 2006), and the beds of former surging glaciers commonly consist of smooth and streamlined sediment (e.g. Ottesen et al., 2008a). A general model of temperate glacier surging should also be able to explain surges of soft-bedded glaciers, as well as those with irregular rockbeds. A surge mechanism based on till instability has been sketched by Nolan (2003), but a quantitative model of this process has not been developed.

The hydrologic switch model is appealing because temperate glacier surges are clearly associated with large changes in basal drainage conditions. Existing theory can explain surge termination in terms of a switch from a distributed to a conduit system, following input of surface meltwater to the bed. How the opposite switch triggers surge initiation, and how well the idealized model parameters adopted by Kamb represent the real world, remain open questions (Harrison and Post, 2003).

The idea of a *thermal switch mechanism* for glacier surging has been around for a long time (see Sharp, 1988a for a review), and has recently been developed by Murray et al. (2000) and Fowler et al. (2001). According to this model, surge cycles on polythermal glaciers occur in response to switches between frozen and unfrozen conditions at the bed. During the quiescent period, the glacier (at least in its lower regions) is cold-based and slow-moving. Build-up of ice in the reservoir area increases the driving stress, which leads to higher ice creep rates. In turn, this generates heat and initiates a positive feedback between accelerated ice motion and strain heating. Eventually, part of the bed is raised to the pressure melting point, and further energy dissipated during glacier motion is used to produce meltwater. Cold ice and permafrost located downglacier can prevent this water from escaping, so that rising basal water pressures lead to reduced basal drag and faster sliding. Surge propagation occurs by stress transfer from the surging area (where basal drag is small) to the surrounding ice (where basal drag is high). The local increase in stress at the boundaries of the surging area increase ice deformation rates, leading to a positive feedback between strain rate and strain heating. In addition, injection of pressurized water along the bed and into the ice can also contribute significantly to surge propagation. Surge termination occurs when subglacial water is able to dissipate, either through the bed or via thrust faults extending from the bed to the surface.

A quantitative model of thermally regulated surges was developed by Fowler et al. (2001), relating changes in ice velocity and thickness to thermal evolution and the production and evacuation of meltwater at the glacier bed. Evolution of the system through time was simulated using a 'lumped', essentially one-dimensional, version of the model, which retains key physical processes while ignoring complex spatial phenomena such as surge propagation. The glacier bed was represented as a till layer (which may be frozen or unfrozen) overlying an aquifer, which appears to be a reasonable approximation to conditions below Bakaninbreen and Trapridge Glacier. During the 'quiescent period', the till is completely frozen, velocity is low and ice thickness increases. A surge begins when the upper part of the till layer thaws, and basal motion increases in response to water trapped between the ice base and the deeper, frozen till. Surge termination occurs when the till thaws completely, and water is able to escape through the underlying aquifer. The amplitude and form of velocity and ice thickness cycles were found to vary with till thickness and aquifer permeability. While this model appears to rely on highly specific basal conditions, it shares many essential features with earlier models of temperate glacier surges developed by Fowler (1987, 1989). Although some of the equations used in the 'temperate' and 'polythermal' versions of the model have different physical interpretations, they have equivalent mathematical effects and give rise to similar oscillatory behaviour. Importantly, Fowler's modelling predicts a spectrum of surge behaviour, ranging from brief, violent surges to more prolonged, slower events, depending on the rates that water and heat are generated and dissipated. We suspect that these principles, though manifest in a variety of ways, may underlie all glacier surges, and could point the way to a unified surge theory.

CHAPTER 6
THE GREENLAND AND ANTARCTIC ICE SHEETS

Hot water drilling on Rutford Ice Stream, West Antarctica (Tavi Murray)

6.1 INTRODUCTION

A comprehensive view of the great ice sheets covering most of Greenland and Antarctica has emerged only very recently, due to the immense logistical difficulties involved in obtaining representative data in remote and hostile environments. With the advent of modern transport, geophysical methods and remote-sensing techniques, the mass balance and dynamics of both ice sheets can now be studied in great detail. These recent developments, however, build on a long tradition of research in the polar regions. Reading the older literature, one is frequently struck by the deep insight of the pioneers, who were often working under conditions of extreme hardship. A good review of research on the polar ice sheets between 1930 and 1980 was provided by Robin

(1981), and a longer-term perspective on the development of science in Antarctica can be found in Fogg (2005).

Modern understanding of the Greenland and Antarctic Ice Sheets has developed in parallel with (and has been largely motivated by) an increasing awareness of their importance for global environmental change. The recent discovery that ice streams and outlet glaciers can undergo large variations in speed has led to the realization that the polar ice sheets can respond much more rapidly to climatic change than was hitherto thought (Bamber et al., 2007). Consequently, there is ongoing concern about the possible impact of ice sheet retreat on global sea levels (Shepherd and Wingham, 2007). In this chapter, we review recent research on the Greenland and Antarctic Ice Sheets, focusing on their surface mass balance and the factors that influence their dynamics. The implications of

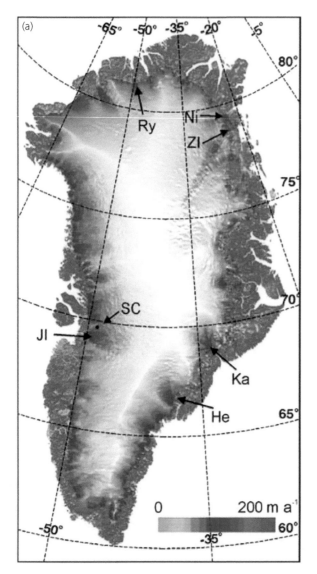

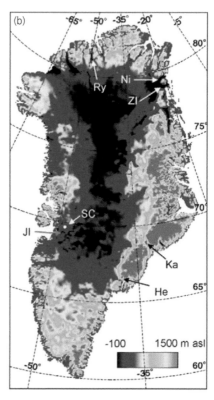

Figure 6.1 (a) Bed elevation, with areas below sea level shown in black. SC: Swiss Camp JI: Jakobshavn Isbrae; He: Helheim Glacier; Ka: Kangerdlugssuaq; ZI: Zachariae Isstrøm; Ni: Nioghalvjerdsbrae ('Seventy-nine Glacier', at 79° N); Ry: Ryder Gletscher (Bamber et al., 2007). (b) The Greenland Ice Sheet: surface topography and balance velocities.

ice sheet fluctuations for sea-level change are discussed in chapter 7.

6.2 THE GREENLAND ICE SHEET

6.2.1 OVERVIEW

The Greenland Ice Sheet is by far the largest mass of glacier ice in the northern hemisphere, extending 2500 km from north to south and up to 1000 km from east to west (fig. 6.1a). With an area of 1,736,000 km^2 and a volume of 2,600,000 km^3, it contains 10 per cent of the Earth's total fresh water. If spread evenly across the world's oceans, this is equivalent to 6.5 m of sea-level rise (Williams and Ferrigno, 2009). The interior of the Greenland Ice Sheet consists of a northern and a southern dome, with elevations of 3200 m and 2850 m, respectively,

linked by a long saddle with an elevation of 2500 m. Bedrock below this central part of the ice sheet is rather flat and close to sea level, but the periphery is almost completely fringed by coastal mountains, through which outlet glaciers drain the interior (fig. 6.1a; Thomas, 2004; Bamber et al., 2007).

The balance velocities shown in figure 6.1b highlight large areas of 'sheet flow' in the interior, with velocities of ~10^1 m yr^{-1}, and fast-flowing outlet glaciers, with balance velocities of ~10^2 to 10^4 m yr^{-1} (Bamber et al., 2000). Many of Greenland's outlet glaciers terminate on land or in lakes, but the largest, fastest-flowing glaciers terminate in tidewater. In recent years, field observations and remote-sensing studies have shown that both sheet flow and outlet glacier velocities deviate from balance velocities on a range of timescales (sections 6.2.3 and 6.2.4). Calving at the termini of tidewater outlet glaciers

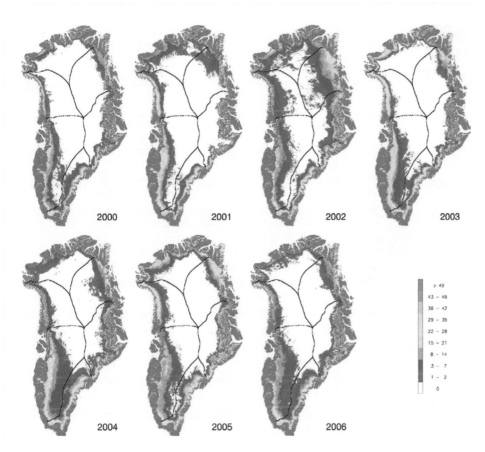

Figure 6.2 Number of melt days on the Greenland Ice Sheet, 2000–6, derived from MODIS land-surface temperature data. Ice-free land is shown in green (Hall et al., 2008).

accounts for over half of the total losses of ice from the Greenland Ice Sheet, so any changes in ice velocity can have important implications for ice sheet mass balance (Rignot and Kanagaratnan, 2005; Joughin et al., 2008b).

6.2.2 CLIMATE AND SURFACE MASS BALANCE

Climatic data for Greenland are available for a number of surface weather stations (Steffen and Box, 2001; Box, 2002), and regional coverage for some variables is provided by satellite observations (e.g. Comiso et al., 2003). Many data gaps exist, however, and regional climatological studies of Greenland commonly make use of synthetic data sets such as ERA-40, derived by reanalysis of global weather data by the European Centre for Medium-Range Weather Forecasts (ECMWF) (Uppala et al., 2005). The climate of Greenland exhibits large inter-annual variability, in terms of both temperature and precipitation (Hanna et al., 2006, 2008). Much of the variability in temperature is associated with shifts in the relative strength of the Icelandic Low and the Azores High, commonly known as the North Atlantic Oscillation (NAO) (Hall et al., 2008). The positive phase of the NAO reflects a deeper Icelandic Low (i.e. more frequent travelling depressions over the north Atlantic) and a stronger high-pressure region over the subtropical Atlantic. These conditions correspond to colder winters over Greenland.

In contrast, the negative phase of the NAO is characterized by a weaker Icelandic Low and Azores High, corresponding with milder winters in Greenland. The relationship between the NAO and variations in precipitation over Greenland is less clear (Hanna et al., 2006). Large, low-latitude volcanic eruptions have also had a significant effect on the climate of Greenland, with particularly cool years following eruptions of Mount Pinatubo, Philippines (1992), El Chichon, Mexico (1982) and Agung, Bali (1963) (Hanna et al., 2005).

Overall, there has been a trend towards higher air temperatures over Greenland since the 1980s. The temperature increase over Greenland is strongly correlated with, but greater than, the temperature trend for the whole northern hemisphere, suggesting that it reflects anthropogenic greenhouse gas concentrations rather than regional climatic variability (Hanna et al., 2008). Warmer summers and longer melt seasons have led to increased ice surface temperatures and runoff, although there is large inter-annual variability, with greater than average melt in 2002 and 2005, and lower than average in 2000, 2001 and 2006 (fig. 6.2; Hall et al., 2008). The average snow accumulation rate over the Greenland Ice Sheet exceeds the average melt rate, so the surface mass balance (i.e. excluding calving losses) is positive (Bales et al., 2001; Hanna et al., 2006). Data from snowpits and climatic modelling indicate that total accumulation averages $299 \pm 23\,\mathrm{kg\,m^{-3}\,yr^{-1}}$, but with significant inter-annual

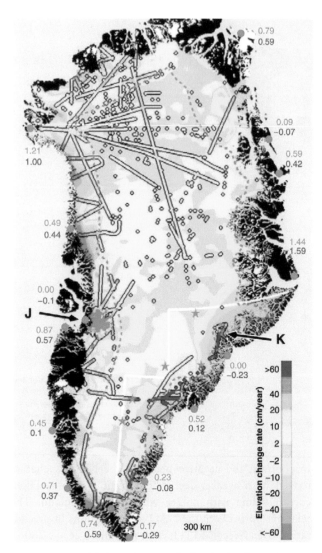

Figure 6.3 Rates of thickness change on the Greenland Ice Sheet. Blue areas indicate thickening at high elevations, whereas yellow and red areas indicate thinning of the ice sheet margins and outlet glaciers (Krabill et al., 2004).

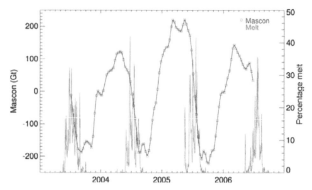

Figure 6.4 Mass balance cycles on the Greenland Ice Sheet. GRACE-derived mass concentration (mascon) in Gt (blue) and percentage area of melt (red) (Hall et al., 2008).

the elevation change on marine-terminating glaciers is the result of dynamic thinning (i.e. longitudinal extension and vertical shortening; section 5.2.3), associated with calving retreat. These dynamical changes, and their implications for the mass budget of the Greenland Ice Sheet, are discussed in detail in section 6.2.4.

Annual mass balance cycles for the Greenland Ice Sheet have been determined using gravimetric data from the Gravity Recovery and Climate Experiment (GRACE; Luthcke et al., 2006; Velicogna and Wahr, 2006; section 2.5.2). Figure 6.4 shows cycles of mass gain in winter and mass loss during summer, with an amplitude of around 300 Gt. Although the seasonal cycle of mass loss clearly coincides with surface melting, a significant proportion of the loss actually relates to increased calving during the summer months (cf. Howat et al., 2008).

6.2.3 ICE SHEET FLOW

Until recently, it was believed that flow of the 'inland ice' in Greenland was unlikely to vary on short timescales, since thick, cold ice appeared to present an impenetrable barrier to surface-to-bed drainage. A radical change in perspective was triggered by a landmark paper by Zwally et al. (2002a), which presented clear evidence for seasonal velocity fluctuations at Swiss Camp, near the equilibrium line of the west-central Greenland Ice Sheet. GPS measurements made over several seasons revealed dramatic speed-ups in early summer, during which ice velocities rose up to 25 per cent higher than the winter average of ~ 32 m day^{-1} (fig. 6.5). The onset of the speed-ups occurs early in the ablation season, and Zwally et al. (2002a) proposed that they reflect increased basal motion in response to surface meltwater reaching the glacier bed. Because the ice at Swiss Camp is over 1200 m thick and mostly below the pressure-melting point, this implies highly efficient formation or reactivation of englacial drainage systems each melt season. Subsequent work has shown that large areas of the ablation zone of the ice sheet are affected by seasonal velocity fluctuations (e.g. Joughin et al., 2008a; van de Wal et al., 2008). In some areas,

variation (Cogley, 2004). In recent decades, there has been a significant increase in accumulation over Greenland, largely offsetting the trend towards increasing ablation. As a result, there is no statistically significant trend in surface mass balance, despite strong warming over Greenland (Hanna et al., 2005; Box et al., 2006).

Several studies have shown that the Greenland Ice Sheet is currently thinning at low elevations (<2000 m) and thickening at higher elevations (>2000 m) (e.g. Krabill et al., 2000, 2004; Zwally et al. 2005; Luthcke et al., 2006; Pritchard et al., 2009; fig. 6.3). This trend is consistent with the evidence for increases in both ablation losses and accumulation totals, although surface mass balance alone cannot explain the magnitude of all of the observed elevation changes. On land-terminating sectors of the ice sheet, thinning rates are comparable to losses by surface melting, but on marine-terminating outlet glaciers, the thinning is greater than can be accounted for by ablation losses (Sole et al., 2008). It is clear that most of

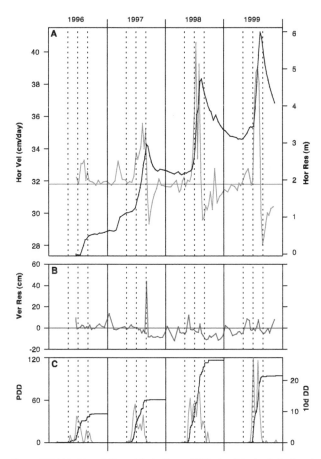

Figure 6.6 Part of the western margin of the Greenland Ice Sheet, showing numerous melt ponds (dark blue) (NASA, dhttp://visibleearth.nasa.gov).

Figure 6.5 Velocity variations at Swiss Camp. (A) Horizontal velocity (red) and cumulative additional motion relative to the wintertime average of 33.33 cm day^{-1} (black). (B) Vertical residual (blue) indicating a short-lived uplift event in summer 1997. (C) Positive degree days (PDDs; red) and cumulative PDDs (black). Vertical dotted lines indicate 1 May, 1 July and 1 September each year (Zwally et al., 2002a).

speed-ups greater than those at Swiss Camp have been recorded, with short-term summer velocity peaks up to 100 per cent greater than the winter average.

In several cases, transient speed-ups have been shown to coincide with the drainage of supraglacial lakes. In gently sloping parts of the ablation zone of the Greenland Ice Sheet, large numbers of shallow lakes form during each melt season, typically ranging between 10^2 and 10^3 m in diameter (Echelmeyer et al., 1991; Box and Ski, 2007; McMillan et al., 2007; fig. 6.6). The presence of extensive ponded water substantially alters ice sheet albedo, leading to more efficient melting (Lüthje et al., 2006). Using MODIS imagery, Box and Ski (2007) showed that along the western ablation zone between 65.76° and 69.87° N, total lake volume is in excess of 1 km^3 by July each year. Lake drainage events begin in the second half of June and continue until August, with activity apparently migrating inland as the ablation season progresses. During the largest single drainage event recorded by Box and Ski (2007), ~97.4 × 10^6 m^3 of water was routed into the glacier, indicating a discharge of at least 16.2 × 10^6 m^3 day^{-1}.

In some cases, lake drainage has been shown to occur by hydrologically assisted fracture propagation, where a combination of tensile stresses and large water supply allows water-filled fractures to penetrate from the surface to the bed (Alley et al., 2005b; Das et al., 2008; section 3.3.2). This process provides a highly efficient means of routing surface water through thick, cold ice, and once fracturing is initiated it can cause lakes to drain in a matter of hours. One event has been documented in detail by Das et al. (2008). Pressure sensors installed in a lake showed that slow drainage began around 00.00 UTC (Universal Time Coordinated) on 29 July. The drainage rate increased dramatically around 16 hours later, whereupon the whole lake drained in ~1.4 hours. The average discharge during this rapid drainage phase was 8700 m^3 sec^{-1}, greater than the average flow of Niagara Falls. Lake drainage coincided with increased icequake activity and acceleration and uplift of the ice, and was followed by ice subsidence and deceleration over the subsequent 24 hours. Following drainage, two large moulins were observed in the lake floor, both located on km-scale fractures. Das et al. (2008) concluded that drainage evolution proceeds in four stages:

(1) slow initial drainage;
(2) connection to the bed through an extensive fracture system;
(3) moulin enlargement by flowing water and closure of the adjacent fractures;
(4) moulin-routed drainage of daily meltwater to the bed.

Ground-penetrating radar surveys by Catania et al. (2008) have revealed numerous vertical structures in areas of extending flow around Swiss Camp, interpreted as hydrologically assisted fractures (fig. 6.7). This recent work thus provides strong support for the hypothesis advanced by Zwally et al. (2002a).

The sudden influx of such large volumes of water to the ice sheet bed must result in locally high basal water

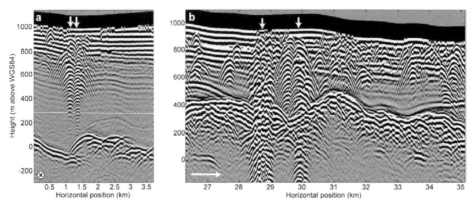

Figure 6.7 Vertically stacked reflectors in ground-penetrating radar profiles, interpreted as hydrologically assisted fractures. Ice flow perpendicular to the page in (a) and to the right in (b). The prominent bed reflectors indicate ice thickness of ∼1000 m (a) and 600 m (b) (Catania et al., 2008).

pressures on timescales of hours to days, reducing ice–bed coupling and increasing basal motion rates. Because reduced basal drag in one area leads to increased longitudinal stresses in adjacent regions, the effects of bed decoupling can be transmitted both upglacier and downglacier, so that surface velocity fluctuations occur over much larger areas than those directly affected by basal water pressure anomalies (Price et al., 2008).

The realization that surface melting can affect the dynamics of the Greenland Ice Sheet has prompted widespread speculation both in the popular press and the scientific literature that climatic warming could increase ice flux to the oceans, accelerating sea-level rise. For example, Parizek and Alley (2004a) modelled the future evolution of the Greenland Ice Sheet under a range of global warming scenarios, assuming a linear relationship between surface melt rates and ice velocity. This appeared to show that dynamic effects could increase rates of ice loss over those by melting by several per cent. However, it is clear that rates of basal motion are not simply a direct, linear function of meltwater inputs, as assumed in the model (see sections 5.4.1 and 5.5.1). On valley glaciers and the Greenland Ice Sheet, speed-ups have been shown to coincide with periods of increasing basal meltwater storage, when water inputs to the bed exceed the capacity of basal drainage systems (e.g. Vieli et al., 2004; Bartholomaus et al., 2008; van de Wal et al., 2008). With increased discharge, however, conduits are enlarged by wall melting, and become more efficient at evacuating basal water (section 3.2). Falling water storage at the bed then leads to increased ice–bed coupling and reduced rates of basal motion. Interestingly, the Swiss Camp velocity records of Zwally et al. (2002a) exhibit 'extra slowdowns' following the early summer velocity peaks. The key issue is whether the effect of enhanced surface meltwater production results in a net increase in mean annual ice velocities. From a 17-year record of ice velocities in western Greenland, van de Wal et al. (2008) found that mean flow

speeds have decreased by around 10 per cent, possibly due to reduced driving stresses in response to ice thinning or surface slope reduction. They found no correlation between annually averaged velocity and ablation totals, suggesting that increasing meltwater inputs will probably have only a limited effect on the response of the inland ice to climate change.

6.2.4 ICE STREAMS AND OUTLET GLACIERS

The outlet glaciers draining the Greenland Ice Sheet are very variable in terms of size, flow speed and ice discharge. A fundamental distinction exists between land-terminating outlet glaciers, such as the Russell Glacier in south-west Greenland, which typically have velocities of tens of metres per year, and tidewater outlet glaciers with velocities of hundreds to thousands of metres per year (e.g. Chandler et al., 2005; Rignot and Kanagaratnam, 2005; Joughin et al., 2008a). The largest outlet glaciers terminate in tide water, and include Jakobshavns Isbrae in the west and Helheim Glacier and Kangerdlugssuaq in the south-east.

Much of the ice in northern Greenland is discharged via the 700 km *North-east Greenland Ice Stream*, which was discovered in the 1990s through analysis of synthetic aperture radar imagery (Fahnestock et al., 1993; Joughin et al., 2001; fig. 6.8). Velocities reach around 500 m yr^{-1} at the downstream end, where fast flow is associated with low driving stresses and low basal drag, similar to the Ross Ice Streams in Antarctica (section 6.3.4). Fahnestock et al. (2001) have shown that basal melting rates are very high in the onset zone of the ice stream, implying a geothermal heat flux of ∼970 mW m^{-2}, 17 times higher than the continental background value. The region of enhanced melting coincides with a free-air gravity anomaly, strongly suggesting the presence of a subglacial volcanic centre.

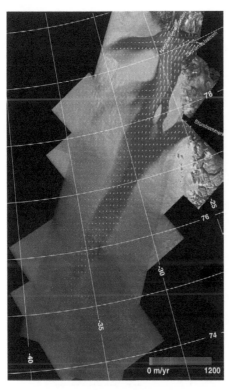

Figure 6.8 The Northeast Greenland Ice Stream velocities, derived from synthetic aperture radar (Joughin et al., 2001).

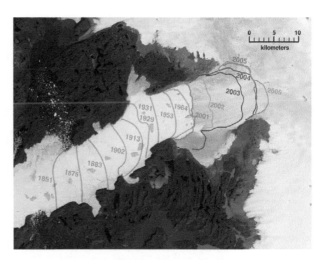

Figure 6.9 Retreat of Jakobshavn Isbrae, 1851–2006. The background image shows the glacier in 2001, prior to the most recent episode of retreat (NASA image by Cindy Starr, based on data from Ole Bennike and Anker Weidick, Geological Survey of Denmark and Greenland, and Landsat data, http://earthobservatory.nasa.gov).

The largest outlet glacier in west Greenland is Jakobshavns Isbrae (also known by its Greenlandic name *Sermeq Kujalleq*), which drains a catchment area of 110,000 km², or around 6 per cent of the whole ice sheet. Since the middle of the nineteenth century, the ice front has retreated intermittently back into Jakobshavns (*Ilulissat*) Isfjord, interspersed with periods of stability (fig. 6.9; Sohn et al., 1998; Csatho et al., 2008). A dramatic phase of retreat commenced in 1998, which has been widely regarded in the media as an ominous symptom of climate change. Scientific attention on this glacier has focused on determining how and why retreat is initiated and maintained.

Prior to 2001, the lowermost ~15 km of Jakobshavns Isbrae formed a floating ice tongue, with partially grounded areas in its southern half. The floating tongue appears not to have been a single, coherent mass, but a heavily fractured melange of blocks restrained by the sides of the fjord (Johnson et al., 2004). Velocities on the floating part of the tongue varied between ~5.7 and ~6.7 km yr⁻¹, declining gradually upstream on the grounded trunk glacier (Echelmeyer and Harrison, 1990; Joughin et al., 2004a). On both floating and grounded parts, seasonal fluctuations in velocity appear to have been small. This situation began to change in the late 1990s, when the glacier tongue began to thin, accelerate and retreat. Rapid retreat commenced in 2001, with the greatest ice loss occurring between 2002 and 2003, and

by 2006, the floating tongue of the glacier had all but disappeared. These changes were accompanied by a dramatic acceleration of the lower part of the glacier. Between May 1997 and October 2000, the velocity of the floating tongue increased to 9.4 km yr⁻¹, then reached 12.6 km yr⁻¹ in spring 2003 (Joughin et al., 2004; Luckman and Murray, 2005). The acceleration of the lower glacier continued until at least 2007, the latest year for which figures were available at the time of writing (2009) (Joughin et al., 2008b; fig. 6.10). As a result of this flow acceleration, calving losses from the ice front increased from 24 km³ yr⁻¹ in 1996 to 46 km³ yr⁻¹ in 2005 (Rignot and Kanagaratnam, 2005). Since the mid 1990s, the glacier has also exhibited marked seasonal velocity fluctuations, which show a very close correspondence with annual fluctuations of the calving front position (Luckman and Murray, 2005; Joughin et al., 2008b). Summer velocities as high as 45 m day⁻¹ (equivalent to 16 km yr⁻¹) have been recorded (Dietrich et al., 2007).

Many other tidewater outlet glaciers in Greenland have undergone recent acceleration and retreat, especially in the southern half of the island (e.g. Rignot and Kanagaratnam, 2005; Howat et al., 2008). The behaviour of the two largest outlet glaciers on the south-east coast, Helheim Glacier and Kangerdlugssuaq, has been particularly well studied (Howat et al., 2005, 2007; Stearns and Hamilton, 2007; Joughin et al., 2008c). Between them, these glaciers carry 35 per cent of the total ice discharge from the Greenland Ice Sheet, draining around 10 per cent of its area. The terminus positions of both glaciers were stable from the mid twentieth century until 2002, after which Helheim Glacier retreated 7 km in the following 3 years (fig. 6.11). The retreat of Kangerdlugssuaq commenced slightly later, but the ice front retreated

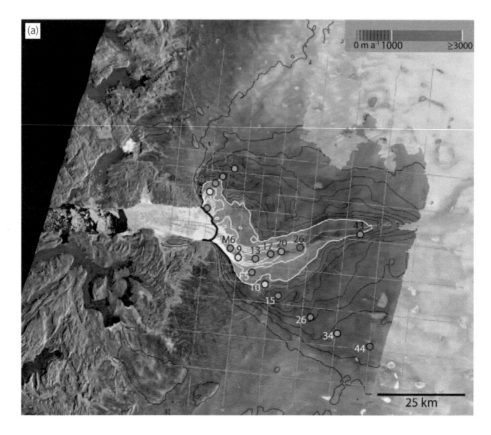

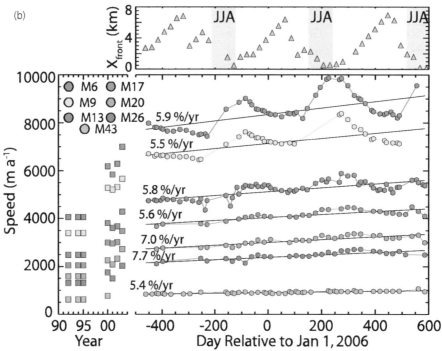

Figure 6.10 Recent changes in the velocity of Jakobshavn Isbrae. (A) Velocity map for 1 January 2006. Isolines are at intervals of 100 m yr^{-1} up to 900, and at intervals of 1 km yr^{-1} for faster speeds. (B) Time series of velocities at several points along the main glacier trunk (locations shown in (A)). The upper panel shows oscillations in ice front position, reflecting greater calving rates in summer (Joughin et al., 2008b).

at over 4 km yr^{-1} in 2004–6. As was the case for Jakobshavns Isbrae, ice front retreat coincided with thinning and acceleration. Unlike that glacier, however, the phase of acceleration and retreat was followed by slowdowns and stabilization of the ice front position. Velocities at the terminus of Helheim Glacier increased from ~23 m day^{-1} in 2000 to ~30 m day^{-1} in 2005, before falling to around the 2000 values in 2006.

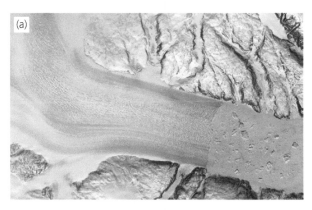

Figure 6.11 ASTER images of Helheim Glacier, east Greenland. (a) 12 May, 2001. (b) 19 June 2005 (NASA, http://earthobservatory.nasa.gov).

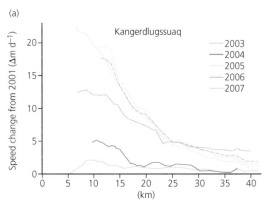
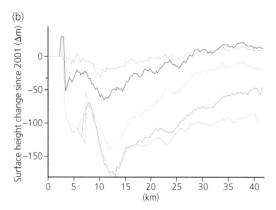

Figure 6.12 Kangerdlugssuaq Glacier, east Greenland: along-flow variations in velocity (a) and surface elevation change (b) since 2001. The horizontal scale is measured upglacier from the 2001 terminus (Howat et al., 2008).

Similarly, velocities at the terminus of Kangerdlugssuaq increased from ~20 m day^{-1} to ~40 m day^{-1} between 2002 and 2005, before falling to ~30 m day^{-1} in 2006 (figs 6.12 and 6.13). The recent behaviour of Helheim Glacier was modelled by Nick et al. (2009), using a higher-order flow-line model coupled to a height-above-buoyancy calving criterion (fig. 6.14).

Large calving events on both Helheim and Kangerdlugssuaq Glaciers coincided with glacial earthquakes, strongly suggesting that calving-related processes are the source of the seismicity (Joughin et al., 2008c; cf. O'Neel et al., 2007). Similar earthquake events in east Greenland have been detected at remote seismic stations for many years (e.g. Ekstrom et al., 2006), allowing temporal variations in calving activity to be inferred. The number of seismic events per year began to increase in 2002, and more events occurred in 2005 than the combined total for 1993–95. The events show a strong seasonal cycle, with almost five times as many in summer than in winter.

In a study of 32 tidewater outlet glaciers in south-east Greenland, Howat et al. (2008) found that most underwent net retreat, thinning and acceleration between 2000 and 2006 (Fig. 6.13). Twenty-seven glaciers increased in velocity over this period, 23 of which underwent

simultaneous retreat. Retreat occurred in all cases where acceleration was by 10 per cent or more. In all cases of ice acceleration, the greatest changes occurred near the terminus and declined in magnitude upglacier. The near-simultaneous behaviour of tidewater outlet glaciers over a large geographical area indicates a common forcing mechanism. The most likely trigger for initial retreat and acceleration is ice-tongue thinning in response to higher summer temperatures and/or warmer waters in the fjords. Where ice is grounded in deep water, ice thinning – either by surface or basal melting – reduces the mean basal effective pressure, lowering resistance at the bed and encouraging acceleration. Increased ice velocities near the terminus result in longitudinal stretching, which in turn leads to both dynamic thinning and calving retreat (section 5.5.3). A positive feedback is therefore set in motion between acceleration, thinning and calving retreat, which continues until the glacier stabilizes at a new pinning point (section 5.5.4; Benn et al., 2007b; Pfeffer, 2007b; Howat et al., 2008; Nick et al., 2009).

The stabilization of Helheim Glacier occurred when the glacier front reached shallower water, following rapid calving retreat through an overdeepening (fig. 6.14). The same may well be true of Kangerdlugssuaq, although bed topography is not known for that glacier. Thinning of a

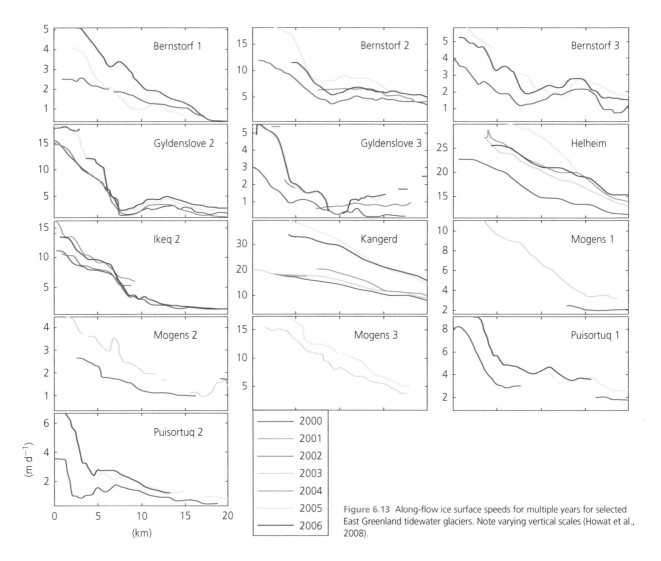

Figure 6.13 Along-flow ice surface speeds for multiple years for selected East Greenland tidewater glaciers. Note varying vertical scales (Howat et al., 2008).

tidewater glacier tongue results in a reduction of basal effective pressure below the whole glacier tongue, which can only be reversed if the glacier thickens once more, or retreats into shallower water. This contrasts with transient changes in effective pressure associated with increased surface-to-bed meltwater drainage, which only lead to short-term glacier speed-ups (Howat et al., 2008).

While the thinning–acceleration feedback can explain much of the observed behaviour, other factors clearly play important roles locally. For example, loss of back-stress appears to have been crucial in modulating the behaviour of Jakobshavns Isbrae following the disintegra-tion of the floating ice tongue (Johnson et al., 2004). Prior to 2000, the floating tongue appears to have acted as an 'ice jam', restrained between the fjord wall to the north and grounded ice to the south. This source of resistance was transmitted upglacier as a backstress, limiting the rate of ice flow. The progressive break-up of the tongue, therefore, resulted in a loss of this backstress, leading to ice acceleration. Certainly the role of varying backstress can be clearly seen in the seasonal behaviour of the glacier. In winter, sea ice binds together masses of calved

icebergs, providing an additional source of resistance at the front, reducing ice velocity. Calving is inhibited and the ice front advances. In summer, calving recommences and loss of ice at the front is accompanied by ice acceler-ation (fig. 6.10). Holland et al. (2008) have argued that the initial break-up of the floating tongue of Jakobshavns Isbrae was triggered by bottom melting, in response to the arrival of relatively warm water from the North Atlantic. However, Luckman and Murray (2005) have pointed out that acceleration of the lower glacier preceded calving retreat, so loss of backstress is unlikely to be the initial trigger. The initial phase of acceleration was coincident with warm summers, suggesting that basal lubrication by surface-derived meltwater was involved. Other factors that may be implicated, locally or regionally, in the recent acceleration and retreat of tidewater glaciers include declining sea ice extent and surface meltwater increasing the depth of penetration of surface crevasses. It is clear, however, that one simple model cannot account for all of the observed behaviour, which likely reflects a complex interplay of many environmental and glaciological controls (Csatho et al., 2008).

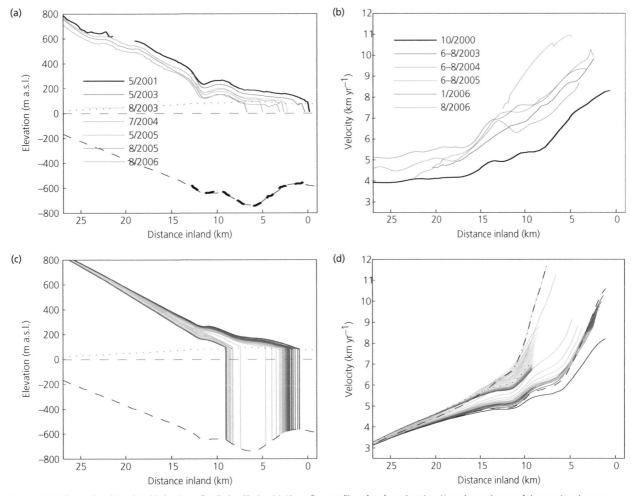

Figure 6.14 Observed and simulated behaviour of Helheim Glacier. (a) Along-flow profiles of surface elevation. Note the readvance of the terminus between 2005 and 2006. (b) Along-flow profiles of surface velocity. Note acceleration of the ice between 2000 and 2005 and the slowdown in 2006. (c) Simulated ice front positions. (d) Velocity (Nick et al., 2009).

6.3 THE ANTARCTIC ICE SHEET

6.3.1 OVERVIEW

The Antarctic Ice Sheet is the world's largest store of fresh water, with an area of ~13.5 million km^2 and a volume (including fringing ice shelves) of 25.4 million km^3. The Antarctic Ice Sheet is divided into two unequal parts by the Transantarctic Mountains, close to the neck between the Weddell and Ross Seas (fig. 6.15). The larger part, the East Antarctic Ice Sheet (EAIS) has a total grounded ice volume of 21.7×10^6 km^3, and the West Antarctic Ice Sheet (WAIS) has a grounded ice volume of 3.0×10^6 km^3. Most of the WAIS and large areas of the EAIS are grounded below sea level (fig. 6.16; Lythe et al., 2001; Bamber et al., 2007). In total, around 8.5 per cent of the volume of the grounded ice sheet lies below sea level. This leaves 22.6 million km^3 of ice above sea level, which, if it were to melt completely, is equivalent to 57 m of sea-level rise (Lythe et al., 2001). The EAIS appears to be a very stable feature, which has remained close to its present volume during several interglacials (Sugden et al., 1995;

Marchant et al., 2002; Siegert, 2008). However, the largely marine-based WAIS may be more vulnerable, and over the years a great deal of research has focused on the possibility that rapid deglaciation might occur in response to anthropogenic climate change (Vaughan, 2008). Complete melting of the WAIS could raise global sea level by an average of 5 m.

Extending northwards from West Antarctica is the Antarctic Peninsula. The ice caps and glaciers in the Peninsula are nourished locally and are not considered to be part of the Antarctic Ice Sheet (van der Veen, 1999a). Some aspects of the glaciation of the Antarctic Peninsula, including recent changes to its fringing ice shelves, are discussed in section 5.6. Independent glaciers also exist in some areas of continental Antarctica, where precipitation is insufficient to maintain large volumes of ice. Among the best known of these areas are the Dry Valleys in Victoria Land, East Antarctica.

The East Antarctic Ice Sheet largely consists of a relatively flat dome, the South Polar Plateau, rising to a maximum elevation of over 4000 m. Ice flows radially towards the coast, and through the Transantarctic

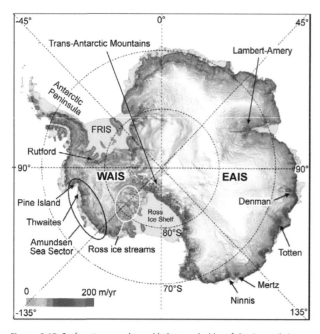

Figure 6.15 Surface topography and balance velocities of the Antarctic Ice Sheet. Major ice streams and other features are shown. FRIS: Filchner-Ronne Ice Shelf (Bamber et al., 2007).

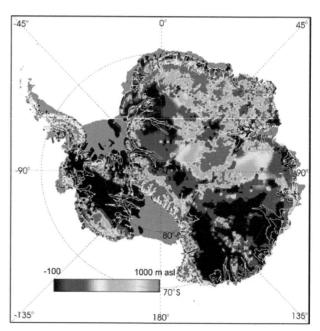

Figure 6.16 Antarctic basal topography with areas below sea level shown in black (Bamber et al., 2007).

Mountains into West Antarctica and onto the Ross Ice Shelf. In the coastal regions, ice discharge becomes focused into outlet glaciers and ice streams, separated by local ice domes and mountain ridges. One of the outlets of the EAIS is popularly known as the 'world's largest glacier', and consists of three confluent ice streams (Lambert, Mellor and Fisher Glaciers) and the Amery Ice Shelf at their seaward end (fig. 6.17). The grounded portion alone is over 1.48 million square kilometres in area, or 16 per cent of the total EAIS (Fricker et al., 2000; Wen et al., 2007). Other ice streams in East Antarctica include the Shirase, Jutulstraumen, Denman, Totten, Ninnis and Mertz Glaciers, the latter two named after men who died during Douglas Mawson's ill-fated traverse of King George V Land in 1912–13 (Bickel, 2000).

The topography of the West Antarctic Ice Sheet is more complex than the EAIS, and consists of three major ice domes and numerous local domes and ridges. Ice flow is divided into three distinct sectors by topographic divides (Fahnestock and Bamber, 2001; Bindschadler, 2006; fig. 6.15). The *Amundsen Sea sector* drains northwards via Pine Island Glacier, Thwaites Glacier and numerous smaller ice streams and outlet glaciers. Several of these terminate in fringing ice shelves, the largest of which are the Getz Ice Shelf and the Abbot Ice Shelf. The *Weddell Sea sector* drains into the vast Ronne Ice Shelf (360,000 km²), from the central plateau and the Ellesworth Mountains. Major ice streams in this sector include the Institute, Rutford and Evans Ice Streams (Doake et al., 2001). The *Ross Sea sector* drains westwards into the Ross Ice Shelf (850,000 km²), via the Ross or Siple Coast ice streams.

Figure 6.17 Surface velocities of the Lambert Glacier – Amery Ice Shelf system, derived by synthetic aperture radar (Canadian Space Agency/NASA/ Ohio State University, Jet Propulsion Laboratory, Alaska SAR facility, http://earthobservatory.nasa.gov).

The Ross Ice Streams were discovered during airborne radar surveys conducted by the Scott Polar Research Institute and the Technical University of Denmark, and were given the names Ice Streams A–F (Rose, 1979). In 2001, Ice Stream B was renamed Whillans Ice Stream, in memory of glaciologist Ian Whillans (1944–2001). In 2003, the American Advisory Committee for Antarctic Names (ACAN) made the decision to rename the remaining ice streams and inter-stream ridges in honour of other glaciologists who have conducted important work in West Antarctica (table 6.1). The decision to replace existing nomenclature with the names of living scientists was controversial, although if this is the start of a trend, the

Former name	New name
Ice Stream A	Mercer Ice Stream
Ice Ridge AB	Conway Ice Ridge
Ice Stream B1	Van der Veen Ice Stream
Ice Stream B2	Whillans Ice Stream
Ice Ridge BC	Engelhardt Ice Ridge
Ice Stream C	Kamb Ice Stream
Ice Ridge CD	Raymond Ice Ridge
Ice Stream D	Bindschadler Ice Stream
Ice Ridge DE	Shabtaie Ice Ridge
Ice Stream E	MacAyeal Ice Stream
Ice Ridge EF	Harrison Ice Ridge
Ice Stream F	Echelmeyer Ice Stream

Table 6.1 Old and new names for the Ross Ice Streams and inter-stream ridges.

authors of this book look forward to being similarly honoured.

6.3.2 CLIMATE AND MASS BALANCE

In contrast with the north polar regions, the Antarctic consists of a large continent surrounded by oceans. At the latitude of Drake Passage between South America and the Antarctic Peninsula (55–63° S), the only land consists of small, widely spaced islands which have negligible influence on either atmospheric or oceanic circulation. In consequence, air flow in the southern mid latitudes has a much stronger zonal (west to east) component and higher average windspeeds than in the northern hemisphere (Barry and Chorley, 2003). Westerly air flow, together with oceanographic factors, drives the Antarctic Circumpolar Current clockwise around the Antarctic continent. As a result of the mainly zonal atmospheric and oceanic circulation in the southern mid latitudes, there is much less poleward atmospheric and oceanic energy flux than in the northern hemisphere, and as a result Antarctica experiences colder and drier conditions than any part of the Arctic.

The cold waters around Antarctica allow rapid sea ice formation in winter. At its maximum extent in September, sea ice cover is around 16×10^6 km^2, shrinking to $\sim 2 \times 10^6$ km^2 in February. This is a much larger seasonal fluctuation than in the Arctic, where the configuration of the continents encourages the retention of ice for longer periods. Also unlike the Arctic, where average sea ice area has been decreasing by around 4 per cent per decade, the average area of Antarctic sea ice has been increasing, at a rate of 1.7 ± 0.3 per cent per decade (Comiso and Nishio, 2008).

Due to a combination of atmospheric circulation patterns and high elevation, the South Polar Plateau is the coldest place on Earth. The mean annual air temperature at the Russian Vostok Station near the centre of the Antarctic continent is $-55.4\,°C$, with summer and winter means of $-36.1\,°C$ and $-66.8\,°C$, respectively (Turner et al., 2004). The lowest temperature ever recorded at the Earth's surface, $-89.2\,°C$, was at Vostok. The coastal regions of continental Antarctica are mild in comparison, with mean annual temperatures around $-10\,°C$, and summer and winter means of around $-2\,°C$ and $-16\,°C$, respectively. Continental Antarctica is also very windy, with mean annual windspeeds of 5–10 m s^{-1} at most stations. Drifting of snow in cold, windy regions makes measurement of precipitation prone to large errors, so evaluations of Antarctic precipitation utilize climate models with observed temperature, wind, pressure and moisture fields as input, tested against snow depth data (Bromwich et al., 2004a; van den Berg et al., 2006). Precipitation is at a maximum close the coast, where travelling depressions advect moisture evaporated from the oceans. As winds are driven upslope against the ice sheet margin or mountain ridges, orographic effects further enhance precipitation. Snowfall also declines with distance from the coast on the large ice shelves. Graf et al. (1999) determined snow accumulation rates along a flow line of the Ronne Ice Shelf, Antarctica, and showed that they decline from around 20 cm yr^{-1} (water equivalent) at the shelf edge to around 10 cm yr^{-1} at the grounding line, 600 km farther south. Very arid conditions prevail on the Polar Plateau, where the thin, cold air contains very little water vapour. Mean annual precipitation averaged over the whole of Antarctica is only ~ 195 mm yr^{-1}, but is typically three to five times higher than this at the coast.

There is no evidence of an Antarctic-wide temperature trend over the last 40 to 50 years (J. Turner et al., 2005; fig. 6.18). Of 18 climate stations in Antarctica (including the Antarctic Peninsula) for which long-term temperature trends could be computed, 11 have experienced warming and 7 have cooled. However, statistically significant trends were found at only 3 stations: Faraday/Vernadski on the Antarctic Peninsula, Novolazarevskya in East Antarctica, both of which have warmed, and Amundsen–Scott at the South Pole, which has cooled. Warming in the Peninsula has been an exceptional $3.7 \pm 1.6°$ per century, but this is not representative of the continent as a whole. In contrast, continental Antarctica has experienced a statistically significant increase in precipitation, by 1.3 to 1.7 mm yr^{-1} for 1979–99 (Bromwich et al., 2004a).

Low air temperatures mean that surface melting is negligible throughout almost all continental Antarctica. Consequently, surface mass balance is positive over the entire Antarctic Ice Sheet, and follows essentially the same pattern as precipitation, modified by snow blow and losses by sublimation. Accumulation rates have been determined in large numbers of shallow cores taken during traverses of the ice sheet (Vaughan et al., 1999; van den Berg et al., 2006). The sampling density is quite

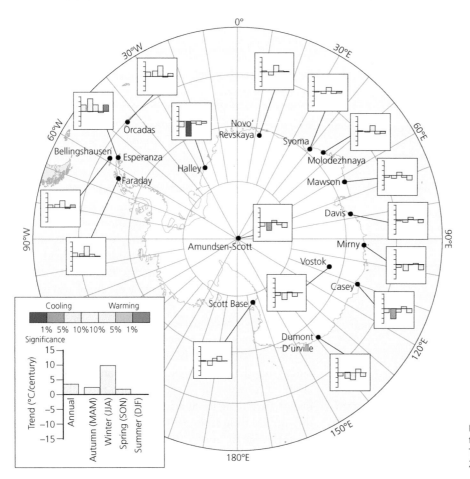

Figure 6.18 Antarctic near-surface air temperature trends for 1971–2000. The stations shown have a minimum of 27 years of data (Turner et al., 2005a).

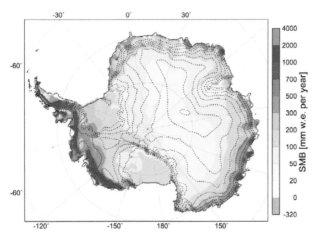

Figure 6.19 Surface mass balance of the Antarctic Ice Sheet, derived from observations and modelling (van de Berg et al., 2006).

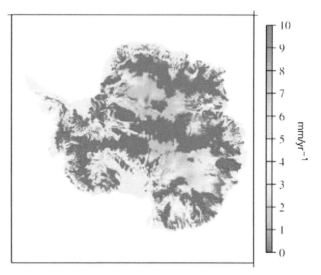

Figure 6.20 Calculated subglacial melt rates below the Antarctic Ice Sheet. The results are sensitive to the assumed geothermal heat flux and other variables (Llubes et al., 2006).

high in some regions but very low in others, complicating efforts to determine continent-wide accumulation patterns. Using a combination of observations and modelling, van den Berg et al. (2006) calculated mean accumulation rates of 170 ± 3 mm yr^{-1} over the grounded portion of the ice sheet, and a total surface mass balance of around $+2 \times 10^{15}$ kg yr^{-1} (fig. 6.19).

Large areas of the bed of the Antarctic Ice Sheet are at the pressure-melting temperature and basal melt rates can be quite high, depending on the balance between geothermal heat flux, strain heating, conduction and heat advection (fig. 6.20; Llubes et al., 2006; cf. section 2.3.2). Calculated basal melt rates are highest below fast-flowing outlet glaciers and the onset zones of ice streams (see section 6.3.4), with an overall mean rate of ~3.5 mm yr^{-1}.

Figure 6.21 A massive tabular iceberg (B-15) calved from the Ross Ice Shelf in mid March 2000. The iceberg is ~300 km long and ~40 km wide, and is one of the largest ever observed (MODIS image from NASA, http://veimages.gsfc.nasa.gov/248/MODIS1000032.jpg).

This is two orders of magnitude smaller than the average surface accumulation rate, but nevertheless constitutes a non-trivial mass balance term. Additionally, the production and storage of basal meltwater plays a central role in dynamic processes, modulating rates of ice flux to the oceans (Bell, 2008; section 6.3.4).

The majority of ice loss from the Antarctic Ice Sheet is at the ice–ocean interface, either by calving or bottom melting on ice shelves. Although infrequent, the release of tabular icebergs is by far the most important process in terms of the volume calved from continental Antarctic ice shelves (fig. 6.21). Some events can be very large indeed. For example, iceberg B-15, which calved from the Ross Ice Shelf in March 2000, measured 295 km × 40 km (Arrigo and van Dijken, 2003). Over the 4-year period 1998–2002, the Fichner–Ronne and Ross Ice Shelves calved a total of six large tabular bergs, with a combined area of >37,000 km². For stable ice shelves, calving of tabular bergs forms part of a cycle, with loss of large areas of the shelf followed by an advance of the ice edge to around its former position. The last major calving event from the Amery Ice Shelf occurred in austral summer 1963–64, since which time new longitudinal rifts have formed along suture zones where neighbouring flow bands have become separated during ice shelf spreading (Fricker et al., 2002). Calving will occur when these rifts are linked by a growing transverse rift, and the likely timing of this event suggests a calving cycle of 60–70 years.

Once released, Antarctic bergs drift around the polar seas, breaking up and melting (Scambos et al., 2005). Large bergs can have very large impacts on marine ecology, by increasing sea ice cover and reducing the productivity of phytoplankton (Arrigo and van Dijken, 2003).

Reduced primary productivity can cascade up to higher trophic levels, affecting, for example, penguin colony size and feeding habits.

Mass balance beneath Antarctic ice shelves (melting and freezing) exhibits considerable spatial variability, often with accumulation and ablation occurring simultaneously beneath different parts of the same shelf. Average bottom freezing rates for the base of Antarctic ice shelves are in the region of 30–50 mm yr^{-1}. These figures, however, mask considerable spatial variability, both between and within ice shelves. Rates of basal accretion by freeze-on have been calculated as 10–35 mm yr^{-1} for the Ross Ice Shelf and 300–600 mm yr^{-1} for the Amery Ice Shelf, where accumulation by basal freezing exceeds that attributable to snowfall (fig. 6.22; Budd et al., 1982). The Ross Ice Shelf can be subdivided into three zones, based on patterns of basal melting and freezing (Souchez and Lorrain, 1991). *Zone 1* is an area of enhanced bottom freezing near to the grounding line, due to the presence of fresh meltwater draining from beneath terrestrial ice streams. Localized basal melting also occurs in this zone in association with vigorous meltwater efflux points. *Zone 2* is an area of slow bottom freezing caused by effective upward heat transport through the ice shelf and the presence of cold isothermal water of low salinity. This gives way to an outer *Zone 3*, where stronger circulation and greater heat exchange produces net basal melting.

Using a combination of satellite radar altimetry and airborne radio echo sounding, Fricker et al. (2001) found that marine ice accounts for approximately 9 per cent of the volume of the Amery Ice Shelf. Marine ice is concentrated in the north-west of the shelf, reflecting clockwise water circulation. In a remote-sensing study of 23 ice

(a)

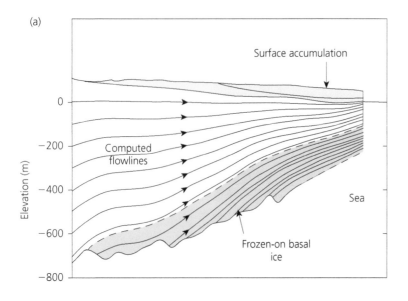

(b)

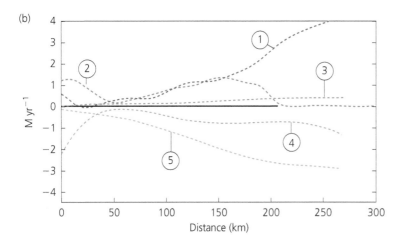

Figure 6.22 (a) Longitudinal profile of the Amery Ice Shelf, East Antarctica, showing ice flow lines, and surface and basal ice accumulation. (b) Mass balance components for the ice shelf: (1) surface accumulation rate; (2) basal ice accretion rate; (3) horizontal ice flux; (4) longitudinal strain thinning; (5) transverse strain thinning (Drewry, 1986, after Budd et al., 1982).

shelves around continental Antarctica, Rignot and Jacobs (2002) found that the highest bottom melt rates are in the Amundsen Sea Embayment (~10–40 m yr^{-1}), reflecting incursion of relatively warm ocean water (section 6.3.7). Average basal melt rate over all ice shelves around continental Antarctica is estimated to be around 0.4 m yr^{-1}.

6.3.3 FLOW OF INLAND ICE

Flow of the inland ice in East Antarctica is primarily by internal ice deformation (van der Veen, 1999a). Large numbers of local flow anomalies have been detected, however, marking transitions from high drag to low drag and vice versa (Rémy et al., 1999). Many of these ice flow anomalies are associated with subglacial lakes, where basal drag is zero (Bell et al., 2002; Wendt et al., 2006). Because such lakes are limited in extent compared to the area and thickness of the surrounding ice sheet, variations in ice velocity are of local importance and probably do not significantly influence the large-scale dynamics of the ice sheet. Mountain ranges and cold basal ice around the periphery of the EAIS act as a barrier to ice flow,

preventing large basal velocities except where ice flow is focused in outlet glaciers or ice streams.

In West Antarctica, drainage of the ice sheet interior is focused in ice streams separated by almost stagnant ice domes and ridges. The position of ice divides is not constant in time, and divides migrate on millennial timescales in response to changing ice stream dynamics and other factors (e.g. Siegert et al., 2004b). Ice divide migration can be inferred from the structure of internal layering detected using ice-penetrating radar. At ice divides, ice layers at depth commonly form upwarped arches termed *Raymond bumps*, which reflect patterns of vertical strain in the ice (Raymond, 1983; fig. 6.23). Horizontal velocity is zero at the divide, and increases with distance on either side. The resulting velocity gradients result in horizontal stretching and vertical thinning, and the patterns of cumulative strain are reflected in variations in ice layer thickness. Layers become thinner with depth below the surface (where strain rates are higher and the ice is older) and with distance from the divide, creating a characteristic 'bump', which increases in amplitude with depth. Under steady-state conditions, Raymond bumps are

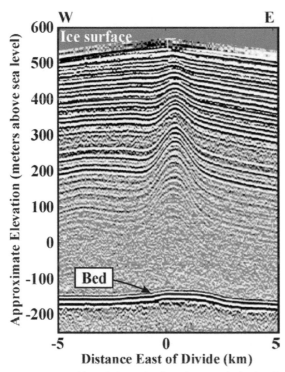

Figure 6.23 Raymond bump in internal ice layers in Roosevelt Island, an ice rise in the Ross Sea (Martin et al., 2006, data from H. Conway and A. Gades).

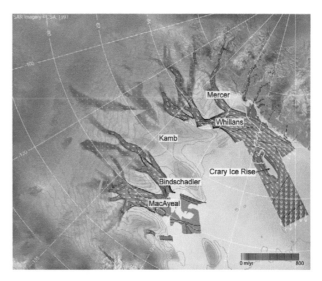

Figure 6.24 Velocity of the Ross Ice Streams and inter-stream areas, derived from RADARSAT data (Joughin et al., 2002).

symmetrical, but will deviate from this simple form due to various factors, such as vertical changes in ice rheology, surface accumulation rates, temperature, ice divide migration and basal sliding (Pettit et al., 2003; Martin et al., 2006; Price et al., 2007). Inverse modelling of bump evolution can be used to reconstruct the history of divide migration and the changing structure of the ice sheet. Knowledge of ice divide history is important for understanding long-term ice sheet dynamics and choosing optimum sites for drilling deep ice cores (Jacobel et al., 2005).

6.3.4 ICE STREAMS

Like the outlet glaciers draining the Greenland Ice Sheet, Antarctic ice streams exhibit a wide range of characteristics. Truffer and Echelmeyer (2003) have defined a spectrum of ice stream types, with topographically confined outlet glaciers (which they termed *isbrae*) as one end member, and the Ross Ice Streams as the other (fig. 6.24). (Truffer and Echelmeyer somewhat confusingly referred to the latter as 'ice-stream type' ice streams, but here we adopt the less ambiguous term *Ross-type ice streams*.) Ross-type ice streams are typically many tens of kilometres wide and around 1 km thick. Ice stream location is not strongly controlled by bed topography, and the ice streams are embedded within regions of cold-based ice. They have low surface slopes and driving stresses ($\sim$10 kPa), but flow with speeds of many hundreds of metres per year. Fast flow is possible because the ice stream has a weak bed, associated with low effective

pressures and soft sediments. Ross-type ice streams are bounded by narrow lateral shear zones which support most of the driving stress. In consequence, the ice deforms almost exclusively by transverse shear. In contrast, isbrae-type ice streams are confined within bedrock channels. They have steep surface slopes and very high driving stresses, often exceeding 200 kPa. High basal drag means that ice deformation occurs by vertical as well as lateral shear. Antarctic streams and outlet glaciers span the entire spectrum, from steep, topographically confined glaciers, such as Byrd Glacier in the Transantarctic Mountains, to Ross-type ice streams embedded within areas of sheet flow, such as Whillans Ice Stream. Byrd Glacier in the Transantarctic Mountains attains speeds of up to 800 m yr^{-1} apparently entirely by vertical shearing of the ice, whereas this process contributes almost nothing to similarly high surface velocities on Whillans Ice Stream (Whillans et al., 1989; Whillans and van der Veen, 1997).

Because of the central role they play in controlling transfer of ice from the interior of West Antarctica to the ocean, Ross-type ice streams have been the focus of a major research effort since the 1980s. Understanding how these ice streams function and evolve requires knowledge of four key areas:

(1) ice stream onset zones;
(2) ice stream beds;
(3) lateral shear margins;
(4) grounding zones between ice streams and floating ice shelves.

Ice stream onset zones mark the transition between two distinct flow regimes: tributary areas of slowly moving inland ice and areas of streaming flow. In the former, vertical shear is an important component of ice flow (section 6.3.3) and velocity increases with driving stress, whereas in areas of streaming flow, vertical shear is much less important than lateral shear and velocities increase as

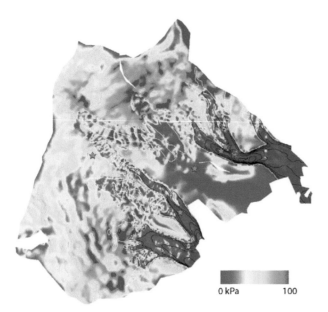

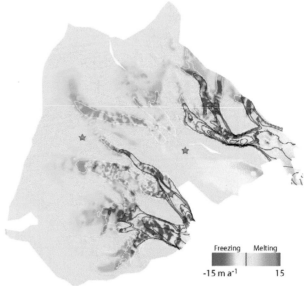

Figure 6.25 Modelled basal drag below the Ross Ice Streams. The data are derived from force budget analysis, in which basal drag is taken to equal the residual after driving stresses, lateral drag and longitudinal stress gradients have been calculated (Joughin et al., 2004c).

Figure 6.26 Modelled basal melt rates below the Ross Ice Streams (Joughin et al., 2004c).

driving stress decreases (Bindschadler et al., 2001; Hulbe et al., 2003). Ice stream onset zones can be identified on satellite imagery from a distinctive suite of morphological features, including transverse crevasses formed in response to extending flow, and the appearance of *flow stripes* or sets of longitudinal ridges and troughs (Bindschadler et al., 2001). Flow stripes appear to be the surface expression of flow perturbations originating at the glacier bed (Gudmundsson et al., 1998). A mismatch between surface flow stripes and folding of internal ice layers was observed by Campbell et al. (2008), who proposed that flow stripes originate near the ice stream onset by transverse shortening and folding of the ice, then become progressively modified by aeolian processes as they are advected downstream. Flow stripes have been widely used as ice flow markers, allowing past dynamic changes to be reconstructed (e.g. Hulbe and Fahnestock, 2007; section 6.3.6).

Morphologically, ice stream onset zones occur at the boundary between the convex-up profile of inland ice and the concave-up profiles typical of ice streams, and have comparatively high surface gradients. Higher than average surface gradients are associated with high driving stresses (Bindschadler et al., 2001; Whillans et al., 2001; equation 4.10). Figure 6.25 shows patterns of basal shear stress below Mercer and Whillans Ice Streams, derived using inverse modelling methods (Joughin et al., 2004b). Regions of high basal stress at the ice stream onsets are clearly visible, particularly in the two tributary branches of Whillans Ice Stream, and contrast sharply with the low basal stresses below the ice streams and the grounding zone between the ice streams and the Ross Ice Shelf.

The increase in velocity in ice stream onset zones is largely accomplished by basal motion, associated with increased production of meltwater at a temperate bed. In section 2.3.2, we saw that strain heating is directly proportional to the product of the shear stress and sliding speed/strain rate (equations 2.10 and 2.11). That is, large values of basal drag and high velocities favour strain heating and meltwater production, which in turn promote more rapid basal motion. Increasing velocities, strain heating and basal meltwater production are therefore mutually reinforcing at ice stream onsets (Bindschadler and Choi, 2007). The high basal melt rates under onset zones contrast with minimal melt rates under the fast-flowing parts of ice streams, where basal shear stresses are typically very low (fig. 6.26; Whillans et al., 2001; Price et al., 2002; Vogel et al., 2003; Joughin et al., 2004c). Basal melting below ice stream onsets is therefore critical for the water balance of the system downstream, and the maintenance of conditions conducive to fast flow (Raymond, 2000; Whillans et al., 2001).

Ice stream beds in West Antarctica are typically underlain by layers of soft, unconsolidated sediments, metres thick (Blankenship et al., 1987; Doake et al., 2001; Kamb, 2001; Vaughan et al., 2003; Murray et al., 2008). Samples of bed materials obtained from boreholes consist of massive, clay-rich till derived from Tertiary marine sediments and crystalline bedrock (Tulaczyk et al., 1998; Kamb, 2001). Unremoulded samples generally have very high porosities and low shear strength, particularly near the top of the till layer. When in situ, the pore space of the till is fully water saturated, and water levels in boreholes indicate that pore-water pressures are close to (and sometimes in excess of) ice overburden pressures.

When the existence of a thick layer of weak sediment below Whillans Ice Stream was first inferred from seismic evidence, it was thought that the entire layer must be undergoing pervasive deformation (Alley et al., 1986, 1987; Blankenship et al., 1987b). Indeed, the conclusion that fast flow of the ice stream was maintained by till deformation did much to establish the 'deforming bed paradigm' discussed in section 4.5.1. Subsequent work, however, has shown that till deformation is probably discontinuous in both time and space. Down-borehole drag-spool measurements conducted on Whillans Ice Stream showed that 80–100 per cent of the total surface velocity was accommodated by shear within 3 cm of the bed, either by sliding or till deformation in a narrow shear zone (Engelhardt and Kamb, 1998; Kamb, 2001). This result also shows that vertical shearing within the ice is negligible, reflecting the fact that very little of the driving stress is opposed by traction at the bed. In contrast, drag-spool measurements on Bindschadler Ice Stream appeared to show that most ice stream motion is accommodated by till deformation deeper than the spool anchor (34–63 cm beneath the top of the till). At the start of the measurement period, however, 80 per cent of the ice motion appeared to be by sliding, and the conclusion that deep deformation became increasingly important during the measurement period is based on the assumption that the anchor remained securely tethered in the till. Till porosity profiles from Bindschadler Ice Stream and shear strength profiles from Whillans Ice Stream are consistent with a thin shear zone at the top of the till, rather than deep, pervasive deformation (Kamb, 2001). Furthermore, the properties of till samples recovered from ice stream beds are consistent with a Coulomb-plastic rheology, which again should favour failure near the ice–bed interface rather than deep, pervasive deformation (Tulaczyk et al., 1998, 2000a; section 4.5.2). Regardless of the exact mechanism of basal motion, however, the key finding is that ice stream beds are very weak and generally only able to support basal shear stresses in the range 1–5 kPa (Engelhardt and Kamb, 1998; Kamb, 2001; Joughin et al., 2002).

Geophysical studies indicate that conditions at the beds of Antarctic ice streams are both spatially variable and rapidly changing (e.g. Vaughan et al., 2003; Smith, 2006; Murray et al., 2008). Repeat imaging of the bed of Rutford Ice Stream has revealed rapidly changing bed forms, probably mega-scale glacial lineations or drumlins (Smith et al., 2007). Dynamic subglacial landforms have also been detected by highly detailed radar and seismic imaging in the onset zone of Rutford Ice Stream (fig. 6.27; King et al., 2007, 2009).

Lateral shear margins play an important role in the force balance of ice streams where the bed offers little resistance to flow (Whillans and van der Veen, 1997; van der Veen, 1999a; Raymond et al., 2001; Whillans et al., 2001). The lateral boundaries of ice streams mark the point where basal conditions transition from a low drag, well-lubricated bed to a high-drag, frozen bed. Any driving stress that is not supported beneath the ice stream must be transferred via the lateral margins to the surrounding high-drag regions of the ice sheet bed. Raymond et al. (2001) showed that, where there is negligible drag below an ice stream, there is a zone of elevated basal shear stresses just outside the ice stream shear margins, where traction at the bed supports almost all of the driving stress across the whole ice stream (fig. 6.28). Stress is transferred from the ice stream to this zone predominantly via the lateral shear stress τ_{xy}. Consequently, the shear margin is a region of very high stresses (fig 6.29; Raymond et al., 2001).

The transverse velocity profiles of ice streams reflect these stress patterns. Velocity gradients ($\partial u/\partial y$) are very high near the margins, with more plug-like flow near the centreline, especially where basal drag is very small (Whillans and van der Veen, 1997; Whillans et al., 2001). Figure 6.30 shows selected velocity profiles across Ross Ice Streams (Joughin et al., 2002), which match quite closely the form of the theoretical velocity profiles discussed in section 4.6.2. Also shown in the figure is the value of the lateral shear stress τ_{xy} across the ice stream. The lateral shear stress is zero at the centreline and rises to values of 100–200 kPa at the margins (note that the sign of τ_{xy} reverses at the centreline because the direction of shear, relative to the coordinate system, switches at that point). If all of the driving stress were supported by the margins and the ice were uniformly thick, then τ_{xy} would vary linearly across the glacier. This is because the total downflow-directed force exerted by the ice (which must be supported by the lateral shear stress) is directly proportional to ice stream width. In reality, τ_{xy} deviates from this simple pattern, primarily due to transverse variations in ice thickness or bed properties.

As a result of very high strain rates, shear margins are intensely crevassed. A characteristic zonation has been recognized, reflecting varying strain rates across the shear margin (Raymond et al., 2001; figs 2.29 and 6.31). The outermost zone is several hundreds of metres wide, and is marked by relatively evenly spaced arcuate crevasses. Closer to the ice stream is a central zone several kilometres wide where strain rates are highest, with multiple intersecting crevasses giving the ice a chaotic, granular appearance. The innermost zone, on the ice stream side of the shear zone, again has relatively low strain rates, and crevasse patterns are similar to those in the outer zone.

Migration of ice stream margins can occur in response to advection of cold ice from inter-stream ridges, which narrows the ice stream, or strain heating of the margins, which widens the stream as the ridge ice is warmed and

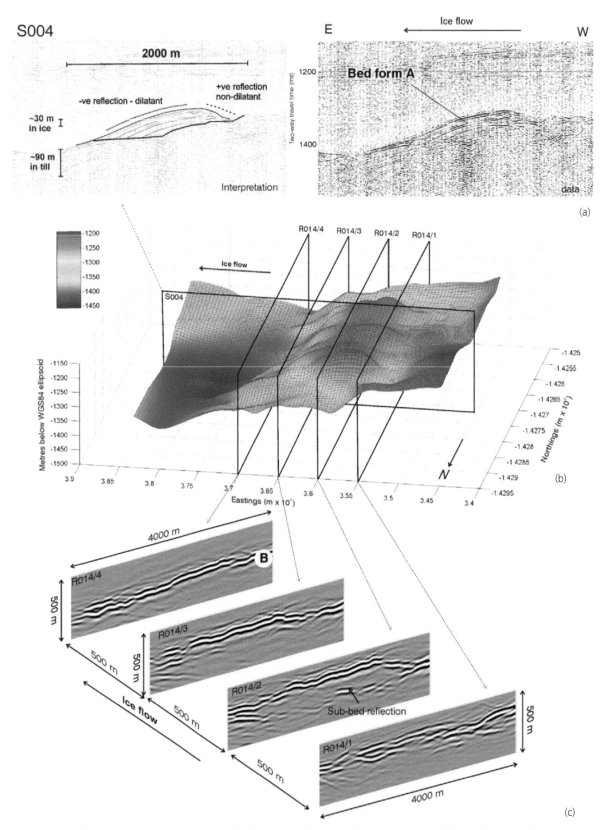

Figure 6.27 Subglacial bedforms below the onset zone of Rutford Ice Stream. (a) Seismic profile and interpretation. (b) Three-dimensional view of bedform morphology, showing the locations of seismic and radar profiles. (c) Transverse radar profiles of the bedforms (King et al., 2007).

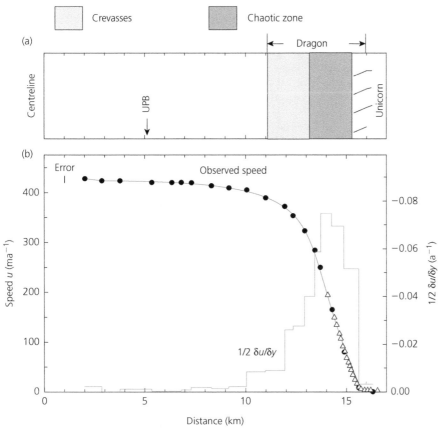

Figure 6.28 Characteristics of ice stream shear margins. (a) Zones in marginal shear patterns. 'Dragon' and 'Unicorn' are features at the margin of Whillans Ice Stream. (b) Observed speed and strain rate across part of the shear margin of Whillans Ice Stream (Raymond et al., 2001).

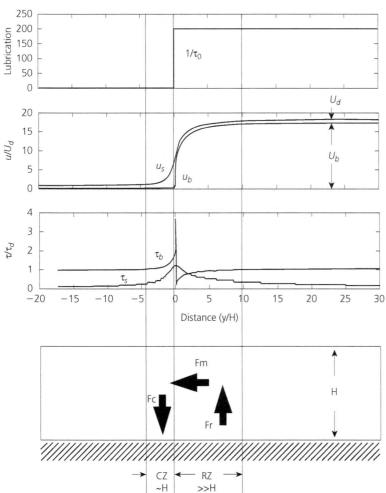

Figure 6.29 Modelled variation in velocity at the surface (u_s) and bed (u_b) and associated basal (τ_b) and lateral (τ_s) shear stress (middle panels) in response to a jump in lubrication (top panel). The bottom panel is a schematic representation of force redistribution across the shear margin. Forces near the margin of the ice stream are supported by the surrounding ice. Forces in the *relaxation zone* (Fr) are transferred via the margin (Fm) to the surrounding *concentration zone* (Fc) (Raymond et al., 2001).

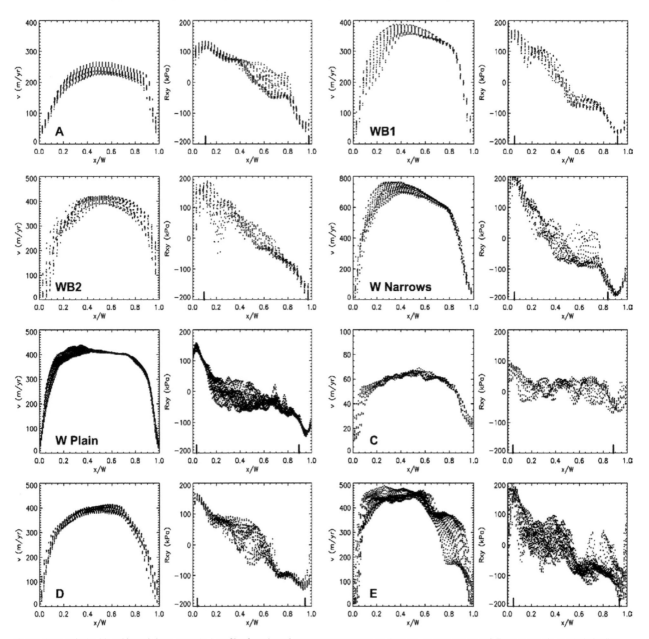

Figure 6.30 Velocity (v) and lateral shear stress (R_{xy}) profiles for selected transects across Ross Ice Streams. A: Mercer; W: Whillans; C: Kamb; D: Bindschadler; E: MacAyeal (Joughin et al., 2002).

entrained (van der Veen and Whillans, 1996; Raymond et al., 2001). Where most of the driving stress is supported by lateral drag, centreline velocities are proportional to the fourth power of the ice stream width (van der Veen, 1999a; equation 4.25). It might be expected, therefore, that even a slight increase in ice stream width could lead to a large increase in velocity, with important consequences for ice stream stability. Stearns et al. (2005), however, have found that recent changes in width of Whillans Ice Stream do not have a simple relationship to ice velocity. In some places, the ice stream has been decelerating and narrowing, as might be expected from decreased heat production under the margin. At other places, the ice stream is decelerating and widening. This is most likely due to a

reduction in water supply from the onset zone, decreasing pore-water pressures in the till and strengthening the bed (Joughin et al., 2002; Vogel et al., 2003). Increased shear stress at the bed could lead to locally increased heat production near the margins (equation 2.11), accounting for concomitant deceleration and widening.

Grounding zones (or perhaps more accurately, *ungrounding zones*) of ice streams mark the transition between lightly grounded ice and fully floating ice shelves. Where the bed is relatively flat, this transition zone may extend over ~10^2 km, forming a complex mosaic of lightly grounded and ungrounded ice. The surface expression of this zone is an *ice plain* (Bindschadler et al., 2005), with surface slopes intermediate between those of an active ice stream (order 10^{-2}) and ice shelf (order 10^{-4}). *Ice rises*, or shallow domes rising above the ice plain surface occur over areas of more firmly grounded ice.

Ice flow in grounding zones and adjacent areas of ice streams is very sensitive to small changes in force balance, and is most clearly illustrated by the way that velocities are modulated by tidal variations. Bindschadler et al. (2003a, b) showed that the lower reaches of Whillans Ice Stream exhibit stick-slip motion, in which brief episodes of forward movement are interspersed with longer periods of little or no motion (fig. 6.32). During spring tide periods, rapid motion events regularly occur near high tide and during falling tide, but the correlation is weaker at neap tides, when tidal magnitudes are lower. Because effective pressures are so low below the ice plain of Whillans Ice Stream, even the modest tidal range of ~2 m is sufficient to cause substantial changes of stress at the bed. Higher normal effective stress at low tide will increase the shear strength of a Coulomb-plastic material (cf. section 4.2.3), possibly causing the bed to 'stick'. Accumulated elastic strain in the

Figure 6.31 Shear margin of Whillans Ice Stream, photographed from an altitude of 3000 m in February 1947 (US Navy, from Swithinbank, 1988).

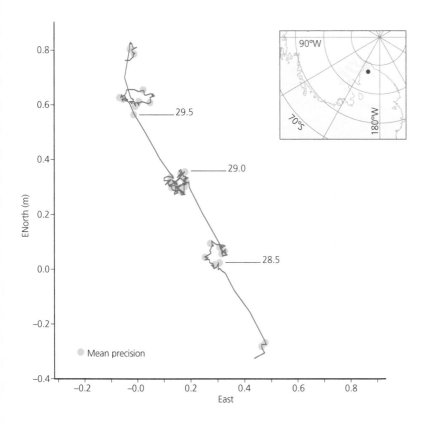

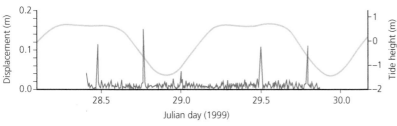

Figure 6.32 Stick-slip motion of Whillans Ice Stream. (Top) Horizontal position of marker. Grey dots indicate the beginning of each hour with Julian day indicated by arrows. (Bottom) Horizontal displacement between successive 5-minute positions of the marker (solid line), and modelled ocean tide (dashed) (Bindschadler et al., 2003a).

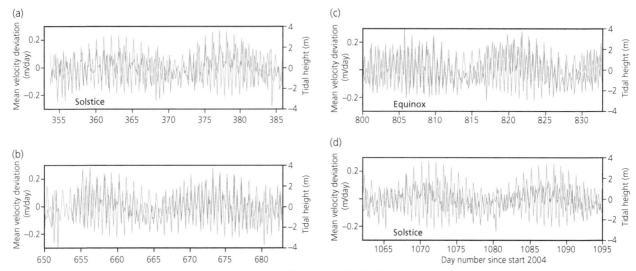

Figure 6.33 Deviation from the mean velocity of Rutford Ice Stream (red line) and tidal forcing (green line). Note the contrasting response when the two semi-diurnal tides have very different amplitudes (a, d) and when they have similar amplitudes (b, c) (After Murray et al., 2007).

till could then be released when the normal effective stress is low, resulting in sudden forward motion.

Tidally modulated velocity fluctuations have also been observed on Rutford Ice Stream (Gudmundsson, 2006; Murray et al., 2007; Aðalgeirsdottir et al., 2008). The tidal influence extends at least 40 km upstream of the grounding line, where velocity fluctuates by up to 20 per cent on a range of frequencies. The maximum fluctuation in forcing is on semi-diurnal frequencies (i.e. over the course of a single tidal cycle), but velocity fluctuations are actually greatest at 2-week, annual and semi-annual frequencies (Murray et al., 2007). Ice velocities are highest at the equinoxes, when there are two semi-diurnal tides of equal magnitude, and lowest at the solstices, when one of the semi-diurnal tides has a lower magnitude (fig. 6.33). This intriguing behaviour is still not fully understood, but it is thought to reflect two distinct processes acting in tandem. First, as appears to be the case at Whillans Ice Stream, the yield strength of the basal till may be non-linearly related to basal effective pressure, and hence will vary with tidal range. Second, at high tides the water column will exert a greater backstress on the ice stream, increasing resistance to flow (cf. section 5.5.1). These two processes act in opposition, potentially giving rise to a complex range of responses to tidal forcing.

The sensitivity of ice streams to small changes in force balance in the grounding lines implies that changes at the marine boundary of ice streams could propagate upstream and influence ice discharges, possibly with major implications for ice sheet mass balance. This important topic is discussed in section 6.3.6.

6.3.5 HYDROLOGY AND SUBGLACIAL LAKES

Until recently, almost nothing was known about the hydrology of the Antarctic Ice Sheets. This situation has changed dramatically with the advent of improved geophysical techniques and accurate repeat measurements of surface elevation, which have revealed varied and dynamic subglacial drainage systems below both East and West Antarctica. These drainage systems are recharged entirely by basal melting, and water storage and flux are controlled largely by patterns of basal melting and potential wells at the bed. This contrasts with the situation in much of Greenland, where subglacial water storage and flux are modulated by seasonal variations in water supply from the surface (section 6.2.3).

Subglacial lakes were first discovered beneath the Antarctic Ice Sheet in the late 1960s, during airborne radio echo sounding surveys to determine ice thickness (Robin et al., 1970). Since that time, around 150 examples have been identified, with new ones frequently being added to the inventory (Siegert, 2000; Dowdeswell and Siegert, 2002; Carter et al., 2007). Subglacial lakes have very distinctive radar signatures, showing up as bright sub-horizontal reflectors contrasting with adjacent areas of the bed (fig. 6.34; Carter et al., 2007). Lake extent can also be determined from surface topography, because the ice sheet surface is exceptionally flat above areas where basal traction is zero (Bell et al., 2006). On a continental scale, lake distribution is determined by pressure-melting conditions at the bed. Within these areas, lakes occur in three main settings:

(1) close to ice divides, where both ice surface slope and velocity are small;
(2) at ice stream onsets;
(3) in prominent subglacial troughs.

In addition, several 'fuzzy lakes' and 'indistinct lakes' have been identified from radar data, which may be areas of saturated sediments or flowing water (Carter et al., 2007).

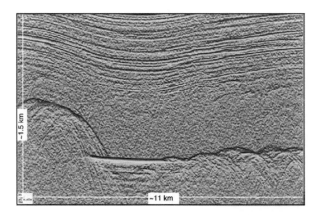

Figure 6.34 Subglacial lake imaged on a radar profile. The lake surface is picked out by the bright subhorizontal reflector left of centre, and contrasts strongly with the adjacent rough glacier bed. Note the layering in the overlying ice (Carter et al., 2007).

The largest known subglacial lake is Lake Vostok near the centre of the East Antarctic Ice Sheet (fig. 6.35; Siegert, 2000; Siegert et al., 2003; Studinger et al., 2003). Measuring ~240 km × ~50 km and with an area of 14,000 km^2, it is 20 times the size of Lake Geneva. Lake Vostok is located on a major geological boundary, and occupies a large basin which is probably tectonic in origin. The ice is thicker over the northern part of the lake (4300 m) than the south (3700 m), and the lake–ice interface slopes upwards from north to south, in the opposite direction to the ice surface. The lake has maximum depths of around 1200 m in the southern part, shallowing to the north. Ice flows over the lake approximately west to east, at speeds of around 3 m yr^{-1}. The thicker ice above the northern end of the lake means that the pressure-melting point is lower there, and melting occurs at the base of the ice. In contrast, ice freezes onto the base of the ice at the southern end, and has built up a layer of accretionary ice up to 210 m thick. Water circulates through the lake to balance melting and freezing.

Subglacial Lake Ellesworth in West Antarctica also occupies a subglacial trough, one of several subglacial valleys that terminate on flanks of Byrd Subglacial Basin (Vaughan et al., 2007). In this case, however, the trough appears to be a glacial erosional feature formed when the West Antarctic Ice Sheet was smaller than at present. Unlike Lake Vostok, which has a closed hydrological system, Subglacial Lake Ellesworth probably receives water from a large catchment of warm-based ice via some kind of subglacial drainage system. The lake has been identified as a possible target for sampling for microbial life (Siegert et al., 2004a). The possibility that Antarctic subglacial lakes may harbour unique microbial ecosystems provides a strong motivation to drill into them, but also reason to proceed with great care to avoid contamination (Siegert, 2000; Zotikov, 2006; section 3.8.2).

Lake formation beneath ice streams is favoured by enhanced meltwater production below onset zones

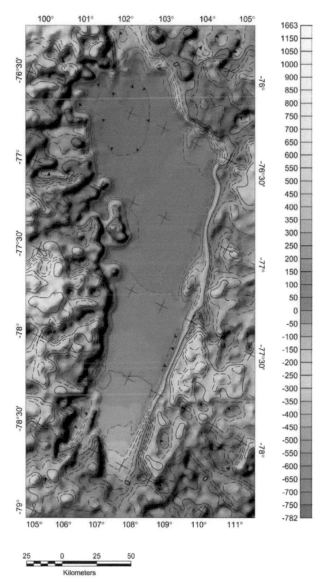

Figure 6.35 Subglacial basin occupied by Lake Vostok. Elevations in metres above sea level (Studinger et al., 2003).

(section 6.3.4) and very low hydraulic gradients (Bindschadler and Choi, 2007; Carter et al., 2007). Recent analyses of ice elevation data have identified coherent areas of uplift and subsidence in West Antarctic ice streams, indicating episodic transfer of water between subglacial lake basins (Gray et al., 2005; Fricker et al., 2007). Elevation changes are typically several metres, affecting many tens of square kilometres, implying the transfer of very high volumes of water (fig. 6.36). One subglacial lake near the grounding line of Whillans Ice Stream drained ~2 km^3 of water into the ocean over 3 years. Continuous water flow beneath West Antarctic ice streams appears to be discouraged by low potential gradients, and water tends to accumulate in areas of low hydraulic potential (section 3.7.4). Water accumulation will increase the local potential, until it reaches the threshold required to push it further through the system.

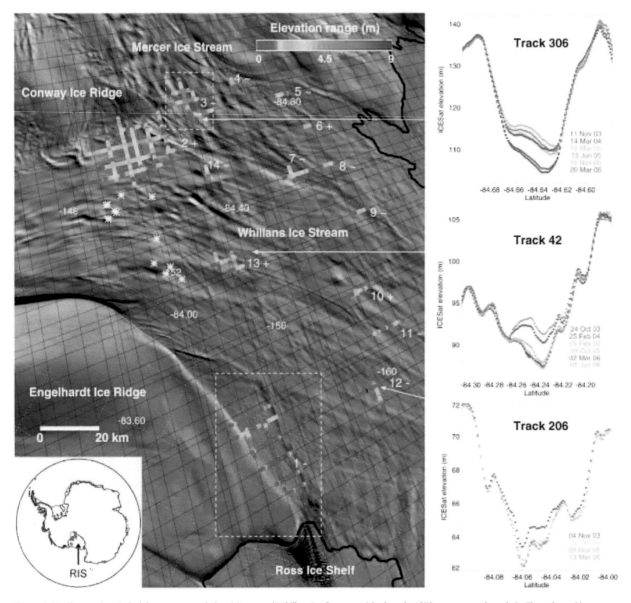

Figure 6.36 Changes in subglacial water storage below Mercer and Whillans Ice Streams, picked out by ICESat repeat-track analysis. The coloured bars represent the amount of observed elevation change, with the magnitude and sign indicated by the numbers. The panels on the right show selected profiles, with patterns of uplift and subsidence (Fricker et al., 2007).

Using seismic data, King et al. (2004) identified an extensive area of water at the ice–bed interface in the onset zone of Rutford Ice Stream. The ponded water extends over an area at least 1 km × 0.2 km, elongated in the direction of ice flow, and appears to be 0.4–0.6 m thick. King et al. argued that this likely forms part of a canal system, although it was not possible to determine whether the ponded water forms a single body of water ~200 m wide, or a number of narrower sub-parallel canals. Evidence for canal systems below Rutford Ice Stream has been presented by Murray et al. (2008).

Episodic lake drainage and filling has also been detected beneath the East Antarctic Ice Sheet (Wingham et al., 2006a). The drainage of one lake over a distance of

290 km into at least two others was indicated by paired subsidence and uplift events, in which 1.8 km³ of water was transferred during a period of 16 months.

6.3.6 ICE STREAM STAGNATION AND REACTIVATION

Several lines of evidence indicate that some of the Ross Ice Streams have switched between active and stagnant modes on centennial timescales. Rose (1979) drew attention to the 'enigma' of Kamb Ice Stream (then known as Ice Stream C), which is a region with a smooth, unbroken surface but a radar signature typical of adjacent fast-flowing ice streams. The lower trunk of Kamb Ice

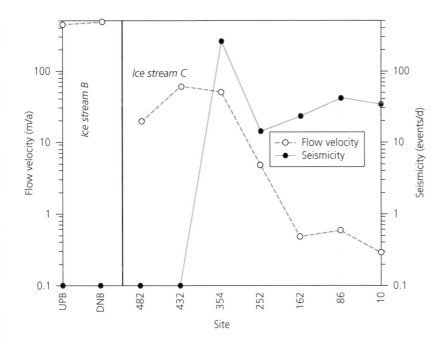

Figure 6.37 Plots of ice velocity (dashed red line) and seismicity (solid blue line) for Kamb Ice Stream (Anandakrishnan and Alley, 1997).

Stream is almost stagnant, flowing at similar speeds to the inter-stream ridges ($\sim 1\,\mathrm{m\,yr^{-1}}$), while its tributaries flow at rates of 60–$80\,\mathrm{m\,yr^{-1}}$, about half that of equivalent parts of adjacent ice streams (fig. 6.24; Joughin et al., 2002). Microearthquake activity is high throughout the trunk of Kamb Ice Stream, even where velocities are very low (fig. 6.37; Anandakrishnan and Alley, 1994; Anandakrishnan et al., 2001). Seismicity is highest in areas of decelerating or slow flow, and is consistent with intermittent slip events at sticky spots on the bed. Borehole video imaging has provided clear evidence that basal freezing occurs under Kamb Ice Stream (fig. 6.38; Carsey et al., 2002; Vogel et al., 2005; Christoffersen et al., 2006). Near the junction between the active upper part and the stagnant lower trunk, a stratified ice layer 10–14 m thick occurs above the bed, with isotopic and structural characteristics consistent with basal freezing in the presence of saturated sediments. Sediment content increases downwards, except for a layer immediately above the ice base, which is sediment poor. The presence of this relatively clear ice and water-filled gaps at the ice–bed interface suggest that basal water is more abundant now than in the last few centuries.

Detailed radar surveys of the lower part of Kamb Ice Stream have revealed buried zones of marginal crevasses, similar to those found at active ice stream margins, at depths of 7–20 m (Shabtaie and Bentley, 1987; Anandakrishnan et al., 2001). Snow accumulation rates imply that the lower section of the ice stream was in a state of fast flow until about the mid nineteenth century, after which it stagnated. Other relict flow features adjacent to the margins of the ice stream indicate flow reorientation and narrowing of the stream around 200

Figure 6.38 Stratified basal ice imaged by borehole video camera, Kamb Ice Stream. The alternating layers of debris-rich and clear ice are a few millimetres thick (Up in the figure is down in the ice.) (Vogel et al., 2005).

years prior to complete stagnation (Catania et al., 2006). Through analysis of deformed internal layering, Ng and Conway (2004) concluded that average pre-stagnation velocities were $\sim 350\,\mathrm{m\,yr^{-1}}$ in the trunk of the ice stream.

It is possible that a comparable shutdown is currently in progress in neighbouring Whillans Ice Stream, which has been slowing down progressively since the first velocity measurements were made in the early 1970s (Whillans et al., 2001; Joughin et al., 2002, 2005; Stearns et al., 2005). At present, the ice stream is flowing at close to its balance velocity, but until recently it was evacuating ice from the interior faster than it was being replenished by snowfall. Ice velocity decreased by 23 per cent between 1973 and 1997, since which time slowdown has been continuing at a rate of $0.6\%\,\mathrm{yr^{-1}}$. Extrapolation of this trend suggests that flow could cease as early as the middle of the present century, although of course there is no

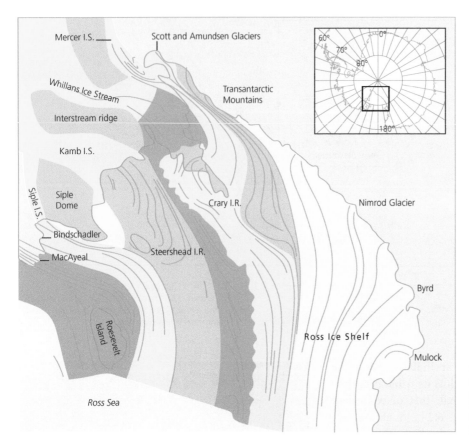

Figure 6.39 Flow units and streak lines on the Ross Ice Shelf. Note the contorted patterns in flow units derived from the Ross Ice Streams compared with those originating in the Transantarctic Mountains (Hulbe and Fahnestock, 2007).

guarantee that the current trend will continue. The changes are not spatially uniform, with the most dramatic occurring in the distal part of the ice stream, where deceleration has been accompanied by inward migration of the northern ice stream boundary. The combination of deceleration and narrowing has led to thickening of this region by around 0.2 m yr^{-1}. In contrast, further upglacier the ice stream has been slowing and widening, with associated thinning.

Hulbe and Fahnestock (2004, 2007) have used flow indicators on the Ross Ice Shelf to show that there have been additional episodes of ice stream stagnation and reactivation in the last 1000 years. Patterns of ice flow are recorded by trains of streak lines and shear margin crevasses which originate at ice stream onset zones and lateral margins, respectively (section 6.3.4). These features can be traced from the ice streams right across the shelf, and exhibit complex patterns somewhat like the looped moraines on surging glaciers (fig. 6.39; section 5.7.1). Distortion of streak lines can result from the development of an obstacle to the flow path across the shelf – such as an ice rise – or from oscillations in the discharge of adjacent ice streams. Hulbe and Fahnestock (2007) used a forward modelling approach to reconstruct the pattern and timing of ice flow variations, and concluded that Whillans Ice Stream ceased rapid flow about 850 years ago and restarted about 400 years later, whereas MacAyeal Ice Stream either stopped or slowed considerably between 800 and 700 years ago, restarting about 150 years later. In addition, subsurface crevasse patterns suggest the existence of a former active ice stream (Siple Ice Stream) on the northern flank of Siple Dome, which stagnated between 250 and 500 years ago (Nereson, 2000). There is also some evidence that ice streams in the Weddell Sea sector of the West Antarctic Ice Sheet can switch from active to stagnant modes (Vaughan et al., 2003).

Several alternative hypotheses have been proposed to explain long-term changes in ice stream velocity (Anandakrishnan et al., 2001), but in recent years two main ideas have received most attention. (1) Ice streams may be subject to cyclic, internally regulated switches of conditions at the bed, similar to surge cycles; or (2) changes in ice configuration or water flow may occur in the ice sheet interior in response to climate change or other factors, and the effects then cascade downstream to affect ice streams.

The idea behind *internally regulated velocity cycles* is that both fast and slow modes of ice stream flow lead to progressive changes in ice geometry and basal energy balance, which eventually trigger a switch from one mode to the other. Fast ice stream flow thins the ice, reducing the distance between the bed and the cold surface, and also

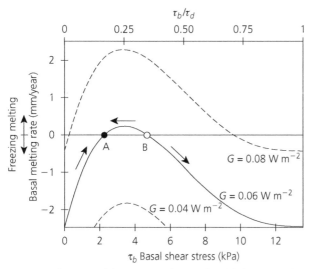

Figure 6.40 Illustration of the undrained plastic bed model, showing basal melt rate as a function of shear stress and geothermal heat flux (G) (Modified from Tulaczyk et al., 2000b).

advects cold ice from the ice sheet interior. The resultant chilling of the bed encourages basal freezing and ice stream deceleration. Conversely, when the bed is frozen, slow flow leads to ice thickening, reducing conductive heat losses and warming the bed. This scenario is essentially the same as the model of thermally regulated surges discussed in section 5.7.4, but with a much longer cycle length than those of surging valley glaciers.

A time-evolving numerical model of ice stream velocity cycles has been developed by Bougamont et al. (2003b), which couples equations for ice flow (incorporating both basal and lateral drag terms), subglacial energy balance and bed properties. The model builds on the *undrained plastic bed* model proposed by Tulaczyk et al. (2000b), which represents the ice stream bed as a Coulomb-plastic material with a yield strength inversely proportional to pore-water pressure (sections 4.2.3 and 4.5.4). Tulaczyk et al. (2000b) argued that, beneath most of West Antarctica, conductive heat fluxes probably exceed the geothermal heat flux, so that production of meltwater at the bed depends crucially on the magnitude of the strain heating term. Now, for ice streams, basal velocity is inversely proportional to the basal shear stress, so that sliding speeds are greatest when traction at the bed is low. Therefore, strain heating does not simply increase with either τ_b or U_b, but is highest at the combination of values where their product is maximized. Interactions between basal melt rates, bed properties, and ice velocity creates important feedbacks within this system, as illustrated in figure 6.40. To the left of point A, basal shear stress is low and basal freezing will remove water from the till layer, reducing the pore-water pressure and increasing till yield strength. This will increase strain heating and reduce the freezing rate, introducing a feedback that will move the bed condition towards point A. On the other hand, to

the right of point B, basal sliding rates are low and basal freezing will strengthen the bed and reduce strain heating. Thus, the bed will evolve towards a frozen state, with high basal shear stress and low velocities. Finally, between points A and B, melting at the bed produces excess water which will decrease till strength, so that the bed will evolve towards point A. Thus, point A is an *attractor*, representing a stable ice stream condition with low basal shear stress and high velocities. A second stable state exists to the right of point B, where the bed is frozen; the strength of the bed is greater than the driving stress and the ice moves slowly by ice creep alone. Processes of basal freeze-on were investigated in more detail by Christoffersen and Tulaczyk (2003a), who showed that migration of pore water towards the freezing interface results in overconsolidation of the dewatered till, increasing its shear strength.

The undrained plastic bed model thus predicts two stable modes, a low bed-strength, high-velocity *ice stream* mode, and a high bed-strength, low-velocity *ice sheet* mode. The time-evolving model of Bougamont et al. (2003b) shows how an ice stream bed could cycle between these modes due to the effect of ice stream thickness on subglacial energy balance. The model results are consistent with the existence of sediment-rich basal ice below Kamb ice stream, the continuing motion of its tributaries after shutdown of the trunk, and evidence for long-term cycles of ice stream shutdown and reactivation (Hulbe and Fahnestock, 2004, 2007; Christoffersen et al., 2006).

A major shortcoming of the undrained plastic bed model is that it considers only the local balance between melting and freezing, and ignores the effect of water transport along the bed (i.e. the bed is *undrained*). There is widespread agreement that ice stream onset zones play a crucial role in the water balance of ice streams (section 6.3.4; Whillans et al., 2001; Vogel et al., 2003; Joughin et al. 2004c; Bindschadler and Choi, 2007), and strong evidence for long-distance transport of water from regions of net melting towards the grounding zone (Gray et al., 2005; Fricker et al., 2007). Thus, although the undrained plastic bed model does represent one end member of possible ice stream states, it does not, as yet, incorporate all factors known to be important in the functioning of the Ross Ice Streams.

Long-distance transport of water is emphasized in alternative explanations of ice stream stagnation and reactivation, which invoke diversion of basal water or ice from one catchment into another. A *water piracy* hypothesis has been advocated by Alley et al. (1994), Anandakrishnan and Alley (1997) and Anandakrishnan et al. (2001), who argued that water flow paths in the onset zone of Kamb Ice Stream are sensitive to changes in ice surface topography, due to a strong transverse bed slope. According to this view, meltwater produced in the Kamb Ice Stream onset zone was captured by Whillans Ice Stream in response to Holocene ice sheet thinning, starving the former of water and triggering shutdown.

This view has found some support in recent studies of the water balance of the Ross Sea sector of the West Antarctic Ice Sheet. Modelling by Joughin et al. (2004c) suggests that 0.48–$0.63\,km^3$ of meltwater is produced annually in the catchments of the active Whillans, MacAyeal and Bindschadler Ice Streams, whereas only $0.32\,km^3\,yr^{-1}$ is produced in the Kamb Ice Stream catchment, although its area is similar. Most of the melting occurs in the onset zones, from where it is evacuated beneath the ice streams towards the coast. Large areas of Kamb and Whillans Ice Streams are subject to basal freezing rates of a few $mm\,yr^{-1}$, and hence are inferred to be more sensitive to changes in water balance than MacAyeal and Bindschadler Ice Streams, which have more restricted areas of freezing. Joughin et al. (2004c) concluded that the shutdown of Kamb Ice Stream could have been triggered by diversion of basal water to the Whillans Ice Stream catchment. Analysis by Wright et al. (2008) supports the idea that, in some parts of Antarctica, subglacial water flow routing is very sensitive to changes in surface elevation of only 5–15 m.

Conway et al. (2002) have presented evidence for reorientation of ice flow in Engelhardt Ice Ridge (between Kamb and Whillans Ice Streams), which could have contributed to ice stream shutdown. Flow stripes and internal radar reflectors show that ice formerly streamed across the ridge into Kamb Ice Stream, but stopped about 250 years ago, apparently in response to thinning and inland migration of the Whillans Ice Stream onset zone. Thinning of Whillans Ice Stream and thickening of Kamb Ice Stream since that time reversed both the direction of ice flow and the subglacial hydraulic gradient. The shutdown of Kamb Ice Stream about 150 years ago thus appears to have been linked to a combination of ice and water piracy in the upper catchment.

While it is undoubtedly true that ice streams are sensitive to changes in ice flow and meltwater fluxes in the interior, this does not rule out the possibility that flow oscillations could occur in response to internal feedbacks. Ice stream dynamics almost certainly reflect complex interactions between external forcing factors and internal feedback processes, and will no doubt continue to vex and intrigue glaciologists for some time to come.

6.3.7 STABILITY OF THE WEST ANTARCTIC ICE SHEET

Ever since radio echo soundings revealed that much of the bed of the West Antarctic Ice Sheet lies below sea level, glaciologists have speculated that ice sheet thinning might

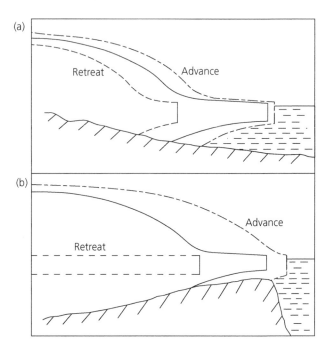

Figure 6.41 Influence of bed topography on the behaviour of two-dimensional ice sheets. (a) Bed sloping away from the ice, tending to damp ice sheet response to an initial perturbation. (b) Bed sloping towards the ice, amplifying ice sheet response.

lead to irreversible grounding line retreat and rapid deglaciation (e.g. Weertman, 1974; Oppenheimer, 1998; Vaughan, 2008). The possibility that rapid retreat of the West Antarctic Ice Sheet might be triggered by anthropogenic climate change was first mooted in a landmark paper by Mercer (1978). The *marine ice sheet instability* hypothesis is based on the idea that ice flow rates and changes in grounding line position are strongly coupled, and potentially subject to destabilizing feedbacks. By definition, the grounding line is located where ice thickness equals the flotation thickness. The grounding line will therefore advance or retreat if the ice thickens or thins in response to changes in ice flux from inland, dynamic thinning and/or local mass balance (section 5.2.3). Grounding line stability depends critically on how the system responds to an initial perturbation to the ice thickness, particularly whether feedback processes amplify or damp the initial signal.

The key feedback in the marine ice sheet instability hypothesis is between dynamic thinning rates and ice thickness at the grounding line. For an unconfined ice shelf, strain rates are a function of the cube of the ice thickness (equation 5.18). So, if it is assumed that ice flux through the grounding line is governed by this relationship, the ice flux will increase with ice thickness. If the ice sheet bed slopes down towards the sea, any increase in ice thickness (due, say, to increased snowfall) will lead to a grounding line advance (fig. 6.41a). The greater ice thickness at the new grounding line, however, will increase strain rates and ice losses, offsetting the increase in surface mass balance and limiting the advance. Similarly, if an initial thinning causes the grounding line to retreat into shallower water, reduced strain rates will tend to off-set the thinning trend and limit the amount of retreat. In this case, a negative feedback between strain thinning and ice thickness encourages grounding line stability. The opposite is true if the bed slopes down towards the ice sheet interior, as it does below the West Antarctic Ice Sheet (fig. 6.41b). An initial thickening will cause the ice to advance into shallower water, reducing the ice thick-ness at the new grounding line. Lower strain rates will encourage further advance until the bed slope changes direction. Conversely, an initial thinning will cause the grounding line to retreat into deeper water. Higher strain rates at the new grounding line will increase losses from the ice sheet, leading to further retreat. In this case, a strong positive feedback means that there can be no stable ice position in an area of reverse bed slope.

Backstress from laterally confined ice shelves could help to stabilize the grounding line, by providing an addi-tional source of resistance to ice flow (section 5.6.3). If the ice shelf were to break up, the loss of backstress might precipitate irreversible grounding line retreat. Mercer (1978) argued that rapid deglaciation of the West Antarctic Ice Sheet might be relatively easy if its fringing ice shelves were removed, and suggested that ice shelf col-lapse in the Antarctic Peninsula could provide an early warning sign that the process had begun.

Many attempts have been made to model the transi-tion zone between grounded ice sheets and floating ice shelves. Until the development of higher-order models (section 5.3.3), this could not be done rigorously because incompatible reduced models were used to represent the grounded and floating parts of the ice (see van der Veen, 1999a). In such models, it is assumed that all resistance in the grounded part is from basal drag, and the ice moves entirely by vertical shearing (the 'shallow ice approxima-tion'). In contrast, in the floating part, basal drag is zero and the ice is assumed to flow entirely in response to lon-gitudinal stresses. This means that, at the grounding line, one set of assumptions must be replaced by another, cre-ating serious problems in modelling flow in the transition zone and limiting the usefulness of any results.

Solutions of this 'grounding-line problem' only became possible with the development of higher-order models. A variety of numerical approaches have been taken to solving the Stokes equations, with varying degrees of suc-cess (e.g. Hindmarsh and LeMeur, 2001; Hindmarsh, 2006; Pattyn et al., 2006). Initial attempts proved sensi-tive to factors such as grid spacing, indicating that the results may reflect numerical artefacts as well as model physics (Vieli and Payne, 2005). An exact solution was obtained by Schoof (2007a, b), and confirms that, in the absence of lateral drag, marine ice sheets are susceptible to the kind of instability envisaged by Weertman (1974), and that no stable grounding line position is possible in areas of reversed bed slope. Schoof's model was not intended to represent a realistic ice sheet, but is an exact solution of a classic problem that can be used as a bench-mark against which more complex models can be compared.

In reality, Antarctic ice streams are bounded by cold, grounded ice at their lateral margins, and the transition between grounded and floating ice is largely mediated by lateral drag rather than longitudinal stresses (section 6.3.4; van der Veen, 1999a). Furthermore, thermal effects may introduce stabilizing mechanisms, because basal freezing can occur in response to thinning (Bougamont et al., 2003b). Incorporation of these effects into a rigor-ous model presents formidable difficulties, ensuring that the question of marine ice sheet stability is likely to remain 'glaciology's grand unsolved problem' for some years to come (cf. Weertman, 1976).

Observational data are as yet insufficient to decide the question, although recent studies have shown that parts of the West Antarctic Ice Sheet can respond rapidly to climatic and oceanic forcing. As Mercer (1978) pre-dicted, many ice shelves around the Antarctic Peninsula have indeed retreated or collapsed, and this has led to increased glacier discharge from the interior (section 5.6.3). Until recently, it was believed that these events would not be repeated farther south, since the recent climatic warming on the Peninsula is not seen in most of continental Antarctica, and the large Ross and Filchner–Ronne Ice Shelves are unaffected by surface melting. Indeed, observations on the Ross and Weddell Sea sec-tors of the ice sheet show a complex but apparently not unstable system (section 6.3.4). However, recent studies of the Amundsen Sea sector have revealed a different picture.

Logistically, the Amundsen Sea sector is harder to reach than other parts of the West Antarctic Ice Sheet, and fieldworkers have tended to neglect it in favour of the more accessible Ross and Weddell Sea sectors. Within the last decade, however, our view of this remote region has been transformed by remote-sensing studies, which have revealed dramatic evidence of ice stream thinning and acceleration (Wingham et al., 1998; Rignot et al., 2002; Shepherd et al., 2002; Joughin et al., 2003). The most dramatic changes have occurred on the largest ice stream, Pine Island Glacier, but they have also affected the neighbouring Thwaites, Smith and Kohler Glaciers (fig. 6.42). Large parts of the catchments of these glaciers are grounded well below sea level (fig. 6.43).

Thinning has been greatest on the ice shelves in front of the glaciers, with rates of up to $\sim$4 m yr^{-1}, although the thinning extends well inland along the ice streams at lower rates ($\sim$10 cm yr^{-1}). As a result of this thinning, the

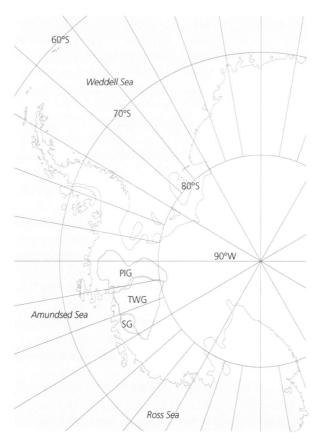

Figure 6.42 Map showing the catchments of Pine Island Glacier (PIG), Thwaites Glacier (TWG) and Smith Glacier (SG). These three glacier systems account for over 90 per cent of the ice discharge from the Amundsen Sea sector of the West Antarctic Ice Sheet (Shepherd et al., 2002).

grounding line of Pine Island Glacier retreated almost 1 km per year between 1992 and 1998 (Rignot, 1998). As has been the case in Greenland (section 6.2.4), glacier thinning has been accompanied by flow acceleration (Rignot et al., 2002; Joughin et al., 2003; Rignot, 2008). Altogether, the glacier has accelerated by up to 42 per cent between 1996 and 2007, and 73 per cent since 1974 (fig. 6.44). Smith Glacier accelerated 27 per cent in 1996–2006, and 8 per cent in 2006–7 alone. The total ice flux from the Amundsen Sea sector increased from $\sim$184 GT yr^{-1} in 1974 to $\sim$215 GT yr^{-1} in 1996, $\sim$237 GT yr^{-1} in 2000 and $\sim$280 GT yr^{-1} in late 2007.

These changes were triggered by increased basal melting below fringing ice shelves (Rignot and Jacobs, 2002; Shepherd, 2004; Payne et al., 2007). Around most of Antarctica, relatively warm Circumpolar Deep Water remains beyond the continental shelf, and does not come into contact with the ice sheet. However, incursions do occur into the Amundsen Sea, increasing melt rates below the relatively small shelves (Jenkins et al., 1997; Hellmer et al., 1998). Ice-shelf thinning causes grounding line retreat, reduced basal drag and loss of backstress, leading to acceleration of the adjacent ice stream. Acceleration and thinning are then propagated up-glacier by longitudinal stress gradients (Payne et al., 2007). Changes in energy flux at the marine boundary, therefore, can be communicated deep into the ice sheet, increasing ice flux from the interior. Whether or not this process continues into the future will have significant implications for future sea-level change (section 7.5.2).

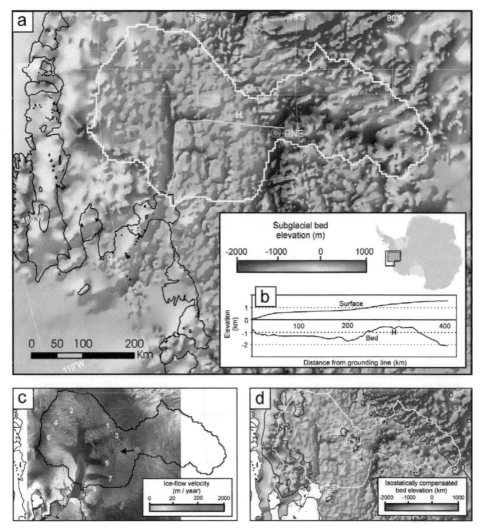

Figure 6.43 (a) Subglacial topography of the Pine Island Glacier basin. (b) Ice surface and bed elevation along the transect indicated in (a). (c) Ice velocity. (d) Bed topography after isostatic adjustment to removal of ice cover (Vaughan et al., 2006).

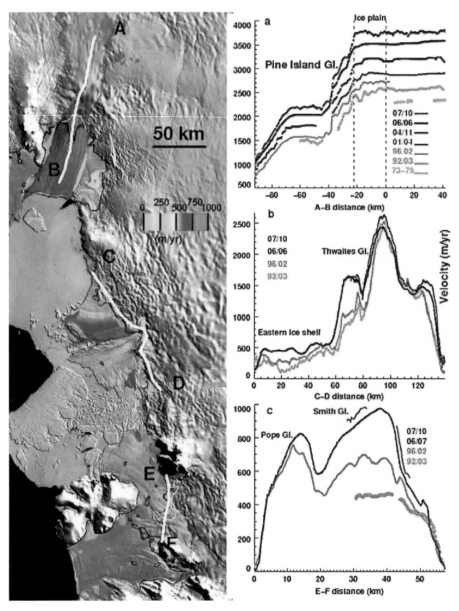

Figure 6.44 Acceleration of Amundsen Sea sector ice streams. Main panel: Velocity increase between 1996 and 2006. Right-hand panels: changes in velocity through time, along profiles shown in map; (a) Pine Island Glacier; (b) Thwaites Glacier; (c) Smith Glacier (Rignot, 2008).

CHAPTER 7

GLACIERS AND SEA-LEVEL CHANGE

Calving glaciers and icebergs, Hornsund, Svalbard (Doug Benn)

7.1 INTRODUCTION

Sea-level rise is widely considered to be among the most potentially damaging effects of climate change (Church et al., 2008). In a warming world, melting glacier ice and thermal expansion of ocean water can both cause rising sea levels, and there is concern that anthropogenic climate change could have a severe impact on many coastal regions. Two-thirds of the world's megacities are located on coasts, and population growth and coastal development are exposing increasing numbers of people to the potential effects of high sea levels. The most immediate effects of rising sea level are submergence of low-lying coastal areas and increased susceptibility to extreme events such as storm surges. Other potentially damaging effects include increased coastal erosion and saltwater intrusion into coastal groundwaters.

In recent times, sea-level change has been modest, although this could change if the Earth experiences higher temperatures in the future. The geological record shows that the growth and decay of glaciers and ice sheets has had a profound effect on global sea level, causing very large regional and global sea-level changes over the course of glacial cycles. There is some evidence that, in the geologically recent past, global ice volumes were significantly less than they are now, suggesting that relatively small temperature increases could have wide-reaching impacts on coastal regions.

In this chapter, we discuss the links between glaciers and sea-level change, then go on to review evidence for changing sea levels over glacial–interglacial cycles. We then examine the evidence for sea-level change in recent times, focusing on the role played by changes in the world's glaciers and ice sheets. We conclude by considering the

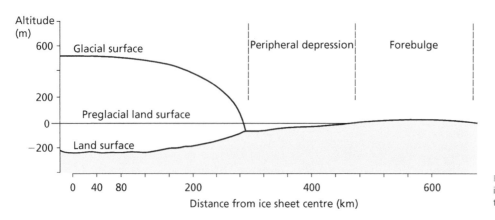

Figure 7.1 The principle of glacio-isostasy, showing the depression of the crust below an ice sheet.

outlook for the future, and how ice volume and sea level could change within our lifetimes and those of our descendants.

7.2 CAUSES OF SEA-LEVEL CHANGE

7.2.1 OVERVIEW

A change in sea level at a particular coastal location can occur in response to a change in elevation of either the ocean surface or the land surface. In either case, local sea-level change is measured relative to the land, and is therefore properly known as *relative sea-level change*. Many processes can cause a relative rise (transgression) or fall (regression) of sea level. On geological timescales, *tectonic processes* alter the configuration and capacity of the ocean basins, and even in the shorter term can cause dramatic uplift or subsidence of coastal areas.

Sea level can also change in response to variations in the density of ocean water (*steric* sea-level change). Fresh water at atmospheric pressure is most dense at $3.97\,°C$, and expands when it is heated or cooled away from that temperature. In the oceans, this is complicated by salt content, and an increase in salinity shifts the density maximum to a lower temperature. At a salinity of 24.7‰ (parts per thousand), the density maximum is at the freezing point ($-1.22\,°C$ for water with this salinity). These effects complicate the effect of temperature changes on ocean volume, but as a general principle a net warming of global sea temperature will cause ocean expansion and sea-level rise, while a net cooling will cause ocean contraction and sea-level fall.

Glaciers and ice sheets can influence sea-level change in several ways.

(1) *Glacio-eustasy.* During glacier and ice sheet build-up, moisture evaporated from the oceans falls as snow and is stored in the form of ice. Moisture is thus removed from the oceans and not replaced by runoff. When global land ice volume decreases, stored water is returned to the oceans. Sea-level change resulting from variations in land ice volume are termed *glacio-eustatic* variations.

(2) *Glacio-isostasy.* In regions occupied by ice sheets, the load placed by the ice on the Earth's crust causes it to displace the underlying mantle (fig. 7.1). This affects relative sea level in and around the areas occupied by ice sheets, but does not have a global impact.

(3) *Hydro-isostasy.* The force balance on the Earth's crust is influenced by the distribution of water in the oceans. Changes in the amount of water in the oceans affects patterns of crustal loading, causing regional and global sea-level changes.

(4) *Geoidal eustasy.* The surface of the world's oceans is not a smooth sphere or ellipsoid, but has bulges and troughs produced by local variations in the Earth's gravitational field (fig. 7.2). This equipotential surface is known as the *geoid*, and will change shape as the Earth's gravitational field adjusts to changing ice sheet configurations, causing sea-level changes which may vary in magnitude and sign.

(5) *Isostatic response to erosion and deposition.* Local crustal loading changes through time as a result of glacial erosion and deposition, which remove mass from continental interiors and add it to continental margins. This process may result in large, local sea-level changes, especially over long timescales.

Relative sea-level change at any particular location is influenced by global eustatic and regional isostatic factors, plus hydro-isostatic, geoidal, steric and tectonic influences. The impact of glacier and ice sheet growth and decay will therefore vary greatly in magnitude and sign depending on location. The idea of 'global' sea-level change, therefore, is a simplification and has to be treated with caution.

7.2.2 GLACIO-EUSTASY AND GLOBAL ICE VOLUME

The main control on global ocean levels on Quaternary timescales is the uptake and release of water by evolving ice sheets over glacial cycles. As a first approximation

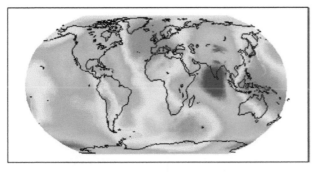

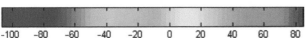

Figure 7.2 Deviations of the geoid from a perfect ellipsoidal surface. In the North Atlantic and western Pacific Oceans, the geoidal surface is higher than the ellipsoid, whereas in the Indian Ocean it is much lower (www.mathworks.com/access/helpdesk/help/toolbox/map/f5-6923.html).

over short intervals of time, mean sea-level change (ΔSL) is related to ice volume changes (ΔV_i) as follows:

$$\Delta SL = (\rho_i/\rho_w)(\Delta V_i/A_{ocean}) \qquad (7.1)$$

where ρ_i and ρ_w are the mean densities of glacier ice and water, respectively, and A_{ocean} is the area of the global ocean. At the present time, the area of the world's oceans is 361.6 million km^2, so an ice volume of 4018 km^3 is equivalent to 1 cm of ocean-depth change, averaged over the entire ocean area. Changes in global sea level through Quaternary time can therefore be used to estimate changes in the volume of ice sheets (Fairbanks, 1989; Lambeck and Chappell, 2001; Milne et al., 2002; Cutler et al., 2003). However, it must be emphasized that water released from melting glaciers and ice sheets will *not* be evenly distributed around the world, due to hydro-isostatic, glacio-isostatic, geoidal and other factors. Sea-level change estimates calculated from equation 7.1 are global averages, and cannot be used to predict sea-level change at any given locality.

Reconstructions of past glacio-eustatic sea-level change and the corresponding ice volume fluctuations have been made using two complementary approaches. First, records of sea-level change can be obtained by dating raised or submerged features that have a known or inferred relationship to former sea level. Fossil coral reefs provide excellent records because they are directly datable, and are located sufficiently far from former ice sheets to be unaffected by glacio-isostasy (although local tectonic factors can be important). Living corals only occur in the photic zone close to the sea surface, and some taxa have a narrow range of depth tolerance. For example, *Acropora palmata* has a present-day depth range of 5 m, so the age and elevation of this coral in fossil reef complexes allows former sea levels to be reconstructed to that level of accuracy. The first detailed eustatic curves for

the late Quaternary were obtained from fossil reefs on Barbados and Tahiti (Fairbanks, 1989; Bard et al., 1990, 1996); and eustatic sea-level curves spanning the whole of the last glacial cycle have been constructed from tectonically raised coral reefs around the Huon Peninsula, New Guinea (section 7.3; Bloom et al., 1974; Chappell et al., 1996; Cutler et al., 2003). High-resolution sea-level records for parts of the period since the LGM have also been obtained from the depth of submerged sediments on stable, low-latitude continental shelves (Hanebuth et al., 2000; Yokoyama et al., 2000). The precision of such reconstructions is limited by uncertainties about the depth at which the sediments were deposited or corals originally grew, the history of tectonic uplift or subsidence at the sites, and errors in radiometric age determinations. Additionally, local hydro-isostatic and geoidal factors may compromise the degree to which local sea-level histories represent global ice volume changes. The amount of local sea-level change attributable to hydro-isostatic loading and geoidal changes cannot be measured directly, but must be derived from geophysical models (e.g. Peltier, 1987, 1999; Lambeck and Chappell, 2001; Milne et al., 2002; Lambeck et al., 2003). Depending on location, the hydro-isostatic and geoidal components may be positive or negative for any given time period, although models disagree over details (e.g. Milne et al., 2002; Peltier, 2002).

The second approach to reconstructing global ocean volume changes employs the oxygen isotope record from deep ocean sediments. The proportion of the heavy isotope of oxygen (^{18}O) in the tests of marine microorganisms depends on the isotopic composition of the water, together with water temperature, ocean circulation and other factors (Bradley, 1999; Waelbroeck et al., 2002; cf. section 2.7). The bulk isotopic composition of the world's oceans is modulated by the storage of light oxygen isotopes in glaciers and ice sheets, so that fluctuations in oxygen isotope ratios (expressed as $\delta^{18}O$) in microfossils within deep ocean sediments can provide a proxy for ice sheet volume and glacio-eustatic sea-level change, provided the influence of other factors can be determined (e.g. Shackleton, 1967, 1987; Waelbroeck et al., 2002). Early attempts to remove the influence of water temperature focused on the $\delta^{18}O$ record from benthic (deep-water) foraminifera, because deep-ocean temperatures change little compared with those close to the ocean surface (Duplessy, 1978; Chappell and Shackleton, 1986; Labeyrie et al., 1987). However, deep-water temperatures are now known to vary by as much as $\pm 2\,°C$ over the course of a full glacial cycle, and uncertainties in deep-sea temperatures of only $\pm 1\,°C$ correspond to eustatic sea-level variations of ± 30 m. Furthermore, local hydrological influences on foraminiferal $\delta^{18}O$ remain unquantified in this approach.

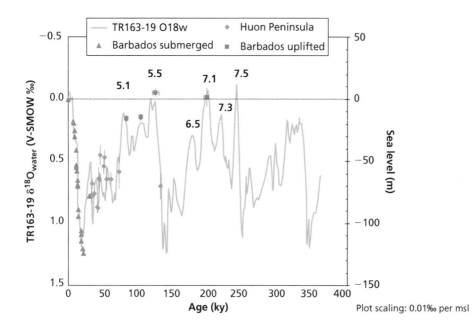

Figure 7.3 Reconstructed glacio-eustatic sea-level change over the last three glacial–interglacial cycles, based on oxygen isotope and Mg/Ca analysis of deep-sea core TRC163-19 from Cocos Ridge, eastern Pacific Ocean (solid line), compared with estimates derived from fossil coral reefs at Huon Peninsula, New Guinea, and Barbados (point symbols) (Lea et al., 2002).

Recently, new theoretical and empirical methods have been developed to identify the thermal, hydrological, and ice volume components of foraminiferal $\delta^{18}O$ records. These include the direct measurement of $\delta^{18}O$ values in pore fluids in ocean cores (Schrag et al., 2002), and the reconstruction of water temperatures from calcium/magnesium ratios in foraminiferal calcite (e.g. Elderfield and Ganssen, 2000; Lear et al., 2000; Lea et al., 2002; Martin et al., 2002). When used in combination, these new methods yield eustatic curves that extend further back in time, and with higher resolution than those derived from corals or other sea-level indicators, which are limited by the availability of suitable datable materials (Lea et al., 2002; Billups and Schrag, 2003; Siddall et al., 2003). Data from submerged and raised coral reefs provide striking confirmation of eustatic curves derived from oxygen isotope ratios, increasing confidence that they provide accurate records of global ice volume changes (e.g. Chappell et al., 1996; Fleming et al., 1998; Lea et al., 2002; Cutler et al., 2003; fig. 7.3).

7.2.3 GLACIO-ISOSTASY AND ICE SHEET LOADING

The Earth's lithosphere (with a mean density of approximately 2800 kg m^{-3}) tends to be in buoyant equilibrium with the underlying mantle (with a mean density of around 3300 kg m^{-3}), somewhat like a boat floating on water. The addition of mass to part of the crust will cause it to sink into the asthenosphere (the plastic upper mantle), just as a boat does when cargo is loaded. This principle is reflected in the origin of the word *isostasy*, which is derived from the Greek *isostasios* meaning 'equal standing' or 'in equipoise'. Ignoring for the moment the influence of crustal rigidity, the equilibrium amount of crustal

depression (ΔZ) resulting from ice sheet loading is a function of ice sheet thickness (H) and the ratio between the densities of ice and mantle:

$$\Delta Z = (\rho_i/\rho_m)H \qquad (7.2)$$

Taking mantle density as ~3300 kg m^{-3}, the depression created beneath an ice sheet will be approximately 0.27 times the ice thickness, so the amount of depression will decrease from the centre of the ice sheet towards the margins.

This simple picture, however, is complicated by two main factors. First, rigidity of the lithosphere means that the load of an ice sheet is partly spread beyond the margin; and second, the crust–asthenosphere system takes time to respond to changes in loading, so isostatic depression lags behind ice sheet growth, and isostatic uplift lags ice sheet decay.

The effects of crustal rigidity mean that a *peripheral depression* develops up to 150–180 km beyond the ice sheet margin (fig. 7.1). This means that, if eustatic sea-level fall is not too great, the sea can flood coastal sites in unglaciated areas due to crustal loading by an adjacent ice sheet. The lateral displacement of asthenospheric material from below the ice sheet results in the formation of a *forebulge*, or area of positive vertical displacement beyond the peripheral depression. The distance between the ice sheet edge and the forebulge depends on the *flexural parameter* of the crust, which represents the amplitude of bending of the lithosphere and is calculated from lithospheric density, thickness, elasticity and other factors (Walcott, 1970). The equilibrium vertical displacement in the region of the forebulge is not large, only about 4 per cent of the depression beneath the ice sheet centre (van der Veen, 1999a). However, the forebulge is considerably higher

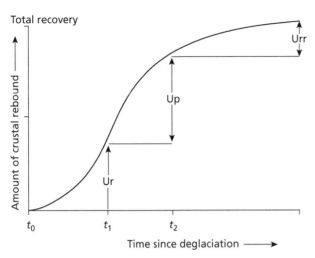

Figure 7.4 Three periods of glacio-isostatic recovery. Initial restrained rebound (Ur) is followed by a period of rapid postglacial rebound (Up). Future residual rebound (Urr) is expected to decrease through time. Initial ice retreat and crustal unloading occurs at t_0, but some ice cover persists until t_1. Time t_2 denotes the present day (Modified from Andrews, 1970).

during ice sheet build-up, because the diffusive, viscous nature of flow in the asthenosphere means that material initially 'piles up' near the ice sheet margins. Because the flexural parameter is not affected by ice sheet thickness, the centre of the forebulge is located 250–280 km from the ice margin, regardless of ice sheet size. During deglaciation and unloading, the lithosphere rebounds beneath the ice sheet, and the forebulge collapses and migrates slowly towards the former ice centre (Walcott, 1970).

Relative sea levels in areas occupied by ice sheets exhibit a lagged isostatic response to ice sheet loading and unloading. Complete adjustment of the lithosphere and asthenosphere to changes in loading may take in excess of 10,000 years, so equilibrium may never be attained over the course of a glacial–interglacial cycle. Response times depend on the rheology of the mantle. Glacio-isostatic recovery in response to deglaciation has been subdivided into three periods by Andrews (1970; fig. 7.4).

(1) *Restrained rebound* takes place beneath a thinning ice sheet. Uplift rates increase through time as the load is progressively removed. Due to occupancy by ice, this period is not recorded by sea-level histories, but must be inferred using geophysical models (e.g. Peltier and Andrews, 1983; Lambeck and Chappell, 2001). For sites close to former ice sheet centres, restrained rebound can represent a considerable proportion of the whole.

(2) *Postglacial uplift* is the phase of rebound recorded in an area once ice withdraws, and can be reconstructed from direct evidence.

(3) *Residual uplift* is the rebound still to take place. Due to the long response time of lithospheric recovery, regions such as central Scandinavia and Hudson Bay in Canada are still rising in response to the disappearance of the great Pleistocene ice sheets.

Rates of unrestrained (i.e. postglacial and residual) rebound decline exponentially through time, with the total amount of rebound R in time t given by:

$$R = a.e^{\lambda t} \tag{7.3}$$

where e is the base of the natural logarithms and a and λ are empirical constants. The half-life of this function, or the time required to accomplish half of the remaining rebound, is $0.693/\lambda$, where 0.693 is the natural logarithm of 2. Dyke and Peltier (2000b) found that for the Laurentide Ice Sheet, rebound half-lives lie in the range 1.2–1.4 ka at the uplift centre, rising to 1.7–2.0 ka closer to the ice limit, then falling to <1 ka at the ice limit itself. These values are about half of those for the Fennoscandian Ice Sheet, which has a rebound half-life of 4.6 ka near the centre of loading (Peltier, 1996), the difference reflecting the contrast in ice sheet size.

A detailed picture of current uplift rates within the limits of the Pleistocene ice sheets has been obtained using data from GPS networks. In Fennoscandia, the area of most rapid uplift is around the Gulf of Bothnia (fig. 7.6, Fjeldskaar et al., 2000), whereas in Canada it is centred on Hudson Bay (fig. 7.5; Sella et al., 2007), in both cases reflecting the areas where ice cover was thickest. More rapid rates of isostatic uplift have been measured in parts of southern Alaska, which are currently rising at up to 3.2 cm yr^{-1} in response to ice loss since the Little Ice Age maximum (fig. 7.7; Larsen et al., 2005). Ice losses in this region have been very large, amounting to over 3000 km^3 in the Glacier Bay area alone. This is equivalent to 8 mm of global sea-level rise, although of course this region is currently experiencing a fall in relative sea level due to an excess of rebound over the eustatic component.

Attempts have been made to quantify amounts of residual uplift from *free air gravity anomalies*, or regional variations in the force of gravity. Negative gravity anomalies should be expected where rebound is incomplete and the lithosphere remains out of isostatic equilibrium (Walcott, 1970). Analyses of GRACE data for North America have highlighted a large negative gravity anomaly over northern North America (Ivins and Wolf, 2008; van der Wal et al., 2008). However, not all of this anomaly is attributable to residual rebound. Ice melt in Greenland and Alaska accounts for around 25 per cent of the signal, whereas changes in water storage elsewhere on the continent introduce further uncertainties. Once these factors are taken into account, the isostatic component appears to exhibit two distinct peaks that may reflect former ice domes of the Laurentide Ice Sheet.

This is not known directly, but must be derived by tuning geophysical models to fit empirical data (e.g. Lambeck, 1993a, b, c, 2004). Indeed, geophysicists have been able to learn much about the characteristics of the lithosphere and the underlying asthenosphere from chronologies of sea level and ice sheets (e.g. Wu and Peltier, 1983; Peltier, 1987, 2002b; Lambeck, 1990, 1993a). Geophysical models contribute greatly to our understanding of the relationships between ice sheets and sea-level change. For regions where former ice extent and isostatic unloading history are reasonably well known, models can be used to derive

best-fit values for asthenosphere viscosity and the crustal flexural parameter, and so represent important sources of knowledge about geophysical processes which cannot be observed directly (e.g. Lambeck, 1993b, c; Shennan et al., 2000). Alternatively, where unloading history can be partially reconstructed from shoreline isobases, but evidence

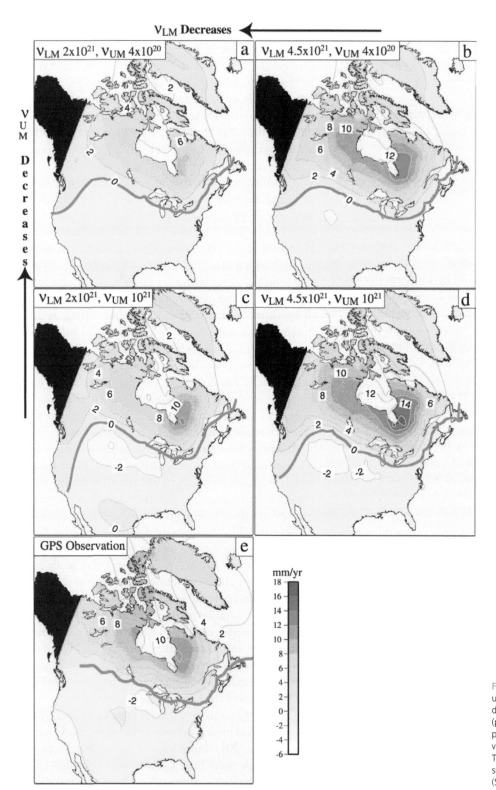

Figure 7.5 Current rates of isostatic uplift in Canada and the USA determined from GPS networks (panel e), compared with model predictions using four different viscosity structures (panels a–d). The hinge line separating uplift from subsidence is shown in green (Sella et al., 2007).

for ice sheet extent is fragmentary, models can be used to reconstruct ice distribution. For example, Elverhøi et al. (1993), Lambeck (1995) and Siegert and Dowdeswell (1995) have used the distribution of raised shorelines as input for models of the Late Weichselian Ice Sheets. In turn, such models can be tested against independent evidence for ice extent in the region based on glacigenic landforms and sediments (e.g. Vorren and Kristoffersen, 1986; Vorren et al., 1989, 1990; Solheim et al., 1990; Sættem, 1994; Landvik et al., 1998). Such modelling exercises are complicated by the fact that ice sheet extents (particularly vertical dimensions) and chronologies are often poorly constrained, data on relative sea-level change may be incomplete, and Earth-rheological parameters derived for one region may not be applicable in another. In cases where empirical reconstructions and modelling results are in disagreement, it may not be easy to tell where the error lies. Nevertheless, when used in combination, empirical observations and geophysical models yield powerful insights into feedbacks between ice sheet evolution and sea-level change.

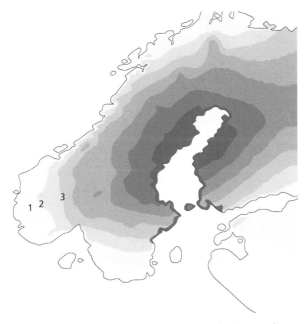

Figure 7.6 Current rates of isostatic uplift in Fennoscandia, in mm yr^{-1} (Fjeldskaar et al., 2000).

7.3 SEA-LEVEL CHANGE OVER GLACIAL–INTERGLACIAL CYCLES

7.3.1 ICE SHEET FLUCTUATIONS AND EUSTATIC SEA-LEVEL CHANGE

Over the course of the last full glacial cycle, changes in global land ice volume have caused eustatic sea levels to vary by up to 120 m. At the peak of the last interglacial

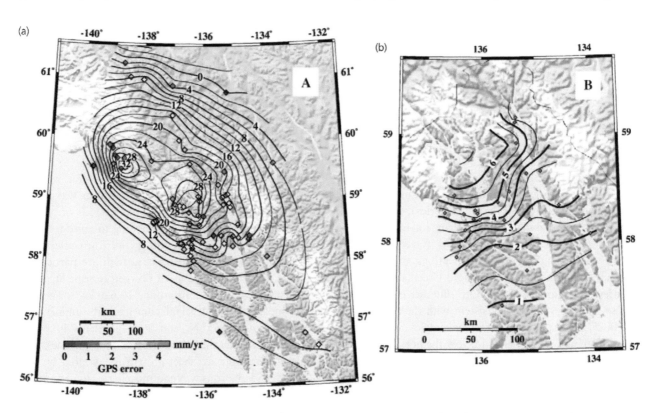

Figure 7.7 Uplift rates in south-east Alaska. (a) Uplift rates (mm yr^{-1}) measured at GPS stations, colour-coded to indicate measurement errors. Contour interval is 2 mm yr^{-1}. The two uplift peaks correspond to the Glacier Bay (right) and Yakutat (left) Ice Fields. (b) Relative sea-level change (m) since AD 1770 (Larsen et al., 2005).

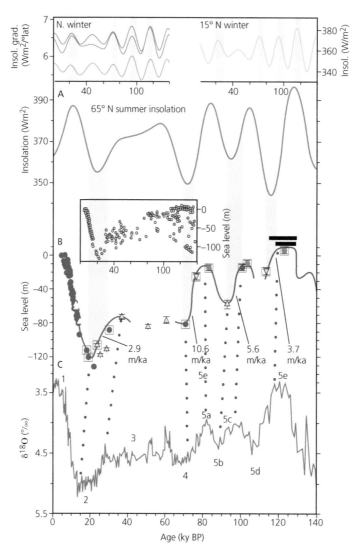

Figure 7.8 Sea-level record and insolation variations over the last full glacial cycle. (A) Summer insolation at 65° N. Grey bars indicate periods of rapid sea-level fall and ice sheet growth. Left inset: winter insolation gradients. Top curve is 35–50° N, middle curve is 50–65° N and bottom is 20–35° N. Right inset: winter insolation curve at 15° N. (B) Coral sea-level record from New Guinea and Barbados. (C) Benthic oxygen isotope curve from core VP19-30, Carnegie Ridge, east Pacific Ocean. Numbers are isotope stages (Cutler et al., 2003).

period, around 125,000 years ago (Oxygen Isotope Stage 5e, known as the Eemian Interglacial in Europe and the Sangamon Interglacial in North America), mean global sea levels were 4–6 m higher than today, indicating smaller land ice volumes than now (Overpeck et al., 2006). According to model simulations, most of this amount is attributable to a reduced Greenland Ice Sheet (Cuffey and Marshall, 2000; Otto-Bliesner et al., 2006). These model results are consistent with the age of basal ice in different parts of the Greenland Ice Sheet, which dates from before the Eemian below the north dome, but not the south (NGRIP, 2004). This evidence for a much reduced Greenland Ice Sheet in the Eemian is important because it highlights the possibility that the ice sheet could undergo similar shrinkage in response to future climate change (section 7.5.2).

Following the high sea levels of the Eemian, repeated sea-level fluctuations have occurred, reflecting waxing and waning of the world's land ice. Four periods of particularly rapid sea-level fall, corresponding to episodes of ice sheet growth, have been identified (fig. 7.8; Cutler et al., 2003). Implied rates of ice sheet growth are particularly high for the transition between Oxygen Isotope Stages 5a and 4, when sea level fell by around 10.6 m ka^{-1}. The periods of most rapid sea-level fall coincide with summer insolation minima at 65° N, and large gradients in winter insolation between the subtropics and mid latitudes. This suggests that rapid ice sheet growth is favoured by cool summers and vigorous winter atmospheric circulation (i.e. low ablation and high snow accumulation).

At the LGM, approximately 21,000 years ago, eustatic sea level was 125 ± 5 m lower than the present day,

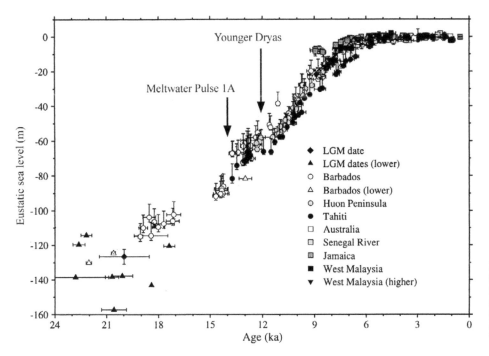

Figure 7.9 Eustatic sea-level change from the LGM to the present day. Note the rapid rise in sea level approximately 14,600 years ago (Meltwater Pulse 1A) and the effect of glacier regrowth during the Younger Dryas (Fleming et al., 1998).

equivalent to a land-based ice volume of around $4.7 \times 10^7 \, \text{km}^3$ (Fleming et al., 1998). Between 20,000 and 7000 years ago, eustatic sea level rose around 120 m, at an average rate of 1 m per century. Superimposed on this trend are short-lived periods of very rapid sea-level rise, with rates up to 4 m per century (Fairbanks, 1989; Fleming et al., 1998; Lambeck et al., 2002; fig. 7.9). Some of these events feature in some sea-level records but not others, and may be artefacts of various uncertainties in the data. Others, however, appear in several records and are interpreted as the signature of major *meltwater pulses*. The largest such event (Meltwater Pulse 1A) was approximately 14,600 years ago, when sea level rose by around 20 m in 500 years (Fairbanks, 1989; Clark et al., 2002). An earlier meltwater pulse apparently occurred around 19,000 years ago (Yokoyama et al., 2000), probably mostly from the large northern hemisphere ice sheets (Clark et al., 2004). Most of the decay of the large ice sheets was completed by 7000 years ago, although 3–5 m of water has been added to global sea level since that time. Eustatic sea level appears to have changed little from 2000 years ago to the start of the nineteenth century (Lambeck et al., 2004).

7.3.2 SEA-LEVEL HISTORIES IN GLACIATED REGIONS

Within areas occupied by the great Pleistocene ice sheets, glacial erosion and deposition tends to destroy or obscure evidence for former sea levels, so empirical evidence for relative sea-level change is mainly confined to the period since ice withdrawal. For this period, however, a wide range of evidence is available. Geomorphological features such as deltas, beaches, shingle ridges and erosional platforms mark former shorelines, which may be either submerged or raised relative to present sea level. The altitude of such features does not coincide exactly with the contemporary sea level, however. Shingle ridges, for example, are piled up by extreme storm events and can form several metres above the normal high-water mark (Otvos, 2000). More accurate sea-level data can be obtained from the distribution of marine sediments, which can be identified using microfossils such as diatoms. Estuarine flats and closed basins reached by the sea only at times of high sea level (isolation basins) are particularly suitable areas for stratigraphical studies of sea-level change (e.g. Smith and Dawson, 1983; Shennan et al., 2000). Submerged forests and peats also provide evidence for sea-level change. On many coasts, however, fossiliferous marine sediments are absent, and sea-level reconstructions must rely on geomorphological evidence. Until recently, dating raised beaches has been problematical, but new opportunities have opened up with the advent of optically stimulated luminescence and cosmogenic nuclide exposure dating (Aitken, 1998b; Gosse and Phillips, 2001).

Relative sea-level histories in areas within or near the limits of the Pleistocene ice sheets exhibit large variations over distances of hundreds or even tens of kilometres, because the glacio-isostatic component of sea-level change varies according to local ice loading history. Where ice thicknesses were substantial, and the amount of isostatic depression was greater than eustatic sea-level lowering, the history of sea-level change following

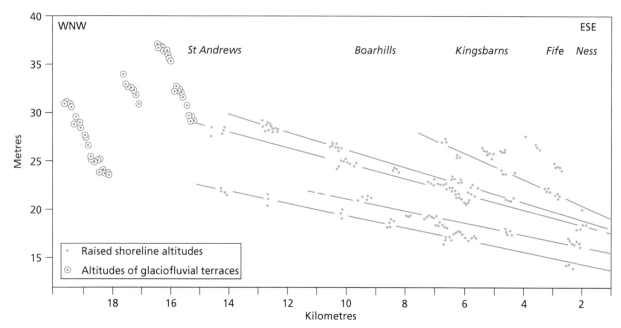

Figure 7.10 Height–distance diagram for shoreline fragments in part of eastern Scotland. The glacio-fluvial terraces WNW of St Andrews connect the shorelines to former ice margin positions. There is also evidence for an ice limit at Boarhills (not shown) (Lowe and Walker, 1997, using data from Robin Cullingford and David Smith).

deglaciation is dominated by progressive emergence of the land (relative sea-level fall). Conversely, close to ice sheet margins, where isostatic depression was smaller than eustatic lowering, postglacial sea-level histories are dominated by submergence (relative sea-level rise). At intermediate localities, relative sea-level change may exhibit periods of emergence and submergence.

In formerly glaciated regions, raised shorelines are typically tilted because they have experienced varying amounts of isostatic uplift depending on proximity to the centre of ice loading. Younger shorelines are tilted less steeply than older ones, due to decreasing amounts of differential uplift through time. In areas of continuously falling sea level, younger shorelines occur at progressively lower altitudes, but this is not the case in areas with more complex histories of relative sea-level change. The extent of shorelines is also controlled by contemporary ice limits, because the sea may be prevented from occupying some low-lying areas by glacier ice. This means that, although the altitude of a particular raised shoreline increases towards the ice sheet centre, it will not penetrate as far towards the loading centre as a younger and lower shoreline if glacier ice becomes less extensive in the intervening period (fig. 7.10).

The maximum elevation attained by the sea on any particular stretch of coast is known as the *marine limit* (Andrews, 1970). From figure 7.11, it can be seen that there are two main controls on spatial variations in the altitude of the marine limit. First, the regional tilt of a raised shoreline will mean that the marine limit will increase towards the centre of ice loading; and second, later persistence of ice closer to the centre of the ice

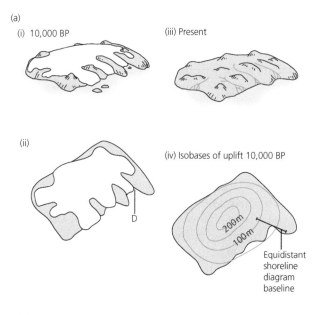

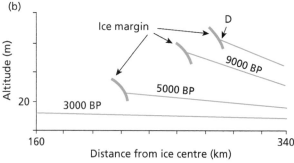

Figure 7.11 (a) The glacio-isostatic signature of a hypothetical island ice sheet. (i, ii) Ice sheet maximum at 10,000 BP. (iii, iv) After complete deglaciation and isostatic rebound. Isobases of raised shorelines show greatest amount of uplift near the centre of the island. (b) Hypothetical height–distance diagram drawn along the baseline shown in (a).

sheet will cause the marine limit to decrease towards the loading centre. The actual pattern will depend on the balance between these two factors. Large falls in the marine limit over short distances reflect pauses in ice sheet retreat while significant isostatic rebound occurred, whereas a small height difference between a shoreline and one that extends much further towards the former ice sheet centre reflects rapid retreat of the ice relative to the rate of isostatic uplift.

Contours joining points of equal elevation, and therefore uplift, on contemporaneous shorelines are known as *isobases*. The form of isobases represents the pattern of differential uplift relating to former ice loads, providing valuable information on the positions of ice sheet dispersal centres during glaciations (fig. 7.12). In detail, however, isobase maps are often complex, and show that uplift patterns rarely form simple domes. Isobases may exhibit abrupt discontinuities, suggesting that rates of uplift were different on either side of fault zones. An example from arctic Canada is shown in figure 7.13, where isobase patterns point to the presence of a fault in the channel between Prince of Wales Island and Somerset Island (Peel Sound), and a block of untilted terrain in northern Prince of Wales Island (Dyke et al., 1991). Further evidence for differential uplift and faulting during isostatic recovery has been discovered in the Forth Valley and the Glen Roy area, Scotland (Sissons and Cornish, 1982). Indeed, faulting may be the norm where large blocks of the Earth's crust undergo rapid uplift. Generally, however, the measured amount of displacement across faults is small, so that faulting and shoreline displacement may go unnoticed unless very accurate surveys are undertaken.

Temporal patterns of glacio-isostatic uplift can be calculated from dated series of raised shorelines by subtracting the eustatic component of sea-level change (Klemann and Wolf, 2005). This approach neglects hydro-isostatic

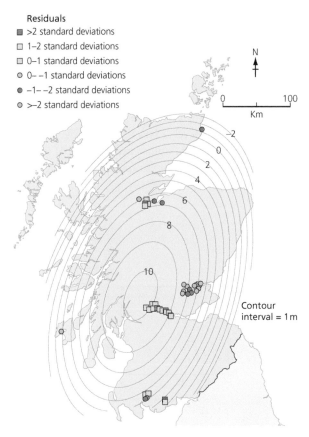

Residuals
■ >2 standard deviations
□ 1–2 standard deviations
□ 0–1 standard deviations
○ 0–−1 standard deviations
● −1–−2 standard deviations
● >−2 standard deviations

Contour interval = 1 m

Figure 7.12 Isobases of the Main Postglacial Shoreline in Scotland, as calculated by fitting a quadratic trend surface to shoreline elevation data (Smith et al., 2000).

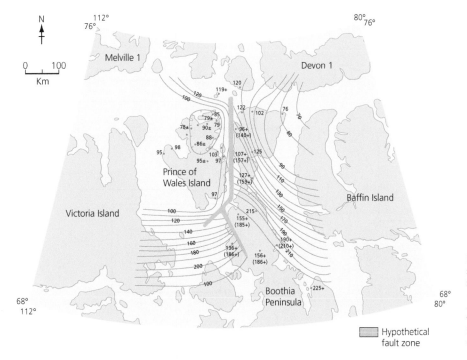

Hypothetical fault zone

Figure 7.13 Isobases for the 9.3 ka BP shoreline in the central Canadian Arctic, showing evidence for a fault zone between Prince of Wales Island and Somerset Island, and the Prince of Wales isobase plateau, which is in an area that has undergone isostatic uplift without significant tilting (Dyke et al., 1991).

and geoidal components of sea-level change, introducing a level of uncertainty, but is still superior to uplift patterns inferred from 'uncorrected' shoreline elevations.

The varying influence of glacio-eustatic and glacio-isostatic factors can be seen in figure 7.14, which shows representative sea-level curves from around the British Isles. In the south, beyond or close to the limits of the last ice sheet, Holocene sea-level change has been dominated by submergence. Although isostatic rebound did occur in this area, it was exceeded by the glacio-eustatic component resulting from ice sheet melting, giving sustained sea-level rise. In Scotland, where the ice sheet was thickest, the early parts of the sea-level curves record relative sea-level fall because local isostatic uplift exceeded global eustatic sea-level rise at that time. By around 8000 BP, however,

isostatic uplift rates had fallen to low values, but global eustatic sea level was rising rapidly. As a result, a period of relative sea-level rise occurred around the coasts of Scotland – an event known as the Main Postglacial Transgression (Sissons, 1983). By around 5000 BP, the great ice sheets had vanished and the glacio-eustatic contribution to sea-level change was very small. Isostatic uplift (which had continued throughout the Main Postglacial Transgression) was then able to take over once more as the main control on relative sea-level change. Thus, over the past few thousand years, relative sea level has fallen around most of the Scottish coast – a trend that continues today. Detailed accounts of Late Glacial and Postglacial sea-level change in Britain have been provided by Sissons (1983) and Shennan et al. (2000).

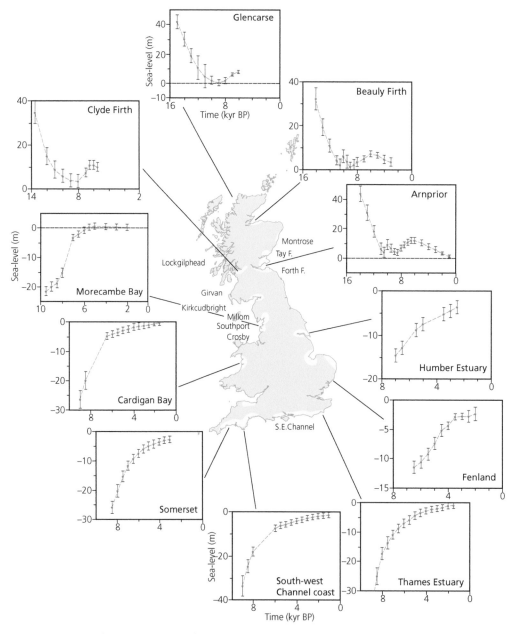

Figure 7.14 Postglacial sea-level curves from around the coast of Britain, demonstrating the influence of glacio-isostasy in the north (emergence) and glacio-eustasy in the south (submergence) (Lambeck, 1995).

7.4 GLACIERS AND RECENT SEA-LEVEL CHANGE

7.4.1 RECORDED SEA-LEVEL CHANGE

Tidal gauge records provide the primary data source for sea-level change over most of the last century (Douglas, 2001). While these provide valuable information for many areas, it is very difficult to derive accurate 'global' figures from tidal gauge data, due to very uneven spatial sampling and the influence of vertical movements of the land. In addition to the tectonic and glacio-isostatic factors discussed above, coastal areas can also change in elevation due to local factors, such as groundwater withdrawal, compaction of sediments or changing patterns of sedimentation, making it difficult to extract the eustatic and steric signal.

Since 1993, near-global measurements of sea level between 66°N and S have been made by high-precision satellite-borne altimeters (TOPEX/Poseidon and Jason-1, launched in 1992 and 2001, respectively; Cazenave and Nerem, 2004). In studies based on tide gauges, most investigators assumed a uniform 'global' signal, but the altimeter data have revealed a complex pattern of large regional variations (fig. 7.15). Parts of the western Pacific and eastern Indian Oceans have experienced rates of sea-level rise ten times the global mean, whereas other regions have experienced a fall. To a large degree, these patterns reflect decadal-scale dynamic factors, such as El Niño–Southern Oscillation in the tropical Pacific. During El Niño events, sea levels and water temperatures are elevated in the eastern Pacific, while during La Niña the opposite is true. Over the 1993–2003 measurement period, there was a transition from El Niño-like conditions at the beginning to more La Niña-like conditions at the end, explaining much of the pattern of falling sea level in the eastern Pacific and rising sea level in the west.

Overall, the satellite altimeter data show a clear net increase in mean global sea level (fig. 7.16; Cazenave and Nerem, 2004; Bindoff et al., 2007). The mean rate of sea-level rise between 1993 and 2003 was 3.1 ± 0.7 mm yr^{-1}. When compared with earlier trends reconstructed from tide gauge records, a shift in the rate of sea-level rise is evident around 1935 (fig. 7.17). From 1870 to 1935, the mean rate was 0.71 ± 0.40 mm yr^{-1}, whereas from 1936 to 2001, it was 1.84 ± 0.19 mm yr^{-1} (Church and White, 2006). An acceleration in the rate of sea-level rise is therefore apparent.

This rise in global mean sea level is almost equally attributable to steric sea-level change (thermal expansion) and glacio-eustasy. Figure 7.18 shows global patterns of sea-level change for the period 1993–2003, compared with the steric component calculated from ocean temperature changes. The overall patterns are similar, but some areas have experienced a greater rise, or a switch from fall to rise, as a result of the glacio-eustatic component. According Intergovernmental Panel for Climate Change (IPCC) estimates (Bindoff et al., 2007), the current glacio-eustatic component is around 1.2 mm yr^{-1}, with the contributions due to the ice sheets and smaller ice masses, as shown in table 7.1. Estimating the glacio-eustatic components of sea-level rise is far from straightforward, however, and is discussed in detail in the following section.

7.4.2 GLOBAL GLACIER MASS BALANCE

Quantifying the contribution made to sea-level rise from the world's $\sim$160,000 glaciers and the two great ice sheets is currently the focus of a major research effort (Bamber and Payne, 2004; Dyurgerov and Meier, 2004).

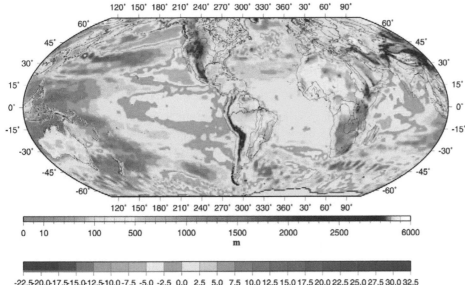

Figure 7.15 Sea-level trends for 1993–2003, from the TOPEX/Poseidon mission (Cazenave and Nerem, 2004).

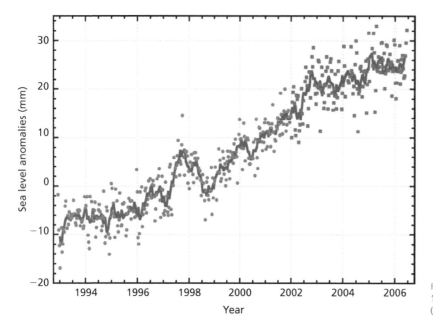

Figure 7.16 Trends in global mean sea level, 1993–2007, relative to the 1993–2001 mean (Bindoff et al., 2007).

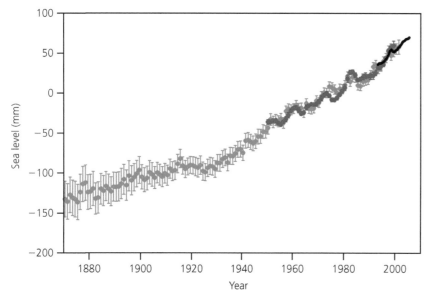

Figure 7.17 Trends in annual global mean sea level 1870–2001, based on adjusted tide gauge measurements (red points and blue curve) and satellite altimeter data (black curve), relative to the 1993–2001 mean (Bindoff et al., 2007).

Obtaining reliable estimates, however, presents considerable challenges. For glaciers and ice caps, the difficulty lies in upscaling information from the tiny fraction of glaciers that have mass balance records, and involves chains of assumptions that require careful testing. On the other hand, the immense size of the great ice sheets meant that it was impossible to obtain representative figures for their mass balance until the advent of modern remote-sensing techniques. In this section, we review the current best estimates of glacier and ice sheet mass balance trends, and their implications for sea-level change.

In the literature, changes in amount of glacier ice are variably quoted in terms of volume of water, volume of ice or mass. For consistency, we quote all figures in terms of mass, in gigatonnes, assuming $1\,\text{km}^3$ of fresh water has a mass of 1 Gt and $1\,\text{km}^3$ of glacier ice has a mass of 0.9 Gt. For converting glacier mass changes into sea-level change, 362 Gt of meltwater is considered equivalent to 1 mm of global sea level, given that the area of the world's oceans is ~362 million km^2 (Dyurgerov and Meier, 2004).

Glaciers and ice caps

Several research groups have attempted to estimate the total current mass balance of all the world's glaciers and ice caps (i.e. excluding the Greenland and Antarctic Ice Sheets), by extrapolating from available measurements (e.g. Cogley and Adams, 1998; Church and Gregory, 2001; Dyurgerov, 2002, 2005; Raper and Braithwaite, 2005). This procedure is not straightforward, and the final results are subject to many uncertainties (Dyurgerov and Meier, 2004). First, for practical reasons, glaciers with long mass balance records tend to be of small size and easy of access, and are not a representative sample.

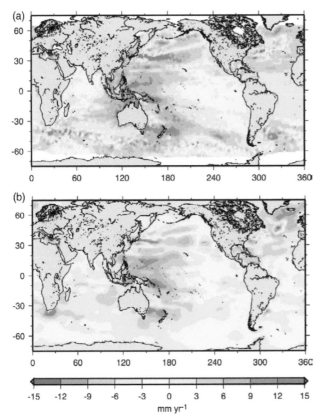

Observed sea-level rise	
(mm yr⁻¹ 1993–2003)	3.1 ± 0.7
Estimated contribution	
Glaciers and ice caps	0.77 ± 0.22
Greenland Ice Sheet	0.21 ± 0.07
Antarctic Ice Sheet	0.21 ± 0.35
Thermal expansion	1.6 ± 0.5
Total	2.8 ± 0.7
Difference between observed	
rise and sum of estimates	0.3 ± 1.0

Table 7.1 Estimated contributions to observed recent sea-level change (Bindoff et al., 2007).

Figure 7.18 (a) Linear trends in mean sea level, 1993–2003, from the TOPEX/Poseidon mission. (b) Steric component of mean sea-level change over the same period (Bindoff et al., 2007).

Mass balance records therefore have to be adjusted for glacier size before they can be extrapolated to larger areas. Second, glacier size distributions are very poorly known, and until complete inventories are available, must be estimated from empirical functions. Third, many glacierized areas are poorly represented in the mass balance records. Around 70 per cent of glaciers with long records are located in Scandinavia, the European Alps, North America and the Former Soviet Union, and many important glacierized regions lack any observations at all. The existing data set is biased towards maritime regions, and glaciers in continental settings are under-represented. Fourth, the total area occupied by glaciers and ice caps has not been measured accurately. For example, estimates of the area of independent glaciers around the periphery of the Antarctic Ice Sheet vary by an order of magnitude (Dyurgerov and Meier, 2004). Finally, to estimate the total sea level potential in the world's glaciers and ice caps, the total volume of glacier ice must be estimated from area-volume functions (Bahr et al., 1997).

The most comprehensive attempt to estimate global glacier and ice cap mass balance is that of Dyurgerov and Meier (2005). They estimate that the current total glacier area (excluding the two ice sheets) is 785,000 ± 100,000 km², with a volume of about 260,000 ± 65,000 km³. This is equivalent to about 0.65 ± 0.16 m of

sea-level change. Figure 7.19 shows calculated cumulative glacier mass balances for seven regions for the period 1960–2003. Most regions show a persistently negative mass balance, although glaciers in the Andes and Europe had positive mass balances for part of the time. There is also considerable variation within each region. In Europe, for example, Alpine glaciers generally had positive mass balances during the late 1970s to mid 1980s, after which they mainly had negative balances, whereas glaciers in Scandinavia mostly had more positive balances for most of the measurement interval (cf. fig. 2.33). Taken together, the world's glaciers and ice caps are losing mass. For the period 1961–2003, Dyurgerov and Meier (2005) calculated that the mean contribution to sea-level rise was +0.51 mm yr⁻¹, rising to +0.93 mm yr⁻¹ in the decade 1994–2003.

The regions with most negative cumulative balances are Patagonia, Alaska and the north-west USA and Canada. Comparison of airborne laser altimetric measurements with older data for a large sample of glaciers in Alaska and adjacent parts of Canada by Arendt et al. (2002) indicated a mean thickness change of −0.52 m yr⁻¹ from the mid 1950s to the mid 1990s, accelerating to −1.8 m yr⁻¹ from the mid 1990s to 2000–1. Extrapolation to all glaciers in the region yielded a total contribution to mean sea-level change of +0.14 ± 0.04 mm yr⁻¹ for the earlier period, rising to +0.27 mm yr⁻¹ during the past decade. Rignot et al. (2003) found that the North and South Patagonian Ice Fields have been thinning at similar rates, losing ice equivalent to sea-level rise of 0.042 mm yr⁻¹ from 1975 to 2000, and 0.105 mm yr⁻¹ from 1995 to 2000 (the area of the Patagonian Ice Fields is around 10 per cent that of the Alaskan and Canadian glaciers). Many of the Patagonian outlet glaciers terminate in large proglacial lakes, and a combination of dynamic processes and iceberg calving has contributed to rapid ice loss. However, Arendt et al. (2002) found that some Alaskan tidewater glaciers advanced, probably reflecting the influence of calving dynamics on mass balance (section 5.5.4).

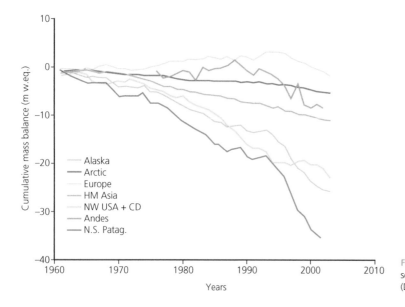

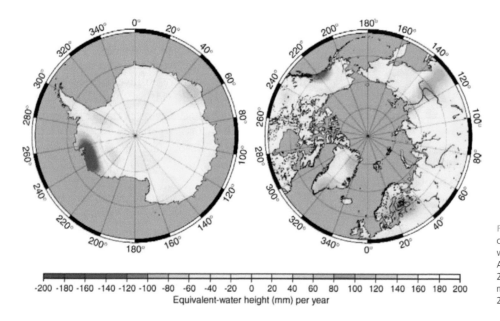

Figure 7.19 Cumulative glacier mass balances calculated for seven regions. Note the increased mass loss in the 1990s (Dyurgerov and Meier, 2005).

Figure 7.20 Maps of ice volume change (expressed as equivalent water thickness change) for Antarctica and Greenland, 2002–2005, from GRACE gravity measurements (Ramillien et al., 2006).

The polar ice sheets

If the Greenland and Antarctic Ice Sheets were to melt completely, glacio-eustatic sea-level rise would be around 63.5 m, resulting in huge loss of habitable land. Trends in their mass balance, therefore, will have a correspondingly large influence on intermediate- to long-term sea-level change. The immense size of the polar ice sheets, however, made it impossible to obtain meaningful mass balance estimates using traditional glaciological methods, and a comprehensive picture has only begun to emerge very recently (Bentley, 2004; Thomas, 2004). The rapid development of geodetic and gravimetric remote-sensing techniques has revolutionized our understanding of the mass balance of the polar ice sheets, and has revealed them to be highly dynamic systems capable of very rapid response to climate change. Although uncertainties remain over the precise magnitudes, it is now clear that the mass balance of the Greenland Ice Sheet is currently strongly negative, and that the Antarctic Ice Sheet as a whole is close to zero net balance.

Using gravimetric data from the GRACE mission, Ramillien et al. (2006) calculated that for the period July 2002 to March 2005, the mean annual mass deficit of the Greenland Ice Sheet was 129 ± 15 Gt yr^{-1} (fig. 7.20). The mass deficit calculated by Ramillien et al. is rather higher than that derived from GRACE data by Velicogna and

A

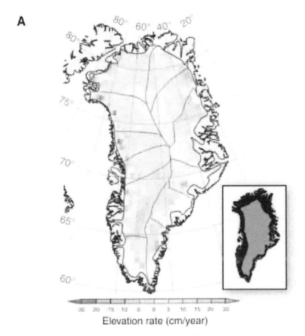

Elevation rate (cm/year)

B

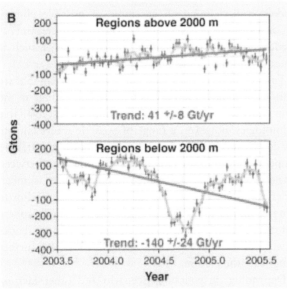

Figure 7.21 (A) Rate of elevation change of the Greenland Ice Sheet, 1992–2003, from satellite radar altimetry. Inset: areas above (grey) and below (black) 2000 m. (B) Time series of elevation change for areas above and below 2000 m. Note the large annual cycles in the lower panel (Shepherd and Wingham, 2007).

sheet, particularly in southern Greenland (Krabill et al., 1999, 2000; Velicogna and Wahr, 2006; Shepherd and Wingham, 2007; fig. 7.21). For example, the coastal portions of Kangerdlugssuaq and Helheim Glaciers – the two largest outlet glaciers in eastern Greenland – lost a combined $46 \pm 7 \, \text{Gt yr}^{-1}$ in the period 2001–6 (Stearns and Hamilton, 2007; section 6.2.4). The very high mass deficit from this and other coastal regions is largely due to increased glacier velocities and associated calving rates, and only partly to increased surface melting (Howat et al., 2005; Rignot and Kanagaratnam, 2005; Luckman et al., 2006; Sole et al., 2008).

In Antarctica, altimetric and gravimetric measurements reveal complex patterns of mass change (Ramillien et al., 2006; Wingham et al., 2006b; Shepherd and Wingham, 2007). For the period 2002–5, the East Antarctic Ice Sheet (accounting for almost 90 per cent of the ice mass in the continent) exhibited an increase of around $67 \pm 28 \, \text{Gt yr}^{-1}$. Most of the increase was concentrated in coastal areas, in response to greater amounts of snowfall (Wingham et al., 2006). In contrast, the West Antarctic Ice Sheet (WAIS) lost mass during the same period, with mean annual losses amounting to $107 \, \text{Gt yr}^{-1}$. Almost all of the observed change relates to changes in ice discharge, as melting is negligible in this region (Rignot, 2006). The Siple Coast sector of WAIS, which drains into the Ross Ice Shelf, shows a complex pattern of mass change associated with changes in the velocity of major ice streams (Joughin and Tulaczyk, 2002). The most dramatic mass losses, however, are in the Amundsen Sea sector of WAIS, especially the Pine Island, Thwaites and Smith Glaciers (Shepherd et al., 2002; Rignot, 2008; section 6.3.6; figs 7.20 and 7.22). In the Antarctic Peninsula, accelerated glacier flow has increased ice discharge to the oceans (De Angelis and Skvarca, 2003; Pritchard and Vaughan, 2007). However, most glacier ice in the Antarctic Peninsula is in the continental ice cap of Dyer Plateau, which is experiencing snowfall-driven growth sufficient to offset the coastal dynamic changes (Shepherd and Wingham, 2007).

Expressed in terms of sea-level change, the calculated contributions of the polar ice sheets for the period 2002–5 are $+0.36 \pm 0.04 \, \text{mm yr}^{-1}$ (Greenland), $+0.3 \pm 0.06 \, \text{mm yr}^{-1}$ (West Antarctica) and $-0.19 \pm 0.07 \, \text{mm yr}^{-1}$ (East Antarctica) (Ramillien et al., 2006). The overall contribution of the great ice sheets to sea-level rise is therefore around $0.47 \, \text{mm yr}^{-1}$, with the Greenland contribution somewhat more than the IPCC estimate, and the Antarctic contribution rather less. Combining this figure with that calculated by Dyurgerov and Meier (2004) for glaciers and ice caps yields a total current glacio-eustatic contribution to sea-level change of $\sim 1.4 \, \text{mm yr}^{-1}$. This figure is certain to be revised as improved methods of calculating regional mass balance become available.

Wahr (2005) ($82 \pm 28 \, \text{Gt yr}^{-1}$ for 2002–4), and from altimetric data by Krabill et al. (2004) ($80 \pm 12 \, \text{Gt yr}^{-1}$ for 1997–2003). In part, the discrepancy may arise from the different measurement periods used, and an increase in mass deficit through time. According to Rignot and Kanagaratnam (2005), mass losses from Greenland increased from $82 \pm 29 \, \text{Gt yr}^{-1}$ in 1996 to $202 \pm 37 \, \text{Gt} \, \text{yr}^{-1}$ in 2005. Using GRACE data, Velicogna and Wahr (2006) found a similar rapid increase in mass losses, from $104 \pm 54 \, \text{Gt yr}^{-1}$ in 2002–4 to $246 \, \text{Gt yr}^{-1}$ in 2004–6. Most of the mass loss is from lower elevations on the ice

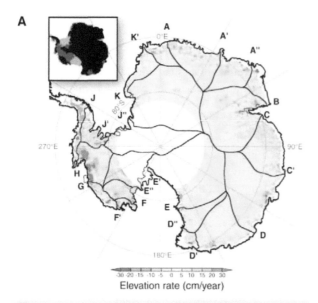

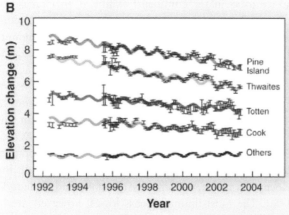

Figure 7.22 (A) Rate of elevation change of the Antarctic Ice Sheet, 1992–2003, from satellite radar altimetry. The inset shows continental-based (black), marine-based (dark grey) and floating (light grey) parts of the ice sheet. Note the very large reduction of surface elevation in the Amundsen Sea sector of West Antarctica. (B) Elevation change of the trunks of selected outlet glaciers (Shepherd and Wingham, 2007).

7.5 FUTURE SEA-LEVEL CHANGE

7.5.1 IPCC CLIMATE AND SEA-LEVEL PROJECTIONS

Any predictions of future glacio-eustatic sea-level change are subject to large uncertainties, including future levels of greenhouse gas emissions, how such emissions translate into gas concentrations in the atmosphere, how greenhouse gases together with other anthropogenic and natural forcings translate into climate change, and the resulting response of diverse glacier systems throughout the world.

Various strategies have been developed to address these uncertainties. The IPCC has defined a suite of 'emissions scenarios', in which greenhouse gas emissions follow different trajectories, depending on alternative demographic, economic and technological development

paths (Fisher et al., 2007). Some scenarios assume ever-increasing greenhouse gas emissions due to high population growth, slow economic development and slow technological change (Scenario A2), while others assume widespread economic restructuring and a reduction in emissions by the mid twenty-first century (Scenario B1). It is impossible to know how greenhouse gas emissions will actually evolve in the twenty-first century and beyond, but it is likely that they will fall within the envelope defined by these scenarios.

Greenhouse gas concentrations depend on both emission levels and the rate at which greenhouse gases are removed from the atmosphere by natural processes, and future levels must be estimated by coupling emissions scenarios to geochemical models (Denman et al., 2007). Alternative greenhouse gas concentrations (and other factors such as aerosol loading) can then be used as input for climate simulations using General Circulation Models (GCMs). Uncertainties at the modelling level are addressed by considering the results of different climate models, and defining an ensemble of possible climate responses (Randall et al., 2007). Yet more uncertainty is introduced by unknown natural climate forcings, such as large volcanic eruptions and solar variability.

Taken together, all of these uncertainties mean that we cannot know how the Earth's climate will change in the coming decades and centuries. By using the best available technologies, however, a range of possible future climate trends can be defined (fig. 7.23). The 'best estimates' of temperature rise in the twenty-first century vary between 2 and 4 °C, but the full range of uncertainty is much wider. Future sea-level rise depends on which climate scenario unfolds, and the ways in which the hydrosphere and cryosphere respond to the climate signal. Feedbacks between different components of the Earth system further complicate the picture.

Depending on emissions scenario and other factors, the IPCC estimates that, by the end of the century, mean global sea level will be around 0.2–0.6 m higher than 1980–99 levels (Meehl et al., 2007). Loss of glacier ice accounts for 30–45 per cent of the total, although this assumes that calving losses remain constant. Ice dynamic processes are a major source of uncertainty in sea-level projections (Bamber et al., 2007; Pfeffer et al., 2008), and since glaciologists' ability to model changing ice-sheet dynamics is not yet up to the task, no attempt was made by the IPCC to explicitly calculate this component. Instead, the possible impact of future dynamic changes was illustrated by scaling up current ice discharges within what seem like reasonable limits. This could add a further 0.09–0.17 m of sea-level rise to the total.

It must be emphasized that these figures do not imply uniform global sea-level rise, and that the changes experienced along the world's coasts will differ substantially in

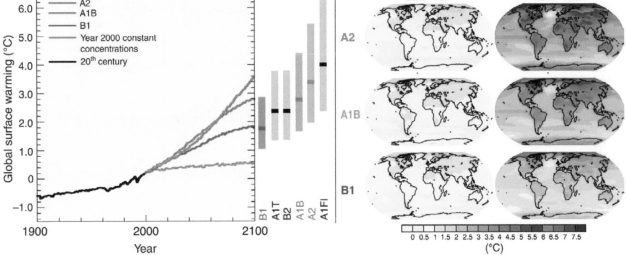

Figure 7.23 Alternative climate projections for the twenty-first century. Left panel: Multi-model averages of surface temperature (relative to the 1980–1999 mean), for emissions scenarios A2, A1B and B1. The vertical bars indicate the likely range and the best estimate for each scenario. Right panels: Projected regional temperature changes for the early and late twenty-first century for three emissions scenarios (IPCC, 2007).

magnitude and sign. Tectonic and glacio-isostatic uplift or subsidence can have large effects on relative sea-level change, while hydro-isotatic, geoidal and other dynamic influences can also introduce large regional variations. The effect of elevated landmasses on the shape of the geoid means that the magnitude of glacio-eustatic sea-level rise will also vary depending on the origin of the meltwater. Loss of mass from an ice sheet reduces its gravitational pull on the surrounding ocean, locally lowering the geoidal surface. This means that sea level can actually fall adjacent to shrinking ice sheets (fig. 7.24; Mitrovica et al., 2001; Tamisiea et al., 2001). A reduction of the mass of the Antarctic Ice Sheet produces the greatest amount of sea-level rise in the northern hemisphere and vice versa.

There is now consensus that continuously rising concentrations of anthropogenic greenhouse gases will lead to further atmospheric warming and reduction of global ice volume. The sea levels that the people of the late twenty-first century will actually experience depends on the way that natural systems respond to anthropogenic and other forcings, as well as the decisions made by people at many levels of society. The task facing glaciologists, therefore, is not to predict what *will* happen to the world's glaciers and ice sheets, but to address the question: what will happen if…?

7.5.2 PREDICTING THE GLACIAL CONTRIBUTION TO SEA-LEVEL CHANGE

The most important tools used by glaciologists to predict how glaciers and ice sheets might respond to climate

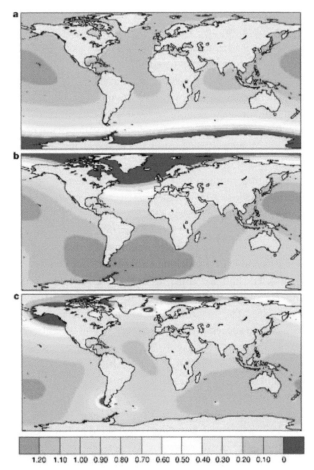

Figure 7.24 Modelled patterns of sea-level change resulting from ice melt in (a) Antarctica, (b) Greenland and (c) mountain glaciers and ice caps. The units are mm yr^{-1} for a 1 mm yr^{-1} ice melt contribution (Mitrovica et al., 2001).

change are numerical models (section 5.3; Huybrechts, 2004; van de Wal, 2004). Modelling strategies vary widely depending on the particular aims of the study, but can be broadly classified into two main types. First, *static models* focus on surface mass balance, and calculate ice mass gains or losses without considering ice flow processes (e.g. ACIA, 2005; Radic and Hock, 2006). Second, *dynamic models* couple equations describing the surface mass balance to others representing ice flow, so that the geometry of the system can evolve in response to changes in mass balance and vice versa (van der Veen, 1998a; Marshall et al., 2004; van der Veen and Payne, 2004). Both approaches have advantages and disadvantages. Although clearly less realistic, static models can represent many important surface processes very well, and allow changes in surface mass balance to be calculated for large regions (Oerlemans, 2001; Braithwaite and Zhang, 2003; Bougamont et al., 2005, 2007). Consequently, static models have provided useful information on likely short-term rates of sea-level change. In the longer term, however, static models tend to overestimate mass losses because the area subject to melting is held constant (Schneeberger et al., 2003). Dynamic models represent glacier systems more realistically, but require many more inputs. In addition, small grid spacing is required to adequately represent the flow of valley glaciers and ice streams, forcing a compromise between model resolution and geographical area. This means that modellers must decide whether to model small regions in detail, or large regions at a much coarser resolution. The first option is most appropriate for valley glaciers and ice caps, whereas the second is necessary for modelling the evolution of ice sheets.

Dynamic models have been used to assess possible future mass losses from valley glaciers by Oerlemans et al. (1998) and Schneeberger et al. (2003). A key issue is finding ways to upscale the results of modelling a small sample of glaciers to large regions. Oerlemans et al. (1998) conducted modelling experiments for a sample of 12 glaciers and ice caps, to determine volume changes under a range of temperature and precipitation forcings (fig. 7.25). The magnitude of glacier response is highly variable, so successful upscaling depends on weighting the results appropriately. Figure 7.26 shows the results of two alternative weighting procedures. Although the absolute values of volume change differ, the results imply that with a warming rate of 0.04 °C yr^{-1} (4 °C per century) and no increase in precipitation, little glacier ice would be left by 2100, whereas if warming is restricted to 0.01 °C yr^{-1} and precipitation increases by 10 per cent per degree of warming, then ice losses will be restricted to 10–20 per cent of the 1990 value.

In Greenland, surface ablation and accumulation are very important components of the total mass budget, and

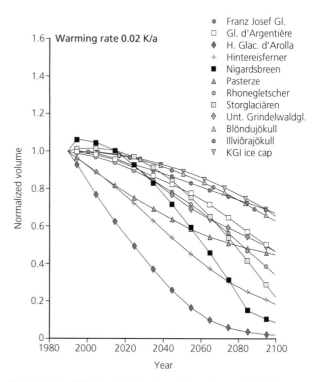

Figure 7.25 Modelled twenty-first-century volume change of 12 glaciers, for a warming rate of 0.02 °C yr^{-1}, and no change in precipitation (Adapted from Oerlemans et al., 1998).

static models have been widely used to determine how surface mass balance will change under different climate scenarios (sections 2.2–2.4). An important threshold is the point at which surface mass balance becomes negative. At the present time, the surface mass balance (i.e. exclusive of calving losses) is positive, and while this is the case the Greenland Ice Sheet will continue to occupy most of the island. Calving losses may cause the outlet glaciers to retreat to the coast, but once the ice sheet is entirely land-based it will stabilize, provided surface mass balance remains zero or positive. Gregory and Huybrechts (2006) modelled the effect of different combinations of temperature and precipitation change on the surface mass balance of Greenland expressed in terms of sea-level rise (fig. 7.27). Then, they used a GCM to model climate changes under a range of greenhouse gas scenarios and concluded that the surface mass balance will become negative if mean annual air temperature increases by more than 4.5 °C (equivalent to a global average increase of 3.1 °C). This is somewhat higher than earlier estimates (e.g. Gregory et al., 2004), but could still be crossed by the end of the present century under some IPCC warming scenarios. While the concept of a temperature threshold is robust, its precise value depends to a large degree on model physics. Bougamont et al. (2007) found that a positive degree-day model (as used by Gregory and Huybrechts) predicted almost twice as much runoff from the Greenland Ice Sheet than an energy balance model, about half of the difference being due to

(a)

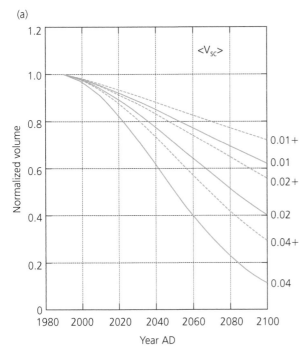

(b)

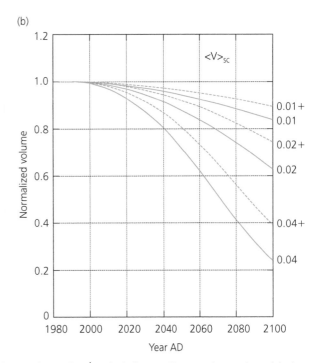

Figure 7.26 Scaled glacier volume changes for a range of climate scenarios. Labels are in degrees C yr^{-1}, and + indicates a 10 per cent increase in precipitation per degree warming. The left-hand panel ($<V_{SC}>$) shows the average of all model results, and the right-hand panel ($<V>_{SC}$) shows results weighted by glacier volume (After Oerlemans et al., 1998).

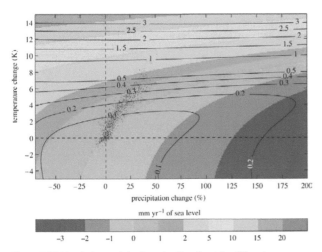

Figure 7.27 Mass balance of the Greenland Ice Sheet for different combinations of temperature and precipitation change, expressed as sea-level equivalent. The colours represent model averages and the isolines the standard deviation. The dots show decadal mean climates in GCM simulations of climate under a range of greenhouse gas scenarios (Gregory and Huybrechts, 2006).

differences in refreezing in the snowpack (see section 2.4.3). These results indicate that modelled snowpack processes have a potentially large effect on predicted ice sheet mass balance, and associated rates of sea-level rise.

Although calving losses from Greenland will cease if the ice retreats inland, calving glacier dynamics introduce large uncertainties into Greenland's contribution to sea-level rise in the short term. Acceleration of many of Greenland's tidewater outlet glaciers in the first years of the twenty-first century prompted widespread concern

that increased calving losses could lead to very rapid rates of sea-level rise (e.g. Rignot and Kanagaratnam, 2005). More recent research, however, strongly suggests that the acceleration is a transient phenomenon related to local fjord overdeepenings, and that flow speeds decrease again after glaciers retreat to stabilizing pinning points (Howat et al., 2008; section 6.2.4). Newly developed numerical models (Nick et al., 2009) open up the possibility of predicting changing calving fluxes, provided bed topography is known.

The possible influence of subglacial hydrology on the evolution of the Greenland Ice Sheet was addressed by Parizek and Alley (2004a), who conducted experiments with a thermo-mechanical flow-line model that coupled basal sliding to surface meltwater production. Parizek and Alley assumed a linear scaling between melt rates and basal velocity based on the data set of Zwally et al. (2002a), which greatly increased mass losses from the ice sheet in response to climate warming. As discussed in sections 5.4.1 and 6.2.3, however, the relationship between basal lubrication and sliding is not straightforward, because subglacial drainage systems can adjust to evacuate increased water inputs. Indeed, there is evidence that ice velocity in some sectors of the ice sheet has actually decreased in parallel with increased melting. Coupling models of subglacial hydrology to realistic treatments of basal sliding remains one of the greatest challenges facing glacier modellers (section 5.3.3).

The long-term evolution of the Greenland Ice Sheet has been modelled by Ridley et al. (2005), who coupled a

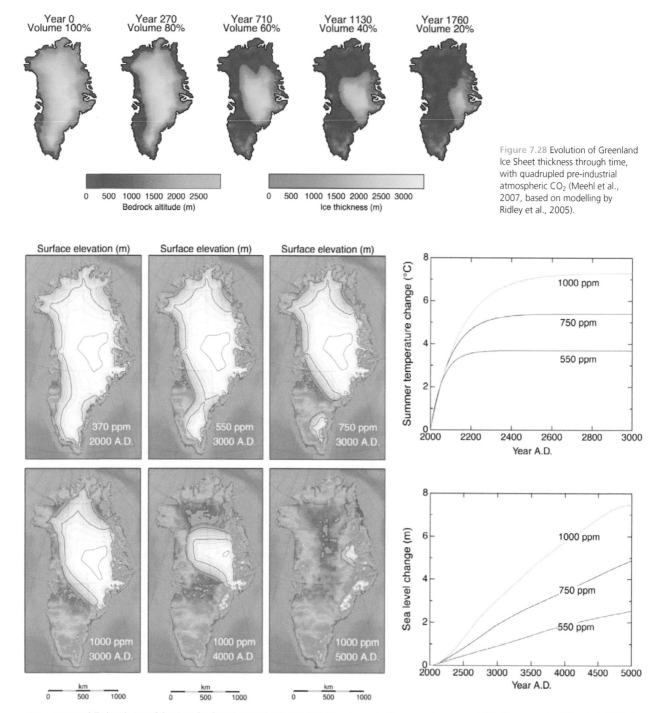

Figure 7.28 Evolution of Greenland Ice Sheet thickness through time, with quadrupled pre-industrial atmospheric CO_2 (Meehl et al., 2007, based on modelling by Ridley et al., 2005).

Figure 7.29 Modelled evolution of the Greenland Ice Sheet and associated sea-level rise under three greenhouse gas stabilization scenarios (Alley et al., 2005a).

three-dimensional, thermo-mechanical ice sheet model and the Hadley Centre Coupled Ocean–Atmosphere general circulation model (HadCM3). The model was forced with an increase of atmospheric CO_2 to four times pre-industrial levels, which was held constant for a model run of 3000 years. The ice sheet retreated towards the eastern mountains (fig. 7.28), and by the end of the model run had almost completely vanished. The total direct contribution to global mean sea level was approximately 7 m, with peak rates of sea-level rise of 5 mm yr^{-1} early in the simulation. Modelled retreat of the Greenland Ice Sheet, and associated rates of sea-level rise, for a range of greenhouse gas scenarios are shown in figure 7.29 (Alley et al., 2005a). Note that the modelled patterns of retreat differ from those in figure 7.28; this reflects different parameterizations of climate and ice sheet processes.

Charbit et al. (2008) studied the effect of cumulative CO_2 emissions and natural CO_2 drawdown processes on long-term ice sheet behaviour. They found that total atmospheric loadings of 1000–2500 Gt of carbon cause

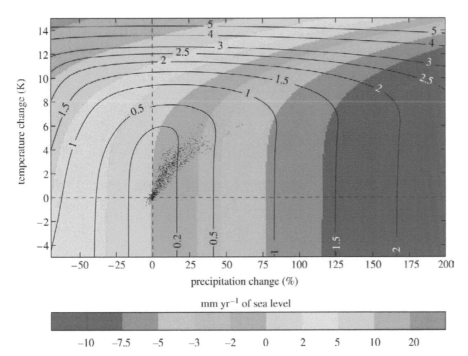

Figure 7.30 Mass balance of the Antarctic Ice Sheet for different combinations of temperature and precipitation change, expressed as sea-level equivalent. The colours represent model averages and the isolines the standard deviation. The dots show decadal mean climates in GCM simulations of climate under a range of greenhouse gas scenarios (Gregory and Huybrechts, 2006).

rapid shrinkage of the Greenland Ice Sheet, but slight recovery took place after several millennia. The greatest ice loss was in the south, where higher albedo prevented regrowth even when CO_2 levels fell once more. Higher loadings (3000–3500 Gt) caused irreversible melting. (In 2007, total CO_2 emissions from fossil fuels amounted to 8.5 Gt.)

Ice sheets also interact with other parts of the Earth system, potentially changing the magnitude and sign of ice sheet response to an initial signal. For example, much attention has been focused on the possibility that ice sheet melting could affect the North Atlantic Meridional Overturning Circulation (NAMOC), by suppressing the formation of dense waters off Greenland (see Broecker et al., 1988; Fichefet et al., 2003). Simulations with an Earth system model with coupled atmosphere, ocean, and ice sheet components, by Vizcaino et al. (2008), found that NAMOC shutdown occurred in the most extreme warming scenarios ($4 \times CO_2$) in response to increased freshwater fluxes into the oceans. Most of this water reflected increased precipitation and river discharge, and meltwater from Greenland did not contribute significantly to the overall freshwater flux. On the other hand, NAMOC shutdown drastically reduced the rate of decay of the Greenland Ice Sheet, so that its long-term contribution to sea-level rise is actually less than in a $3 \times CO_2$ simulation.

For the Antarctic Ice Sheet, surface melting is not a big issue in the immediate term. Gregory and Huybrechts (2006) found that the surface mass balance of the Antarctic Ice Sheet is likely to become increasingly positive under a warming climate, due to increased precipitation (fig. 7.30). Thus, if calving rates remain constant, increasing snowfall over Antarctica will partially offset sea-level rise from other sources. Furthermore, modelling by Swingedouw et al. (2008) suggested that significant runoff from the Antarctic Ice Sheet could moderate warming in the Southern Hemisphere, by introducing cold water to the surrounding ocean. This would slow the retreat of sea ice, helping to maintain low temperatures due to low surface albedo.

The greatest uncertainty about the future of the Antarctic Ice Sheet concerns ice dynamics, particularly in the Amundsen Sea sector of the marine-based West Antarctic Ice Sheet (Alley et al., 2005a; Bamber et al., 2007; Meehl et al., 2007; Shepherd and Wingham, 2007; section 6.3.7). Glaciological models are not yet capable of dealing with the key processes governing the stability of marine ice sheets, and it is not known whether the recent dramatic acceleration and thinning of Pine Island Glacier and adjacent ice streams will be sustained in the long term, and whether feedbacks will stabilize or destabilize the system. The Amundsen Sea sector alone contains enough ice to raise global sea level by an average of 1.5 m, and of all the world's ice masses, this region has the potential to be the single largest contributor to sea-level change in the coming century. As thinning glaciers unground into the smooth, overdeepened beds inland, mass loss from this region appears set to increase for years to come (Rignot, 2008). By how much remains to be seen.

Given these great uncertainties, knowledge about past sea levels and ice sheet extent may provide useful insights into the likely magnitude of future changes. Overpeck et al. (2006) and Otto-Bliesner et al. (2006) compared future climate scenarios with modelled climate of the last (Eemian) Interglacial period (section 7.3.1).

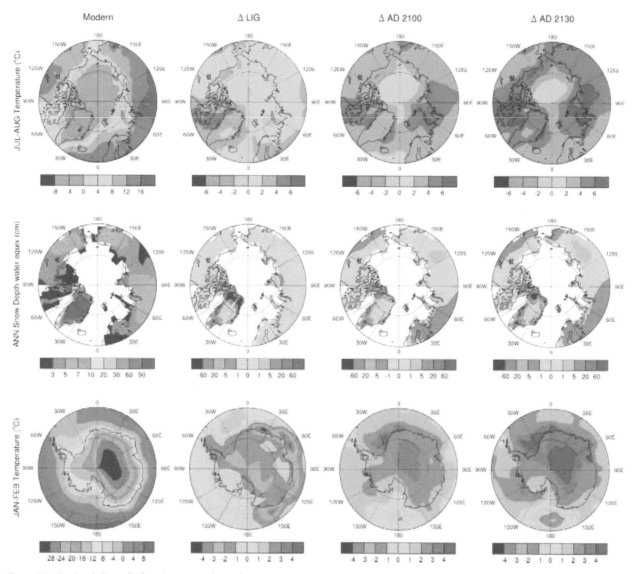

Figure 7.31 Simulated climate for four time periods, from left to right: modern; difference between modern and 130,000 years ago (ΔLIG); difference between modern and 3 × pre-industrial CO_2 levels (ΔAD 2100); difference between modern and 4 × pre-industrial CO_2 levels (ΔAD 2130). The top and bottom panels show summertime temperatures, and the middle panels show snow depth (Overpeck et al., 2006).

They concluded that, by 2100, climate in the polar regions may be warmer than during the Eemian (fig. 7.31). Given that Eemian sea levels appear to have been 4–6 m higher than today, these results imply that parts of the Greenland and Antarctic Ice Sheets could contribute substantially to sea-level rise in the next hundred years. Although these results are far from certain (Oerlemans et al., 2006), they lend some weight to the idea that the polar ice sheets could undergo significant changes within a few human generations, with wide-reaching effects.

PART TWO
GLACIATION

Flaajökull Foreland, Iceland (D.J.A. Evans)

CHAPTER 8
EROSIONAL PROCESSES, FORMS AND LANDSCAPES

Mount Cook & Lake Pukaki, New Zealand (D.J.A. Evans)

8.1 INTRODUCTION

Erosion in glacial environments results in some of the most familiar and characteristic forms and landforms, such as striated rock surfaces, roches moutonnées, cirques and troughs, and creates some of the most spectacular landscapes on Earth. The presence of glacier ice in a catchment also strongly influences the evolution of erosional forms beyond the ice margin, such as proglacial meltwater channels.

Erosional forms exist on a wide variety of scales, reflecting the operation of glacial processes over different amounts of time. In this chapter, we discuss the characteristics and origins of erosional forms at four scales:

(1) *Small-scale forms.* These are superficial erosional marks such as striae and friction cracks, which commonly record single rock failure events.

(2) *Intermediate-scale forms.* These include bedforms, depressions or channels, which, although locally impressive, are small compared with the ice flow unit responsible for their formation.

(3) *Large-scale forms.* These are erosional forms which are comparable in scale with the associated glacier or ice stream, such as cirques and troughs.

(4) *Landscapes of glacial erosion.* At the largest scale, erosional landscapes are distinctive assemblages of landforms that record long-term, regional patterns of glaciation.

Some relationships between these forms, and characteristic dimensions, are shown in table 8.1. Note that the dimensions of forms at different 'scales' may overlap. The important point to bear in mind, however, is that small-scale erosional marks are commonly superimposed on intermediate-scale bedforms and channels, which in turn are superimposed on large-scale ice-eroded channels or surfaces, which combine to form erosional landscapes, in a spatial hierarchy of forms.

At all scales, it is useful to bear in mind the sets of variables that control or influence processes and patterns of erosion, and the form, size and distribution of erosional forms. Sugden and John (1976) grouped these variables into four categories:

(1) *Glaciological variables* encompass the characteristics of the ice, particularly conditions at the bed. They include basal shear and normal stresses, subglacial water pressures and drainage system configuration, flow direction and basal velocity, thermal regime and the amount of debris held in the basal ice. Basal thermal regime is particularly important, because sliding and significant erosion can only occur where the basal ice is at the pressure-melting point. Cold-based ice tends to protect the underlying substratum, although locally impressive quarrying can occur below cold-based ice margins. Significant debris entrainment is possible in zones of net freezing, but not in zones of net melting.

(2) *Substratum characteristics* incorporate the physical characteristics of the bed, and include the structure, lithology, joint distribution and degree of weathering of hard rockbeds, and the thickness and composition of unconsolidated sediments. The erosion of hard rockbeds is strongly influenced by lithology and the degree of preglacial weathering. Thus, glacial erosion is thought to have been particularly effective in the early stages of the present, Quaternary ice age, due to the widespread availability of deep-weathered regolith (saprolites) formed during the Tertiary period. After the removal of large areas of this regolith during early glaciations, the erosional capability of later glaciations was reduced due to the occurrence at the ground surface of unweathered bedrock. Another important factor is the permeability of the substratum, which influences the efficiency of drainage at the bed.

(3) *Topographic variables* encompass the morphology of the glacier bed at a wide range of scales, from small-scale roughness elements up to the relative relief of the whole glacierized catchment. At the smallest scale, topographic variables influence local patterns of glacier flow and determine the location of stress concentrations at the bed, whereas at the largest scale, relief influences the location of glacier masses, and their morphology, dynamics and efficiency as agents of erosion. In high-relief settings, the various modes of rockslope failure (RSF) that take place through time may help to initiate or accentuate glacial erosional forms (e.g. Hewitt, 1999; Jarman, 2002, 2003, 2006; Turnbull and Davies, 2006).

(4) *Temporal variables* include the duration of glaciation and changes in any of the above variables over time.

In combination, these variables influence the modification of glacier beds in such a way that they become more efficient pathways for the evacuation of glacier ice or meltwater. This framework focuses attention on the links between process and form, and encourages recognition of equilibrium forms (Sugden and John, 1976).

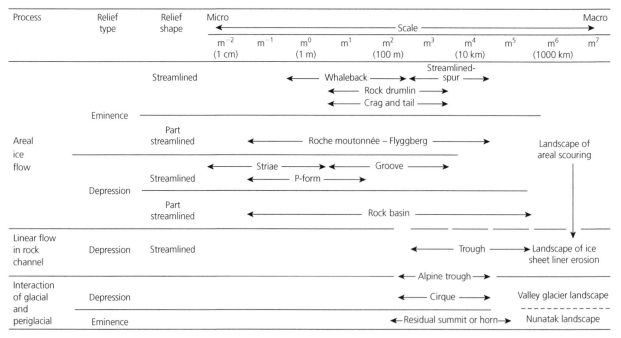

Table 8.1 Size classification of glacial erosional forms (after Sugden and John, 1976)

8.2 SUBGLACIAL EROSION

As discussed in chapter 4, glaciers can be underlain by rock ('hard beds') or unlithified sediments ('soft beds'). Subglacial erosion of rock surfaces can be subdivided into two distinct processes of *abrasion*, or the grinding of fine-grained material, and *quarrying*, or the failure of larger pieces or rock. Although these two processes operate at very different scales, they can both be understood within the framework of fracture mechanics. Rock fracture is not required to dislodge fragments from soft beds, although the release of particles or particle aggregates from unlithified glacier beds involves similar physical processes to hard-bed erosion.

8.2.1 ROCK FRACTURE: GENERAL PRINCIPLES

All rocks are traversed by cracks on a wide range of scales, from large, readily visible joints and bedding surfaces, down to microscopic flaws and cleavage planes in mineral grains (fig. 8.1). The size and density of such cracks is highly variable, and is a major factor determining the strength of different rocks. For example, shale has closely spaced bedding planes, whereas joints in some granites are up to several metres apart. In some rocks, cracks can be continuous enough to completely isolate volumes of rock, such as joint-bounded blocks. When this is the case, erosion is simply a matter of removing fragments defined by existing fractures, a process known as *discontinuous rock mass failure* (Addison, 1981). More usually, cracks are not continuous, but must be extended by imposed stresses before failure can occur. The mechanisms of *crack growth* (or *crack propagation*) have been intensively studied by engineers seeking an understanding of how and why structures, such as bridges, skyscrapers and aircraft, fail in certain circumstances. A good introduction to the science of *fracture mechanics* has been provided by Janssen et al. (2004), outlining principles of crack growth that apply equally well to engineering structures and rock below glaciers.

Cracks in rocks can grow in each of the three modes described in section 4.7.1, or by a mixture of two or more modes. As explained in section 4.2.1, both tensile and shear stresses arise in a material in response to externally applied shear stresses and/or spatial variations in normal stress. Since the magnitude of both shear and normal stresses varies with orientation, the likely mode of crack growth depends on the orientation of the crack relative to the stress field.

Stress concentrations around the crack tip magnify the externally applied force, an effect that increases with crack length (fig. 4.34; Janssen et al., 2004). Crack growth releases stored (elastic) strain energy from the surrounding rock, which adds to the stress concentration at the crack tip. The longer the crack, the more strain energy is released and the easier it will be to make the crack grow. For each material, there is a threshold crack size, dividing two very different types of behaviour. Below the threshold, the small amount of strain energy released is insufficient to cause additional fracture, and further crack growth is unlikely. Above the threshold, the strain energy released by crack growth adds substantially to the energy available for fracturing, and the crack can grow in an explosive manner. The threshold length is called the *critical Griffith length* after the aircraft engineer A.A. Griffith (1893–1963), who laid the foundations of modern fracture mechanics in the 1920s. Cracks longer than the threshold length are known as *critical cracks*, and those below it are *subcritical cracks* (fig. 8.2). Although final rock failure involves the growth of critical cracks, rock masses lacking critical cracks can be weakened over time by a process of *subcritical crack growth*, whereby cracks below the critical Griffith length grow slowly until they reach the threshold length and become susceptible to rapid failure. Subcritical crack growth can occur under static deviatoric stresses as a result of stress–corrosion reactions that weaken atomic bonds at crack tips (Atkinson, 1984, 1987).

Figure 8.1 Joints on former glacier beds. (a) Deep and widely spaced joints in abraded granite, Makinson Inlet, Ellesmere Island. (b) Shallow and densely spaced joints on a limestone whaleback, Svalbard (D.J.A. Evans).

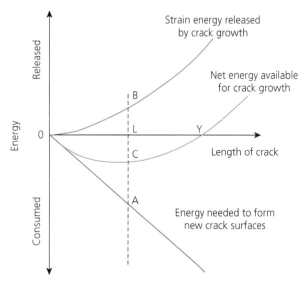

Figure 8.2 Griffith model of crack growth. Line 0-A represents the energy requirement to grow cracks of length L; 0-B is the strain energy released by the progress of the crack; and 0-C is the net available energy in the system. For cracks shorter than length Y, more energy is needed for crack growth than is released; these subcritical cracks will not tend to grow. In contrast, for cracks longer than length Y, more energy is released by crack growth than is required for crack extension – these are critical cracks that will grow catastrophically (Modified from Gordon, 1976).

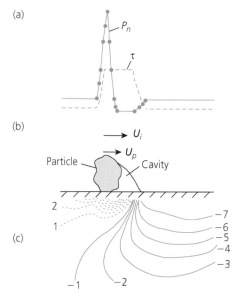

Figure 8.3 Calculated stress patterns in a horizontal rockbed associated with a rock particle held in sliding ice. (a) Distribution of normal (P_n) and shear (τ) stresses at the rock surface. (b) The velocity of the particle (U_p) is less than that of the ice (U_i) owing to frictional retardation. As a result, a low-pressure cavity opens up in the lee of the particle. (c) Contoured values of the maximum stress in the rock. Positive values are compressive, negative values are tensile (Redrawn from Drewry, 1986, after Ficker et al., 1980).

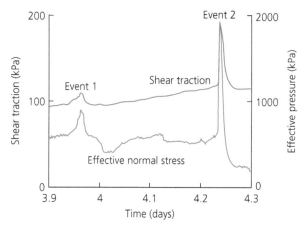

Figure 8.4 Transient increases in shear traction and normal stress measured on pressure sensors installed below Engabreen, reflecting the passage of large clasts over the bed (After Cohen et al., 2005).

On hard glacier beds, larger than average stresses occur:

(1) on bumps and the edges of steps, which support the stress transferred from low-pressure cavities or other weak areas of the bed;
(2) at the contacts between the bed and particles held in the ice;
(3) at the contacts between debris particles.

Elevated stresses near the crest of bumps and the edge of steps result in large stress gradients in their immediate vicinity (Morland and Boulton, 1975; fig. 3.21), and the associated pattern of deviatoric stresses will encourage growth of suitably oriented cracks. In addition to spatial variations in stress, subglacial rock surfaces are subject to stress fluctuations in time, associated with variations in water pressure and the passage of rock particles dragged along by the ice. The extent of subglacial cavities increases with sliding speed and is inversely proportional to effective pressure ($N = P_i - P_w$; section 3.2.4), so loading on bedrock surfaces will vary through time in response to water pressure fluctuations. Stress gradients in the vicinity of cavities will be highest at times of falling water pressure, because water pressure in cavities can decrease more rapidly than cavity closure can reduce the load on adjacent rock surfaces (Iverson, 1991a; Cohen et al., 2006).

Stress concentrations below particles held in basal ice are also closely associated with ice flow and the development of lee-side cavities (section 4.4.4). Cohen et al. (2005) have shown that water-filled cavities play a very important role in enhancing stress concentrations beneath particles. Figure 8.3 shows modelled stress patterns in a flat glacier bed associated with a particle with a low-pressure cavity in its lee (Ficker et al., 1980). Maximum normal stresses (P_n) are developed below the point of contact, and high shear stresses (τ) occur below and for some distance in front of the particle. Crack growth and rock fracture are most likely in the zone of maximum stress gradient below and in front of the leading edge of the particle. Pressure sensors installed below Engabreen, Norway, have recorded large fluctuations of normal and shear stresses associated with the passage of rock particles and variations in water pressure (Cohen et al., 2005; fig. 8.4).

8.2.2 ABRASION

Abrasion is the wear of rock surfaces by the processes of *striation* (the scoring of bedrock) and *polishing*, in which the roughness of rock surfaces is reduced by the removal of small protuberances. Striations (or *striae*) are formed when asperities protruding from rock particles are dragged over bedrock or other clasts, scouring out thin grooves. As an asperity passes over a rock surface, transient stress concentrations below the asperity promote crack growth and brittle failure. As the asperity moves on, a striation is left behind as a trail of damage. Striae are therefore the cumulative effect of small brittle failure events marking the passage of overriding particles (fig. 8.5). Drewry (1986, p. 51) has described how the striation process is not continuous, 'but comprises jerky steps. As grooving commences there is a build-up of elastic strain at the asperity tip. This is released giving rise to the impact of the asperity against the rock surface with subsequent production of rock chips ... The sequence then recommences.' Evidence for this type of jerky abrading motion comes from studies of industrial rock cutting (Drewry, 1986), and from the micromorphology of natural striae. Although striae seem continuous with the naked eye, through the microscope they are seen to consist of numerous crescent-shaped fractures, each marking a discrete failure event (fig. 8.6; Iverson 1990, 1995).

In this book, we define *polishing* as smoothing of a rock surface resulting from the removal of small asperities by overriding rock particles and ice. Stress concentrations in asperities locally increase the likelihood of critical or subcritical crack growth, so that brittle failure is more likely to occur at these locations than elsewhere, reducing the relief of the surface. A similar process will wear down asperities on striating clasts; as a striation is cut, the cutting point is worn down. Evidence for this is provided by the common observation that striations become wider and shallower downglacier from their point of initiation. Striation and polishing are thus complementary parts of the same process: the act of cutting a striation in bedrock polishes the striating clast, and vice versa.

Stress concentrations can be expected to occur around protuberances on the bed even below flat, non-striating clast surfaces and masses of compressed rock flour, so that the breakage and removal of asperities – or polishing – can be expected to be independent of the striation process, and may occur in its absence. This may explain the otherwise puzzling evidence that clean ice can achieve some degree of subglacial abrasion, even though ice is much softer than rock. Budd et al. (1979) conducted laboratory experiments that demonstrated that clean, sliding ice could erode rough granite slabs. It is likely that the main mechanism of abrasion below clean ice is polishing, since all of the drag below clean, wet-based ice is the form drag supported by protuberances, which will therefore be the locus

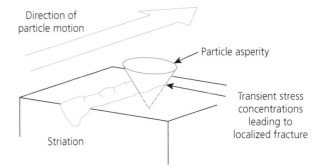

Direction of particle motion

Particle asperity

Transient stress concentrations leading to localized fracture

Striation

Figure 8.5 Simple schematic representation of the striation process (Modified from Drewry, 1986).

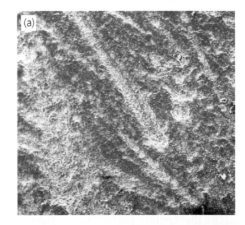

(a)

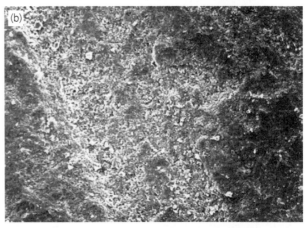

(b)

Figure 8.6 Scanning electron micrographs of striae at a range of scales. Note the abundant evidence for brittle fracture within striae compared with the smooth appearance of the neck between striae (Courtesy of L.A. Owen).

of high stress concentrations (see Section 4.4). Budd et al. (1979) reported extrapolated erosion rates of up to 55 mm yr^{-1}, although it is probable that such high rates will not be sustained as the roughness of the bed decreases and the stress concentration effect is reduced.

Theories of subglacial abrasion have been developed by Boulton (1974, 1979) and Hallet (1979b, 1981), allowing abrasion rates to be predicted from *wear laws*. For example:

$$\dot{A} = \alpha^* C_d v_p N \qquad (8.1)$$

which predicts abrasion rate ($\dot{A}$) from the concentration of particles in the basal ice (C_d), particle velocity (v_p), the normal stress pressing the particle against the bed (N) and an empirical factor α^* incorporating the effects of rock hardness (Hallet, 1979b). The influence of normal stress and velocity on abrasion rates can be readily understood using the analogy of sandpaper on wood: the harder the sandpaper is pressed down and the faster it is moved along, the more effective the sanding process. Empirical validation of this type of wear law has been provided in a series of laboratory experiments conducted by Lee and Rutter (2004), who measured progressive abrasion of a range of rock types using counter-rotating rock wheels forced together under variable loads. The lower wheel was half-submerged in a bath of water, so that abrasion products were continually removed from the abrading surface. Lee and Rutter found that the data could be described by the following function:

$$\dot{A} = k_w N^n n_{rock}^{*m} \qquad (8.2)$$

where n_{rock}^* is the rock porosity, and k_w, n and m were empirically found to be $10^{-8.11}$, 8.33 and 4.50, respectively. The importance of normal force is consistent with Coulomb-like friction, where shear stress is directly proportional to the applied normal stress (section 4.2.3). Equation 8.2 scales abrasion rates to the amount of shear displacement, or the relative motion of the two rock wheels, whereas in equation 8.1, abrasion rates refer to a fixed point on a glacier bed, so the amount of shear displacement is represented by the velocity of overriding debris (v_p).

In equation 8.2, rock porosity is used as a proxy variable for rock strength, and plays the same role as the prefactor α in equation 8.1. In Lee and Rutter's experiments, wear was most rapid for high-porosity sandstones, and commonly involved whole grains being pulled out of the surrounding rock. Much lower rates of wear were recorded for low-porosity granite and metamorphic rocks, which developed striated and polished surfaces. The importance of rock strength had been demonstrated many years earlier in a series of experiments conducted beneath Breiðamerkurjökull, Iceland, by Boulton (1979), in which plates of different types of rock were bolted onto the glacier bed and left to be striated by the overriding ice (fig. 8.7). Measured abrasion rates were found to be inversely proportional to rock hardness, and were several times higher for limestone (hardness = 180–210 kg mm^{-2}) than for basalt (hardness = 865–905 kg mm^{-2}).

During the striation process, large numbers of small particles are created, which tend to accumulate around asperities in the striating clasts (Boulton, 1974; Iverson, 1990). If these wear products are not removed, they will tend to increase the effective area of contact between the clast and the bed, reducing clast–bed friction and the

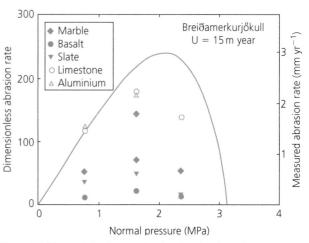

Figure 8.7 Measured abrasion rates for three locations beneath Breiðamerkurjökull, Iceland. Rates of abrasion were measured for platens of different materials bolted on to bedrock. Material hardness was as follows: marble = 450–510 kg mm^{-2}; basalt = 865–905 kg mm^{-2}; slate = 605–660 kg mm^{-2}; limestone = 180–210 kg mm^{-2}; aluminium = 50–60 kg mm^{-2}. The inverse relationship between hardness and abrasion rates is clearly apparent. Dimensionless abrasion rates predicted by Boulton (Coulomb friction) theory are also shown. The general pattern of abrasion rates, first rising then falling with normal pressure, is also apparent in the measured rates (Drewry, 1986, after Boulton, 1979).

efficiency of the striating process. Beneath glaciers, fine wear products ($<50\,\mu m$) can be removed by water flowing in thin films at the bed (section 3.4.2; Hallet, 1979a; Cohen et al., 2006).

8.2.3 QUARRYING

The fracture of large fragments (greater than about 1 cm) from rockbeds occurs by essentially the same process as abrasion. Temporary stress concentrations below overriding clasts tend to enlarge cracks in the rock, which can then lead to the isolation of fragments from the parent mass. The distribution of joints and other macroscopic cracks in rock exert a strong influence on quarrying processes, although the form of fractures on flat bedrock surfaces indicate that microcracks can grow rapidly in response to stress concentrations below clasts.

Recent work on quarrying has focused on fracturing at the lip of rock steps, where large stress gradients can develop at times of falling water pressure in lee-side cavities (fig. 8.8; Iverson, 1991a; Cohen et al., 2006). For a steady-state water-filled cavity, in which water pressure is equal to the ice overburden pressure, stresses are at a maximum adjacent to the cavity. If water pressure drops, bedrock stresses increase dramatically while the ice adjusts towards a new equilibrium. Stress in the bedrock is generally vertically oriented and compressive, leading to the development of tensile deviatoric stresses in the direction of glacier flow, so that fragments will spall off approximately parallel to the back of the step. If the cavity is connected to a basal meltwater system fed from the surface, large diurnal fluctuations in water pressure can occur during the ablation season (section 3.5).

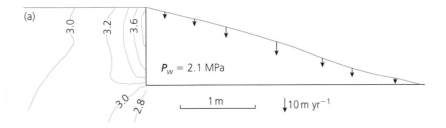

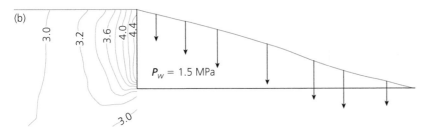

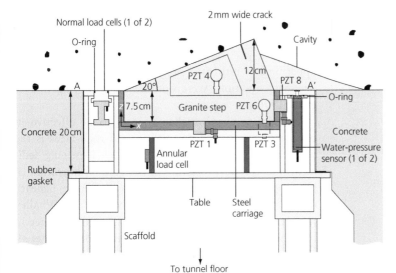

Figure 8.8 Modelled principal stresses in bedrock upstream of a step cavity. (a) Steady-state case, where water pressure in the cavity (P_w) = 2.1 MPa. Principal stresses are at a maximum adjacent to the step. Downward-pointing arrows show the vertical component of ice flow in the cavity roof. (b) Stress pattern associated with a sudden drop of water pressure to P_w = 1.5 MPa. Note the dramatic increase in principal stresses and vertical ice velocities (Modified from Iverson, 1991a).

Figure 8.9 Diagram of artificial rock step used in subglacial quarrying experiments (Redrawn from Cohen et al., 2006).

Iverson (1991a) also argued that water located in joints and microfractures plays an important role in the fracture of bedrock in the vicinity of subglacial cavities. He showed that water pressure in rock fractures cannot change as rapidly as that in larger cavities, because there is a limit to the rate at which water can flow into or out of narrow fissures. This means that if the water pressure at the bed fluctuates, there will be a delay during which the water pressure in rock fractures will differ from that in adjacent cavities. This delay will be greatest for massive rocks with narrow fractures, and least for porous, well-jointed sedimentary rocks. If the water pressure at the ice–bed interface falls, water pressure in rock fractures will remain relatively high while the system adjusts. This will tend to weaken the rock, because the water will support some of the confining pressure, reducing the rock-to-rock friction across the fracture (this effect is summarized in the Coulomb equation; section 4.2.3). At the same time, the reduced pressure in the adjacent cavity will mean that the proportion of the weight of the glacier

formerly supported by pressurized water will be shifted to the bedrock surface, increasing the local shear stress. The combination of reduced rock–rock friction, higher shear stresses and the presence of low-pressure cavities will create ideal conditions for rock fracture. In addition, high-pressure water in rock joints will also encourage long-term subcritical crack growth, further weakening the rock. Quarrying is even more likely to occur if falling cavity water pressures coincide with the passage of a large particle over the rock step.

Iverson's model of the quarrying process has been very elegantly confirmed by experiments conducted in Svartisen Subglacial Laboratory by Cohen et al. (2006). A protruding rock step, with a shallow crack cut parallel to the lee face, was installed at the glacier bed and allowed to be enveloped by the ice (fig. 8.9). A cavity in the lee of the step could be pressurized by water pumped from the tunnel below, and allowed to contract when pumping ceased. Effective pressure on the rockbed and water pressure in the cavity were continuously monitored. Propagation of the

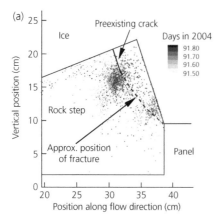

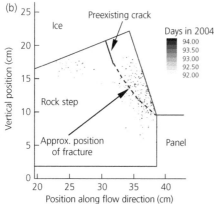

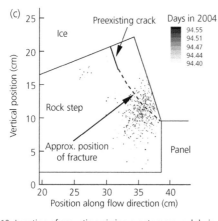

Figure 8.10 Location of acoustic emission events measured during successive pump tests. Shading indicates timing of events (Cohen et al., 2006).

crack was detected using an acoustic emission monitoring system, which picks up high-frequency waves generated by rapid release of stored strain energy. The monitoring system allowed the source of emission events to be located precisely, and showed that the crack propagated downwards during periods of falling cavity water pressure and increased loading on the step (fig. 8.10).

The experimental results of Cohen et al. (2006) add weight to the conclusions of Hallet (1996), who modelled quarrying rates for a glacier bed composed of step cavities and rock ledges. According to the model, quarrying rates are small when there are few cavities at the bed, because average stresses on the area of ice–rock contact are relatively small. When cavities form on larger areas of the bed

(i.e. when higher volumes of pressurized water cause more extensive ice–bed separation), the basal stresses have to be supported on a smaller area of ice–rock contact, elevating the stress concentrations at these points and increasing the likelihood of crack growth and rock fracture. For very extensive cavity formation, quarrying rates diminish, because the area affected by high loads is very small. There are few empirical data on subglacial quarrying rates that would allow testing of this model, but Hallet has argued that conditions favourable for rapid quarrying existed below Variegated Glacier during the 1982–83 surge, when effective pressures were small and an extensive system of linked cavities is thought to have existed beneath the glacier (section 5.7.3; Kamb et al., 1985; Kamb, 1987; Humphrey and Raymond, 1994). During the surge, the sediment flux in the proglacial stream was very large, equivalent to basal erosion rates of about 0.5 m per 17-year surge cycle. Since erosion rates during quiescent phases are likely to have been small, it was concluded that quarrying rates may have been as high as 100 mm yr^{-1} during the surge. Such rates are at the upper limit of estimated subglacial erosion rates, indicating that conditions were optimal for quarrying (Hallet et al., 1996).

Another mechanism that appears to be effective in removing fragments from well-jointed rock faces is the squeezing of glacier ice or subglacially deforming sediment into crevices, resulting in the gradual liberation of blocks. Ice injection has been reported from beneath Øksfjordjøkelen in Arctic Norway by Rea and Whalley (1994), and soft sediment injection has been implicated in bedrock plucking at a variety of locations by Evans et al. (1998).

8.2.4 EROSION BENEATH COLD ICE

Sliding is negligible or non-existent below cold-based ice, due to the high adhesive strength of ice and rock below the pressure-melting point. Therefore, abrasion will not be an effective erosive mechanism beneath predominantly cold-based glaciers. Drewry (1986), however, has pointed out that clasts embedded in the base of cold-based ice will be subject to some drag force because the ice above the bed will be shearing by internal deformation. This drag force may be sufficient to cause rotation and slip of the clast over the bed, possibly resulting in a limited amount of abrasion. In comparison with predominantly wet-based glaciers, however, the amount of abrasion will be negligible.

In contrast, plucking processes can be effective under certain conditions below cold ice. The main requirement is that the shear strength of at least part of the substratum should be low enough to allow deformation or failure under glacially imposed stresses. This condition is most likely to be met for well-jointed rocks or frozen sediments. Boulton (1979) has described glacial plucking of frozen dune sands below Wright Lower Glacier in the Dry Valleys region, East Antarctica, at temperatures of −15 °C. The sands are partly cemented by ice, but also contain ice-free pores and ice lenses derived from blown snow. The ice–sand interface is relatively smooth, but with a few hummocks reflecting the original dune morphology. In several places, blocks of sand have been displaced and incorporated within the ice, with varying amounts of disaggregation. There was evidence for both ductile deformation and brittle failure of the sands. Blocks were found up to 2 m above the bed, and appear to have been incorporated within the ice by the differential ice flow around the flanks of hummocks.

Fitzsimons et al. (1999, 2001) reported very thorough investigations of basal ice below Suess Glacier, also in the Dry Valleys. Layers of debris-rich ice, boulders and blocks of frozen sediments occurred over 2 m above the glacier bed, clearly indicating subglacial erosion and debris entrainment. Laboratory shear tests on the basal ice facies showed that, except for a distinctive layer of amber ice at the very top of the sequence, the ice was stronger than the overlying clear ice. Measured velocity profiles in the basal ice layers, however, revealed that much of the displacement was focused in thin horizons, generally between layers with contrasting physical characteristics or debris content. In some cases, slickensides and offset structures provided clear evidence of slip between layers. An interesting feature of the basal ice is the presence of gas-filled cavities between boudinaged blocks or at interfaces between layers. The cavities indicate failure in association with rheological contrasts between adjacent layers – for example, fracturing within stiff frozen sediment sandwiched between ductile ice layers, or failure along the boundary between layers with different rheology. It thus appears that differential strain response within heterogeneous basal materials can cause high enough stress gradients at rheological boundaries to cause failure. In contrast, homogeneous materials appear to undergo only slow creep under the same applied external force. If rheological contrasts exist within the substrate below the glacier (due, for example, to variations in grain size or ice content), differential strain response to imposed stresses could induce failure, leading to plucking and debris entrainment.

8.2.5 EROSION OF SOFT BEDS

As we saw in section 4.5, unlithified bed materials can be mobilized by glacial stresses due to the failure of sediment at its yield stress. This can occur by the development of failure planes below the ice–bed interface, resulting in intact fragments of unconsolidated sediment or weak rock being detached along a failure plane or décollement surface. Detached fragments will not be disaggregated during transport if their yield strength exceeds the imposed

stresses. The fragments can have greater frictional or cohesive strength than the surrounding material due to differences in consolidation or grain size. The grain size of the material is crucial to its drainage characteristics and therefore its resistance to disaggregation. Sands, gravels and diamicts tend to be more resistant to disaggregation than silts and clays, owing to higher coefficients of friction and/or higher permeabilities (Boulton and Hindmarsh, 1987; Benn and Evans, 1996). Rafted fragments (or rafts) vary greatly in size, from the centimetre scale (soft clasts, lenses, inclusions or boudins) up to several kilometres (mega-blocks) (see sections 10.3.3 and 11.3.2.6). In some cases, rafts and other fragile inclusions such as mollusc shells can be transported subglacially without disaggregation, although they may be eroded at their edges. In other situations, rafts can undergo some deformation, but with lower cumulative strains than the surrounding material, behaving like augen or boudins in metamorphic rocks and giving rise to *tectonic lamination* (Hart and Boulton, 1991; Hart and Roberts, 1994; Boulton et al., 2001a). The erosion of soft beds results ultimately in the creation of overdeepened sections of the glacier substrate, either by incremental removal of rafts (*excavational deformation*; Hart et al., 1990) or by wholesale removal of thrust blocks – concepts that we visit in more detail in sections 10.3.3 on the production of glacitectonite and 11.3.2 on glacitectonic processes and landforms. The erosional signatures of such processes are covered in section 8.5.2.

8.3 SMALL-SCALE EROSIONAL FORMS

A very wide range of small-scale erosional forms have been recognized, and a variety of descriptive terms are used in the literature (e.g. Laverdiere et al., 1979, 1985). In this book, we have adopted the most widely used terminology, and describe small-scale forms under four headings:

(1) striae and polished surfaces;
(2) rat tails;
(3) chattermarks, gouges and fractures;
(4) P-forms.

The general characteristics of many small-scale forms are depicted in figure 8.11.

8.3.1 STRIAE AND POLISHED SURFACES

Striae (singular: *striation*) are scratches incised into bedrock or clast surfaces, and have long been recognized as evidence for scoring by particles embedded in glacier ice. They are direct results of subglacial abrasion and, according to experimental studies and field observations, can be eroded rapidly if optimum conditions are satisfied (Boulton, 1974; Atkinson, 1984; Iverson, 1990). Striated

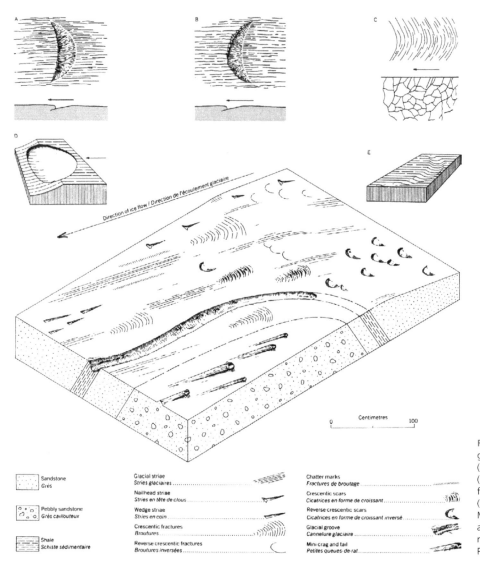

Figure 8.11 Small-scale forms of glacial erosion. (A) Lunate fracture (plan and section). (B) Crescentic gouge (plan and section). (C) Crescentic fractures (plan and section). (D) Conchoidal fracture. (E) Sichelwanne. Main diagram: striae, fractures, grooves and associated features and their relationships to bedrock lithology (After Prest 1983; Embleton and King, 1975)

rock surfaces often display areas of polishing which when viewed under the microscope comprise a surface of densely spaced micro-striae.

Close inspection of striae reveals that their edges and bases are rough and composed of numerous arcuate fracture surfaces, attesting to their excavation by indentation fracture beneath asperities in overpassing clasts (section 8.2.2; fig. 8.6). Rea (1996) has also demonstrated how polished surfaces may occur in striae. Indentation theory suggests that the width and depth of striae depend on the shape of the asperity, and that they increase with the load pressing the particle against the bed (Drewry, 1986). Striae may widen downglacier due to the progressive blunting of the indenting asperity as it is dragged across the bed. The widening may be gradual (*wedge striations*) or abrupt (*nail-head striations*), depending on the rate of asperity blunting relative to the velocity of the abrading clast (fig. 8.11). Rarely, striae become narrower downglacier. The reasons for this are unclear, although one possibility is that they reflect reductions in the normal force pressing the indenting asperity against the bed.

Striae also commonly terminate abruptly at a deep, blunt end next to the narrow up-ice end of another striation. This *en echelon* pattern is thought to relate to the rotation or 'flip-out' of the striating clast, which lifts one asperity clear of the substrate, but brings a new, sharper asperity in contact along an adjacent flow line. Although individual striae may maintain a straight line for more than a metre, they sometimes deviate from the mean flow direction if the striator clast has been rotated while still in contact with the substrate.

The orientations of striae may vary considerably on a single rock outcrop. While striae on the flat upper surface of an outcrop are commonly parallel with each other and deviate only slightly from the average ice flow direction, striae on uneven surfaces generally deviate markedly due to the irregularities in the basal flow of the glacier (fig. 8.12; Rea et al., 2000).

Although multiple striae directions can be produced contemporaneously by complex flow over a rough bed (Rea et al., 2000), bedrock outcrops often display two or more sets of striae relating to separate ice flow events

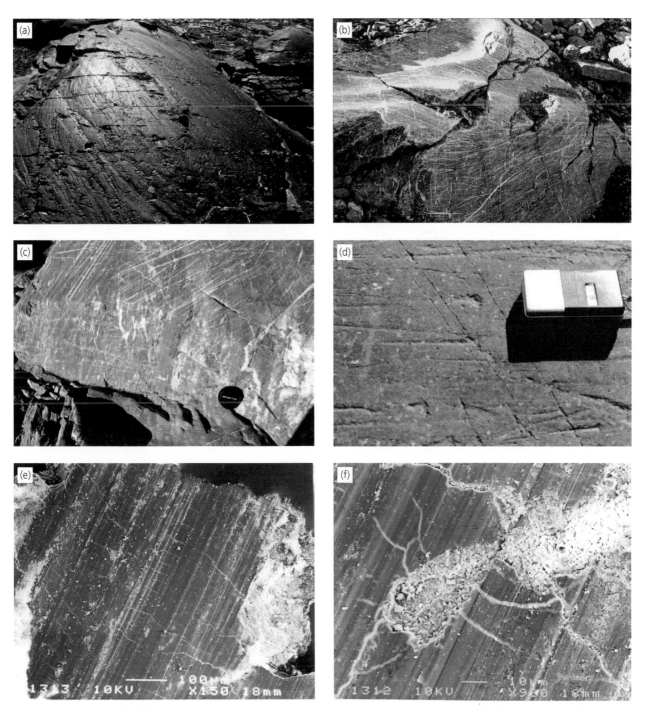

Figure 8.12 Glacial striae and polish. (a) Cross-cutting striae on moulded bedrock, Chandra Tal, Lahul Himalaya. (b) Variable striation directions reflecting complex ice flow around a bedrock bump, St Jonsfjorden, Svalbard. (c) Multiple striation directions on a smoothed surface and lee-side cavity face, St Jonsfjorden, Svalbard. Note three ice flow directions on smoothed surface and transverse ice flow on the lee-side face. (D.J.A. Evans) (d) Nail-head striae, Quebec, Canada. (C. Laverdiere) (e) SEM photomicrograph of part of a polished surface within a striation, showing numerous micro-striations (<10 µm wide). (Rea, 1996) (f) SEM photomicrograph of a polished surface from the study of Rea (1996), demonstrating that the polish is 'plastered' onto the bed, because it has peeled off to reveal rock flour below (B.R. Rea).

(fig. 8.12). The sequence of striating events cannot always be determined, except where shallow striae cut across deeper striae produced by older ice flows. Cross-cutting relationships may result from shifts in ice divides or dispersal centres during a single glaciation, or from separate glacial episodes. Many glacial geomorphologists have employed multiple striae directions in regional reconstructions of former glacier flow (e.g. Veillette et al., 1999;

Jansson et al., 2002). For example, well-preserved records of cross-cutting striae in Scandinavia allow shifts in ice dispersal centres during the Weichselian and Younger Dryas to be reconstructed (Anundsen, 1990; Kleman, 1990). Complex patterns of multiple ice flow directions have been reported from the Canadian Atlantic provinces by Stea (1994), who combined striae patterns with data on till provenance and glacially streamlined landforms.

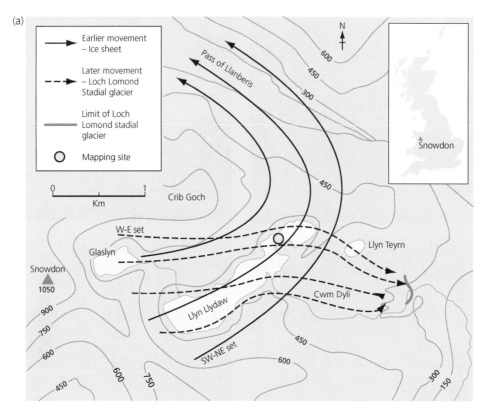

Figure 8.13 Cross-cutting striae on a bedrock outcrop in the Snowdon massif, Wales. (a) Location map. (b) Maps of bedrock outcrop, showing: (i) pattern of striae produced by ice sheet flow; (ii) pattern of striae produced by local, Younger Dryas valley glacier; (iii) flow pattern inferred from striae in lee-side cavities (Redrawn from Sharp et al., 1989a).

Two separate sets of cross-cutting striae have been documented by Sharp et al. (1989a) in the Llyn Llydaw cirque complex, North Wales. The first set records former ice flow from south-west to north-east at the base of the last (Dimlington Stadial) ice sheet, whereas the overlying set were cut by a smaller Loch Lomond Stadial glacier flowing from west to east (fig. 8.13). The preservation of earlier striae attests to the lower sliding velocity and lower erosional capacity of the Loch Lomond Stadial cirque glacier. Sharp et al. (1989a) have used these cross-cutting striae in an attempt to quantify the amount of erosion by the Loch Lomond Stadial glacier. They argued that the almost total absence of striae of <0.4 mm width in the Dimlington Stadial set suggests that they have been erased by the Loch Lomond Stadial ice, placing a limiting value on the amount of erosion that occurred during that event.

Polished surfaces appear as a shine, gloss or lustre (fig. 8.12a) on abraded rock outcrops. Some apparently polished surfaces may reflect precipitation of solutes from subglacial meltwater (Hallet, 1976a). Support for a precipitate origin comes from the fact that some parts of polished surfaces actually lift off after prolonged periods of subaerial exposure.

8.3.2 RAT TAILS

Rat tails are small, residual longitudinal ridges extending down-ice from resistant rock knobs or nodules (figs 8.11

and 8.14). They are essentially small-scale equivalents of crag and tails (section 8.4.3), created by the removal of less resistant material to either side. In some examples, elongate scalloped troughs curve round the up-ice side and flanks of the rat tail in an elongate sickle-shaped depression. Most researchers regard rat tails as the product of differential abrasion of bedrock surfaces, and the presence of lateral troughs as evidence for small-scale streaming of ice around the obstacle.

8.3.3 CHATTERMARKS, GOUGES AND FRACTURES

Fracture marks or cracks in bedrock record the removal of rock flakes by subglacial quarrying (fig. 8.11; section 8.2.3). These are variably known as chattermarks, crescentic gouges or crescentic fractures (*sichelbruche* or *parabelrisse*), conchoidal fractures (*muschelbrüche*) and lunate fractures. *Chattermarks* are crescent-shaped fractures, usually a few centimetres across, with their open or concave sides facing down-ice. They commonly occur as a series of closely spaced fractures nested one inside the other, resulting from repeated fracture events beneath a single overpassing clast, possibly in association with 'stick-slip' motion by the overriding ice (figs 8.11 and 8.15). The spacing between chattermarks is commonly remarkably consistent, suggesting that fracture events

(b) (i)

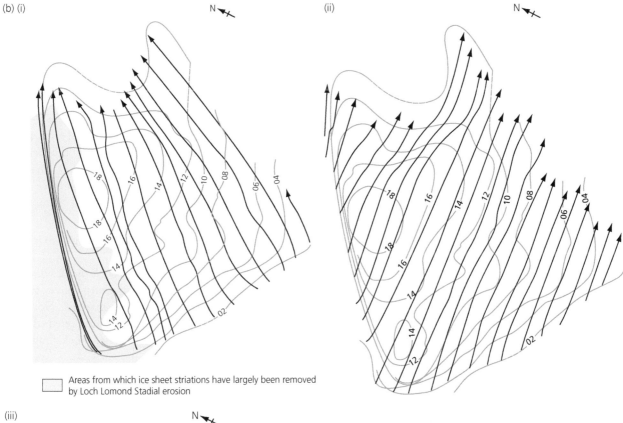

Areas from which ice sheet striations have largely been removed by Loch Lomond Stadial erosion

(iii)

Figure 8.13 (*Continued*)

Figure 8.14 Rat tails in Quebec, Canada, showing ice flow from left to right (C. Laverdiere).

occur at regular intervals, possibly in association with diurnal variations in subglacial water pressure.

Larger arcuate gouges and fractures may measure from a few centimetres to more than a metre across. *Crescentic gouges* and *lunate fractures* are similar except that their horns are aligned in opposite directions; the horns of lunate fractures point down-ice, whereas those of crescentic gouges point up-ice (fig. 8.15). Both may be present together on the same rock outcrop. Crescentic gouges and lunate fractures are bounded by two fracture planes: one that dips in a down-ice direction (principal fracture) and one that is vertical and constitutes the down-ice termination of the principal fracture (figs 8.11a, b). The principal fracture plane may be gently or steeply dipping, but it always dips down-ice, providing a useful method of reconstructing former glacier flow direction. *Conchoidal fractures* are produced when the fracture plane is concave upward, the final form telling us very little about

Figure 8.15 Cracks and gouges. (a) Crescentic fractures and striae showing ice flow from left to right. (C. Laverdiere) (b) Crescentic gouges, Nigardsbreen, Norway, showing ice flow from right to left. (c) Chattermarks in grooves across striae, St Jonsfjorden, Svalbard, showing ice flow from left to right (D.J.A. Evans).

former ice flow direction. *Crescentic fractures* are arcuate fracture planes which may dip either up-ice or down-ice and may have their horns pointing upglacier or downglacier. Fractures with up-ice pointing horns commonly occur in isolation, whereas those with down-ice pointing horns generally occur in a series, the width of individual fractures decreasing in a down-ice direction (fig. 8.15). Series of crescentic fractures therefore provide a clear sense of former ice flow direction.

8.3.4 P-FORMS

Dahl (1965) introduced the term *P-forms* or *plastically moulded forms* to refer to smoothed depressions eroded into bedrock, reflecting his impression that they are formed by a plastic medium or media. Kor et al. (1991) introduced the alternative, non-genetic term *S-form* or *sculpted form*, but although we recognize that P-forms probably originate by a variety of glacial and fluvial erosional processes, we continue to use the original, widely accepted term. P-forms exhibit a wide variety of shapes and sizes, and may be classified into three broad types, according to whether they are parallel or transverse to ice flow, or non-directional (fig. 8.16).

(1) *Transverse forms* are aligned at right angles to ice flow and include:
  * *muschelbrüche* (singular: *muschelbrüch*), which are shaped like mussel shells, with sharp convex-upflow rims and indistinct downflow margins. Note that muschelbrüche have also been interpreted as scars left by the removal of quarried rock fragments (section 8.3.3);
  * *sichelwannen* (singular: *sichelwanne*), which are sickle-shaped depressions, often containing striae, with horns pointing downglacier (fig. 8.17a);
  * *comma forms*, which are similar to sichelwannen, but with one horn missing or less well-developed than the other;
  * *transverse troughs*, which are elongate hollows with a relatively steep, planar upflow slope and a gentler downflow slope which is commonly scalloped.
(2) *Longitudinal forms* are aligned parallel to ice flow, and include:
  * *spindle flutes* (fig. 8.17e), which are spindle-shaped depressions with a pointed end oriented upflow and broadening downflow to a distinct, rounded termination (closed spindles), or a smooth, gradual ramp (open spindles);
  * *cavettos*, which are curvilinear, undercut channels eroded into steep or vertical rockfaces; they may contain striae and crescentic gouges;
  * *furrows*, which are elongate, flow-parallel grooves which may be straight, curved or winding in planform (figs 8.17b, c); they may have smaller P-forms or striae on their floor and walls.
(3) *Non-directional forms* are of two main kinds:
  * *undulating surfaces*, which are low-amplitude undulations found on the lee sides of bedrock humps;
  * *bowls and potholes*, which are near-circular, deep depressions a few centimetres to several metres deep (fig. 8.17d).

Although P-forms vary widely in size and collectively may cover very large areas (fig. 8.18), they are mostly small-scale features. The exception to this is the bedrock *mega-groove* (figs 8.17 f, g), which may exceed 10 km in

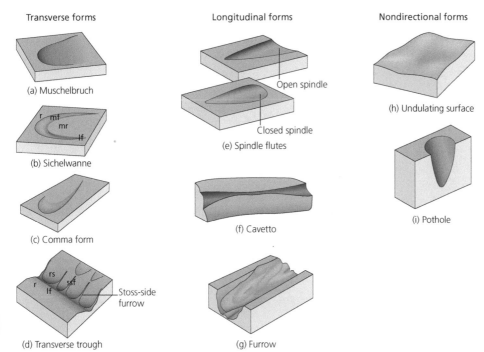

Transverse forms

(a) Muschelbruch

(b) Sichelwanne

(c) Comma form

(d) Transverse trough

Longitudinal forms

Open spindle

Closed spindle

(e) Spindle flutes

(f) Cavetto

(g) Furrow

Nondirectional forms

(h) Undulating surface

(i) Pothole

Figure 8.16 Classification scheme for P-forms (after Kor et al., 1991)

Figure 8.17 Examples of P-forms. (a) Sichelwanne at Loch na Keal, Isle of Mull, Scotland. (b) Curved channels at Loch na Keal. (Landform Slides) (c) Groove with chattermarks, Maringouin Cape, Bas Saint-Laurent, Canada. (C. Laverdiere) (d) Large pothole, Finland. (J. Väätäinen, from Andersen and Borns, 1994, reproduced with permission of Scandinavian University Press) (e) Large bedrock flutes with lateral furrows accentuated, Wilton Creek, Ontario. (Shaw, 1988a, reproduced with permission of the *Canadian Journal of Earth Sciences*) (f) The Kelleys Island bedrock grooves, Lake Erie, Ohio, USA. Ice flow was away from the viewer. (g) Detail of the Kelleys Island bedrock grooves, Lake Erie, Ohio, USA. Ice flow was towards the viewer. (T.G. Fisher) (h) Bedrock mega-grooves east of Loch Broom, north-west Scotland, as depicted by NEXTMap land surface elevation data. Inset view is a ground photograph of the mega-grooves (After Bradwell et al., 2008a).

Figure 8.17 (*Continued*)

length and tens of metres wide and deep. Bedrock mega-grooves are all the more impressive in that they are linear features that run sub-parallel, often across the grain of the bedrock into which they have been eroded (e.g. Goldthwait, 1979; Bradwell, 2005; fig. 8.17 h).

Four media have been invoked to explain the erosion of P-forms (Gjessing, 1965; Gray, 1981): (1) *debris-rich basal ice*; (2) *saturated till* flowing between the ice base and bedrock; (3) *subglacial meltwater* under high pressure; and (4) *ice-water mixtures*. Gjessing (1965) concluded that most forms reflect erosion by saturated till, which he envisaged moving as a viscous liquid, driven by pressure differences at the glacier bed. Subsequent studies have tended to stress the importance of either debris-rich ice or flowing water, media with strikingly different viscosities and flow behaviour.

Several early research papers on P-forms (e.g. Dahl, 1965) concluded that they are the product of subglacial meltwater erosion, and this was an especially attractive interpretation wherever P-form assemblages include potholes (e.g. Gjessing, 1967). Potholes on former glacier beds are morphologically similar to those scoured out by turbulent subaerial streams, and have detailed features strongly suggestive of fluvial flow. For example, some potholes have helical grooves incised on their walls (Kor et al., 1991). Some subglacial potholes might mark the position of *plunge pools* at the lower end of moulins.

Some workers have argued that all P-forms are eroded by subglacial meltwater, due to their morphological similarity to scours eroded by fluids (Shaw and Kvill, 1984; Shaw and Sharpe, 1987; Bradwell, 2005;

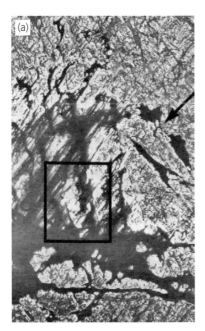

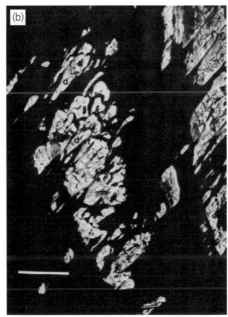

Figure 8.18 Aerial photograph of the Henvey Inlet area, Ontario, showing widespread P-forms. (a) General view showing furrows and intervening rock drumlins. Ice flow was towards the SW (arrow), cutting across the NW–SE trending rock structure. Arrow length equivalent to 0.5 km. (b) Detail of boxed area showing a group of sichelwannen (s), furrows (f) and rock drumlins (d) (Kor et al., 1991).

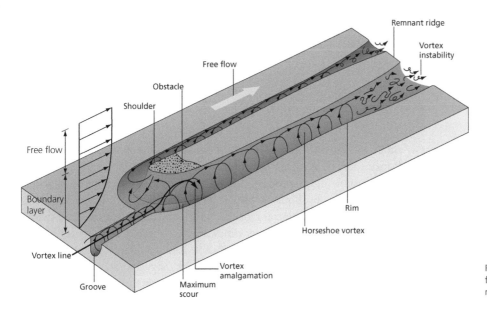

Figure 8.19 Proposed mechanism for the erosion of sichelwannen by meltwater (After Shaw, 1994)

Munro-Stasiuk et al., 2005). The possibility that sichelwannen could be eroded by flowing water was demonstrated experimentally by Hjulstrom (1935), who reported cavitation erosion by turbulent water flowing around a bolt in a metal plate. Shaw (1988b, 1994), Sharpe and Shaw (1989) and Kor et al. (1991) argued that P-forms are eroded by vortices in turbulent subglacial meltwater streams which impinge on the rockbed over short distances. Sichelwannen and comma forms were explained as the products of erosion by horseshoe vortices set up in subglacial meltwater by bedrock obstacles in the boundary layer (fig. 8.19). Even bedrock mega-grooves in Scotland have been explained as Nye channels (Bradwell, 2005) based predominantly on their similarity to fluvial rock channels (Wohl, 1998), although the linearity and sub-parallel alignments of the mega-grooves are unlike the forms of previously reported tunnel channels and valleys (see below). Indeed Bradwell et al. (2008a) have more recently reassessed the mega-grooves of Scotland as features likely eroded by flow-parallel debris bands in ice streams (cf. Catania et al., 2005), due to the fact that they are part of a landform suite that can be used to demarcate the beds of palaeo-ice streams in the region. The fluvial hypothesis is based almost entirely on a *form analogy* between P-forms and non-glacial forms scoured by fluid flows, and the similarity between such forms and scours eroded around obstacles by wind and water are indeed very striking.

The presence of striae in P-forms is often cited as evidence of their erosion by glacier ice (e.g. Goldthwait, 1979),

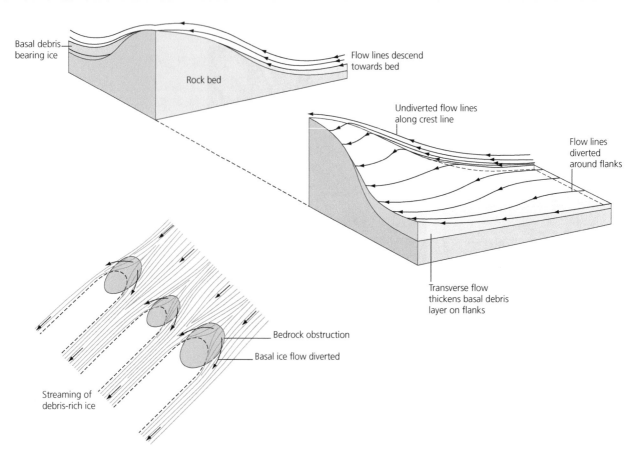

Figure 8.20 Lateral and vertical migration of debris caused by the streaming of basal ice around bed obstructions (Redrawn from Drewry, 1986, after Boulton, 1979).

but opponents of the ice-erosion hypothesis allude to the fact that striae occur only on parts of many P-forms and often do not occur at all, suggesting that glacial erosion merely ornamented the P-forms after they were cut by meltwater (Shaw 1988a, 1994; Munro-Stasiuk et al., 2005). It has also been argued that many P-forms are too tortuous to have been eroded by basal ice. However, Boulton (1974) observed P-forms in contact with debris-rich basal ice beneath Breiðamerkurjökull and Glacier d'Argentière. Boulton (1974, 1979) argued that P-forms are eroded where basal debris concentrations are higher than average due to differential ice flow around bedrock protuberances. Ice flows more rapidly around the sides of obstacles than over the tops, concentrating debris-rich ice in the troughs between obstacles, providing conditions favourable for the erosion of cavettos and grooves on the flanks of roches moutonnées (fig. 8.20). In addition, Boulton (1974) observed that sichelwannen in subglacial positions occur near points of cavity closure, which he considered to reflect enhanced basal ice pressure against the bed and streaming of debris-rich ice around bedrock obstructions. In Boulton's view, it is the enhanced plastic flow of basal ice around obstructions and its separation into debris-rich and debris-poor streams that is responsible for the differential abrasion of bedrock substrates producing most P-forms. He accepts that potholes and other fluvial forms may coexist, but they relate to phases

of fluvial erosion at the bed and should not be considered as glacially abraded P-forms merely because they are juxtaposed; potholes within other P-forms are explained by Boulton (1974) as the products of alternating fluvial and glacial erosion, which may take place on a yearly basis or over longer timescales. Furthermore, where ice streams over the top of a pothole it may undergo pressure fluctuations and therefore P-forms may emanate from potholes.

If glacier ice is responsible for the erosion of P-forms, then abrasion forms like striae should occur not just on the walls of individual forms, but also in the tightest corners. It is sometimes claimed that ice is unable to stream in tight folds and in different directions over small distances, presenting a major flaw in the ice abrasion hypothesis. However, observations made by Rea and Whalley (1994) in a large cavity beneath an outlet of Øksfjordjøkelen showed that ice can be squeezed through narrow openings and can turn acute corners into cavities, resulting in a wide range of striae alignments over small distances (fig. 8.21). Furthermore, Rea et al. (2000) have documented striae cut at 90° to the main glacier flow direction due to localized ice deformation into transverse trenches. Striae cut vertically into the lee-side faces of rock steps in the proglacial area of Øksfjordjøkelen attest to the abrasion of widening joint systems by small ice streams moving in a totally different direction to the main glacier flow (Rea, 1994; Rea and Whalley, 1994; Rea et al., 2000).

Figure 8.21 Striae conforming to a P-form on the side of a whaleback, St Jonsfjorden, Svalbard. Ice flow from right to left (D.J.A. Evans).

Figure 8.22 Roche moutonnée at the head of Moraine Fjord, South Georgia (D.J.A. Evans).

A hybrid origin for at least some P-forms was proposed by Boulton (1974), who grouped them into (1) glacially abraded forms such as cavettos, sichelwannen and troughs; (2) fluvially eroded forms such as potholes; and (3) intermediate forms, which are less easy to classify, but have been affected by fluvial and glacial processes. Boulton's approach accepts the role of both subglacial fluvial erosion and glacial abrasion, but assigns different P-forms to those processes. The fact that some of the forms described by Shaw (1988a, 1994) carry a strong fluvial signature is not in doubt, but those who advocate fluvial erosion exclusively for all P-forms (e.g. Munro-Stasiuk et al., 2005, who go so far as to propose a fluvial origin for striae) are rejecting some of the rare and invaluable subglacial observations and laboratory experiments available to glacial research. Because the P-forms of the Isle of Mull, Scotland, contain fluvial and glacial signatures, Gray (1992) proposed a two-phase evolution regardless of the specific P-form type: fluvial erosion by corrasion and cavitation followed by glacial abrasion. Such interpretations are prompted by the fact that striae commonly only partially cover the surfaces of P-forms, possibly recording partial ornamentation of fluvial forms by glacier ice. However, the distribution of striae in grooves is perhaps better explained by the fact that the glacially eroded grooves acted as drainage routeways during deglaciation and underwent cosmetic changes by limited meltwater erosion.

8.4 INTERMEDIATE-SCALE EROSIONAL FORMS

Intermediate-scale erosional forms reflect the interaction between geology, topography and patterns of ice and water flow. As such, they can yield important insights into former glacial conditions, particularly when morphological studies are combined with modern glaciological theory. In this section, we consider roches moutonnées, whalebacks and rock drumlins, crag and tails, and channels.

Figure 8.23 Giant roche moutonnée on Deeside, Scotland. Note trees for scale (D. Sugden).

8.4.1 ROCHES MOUTONNÉES

Roches moutonnées (singular: *roche moutonnée*) are asymmetric bedrock bumps or hills with abraded up-ice or stoss faces and quarried down-ice lee faces (fig. 8.22). The name was first introduced by Horace-Bénédict de Saussure (1740–99), based on a fancied resemblance to the wavy wigs of the eighteenth century, which were called moutonnées after the mutton fat used to hold them in place. Roches moutonnées range in size from less than 1 m to several hundreds of metres across. Some researchers have regarded roches moutonnées as one part of an erosional continuum, ranging from small *asymmetrical rocks* to major *bedrock steps*, although the term roche moutonnée usefully describes asymmetric erosional forms at all scales. Examples of large roches moutonnées are the asymmetrical granite hills of New England (up to 1.3 km long and 50 m high; Jahns, 1943) and Deeside, north-east Scotland (up to 150 m high; Sugden et al., 1992; Glasser, 2002a; fig. 8.23). Very large asymmetrical hills, known as *flyggbergs*, occur in parts of Sweden and are up to 3 km long and 350 m high. Large asymmetric hills may have smaller roches moutonnées superimposed on their surfaces (Glasser and Warren, 1990).

The distribution of small-scale erosional forms on the surfaces of roches moutonnées in part of Finland has

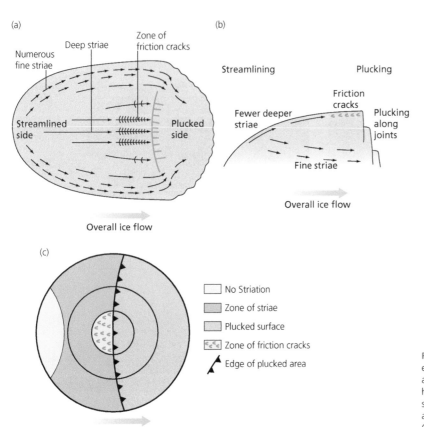

Figure 8.24 Idealized distribution of small-scale erosional forms on roches moutonnées in plan (a) and profile (b) view. (c) Stereographic model (upper hemisphere projection) of the distribution of small-scale erosional features on roches moutonnées, with arrow indicating ice flow direction (Redrawn from Chorley et al., 1984, after Rattas and Seppälä, 1981).

been studied by Rastas and Seppälä (1980). Striae are widespread on the stoss sides, except for steep, upglacier-facing surfaces, and striae and friction cracks occur together on gently sloping stoss-side surfaces (fig. 8.24). Polished facets are confined to the flanks and gently sloping surfaces of the lee sides, and plucked surfaces are found on steep, downglacier-facing parts.

The large-scale form and surface morphology of roches moutonnées reflect the distribution of stresses in bedrock humps beneath sliding ice, and the associated processes and patterns of rock failure. On the stoss side of humps, normal stresses are higher than average and particles held in basal ice are brought towards the bed, increasing the effectiveness of abrasion (section 8.2.2; Boulton, 1974; Hallet, 1979a). In contrast, normal pressures on the lee side are lower than average, encouraging the formation of cavities and suppressing abrasion. Instead, the presence of cavities on the lee side of bumps promotes crack propagation and fracture in the bedrock just upstream from the cavity, particularly under conditions of fluctuating water pressures (section 8.2.3; Iverson, 1991b; Sugden et al., 1992; Hallet, 1996). The preferential action of abrasion and quarrying on the stoss and lee sides, respectively, of bedrock bumps leads to the evolution of the classic roche moutonnée form. Study of preglacial joint structures in large roches moutonnées in New England by Jahns (1943) indicated that 33 m of rock had been removed by quarrying from the lee sides, com-

pared to a maximum of only 4 m on their abraded stoss ends. In this case, therefore, quarrying was a far more effective process of subglacial erosion than abrasion. Recent work suggests that quarrying is generally the more important process (Boulton, 1979; Drewry, 1986; Iverson, 1995), thus explaining why roches moutonnées are such widespread forms in glaciated hard-rock terrain.

The important role of cavities in the quarrying process suggests that roches moutonnées will tend to form where low-pressure cavities exist at the glacier bed. Such cavities are most likely below thin ice, where the average ice overburden pressure p_i is low, and in areas where subglacial water pressures undergo large fluctuations due to variations in the supply of meltwater from the glacier surface (section 8.2.3; Iverson, 1991b; Sugden et al., 1992). It may be expected, therefore, that roche moutonnée formation will be encouraged below thin, temperate valley glaciers and near the melting margins of ice sheets. Empirical support for this idea has been provided by Sugden et al. (1992), who noted that boulder trains originating from the lee sides of roches moutonnées in north-east Scotland extend only a few hundred metres downflow. Such short transport distances indicate that the trains relate to a pulse of erosion at the end of the last glacial cycle, when ice was thin and meltwater was abundant. A similar conclusion has been reached by Roberts and Long (2005), based on their study of roches moutonnées and whalebacks on the former bed of

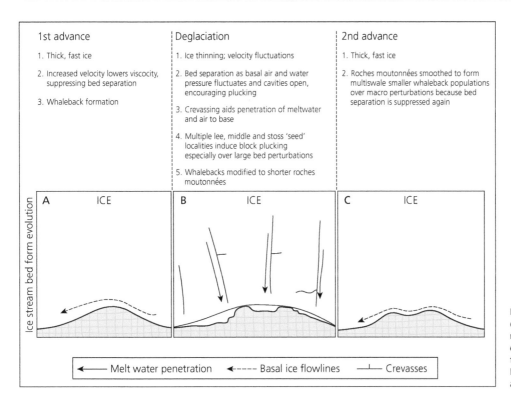

Figure 8.25 A model for the development of whalebacks and roches moutonnées based on erosional bedform distribution on the former bed of Jakobshavns Isbrae, Greenland (After Roberts and Long, 2005)

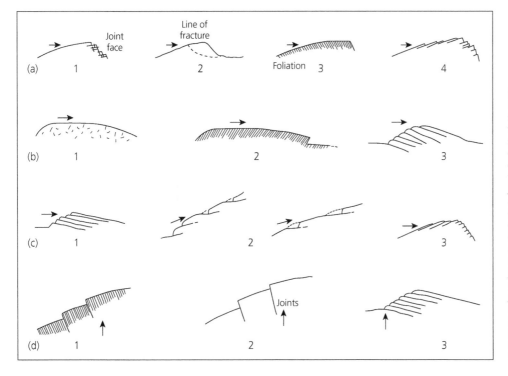

Figure 8.26 Schematic relationships between bedrock structure and glacial erosional forms. (a) Quarried lee slopes associated with (1) jointing; (2) massive rocks; (3) steep down-ice dipping foliation; (4) up-ice dipping joints. (b) Patterns of abrasion associated with (1) massive rocks; (2) steep up-ice dipping foliation; (3) down-ice dipping joints. (c) Quarried stoss slopes associated with (1) down-ice dipping joints; (2) up-ice dipping joints with layers truncated on up-ice sides; (3) up-ice dipping joints with layers truncated on down-ice sides. (d) Quarried lateral slopes (transverse cross-sections) associated with three structural patterns (After Gordon, 1981).

Jakobshavns Isbrae, Greenland. They conclude that roches moutonnées form under thin ice during deglaciation and that whalebacks (see section 8.4.2) indicate thicker ice where bed separation and cavity production is suppressed. Roberts and Long (2005) also identified a relationship between increased erosional bedform density and thick, fast-flowing ice (fig. 8.25), implying that zones of former fast glacier flow may be mapped

using erosional bedform characteristics (cf. I.S. Evans, 1996; Hall and Glasser, 2003).

The morphology of roches moutonnées is also strongly influenced by bedrock structure (Gordon, 1981; Rastas and Seppälä, 1981; Glasser and Warren, 1990; Sugden et al., 1992). Quarrying and plucking are encouraged by the presence of favourably oriented joint systems and other pre-existing fracture surfaces (fig. 8.26). In some situations, limited plucking can even occur on stoss-side surfaces. The influence of jointing on the evolution of a granite roche moutonnée was reconstructed by Sugden et al. (1992) and is shown in figure 8.27. The initial stages of quarrying exploit sheet joints developed parallel to the original ground surface, yielding arcuate slabs. As erosion progresses, deeper vertical and horizontal tectonic joints are exposed, leading to the creation of a stepped lee-side profile. Olvmo and Johansson (2002) provided evidence for the influence of preglacial etching patterns on the shape of roches moutonnées and whalebacks, and proposed that glacial erosion is most effective when ice flow is coincident with etched structures, suggesting that deeply weathered rock structures play a significant role in the shaping of glacial erosion features (including P-forms; Olvmo et al., 1999; see also section 8.5 for influences on larger landforms).

Some researchers go somewhat further, by suggesting that roches moutonnées are essentially remnants of preglacial weathering which have been only been slightly modified by glacial erosion (e.g. Lidmar-Bergström, 1988; Lindstrom, 1988; Patterson and Boerboom, 1999). According to this view, roches moutonnées are immature glacial erosional forms that would be completely removed by lee-side cliff retreat under prolonged glaciation. Lindstrom (1988) regarded this 'weathering hypothesis' as an alternative to the 'classical glacial theory', but the two models are in fact complementary. Some roches moutonnées do appear to be slightly modified preglacial hills, but in many areas, such as the floors of cirques and troughs, they are clearly due entirely to differential glacial erosion. Between these end-members there is likely to be a continuum of forms with varying degrees of inherited topography.

8.4.2 WHALEBACKS AND ROCK DRUMLINS

Whalebacks and *rock drumlins* are elongate, smoothed bedrock bumps which lack the quarried lee faces characteristic of roches moutonnées (Sugden and John, 1976; I.S. Evans, 1996). Whalebacks are approximately symmetrical, looking rather like the backs of whales breaking the ocean surface (fig. 8.28), whereas rock drumlins (also known as *tadpole rocks*; Dionne, 1987) are asymmetrical, with steeper stoss faces and gently tapering lee sides. Both are commonly ornamented with abundant striae, friction cracks and P-forms. Whalebacks up to 1 km long have been described by I.S. Evans (1996) from the Coast Mountains of British Columbia, Canada.

The absence of quarried lee faces on whalebacks and rock drumlins is thought to imply that low-pressure cavities did not exist at the glacier bed during their formation (I.S. Evans, 1996). Cavity formation is suppressed below thick ice where average ice overburden pressures are high, and in such situations abrasion may take place

over most of the bed, creating smoothed, symmetrical whalebacks. According to Evans, whalebacks can form below ice a few hundred metres thick, but are best developed where ice is 1–2 km thick. Asymmetrical rock drumlins are thought to occur where abrasion is focused on

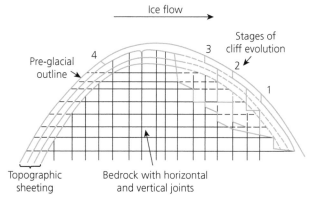

Figure 8.27 Conceptual model of evolutionary stages of the formation of roches moutonnées, based on examples in eastern Scotland. Stage 1 involves the removal of lee-side surface slabs and the exploitation of topographic sheeting. Stage 2 involves the up-ice migration of vertical faces with a limitation on the depth of crack propagation. Stage 3 involves continued back-wearing and successive cycles of deepening. Stage 4 involves the production of the typical roche moutonnée profile, with a staircase cliff and an abraded upper surface (Modified from Sugden et al., 1992).

Figure 8.28 Erosional bedforms. (a) Whaleback with the traces of minor plucked blocks on its surface, Konowbreen, St Jonsfjorden, Svalbard. (b) Rock drumlin on the foreland of Nigardsbreen, Norway (D.J.A. Evans).

the stoss side, and both abrasion and plucking are suppressed on the lee side. Such conditions may exist where either: (1) normal stresses over the lee side of bumps are low enough to inhibit abrasion, but too high to allow cavity formation; or (2) cavities form, but are not subject to the fluctuations in water pressure that encourage quarrying. The latter condition is most likely to occur where surface meltwater is prevented from reaching the bed, such as beneath ice streams in polar areas which have surface layers of cold ice. Whalebacks and rock drumlins are widespread on the floors of troughs in the Coast Mountains of British Columbia, which channelled ice streams draining the Cordilleran Ice Sheet during the last glaciation (I.S. Evans, 1996). Such forms may be characteristic of erosion beneath thick, fast-flowing ice streams and outlet glaciers, as proposed by Roberts and Long (2005) to explain the distribution of whalebacks and roches moutonnées on the former bed of Jakobshavns Isbrae, Greenland.

Several other ideas have been proposed to explain the formation of whalebacks and rock drumlins. These include:

(1) the survival of preglacial bedrock hills (Lindstrom, 1988), an explanation that complements theories of larger-scale landscape inheritance (e.g. Lidmar-

Bergström, 1997; Olvmo et al., 1999; Johansson et al., 2001a, b; Olvmo and Johansson, 2002);

(2) the remodelling of roches moutonnées and the removal of quarried faces due to changing ice flow directions (Anundsen, 1990);

(3) bedrock structure which is unfavourable to the development of plucked lee faces (Gordon, 1981).

Subglacial fluvial erosion was also considered by Kor et al. (1991), because rock drumlins are often ornamented with what they interpret as fluvially cut P-forms. The occurrence of P-forms, however, may simply reflect efficient abrasion due to high overburden pressures, high sliding velocities and large areas of intimate ice–bed contact beneath thick ice (sections 8.2.2 and 8.3.1).

8.4.3 CRAG AND TAILS

Erosional crag and tails are elongate, streamlined hills consisting of a resistant bedrock crag at the up-ice end, and a tapering tail of less resistant rock extending down-ice. They are produced by the streaming of ice around the obstacle, and the protection of the 'tail' from erosion. The classic example of a crag and tail is Edinburgh Castle and the Royal Mile (fig. 8.29a; Sissons, 1967; Evans and Hansom, 1996). In this case, the 'crag' is a volcanic

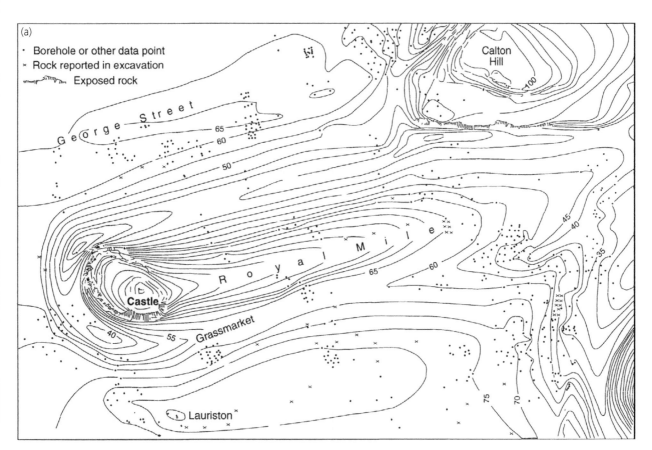

Figure 8.29 Crag-and-tail features. (a) Map of the bedrock surface around Edinburgh Castle and the Royal Mile, showing a classic crag-and-tail morphology. (Evans and Hansom, 1996, after Sissons, 1971). (b) Plan views of typical horned crag and tails from Quebec-Labrador, Canada. More immature features possess poorly developed horns (After Jansson and Kleman, 1999).

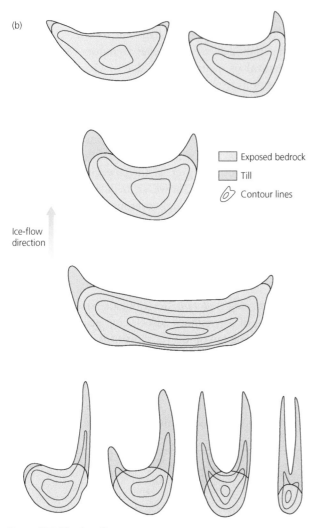

(b)

Exposed bedrock

Till

Contour lines

Ice-flow direction

Figure 8.29 (Continued)

8.4.4 CHANNELS

Channel forms, cut by glacial meltwater, are very distinctive features of many glacial landscapes. Indeed, in polar environments where glacier ice can carry a relatively small debris load and is commonly largely frozen to its bed, the erosional impact of meltwater is often the only imprint left by past glaciations. Some of the most spectacular meltwater drainage systems often evolve over several glaciations, giving rise to complex local and regional patterns of meltwater incision juxtaposed with evidence of buried interglacial and glacial channels (e.g. Evans and Campbell, 1995; Glasser et al., 2004; Kehew and Kozlowski, 2007). In this section, we describe channels formed in subglacial, ice-marginal and proglacial environments, and discuss their formation in terms of the theory of meltwater flow and the erosional processes detailed in chapter 3.

Subglacial (Nye) channels and meltwater corridors

Nye channels are erosional features cut in bedrock and consolidated sediments by subglacial drainage. Most extend for a few tens to a few thousands of metres, and are up to a few tens of metres wide. The very largest features, however, form major valley systems up to 100 km in length, known as *tunnel valleys*. The origin of these large, complex forms is discussed in the next section. Channels can occur as isolated features or as part of complex, many-branched systems extending over large areas. Such systems can take the form of *dendritic networks*, similar to subaerial channel patterns, recording efficient subglacial drainage along discrete conduits, or *anastomosing systems* in which multiple channels split and rejoin. Anastomosing systems may represent former distributed drainage in linked cavity networks (section 3.4.3), but may also be time-transgressive patterns resulting from channel migration and switching. Thus, careful work is needed in the interpretation of former subglacial channel networks, to determine whether all parts of the system were occupied contemporaneously (e.g. Sissons, 1963; Booth and Hallet, 1993).

In water-filled subglacial channels, flow is driven by gradients in water pressure as well as elevation, so water can flow uphill (section 3.2.2). Undulatory long profiles of such channels as they cross topographic barriers is one of the clearest criteria for distinguishing subglacial from subaerial channels. At a local scale, subglacial channel pathways are influenced by bed topography. In particular, channels will tend to avoid high points on the bed, which are areas of high elevation head, and commonly cross topographic barriers at low points or cols (Booth and Hallet, 1993). The influence of both ice thickness and basal topography on subglacial drainage has been examined by Booth and Hallet (1993), who compared channel networks cut below the Late Glacial Puget Lobe of the

plug of hard, resistant basalt, and the 'tail' is composed of less resistant sedimentary strata. In this part of Scotland, soft strata have been preferentially removed over large areas, leaving the more resistant volcanic plugs and their associated residual tails standing prominently on the urban skyline. The great length of the Royal Mile 'tail' indicates a long 'pressure shadow' in the lee of the crag, possibly reflecting high sliding velocities below this part of the last ice sheet. Erosional tails may also be cut in pre-existing Quaternary sediments. However, good exposures are necessary to differentiate such features from large flutings and cavity fills deposited in the lee of basal obstructions (section 11.2.3). Horned crag and tails have been reported by Jansson and Kleman (1999) from the Quebec-Labrador region of Canada. These features are characterized by two till ridges, emanating from the lateral flanks of a bedrock crag and extending downflow (fig. 8.29b), and were interpreted by Jansson and Kleman as the products of ice that is frozen to the summit of the crag but melting lower down, thereby preventing till accumulation in the lee side of the protuberance.

Cordilleran Ice Sheet with theoretical patterns predicted by Shreve's (1972) theory (section 3.2). Subglacial channels are particularly well developed in this area, and are typically 10–100 m deep, 50–150 m wide, and up to 8 km long. Furthermore, the former ice surface can be reconstructed with considerable confidence, due to abundant ice-flow indicators and well-constrained ice limits, allowing basal equipotential contours to be drawn. There is generally remarkably good agreement between observed and predicted patterns, with many of the major channel systems coinciding with or closely similar to reconstructed flow paths. Some degree of uncertainty is inevitable in such reconstructions because: (1) the present distribution of channels may not represent the former drainage system at a single point in time, but is probably time-transgressive; (2) the true basal topography is commonly partially obscured by later deposits; and (3) Shreve theory provides only an approximate picture of subglacial conditions. The success of the reconstruction, despite these problems, strongly supports the idea that drainage below the Late Glacial Puget Lobe was through a dendritic system of water-filled Nye channels directed by ice pressure and local topography. Such reconstructions, therefore, add considerably to our understanding of past glacier dynamics.

Below valley glaciers, high relative relief exerts an important control on the location of subglacial channels, which commonly follow valley axes. Focused meltwater erosion can produce deep, narrow *slot gorges*, such as the gorge at Berekvam, Norway, which is 2 m wide and 30 m deep, or the Corrieshalloch Gorge in Northern Scotland, which is 10–20 m wide and 60 m deep. However, it is often difficult to determine to what extent pre-existing fluvial gorges were influential in guiding subglacial meltwater, and how effective deglacial and postglacial subaerial streams have been in deepening such forms.

Subglacial meltwater flow is also strongly influenced by local topography if channels are only partially filled with water for much of the time (section 3.2.2). In this case, water pressure is atmospheric, and hydraulic potential is governed solely by elevation. Such conditions are most likely close to glacier margins, where ice is thin and tunnel creep closure rates are low, and where most meltwater is derived from the surface and is thus subject to large diurnal fluctuations. Fossil channel systems may exhibit increasing conformity with local topography with decreasing elevation, with predominantly ice-directed channels on the upper slopes giving way to downslope-oriented channels near the valley floors. Such networks are commonly time-transgressive systems consisting of higher, older elements controlled by the ice surface slope, and lower, more recent slope-directed elements cut below thinning ice in the later stages of deglaciation (e.g. Gordon, 1993).

Anastomosing systems of Nye channels record several generations of channel incision or former linked cavity

drainage network (section 3.4.3). Very clear examples of former linked cavity systems have been exposed on limestone bedrock by recession of the Glacier de Tsanfleuron, Switzerland (Sharp et al., 1989b), Castleguard Glacier, Alberta, Canada (Hallet and Anderson, 1980), and Blackfoot Glacier, Montana, USA (Walder and Hallet, 1979). Such systems can be subdivided into five geomorphic units, each associated with a different set of subglacial processes:

(1) areas of intimate ice–bedrock contact;
(2) Nye channels incised into bedrock;
(3) lee-side cavities on the down-ice side of bedrock obstacles;
(4) surface depressions filled with calcite precipitates;
(5) karst sinkholes (fig. 8.30).

Areas of intimate ice–bedrock contact are striated, but also display solutional hollows and calcite precipitates formed during regelation sliding (section 4.4.2; Hallet, 1976a; Sharp et al., 1989b). The *Nye channels* are aligned largely parallel to former glacier flow and form an anastomosing rather than arborescent drainage pattern. Many channels have blind terminations, although it is possible to detect evidence of limited water flow from one channel to another, or from channels into cavities. Therefore, the Nye channels act as links between lee-side cavities, and together the channels and cavities act as interconnected drainage routeways. Three types of *lee-side cavity* were recognized by Sharp et al. (1989b):

(1) large cavities aligned oblique to former ice flow direction;
(2) elongate channels aligned parallel to former ice flow and continuous with Nye channels;
(3) small cavities.

The large cavities were apparently fully integrated into the subglacial drainage system, whereas the small cavities were poorly connected. *Surface depressions* or hollows that are almost totally filled by precipitates are located at the heads and margins of Nye channels. They are differentiated from other channels and cavities by their shallowness and lack of elongation, and are thought to document cavity closure and meltwater freezing during periods of low discharge. *Karst sinkholes* occur in association with Nye channels and elongate cavities, indicating a connection between the subglacial and subterranean drainage systems. Sinkholes occur at both the upstream and downstream ends of Nye channels, suggesting that they feed meltwater into as well as receive meltwater from the channels; a sinkhole may provide meltwater during periods when the subterranean system is full of water or is blocked by glacier ice.

Very impressive fossil anastomosing and dendritic channel networks, referred to as The Labyrinth, have been described by Sugden et al. (1991) and Denton

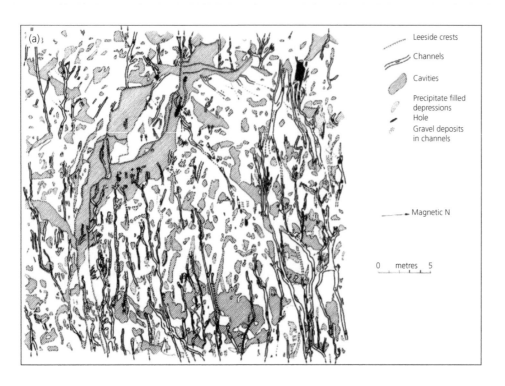

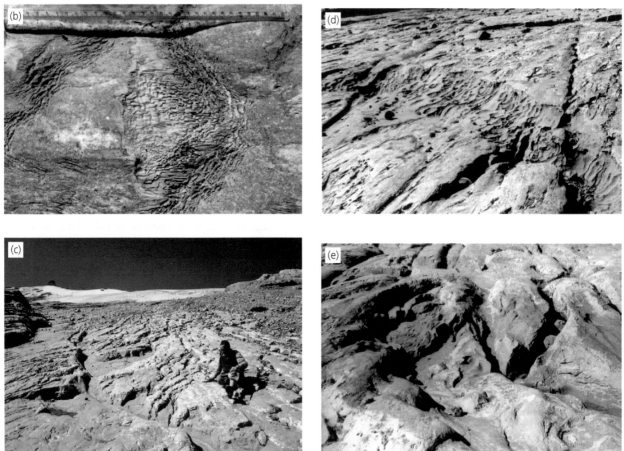

Figure 8.30 Former linked-cavity drainage system exposed on the foreland of Glacier de Tsanfleuron, Switzerland. (a) Map of part of the proglacial area, showing the distribution of erosional features. White areas represent areas of former intimate glacier–bed contact. (Sharp et al., 1989b, reproduced with permission of Wiley) (b) Detail of an area of intimate glacier–bed contact, showing striated bedrock, solution furrows and subglacially precipitated calcite. Ice flow was from left to right. (c) Network of anastomosing Nye channels. (d) Former lee-side cavity, showing water eroded surfaces. (e) Precipitate-filled depression at the margin of a Nye channel ((b–e) M.J. Sharp).

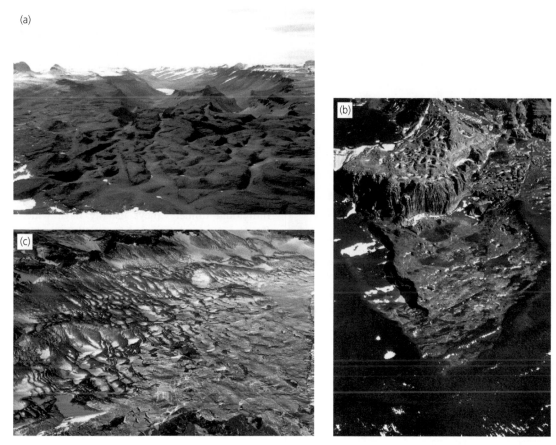

Figure 8.31 Landforms of subglacial fluvial erosion by catastrophic outburst floods in the Transantarctic Mountains. (a) The Labyrinth, a spectacular series of subglacial meltwater channels exposed in front of the Upper Wright Glacier, Dry Valleys. (D.R. Marchant) (b) Spatulate-shaped basin in meltwater-stripped sandstone at the base of a 200 m cliff (Battleship Promontory). Plunge pools occur at the cliff base and giant potholes cover the area above the cliff top. (c) Corrugated bedrock in sandstone in the Coombs Hills. This view shows a transition from stripped bedrock with plunge pools and patches of regolith at lower right, through corrugated bedrock with no regolith, corrugated bedrock with residual ripples comprising dolerite boulders up to 35 cm diameter on the corrugation crests, corrugations in regolith, to undisturbed regolith at top left. This transition attests to the varying intensity of meltwater erosion along a subglacial flood track (Denton and Sugden, 2005; images provided by D.E. Sugden).

et al. (1993), from the Upper Wright Valley, Antarctica (fig. 8.31). Individual channels are up to 50 m deep, and in several places are punctuated by huge potholes tens of metres deep and across. Denton and Sugden (2005) considered these channels to be part of a network of subglacially eroded forms, including scablands and giant scour holes and scallops, that cover the high plateaux of the Transantarctic Mountains in Victoria Land. They infer from the distribution of the landforms, some of which occur juxtaposed to regolith preserved by cold-based ice coverage, that erosion was affected by subglacial meltwater outbursts that breached an otherwise predominantly frozen-bedded ice sheet that overtopped all but the highest summits of the region (see catastrophic meltwater floods in section 8.4.4). The huge volumes of water necessary for the erosion of the landforms were likely released catastrophically from subglacial or surficial lakes, an event that has been dated to sometime during the period 14.4–12.4 Ma ago by Lewis et al. (2006). Further examples of huge subglacial drainage networks are being discovered regularly by ongoing SWATH bathymetric surveys of the Antarctic

shelf, where they are interpreted as the products of numerous catastrophic outbursts that likely triggered periods of fast glacier flow (e.g. Wellner et al., 2001; Lowe and Anderson, 2003; section 12.4.2).

The occurrence of meltwater corridors on the former bed of the north-west Laurentide Ice Sheet (St Onge, 1984; Rampton, 2000) clearly shows that subglacial meltwater discharges do not always result in deep incision. Meltwater corridors are characterized by zones or corridors of scoured bedrock that run through large areas of till blanket (fig. 8.32). Their boundaries with the surrounding till appear sharp, but are often marked by narrow zones of modified till where meltwater discharges were insufficient to remove all of the till cover. In addition to the fluvially scoured bedrock, the corridors contain discontinuous glacifluvial sediments and eskers, and occasional tunnel channels where erosion clearly became more focused.

Tunnel valleys and tunnel channels

Tunnel valleys or *channels* (also known as *rinnentaler* or *tunneldale*) are large, overdeepened channels cut into

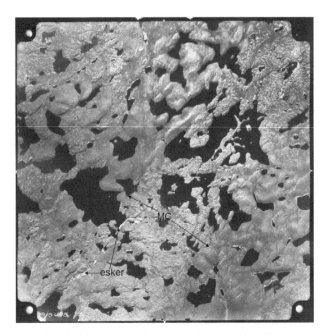

Figure 8.32 Aerial photograph of a typical meltwater corridor in Slave Province, Nunavut, north-west Canada. The corridor is marked by the meltwater-scoured bedrock (MC). Former subglacial water flow direction was towards the top right of the image (National Aerial Photograph Library, Ottawa, photograph A15495-113, after Rampton, 2000).

bedrock or sediment, which can reach >100 km long and 4 km wide (figs 8.33 and 8.34; Ó Cofaigh, 1996). They can occur in isolation or as parts of dendritic or anastomosing patterns extending over very large areas, and have been recognized in many areas formerly covered by Pleistocene ice sheets, including North America, north Germany, Denmark, Poland, Byelorussia, the floor of the North Sea, Britain and Ireland (Wingfield, 1990; Praeg, 2003; Kristensen et al., 2008). They share many characteristics with Nye channels, including undulatory bed long profiles, overdeepened basins along their floors, and hanging tributary valleys (e.g. Ó Cofaigh, 1996). Individual tunnel valleys usually have wide, relatively flat bottoms and steep sides, and the numerous troughs that occur along their lengths may be occupied by lakes (e.g. Clayton et al., 1999; fig. 8.34). These characteristics provide clear evidence that they were excavated by subglacial meltwater flowing under pressure, a process that was spectacularly confirmed by the excavation of a tunnel channel beneath the outer snout of Skeiðarárjökull during the 1996 jökulhlaup (Russell et al., 2007). Further geomorphic evidence for a subglacial origin is the tendency

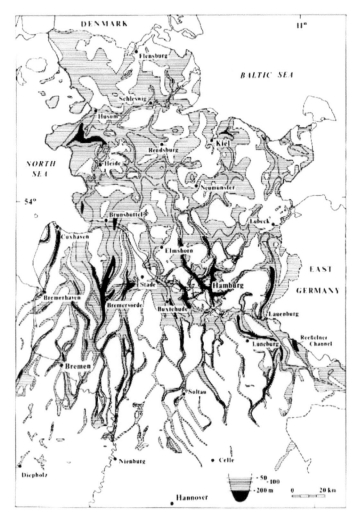

Figure 8.33 Elsterian age buried tunnel valleys, north-west Germany (Ehlers et al., 1984, reproduced with permission of Elsevier).

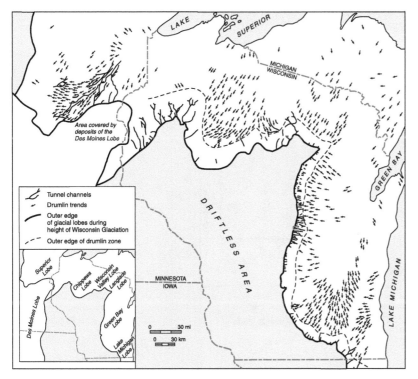

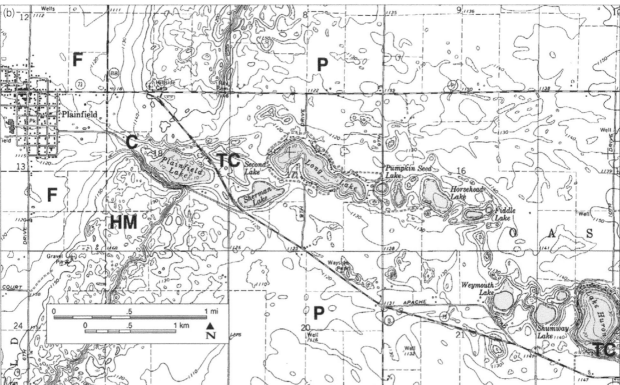

Figure 8.34 Maps and photographs of tunnel valleys and channels in the Midwest, USA. (a) Map of Laurentide Ice Sheet lobe margins and associated features (Clayton et al., 1999, after various authors). Location of (b) is marked by a small box on the margin of the Green Bay lobe. (b) Topographic map showing a tunnel channel at the margin of the Green Bay Lobe of the Laurentide Ice Sheet in Wisconsin (Clayton et al., 1999). Note the water-filled depressions that occur along the bottom of the feature and the breaching of the Hancock end moraine (HM). A subaerial outwash fan (F) has been deposited beyond the moraine and this has been channelled (C) at its apex. Pitted outwash (P) dominates the land surface on the ice-proximal side of the Hancock moraine. Ice flow was from east to west. (c) Tunnel valleys, valley mouth fans and the St Croix Moraine (shaded) in east central Minnesota and Wisconsin, USA. A = Comfort Lake tunnel valley and fan; B = Buffalo Lake tunnel valley and fan; C = Goose Lake tunnel valley and fan; D = Arsenal fan and Lino Lakes tunnel valley system; E = Elk River fan complex. (d) Aerial photograph of Comfort Lake fan and associated tunnel valleys partially delineated by elongate lakes (Patterson, 1994 and aerial photograph by Mark Hurd Aerial Surveys). (e) Classification scheme for the morphology of tunnel valleys (channels) in deglaciated landscapes (Kehew and Kozlowski, 2007), based on Michigan case studies. In each case, panel A depicts erosion of the initial channel. From left to right: Type I is produced by a small infill of sediment and ice (B) followed by ice melt-out; Type II is produced by infill by large volumes of sediment followed by glacier readvance and moraine construction (B) and then ice melt-out to produce a valley that continues through the moraine; Type III is produced where no buried ice occurs and glacier readvance results in deposition of till (B) followed by outwash (C), leaving a chain of lakes; Type IV is produced where large volumes of ice and debris choke the channel and outwash is prograded over the sequence by an ice readvance from a different direction (B), followed by melt-out and channel collapse which truncates the fan surfaces; Type V is produced where subglacial drainage persists throughout deglaciation and so the channel hosts an esker.

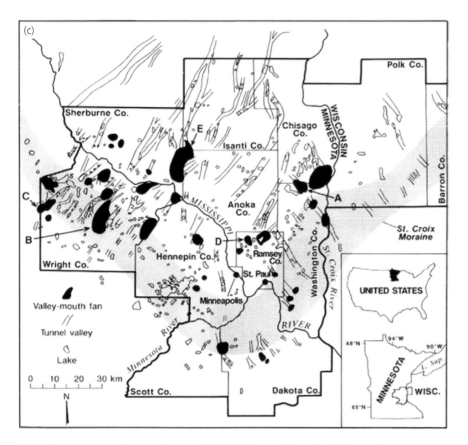

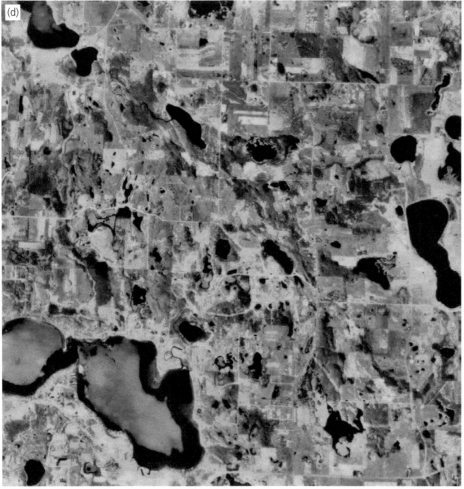

Figure 8.34 (Continued)

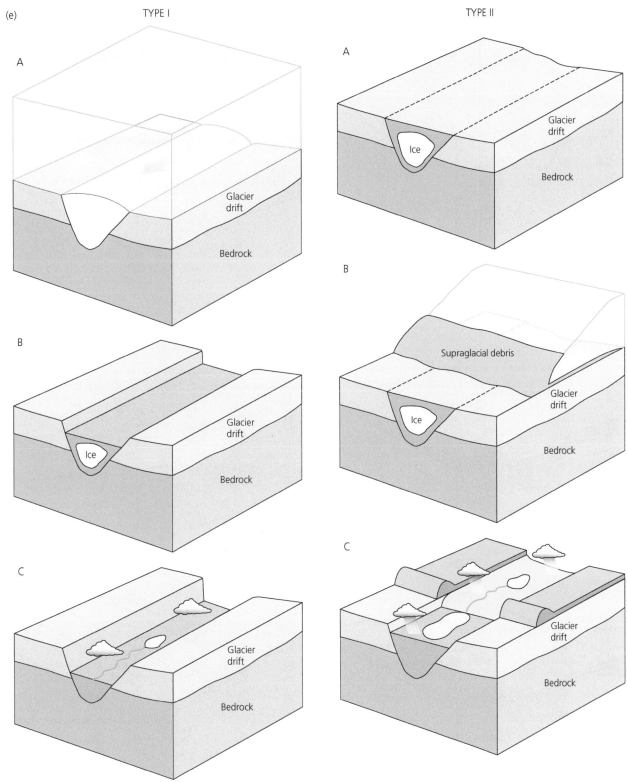

(e)

TYPE I

TYPE II

Figure 8.34 (*Continued*)

for tunnel valleys to terminate abruptly at major moraines, often after ascending a reverse slope, where they may grade into large subaerial ice-contact fans (e.g. Patterson, 1994; Clayton et al., 1999). The surfaces of these fans may lie up to 100 m above the tunnel valley bottom, reflecting deposition from pressurized meltwater emerging from beneath the ice. More recently, the term *tunnel channel* has been applied to channels cut by rivers that fill the channel, in contrast to tunnel valleys, which are valleys cut by lateral planation of a river that is narrower than the valley (Mooers, 1989a; Clayton et al., 1999; fig. 8.35). However, it is apparent that some researchers

TYPE III   TYPE IV

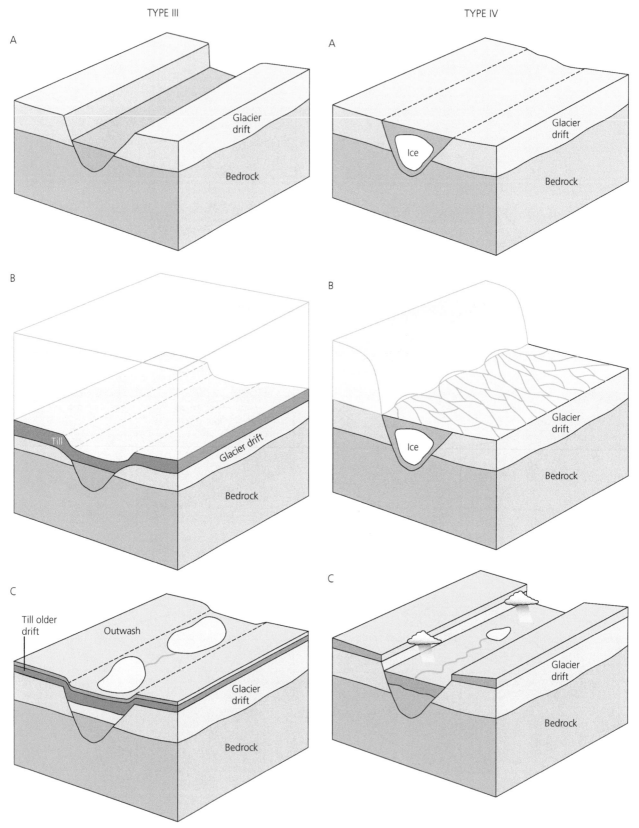

Figure 8.34 (*Continued*)

now regard all glacifluvially eroded valleys in their study areas as subglacial 'channels' in the hydrologic sense and have modified their terminology accordingly (e.g. Beaney, 2002; Rains et al., 2002; Sjogren et al., 2002; Fisher et al., 2005). Because *tunnel channel* is a genetic term, we use the more descriptive term *tunnel valley* from hereon in this section, unless a channel origin has been convincingly inferred.

TYPE V

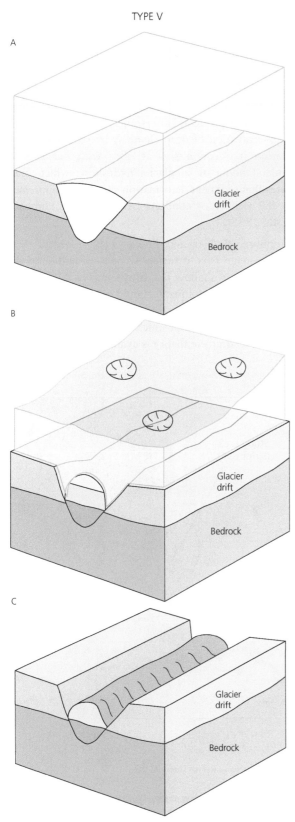

Figure 8.34 *(Continued)*

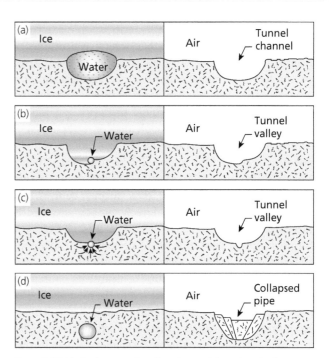

Figure 8.35 Idealized cross sections through tunnel channels and valleys, showing their formation in soft sediment (hatched pattern): (a) Tunnel channel eroded by subglacial river that fills the channel. (b) Tunnel valley eroded by subglacial river that is smaller than the valley. (c) Tunnel valley formed by flowage/creep of subglacial material into river that is smaller than the valley. (After Boulton and Hindmarsh, 1987) (d) Subglacial pipe where the tunnel is sapped into subglacial material and enlarged by meltwater (see section 3.4). Surface expression is due to collapse of overlying material (After Clayton et al., 1999).

Tunnel valleys may be completely infilled by thick sedimentary successions, including glacigenic, glacifluvial, glacilacustrine, glacimarine and non-glacial deposits, and may not have any clear topographic expression on the surface (Ó Cofaigh, 1996; Kehew and Kozlowski, 2007) except for linear chains of depressions (*kettle chains*) in some cases. Therefore, detailed studies of borehole logs, geophysical data and sedimentary exposures are necessary to determine their morphology and extent (e.g. Jørgensen and Sandersen, 2006). Some care needs to be taken wherever old subaerial drainage networks have been re-excavated or partially cross-cut by later subglacial tunnel valleys. For example, Early Pleistocene (preglacial) river valleys lie buried beneath later glacial sediments on the southern Alberta prairies in Canada, and the occurrence of chains of elongate depressions demarcate the positions of subglacial channels, which are often centred over the old buried valleys (Evans and Campbell, 1995). In such situations, the long profiles of the buried valley bottoms reveal that they are normal fluvial forms, whereas the superimposed chains of depressions on the prairie surface indicate that subglacial meltwater excavated some of the glacial sediments filling the valley. Kehew and Kozlowski (2007) have classified the various geomorphic forms representative of tunnel valleys (channels) according to their post-erosional evolution (Types I–V; fig. 8.34e), demonstrating that they may be partially masked by depositional processes (Types II–IV), or could be clearly demarcated in the landscape due to minimal burial (Type I), or host esker ridges along their floors (Type V).

Lykke-Andersen (1986) suggested that some Danish tunnel valleys may in fact be old river valleys excavated by proglacial meltwater or even glacier ice. Jørgensen and Sandersen (2006) proposed that several generations of Danish subglacial tunnel valley are superimposed, based on the fact that tunnel valleys can be age-differentiated according to their preferred orientations. Clearly, ice sheet sub-marginal and marginal processes can modify landforms considerably through one or several glacial cycles, and therefore many tunnel valley sections could be polygenetic (e.g. Huuse and Lykke-Andersen, 2000b; Kozlowski et al., 2005). Despite the tendency for subglacial streams to reoccupy their former courses, the north German tunnel valleys largely cut across the Early Pleistocene channels and river networks. Polygenetic forms are often very difficult to decipher. For example, Clayton et al. (1999), Kehew et al. (1999, 2005) and Kehew and Kozlowski (2007) identified palimpsest forms, where glacier readvance led to the modification of pre-existing glacifluvial erosional features. These include *palimpsest tunnel valleys*, where later sediments cover the erosional feature, and *palimpsest spillways*, where former proglacial lake spillways (see page 299) are partially masked by glacial readvance sediments.

A significant research problem in tunnel channel genesis is the large size of the channels, which, if they ever experienced bankfull conditions, would imply water discharges far in excess of those that could be maintained by steady-state basal melting. Two main theories have been advanced to explain this:

(1) tunnel valleys could result from progressive excavation of sediment by normal discharges, in conjunction with subglacial sediment deformation;
(2) they could be excavated by extremely large, transient discharges associated with catastrophic drainage events.

These ideas are discussed in turn below.

Drainage over subglacial deforming sediment

Tunnel valley genesis has been explained by Shoemaker (1986a) and Boulton and Hindmarsh (1987) as the result of steady-state meltwater drainage over subglacial deforming sediment. They argued that where subglacial discharges cannot be accommodated by Darcian groundwater flow, piping failure will initiate subglacial drainage conduits, which will help to maintain glacier stability. According to this model, deforming sediment will tend to creep into the conduit, whereupon it is flushed out by meltwater. Continued sediment excavation by this process is argued to result in the formation of a valley system much larger than the active conduit (fig. 8.35c). The final stage of this process occurs when discharges fall, and the conduit fills with sediment and contracts by ice creep. Piotrowski et al. (1999) reported stratigraphic evidence for the squeezing of soft sediment into a small subglacial meltwater channel during its excavation into stratified outwash.

There are two main problems with the deforming bed/steady-state drainage model (Ó Cofaigh, 1996):

(1) There is little field evidence in support of the idea that sediment deformation occurs in conjunction with tunnel valley excavation, although many tunnel valleys do occur in areas where there are extensive deformed tills (Mooers, 1989a; Patterson, 1994).
(2) The excavation of deep channels by sediment deformation into conduits is at odds with our present understanding of subglacial hydrology, which indicates that sediment deformation should be suppressed in the vicinity of conduits (Alley, 1992b; Hubbard et al., 1995).

Furthermore, Walder and Fowler (1994) argued that excess drainage over deformable beds should take the form of broad, shallow anastomosing canal systems, rather than major conduits. It should be emphasized however, that little is known about drainage conditions on soft glacier beds, and that much more research is necessary before a complete theory is available.

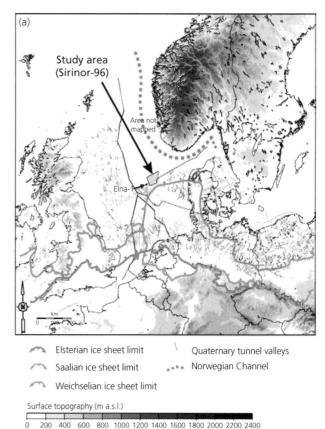

Elsterian ice sheet limit   Quaternary tunnel valleys
Saalian ice sheet limit   Norwegian Channel
Weichselian ice sheet limit

Surface topography (m a.s.l.)

0   200 400 600 800 1000 1200 1400 1600 1800 2000 2200 2400

Figure 8.36 Major tunnel valleys in the North Sea. (a) Location map of tunnel valleys and their relationships to major ice sheet margins. The Sirinor study area is the target of a 3D seismic survey (depicted in (b) and Elna 1 is a borehole location. (Kristensen et al., 2007) (b) A 3D map of the cross-cutting tunnel valleys in the Sirinor study area located in (a). Salt diapirs (SD) are indicated. (c) Interpretations of west–east seismic profiles in the west-central North Sea, showing three generations of incisions (After Ehlers and Wingfield, 1991).

Catastrophic meltwater floods

Several researchers have invoked catastrophic drainage events to account for the excavation of tunnel valleys. Extensive tunnel valley systems in northern Germany (fig. 8.33) have been interpreted as the products of outburst floods below the margins of the Scandinavian Ice Sheet (Ehlers and Linke, 1989). There is evidence for several generations of erosion, and a complex excavation history was envisaged involving outburst floods and intervening periods of resculpturing and widening by glacier ice. Catastrophic floods have also been implicated by

Wingfield (1990) to account for extensive networks of elongate, blind-ended depressions more than 100 m deep on the floor of the North Sea in the marginal zone of the former Scandinavian and British Ice Sheets, where seismic profiling has enabled the identification of three generations of incision (fig. 8.36c). Although there is controversy surrounding the interpretation of the North Sea incisions (see review in Ehlers and Wingfield, 1991), their similarity to the tunnel valleys of northern Europe has prompted a genetic comparison with those terrestrial examples. Wingfield (1990) proposed that the incisions were excavated by rapid headward erosion of channels during jökulhlaups from large subglacial lakes. Alternatively, Kristensen et al. (2007) regarded some of the North Sea tunnel valleys as polyphase features that developed during ice-marginal oscillations over several glaciations. This explains a range of features in their study area, such as the lack of branching and anastomosing patterns, the undulatory bottom profiles, adverse bed slopes at termini and cross-cutting relationships (figs 8.36a, b).

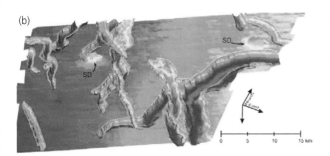

(b)

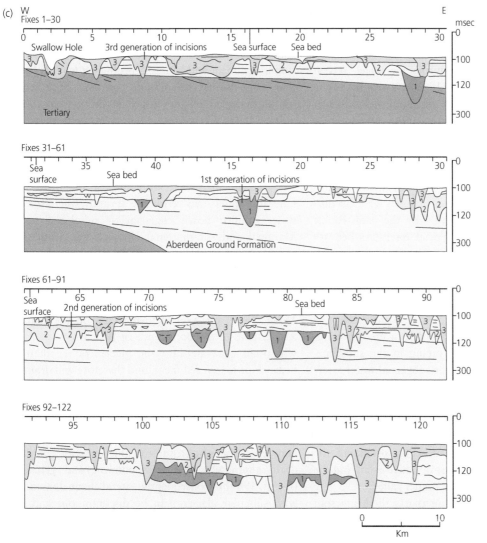

(c)

Figure 8.36 (*Continued*)

Extensive tunnel valley networks in south-central Ontario, Canada, have been interpreted by Brennand and Shaw (1994) as the products of exceptionally large catastrophic lake drainage beneath the Laurentide Ice Sheet. These networks consist of multiple, large anastomosing channels, with undulating long profiles and deep, overdeepened basins, and commonly contain eskers running along their floors. The case for catastrophic lake drainage in this region is highly controversial, and is based on radical interpretations of many landform assemblages, including drumlins, flutings and bedrock erosional marks, as the products of huge subglacial sheet floods (section 11.2.8). Because the inferred drainage events are thought by many researchers to be too large and extensive to have been supplied by subglacial meltwater, Shoemaker (1992) has suggested that they originated by reversed subglacial drainage from proglacial lakes, supplemented by supraglacial melting. There is no independent field evidence for this scenario, and there are strong reasons for regarding it as physically implausible (Ó Cofaigh, 1996; Clarke et al., 2005).

A radial network of tunnel valleys formed beneath the Superior Lobe of the southern Laurentide Ice Sheet in Minnesota, USA (figs 8.34a, c, d), has been interpreted as resulting from either simultaneous incision by catastrophic discharges, or multigenerational incision during steady-state, stable drainage during ice sheet retreat (Mooers, 1989a; Patterson, 1994). A catastrophic origin for the tunnel channels beneath the snout of the former Green Bay Lobe in Wisconsin has been proposed by Clayton et al. (1999), based on their location in a narrow marginal zone thought to have been frozen to its bed. This frozen toe zone explains the lack of drumlins beneath a glacier that otherwise produced widespread streamlined landforms (fig. 8.34a). The tunnel channels were cut during short, widely spaced intervals by the catastrophic release of meltwater from subglacial reservoirs beneath the warm ice zone located upglacier. This catastrophic drainage origin has been modelled by Hooke and Jennings (2006). They suggested that subglacial lakes trapped behind glacier sub-marginal permafrost can repeatedly drain rapidly in response to high water pressure gradients in the substrate. Seepage of the meltwater at the glacier margin initiates piping in the substrate, which catastrophically siphons the subglacial water reservoir. Towards the end of rapid reservoir drainage, ice creep and till deformation close the conduit, and permafrost re-aggradation closes the seal once more. In the presence of sub-marginal permafrost, a subglacial reservoir could re-form in a matter of decades, providing the impetus for a further catastrophic drainage event and the incision or re-incision of another tunnel valley.

Like their north European counterparts, the Minnesota and Wisconsin tunnel valleys/channels terminate at sand and gravel fans, which are up to 85 m higher than the valley floors and form part of end moraine complexes, documenting the deposition of outwash by subglacial meltwater escaping from the ice margin under pressure. The distribution of the fan complexes and other glacial landforms in the area lends strong support for their formation in several stages during ice retreat, with different glacier lobes producing glacifluvially eroded tunnels in different ways (Clayton et al., 1999; Cutler et al., 2002). The polygenetic origin of some deeply incised valleys, such as those reported from the former marginal zones of the Lake Michigan and Saginaw lobes of the southern Laurentide Ice Sheet by Kozlowski et al. (2005) and Kehew and Kozlowski (2007), require us to entertain the concept of combined subglacial and proglacial incision due to repeated outburst floods from receding glacier margins. Resolution of the debate concerning the origin of tunnel valley complexes is clearly important, and has wide-reaching implications for the nature of drainage below large ice sheets and the relative contribution of catastrophic and steady-state conditions to landscape evolution and environmental change.

Ice-marginal (lateral) channels

Water draining along the margins or in sub-marginal zones of glaciers can be responsible for considerable incision of sediment and bedrock, producing *lateral channels* that mark former ice margin positions. Such features are particularly well developed at the margins of subpolar glaciers where large volumes of meltwater cannot penetrate to their frozen beds. Following deglaciation, lateral meltwater channels are left perched above valley sides, and are usually distinguishable from subglacial and subaerial channels by their planform and distribution. Lateral channels may terminate abruptly where the meltwater drained down englacial or subglacial tunnels. Marginal and sub-marginal channels can form nested *inset sequences*, which can be used to reconstruct glacier recession patterns.

The evolution of lateral meltwater channels at the margins of subpolar glaciers has been studied intensively by Maag (1969) on Axel Heiberg Island in the Canadian Arctic. Due to the occurrence of permafrost, meltwater will often excavate into the glacier rather than into bedrock or frozen sediment, and therefore will drain submarginally. In successive summer seasons, streams may reuse old channels even though they may fill up with snow or partially close due to ice creep during the winter (fig. 8.37). In addition, meltwater routes may alternate between marginal and sub-marginal positions, and may drain alternately over rock/sediment and glacier ice (Evans, 1989b). Consequently, lateral channels associated with subpolar glaciers can be discontinuous and often document sub-marginal as well as ice-marginal drainage.

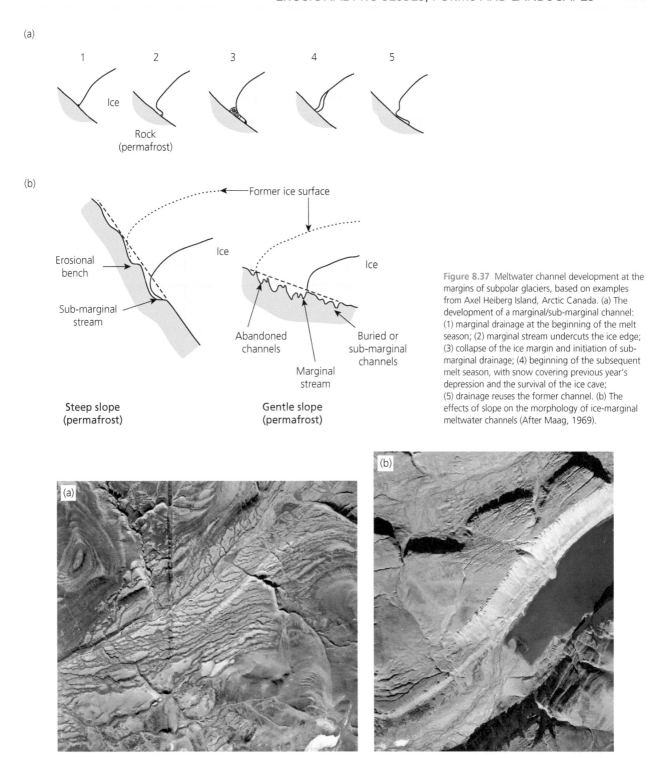

Figure 8.37 Meltwater channel development at the margins of subpolar glaciers, based on examples from Axel Heiberg Island, Arctic Canada. (a) The development of a marginal/sub-marginal channel: (1) marginal drainage at the beginning of the melt season; (2) marginal stream undercuts the ice edge; (3) collapse of the ice margin and initiation of sub-marginal drainage; (4) beginning of the subsequent melt season, with snow covering previous year's depression and the survival of the ice cave; (5) drainage reuses the former channel. (b) The effects of slope on the morphology of ice-marginal meltwater channels (After Maag, 1969).

Figure 8.38 Lateral meltwater channels cut at the margins of cold-based glacier snouts during their recession from LGM limits in the Canadian Arctic. (a) Eastern Borden Peninsula, Baffin Island. (b) The head of Flagler Bay, Ellesmere Island (National Air Photograph Library, Ottawa).

Maag (1969) also stressed the importance of: (1) slope angle in the production of lateral channels, steep slopes being eroded into benches rather than channels; and (2) the nature of the substrate. Incision rates as high as 1 m per 24 hours, due to the combined effect of fluvial and thermal erosion, were recorded in frozen gravels on Axel Heiberg Island. Incision rates and channel depths in

bedrock will obviously vary according to lithology and discharge magnitudes during channel occupancy.

Lateral meltwater channels cut along the cold-based northern margins of the Laurentide Ice Sheet have been mapped by Dyke (1993) and Dyke et al. (1992), who used them to reconstruct regional patterns of ice recession (fig. 8.38). Extensive flights of inset meltwater channels with very shallow gradients are evident on the upper slopes of the fjords and intervening plateaux of southeast Ellesmere Island, Canadian Arctic, documenting meltwater drainage along the margins of low-gradient fjord glaciers. Retreat patterns of subpolar glaciers have been reconstructed for large areas of Ellesmere Island, using lateral meltwater channels in conjunction with other ice-marginal accumulations (e.g. England, 1990; Evans, 1990b). The widespread production of marginal meltwater channels at the edge of the receding British Ice Sheet has been recognized only recently by Clark et al. (2006) and Greenwood et al. (2007), even though some local mapping projects have identified dense channel networks (e.g. Arthurton and Wadge, 1981).

Although lateral meltwater channels are best developed along cold-based glacier margins, they have been reported from regions where glacier snouts are temperate – for example, in Alaska (Syverson and Mickelson, 2009). In the Canadian Cordillera, lateral meltwater channels demarcate the receding margins of valley glaciers debouching from the mountain ice field, explained by Dyke (1993) as the product of either: (1) high subglacial water pressures deflecting some surface water along the glacier margins; (2) spring snow melt on surrounding slopes and meltwater drainage along the glacier margin while the bed was still frozen after the penetration of the winter cold wave; or (3) cold-based conditions in the outer marginal zone of otherwise temperate glaciers. Dyke (1993) proposed that the spectacular lateral meltwater channels of the central Canadian Arctic display a very fine temporal resolution, perhaps of 'melt event' scale. Caution is warranted with such interpretations in situations where meltwater may have originated from ice-contact lakes or from unusual flood events, where a catastrophic origin can be determined based on the occurrence of ice block melt-out craters and dissected accumulations of very coarse, poorly sorted glacifluvial sediment (e.g. Roberts et al., 2003).

The diversion of large rivers by advancing glaciers can result in the excavation of large-scale channels along or just beyond the glacier margins. Such channels, known as *ürströmtaler*, mark successive marginal positions of the Scandinavian Ice Sheet in Poland, Germany and the Netherlands (e.g. ter Wee, 1983), with continuations on the German Bight sector of the North Sea (Figge, 1983). The ürströmtaler mark the positions of the main European rivers after they were diverted from their normal south–north drainage direction by the advancing ice sheet to flow in a westerly direction along an ice margin.

Proglacial channels and flood tracks

Channels and gorges cut by proglacial streams can achieve impressive dimensions due to the erosive power of high discharges and sediment loads during peak flows. In areas formerly occupied by glacier ice, such channels can

originate as subglacial Nye channels and undergo subsequent subaerial enlargement after deglaciation, and the relative contribution of subglacial and subaerial erosion can be unclear. In most glaciated terrains, anomalously located fluvial channels can be interpreted only as the products of glacially induced drainage diversion when ice filled valleys and fed meltwater down minor topographic depressions, enlarging them to sizes that cannot be explained by fluvial erosion in non-glacial conditions. The impressive 20 km long and 300 m deep York Canyons that cut across Meta Incognita Peninsula on Baffin Island are excellent examples of such a drainage diversion, produced by the Noble Inlet advance of Quebec-Labrador ice onto the island at the end of the last glaciation (Kleman et al., 2001).

The very largest proglacial channels, however, are cut by glacier lake outburst floods (GLOFs; section 3.7), as documented by studies of the areally scoured and pothole-covered erosional tracts of Icelandic proglacial meltwater streams (e.g. Carrivick et al., 2004; Alho et al., 2005). The immense erosive capacity of such floods is perhaps most clearly demonstrated by the *Channelled Scablands* of Washington state, USA (Baker, 1981; Baker et al., 1987; Alt, 2001), which are the deeply dissected remnants of the formerly extensive loess and basalt deposits of the Columbia Plateau. This landscape was first attributed to erosion by a large GLOF, known as the Spokane Flood, by J. Harlen Bretz, in a series of controversial publications (e.g. Bretz, 1923; Bretz et al., 1956; see Evans, 2004 for review and further references). However, it was not until the 1960s that Bretz's ideas of a proglacial flood were finally accepted by the geological community, largely because of the absence of an obvious source for a flood large enough to accomplish such massive erosion. Sedimentary records have revealed that proglacial floods have been regular occurrences in the region during several glaciations (Waitt, 1980; Bjornstad et al., 2001). Comprehensive coverage of the controversy surrounding the acceptance of Bretz's flood hypothesis has been provided by Gould (1980), Baker (1981) and Baker et al. (1987), who regarded the Spokane Flood as one of the most valuable of what William Morris Davis endearingly entitled 'outrageous geological hypotheses', which challenge accepted thinking and offer a new paradigm.

The Channelled Scablands were created by catastrophic drainage of Glacial Lake Missoula, a large, ice-dammed lake ponded up by the margin of the late Pleistocene Cordilleran Ice Sheet, and for this reason the Washington floods are sometimes referred to as the Lake Missoula floods (fig. 8.39). Glacial Lake Missoula in Montana had a maximum volume of approximately 2500 km^3, and on several occasions drained catastrophically through the ice dam, releasing up to 2184 km^3 of water (comparable to the volume of Lake Ontario;

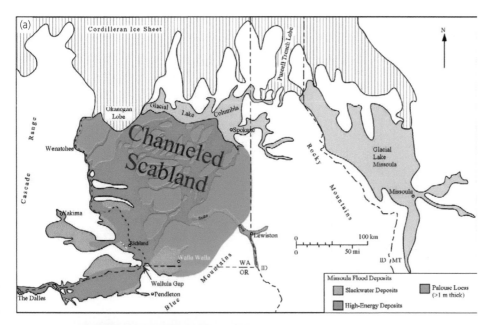

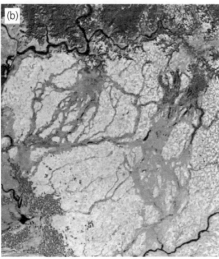

Figure 8.39 (a) Map showing the extent of the Missoula Floods and the location of the feeder lakes Missoula and Columbia (Provided by B.N. Bjornstad) (b) Landsat false-colour image, showing the Missoula flood tracks in darker green.

Clarke, 1986). The flood waters scoured across the northern part of the Columbia Plateau towards the Pasco Basin, and then down the Columbia river valley towards the Pacific Ocean. The awesome erosional capacity of these floods is manifest in the Channelled Scablands, which comprise anastomosing flood channels and gorges that contain erosional and depositional landforms, such as rock basins, giant cataract alcoves, large residual 'islands', mega-bars of gravel and giant current dunes covering an area of approximately 40,000 km² (Baker et al., 1987; fig. 8.40). The escaping waters stripped a cover of loess up to 60 m thick and plucked out basalt blocks from channel walls and floors, in many cases leaving isolated mesas or tablelands, known locally as *scabs*, between channels. In confined channels, water depths attained 100–200 m, reaching almost 300 m in the

narrow eastern entrance to the Columbia river gorge. Some channels were enlarged by the headward retreat of major waterfalls; one of the largest, Dry Falls, was over 5 km wide and 120–130 m high (fig. 8.40b). Baker (1981) and Baker et al. (1987) concluded that the majority of this erosion was achieved in one or a few exceptionally large floods, perhaps over a period of hours, although Waitt (1980, 1985) has clearly demonstrated that there were up to 40 flood events. Various calculations of the flood discharges from Glacial Lake Missoula have been attempted and range from 2.7 to 21×10^6 m³ sec⁻¹, astounding figures equivalent to between 2 and 20 times the mean flow of all the world's rivers into the oceans (Clarke, 1986; Baker et al., 1987). The most recent modelling by Miyamoto et al. (2007) has identified that the largest volumes of water required for the production of the Channelled Scablands could not have been sourced from Lake Missoula alone, highlighting the fact that other linked glacial lake systems in the region, like Glacial Lake Columbia, and subglacial sources of meltwater, are also likely to have contributed to the most erosive flows.

The erosional landforms of the Channelled Scablands, all of which have been identified at a range of scales in other proglacial and subglacial flood tracks, are grouped into three categories by Baker et al. (1987):

(1) *Scabland erosion complexes* are formed by the fluvial incision of bedrock. Experiments show that forms such as longitudinal grooves, potholes and transverse erosional ripples eventually become incised by inner channels, which migrate upstream by knickpoint recession. This has been modelled for the Channelled Scablands by Baker et al. (1987), who inferred an evolutionary sequence involving: (a) early stages of streamlining and incision of the loessic mantle;

Figure 8.40 Various features of the Channelled Scablands of Washington and Idaho, USA. (a) West Potholes cataract. The water spilled from a lip 3 km wide, over cliffs 100 m high and into plunge basins 40 m deep. Boulders 30 m in diameter were carried from the cataract and deposited in huge bars downstream. (b) Dry Falls cataract complex. (c) Butte-and-basin scabland development in Lenore Canyon. (d) Giant current ripples near Spirit Lake, Idaho. The gravel ripples are spaced 85 m apart and are 4 m high on average (Baker et al., 1987, reproduced with permission of the Geological Society of America).

(b) plucking of the underlying well-jointed basalt lava flow by turbulent vortices or *kolks*; (c) pothole enlargement and coalescence, leading to the development of butte and basin topography; and (d) the growth of inner channels and cataract recession.

(2) *Streamlined erosional residuals* are isolated loess hills with approximately lemniscate planforms, interpreted as erosional islands, which have been streamlined to reduce the drag or resistance to flowing water.

(3) *Scour marks* are produced wherever obstacles exist within scabland channels. Such obstacles will initiate a horseshoe-shaped vortex and scour hole at their upflow boundaries, and a wake vortex in the downstream zone of flow. Analogous small-scale forms can be observed in wind-blown sand or snow in the vicinity of fence posts or other obstacles. Scour marks may be associated with lee-side depositional forms known as *pendant bars*, depending on flow conditions.

Giant depositional features also occur in the Channelled Scablands, including *bars* tens of metres high on the floor of channels, and *giant current ripples* 1–15 m high and spaced 20–200 m apart (Bretz et al., 1956; Baker et al.,

1987). When viewed from above, the crest forms of the gravel waves are similar to those of normal sand ripples or dunes, but are composed of cross-bedded gravel and boulders (fig. 8.40d; see sections 10.4.4 and 10.4.5). The large size and coarse composition of such bedforms makes the immense size of the Missoula floods all the more impressive.

The range of landforms in the Channelled Scablands has been employed by Benito (1997) to calculate threshold values of stream power for different parts of the flood tracks. These values range from $500 \, \text{Wm}^{-2}$ for the streamlined hills up to $4500 \, \text{Wm}^{-2}$ to produce the major channels. The implications of Benito's calculations also echo the views of Miyamoto et al. (2007) in that major erosion appears possible during only a few of the multiple floods, specifically those that produced the required magnitude to exceed the threshold values for significant geomorphic work.

Since the acceptance of the Missoula Flood hypothesis, proglacial flood tracks have been identified in a number of locations. Cataclysmic flood tracks of a size comparable with those produced by the Missoula floods have been

Figure 8.41 Erosional features associated with a GLOF from moraine-dammed Queen Bess Lake, British Columbia, Canada, in 1997. (a) Flood-scoured bedrock directly below the breached moraines. (b) View down the west fork of the Nostetuko River, showing erosional features cut by the flood waters (Kershaw et al., 2005).

recognized at the margins of proglacial lakes in Siberia by Rudoy et al. (1989), Rudoy (2002), Baker et al. (1993) and Rudoy and Baker (1993). Large channels identified on the floor of the North Sea basin and the English Channel have been interpreted as the erosional products of large meltwater discharges, either beneath or in front of the retreating Scandinavian/British Ice Sheet (Wingfield, 1990; Ehlers and Wingfield, 1991; Gupta et al., 2007). Other channel systems and streamlined residuals thought to have been cut by GLOFs have been described from the Great Plains of North America (e.g. Kehew and Lord, 1987), Sweden (Elfström and Rossbacher, 1985) and Norway (Longva and Thoresen, 1991). At smaller scales, but nonetheless devastating in impact, GLOFs produce narrow corridors of erosion in valley bottoms in mountain landscapes that emanate from breaches in moraine dams. Although these flood tracks are usually characterized by large volumes of debris flow and glacifluvial material, they are also the locations of some considerable incision of alluvial valley fills and stripping of sediment down to bedrock (Clague and Evans, 2000; Cenderelli and Wohl, 2003; Kershaw et al., 2005; fig. 8.41).

Spillways are distinctive types of proglacial channels, produced where water decants from ice-dammed or proglacial lakes over cols or low points on the watershed. Spillways can be cut over a long period of time by regulated overflow water, although catastrophic jökulhlaup-like release of water down spillway courses can produce scabland topography (Kehew and Lord, 1986). An impressive spillway at Newtondale on the North York Moors, England, was cut by water overflowing from an ice-dammed lake on the north side of the Moors (fig. 8.42). The spillway carried the water through the unglaciated highlands of the Moors into Glacial Lake Pickering, which was located to the south of the high terrain and dammed by a glacier lobe on the east Yorkshire coast. Newtondale has been at the centre of prolonged

Figure 8.42 Oblique aerial photograph of Newtondale, North Yorkshire, England. Although its precise origin is a matter of some controversy, the simplest interpretation of this feature is that it is a proglacial lake spillway cut across the unglaciated uplands of the North Yorkshire Moors by meltwater draining from a glacier margin to the north. Note the North Yorkshire Moors railway line in the valley bottom (Aerial photograph BA30, Cambridge University Collection of Aerial Photography).

debate since the early twentieth century, but alternative interpretations have yet to be proven as more appropriate (cf. Glasser, 2002b). An extensive network of undisputed deglacial spillways occurs throughout the Canadian provinces of Alberta, Saskatchewan, Manitoba and Ontario, and the US states of North Dakota, Minnesota, Michigan and Wisconsin, documenting overflow from proglacial lakes during the retreat of the Laurentide Ice Sheet (e.g. Kehew and Lord, 1986, 1987, Clayton and Knox, 2008). Quite often, the release of water from one lake caused receiving lakes downstream to rapidly incise their outlets, thus producing a chain reaction of catastrophic drainages. Some narrow bedrock canyons, like the Ouimet Canyon in Ontario (100 m wide and 100 m deep), and much larger channels, like the Souris Spillway of Saskatchewan and North Dakota (1 km wide and 45 m deep), were excavated over short periods of time during catastrophic drainage. In some situations,

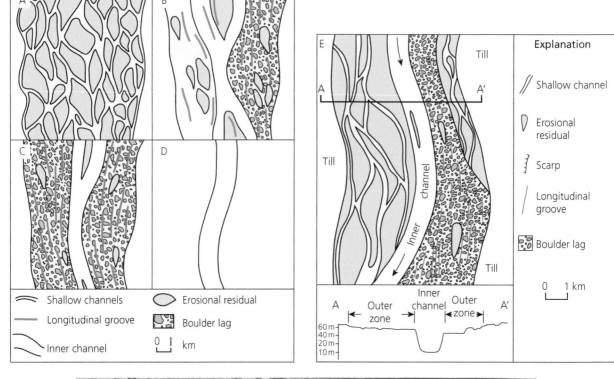

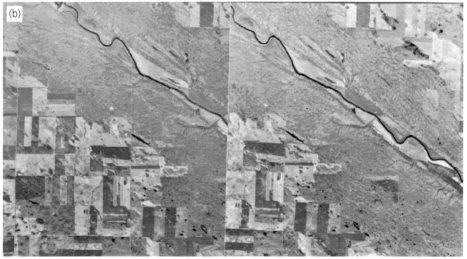

Figure 8.43 Spillways of the North American Great Plains. (a) Schematic diagram of four stages in spillway development. A–D shows a sequential development of: (A) shallow anastomosing channels with irregular inter-channel residuals and no inner channel; (B) fewer, more streamlined residuals, with an anastomosing channel pattern still evident in places and with a poorly developed inner channel; (C) well-developed inner channel, lemniscate residuals and longitudinal grooves; (D) larger inner channel with terrace remnants, few residuals and no outer zone. E shows the final assemblage of landforms and a cross profile at A-A' (Kehew and Lord, 1986, reproduced with permission of the Geological Society of America) (b) Aerial photograph stereopair of the Souris spillway near Hitchcock, Saskatchewan, showing the outer zone of boulder lag, not in use for crop production (Kehew and Lord, 1989). (c) Characteristics of the inner channels of spillways: (i) Aerial photograph mosaic of erosional residuals in the Souris spillway between glacial Lakes Souris and Hind, Manitoba. (Aerial photographs A24965-214, A24966-09 and A24966-40, Manitoba Department of Natural Resources) (ii) Streamlined erosional hill (1.6 km long, 0.5 km wide, 20 m high) in the Souris spillway. (A. Kehew) (d) Characteristics of the outer channels of spillways: (i) The Souris spillway near Hitchcock, Saskatchewan. The outer zone (10 km wide) extends from the foreground to the cultivated land in the distance, contains conspicuous longitudinal grooves and is covered by boulder lag deposits. The inner channel in the centre of the photograph is 1 km wide and 35 m deep. (A. Kehew) (ii) Stereopair of erosional anastomosing channels near Minot, North Dakota. The channel incisions have isolated a large number of erosional residuals (Kehew and Lord, 1986, Aero Service Corporation photographs 340-2471 and 2472).

spillways can resemble subglacial tunnel valleys, where subsequent glacier advance has resulted in the deposition of glacial sediment over the landform (*palimpsest spillways*; Clayton et al., 1999; Kehew and Kozlowski, 2007).

Figure 8.43a shows a fourfold evolutionary sequence of spillway development (Kehew and Lord, 1986, 1987). (A) In the early stages of a flood, no channel exists to carry the outburst discharge, and a system of anastomosing

(c)(i)

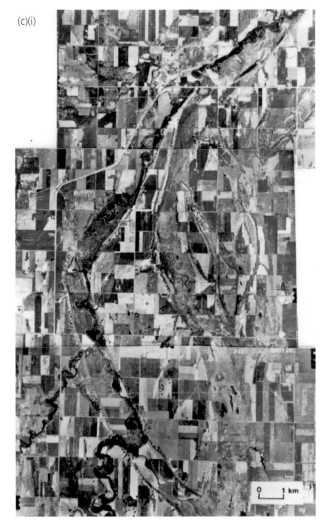

(d)(i)

(d)(ii)

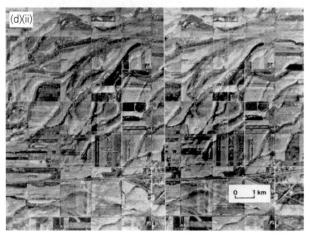

(c)(ii)

Figure 8.43 (Continued)

in the inner channel, causing deep incision. The final form of inner channels of flood spillways is characterized by: (1) a trench-like shape; (2) uniform width and side slopes; (3) regular meander bends; (4) occasional bifurcation to form parallel or anastomosed channels separated by linear ridges or streamlined erosional residuals; (5) occasional isolated and often streamlined erosional residuals. Such channels can be 1–3 km wide and 25–100 m deep. Deposits formed by catastrophic floods are described in section 11.5.1.

Icebergs carried by GLOFs can exert a considerable impact on the landscape by scouring out grooves and depressions. Longva and Thoresen (1991) have identified a variety of iceberg scours and gravity craters formed during a GLOF from Glacial Lake Nedre Glamsjo in Norway (fig. 8.44). Peak discharges were responsible for the erosion of anastomosing channels and intervening streamlined mounds with a maximum relief of 2 m, together with iceberg scour marks up to 1 m deep and 2 km long. As current velocities gradually decreased, the icebergs were arrested while still scouring, and where they finally came to rest they produced *gravity craters*, consisting of shallow, semicircular depressions up to 30 m wide. Some icebergs were refloated after making one gravity crater and thus produced *crater chains*; this may have been achieved either by a fast drifting iceberg with a

channels is eroded. (B) As discharges rise, a central channel is cut, leaving erosional remnants of the earlier anastomosing channel system in an outer zone. Scouring of the outer zone produces longitudinal grooves and streamlined erosional hills, and boulder lags represent particles too large to be carried along by the flow. (C) Channel incision and scouring of the outer zone continue. (D) Flow gradually becomes increasingly focused rocking motion, by reflotation after berg break-up or by short pulses of increased discharge. Some icebergs appear to have been reoriented without being refloated, producing a close cluster of gravity craters. The association of the Norwegian craters with an obvious flood track appears to support their origin by iceberg impact. Kettle holes created by the melt-out of buried ice blocks are described in section 11.4.6. A particularly diagnostic type of iceberg scour on the floor of Hudson Bay has been used in conjunction with channels and large sand waves as the final drainage pathway for Glacial Lake Agassiz (Clarke et al., 2004; Dyke, 2004). As the Laurentide Ice Sheet shrank to a narrow ridge separating Lake Agassiz in the south from the Tyrrell Sea to the north at around 7.7 ka^{14}C years BP, the lake drained catastrophically beneath the ice sheet. The rapidly flowing water mobilized large, multi-keeled tabular icebergs in the Tyrrell Sea, causing them to swirl in the current and pivot in turn on different keels, while other icebergs inscribed circular ridges on the sea floor. The unidirectional convexity of the scours reflects a consistent current flow direction.

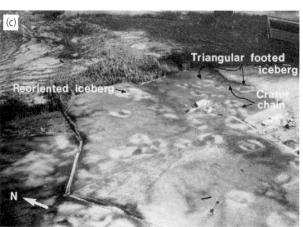

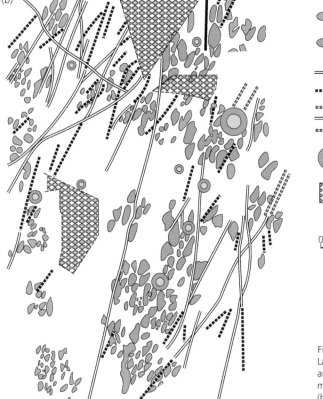

	Elongate mounds
	Elongate mounds used for statistics
	Iceberg scours
	Diffuse iceberg scours
	Youngest scour / Oldest scour / Berm
	Iceberg gravity crater with pressure ridge
	Bushes

0 ⟶ 500 M

Figure 8.44 Iceberg scours and gravity craters in flood tracks from Glacial Lake Nedre Glamsjo, Norway. (a) Aerial photograph (Fjellanger Widerøe) of an area near the junction of the Vorma and Glomma rivers, showing elongate mounds (light patches) and iceberg grounding and scour features (dark lines). (b) Geomorphological interpretation of the area shown in (a). (Longva and Thoreson, 1991) (c) Iceberg gravity craters and crater chains (After T.H. Bargel).

Smaller-scale features produced by catastrophic flooding by glacial meltwater include scratches or 'striations' on bedrock surfaces. Although normally attributed to direct glacial abrasion, striations have been associated with GLOFs by McCarroll et al. (1989). Such scratches are distinguishable from glacially abraded striations because they have low length: width ratios and a tapering shape, indicating a striking blow rather than steadily applied pressure. Furthermore, they possess numerous cross-cutting directions, which should not be misinterpreted as the products of directional changes in glacier flow.

8.5 LARGE-SCALE EROSIONAL LANDFORMS

Large-scale erosional landforms constitute some of the most impressive geomorphological features on the Earth's surface, providing striking testimony to the immense erosive potential of glacier ice over long timescales. In this section, we discuss the characteristics and genesis of basins and overdeepenings in bedrock and unconsolidated sediments, troughs and fjords, cirques and strandflats.

8.5.1 ROCK BASINS AND OVERDEEPENINGS

Basins can be eroded into bedrock by glaciers at a wide variety of scales, from small hollows between roches

Figure 8.45 Rock basins separated by a rock bar (riegel) in a cirque in Kananaskis Country in the Canadian Rockies (D.J.A. Evans).

moutonnées to large overdeepenings occupying the full width of cirques and troughs. Large overdeepenings are common in deglaciated valleys, where they are generally occupied by lakes (fig. 8.45), and have been identified beneath many existing glaciers in the course of radio echo-sounding studies (fig. 8.46). In addition, numerous rock basins can occur together in ice-scoured *knock and lochan* topography (Rea and Evans, 1996; section 8.6.1). The formation, size and shape of rock basins is controlled by glaciological variables, such as the thermal regime and stress conditions at the glacier bed, and substratum characteristics, particularly bedrock structure and lithology. The main processes of rock basin erosion are quarrying and abrasion (sections 8.2.2 and 8.2.3). Effective quarrying and abrasion can only occur where the glacier bed is predominantly wet-based, such as the beds of temperate glaciers and the base of fast-flowing ice streams. Additionally, quarrying is most effective where there are large water-pressure variations at the bed, which encourage the large transient stress gradients required for rock failure (Iverson, 1991a; Hallet, 1996). The role of water-pressure variations on the erosion of large overdeepenings below valley glaciers has been examined by Hooke (1991), using data from Storglaciären. He argued that quarrying is focused on the lee side of steps on the bed, because these locations favour the formation of surface crevasses, which allow surface meltwater to reach the bed (fig. 8.46). Fluctuating water pressures will encourage rapid erosion downglacier from rock steps, overdeepening the bed profile. In turn, the evolving glacier bed topography favours continued surface crevassing, so that overdeepenings of the glacier bed will be amplified over time by a positive feedback. A similar process may operate on a small scale, whereby cavities become enlarged by differential erosion.

Bedrock structure and lithology control bed resistance to quarrying and abrasion, and thus exert a strong influence on the location and morphology of rock basins (Sugden and John, 1976; Gordon, 1981). In addition, rock weaknesses may have been partially excavated by non-glacial processes prior to glacier advance. Preglacial weathering can isolate resistant bedrock protuberances and weaken well-jointed rock in the intervening areas,

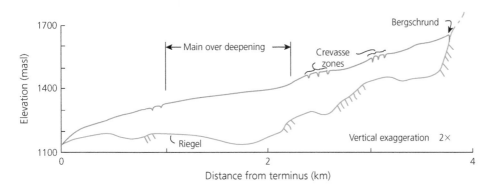

Figure 8.46 Longitudinal section through Storglaciaren, Sweden, showing the location of surface crevasse zones, the bedrock profile and inferred zones of enhanced quarrying (Modified from Hooke, 1991)

preparing the way for effective subglacial erosion (Johansson et al., 2001a, b). Lithological controls on rock basin development were emphasized by I.S. Evans (1994), based on studies of cirques in the English Lake District and the Cayoosh Range of British Columbia, Canada. A very strong relationship between regional fault patterns and the orientation of the proglacial and deglaciated lakes of Patagonia has been demonstrated by Gonzales and Aydin (2008). They infer that erosion will have been concentrated in the areas of faults and their damage zones, resulting in glacially overdeepened and widened valleys, comprising segments and tributary arms that are aligned obliquely due to the control of bedrock structure.

Shallow basins are commonly elongated because: (1) they are guided by lithological changes or joint/fault systems in exposed strata; (2) glacier ice excavates preferentially along those zones of weakness lying sub-parallel to ice flow direction; and (3) preglacial fluvial relief may be preferentially overdeepened. Large volumes of relatively soft bedrock can be removed when glaciers exploit fluvially dissected terrain. For example, Clayton (2000) has estimated the volume of Jurassic mudrock and Cretaceous chalk removed by the Anglian Ice Sheet margin to produce the English Wash and Fen basins once the ice had penetrated the chalk escarpment. The volume of $555 \, km^3$, if it was all excavated during 20 ka of Anglian glaciation, converts to an erosion rate of up to $4000 \, mm \, ka^{-1}$. Much of this material ($363 \, km^3$) can be accounted for in the estimated volume of the regional chalky till. The efficacy of glacial erosion in this case study is illustrated by averaging out the total erosion rate over the 440 ka since the Anglian ($167 \, mm \, ka^{-1}$) and comparing it to the same rate for the nearby and unglaciated Thames basin, which is underlain by less resistant Oxford clay ($45 \, mm \, ka^{-1}$).

8.5.2 BASINS AND OVERDEEPENINGS IN SOFT SEDIMENTS

Erosional basins also occur in areas underlain by soft sediments, and are thought to result from three main processes:

(1) Where the sediments form a rigid bed, erosion may occur by *plucking*, as for rockbeds. Other factors being equal, erosion rates should be higher than for hard rockbeds, so that the location and form of basins may simply reflect the distribution of pre-existing sediments.

(2) *Glacitectonic thrusting* near glacier margins can excavate large basins from sub-marginal positions, the eroded material being repositioned beyond the ice margin as ice thrust masses (Aber et al., 1989; Aber and Ber, 2007). Basins formed in this way commonly occur immediately up-ice of thrust-block moraines, marking major ice margin positions. Thrust masses and basins occurring together are known as *hill-hole pairs*, and in some cases the dimensions of the thrust masses correspond closely to those of the basin. Hill-hole pairs are reviewed in detail in section 11.3.2. A conceptual model of glacitectonic disturbance and basin overdeepening based on structural studies of thrust moraines in Germany and Spitsbergen is shown in figure 8.47. The lateral margins of basins formed by thrusting are commonly straight and parallel to the former ice flow direction, reflecting the location of *tear faults* at the boundaries of the thrust mass. Where ice-marginal thrust belts are overridden by continued glacier advance, excavational basins and thrust masses may appear smoothed or even streamlined.

(3) On a *deformable bed*, erosion may occur by the net advection of deforming sediment away from the site, leading to the production of overdeepenings beneath

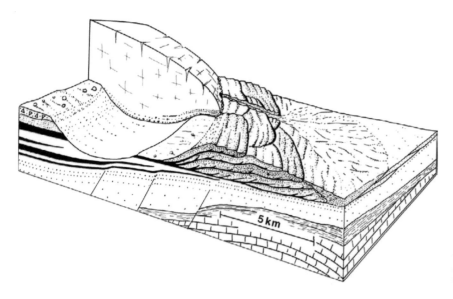

Figure 8.47 Basin excavation associated with thrust moraine formation at the margins of an ancient ice lobe, Dammer Berge, Germany (van der Wateren, 1995).

glacier snouts. Boulton (1996a) has argued that erosion by this process should be most rapid just upglacier from glacier equilibrium lines, where the balance velocity is rising downglacier and glacier flow is extensional (see section 5.2.1). According to this view, therefore, erosional basins or overdeepenings should occur beneath the lower part of the glacier accumulation area and will be deepest where glacier snouts have occupied the same position for a prolonged period, transporting material by a variety of subglacial processes to marginal till stacks (Krüger, 1994; Evans and Hiemstra, 2005). Several variables will complicate this simplified cause-and-effect relationship,

including the influence of sediment properties and water pressures on deformation rates, and the influence of topography on glacier thicknesses and velocities. Additionally, subglacial deformation may preferentially enlarge pre-existing basins produced by ice-marginal tectonics and other processes.

8.5.3 TROUGHS AND FJORDS

The most spectacular manifestations of glacial erosion are troughs and fjords carved by ice flow through major rock channels (fig. 8.48). The differences between glacial and fluvial valleys are often expressed in visual terms, but

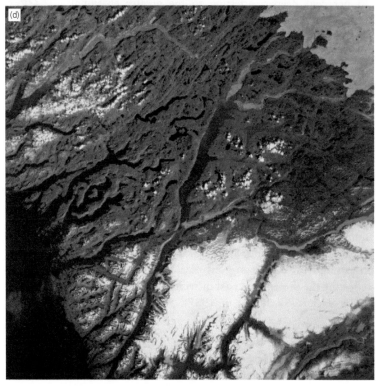

Figure 8.48 Troughs formed by deep erosion by outlet glaciers. (a) Yosemite Valley, California. (Landform Slides) (b) Western Brook Pond, a large lake occupying an overdeepened trough in Gros Morne National Park, western Newfoundland, Canada. The sea can be seen in the distance, the valley having been a fjord until glacio-isostatic uplift raised its threshold above sea level. (c) Sognefjord, Norway. (D.J.A. Evans) (d) Landsat image of part of the coast of west Greenland, with the Greenland Ice Sheet visible at top right. The longest fjord in this view is Sondre Stromfjord. Note the dendritic pattern of the fjords and the numerous lakes, indicative of areal scour on the highlands (Landsat image 21654-14185 (C), 8-3-79, Canadian Centre for Remote Sensing).

Fjord district	Fjord stage*	Tidal range[†]	River discharge[‡]	Climate	Sedimentation rate[#]
Greenland	1, 2	Medium to low	High	Subarctic to arctic	Medium to high
Alaska	1, 2, 3, 4	High	Low to high	Subarctic maritime	Medium to high
British Columbia	3, 4	High	Medium to high	Temperate maritime	Medium to high
Canadian maritimes	4, 5	Low to medium	Low to high	Subarctic to temperate maritime	Low
Canadian Arctic archipelago	1, 2, 3, 4	Low to high	Low to medium	Arctic desert to maritime	Low to medium
Norwegian mainland	3, 4	Low	Low to medium	Subarctic to temperate maritime	Low
Svalbard	2, 3	Low	Low	Arctic island	Medium
New Zealand	4, 5	Medium	Low to medium	Temperate maritime	Low to medium
Chile	2, 3, 4	Low	Low to high	Temperate to subarctic maritime	Medium to high
Scotland	4, 5	Low to high	Low	Temperate maritime	Low

*Stage 1 = glacier filled; stage 2 = retreating tidewater glaciers; stage 3 = hinterland glaciers; stage 4 = completely deglaciated; stage 5 = fjords infilled
[†]low = <2 m mean range; medium = 2–4 m mean range; high = >4 m mean range
[‡]low = <50 m^{3}s^{-1} mean annual discharge; medium = 50–200 m^{3}s^{-1}; high = >200 m^{3}s^{-1}
[#]low = <1 mm yr^{-1} average over entire fjord basin; medium = 1–10 mm yr^{-1}; high = >10 mm yr^{-1}

Table 8.2 Typical characteristics of major fjord coastlines (Syvitski et al., 1987)

have been quantified by Amerson et al. (2008) based on a sample set from Idaho, USA. They demonstrated that the cross-sectional areas of glaciated valleys are 80 per cent greater than their fluvial counterparts and also tend to be 30 per cent deeper, attesting to the greater erosional capacity of glacial systems in valley settings. The largest troughs on earth, the Thiel and Lambert troughs, are presently occupied by outlet glaciers of the Antarctic Ice Sheet and are approximately 1000 km long, >50 km wide and up to 3.4 km deep (Drewry, 1983; BEDMAP, 2002). Very long and deep open fjords (>100 km long and up to 3 km deep from mountain crest to sea bottom) occur along the coasts of Greenland, Norway and the Canadian Arctic islands of Ellesmere and Axel Heiberg, the longest in each region being Nordvestfjord/Scoresby Sund (300 km), Sognefjord (220 km) and Greely Fiord/Nansen Sound (400 km), respectively (see table 8.2 for a summary of fjord characteristics). Fjords are also well developed along the glaciated coasts of British Columbia in Canada, southern Chile, Iceland, Svalbard, western Scotland and the south-west corner of New Zealand's South Island, reflecting the present and past distribution of low-altitude outlet glaciers (Syvitski et al., 1987).

Trough cross profiles

The cross profiles of troughs and fjords are often referred to as U-shaped, although in reality they tend to be asymmetric, with one steep and one gentler slope, or approximately parabolic (Graf, 1970). Glacial valley cross profiles have been the subject of some advanced mathematical analysis (Hirano and Aniya, 1988, 1989, 1990,

2005; Harbor, 1990; Morgan, 2005), but the form of many troughs and fjords can be approximated by a power law model (e.g. James, 1996; Li et al., 2001):

$$D_V = aW_V^b \tag{8.3}$$

where W_V = valley half-width, D_V = valley depth, and a, b = constants.

For many troughs, the exponent b is approximately equal to 2. A complete description of trough form also requires the relative dimensions to be defined. Graf (1970) suggested using a *form ratio*:

$$F_R = D_V/w_v \tag{8.4}$$

where w_v = valley top width, and obtained form ratios of 0.242–0.445 for troughs in the Beartooth Mountains of Montana and Wyoming. Increases in F_R values indicate valley narrowing. A widely cited statistical analysis that attempts to identify trends in the development of glacial valley morphology is the b-F_R model of Hirano and Aniya (1988). This is depicted by the b-F_R diagram, which plots the exponent b from the power law model (equation 8.3) against F_R from the from ratio (equation 8.4). A b-F_R diagram containing plots from studies on glacial valleys in a variety of settings is depicted in figure 8.49. The Antarctica, Patagonia and Canadian Cordillera data on this diagram were used by Hirano and Aniya (1988) to propose two models of valley development: 1) the Rocky Mountain or alpine model, in which valleys deepen without widening; and 2) the Antarctica–Patagonia or ice sheet model, in which valleys widen without deepening.

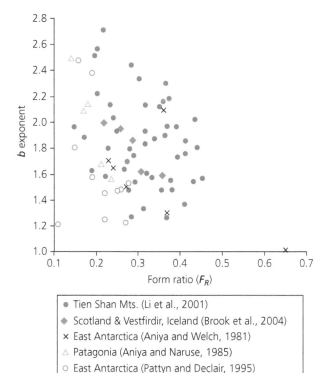

Figure 8.49 The 'b-F_R' diagram of Hirano and Aniya (1988), with data from various glacial valleys (Data from Aniya and Welch, 1981; Aniya and Naruse, 1985; Pattyn and Decleir, 1995; Li et al., 2001; Brook et al., 2004).

Although subsequent studies have delivered data that appear to substantiate the b-F_R concept, attention has been drawn towards the role of bedrock characteristics in glacial valley profile development (Augustinus, 1992a; Pattyn and Decleir, 1995; Brook et al., 2004). For example, Brook et al. (2004) demonstrated clear statistical relationships between rock mass strength and various measures of glacial valley cross-profile morphology, including F_R and exponent b.

In reality, trough cross profiles may depart significantly from the ideal parabolic form. For example, Sognefjord exhibits a major break in slope at present sea level, with relatively gentle slopes above the water line and steep slopes below sea level. Nesje and Whillans (1994) have argued that the upper slopes have been modified by subaerial weathering and denudation, whereas the lower slopes preserve their glacial form due to the support provided by the water.

The evolution of trough cross profiles has been modelled by Harbor (1992), using theoretical velocity distributions in a valley glacier cross section (fig. 8.50a). For a V-shaped valley, modelled basal velocities are highest partway up the valley sides, and lowest below the glacier margins and centreline. If it is assumed that erosion rates are proportional to the sliding velocity, the most rapid erosion will occur on the valley sides, causing broadening and steepening of the valley. Eventually, an equilibrium profile is attained, which continues to deepen over time. Pulsed erosion over multiple glacier cycles produce a

series of breaks in slope and overhangs on the trough walls above the downcutting glacier (fig. 8.50b). Such slopes are likely to be unstable and prone to collapse, thus modifying the form of the trough by paraglacial reworking. For this reason, oversteepened, ice-free slopes in troughs are termed the *zone of glacial influence*, in which slope evolution is strongly influenced by the contemporary or former presence of glacier ice (fig. 8.50c). Verification of the patterns or erosion predicted by Harbor's model has been provided by calculations of erosion rates using terrestrial cosmogenic nuclides (Fabel et al., 2004).

Parabolic cross profiles may develop with the aid of pressure-release mechanisms in the underlying bedrock. It has long been recognized that freshly exposed rock will undergo expansion (*dilation* or *dilatation*) due to removal of the confining pressure, and engineers involved in road and rail construction are well aware of the problems of pressure release in freshly cut rockfaces. Rock dilation results in the development of fractures parallel to the ground surface known as dilation joints or sheeting, particularly on massive rocks where there are few pre-existing discontinuities along which strain may be accommodated. Dilation critically weakens rock masses, facilitating subsequent subglacial erosion. The most obvious time for dilation to take place is after deglaciation, when the ice overburden is removed and freshly eroded rock surfaces are exposed, although it is probably also important beneath active ice, once glacial erosion has removed rock slabs that have been fully released along dilation joints during a previous non-glacial period. Thus, pressure release may serve to accelerate subglacial erosion in favourable locations.

Erosion rates in troughs that carry major outlet glaciers are likely to be greater than those below tributary glaciers, resulting in *hanging valleys* perched above the main trunk floor. Hanging valleys may exhibit little or no evidence of glacial erosion and may retain preglacial profiles, particularly in areas where the ground adjacent to troughs was occupied by thin, cold-based ice (Sugden and John, 1976).

Trough long profiles

Troughs have been classified by Linton (1963) as:

(1) *alpine*, cut by valley glaciers emanating from high ground;

(2) *Icelandic*, which have a closed *trough head* at the upper end and have been cut by ice spilling over from a plateau ice cap or ice sheet;

(3) *composite*, which are essentially *through troughs* or *through valleys* open at both ends and cut beneath an ice sheet;

(4) *intrusive* or *inverse*, cut against the regional slope where ice impinges on the downstream ends of highland valleys (e.g. New York Finger Lakes).

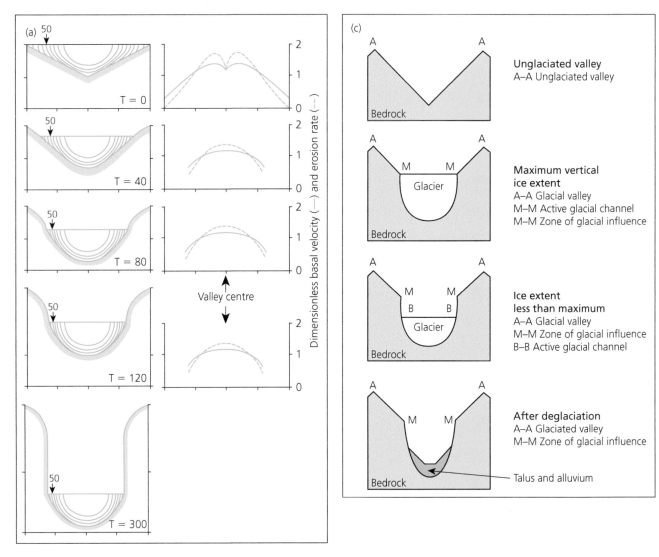

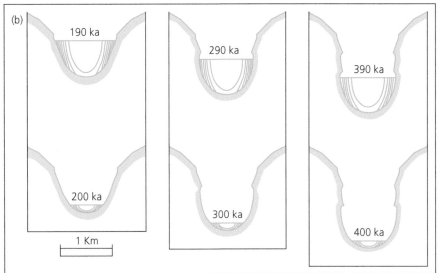

Figure 8.50 Evolution of trough cross-profiles, as modelled by Harbor et al. (1988) and Harbor (1992). (a) Evolving valley cross-sections at different time steps (T = 0–300) and corresponding basal velocities and erosion rates. (b) The development of irregular cross profiles due to glacier expansion and contraction on 100 ka cycles. (c) Schematic evolution of troughs by erosion in the active glacial channel (B-B) and subaerial slope modification in the zone of glacial influence (M-M). At the maximum vertical ice extent, the active glacial channel and the zone of glacial influence coincide (Harbor, 1992b, reproduced with permission of the Geological Society of America).

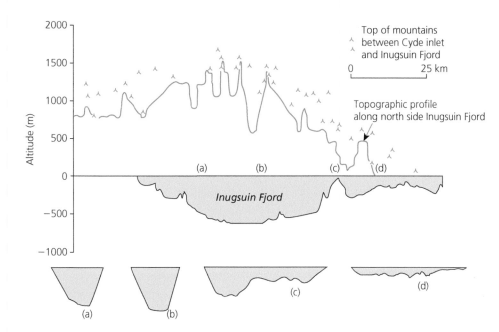

Figure 8.51 Relationship between the depth of Inugsuin Fjord, Baffin Island, and surrounding topography. The fjord is deep and narrow in the vicinity of high mountains, but wide and shallow towards the coast (Sugden and John, 1976, after Loken and Hodgson, 1971).

Trough heads represent important process thresholds in the landscape, and may form beneath ice sheets at the upper end of ice streams where erosion rates suddenly increase, or at the head of valley glaciers. The location of trough heads may be influenced by several factors, including: (1) a pre-existing valley step; (2) increases in ice discharge due to the convergence of two or more tributary glaciers; or (3) lithological or structural variation in the bedrock leading to a downglacier increase in erodibility.

In contrast with river valleys, troughs and fjords commonly have *overdeepenings* on their floors (section 8.5.1). Overdeepened basins commonly form where ice discharges are relatively high, such as at the junctions of tributary valleys, or at narrowings in the valley profile. Overdeepened sections terminate at a *sill* or *threshold*, which are commonly located where ice flow becomes less constrained and velocities decrease. This is clearly illustrated by the relationship between topography and basin depth along Inugsuin Fjord on Baffin Island, Canada, which has been overdeepened by glacier ice converging on it from the surrounding mountains (fig. 8.51; Løken and Hodgson, 1971). The area of deepest erosion in troughs and fjords marks the long-term average position of maximum ice discharge, which may broadly coincide with the equilibrium line on the ice surface. In coastal mountains, the shallowing of fjords and the occurrence of thresholds may be associated with the increased bouyancy and eventual flotation of glacier ice in the marine environment, in addition to the flow divergence induced by the more open topography (Shoemaker, 1986b). The glacio-isostatic rebound that accompanies deglaciation can result in the emergence of fjord thresholds to produce numerous low-lying bedrock islands (*skerries*), such as those at the mouths of the Norwegian fjords, or landlocked lake

basins, such as those in western Newfoundland, Canada. Some fjords have simple, single basin forms (e.g. Milford Sound, New Zealand), whereas others have several basins and multiple sills (e.g. Hardangerfjord, Norway). These contrasting forms give rise to profoundly different postglacial sedimentation patterns.

Alley et al. (2003a, b) have argued that the gradient of the reverse slope at the downglacier end of overdeepenings is limited by glacio-hydraulic processes. Water flowing up a reverse slope at the glacier bed experiences an increase in elevation head, and because the total hydraulic potential must decrease downflow, water pressure must drop by at least as much (section 3.2.2).

If water at the pressure-melting point undergoes a drop in pressure, it becomes supercooled with respect to the new conditions, and tends to freeze. If other factors are equal, the supercooling effect increases with the gradient of the reverse slope. That is, the steeper the reverse slope at the downglacier end of a trough, the greater the amount of basal freezing. Effective erosion of the bed requires temperate basal conditions, so glacio-hydraulic supercooling will act as a check on erosion rates. This will therefore create a stabilizing feedback on patterns of erosion. Subglacial erosion deepens the trough, which increases the gradient of the reverse slope at its downglacier end. Conversely, steepening of the reverse slope will reduce erosion. Through time, therefore, the slope will assume a form in which these effects are in balance.

Using a simple, equilibrium model of subglacial hydrology, Alley et al. (2003a, b) argued that the gradient of the reverse slope should be ~50 per cent greater than (and in the opposite direction to) the ice surface slope. As they acknowledged, however, this conclusion is based on the assumption of steady state on both short and long timescales. It is more realistic to regard the form of overdeepenings as the cumulative effect of non-steady processes acting over long periods of time.

Another aspect of trough/fjord form critical to understanding glacial erosion patterns is the stepped long profile, which comprises basins and intervening *riegels*, or transverse rock bars. Irregularities in the glacier bed may arise from several factors, the most important of which are: (1) spatial variations in bedrock lithology and structure; (2) changing glacier discharge as a result of ice confluence or diffluence; and (3) pre-existing aspects of the relief. Anderson et al. (2006) have modelled the evolution the glacial valleys in order to elaborate on Sugden and John's (1976) explanation of stepped long profiles, specifically that the deepest erosion takes place below the long-term ELA (MacGregor et al., 2000; Tomkin and Braun, 2002). Their results indicate that additional overdeepening and the production of trough heads takes place at the points where tributary glaciers join the main trunk ice, echoing the point made above about Inugsuin Fjord. Therefore, the pattern of erosion intensity is inherited from preglacial fluvial networks.

The influence of geological structure and topography

Several researchers have argued that the location and planform of troughs and fjords reflects pre-existing topography or geological structures. The sinuous or meandering courses that some troughs and fjords display are often interpreted as preglacial legacies, where glacier ice has excavated pre-existing fluvial valleys (e.g. Augustinus, 1992b). Such interpretations are supported where traces of largely unmodified fluvial valleys occur in the same landscape (e.g. Linton, 1963), or fjord systems display dendritic patterns, such as in the Canadian Arctic and Fiordland, New Zealand (Augustinus, 1992b). Numerical modelling of landscape evolution due to glaciation suggests that fluvial forms will begin taking on glacial signatures after approximately 100 ka of glaciation (Jamieson et al., 2008). The juxtaposition of glacially overdeepened and unmodified valleys is a product of differential modification according to valley alignment and former ice flow direction. Those valleys orientated parallel to former ice flow are most likely to carry large volumes of ice and experience the most intense glacial erosion.

Open-ended or *through troughs*, especially those which have breached pre-existing watersheds, are more difficult to explain, but probably result from the streaming of ice down preglacial fluvial valleys and the headward erosion of trough heads. Such a change in the basal topography over time can result in the capture of glacier ice and a radical change in basal flow directions. Alternatively, glacier ice may exploit regional zones of structural weakness in the bedrock, such as fault systems, and therefore bear no resemblance to preglacial drainage patterns; troughs excavated along such zones are entirely glacial in origin.

The role of geologic structures and pre-existing structural forms (e.g. grabens, regional strike-slip faulting) in the funnelling of glacier ice has been stressed by numerous researchers. The close affinity between major ice streams and geologic features is certainly obvious at some locations; for example, Lambert Glacier in Antarctica occupies the Lambert Graben, a major structural feature. Just as the sinuosity of troughs and fjords has been linked to pre-existing fluvial valleys, so trough and fjord alignments have been linked to bedrock lineaments such as faults and intrusions. The rectilinear pattern of some fjord systems has been linked to intersecting lines of fracture at a regional scale. Careful mapping of structural lineaments and intrusions in Norway (Nesje and Whillans, 1994), Ellesmere Island in the Canadian Arctic (England, 1987) and Chile (Glasser and Ghiglione, 2009) has revealed a close relationship between the orientation of such bedrock features and fjords/valleys. Furthermore, the remarkably straight and parallel cliffs of Somerset, Devon, and the north-west Baffin Islands in the Canadian Arctic, which form virtually unbroken coastlines, in places up to 200 km long, have been cited as evidence for large-scale block faulting by England (1987). This interpretation for the inter-island channels of the central Canadian Arctic is supported by the occurrence of remnants of formerly continuous Tertiary river valleys on the islands (horsts) and on the floors of the channels (grabens; Dyke et al., 1992). Rift valleys in non-glaciated regions provide clear examples of tectonically controlled landforms and it is not unreasonable to suggest that such landforms exist in glaciated regions, especially where geological evidence provides support for a tectonic origin. However, it is most likely that a continuum of landforms exists, ranging from tectonically controlled grabens, through glacially modified river and fault systems, to entirely glacially eroded troughs and fjords. The fjords and troughs of northern Ellesmere Island contain several contrasting elements, such as straight walls, rectilinearity and sinuous reaches, indicating that a combination of factors (some perhaps more dominant than others) is critical to the glacial excavation of such landscapes (fig. 8.52).

An argument in favour of the tectonic origin of some inter-island channels, such as those of the central Canadian Arctic archipelago, is the lack of an integrated upland glacier source area, like those surrounding the fjord coastlines of Norway, British Columbia, Chile and New Zealand, and the absence of a nearby continental interior that could supply large volumes of ice. In addition, some of the largest channels, such as Lancaster Sound, are oriented transverse to the former ice flow of the Laurentide Ice Sheet. In contrast, the troughs and

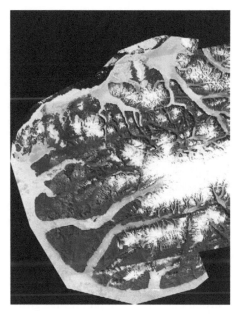

Figure 8.52 Landsat image of the sea ice-filled fjords of north-west Ellesmere Island, Canadian High Arctic, showing remarkably long, straight fjord walls and both rectilinear and sinuous fjord patterns.

fjords presently occupied by the outlet glaciers of the Greenland and Antarctic Ice Sheets lie parallel to ice flow and are much more likely to have been eroded by the ice draining through them. Similarly, the evolution of the Baffin Island fjords, which are oriented parallel with former ice flow at the margins of the Laurentide Ice Sheet, has been numerically modelled by Kessler et al. (2008), and found to be consistent with glacial erosion of pre-existing fluvial topography due to a positive feedback on accelerated ice flow. Their model implies that the present topography was largely eroded by ice in a million years. Nonetheless, an exhaustive study by Glasser and Ghiglione (2009) has revealed a striking similarity between fjord orientations and structural lineaments in Chile, prompting the conclusion that geological structure is the primary control on fjord development in this region.

8.5.4 CIRQUES

Cirque morphology

Cirques (also known as *corries or coires* in Scotland, and *cwms* in Wales) are among the most characteristic forms of glacial erosion in mountainous terrain (fig. 8.53). Evans and Cox (1974) defined a cirque as:

> a hollow, open downstream but bounded upstream by the crest of a steep slope (headwall), which is arcuate in plan around a more gently sloping floor. It is 'glacial' if the floor has been affected by glacial erosion while part of the headwall has developed subaerially, and a drainage divide was located sufficiently close to the top of the headwall for little or none of the ice that fashioned the cirque to have flowed in from outside.

The typical cirque profile (flat-floor/overdeepened basin connected to a steep backwall by a concave slope) is a product of subglacial erosion on the cirque floor and lower backwall, and subaerial frost action on the upper backwall, the exact form being controlled by bedrock structure. Cirques are rarely simple features; their headwalls may be poorly developed or missing, they may have rectilinear planforms, or they may occur nested one within the other (e.g. Gordon, 1977; Bennett, 1990). The following types can be defined:

(1) *simple cirques*, which are distinct, independent features;
(2) *compound cirques*, in which the upper part consists of two subsidiary cirques of approximately equal size;
(3) *cirque complexes*, in which the upper part consists of more than two subsidiary sidewall or headwall cirques;
(4) *staircase cirques*, where two or more cirques occur one above the other;
(5) *cirque troughs*, where the cirque marks the upper end of a trough.

Some cirques may have attributes of more than one category – for example, compound or staircase cirques forming trough heads. A wide range of cirque sizes exists, ranging from valley-side niches less than 50 m from backwall to lip, to cirques with floors several kilometres long and forming *alpine valley heads* (Gordon, 1977; Haynes, 1995).

Attempts to provide general classifications of cirque form have employed a variety of dimensions (e.g. Gordon, 1977). Some commonly used measurement parameters are: *length* from headwall to threshold; *width* between the sidewalls; *depth* from headwall crest to floor; *threshold elevation*; *area*; *volume*; and *aspect*. The long profiles of cirques have been generalized mathematically. Haynes (1968), for example, used logarithmic curves of the form:

$$y = k \, (1 - x) \, e^{-x} \tag{8.5}$$

where y is altitude, x is distance along the profile and k is a constant describing the concavity of the long profile. Where $k = 2$, the basin is deep with a steep headwall, and where $k = 0.5$, the cirque is shallow with a flat or outward-sloping floor. Thus, high k values correspond with well-developed, overdeepened cirques (fig. 8.54).

Several attempts have been made to reconstruct sequences of cirque evolution using *ergodic models*, in which a continuum of forms in space is used as a substitute for evolutionary development through time. This approach is common in geomorphology, where long-term processes cannot be measured, and is based on the idea that in areas with evolving landforms there will be examples at various stages of development, which can be arranged in a hypothetical developmental sequence (e.g. Schumm and Lichty, 1965). This sequence is reflected in the scale of cirque development devised by

Evans and Cox (1995), based on morphology. They identified five 'grades' (cf. I.S. Evans, 2007), defined as:

- Grade 1 – classic; with all textbook attributes.
- Grade 2 – well defined; headwall and floor clearly developed, headwall curves around floor.

- Grade 3 – definite; no debate over cirque status, but one characteristic may be weak.
- Grade 4 – poor; some doubt, but well-developed characteristics compensate for weak ones.
- Grade 5 – marginal; cirque status and origin doubtful.

In general, as cirque size increases, the degree of enclosure of the planform and long profile increases (Gordon, 1977; Evans and Cox, 1995), indicating that cirques develop from hollows by progressive retreat of the backwall and flattening of the floor, the latter resulting in slope reversal and overdeepening (fig. 8.55a). Rates of increase in length, width and depth vary, but it is generally cirque length that develops fastest. This has been termed *allometric* cirque development by I.S. Evans (2006), illustrated clearly by allometric plots of cirque length, width and height, or amplitude against size, for Wales and the English Lake District (fig. 8.55b). It has been argued that this evolution reflects the progressive growth of the

Figure 8.53 Typical cirque forms. (a) A cirque presently occupied by a small glacier below the aptly named Cirque Mountain, Torngat Mountains, Labrador, Canada. Several neoglacial moraines can be seen in the foreground (D.J.A. Evans) (b) Map of the cirques surrounding Cader Idris, North Wales. The lakes Llyn Cau and Llyn y Gadair occupy rock basins, but are also partially moraine-dammed (Embleton and King, 1975).

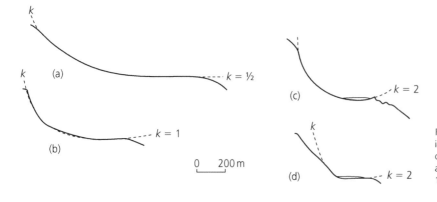

Figure 8.54 'k-curves' fitted to cirque long profiles in Scotland. Eighty-one per cent of the sampled cirque long profiles closely resemble k-curves, although some, such as (d), do not (After Haynes, 1968, reproduced).

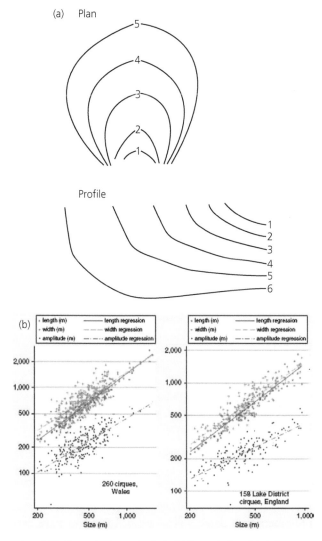

Figure 8.55 Cirque development criteria. (a) Conceptual model of cirque development by the progressive enlargement and deepening of a hollow (Gordon, 1977, reproduced with permission of Wiley-Blackwell). (b) Allometric plots of length, width and amplitude (height) against size for cirques in Wales and the English Lake District. Note that the gradient for amplitude is lower than those for length and width, attesting to slower rates of deepening over time (I.S. Evans, 2007).

occupying snow and ice body, from a snow patch in a *nivation hollow*, through a larger snowpatch in a *nivation cirque*, to a glacier in a true cirque. However, this view is probably highly simplistic, and cirques are likely to evolve from a variety of pre-existing hollows in several episodes of glacier occupancy (Evans and Cox, 1995).

Processes of cirque formation

The enlargement of mountainside hollows can occur by snowpatch erosion or *nivation*, which embraces a range of weathering and transport processes accelerated by late-lying or permanent snow (Ballantyne and Harris, 1994). Nivation processes include intensive freeze–thaw activity, enhanced chemical weathering, slopewash, debris transport by snowcreep or slip, and solifluction, which are encouraged by repeated freeze–thaw cycles

and/or abundant meltwater during the summer months. The relative importance of these processes, and their variation with climate and snowpatch thickness, are very poorly known. Mountainside hollows can also originate by fluvial erosion or large-scale slope failures. True cirque formation occurs once hollows are occupied by glacier ice – that is, where the ice is sufficiently thick and steep to generate the stresses required for internal creep deformation and basal sliding. Ballantyne and Benn (1994a) argued that the threshold length between snowpatches and glaciers is approximately 30–70 m from backwall to toe and it has been demonstrated that small mountainside niches can host eroding glacierets (Carr, 2001a) and even eroding snowpatches (Shakesby et al., 1999).

Glaciers enlarge cirques by subglacial quarrying and abrasion of the cirque floor (e.g. Bennett et al., 1999b), while subaerial slope retreat of the bounding rockwalls due to oversteepening supplies debris to the glacier surface. Effective subglacial erosion and cirque downcutting requires that the ice is at least partially wet-based, and overdeepened rock basins mark sites where subglacial erosion was particularly effective. Possible scenarios include:

(1) cirque occupance by small, subpolar glaciers, which are wet-based where the ice is thickest, but cold-based in the thinner marginal zone (Richardson and Holmlund, 1996);

(2) occupance by temperate glaciers, with overdeepening occurring in areas where sliding velocities were highest, in the region of the glacier equilibrium line;

(3) enhanced erosion due to the availability of surface meltwater via bergschrunds or crevasse fields (Hooke, 1991).

Backwall retreat occurs mainly by a combination of mechanical weathering and mass movements (section 9.3.1). Turnbull and Davies (2006) have speculated that rockslope failures contribute significantly to the evolution of cirque morphology, and Jarman (2006) provided striking evidence for the early stages of mountainside niche evolution by rockslope failure (fig. 8.56). Debris delivered to the glacier surface, both by steady-state slope processes and catastrophic rockslope failure (e.g. Arsenault and Meigs, 2005) is transported away by glacier flow.

The least-known aspect of cirque erosion is what happens in the transition zone between subglacial and subaerial process domains, at the upper margin of the glacier. It has been argued that mechanical weathering may be particularly effective in the *randkluft* (the crevasse between a cirque glacier and the backwall). Measurements taken by Gardner (1987) in the randkluft of the Boundary Glacier in the Canadian Rockies indicate that considerable frost shattering takes place on the lower backwall rather than in the bergschrund, where summer temperatures do not rise above freezing. During the ablation season, the randkluft migrates downslope

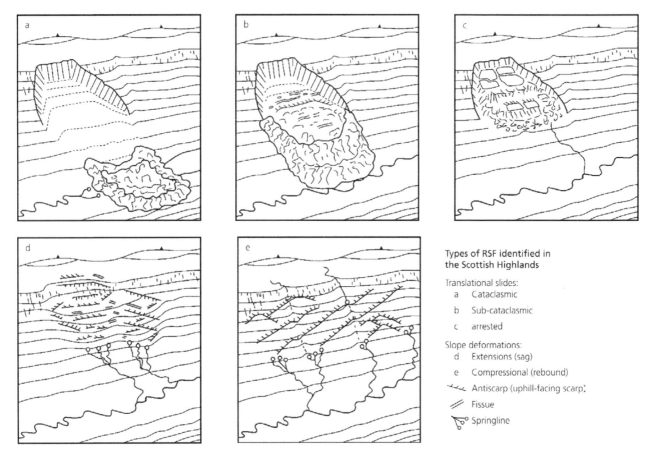

Figure 8.56 Types of rockslope failure identified by Jarman (2006) in the Scottish Highlands. The depressions produced by such rockslope failures constitute ideal mountain side niches for the future development of cirques (reproduced with permission of Elsevier).

and thus extends the rock surface area exposed to freeze–thaw weathering. Over longer periods of glacier thickening and thinning, large areas of cirque backwalls can be exposed to mechanical breakdown in this way. Furthermore, during the ablation season, subglacial quarrying may be facilitated downglacier of the randkluft or bergschrund, due to large diurnal variations in the supply of meltwater from the surface (Hooke, 1991; Iverson, 1991a).

The definition of cirques cited at the beginning of this section excludes those features fed by ice from outside the cirque basin. However, this is difficult to demonstrate in many situations, such as where cirques are incised into the margins of plateaux. Ice is likely to form first on the high plateau at the onset of glacial conditions and spill over cirque headwalls, so modifying their form (Rea et al., 1998; Rea and Evans, 2003). Thus, in such settings, long-term cirque evolution over several glacial cycles may reflect varying conditions, a possibility that presents problems for ergodic models.

Some calculations of cirque erosion rates and development cycles have been attempted in a variety of settings. The erosion rates of cirques cut into the sides of radiometrically dated volcanoes in Antarctica have been calculated by Andrews and LeMasurier (1973) as 360–460 mm ka^{-1}. For these cirques, rates of backwall erosion were far greater (5800 mm ka^{-1}) than sidewall erosion rates (800 mm ka^{-1}), indicating that these cirques will tend to become more elongated over time. This pattern is not true of all cirques, however, as relative backwall and sidewall erosion rates will be strongly dependent on geological and microclimatic factors (Bennett, 1990; Bennett and Glasser, 1996). Larsen and Mangerud (1981) calculated erosion rates of 500–600 mm ka^{-1} for a cirque glacier in western Norway, indicating that up to 125 ka would be required to erode the cirque. Somewhat lower erosion rates of 60 mm ka^{-1} were calculated by Andrews and Dugdale (1971) for cirques on eastern Baffin Island, Canada, by assuming that the cirque was occupied by ice for 1 million years. Anderson (1978) calculated erosion rates of 8–76 mm ka^{-1} for cirque glaciers on Baffin Island, indicating that 2–14 million years were required for cirque production. Studies of sediment concentration in meltstreams indicate present-day erosion rates of around 200–600 mm ka^{-1} for cirque and valley glaciers in Norway.

Figure 8.57 Examples of strandflats. (a) A strandflat linked to the floors of deep troughs at the margin of the Eggum massif, Vestagoy, northern Norway. Beaches on the strandflat have been constructed from reworked moraines (A. Guilcher). (b) A narrow strandflat near Cape Sabine, Ellesmere Island, Arctic Canada, the scene of the grim ending to Lieutenant Adolphus Greely's Lady Franklin Bay Expedition in 1884, when only 6 of a 25-man crew survived starvation (D.J.A. Evans).

8.5.5 STRANDFLATS

Named after the prominent examples off the Norwegian coast, *strandflats* are extensive, undulating rock platforms located close to sea level around the coasts of high-latitude landmasses, including Norway, Greenland, Spitsbergen, Iceland, Scotland, Ireland and Antarctica. Where partially submerged, they appear as a zone of low rocky islands or skerries, known as *skjaergard* in Norway. Strandflats are up to 50 km wide, cut across geological structures, and usually end abruptly at an inland cliff-line or break of slope (fig. 8.57). In some instances they may extend short distances up fjords, but are generally associated with the shallow water depths of fjord mouths. When viewed parallel to the coast, strandflats are remarkably horizontal, but they may possess very small seaward slopes, partially due to postglacial glacio-isostatic displacement. In some locations, several strandflats may occur at various heights above and/or below present sea level.

Strandflat formation has been explained by four main mechanisms, acting in combination: (1) frost action, combined with active sea ice rafting; (2) marine erosion; (3) subaerial erosion; and (4) subglacial erosion. *Frost action* is considered to be an effective erosive agent on periglacial shores, due to a potent combination of frequent cold temperatures and abundant water (Dawson et al., 1987). Furthermore, frost-shattered debris can be readily removed by sea ice which freezes onto the nearshore sea floor in winter (the *ice foot*), then breaks up and floats away in spring and summer. Together, therefore, frost shattering and ice rafting are capable of rapid coastal erosion. *Marine erosion* includes quarrying and abrasion by large, high-energy storm waves. Some strandflats exhibit striated and smoothed bedrock outcrops, providing evidence for *glacial erosion.*

Abrupt inner margins and cliff lines are consistent with marine erosion, although strandflats tend to be very wide, with small or no seaward slopes, unlike marine platforms, which have prominent seaward slopes. It is difficult to reconcile the large width and low slope of strandflats with a purely marine origin, because the energy of large waves is dissipated as they travel through shallow water, severely reducing their erosive capability at the shoreward margin of a wide platform. A worldwide review of strandflats by Guilcher et al. (1986) highlighted their probable polygenetic nature, but stressed the impact of glacial erosion in the evolution of the more extensively developed and wider forms. They provided evidence for the modern-day frost action and coastal erosion of the strandflats of Spitsbergen and the South Shetland Islands, Antarctica, but suggested that the Spitsbergen examples are so wide that they must have been initiated either by glacial erosion by piedmont lobes or as Neogene erosion surfaces. Similarly, Guilcher et al. invoked a preglacial planation surface as a precursor for the small strandflats of western Ireland.

It appears from the literature that enough evidence is available to support a combined frost action and marine genesis for narrow strandflats with seaward tilts, which should therefore be termed *marine platforms* (e.g. Hansom, 1983a). The wider, often horizontal platforms of glaciated coasts, which commonly merge with glaciated cirque, valley and fjord floors, and which contain direct evidence of glacial erosion, are more likely to have been at least partially shaped by glaciation. The fact that the strandflats of Norway merge inland with the preglacial (or *paleic*) surface lends strong support to the theory that they are glacially modified erosion surfaces (*etchplains* and *peneplains*), which have escaped deep ice erosion because of the reduction in erosive capacity at the ice sheet margins. One intriguing possibility is that strandflat erosion may occur by planation beneath partially floating glacier tongues near marine ice sheet margins. However, the abrupt inner margins of many strandflats, which are marked by bedrock cliffs extending over long distances and are transverse to former glacier flow, are difficult to explain by glacial erosion and therefore probably document at least a partial marine influence in strandflat genesis. Porter (1989) has argued that strandflats formed during 'average' glacial conditions prevailing throughout most of the Quaternary, when relative sea level along many glaciated coasts was similar to that of today, and coastal zones were subject to repeated alternations between periglacial, marine and subglacial erosive processes during glacier advance and retreat cycles.

8.6 LANDSCAPES OF GLACIAL EROSION

Sugden and John (1976) defined distinctive *landscapes of glacial erosion*, consisting of regional associations of erosional landforms. Their classification is based on both morphology and processes of erosion, and therefore sheds light on the links between glacier dynamics and landscape evolution. Five landscapes of glacial erosion were recognized:

(1) *landscapes of areal scouring*, which everywhere bear signs of glacial erosion;
(2) *landscapes of selective linear erosion*, in which erosion is confined to troughs between unmodified plateau remnants;
(3) *landscapes of little or no glacial erosion*, which are essentially unmodified preglacial landscapes that have survived one or more periods of glacial occupancy;
(4) *alpine landscapes*, consisting of dendritic networks of troughs separated by ridges;
(5) *cirque landscapes*, in which essentially discrete cirques are set in a hill or mountain massif.

These landscapes reflect the long-term influences of pre-existing topography, glacier and ice sheet morphology, basal thermal regime and mass flux, and develop cumulatively over multiple glacial cycles. They are therefore more likely to relate to 'average' glacial coverage during the Quaternary than to 'instantaneous' ice configurations at glacier maxima (Porter, 1989).

8.6.1 AREAL SCOURING

Landscapes of areal scouring are extensive tracts of subglacially eroded bedrock, consisting of rock knobs, roches moutonnées and overdeepened rock basins (figs 8.58 and 8.59). In reference to this characteristic morphology, Linton (1963) referred to such terrain as *knock and lochain* topography, from the Scots Gaelic words *cnoc*, meaning 'knoll' and *lochain*, meaning 'small lake'. The location of high and low points in areally scoured terrain reflects bedrock lithology and structure. Hollows and rock basins occur where joint density is high or where less-resistant rocks crop out, whereas knolls are underlain by more resistant rocks. However, relief amplitude is generally low, and is typically less than 100 m.

Figure 8.58 View across areally scoured terrain towards the southern margin of the Greenland Ice Sheet. To the right of this view, areal scour has only partially modified an older cirque terrain. A landscape of selective linear erosion can be seen near the ice sheet margin. Austmannadalen is the valley descended by Fridtjof Nansen and his team after their first crossing of the ice sheet in 1888 (Danish National Survey and Cadastre; photo annotated and provided by R.S. Williams, USGS)

Spectacular examples of areally scoured landscapes occur on the Canadian Shield, in West Greenland and in north-west Scotland (Gordon, 1981; Rea and Evans, 1996).

The widespread evidence for abrasion and plucking in landscapes of areal scouring indicates that they develop below wet-based ice, in situations where flow is laterally extensive and not focused into narrow channels. The extent of areal scouring, however, does not necessarily reflect the distribution of warm basal conditions at any one time, due to the distinct possibility that the present landscape is a palimpsest produced during several glacial stages. The impression of deep glacial erosion in Precambrian shield areas can be misleading. For example, the classic landscapes of areal scouring in north-west Scotland form part of a land surface that can be traced beneath Late Precambrian sedimentary rocks (Stewart, 1991), showing that in this area glacial erosion has exposed and modified an extremely ancient erosion surface. Similarly, the deep basin on the Hudson Bay area is a very ancient feature which contains Palaeozoic sedimentary rocks, showing that the Canadian Shield must have been eroded to something like its present level by Palaeozoic times.

Figure 8.59 (a) Landscape of areal scour near Loch Laxford, north-west Scotland. (RCAHMS aerial photograph, Crown Copyright). (b) A ground view of the knock-and-lochan topography in this area (D.J.A. Evans).

8.6.2 SELECTIVE LINEAR EROSION

Landscapes of selective linear erosion are characterized by deep troughs separated by essentially unmodified plateau surfaces (fig. 8.48; Sugden, 1974; Sugden and John, 1976). Striae or polished surfaces may cover the trough sides right up to the cliff top, showing that the troughs were once completely filled with erosive ice, whereas periglacial forms such as blockfields and tors may be preserved on the plateaux a few tens of metres from the trough edge. In planform, the troughs may form a dendritic network or a more complex interconnected pattern. Landscapes of selective linear erosion are well represented in east and north Greenland, the Allegheny plateau area of North America, Scandinavia, the Cairngorms of Scotland, and some Arctic islands; such landscapes are often well developed in areas removed from the maritime fringes of ice sheets, reflecting the relatively low mass turnover at great distances from oceanic moisture sources, where rapid, erosive flow can only occur in favourable locations (Sugden, 1974; Sugden and John, 1976; Hall and Glasser, 2003). It is important to remember that bedrock will exert some influence over the regional patterns of glacial erosion. For example, the role of bedrock lithology in the longer-term development of troughs and fjords over the last 75 million years has been demonstrated by Swift et al. (2008) using the fjord landscape of east Greenland. They showed that the more resistant Caledonian Basement rocks have restricted glacial valley widening, and that instead preglacial valleys have been overdeepened. In contrast, the less resistant Devonian and Mesozoic bedrock has been eroded into a system of shallower and wider fjords.

Landscapes of selective linear erosion develop beneath ice sheets, with the troughs marking former ice streams and the intervening plateaux marking areas of slowly moving or cold-based ice. The presence of troughs is thought to reflect positive feedbacks between subglacial topography, ice velocity, basal temperature and erosion rates. Numerical modelling of landscape development under a hypothetical ice sheet appears to confirm such theoretical notions of feedback relationships. For example, Jamieson et al. (2008) showed that valley overdeepening through time eventually stabilizes the thermal regime of an ice sheet in that the valleys become the permanent locations of warm-based ice, thereby creating a positive feedback in terms of increased sliding and erosion rates. Where ice sheets occupy irregular topography, basal melting will be encouraged within valleys or troughs, where the ice is thickest. Basal sliding provides another source of heat at the glacier bed, promoting further basal melting and increased sliding rates. Sliding conditions may also propagate up onto the valley heads and margins of plateaux surrounding the troughs (Hall and Glasser, 2003). As a result, erosion rates will be highest within valleys, deepening and enlarging them, and promoting more basal melting, fast sliding and erosion. Thus, pre-existing valleys in a landscape will tend to be exploited by ice flow and transformed into troughs. In contrast, high points beneath the ice sheet will tend to be occupied by thin, cold-based ice, which does not erode but protects the pre-existing land surface. Pre-existing valley systems will also encourage faster-than-average ice flow by collecting ice from large catchment areas, thus increasing the balance velocity (section 5.2.1).

It is difficult to derive quantitative assessments of the amount of glacial erosion represented by troughs, due to the unknown depth of preglacial valleys and in some cases an incomplete understanding of the tectonic contribution to relief production. By using the reconstructed *paleic* or preglacial surface in the Sognefjord drainage basin and subtracting the present-day topography of the fjord, Nesje et al. (1992) calculated that $7610 \, \text{km}^3$ of material has been removed during successive Quaternary glaciations (fig. 8.60). Depending on the amount of Quaternary time over which glaciations have been eroding Sognefjord, this yields a range of values from $102 \, \text{cm ka}^{-1}$ to $330 \, \text{cm ka}^{-1}$ for ice stream erosion rates. Their reconstruction of the paleic surface allows us to view the nature of the preglacial relief, and provides evidence for the channelling of ice and the concentration of glacial erosion by pre-existing topography. Nesje and Whillans (1994) proposed the following evolutionary development of Sognefjord as a representative example of the Norwegian fjords (fig. 8.61):

- *Phase 1*: deep chemical weathering and erosion during the Mesozoic and early Tertiary to produce the paleic surface just above sea level.
- *Phase 2*: a preglacial phase of Tertiary uplift, which was greater towards the west coast, and predominantly fluvial incision of valleys guided by rock fracture patterns.
- *Phase 3a*: interglacials and interstadials when subaerial processes erode the fractured rock, especially at valley heads, and fill valley floors with the debris (the paleic surface not greatly affected).
- *Phase 3b*: repeated glaciation throughout the Quaternary, resulting in overdeepening of pre-existing valleys and a near uniform but very restricted net erosion of the paleic surface.

This reconstruction emphasizes the role of slope mass wasting during interglacials and interstadials, based on the observation that the largest accumulations of coarse debris today occur in fans at valley heads and in deltas, both located below bedrock fracture zones.

In a similar terrain, the Torngat Mountains of northern Labrador, Staiger et al. (2005) have used cosmogenic nuclide dating to calculate fjord and valley incision rates. The results indicate that >2.5 m of erosion occurs in glaciated valleys during a single glaciation, contrasting

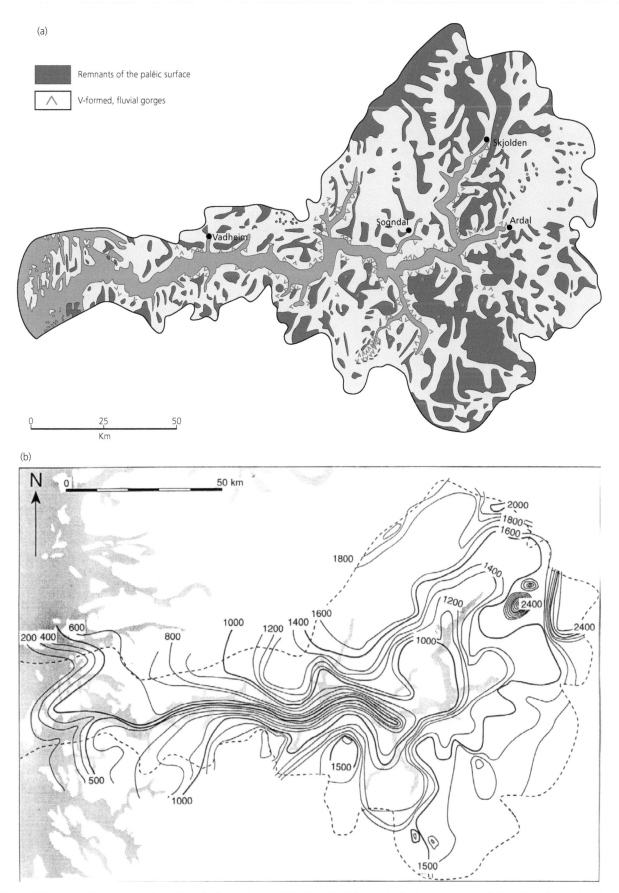

Figure 8.60 Impacts of glacial erosion in the Sognefjord area, Norway. (a) The Sognefjord drainage basin, showing remnants of the paléic surface, V-shaped valleys and gorges (V symbol) and Tertiary soil remnants (red dots). (b) Reconstruction of the paléic surface based on interpolation of paléic surface remnants with elevations in metres. The dashed line marks the edge of the modern drainage basin (Nesje and Whillans, 1994, reproduced with permission of Elsevier).

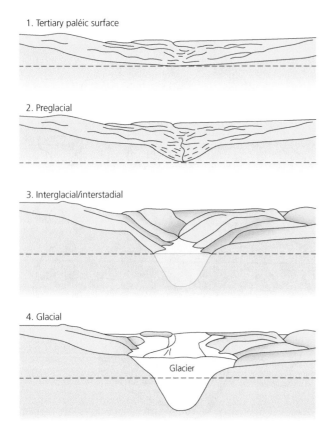

1. Tertiary paléic surface

2. Preglacial

3. Interglacial/interstadial

4. Glacial

Glacier

Figure 8.61 Schematic reconstruction of the main phases of development of Sognefjord, Norway. The dashed line represents sea level (Nesje and Whillans, 1994, reproduced with permission of Elsevier).

with erosion rates of <1.4 m Ma^{-1} (i.e. over numerous glacial–interglacial cycles) on summit plateaux where ancient blockfields are preserved (see section 8.6.3). Despite the fact that such calculations are ballpark figures, it is interesting to note that the incision rates in Labrador are much lower than those in an area of similar bedrock in Norway.

Recent acquisitions of Antarctic subglacial topography through BEDMAP (2002) have enabled Jamieson et al. (2005) to calculate erosion rates beneath the East Antarctic Ice Sheet, based on a reconstruction of the preglacial topography of the Lambert basin. By taking into account the volume of sediment eroded, as recorded by offshore stacks of fluvial and glacigenic deposits, they demonstrated that fluvial and glacial erosion have been focused on the Lambert Graben over the last 118 million years, and have been in the order of 1–2 m Ma^{-1} and 0.89–1.79 m Ma^{-1} respectively. This study provides clear indications that the tectonic history and physiography of the region control the denudation rates, and that selective linear erosion by ice sheets accentuates the preglacial fluvial drainage patterns. In comparison to the vastly greater erosion rates for Sognefjord (1–3.3 m ka^{-1}), linear erosion rates beneath the Antarctic Ice Sheet are extremely low and are compatible with the rates of <1.4 m Ma^{-1} calculated for summit plateaux in Labrador (Staiger et al., 2005).

The dimensions of troughs appear to be scaled to the amount of ice that discharged through them, in the same way as river channels are adjusted to bankfull discharges. A relationship between trough size and ice discharge was first noticed by the pioneering German geomorphologist Albrecht Penck (1858–1945), who proposed the *law of adjusted cross sections*. This relationship has been tested for Greenland (Haynes, 1972), British Columbia (Roberts and Rood, 1984) and New Zealand (Augustinus 1992b), by correlating parameters such as trough width, depth, length and cross-sectional area with the ice catchment area. Ideally, the first four parameters should be measured relative to preglacial erosion surfaces, if they are identifiable, between the troughs being studied. The assessments undertaken in these studies confirm that relationships exist between trough size parameters and glacier contributing area, which is the equivalent of the drainage basin in fluvial systems. Roberts and Rood (1984) proposed that trough length is proportional to glacier contributing area, following a power function similar to that used for fluvial systems:

$$L_T = aA_G{}^b \qquad (8.6)$$

where L_T = trough length, A_G = glacier contributing area and a, b = constants. Significant correlations between various trough parameters and glacier contributing area indicate that the erosion of troughs and fjords increases in direct proportion to the volume of ice discharged through them. Making comparisons of erosion rates, Augustinus (1992b) concluded that, because fjords were deeper and longer in British Columbia than in New Zealand, more intense glacial erosion had taken place on the British Columbia coast, even though glacier contributing areas were the same. However, Brook et al. (2003), comparing new data from western Scotland with existing data from the British Columbia and New Zealand fjords, subsequently pointed out that power law functions had been misapplied and the sensitivity of outlet valley dimensions to increases in glacier contributing area is essentially similar between all three geographical settings (fig. 8.62).

Problems arise where troughs or fjords have been accessed by glacier ice flowing from ice sheet interiors and overtopping watersheds, such as occurred in the through troughs and fjords of the Torngat Mountains of Labrador and eastern Baffin Island during Laurentide Ice Sheet glaciations. Such patterns of ice flow will result in disproportionate trough sizes relative to local drainage areas. Haynes (1972) suggested that trough size-drainage area assessments can therefore be used to differentiate those features eroded by local versus regional ice. However, even within local ice sheet centres, ice streams

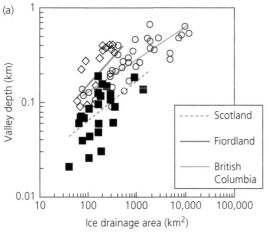

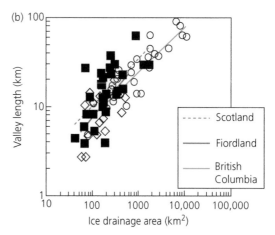

Figure 8.62 Graphs showing the relationships between glacier drainage basin area and outlet valley dimensions (a = valley depth below sea level; b = valley length), based on data from British Columbia (circles; Roberts and Rood, 1984), New Zealand (diamonds; Augustinus, 1992b) and Scotland (black squares), with power law curves fitted to each data set (Brook et al., 2003, reproduced with permission of Wiley).

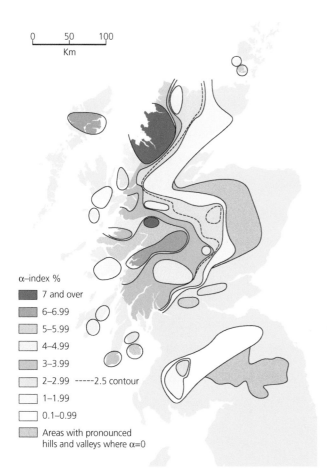

Figure 8.63 Valley connectivity indices for Scotland, showing areas of high trough connectivity for the former ice dispersal centres of the western Highlands (After Haynes, 1977).

may cross from one drainage basin to another. The amount of watershed breaching by an ice mass has been quantified by Haynes (1977) and Augustinus (1992b), using the interconnectivity of troughs (fig. 8.63). Connectivity is high at former ice dispersal centres where glacier ice was capable of overtopping and incising through preglacial watersheds, and low in peripheral areas.

8.6.3 LANDSCAPES OF LITTLE OR NO GLACIAL EROSION

Numerous examples exist of preglacial landforms, sediments and weathering horizons that have survived intact despite occupation of an area by glacier ice, perhaps on numerous occasions. Such *landscapes of little or no glacial erosion* owe their preservation to protection by cold-based ice caps or sectors of ice sheets (e.g. Sugden, 1974; England, 1987; Kleman, 1992, 1994; Dyke, 1993; Kleman and Bergstrom, 1994; Fabel et al., 2002; Stroeven et al., 2002a, b; Sugden et al., 2005).

For example, land surfaces of probable Neogene age are preserved within the limits of Late Wisconsinan ice caps in the central Canadian Arctic archipelago (Dyke et al., 1992; Dyke, 1993; fig. 8.64). The land surfaces display dendritic fluvial valleys, tors, blockfields, patterned ground and gelifluction terraces – features characteristic of subaerial weathering and erosion in a periglacial climate. Similar rock types which are known to have been ice scoured during the last glaciation bear little or no weathered regolith, indicating that the weathered landscapes represent tens of thousands of years of preglacial subaerial modification. In such areas, the only signs of glacial modification may be marginal meltwater channels cut during deglaciation (see sections 8.4.4 and 12.3.3). Palaeo-surfaces, often containing Neogene weathering residues and etchforms, exist in numerous highland terrains which are dissected by troughs and fjords (Sugden and John, 1976; Nesje and Whillans, 1994; Johansson et al., 2001a, b; Sugden et al., 2005; Goodfellow, 2007; Goodfellow et al., 2008). In such areas, selective linear erosion marks the position of former wet-based ice

streams, whereas the intervening remnant preglacial land-scapes record the location of cold-based ice. Gellatly et al. (1988) and Rea et al. (1996a) have described blockfields melting out from beneath cold-based plateau ice caps in the mountains of Arctic Norway. The plateaux contain no evidence for glacial modification, but rise above deeply dissected, ice-moulded terrain. The presence of protective, cold-based ice caps on high plateaux has important implications for glacier reconstructions, because blockfields on high surfaces could be mistaken for evidence of former nunataks (section 12.3.5).

At lower latitudes, deep-weathered bedrock or sapro-lites exist in some areas that have been covered by ice sheets on a regular basis throughout the Pleistocene (e.g. Hall and Sugden, 1987; Patterson and Boerboom, 1999; Bonow, 2005; Olvmo et al., 2005), attesting to erosional selectivity in glacial systems. Based on studies in Scotland, Glasser (1995) and Hall and Glasser (2003) argue that erosional selectivity reflects preferential ice streaming along preglacial valleys, and correspondingly minor erosion of interfluves. This shows that the most intense

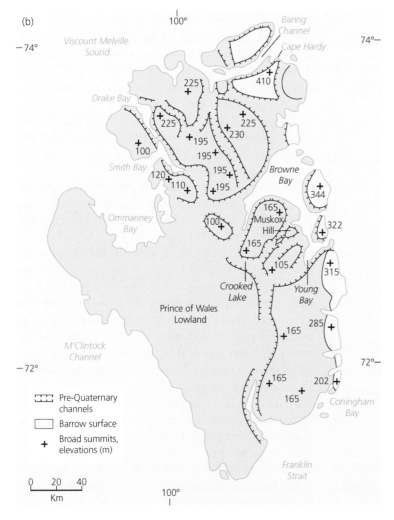

Figure 8.64 A landscape of restricted glacial erosion, Prince of Wales Island, Canadian central arctic archipelago. (a) Landsat image of part of the island, showing fragments of a Tertiary fluvial valley network. (b) Map of the island, showing the pre-Quaternary physiographic elements, which have survived numerous ice sheet glaciations (After Dyke et al., 1992).

erosion occurs where the topography favours convergence of flow, implying that preglacial relief is critical to patterns of erosion by glacier ice during multiple glaciations.

The occurrence of ancient landforms such as tors, saprolites and etchforms, in addition to the preservation of blockfields on high summits in formerly glaciated regions, has been at the centre of debates about the extent of former glaciations for decades. Two schools of thought have emerged over that time, one maintaining that such features could not have survived ice coverage and therefore represent nunataks, and the other proposing that preservation was possible beneath cold-based ice. Resolution of this issue has been brought closer recently by the advent of cosmogenic radionuclide (CRN) surface exposure dating, which has provided ages on these supposed ancient land surfaces. Rather than deliver either very old or very young dates, thereby verifying one or the other of the two schools of thought, the results have indicated complex erosion histories that can be explained most satisfactorily by temporally variable glacial erosion rates beneath cold-based ice, interspersed with periods of periglacial weathering during interglacials (e.g. Sugden et al., 2005). Of great significance in this explanation is the acknowledgement that cold-based ice can achieve localized but nonetheless significant erosion by block removal (e.g. Atkins et al., 2002). An excellent example of the application of terrestrial cosmogenic nuclide (TCN) to the dating of minimally eroded landscapes is that of Phillips et al. (2006) and Hall and Phillips (2006) on the Cairngorm tors in Scotland, features previously cited as evidence for the preservation of pre-Quaternary landforms (cf. Hall and Sugden, 2007). The conclusions of this case study are that tors can survive multiple glaciations and gradual modification can be accomplished by cold-based ice through the plucking of blocks detached by interglacial/interstadial weathering (fig. 8.65; e.g. Johansson et al., 2001). Similar conclusions have been proposed by André (2004), based on tors in northern Sweden. Considerable evidence of preglacial landform survival and the occurrence of clear trimlines between glacially modified and relict terrain, supported by TCN dating, has been presented widely from areas formerly covered by ice sheets (fig. 8.66; e.g. Clarhäll and Kleman, 1999; Fabel et al., 2002; Stroeven et al., 2002a, b; Harbor et al., 2006; Staiger et al., 2006). Such studies imply that glacial thermal regimes have a significant role to play in the efficacy of glacial erosion, and that the supply of sediment to the global oceans is extremely uneven within regions (Stroeven et al., 2002c; Staiger et al., 2005, 2006), as well as between them (Jamieson et al., 2005).

Some remarkable evidence for the non-erosive, protective role of cold-based ice has been reported from Arctic locations where Quaternary sediments have escaped

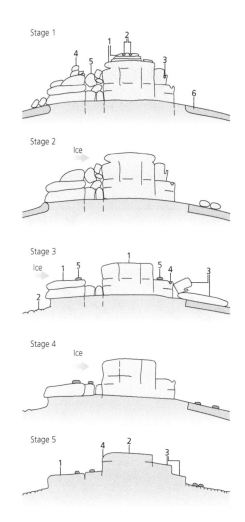

Figure 8.65 A model of tor evolution based on a Cairngorm Mountain case study. Stage 1 is an unmodified tor with the following features: (1) widened exfoliation joints and rounded block edges; (2) weathering pits; (3) weathering pit spillway; (4) perched block superstructure; (5) toppled blocks; (6) regolith. At Stage 2, the tor is covered by cold-based ice with internal flow represented by the arrow. The superstructure is displaced a short distance downflow. At Stage 3, continued modification produces: (1) unweathered surface with no weathering pits; (2) regolith removal; (3) boulder train; (4) locally preserved weathered surface and weathering pits; (5) erratics. At Stage 4, the tor has been reduced to an upstanding plinth that can resist further erosion by cold-based ice. At Stage 5, wet-based and sliding ice reduces the tor to a low slab with: (1) erratics on slab surface lacking open expansion joints; (2) abraded surfaces; (3) lee-side plucking; (4) stoss-side abrasion (Phillips et al., 2006).

removal and thereby featured centrally in debates about ice sheet extent during the LGM. On Baffin Island, Davis et al. (2006) have provided compelling evidence for the near-perfect preservation of a pre-LGM raised marine delta, complete with in situ marine shells dating to >54 ka BP, despite inundation by the north-east margin of the Laurentide Ice Sheet. Although this delta had been traditionally cited as evidence of ice-free conditions during the LGM, TCN dating of erratic boulders draping the surface of the feature reveal that it has been overridden. A similar interpretation of preservation of pre-LGM raised beaches beneath cold-based ice appears to be the only way to reconcile maximalist and minimalist

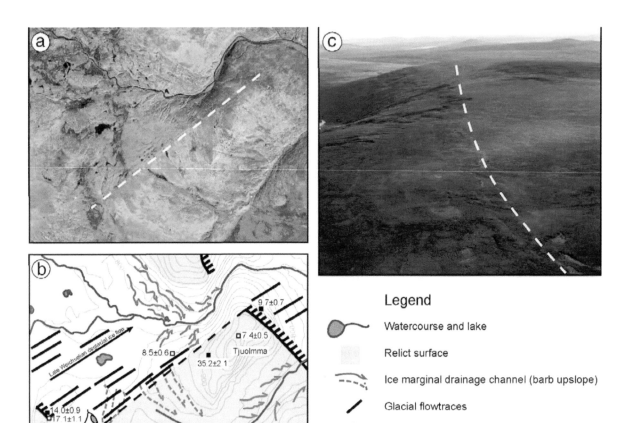

Legend

🫧 Watercourse and lake

⬚ Relict surface

⤳ Ice marginal drainage channel (barb upslope)

／ Glacial flowtraces

𝗘𝗘 Glacial transverse scarp

⸓ Lateral sliding boundary

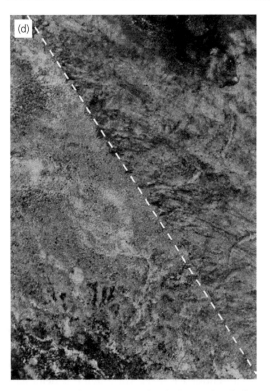

Figure 8.66 Examples of the evidence used in the demarcation of a former subglacial thermal/sliding boundary beneath the Fennoscandian Ice Sheet, northern Sweden. The infrared aerial photograph (a), the geomorphology map (b) and oblique aerial view (c) show the preservation of pre-Late Weichselian meltwater channels on Mount Tjuolmma. The channels have been cross-cut by subglacial till lineations (flutings). Photograph (c) is taken from the position of the eye symbol in (b). Aerial photograph (d) shows the sharp boundary (dotted line), interpreted as a glacier sliding boundary, between pre-Late Weichselian patterned ground and subglacial flutings on Mount Sjuva (Harbor et al., 2006).

reconstructions of the Barents Sea Ice Sheet in western Svalbard (Landvik et al., 2005; Evans and Rea, 2005 and references therein).

8.6.4 ALPINE LANDSCAPES

Alpine landscapes are among the most instantly recognizable landscapes of glacial erosion. They form in areas of repeated valley glaciation, and comprise networks of troughs and cirques deeply incised into mountain massifs. During glacial occupancy, the interfluves are commonly ice free, and are typically reduced to narrow *arêtes* and sharp rock peaks or *horns* (fig. 8.67). An ergodic model of the evolution of an alpine landscape is shown in figure 8.68. *Stage A* represents a fluvial landscape, with rounded interfluves, sinuous valley floors, interlocking spurs and tributaries graded to the main valley floor. In *Stage B*, small cirque and niche glaciers occupy favourable locations, and accomplish limited erosion. *Stage C* represents occupation of the landscape by confluent valley glaciers; and *Stage D* represents the mature alpine landscape, consisting of troughs, hanging valleys, truncated spurs and narrow interfluves. At times of maximum glacier cover, ice may spill from one catchment to another over low points on the watershed, creating glacial breaches (fig. 8.69). The intensification of glacial erosion in mountain settings during the Quaternary may even be the mechanism through which climate change modulates orogenesis (e.g. Berger et al., 2008). Although alpine

Figure 8.67 Aerial photograph of a low-relief alpine landscape occupied by a complex of valley and cirque glaciers, Wootton Peninsula, north-west Ellesmere Island, Arctic Canada (National Air Photograph Library, Ottawa).

landscapes are traditionally associated with long periods of glacier occupancy, it is the occupancy of alpine styles of glacier cover rather than ice sheets that develop alpine topography. This has been equated to Porter's (1989) concept of 'average' glacial conditions by McCarroll (2006), using the celebrated alpine glacial landscape of Snowdonia in North Wales as an example. This landscape has most likely been produced more by average glacial conditions during periods in northern Europe when conditions were not cold enough to produce ice sheets. Such periods cumulatively equate to far longer periods of glacier occupancy than those associated with maximum ice sheet coverage.

Alpine landscapes are most spectacularly developed in high-relief, tectonically active mountain environments, such as the Himalaya, the North American Cordillera, the Andes and the New Zealand Alps. In such areas, relative relief from the valley floors to the summits can be in excess of 3000 m, attesting to very high erosion rates linked to rapid uplift, slope instability, and high throughput of water and ice. Relative relief is generally lower for tectonically stable alpine environments, and is least for old, relict mountain terrains, such as the Highlands of Scotland and the mountains of western Scandinavia. The relationship between relative relief and tectonic activity suggests that alpine environments may undergo a long-term evolutionary sequence mirroring that of the mountain belt in which they occur. Glaciation will be initiated during mountain uplift, when the mountain mass intersects the regional snowline. Glacial incision and alpine landscape development will reach a maximum when rapidly uplifting, high altitude terrain extends high above the snowline. Once uplift ceases, relative relief is reduced by erosion at an exponentially decreasing rate. Thus, mountain belts that have been tectonically active in the recent geological past (such as the European Alps) still have high-relief alpine terrain, whereas ancient mountain belts (such as the Scottish Highlands) exhibit much more subdued alpine

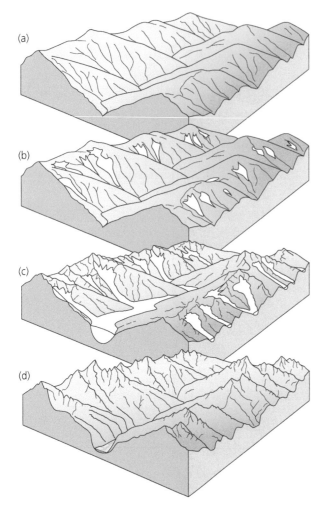

Figure 8.68 Ergodic model of the development of an alpine landscape from a fluvial precursor (Flint, 1971, reproduced with permission of Wiley).

Figure 8.69 Deep glacial breach, Cuernos del Paine, Chile. The breach is cut through a granite batholith (pale colour) and its shale capping (dark colour) (C.M. Clapperton).

landscapes. This simple evolutionary sequence will, of course, be interrupted by global and regional climatic change, inundation of the massif by ice sheets, and other factors, but serves as a useful basic framework for understanding alpine landscapes.

The role of alpine glaciation in relief development has been the subject of a number of recent studies that have utilized digital elevation models (DEMs), rock exhumation rates and landscape evolution models. For example, Whipple et al. (1999) and Brocklehurst and Whipple (2002) identified the prominent processes of headward erosion into mountain ranges, and the overdeepening of valleys above the ELA at rates that outstrip those of fluvial erosion downvalley. The net result is the overall reduction in relief so that area–altitude relationships can be correlated with the extent and long-term impact of glaciation (Brocklehurst and Whipple, 2004). Moreover, a landscape evolution model developed by Braun et al. (1999) shows that the evolution of glacially shaped valleys gradually produces a landscape that can hold more ice, thereby maintaining greater rates of glacial erosion in the upper, glaciated stretches of mountain valleys

through positive feedback. A further potential feedback, the uplift of mountain masses in response to glacial incision (Molnar and England, 1990), remains unproven (Brocklehurst and Whipple, 2002), although calculations of erosion rates by Spotila et al. (2004) have shown that glaciers can remove rock at the same rate that it is tectonically uplifted, thereby limiting the altitude to which mountains may rise (cf. Oskin and Burbank, 2005). This 'glacial buzzsaw' effect has been verified by Mitchell and Montgomery (2006), based on their assessments of the glacial topography in the Cascade Range of the USA, and supported by Tomkin (2007), using numerical modelling that couples rock uplift and erosion rates. Particularly informative in this respect are altitude–area distributions in alpine landscapes; more specifically, the proportion of land surface lying above the zone of cirques. Only 10 per cent or less total land area lies above the zone of cirques (the altitude of the long-term ELA) in the studies undertaken by Mitchell and Montgomery (2006) in the US Cascades and by Montgomery et al. (2001) in the Andes, demonstrating the dominance of glacial erosional processes in removing rock at or near the ELA at high altitude (MacGregor et al., 2000; Tomkin and Braun, 2002). Moreover, the paraglacial slope failure and debuttressing that occurs during non-glacial conditions is focused above the long-term ELA and helps to maintain the 'peak concordance' that is correlative with cirque altitudes across mountain ranges.

An assessment of the long-term development of an alpine landscape has been made by Kirkbride and Matthews (1997) for the Ben Ohau Range, South Island, New Zealand. This north–south trending mountain range trends obliquely to the regional north–west to south–east rise in glacier ELAs, which resulted in a marked gradient in past glacier distribution along the range. As a consequence, the south end of the range is a non-glacial landscape of fluvial incision, whereas the north end is a classic alpine glacial landscape of troughs and narrow, pinnacled intervening ridges. By combining detailed morphometric analyses of these landscapes with the reconstructed climatic and tectonic history of the region, Kirkbride and Matthews were able to demonstrate that valley forms develop recognizable glacial profiles after approximately 70,000 years of glacier occupancy; cirques become established after 200,000 years of occupancy; and fully developed troughs require at least 320,000 years of occupancy. It was not possible to determine absolute rates of erosion, due to uncertainty surrounding the amount of fluvial incision of the valleys, and spatial variations in glacier mass throughput and erosiveness. The continued evolution of the tectonically active New Zealand alpine glaciated terrain has been numerically modelled by Herman and Braun (2006, 2008), who demonstrated that a further 360,000 years of glacial erosion will likely result in the overall smoothing of the topography and the widening and deepening of basins. Their model also highlights the downstream variability in erosion patterns through time:

(1) Downstream sections of valleys are affected by erosion mostly at glacial maxima, and hillslope processes are dominated by fluvial incision and landsliding.

(2) Central sections of valleys are subject to oscillating patterns of fluvial and glacial erosion, and hillslopes are subject to fluvial incision, landsliding and glacial abrasion.

(3) Upper parts of catchments remain glaciated, and hillslopes are therefore subject to glacial and periglacial processes. The very dynamic nature of hillslope evolution in the region is driven by high tectonic uplift rates and large precipitation totals, making a steady-state glacial morphology difficult to achieve.

8.6.5 CIRQUE LANDSCAPES

Cirque landscapes are characterized by independent cirques incised into upland terrain. The appearance of the landscape is strongly influenced by cirque density, and may range from isolated cirque basins cut into the edges of plateaux, to densely packed cirques separated by narrow arêtes (fig. 8.70).

Because of their assumed association with small glaciers, cirques have been used as indicators of long-term average palaeoclimatic conditions (e.g. I.S. Evans, 1999). Two principal cirque attributes have been employed: *orientation* and *altitude*. Cirque orientations for specific regions can be portrayed effectively in *cumulative vector diagrams* (fig. 8.71). Successive legs, proportional to the number of cirques with particular orientations, are drawn on the diagram, beginning with the orientation approximately opposite to the vector mean. The straight line joining the first and last legs is the *resultant vector*, which can be used to summarize the distribution. A measure of cirque preferred orientation or *cirque asymmetry* can be calculated by expressing the length of the resultant vector

Figure 8.70 Cirque landscape, verging on alpine landscape, Torngat Mountains, Labrador, Canada (National Air Photograph Library, Ottawa).

as a percentage of total length of all the vectors, a quantity known as *vector strength*.

Such diagrams and statistics give an immediate impression of the preferred orientation of cirques in any particular region. Strong to extreme asymmetry may reflect structural and topographic controls, but such factors are unlikely to explain marked cirque asymmetry over large regions unless combined with climatic variables. The tendency for cirques in mid latitudes to face eastwards reflects the accumulation of snow and glacier growth in the lee-side hollows of upland terrains located in the westerly wind belts (e.g. Evans, 1977; Benn, 1989b). The poleward aspects of many cirques (north- and south-facing in the northern and southern hemispheres respectively) result from reduced ablation and preferential glacier survival in such locations, where direct solar insolation is lowest. Low cirque asymmetry may indicate that several generations of cirques, each relating to different palaeoclimates, may exist in the same area. For example, Haynes (1995) has argued that in the Antarctic Peninsula, the largest alpine valley heads with predominantly lee-side and poleward aspects (south-east and south) date to early marginal glacial conditions, whereas the smaller cirques, which face predominantly windward (north-west), formed during more recent, less marginal glacial conditions. Therefore, weak vectors may arise in locations where glaciation is less marginal and cirques are well developed on the windward sides of mountain terrains, where the highest direct precipitation occurs. In some regions, strong windward vectors may result from this precipitation control (e.g. Andrews and Dugdale, 1971; Aniya and Welch, 1981). The tendency for cirque asymmetry to be

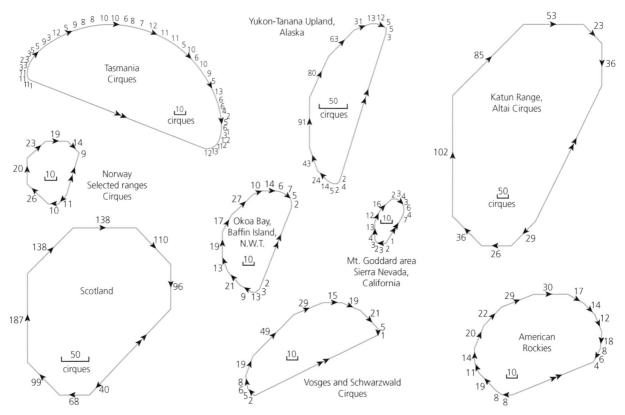

Figure 8.71 Cumulative vector diagrams of cirque orientations. All vector diagrams are at the same scale, except for Scotland and Yukon-Tanana (half scale) and Katun (one-fifth scale). North is towards the top of the page. Resultant vectors are indicated by double arrows (Modified from I.S. Evans, 1977).

influenced by total glacier ice cover of a region (intensity of glacierization) was termed the *law of decreasing glacial asymmetry with increasing glacier cover* by Evans (1977). The greater asymmetry in cirque orientation is thus conceivably a measure of the marginality of a region to glacierization over the long glacial history of the Earth, and may explain patterns of greater asymmetry in mid- and low-latitude upland terrains.

Regional trends in *cirque elevation* have also been interpreted in terms of palaeoclimate. For example, cirque floor altitudes in Scotland and Wales rise from south-west to north-east, suggesting that there was a similar altitudinal distribution of cirque glaciers at various times during the Pleistocene (Sissons, 1967; Bennett, 1990; I.S. Evans, 1999). This pattern most probably reflects former precipitation gradients, with conditions being most favourable for glacier growth near oceanic moisture sources. Palaeoclimatic reconstructions based on cirque floor altitudes must, however, be treated with caution. At best, they can only offer information on average conditions during the Pleistocene, as cirques must have been initiated and enlarged on several occasions under varying climates. Additionally, cirque floor altitudes may have been influenced by regional uplift or subsidence since their formation, particularly in tectonically active regions such as New Zealand, the Himalaya and the American Cordillera (Holmes, 1993). Finally, the altitude of a cirque basin

does not necessarily bear a close correspondence to the ELA of the glacier which formed it. For example, Richardson and Holmlund (1996) found that in northern Sweden, the cirque occupied by Passglaciären probably underwent most rapid erosion on occasions when it was occupied by the head of an extended valley glacier, rather than a small cirque glacier as at present. Thus, in this case, the cirque floor is probably a perched rock basin eroded below the upper reaches of a large valley glacier, and at times of more restricted cirque glaciation it is an essentially fossil form, bearing no clear relationship to the prevailing climatic conditions. Similarly, I.S. Evans (1999) has demonstrated that cirques in Wales developed during the earlier and later stages of glacial cycles, when localized mountain ice fields predominated; during full glacial conditions the landscape was submerged by an ice sheet.

8.6.6 CONTINENT-SCALE PATTERNS OF EROSION

Benchmark work on reconstructing the large-scale distribution of landscapes of glacial erosion in terms of ice sheet thermal regime was conducted by Sugden (1977, 1978), who used a simple glaciological model of the Laurentide Ice Sheet. His reconstruction identified a thermal pattern consisting of a wet-based inner area

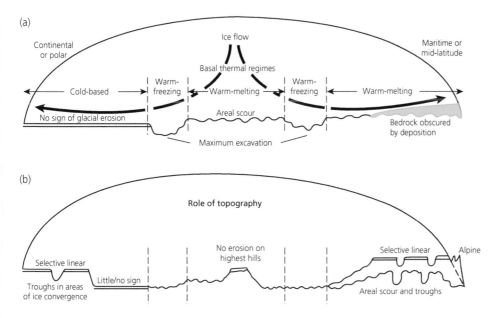

Figure 8.72 Conceptual model of erosion by ice sheets. The left-hand side represents polar or continental conditions and the right-hand side mid-latitude or maritime conditions. (a) Thermal regimes and idealized erosional effects. (b) Schematic effect of topography, showing how uplands experience minimal erosion or selective linear erosion. Massifs protruding above the ice sheet surface near the margin are sculpted into alpine landscapes. In cold-based zones, troughs will form in areas of flow convergence (Chorley et al., 1984).

surrounded by an outer zone of cold-based ice, which in turn was followed by a zone of wet-based ice in some areas (fig. 8.72). This pattern was then compared with the distribution of landscapes of areal scouring, selective linear erosion and little or no glacial erosion, with the intensity of erosion being estimated from lake basin density. A general correspondence between thermal zones and patterns of erosion was very clear; landscapes of areal scour coincide with inferred wet-based areas, landscapes of little or no glacial erosion occur where the ice was calculated to be cold-based; and the zone of most intense erosion occurs in the transition zone between wet-based and cold-based ice, where plucking and debris entrainment is likely to be widespread (fig. 8.72a). We now understand that erosional landforms are extremely unlikely to have formed synchronously during the ice sheet maximum, and that basal thermal zones must have migrated during glacier expansion and retreat, resulting in palimpsest erosional landscapes (Holmlund and Naslund, 1994; Haynes, 1995). This leads to the preservation of warm bed features under cold-based ice (e.g. Dyke et al., 1992; Clark et al., 2000; Fredin and Hättestrand, 2002; Kleman et al., 2002; Clarhäll and Jansson, 2003; section 12.4.4) and the superimposition of subglacial landforms (section 12.4.3), but nonetheless Sugden's model highlights the predominant locations of erosional zones through numerous glaciations, resulting in cumulative imprints of erosional intensity. Complications are introduced by topography and substrate geology. High massifs will experience minimal erosion, selective linear erosion, or areal scour and trough formation, depending on their position relative to the pressure melting point at the bed and the degree of flow convergence (fig. 8.72b). Massifs protruding above the ice sheet surface close to the margin will develop into alpine landscapes, which will be more

rugged in maritime than continental environments. It should be noted, however, that alpine landscapes can also develop during local glaciations, at times when the large ice sheets were more restricted (e.g. I.S. Evans, 1999).

An important factor influencing large-scale patterns of glacial erosion is former glacier mass balance, particularly the availability of precipitation (Sugden and John, 1976). Other things being equal, higher precipitation totals will increase the mass turnover, and therefore the erosive capability, of the glacier system (section 5.2.1). For example, the amount of glacial erosion in Scotland is greater in the west than the east, due to the greater proximity of the western Highlands to oceanic moisture sources (fig. 8.73; Linton, 1963; Hall and Sugden, 1987). In the eastern part of the country, successive ice sheets probably had low mass turnover, allowing the survival of preglacial landscape elements. The zonation of such landscapes has provided estimates of the intensity of subglacial erosion associated with ice sheets or phases of alpine glaciation (e.g. Sugden 1974, 1978; Haynes 1977). Calculations of continent-scale glacial erosion require numerical models and some bold generalizations on aspects such as bed properties, palaeoclimatic drivers and subglacial hydrology. Such an approach has been undertaken by Hildes et al. (2004) for the North American ice sheets. Their models yield erosion rates of 0.41–0.58 m over a single glacial cycle, a figure that factors up to much lower total glacial erosion than empirical estimates. This may be explained by progressively lower rates of erosion by the Quaternary ice sheets, as they gradually exhausted the supply of weathered regolith (Clark and Pollard, 1998; Roy et al., 2004), thereby highlighting a significant problem in assessing erosion in landscapes with complex erosional histories.

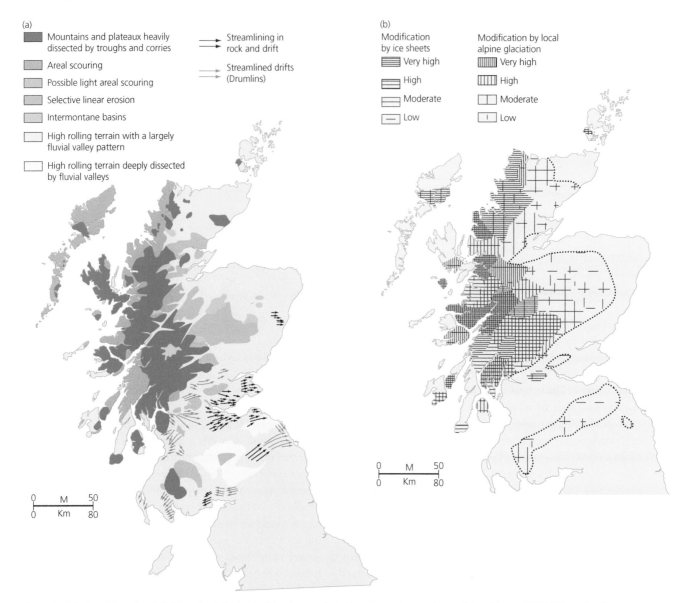

Figure 8.73 Style and intensity of glacial erosion in Scotland. (a) Landscapes of glacial erosion and areas of streamlining. (Haynes, 1983) (b) Intensity of landscape modification by ice sheets and alpine glaciers. The intensity of ice sheet glaciation was measured by a valley connectivity index, and the intensity of mountain glaciation is reflected in the proportion of valleys which terminate in cirques. (After Haynes, 1977) (c) Idealized examples of the different types of glacial erosion: (i) preglacial landscape preserved; (ii) mountain glaciation; (iii) interconnected valleys of ice sheet glaciation; (iv) areal scouring (After Haynes, 1983).

Long timescale changes in glacial erosional patterns and rates over Iceland have been calculated by Geirsdóttir et al. (2007), based on reconciliation of offshore sedimentation rates and volcanically controlled topographical evolution. From 5.0–4.0 million years BP, erosion rates were around $5\,cm\,ka^{-1}$, which contrasts markedly with a rate of $50–175\,cm\,ka^{-1}$ for the last 0.5 million years. This increasing trend in erosion rate is thought to reflect amplified glaciations in Iceland, coincident also with changes in ice-rafted detritus and the increased amplitude of oxygen isotope excursions in the ocean core records. Increased topographic relief due to constructive volcanic activity after 2.5 Ma also acted as a positive feedback on glacial erosion rates over the island.

One of the most spectacular effects of glacial erosion is the diversion of major river systems. Many of the river systems of northern Europe were diverted from their preglacial courses by the ice sheet advances during the middle Pleistocene Saalian glaciation (e.g. Arkhipov et al., 1995. For example, the river Thames, England, was diverted southwards from its preglacial course by the advance of the Anglian Ice Sheet (Gibbard et al., 1988; Rose, 1994). Figure 8.74 shows the reconstructed Tertiary drainage system of the Canadian Shield, and the present drainage basin of the Mackenzie river, resulting from glacial diversion of the original west-to-east flow direction of the precursor Bell River (Duk-Rodkin and Hughes, 1994). Similarly spectacular are cases of drainage

(c) (i)

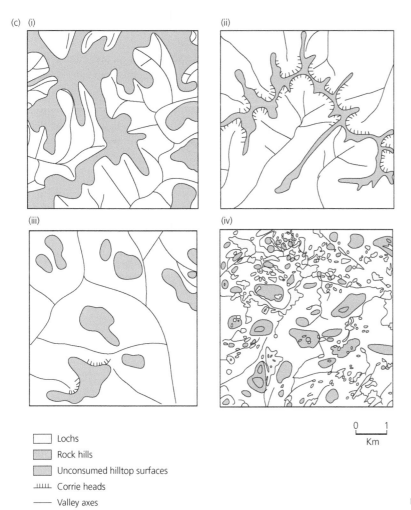

Lochs

Rock hills

Unconsumed hilltop surfaces

⊥⊥⊥⊥ Corrie heads

——— Valley axes

0   1
Km

Figure 8.73 (*Continued*)

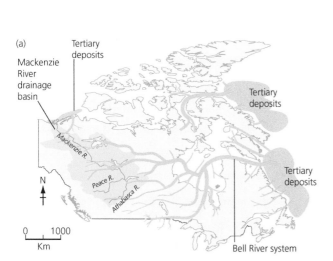

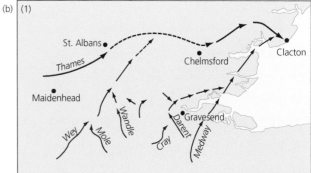

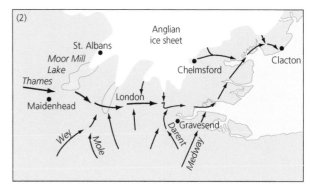

Figure 8.74 Examples of major drainage diversions by ice sheets.
(a) Reconstruction of the preglacial Bell River system, Canada, and the glacial
diversion of the headwaters to form the modern Mackenzie river drainage
basin. (Duk-Rodkin and Hughes, 1994). (b) Maps depicting the (1) pre-Anglian
drainage route and (2) the Anglian Glaciation diversion of the river Thames
(After Bridgland, 1994).

divide modification over numerous glaciations in mountain landscapes. Riedel et al. (2007) provided evidence of drainage divide elimination and migration in the North Cascade Mountains of western North America due to continental-scale glaciation, whereby an ice sheet encroached on a mountain landscape. This involved the reversal of drainage over watersheds and resulted in a divide migration of tens of kilometres. Clear indicators of this regional-scale impact of glacial diversion of river systems are relict dendritic drainage patterns and 'barbed tributaries' (rivers intersecting at acute angles because drainage basins now drain in the opposite direction); enhanced rectangular drainage patterns; the erosion of long canyons downstream of breached divides; beheading and bisection of major mountain valleys; and the deposition of large alluvial fans at canyon mouths.

CHAPTER 9
DEBRIS ENTRAINMENT AND TRANSPORT

Younger Dryas moraines in Llyn Llydaw, Snowdonia, Wales (D.J.A. Evans)

9.1 INTRODUCTION

Glacial sediments can be bewildering in their variety and complexity, reflecting the enormous diversity of ways in which debris can travel through glacial systems from source to the place of final deposition. Debris is subject to deformation, reworking and resedimentation by active and stagnant ice, flowing water, gravity and wind, and can undergo several cycles of reworking before final deposition and preservation in the geological record. Active depositional centres constantly shift position on a daily, seasonal and often random basis, as glaciers advance and retreat. In consequence, glacigenic and associated sediments exhibit enormous variability, and include sediments found in many other sedimentary environments, as well as several unique to glacial settings. Additionally, many glacial sedimentary processes are difficult or impossible to observe directly due to the inaccessible or highly dangerous environments in which they operate.

Understanding glacigenic sediments and depositional landforms has therefore provided glacial geologists with major problems, and a seemingly endless source of controversy and argument. Many solutions have been proposed to the problems of description and classification, and no overall consensus has been reached (e.g. Dreimanis, 1989; Brodzikowski and van Loon, 1991; Hambrey, 1994). Since the 1970s, powerful systematic approaches to the study of glacial sediments have been developed as part of a wider movement in sedimentology as a whole (Boulton, 1972; Dreimanis, 1989; Walker, 1992; Evans, 2003a). These approaches recognize that glacial depositional systems exhibit order on many different levels, either as a series of steps in *time*, or on a range of scales in *space*.

The time dimension is important for understanding the origin of deposits, because different sediment properties develop at different points during the erosional, transport and depositional history of the debris as it travels through the glacial system. The sequence of steps can be regarded as a type of *debris cascade system* (Chorley et al., 1984).

In this chapter, we begin with the concept of the glacial debris cascade, and discuss how it can be used as the basis for understanding the physical properties of glacial sediments. We then discuss the importance of the spatial dimension in understanding glacial sediments, and outline a hierarchical approach to the study of sediments, landforms and landscapes. The remainder of the chapter focuses on processes of debris entrainment and transport, and how these impart distinctive signatures on glacial sediments. Section 9.3 reviews processes of debris entrainment at the surface and beds of glaciers, and the formation of debris-rich basal ice. In section 9.4, we examine how debris is transported through glaciers by flowing ice and water, while section 9.5 focuses on the effects of different transport processes on debris properties, particularly grain size distributions and particle morphology.

9.2 APPROACHES TO THE STUDY OF GLACIAL SEDIMENTS

9.2.1 THE GLACIAL DEBRIS CASCADE

Glacial sediments can be viewed as outcomes of a series of processes extending back in time. In figure 9.1, this *debris cascade system* is broken down into three stages: (1) debris source, (2) transport path and (3) deposition. The *debris source* is the primary input to the system, and may be subglacial (e.g. plucked and abraded bedrock, or overridden sediments) or extraglacial (e.g. rock walls and valley-side regolith). The debris is then carried along one or more *transport paths*, including active or passive glacial transport, rivers and streams, and subaerial or subaqueous slopes. Debris may pass between transport paths many times prior to deposition. *Depositional*

processes refer to the mechanisms which lay down the final deposit, and include glacial, fluvial, gravitational, subaqueous and aeolian processes. Of these, only glacial depositional processes are unique to glacial environments, whereas processes from the other categories can operate in non-glacial environments as well as on glaciers. For example, fluvial processes can deposit sediment in subglacial tunnels, on proglacial braid-plains or in unglaciated valleys.

Sediment properties can be altered at all stages of the debris cascade, such that glacigenic deposits retain a 'memory' of their history of entrainment, transport and deposition. Examples of properties acquired during each of the three stages are as follows.

(1) The *debris source* controls the lithology of the particles in a sediment, but can also influence particle morphology and grain-size distribution (Gomez et al., 1988; Benn and Ballantyne, 1994; Benn, 2004a; Hoey, 2004; Walden, 2004).

(2) The *transport path* determines the wear processes experienced by particles as they move through the system, so exerts a strong influence on particle morphology (Boulton, 1978; Dowdeswell et al., 1985; Benn and Ballantyne, 1994; Benn, 2004a). Grain-size distributions are also modified during transport as the result of progressive wear and the preferential transport of particular size grades by water, wind and gravitational processes (Church and Gilbert, 1975; Humlum, 1985a; Werritty, 1992; Hoey, 2004). Particle lithology can also be influenced through contrasts in the durability of rock types under different types of transport (Slatt and Eyles, 1981). For glacially and gravitationally deformed sediment, aspects of the strain history can be reflected in sediment fabric and structure (Banham, 1977; Benn, 1994a, 2004b; Hart and Roberts, 1994; Benn and Evans, 1996).

(3) *Depositional processes* determine the geometry and extent of sediment units, sedimentary structures (e.g. lamination, cross-bedding and grading), grain-size distributions, particle morphology, fabric, geotechnical

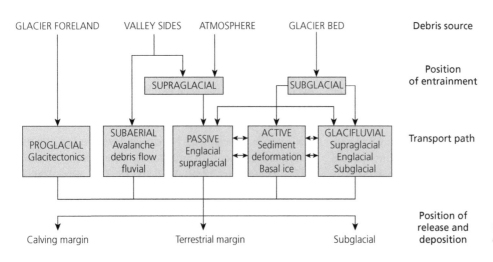

Figure 9.1 The glacial debris cascade system.

properties (e.g. porosity, shear strength and permeability) and other properties. Sediment properties are also influenced by *post-depositional processes*, such as diagenesis, faulting and folding, frost-heaving, and winnowing and leaching of fine material by wind and water (Evans and Benn, 2004b).

9.2.2 SPATIAL HIERARCHIES OF SEDIMENTS AND LANDFORMS

Sediments are typically laid down in assemblages, the characteristics of which reflect the range of processes active in particular environments. Such assemblages can be recognized at a wide range of scales, from the very small to that of a whole continent or ocean basin. For example, cross-bedded sands may be part of an assemblage of sand and gravel infilling a fluvial channel; in turn, the channel fill could be part of an assemblage of channel and bar deposits in a braided river system; and the braided river system part of a yet larger assemblage of deposits laid down during a glaciation. The environmental context of a sediment can therefore be defined at different levels of a spatial hierarchy, beginning with the immediate locality and panning out to ever wider horizons. At each successive level, the controls on the sedimentary system become larger in scale and longer-lasting in effect. For our example of fluvial sands, at the local level the main controls on deposition are the shape of the immediate riverbed and the short-term flow conditions as determined by rainfall, glacier melt and release of stored water. At the largest scale, the formation, location and extent of a braided river system is controlled by global factors such as long-term climatic cycles, relative sea-level fluctuations and tectonics.

The idea of nested spatial scales provides a powerful means of describing how sediments, landforms and landscapes fit together, and of determining how organization in the landscape reflects the organization of depositional processes and external controls in the environment. This hierarchical approach to sedimentology forms the basis of *facies modelling*, or constructing descriptive and predictive models of relationships between different deposits (Walker, 1990, 1992; Reading, 1996).

Four levels of organization can be recognized (Walker, 1992; fig. 9.2). From the smallest to the largest scale, these are: (1) *facies*, or individual deposits; (2) *sediment-landform associations*; (3) *depositional systems*, or *landsystems*; and (4) *glacial systems tracts*, or large-scale linkages of depositional systems. The boundaries between these levels are to some extent arbitrary, and differ slightly from those recognized by other authors. Those defined below have been tailored to the particular problems of description of glacial sediments, landforms and environments.

Facies

The fundamental unit of this spatial hierarchy is the individual sedimentary deposit or *facies* (from the Latin word

meaning 'aspect;' or 'appearance' of something). In this context, facies refers to a body of sediment with a distinctive combination of properties that distinguish it from neighbouring sediments (Walker, 1992; Reading and Levell, 1996). Facies are distinct sediment types deposited by single processes or groups of processes acting in close association. When we wish to refer only to the objective, physical characteristics of a deposit, with no reference to depositional processes, the term *lithofacies* is used. Examples of lithofacies are *cross-bedded sand* and *silt/clay laminae*. *Genetic facies* are named sedimentary units which state or imply a specific mode of formation. The genetic facies equivalents of the previous lithofacies examples might be *dune-bedded sands* and *varves*, which imply an origin by bedform migration and annual lacustrine sedimentation cycles, respectively. Note that genetic facies carry varying amounts of environmental information because some can occur in an extremely wide variety of environments, whereas others are formed in a very limited number of settings. For example, the presence of subglacial till carries more information about the former distribution of glacier ice than a debris flow deposit, because the former requires uniquely glacial processes for its deposition. The sediment facies commonly found in glacial environments are discussed in detail in chapter 10.

Sediment–landform associations

The next level in the hierarchy is the *sediment–landform association* or *facies association*, defined as an assemblage of genetically related facies (Walker, 1992). Facies associations consist of sediments that were laid down beside each other or in an unbroken vertical succession. An unbroken *vertical* transition between two facies reflects a *lateral* shift in depositional processes, such as where fluvial bars migrate downstream and bury older channel deposits, or where glacier retreat causes ice-marginal

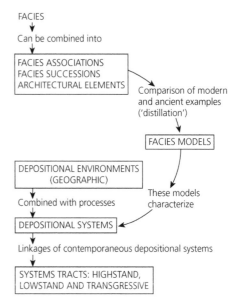

Figure 9.2 Hierarchical sediment classification (Walker, 1992).

deposits to be superimposed on subglacial sediments. The use of vertical facies associations to infer the spatial patterns of depositional processes was first proposed by the German geologist Johannes Walther (1860–1937), and is now known as *Walther's Law of Facies*.

A number of glacial researchers present their stratigraphic sequences using non-genetic *lithofacies associations*, or LFAs, which allow the grouping of lithofacies based on their physical similarities and the nature of their contacts with each other. The boundaries of facies associations are marked by *surfaces of erosion or nondeposition* or *bounding discontinuities*, which represent breaks in deposition of varying duration. Similar, more extensive discontinuities are also used to define depositional systems and systems tracts at higher levels in the hierarchy (van Wagoner et al., 1988; Walker, 1990, 1992). *Sediment–landform associations* are facies associations that have surface expression in a landform which is genetically related to the underlying facies. A related concept is the *architectural element*, which is a facies association with characteristic three-dimensional geometry, which may be buried by younger deposits (Miall, 1985; Walker, 1992; Boyce and Eyles, 2000; see section 11.2.1). An example of a sediment–landform association is a streamlined ridge of deformed sediments aligned parallel to glacier flow, characteristic of a fluted till surface; and an example of an architectural element is a buried river channel filled with cross-bedded sands and gravels. In glacial environments, sediment–landform associations can also be defined on the basis of distinctive deformation structures, such as thrust-block moraines, which are composed of dislocated wedges of sediment separated by low-angle reverse faults.

Facies associations and sediment–landform associations place facies within a *context*, indicating the local stratigraphic, structural and genetic relationship between closely associated facies. They therefore imply particular depositional environments or sub-environments, such as a proximal reach of a braided river or an advancing ice margin, in a way that lithofacies or genetic facies seldom can. Subglacial, ice-marginal, glacifluvial, glacilacustrine and glacimarine sediment–landform associations are discussed in chapter 11.

Depositional systems and landsystems

Facies associations and sediment–landform associations can be grouped together into *depositional systems* or *landsystems*. These are large-scale groupings of facies deposited in the same overall environment, such as the infill of a major proglacial lake or the assemblage of sediments and landforms deposited along the margins of a retreating glacier. Glacial landsystems also include erosional forms that were created broadly contemporaneously with sediments in the same environment. Early examples of this approach

included *subglacial landsystems*, consisting of areas of ice-scoured bedrock, subglacial tills, fluted moraines and eskers; and *glaciated valley landsystems*, consisting of troughs, rockwalls, lateral moraines and fluvial forms (e.g. Boulton and Eyles, 1979; Eyles, 1983b; Powell, 1984; Eyles et al., 1985; Eyles and Eyles, 1992). In the first edition of this book (Benn and Evans, 1998), we described glacial landsystems under the four categories of (1) subglacial; (2) ice-marginal and supraglacial; (3) subaqueous; and (4) glaciated valleys, but based on the considerable development and refinement of the glacial landsystems concept, as reported in Evans (2003a), we present landsystems in chapter 12 under eight categories: (1) ice-marginal terrestrial; (2) ice sheet beds and ice streams; (3) surging glaciers; (4) lowland supraglacial; (5) subaquatic; (6) glaciated valley; (7) plateau ice field; and (8) paraglacial.

Systems tracts

Petroleum geologists have developed and now routinely employ the concept of *sequence stratigraphy* as a means of predicting the location of hydrocarbon reservoirs (Posamentier et al., 1988; van Wagoner et al., 1988; Emery and Myers, 1996; Catuneanu, 2006). Sequence stratigraphy links depositional systems into genetically related *systems tracts*, and provides a powerful means of analysing the complex relationships between facies in a depositional basin, and unravelling the history of environmental changes contained in the geologic record. Sequence stratigraphic concepts have been applied to glacial sediment successions by Boulton (1990), Eyles and Eyles, (1992) and Martini and Brookfield (1995), and are discussed in chapter 12.

The study of the large-scale distribution of depositional systems and landsystems is one of the most exciting branches of glacial geology, because it is at this scale that the impact of glaciation is revealed at its most impressive (cf. Evans, 2003a). Large-scale studies allow long-term cycles of landscape change to be reconstructed, and the links between glacial cycles and global oceanographic and climatic change to be identified (Andrews, 2006; Clark et al., 2006; Kleman et al., 2006).

9.3 GLACIAL DEBRIS ENTRAINMENT

Debris can enter glacial systems either at the upper surface of the ice or at the glacier bed (fig. 9.1).

9.3.1 SUPRAGLACIAL DEBRIS ENTRAINMENT

On valley and cirque glaciers where ice surfaces are overlooked by valley walls, and where isolated peaks or *nunataks* protrude through ice sheet or ice cap surfaces, sediment can be delivered to glacier surfaces by gravitational processes. Elsewhere, supraglacial debris entrainment is usually negligible, except in some volcanically active areas such as Iceland, where far-travelled ashfalls (tephra) can provide a major, but intermittent source of sediment. On ice sheet surfaces far from nunataks and volcanic sources, meteorites can constitute the most important source of large particles. Although negligible in terms of glacial debris transport, meteorites found on the Antarctic Ice Sheet provide a valuable resource for planetary scientists, and many thousands of specimens have been recovered since systematic collection began in the 1970s (Harvey, 2003).

Rocks falling from cliff faces are among the most important sources of debris to valley glaciers (Boulton and Eyles, 1979; Small, 1987; Benn et al., 2003). Debris is released from rockslopes when pre-existing weaknesses are enlarged by mechanical weathering. In glacial environments, frost action is the principal mechanism, acting at the granular scale (microgelivation) or exploiting bedding planes or joints (macrogelivation) (fig. 9.3; Matsuoka, 2001; Matsuoka and Murton, 2008). Two main processes are responsible for frost damage to rocks. First, water undergoes a 9 per cent expansion of volume on freezing, which can force apart the sides of a water-filled pore or crack, causing crack extension. To be most effective, this process requires a high degree of rock saturation and rapid freezing to temperatures around $-5\,^{\circ}C$. Second, liquid water may migrate under capillary pressure towards freezing centres to feed expanding lenses of *segregation ice*, which gradually wedge apart the rock (Walder and Hallet, 1986; Hallet et al., 1991; Matsuoka and Murton, 2008). This process is likely to be most effective during sustained low temperatures of -4 to $-15\,^{\circ}C$. Particularly effective frost weathering is known to occur in randklufts at the margins of glaciers and below snowpatches, due to the microclimate and ready availability of water in these locations (Ballantyne et al., 1989).

Figure 9.3 Frost-shattered sandstone on Adventtoppen, Svalbard (Doug Benn).

Rockfalls associated with periglacial weathering tend to follow annual and diurnal temperature cycles. They are rare in cold weather and at night, when loose rock is cemented by ice, and are most common in warm conditions, particularly when cliffs receive direct sunlight. This fact is well known to mountaineers, who aim to be out of range of active slopes before sunrise. The location of rockfalls is mainly controlled by rockslope form, aspect, lithology and geological structure (André, 1997).

Large-scale rockslope failures are less closely controlled by freezing cycles, although frost action can contribute to the weakening of slopes. Periods of heavy rainfall or rapid snowmelt can promote failure by increasing the water pressure in joints and reducing the frictional strength of the rock. In tectonically active areas, earthquakes can trigger rockslope failures, delivering large amounts of debris to glacier surfaces over extensive areas (Jibson et al., 2006). Glacier retreat also encourages large failures due to the exposure of oversteepened slopes, release of glacier-induced stresses and changes to drainage conditions (André, 1997; Ballantyne, 2002a; Geertsema et al., 2006; Ballantyne et al., 2008). During periods of climatic warming, rockfall activity can be accelerated due to degradation of permafrost which formerly cemented rock materials (Harris et al., 2003; Noetzli et al., 2003; Gruber et al., 2004; Gude and Barsch, 2005).

Long-term rates of debris delivery to glaciers from rockfalls are poorly known. Measured and estimated rockwall retreat rates in present-day mountain environments range between 0.05 and 6.00 mm yr^{-1} (Ballantyne and Harris, 1994; Arsenault and Meigs, 2005). However, the amount of debris delivered on to glacier surfaces by individual rockfalls varies considerably, ranging from single particles (low-magnitude events) to huge rock avalanches (high-magnitude events). Low-magnitude events occur relatively frequently, providing a steady input of debris over long time periods, whereas high-magnitude events tend to be rare. The relative importance of low- and high-magnitude events in the long-term debris budget of valley glaciers is very poorly known, and is likely to be highly variable from place to place. Gordon et al. (1978) calculated that a single rock avalanche on to Lyell Glacier, South Georgia, in September 1975 supplied the same amount of debris as 93 years of 'normal' low-magnitude rockfalls on to the glacier. The contribution of high-magnitude events to rockwall retreat in a glacierized catchment in Alaska has been determined by Arsenault and Meigs (2005).

When they occur, however, rockfalls and rockslides are major events that can profoundly influence local topography and glacier behaviour. A huge rock avalanche in December 1991 removed 20 m from the summit of Mount Cook (formerly 3764 m) in the New Zealand

Alps, sending an estimated 14 million m^3 of rock and ice onto the Tasman Glacier below and obliterating the Hochstetter Icefall en route (fig. 9.4; Kirkbride and Sugden, 1992; Owens, 1992). By blanketing large areas of glacier surfaces with debris, large avalanches protect the ice from ablation, altering the mass balance. Several glaciers have been known to readvance after major rockfalls, including the Brenva Glacier, Italian Alps, following a rockfall in 1922 (Porter and Orombelli, 1981); the Bualtar Glacier, Pakistan, after a major slope failure in 1986 (Gardner and Hewitt, 1990); and several glaciers in Alaska following the magnitude 7.9 Denali Fault earthquake of November 2002 (Jibson et al., 2006).

Snow and ice avalanching can be an extremely important mechanism for transporting debris onto glacier surfaces, especially in high mountain environments such as the North and South American Cordillera, European Alps and the Himalaya, where large quantities of snow can accumulate on unstable slopes (Owen and Derbyshire, 1989; Benn and Owen, 2002; Benn et al., 2003). The collapse of snowslopes or ice seracs can result in very large, destructive avalanches, which can scour and pluck debris from underlying rock surfaces (fig. 9.5). Repeated avalanches from the same slope build out dirty avalanche cones at glacier margins, composed of a mixture of snow and crushed ice, ice blocks and rock (Humlum, 2005). In high mountain environments such as the European Alps and the Himalaya, snow avalanching provides the single most important source of supraglacial debris (fig. 2.9).

Flowage of wet, unconsolidated sediments (*debris flows*) can deliver large quantities of sediment to the margins of some glaciers, especially where ice is overlooked by older glacial deposits. This situation commonly occurs during deglaciation in mountainous terrain, when the oversteepened inner slopes of lateral moraines are exposed above ablating glaciers. Such slopes are generally composed of poorly sorted debris with a fine-grained matrix and can degrade very rapidly, yielding numerous debris

flows during times of snowmelt or heavy rainfall (fig. 9.6; Ballantyne and Benn, 1994a). *Flowing water* from valley sides tends to be a relatively unimportant source of supraglacial debris, but can be locally significant in some glacier ablation areas (Evenson and Clinch, 1987).

Debris that accumulates on glacier surfaces can be incorporated into the body of the glacier in two main ways: (1) burial by snow and ice, and (2) by falling down crevasses or other holes in the glacier surface (fig. 9.7). Burial by snow and ice occurs mainly in glacier accumulation areas, where more snow falls each year than melts during the ablation season. In this case, debris on the glacier surface is progressively buried by snowfall and/or snow and ice avalanches (fig. 9.8), and is incorporated

Figure 9.5 Full-depth snow avalanche, Larsbreen, Svalbard, June 2005 (Doug Benn).

Figure 9.4 Large lobate rock avalanche deposit on the Tasman Glacier, New Zealand, following the catastrophic slope failure of December 1991 (David Evans).

Figure 9.6 Debris flows from sediment-covered valley sides onto a snow surface, Fåbergstollsdalen, Norway (Doug Benn).

within the snowpack. With continued accumulation, the debris is carried to progressively deeper levels within the glacier. Descent of the debris towards the glacier bed can also be encouraged by patterns of ice flow and basal melting. A limited amount of debris burial can also occur at the lateral margins of glacier ablation zones, at the foot of avalanche cones. Such cones represent localized areas of snow or ice accumulation, but in this case deep burial is unlikely, because the debris will be exhumed by melting once the surrounding snow is transported away from the active

cone by glacier flow. Crevasses can swallow up debris in any part of the glacier, the only constraints being the presence of open crevasses and a source of debris (fig. 9.9). This mechanism can occur either where fresh rockfalls deliver debris into open fractures, such as lateral crevasses or the junction of confluent glaciers, or where crevasses open up beneath a surface veneer of debris, such as in icefalls. The randkluft at the upper edge of a glacier can be a very important routeway for supraglacially entrained debris to pass into englacial and subglacial transport.

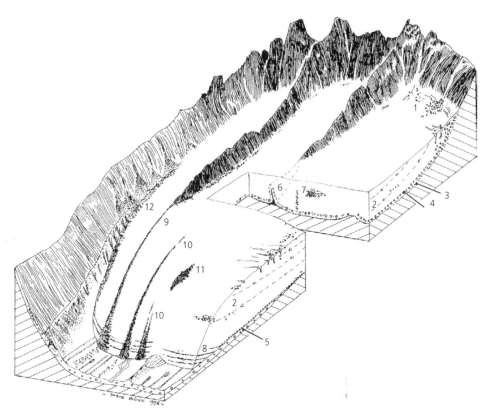

Figure 9.7 Debris transport through a valley glacier. (1) Rockfall debris entrained supraglacially and buried by snow or ingested in crevasses. (2) Ice flow transports debris away from the surface in the accumulation area and towards the surface in the ablation area. (3): Basal tractive zone. (4) Suspension zone. (5) Basal till. (6) Debris septum elevated from the bed below a confluence. (7) Diffuse septum and cluster of rockfall debris. (8) Debris elevated from the bed by shearing. (9) Ice–stream interaction medial moraine. (10) Ablation-dominant medial moraines. (11) Avalanche-type medial moraine. (12) Supraglacial lateral moraine. Glacifluvial transport not shown.

Figure 9.8 Debris entrainment in ice avalanche cones below the 'Spider Wall', south face of Nuptse, Nepal (Doug Benn).

Figure 9.9 Debris-filled surface crevasses, Belvedere Glacier, Italy (Doug Benn).

9.3.2 INCORPORATION OF DEBRIS INTO BASAL ICE

Debris at glacier beds, including freshly released rock fragments, till and pre-existing sediments, can be incorporated into basal ice and enter glacial transport. Basal ice tends to be distinctly different from the ice that makes up the rest of the glacier as the result of intense deformation, melting, refreezing and debris entrainment processes operating near the bed. Basal ice can have a very high debris content (>75 per cent by volume in some cases), and distinctive chemical and isotopic composition and structural characteristics (Hubbard and Sharp, 1989; Alley et al., 1997; Knight, 1997; Hubbard et al., 2009; fig. 9.10). Knowledge of the formation and characteristics of basal ice is important for several reasons. First, the concentration of debris in basal ice controls ice rheology and subglacial friction, both of which influence glacier motion (sections 4.3.2 and 4.4.3). Second, the presence of debris in basal ice has a large influence on rates of subglacial erosion (section 8.2). Third, debris can be transported large distances in basal ice, not only within glaciers and ice sheets, but also far beyond glacier margins in rafted icebergs.

A number of processes can lead to the formation of debris-rich basal ice, the most important of which are: regelation, glaciohydraulic supercooling, net adfreezing, ice deformation around rock fragments, and the entrainment of pre-existing ice. These are reviewed in turn below.

Regelation

At temperate glacier beds, debris can become incorporated within basal ice as a result of regelation processes, either by (1) refreezing on the downglacier side of rock bumps, or (2) infiltration of ice into subglacial sediments.

Freeze-on at the lee side of bedrock bumps was termed the *heat pump effect* by Robin (1976). As discussed in section 4.4.2, glacier sliding over a rough bed involves large fluctuations in basal pressure in the vicinity of rock knobs. On the upglacier side of obstructions, where the

glacier is flowing towards the bed, pressure is higher than average, causing spontaneous melting of the ice. Robin argued that melting occurs at the bed and at grain boundaries within the basal ice, from which water is expelled through veins. This process removes heat from the ice and lowers its temperature to the depressed pressure-melting point. As the ice flows over the lee side of the obstruction the pressure falls, but the temperature cannot rise to the new, elevated pressure-melting point because the lost meltwater is not available to heat the interior of the basal ice. Therefore, cold patches will develop, causing the ice to locally freeze onto the bed. As a result, rock fragments can adhere to the ice and be removed from the bed.

The heat pump effect has been demonstrated experimentally by Röthlisberger and Iken (1981), who also considered the role of cavity formation on the plucking mechanism. The opening of basal cavities beneath a glacier is encouraged by locally high basal water pressures, which tend to inhibit the formation of cold patches. The heat pump effect, therefore, will tend to operate either in 'passive cavities' that experience only low water pressures, or during the closure of 'active cavities' as water pressures fall (cf. Iverson, 1991a). This emphasizes the importance of non steady-state conditions for subglacial erosion.

Where temperate ice overlies unconsolidated sediments, debris can be incorporated into the basal ice by *regelation infiltration* (section 3.4.6; Iverson and Semmens, 1995; Iverson et al., 1995, 2007; Clarke, 2005). This will occur if the ice pressing down on sediment grains is at higher pressure than the water in the pore space below, in which case pressure-melting and refreezing allows the ice to penetrate downwards into the sediment. The rate of ice infiltration v_r is directly proportional to the difference between the ice pressure P_i and the pore-water pressure P_w, and inversely proportional to the thickness of the ice-cemented sediment layer at the glacier sole H_s:

$$v_r = \frac{K_s(P_i - P_w)}{H_s} \tag{9.1}$$

where K_s is a constant (Iverson, 2000). Equation 9.1 shows that the infiltration rate slows down as basal ice accumulates, so it is unlikely that this process alone will produce basal ice layers more than a few centimetres in thickness. Regelation infiltration was elegantly demonstrated in experiments conducted in the Svartisen Subglacial Laboratory by Iverson et al. (2007), who were able to control the water pressure in a prism of sediment installed at the glacier bed (see p. 128). When the sediment prism was excavated at the end of the experiment, ice had penetrated tens of millimetres into the sediment, in good agreement with theoretical predictions (fig. 9.11).

Regelation ice is commonly banded, with closely spaced laminae of clear and debris-rich ice (Souchez and

Figure 9.10 Debris-rich basal ice below Von Postbreen, Svalbard (Doug Benn).

Lorrain, 1987; Hubbard and Sharp, 1993, 1995). Individual laminae are commonly 0.1–1 mm thick and form layers broadly parallel to the bed, although they may pinch out or merge laterally. Each lamina represents one refreezing event. Hubbard and Sharp (1995) also identified a distinct type of *clear ice* below temperate Alpine glaciers. The ice is devoid of layering, but contains dispersed smears of debris and occasional clouds of deformed and flattened bubbles. They argued that this ice forms by internal melting, freezing and ice recrystallization during enhanced creep near the bed, as described by Lliboutry (1993).

The accretion of regulation ice results in isotopic fractionation, because heavy isotopes of oxygen and hydrogen (^{18}O and deuterium) preferentially bond into growing ice crystals (section 2.7.2). During the initial stages of refreezing, therefore, regulation ice will be isotopically heavier than the remaining meltwater, but as freezing proceeds, each increment of ice will be isotopically lighter than the one before, due to the depletion of heavy isotopes in the water. As a result, regulation ice often exhibits *compositional banding*, with each freezing cycle being represented by a downward gradation from isotopically heavy to isotopically light ice (fig. 9.12; Souchez and Jouzel, 1984; Hubbard and Sharp, 1993). The bulk composition of regulation ice will be isotopically heavier than the parent meltwater if some of the water is able to drain away before freezing is complete.

Glaciohydraulic supercooling

In recent years, much attention has focused on the formation of debris-rich basal ice layers by the freezing of supercooled water (e.g. Alley et al., 1997, 1998; Roberts et al., 2002; Tweed et al., 2005; Cook et al., 2006). The cause of the freezing process is essentially the same as for regulation, except that it occurs over larger spatial scales. If water flowing along the bed of a temperate glacier is initially at thermal equilibrium and then experiences a pressure drop, it will freeze unless it can be warmed to the new pressure-melting point. Energy for warming can be supplied from viscous heat dissipation from water flow, geothermal heat, or strain heating (section 2.3.2). However, if these heat sources are insufficient to raise the water temperature to the new pressure-melting point, it will become supercooled relative to its surroundings and freeze onto the glacier sole. Supercooling is likely to occur where basal water flows out of overdeepenings, because the along-flow pressure drop must be large enough to more than compensate for the increase in elevation head (by definition, the total hydraulic potential must decrease downflow; section 3.2.2). There is no reason to suppose, however, that supercooling can only occur where water flows up a reverse slope.

Evenson et al. (1999) have described the formation of icings at the margins of Matanuska Glacier, Alaska, where supercooled subglacial water emerges from vents. Openwork aggregates of ice crystals form around vent margins, and sediment is trapped in interstitial voids when through-flowing water freezes. Through time, this process builds up accumulations of debris-rich ice within conduits, fractures and moulins, as well as vent margins. Lawson et al. (1998) argued that freezing of supercooled water can also account for laterally extensive layers of stratified basal ice up to 10 m thick exposed around the glacier margins. This distinctive ice facies consists of alternating sub-horizontal laminae and lenses of debris-rich, debris-poor and debris-free ice, with individual laminae typically 5–20 mm thick. There are large lateral variations

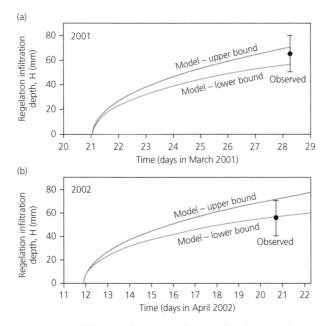

(a)

(b)

Figure 9.11 Modelled rates of regulation infiltration of basal ice into subglacial sediment, compared with observed values (After Iverson et al., 2007).

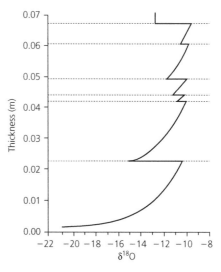

Figure 9.12 Modelled isotopic variation in regulation ice, resulting from the preferential incorporation of heavy isotopes of oxygen in the first ice to form in each cycle (After Hubbard and Sharp, 1993).

in sediment concentration, texture and sorting, and in places sedimentary structures such as channel forms and cross-bedding can be seen. Freezing of supercooled silty water could also account for the formation of *clotted ice*, which is generally clear, bubble-free ice containing dispersed, elongated 'clots' of silt-sized debris. Clotted ice occurs in a wide variety of environments, however, and might form wherever turbid water undergoes bulk freezing (Knight, 1994; Christofferssen et al., 2006).

Ice in the stratified facies is isotopically heavier than englacial ice, reflecting fractionation processes during refreezing (Titus et al., 1999). If the ice also contains tritium (^{3}H), which was produced during atmospheric nuclear weapons tests in the 1950s and 1960s, this shows that the parent water was of more recent origin than the overlying uncontaminated englacial ice.

In Iceland, debris-rich ice has been shown to form from supercooled water during both normal, ablation-driven flows and glacier lake outburst floods (GLOFs; section 3.7; Roberts et al., 2002; Tweed et al., 2005). The process may be particularly important during GLOFs, which entrain copious volumes of sediment, and potentially explain anomalously thick sections of debris-rich ice. Other researchers have concluded that glaciohydraulic supercooling can account for some, but not all occurrences of debris-rich basal ice in Iceland (Cook et al., 2006, 2007; Swift et al., 2006).

Net adfreezing

Accumulation of basal ice will also occur where meltwater flows from temperate to cold areas of the bed (Hubbard and Sharp, 1989; Hubbard, 1991; Christoffersen and Tulaczyk, 2003a). Freezing will occur wherever a supply of water coincides with a net energy deficit at a glacier bed. Near glacier margins, adfreezing can occur in response to the penetration of a winter cold wave through the ice or the circulation of cold air via cavities, crevasses and tunnels. Hubbard and Sharp (1995) described a distinctive type of basal ice formed by the freezing of standing water in bedrock hollows or cavities. This *interfacial ice* consists of alternating bands of ice crystals and debris-rich ice which undulate in parallel with the large-scale roughness of the bed.

Below thicker ice, freezing can occur where more heat is conducted away from the bed than can be provided by geothermal heat or sliding friction, such as below slow-moving ice or at the base of ice streams where sliding rates are high but basal traction is very low (sections 5.2.4 and 6.3.6). The debris content of basal ice formed by net adfreezing is highly variable, depending on the conditions at the ice–bed interface. Very high debris concentrations can result from the downward migration of a freezing front into saturated sediments, so that the ice merely forms an interstitial cement between mineral grains (Harris and Bothamley, 1984). In such cases, masses of pre-existing sediment frozen onto the base of the glacier can preserve delicate sedimentary structures such as cross-bedding and lamination. Adfreezing above a bedrock interface will usually result in basal ice with less debris, with the concentration depending on the proportions of water and debris that are available (Hubbard and Sharp, 1989). As is the case for regelation ice, ice formed by net adfreezing is generally isotopically heavier than meltwater derived from meteoric ice, due to fractionation and incomplete freezing.

Net adfreezing over till beds has been studied in detail by Christoffersen and Tulaczyk (2003a) and Christoffersen et al. (2006), to explain the characteristics of stratified basal ice below Antarctic Ice Streams (section 6.3.6). Freezing at the ice–till interface creates a hydraulic gradient within the till, actively drawing pore water towards the freezing front (fig. 9.13). Initially, sediment is rejected during freezing, forming a layer of clean *segregation ice*. The freezing front then advances downwards into the till, and a new layer of segregation ice is formed up to a few centimetres below the first. Repeated cycles of freezing produce stratified basal ice which can total several metres in thickness. The physics of the freezing process is complex, and involves local regelation infiltration and supercooling.

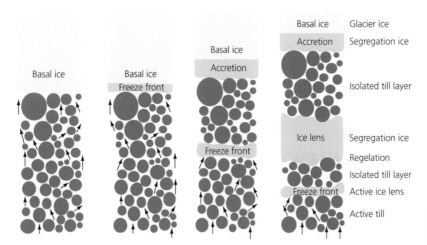

Figure 9.13 Development of stratified basal ice from ice segregation and freezing-front migration (Christoffersen et al., 2006).

Deformation of ice around rock fragments

Ice can deform rapidly under high stresses, and in some circumstances is able to completely or partially encase rock fragments at glacier beds. This process has been observed under Øksfjordjøkelen, Norway, by Rea and Whalley (1994). The bed of this glacier is at the pressure-melting point and consists of a series of abrupt, well-jointed rock steps aligned transverse to ice flow. Rea gained access to the glacier bed via a large cavity between two rock steps, and observed basal ice being squeezed under pressure into rock joints. In one case, the ice had flowed into the joints around and beneath a large block on the bed, completely surrounding it (fig. 9.14). The presence of this ice dramatically reduced the friction between the block and the bed, facilitating removal of the block by the overriding ice. Rea and Whalley suggested that such injection of plastic ice into rock joints may be a very important plucking mechanism, particularly where basal shear stresses are high and sliding rates low. It is probable that this process also requires low basal water pressures to be effective.

Entrainment of pre-existing ice

The retreat of glaciers containing high debris concentrations can result in the preservation of very large quantities of buried glacier ice (e.g. Moorman and Michel, 2000;

Everest and Bradwell, 2003). This is particularly important in permafrost regions, where melting ceases once ice is covered by an overburden of debris that exceeds the active layer thickness (see section 2.4.5). In some cases, buried glacier ice has survived since the LGM (e.g. St Onge and McMartin, 1995; Murton et al., 2005). Several glaciers on Ellesmere Island in the Canadian Arctic have recently readvanced over buried glacier ice and reincorporated it and its overburden (fig. 9.15; Evans, 1989b). This reincorporation results in the reactivation of the buried glacier

Figure 9.14 Deformation of ice around a joint-bounded block, Oksfjordjøkelen, Norway (Brice Rea).

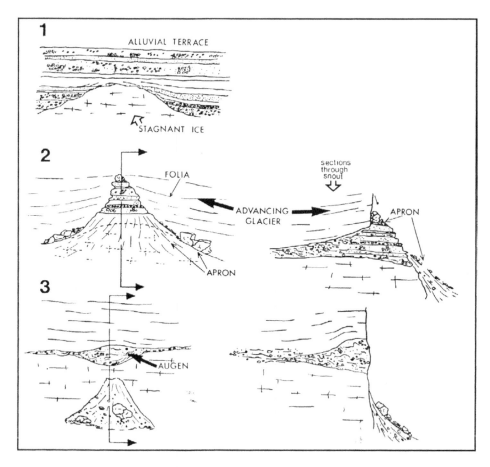

Figure 9.15 An explanation of the formation of gravel pods in the fronts of polythermal glaciers. (1) Progradation of glacifluvial sediments preserves dead ice. (2) Glacier overrides and deforms the debris-covered ice. (3) Complete overriding results in an isolated gravel clot at the boundary between old and new glacier ice (Evans, 1989b).

ice, and the former debris cover becomes an englacial debris band which attenuates during further glacier flow. In many cases, the buried glacier ice on Ellesmere Island has remained continuous with the glacier snout and so forms a supraglacial 'ramp'. This ramp is subject to fluvial incision and collapse, which allows the repositioning of former surface debris within the buried ice. In this way, further prospective englacial debris bands are initiated prior to ramp reactivation.

Advancing glaciers in cold regions often terminate in steep ice cliffs, below which frontal aprons of ice blocks and debris accumulate. Overridden and incorporated aprons are recognizable in the basal ice facies of many polythermal glaciers on Ellesmere Island, where mosaics composed of individual ice blocks are still visible (fig. 9.16; Evans, 1989b). The overriding of aprons by internally deforming glaciers leads to the incorporation and gradual attenuation of ice and sediment accumulations to produce complex basal ice layers of alternating debris- and ice-rich folia.

Deformation of basal ice

The structure and thickness of basal ice layers can be strongly influenced by the deformation history of the ice. In zones of extending glacier flow, the ice is stretched and the basal ice layer will become thinner unless it is renewed

by some other process, such as freezing-on. The effects of stretching can be observed where basal ice consists of layers with different flow responses to stress. Clean, coarse-grained ice tends to be quite ductile, and will stretch readily without breaking, whereas fine-grained or debris-rich ice tends to be stiffer and more susceptible to brittle deformation. When stretched, bands of fine or dirty ice will often break up, forming discontinuous lenses called *boudins*, after the French sausages they resemble (fig. 9.17; Hambrey and Milnes, 1975; Hubbard and Sharp, 1989; Hambrey, 1994). Boudins can provide information on the relative competence of different components of basal ice layers (Fitzsimons et al., 2000). The stretching of basal ice layers undergoing simple shear can also serve to rotate and lengthen lenses within the ice, creating new stratification (Hooke and Hudleston, 1978).

Compressive flow will tend to shorten and thicken basal ice layers, and is thought to be responsible for many of the thick sequences of basal ice observed at glacier margins, particularly on polythermal and surging glaciers. Thickening of basal ice sequences is commonly accompanied by folding and the development of thrust faults, often resulting in complex stratigraphies (fig. 9.18; Hambrey and Müller, 1978; Sharp et al., 1988; Knight, 1989, 1997; Fitzsimons, 1990; Alley et al., 1997; Hambrey and Lawson, 2000). Impressive stratified sequences can be built

Figure 9.16 (a) Frontal apron comprising dry calved ice blocks, Ellesmere Island. (b) Entrained ice blocks beneath foliated debris-rich ice, Eugenie Glacier, Ellesmere Island (David Evans).

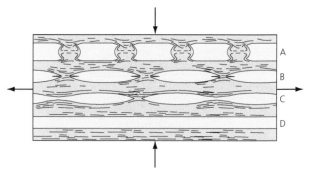

Figure 9.17 Boudins resulting from the extensional deformation of glacier ice with variable rheology (After Hambrey and Milnes, 1975).

Figure 9.18 Folded basal debris-rich ice, Ellesmere Island (David Evans).

up by the repetition of layers in recumbent folds, and by faulting. Indeed, where exposures of basal ice are limited, it may be difficult to distinguish stratification created by ice deformation from primary layering caused by cyclic adfreezing by using field evidence alone. Careful studies of the isotopic composition of the ice may be required to determine its true origin. Layered ice sequences formed by complex folding at the margin of the Greenland Ice Sheet have been termed the *banded ice facies* by Knight (1994).

9.4 DEBRIS TRANSPORT AND RELEASE

Debris is transported through glacier systems by ice flow, in supraglacial, englacial and/or subglacial positions. On many glaciers, however, the greatest volumes of sediment are actually transported by flowing water (Kirkbride, 1995a; Alley et al., 1997). In this section, we review processes of glacial and glacifluvial debris transport, and associated glacial structures. Processes that transport sediment beyond glacier margins are reviewed in chapter 10.

9.4.1 SUBGLACIAL TRANSPORT

Because most debris input to glaciers occurs either at the bed or at the margins below valley walls or nunataks, glacial debris load tends to be initially concentrated either at the bed or within a few metres of it (fig. 9.7). Boulton and Eyles (1979) introduced the term *bed-parallel debris septum* (a *septum* is a partition) to refer to the concentrated zone of debris close to the margins and bed of glaciers. The bed-parallel septum consists of two zones: the *basal tractive zone* in contact with the bed, and the *suspension zone* close to the bed but not in contact with it. Debris in the basal tractive zone is sheared between the glacier sole and the immobile substrate at depth (fig. 9.19). Where the bed consists of hard bedrock, the tractive zone may be only one clast thick, but it may be considerably thicker where

the bed consists of till or other unlithified sediments. Debris transport rates will be directly proportional to the strain rate and thickness of the tractive zone, and inversely proportional to sediment porosity. The deep, rapidly shearing deforming layers envisaged by some researchers, following the introduction of the 'deforming layer paradigm', implied very high rates of sediment transport (Boulton, 1979, 1987; Alley et al., 1987; Hooke and Elverhøi, 1996). Recent observations of till deformation, however, indicate that intense shearing tends to be concentrated in a zone up to a few centimetres thick immediately beneath the glacier sole, with lower strain rates at greater depth (section 4.5.3), implying lower, but still geologically significant sediment fluxes (Alley, 2000). The data presented by Boulton (2006), for example, imply till fluxes below Breiðamerkurjökull, Iceland, equivalent to $\sim 17\,\mathrm{m}^3$ yr^{-1} per unit width of the glacier.

The lower part of the suspension zone typically consists of debris-rich basal ice (section 9.3.2). In valley glaciers, this will pass upwards into debris structures of supraglacial origin, described in the following section. Melting or freezing at the glacier bed will cause debris to pass between the suspension zone and the tractive zone. Rates of sediment transport in the suspension zone will depend on debris-layer thickness and concentration, basal sliding speeds and strain rates in the ice.

9.4.2 HIGH-LEVEL DEBRIS TRANSPORT

Debris bands and septa

In the accumulation zones of valley glaciers, glacier ice typically contains a variety of debris structures reflecting the style and location of supraglacial debris entrainment. Debris bands can be either parallel to the primary stratification of the glacier, reflecting deposition on the glacier surface followed by burial in the snowpack, or cut across the stratification, reflecting deposition in crevasses (fig. 9.20; Hambrey et al., 1999; Goodsell et al., 2005a, b; Gulley and Benn, 2007). The composition of such debris bands is essentially the same as that of the parent material; debris bands derived from rockfalls, for example, commonly consist entirely of coarse, angular clasts. Supraglacially derived debris bands are not randomly distributed through glaciers, but are commonly arranged in longitudinal sets, reflecting the location of persistent debris sources and ice flow lines.

As a result of patterns of ice flow, debris bands are progressively deformed and rotated as they are advected downglacier. Debris from bed-parallel debris septa can be elevated from the bed where ice flow units converge around bedrock obstructions or at valley confluences. Figure 9.21 shows how basal ice can be elevated from the bed at a glacier confluence, forming a *medial debris septum*. Such structures can be vertical or inclined, depending

Figure 9.19 The basal tractive zone: sheared till and basal ice, Scott Turnerbreen, Svalbard (Jason Gulley).

Figure 9.20 Debris bands parallel to ice stratification, which originated as supraglacial debris flows buried within the snowpack. Nick Hulton for scale (Doug Benn).

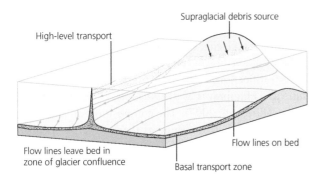

Figure 9.21 Elevation of basal debris into high-level transport at a glacier confluence (Modified from Boulton, 1978).

on the relative strength of the flow units (Eyles and Rogerson, 1978; Boulton and Eyles, 1979; Goodsell et al., 2005a, b). In the case of converging flow around a basal obstruction or a glacier confluence in the accumulation area, the medial septum will not extend to the surface until exposed by melting in the ablation zone (fig. 9.7). In contrast, if a glacier confluence is in the ablation zone,

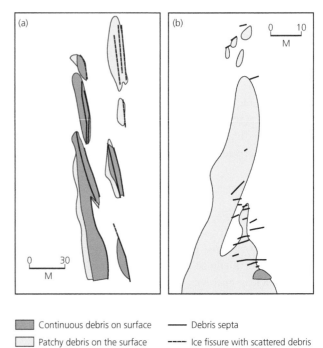

Continuous debris on surface — Debris septa
Patchy debris on the surface ---- Ice fissure with scattered debris

Figure 9.22 Maps of debris septa. (a) Longitudinal septa exposed on the eastern medial moraine, Haut Glacier d'Arolla, Switzerland. (b) Transverse septa, western medial moraine, Bas Glacier d'Arolla (Redrawn from Small, 1987).

the resulting medial debris septum will usually extend from bed to surface. Medial debris septa derived from elevated basal tractive zones typically consist of sharply defined, concentrated bands of debris-rich ice (Vere and Benn, 1989).

Medial debris septa can also form downglacier of persistent supraglacial debris sources, as successive generations of debris are carried downglacier along the same flow line (Small et al., 1979; Gomez and Small, 1985; Small, 1987). In this case, the debris septa are not necessarily sharply defined, but can consist of coarse clasts dispersed through the ice or numerous individual debris bands grouped together in sub-parallel chains (Gomez and Small, 1985; Small, 1987; Goodsell et al., 2005a, b; fig. 9.22). Detailed structural analysis has shown that debris structures are commonly deformed into tight folds as the surrounding ice is stretched and sheared (Hambrey et al., 1999; Hambrey and Glasser, 2003; Goodsell et al., 2005a, b; fig. 9.23).

Medial moraines

In glacier ablation zones, debris from medial debris septa accumulates on the ice surface, forming medial moraines. These are among the most striking features of valley glaciers, providing a graphic picture of the movement of both ice and debris (fig. 9.24). However, medial moraines can give a misleading impression of the quantity of debris in transport, as the debris on the surface is generally much more concentrated and laterally extensive than that

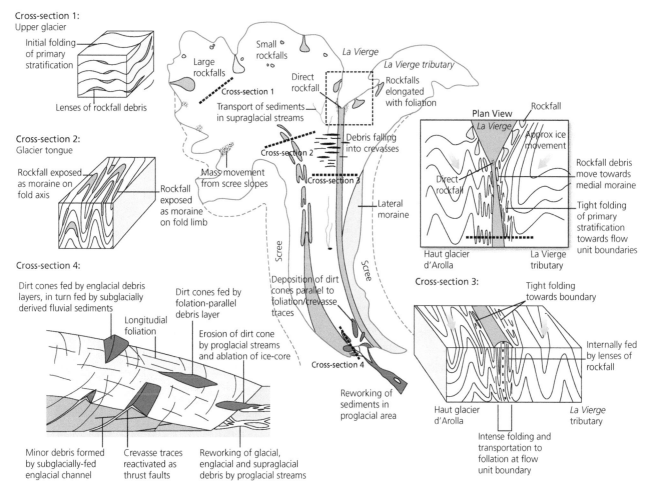

Cross-section 1:
Upper glacier

Initial folding of primary stratification

Lenses of rockfall debris

Cross-section 2:
Glacier tongue

Rockfall exposed as moraine on fold axis

Rockfall exposed as moraine on fold limb

Cross-section 4:

Dirt cones fed by englacial debris layers, in turn fed by subglacially derived fluvial sediments

Dirt cones fed by foliation-parallel debris layer

Longitudial foliation

Erosion of dirt cone by proglacial streams and ablation of ice-core

Minor debris formed by subglacially-fed englacial channel

Crevasse traces reactivated as thrust faults

Reworking of glacial, englacial and supraglacial debris by proglacial streams

Large rockfalls

Small rockfalls

La Vierge

La Vierge tributary

Direct rockfall

Rockfalls elongated with foliation

Transport of sediments in supraglacial streams

Cross-section 1

Mass movement from scree slopes

Cross-section 2

Debris falling into crevasses

Cross-section 3

Lateral moraine

Scree

Scree

Deposition of dirt cones parallel to foliation/crevasse traces

Cross-section 4

Reworking of sediments in proglacial area

Plan View

La Vierge

Rockfall

Approx ice movement

Direct rockfall

Rockfall debris move towards medial moraine

Tight folding of primary stratification towards flow unit boundaries

Haut glacier d'Arolla

La Vierge tributary

Cross-section 3:

Tight folding towards boundary

Internally fed by lenses of rockfall

Haut glacier d'Arolla

La Vierge tributary

Intense folding and transportation to follation at flow unit boundary

Figure 9.23 Schematic summary of debris transport processes at Haut Glacier d'Arolla (Goodsell et al., 2005a).

Figure 9.24 Medial moraines on a calving glacier, east Greenland (Tavi Murray).

within the glacier, due to the effects of glacier ablation in redistributing the debris.

A classification of medial moraines was proposed by Eyles and Rogerson (1978), based on the relationship between debris supply and the morphological development of the moraine. Although fault can be found with this scheme, no alternative comprehensive classification has been proposed, and we retain it as a useful framework. Three broad types of medial moraine can be recognized:

(1) *Ice–stream interaction* (ISI) moraines, which find immediate surface expression downstream from glacier confluences;
(2) *Ablation-dominant* (AD) moraines, which emerge at the surface as the result of the melt-out of englacial debris;
(3) *Avalanche-type* (AT) moraines, which are transient features formed by exceptional rockfall events onto a glacier surface.

These three types are described in turn below.

Figure 9.25 Medial moraines on Storbreen, Norway. (a) Avalanche-type medial moraine reflecting a single large rockfall event. (b) Ablation-dominant medial moraine, formed by melt-out of scattered rockfall debris. (c) Vertical septum of basal debris elevated from the bed. Melt-out has produced an extensive cover of supraglacial debris, which has been cleared to reveal the internal structure (Doug Benn).

Ice–stream interaction (ISI) moraines form at the intersection of confluent valley glaciers below or close to the equilibrium line (fig. 9.7). The moraines find immediate surface expression at the confluence, commonly by the merging of supraglacial lateral moraines. They may be simply superficial features, although debris supply is commonly supplemented downglacier by the melt-out of englacial debris, which may consist of either rockfall material or elevated basal debris (fig. 9.25). Moraine morphology is strongly influenced by the character of ice flow, and lateral compression between merging ice streams can counteract the tendency of moraine debris to spread out laterally at the surface, particularly if the glacier is undergoing longitudinal extension (Eyles and Rogerson, 1978; Smiraglia, 1989). Complex surface forms can also develop if the two ice streams have different velocities or discharges, causing one to shear up over the other.

Ablation-dominant (AD) moraines emerge in ablation areas by the melt-out of englacial debris septa formed further upglacier. Three subtypes can be defined, depending on the mechanism of debris entrainment. *AD1* or *below firn-line type* medial moraines form in glacier ablation zones in the lee of nunataks where ice is heavily crevassed. Rockfall debris entering crevasses comes to occupy shallow englacial positions, and is then exposed downglacier

by ablation. A prominent moraine of this type occurs on Austerdalsbreen, an outlet of the Jostedalsbre Ice Field in south-central Norway. The moraine forms at the junction of two icefalls, Odinsbreen and Thorsbreen, that drop 800 m from the snowfields above. Debris enters the glacier via numerous crevasses and progressively melts out downglacier, forming an upstanding ice-cored ridge. The ridge attains a maximum height of 12 m above the adjacent clean ice, approximately 1200 m from the confluence, at which point the debris supply becomes exhausted. Thereafter, debris spreads laterally, the ridge declines in relief, and moraine width increases from 40 m to over 200 m in the terminal zone.

AD2 or *above firn-line type* medial moraines are fed by debris in the accumulation area, either at the junction of flow units or persistent rockfall sites, such as gullies in the glacier backwall. During transport through the glacier accumulation area, the debris becomes more deeply buried by annual snowfall, then melts out below the firn line as deep ice is exposed by ablation. Several variants of this moraine type can be recognized, according to the mechanism of debris entrainment:

(1) Rock debris falls onto the glacier surface and is buried in the snowpack. The resulting medial moraine consists of longitudinal bands of coarse, angular debris,

sometimes with several parallel bands, which may be of different rock types (Vere and Benn, 1989).

(2) Rock debris falls into marginal crevasses, resulting in transverse debris bands that melt out to form beaded moraines. These may merge downglacier into a continuous cover of superficial debris. A well-studied example of this type occurs on the Glacier de Tsidjiore Nouve, Switzerland (Small et al., 1979; Small and Gomez, 1981; Small, 1987).

(3) Basal debris is elevated from the bed at a glacier confluence, forming a longitudinal septum rising upwards from the bed but not immediately reaching the surface. At higher levels, englacial debris consists of rockfalls buried by snow or ingested in crevasses. As the resulting medial moraine melts out in the ablation zone, subglacial and rockfall debris become mixed together.

The quantity of englacial debris nourishing AD2 moraines is highly variable, leading to very different degrees of morphological development. If the amount of debris is small, the surface expression of the moraine consists of only a diffuse scatter of particles, but if a large amount of englacial debris is present, prominent upstanding ridges will develop (Small et al., 1979).

AD3 or *subglacial rock-knob type* medial moraines form where debris is elevated from the bed in the lee of ice-covered rock knobs by converging ice flow. Basal debris-rich ice is rotated to form a vertical or steeply dipping longitudinal septum, which generally has a high debris content, composed of abraded clasts in a fine-grained matrix. When the septum melts out from the ice, it forms a low, steep-sided ridge of poorly sorted till on the glacier surface, which gradually degrades on exposure to the weather (Vere and Benn, 1989). Debris accumulating on the adjacent ice retards ablation, increasing the relief of the moraine above the rest of the glacier. Debris then flows down the flanks of the moraine, broadening and thinning the cover of debris on the surface. An example of this type of moraine on Storbreen, Norway, has been described by Vere and Benn (1989) (fig. 9.25).

Patterns of ice deformation can also strongly influence the formation of ablation-dominant moraines. Hambrey and Glasser (2003) have shown that, where ice from multiple basins feeds a narrow glacier tongue, convergent flow results in transverse shortening of the ice and the development of tight folds with flow-parallel axial planes. If deformation is sufficiently intense – which is commonly the case at flow-unit boundaries – the original, primary stratification of the glacier is overprinted by a longitudinal foliation, and once-horizontal debris structures are rotated into longitudinal septa. In some cases, foliation containing basally derived debris also shows an axial planar relationship with folding, suggesting that debris-rich

basal ice has been folded within the body of the glacier. Examples of medial moraines in the European Alps formed or influenced by these processes have been described by Goodsell et al. (2005a, b) and Robertson (2008).

Avalanche-type (AT) moraines consist of isolated concentrations of debris passing downglacier (fig. 9.25). Avalanche-type moraines originate from large, low-frequency rockfall events, and can become elongated in the downglacier direction during extensional ice flow. Clearly, a discrete medial moraine will not develop from rockfalls large enough to mask a considerable part of the glacier surface. Eyles and Rogerson (1978) described a multi-ridged example almost 1 km long from the Berendon Glacier.

The morphological expression and downglacier evolution of moraines is influenced by the amount of available debris (Eyles and Rogerson, 1978; Anderson, 2000). Debris thicker than a few centimetres reduces ice ablation rates, so medial moraines tend to form ice-cored ridges standing proud of the surrounding clean ice surfaces. Differential ablation rates cause ridge relief to increase downglacier, and debris rolls, slides and flows down the ridge flanks onto the adjacent clean ice, increasing the width of the moraine. If the ice underlying the ridge contains abundant debris, the surface debris cover can be replenished by melt-out, so that moraines fed by englacial debris septa tend to be long-lived, high-relief features. However, where debris does not extend through the full depth of the glacier, the supply to the moraine will eventually become exhausted. The ridge form will then degrade and the moraine will evolve into a low-relief veneer which is transported without further modification to the terminus.

Englacial thrusting

Basal debris can be elevated to englacial or supraglacial positions by compressive deformation near glacier margins. Earlier in this chapter, we discussed how basal ice layers can be thickened in areas of compressive flow, carrying debris to higher levels within the glacier. In this section we focus on the role of thrust faults in lifting debris from glacier beds.

High longitudinal compressive stresses can develop near glacier termini in a variety of situations, including:

(1) rapid glacier advances, especially the propagation of surge fronts;
(2) where wet-based ice decelerates against cold-based ice, either where a winter cold wave penetrates through thin ice to the bed of a temperate glacier, or year-round at the margins of some polythermal glaciers;
(3) where ice flows against a topographic barrier.

Folding and thrusting of ice has been observed in all these situations, and can elevate debris by two main mechanisms

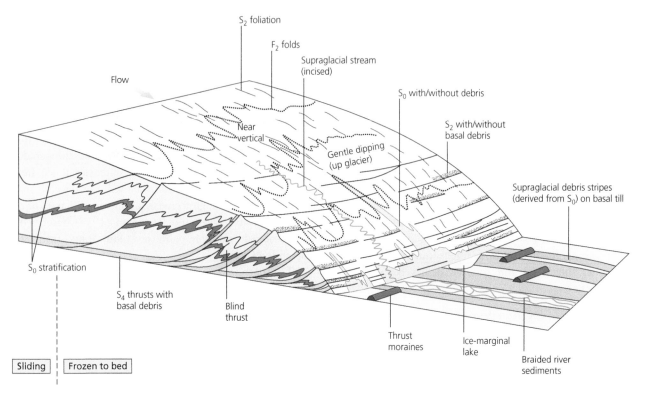

Figure 9.26 Elevation of basal debris by thrusting and folding (Hambrey et al., 1999).

(fig. 9.26). First, englacial debris can be carried within overthrust ice slabs, which climb up and over static or more slowly moving ice downglacier. Second, debris can be elevated along the thrust plane itself, by water flow, shear, till injection or some other process. Over the years, there has been a great deal of debate about whether debris can be elevated along thrust planes (see Glasser and Hambrey, 2003), but observations on several glaciers have left no doubt that it can be an effective process.

Thrusting is particularly important at advancing surge fronts, in both temperate and polythermal glaciers, due to intense longitudinal stress gradients and ice deformation (e.g. Sharp et al., 1988; Clarke and Blake, 1991; Hambrey and Lawson, 2000; Murray et al., 2000). Ice exposures in the margins of now quiescent surging glaciers provide abundant evidence of debris-filled structures extending upwards from the bed (fig. 9.27; Glasser et al., 1998). Although some examples have been interpreted as reoriented crevasse fills (e.g. Evans and Rea, 1999; Woodward et al., 2002; Glasser et al., 2003), the presence of drag folds and other structures shows that thrusting was clearly involved in the formation of many of these features. Debris-filled thrust faults are also widespread near the termini of Svalbard glaciers with no known history of surging. Hambrey et al. (1999, 2005) and Glasser and Hambrey (2003) have argued that englacial thrusting is a key mechanism of debris elevation in the marginal zones of non-surging polythermal glaciers, explaining the thick mantles of debris covering many

glacier termini in Svalbard. Although small Svalbard glaciers commonly do not display any of the 'classic' structural signatures of surges, such as looped moraines (Hambrey et al., 2005), the distinction between surging and non-surging glaciers may not be clear-cut (section 5.7.1) and it is possible that many have experienced surges or surge-like advances during the Little Ice Age (e.g. Hagen et al., 1993; Hansen 2003). Thrust faults and associated structures on at least some 'non-surging glaciers' may, therefore, be inherited from earlier dynamic behaviour. Debris-filled thrust faults have rarely been observed on temperate glaciers, although Glasser and Hambrey (2002) have shown that basal debris is elevated along thrusts near the margin of Soler Glacier, Patagonia, where it flows against a reverse slope. Echelmeyer and Zhongxiang (1987) and Tison et al. (1993) have identified debris within shear planes rising from frozen glacier beds in China and Antarctica.

Most thrust faults are not associated with large amounts of debris, but infills of diamicton several centimetres thick are not uncommon. Larger wedge-like infills can also occur, especially at the intersection of faults. According to Hambrey et al. (1999) and Glasser and Hambrey (2003), thrusts can incorporate metres-thick slices of the glacier bed, including basal till, glacifluvial and glacimarine sediment, while preserving original sedimentary structures (see section 12.3.3). More usually, thrust faults are marked by thin layers of debris-rich ice or films of fine debris.

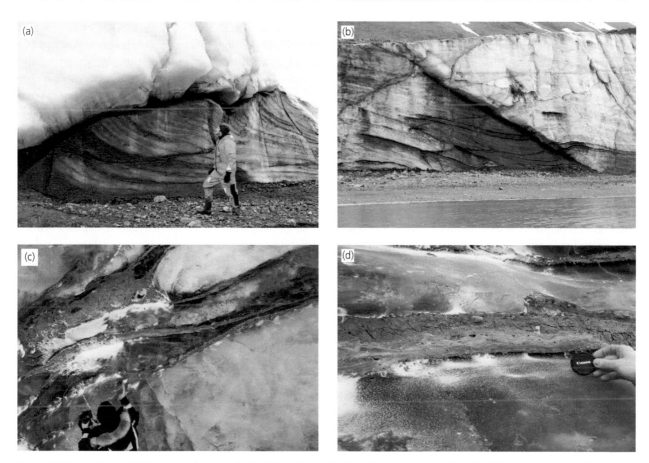

Figure 9.27 Elevated basal debris in surge-type glaciers, Svalbard. (a, b) Debris-filled thrust faults, Nansenbreen. Note drag folds on the hanging wall and foot wall. (c, d) Thrust faults infilled with diamicton and sorted sand, Tunabreen. Note faulting in (d) (Doug Benn).

Debris-mantled glacier surfaces

Where glacier ice contains debris, ice ablation results in the progressive accumulation of superficial debris. Debris-covered glaciers are common in high-relief mountain environments such as the Himalaya and the Southern Alps of New Zealand, where mass wasting processes deliver large volumes of debris to glacier surfaces, in both the accumulation and ablation zones (fig. 9.28). Continuous debris mantles can also result from the accumulation of debris that has been elevated from the bed by folding and thrusting. On most glaciers, melting is the dominant surface ablation process, but in cold, dry environments, ice can also be lost by sublimation (section 2.4.4). Debris accumulation by sublimation has been reported from Antarctica by Marchant et al. (2002) and Kowalewski et al. (2006), who described examples of diamictons that grade downwards into glacier ice displaying debris bands that contain the same material as the surface diamicton. Extremely slow freeze-drying can allow very delicate englacial structures to be inherited from the parent ice, including foliation, augen, laminae bent round clasts, and attenuated folds (Shaw, 1977; Lundqvist, 1989b). However, sediments produced by sublimation tend to be extremely loosely packed, and liable

Figure 9.28 The debris-mantled Rongbuk Glacier, on the north side of Mount Everest, June 2004. The peak of Pumori is enveloped in monsoon cloud (Doug Benn).

to collapse at the slightest touch, so *sublimation till* is very unlikely to survive complete deglaciation (Lundqvist, 1989a). In permafrost environments, however, debris-covered glacier ice can persist for thousands or even millions of years (e.g. Murton et al., 2005). The oldest known surviving glacier ice on Earth is in the Dry Valleys of Antarctica, where the unique climatic conditions are conducive to very gradual loss of ice by vapour transfer through the overlying debris (Sugden et al., 1995; Marchant et al., 2002; Kowalewski et al., 2006).

The presence of thick debris cover profoundly influences patterns of ablation (section 2.4.5), and uneven reworking and deposition of debris during glacier ablation is responsible for highly distinctive landform assemblages underlain by complex sediment successions. Debris variability on the glacier surface is responsible for large differences in the albedo and insulating properties of the glacier over short distances, resulting in *differential ablation.* Supraglacial debris mantles commonly exhibit great spatial variability in thickness, lithology, and grain size. The initial position of thicker debris accumulations reflects the distribution of englacial septa, which melt out of the ice, forming longitudinal or transverse concentrations of debris separated by areas of bare ice. As ablation proceeds, continuing melt-out and lateral spreading of debris gradually results in complete coverage of the glacier surface, although spatial differences in debris thickness and composition commonly persist, at least during the initial stages of ablation. For example, longitudinal bands of granite and schist debris, reflecting variations in catchment lithology, can be traced the full length of the debris-mantled tongue of Khumbu Glacier (fig. 9.29; Fushimi et al., 1980).

On debris-covered glaciers, spatial variations in ablation rates can lead to the development of highly irregular topography during periods of negative mass balance,

forming hills and hollows with a relief of up to 50 m (fig. 9.28). As slopes increase, debris slides, flows or falls into the surrounding low points, exposing ice in the topographic highs and increasing the thickness of insulating debris cover in the depressions. Rapid ablation of exposed ice and reduced ablation in hollows eventually results in *topographic inversion*, with former hollows evolving into new high points, and vice versa (Eyles, 1979; Kirkbride, 1993; Nakawo et al., 1999; Iwata et al., 2000). Topographic inversion may take place several times before the supraglacial debris is finally deposited, resulting in complex sediment assemblages (section 11.4.2). The development of very thick debris mantles reduces ablation rates over the entire glacier surface to such an extent that topographic inversion is inhibited, and a more subdued, low-relief topography develops. This tends to be the case only on longer glaciers with very high debris content, such as Khumbu Glacier (Iwata et al., 2000).

Where debris cover is thick, ablation tends to be focused where the debris cover is perforated by moulins, lake basins and other holes in the glacier surface. The slopes around such holes tend to be too steep to support debris, so that bare ice is exposed and ablation rates are high. Ablation proceeds by *backwasting*, or the parallel retreat of ice slopes around the hole or lake shore (fig. 9.30). Calving also leads to slope retreat, especially where ice cliffs are undercut by thermo-erosional notches (section 5.5.2). Expansion of lake basins can be the most important component of ablation on the lower part of large debris-mantled glaciers (Kirkbride, 1993; Sakai et al., 2000, 2002; Benn et al., 2001; Röhl, 2008).

Debris cover thickness tends to increase downglacier, which can lead to a reversal of the ablation gradient, with higher melt rates in the upper ablation zone than in the terminus region (section 2.4.5). During periods of negative mass balance, surface lowering in the upper ablation zone leads to a reduction in the overall gradient of the glacier, potentially setting in train a self-reinforcing cycle of downwasting and stagnation. Lower ice thicknesses

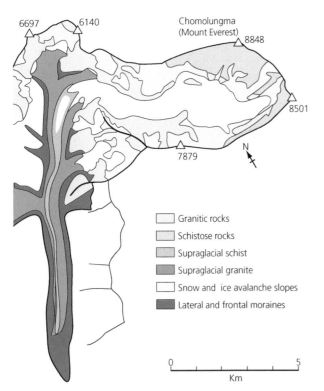

Figure 9.29 Distribution of supraglacial debris types and catchment lithology, Khumbu Glacier, Nepal (Fushimi et al., 1980).

Figure 9.30 Backwasting ice cliffs, Ngozumpa Glacier, Nepal (Doug Benn).

and gradients cause a reduction in driving stresses and ice velocities, and large areas of the glacier terminus can become inactive (Bolch et al., 2008b; Quincey et al., 2009). Reduced ice surface gradients also impede drainage of the glacier, encouraging the formation of supraglacial lakes. Rapid ablation of exposed ice around lake margins increases rates of surface downwasting, and eventually the debris-covered terminus of the glacier can be cut off from the active ice. Reduction of ice velocities greatly reduces the rate of supraglacial debris transport (Kellerer-Pirklbauer, 2008). Rather than being carried downglacier by ice flow, debris is more or less randomly reworked on the glacier surface, or scavenged by supraglacial and englacial streams and transported by the glacifluvial system (see section 9.4.3).

Downwasting of debris-covered glaciers produces a distinctive terrain known as *glacier karst*, in recognition of its close similarity to karst features on limestone terrain. Although glacier karst is commonly associated with stagnant glacier tongues, it can also develop on actively moving glacier tongues if ablation rates are sufficiently high (Kirkbride, 1995b). Near-surface englacial conduits form in very similar ways to limestone caves, when water exploits and enlarges permeable structures within the ice (Gulley and Benn, 2007). The collapse of conduit roofs creates surface hollows, similar to the creation of large sinkholes on limestone when cave chambers collapse. Collapsed conduits are identifiable on glacier surfaces as chains of crater-like depressions filled with water (fig. 9.31; Kirkbride, 1993). If the glacier drainage system has a high base level, due, for example, to the presence of a large moraine barrier, glacier downwasting can lead to the development of large, potentially unstable proglacial lakes (section 3.7.2; Bolch et al., 2008a; Quincey et al., 2009). Detailed descriptions of depositional and erosional processes, topographic development and sedimentology associated with glacier karst at the margins of Kötlujökull, a debris-mantled outlet of the ice cap Myrdalsjökull, Iceland, have been provided by Krüger (1994).

9.4.3 GLACIFLUVIAL TRANSPORT

Glacial drainage systems (sections 3.3 and 3.4) can transport very large volumes of sediment through glaciers, leading to high sediment yields from many glacierized catchments (Gurnell, 1987, 1995; Bogen, 1996; Hodson et al., 1997; Buoncristiani and Campy, 2001; Swift et al., 2002, 2005a). Indeed, sediment evacuation by subglacial meltwater dominates the sediment budget of most temperate glaciers and ice caps, transporting greater amounts of sediment than ice flow (Kirkbride, 1995a; Kirkbride and Spedding, 1996; Alley et al., 1997; Spedding, 2000). Glacifluvial and glacial transport systems are not distinct, separate entities, but form a closely coupled system, with many linkages and interactions (Swift et al., 2002). In this section, we focus on processes of glacifluvial sediment transport, and the linkages between glacifluvial and glacial transport. For more specialized coverage of fluvial processes and forms, the reader should consult texts by Allen (1982), Knighton (1998), Bridge (2003) and Bridge and Demicco (2008).

Glacial streams and rivers can transport sediment in the form of *suspended load* held within the flow and *bedload* carried along the channel floor, in addition to *dissolved ions* (section 3.9). Rates of sediment transport depend on both the availability of sediment and the characteristics of the flow. Flow *competence* is defined as the largest particle a flow can carry, and flow *capacity* as the total quantity of the available sediment that can be transported. Both competence and capacity increase with the boundary shear stress, which is correlated with flow velocity and discharge.

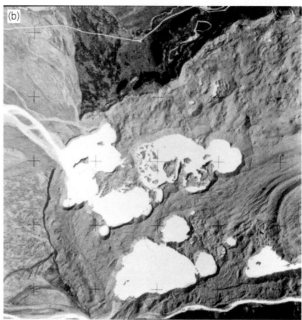

Figure 9.31 Vertical aerial photographs of the terminus of the Tasman Glacier, New Zealand, in (a) 1971 and (b) 1986. In (a), the chain of circular depressions was probably formed by the collapse of an englacial conduit. By 1986, the craters had become enlarged into a series of lakes, following backwasting of exposed ice cliffs (New Zealand Department of Scientific Research).

Suspended sediment

Suspended sediment refers to grains maintained in transport above the bed. In stagnant water, grains will settle under gravity at a rate proportional to their diameter, so for grains to remain in suspension in flowing water, the settling velocity must be balanced by a vertical component of water flow lifting the grains away from the bed. Suspended sediment is therefore a characteristic of turbulent flows in which some of the flow is directed upwards. The maximum size of particles that can remain in suspension will increase with the turbulence and velocity of the flow, but will generally be smaller than fine sand.

Suspended sediment concentrations in glacial streams are not simply a function of flow conditions, however, but also reflect the availability of sediment. For individual flood events, peaks in suspended sediment concentrations often occur before peak discharges, and concentrations are often lower during the falling stage than for similar discharges during the rising stage (fig. 9.32; Gurnell, 1982). This reflects the rapid flushing out of available fine-grained sediment from the glacier bed and channel floor, and the subsequent depletion of the sediment supply. Alternatively, melt streams may display brief pulses or 'slugs' of suspended sediment from storage due to the collapse of a channel wall or changes in channel patterns. The effects of flushing can also be recognized in seasonal variations in suspended sediment concentration (Swift et al., 2002). In winter, when meltwater discharges are low, suspended sediment concentrations are also low, typically only a few milligrams per litre. Sediment concentrations rise rapidly with increasing discharges in the spring, and may reach a few tens of grams per litre. Concentrations tend to fall off later in the ablation season, even though meltwater discharges can remain high, owing to the reduction in available fine-grained sediments.

Where meltwater discharges are high and loose sediment is readily available, suspended sediment concentrations can become very high, forming hyperconcentrated flows (Saunderson, 1977; section 10.4.10). The suspended sediment is usually of mixed grain size, up to and including gravel, and water content is about 20–60 per cent by weight (Pierson and Costa, 1987). Hyperconcentrated flows are therefore transitional between debris flows and normal stream flows. Particles are maintained in suspension by various mechanisms, including fluid turbulence, but where clay is present, turbulence is suppressed and cohesive strength may play an important role. Flows may be homogeneous or consist of two layers, with an upper, low-concentration component and a lower, coarse-grained, high-concentration 'carpet'. Deposition results from falling shear stresses and flow velocities associated with a reduction in stream gradient and/or flow depth, and occurs by partial or wholesale 'freezing' or locking of the suspension.

Bedload

Fluvial bedload refers to grains swept along close to the bed of a stream, in continuous or intermittent contact with the bed. Three basic modes of bedload transport are possible: sliding, rolling and saltation (Allen, 1982, 1985). Sliding particles retain continuous contact with the bed, but undergo negligible net rotation. Sliding is most important for disc-shaped gravel particles. Rolling particles are also in continuous contact with the bed, but move by rotating like a ball or a wheel. Saltation (from the Latin *saltare*, meaning 'to leap') involves particles taking relatively long jumps along the bed, and touching the bed only at the start and finish of each jump. The movement of grains in streams may involve all three of these processes occurring together, but in general a sequence of

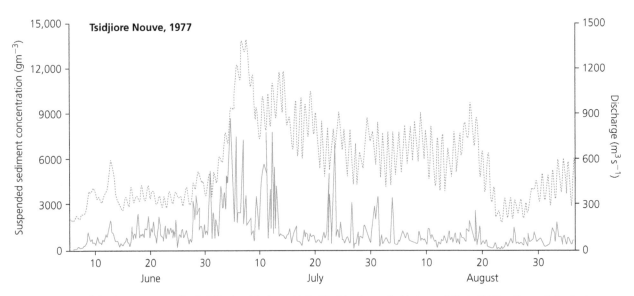

Figure 9.32 Suspended sediment concentration (solid line) and discharge (dashed line) for the proglacial stream of Glacier Tsidjiore Nouve, Switzerland (Modified from Gurnell, 1982).

dominant transport modes from sliding to rolling to salta-tion occurs with increasing driving force. When forces tending to lift the particle exceed the gravitational forces that cause settling, contact with the bed is lost and the particle enters suspension. The forces imposed on parti-cles by a flowing fluid are primarily a function of fluid velocity, although fluid viscosity and the presence or absence of turbulent flow are also important factors. A mathematical treatment of bedload transport was pre-sented by Allen (1982).

The grain size and quantity of sediment that can be maintained in motion as bedload increases with driving stress, and varies between small quantities of silt for low flows, through to heavy loads of all grain sizes, up to large boulders for extreme flood events. At Lyngsdalselva, Norway, Ashworth and Ferguson (1986) showed that bedload transport was an order of magnitude higher during flows of $10\,\mathrm{m^3\,s^{-1}}$ than for flows of $5\text{--}8\,\mathrm{m^3\,s^{-1}}$. The high discharges often associated with glacial melt-streams therefore lead to a huge capacity to transport sediment. For example, Østrem (1975) found that the meltstream of Nigardsbreen, Norway, transported around 400 tonnes of bedload during a 27-day period in the sum-mer of 1969. A more subjective impression of this capac-ity can be gained by standing beside a glacial stream in full flood and listening to the deep rumbling of boulders as they crash together and move past. High discharges,

however, do not necessarily mean high bedload, as sediment fluxes can be strongly supply-limited (Pearce et al., 2003).

Linkages between glacifluvial and glacial transport

Sediment can be entrained into glacial drainage systems at the bed, within the ice or at the glacier surface. On 'clean' glaciers, with small concentrations of supraglacial and englacial debris, the bed is by far the most important of these sediment sources. The configuration of the sub-glacial drainage system is a critical control on the effi-ciency of fluvial sediment transport (Swift et al., 2002; Swift, 2006). Distributed drainage systems can access large parts of a bed, but low flow velocities and the restricted size of flow paths limit sediment transport (Alley et al., 1997; section 3.4). Furthermore, the reten-tion of sediment below glaciers can be expected to sup-press erosion, by allowing thick till layers to build up between ice and bedrock. On the other hand, channelized drainage systems are characterized by much higher flow velocities, and can rapidly strip subglacial sediment from flow paths.

Swift et al. (2002, 2005a) found that the suspended sediment load of meltstreams emerging from Haut Glacier d'Arolla increased very steeply and non-linearly with discharge (fig. 9.33). For example, an increase in

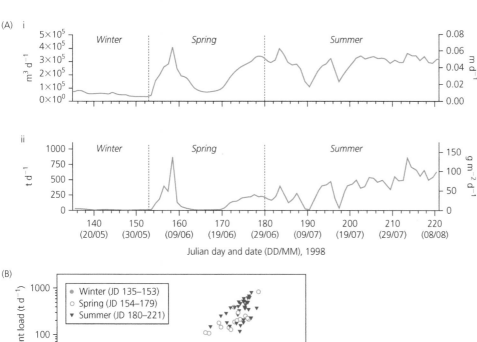

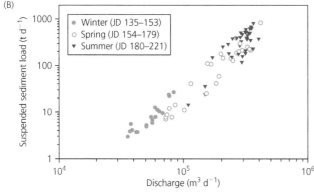

Figure 9.33 (A) Time series of discharge (top) and suspended sediment load (bottom), Haut Glacier d'Arolla, 1998. The right-hand scales show spatially averaged values. (B) Suspended sediment yield versus discharge (Swift et al., 2002).

discharge of 7.5 times led to an increase in suspended sediment load of 125 times – much greater than would be expected from theoretical discharge–capacity relationships (Alley et al., 1997). This pattern reflects the evolution of the glacial drainage system during the melt season, from an inefficient, low-capacity distributed system in the late winter and spring, to an efficient, high-capacity channelized system in summer. Efficient sediment evacuation is also encouraged by large diurnal discharge variations, which increase the exchange of water between channels and the surrounding bed (section 3.4.1).

These data indicate that sediment will tend to be retained below glaciers with predominantly distributed drainage systems, but will tend be flushed out where integrated, channelized systems develop during the melt season (Swift et al., 2002; Swift, 2006). Where rates of meltwater production are high, therefore, sediment can be readily evacuated from beneath glaciers and transported beyond the glacier margin in proglacial streams. In areas of high summer precipitation, such as the monsoon-dominated Himalaya, intense rainfall events can also lead to efficient flushing of sediment (Hasnain and Thayyen, 1999b; fig. 9.34).

Subglacial sediment yields are likely to be very small below debris-covered glaciers where surface melt rates are small. Ice-cored moraines, however, represent large sediment stores that can be progressively entrained by meltstreams over long periods of time (Etzelmüller et al., 2000; Irvine-Fynn et al., 2005; Lukas et al., 2005; Moorman, 2005). Low ice velocities on stagnating, debris-covered glaciers further increase the importance of the glacifluvial system in the total sediment budget.

9.5 EFFECTS OF TRANSPORT ON DEBRIS

Different transport processes can leave contrasting signatures on debris, allowing aspects of transport history to be reconstructed from the characteristics of glacial deposits. Boulton (1978) drew a fundamental distinction between *active transport* in the basal tractive zone and *passive transport* at higher levels within or on the glacier. During active transport, abrasion and crushing processes create distinctive grain-size distributions and progressively modify the *form* or *morphology* of particles. In contrast, during passive transport in supraglacial and englacial positions, inter-particle contact forces are generally small and debris passes through the glacier system with little or no modification (fig. 9.35).

Recent work has shown that this simple picture needs to be modified somewhat, for a number of reasons. First, supraglacial debris transport is not always 'passive'. Significant particle size reduction can occur in response to periglacial and chemical weathering processes, and where supraglacial debris contains abundant large boulders, these can grind together as the underlying ice melts, causing fracture and edge-rounding (Owen et al., 2002a; fig. 9.35). Second, *glacifluvial processes* can transport significant volumes of sediment in supraglacial, englacial and subglacial environments (section 9.4.3), increasing the roundness and sorting of sediments. In this section, we consider the impact of active, passive and glacifluvial transport on particle size distributions and particle morphology.

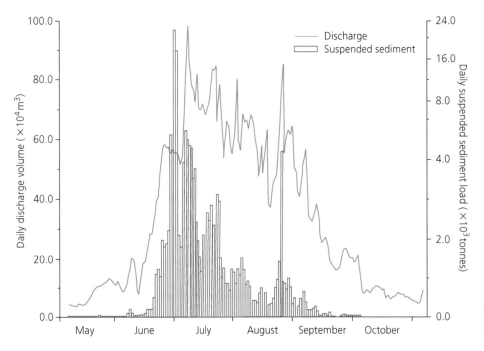

Figure 9.34 Daily discharge and suspended sediment load of the proglacial river, Dokriani Glacier, India (After Hasnain and Thayyen, 1999b).

9.5.1 GRANULOMETRY

In general, supraglacial and englacial debris retains the characteristics of the parent material. If the parent debris is periglacially weathered material that has fallen onto the glacier surface, it tends to be predominantly coarse-grained, with a deficiency in the finer grades (fine sand and smaller; fig. 9.36). However, debris in high-level transport can also have originated at the glacier bed or pre-existing sediments, in which case, finer-size fractions may be present.

Abrasion and breakage of particles during active transport produces grain size distributions that are very different from those resulting from periglacial mechanical weathering or transport by flowing water (fig. 9.36). Distributions commonly exhibit a very broad range of particle sizes, from clay or silt up to pebbles or larger, and are typically *bimodal* or *polymodal* (i.e. with two or more peaks in the grain size distribution). Such distributions reflect progressive particle size reduction or comminution,

as particles are fractured under applied stresses during shear. Boulton (1978) and Haldorsen (1981) attributed bimodal and polymodal distributions to the effects of two processes: *crushing*, or the breakage of two interlocking grains; and *abrasion*, or the production of fine-grained fragments as two grains slide past each other. The products of crushing and abrasion were studied by Haldorsen (1981), who simulated subglacial particle wear using a ball mill, in which rock fragments and steel balls are tumbled in a drum. She found that crushing can produce a range of particle sizes, from 0.016 to 2 mm, whereas experimental abrasion produced fragments in the range 0.002–0.063 mm, both distributions closely resembling modes found in natural tills (fig. 9.37). Haldorsen interpreted coarser clast modes in the tills as residual rock fragments which could have been reduced further by subglacial crushing and abrasion, given sufficiently long transport distances. The identification of separate components in actively transported debris is facilitated by

Figure 9.35 Debris in supraglacial transport. (a) Coarse angular debris on the surface of Batal Glacier, India. (b) Crushing and edge-rounding at the contact between two large boulders on the surface of Miage Glacier, Italy (Doug Benn).

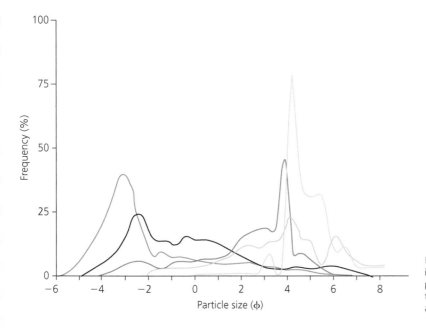

Figure 9.36 Grain size distributions from debris-rich ice, Matanuska Glacier, Alaska. The curves with prominent peaks in the fine grain sizes (2–6 ϕ) are typical of actively transported debris (Drewry, 1986, after Lawson, 1979a).

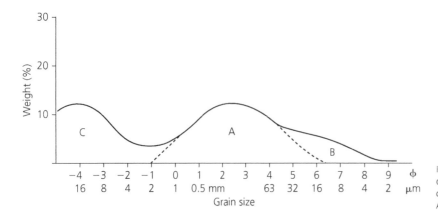

Figure 9.37 Interpretation of the mean grain size distribution of 150 till samples. Area A: resistant crushed fraction. Area B: component from abrasion. Area C: residual clast mode (Haldorsen, 1981).

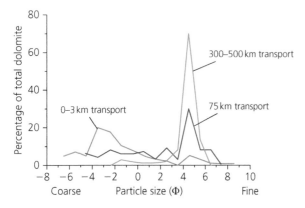

Figure 9.38 Frequency distribution of dolomite in till samples from the Hamilton-Niagara region, Canada, showing the effects of progressive comminution during transport (Redrawn from Dreimanis and Vagners, 1971).

Gaussian component analysis, which statistically resolves polymodal distributions into a series of normal (bell-shaped or Gaussian) curves (Sheridan et al., 1987). This technique was applied by Sharp et al. (1994) to debris from basal ice formed during the quiescent and surging phases of Variegated Glacier (section 5.7.3), allowing the effects of different processes of wear and selective entrainment and deposition to be identified.

Dreimanis and Vagners (1971) conducted seminal research on changes in the granulometry of subglacial debris with distance from known bedrock sources. Coarse modes were well developed close to the debris source, and fine modes increase in importance with increasing transport distance. However, there appeared to be a lower size limit beyond which no further comminution occurred, regardless of transport distance (fig. 9.38). This limit was termed the *terminal grade*, and was interpreted as the particle size at which the available energy during transport is insufficient to cause further fracture. The existence of terminal grades can be partially explained by Griffith crack theory (section 8.2.1), which shows that the energy required for crack growth and rock fracture is inversely proportional to crack length. The fracture of small particles can occur only by the growth of short cracks, and therefore requires greater energy input than the fracture of large

particles by the growth of relatively long cracks (Drewry, 1986). Terminal grades are equivalent to the *limit of grindability* recognized by engineers, and appear to be related to the maximum stress concentrations that occur during shear. As would be expected, softer materials, such as mica or feldspar, have smaller terminal grades than harder, resistant minerals such as quartz. Haldorsen (1983) found that the finer grades in experimentally crushed debris and basal tills are enriched in feldspar and sheet silicate minerals, compared with coarser grades in which quartz is more abundant. However, terminal grades should not be defined on the basis of sediment granulometry alone, as grain size modes may represent transient stages of comminution, which could have been further reduced given more time (Haldorsen, 1981; Drewry, 1986).

Another approach to interpreting particle size distributions has been taken by Hooke and Iverson (1995) and Iverson et al. (1996), using techniques developed by Sammis et al. (1987) for the study of tectonically crushed rocks. This method considers the number of grains per size fraction as shown on double logarithmic plots. Hooke and Iverson noted that, for subglacially sheared tills, the plotted lines tend to be straight, suggesting that the distributions are self-similar at all scales (i.e. they have *fractal* characteristics; fig. 9.39). Sammis et al. (1987) argued that the gradients (*fractal dimensions*) of distributions produced entirely by grain fracture should be around 2.58, reflecting grain fracture near the contact between two similarly sized particles, where stress concentrations are highest. Progressive fracturing in these locations during shear will result in maximum cushioning for each particle and the minimization of stress concentrations. Calculated fractal dimensions for the subglacial samples were close to 3, leading Hooke and Iverson (1995) to propose a modified form of this mechanism for deforming subglacial tills, in which slippage and abrasion between grains occurs in addition to shear by intergranular fracture, producing an excess of fines. However, Benn and Gemmell (2002) showed that apparently 'fractal' grain-size distributions can be produced by mixing of distinct populations of sorted sediments (figs 9.40 and 9.41).

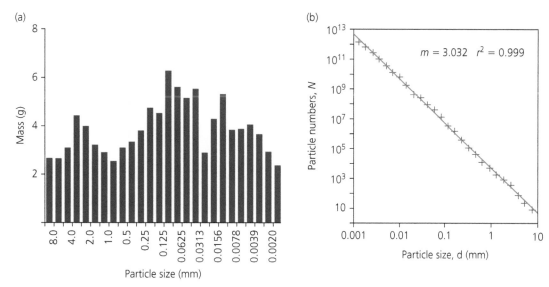

Figure 9.39 Till grain size distribution on (a) a conventional mass vs particle size plot; and (b) a double logarithmic particle numbers vs size plot, showing the calculated 'fractal dimension', m (After Benn and Gemmell, 2002).

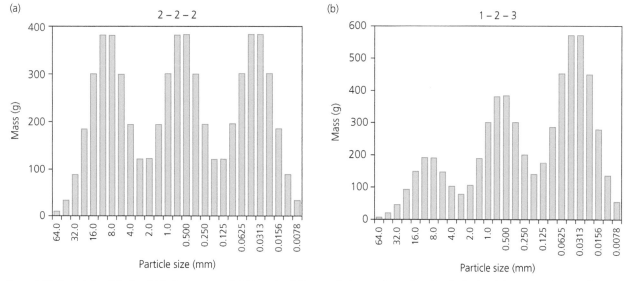

Figure 9.40 Simulated particle size distributions composed of (a) equal and (b) unequal mixtures of three Gaussian distributions (After Benn and Gemmell, 2002).

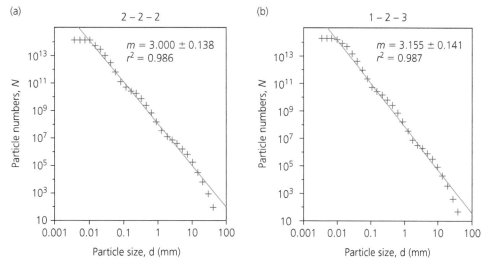

Figure 9.41 Apparently 'fractal' particle size distributions of the simulated sediment mixtures shown in figure 9.40 (After Benn and Gemmell, 2002).

Such mixtures appear to be fractal because the inverse-logarithmic relationship between grain size and particle numbers per unit mass dominates the distribution, so that individual modes introduce only insignificant deviations from a straight line. The relationship between grain size and particle numbers is cubic, hence 'fractal dimensions' will always be close to 3; lower values simply reflect relatively larger amounts of fine-grained sediment in the distribution. Benn and Gemmell concluded that this method cannot be used to infer particular genetic processes. Indeed, field studies have found no significant difference in the 'fractal dimensions' of subglacial and supraglacial sediment matrix (Khatwa et al., 1999).

Some subglacial sediments are produced by the cannibalization of pre-existing sediments, in which case grain size distributions may be inherited from other glacial or non-glacial processes. Subsequent shear of such sediment during active transport can modify the original distribution by particle communion or mixing of the sediment with other debris.

In glacifluvially transported sediments, hydrodynamic processes result in selective entrainment, transport and deposition (Powell, 1998). During rising stages, sorting can arise from the preferential entrainment of the finer fractions from heterogeneous source sediments. During transport, grain size distributions of bedload and suspended load are modified by varying flow conditions, and during falling stages progressively finer sediments settle out of the flow. Consequently, glacifluvial sediments are commonly well sorted, with individual beds exhibiting a single major mode spanning a relatively small range of grain sizes. Grain size distributions are approximately Gaussian when plotted on a logarithmic scale (Hoey, 2004). This simple picture can be complicated by a number of factors. For example, bimodal grain-size distributions can result from mixing of suspended load and bedload fractions, or if finer sediments are trapped in voids between larger grains. The reader is referred to Bridge (2003) for a detailed discussion of sorting processes in fluvial systems.

9.5.2 CLAST MORPHOLOGY

Glacial and glacifluvial transport imprint very distinctive signatures on the morphology of sedimentary particles (e.g. Boulton, 1978; Dowdeswell et al., 1985; Benn and Ballantyne, 1994; Spedding and Evans, 2002; Benn, 2004a; Hambrey and Ehrmann, 2004). Clast morphology, or form, can be regarded as the sum of three characteristics at different scales:

(1) *shape*, or the relative dimensions of the long, intermediate and short axes (a, b and c axes respectively) (fig. 9.42);
(2) *roundness*, or the degree of curvature of clast edges;
(3) *texture*, or the character of the clast surface (Barrett, 1980).

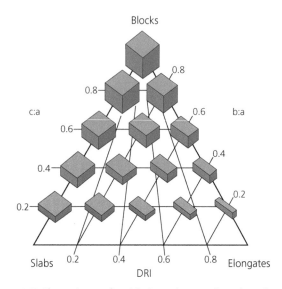

Figure 9.42 The continuum of particle shapes, between the end members of blocks, elongates and slabs. The diagram is scaled using the ratios between the (a) long, (b) intermediate and (c) short axes. DRI is the 'disk-rod index', equal to (a–b)/(a–c) (Benn and Ballantyne, 1993).

In addition, we can assess the *wear patterns*, or the distribution of different types of wear over the clast surface (Benn, 1994a, 1995).

Clast shape is dependent on both process and lithology. For example, periglacially weathered clasts of massive, coarse-grained rock such as granite tend to have elongate and slabby shapes (Ballantyne, 1982; Benn and Ballantyne, 1994; fig. 9.43). In contrast, actively transported clasts of the same lithology have more compact, blocky shapes, which may reflect both the initial shapes of subglacially plucked fragments and preferential breakage of clasts across their long axes during transport at the ice–bed interface (Benn, 1992). For fissile rocks such as shale and some limestones, actively transported clasts generally have more elongate shapes, indicating preferential fracture along bedding planes.

Passively, actively and glacifluvially transported sediments also have contrasting roundness characteristics. Frost-weathered clasts are commonly angular or very angular, and debris transported by snow avalanches tends to have particularly sharp, fragile edges. During active transport, fracture and abrasion processes have opposite effects on clast roundness. Abrasion increases edge-rounding and creates polished facets, whereas fracture creates new, sharp edges and fresh faces. The balance between these two processes means that both angular and well-rounded forms tend to be rare in actively transported debris, and most clasts have intermediate roundness characteristics (i.e. sub-angular and sub-rounded; fig. 9.44). Studies of clast-roundness evolution with distance from a known source have demonstrated that only short distances of active transport are necessary for significant edge-rounding to occur (Humlum, 1985a).

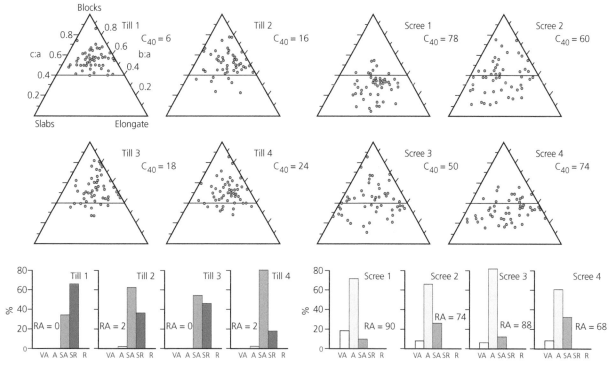

Figure 9.43 Clast shape and roundness characteristics of actively transported debris (till) and scree (representative of passively transported rockfall material) near the margins of Storbreen, Norway. The roundness categories are: VA: very angular; A: angular; SA: sub-angular; SR: sub-rounded; R: rounded. C_{40} is the percentage of clasts with c:a ratios $\leqslant$ 0.4, and RA is the percentage of clasts in the VA and A classes. Note the low C_{40} and RA values for the till samples compared with the scree (Benn and Ballantyne, 1994).

Figure 9.44 Typical wear patterns on actively transported clasts, from tills deposited by an outlet of the LGM Patagonian Ice Field (Doug Benn).

Glacifluvially transported clasts become increasingly well rounded during transport, and on most lithologies fracturing events are rare.

The surface texture of actively transported glacial debris is also very distinctive, especially when compared to clasts transported in fluvial or coastal environments. Interactions between particles, and between particles and the bed, result in polished and striated faces, particularly on fine-grained lithologies such as limestones, slate and basalt (figs 9.44 and 9.45). Coarse-grained rocks such as granite tend not to bear clear striae, because the different mineral constituents offer varying resistance to erosion and strongly influence patterns of surface wear. Striae on clasts may be straight and parallel, indicating stability of the clast relative to the striating medium, or curved and exhibiting a wide variety of orientations, reflecting clast rotation and realignment during wear (Hicock, 1991; Benn, 1995). Straight striations aligned parallel to clast long axes are typical of the upper surfaces of clasts lodged beneath sliding ice (Benn, 1994a; Krüger, 1994). Cross-cutting striae on individual clasts attest to clast rotation or multiple episodes of transport.

Shape and roundness co-vary in systematic ways, allowing different clast populations to be distinguished. In figure 9.46, actively and passively transported clasts form two distinct populations, the former dominantly blocky and edge-rounded, the latter more slabby, elongate and angular. Freshly plucked rock fragments form a third population, with blocky and angular characteristics. The transport history of samples from end moraines in the same catchments can then be interpreted with confidence, using these end members as control samples (Benn, 1992, 2004a). The shape and roundness indices used in these diagrams were chosen to maximize the distinction between actively and passively transported clasts (Ballantyne, 1982; Benn and Ballantyne, 1994). They are much less sensitive, however, to differences between actively and fluvially transported clasts, and different indices should be used to determine their relative proportions (Benn, 2004a).

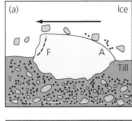
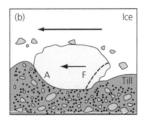
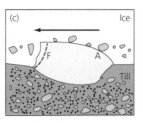
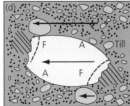
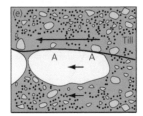
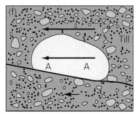

Figure 9.45 Development of asymmetric clast wear patterns beneath sliding ice and within deforming till, showing principal locations of abrasion (A) and fracture (F). Arrow lengths show relative velocities. (a) Lodged clast with stoss-lee form owing to stoss-side abrasion and lee-side fracture below sliding ice. (b, c) Double stoss-lee morphology resulting from a two-stage process of ploughing and lodgement. (d) Double stoss-lee clast resulting from a single-stage process within a deforming layer. Low pressure zones are shaded. (e, f) Flat, polished facets eroded on the upper and lower surfaces of clasts, where there is significant slip between the clast and the adjacent shear plane (Modified from Krüger, 1984, and Benn and Evans, 1996).

An important aspect of clast morphology is the asymmetry of wear patterns. Clasts from fluvial or coastal environments are generally evenly worn on all parts of their surfaces, but this is often not the case for subglacially transported or deposited clasts, owing to the asymmetric distribution of stresses in the subglacial environment (section 10.3.4.2).

9.5.3 PARTICLE MICROMORPHOLOGY

Matrix particles exhibit a range of shapes and surface textures, and several studies have sought diagnostic criteria to identify grains that have undergone active subglacial transport (see Carr, 2004 for a review of techniques and examples). Particular attention has focused on the surface texture of quartz grains, as revealed through the scanning electron microscope or SEM (Whalley, 1996), because quartz is a durable, common mineral which exhibits fracture patterns that were thought to be relatively independent of mineral structure. However, results have been ambiguous, and the identification of distinct 'glacial' quartz textures remains elusive.

Krinsley and Doornkamp (1973) proposed that 'glacial' quartz grains can be recognized on the basis of their sphericity (high c:a ratios), angularity and textures indicative of mechanical fracture – characteristics that were subsequently used to infer the presence of actively transported debris in the geological record (e.g. Drewry, 1975). Features such as conchoidal fractures, sharp angular edges and the presence of 'pre-weathered' and high-relief particle morphology were regarded by Mahaney and Kalm (2000) and Mahaney et al. (2001) as particularly

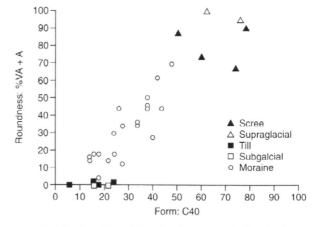

Figure 9.46 Co-plot of C_{40} and RA indices for clast samples from Storbreen, Norway in which moraine samples reflect mixing of control populations.

diagnostic of subglacial environments (cf. Hart, 2006). Similarly, Eyles (1978) proposed that a *microblock texture*, consisting of a series of steps and angular corners, was a reliable diagnostic criteria for subglacially transported quartz grains. However, grains with similar features occur in non-glacial environments characterized by high energy transport or mechanical weathering, and even microblock textures could be the product of the history of quartz grain crystallization, deformation and preglacial fracture rather than its glacial transport history (Gomez et al., 1988). With our present state of knowledge, the only definitive statement that can be made about quartz microtextures is that they are more ambiguous as indicators of active subglacial transport than the morphology of clast-sized particles.

CHAPTER 10
GLACIGENIC SEDIMENTS AND DEPOSITIONAL PROCESSES

Twin Otter landing at Dobbin Bay, Ellesmere Island, Arctic Canada (D.J.A. Evans)

10.1 INTRODUCTION

Sediments can be classified either using lithologic or genetic approaches. A genetic approach is taken here for two main reasons. First, it emphasizes the links between facies characteristics and glaciological processes and depositional environments, providing continuity with part I of this book. Second, genetic facies may have a range of lithofacies characteristics, depending on variations in the source material and the relative importance of various depositional processes. A purely descriptive approach to discussing glacigenic and related sediments would therefore be over-complex and repetitive. For each genetic facies covered in this chapter, we provide a definition, a brief account of the processes of deposition, and typical distinguishing characteristics, such as grain size distribution, particle morphology, fabric, internal structures and

facies dimensions and geometry. Before we deal with sediments, we first need to review the structures imparted on the whole range of glacigenic facies by various styles of tectonics and deformation.

10.2 SEDIMENT DESCRIPTION AND CLASSIFICATION

10.2.1 SEDIMENT DESCRIPTION

A comprehensive description of a sediment should include details of:

(1) its *properties*, allowing aspects of its erosional, transport and depositional history to be reconstructed;

(2) its *geometry*;

(3) its *position* with respect to the adjacent sediments and the land surface at a range of scales.

Sediment properties provide the basis for defining and describing lithofacies. The choice of properties to be described must be guided by the object of the investigation, because different properties convey different types of information (Evans and Benn, 2004b). Properties commonly described in glacial process studies are: grain size distributions, structures, fabric, particle form, matrix properties and density. Some of the techniques for measuring lithofacies properties are described briefly in this book; additional information can be found in texts by Hubbard and Glasser (2005) and Evans and Benn (2004a). A particularly important characteristic is *sediment texture*, defined in terms of the dominant grain size and the grain size distribution. Poorly sorted sediments are termed *diamictons* and their lithified equivalents are termed *diamictites* (Flint et al., 1960). Some researchers use the term *diamict* to cover both diamictons and diamictites (e.g. Eyles et al., 1983).

Sedimentologists have devised various shorthands or *lithofacies codes* for recording lithofacies properties in the field. These codes consist of letters denoting sediment lithology, plus additional letters for internal structures or inclusions. A widely used scheme for the deposits of braided rivers was introduced by Miall (1977, 1978) and later extended by Eyles et al. (1983) to include codes for diamictons. Further refinements have been made by Benn and Evans (1998) to account for laminated sediments and granule size materials in glacial outwash. A more comprehensive set of symbols to depict texture, bedding and structure in stratigraphic diagrams has been compiled by Evans and Benn (2004b); these are shown together with facies codes in figure 10.1. The diamicton code is less satisfactory than that for fluvial sediments because it omits several important diagnostic properties, such as density, particle morphology, fabric and the nature of inter-beds, intra-beds and intra-clasts. Reliance on lithofacies coding alone yields insufficient information for

correct interpretation, so the codes should only be used as a supplement to other observations (e.g. Kemmis and Hallberg, 1984; Hambrey, 1994). Given the complexity of glacigenic sediment sequences, it is wise to include as much detailed observation and data as possible, and detailed coding schemes have been devised for specific purposes. For example, Ghibaudo (1992) proposed a useful scheme for describing subaqueous mass flow deposits, incorporating code letters for textural variation and a range of internal structures. An excellent example of a more intensive documentation of glacigenic sediment sequences is that of Krüger and Kjær (1999), in which seven levels of information are presented in vertical profile logs, including genetic interpretations that identify landform–sediment associations (fig. 10.2).

Diamictons, which are very common in glacially influenced depositional environments, can also be described and interpreted through micromorphology. The wide range of microfabrics and microstructures visible on thin sections of diamictons (fig. 10.3) provides extra criteria for genetic interpretations (see Carr, 2004b for a practical guide). Although many features are common to diamictons deposited by both mass flowage and subglacial traction, micromorphology has nonetheless demonstrated that deformation and pore-water movement is widespread in the production of tills (van der Meer, 1993; Menzies, 2000; Hiemstra and Rijsdijk, 2003; van der Meer et al., 2003; Menzies et al., 2006), and, in combination with other macroscopic descriptive criteria, provides a powerful multifaceted approach to sediment assessment (e.g. Carr, 2001; Lachniet et al., 2001; Phillips et al., 2002; Menzies and Zaniewski, 2003; Hiemstra et al., 2004; Evans and Hiemstra, 2005).

The *geometry* and *position* of lithofacies are usually portrayed graphically in vertical profiles, two-dimensional logs, maps, or some combination of the three. *Vertical profiles* are measured vertical sections on which bed

Primary glacigenic deposits (tills)	Glacifluvial deposits	Gravitational mass movement	Deposits from suspension settling and iceberg activity
Glacitectonite	Plane bed deposits	Debris fall deposits	Cyclopels
Subglacial traction till	Cross-bedded facies	Gelifluction deposits	Cyclopsams
Subglacial melt-out till	(dunes and antidunes)	Slide and slump deposits	Varves
	Gravel sheets	Debris flow deposits	Dropstone mud
	Ripple cross-laminated facies	Scree	Dropstone diamicton
	Hyperconcentrated flows	Turbidites	Silt and mud drapes
			Undermelt diamicton
			Iceberg contact deposits
			Ice-keel turbate

Table 10.1 Classification of glacial deposits and associated facies used in this book

thickness and lithology are shown, alongside additional information such as structural data and the nature of contacts between lithofacies (e.g. gradational, erosional, deformed). Modal grain size of lithofacies is shown on vertical profiles using the width of the bed, providing a rapid impression of the overall character of a succession (fig. 10.4). Vertical profiles are ideal for portraying the lithology of cores, and are also often used for recording outcrop data. However, vertical profiles give very limited information on the geometry of facies and facies associations, and the relative importance and extent of bounding surfaces, so where outcrop data are available, two-dimensional logs are preferable. *Two-dimensional logs* vary from sketches to very detailed measured drawings, and are a good method for showing lithofacies geometry and the larger-scale geometry (*architecture*) of

(a)

Code	Description	Code	Description
Diamictons	*Very poorly sorted admixture with wide range of grain sizes*	*Sands*	*Particles of 0.063–2 mm*
Dmm	Matrix-supported, massive	St	Medium to very coarse and trough cross-bedded
Dcm	Clast-supported, massive	Sp	Medium to very coarse and planar cross-bedded
Dcs	Clast-supported, stratified	Sr (A)	Ripple cross-laminated (type A)
Dms	Matrix-supported, stratified	Sr (B)	Ripple cross-laminated (type B)
Dml	Matrix-supported, laminated	Sr (S)	Rippled cross-laminated (type S)
- - - (c)	Evidence of current reworking	Scr	Climbing ripples
- - - (r)	Evidence of re-sedimentation	Ssr	Starved ripples
- - - (s)	Sheared	Sh	Very fine to very coarse and horizontally/ plane-bedded or low angle cross-lamination
- - - (p)	Includes clast pavement(s)		
		Sl	Horizontal and draped lamination
Boulders	*Particles >256 mm (b-axis)*	Sfo	Deltaic foresets
Bms	Matrix-supported, massive	Sfl	Flasar bedded
Bmg	Matrix-supported, graded	Se	Erosional scours with intraclasts and crudely cross-bedded
Bcm	Clast-supported, massive		
Bcg	Clast-supported, graded	Su	Fine to coarse with broad shallow scours and cross-stratification
Bfo	Deltaic foresets		
BL	Boulder lag or pavement	Sm	Massive
		Sc	Steeply dipping planar cross-bedding (non-deltaic foresets)
Gravels	*Particles of 8–256 mm*		
Gms	Matrix-supported, massive	Sd	Deformed bedding
Gm	Clast-supported, massive	Suc	Upward coarsening
Gsi	Matrix-supported, imbricated	Suf	Upward fining
Gmi	Clast-supported, massive (imbricated)	Srg	Graded cross-laminations
Gfo	Deltaic foresets	SB	Bouma sequence
Gh	Horizontally bedded	Scps	Cyclopsams
Gt	Trough cross-bedded	- - - (d)	With dropstones
Gp	Planar cross-bedded	- - - (w)	With dewatering structures
Gfu	Upward-fining (normal grading)		
Gcu	Upward-coarsening (inverse grading)	*Silts and clays*	*Particles of <0.063 mm*
Go	Openwork gravels	Fl	Fine lamination, often with minor fine sand and very small ripples
Gd	Deformed bedding		
Glg	Palimpsest (marine) or bedload lag	Flv	Fine lamination with rhythmites or varves
		Fm	Massive
Granules	*Particles of 2–8 mm*	Frg	Graded and climbing ripple cross-laminations
GRcl	Massive with clay laminae		
GRch	Massive and infilling channels	Fcpl	Cyclopels
GRh	Horizontally bedded	Fp	Intraclast or lens
GRm	Massive and homogeneous	- - - (d)	With dropstones
GRmb	Massive and psuedo-bedded	- - - (w)	With dewatering structures
GRmc	Massive with isolated outsize clasts		
GRmi	Massive with isolated, imbricated clasts		
GRmp	Massive with pebble stringers		
GRo	Open-work structure		
GRruc	Repeating upward-coarsening cycles		
GRruf	Repeating upward-fining cycles		
GRt	Trough cross-bedded		
GRcu	Upward coarsening		
GRfu	Upward fining		
GRp	Cross-bedded		
GRfo	Deltaic foresets		

Figure 10.1 Lithofacies codes (a) and conventional symbols (b) employed in the recording of lithological information and sedimentary structures and bedding contacts in stratigraphic cross sections and vertical profile logs (Evans and Benn, 2004b).

(b)

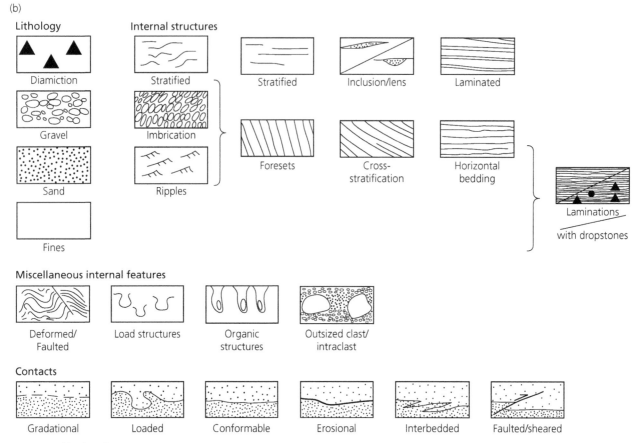

Figure 10.1 (Continued)

associations. Lithological data can also be shown. Facies geometry and architecture can be logged in three dimensions and shown on *fence diagrams*, but data of this quality are very rarely available. *Landform maps* have been important tools for glacial geomorphologists for many years, and when combined with two-dimensional logs can give a very good summary of the three-dimensional form of sediment assemblages. Further details on the various methods of sedimentary data collection and presentation have been provided by Evans and Benn (2004a).

In the following sections, we describe the principal genetic facies commonly found in glacial, proglacial, glacilacustrine and glacimarine environments, using the classification shown in table 10.1. Shoreface and tidal sediments are not described, and interested readers are referred to Allen (1982), Reading and Collinson (1996) and Bluck (1999) for good reviews.

10.2.2 DEFORMATION STRUCTURES

The process–form relationships of glacitectonic disturbance, gravity-induced deformation and subglacial shearing are assessed throughout this and following chapters, but we need to review the range of structures commonly observed in glacigenic facies at this stage in order to appreciate their use in sediment description and classification. A significant contribution in this realm is the overview of McCarroll and Rijsdijk (2003), whose classification scheme for all the potential deformation structures in glacial environments is reproduced in figure 10.5. This scheme identifies the types of structures produced by glaciers in Quaternary sediments according to the four principal styles of deformation, pure shear, simple shear, compressional and vertical deformation (see section 4.2), and provides comparative criteria for undeformed materials. The glacigenic depositional settings in which each style of deformation can occur are summarized in table 10.2.

The various fold and fault types common to glacial sediments have been reviewed by Evans and Benn (2004b), and figure 10.5a shows typical examples of process–form relationships relative to deformation style. For example, pure shear (type P) can be recognized in matrix realignment and brittle fracture where materials are loaded by glacier ice. Simple shear (type S) is common in subglacially deformed materials and debris flow deposits (van der Wateren, 1995). Compressional structures (type C) are common in proglacial situations where there is the accommodation space for the vertical

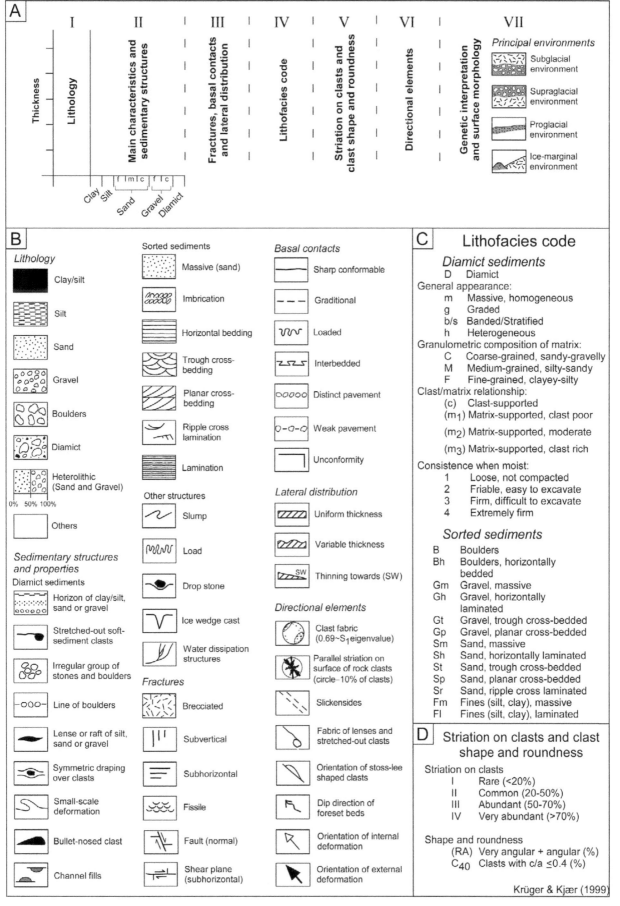

Figure 10.2 Template data chart for describing tills and associated sediments based on the ice-cored morainic landscape of Kötlujökull, Iceland (Krüger and Kjær, 1999).

Microfabrics and Microstructures within the Plasma and S-Matrix of Glacial Sediments

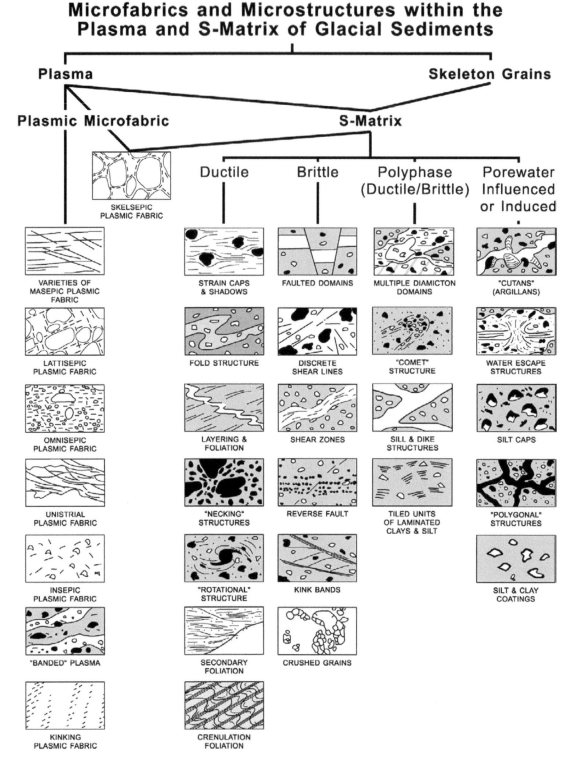

Figure 10.3 Typical deformation structures in thin sections (Carr, 2004, after Menzies, 2000 and van der Meer, 1993)

displacement of folded and thrust strata, and the atmospheric pressures to which pressurized groundwater will migrate and often produce hydrofractures. Vertical deformation structures (type V), such as sags, diapirs, isolated pods, open folds and normal faults are a response to gravity-driven processes, which are very common in proglacial settings where ice is melting out or sediments have high water contents. Undeformed sediments (type U) are identified by the descriptions of their bedding and sedimentary structures and are most common in more distal glacilacustrine and glacimarine environments. However, localized deformation by iceberg dumping and

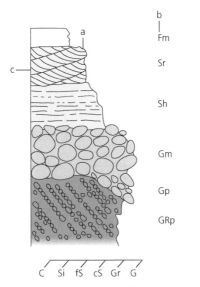

Figure 10.4 Conventions used in vertical profile logs. (a) Log width indicates modal particle size, as depicted on the lower scale, where C = clay, Si = silt, fS = fine sand, cS = coarse sand, Gr = granules and G = gravel. Glacial researchers usually add an extra gradation to the right for diamictons. (b) Facies codes. (c) Graphical symbols indicate sediment structures (e.g. ripples, planar bedding).

keel ploughing (Thomas and Connell, 1985; section 10.6.6) is common in these environments. Additionally, in situations where glacier ice deforms sediments, it is usual to find a range of structures associated with water movement and water-saturated materials, such as hydrofractures, liquefaction and homogenization. Phillips et al. (2007) have demonstrated that such structures are ubiquitous, at least at the micro-scale, in both subglacial and proglacial environments.

Glacial sediments, particularly subglacial deposits, commonly exhibit structures associated with shear zones. Subglacial shearing can produce structures closely similar to *fault gouge* found in geological shear zones, including fault breccias and fluted surfaces (Eyles and Boyce, 1998). The shear zone beneath glaciers is characterized by an upwards increase in deformation intensity from undeformed in situ *footwall* materials to strongly sheared materials at the ice–substrate contact. This can be depicted by typical strain ellipsoids and strain trajectories across the shear zone (fig. 10.5b; van der Wateren et al., 2000). Geologically, a shear zone is recognizable through a variety of structures, as shown in figure 10.5c, which relates to simple shear as depicted in the strain ellipse. The geometry and distribution of these structures vary with the type of shear zone and the distribution of strain within it (fig. 10.5d; van der Wateren, 1995). The hard rock analogues can be further applied with respect to sedimentary structures in shear zones whereby tectonic lamination and homogenization of materials can be compared to mylonitization between fault blocks.

10.2.3 PRIMARY AND SECONDARY DEPOSITS

In classifying sediments deposited in glacial environments, a distinction is sometimes made between *primary deposits*, laid down by uniquely glacial agencies, and *secondary deposits*, which have undergone some form of reworking by non-glacial processes (Lawson, 1981b; Dreimanis, 1989). According to this distinction, primary deposits include tills deposited at the ice–bed interface or the subglacial traction zone by the combined processes of deformation, ploughing, lodgement and melt-out, whereas secondary deposits are composed of glacial debris that has been remobilized and deposited by gravitational flowage, stream flow or other agencies. This may take place on a downwasting glacier surface, or after deglaciation. In the former case, it can be argued that the deposit is till, but this is not true for the latter, especially if a long period of time has elapsed since deglaciation. The boundary between primary and secondary deposits, however, is arguable, and it is not always clear at which point sediments lose their uniquely glacial character and become some other kind of deposit. This difficulty is reflected in the debate over the definition of the word *till* (Dreimanis, 1989). This debate kept the Till Work Group of the International Union for Quaternary Research (INQUA) occupied for more than 15 years in the 1970s and 1980s, and was not resolved to everyone's agreement. Most members of the Till Work Group accepted a broad definition of till as 'sediment that has been transported and deposited by or from glacier ice, with little or no sorting by water' (Dreimanis, 1989). This definition would embrace all primary deposits and some secondary deposits, particularly those formed at or near glacier margins by gravitational reworking processes. In reality, field and laboratory techniques are not actually capable of refined assessments of the exact genesis of tills, making complex classification schemes such as that proposed by the Till Work Group difficult to apply in practice (cf. Evans et al., 2006a); more specifically, such schemes give a false sense that the glacial research community has accomplished a foolproof forensic procedure for the reconstruction of ancient process–form relationships. Therefore, many glacial geologists prefer the more restrictive definition proposed by Lawson (1981b): 'Till is sediment deposited directly by glacier ice, and has not undergone subsequent disaggregation and resedimentation.' According to this definition, only sediments deposited by primary, uniquely glacial processes can be classified genetically as tills, and mass movement deposits from glacial settings are grouped with similar deposits from other environments, rather than with tills. Lawson's definition is adopted in this book, because it emphasizes the importance of depositional processes, rather than position of deposition, in sediment classification.

(a)

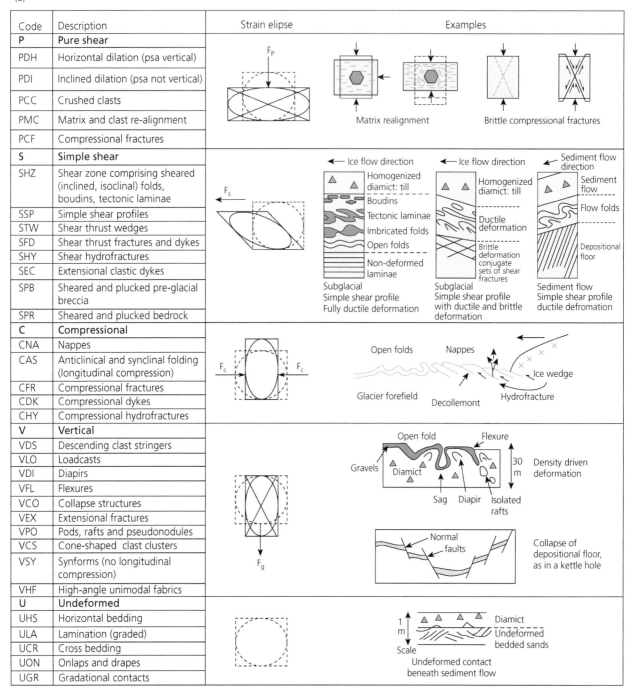

Code	Description	Strain elipse	Examples
P	**Pure shear**		
PDH	Horizontal dilation (psa vertical)		
PDI	Inclined dilation (psa not vertical)		
PCC	Crushed clasts		
PMC	Matrix and clast re-alignment		
PCF	Compressional fractures		
S	**Simple shear**		
SHZ	Shear zone comprising sheared (inclined, isoclinal) folds, boudins, tectonic laminae		
SSP	Simple shear profiles		
STW	Shear thrust wedges		
SFD	Shear thrust fractures and dykes		
SHY	Shear hydrofractures		
SEC	Extensional clastic dykes		
SPB	Sheared and plucked pre-glacial breccia		
SPR	Sheared and plucked bedrock		
C	**Compressional**		
CNA	Nappes		
CAS	Anticlinical and synclinal folding (longitudinal compression)		
CFR	Compressional fractures		
CDK	Compressional dykes		
CHY	Compressional hydrofractures		
V	**Vertical**		
VDS	Descending clast stringers		
VLO	Loadcasts		
VDI	Diapirs		
VFL	Flexures		
VCO	Collapse structures		
VEX	Extensional fractures		
VPO	Pods, rafts and pseudonodules		
VCS	Cone-shaped clast clusters		
VSY	Synforms (no longitudinal compression)		
VHF	High-angle unimodal fabrics		
U	**Undeformed**		
UHS	Horizontal bedding		
ULA	Lamination (graded)		
UCR	Cross bedding		
UON	Onlaps and drapes		
UGR	Gradational contacts		

(b)

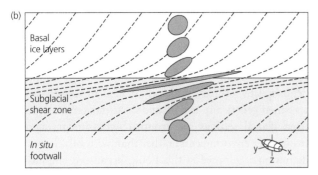

Figure 10.5 Deformation structures typically found in glacial materials. (a) Different types of deformation and their coding, as proposed by McCarroll and Rijsdijk (2003). The forces on the strain ellipses (Fp, Fs, Fc and Fg) refer to pure shear, simple shear, compressional and gravitational. (b) Conceptual diagram of the subglacial shear zone with typical strain ellipses and strain trajectories. (c) Structures produced within a shear zone by simple shear (van der Wateren et al., 2000). (d) Comparisons of structures produced in typical shear zones where top panel is of a soft sediment nappe from a thrust moraine, middle panel is a subglacial deforming layer and bottom panel is at the base of a debris flow. Graphs show finite shear strains (γ) with depth. S_r is the minimal strain horizon with *rooted* structures; S_b contains boudins, detached folds and transposed foliation; and S_h is the area of maximum strain characterized by homogenized materials. The A and B horizons refer to the subglacial deforming till stratigraphy of Boulton and Hindmarsh (1987).

(c)

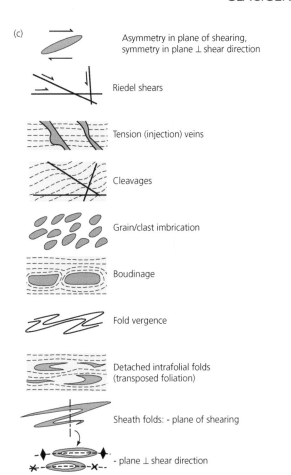

Asymmetry in plane of shearing, symmetry in plane ⊥ shear direction

Riedel shears

Tension (injection) veins

Cleavages

Grain/clast imbrication

Boudinage

Fold vergence

Detached intrafolial folds (transposed foliation)

Sheath folds: - plane of shearing

- plane ⊥ shear direction

10.3 PRIMARY GLACIGENIC DEPOSITS (TILL)

10.3.1 OVERVIEW

Primary glacigenic deposits, as defined in section 10.2.3, form subglacially by some combination of lodgement, bed deformation, ploughing and melt-out. Subglacial deposition usually involves complex interactions between these basic processes, and consequently, subglacial tills are enormously variable, defying attempts to classify them into distinct, exclusive categories. In the first edition of this book (Benn and Evans, 1998), we adopted an approach that regarded tills deposited by each of the three primary processes of lodgement, melt-out and deformation as end members of a continuum. Tills deposited by more than one process therefore occupied intermediate positions between these end member types (Dreimanis, 1989; Hicock, 1990, 1993; Benn and Evans, 1996). It was also recognized that secondary modification could result from gravitational flowage. In a review by Evans et al. (2006a), a revised till classification scheme was proposed, because it was considered extremely unlikely that all subglacial tills are anything other than hybrids produced by a range of processes in the subglacial traction zone (e.g. Alley, 2000; Ruszczynska-Szenajch, 2001; Nelson

(d)

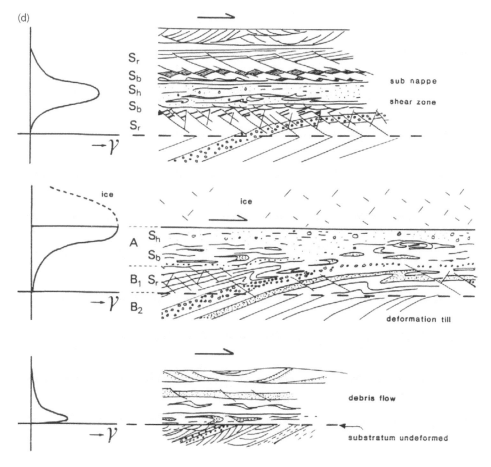

Figure 10.5 (*Continued*)

Deformation style	Subglacial	Ice marginal	Proglacial	Glacilacustrine	Glacimarine
Pure shear	Very common	Common	Absent	Absent	Absent
Simple shear	Dominant	Common*	Rare–common*	Absent	Absent
Compressional	Rare	Dominant	Rare	Absent	Absent
Vertical	Absent	Rare	Common	Very common	Rare
No deformation	Absent	Rare	Common	Dominant	Dominant
Iceberg dump and deformation	Absent	Absent	Absent	Common	Common

* The geometry of simple shear deformation structures relates to ice flow in ice marginal locations and to slope in proglacial settings.

Table 10.2 Typical occurrence of deformation styles in the different glacial depositional environments (McCarroll and Rijsdijk, 2003)

et al., 2005). In this section, we extend and elaborate on the Evans et al. (2006a) scheme.

The coexisting subglacial processes of melt-out, deformation, flow, sliding, lodgement and ploughing effectively mobilize, transport and deposit sediment, forming a wide range of deposits from glacitectonically folded and faulted stratified material to texturally homogeneous diamicton. The dominance of any one subglacial process varies both spatially and temporally, giving rise to the possibility that a till or complex till sequence contains a superimposed signature of several former transportation/deposition processes at the ice–bed interface. Convincing arguments have been made that deformation is involved in the production of all tills (cf. Ruszczynska-Szenajch, 2001; van der Meer et al., 2003), leading Menzies et al. (2006) to propose substituting the label 'tectomict' for the term 'till' because it is a structural rather than a depositional sediment. We prefer, however, to retain the older, widely accepted terminology.

The interpretation of massive tills has prompted seemingly endless debate and controversy. For example, disagreement has arisen over the interpretation of massive tills deposited by the Laurentide Ice Sheet, and some researchers have argued that they provide evidence for widespread subsole deformation (Boulton and Jones, 1979; Alley, 1991; Hicock and Dreimanis, 1992; Evans, 2000b), while others contend that they are melt-out tills (Clayton et al., 1989; Mickelson et al., 1992; Ronnert and Mickelson, 1992). Regardless of their precise origin, however, these tills display unequivocal evidence of transport and deposition in the subglacial traction zone (section 9.4.1). We therefore group all massive subglacial tills under the umbrella term *subglacial traction till*, while recognizing that finer genetic interpretation and classification may be possible in some cases. Subglacially modified materials that retain some of the structural characteristics of the parent rock or sediment body are termed *glacitectonite*. In the following sections we summarize the processes involved in the release and deposition of material from debris-rich glacier ice, then describe the characteristics of glacitectonites and subglacial traction till.

10.3.2 PROCESSES OF SUBGLACIAL TILL FORMATION

Subglacial deposition happens through the complex interaction of several factors, including the basal melt rate, the balance between frictional and driving forces acting on particles, debris grain size distribution, the configuration of the glacier bed, and the rate at which water can drain away from the site of deposition. We can recognize four basic processes of debris deposition: (1) *lodgement*, or deposition directly from sliding ice by frictional processes; (2) *frictional retardation* in a subglacial deforming layer; (3) *melt-out*, or deposition due to the melting of stagnant or slowly moving ice; and (4) *deposition by gravity*.

Lodgement

Lodgement is the plastering of glacial debris from the base of a sliding glacier onto a rigid or semi-rigid bed (Dreimanis, 1989). In effect, this occurs where the frictional drag between the debris and the bed is greater than the shear stress imposed by the moving ice, and is therefore sufficient to inhibit further movement of the debris (Boulton, 1975, 1982). The process of lodgement can occur for single particles or for masses of debris-rich basal ice.

Different situations where lodgement can occur are shown in figure 10.6. Where the glacier bed is rigid, friction between an overriding particle and the bed is mainly due to the interlocking of asperities beneath the particle and small roughness elements on the bed, and the resisting forces arise from the friction between these essentially rigid elements (Drewry, 1986). In contrast, where the glacier bed consists of deformable materials such as sand, gravel or till, particles protruding from overriding ice can plough through the bed. Ploughing particles will lodge on the bed if the bed resistance is high enough, which depends on the rate at which excess pore-fluid pressures can be dissipated (Thomason and Iverson, 2008). Particles may also be lodged against other particles already protruding from the bed, forming clusters (Boulton, 1975,

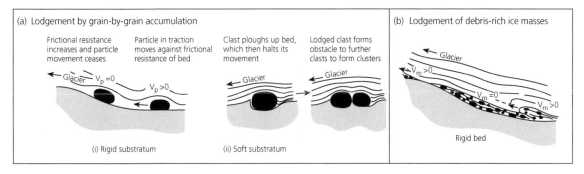

Figure 10.6 The lodgement process. (a) Particles lodging (i) by frictional retardation against a rigid bed and (ii) against obstacles on a soft bed. Particles can plough up the bed, which then halts movement, or lodge against other particles. (b) Lodgement of debris-rich ice masses. The whole assemblage lodges against the bed, and melts out in situ (Modified from Boulton, 1982).

1982). Additionally, entire masses of debris-rich ice may lodge against the bed.

Lodgement is unlikely to result in the grain-by-grain accumulation of rigid deposits. Accumulating sediment below a wet-based, sliding glacier is likely to consist of some combination of immobile lodged particles, ploughing particles maintained in traction by the glacier sole, and shearing sediment masses (Krüger, 1994). In a study of basal tills at Slettmarkbreen, Norway, Benn (1994a) found that large clasts tended to lodge on the bed, while smaller clasts were more liable to remain in transport by shearing in a thin subglacial deforming layer. This probably reflects the high drag force beneath large, heavy clasts (section 4.4.3) and the tendency for small particles to be carried along in the deforming layer, whereas larger clasts can penetrate the deforming layer and lodge against deeper, more rigid parts of the substratum.

Slabs of debris-rich ice will lodge against the bed if the frictional drag at the base of the slab exceeds the strength of the ice overlying the slab. Rheological differences between clean and debris-rich ice may encourage the development of a décollement plane immediately above the debris-rich ice. This plane can evolve into the new sliding base of the glacier, the ice below becoming part of the substratum. Once the slab has lodged, the interstitial ice will gradually melt out, releasing the debris, which will either become part of the immobile substratum or be remobilized as a subglacial deforming layer.

Frictional retardation in a deforming layer

In section 4.5 we saw that subglacial sediment deformation will occur where there is coupling between the ice and the substrate, and the shear traction imposed by the ice equals the yield strength of the subglacial sediment. Sediment in a subglacial deforming layer, therefore, will come to rest if the ice decouples from the bed or the sediment yield strength is greater than the applied shear traction. Both of these situations can arise in response to changing pore-water pressures in the substrate. Decoupling tends to occur at high pore-water pressures,

when a film of water forms at the ice–sediment interface. Since the water cannot support a shear stress, the underlying sediment is no longer subject to glacially imposed stresses and will cease to deform (section 4.5.3). Sediment yield strength is inversely proportional to pore-water pressure (section 4.2.3), so will increase if water is able to drain away from the bed. Therefore, deformation will also cease when pore-water pressures are low.

Where glacier beds are connected to surface-fed drainage systems, pore-water pressures can fluctuate on a diurnal basis (e.g. Boulton, 2006; Kavanaugh and Clarke, 2006), and it is likely that sediment will be deposited and remobilized many, many times. In unconnected regions, cycles of deposition and mobilization could occur on longer timescales. Hart and Boulton (1991) and Hart (1995a) visualized such cycles in terms of upward or downward migration of the base of the deforming layer. According to this model, thickening of the deforming layer erodes and mobilizes deeper sediments (*excavational deformation*), whereas thinning of the layer results in sediment deposition.

Melt-out

Deposition by melt-out refers to direct sediment deposition through the melting of stagnant or very slowly moving debris-rich ice. Recent developments in the understanding of glacio-hydraulic supercooling (sections 3.2.4 and 9.3.2; Alley et al., 1998, 1999; Lawson et al., 1998; Evenson et al., 1999; Roberts et al., 2002) have provided fresh impetus for the concept of till production by subglacial melt-out, by recognizing that thick sequences of basal debris-rich ice can be more widespread in glacier systems than previously thought (Larson et al., 2006). Melt-out can occur in supraglacial or subglacial positions. Only subglacial melt-out is discussed here, and supraglacial melt-out is described in section 9.4.2. The energy required for subglacial melt-out can be supplied from geothermal sources and sensible heat from incoming meltwater. The amount of heat provided by these sources is generally not large,

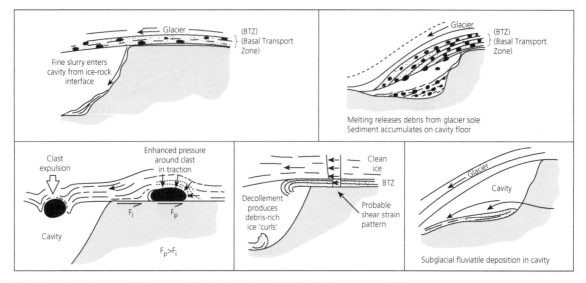

Figure 10.7 Observed mechanisms of debris accumulation in subglacial cavities (Modified from Boulton, 1982).

and typical rates of basal melt-out are in the range 5–12 mm yr^{-1} (Drewry, 1986; Paul and Eyles, 1990), although they may be much higher in volcanically active areas.

During the melt-out process, debris-rich ice undergoes *thaw consolidation*, or volume reduction consequent on ice melt and meltwater drainage. The amount of consolidation depends on the original debris content of the ice, and will be large if the debris content is low. Where debris content is high, the volume of the deposited debris may be only slightly less than that of the parent ice, and quite delicate englacial structures may be preserved. The behaviour of released debris during thaw consolidation also depends on the balance between meltwater production and drainage away from the site (Paul and Eyles, 1990). If the meltwater produced during the melt-out process is free to drain away at a rate equal to or higher than the rate of production, debris will be deposited with little or no disturbance, other than that due to consolidation. However, where meltwater cannot drain away freely, pore-water pressures will tend to rise during melt-out, decreasing the frictional strength of the deposited debris and increasing the likelihood of failure and remobilization. Paul and Eyles (1990) have argued that such conditions are very common where the debris content of basal ice is predominantly fine grained. Nevertheless, Carlson (2004) has demonstrated that till with low hydraulic conductivity can accommodate the transport of 1.3 m^3 water yr^{-1}m^{-2}, up to three orders of magnitude more than is typically produced at the base of a glacier. Remobilization of melted-out debris can also occur if the glacier bed is sloping, and according to Paul and Eyles (1990), failure can occur on slopes as low as 8°. The implications of such processes for the preservation potential of melt-out deposits are discussed in section 10.3.4.

Figure 10.8 Till curls at base of Øksfjordjøkelen, Arctic Norway (B.R. Rea).

Deposition by gravity

Basal debris may also be deposited by gravity into subglacial cavities, either in the lee of obstructions or below ice overhangs at the glacier margin (fig. 10.7; Boulton, 1982). If basal melting rates are high, debris may simply melt out from the roof of a cavity and fall to the floor. This process can be very important in ice marginal cavities during the ablation season, as anyone who has crawled into one will be aware. Additionally, slurries of saturated debris can flow into cavities from the ice–bed interface. Below thicker ice, clasts can be actively extruded from the basal ice into cavities by excess ice pressure (Boulton, 1982). A related phenomenon is the formation of *till curls* (fig. 10.8), or curved masses of debris-rich ice, which spall off from the bed of a glacier as it enters a cavity due to the different strain response of debris-rich and clean ice to stress release. Debris from all cavity-fill mechanisms depicted accumulates on the bottom of the cavity and may ultimately re-couple with the glacier base, either frozen into the basal ice or as a subglacial deforming layer.

10.3.3 GLACITECTONITE

The term *glacitectonite* refers to subglacially sheared rocks and sediments, and was introduced by Banham (1977) as an analogous term with tectonite, a metamorphic rock with structures and fabrics imprinted by its tectonic history (cf. Dreimanis, 1989; Pedersen, 1989). The term thereby embraced sheared materials that retain some primary structures (*exodiamict glacitectonite*), and diamictons, in which all primary structures have been destroyed by shear (*endiamict glacitectonite*). We here employ a refined version of the definition of glacitectonite: *rock or sediment that has been deformed by subglacial shearing (deformation) but retains some of the structural characteristics of the parent material, which may consist of igneous, metamorphic or sedimentary rock, or unlithified sediments* (Benn and Evans, 1996).

This definition is close to that of Banham's (1977) 'exodiamict glacitectonite'. Because pre-existing structures are still recognizable, they can be used as strain markers; examples include lamination, cross bedding or even overridden trees (Fleisher et al., 2006). Glacitectonites can be subdivided into *Type A*, which display evidence of penetrative deformation, and *Type B*, which have undergone non-penetrative deformation, so pre-deformational sedimentary structures are merely folded and faulted. Diamictic products of subglacial shear (Banham's 'endiamict glacitectonite') have previously been widely referred to as *deformation till*, but in this book they are incorporated within the category of *subglacial traction till* (section 10.3.4; Evans et al., 2006a).

Glacitectonites can display evidence for brittle or ductile deformation, or a combination of the two. Some glacitectonites are breccias, broken up and displaced along fault planes (brittle deformation; e.g. Lønne, 2005; Hiemstra et al., 2007), whereas others are folded by pervasive ductile deformation (figs 10.9 and 10.10).

The most complex deformation structures tend to occur in inhomogeneous materials with widely varying strengths, such as inter-bedded gravels, sands, silts and clays. In such cases, the fine-grained units can be highly deformed, with complex elongated folds, whereas coarse-grained units (e.g. sands or gravels) commonly form relatively undeformed pods or lenses (e.g. Berthelsen, 1979; Hart and Roberts, 1994; Benn and Evans, 1996; Evans et al., 1999a; Evans, 2000a; Roberts and Hart, 2005). This reflects variations in material response to stress, primarily due to contrasts in sediment frictional strength and permeability (section 4.2.3). Pods of stiff sediment commonly have streamlined forms, recording erosion and deformation of their edges during shear within more rapidly deforming material, and are analogous to *augen* and *boudins* in metamorphic rocks (figs 10.9 and 10.10c). Glacitectonites that have undergone high cumulative strains can exhibit a distinctive sub-horizontal banding,

leading ultimately to *banded* or *laminated tills*. Individual bands are commonly lithologically distinct, and represent different rock and sediment types that have been highly attenuated but not mixed during strain. Roberts and Hart (2005) recognized two types of laminae: Type 1 laminae, produced by intergranular shear, and Type 2 laminae, which are the attenuated remnants of pre-existing sedimentary bedding (a highly deformed glacitectonite). Type 1 laminae are the result of the syn-deformational development of matrix variability; it is conceivable that this may be driven by the constant vertical migration of shear zones through the till (e.g. Tulaczyk et al., 2000a, b) or *strain blocking* by large clasts (Boulton and Dobbie, 1998). Banded glacitectonites with scattered pebbles can be mistaken for laminated glacilacustrine and glacimarine deposits with dropstones, but can be distinguished by the presence of lithological banding, asymmetric folds around pebbles, and attenuated folds among the banding (Hart and Roberts, 1994; Ó Cofaigh and Dowdeswell, 2001; Roberts and Hart, 2005). Banded glacitectonites can also be mistaken for other types of deposit, such as debris flows and melt-out tills. Illustrations of numerous examples of glacitectonites have been provided by Evans and Benn (2004a).

Vertical sequences in glacitectonites commonly display increasing deformation up-section, as the result of patterns of cumulative strain in the substrate (Banham, 1977; Pedersen, 1989; Benn and Evans, 1996; Hiemstra et al., 2007). An idealized full sequence might consist (from the base up) of:

(1) undisturbed bedrock or sediment;
(2) glacitectonite with mildly distorted primary structures;
(3) glacitectonite with widespread, penetrative shear structures;
(4) subglacial traction till.

This sequence reflects the fact that overburden pressures, and hence sediment frictional strength, decrease upwards, so that the most intense deformation and the highest cumulative strains will be found at the top of the sequence, as indicated by measurements of deformation in modern glaciers (section 4.5.3; Boulton et al., 2001a; Iverson et al., 2007). It should be emphasized, however, that vertical patterns of deformation will also vary with sediment grain size, consolidation and other properties (fig. 10.11; Benn and Evans, 1996; Evans, 2000a; Roberts and Hart, 2005; Evans et al., 2006a).

Patterns of cumulative strain in glacitectonites have been elegantly demonstrated in microstructural studies by Phillips et al. (2007). Most, if not all, of the strain in the subglacially sheared sediments they studied was accommodated by strain in narrow, water-lubricated detachment zones. Liquefaction and hydrofracturing can lead to local homogenization of sediments and the destruction of older deformation structures.

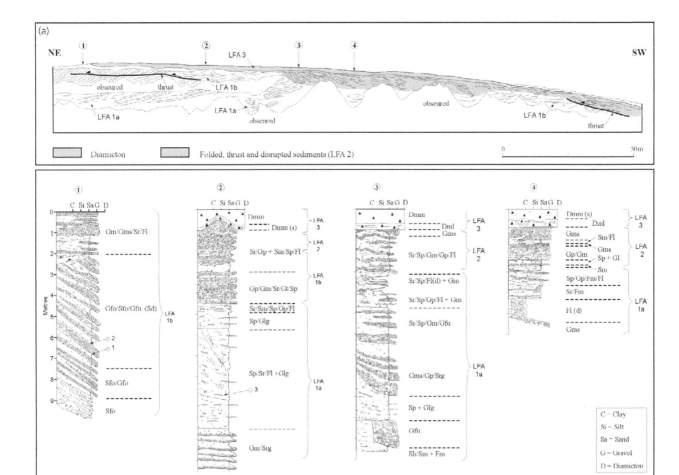

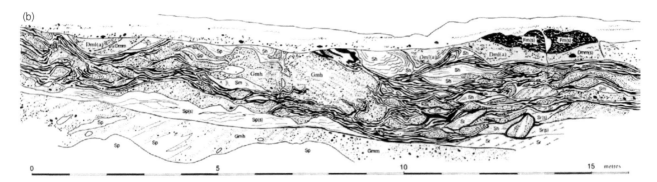

Figure 10.9 (a) Two-dimensional section sketch of the 1995 quarry exposure at Drumbeg, south Loch Lomond, showing the large-scale architecture of the major lithofacies and the impact of glacitectonic disturbance towards the top of ice-contact delta foresets. (b) Detail of glacitectonite exposed at the far right-hand side of the exposure depicted in (a) and composed of sand and gravel pods separated by attenuated silt and clay beds. Note the erosional contact with underlying sand and gravel foreset beds, which have been cannibalized to form the glacitectonite.

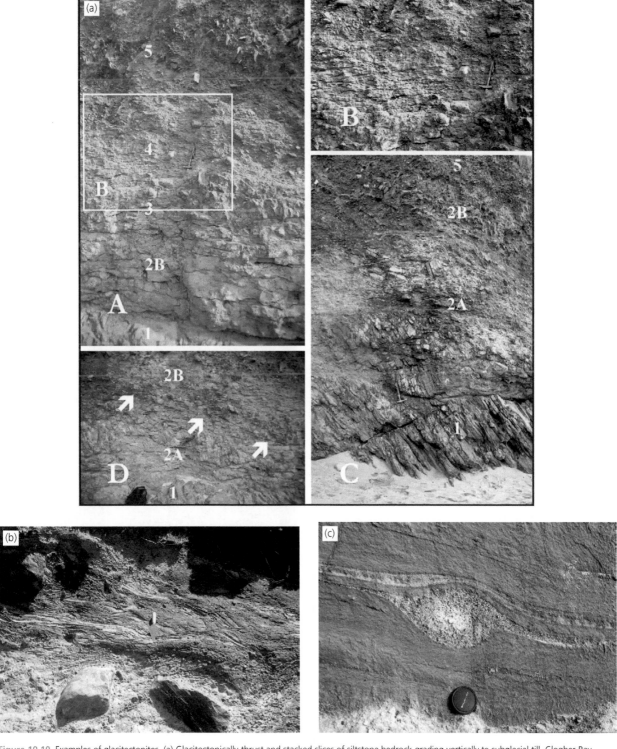

Figure 10.10 Examples of glacitectonites. (a) Glacitectonically thrust and stacked slices of siltstone bedrock grading vertically to subglacial till, Clogher Bay, Ireland. (Hiemstra et al., 2007). (A) The sequence of units 1, 2B, 3, 4 and 5. (B) Fissile character of unit 4. (C) Sequence of units 1, 2A, 2B and 5. (D) Sequence of units 1, 2A and 2B in detail. White arrows indicate the contact between subunit 2A and 2B. Unit 1 is intact (i.e. undeformed, but jointed), bedded siltstone bedrock. Unit 2 is a bed of deformed siltstone, the deformation signature consisting of straight, parallel and anastomosing fractures interpreted as shear planes. Unit 3 is more heavily deformed, brecciated siltstone in which primary bedrock structures are largely unidentifiable. Unit 4 is a zone of heavily sheared siltstone mixed with diamictic sediment. Unit 5 is a massive, predominantly matrix-supported diamicton (Dmm), with dispersed clast-supported zones (Dcm). Vertical changes in the internal characteristics of the stacked slices represent a gradual upward transition from bedrock to subglacially sheared material, thereby recording thrusting, stacking, shearing and pulverizing of soft bedrock and its incorporation into the base of a subglacial till layer. (b) Laminated diamicton containing recognizable sedimentary structures forming part of a glacitectonite carapace over glacilacustrine sediments, Loch Quoich, Scotland. (c) Gravel augen structure or boudin in glacitectonized sands and gravels (highly laminated glacitectonite) at Funen, Denmark. (d) Laminated diamicton produced by the subglacial deformation of periglacial slope deposits, southern Ireland (Photos: D.J.A. Evans).

Figure 10.10 (*Continued*)

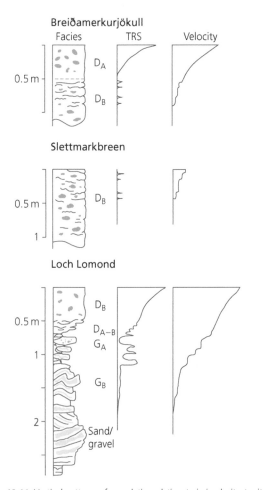

Breiðamerkurjökull

Facies   TRS   Velocity

0.5 m   D_A
         D_B

Slettmarkbreen

0.5 m   D_B

1

Loch Lomond

0.5 m   D_B
         D_A–B
1   G_A

         G_B

2   Sand/
         gravel

Figure 10.11 Vertical patterns of cumulative relative strain in glacitectonites and subglacial traction tills in Iceland, Norway and Scotland. D_A = ductile deformed till; D_B = brittle deformed till; G_A = penetrative glacitectonite; G_B = non-penetrative glacitectonite (Modified from Benn and Evans, 1996).

Figure 10.12 Subglacial traction till features. (a) Fissile, massive matrix-supported diamicton with striated and facetted clasts, Isle of Skye, Scotland. (b) Small grooves eroded into laminated silts by an overlying till base, Lethbridge, Alberta, Canada. The grooves record ice flow at right angles to the pencil. (c) Fold structure in the Skipsea Till, East Yorkshire, England, defined by the weathering out of intra-till sand wisps that give the till a laminated appearance at its base. (d) Deformed rafts (intra-clasts) of shale in matrix-supported diamicton, derived from the cannibalization of underlying strata, southern Alberta, Canada. (D.J.A. Evans). (e) Subglacially deformed till on the Breiðamerkurjökull foreland, showing the massive, friable structure of the formerly dilatant A horizon (above the penknife) overlying the fissile, denser B horizon. (A. Snowball) (f) Boulder pavement at the base of a till in a multiple till sequence, Tierra del Fuego, Chile (D.I. Benn).

10.3.4 SUBGLACIAL TRACTION TILL

Definition

Evans et al. (2006a) introduced the term *subglacial traction till* for sediments previously separately labelled as lodgement till, deformation till and comminution till (fig. 10.12). Here, we extend the scope of the traction till concept to include melt-out tills, in recognition of the fact that debris can pass between basal ice and the substrate many times prior to final deposition. Subglacial traction till is defined as: *Sediment deposited at a glacier sole, the sediment having been released directly from the ice and/or*

Figure 10.12 (Continued)

liberated from the substrate and then disaggregated and completely or largely homogenized during transport.

This definition also encompasses localized lee-side cavity fills beds and stratified sediments deposited by water films and channels. We now review the geologic signatures of the different processes that contribute to the formation of subglacial traction till.

Evidence for ploughing and lodgement

Evidence for clasts dragged along by sliding ice (ploughing) is provided by linear grooves (*sole casts*), and sediment prows in front of clasts. Lodgement takes place when ploughing clasts are arrested by the build-up of

prows. Grooves and prows are commonly clearly visible on freshly exposed glacier beds (e.g. Boulton, 1982; Benn, 1995) and can also be recognized in till exposures (Ehlers and Stephan, 1979; Clark and Hansel, 1989; Piotrowski et al., 2001; Jorgensen and Piotrowski, 2003; fig. 10.12b).

Clasts lodged beneath sliding ice tend to develop asymmetric 'stoss-and-lee' forms, owing to in situ abrasion of their upglacier (stoss) sides and plucking on their downglacier (lee) sides, similar to the erosion of a roche moutonnée in bedrock (fig. 9.45; Boulton, 1978; Krüger, 1984; Benn, 1994a). Krüger (1984) identified 'double stoss-and-lee' forms, which he attributed to a two-stage process of clast wear at the ice–till interface. Prior to deposition, the clast is ploughed through the underlying till, causing abrasion of the lower leading edge and fracture of the lower trailing edge. The clast then lodges against the bed and the stoss side is abraded while the lee side is fractured. Benn (1995) has argued that double stoss-lee forms could also develop on clasts rafted along within deforming till. If the clast maintains a quasi-stable orientation within the till, abrasion will be focused at the lower leading and upper trailing edges, and fracture at the lower trailing and upper leading edges, as till deforms around clasts.

Deformation structures

Since the introduction of the 'deforming bed paradigm' in the 1980s, many criteria have been proposed for the recognition of former deforming layers (e.g. Boulton, 1987; Hart et al., 1990; Hooke and Iverson, 1995; Benn and Evans, 1996). Many of these criteria can be questioned, and indeed, characteristics regarded by some as clear evidence for deformation are interpreted by others as quite the opposite (Piotrowski et al., 2001, 2002; Benn, 2006).

The clearest evidence for deformation is the presence of *deformed inclusions*, such as streaked-out or folded pods of sand or soft rock. Inclusions can highlight patterns of strain in the surrounding till (fig. 10.12c), or exhibit pressure-shadow effects and boudinage (Berthelsen, 1979; Hart and Roberts, 1994; Evans et al., 1995; Benn and Evans, 1996). Piotrowski and Kraus (1997) and Piotrowski et al. (2001) have argued that 'dispersion tails' extending from soft-sediment clasts can be used to differentiate between the effects of ploughing and deformation in tills (fig. 10.13). A soft clast that has been partially disaggregated within a pervasively deforming bed should develop two dispersion tails extending up- and downglacier, reflecting velocity gradients within the till (Hart and Boulton, 1991), whereas one affected by ploughing at the ice–bed interface should be expected to have only one tail extending downglacier from its upper surface. Deformed inclusions may be particularly abundant near the base of till units, where cumulative strains

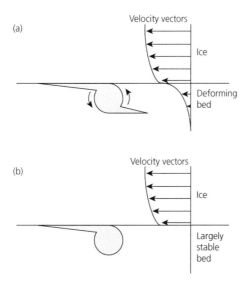

Figure 10.13 Conceptual diagram showing the development of erosion tails from a soft-sediment clast in (a) a deforming bed where the vertical distribution of velocity causes the clast to rotate and dispersion tails develop on its up-ice lower side and down-ice upper side; (b) a stable bed subject to ice sliding where the dispersion tail is only at the sliding interface (Piotrowski and Kraus, 1997).

will be lower and the material less homogenized, and there may be a gradational contact with underlying glacitectonite (Hart, 1995a; Benn and Evans, 1996; Evans and Twigg, 2002). In some cases, it is possible to trace inclusions or rafts back into glacitectonite or undisturbed parent materials (e.g. Benn et al., 2004). The process of detaching rafts of underlying material and then attenuating them in the deforming layer (fig. 10.12d) has been referred to as *cannibalization* by glacial sedimentologists (e.g. Hicock and Dreimanis, 1992; Evans, 2000a).

When strain markers are not available at the macro-scale, evidence of deformation of subglacial materials can be found in a wide variety of microstructures and fabrics, depending on granulometry and strain history (Owen and Derbyshire, 1988; Menzies and Maltman, 1992; van der Meer, 1993; Owen, 1994; Menzies, 2000; van der Meer et al., 2003; Carr, 2004). Faulting or folding is easily recognizable where it affects sorted sediments such as laminated silts and clays (e.g. van der Meer, 1993), but former deformation of diamictons can also be identified from more subtle features (see Carr, 2004 for illustrations). Structures indicative of small-scale brittle deformation include discrete microshears (particularly in clay-rich sediments), brecciation, boudins or augen, and crushed quartz grains (Owen and Derbyshire, 1988; Menzies and Maltman, 1992; van der Meer, 1993).

Under the microscope, clay-rich parts of a deformed till may show strong birefringence in cross-polarized light. This phenomenon is referred to as *plasmic fabric* and is the direct result of clay platelets having aligned parallel to each other under the influence of (subglacial) shear stresses (see van der Meer, 1993; Carr, 2004). If deformation continues to be partitioned into the clay-rich

parts of the till, it could eventually result in strain hardening and brittle failure (faulting). Further microscopic evidence of non-uniform responses by heterogeneous subglacial sediment to deformation is the occurrence of rounded, soft sediment inclusions/intraclasts in till, known as *till pebbles* (van der Meer, 1993). They form where small-scale variations in composition, grain size and/or water content lead to the solidification of patches or fragments in a matrix that is otherwise in a dilated or liquefied state (fig. 10.14).

Microstructural studies of artificially deformed clays by Hiemstra and Rijsdijk (2003) have shown that distinctive grain arrangements develop during shearing. Focused strain along shear zones produces grain lineaments, and the rotation of particles or particle aggregates within shear zones produces 'turbate structures' (fig. 10.15a). These structures consist of small particles arranged concentrically around a larger mass, and have been recognized in many ancient tills (van der Meer, 1993, 1997; van der Meer et al., 2003). Hiemstra and Rijsdijk (2003) showed that both grain lineaments and turbate structures increase in frequency at higher cumulative strains, and pointed out that planar movements are necessary to create the torques required to rotate grains. This is a much more sophisticated view of grain rotation than earlier, mechanically impossible conceptions of deforming layers as a mass of interconnected 'wheels' all rotating in the same direction (cf. van der Meer, 1997).

Deformation of tills in ring-shear experiments by Thomason and Iverson (2006) showed that most strain was accommodated by microshears (Riedel shears) aligned at low [R1] and high [R2] angles to the shearing direction (figs 10.15b, c). The degree to which low-angle shears accommodate strain was quantified using the dimensionless I_L index:

$$I_L = \frac{\sum_{i=1}^{n} L_i \cos \psi_i}{\sum_{i=1}^{m} H_i \cos \theta_i + \sum_{i=1}^{n} L_i \cos \psi_i} \qquad (10.1)$$

where n and m are the numbers of low-angle (L) and high-angle (H) shears respectively, and ψ and θ are the acute angles of low- and high-angle shears respectively. The division between high- and low-angle shears is regarded as $25°$. The I_L index varies from 0 to 1, with high values being indicative of a predominance of shears parallel to the shearing direction. This works on the principle that microshears become aligned more in parallel to the principal zone of displacement under progressively higher strain. An interesting development in this area of research was reported by Larsen et al. (2006), who subjected 40 mm of diamicton and 40 mm of overlying sand to shear strains up to 107. Displacement, as measured from deformation indicators (S-matrix microstructures) in thin

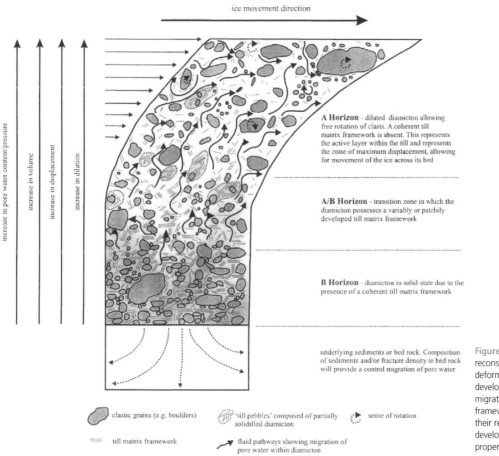

Figure 10.14 Idealized reconstruction of the subglacial deforming till layer and the development of pore-water migration pathways, till matrix framework and till pebbles, and their relationship to horizon development and geotechnical properties (Evans et al., 2006a).

sections, took place only in a 14–20 mm thick zone within the diamicton. By plotting I_L indexes from Pleistocene tills together with the results of their laboratory experiments, Larsen et al. (2006) identified a progressive I_L index increase during shearing (fig. 10.15d), demonstrating the potential of the index for estimating strain in ancient tills. However, the homogenized tills from their Pleistocene sites record strain values of only 8–13, suggesting that most of the microstructural changes take place in tills at low strains.

Subglacial meltwater and ice–bed decoupling

Glaciological evidence for glacier sliding and water flow over till beds was discussed in section 4.5.3. Some sediment facies and structures have been interpreted as the products of intermittent flow of a water film at the ice–bed interface and ice–bed decoupling (e.g. Eyles et al., 1982; Brown et al., 1987; Piotrowski and Kraus, 1997; Piotrowski and Tulaczyk, 1999; Piotrowski et al., 2001). Clark and Walder (1994) and Evans et al. (1995) introduced the concept of *canal fills*, or the infills of former braided canal systems that develop at the ice–bed interface during periods of decoupling. They are recognizable as stratified lenses with flat upper surfaces and concave-up bases, containing primary depositional structures or bedforms, in addition to thin sand layers and stringers

(fig. 10.16). Intra-till stratified sediments may be un-deformed, implying that overlying tills were deposited by passive melt-out. In many cases, however, intra-till melt-water deposits have been glacitectonically deformed, implying ice–bed re-coupling and shearing of the bed following deposition (Eyles et al., 1982; Boyce and Eyles, 2000). Other structures associated with meltwater deposits that are indicative of localized ice–bed coupling include slickensides and flat-topped striated clasts.

It is possible that some intra-till stratified sediments record water flow *within* a till, either during or immediately after till deposition. Some researchers have interpreted subtill stratified units as evidence for water films at sliding till bases, rather than former ice–till interfaces (Hindmarsh, 1996; Kjær et al., 2006). In addition, during till deposition by deformation, meltwater may drain through the till by pipe flow, as has been observed at the base of the Rutford Ice Stream, Antarctica, by King et al. (2004). Potential sedimentary signatures of this process have been proposed by Alley (1991), who suggested that mini eskers could develop in the till. Such features are very often difficult to differentiate from rafts cannibalized from older sediments, especially when they have been folded and/or attenuated after their emplacement. However, differentiation may be possible where intra-till channel fills possess different lithological and grain size

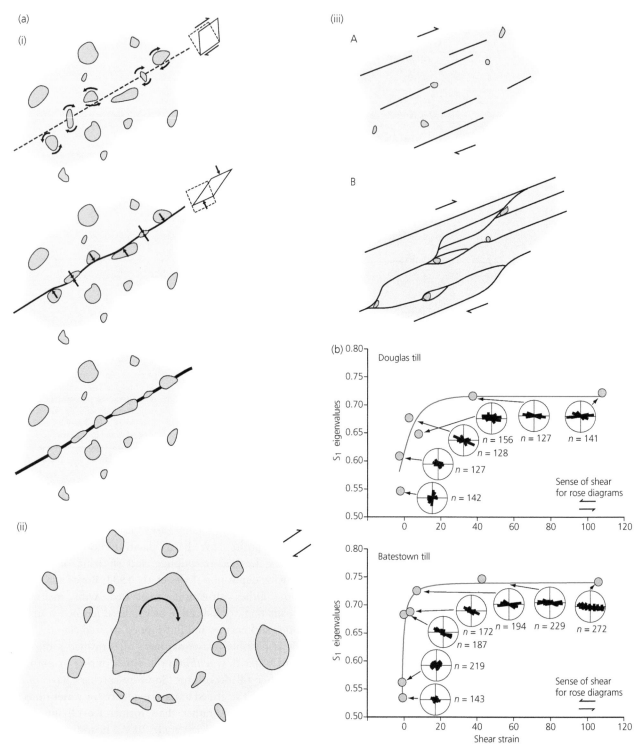

Figure 10.15 Micromorphological evidence of artificially induced strain. (a) Grain alignments and plasmic fabrics developed during a shearing experiment on potters clay where strain was <0.1 (Hiemstra and Rijsdijk, 2003). (i) Schematic illustration of the development (sequentially from top to bottom) of grain lineaments and unidirectional plasmic fabrics. Dashed line in upper sketch represents the position of a developing shear plane. Elongate grains near the shear plane rotate until they are aligned plane-parallel. They also move towards the shear plane due to contraction of the sediment (see strain boxes). (ii) Sketch to illustrate the relationship between unidirectional plasmic fabrics, skelsepic plasmic fabrics and turbate structures. (iii) Schematic diagram to show the development of branching and merging characteristics in unistrial plasmic fabrics – short, discontinuous unistrials in (A) grow to form continuous features in (B), and where unistrials meet they split or bifurcate. (b) Microfabric strength related to strains up to 108 produced by shearing experiments on two Laurentide tills (Douglas and Batestown). Rose diagrams record the alignments of sand grain long axes in two dimensions and therefore S1 eigenvalues will not fall below 0.5. (Thomason and Iverson, 2006). (c) (i) Photomicrograph under cross-polarized light of Douglas Till showing microshears at a shear strain of 39 and (ii) schematic reproduction of Riedel shear alignments and strain ellipsoids representing steady state V1 eigen vector. (Thomason and Iverson, 2006). (d) Graph showing the relationship between shear strain and I_L index based on ring shear experiments and Pleistocene tills, the latter marked by dashed lines. (Larsen et al., 2006a). More detail on micromorphological features is available in Carr (2004).

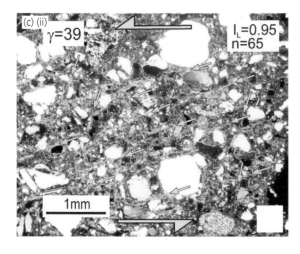

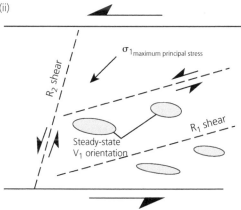

Figure 10.16 Examples of proposed ice–bed separation features. (a) Horizontally stratified till consisting of sand layers mm thick, intercalated with till matrix. (b) Single horizontal stringer of stratified sand in till matrix and draping a lodged clast (Piotrowski and Tulaczyk, 1999, reproduced with kind permission of Jan Piotrowski).

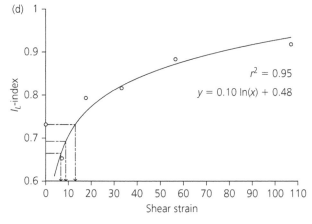

Figure 10.15 (*Continued*)

signatures to the substrate from which the cannibalized rafts originated.

Lee-side cavity fills

The opening and closure of lee-side cavities may be recorded by cavity-fill deposits within till successions. Sedimentological evidence for ancient lee-side tills in the Canadian Rocky Mountains has been provided by Levson and Rutter (1989a, b), who described massive diamictons with rare, steeply dipping sand and gravel lenses located on the lee side of bed obstructions. The clasts within the diamicton are striated and possess a strong fabric with a dominant downvalley plunge of a-axes, and the sand lenses also dip consistently in a downvalley direction. The sediments are truncated by an overlying subglacial till, suggesting that the glacier eventually contacted the lee-side fill once the cavity was full.

Possible cavity fill deposits have been observed in the lee side of drumlins (e.g. Dardis et al., 1984; McCabe and Dardis, 1989), and large examples associated with tuyas (subglacial volcanic edifices; section 11.2.11) have been described by Bennett et al. (2006).

Clast (boulder) pavements

Clast or *boulder pavements* are distinctive features that occur within some till successions, consisting of horizontal layers of clasts one or more particles thick, commonly with planar, striated upper surfaces (fig. 10.12f). Perhaps the simplest explanation for layers of large clasts within a till succession is that they are lag deposits marking former positions of the ice–till interface. Preferential removal of the matrix of the underlying till by the combined effects of subglacial meltwater flushing and glacier sliding would isolate larger clasts at the tops of till sheets, which could then be buried by renewed till deposition (e.g. Boyce and Eyles, 2000). Careful study of clast fabric in boulder pavements by Hicock (1991) has shown that some have likely been disturbed by till deformation after formation.

Alternative mechanisms for the formation of clast pavements at the base of thick deforming layers have been proposed by Clark (1991) and Boulton (1996a). Clark's idea was that pavements represent clasts that have sunk to the base of a dilatant till layer, as can be observed in debris flows. In contrast, Boulton (1996a) argued that subtill boulder pavements are formed during excavational deformation at the base of a deforming layer. According to this model, downward movement of the base of a deforming layer into an older till unit will result in the preferential mobilization of fine material and the accumulation of large particles which resist entrainment. The Boulton model therefore predicts that the boulders in subtill pavements will be lithologically similar to those in the underlying sediment, whereas Clark's model predicts that they should be similar to those in the overlying till.

Evidence for melt-out

Subglacial melt-out till has been defined as: *sediment released by the melting of stagnant or slowly moving debris-rich glacier ice, and directly deposited without subsequent transport or deformation.*

However, it may be difficult or impossible to distinguish between sediment that has been transported within basal ice, then subsequently deposited by passive melt-out, and sediment that has been transported by ploughing and subsole deformation, then deposited by frictional retardation. Both sets of processes can produce massive tills with varying degrees of evidence for shearing, dewatering and remobilization. In some cases, however, it may be possible to recognize signatures diagnostic of passive melt-out from debris-rich basal ice.

There have been very few studies in modern environments where subglacial melt-out has been observed. The most detailed and influential study is that of Lawson (1979a, b), who compared the characteristics of glacier ice and associated deposits at the margin of Matanuska Glacier, Alaska. Important observations have also been made in Svalbard by Boulton (1970a, b, 1971) and at the margin of Burroughs Glacier, Alaska, by Mickelson (1973, 1986), Ronnert and Mickelson (1992), and Ham and Mickelson (1994). Tills deposited by the passive melt-out of debris-rich, foliated ice may inherit foliation as well as the grain size, particle morphology and fabric characteristics from the parent ice. Structures may include discontinuous layers and lenses, textural and compositional banding, and flow structures (Lawson, 1979a, b; fig. 10.17). The dip of the foliation in melt-out till should

be less than that in the ice source, due to the volume reduction and compaction consequent on ice melting and dewatering (fig. 10.18; Boulton, 1971). Where the original debris concentrations are high, the thaw consolidation ratio will be low, so that, theoretically at least, delicate englacial structures may be preserved, especially where drainage away from the site is efficient. Some photographs of modern melt-out tills, such as that reproduced in figure 10.17e, reveal that the parent ice is so heavily debris-charged that it appears to be frozen stratified sediment; as these sediments have been in basal glacier ice, presumably they have been incorporated by apron overriding (section 5.4.3) and then released without being disaggregated by the glacier, suggesting that they might be more accurately described as glacitectonites. Some melt-out tills are structureless and massive, and lack any visible foliation (Lawson, 1979a).

The porosity of freshly deposited melt-out till is variable. Lawson (1979a) found that subglacial melt-out till at Matanuska Glacier was much more compact than debris melted out on the glacier surface. In contrast, Ronnert and Mickelson (1992) described fresh subglacial melt-out till with high porosity. The main controls on porosity are the debris concentration in the parent ice, effective overburden pressures and drainage conditions.

Many of the proposed characteristics of melt-out deposits have been inferred from the geological record (e.g. Haldorsen and Shaw, 1982; Shaw, 1982, 1983a; Piotrowski, 1994; Munro-Stasiuk, 2000). However, the criteria regarded as diagnostic of melt-out by some researchers are interpreted as the products of deformation or glacilacustrine deposition by others. For example, Evans et al. (2006b) have provided alternative interpretations of supposed melt-out tills identified in southern Alberta, Canada, by Munro-Stasiuk (2000); the sediments in question have characteristics that may be explained as glacitectonically disturbed glacilacustrine sediments. Similarly, possible Pleistocene examples of foliated melt-out tills have been described by Shaw (1982, 1983a), although many of these examples appear to be very similar to glacitectonites, and the foliation could have originated by subglacial deformation rather than shear of debris-rich ice. Indeed, it may be very difficult to distinguish between these two origins without very detailed study.

One feature that may be diagnostic of melt-out is the occurrence of scour fills beneath clasts (fig. 10.17b). Shaw (1983a) suggested that these structures are produced by subglacial meltwater scouring beneath clasts held in the overlying ice. Clasts with scour fills have been reported from stratified tills by Munro-Stasiuk (2000). An alternative origin for such clasts and scours has been proposed by Evans et al. (2006b), who invoked the role of anchor ice in glacilacustrine sedimentation (see section 10.6).

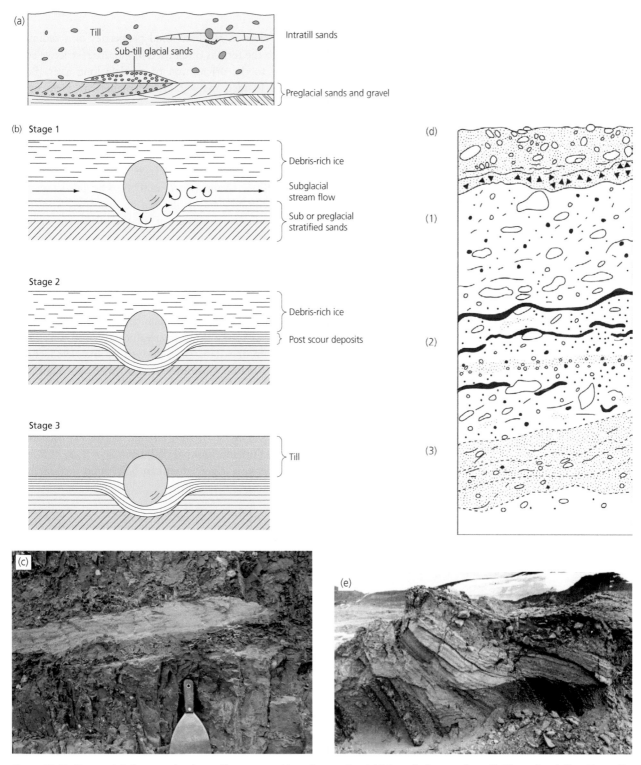

Figure 10.17 Characteristic features of melt-out till, as proposed by various studies. (a) Schematic diagram of stratified lenses in subtill and intra-till positions (Shaw, 1983a) (b) Explanation of the development of scour fills beneath clasts held in basal ice during melt-out (Shaw, 1983a) (c) Soft sediment intra-till lens, Alberta, Canada (D.J.A. Evans) (d) Idealized vertical profile of typical features of proposed melt-out at the Matanuska Glacier, Alaska. (1) Structureless, pebbly, sandy silt. (2) Discontinuous laminae, stratified lenses and pods of texturally distinct sediment in massive pebbly silt. (3) Layers of texturally, compositionally or colour-contrasted sediment. Laminae or layers may drape over large clasts. (Lawson, 1981a) (e) Sublimation till from the Antarctic Peninsula (S. Rubulis)

The drainage of meltwater from sites of subglacial melt-out may result in the deposition of inter-till stratified lenses and beds (e.g. Shaw, 1982; Munro-Stasiuk, 2000), or the formation of dewatering structures (Carlson, 2004). Some till sequences are penetrated by numerous vertical pipes, which are sometimes filled with sorted sediments. These might indicate dewatering of sediment during and following melt-out, although other origins (such as hydrofracturing) should be considered.

It has been argued that unremoulded melt-out deposits are probably very rare, due to the high chance that original structures will be destroyed during dewatering or gravitational reworking (section 10.3.2; Paul and Eyles, 1990). If this is true, however, *remoulded* melt-out deposits could still be very common, and could be difficult to distinguish from those deposited by other subglacial processes.

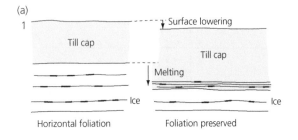

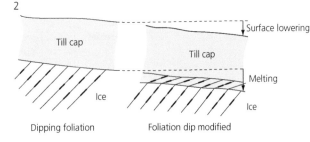

Till fabric

Fabric refers to all directional properties of a sediment, although the term is often used in a more restrictive sense to refer to the orientation of particle long axes (Benn, 2004b). In fact, many fabric elements can yield useful information on subglacial processes. The orientations of clast long a-axes, the alignment of a–b planes, polished facets, and stoss-lee forms can reveal much about the deformational and depositional history of tills (Hicock, 1991; Benn, 1994a, 1995; Li et al., 2006; Piotrowski et al., 2006; Evans et al., 2007a). The fabric of fine-grained components in tills can be measured from their *natural remanent magnetism*, or the alignment of magnetic particles. Magnetic grains in subglacial deposits tend to be aligned parallel to the direction of shear, and commonly show close agreement with clast fabric orientations (Boulton, 1976; Hooyer et al., 2008; Shumway and Iverson, 2009).

Fabric 'shape' can be quantified using the relative magnitudes of the three eigenvalues S_1, S_2 and S_3 calculated from the data (Dowdeswell et al., 1985; Benn, 1994a, 2004b). Isotropic fabrics (with no significant preferred orientation in any direction) have $S_1 \approx S_2 \approx S_3$; girdle fabrics (with most clast orientations confined to a plane, but no significant preferred orientation within the plane) have $S_1 \approx S_2$, $S_3 \approx 0$; and cluster fabrics (with most clast orientations parallel) have $S_1 \approx 1$, $S_2 \approx S_3 \approx 0$. Ratios between eigenvalues can be plotted on triangular

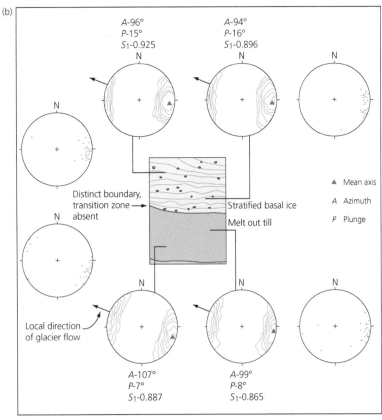

Figure 10.18 Melt-out characteristics. (a) The preservation and modification of melt-out tills from debris-rich ice, showing (1) horizontal ice surface and folia and (2) dipping ice surface and folia and the resulting modification of fabric (Sugden and John, 1976, after Boulton, 1971) (b) Stratigraphic and fabric characteristics of subglacial melt-out tills from Matanuska Glacier, Alaska, showing virtually identical clast fabrics in till and in glacier ice (Modified from Lawson, 1979b)

diagrams analogous to particle shape diagrams, and samples from particular types of sediment tend to cluster in predictable fields (fig. 10.19). The technique is not intended to define unique 'genetic fingerprints' (cf. Bennett et al., 1999a), but provides a valuable aid to visualization and interpretation of fabric data.

Two fundamental mechanisms have been proposed for the orientation of particles in deforming media, both of which have been invoked to explain till fabric characteristics. First, in *Jefferey-type rotation*, particles roll continuously in response to velocity gradients in a shearing viscous medium (Jefferey, 1922). Second, *March-type*

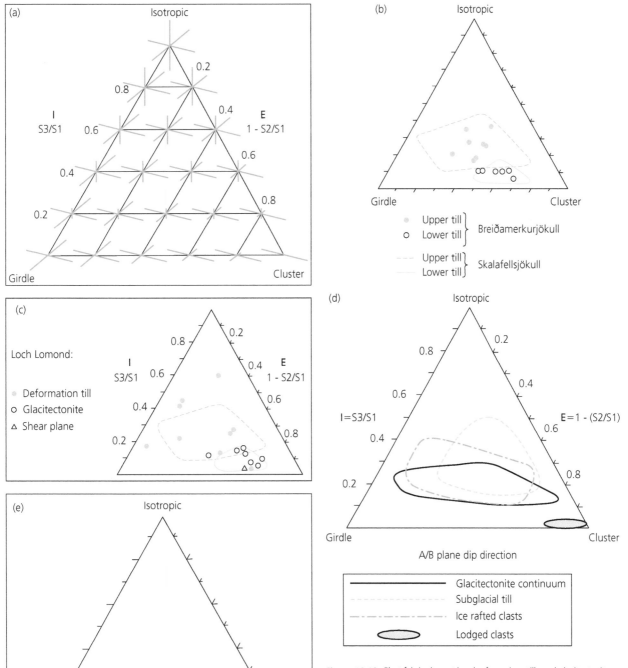

Figure 10.19 Clast fabric shape triangles for various tills and glacitectonites. (a) Key to fabric shape triangle, in which sample populations are plotted according to their isotropy and elongation (Benn, 1994a). (b) A-axis samples from Breiðamerkurjökull, with envelopes for similar A horizon (dashed line) and B horizon (solid line) tills from Skalafellsjökull (after Benn and Evans, 1996). (c) A-axis samples from the Loch Lomond study of Benn and Evans (1996), including envelopes for Icelandic A (dashed line) and B (solid line) horizons. (d) A/B plane samples from glacitectonites, Icelandic subglacial tills, ice-rafted sediments and subglacially lodged clasts (Evans et al., 2007a). (e) A-axis samples from melt-out till and debris-rich ice (data from Lawson, 1979b).

March rotation

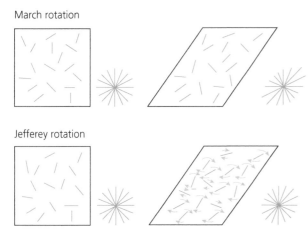

Jefferey rotation

Figure 10.20 Jefferey and March models of particle orientation in a deforming medium. The March model shows that particles rotate passively and the fabric ellipsoid reflects the deformation ellipsoid, whereas the Jefferey model implies that particles continuously roll so that the fabric ellipsoid may be more or less elongate than the deformation ellipsoid (after Benn and Evans, 1996).

rotation occurs when particles act as passive strain markers and rotate towards parallelism with the principal axis of extensional strain (March, 1932; fig. 10.20). This behaviour is consistent with Coulomb-plastic rheology, in which strain is accommodated by slippage between grains (section 4.2.3; Ildefonse and Mancktelow, 1993; Hooyer and Iverson, 2000). Jefferey-type rotation results in 'weak' fabrics, with many particle axes aligned transverse or oblique to the flow direction. In contrast, March-type rotation produces 'strong' cluster fabrics, with a single mode parallel to the direction of shear.

Laboratory experiments and some field studies support the idea that March rotation is the dominant mechanism for orientating particles in deforming till. Ring-shear experiments by Hooyer and Iverson (2000) have shown that particle long axes rapidly rotate into parallelism with the principal strain axis and remain there. Stable cluster fabrics are attained at cumulative strains of around 2.0, with no further modification up to strains of at least 370. Thomason and Iverson (2006) also identified a clear strengthening of sand grain microfabric with increased strain; fabric strength attained S_1 values of 0.71–0.74 at shear strains of 7–39, but then remained steady up to shear strains of 108 (fig. 10.15c). Detailed studies of clast fabrics in recently exposed subglacial tills at Breiðamerkurjökull, Iceland, by Benn (1995), also indicated that March rotation plays a major role in fabric development. The sampling site was located close to where Boulton conducted his measurements of subglacial till deformation in the 1980s, having been exposed by glacier retreat in the intervening period (Boulton and Hindmarsh, 1987; section 4.5.1). The till has a two-tiered structure, with a low-strength, high-porosity upper layer (A horizon) overlying a stronger, higher density lower till (B horizon). Clasts in both horizons had a–b planes close

to horizontal, with striated, polished facets on their upper and lower surfaces, indicating clast 'gliding' in quasi-stable orientations. In the lower till and fluted parts of the upper till, clasts had strong cluster a-axis fabrics parallel with the direction of ice flow (fig. 10.19b). In unfluted parts of the upper till, however, a-axis fabrics were weaker, which was interpreted as evidence of unsteady strain conditions in response to pore-water pressure fluctuations and clast interactions. Larsen and Piotrowski (2003) reported clast fabrics from stacked till units in Poland which are uniformly strong throughout the sequence, interpreted as evidence for March rotation during Coulomb-plastic till deformation.

Lodged clasts also tend to have strong preferred orientations, reflecting tightly constrained shear at the ice–till interface. When particles are bridged between ice and till, movement of the ice will tend to rotate the particle until its a-axis and a–b plane are parallel to the plane of shear, although an upglacier imbrication will develop due to the tendency for till matrix to plough up in front of the particle. Additionally, the a-axes will rotate into parallelism with the direction of shear due to the drag imposed by the surrounding ice and till matrix (Benn 1994a, 1995; Benn and Evans, 1996). Evans et al. (2007a) presented a–b plane fabrics of clasts lodged on the forelands of Icelandic glaciers. The clasts were partially buried in a fluted till surface, with their upper exposed faces striated parallel to the surrounding flutings, and exhibited strong cluster fabrics with near horizontal a–b planes.

Clasts in glacitectonites commonly have strong cluster fabrics, reflecting the strongly constrained shear that takes place in such materials (fig. 10.19c). There is some variation, however, depending on the degree of modification of the parent material. Using a–b plane clast macrofabrics from glacitectonites that have undergone various degrees of shearing, as indicated by their structural appearance, Evans et al. (1998, 2007a) provided a range of strain signatures indicative of a glacitectonite continuum (fig. 10.19d). Weaker fabrics are typical of the more immature glacitectonites that have been subject to lower cumulative strain.

Strong clast macrofabrics have also been reported for melt-out tills (fig. 10.18b). This is thought to reflect the original englacial fabrics, but with some overprinting during dewatering and consolidation (Boulton, 1970b; Lawson 1979a, b; Lundqvist, 1989b; Murton et al., 2005). Englacial fabrics often have strong preferred orientations parallel to the direction of shear, due to the rotation of clasts by the surrounding deforming ice (fig. 10.19e; Lawson 1979a; Ham and Mickelson, 1994). Preferred orientations of pebbles in ice are commonly parallel to ice flow directions, but may be transverse in zones of compressive flow (Boulton, 1970a). Melt-out till fabrics collected by Lawson (1979a, b) faithfully reflect

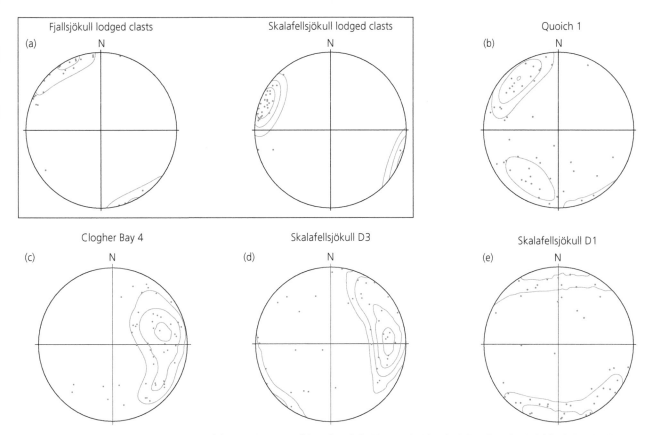

Figure 10.21 Examples of contoured stereonets of clast A/B plane macrofabrics for subglacial materials. (a) Lodged clasts from the Skalafellsjökull and Fjallsjökull fluted forelands, Iceland. (b) Immature glacitectonite from glacially overridden glacilacustrine sediments, Loch Quoich, Scotland. (c) Mature glacitectonite from sheared siltstones at Clogher Bay, SW Ireland. (d) Icelandic A horizon. (e) Icelandic B horizon (data from Evans et al., 1998, 2007a; Hiemstra et al., 2007; Evans and Twigg, 2002).

former ice flow directions, unlike those of glacigenic debris flow deposits in the same area. The melt-out till fabrics, however, display a reduction in dip values and an increase in dispersion relative to englacial fabrics. Compaction during the melt-out process reduces the range of dip values and 'flattens' the fabric, reducing its isotropy, and clast interactions during settling weaken the preferred orientation, reducing the elongation (fig. 10.19e; Lawson, 1979a, b; Benn, 1994a). Lawson's data set has been widely used as a reference for the interpretation of ancient tills, but it would be unwise to assume that all tills deposited by melt-out will share the same fabric characteristics. Indeed, given that such deposits will experience varying degrees of remoulding during and following deposition, it is likely that their fabric characteristics will be highly variable.

Many massive Pleistocene and modern subglacial tills have 'weak' clast fabrics (e.g. Hicock, 1992; Hart, 1994; Benn and Evans, 1996; Benn, 2006; figs 10.19, 10.21). Some researchers have argued that this is evidence for former thick, subglacial deforming layers, in which clasts underwent continuous rotation by Jefferey rotation, (e.g. Hicock and Dreimanis, 1992; Hart, 1994; Hicock et al., 1996). On the other hand, others have argued that only March rotation can occur in subglacial deforming layers,

so any till with a 'weak' fabric cannot have formed by subglacial deformation (e.g. Hooyer and Iverson, 2000; Piotrowski et al., 2001; Thomason and Iverson, 2006; Iverson et al., 2008). Thus, the same evidence – weak clast fabrics – has been adduced as evidence both for and against subglacial deformation.

Several possible mechanisms might account for 'weak' fabrics in basal tills. First, unsteady patterns of deformation (in space and time) could 'weaken' fabrics, so that March rotation need not necessarily cause monotonic development of strong cluster fabrics (Benn, 1995; Benn and Evans, 1996). Fabric strength may also be affected by clast collisions in coarser grained tills (Ildefonse et al., 1992). For example, the tendency for smaller particles to display weaker or transverse a-axis fabrics than larger particles in the same till (Kjær and Krüger, 1998; Carr and Rose, 2003; Carr and Goddard, 2007) may be due to smaller particles being more susceptible to collisions with equal-sized or larger neighbours (Thomason and Iverson, 2006). Inhomogeneous or unsteady deformation may therefore produce a wide range of clast fabric strengths, with localized fabric patterns reflecting local strain conditions.

Alternatively, some degree of particle rolling might occur in conjunction with March rotation, and exert

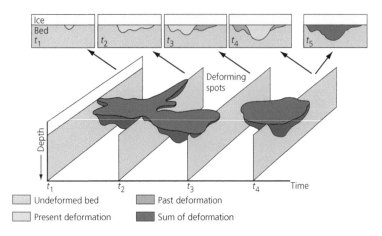

Figure 10.22 Schematic reconstruction of the concept of the stable/deforming subglacial bed mosaic concept of Piotrowski et al. (2004). The position of the deforming spots on the bed changes spatially and temporally, producing overprinted deformation signatures in the subglacial material.

some influence on fabric development. Some evidence for this is provided by pioneering experiments conducted by Hart et al. (2009), in which autonomous probes were inserted into till at the bed of Briksdalsbreen, Norway, via boreholes. Data from tilt sensors in the probes was transmitted to the glacier surface, providing a continuous record of probe movement within the till. The probes tended to exhibit progressive reduction of dip (i.e. March rotation), on which was superimposed short-term dip oscillations and rotation about the a-axis. These data indicate that clast behaviour beneath glaciers is likely to be rather more complex than simple idealized models would suggest.

Finally, it is possible that massive tills with 'weak' fabrics either did not form part of subglacial deforming layers or, if they did, their characteristics have been overprinted by some other process. For example, repeated ploughing of soft sediment by clasts and ice keels might produce a well-mixed, massive sediment with 'weak' clast fabrics. Partial sediment remoulding during melt-out of debris-rich ice may have a similar effect. Testing these alternative ideas will require very careful research design and painstaking data collection.

Subglacial till mosaic

Observations below modern glaciers indicate that many glacier beds may consist of mosaics of deforming and sliding bed conditions that vary in space and time (section 4.5.3). Sedimentological evidence for spatially and temporally varying bed conditions has been presented by a number of researchers, including Dreimanis et al. (1986), Hicock (1990), Evans et al. (1995), Benn and Evans (1996), Piotrowski and Kraus (1997), Piotrowski et al. (2001, 2002) and Shumway and Iverson (2009). The sedimentary signature of such *subglacial till mosaics* is characterized by changes in till properties over short distances, depending on the predominant process at the time the glacial depositional system shut down. Stratified sediments record periods of subglacial meltwater sheet flow between

phases of accretion of till. Such inter-bedded sequences could then be glacitectonized during later periods of ice–bed coupling. This temporally and spatially variable impact of deformation on glacier substrates is illustrated in figure 10.22, which shows how deformation signatures may be superimposed on parts of the subglacial till layer (Piotrowski et al., 2004). This concept has been elucidated by van der Meer et al. (2003), with reference to the migration of the hard, mobile and quasi-mobile zones (H, M and Q beds of Menzies, 1989) of active glacier beds.

The properties of many subglacial tills suggest that they have complex histories, involving various combinations and cycles of lodgement, deformation and melt-out. Examples include till complexes formed by all three processes at Bradtville, Ontario (Dreimanis et al., 1986; Hicock, 1990); and hybrid lodgement and deformation tills described by Hicock and Dreimanis (1992) and Benn (1994a). The difficulty of reconstructing till history has been illustrated by Ham and Mickelson (1994), who described subglacial tills exposed by recent retreat of Burroughs Glacier, Alaska. The tills are known to have been frozen into the glacier sole, and to have subsequently melted out. Some characteristics of the till, however, indicate that it is not a simple 'melt-out till'. First, slickensided shear surfaces are present throughout, particularly in the uppermost 0.5 m, which has a pronounced subhorizontal fissile structure, possibly indicative of subsole deformation. Second, stoss-and-lee boulders and lee-side flutes are common at the upper surface of the till, indicating modification of the till below sliding ice. Third, clast fabrics do not identify any particular mode of deposition. These characteristics suggest that the till may have undergone several cycles of erosion, transport and deposition beneath the glacier, involving repeated melt-out and freezing. A similar complex evolutionary cycle has been proposed by Evans and Hiemstra (2005) for tills deposited at the snouts of some Icelandic glaciers, in order to account for a range of sedimentary features. A very useful model for the complex evolution of subglacial till

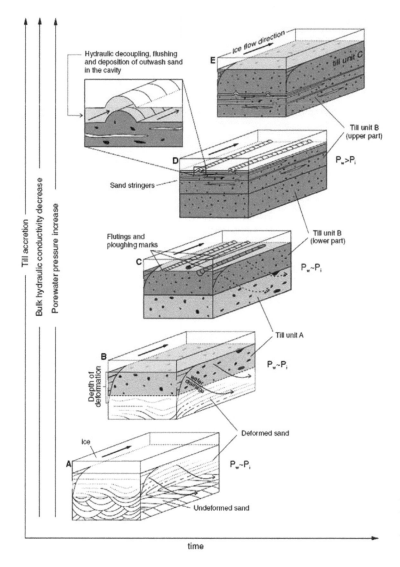

Figure 10.23 Schematic model of the time-transgressive evolution of a till succession and its overprinted signatures of deformation, lodgement, ploughing and sliding. The major processes are marked as: (A) and (B) bed deformation; (C) lodgement and ploughing; (D) basal decoupling and meltwater flushing (Piotrowski et al., 2006)

was presented by Piotrowski et al. (2006), based on a till sequence in Poland (fig. 10.23).

Subglacial traction till: summary

In the subglacial tractive zone, debris is made available for till production via melt-out from the ice base, quarrying and abrasion of hard rock surfaces, and the excavation of particles or particle aggregates from weaker substrates. Sediment is also reworked in water films and canals at the ice–sediment interface or within the till. Subglacial deformation can alternate with ice–bed decoupling and sliding, debris freeze-on and melt-out. Deformation typically takes place close to the bed, and may be partitioned according to sediment rheological properties. The deformation history of a sediment is recorded in the total strain signature, although this may be partially or completely overprinted during dewatering, mass movement or other processes. In some cases, depositional and deformational processes can be reconstructed in considerable detail, although in others it may not be

possible to say much beyond the fact that a till has been transported and deposited in the subglacial tractive zone.

Zones of maximum strain migrate vertically in subglacial materials in response to changing pore-water pressures, even on a diurnal basis (section 4.5.3), potentially resulting in complex cumulative strain patterns (fig. 10.24). Many subglacial tills are heterogeneous, because they have either inherited textural variability from the substrate or have reworked sliding bed deposits or other stratified materials following a phase of ice–bed decoupling. Such compositional variations will affect the response of the till to changes in pore-water pressure and/or any imposed shear stress. For example, the relatively more sandy parts of the matrix will dilate more rapidly than adjacent clay-rich areas, resulting in vertical and/or lateral variations in till dilation. Clay-rich areas of the till will possess a relatively higher cohesive strength and also impede pore-water flow through the diamicton. This results in deformation partitioning, where dilation, liquefaction and therefore flow displacement occur in a

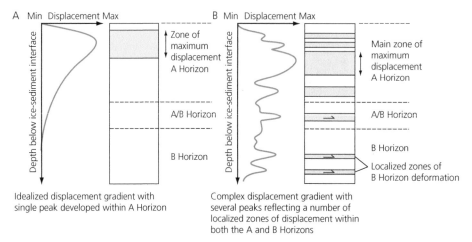

Figure 10.24 Theoretical cumulative displacement curves through subglacially deforming till. (A) A simple vertical decrease in displacement as portrayed in the Breiðamerkurjökull case study, but incorporating field evidence of maximum displacement at depth in the till. (B) A complex curve depicting multiple failure loci due to deformation partitioning (Evans et al., 2006a).

number of discrete zones within the till (Benn and Evans, 1996; Evans, 2000a). Therefore, subglacial till could be composed of an anastomosing network of dilated/liquefied zones enclosing domains of partially solidified diamicton. In contrast, the higher frictional strength of sand-rich lithologies means that when pore-water pressures are relatively low, they are more competent and resistant to deformation. Consequently, any deformation will be partitioned into the less competent, weaker, clay-rich parts of the sediment. As a result, the clay-rich parts of the till may exhibit an apparently higher intensity of deformation compared to the adjacent sandier horizons, which may lack any obvious signs of deformation.

The grain size distribution of subglacial traction till can vary widely. Some distributions simply reflect mixing of the parent materials, with little or no modification by crushing or abrasion (Elson, 1989), whereas others show clear evidence for grain fracture and comminution during shear (e.g. Boulton et al., 1974; Benn, 1995, 1996; Hooke and Iverson, 1995). The amount of grain size modification depends on the relative importance of dilation and grain fracture in the shearing process, which in turn reflects till strength and pore-water pressures (section 4.4.2; Hiemstra and van der Meer, 1997). Dilatant, low-strength tills can deform with little or no grain crushing because grains can climb over each other during shear. In contrast, stiff, high-strength tills may deform mainly by the fracture of interlocking grains, leading to distinctive grain-size distributions (Hooke and Iverson, 1995). Where tills have been completely homogenized by deformation of pre-existing sediments, they are expected to have rather uniform grain-size distributions over large areas, due to thorough mixing of materials under high cumulative strains (Alley, 1991).

Till sequences in the geological record are often very complex. Architectural element analysis has been conducted by Boyce and Eyles (2000) on a till sequence in Ontario, Canada, in order to assess the impacts of the subglacial process–form mosaic on sediments and structures over a single glacial cycle. They recognized three architectural elements with characteristic internal structures, separated by and including seven orders of bounding surface (fig. 10.25a). The oldest element is a deformed zone comprising glacitectonized pre-existing sediment (glacitectonite). This is overlain by a series of 'diamict elements', occasionally separated by erosional surfaces, clast pavements and 'inter-bed elements' of stratified sediment. Their explanation of the till sequence involves a combination of subglacial deformation events punctuated by phases of ice–bed separation and subglacial meltwater sheet flow (fig. 10.25b).

A conceptual model for the production of subglacial traction till is presented in figure 10.26. The shaded area of the glacier bed represents the subglacial traction till (STT), the final thickness of which does not reflect the thickness of the subglacial traction/deforming layer at any one point in time. The likely relationships between the STT and other deposits in the geological record are represented in the profile logs. Not all the process–form products would necessarily be coeval, and signatures may be overprinted due to glacier advance-and-retreat cycles. The graphs to the right of each vertical profile log indicate the likely relative pattern of cumulative strain. The likely process–form products are presented in the labelled boxes as follows:

(a) Pre-advance lake sediments capped by STT. (b) Lee-side cavity fill diamicts and stratified sediments capped by STT where ice has re-coupled with the bed due to cavity closure. In A and B, the tops of the lake sediment and cavity fill sequence are classified as glacitectonite which is locally cannibalized and dragged into the base of the deforming till layer after folding/thrusting is induced

(a)

ARCHITECTURAL ELEMENTS IN NORTHERN TILL		CODE	OUTCROP (2-D) GEOMETRY	APPROX. SCALE	L/T	LITHOFACIES ASSEMBLAGE	INFERRED PROCESSES
Diamict element		DE	Tabular diamict beds, planar to gently undulating bounding contacts; boulder pavement often marking upper surface	> 100 m (L) < 10 m (T) > 10³ m² (A)	> 10	Dmm, Dms Dms, Sm, Sr	Subglacial aggradation of deformation till
Interbed	Coarse	I-c₁	Laterally continuous, sheet-like sands and gravels separating diamict elements	> 100 m (L) < 5 m (T) 10³ m² (A)	> 25	Sm, Sr, Sp Gm, Gp, Dmm	Ice-bed separation; erosion and deposition by subglaciofluvial meltwater sheet-flow
		I-c₂	Laterally discontinuous sand gravel body, with pinch and swell geometry	< 10 m (L) < 1 m (T) 10² m² (A)	< 10	Sm, Sp, Gm, Gcm	Ice-bed separation; localized incision by subglaciofluvial meltwater sheet-flow
	Fine	I-f	Laterally continuous tabular silt and mud units separating diamict elements	> 10 m (L) < 1 m (T) 10² m² (A)	> 10	Fl, Fm, Flr, Fl	Ice-bed separation; low energy sedimentation in subglacial water body
Deformed zone		DZ	Undulatory zone of deformed till and thrusted sediments at base of till sheet; variable thickness and spatial extent	Variable (L) > 10 m (L) 10² m² (A)		Dmm, Dms + included sub-till sediments	Subglacial deformation of pre-existing strata

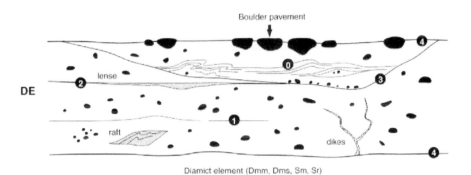

DE

Diamict element (Dmm, Dms, Sm, Sr)

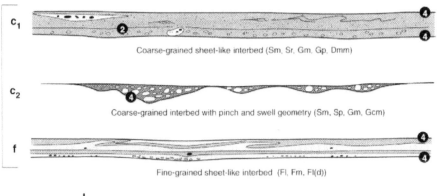

c₁

Coarse-grained sheet-like interbed (Sm, Sr, Gm, Gp, Dmm)

c₂

Coarse-grained interbed with pinch and swell geometry (Sm, Sp, Gm, Gcm)

f

Fine-grained sheet-like interbed (Fl, Fm, Fl(d))

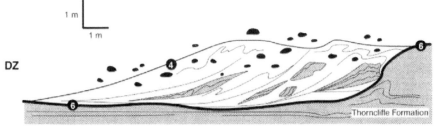

1 m

1 m

DZ

Thorncliffe Formation

Deformed zone (Dmm, Dms, Sd)

Figure 10.25 Application of architectural element analysis to a complex till sequence in Ontario, Canada. (a) (above) Architectural element types and (below) typical bounding surfaces. DE = diamict element, where tabular beds are separated by planar to gently undulating bounding surfaces with occasional clast pavements; I = inter-beds comprising coarse (I-c) and fine (I-f) ice–bed separation deposits; DZ = deformed zone comprising glacially deformed pre-existing strata. (b) (overleaf) A conceptual model to explain the deposition of the till sequence– (A) Erosion and deformation of pre-existing unlithified sediments to form drumlinized surface. (B) Conformable aggradation of deformation tills on drumlinized surface. (C) Subglacial fluvial reworking of diamicton to form sheet-like inter-beds and boulder pavements. (D) Continued aggradation of deformation tills and stratified inter-beds (Boyce and Eyles, 2000)

(b)

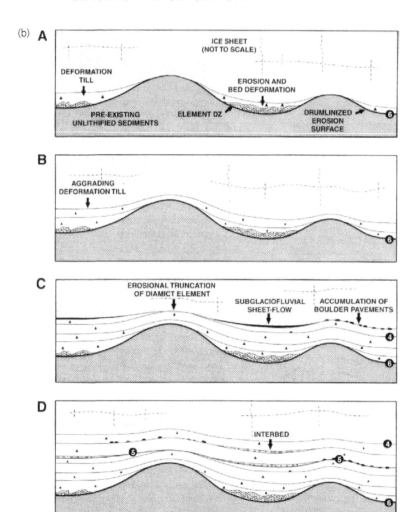

Figure 10.25 (*Continued*)

by ice-keel ploughing, clast ploughing and/or migrations in the locus of failure planes in subglacial sediment, aided by cyclical changes in pore-water pressure. This produces a crudely laminated diamicton; (C) Sheared tills with partially to intensively glacitectonized stratified intra- and inter-beds produced over hard beds where (i) STT and stratified inter-beds (canal fills) are continually aggraded due to continuous replenishment of deformable sediment from up-ice; (ii) large clasts are lodged to produce a clast pavement at the base of a thin STT in which finer-grained material is continuously advected down-ice; (iii) a single STT with associated glacitectonized canal fill at a site where aggradation is restricted by poor till continuity; (iv) regelation of STT (dark grey shade) leads to the stacking of tills which later melt out (note the similarity with sequences produced by marginal incremental thickening in box (g); (D) Pre-advance or remobilized melt-out till overlain by STT. The top of the melt-out till is classified as glacitectonite and is locally cannibalized and dragged into the base of the deforming till layer after folding/thrusting is induced at the STT/melt-out till interface,

and/or the loci of failure planes migrate through the subglacial sediment pile to produce a crudely laminated diamicton; (E) Glacitectonized soft bedrock grades upwards into STT with bedrock rafts ('comminution till'); (F) Glacitectonite (glacially overridden lake sediments) overlain by deforming STT where (i) the glacitectonite has been locally cannibalized and dragged into the base of the deforming layer after folding/thrusting is induced at the STT/stratified sediment interface, and/or the loci of failure planes migrate through the subglacial sediment pile to produce a crudely laminated diamicton (this may be assisted by seasonal migration of the freezing front or cyclical changes in pore-water pressures); (ii) the top of the glacitectonite is eroded by the base of the STT. The contact between the two sediment bodies is essentially a fault gouge; (G) Glacifluvial outwash or lake sediment, glacitectonized in its upper layers and overlain by a stacked sequence of macroscopically massive STTs. Incremental stacking occurs where the glacier margin is stationary and slabs of STT freeze on and melt out every year.

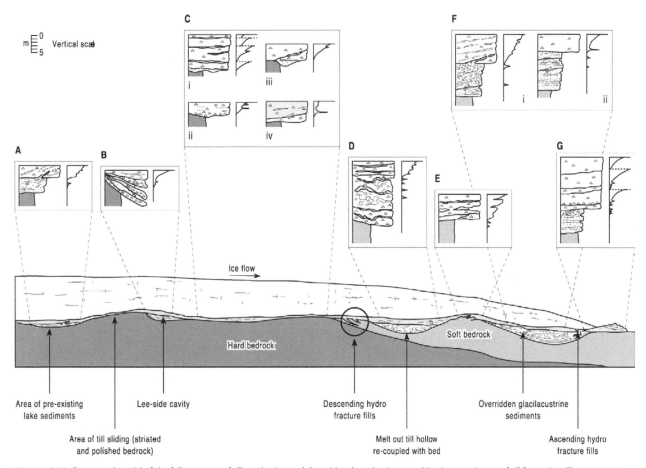

Figure 10.26 Conceptual model of the full spectrum of till production and depositional mechanisms resulting in a continuum of till formation. The processes depicted are those that would be expected beneath a large, predominantly temperate outlet glacier with a seasonally frozen snout (Evans et al., 2006a).

10.4 GLACIFLUVIAL DEPOSITS

10.4.1 TERMINOLOGY AND CLASSIFICATION OF GLACIFLUVIAL SEDIMENTS

Sediments deposited from flowing water are ubiquitous in many glacial environments because of the seasonal abundance of meltwater. Fluvial facies can be deposited in subglacial and englacial conduits, in supraglacial and proglacial streams, and near subaqueous meltwater portals, and are therefore important constituents of many subglacial, ice-marginal, proglacial, and subaqueous sediment–landform associations. Individual lithofacies reflect local sediment supply and water flow conditions rather than the position of deposition, so that some facies may be deposited in all of the above environments. In each environment, however, facies occur in different combinations, forming distinctive vertical successions and architectural elements that reflect larger-scale channel patterns and their change through time.

In fluvial systems, it is possible to recognize a hierarchy of depositional units, representing the storage of sediment at different scales (Ashley, 1990; Miall, 1992). This is manifest on channel floors as *bedforms*, of which there are four basic types: plane beds, ripples, dunes and antidunes (Allen 1983, 1985; Ashley, 1990). These are patterns in the bedload which represent a dynamic equilibrium response of the bed to prevailing flow conditions. Bedforms are repetitive, mobile structures that migrate down-current in response to the erosion of sediment from their upstream faces (where shear stresses are highest) and the deposition of sediment on their lee faces (where shear stresses are low). The form, wavelength and height (amplitude) of bedforms reflect specific combinations of sediment grain size and flow velocity, as shown in phase diagrams based on flume experiments (fig. 10.27). Such phase diagrams depict *existence fields* for the bedform types, the boundaries of which may be sharp or gradational, depending on whether or not changes in flow conditions bring about a sudden jump between bedforms that are stable for any given particle size. The smallest bedforms (ripples) reflect flow conditions very close to the bed, in the viscous sub-layer, and are classed as *microforms*; whereas the larger bedforms (dunes) reflect conditions in the outer part of the boundary layer of the flow, and are termed *mesoforms* (Allen, 1982, 1985).

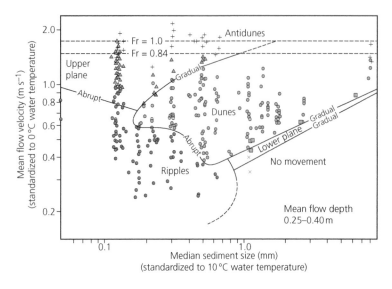

Figure 10.27 Bedform existence fields shown as a function of median sediment size and mean flow velocity (Ashley, 1990, reproduced with permission of the *Journal of Sedimentary Research*).

Microforms can be superimposed on the upstream faces of mesoforms. In turn, bedforms occur within larger depositional units such as longitudinal and bank-attached bars (*macroforms*), and channel fills. These units respond to large-scale flow patterns and long-term fluvial regime and tend to be active only during peak discharges. At yet larger scales, macroforms and channels fills are nested within channel systems, valley fills and basin fills (Miall, 1992).

Of significance with respect to bedform production is the distinction between *subcritical* and *supercritical flow*. This is determined by the *Froude number (Fr)*:

$$Fr = U / \sqrt{gd}$$

where U is depth-averaged flow velocity, g is gravity and d is flow depth. A value of $Fr > 1$ represents a supercritical or fast and shallow flow, whereas a value of $Fr < 1$ defines subcritical flows in relatively deeper and slower-moving water. Bedforms indicative of supercritical flow include antidunes (section 10.4.5), as indicated by their presence on the phase diagram at Fr values approaching and above 1.0. A lateral transition between the two flow regimes produces a step in the water surface, referred to as a *hydraulic jump*, resulting in significant changes in sedimentary processes and forms (Alexander et al., 2001).

Fluvial sediments are characterized by stratification and other sedimentary structures at several scales, which record the migration and accretion of bedforms and larger depositional units (Allen, 1983; Miall, 1985, 1992). In the following sections, we describe genetic facies formed by the accretion and migration of bedforms below flowing water. We also consider facies formed by settling of suspended particles from portions of flows that have come to a halt, and facies deposited from hyperconcentrated flows, or sediment-water mixtures that are transitional between fluvial flows and debris flows. The ways in which genetic facies combine in larger units (architectural

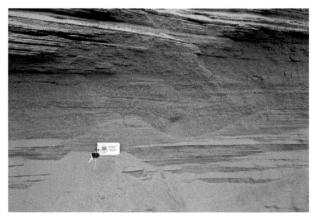

Figure 10.28 Horizontally bedded sand facies containing erosional scours, Drumheller, Alberta, Canada (D.J.A. Evans)

elements) representing channel fills and bar forms are discussed in section 11.5.1. Examples of fluvial facies and their structures as they occur in glacier-fed systems are presented in Evans and Benn (2004b).

10.4.2 PLANE BED DEPOSITS

Net deposition of sediment on flat, sandy stream beds produces beds of horizontally bedded or laminated sand, with the laminae recording minor fluctuations in flow velocity or sediment supply. Lamination can be strong or faint, depending on depositional conditions (fig. 10.28). Grain size tends to be very fine to coarse sand (0.0625–1.00 mm). For units deposited under upper flow regime conditions, bedding planes display thin, linear grooves and ridges aligned parallel to the former flow direction, called *parting lineation*. This lineation forms in response to instabilities in the boundary layer of the water flow (Allen, 1982, 1985), and is often observed in sandstones that split along bedding planes. It is much harder to detect in unlithified Quaternary sediments.

Sedimentation on plane beds under the lower flow regime also results in horizontally bedded and laminated sands. In this case, the sand tends to be coarse to very coarse (0.8–2.00 mm), and parting lineation is absent.

10.4.3 RIPPLE CROSS-LAMINATED FACIES

Sand facies deposited by the migration and vertical accretion of current ripples display distinctive internal structures called *ripple cross lamination*, the form of which depends on the balance between down-current migration and suspension sedimentation during ripple formation (Jopling and Walker, 1968; Ashley et al., 1982; Allen, 1985). Three basic types of cross lamination can be recognized:

- *Type A* develops when more sediment is eroded from the stoss sides during ripple migration than is added from suspension. Successive lee-side positions of the ripples are preserved as cross laminae that dip down-current, but the stoss positions are not preserved, due to net erosion. Instead, individual sets of lee-side laminae are separated by diachronous erosion surfaces that dip up-current (fig. 10.29). The *angle of climb* of these surfaces is variable, and reflects the net amount of deposition from suspension averaged over the entire bedform. There are two limiting cases for Type A cross lamination, based on the angle of climb. In the first case, the angle of climb is zero relative to the base of the bedform, recording ripple development by migration only, with no addition of sediment from suspension. The ripples form isolated sets sometimes known as *starved ripples*. In the second case, the angle of climb is identical to the angle of the original stoss faces, recording an exact balance between deposition from suspension and erosion on the stoss face of the bedform. This type is also known as critical cross lamination (Allen, 1985).
- *Type B* cross lamination displays successive stoss face positions as well as the lee face positions, and records net deposition from suspension over the whole bedform (fig. 10.29). Net deposition is still highest on the lee faces, however, due to the redistribution of sediment by the water current. The angle of climb is now greater than the slope angle of the stoss faces, and is defined as the angle between successive ripple crest positions.
- *Type S*, also known as *sinusoidal cross lamination* or *draped lamination*, is characterized by only small variations in thickness between stoss-side and lee-side laminae and a very steep angle of climb (60–90°). Ripple profiles are only weakly asymmetric or symmetrical. This type of cross lamination records small or no ripple migration (i.e. weak or zero current flow) and a dominance of suspension sedimentation.

Vertical gradations between two or more of these ripple types are often observed in sandy fluvial and glacilacustrine

sediments, recording temporal changes in current flow and suspended sedimentation rates. Upward transitions from A–B–S, A–B and B–S are characterized by an upward increase in the angle of climb, paralleled by a decrease in mean sediment size, and are formed during waning flows and an increase in deposition from suspension. Upward transitions from S–B–A, S–B or B–A have a decreasing angle of climb and an increase in mean sediment size, and record rising flows and a decrease in the importance of suspension sedimentation. Excellent examples of different types of transition are illustrated by Jopling and Walker (1968) and Ashley et al. (1982).

10.4.4 DUNES

Sand and gravel facies produced by migration and vertical accretion of dunes have a cross-bedded internal structure, resembling large-scale ripple cross lamination. The dominance of cross bedding indicates that dunes form by the migration of gravel and sand sheets (Carling, 1996). Two basic types can be recognized, although more exhaustive classification schemes are used in detailed sedimentological studies (Allen, 1982).

Planar cross-bedded gravels and sands originate by the migration of transverse, two-dimensional dunes with approximately planar lee surfaces. The thickness of sets gives a minimum height for the original bedform, and generally ranges between 0.25 m and 4 m or more for

Figure 10.29 Rippled sand facies. (a) Climbing ripple forms in glacial lake sediments, Drumheller, Alberta, Canada. (D.J.A. Evans) (b) The production of cross-laminated rippled sands and their relationship with rates of bedform climbing (Fritz and Moore, 1988, reproduced with permission of Wiley)

(b)

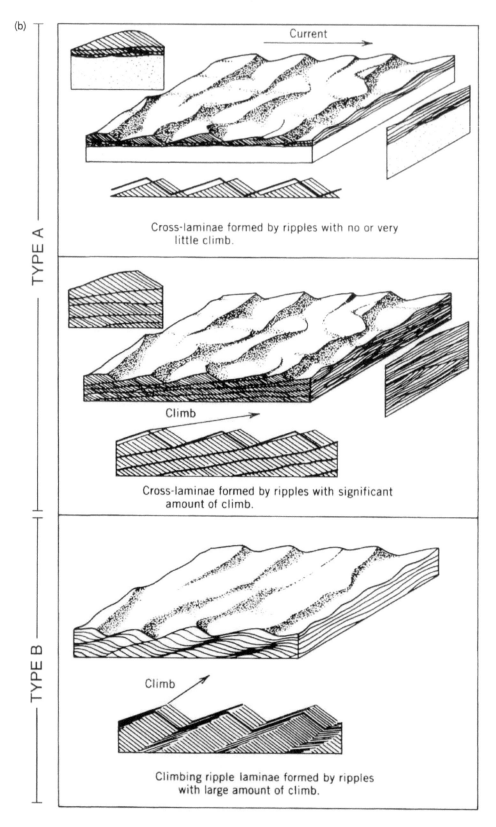

Current

Cross-laminae formed by ripples with no or very little climb.

TYPE A

Climb

Cross-laminae formed by ripples with significant amount of climb.

Climb

Climbing ripple laminae formed by ripples with large amount of climb.

TYPE B

Figure 10.29 (Continued)

gravels, and 0.05 to at least 5 m for sands (typically less than 1 m; Miall, 1977). The thickness of lee-side beds may be as much as 0.4 m in larger gravel sets, recording a considerable amount of sediment transport over the original dune surface. Planar cross-bedded gravels and sands can also form by the growth of small deltas from bars into deep channels, but this is thought to be rare. Gravel foresets are common in the *megadunes* commonly produced in river channels fed by the large discharges associated with glacier lake outburst floods. For example,

Carling et al. (2002) described gravel dunefields in the Chuja-Katun river system in the Altai Mountains of southern Siberia, in which individual dunes are 20 m high and 300 m long in the direction of former flow, and comprise cross sets of well-sorted open-framework cobbles and bimodal fine gravels. The attitude of the bedding provides information on depositional process within the dunes, with steeper beds produced by lee-side avalanching and more shallow-angled beds representing deposition in bedload sheets (cf. Carling, 1996).

Trough cross-bedded gravels and sands form by the migration of three-dimensional dunes or two-dimensional dunes with scalloped lee faces (Miall, 1977; Ashley, 1990). The cross beds have curved surfaces, concave-up and concave downstream, reflecting deposition in lee-side hollows. The maximum dip of cross beds may reach 30°. Sets typically range from 0.2 m to 3 m in thickness for gravels, and 0.05 m to 0.6 m for sands (Miall, 1977). Trough cross bedding may occur in solitary scoops eroded into other facies, or form cosets (multiple sets) of cross-cutting units called *festoon cross bedding*.

Planar and trough cross beds can be cross-cut by low-angle erosion surfaces called *reactivation surfaces* (fig. 10.30; Boothroyd and Ashley, 1975). These surfaces, which may be planar, curved or irregular, record reworking of dune slipfaces, often during lower stages of flow. Renewed cross-bed deposition above the reactivation surface indicates a re-establishment of downcurrent dune migration. Gravel foresets separated by reactivation surfaces are indicative of repeated high discharges in a river system (Russell and Marren, 1999).

Measurement of the direction of dip of cross beds in planar and trough cross-bedded gravels and sands is a useful indicator of palaeoflows. It should be remembered, however, that in braided or meandering rivers, bedforms migrate at a variety of angles to the main trend of the river, and that reconstructions based on a few measurements may give a misleading view of former river flow directions. Large samples typically show a broad scatter of palaeoflow directions, the centre of which gives a good indication of the general direction of flow.

10.4.5 ANTIDUNES

Antidunes occur on subaqueous beds as low-amplitude sinusoidal waves which are in phase with water surface waves. This is because the surface of supercritically flowing water directly affects bed shape (Alexander et al., 2001). Sediment facies deposited by the migration of antidunes can form *foresets* or *backsets*, depending on the direction of bedform migration (Allen, 1982, 1985; Rust and Gibling, 1990). Backsets typically dip upstream at a shallow angle, and provide the clearest diagnostic feature of antidunes. Additionally, downstream dipping surfaces may have smaller bedforms superimposed on them, a situation which cannot occur on dunes. The term 'antidune' is not strictly appropriate because it implies that the bedforms migrate upstream; moreover they are not really dunes, but the term seems to be firmly established in sedimentology.

10.4.6 SCOUR AND MINOR CHANNEL FILLS

Scours are overdeepened parts of river beds related to disturbances of the current around dunes or other obstructions. They are elongate parallel to flow, typically asymmetric transverse to flow, and may be up to 0.45 m deep and 3 m wide (Miall, 1977). The subsequent infilling of scours with sand, gravel or pebbly sands produces cross-bedded facies similar in some respects to trough cross beds formed by dune migration. However, scour fills show more internal variation, and sometimes display minor sedimentary structures such as ripple cross

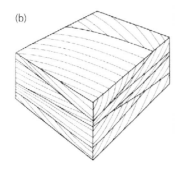

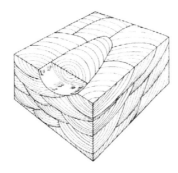

Figure 10.30 Sand and gravel cross-bedding. (a) A complex sequence of cross-bedded sands and gravels in glacifluvial outwash, Kells, Ireland. (D.J.A. Evans) (b) Block diagrams showing (left) planar and (right) trough cross-bedding. Bold lines denote reactivation surfaces (Reinick and Singh, 1980, reproduced with permission of Springer-Verlag).

lamination and parting lineation on bedding planes, indicating that the infill does not consist of lee-side slipfaces. Scour erosion and infill may be quite widely separated in time, unlike the development of erosional and depositional surfaces in festoon cross bedding.

10.4.7 GRAVEL SHEETS

Gravel sheets form by clast-by-clast accretion on low-relief parts of riverbeds, such as longitudinal and bank-attached bars (Miall, 1977; Collinson, 1996). Sediment movement and deposition are strongly episodic, occurring during floods and subsequent waning flows; between floods, gravel sheets are commonly exposed above water level. Typical dimensions for individual depositional units range up to 1 m in thickness and tens to hundreds of metres in lateral extent.

Gravel sheets consist of massive to crudely bedded clast-supported gravels (fig. 10.31; Miall, 1977). The clast-supported framework reflects grain-by-grain deposition from bedload. During this process, fine gravel particles can become trapped in the spaces between large clasts, resulting in a poorly sorted deposit, although crude horizontal stratification can develop during fluctuating flows. Clusters of large clasts sometimes occur, due to the trapping of clasts upstream of obstructions, in a process similar to the development of clast clusters in lodgement tills (Brayshaw, 1984). Clast accumulation in gravel sheets can result in fining-up units as the deposit builds to shallower water levels, but coarsening-up units can also occur as the result of downstream bar migration (Miall, 1985). Gravel sheets can be openwork, but fine matrix commonly occupies the spaces between clasts, due to filtering of sand or silt into the framework following deposition. Clast fabrics typically show upstream imbrication of a–b planes, and a preferred orientation of a-axes transverse to flow (Boothroyd and Ashley, 1975; Rust, 1975). This fabric represents the most stable alignment of pebbles beneath flowing water, maximizing the strength of the bed and minimizing the traction, and is widely regarded to be a good palaeocurrent indicator.

Giant bars are common in landscapes affected by glacier lake outburst floods, where they display large-scale gravel cross bedding or even foresets on their slipfaces. Upward-coarsening gravel sequences, documenting increasing flow competence during the rising limbs of floods, are also common in GLOF-fed rivers (Marren, 2005). Carling et al. (2002) described sheets of pebble gravels interspersed with cobble and boulder beds in the proximal parts of bars in the Chuja-Katun river system in the Altai Mountains of southern Siberia. Where giant bars have been constructed in the upper parts of tributary valleys to the main flood course, they comprise self-similar stacks of down tributary thinning 'rhythms'. These *gravel rhythmites* comprise proximal pebble and

Figure 10.31 Gravel sheets in Late Devensian glacial outwash, Lackford, south-east England. Note the intact mammoth tusk protruding from the gravel and to the right of the trowel, attesting to short travel distances (D.J.A. Evans).

Figure 10.32 A gravel rhythmite (up to 8 m thick) in the Komdodj giant bar, produced by catastrophic glacial lake flooding at the confluence of the Chuja–Katun river system, Altai Mountains, Siberia (P.A. Carling)

granule sheets that fine distally into cross-bedded coarse sands and granules (fig. 10.32).

10.4.8 SILT AND MUD DRAPES

During waning flows, pools of stagnant water are commonly left in abandoned channels, allowing fine suspended sediment to settle out. This sediment forms drapes of mud, silt and sometimes very fine sand blanketing the underlying deposits (fig. 10.33; Miall, 1977). Drapes will tend to be thickest in the middle of the pool, where the water is deepest, forming thin, concave-up lenses of variable lateral extent. Typical thicknesses range from a few millimetres to a few centimetres. Beds are commonly massive, but can be laminated if sediment delivery occurred in pulses. Fluvial mud drapes should not be confused with glacimarine mud drapes (discussed in section 10.6).

Figure 10.33 Sandy and silty drapes in a vertical sequence of gravel sheets, Alberta, Canada (D.J.A. Evans).

Figure 10.34 Hyperconcentrated flow deposit with intra-clasts, Skeiðarársandur, Iceland (A.J. Russell).

10.4.9 HYPERCONCENTRATED FLOW DEPOSITS

The deposits of hyperconcentrated flows (section 9.4.3) are intermediate between stream flow and debris flow end members (Mulder and Alexander, 2001; Benvenuti and Martini, 2002). They form sheets or channelized lenses, commonly tens of centimetres to a few metres thick and several metres to tens of metres across. Particularly extensive deposits, hundreds or thousands of metres across, can be formed by GLOFs (Lord and Kehew, 1987; Maizels, 1989a, b; Russell and Knudsen, 2002; Marren, 2005). In general, deposits consist of matrix-supported gravels or diamictons often containing *rip-up clasts*, reflecting mass transport and rapid deposition. Rip-up clasts may be composed of a variety of materials derived from materials eroded by flood waters, and may be transported and deposited in a frozen or unfrozen state; diamicton intra-clasts, ripped up from local tills, are particularly common in glacifluvial flood deposits (e.g. Russell and Marren, 1999; Russell and Knudsen, 2002). Hyperconcentrated flow units may be entirely massive or display internal structures, such as crude stratification and basal inverse- or inverse-to-normal grading (fig. 10.34; Maizels, 1989a, b; Todd, 1989). Stratification can consist of crude horizontal or cross bedding, and probably records pulsed deposition from surges in the flow (Maizels, 1989b). There is commonly an upward increase in textures, structures and sediments indicative of fluvial processes, such as better sorting, well-developed bedding and cross bedding, and winnowed gravels. These characteristics are thought to result from the transition from hyperconcentrated to normal stream flow during falling discharges, whereupon formerly suspended sediment is reworked as bedload (Maizels, 1989a, b). Clast fabrics in the massive parts of hyperconcentrated flow deposits tend to show preferred a-axis orientations parallel to flow and up-flow a-axis imbrication, characteristics typical of sheared, concentrated sediment-water mixtures (Todd, 1989).

10.5 GRAVITATIONAL MASS MOVEMENT DEPOSITS AND SYN-SEDIMENTARY DEFORMATION STRUCTURES

10.5.1 OVERVIEW

The abundance of steep, unstable slopes in glacial environments makes gravitational mass movements very important agents of sediment reworking and deposition. Gravitational processes operate in subglacial cavities, on the surface and around the margins of glaciers, and on proglacial and subaqueous debris slopes. Deposits formed by gravitational mass-wasting processes are usually classified according to the geometry of the moving mass, the velocity and mechanism of movement, water content during transport and deposition and the location (subaerial or subaqueous) of deposition (Prior and Bornhold, 1990; Mulder and Alexander, 2001; fig. 10.35). Five basic types of deposit are described in the following sections:

(1) *fall deposits*, composed of particles that have fallen, rolled, bounced or slid down steep slopes;
(2) *gelifluction deposits*, formed by the slow downslope movement of seasonally frozen debris;
(3) *slump and slide deposits*, formed by the sliding of intact blocks of sediment or rock, with varying amounts of internal deformation;
(4) *debris flow deposits/sediment gravity flows*, resulting from the flowage of concentrated sediment-water mixtures;
(5) *turbulent flow deposits* or *turbidites*, deposited from turbulent underflows below standing water.

It should be emphasized that, in nature, mass movements commonly undergo transformations from one type to another during transport. For example, slumps can evolve into debris flows if the component debris becomes liquefied and disaggregated; and subaqueous debris flows can partially or completely transform into turbulent

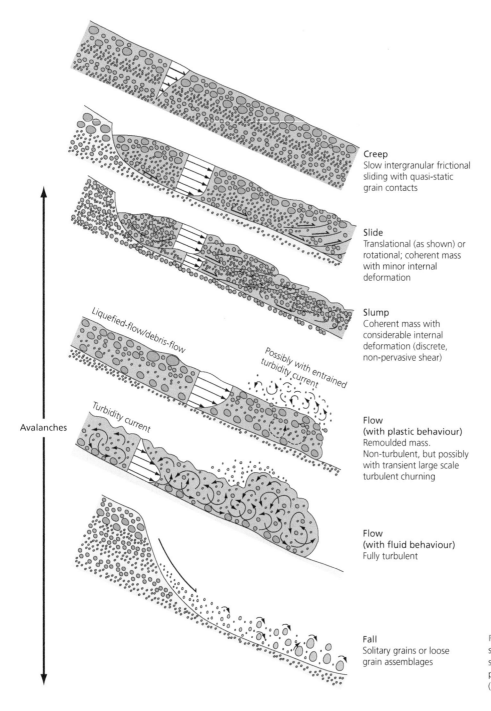

Creep
Slow intergranular frictional
sliding with quasi-static
grain contacts

Slide
Translational (as shown) or
rotational; coherent mass
with minor internal
deformation

Slump
Coherent mass with
considerable internal
deformation (discrete,
non-pervasive shear)

Flow
(with plastic behaviour)
Remoulded mass.
Non-turbulent, but possibly
with transient large scale
turbulent churning

Flow
(with fluid behaviour)
Fully turbulent

Fall
Solitary grains or loose
grain assemblages

Avalanches

Liquefied-flow/debris-flow

Possibly with entrained
turbidity current

Turbidity current

Figure 10.35 Range of gravity-driven
sediment transport processes on
subaqueous slopes. The velocity
profiles are schematic and not to scale
(Modified from Nemec, 1990).

underflows due to sediment mixing with the overlying water column. As a result of transformations during movement, mass-movement deposits may exhibit lateral and vertical changes in structure, grain size, sorting and other properties, which can be used to reconstruct the sequence of events during transport and deposition. Another important point is that debris can undergo several cycles of movement and redeposition by gravitational and other processes. This is particularly true in ice-contact environments where the topography is subject to major changes due to ice melt and advance, and for steep subaqueous slopes which may undergo alternating phases of oversteepening and failure. Such multiple reworking episodes can form complex successions that require careful study for their correct interpretation; some examples are described in chapter 11.

10.5.2 FALL DEPOSITS

Fall deposits form by the accumulation of debris at the foot of slopes, following transport by falling, rolling, sliding and bouncing down the slope. Fall deposits can be predominantly coarse-grained or diamictic, and can accumulate by numerous low-magnitude events or single high-magnitude events. In this book, diamictic deposits are termed *debris fall deposits*, coarse fall deposits constitute *rockfall deposits* or *talus*, and very coarse deposits from single high-magnitude rockslope failures are *rock avalanche deposits*. Fall deposits can be deposited in subaerial and subaqueous environments, and to a much more limited extent in subglacial cavities. The characteristics of fall deposits tend to differ in each of these three types of environment, and are discussed in turn below.

Subaerial fall deposits

The extent and geometry of subaerial fall deposits is very variable, depending on debris supply and the position of deposition. At the base of ice and snow slopes, deposits tend to accumulate in cones and wedge-shaped bodies, which may be only tens of centimetres thick where debris supply is limited. Where debris supply is high, talus deposits below ice and snow slopes can be several metres thick and coalesce into ramparts tens or hundreds of metres across (Lawson, 1981b, 1989; Evans, 1989b; Ballantyne and Harris, 1994). Extensive fall deposits can also form below debris slopes, such as steep-sided moraines, recently deglaciated slopes and unglaciated hillsides, and take the form of cone-shaped or sheet-like bodies, depending on the nature of the source area (Ballantyne and Harris, 1994). *Talus* is a loose, clast-dominated deposit, with varying amounts of void space and interstitial matrix (Hinchliffe and Ballantyne, 1999; fig. 10.36). In *openwork* deposits, matrix material is completely absent, although this is typically only the case for near-surface parts of talus accumulations. Deposits may be massive and structureless, but crude stratification and vertical size-sorting may be present. Particle morphology is commonly typical of periglacially weathered or passively transported debris, with high percentages of angular and slabby or elongate clasts, although this is not exclusively the case, and talus derived from glacier ice or pre-existing sediment can display a variety of clast morphologies. Fabrics can be isotropic, particularly in small talus bodies (Lawson, 1979b, 1989), but more commonly have some degree of preferred orientation, parallel or transverse to the depositional slope. Strong downslope preferred orientations develop where clasts slide down the depositional surface prior to deposition, or where post-depositional shearing of the deposit occurs; whereas more isotropic fabrics arise where particles interlock with a rough talus surface at varying angles (fig. 10.37; Benn, 1994b). Talus and avalanche deposits often display marked *fall sorting* or downslope coarsening, due to the greater momentum of large particles and their consequent tendency to travel farthest. *Debris fall deposits* have been described by Lawson (1979a, 1981a, 1989), who called them 'slope colluvium' and 'ice-slope colluvium'. The deposits are typically chaotic, consisting of unsorted and

Figure 10.36 A talus slope in the Torngat Mountains, northern Labrador, Canada (D.J.A. Evans).

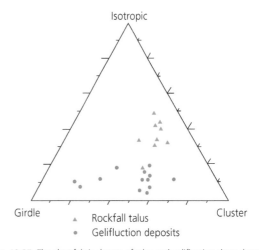

Figure 10.37 The clast fabric shapes of talus and gelifluction slope deposits (data provided by C.K. Ballantyne)

texturally diverse materials, sometimes enclosing intact blocks of pre-existing sediment. The action of running water between fall events can form inter-beds and lenses of sorted sediments and winnowed lag horizons.

Subaqueous fall deposits

Wherever subaqueous slopes are very steep, sediment may be subject to rapid downslope movement by *debris fall*, similar to rockfall in terrestrial environments. The term *grain fall* is also used in the literature, although Nemec (1990) has suggested that it be avoided because of its use in reference to suspension sedimentation. Debris may fall as single particles or as masses of strongly dispersed particles; each particle is driven in a downslope direction by its own momentum, and will bounce, slide or roll when impacting the bottom. Therefore, debris fall sediments will display a downslope coarsening similar to the fall sorting that occurs on terrestrial scree slopes (fig. 10.38). In addition, a fining-upward or normally graded sequence may be produced by a debris fall due to the tendency for larger particles to be deposited first and then to be overrun by the finer-grained tail of the avalanche.

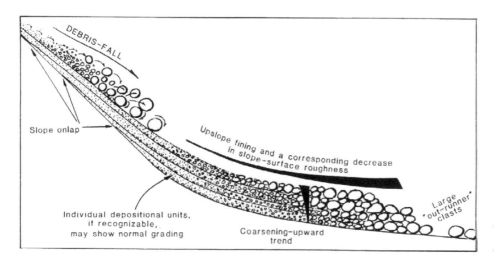

Figure 10.38 Schematic depositional characteristics of debris falls (Nemec, 1990, reproduced with permission of Blackwell)

Fall deposits can show up very clearly on side-scan sonar images of present-day subaqueous slopes as lobate, fall-sorted spreads of coarse debris (Nemec, 1990; Prior and Bornhold, 1990). They can be difficult to identify in section in ancient deposits, mainly because debris falls tend to become enveloped by later fine-grained sediments, causing them to resemble diamictic debris flows. Diagnostic criteria might be isolated clasts overlain by layered drapes of fine sediment, indicating that the fine drape was deposited after the clasts, or evidence that large clasts have ploughed through the underlying sediment before coming to a halt. Large subaqueous fall deposits are often referred to as *olistostromes*.

10.5.3 SLIDE AND SLUMP DEPOSITS

As is the case in subaerial environments, sediment on subaqueous slopes can fail along internal shear planes, and undergo downslope transport as *slumps* or *slides* (Prior and Bornhold, 1988; Nemec, 1990). Slide sheets undergo variable amounts of internal deformation, and may disaggregate during transport and evolve into debris flows. They are identifiable on images of present-day sea and lake floors from transverse compressional ridges on the slide surface, which may mark the surface trace of dipping shear faults within the slope.

Slides and slump deposits form sheet-like masses around 10 cm to >100 m thick. In slides, the internal beds are mainly undisturbed, but may exhibit compressional or tensional deformation near the toe and head zones, respectively. Slumps display more extensive evidence of internal deformation, consisting of overfolds, box folds, imbrication of soft clasts and branching thrust faults (Allen, 1982). In large slumps and slides, the style of internal deformation can be very similar to that caused by proglacial glacitectonics, and correct interpretation will require consideration of the surrounding facies, palaeo-slope and surface topography. Slumps commonly generate

debris flows, either in the form of local, small-scale flows, or large flows resulting from the large-scale transformation of the whole mass. Where subaqueous mass movement disrupts sediment but does not result in its long-distance transport and disaggregation, it produces what is known as gravity-induced soft sediment deformation. This can lead to the production of sediment melange, consisting of flow structures and attenuation features such as boudins, similar to those found in glacitectonite. Internal deformation structures in subaqueous slides and subglacially deformed materials have been compared by Hiemstra et al. (2004), based on the analysis of thin sections through Alaskan fjord sediments. In slide deposits, cumulative strains decrease upwards from the base, as recorded by decreasing angles in conjugate shear planes. In contrast, in a typical subglacial situation, the highest cumulative strains are towards the top of the sediment package (fig. 10.39).

10.5.4 DEBRIS (SEDIMENT-GRAVITY) FLOW DEPOSITS

Debris flow deposits are formed by the gravitational flowage (*sediment-gravity flows*, *density flows* or *mass flows*) of sediment-water mixtures. They are important in glacial sedimentological terms because they produce diamictons that are often difficult to differentiate from subglacial tills, especially as the flowage process can mix heterogeneous sediment sources to create a continuum from stratified to massive diamictons. This process of homogenization has been likened to a cement mixer by Eyles and Eyles (2000; fig. 10.40). Flowage can take place in both subaerial and subaqueous environments. Subaqueous flows can consist of debris introduced directly into standing water from ice margins or melt-streams, or material remobilized by subaqueous slope failures. In subaerial flows, movement occurs when the stresses imposed by the weight and surface gradient of the

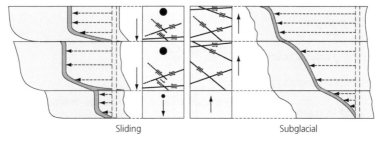

Sliding   Subglacial

Figure 10.39 Conceptual model of the deformation patterns in subaqueous slides compared to subglacial deforming layers. The strain marker prior to sliding/deformation is outlined by dashed lines and yellow shade. Dashed arrows point to its post-sliding/deformation shape and position (brown solid line). The resulting conjugate shear planes are presented in the adjacent columns, with the increasing trend in angles between planes, and therefore trends in increasing deformation intensity, depicted by solid black arrows. Note the reversal of trends. Two discrete failure planes have been extended into the subglacial case to facilitate comparison (Hiemstra et al., 2004)

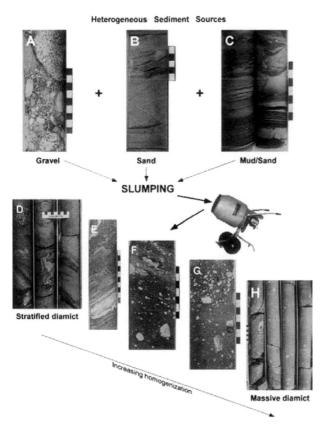

Figure 10.40 The 'cement-mixer' analogy for the generation of diamictons by downslope debris flow, based on the interpretation of Lower Permian glacigenic facies, Western Australia. (A–C) Various sediment sources feeding material to slumps. (D) Well-stratified diamicton of crudely intermixed and folded gravel, sand and mud. (E) Well-stratified diamicton with folding of sand- and mud-rich units. (F) Stratified diamicton with greater degree of homogenization of sand- and mud-rich sediment. (G) Weakly stratified diamicton with clast-rich horizon. (H) Massive diamictons in which complete homogenization of source materials has occurred (Eyles and Eyles, 2000)

material exceed its yield strength. High stresses and flow mobilization may be triggered by high deposition rates, slope oversteepening by ice-push or erosion, iceberg calving, wave impacts or even earthquake shocks. In glacilacustrine and glacimarine settings, mass flows result in the rapid, strongly episodic transfer of large amounts of debris, which can build up thick sediment successions. Flows can also be erosive, scouring out pre-existing sediment and depositing it further downslope.

In many texts, the term *flow till* is used to refer to subaerial sediment flows deposited in direct association with glacier ice or from freshly deposited till (e.g. Sugden and John, 1976; Dreimanis, 1989; Hambrey, 1994). Similarly, the terms *subaquatic flow till* and *submarine flow till* (Dreimanis, 1989) have been employed for subaqueous flow deposits. However, because processes and products of debris flow can be identical in glacial and non-glacial environments, it is preferable to employ terms that emphasize mechanisms of flow transport and deposition, rather than the origin of the sediment or the location of the flow (Lawson, 1979a, 1981a, 1989; Nemec, 1990). In this book, the general term *debris flow deposit* is used in preference to 'flow till' with its environmental implications. The prefix *glacigenic-* may be added where direct association with glacier ice can be demonstrated (Lawson, 1979a).

Subaqueous mass flows can exhibit a range of flow behaviour depending on their grain size distribution, density, matrix strength, buoyancy, pore pressure and turbulence (Nemec, 1990; Mulder and Alexander, 2001). An important distinction can be made between *debris flows*, which are high-concentration plastic slurries, and *turbidity currents* (or *fluidal flows*), which are rapidly moving turbulent underflows. These two types can be further subdivided according to the dominant flow mechanism, although it should be recognized that one flow type can evolve into another due to changes in water content and strength during flow (Prior and Bornhold, 1989; Nemec, 1990).

Because subaerial and subaqueous debris flow deposits commonly have distinct characteristics, due to the role of the water column in modifying subaqueous flows (fig. 10.41), they are described separately below. We also subdivide subaqueous flows and their deposits into cohesive debris and cohesionless varieties, and provide separate coverage of turbidity currents and turbidites. The distinction between *cohesive* and *cohesionless* subaqueous flows is important, because the presence or absence of matrix cohesion strongly influences flow behaviour and the character of the resulting deposits (Nemec, 1990; Mulder and Alexander, 2001).

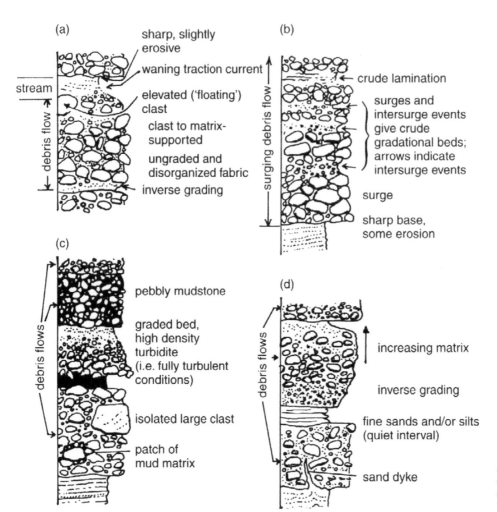

Figure 10.41 Examples of different structures resulting from debris flows in subaerial (a, b) and subaqueous (c, d) settings (after Nemec, 1990)

Subaerial debris flow deposits

Subaerial sediment flow deposits are usually, but not always, diamictons with a fine-grained matrix (clay, silt or sand) and variable clast content. Bulk grain size distributions are essentially those of the parent debris, due to the unimportance of sorting and winnowing during transport. Because inter-particle contact forces are low during flowage, particle morphology is inherited with little or no modification from the source material, except for fragments of very weak rocks such as shale. Particle morphology may be characteristic of actively or passively transported glacigenic debris, water-worn sediment or some mixture, depending on the source.

Deposits form lobate sheets or channelized lenses, generally a few centimetres to 2 m or more thick, and a few square metres to hundreds or even thousands of square metres in area. Sequences of flows can build up to over 10 m thick and cover areas of several tens of thousands of square metres (Lawson, 1979a). In section, individual flows form tabular or lens-shaped units, and if flows are channelized, units can have concave-up, erosional bases and flat tops (fig. 10.42). Where successive flows are similar in texture, individual units may be difficult to distinguish, but boundaries are commonly marked by basal concentrations of clasts, upper washed horizons, inter-beds of silt, sand or gravel, or more subtle bedding structures. At first acquaintance, sediment flow deposits can appear totally chaotic, but close examination often reveals a surprising amount of internal organization which can be used to reconstruct the processes of deposition in some detail. This has been demonstrated at the micro-scale by Lachniet et al. (2001), Menzies and Zaniewski (2003) and Phillips (2006), who have identified many features similar to those produced in tills, but also some structures that are particularly diagnostic of subaerial debris flows. For example, normal faults and dewatering structures are very common in debris flow thin sections, particularly *tiled structures* produced by the pulsed movement of fine silts and clays (cf. Menzies, 2000; Carr, 2004). A certain amount of caution needs to be emphasized here, however, as some thin section analyses of tills have highlighted a dominance of water-escape structures over shear indicators (Benn et al., 2004; Evans and Hiemstra, 2005).

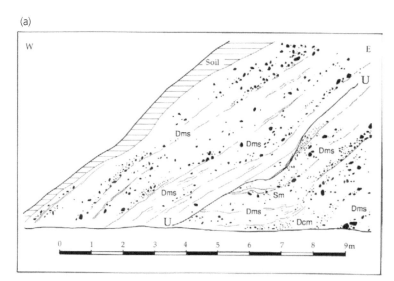

Figure 10.42 Subaerial debris flow deposits, Isle of Skye, Scotland. (a) Sketch of a section through a moraine, showing stacked debris flows. Note the basal clast clusters, planar and channelized contacts and unconformity 'U'. (b) Photograph of part of the same section (D.I. Benn).

The internal structure of sediment flow deposits is strongly related to the water content during flow and deposition. Four basic types of debris flows were defined by Lawson (1979a, 1982). *Type I flows* have a low water content (8–14 per cent by weight) and tend to behave as a rigid or semi-rigid 'plug' rafted along on a thin basal shear zone. They form clast-rich, poorly sorted deposits that can incorporate blocks of pre-existing sediments (fig. 10.41). The only visible structures may be zones of short, curving shear planes near the base, and thin silt or sand horizons near the top, reflecting post-depositional washing by water. Concentrations of large clasts may be present at the margins of the flow.

Type II flows have a water content of around 15–20 per cent, and consequently have lower shear strength. Active flows tend to occupy channels that are partly eroded by the flow and partly bounded by levees of coarse, bulldozed sediment, and run out into pronounced depositional lobes. The resulting deposits typically have concave-up lower surfaces in transverse section, with a massive, poorly sorted interior underlain and flanked by more organized, sheared sediment. The sheared horizons tend to show distinct size sorting, with concentrations of clasts at the base due to sinking of coarse particles through saturated, liquefied matrix, and their subsequent transport as bedload (fig. 10.41). The sediment immediately above such clast concentrations can be depleted in coarse material relative to the rest of the deposit, and may be inversely graded. Curving shear surfaces and streaked out smears of silty clay may be present. Washed horizons

commonly occur at the flow top, as the result of surface water flow and ponding following deposition.

Type III flows have yet higher water content during flow (18–25 per cent) and still lower shear strength. Shear occurs throughout all or most of the sediment mass, with only poorly developed plug zones. Due to the low sediment strength, flows tend to be thin (generally <0.5 m), rapidly moving and very erosive, cutting channels in pre-existing sediments. Flow is often turbulent, with sediment pulses sloshing irregularly down channels and feeding lobate terminal fans. Type III deposits resemble the lower, sheared part of Type II flows, with pronounced basal clast concentrations. There are often marked lateral and longitudinal variations in the thickness of deposits and basal clast concentrations, reflecting discharge fluctuations or *surges* in the parent flows (fig. 10.41). Pulsed flow can also result in a crudely stratified deposit, with inter-bedded diamicton, clast concentrations, and washed horizons. Individual particles can be thicker than the whole deposit, projecting above the flow surface.

Type IV flows have the highest water content (>25 per cent) and very low strength. Active flows are rapidly moving slurries occupying narrow channels and feeding broad, thin terminal lobes, but because water is abundant during their formation, they are often reworked by stream flow. The resulting deposits consist mainly of silt and sand, sometimes with thin basal layers of coarse sand or fine gravel, because flows with high water content cannot support clasts due to their very low matrix strength.

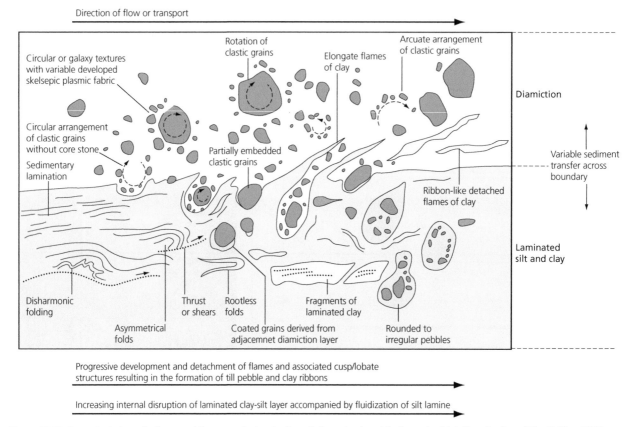

Figure 10.43 Conceptual schematic diagram of the range of microstructures that can develop at the base of a debris flow diamicton (After Phillips, 2006).

Individual depositional units are thin (0.02–0.1 m), but can build up thick sequences of thin laminae.

A micro-morphological study of the basal structures from a debris flow deposit developed in glacifluvial materials has been undertaken by Phillips (2006), who provided an idealized diagram of the range of microstructures that can be visible in thin section (fig. 10.43). Common structures in the substrate are folds, thrusts and shears, and these can be associated with 'rotated' to slightly attenuated diamicton pebbles, which are derived from the overlying debris flow. The contact between the debris flow and the substrate can be marked by elongate 'flames' of the substrate material separating lobate or pendant structures of the debris flow diamicton, which are progressively tilted downflow. The base of the debris flow contains detached 'flames' or ribbons of the substrate material and indicators of rotational deformation, such as circular, arcuate and galaxy-like grain arrangements. The latter have frequently been cited as diagnostic criteria of subglacial deformation (van der Meer, 1993; Menzies, 2000; Khatwa and Tulaczyk, 2001), a concept clearly disputed by Phillips's (2006) findings.

Pebble fabrics in sediment flow deposits are very variable, and display a wide range of isotropy and elongation values (fig. 10.44; Lawson, 1979a, b; Mills, 1991; Benn, 1994b). Fabric shape and preferred orientations are dependent on patterns of strain within the parent flow.

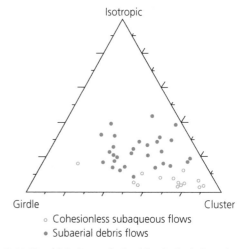

Figure 10.44 Clast fabric shapes of subaerial and cohesionless subaqueous debris flow deposits (Subaerial data from Lawson, 1979a; Mills, 1991; subaqueous data provided by F. Massari).

Fabrics in non-deforming plug zones can be isotropic, but well-organized fabrics can develop in zones of intense shearing as particles are rotated by the deforming matrix (Lawson, 1979a). Particle a-axes and a–b planes can be aligned parallel to the frontal and lateral margins of flows, where flow is compressive, whereas a-axis fabric maxima will tend to be parallel to flow below the central parts, where flow is extending (fig. 10.45; Boulton, 1971; Owen, 1991). In all cases, preferred orientations will

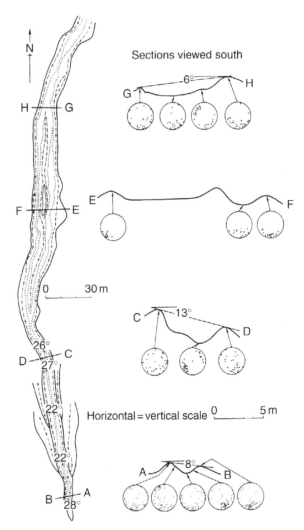

Figure 10.45 Clast fabric patterns in a subaerial debris flow, near the Batura Glacier, Pakistan (Owen, 1991, reproduced with permission of *Zeitschrift fur Geomorphologie*).

The range of deposits associated with each flow type is depicted on the right-hand side of this diagram, and we now review the main process–form relationships under the headings of cohesive debris flows (debris flow in fig. 10.46) and cohesionless or non-cohesive debris flows (hyperconcentrated density flows and concentrated flows in fig. 10.46). The turbidity type flows of figure 10.46 are then covered in section 10.5.5.

Cohesive debris flows

Cohesive subaqueous debris flows contain some clay matrix material which, when mixed with water, acts as a fluid with cohesive strength (Johnson and Rodine, 1984). Large clasts, boulders and blocks of pre-existing sediment can be rafted along at high levels of the flow, partly because of the cohesive strength of the matrix and partly because particles enveloped in a dense slurry have low buoyant weight. The upper part of the flow commonly consists of a semi-rigid 'plug' riding passively on a thin basal shear zone, as for Type I subaerial flows (see page 407; Lawson, 1981b, 1982). Unlike subaerial flows, however, the uppermost parts of subaqueous cohesive flows can become diluted during flow due to turbulent mixing with the overlying water column. This can cause transformation of part of the flow into a low-density, turbulent underflow, altering the characteristics of the final deposit.

The velocities of cohesive debris flows increase non-linearly with shear stress. Increasing shear stresses also cause thickening of the basal shear zone at the expense of the plug zone; and for high stresses (or low-strength flows) the plug zone may disappear, as for Type IV subaerial flows. Deposition occurs once the shear stress falls below the shear strength again and the basal shear zone thickness shrinks to zero, a process known as *freezing*. This has nothing to do with the transformation of water to ice, so when describing flow behaviour in glacial environments, the term 'freezing' should be used with caution.

Deposits of cohesive debris flows form sheet-like or lobate beds, with planar or slightly scoured bases. Individual units are typically decimetres to several metres thick, and a few metres to hundreds of metres in lateral extent. Very extensive debris flow deposits, thousands of metres long, have been identified on glacially influenced submarine slopes (Laberg and Vorren, 2000; Dowdeswell et al., 2008c; Ottesen et al., 2008a; Leynaud et al., in press). Such large-scale debris flows are often produced by the downslope disaggregation of subaqueous slumps and slides, and can be recognized by the presence of blocks of the substratum in the deposits.

Former plug zones in cohesive debris flow deposits are typically composed of massive, matrix-supported diamicton or muddy sands and gravels (fig. 10.47). Very large clasts and intact or remoulded soft-sediment rafts may be present (Ghibaudo, 1992). The upper parts of flows may pass vertically into normally graded and stratified

show no systematic relationship to regional ice flow directions (Lawson, 1979a, b).

Subaqueous debris flow deposits

For subaqueous debris flows, a fundamental distinction is made between *cohesive* and *cohesionless* flows, because the presence or absence of matrix cohesion exerts an important control on flow behaviour and the character of the resulting deposit (section 8.4.4; Nemec and Steel, 1984; Postma, 1986; Nemec, 1990). A similar distinction can also be made for subaerial flows, but flows that behave as cohesionless materials are unimportant in terrestrial glacial environments. It is important to recognize that there is a continuum of subaqueous debris flow types, ranging from cohesive debris flows through hyperconcentrated flows (see section 10.4.9) to turbidity currents (see section 10.5.5). A detailed classification scheme for this continuum has been presented by Mulder and Alexander (2001), based on physical flow properties and grain support mechanisms, and is replicated in figure 10.46.

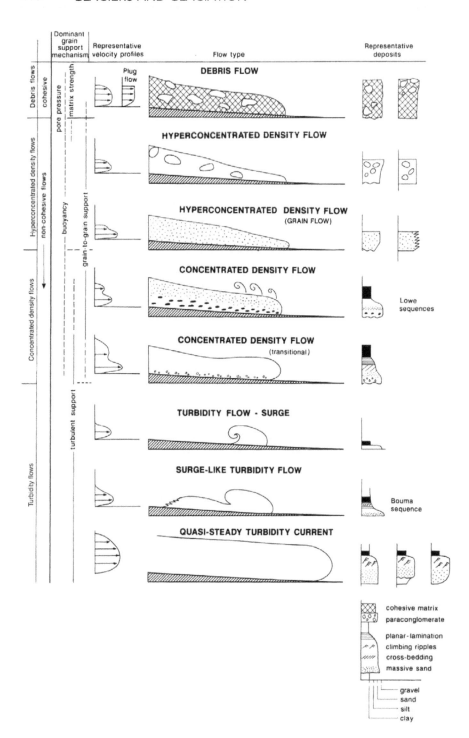

Figure 10.46 Classification of subaqueous sedimentary density flows, based on the criteria of dominant grain support mechanism, velocity profile, flow shape and schematic sedimentary logs (Mulder and Alexander, 2001).

gravels, sands and silts, deposited from turbulent flows derived from the debris flow (fig. 10.41). The relative thickness of the debris flow and turbulent flow components of such two-layer deposits is variable, and depends on the extent to which flow modification occurred. The deposits of turbulent flows (turbidites) are described in detail in section 10.5.5.

Basal sheared zones of cohesive debris flow deposits can be identified by the presence of attenuated wisps of clay or silt, shear planes and normal or inverse grading. Normal grading reflects the sinking of large particles

through weak, shearing sediment, whereas inverse grading is generally attributed to some combination of upward clast migration due to shearing and a sieving effect (these mechanisms are explained under *cohesionless debris flows*, below). Another possible mechanism for the development of inverse grading was proposed by Broster and Hicock (1985), who argued that normally graded beds on the upper surfaces of cohesive flows (formed by turbulent mixing) could be rolled over the nose of the flow, conveyor-belt fashion, to be redeposited upside down beneath the flow as an apparently inversely graded layer.

Figure 10.47 Subaqueous mass flow deposits. (a) Thin lenses of diamicton with silt partings, Loch Quoich. (b) Subaqueous graded mass flow deposits in the foresets of a glacilacustrine delta, Drymen, Scotland (D.J.A. Evans).

Cohesionless debris flows

Cohesionless debris flows are high-concentration, non-turbulent flows in which cohesion contributes little or nothing to sediment strength; they tend to occur in sandy and gravelly materials (Nemec and Steel, 1984; Postma, 1986; Nemec, 1990; Nemec et al., 1999). Grains are supported during transport by one or more mechanisms, including:

(1) *dispersive pressure*, or the upward component of the pressure exerted by neighbouring shearing grains;
(2) *buoyancy effects*, due to the density of the surrounding matrix which makes grains lighter than they would be in water;
(3) *liquefaction and fluidization*, or the upward movement of pore fluid during dewatering (Lowe, 1976a, b; Nemec, 1990).

Various subtypes of cohesionless debris flows have been defined on the basis of the dominant particle support mechanism during transport, including *grain flows*, *fluidized flows* and *liquefied flows*, although these mechanisms rarely work in isolation (Nemec and Steel, 1984). Cohesionless debris flows can originate by the failure of cohesionless materials on subaqueous slopes, or by the gravitational transport of sediment delivered to subaque-

ous slopes by stream flow or dumping from glacier margins (e.g. Nemec et al., 1999).

Liquefied flows are triggered by elevated pore-water pressures associated with transient shocks, such as iceberg calving events, earthquakes and sudden loading by overpassing sediment. These shocks will cause the grains to be suspended in fluid for a short period during which flow is possible. As grains begin to settle out and regain contact with each other after the initial shock, they displace pore water upwards. This means that a liquefied flow will 'freeze' from the bottom upwards. Commonly, the upward flow of water is not uniform and is concentrated in pipes, producing sand volcanoes on the surface, and ball and pillow structures within the sediments. The temporary suspension or upward movement of grains by escaping fluid is called *fluidization* or *dewatering* and is not synonymous with liquefaction, which refers to the downward settling of grains through the fluid (Lowe, 1976b).

The morphology of cohesionless flows generally consists of an upper erosive channel and a lower depositional lobe (Prior and Bornhold, 1990). On steep slopes, cohesionless flows may grade into debris falls (see page 575) and exhibit downslope coarsening or fall sorting due to the tendency for the larger particles with most momentum to travel furthest. Liquefaction and hydroplaning of mass flows on submarine continental margins can result in long run-out distances and the production of *contourites*, which are the deposits of bottom currents that hug lower contours.

Cohesionless debris flow deposits typically form dipping sheet-like or lobate masses, centimetres to metres thick. Depositional dips are variable, but are usually in the range of 10–37°, but can be as low as 3° for liquefied or fluidized flows. Bases are commonly erosive, and can form deep, channelized scours (chutes) aligned downslope (Prior and Bornhold, 1990; Carlson et al., 1992; Lønne, 1993). Incorporation of underlying material into eroding flows can result in gradational 'welded' basal contacts.

Deposits, often termed *debrites*, typically consist of gravel, sand or pebbly sand, and may be well or poorly sorted, depending on the source material (figs 10.41, 10.47). Flows fed by subaerial streams tend to be well sorted, reflecting the grain size characteristics of fluvial bedload, whereas flows fed directly by subglacial conduits or derived by reworking of heterogeneous sediments can contain a wide range of grain sizes (Mastalerz, 1990; Lønne, 1993). Torrential subaerial streams can also introduce poorly sorted material to subaqueous slopes (Martini, 1990). Ice-proximal flow deposits can contain an abundance of boulder-sized material.

Units may be massive or exhibit inverse or inverse-to-normal grading. *Inverse grading* is particularly common at unit bases, and is widely thought to result from two main processes. First, *dispersive pressure* generated by colliding grains tends to be greatest around large grains, which preferentially migrate upwards towards zones of lower strain rate (Bagnold, 1956). This process is still poorly understood, however, and disagreement exists about its effectiveness (Legros, 2002; Le Roux, 2003). Second, *kinetic sieving* causes small particles to fall down through the spaces between large ones. Shake a box of popcorn, a bowl of sugar or a large packet of crisps to see this effect in action. *Normal grading* is typically only weakly developed, because strong normal grading implies that grains were able to move freely enough relative to one another to allow differential gravitational settling, conditions usually associated with cohesive or turbulent flows. Particle fabrics in gravelly cohesionless debris flows commonly show preferred a-axis orientations

parallel to flow, and up-flow a-axis imbrication, resulting from particle rotation in a shearing medium (fig. 10.44; Rust, 1977). Imbrication angles are variable, and can increase or decrease upwards depending on patterns of strain in the flow during final deposition.

Density-driven deformation structures

Some sedimentary units are the product of density-driven flowage; their sedimentology, in addition to their deformation structures, reflecting the processes by which they have been emplaced in surrounding materials. Although such features as *ball and pillow, flame* and *dish* structures are routinely identified in dewatered subaqueous sediments (see section 10.2.2 and Evans and Benn, 2004b),

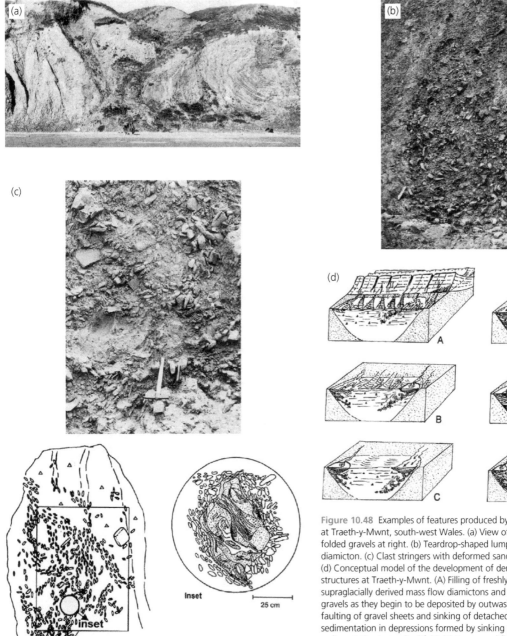

Figure 10.48 Examples of features produced by density-driven deformation at Traeth-y-Mwnt, south-west Wales. (a) View of cliff section with syn-formally folded gravels at right. (b) Teardrop-shaped lump of gravels in underlying diamicton. (c) Clast stringers with deformed sand raft (inset) in diamicton. (d) Conceptual model of the development of density-driven deformation structures at Traeth-y-Mwnt. (A) Filling of freshly deglaciated rock basin with supraglacially derived mass flow diamictons and onset of deformation of gravels as they begin to be deposited by outwash streams. (B) Density-driven faulting of gravel sheets and sinking of detached gravel rafts. (C) Lacustrine sedimentation in depressions formed by sinking gravels. (D) Syn-deformational sedimentation in synclinal fold. (E) Downward flexuring of gravels. (F) Distal sedimentation over the top of the sequence (from Rijsdijk, 2001).

few studies have been conducted on examples of large-scale loading and flowage in Quaternary glacial sediment sequences, and many have been misinterpreted, despite a substantial literature on the process–form relationships of density-driven structures (e.g. Anketell et al., 1970; Brodzikowski et al., 1987; Nichols et al., 1994; Hindmarsh and Rijsdijk, 2000). Such structures are initiated when high-density fluids overlie low-density fluids, setting up a *Rayleigh-Taylor instability*. The reversed density is essentially a store of potential energy that is expended by the sediments as they generate deformation across their boundary. This density-driven deformation results in the production of features such as diapirs, load casts, synclines and detached lumps.

An excellent example of such activity in ancient glacigenic sediments is reported by Rijsdijk (2001) and Hiemstra et al. (2005) from the site of Traeth-y-Mwnt in south-west Wales. At this site, a lower silty diamicton has been loaded by silt, sand, gravel and sandy diamicton to produce large-scale folds, flexures and faults, and smaller-scale rafts, pods, diapirs, sub-vertically inclined clast stringers, clast clusters and teardrop-shaped lumps (fig. 10.48). The conceptual model in figure 10.48d provides an explanation of the features at Traeth-y-Mwnt as density-driven deformation structures produced when gravels and stratified sediments were deposited over supraglacially derived mass flow diamictons that had accumulated in a bedrock basin.

10.5.5 TURBIDITES

In turbidity currents (also known as *underflows* and *fluidal mass flows*), particles are kept aloft within the flow body by turbulent suspension (Kneller and Buckee, 2000; Mulder and Alexander, 2001). The suspended particles render the flow denser than the surrounding water, causing it to move downslope along the bed. Turbidity currents are

subdivided into head and tail regions, the head being 1.5–2 times the thickness of the tail (fig. 10.49). The turbulent head moves downslope, mixing with the ambient water, and requires constant transfer of denser fluid from the tail to maintain momentum. This can be sustained for hours where the turbidity current is fed by dense underflowing meltstreams, but most flows eventually dissipate once the influx of suspended sediment from the initiating disturbance is exhausted.

Sediment is transported by turbidity currents either as *suspended load* or as *bedload* swept along by the over-passing density current. As flows decelerate, particles settle out of suspension and become part of the bedload or the static bed. The higher-settling velocities of larger particles mean that they will be deposited first, while finer material remains in suspension, resulting in *normal grading* or upward fining. Turbidity currents can also be erosive, scouring bottom sediments to form subaqueous channels, often bounded by levees.

Turbidity currents can be initiated in at least three different ways (Nemec, 1990):

(1) They may form directly from sediment-laden melt-water discharging from meltstreams or submerged conduit mouths, in the form of hyperpycnal underflows. Such underflows are important agents for transporting and depositing sediment in glacier-fed lakes, but are less common in glacimarine environments, because suspended sediment concentrations in meltwater inflows are not usually large enough to overcome buoyancy effects. In glacier-fed lakes, the sediment-laden inflowing water is often considerably denser than the lake water, and quasi-continuous underflows occur as a result. Such underflows are generated during maximum melt periods, when the thermal stratification of the lake water can be totally overwhelmed.

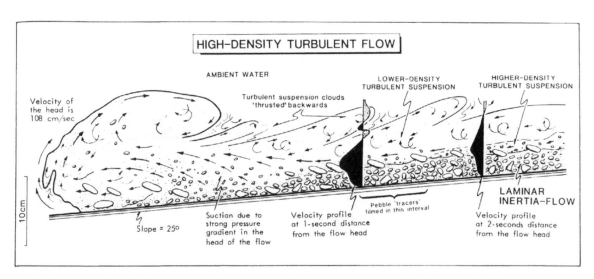

Figure **10.49** Main characteristics of a high-density turbidity current, drawn from a laboratory-generated flow (Postma et al., 1988, reproduced with permission of Elsevier).

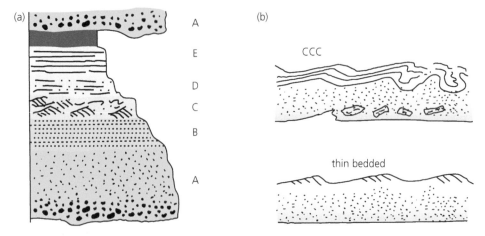

Figure 10.50 Subaqueous underflow deposits (turbidites). (a) Classic 'Bouma' turbidite sequence. Letter codes are explained in the text. (b) Examples of 'base cut-out' distal turbidites. Unit 'ccc' consists of climbing ripples, convolute lamination and rip-up clasts. The 'thin-bedded' unit consists of graded sands and starved ripples. (Walker, 1992). (c) Normally graded, ripple cross-laminated and convolute laminated gravelly sands and sands between gravelly mass flows from glacilacustrine bottomsets, Iceland (D.J.A. Evans).

(2) Turbidity currents may evolve less directly, by intense sediment fallout from suspension plumes, which can blanket large areas of subaqueous slopes with dense, mobile suspensions (Syvitski and Farrow, 1989). This situation is most common in proximal glacimarine environments where there is abundant material in high-level suspension (e.g. Nemec, 1995).

(3) They may evolve from the progressive downflow dilution of debris flows, due to turbulent mixing with the overlying water column. Such sediment failures are usually associated with high sedimentation rates at subaqueous ice-proximal or ice-contact depo-centres, such as fans and deltas.

Flow in case (3) will tend to occur in short-lived surges, whereas cases (1) and (2) range from brief surges to more or less continuous 'streams' flowing for hours or days (Nemec, 1990).

Deposition from turbidity currents happens as the flow slows down or becomes less turbulent, by a combination of settling from suspension and reworking as bedload beneath higher, still moving parts of the flow. These processes give rise to a general proximal-to-distal fining of turbidite beds, reflecting the rapid deposition of coarse material and the transport of finer material into deeper parts of the basin. They can also produce a characteristic graded vertical sequence within the bed, known as the *Bouma sequence* (Bouma, 1962; Shanmugam, 1997; Kneller and Buckee, 2000). The full Bouma sequence is, from the bottom up (fig. 10.50):

- (A) massive or normally graded sand or gravel.
- (B) planar-laminated sand;
- (C) ripple cross-laminated sand and silt;
- (D) interlaminated silt and/or clay;
- (E) massive clay or silt;

Division (A) is the product of rapid deposition from suspension of the coarsest material with little or no subsequent reworking. Parallel lamination (Division B) forms as sand is transported and deposited as bedload in the *upper flow regime* (section 10.4.1), driven by the flow above. The sand is derived either from the overlying current or reworking of Division A. Continued deceleration of the flow results in a transition from the upper to the lower flow regime, and a consequent change in bedforms developed at the base of the flow. Ripple cross lamination reflects the downflow migration of ripples during the reworking of Division B and/or the influx of new sediment from suspension. *Climbing ripples* (section 10.4.2) record net deposition of sediment, and *convolute lamination* can develop by the deformation of laminae during loss of excess pore water. Division D forms by the settling of suspended fines during the last stages of the flow, with lamination reflecting pulses in flow velocity. Division E represents the gradual settling of residual suspended load in quiet water conditions.

Many variations on this basic sequence can exist (Kneller and Buckee, 2000). Parts of the sequence can be missing or repeated, depending on flow conditions. High-concentration flows produce thick A and B units,

Figure 10.51 Coarse-grained rhythmites (turbidites) deposited in a slack-water embayment at the snout of Skeiðarárjökull, Iceland, during the 1996 jökulhlaup. Note the outsized cobbles deposited from suspension (A.J. Russell).

with subordinate or missing C to E units, whereas low-concentration flows can produce base cut-out sequences consisting of Divisions C to E only. Repeated divisions can form by pulsed flows or flows reversing their direction of travel after rebounding from topographic highs. 'Floating' clasts, much larger than the surrounding material, can occur in Divisions A and B. These are transported in the flow by gliding along boundaries between regions with contrasting density, in much the same way as a waterskier skims along the boundary between water and air (Postma et al., 1988). Outsized clasts are deposited at high levels if the surrounding flow rapidly immobilizes or 'freezes'.

Several studies of flood deposits in proglacial fluvial systems have identified rhythmic sequences that are best explained as coarse-grained turbidites, due to their similarity to Bouma sequences (e.g. O'Connor, 1993; Smith, 1993; Russell and Knudsen, 1999; Carling et al., 2002). In these settings, deposition is from sediment transport pulses that feed into slack-water areas during cataclysmic floods (fig. 10.51); they have been reported from giant fluvial bars, where they are more often referred to as *gravel rhythmites* (see section 10.4.7).

10.5.6 CLASTIC DYKES AND HYDROFRACTURE FILLS

Some sediment bodies may be injected with materials bursting out from confined aquifers (Nichols et al., 1994; Boulton and Caban, 1995; van der Meer et al., 1999; Le Heron and Etienne, 2005). In glacial environments, the discharge of subglacial groundwater can be impeded by

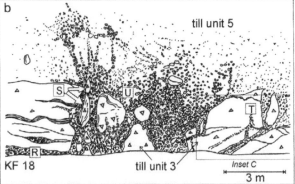

Figure 10.52 Example of hydrofracture fill features at Killiney Bay, Ireland. (a) Photograph and (b) sketch of section KF18, showing gravel dykes and burst-out features. R = lower gravels; S = streamlined till intra-clast; T = offshoot dyke or branch, running initially at low angle parallel to till joints, and then vertically; U = rotated till intra-clast). (c) Inset box marked in (b) showing details of offshoot dykes (Rijsdijk et al., 1999).

permafrost near the glacier margin, allowing artesian pressures to build up. Large pressure gradients can lead to *hydrofracturing* of pre-existing sediments, and the emplacement of *clastic dykes* and associated features (e.g. Dreimanis and Rappol, 1997; Evans and Ó Cofaigh, 2003). An illuminating study of complex gravel-filled clastic dykes injected into till at Killiney Bay, Ireland, was conducted by Rijsdijk et al. (1999). The dykes are rooted in an underlying gravel layer and extend vertically into overlying till, often ending in plumes or *burst-out structures* (fig. 10.52). They are aligned sub-vertically in the till

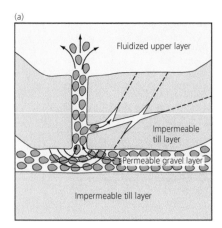

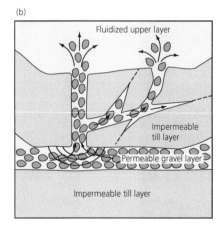

Figure 10.53 Explanation of processes and forms in hydrofracture filling. (a) Production of vertical fracture and dyke filling, with burst out and initiation of a branch by the dilation of pre-existing joints, due to hydraulic pressure along dyke walls. (b) Joint filling leads to tensile strength of overlying till being exceeded and the formation of vertical offshoots and small burst-out plumes (Rijsdijk et al., 1999).

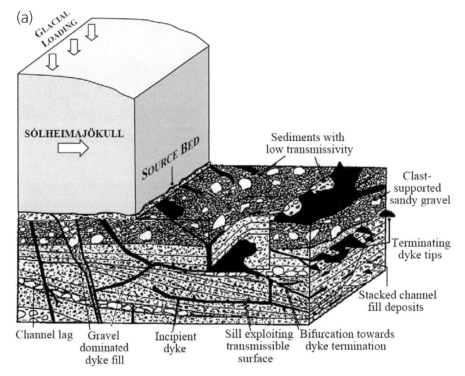

Figure 10.54 Clastic dyke swarm produced by the advance of the Icelandic glacier Sólheimajökull. (a) Schematic diagram showing the relationships between clastic dykes, host sediments, source sediments and glacier snout. (b) Geometry of cross-cutting gravel- and sand-rich dykes, illustrating the reduction in width with distance from source bed. (c) Sedimentology of a clastic dyke, revealing the laminated nature of the sediment fills and the occurrence of laminae symmetry, particularly in logs IV and VI. (d) Alternative models of clastic dyke formation when materials are non-fluidized (Le Heron and Etienne, 2005).

and may show evidence of infilled branches where the pressurized dyke fill has exploited pre-existing joints (fig. 10.53). The sediment infill of the dykes comprises poorly sorted coarse gravels with clast a-axes and a–b planes aligned parallel to the dyke walls. Till intraclasts also occur within the gravel infills; these are commonly streamlined and aligned parallel with the dyke walls. The tops of dykes and branches often end in funnel-shaped clusters of clasts in the overlying till. Rijsdijk et al. (1999) explained the features at Killiney Bay as the products of hydrofracturing of the till aquitard, due to rising water pressure in the underlying gravel aquifer. Fluidized gravels were then injected up to 7 m vertically along the fractures, ending up as burst-out structures in the

uppermost till; the burst-out features, particularly the fan-shaped gravel clusters, indicate that the uppermost till was initially supersaturated and clasts settled through it after bursting out.

Clastic dykes can also be produced by the downward injection of material, for example when a glacier advances over stratified deposits (Boulton and Caban, 1995). Excellent examples of such features were reported by Le Heron and Etienne (2005) from the foreland of Sólheimajökull, Iceland, where a three-dimensional, reticulate clastic dyke swarm was produced in proglacial sediment during a recent glacier advance. The dykes form a bifurcating system, dipping predominantly away from the former glacier terminus, which supplied the

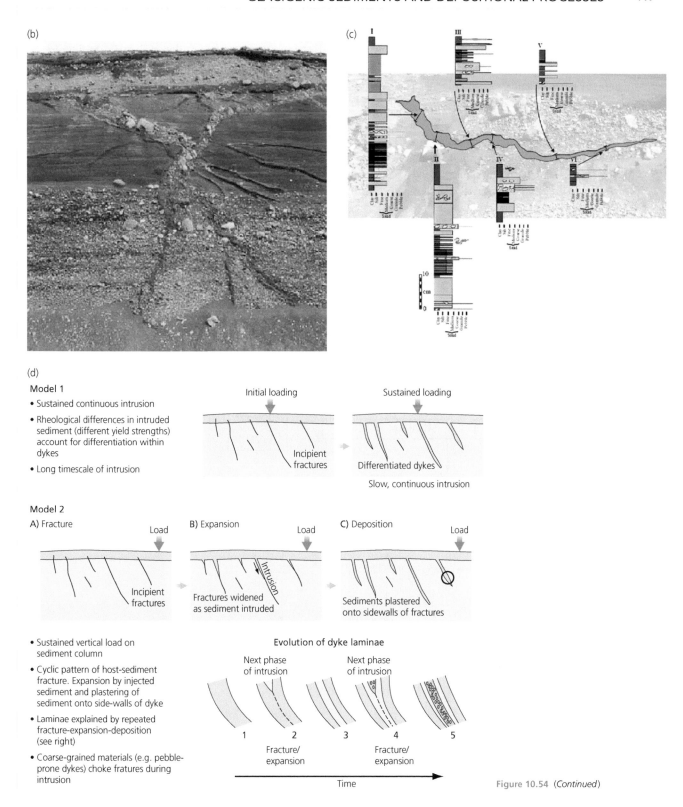

Figure 10.54 (*Continued*)

hydraulic head (fig. 10.54a). Wherever the dykes reach a transmissible bed, they branch laterally to form sills. The source bed for the dyke fills was the sandy subglacial till, which was driven into the hydraulic fractures as they opened up in the proglacial sediments, propagating away from the advancing load. The sediment infills were laminated and locally folded, which may indicate multi-phase hydrofracturing and accretion of injected sediment (fig. 10.54).

Given the important role of pressurized ground-water in glacial systems, it is likely that many more examples of hydrofracture fills exist in ancient glacial sediments, but most have been misinterpreted or simply overlooked.

10.6   GLACIMARINE AND GLACILACUSTRINE DEPOSITS

Sediment can enter glacimarine and glacilacustrine environments from several sources (fig. 10.55; Powell, 1991; Powell and Domack, 1995):

(1) *Supraglacial, englacial and subglacial debris* can be released directly into the water in ice-contact environments, either from glacier margins or calved icebergs.

(2) *Unfrozen subglacial sediment* can emerge from beneath glaciers at the grounding line.

(3) *Glacier meltstreams* can carry sediment directly into lakes or the sea from englacial and subglacial portals or supraglacial channels. Alternatively, meltwater can pass through proglacial channels before entering standing water.

(4) Sediment can also be delivered from non-glacial sources by rivers and slope processes. In this case, the sediment load may consist of paraglacially reworked glacigenic sediment.

The relative contribution of sediment from these sources varies from basin to basin, depending on the lithology, topography and tectonic setting of the catchment, glacier activity and morphology, and the climatic regime. Glacier meltstreams are generally by far the most important sediment source at temperate glacier margins (Powell and Domack, 1995; Hunter et al., 1996). In such settings, thick basal debris sequences are rare and basal sediment is effectively flushed out by subglacial streams, so that direct sediment input from debris-rich ice or icebergs tends to be of minor importance. In contrast, ice shelves may have thick sequences of basal debris-rich ice, much of which can directly enter the water following basal melting. Subpolar tidewater glaciers also tend to release most sediment by the direct melting of the debris-charged basal ice along the submerged ice front, or by the melting of calved ice blocks (Ashley and Smith, 2000). Glaciers originating in mountain terrain commonly have extensive covers of supraglacial debris, which can enter lakes and the sea by gravitational or glacifluvial processes.

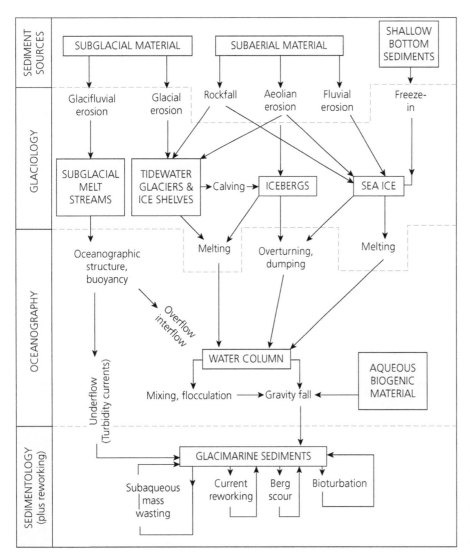

Figure 10.55 Debris cascade for the glacimarine sedimentary system, illustrating the sediment sources, glaciological and oceanographic constraints, and subaqueous processes. Boxes represent sediment stores, and unboxed terms represent processes (Dowdeswell, 1987).

The overall magnitude of sediment supply is also highly variable. Calculated sedimentation rates for proximal glacimarine environments range from ~3 mm yr^{-1} for Antarctica (Ashley and Smith, 2000), to 30–100 mm yr^{-1} in Svalbard (Elverhøi et al., 1983), to 2000–9000 mm yr^{-1} for Alaskan fjords (Powell, 1983). Extreme values of 13,000 mm yr^{-1} were reported by Cowan and Powell (1991) for an ice-proximal basin in front of McBride Glacier, Alaska. Extreme variability in sedimentation rates in single lake systems can arise from periodic glacier surging, as demonstrated by Fleisher et al. (2003) for lakes in front of the Bering Glacier, Alaska. Prior to a glacier surge in 1993–95, rates of suspension settling ranged from 0.6 to 1.2 m yr^{-1}, but during the surge this increased to 2.2–3.1 m yr^{-1}.

10.6.1 WATER BODY CHARACTERISTICS AND SEDIMENT INFLUX

The input of meltwater and sediment from glaciers profoundly influences the characteristics of glacier-fed lakes and marine environments, including the vertical distribution of temperature and density, and the orientation, strength and depth of currents. In turn, these water body characteristics exert strong controls on processes and patterns of sedimentation. Although there are many similarities in the water characteristics of deep lakes and the sea, the influence of salinity and tidal movements in glacimarine environments results in very distinctive differences.

Lake stratification

Patterns of sediment transport and water flow in lakes are strongly dependent on the vertical temperature profile, or *thermal structure*, of the lake water. Particularly important is the relationship of water density to temperature. Pure water at atmospheric pressure has a maximum density of 1000 kg m^{-3} at a temperature of 3.98 °C. Above and below this temperature, the density decreases (fig. 10.56a). The rate of change in density increases as the temperature moves further away from 3.98 °C, so that the density difference caused by a change from 19 to 20 °C is 20 times greater than that caused by a change from 4 to 5 °C. The influence of temperature on water density means that water bodies with different temperatures will have different buoyancies. If less dense water

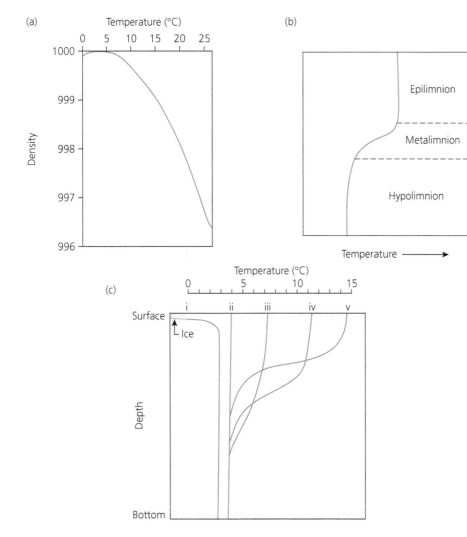

Figure 10.56 Temperature characteristics of lake water. (a) Relationship between temperature (°C) and density (kg m^{-3}) for pure water. (b) Temperature profile resulting from surface heating and wind mixing. (c) Hypothetical evolution of the thermal structure of a deep lake between (i) winter and (v) midsummer. Curve (ii) shows isothermal conditions resulting from lake overturn in autumn and spring (Redrawn from Smith and Ashley, 1985)

overlies denser water, this results in a stable layering known as *thermal stratification*.

In early summer, surface heating of a lake can create a warm upper layer overlying cold, denser water. The form of the temperature/depth curve, however, virtually never has a smooth form due to vertical mixing processes caused by several factors. First, the water surface can be cooled by evaporation, outgoing longwave radiation and sensible heat loss, setting up transient convection currents in the surface water layer according to the time of day and weather conditions. Second, winds will generate turbulence in the water column, and this initiates mixing and the downward transfer of heated surface waters.

A combination of surface heating and vertical mixing produces a characteristic three-layer water column (fig. 10.56b; Smith and Ashley, 1985). The upper layer (the *epilimnion*) consists of warm, low-density water, which is generally isothermal due to efficient vertical mixing. The lower layer, or *hypolimnion*, is made up of cold, dense, and relatively undisturbed water. These two layers are separated by the *metalimnion* or *thermocline*, which is a zone of rapid temperature change. The depth of the thermocline is dictated by factors such as intensity and duration of solar radiation, the temperature and location of inflowing water, the presence of chemically stratified layers and, usually most importantly, wind strength. It has been shown by Ragotzkie (1978) that lakes with larger fetches have thicker midsummer epilimnions. Because of variability in wind activity, heating and cooling, the epilimnion may contain several smaller and short-lived thermoclines.

The thermal stratification depicted in figure 10.56b represents the midsummer situation when maximum stability is attained in the water column. The thermal structure of deep temperate lakes, however, changes through the year in response to variations in the energy balance (fig. 10.56c). The midsummer curve (i) begins to change shape in late summer/early autumn, when the lake surface waters begin to cool, deepening the thermocline (curves ii and iii). Sinking cool water sets up convection, progressively destabilizing and cooling the whole epilimnion. Curve iv represents the situation in late autumn when full-depth mixing has occurred and epilimnion temperatures have fallen to those of the hypolimnion. This process is known as *overturning* and it produces an isothermal water body. Continued heat loss and mixing cools the lake to 4 °C or less, but this rarely creates a reverse thermocline because of the very small density changes in the 0–4 °C temperature range (fig. 10.56a), which are easily broken down by the wind. Ice formation (v) will occur once the surface layer reaches 0 °C, but only if vertical mixing is suppressed during calm weather. Once this ice disappears in the following spring,

the surface water will warm to hypolimnion temperatures and the lake will once again experience isothermal mixing, followed by thermal stratification. This means that temperate lakes go through two overturning periods in spring and autumn, and are thus referred to as being *dimictic*.

Monomictic lakes circulate only once per year, and can be subdivided into warm and cold varieties. Warm monomictic lakes experience winter circulation above 4 °C and have been reported from New Zealand, where the mild maritime climate prevents surface ice formation, but allows the development of deep thermoclines (Pickrill and Irwin, 1982). Cold monomictic lakes undergo summer circulation below 4 °C and this is thought to be the situation in some ice-contact lakes. Some lakes at high latitudes may be subject to different thermal conditions from year to year, because of the variability of duration and timing of the short ice-free season. Some lakes may circulate at various times because of a lack of thermal stratification and are referred to as *polymictic* (Gilbert and Church, 1983).

All of the above examples are known collectively as *holomictic lakes*, which circulate completely to the bottom during at least part of the year. In contrast, *meromictic lakes* contain bottom water which is stabilized by dissolved material and which remains partly or wholly unmixed with overlying water, producing chemical stratification. As glacial meltwater generally contains low dissolved contents, meromictic lakes are not well represented among glacier-fed lakes. However, meromictic lakes are common in basins recently isolated from the sea by glacio-isostatic rebound or still maintaining some contact with seawater.

In glacier-contact lakes, the continuous supply of 0 °C water by channelled inflow, direct glacier melting or melting icebergs will at the very least modulate the temperature. Quite often, the cold water influx will inhibit or disrupt the formation of a thermal stratification of the water column, as outlined above. Churski (1973) has reported on the effects of subaqueous injections of cold water and ice-marginal melting into ice-contact lakes in Iceland (fig. 10.57). These disruptions may grade down-lake into thermally stratified water if the lake body is large enough.

Lake stratification also occurs as a result of variations in suspended sediment concentration, termed *sediment stratification* by Smith and Ashley (1985). Studies of Peyto Lake, Alberta, and Malaspina Lake, Alaska, have revealed gradual increases in suspended sediment with depth (Gustavson, 1975). The processes responsible for this density gradient are unclear, but are thought to relate to a combination of the upward diffusion of sediment from underflows and the downward settling of sediment from interflows, both of which are known to occur in Peyto and Malaspina lakes.

(a)

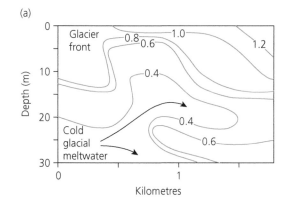

(b)

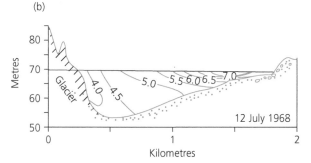

Figure 10.57 Examples of the influence of glacier proximity on the thermal structure of glacier-contact lakes in Iceland. (a) Disruption of the thermal structure of Jökulsarlon by the injection of cold, dense meltwater below the lake surface. The glacier snout was 450 m to the left of the profile (after Harris, 1976). (b) Water cooling adjacent to the margin of Skeiðarárjökull (After Churski, 1973).

Water inflow to glacier-fed lakes

In glacierized catchments, water will flow into lakes directly from glacier ice (via englacial and subglacial portals) and from glacially fed and/or non-glacial rivers. In regions with marked seasonal temperature variations, maximum discharges of both water and sediment load occur during the summer melt season, with peak events coinciding with extreme weather or sudden releases of stored water (sections 3.6 and 3.7).

As water enters a lake, turbulent eddies along the inflow margins cause it to mix with lake water, producing a two- or three-dimensional expansion of the flow. This is coupled with an exponential decrease in velocity of the inflow to produce an expanding and decelerating *plume*. The shape and velocity patterns within the plume are governed by the discharge and velocity of the incoming water, channel shape, suspended sediment concentration and, most importantly, the density differences between the inflow and lake waters and the density stratification of the lake (Bogen, 1983). When viewed from the air, the outer edge of a turbid interflow plume can be seen to possess a very sharp contact with the lake water into which it is discharging. This marks the position of the 'plunge line', where the plume sinks to reach lake water of similar density. Between the inflow point and the plunge line, the

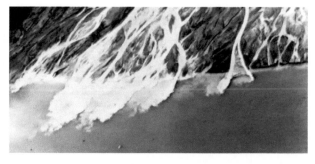

Figure 10.58 Turbid inflow plumes in Peyto Lake, Alberta, Canada, showing sharp contacts between lake water and lakeward edges of plumes (plunge points) (Smith and Ashley, 1985, reproduced with permission of SEPM).

plume spreads out two-dimensionally and decelerates as it mixes with the lake water. The turbulence in the plume entrains lake water as the plume sinks. This increases the volume of the plume and sets up compensatory surface counter-currents to replace the water lost to the sinking plume. The plunge line essentially represents the convergence point of the lakeward-moving plume and the shoreward-moving counter-currents (fig. 10.58). Once the interflow plume reaches the depth at which its density equals that of the surrounding water (usually around the level of the metalimnion), it spreads out horizontally. Clay-sized material can remain suspended in the interflow due to upward-directed currents, but the coarser fractions will drop out progressively in a down-lake direction.

If the density of the diluted inflowing water is less than the density of the epilimnion, it will rise to the surface as an *overflow*. In contrast, if the incoming water is denser than the hypolimnion, it will flow along the bottom of the lake as an *underflow* (fig. 10.59). Overflows and underflows are referred to as *hypopycnal* and *hyperpycnal* inflow, respectively. *Interflows* occur where the density of the inflowing water is intermediate between the dense, cold lake bottom water and the low-density, warm lake surface water, and will move along the thermocline.

The larger the density differences between the inflow and the lake water, the greater the energy required to mix them, so that turbulent diffusion is reduced. This usually results in the maintenance of a discrete density current through the lake. According to field measurements, underflow velocities vary greatly and the flow may persist for hours or even days, especially in areas of high river discharges. Velocities ranging from a few cm sec^{-1} to a few tens of cm sec^{-1} are common (Gustavson, 1975; Gilbert and Shaw, 1981; Smith et al., 1982; Bogen, 1983), but values over 100 cm sec^{-1} have been reported (Weirich, 1984). Interflow velocities have been measured at 2–15 cm sec^{-1} (Smith and Ashley, 1985).

In unstratified or weakly stratified lakes where the inflowing water is of the same density as the lake water, a *homopycnal inflow* will result, in which the incoming plume quickly loses its identity by three-dimensional

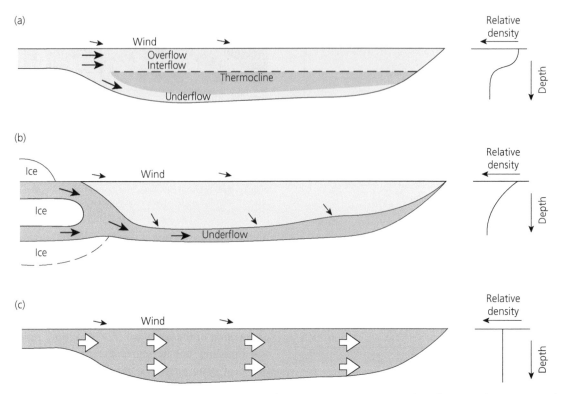

Figure 10.59 Density stratification in glacier-fed lakes. (a) Thermal stratification, showing the positions of overflows, interflows and underflows. This situation is most common in non glacier-contact lakes. (b) Sediment stratification, typical of glacier-contact lakes. (c) No stratification (Adapted from Smith and Ashley, 1985).

mixing (fig. 10.59c). Homopycnal inflows are typical of shallow, freely circulating lakes which receive inflowing water of constant density – conditions which are not common in glacier-fed lakes. However, lakes fed by the lower stretches of glacier meltstreams, especially when further lakes act as sediment traps upstream, may receive homopycnal inflows. Quite often in these situations, more than 80 per cent of the sediment entering the lake leaves it again at its outlet, because the inflowing water is of approximately the same temperature and the same suspended sediment concentration as the lake water, so mixing occurs quickly and water stratification is inhibited (Smith et al., 1982). As the sedimentation rates in such lakes are very low, the bottom sediment is commonly mixed by the action of organisms living on the lake bottom (*bioturbation*).

The density of inflowing water is governed by temperature, dissolved solute content and suspended sediment concentrations. The suspended sediment concentration is the most important source of density variations in inflow water, because it varies over a wider range and changes over shorter periods of time than temperature, which varies over longer periods of days to months. Variability of inflow type and mixing patterns will relate to seasonal and weather-dependent stream discharges, stream temperatures, suspended sediment concentrations and seasonal changes in lake thermal stratification. In some lakes, this may result in the production of different inflow types at different times throughout the year (Pickrill and Irwin, 1982). Where discharge and density vary rapidly, overflows, interflows and underflows may coexist for short periods (Gilbert, 1975; Smith et al., 1982).

Water stratification and inflow in glacimarine environments

Overflows are much more common in glacimarine environments due to the high density of seawater (approximately $1030 \, \text{kg m}^{-3}$) compared with fresh water (approximately $1000 \, \text{kg m}^{-3}$). This means that incoming water can have suspended sediment concentrations as high as $30 \, \text{kg m}^{-3}$ ($30 \, \text{g litre}^{-1}$) and still remain buoyant enough to form overflows. Water upwelling from submerged englacial and subglacial conduits can be observed at the termini of many tidewater glaciers in the form of turbid sediment plumes, surface water disturbances and currents (fig. 10.60; Cowan and Powell, 1990; Powell, 1990; Powell and Domack, 1995).

Where overflows enter the sea, the water column commonly has a two-layer density structure separated by the *pycnocline*, although this can be complicated by undercurrents and influxes of water from sub-basins separated by topographic barriers to produce multilayer circulation (Syvitski et al., 1987). Underflows are rare in the glacimarine environment, because they require suspended sediment concentrations greater than $30 \, \text{kg m}^{-3}$ in order to exceed the density of saline ocean water.

Figure 10.60 Turbid overflow upwelling at a floating glacier margin, Leffert Glacier, eastern Ellesmere Island, Arctic Canada (D.J.A. Evans).

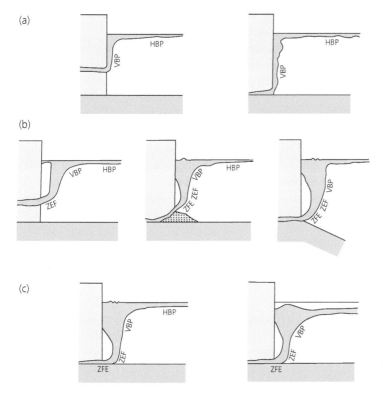

Figure 10.61 Patterns of meltwater discharges from tidewater glaciers. (a) Forced plume dominated by buoyant forces. (b) Axisymmetric jets developing into axisymmetric plumes as momentum forces give way to buoyant forces. (c) Plane jets developing into axisymmetric jets and plumes. ZFE = zone of jet flow establishment; ZEF = zone of established jet flow; VBP = vertical buoyant plume; HBP = horizontal buoyant plume (Adapted from Powell, 1990)

Water discharging into the sea from subglacial and englacial tunnels are subject to two sets of forces: *buoyancy forces*, encouraging upwelling, and *momentum forces*, tending to carry the water horizontally through the water column (Powell, 1990). When discharges are low, buoyancy forces are dominant and the meltwater rises rapidly to the surface as a *forced plume* (fig. 10.61a), but at higher discharges, the momentum forces dominate and the efflux will issue as a sub-horizontal *jet* before rising (figs 10.61b, c). *Axisymmetric jets* are formed where the efflux travels through the free water column before rising, whereas *plane jets* travel along the sea floor before upwelling to form overflow or, more rarely, interflow plumes.

The behaviour of a turbulent jet once it is beyond the tunnel mouth depends largely on the rate of mixing with the surrounding seawater (Powell, 1990). Three zones have been identified (fig. 10.61):

(1) The *zone of flow establishment* (ZFE) occurs close to the tunnel mouth, where the centreline velocity equals the outlet velocity and the centre of the jet acts as a plug flow.

(2) The *zone of established flow* (ZEF) begins at a distance of $6.2 \times D$ from a circular outlet (where D is tunnel or pipe diameter), where the core velocity decelerates. In this zone, turbulent mixing occurs throughout the jet and this can extend to beyond $200 \times D$ for axisymmetric jets and beyond $2000 \times D$ for plane jets.

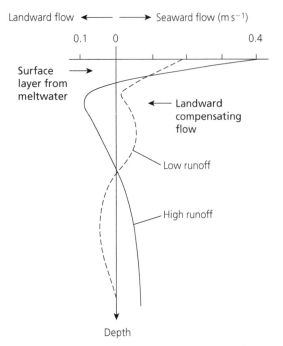

Figure 10.62 A vertical velocity profile through the upper levels of a typical fjord. Meltwater is added to the surface layer, which flows seawards (left to right), and is compensated by a return flow at depth. Return flow occurs at deeper levels at times of low runoff (Drewry, 1986).

Clearly, bottom friction and suspended sediment concentration will cause these distances to vary. At the end of the ZEF, the large amount of turbulence forces the residual axial velocity of the jet to decay to zero and, therefore, buoyancy forces take over.

(3) The efflux rises in a *vertical buoyant plume* (VBP), and once at neutral buoyancy it becomes a *horizontal buoyant plume* (HBP).

General circulation in glacimarine environments is dominated by the seaward flow of surface meltwater plumes from both glacier snouts and river mouths, and compensatory landward flows of deeper saline water. The strength of this circulation is obviously linked to the intensity of meltwater inflow (fig. 10.62). In Arctic fjords, surface plumes of anywhere between 5 and 31‰ salinity and temperatures of 2–4 °C have been observed to overlie more saline, colder water (Elverhøi et al., 1983), but stratification due to subglacial meltwater efflux is uncommon around Antarctica.

Although surface plumes are readily visible near the efflux point due to their high suspended sediment concentration, they become less pronounced with distance due to mixing of the plume with the ambient water. Equal quantities of water are exchanged between layers, resulting in a reduction of the salinity of the ocean water. The amount of mixing and entrainment that takes place is a function of the balance between the *stabilizing forces* associated with density stratification and the *destabilizing forces* associated with the velocity gradient across the interface

of the two water bodies. This is defined by the Richardson number (R_n):

$$R_n = \frac{g}{\rho_w}\left(\left(d_{wc}\frac{\partial \rho_w}{\partial z}\right) \Big/ \left(d_{wc}\frac{\partial U_w}{dz}\right)\right)^2 \qquad (10.2)$$

where g is gravitational acceleration, ρ_w is water density, U_w is water velocity, d_{wc} is depth in the water column, ($\partial \rho_w/\partial z$) is vertical density gradient and ($\partial U_w/\partial z$) is vertical velocity gradient.

A Richardson number of <0 indicates that the water column is unstable and mixing will occur, whereas $R_n > 0$ indicates stable conditions. Moreover, an R_n value >0.25 would suggest that mixing due to shear instability is not significant. In the absence of strong currents and circulation, the melting of glacier ice in contact with ocean water can result in dilution, producing ice-proximal brackish water (Matthews and Quinlan, 1975).

In glacilacustrine settings, overflows only occur when the inflowing water is less dense than the lake water; therefore overflows cannot carry a large amount of suspended sediment. This means that inflowing meltstreams with high suspended sediment concentrations will generally produce underflows rather than overflows. Furthermore, overflows will tend not to be produced in thermally stratified lakes, but will form during the late spring and winter, when incoming water is more likely to contain small suspended sediment loads and be more buoyant than the lake water. Because they occur at the lake surface, overflows are susceptible to wind stress, and due to the effects of the wind are often very difficult to differentiate from shallow interflows.

The role of surface ice

In glacilacustrine environments, the formation of surface ice in winter usually results in an abrupt change in surface temperatures in an otherwise stable temperature profile (fig. 10.57). This combines with reduced winter meltwater discharge and sedimentation rates and negligible wind mixing of surface water, to produce a largely unstratified water column. In marine settings, sea ice formation has a similar considerable impact on seawater stratification, but for slightly different reasons.

Sea ice formation during the winter will eliminate wind mixing of surface waters and may act to dampen the effects of tides and currents. Furthermore, as seawater is frozen, salt is rejected and this produces a layer of high-salinity water below the ice. This high-salinity layer increases in thickness throughout the winter as more salt is released by ice growth and there is vertical mixing of deeper waters (fig. 10.63; Syvitski et al., 1987). At the same time, surface water temperatures are reduced, increasing the depth of the temperature maximum. In this way, marine water stratification caused by meltwater

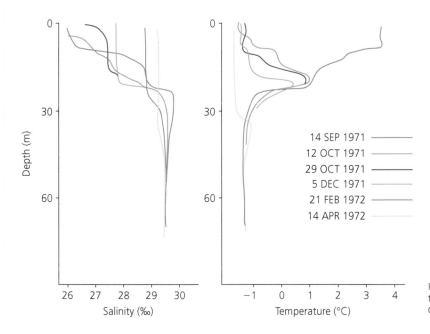

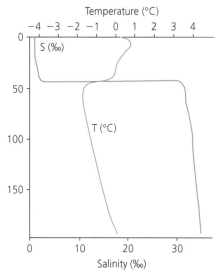

Figure 10.63 Seasonal changes in water column temperatures and salinities in Cambridge Bay, Arctic Canada (Syvitski et al., 1987).

Figure 10.64 Temperature and salinity profiles in Disraeli Fjord, Arctic Canada, due to the influence of the Ward Hunt Ice Shelf, which blocks the fjord mouth, encouraging the survival of a warm lid of fresh water (epishelf lake) (Syvitski et al., 1987, after Keys, 1978).

efflux essentially breaks down during the winter in areas where sea ice forms. The thickness and duration of sea ice obviously varies according to climate and local geography and oceanography, but generally they both increase with latitude. Some sheltered re-entrants, like the fjords of the Canadian and Greenland High Arctic, are termed *frigid*, in that they possess perennial sea ice.

Water stratification in glacimarine environments is further complicated where glaciers or ice shelves block the mouths of fjords. Russell Fiord, Alaska, was transformed from an arm of the sea into a freshening ice-dammed lake by an advance of Hubbard Glacier in 1986, with profound implications for water body characteristics and marine life (Krimmel and Trabant, 1992). Until 2003, the Ward Hunt Ice Shelf, Ellesmere Island, dammed Disraeli Fjord creating a 'lid' of freshwater 44m thick on top of relict ocean water (Keys, 1978). Breakup of the ice shelf caused the lake to drain, and the fjord now functions as an estuarine system (Veillette et al., 2008).

Tides, currents and waves

Water column stratification and circulation in glacimarine environments and large glacier-fed lakes are affected by the actions of tides, currents and waves. In general terms, tides and currents will enhance turnover in the water column and so minimize water stratification. With respect to sedimentation processes, they also influence the transport of suspended sediment and rework sediment on the ocean bottom. The impact of tidal forcing on currents and circulation in glacimarine situations is often more important during the winter, when the inputs of meltwater from glacial and fluvial systems are at their lowest.

Glacimarine environments are affected by tides in a number of ways. Below the surface, tides set up *barotropic* currents which are of more or less constant velocity with depth. The velocity may be affected by bathymetry in some marine re-entrants, such as fjords, where sills may constrict the flow and increase the velocity. However, tidal current velocity is predominantly related to tidal range. For example, tidal current velocities up to $0.31\,\mathrm{m\,sec^{-1}}$ have been measured in Glacier Bay, Alaska, which has a tidal range of 5 m (Matthews and Quinlan, 1975), compared with only $0.0027\text{–}0.0029\,\mathrm{m\,sec^{-1}}$ for Baffin Island fjords, with a tidal range of 1.1 m (Gilbert, 1982).

Tides also affect the nature of surface plumes. Ebb tides, which reinforce flow away from the shore, will lead to plume velocity increases and decreasing plume thickness, whereas flood tides counteract seaward flow and

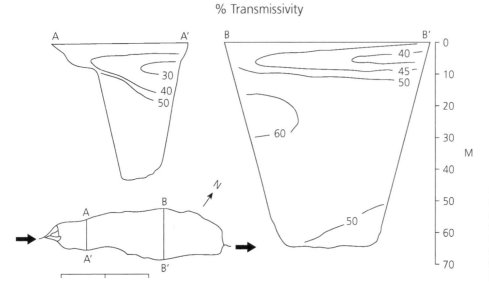

will produce thicker but slower-moving surface layers. Water mixing is also enhanced by flood tides, which can hold surface plumes near to the terminus of tidewater glaciers. Further variability in the behaviour of surface plumes is brought about by various interactions between periods of maximum stream discharge and ebb and flood tides. Tidal impacts on cyclic deposition of laminated sediment are reviewed in section 10.6.4.

Surface currents are generated by wind shear. In glacial environments, katabatic winds blowing offshore will influence the flow direction and the longevity of overflow plumes and will also determine the drift paths and drift speeds of icebergs. Larger-scale oceanic currents, such as long-shelf currents, upwellings, geostrophic currents and subaqueous slope failures, are also important in glacimarine settings, and are a function of landmass positioning, global ocean currents and submarine mega-topography. Subaqueous slope failures are temporally discontinuous and relatively unpredictable, but they are capable of producing considerable impacts through current activity and wave action, and very large failures can initiate devastating tidal waves or *tsunamis*.

The Coriolis effect

Near-surface inflow plumes discharging into water bodies are also affected by the Earth's rotation. Because of the curvature of the Earth, northward- and southward-flowing surface currents will be deflected in order to conserve angular momentum, an effect which is most pronounced close to the Poles. This *Coriolis effect* causes plumes to be deflected to the right in the northern hemisphere and to the left in the southern hemisphere, similar to atmospheric wind systems (Barry and Chorley, 2003). Plumes will be deflected until they impinge on a shoreline, along which they then flow (Hamblin and Carmack, 1978). Transmissivity profiles from Lake Louise, Alberta, which

provide a measure of the amount of light transmitted through the water column according to the suspended sediment concentration, provide evidence of such a deflected interflow (fig. 10.65; Smith and Ashley, 1985). Such inflow patterns have considerable effects on lake bottom sedimentation patterns in a cross-lake direction which are superimposed on the normal proximal-distal downlake patterns.

Bathymetry

The bottom topography of both marine and lacustrine basins may have considerable impact on water circulation and associated sedimentation patterns. Common topographic elements are submerged bedrock hills or depositional features such as moraines or morainal banks, subaqueous channels, and enclosed basins created by erosional overdeepening or uneven sedimentation. Local topography creates sub-basins within larger depositional lake or marine basins, and may have distinct circulation and deposition patterns (fig. 10.66). The transfer of sediment from one sub-basin to another will be impeded to a greater or lesser extent depending on the depth of the topographic barrier relative to dominant current flow (e.g. underflows, interflows or overflow plumes). Bottom topography has the greatest influence on the flow direction of underflows, which are deflected away from topographic high points and steered towards the deepest parts of basins (Smith et al., 1982).

Water stratification in fjords can be modified by the presence of deep, stagnant bottom water behind bedrock or sediment sills at the seaward end. The salinity of this water will be related to the magnitude of the deep water renewal, and the strength and mixing depths of freshwater plumes. Where a shallow sill restricts circulation, fjord bottom waters are isolated from the open sea and deep saline water only rarely penetrates the fjord, thus

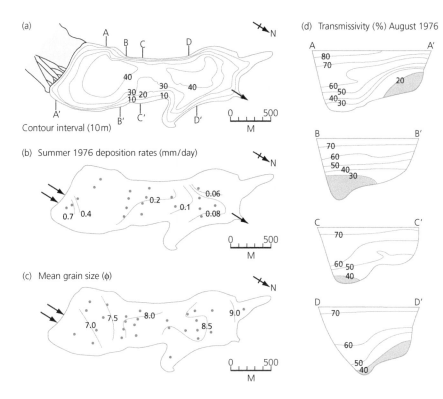

Figure 10.66 The effects of bottom topography on sedimentation in the underflow-dominated Peyto Lake, Alberta, Canada (after Smith et al., 1982). (a) Bathymetric map showing two sub-basins and cross profiles A-A' to D-D'. (b) Summer sedimentation rates measured at points marked by dots. (c) Mean grain size distribution showing correspondence with bathymetry. (d) Transmissivity profiles along the sections A-A' to D-D', showing that the lowest values (highest suspended sediment concentrations = brown) are controlled by bathymetry and positions of inlet and outlet.

producing *euxinic* conditions. For a thorough review of fjord circulation patterns, see Syvitski et al. (1987). Where shallow basins have been glacio-isostatically lifted above present sea level, the sill will help to isolate seawater within the depression as a terrestrial lake is produced by uplift. This results in stratified lakes in which saline bottom water can be preserved (Hattersley-Smith et al., 1970; Retelle, 1986).

10.6.2 DEPOSITIONAL PROCESSES

Deposition from jets

Where subglacial or englacial streams discharge directly into the lacustrine or marine environment, large quantities of fluvial bedload and suspended load can be transported and deposited by jets (Powell and Domack, 1995). The highest sediment concentrations in jets occur close to the efflux point in the ZFE. Bedload and suspended sediment loads can be very high, and sediment can be transported out onto the lake or sea floor as *hyperconcentrated flows* or as *bedload traction carpets* similar to those in turbid subaerial streams. The coarsest sediment is deposited rapidly and close to the efflux point as flow decelerates, and the finer material is transported further by the turbid jet and plume, which maintain some momentum (Powell, 1990). In glacimarine environments, fine suspended sediment is commonly uplifted from the bed in buoyant plumes, promoting more rapid deposition of the coarser material. Sediment deposited from jets forms sheets and channelized lenses of gravel and sand, which typically display rapid downstream fining.

The distance over which such fining takes place relates to the magnitude of the meltwater discharge from the tunnel, being greater during the melt season. Flow can also be erosive, scouring out channels which are later filled by renewed sediment accumulation (Rust and Romanelli, 1975).

Settling from suspension

Suspended sediment in overflows, interflows and underflows can be transported for considerable distances in lakes and the sea, and then gradually settles out through the water column. The processes of transport and suspension settling are essentially similar in glacilacustrine and glacimarine processes, although the salt content of seawater and tidal effects introduce some important differences.

In still water, suspended particles will fall down through the water column with a velocity that reflects the buoyant weight of the particle and the drag forces resisting motion. The buoyant weight and drag forces are influenced by several factors, particularly the size, density and shape of the particle, and the density and viscosity of the water. For spherical particles, the settling velocity is given by Stokes's Law:

$$\nu_f = \frac{g}{18\mu}(\rho_p - \rho_w)D^2 \qquad (10.3)$$

where ν_f is fall velocity, g is gravitational acceleration, μ is water viscosity, ρ_p is particle density, ρ_w is water density and D is grain diameter.

This shows that fall velocity increases with particle size and the density difference between the particle and the water, and decreases with water viscosity. Fall velocities are large for large particles, and very small for small particles, as is strikingly demonstrated by considering the settling of quartz spheres through salt water (Allen, 1985). Terminal settling velocities are $0.00979 \, \text{m sec}^{-1}$ for a $100 \, \mu\text{m}$ particle (fine sand), $0.0000979 \, \text{m sec}^{-1}$ for a $10 \, \mu\text{m}$ particle (silt), and only $0.000000979 \, \text{m sec}^{-1}$ for a $1 \, \mu\text{m}$ clay particle! At these rates, settling times through a 100 m column of salt water are approximately 3 hours (fine sand), 11 days (silt) and 3 years (clay). Since particles will remain in suspension if vertical motions in the water equal or exceed the settling velocity, it is easy to see that fine particles can be transported for considerable distances in well-mixed waters.

The density term in Stokes's Law shows that particles will settle more rapidly in fresh water than in salt water. Additionally, particles with high density, such as iron-rich minerals, will settle out more rapidly than low-density particles of the same size. Settling velocities for non-spherical particles will differ from those calculated by Stokes's Law. Platy particles, such as shell fragments, mica flakes and single clay particles, settle more slowly due to their high surface area relative to their volume (Allen, 1985).

Flocculation and pelletization

The behaviour of suspended clay particles in sea water is strongly influenced by the process of *flocculation*, or the bunching of clay particles into aggregates called *floccules* or *flocs*. Flocculation takes place when the repulsive charge that normally exists between clay particles is reduced by the electrolytic action of salt water, allowing particles to stick together if they collide (Drewry, 1986; Cowan and Powell, 1991). A salinity of only 3–4‰ will promote flocculation. Curran et al. (2004) found that median floccule sizes increased with depth, from around 1 mm to around 1.5 mm. Flocculation, therefore, can create sand-sized particle aggregates, which sink much faster than isolated clay particles, causing them to be deposited alongside fine silts.

Aggregation of clay particles also occurs by the production of *fecal pellets* by zooplankton (Syvitski et al., 1987). This pelletization can produce agglomerations up to $100 \, \mu\text{m}$ across (Syvitski and Murray, 1981), and fecal pellets comprise up to 18 per cent of the suspended particulate matter in the waters just offshore of the Ross Ice Shelf, Antarctica (Carter et al., 1981). Similar processes operate in glacilacustrine environments, and Smith and Syvitski (1982) have reported fecal pellets produced by sediment-ingesting zooplankton in Bow Lake, Canada. These pellets promote the deposition of fine silts and clays that would otherwise remain in suspension until they are flushed out of the lake system.

A final process of particle aggregation is *agglomeration*, or the attachment of mineral grains to organic detritus by cohesion (Syvitski et al., 1987). This process is unimportant in ice-proximal environments, where suspended sediment concentrations are high and organic content is low, but it may be important for the removal of fine clay particles from the water column in distal settings.

Melt-out from floating ice, ice shelves and icebergs

Debris ranging in size from clay and silt up to boulders can be introduced directly to ice-contact glacilacustrine and glacimarine environments by melt of debris-rich ice beneath ice shelves or dumping from icebergs. In the case of iceberg dumping, debris may be transported hundreds or even thousands of kilometres from the ice margin prior to deposition, thus providing a very efficient form of debris dispersion.

Ice rafting

Debris can be rafted out into lakes and the sea by floating ice, to be released into the underlying water column as the ice melts. Clastic material in ocean sediment cores is often interpreted as an indicator of expanded glaciers, but other forms of rafting should be considered (Reimnitz and Kempema, 1988; Gilbert, 1990; Bennett et al., 1996a). Clasts can also be transported by sea and lake ice, in the stomachs of marine mammal (gastroliths) and by floating seaweed.

Sea and lake ice rafting can be subdivided into active and passive types (fig. 10.67). *Active sea/lake ice rafting* involves either: (1) freeze-on of bottom sediment directly onto thickening surface ice (*anchor ice*); (2) the incorporation of suspended sediment directly into forming sea or lake ice; or (3) the formation of anchor ice on the sea or lake floor (Reimnitz and Kempema, 1988). In contrast, *passive sea/lake ice rafting* involves the deposition of sediment onto existing ice surfaces, by: (1) rockfalls and avalanches from overlying slopes; (2) fluvial deposition onto shore-ice surfaces before they break up in the spring; (3) aeolian deposition; and (4) littoral deposition whereby material is moved up through cracks in the shore-ice during high tide (Gilbert, 1990).

One other unusual form of rafting in the marine environment involves the lifting of clasts by marine algae, popularly known as seaweed. Gilbert (1990) has demonstrated that once the alga *Fucus vesiculosus* has grown to three times the weight of the clast to which it is attached, it will float the clast. This is because the alga possesses gas-filled vesicles in its thalli, producing buoyancy. A large number of algae are also found frozen into sea ice and anchor ice in littoral areas. Although the relative importance of algal rafting is difficult to assess, it is responsible for the introduction of individual dropstones

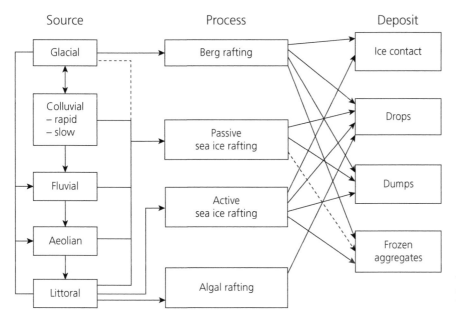

Figure 10.67 Debris sources, transport processes and deposits involved in rafting in the glacimarine environment (excluding extraterrestrial and volcanic sources) (Adapted from Gilbert, 1990).

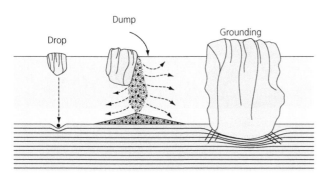

Figure 10.68 Dropping and dumping of debris from icebergs, and sediment deformation by grounded bergs (Redrawn from Thomas and Connell, 1985).

to fine-grained sediments far from their source area and this is normally regarded as indicative of iceberg or at least sea ice rafting. The problem of equifinality here is exacerbated by the fact that the subpolar zone of ice rafting, at least in the northern hemisphere, overlaps a large part of the zone of algal rafting.

Bergs calved from glaciers may contain englacial or subglacial debris bands or spreads of supraglacial debris (Powell and Domack, 1995). Abundant debris is typical of bergs derived from valley and outlet glaciers in high-relief terrain, surging glaciers and some subpolar glaciers. In contrast, bergs from ice shelf margins generally have low debris concentrations, because much of the basal debris may have been lost by basal melting prior to calving (Drewry and Cooper, 1981).

Drifting icebergs gradually melt as they are carried by ocean and lake currents, releasing their sediment load into the underlying sediment column (fig. 10.68). Debris release from individual bergs is typically episodic, reflecting fragmentation and overturning events triggered by changes in the shape and centre of gravity of the melting

berg. As icebergs overturn, any debris that has accumulated on the surface is suddenly dumped into the water. The amount of fragmentation and overturning decreases with distance from the coast, as icebergs slowly attain equilibrium positions with low centres of gravity (Orheim, 1980). A typical debris release history for a tabular iceberg is shown in figure 10.69 (Drewry and Cooper, 1981; Dowdeswell, 1987). Most of the sediment is released by the berg within the first 200 km of the ice margin and the berg is essentially clean after 400 km. This model is based on observations of icebergs at varying distances from the ice margin, as well as laboratory and theoretical analyses of iceberg melting rates in 0 °C water. Melting rates vary between $0.1 \, \mathrm{m \, yr^{-1}}$ and $100 \, \mathrm{m \, yr^{-1}}$, depending on debris content (Russell-Head, 1980; Drewry, 1986).

Debris can be released from icebergs and deposited in four different ways (fig. 10.68; Thomas and Connell, 1985; Gilbert, 1990):

(1) *Iceberg drop* involves the deposition of a single particle, usually after release from the berg by melting.

(2) *Iceberg dumps* refer to the simultaneous deposition of a number of particles in a single event, usually associated with the overturn of an iceberg and the tipping of surface material into the water. Alternatively, concentrations of sediment from former englacial tunnels or crevasse fills can be released during melt-out and/or iceberg overturn.

(3) *Frozen aggregates* are sediments held together by interstitial ice which are released when ice rafts are broken apart.

(4) *Ice contact* rafted sediments are deposited by an iceberg that is grounded on the sea floor.

Dowdeswell and Murray (1990) have suggested that five factors govern the amount of sediment deposited by

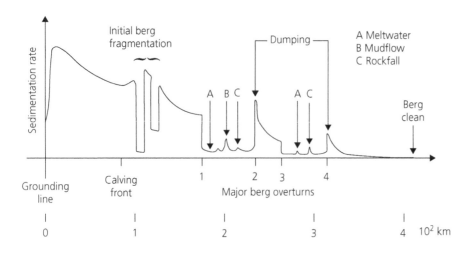

Figure 10.69 Hypothetical debris release history of a drifting and melting iceberg, showing the effects of calving, fragmentation and overturn (Drewry, 1986).

iceberg rafting where the parent ice mass is a tidewater glacier:

(1) the distribution and concentration of sediment in the parent glacier;
(2) the rate of iceberg calving;
(3) the rate of melting at the iceberg base;
(4) the temperature of the ocean water;
(5) the velocity, stability and drift track of the iceberg.

Some of these factors may counteract each other. For example, where sea ice acts as a buttress to a calving glacier, it will reduce the rate of calving and therefore reduce the iceberg sedimentation rate. This will be counteracted by the restriction placed on the drift speed and distance by the sea ice buttress, which will increase ice-proximal sedimentation rates. Where icebergs are derived from ice shelves, undermelt and bottom freezing must be taken into account. Before calving takes place, undermelt may remove basal debris, whereas bottom freezing may protect it. Studies of Holocene glacimarine deposits in Scoresby Sund, East Greenland, by Ó Cofaigh et al. (2001) have revealed a contrast in sedimentation styles that reflect glacier activity through meltwater generation. The inner part of Scoresby Sund is dominated by fine-grained laminated muds deposited by turbidity currents and suspension settling from overflow plumes, reflecting high rates of sedimentation and meltwater flux to the ocean. The outer part of Scoresby Sund, on the other hand, is dominated by coarse-grained iceberg-rafted sediment, interpreted to be the result of the distal location from fast-flowing glacier termini (i.e. further from the dominant influence of turbidity currents and suspension sedimentation) and the high calving flux of the glaciers.

Iceberg scouring

As grounded icebergs are dragged over the lake or seabed by currents, they plough curved, flat-bottomed troughs or furrows which can heavily modify or even destroy sub-glacial lineations produced prior to the break-up of marine-based glaciers (e.g. Polyak et al., 2001; Syvitski

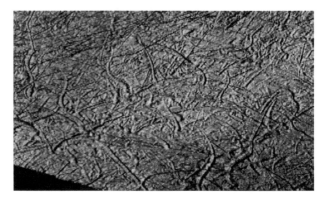

Figure 10.70 An illuminated time-structure map compiled from 3D seismic data from the Barents Sea, showing multiple criss-crossing, curved iceberg furrows developed since the deglaciation of the shelf (Rafaelsen et al., 2002).

et al., 2001; Rafaelsen et al., 2002; Shipp et al., 2002; Vorren, 2003; fig. 10.70). Large examples can be incised up to 30 m deep and up to 1 km wide and stretch for several kilometres, but they are generally much smaller. Spectacular iceberg furrows are also produced in proglacial lakes, some of the best examples being preserved on the former floor of Glacial Lake Agassiz in North America (Woodworth-Lynas and Guigné, 1990; Teller, 2003). The trough bottoms often exhibit a microtopography of ridges and grooves, with a relief of up to 30 cm. These are thought to be the product of dragging of bed material which has been mechanically incorporated into the iceberg keel. Sand volcanoes on top of this microtopography attest to the liquefaction and dewatering of bed material during and immediately after scouring (Hodgson et al., 1988). On either side of the main furrows, linear scour *berms* of displaced blocky material can reach heights of 6 m. The chaotic criss-crossing of iceberg furrows reflects the varied drift patterns of icebergs on the lake or ocean surface (fig. 10.71). These can be affected by winds as well as dominant current patterns. In situations where glacial lakes drain catastrophically, iceberg drift is arrested and the final resting place becomes marked by 'wallow' craters (fig. 10.71). The deformation

Figure 10.71 Iceberg 'wallow craters', produced where catastrophic drainage of an ice-dammed lake occurred along the side of Heinabergsjökull, Iceland (D.J.A. Evans).

Figure 10.72 A trace fossil from Holocene glacimarine silts in Clements Markham Inlet, north-east Ellesmere Island, Arctic Canada. The markings are on a bedding plane that has been opened like a book (left surface is imprint, right surface is cast). The central, elongate marking was made by the side-to-side motion of a fish's tail and the outer imprints were made by the pushing movement of its fins. A further unidentified trail crosses the fish trail from top to bottom (T.G. Stewart).

structures created by iceberg ploughing are described in section 10.6.6.

Reworking by currents and biological activity

Current reworking

Lake and sea currents are generated by a number of mechanisms, including wind shear, tides, storm waves, long-shelf currents, upwellings, geostrophic currents, glacier and iceberg melting, and subaqueous sediment gravity flows. Generally, currents that reach the bottom of a water body tend to winnow the fine-grained materials and leave coarse lag deposits, termed a *palimpsest lag* by Powell (1984). They may also erode pre-existing material to produce erosional contacts or cross-cutting relationships between sedimentary units. Particularly erosive sediment gravity flows may be capable of scouring small-scale bedforms or flute marks similar to those produced in fluvial environments.

Bioturbation

Disturbance of sediment by surface or near-surface dwelling marine organisms is referred to as *bioturbation*. The wide range of structures produced by bioturbation are usually called *trace fossils*, and where they can be recognized and assigned to species they are often termed *lebensspuren* (Bromley, 1996). Five main types can be recognized:

(1) *resting traces*, where creatures rest in one place for a period of time;
(2) *crawling traces*, made by animals making their way across the bed (fig. 10.72);
(3) *browsing traces*, made by animals searching for food on the sediment surface;
(4) *feeding structures*, which include burrows and similar structures created by organisms moving through the sediment for food;
(5) *dwelling structures*, where organisms create semi-permanent burrows.

Due to the high sedimentation rates in proximal glacilacustrine and glacimarine environments, bioturbation is not usually intensive. Bioturbation can, however, result in the partial or complete destruction of primary depositional structures in more distal positions where sedimentation rates are low (Gilbert, 1982).

10.6.3 VARVES AND OTHER GLACILACUSTRINE OVERFLOW/ INTERFLOW DEPOSITS

As explained above, suspended sediment carried into lakes as high-level overflows and interflows spreads laterally and gradually settles out through the water column, with velocities proportional to particle size. The resulting sediments form blanket-like drapes which typically display proximal to distal fining and thinning (Smith et al., 1982; Drewry, 1986). Deposits will also tend to be thicker towards the right or left shore in the northern and southern hemispheres, respectively, due to Coriolis deflection of the circulating plumes (Smith and Ashley, 1985).

The deposits will tend to exhibit different structures below shallow and deep water. Below shallow water (generally above the thermocline), overflow and interflow deposits will tend to form massive muds and silts, due to sediment mixing by bioturbation, wave disturbance and wind-generated currents (Smith and Ashley, 1985). Below deeper water, the deposits typically form laminated fining-up sequences of silt and clay (figs 10.73, 10.74). The laminae reflect fluctuations in the grain size and quantity of incoming sediment, due to daily, meteorological or annual water and sediment discharge cycles (Church and Gilbert, 1975). Short-term cycles produce thin, normally graded laminae with sharp basal contacts and gradual fining-up within each unit. In contrast, annual cycles can produce distinct silt-clay couplets known as *varves*, with sharp contacts between the coarse and fine components, due to marked differences in sediment supply between summer and winter (Ashley, 1975; Ohlendorf et al., 1997). The coarse component is deposited from overflow and interflow plumes generated during the ablation season, and the fine component records gradual settling of the finest material during winter when there is little or no incoming water to the lake

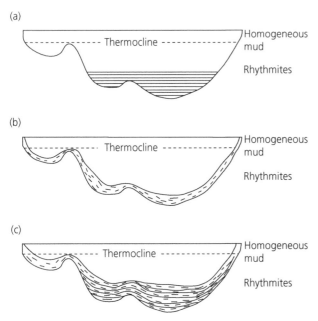

Figure 10.73 Spatial variations in lake sediment characteristics under different dispersal mechanisms. (a) Overflow–interflow. (b) Underflow. (c) Overflow–interflow and underflow (After Smith and Ashley, 1985).

(fig. 10.75). Winter sedimentation commonly takes place below a cover of surface ice. Thin layers of coarser material can occur within winter clay layers in varves, deposited during winter storms or due to sediment reworking on the lake floor. Varves show a proximal to distal decrease in the thickness of the silt component relative to the clay, and also an overall reduction in mean grain size (Ashley, 1975). They also display marked year-to-year variations in thickness, which are valuable in establishing correlations between different parts of a basin (Andren et al., 1999).

Laminated glacilacustrine sediments can be deposited by the combined action of overflows, interflows and underflows (turbidity currents). It is relatively straightforward to distinguish turbidites and overflow/interflow deposits because the latter are regularly and thinly bedded and do not contain current structures such as ripple cross lamination (cf. section 10.5.5). The relative thicknesses of the underflow and overflow/interflow components of composite sequences varies with position with respect to sediment sources. In locations close to sediment influx points, turbidites tend to predominate, with thin clay layers representing short-term pauses in sedimentation (Smith and Ashley, 1985). In contrast, sedimentation in distal locations tends to be dominated by settling from overflows and interflows, with only minor input from turbidity currents. The underflow and overflow/interflow components in composite laminated deposits represent very different rates of sedimentation. Turbidity currents in glacial lakes may last only a few minutes, but deposit several centimetres of sediment; whereas quiet-water settling of clays may only deposit a few millimetres or less

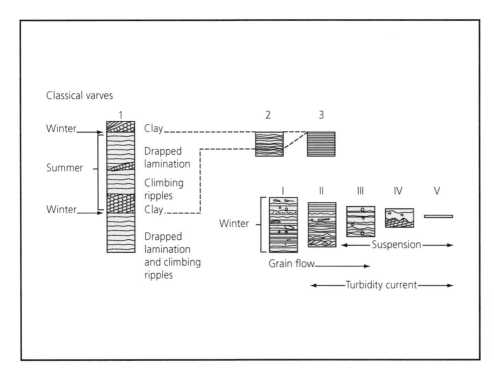

Figure 10.74 Schematic diagram to show the internal characteristics of laminated sediments (varves vs other rhythmites) and their positions of deposition (After Eyles and Miall, 1984).

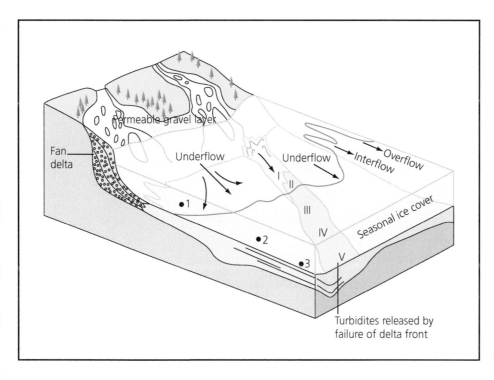

Figure 10.74 (*Continued*)

Rhythmite type	Surge deposit	Varve	
Sediment dispersal mechanisms	Slump-generated Surge currents	Suspension setting	Winter
		Overflow-unterflows Underflows Surge currents	Summer
Time for depositing each rhythmite	Minutes	1 year	

Figure 10.75 A comparison of surge deposits and varves (After Smith and Ashley, 1985).

over many months. Thus, laminated sediments may represent depositional cycles on many different timescales. The term 'varve' should only be used where a seasonal depositional cycle can be clearly demonstrated (Eyles and Eyles, 1992). Recent micro-scale examinations of rhythmically laminated lacustrine sediments have revealed facies characteristics typical of seasonal varve deposition, such as regular alternations between silt-dominated and clay laminae, sharp contacts between clay lamina and overlying silt-dominated lamina, and the normal grading or massive appearance of sand/silt laminae in the coarser parts of sequences (Palmer et al., 2007). In the Canadian Arctic, varve records have been linked to climatic and meteorological events (Lamoureux and Gilbert, 2004a; Hambley

and Lamoureux, 2006; Lamoureux et al., 2006; Chutko and Lamoureux, 2008), and are increasingly being used in conjunction with other palaeolimnological data for high-resolution palaeoenvironmental reconstruction in both glaciated and non-glaciated or nival catchments (e.g. Lamoureux, 2001; Lamoureux and Gilbert, 2004b; Forbes and Lamoureux, 2005; Tomkins and Lamoureux, 2005; Gilbert et al., 2006; Lamoureux et al., 2006).

10.6.4 LAMINATED GLACIMARINE SEDIMENTS

Overflow and interflow plumes of suspended sediment tend to be more important in glacimarine environments

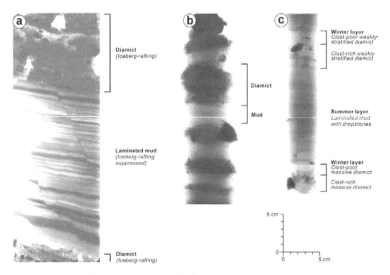

Figure 10.76 Laminated glacimarine sediments identified on X-radiographs of offshore sediment cores. Each example shows the products of alternating suspension sedimentation from meltwater plumes and iceberg rafting. (a) Laminated mud bracketed by diamicton, reflecting the change in importance of iceberg rafting in Scoresby Sund, East Greenland. (b) Alternating stratified diamict and mud that is predominantly massive, but occasionally weakly laminated, Nansen Fjord, East Greenland. Note the sparsity of IRD in the mud units. Muddy facies in both (a) and (b) are interpreted as the product of suspension sedimentation from meltwater plumes during intervals when iceberg delivery was suppressed by the development of shorefast sea ice. (Ó Cofaigh and Dowdeswell, 2001) (c) Glacimarine varves from Disenchantment Bay, Alaska (After Cowan et al., 1997).

than glacial lakes, due to the buoyancy of fresh meltwater in dense, saline seawater. Turbid meltwater plumes emerging from efflux points below grounded ice can rise to the surface or to intermediate depths, and can be transported for many tens of kilometres from the ice front (Syvitski et al., 1987). The buoyancy of freshwater plumes in seawater also means that coarser grains, up to sand size, can be transported in suspension, and plumes are often very prominent at the mouths of glacially fed rivers (Nemec, 1995). Consequently, laminated sediments are common in deglaciated fjords and occur sporadically on high-latitude continental shelves, wherever reworking by iceberg scouring is ineffective (see Ó Cofaigh and Dowdeswell, 2001 for a thorough review). The lamination reflects the cyclicity of suspension sedimentation processes, resulting in the deposition of couplets of a coarse lamina of fine sand or coarse silt fining upwards into a silt or clay lamina (e.g. Cowan et al., 1999). The number of dropstones varies according to the sea ice conditions prevalent at the time of deposition. Some glacimarine laminations may contain no ice-rafted debris (IRD), due to the dominance of extensive shorefast sea ice (Dowdeswell et al., 2000). A laminated appearance can also be imparted by the alternations between IRD-poor, fine-grained muds and stratified, matrix-supported diamictons (fig. 10.76; Cai et al., 1997; Jaeger and Nittrouer, 1999; Smith and Andrews, 2000). The laminated muds in such sequences are interpreted as the products of suspension sedimentation from turbid meltwater plumes (termed *plumites* by Hesse et al., 1997), whereas the stratified diamictons reflect rain-out of IRD combined with suspension settling when iceberg melting is dominant. Such deposits have been referred to as *glacimarine varves* by Cowan et al. (1997), who regard the diamictons as winter layers, introduced by iceberg sedimentation, when plumes are suppressed due to reduced meltwater production (fig. 10.76).

The terms *cyclopel* and *cyclopsam* (with a silent 'p') were introduced by Mackiewicz et al. (1984) for the couplets deposited from overflow and interflow plumes in glacimarine environments. This terminology uses the Greek root *cycl* to refer to the cyclic nature, and the Greek words *pel* and *psam* to denote muddy and sandy units, respectively. Deposition from turbid plumes typically produces couplets of silt and mud (*cyclopels*) and sand and mud (*cyclopsams*). Each couplet has a sharp lower contact, recording the sudden onset of deposition of the coarsest material, and is normally graded, recording the gradual settling of progressively finer material (fig. 10.77; Mackiewicz et al., 1984; Cowan and Powell, 1990). Sedimentation rates can be very high, as much as 15.4 cm of sand in 19 hours near the margins of temperate tidewater glaciers (Powell and Molnia, 1989).

The grading in cyclopels and cyclopsams records variations in sediment supply and settling rates, controlled by fluctuating meltwater stream discharge, tides and wind shear (Cowan et al., 1988). Tidal influence was clearly demonstrated by Cowan and Powell (1990) in a study of sedimentation in McBride Inlet, Alaska, near the tidewater snout of the McBride Glacier. Sediment traps positioned on the sea floor collected the largest amounts of sand within the 2 hours following each low tide, and the least amount around 1 hour after high tide, when silt and mud deposition was more important. This pattern reflects the dynamics of sediment movement during the tidal cycle. When the tide is rising, water velocities are high, maintaining coarse material at high levels. This material

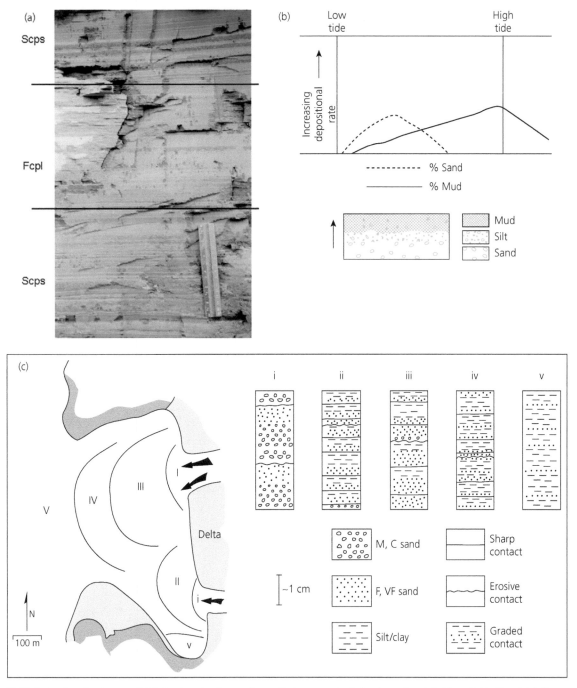

Figure 10.77 Cyclopels and cyclopsams. (a) Cyclopels and cyclopsams in a vertical sequence in glacimarine sediments in Clements Markham Inlet, Ellesmere Island, Arctic Canada. (T.G. Stewart) (b) The relationship between cyclopel/cyclopsam deposition and tides. (Cowan and Powell, 1990, reproduced with permission of the Geological Society of London) (c) The distribution of sequences of sedimentary structures in Riggs Embayment, Glacier Bay, Alaska. The pattern of sedimentation is directly related to distance from stream mouths, Zone i being dominated by proximal downslope processes, and Zones ii–v being characterized by more distal suspension settling (Phillips et al., 1991, reproduced with permission of the Geological Society of America).

gradually sinks towards the bottom during high and ebb tide, finally settling out in the slack water around low tide. This cycle happens against a background of continuous settling of fine silt and clay particles. Thus, each day, two laminae will be produced (one for each tidal cycle), plus additional laminae related to fluctuations in sediment and meltwater influx, wind patterns and sediment reworking on the sea floor (Cowan et al., 1988).

Cyclopsams are generally deposited within 1 km of the point where the meltwater plume enters the sea. Cyclopels may also be deposited in proximal environments during periods of low discharge, but are more common between one and several kilometres of the efflux point (Mackiewicz et al., 1984; Powell and Molnia, 1989). Both cyclopsams and cyclopels can be inter-bedded with turbidites, other mass flow deposits and IRD, particularly in proximal environments (fig. 10.76; Powell and Molnia, 1989; Ó Cofaigh and Dowdeswell, 2001).

10.6.5 ICE-RAFTED DEBRIS AND UNDERMELT DEPOSITS

Material rafted into marine or lacustrine environments by floating ice, in the form of icebergs or ice shelves, ranges in size from clay to boulders (Gilbert, 1990). Collectively, such material is known as *ice-rafted debris* or *IRD*, and can be very prominent in glacimarine and glacilacustrine depositional sequences. Isolated clasts dropped onto a lake or sea bed from floating ice are termed *dropstones*,

and may occur in both massive and laminated sediments. They also occur as outsized clasts in current-bedded bottom sediments, but can be confused with isolated particles that have rolled or bounced out beyond coarse mass flows in more proximal environments. Where the surrounding sediments are laminated, dropstones can be identified by the deformation or penetration of underlying laminae, as described by Thomas and Connell (1985; fig. 10.78).

Sustained rain-out from below ice shelves or multiple icebergs can produce thick deposits of mud, pebbly mud or diamicton. Many terms have been proposed for deposits formed by rain-out below ice shelves, including *undermelt till, subaquatic melt-out till, subaqueous basal*

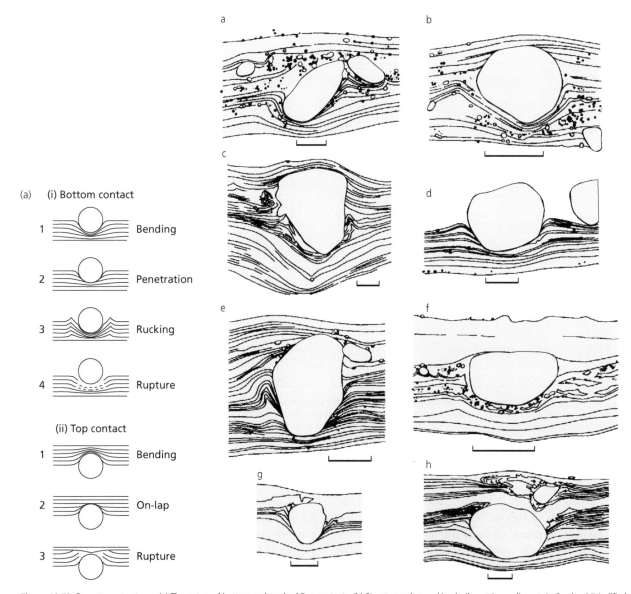

Figure 10.78 Dropstone structures. (a) The nature of bottom and top bedding contacts. (b) Structures observed in glacilacustrine sediments in Scotland (Modified from Thomas and Connell, 1985).

till, grounding-line till and *dropped para-till* (Dreimanis, 1989). In the latter term, the Greek prefix *para-* indicates that the material is closely related to till, but by strict definition is not a till. However, because deposition by rain-out below ice shelves is by settling under gravity, we prefer to avoid the term 'till', and recommend *undermelt diamicton*, as proposed by Gravenor et al. (1984). Where dropstones can be identified, the terms *dropstone diamicton* and *dropstone mud* can be employed, depending on their grain size characteristics. A cut-off point of 10 per cent clasts per unit area is a convenient division between dropstone diamictons and muds (Powell, 1984). In practice, it may not be possible to distinguish undermelt diamicton from dropstone diamicton deposited from multiple icebergs using internal characteristics alone.

Dropstone diamictons and muds form extensive, blanket-like sheets draped over the pre-existing topography, and can be massive or weakly stratified (fig. 10.79). Stratification can be particularly pronounced if ice-rafting occurs alongside other depositional processes, such as the settling of suspended plumes, bottom tractive current activity or sediment gravity flow (Powell, 1984). Stratification will tend to be destroyed by bioturbation, resulting from the activities of bottom-dwelling invertebrates. Other signs of biological activity in dropstone muds can include trace and body fossils, including shells and microfossils.

The rate of deposition of IRD is a function of the debris content of the ice (debris concentration and debris-layer thickness) and the frequency of iceberg passage, which reflects the calving rate and distance from the glacier margin (Dowdeswell and Dowdeswell, 1989; Dowdeswell and Murray, 1990). Sequences of dropstone diamictons and muds may be only a few centimetres thick below small glacier-contact lakes, but can build up to tens of metres thick on glacially influenced coasts.

The grain size distributions of dropstone diamictons and dropstone muds are highly variable, depending on the character of the parent debris and the degree of fall sorting, water column winnowing and bottom winnowing that occurs prior to deposition (e.g. Anderson et al., 1980). Clast-rich dropstone diamictons are typically associated with proximal environments, where they form by debris melt-out below floating glacier ice or frequent icebergs (Hambrey, 1994). If fine-grained sediment is winnowed out in the water column, the deposited sediment can be a clast-supported diamicton or gravel (Powell, 1984). Dropstone muds are matrix-supported diamictons with strongly bimodal particle-size distributions, reflecting dominant suspension sedimentation and minor quantities of dropstones.

Particle morphology is inherited from the parent glacial debris. Characteristics typical of actively transported debris (sub-rounded and sub-angular clasts, faceting and

Figure 10.79 Dropstone muds/silts, diamictons and iceberg dump structures. (a) Dropstone-rich, laminated to massive silts, sands and gravels interbedded with subaqueous mass flow diamictons, Loch Quoich, Scotland. (b) Dropstone in glacimarine silts and clays, illustrating bending of lower bedding contact and on-lap of upper beds, Phillips Inlet, north-west Ellesmere Island, Arctic Canada. (c) Iceberg dump structure in glacilacustrine shallow foreset sands, Drymen, Scotland (D.J.A. Evans).

striations, and crushed quartz grains) are common, indicating the importance of basal debris; but on a local scale, passively transported debris may predominate, reflecting supraglacial debris sources or debris eroded beneath cold ice. No systematic differences in particle morphology are evident between ice-proximal and ice-distal environments (Hambrey, 1994).

Dropstone a-axis fabrics will be influenced by the character of the bottom sediment. Where bottom sediments are relatively stiff, but allow penetration by clasts, they will tend to preserve the high dip angles of clasts that fall vertically through the water column and hit the bottom nose-first, producing a weak cluster fabric with a vertical preferred orientation (Smith, 2000). Conversely, where bottom sediments are either very soft or compacted, vertically impacting clasts will tend to fall sideways because the sediment either cannot hold the clast upright or the clast cannot penetrate the sediment, resulting in a girdle fabric (Domack and Lawson, 1985). Preferred orientations can also develop in response to bottom current activity and reorientation by mass movements. More isotropic fabrics can reflect a combination of these processes and the varying influence of particle shape.

Iceberg dump mounds are formed where icebergs roll over and rapidly dump large quantities of debris in a single location (Thomas and Connell, 1985; Benn, 1989a). These are planar-based mounds of coarse materials, which can occur in isolation, offlap each other or be stacked vertically and inter-stratified with finer-grained sediments (fig. 10.80). They are typically symmetrical cones with a maximum height of 2 m and a width-to-height ratio of 6:1. Asymmetrical cones are produced by multiple dumping events due to repeated overturning of, or basal melt-out from, the same iceberg. Mound bases can be horizontal or inclined, depending on the form of the substratum on which they lie. Clasts at the base of the mounds will commonly deform underlying fine-grained sediments. The mounds themselves consist of poorly sorted gravels with bedding dipping outwards from the cone centre, or comprise a central core of clast-rich diamicton grading upwards and outwards to poorly sorted gravel. These coarse sediments can be inter-stratified with fine sediments towards the top flanks. Iceberg dump events can introduce a wide range of sediment sizes with different settling velocities to the water column, and turbid underflows can transport suspended sediments laterally over a wide area, forming graded turbidites (Gilbert, 1990). The open texture of some diamictons in iceberg dump mounds led Thomas and Connell (1985) to suggest that rafted blocks of debris-rich ice had dropped, then melted out on the bottom. A related process is the rafting of frozen pre-existing sediments by sea ice (Gilbert, 1990). Thomas and Connell (1985) proposed the term *iceberg dump till* for the sediments in iceberg dump mounds, but we would include these sediments in the category of iceberg *dropstone diamictons*, in line with Powell (1984), because they are not tills in the strict sense. The dump structures described above do not display any evidence for faulting, scour or other deformation associated with iceberg contact, so are not produced directly by ice.

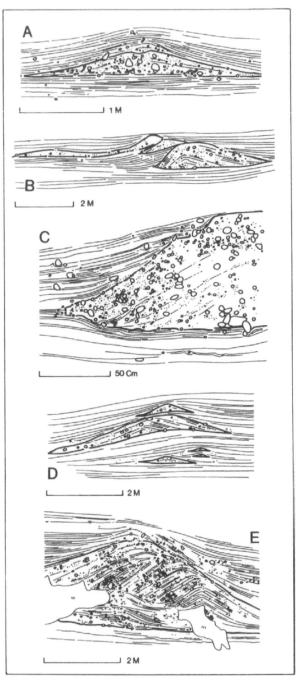

Figure 10.80 Iceberg dump and grounding structures observed in glacilacustrine sediments in Scotland. (A) single symmetric structure. (B) Stacked asymmetric, laterally overlapping structure. (C) Detailed marginal contact. (D) Vertically stacked, symmetric structure. (E) Compound, stacked asymmetric structure (Thomas and Connell, 1985, reproduced with permission of SEPM).

A particular style of lake ice deposition has been proposed by Hall et al. (2006) to explain the sediment–landform assemblages of former lake beds in the Antarctic Dry Valleys. They identify a transport pathway for glacial debris that entails melt-out onto fringing lake ice, subsequent reworking on the lake ice surface and final deposition through moats formed at the terrestrial lake

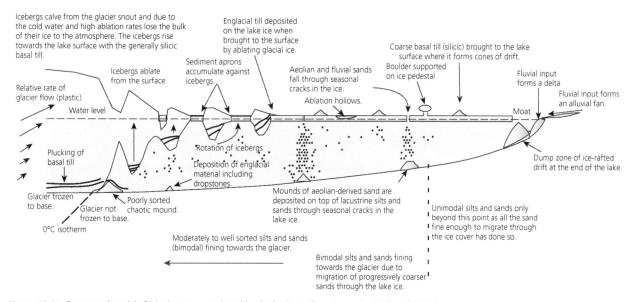

Figure 10.81 Conceptual model of lake ice conveyor deposition in the Dry Valleys, Antarctica (Hall et al., 2006).

margin (fig. 10.81). The resulting mounds of sediment or *lake ice conveyor deposits* are similar to iceberg dump mounds, except that they occur as linear ridges marking the former locations of moats.

10.6.6 ICEBERG GROUNDING STRUCTURES AND SEDIMENTS

Diamictons and clusters of clasts can be deposited directly from icebergs without intervening gravitational transport if bergs lodge on the sea or lake bed for long periods and debris melts out in situ. Diamicton produced by this process has been termed *iceberg till* (Powell, 1984; Dreimanis, 1989), on the grounds that the sediment is deposited directly from glacier ice and is therefore a till. However, although glacier ice is involved, it is no longer part of a glacier, and this usage is not adopted in this book. Instead, we use the term *iceberg contact deposits*.

The frequency of iceberg scouring is so intense in some locations that all primary depositional structures can be destroyed, producing massive, structureless diamicton from pre-existing sediments (Vorren et al., 1983; Dowdeswell et al., 1994b). Fine bottom sediment can also be re-suspended during this process, and the remaining sediment can be depleted of fines depending on currents (Marienfeld, 1992). Massive diamictons produced by ice-keel scouring were termed *ice-keel turbate* by Woodworth-Lynas and Guigné (1990), and can be very difficult to differentiate from non-stratified dropstone diamictons, as well as some subglacial tills, unless some identifiable deformation structures are preserved. Because the exact origin of ice-keel turbates is often difficult to determine, the term is associated with the deformation of lake or sea floor sediments by all forms of drifting ice,

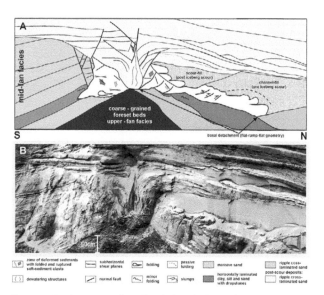

Figure 10.82 An exposure through an iceberg keel scour in a subaqueous ice-contact fan in north-west Germany. The sketch provides a summary of the faulting, folding and mass flow features produced during scouring. Iceberg drag direction was out of the page (Winsemann et al., 2003).

including icebergs and perennial or seasonal sea and lake ice. Post-depositional modification and burial of ice-keel turbate features often makes their identification difficult; Eden and Eyles (2001) provided a good illustration of the modification of the berms and ice-keel turbates in glacilacustrine sediments, where storm waves have reworked the berms of individual scours and deposited hummocky and swaley cross-stratified sands over the remaining deformation structures.

Distinctive faulting structures can be created by single iceberg grounding or ploughing events (Thomas and Connell, 1985; Woodworth-Lynas and Guigné, 1990; Winsemann et al., 2003; Eyles et al., 2005; fig. 10.82). An idealized sequence of events was presented by

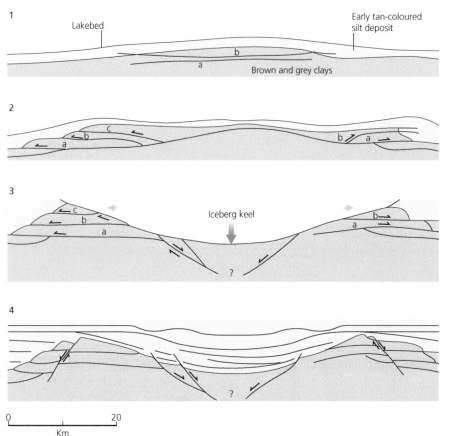

Figure 10.83 Conceptual sequence of the progressive development of fault structures in lake or marine sediments during scouring by a drifting iceberg. (1) Sediment is compacted in front of the scouring iceberg keel and as a result the sediment surface bulges and develops horizontal faults. (2) As scouring continues, the dominant motion of the displaced sediment is radially away from the scour axis, and berms are produced by the horizontal displacement of clay slabs. (3) As the iceberg keel passes over the site, the final emplacement of stacked clay slabs takes place at the scour margins. Vertical loading by the keel causes shallow foundation failure, and shallow faults develop beneath the scour trough. (4) As the keel moves past the site, the stress release causes high-angle normal faulting of the clay slabs on the outer berm margins. Sediment then fills the iceberg scour trough, and faults may be reactivated by the resultant overburden pressure (After Woodworth-Lynas and Guigné, 1990).

Woodworth-Lynas and Guigné (1990), shown in figure 10.83. As the iceberg ploughs forward, slices of sediment are thrust forward and to the sides along low-angle thrust faults. The fault surfaces are highly polished and slickensided, indicating the direction of thrusting. The outward movement of the thrust slices is accompanied by downward movement of a central wedge of sediment, which is typically folded and seamed with microfaults. Downward normal faulting also occurs at the outside margins of the ridges (berms) of thrust sediment. Other features produced by floating lake or sea ice are *soft sediment striations*, which are grooves and intervening flutes with a relief of about 1 cm, superimposed with millimetre-scale striae (Eyles et al., 2005). Such forms are difficult to differentiate from subglacial sole casts unless they are clearly associated with laminated muds in a glacilacustrine sequence.

10.6.7 FOSSILIFEROUS DEPOSITS AND BIOGENIC OOZES

Biological activity can be very important in glacimarine and some glacilacustrine environments, and body or trace fossils can constitute an important component of some facies. Where input of clastic particles is high, such as in proximal environments, fossils may make up only a small volume of the sediment, but in other areas, fossils can make up the bulk of deposits. Sediments made up mainly or entirely from microfossils are termed biogenic oozes. The range of species present in fossiliferous sediments is an important source of palaeoenvironmental data, such as water depth, temperatures, turbidity and salinity (Andrews et al., 1996; Lowe and Walker, 1997; Bradley, 1999).

10.7 WINNOWING STRUCTURES (LAGS, COQUINAS AND BOULDER PAVEMENTS)

Currents flowing in both marine and lacustrine environments can rework bottom materials by transporting away fine-grained sediments, leaving coarse *lags*, especially during periods of non-deposition or reduced sediment influx to the deep water. These have been termed *palimpsest lags* by Powell (1984), because each lag surface in a vertical sedimentary succession represents an incomplete record of former depositional processes. Powell (1984) went on to suggest that glacier readvance over a palimpsest lag would produce a striated clast pavement similar to those produced in subglacial situations (section 10.3.4).

Coquinas are a type of lag deposit consisting of shell-rich bands within water-lain muds and sands. Examples have been described from the early Pleistocene Yakataga Formation in Alaska, which contain whole and broken bivalves, gastropods, brachiopods, worm tubes and barnacles in a matrix of sand or mud, and form extensive planar or concave-up features (Eyles and Lagoe, 1990). The fauna represented in the coquinas prefer shallow-water environments with little or no sediment accumulation, suggesting that the coquinas formed when increased current energy prevented mud deposition. Coquinas within channel fills probably represent shell-rich lags produced by channelized currents, which scoured and winnowed shell-bearing sediments.

Striated *clast* or *boulder pavements* which may have originated partly by current winnowing have been described by Eyles (1988) and Eyles and Lagoe (1990) from the Yakataga Formation, and by McCabe and Haynes (1996) from Late Pleistocene sediment sequences in Ireland. In the Alaskan example, the clasts making up the pavements do not have the attributes normally associated with subglacial lodgement or deformation, such as stoss and lee forms and consistent a-axis fabrics

(sections 10.3.2 and 10.3.4), although their upper surfaces do bear consistently oriented striae, unlike striae formed by sea or shore ice (Hansom, 1983a; Dionne, 1985). The clast pavements were interpreted by Eyles (1988) as the products of winnowing of diamictons by storm waves and currents, followed by glacial abrasion of the coarse lag by later ice advance. Because there is a lack of glacitectonic disturbance of surrounding sediments, Eyles and Lagoe (1990) suggested that the ice margin was partially floating and touched down only locally on high points on the seabed. The Irish boulder pavements described by McCabe and Haynes (1996) occur within waterlain muds and sands. The boulders are closely packed in a mosaic-like pattern, which is generally one boulder thick. Most boulders have striated surfaces, and about 50 per cent bear two or more sets of striae, with no clear dominant striation direction. The boulders are thought to have been delivered by ice rafting, and organized into a pavement by winnowing in the near-shore zone. The inconsistent direction of the striae suggests that they were incised by freely floating ice floes.

CHAPTER 11

SEDIMENT–LANDFORM ASSOCIATIONS

Dave Twigg setting up base station during 1998 survey of Breiðamerkurjökull (D.J.A. Evans)

11.1 INTRODUCTION

Glacial depositional landforms and facies associations are often bewildering in their variety, reflecting a very wide range of processes, topographic settings and glacial environments. Most landforms and sediments forming in modern glacial environments are relatively easy to classify and interpret because we can observe their formation. But some landform–sediment associations are known only from ancient glaciated terrains, such as drained lake or shelf-sea floors, or the former beds of the great mid-latitude ice sheets, and may be of a scale, morphology and structure that has never been observed in active depositional settings. As a result, genetic models for such associations must be based on

interpretation of the geomorphological and geological record. Interpretation of ancient sediments and landforms should be based on detailed, systematic observations, and guided, wherever possible, by appropriate modern analogues and sound theory. Unfortunately, there has always been a tendency in glacial geology for some researchers to indulge in unconstrained speculation, and many weird and wonderful ideas have appeared in the literature to account for certain landforms. In this chapter, we concentrate on those interpretations which are well supported by observations or theory, and generally mention less plausible or outmoded ideas only in passing. This allows us to illustrate the broad range of glacigenic sediment–landform associations and their origins without becoming too distracted by the many curiosities that clutter the literature.

We classify glacial sediment–landform associations according to depositional environment:

(1) subglacial associations;
(2) ice-marginal moraines;
(3) supraglacial associations;
(4) proglacial associations;
(5) glacilacustrine and glacimarine associations.

A problem with this approach is that some glacial depositional landforms are shaped in more than one environment. Cupola hills (see page 503), for example, are ice-marginal glacitectonic moraines that have been modified by subglacial processes; and ice-walled lake plains (section 11.4.4) are supraglacial glacilacustrine associations. We have placed such awkward cases alongside the associations with which they have the closest genetic affinity.

11.2 SUBGLACIAL ASSOCIATIONS

Subglacial associations are among the most enigmatic products of glaciation, and remain central to probably the most contentious debate in geomorphology, the evolution of subglacially streamlined landforms. Advances in our understanding of subglacial process–form relationships have been made, however, particularly by laboratory research into till mechanics, direct observations of ice–bed processes and assessments of landform patterns and distributions. This section begins with a discussion of subglacial facies associations that lack surface expression in landforms, either due to erosion or burial under younger sediments. Such associations may constitute the only record of glaciation in early Quaternary or older strata, and their correct interpretation is very important for reconstructions of former glacier behaviour. We go on to describe subglacial bedforms, such as drumlins, flutings, Rogen moraine (ribbed terrain) and subglacial hummocky terrain, which are distinctive sediment–landform associations produced beneath active glaciers. Lee-side cavity fills and crevasse-squeeze ridges are then described, followed by a discussion on eskers. The section concludes with consideration of subglacial volcanic forms.

11.2.1 SUBGLACIAL FACIES ASSOCIATIONS

Sedimentation and tectonic processes in subglacial environments can result in complex, highly variable associations of diamicton and sorted sediments (see section 10.2). Many facies models have been proposed in the literature, which explain such associations in terms of shifting subglacial environments and processes. Such models, however, are of varying usefulness, and may be strongly influenced by ruling hypotheses and untested assumptions about till genesis and depositional environments. Furthermore, the quantity and quality of the sedimentological data employed in the construction of such models is very variable. As a result, any one subglacial facies association can be interpreted in very different ways by different researchers.

An informative example of changing interpretations of a glacigenic facies association is that of the Catfish Creek Drift Formation, exposed along the northern shore of Lake Erie, Canada. The Catfish Creek Drift Formation is an extensive association of interstratified diamictons, sands, gravels and laminated muds laid down during the late Wisconsinan glaciation (fig. 11.1; Dreimanis et al., 1986). The diamicton units are variably laminated and massive, and some units exhibit isoclinal fold structures or augen-shaped sand and silt inclusions. Sand inter-beds are common, inter-fingering with diamicton units. The Catfish Creek Drift Formation was interpreted by Evenson et al. (1977) as an ice-contact subaqueous association consisting of subaqueous debris flows and water-sorted sediments. Folding and augen structures within diamicton units were regarded as the result of remobilization of pre-existing sediments by subaqueous flowage. This model was challenged by Gibbard (1980), who reinterpreted the association as the result of rain-out or 'undermelt' beneath a floating ice shelf, combined with limited post-depositional flowage and intermittent meltwater discharge. Subsequently, Dreimanis et al. (1986) and Hicock (1992, 1993) conducted detailed analyses of some exposures, including systematic studies of clast morphology and fabric, and concluded that the Catfish Creek Drift Formation records complex switching between subsole deformation, lodgement, flowage and deposition by meltwater in subglacial cavities. (A similar conceptual model of a subglacial till mosaic was reviewed

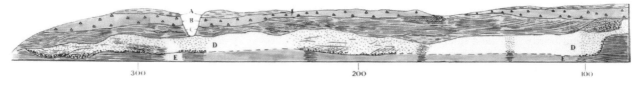

Figure 11.1 The stratigraphic context for the Catfish Creek till, Ontario, Canada. A = Lake Maunee silt; B = Port Stanley till; C = stratified Catfish Creek till; D = sand and gravel; E = laminated clays and silts (Modified from Gibbard, 1980).

on page 395.) Architectural element analysis of complex till sequences in the Great Lakes region by Boyce and Eyles (2000) also suggested deposition by alternating phases of subglacial deformation and ice–bed separation with meltwater sheet flow. The most recent assessment of the Catfish Creek Drift Formation, by Dreimanis and Gibbard (2005), acknowledged the role of multiple subglacial till deposition by competing glacier lobes and the influence of meltwater deposition in subglacial cavities, but retained an 'undermelt' interpretation for some stratified parts of the depositional sequence. The changing interpretations of the Catfish Creek Formation and other complex till stratigraphies in the Great Lakes region reflect changing fashions in glacial geology, and emphasize the need for rigorous analyses using a wide range of descriptive criteria when attempting to establish the origin of complex sediment successions (Hicock, 1992; Hart and Roberts, 1994; Benn and Evans, 1996; Boyce and Eyles, 2000; Evans and Benn, 2004a).

Several scenarios have been proposed to account for vertical and lateral facies variability in subglacial sediment associations. These include:

(1) rheologic superposition of different till facies;
(2) intermittent subglacial drainage during till deposition;
(3) lobal interactions between adjacent ice flow units;
(4) alternations between subglacial and ice-marginal, glacilacustrine or glacimarine environments at oscillating ice margins.

Sedimentary successions resulting from cases (3) and (4) do not comprise single facies associations, but represent two or more superimposed associations formed in different environments or events. As such, they are more appropriately discussed in section 11.6. Successions formed by rheologic superposition and alternate till deposition and subglacial drainage are described in turn below.

Rheologic superposition

Rheologic superposition can be defined as the successive formation of subglacial tills with different deformational histories (Hicock, 1992; Hicock and Dreimanis, 1992; Hicock and Fuller, 1995), and has been inferred from till successions that exhibit systematic vertical changes in fabric characteristics and structures (e.g. Hicock and Lian, 1999). For example, switches between ductile and brittle deformation at a glacier bed may result in partial overprinting of existing till characteristics with new fabrics and structures (section 10.3). This can take the form of the collapse and brittle shear of a ductile horizon, producing boudinage and fracturing of soft-sediment deformation structures (Hicock and Dreimanis, 1992; Hicock, 1992), or the dilation and ductile deformation of hitherto stiff tills, resulting in the rotation of striated, faceted particles away from their original orientations (e.g. Hicock, 1991,

Figure 11.2 Sedimentary and structural features in subglacial till stratigraphy at Kurzetnik, central Poland. (a) A vertical transition from undeformed sands to deformed sands and then to a coarse-grained till (unit A) overlain by a fine grained till (unit B). (b) Sand stringers deposited by meltwater during intermittent ice–bed decoupling between phases of subglacial till (unit B) deposition. (c) Lineaments or fluted surface at the interface between till and sand stringers in unit B, documenting phases of lodgement and ploughing. See figure 10.23 for a conceptual model of time transgressive till formation in which the features illustrated here are used to reconstruct subglacial processes as they are related to phases of till accretion and associated parameters such as hydraulic conductivity and pore-water pressure (Piotrowski et al., 2006).

1992; Benn, 1994a, 1995; Benn and Evans, 1996; Lian and Hicock, 2000). Preservation of a vertical succession of till types, recording a full history of deformation and associated subglacial processes at the site, clearly requires net accretion of till (e.g. Larsen and Piotrowski, 2003; Larsen et al., 2004; figs 10.23 and 11.2), or *constructional deformation* (Hart and Boulton, 1991; Hart, 1995a).

Another example of rheologic superposition is till formed by melt-out of debris-rich basal ice, followed by remobilization in a deforming layer. In this case, structures and textures formed during the shear of debris-rich ice are overprinted during subglacial shear (e.g. Ham and Mickelson, 1994). Shaw (1987) described facies associations in the Edmonton area of Alberta, Canada, which he interpreted as inter-bedded melt-out tills, deformation tills and subglacial channel fills, and Evans (1994) invoked multiple melt-out and deformation events to explain complex till stratigraphies in southern Alberta. However, great care must be taken when interpreting such complex successions, because many till characteristics previously assumed to represent inherited englacial structures are now known to be characteristic of glacitectonites (sections 10.3.2 and 10.3.3), re-enforcing our cautionary comment above about ruling hypotheses and untested assumptions. It is because of our inability to isolate unique diagnostic sedimentological criteria for specific subglacial processes that we have proposed the broader term *subglacial traction till* for the range of subglacial, often stacked sequences of till and associated sediments in section 10.3.

Intermittent subglacial drainage

There are numerous descriptions of till successions that contain inter-beds and lenses of sorted sediments (e.g. Eyles et al., 1982; Brown et al., 1987; Evans et al., 1995; Benn and Evans, 1996; Piotrowski and Kraus, 1997; Piotrowski and Tulaczyk, 1999; Boyce and Eyles, 2000). Sorted inter-beds commonly take the form of broad lenses of silt, sand or gravel, with concave-up lower contacts and nearly planar upper contacts, in addition to thin sand layers and stringers. Lenses may exhibit well-preserved internal structures such as plane bedding and ripple cross stratification, but varying degrees of tectonic folding and attenuation are common at the contacts with overlying tills (fig. 11.3). Clark and Walder (1994) have argued that such successions represent deformation tills and the infills of former braided canal systems developed at the ice–till interface (section 3.4). According to this model, the till units are formed by subsole deformation when the glacier is coupled to the bed, and the canal systems are formed when discharges are too high to be evacuated through the bed and bed separation occurs over large areas. Clearly, the survival of canal fills and other meltwater deposits in a deformational environment requires that they experience only low cumulative strains, either because (1) deformation persists for a short period following canal formation; (2) strain persists but strain rates are very low; (3) net accretion of deformation till occurs and the base of the deforming layer migrates upward, and deeper layers of the deforming layer become immobile (i.e. there is *constructional deformation*; Hart

and Boulton, 1991); or (4) till accretion occurs by passive melt-out following ice stagnation.

This section on subglacial facies associations has drawn on the discussions presented in section 10.3, where we provided explanations for the widely cited examples of overprinted and/or stacked sequences of glacitectonites, subglacial traction tills and melt-out deposits. Summarizing the review of Evans et al. (2006), we provided an overview of the case made by glacial sedimentologists in support of the concept of a temporally and spatially variable mosaic of subglacial processes and forms (cf. Menzies, 1989; Piotrowski and Kraus, 1997; van der Meer et al., 2003). The existence of such a mosaic helps us to explain the complex stacks of tills and associated sediment bodies often identified in former subglacial environments (fig. 11.4). The most important implication of the subglacial mosaic is that we should not divorce rheologic superposition from intermittent subglacial drainage. This has been ably demonstrated by Larsen et al. (2004) and Piotrowski et al. (2006), based on their study of a multiple till sequence overlying glacitectonically deformed outwash in central Poland (fig. 11.2b). The sedimentology, structure, fabric and thin section data suggest that the till sequence was emplaced time transgressively, initially by deformation of sand and accreting till, then by lodgement and ploughing, and ultimately by intermittent decoupling and re-coupling of the bed due to the progressively less efficient drainage of meltwater through the thickening till layer. It is important to acknowledge and understand such conceptual models of subglacial sediment variability and their manifestation in till stratigraphies before we proceed with our discussion on subglacial bedforms, because our models of landform evolution must be compatible with our till production mechanisms. Implications of till variability for regional palaeoglacier flow patterns are discussed separately in section 12.4.6.

11.2.2 SUBGLACIAL BEDFORMS

Subglacial bedforms are longitudinal, transverse or even non-aligned/hummocky accumulations of sediment formed below active ice (Rose, 1987; Menzies and Rose, 1989; Eyles et al., 1999). Longitudinal forms are streamlined features aligned parallel to ice flow, and can be divided into *drumlins*, *flutings* and *mega-flutings*. In valley settings, elongate till masses often extend downvalley from bedrock spurs to form *drumlinoid drift tails* (Mitchell, 1994). Transverse bedforms are classified as *Rogen* or *ribbed moraines*. Additionally, recent research has identified weakly to non-aligned or hummocky terrain that appears to be best explained as a subglacial phenomenon; such terrain must not be confused with supraglacially derived hummocky moraine (see section 11.4.2). Bedforms occur in fields which are positioned in relation to ice

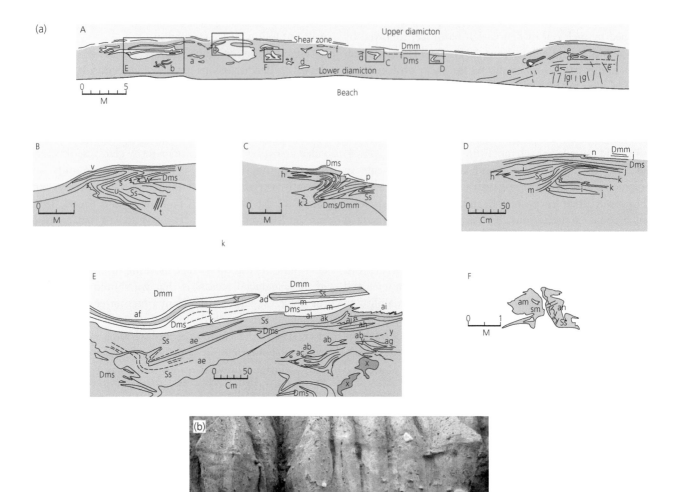

Figure 11.3 (a) Inter-bedded tills, sands and silts at Skipsea, Yorkshire, UK, described by Evans et al. (1995) and Benn and Evans (1996). Tills are variably massive to laminated, and contain clear deformational structures such as isolated folds and smeared out lenses of chalk and unconsolidated sediments. The sorted lenses exhibit both primary depositional structures and tectonic deformation, and commonly grade upward into complexly folded and attenuated inter-beds of silt, sand and diamicton. A = overview of section face showing Skipsea Till and associated intra-beds of stratified sediments; B–F = details of structures outlined by boxes in A with labels locating the following structures: a) deformed stratified, poorly-sorted, coarse and pebbly sands; b) concentrations of chalk clasts; c) concentrations of rounded pebbles; d) deformed sand lenses; e) crude lamination in the till due to subtle grain size variations; f) major discontinuity in the till, marking the top of the zone of deformed sand lenses; g) strongly developed vertical joints; h) interdigitation of till and sand; i) crudely planar stratified sands; j) laminated till comprising red, grey-brown and buff lamina derived from cannibalized soft bedrock; k) chalk stringers; l) laminated till; m) laminated till with chalk stringers; n) chevron fold; o) stratified sand with interstratified minor diamicton beds <10 mm thick; p) till with weak stratification at base and chalk stringers subparallel to lower contact; q) dark brown till with chalk stringers; r) sand with stratification parallel to base of overlying till; s) cross-stratified sands; t) normal faulted sands; u) gentle folds; v) light brown stratified diamicton with sandy intercalations; w) folded stratified diamicton; x) concentrations of subrounded chalk pebbles, some of which are deformed into stringers within the surrounding till; y) sand stringers in till; z) sandy diamicton interstratified with cm thick beds of clayey sand folded into a recumbent fold; aa) deformed stratified clayey sands; ab) stratified sand lenses (dm wide) with convex tops and flat bases; ac) shears in till lined with sand; ad) laminated till grading into massive till; ae) moderately to well sorted, coarse to fine sands; af) rippled medium sands; ag) deformed lenses of rippled sands; ah) fine to very coarse, poorly sorted clayey sands containing pebbles towards the top; ai) stratified diamicton interdigitated with sandy clay; aj) chalk stringer deformed by a diaper; ak) coarse to fine sands with low-angled cross-stratification deformed by small normal faults; al) smooth base of laminated till parallel to the laminations of the underlying sands; am) massive to moderately well-sorted coarse sands; an) faulted, stratified, poor to moderately sorted, medium to fine sands. (b) Isolated channel fills, possibly representing Nye channels on a former glacier bed, in Late Pleistocene sediments exposed on the coast of Tierra del Fuego, southern Chile. The channel is incised downward into compact subglacial traction till and is bounded above by a horizontal erosion surface. The association of the channel and compact till is compatible with the hydraulic theory of Walder and Fowler (1994), which predicts that drainage over rigid or slowly deforming sediments is likely to be via dendritic conduits, in contrast with the braided canals postulated for weak, deformable substrata (Photo: Doug Benn).

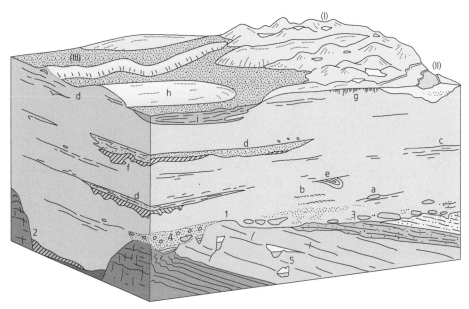

Figure 11.4 Typical subglacial sedimentary sequence and associated surface morphology in a low-relief limestone terrain where stacked tills have been deposited during a single glaciation. The land surface is characterized by: (I) hummocky kame and kettle topography; (II) outwash cut into the till surface and comprising stratified sands and gravels; (III) esker deposited during ice wastage and therefore not truncated by subglacial tills like other channel fills in the subglacial sediment stack below. The base of the glacigenic sedimentary sequence is characterized by: (1) striated rock head; (2) a buried channel/valley with a fill of subglacial sands and gravels and till; (3) glacitectonized rock head, with rock rafts and boulder pavements; (4) the lowermost till, comprising local lithologies and which thickens in the lee sides of rock protuberances as lee-side cavity fills; (5) cold-water karst. The sediments of the subglacial depositional sequence are characterized by: (a) predominantly preferentially aligned, faceted clasts; (b) crude shear lamination produced by the smearing of soft lithologies (glacitectonite/subglacial traction till); (c) slickensided bedding planes (fissility) produced by glacitectonic shear; (d) stratified gravels, sands and clays deposited in subglacial cavities, pipes or canals and truncated by overlying tills (the base of each till unit may be fluted) – they constitute lenses which are elongated in the direction of ice flow and typically internally disturbed by folding and faulting due to post-depositional deformation by glacier/till overriding; (e) folded and sheared-off channel fill; (f) diapiric intrusion of till squeezed up into subglacial cavity; (g) vertical joints produced by post-depositional pedogenic processes; (h) drumlinized surface of upper till sheet; (i) inter-drumlin depressions filled with postglacial solifluction debris and peat. The horizontal scale may range from 10 m to 10 km, and the vertical scale may be 10 cm to 100 m high (Modified from Eyles, 1983a).

divides and ice streams at various times during active glaciation. This means that they are related not just to substrate morphology, local stress variations and sediment supply, but also to ice flow and sediment deformation histories. Longitudinal forms such as drumlins and flutings are clearly very similar to whalebacks and rock drumlins (section 8.4.2), suggesting that some common factors might underlie their formation.

The distinction between drumlins, flutings and mega-flutings is based on the *length* and *elongation ratio* of the bedform. The elongation ratio is defined as:

$$E = l_b/w_b \qquad (11.1)$$

where E is the elongation ratio, l_b is the maximum bedform length and w_b is the maximum bedform width. Elongation ratios and lengths of some drumlins, flutings and mega-flutings are shown in figure 11.5 (Rose, 1987), which depicts drumlins as having axes >100 m long and elongation ratios up to 7:1; flutings are less than 100 m long, with elongation ratios in the range 2:1 to 60:1, or even more; and mega-flutings are elongate forms with long axes greater than 100 m. Drumlinoid forms longer than about 1000 m are termed *streamlined hills*, and very large elongate forms many tens of kilometres long, hundreds of

metres wide and more than 25 m high have been recognized on satellite images and termed *mega-scale glacial lineations* (MSGL) by Clark (1993). However, the dimensions of 'drumlins' and 'flutings' described in the literature commonly differ from those defined in figure 11.5. In recognition of this, Rose has stressed that drumlins, flutings and mega-flutings form a continuum, and any lines drawn between them can only ever be arbitrary.

We now consider the morphology and composition of flutings, mega-flutings/MSGL drumlins and Rogen (ribbed) moraine, and discuss modern theories for their origin.

11.2.3 FLUTINGS

Flutings, also known as *flutes* or *fluted moraines*, are elongate streamlined ridges of sediment aligned parallel to former glacier flow (fig. 11.6; Boulton, 1976; Rose, 1987; Gordon et al., 1992). They are generally a few tens of centimetres to a few metres high and wide, and occur in groupings of sub-parallel ridges on many modern glacier forelands, and some older glacial landscapes. Flutings have low preservation potential, because they are readily degraded by wind and water, and consequently are much more common on modern forelands than on older terrain

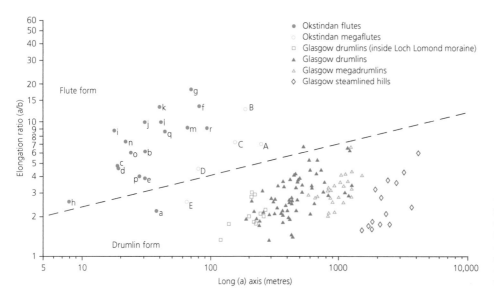

Figure 11.5 The relationship between length and elongation ratio of flutings, mega-flutings, drumlins, mega-drumlins and streamlined hills from the Glasgow area, Scotland, and Okstindan, Norway. The dashed line represents a possible quantitative differentiation of flutings and drumlins (Rose, 1987; reproduced with permission of Balkema).

Figure 11.6 Flutings on modern glacier forelands. (a) Fluting field on the foreland of Isfallsglaciaren, Sweden, the site of Hoppe and Schytt's (1953) seminal study. (b) Degraded flute on the Skalafellsjökull foreland, Iceland, showing its relationship with a lodged and faceted boulder. Note the striae and the perched erratic on the up-ice side of the boulder (D.J.A. Evans).

(Glasser and Hambrey, 2001; Evans and Twigg, 2002). Flutings commonly begin downglacier from lodged boulders or clast clusters, although this is not always the case. They are generally straight or very gently curving, but may swerve abruptly around boulders before resuming a straight course (fig. 11.6b; Gordon et al., 1992). They tend to consist mainly of subglacial till, although cores of deformed, pre-existing sediments have been reported (Paul and Evans, 1974; Boulton, 1976, 1987; Rose, 1989; Eklund and Hart, 1996).

Several models of fluting formation have been proposed (Gordon et al., 1992). The most widely accepted model regards flutings as the product of subglacial sediment deformation in the lee of obstructions on the bed (Boulton, 1976; Benn, 1994a; Eklund and Hart, 1996). According to this interpretation, flutes are initiated when weak, saturated sediment is squeezed under pressure into small lee-side cavities behind obstructions. The sediment flows more readily in response to stress than the ice, so the cavity becomes filled up and prevented from closing. The sediment-filled cavity can then evolve into a fluting by one of two mechanisms:

(1) As sediment flows into the lee-side cavity, the drop in pressure may cause it to freeze and adhere to the basal ice. The frozen sediment is then carried forward by the ice, and new material is added by deformation into the cavity at the upglacier end (Gordon et al., 1992).

(2) The sediment within the cavity can remain unfrozen, and new sediment added at the downglacier end as the cavity is carried forward by ice flow (fig. 11.7; Boulton, 1976; Morris and Morland, 1976; Benn, 1994a). Flutings formed by the second mechanism can become frozen if they propagate into areas where the base of the glacier is below the pressure-melting point (Boulton, 1976; Gordon et al., 1992; Hart, 1995b).

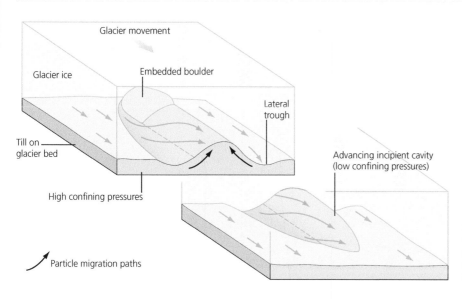

Figure 11.7 A conceptual model of fluting formation by the subsole deformation of sediment into an advancing incipient cavity on the down-ice side of a lodged boulder (Benn, 1994a).

Evidence for sediment deformation in flutings is provided by folds and faults in sorted sediments, magnetic fabrics and clast fabrics (Paul and Evans, 1974; Boulton, 1976; Benn, 1994a; Eklund and Hart, 1996). At several sites, converging *herringbone fabrics* have been reported, in which particles on the flanks of flutings tend to be orientated obliquely downglacier towards the fluting axis (fig. 11.8). Herringbone fabrics in flutings on the forelands of Norwegian glaciers have been used in two alternative interpretations of the relationship between localized ice flow directions and fluting form. Rose (1989) suggested that herringbone fabrics in flutings at Austre Okstindbreen record former patterns of ice flow over the bed, ice flow being convergent at the proximal end of the fluting and divergent at the distal end, due to flow disturbances in the basal ice set up by boulders, bedrock knobs and boulder clusters at the ice–bed interface. However, striation data on boulders in other flutes on the same foreland indicate parallel ice flow over the features (Rose, 1992). At Slettmarkbreen, Benn (1994a) similarly showed that herringbone fabrics do not conform to the ice flow direction recorded by striae on lodged boulders within the flute. His alternative interpretation of these data is that herringbone fabrics record patterns of strain within deforming till rather than shear at the ice–bed interface (fig. 11.7). The herringbone pattern, therefore, records oblique strain patterns within till, reflecting stress gradients set up in the vicinity of boulders.

Benn and Evans (1996) have argued that there may be two contrasting types of fluting: (1) *tapering flutings*, which become lower and narrower with distance downglacier; and (2) *parallel-sided flutings*, which maintain constant cross profiles for considerable distances. Tapering flutings occur on the forelands of Slettmarkbreen and Okstindbreen, Norway (fig. 11.8), whereas examples of parallel-sided flutings occur in front of Breiðamerkurjökull and Skalafellsjökull in Iceland (Evans and Twigg, 2002;

Evans, 2003b) and on the foreland of the Icelandic surging glacier Brúarjökull (Evans and Rea, 1999, 2003). The Brúarjökull flutings are remarkably consistent in morphology over a distance of >1 km and continue to emerge from the downwasting snout. Among the parallel-sided flutings at Brúarjökull are short sediment prows and tapering flutings on the downflow sides of boulders. Evans and Rea (2003) interpreted these forms as incipient flutes formed by boulders that lodged at the glacier bed during surge termination. The implication here is that longer flutings record fast flow events. In contrast with the herringbone fabric patterns in tapering flutings, fabrics from parallel-sided flutes may be either parallel to the flute axis or converge only slightly downglacier (Boulton, 1976; Benn, 1995). However, fabrics from the surge flutings of Brúarjökull are very weak and do not display herringbone or flute-parallel trends, suggesting that there was only weak ice–bed coupling and subglacial sediment may have been emplaced by flowage into grooves in the ice base (Evans and Rea, 2003).

Benn and Evans (1996) have suggested that the contrasting forms of parallel-sided and tapering flutes may reflect differences in till rheology or strain response to stress (section 4.2.3). Parallel-sided flutes at Breiðamerkurjökull are composed of weak, dilatant till which can deform much more readily than ice (Boulton and Hindmarsh, 1987; Benn, 1995). Therefore, pressure differences within grooves in the basal ice are equilibrated by till flow rather than ice creep, so that groove closure is prevented and the fluting can propagate downglacier all the way to the margin. In contrast, the till in the Norwegian tapering flutes was apparently stiff and non-dilatant, and there is likely to have been a lower contrast between the viscosity of the ice and till (Benn, 1994b). Therefore, grooves in the basal ice may have been able to contract by ice deformation, leading to downglacier decay of the flute form. According to this view, the length of tapering flutings is a

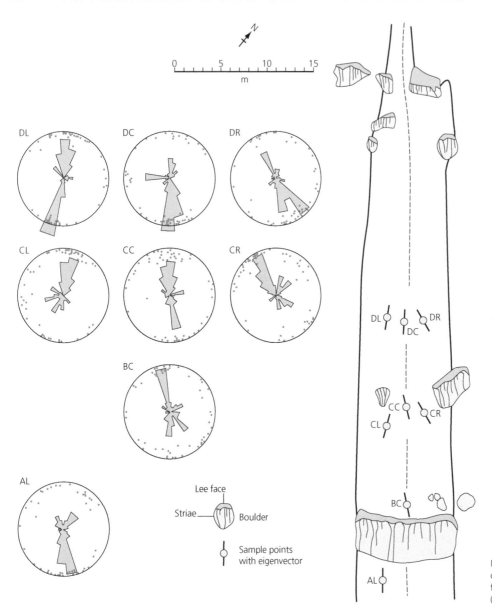

Figure 11.8 'Herringbone' pattern of clast fabrics in a tapering flute on the foreland of Slettmarkbreen, Norway (Benn, 1994a).

function of the relative strain rates of ice and till, whereas the length of parallel-sided flutings is the product of the ice velocity and flow duration. Fluting length, however, is ultimately limited by the position of the ice margin relative to the point of initiation.

The distribution of flutings on glacier forelands may be dictated by the local bed conditions, specifically topography and sediment strength. This has been demonstrated for flutings on the foreland of Saskatchewan Glacier in the Canadian Rockies by Hubbard and Reid (2006). They showed that flutings were produced only where the glacier overran a topographic high immediately down-valley from an overdeepening. This resulted in elevated pore-water pressures in silt/clay-rich diamictons, leading to a reduction in sediment strength and the deformation of the substrate.

Flutings have been observed frozen into the base of glaciers, leading Gordon et al. (1992) to suggest that they

may originate by the melt-out of longitudinal thickenings of basal debris-rich ice. This implies that the characteristics of these flutes are inherited from strain patterns in the basal ice layers rather than from the ice–bed interface (Rose, 1989) or subsole deformation (Benn, 1994a). However, there is the possibility that the flutings described by Gordon et al. (1992) were produced by the deformation of unfrozen sediment beneath thicker temperate ice and then frozen to the glacier sole during ice thinning.

The flutings studied by Rose (1989) at Austre Okstindbreen are composed of a distinctive till unit, allowing the rates of sediment transfer involved in fluting formation to be estimated. Rates range from $0.0077 \, \mathrm{m^3 \, m^{-2} \, yr^{-1}}$ to $0.38 \, \mathrm{m^3 \, m^{-2} \, yr^{-1}}$ (cubic metres per square metre of bed area per year). Application of these figures to larger drumlins and flutings at other locations reveals that subglacial bedforms may be created within a period of between 4 and 400 years irrespective of their size, thus indicating rapid development. It should be noted however, that these sediment transfer rates are for one glacier at one time period, and may not be typical. For example, the parallel-sided flutings on the Brúarjökull foreland in Iceland were produced during a surge that lasted for less than 12 months (Evans and Rea, 2003).

An erosional origin has been suggested for some flutings. Evidence that 'erosional fluting' occurs on a small scale is provided by *sole marks* below the bases of some subglacial tills and on some recently deglaciated surfaces (e.g. Ehlers and Stephan, 1979; Hart, 1995b; Evans et al., 2006a). These sole marks appear to have been ploughed by clasts protruding from the glacier sole, which were dragged over the substratum. Clasts tend to occur at their downglacier ends in contrast with constructional flutings which commonly have clasts at their upglacier ends. A similar erosional model for larger lineations, which invokes ploughing ice keels in the glacier sole, is presented in the next section. Certain crag and tails (section 8.4.3) and flutings with cores of pre-existing stratified outwash have also been regarded as erosional features produced by differential erosion downflow of protecting obstacles such as lodged clasts (e.g. Fuller and Murray, 2000).

11.2.4 DRUMLINS, MEGA-FLUTINGS AND MEGA-SCALE GLACIAL LINEATIONS (MSGL)

Drumlins and MSGL are among the most enigmatic of glacial landforms, and over the last hundred years a very large body of literature and a great many theories of their formation have been published (cf. Menzies, 1984; Menzies and Rose, 1989). In fact, Sugden and John (1976) went so far as to state that 'there are almost as many theories of drumlin formation as there are drumlins'. The problem of determining the origin of large-scale streamlined forms stems from the lack of observations of analogous modern environments, so that researchers must base their arguments on the morphology, internal composition and distribution of features that formed below long-vanished glaciers, where little is known of former subglacial conditions. Recent advances have been made by combining detailed sedimentological studies of drumlin composition with modern glaciological theory and some restricted subglacial processes observations, but many important questions remain. Menzies (1979) provided a very thorough review of the literature up until the late 1970s. Here, we concentrate on the more recent literature, focusing on the most plausible theories of formation.

Morphology

Derived from the Gaelic word *druim* (rounded hill), drumlins are defined as: 'typically smooth, oval-shaped hills or hillocks of glacial drift resembling in morphology an inverted spoon or an egg half-buried along its long axis. Generally the steep, blunter end points in the up-ice direction and the gentler sloping, pointed end faces in the down-ice direction, these two ends being respectively known as the stoss and lee sides' (Menzies, 1979).

Drumlin long axes are oriented parallel to the direction of ice flow, with higher and wider stoss ends which taper down to a pointed lee end. Although our definition of a drumlin gives the impression that this descriptive label may only be applied to features with an oval shape, there is in fact a continuum of drumlinoid forms (fig. 11.9–11.11). Long, narrow drumlins have been termed *spindle* forms, and broader, often asymmetrical drumlins are known as *parabolic* forms. More complex forms also exist, such as *transverse asymmetrical drumlins* or *superimposed drumlins*, consisting of longitudinal ridges superimposed on transverse or oblique hills (Rose and Letzer, 1977; Rose, 1987). Compound spindle drumlins with multiple crests and asymmetrical spindle drumlins with overdeepened hollows along one margin have been observed in Chile by Clapperton (1989). Over the years, numerous indices have been devised to describe drumlin morphology (e.g. Evans, 1987; Coude, 1989; Mitchell, 1994; Smalley and Warburton, 1994). Using tens of thousands of drumlins in Britain and Ireland (fig. 11.12), Clark et al. (2009) found that drumlin length, width and elongation ratios form unimodal distributions, strongly suggesting that they form a single population of landforms. Mean drumlin lengths, widths and elongation ratios are 629 m, 209 m and 2.9 m, respectively.

With increasing elongation ratios, drumlins grade into *mega-flutings* or *megaflutes*. Using Landsat satellite imagery, Clark (1993) identified particularly large *mega-scale glacial lineations* (figs 11.10 and 11.11). The term lineation was proposed in order to accommodate the notion that what we are looking at is potentially a grooved surface rather than depositional ridges/flutings. This has prompted the use of the term *giant glacial grooves* (e.g. Stokes and Clark, 2003a), implying an erosional genesis similar to bedrock *mega-grooves* (section 8.3.4). These lineations are often not easily recognizable on aerial photographs until they are compared with linear features identifiable by a distinct grain on satellite images (fig. 11.10bi). The coverage of individual air photographs is not large enough to view the largest of mega-scale lineations, but they often show up as a series of aligned drift mounds (fig. 11.11b). Critical to a grooving origin for flutings and MSGL (see page 467) are some unique elements of their morphology. First, deep grooves often lie parallel to and alongside individual upstanding ridges (fig. 11.11a), a characteristic interpreted by Clark et al. (2003) as indicative of ploughing of sediment by keels of ice and the piling up of that sediment in adjacent ridges.

Second, flat-bottomed, angular grooves can often be seen adorning the surfaces of other bedforms, giving the impression that an antecedent streamlined till surface has been grooved (fig. 11.11b). Finally, groove widths and depths are often seen to diminish downflow, and ridges will also often peter out in the same direction, characteristics that can be explained by the gradual reduction in size of the ploughing protuberance (this is critical to the ice keel groove-ploughing theory of Clark et al. (2003) (see page 467). Significant advances have been made in the identification of larger subglacial bedforms, using remotely sensed imagery and digital elevation models (Stokes and Clark, 2003a; Jansson and Glasser, 2005b; Smith and Clark, 2005), allowing comprehensive and extensive systematic mapping of former subglacial beds (see section 12.4). Such mapping has progressively

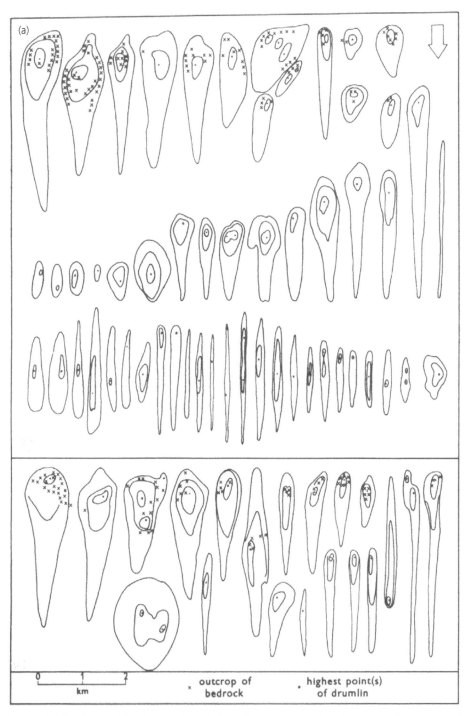

Figure 11.9 Typical drumlin and fluting shapes in (a) central Finland (Gluckert, 1973) and (b) Canada (Shaw, 1983a, reproduced with permission of the International Glaciological Society).

facilitated more representative statistical analyses of sub-glacial bedforms. From such data sources we can measure the dimensions of MSGL; lengths range from 6 to 100 km, widths from 200 to 1300 m, and spacings from 200 m to 5 km. Stokes and Clark (2003) highlighted the remarkable parallel conformity of MSGL, where the standard deviation of lineament orientation amounts to no more than 3.8° over an area of 720 km².

Distribution

Although assessments of individual drumlin shapes are important, the analysis of drumlin distributions and the relationships between drumlins and other landforms perhaps tells us more about the evolution of streamlined shapes and thus the subglacial processes that formed them (e.g. Menzies, 1979; Patterson and Hooke, 1996). We are particularly interested in finding solutions to persistent questions about where under an ice sheet drumlins and associated bedforms are produced and during what phase of a glacial cycle. Together with understanding the mechanisms and implications of bedform superimposition/preservation, once we have answered such questions we can begin complex palaeoglaciological reconstructions.

The distribution patterns of drumlins, mega-flutings and MSGL have been analysed at several scales, and this allows the grouping of similar landforms into *flow sets* (Clark 1994, 1999) according to spatial coherency and general patterns of orientation. Clark (1999) proposed

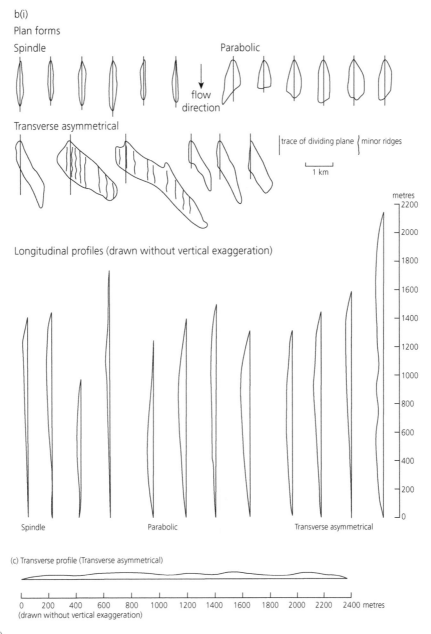

Figure 11.9 (*Continued*)

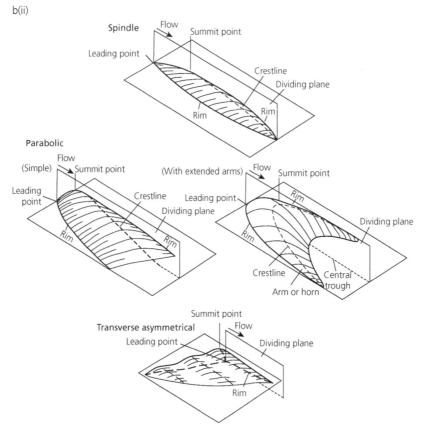

Figure 11.9 (*Continued*)

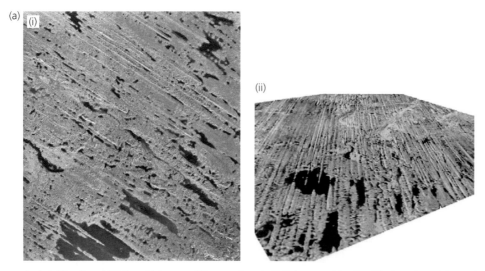

Figure 11.10 Mega-scale glacial lineations. (a) Dubawnt Lake area of Nunavut, Canada. (i) Flutings transitional to ribbed terrain with former ice flow from bottom right. (National Aerial Photograph Library, Ottawa) (ii) Oblique view of a Landsat ETM+ subscene, looking downflow along the Dubawnt Lake mega-scale glacial lineations (MSGL). Note the Rogen/ribbed moraine crossing the grooves. (Stokes and Clark, 2003a) (b) James Bay lowlands, Canada, viewed at satellite image and aerial photograph scales. (After Clark, 1993) (i) Landsat image displaying strong NNW–SSE aligned mega-lineations in the right-hand part, and ENE–WSW aligned ribbed terrain on the left. (ii) Aerial photograph of the central part of the mega-lineations in the Landsat image, showing that the lineations are not easy to detect at this scale, probably due to postglacial dissection (National Aerial Photograph Library, Ottawa).

that the grouping of subglacial bedforms into flow sets must be based on a combination of the following criteria:

(1) *parallel concordance*, whereby each landform has a similar orientation to its neighbour;

(2) *close proximity*, where landforms are closely packed with their neighbours;

(3) *similar morphometry*, whereby adjacent landforms usually display similar forms.

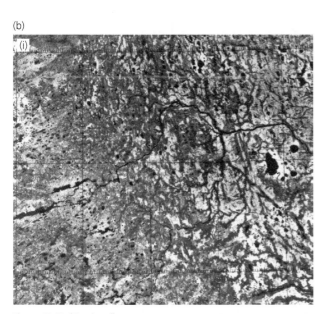

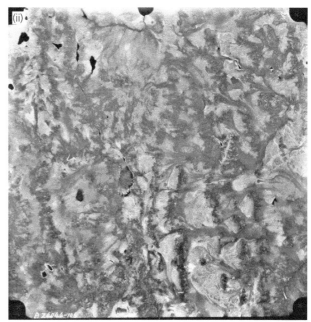

Figure 11.10 (*Continued*)

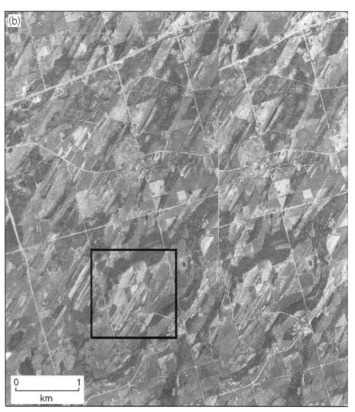

Figure 11.11 Fluting and MSGL characteristics regarded as indicative of substrate grooving by Clark et al. (2003). (a) MSGL in North Dakota, where ridges run parallel to adjacent grooves. (b) Aerial photograph stereopair showing flat-bottomed, angular grooves cutting through the tops of drumlins (best illustrated in boxed area) in the Peterborough drumlin field, Ontario, Canada (National Aerial Photograph Library, Ottawa, photographs A25686–88 and 89).

Drumlins tend to be concentrated in *fields*, often numbering several thousand individuals. Within such fields, drumlins may occur in close association with rock drumlins (section 8.4.2) and Rogen moraine (section 11.2.5; figs 11.10 and 11.13). Drumlin fields commonly form broad bands, aligned either transverse or parallel to former glacier flow directions. Transverse fields may reflect zones of drumlin formation behind former ice margins, whereas longitudinal drumlin fields and/or mega-flutings may mark the position of fast ice streams or glacier lobes

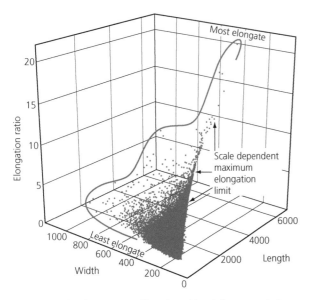

Figure 11.12 Plot of co-variation of length, width and elongation of a large sample of drumlins from the British Isles (Clark et al., 2009, reproduced with permission of Elsevier).

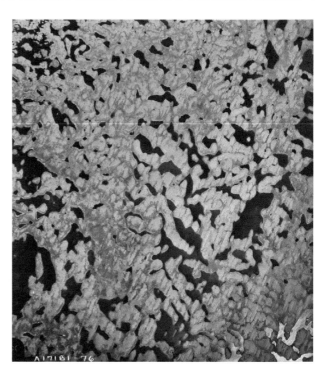

Figure 11.13 Drumlinized ribbed moraine in the area formerly covered by the Keewatin sector of the Laurentide Ice Sheet (Aylsworth and Shilts, 1989, National Aerial Photograph Library, Ottawa).

in former ice sheets (e.g. Patterson and Hooke 1996; see section 12.4), often occurring in lowland areas or in large valleys which acted as conduits for the evacuation of ice from the ice sheet interior (e.g. Mitchell 1994, 2006; Kovanen and Slaymaker, 2004; fig. 11.14). This is not to say that drumlins and mega-flutings are restricted to lowland terrains, and several researchers have stressed that subglacial bedforms generally cannot be correlated with particular topographic settings or substrate lithologies (cf. Patterson and Hooke, 1996). Drumlins occurring in areas of greater relief are often developed on the lee side of spurs and are known as *drumlinoid drift tails* (fig. 11.15); they document confined streaming of glacier ice due to topographic factors.

The most remarkable patterns of coherency and down-ice trends in elongation can be seen in the MSGL of former ice streams (e.g. Canals et al., 2000; Sejrup et al., 2000; Clark and Stokes, 2001; 2003; Stokes and Clark, 1999, 2001, 2003a, b; Wellner et al., 2001, 2006; see section 12.4.2). An excellent illustration of these trends is the former bed of the Dubawnt Lake palaeo-ice stream in northern mainland Canada (Stokes and Clark, 2002a; fig. 11.16). The drumlins and MSGL reveal a pattern of convergence (interpreted as the ice stream onset zone; see section 6.3.4) towards a narrow corridor of highly elongate, closely packed and extremely parallel MSGL (ice stream trunk zone), and then a pattern of divergence towards the former ice margin. Mean bedform lengths and elongation ratios rise through the onset zone up to a maximum in the trunk zone and then tail off again in the divergence zone. The elongation ratios and lengths of the MSGL also increase from the sharply defined boundaries of the trunk zone towards its centre. This pattern of bedform attenuation at the centre of former ice streams has been

noted in several studies (e.g. Dyke and Morris, 1988; Hart, 1999; Briner, 2007) and strongly suggests that patterns of elongation in fields of drumlins, mega-flutings and MSGL can be used as a relative indicator of ice flow velocity.

Like the MSGL at the centre of a palaeo-ice stream bed, drumlin distribution within a swarm may also be regular, but is more likely to be apparently random (e.g. Smalley and Unwin, 1968). The occurrence of bands of drumlins lying abreast of one another within drumlin fields has been used by Smalley and Warburton (1994) to suggest that they reflect a banding of bed properties. The influence of pre-existing variations in small-scale topography and substrate properties on drumlin distribution has been stressed by Boulton (1987) and is discussed on page 463.

Drumlins are found on a wide variety of substrata. In a review of the literature, Patterson and Hooke (1996) found that unconsolidated sediments make up 34 per cent of drumlin sustrata, 18 per cent being till and 16 per cent being stratified sediments. The remaining 66 per cent is rock, divided between shales and slates (33 per cent), crystalline rocks (33 per cent), carbonates (25 per cent), and sandstones, conglomerates and basalts. Menzies (1979) has pointed out that factors such as joint and fracture systems, depth of weathering, lithology and mineralogy need to be assessed when making generalizations about substrates and drumlin occurrence.

Patterson and Hooke (1996) argued that drumlins form in areas where basal shear stresses are low and pore-water pressures are high. They found that, in situations

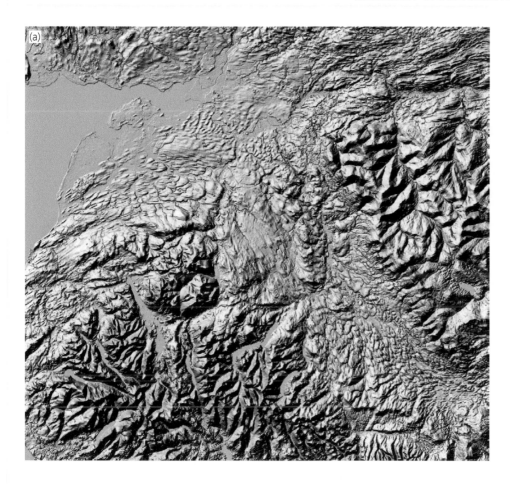

ELEVATION (m)

940   840   740   640   540

Figure 11.14 Drumlin distributions in lowland settings. (a) DEM of north-west England showing the partially overprinted drumlin swarms of the Eden Valley and Solway lowlands, documenting different flow stages beneath the British Ice Sheet during the last glaciation. (NEXTMap Britain data from Intermap Technologies Inc., provided courtesy of NERC via the NERC Earth Observation Data Centre (NEODC)) (b) DTM showing radial pattern of drumlin distribution and outlining the shape and basal flow vectors of the Okanagan Lobe, North American Cordilleran Ice Sheet (Kovanen and Slaymaker, 2004).

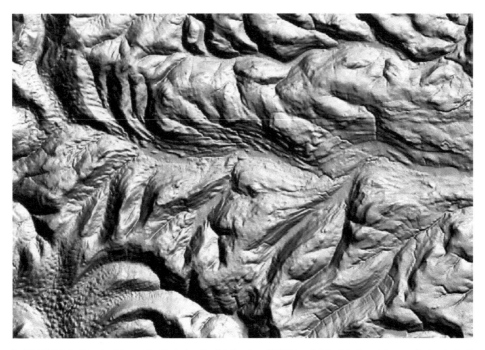

Figure 11.15 Drumlinoid drift tails in Wensleydale and its tributaries, English Pennines (NEXTMap Britain data from Intermap Technologies Inc., provided courtesy of NERC via the NERC Earth Observation Data Centre (NEODC)).

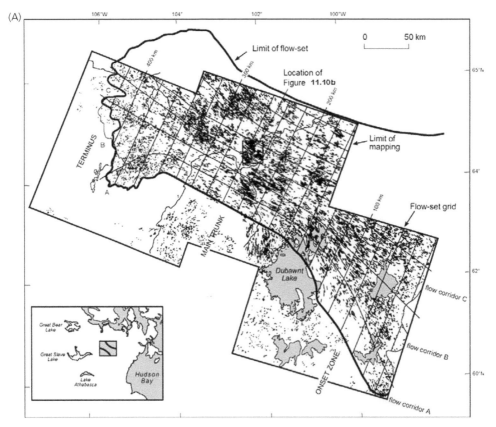

Figure 11.16 Glacial bedforms of the Dubawnt Lake palaeo-ice stream. (A) Map of the flow-set based on mapping of drumlins, flutings and MSGL from Landsat MSS imagery. (B) Map showing the variations in bedform elongation ratios, based on data from individual 20 km grid squares. (C) Down-ice bedform elongation ratios along the three flow corridors, A, B and C in (a) (Stokes and Clark, 2002a).

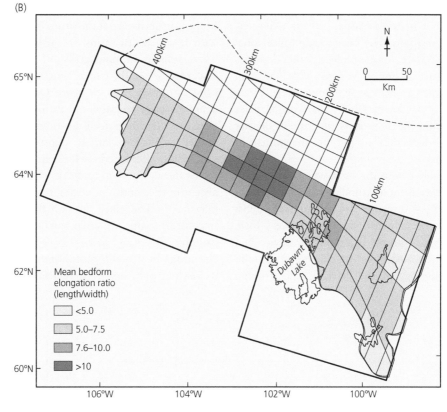

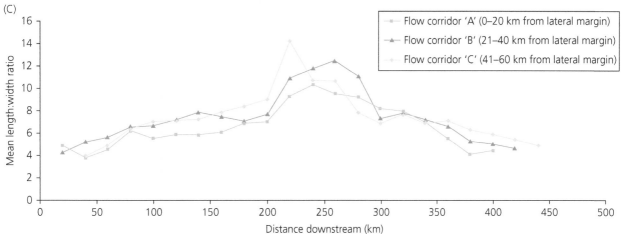

Figure 11.16 (*Continued*)

where it is possible to reconstruct the ice-surface form for particular drumlin-forming events, reconstructed basal shear stresses fall in the range 11–35 kPa. Interestingly, this range is closely similar to that for modern glaciers known to be underlain by deforming substrata. Patterson and Hooke also argued that drumlins commonly occur upglacier from ice-thrust landforms thought to have formed at frozen glacier margins. They argued that the presence of a frozen toe would have blocked subglacial drainage, leading to elevated pore-water pressures. As they pointed out, high pore-water pressures are compatible with low bed strength and low shear stresses (see

section 4.5.4). It should be noted however, that neither former shear stresses nor basal water pressures are known for the vast majority of drumlin fields, and that generalizations are probably premature.

Drumlin field morphometry has been assessed statistically. For example, *drumlin spacing*, defined as the perpendicular distance between adjacent drumlins, may exhibit normal or multi-modal distributions for individual fields, and is highly variable between fields. Such inter-field variability indicates that the glacial processes responsible for drumlin formation differ according to location (Menzies, 1979). *Drumlin density*, the number of drumlins per unit area, varies from 19.3 km^{-2} in Appleby, England, to 1.8 km^{-2} in Nova Scotia, Canada (Menzies, 1979), but great variability may occur within individual fields. In New York State, for example, drumlin densities range from 3.37 to 8.39 km^{-2} (Miller, 1972). Drumlin densities may decrease or increase in a downglacier direction (e.g. Smalley and Unwin, 1968), although Menzies (1979) suggested that measurements of drumlin density are not instructive unless they are put into the context of the local topography.

As highlighted above with the Dubawnt Lake palaeo-ice stream example, changes in drumlin/mega-fluting morphology in a down-ice direction are probably the most instructive spatial criteria when assessing process–form relationships. A good example is a study by Boyce and Eyles (1991), who reported down-ice changes in drumlin shape and sedimentology in the Peterborough drumlin field south of the Dummer Moraine, Ontario (fig. 11.17). In the northern part of the drumlin field, spindle drumlins with length:width ratios of >6 lie directly on bedrock. These spindle drumlins grade into less elongate drumlins towards the south, where ice overrode thick accumulations of outwash. South of the Oak Ridges Moraine, the drumlins have length:width ratios of <3, and are composed of deformation till unconformably overlying thick sequences of older, largely undisturbed sediments. In the northern rock-floored area, eskers record the drainage of meltwater in discrete conduits, whereas eskers are rare or absent in the south, indicating that meltwater evacuation may have taken place through the deforming substrate. In contrast to our earlier Dubawnt Lake palaeo-ice stream example, Boyce and Eyles (1991) considered the down-ice variations in drumlin morphometry to be a function of decreasing duration of subglacial deformation towards

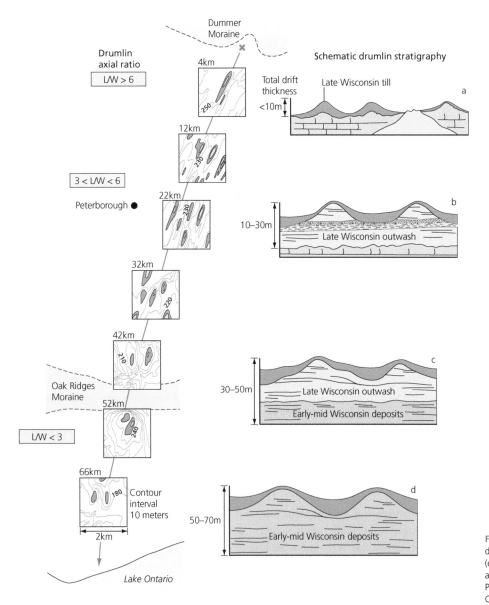

Figure 11.17 The variability in drumlin stratigraphy and morphometry (expressed as length:width ratios, L/W) along a 70 km flow line through the Peterborough drumlin field, Ontario, Canada (After Boyce and Eyles, 1991).

the ice margin. In contrast, in the northern part of the area, pre-existing sediment is thought to have been stripped off by subglacial erosion beneath the ice sheet interior. Similar patterns of downflow change in subglacial bedforms are being reported by a number of studies of subglacial bedforms produced beneath palaeo-ice streams on the Antarctic continental shelf (Wellner et al., 2001, 2006; Ó Cofaigh et al., 2002a; J. Evans et al., 2004). These studies combine swath bathymetry with acoustic data on till properties to illustrate the downflow increase in elongation ratios and the association of MSGL and subglacial till in the lower reaches of palaeo-ice stream trunk zones (fig. 11.18). Additionally, the data reveal that subglacial meltwater erosional features occur only in the upper trunk and onset zone (Lowe and Anderson, 2003), where irregular drumlins are composed largely of crystalline bedrock. This spatial pattern is generally attributed to the gradual accommodation of subglacial meltwater drainage through a deforming till layer that was produced by the ice stream as it traversed the softer bedrock lithologies of the outer shelf (we will pick up on this relationship between substrate characteristics and subglacial bedform generation on page 463).

Finally, it must be pointed out that we cannot assume that juxtaposed drumlins or any subglacially streamlined landforms were synchronously deposited, or that their forms were in equilibrium with the most recent glacial event. Evidence for time-transgressive (diachronous) deposition of drumlins within individual drumlin fields in Ireland is well documented (e.g. Knight et al., 1999; McCabe et al., 1999). Mooers (1989b) has argued that drumlins in central Minnesota, USA, were formed incrementally during the retreat of the Superior Lobe of the Laurentide Ice Sheet, with active drumlin formation occurring only in a 20–30 km wide zone below the margin. Similarly, Patterson and Hooke (1996) found that drumlins commonly occur in transverse belts behind former ice margin positions during glacier retreat. It is now generally accepted that overprinting of one set of drumlins or mega-flutings on top of another can result from ice divide migration, which is a response to changes in the overall shape of the ice sheet. Previous forms may be completely obliterated, may survive intact if they were located in frozen bed zones during later flow phases (Dyke et al., 1992; Kleman, 1992), or may be partially erased or modified (Boulton, 1987; Kleman et al., 1994;

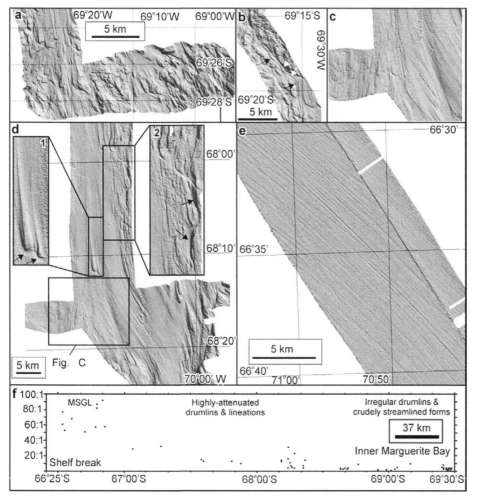

Figure 11.18 Downflow changes in subglacial bedforms in the Marguerite Trough, Antarctic Peninsula, illustrated with EM120 shaded relief images. (a) Short, irregular drumlins and crudely streamlined forms, inner Marguerite Bay. (b) Meltwater channels in bedrock (arrowed), inner Marguerite Bay. (c) Convergence of streamlined bedforms into the deepest part of the trough. (d) Drumlins and lineations formed in sediment and bedrock on the middle shelf. Inset 1 = highly attenuated drumlins up to 120 m high, with crescentic overdeepenings (arrowed) around their stoss sides. Inset 2 = sinuous meltwater channel up to 230 m deep (arrowed). (e) MSGL on the outer shelf. Maximum height of the lineations is 15 m and widths range from 130 to 300 m. (f) Relationship between length:width ratio of subglacial bedforms (y-axis) and distance northwards along Marguerite Trough (x-axis) (Ó Cofaigh et al., 2002a).

Clark, 1999). The remoulding of pre-existing forms and sediments is referred to as *drumlinization*; we will explore the concept of overprinting of subglacial bedforms and the differentiation of complex flow sets in section 12.4.3.

Composition

The composition of drumlins and mega-flutings is extremely varied. Some have rock cores mantled by a superficial *carapace* of till, which surrounds the core like an eggshell (e.g. Hart, 1995b), but most are composed entirely of unconsolidated sediments. Many researchers have reported drumlins with cores of sorted sediments covered by a till carapace (Krüger and Thomsen, 1984; Boulton, 1987; Boyce and Eyles, 1991; Bluemle et al., 1993). In some cases, the sediment core clearly consists of overridden pre-existing materials (e.g. Krüger and Thomsen, 1984; Boyce and Eyles, 1991), but in others the age relationships between sorted sediments and drumlinization are not so obvious. Large exposures through drumlins and mega-flutings are rare, a fine example being that reported by Menzies and Brand (2007) from the New York drumlin field. This exposure clearly documents the deposition of ice-contact delta foreset bedding that has been overridden and drumlinized by a glacier readvance (fig. 11.19). It therefore constitutes a perfect example of a drumlin being seeded at a well-drained subglacial 'sticky spot' (see page 463 onwards). Some researchers have suggested that sorted sediments within drumlins were deposited contemporaneously with drumlin formation (e.g. Dardis and McCabe, 1983, 1987; Dardis et al., 1984; Dardis, 1985, 1987; Hanvey, 1987,

1989; McCabe and Dardis, 1989; Shaw et al., 1989;). Sediment cores may be undeformed, or display complex deformation structures such as overfolds and thrusts and water escape structures (e.g. Krüger and Thomsen, 1984; Stanford and Mickelson, 1985; Boulton, 1987; Bluemle et al., 1993; Wysota, 1994; Shaw et al., 2000). Drumlins and mega-flutings may also consist entirely or predominantly of till (e.g. Newman and Mickelson, 1994). Ice-cored drumlinoid forms have been identified on the foreland of the Icelandic surging glacier, Brúarjökull (Evans and Rea, 2003; Schomacker et al., 2006), where stagnating glacier ice dating from older surges has been overridden by the most recent surge. In such situations, the drumlinoid form is gradually destroyed by the melt-out of the ice core.

Till fabric studies commonly show systematic variations in the preferred orientation of clasts in individual drumlins, revealing former patterns of ice flow or sediment deformation. Fabrics from surface tills tend to be parallel with the former ice flow direction on the top of drumlins (e.g. Schomacker et al., 2006), but parallel to the contours on the flanks, suggesting divergent ice or till flow at the stoss ends and convergent flow at the lee ends or tails of the obstacles (e.g. Andrews and King, 1968; Krüger and Thomsen, 1984). Within till cores, fabrics may bear no relationship to drumlin form, and appear to record till accretion prior to drumlinization (e.g. Stea and Brown, 1989; Newman and Mickelson, 1994). Preferred fabric orientations can vary upsection within drumlin cores, probably as a result of changing ice flow directions during till accretion (Stea and Brown, 1989).

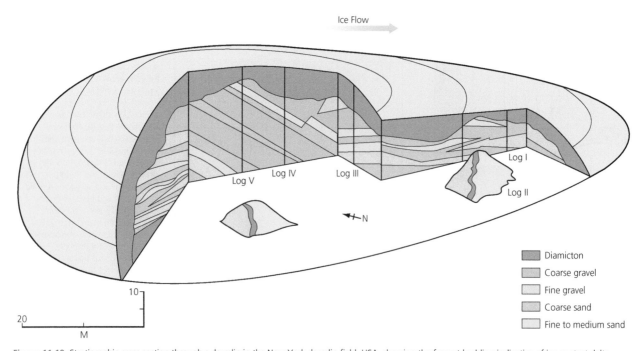

Figure 11.19 Stratigraphic cross section through a drumlin in the New York drumlin field, USA, showing the foreset bedding indicative of ice-contact delta deposition and the till carapace emplaced during drumlinization (After Menzies and Brand, 2007).

Processes of formation

In essence, drumlins and mega-flutings can be explained as the result of either: (1) erosion of the intervening hollows; (2) accretion of sediment in hills, or (3) some combination of the two. *Erosion* of material from hollows and swales could produce drumlins from the remnants of formerly more extensive pre-existing sediments. Alternatively, drumlin *accretion* could occur by the successive additions of sediment in a series of concentric shells.

Boulton (1987) proposed a qualitative model of drumlin formation, based on the principles of sediment erosion and redistribution within subglacial deforming layers (section 4.5). Weaker parts of the bed are assumed to deform relatively rapidly, leaving stronger areas (sticky spots) as static or mobile obstacles. Figure 11.20 shows how drumlins could develop from spatially variable bed materials. Areas underlain by gravel are likely to be well drained, leading to low pore-water pressures and high sediment strength, whereas the finer-grained intervening areas are less well drained and consequently weaker.

There is therefore a spatially variable bed response to glacially imposed stresses: the gravel bodies form rigid or slowly deforming cores, whereas the finer-grained material is more likely to undergo pervasive deformation, forming far-travelled sheaths of highly attenuated glacitectonite or subglacial traction till around the stiffer cores. Streamlining of the residual gravel bodies forms drumlins, which are essentially equivalent to augen in metamorphic rocks. Note how the flow lines around the drumlins come to resemble the till fabric patterns found in drumlin carapaces. Over time, some cores may become *de-rooted* and mobile, although still moving more slowly than the surrounding weak sediment (figs 11.21g, h). When cores have been transported from their original positions, the initial distribution of deformable and non-deformable materials on the substrate may be impossible to reconstruct. De-rooted cores are also characterized by highly deformed stratigraphies, whereas stable cores reveal very little deformation. Kjær et al. (2003b) described drumlins from the recently deglaciated foreland of Slèttjökull in Iceland, where deforming till

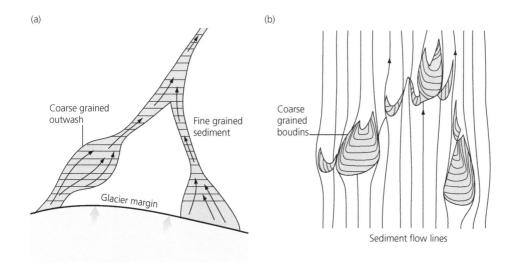

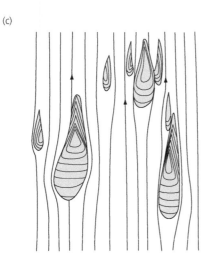

Figure 11.20 Hypothetical example of drumlin formation as a result of the contrasting rheological properties of sediment overridden by a glacier. (a) Initial distribution of coarse-grained proglacial outwash. (b) and (c) The progressive development of drumlins from coarse-grained sediment masses, with rapid deformation of the weaker sediments (After Boulton, 1987).

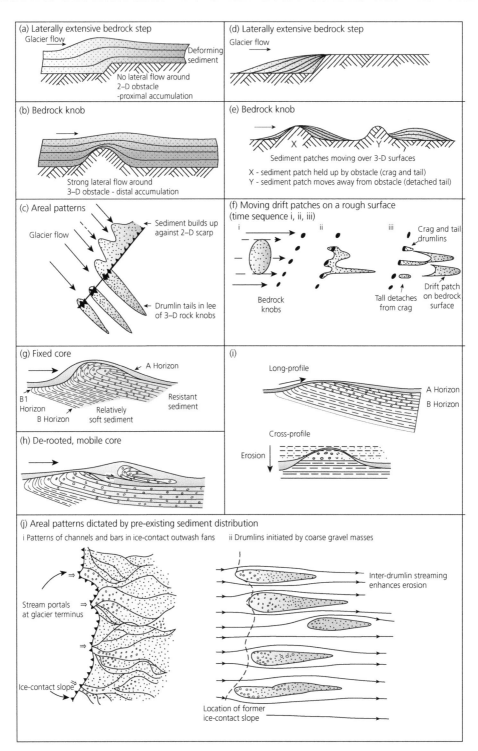

Figure 11.21 Summary schematic diagram showing drumlin forming processes, internal structures and distribution patterns according to the deforming bed model. (a) and (d) Streamlined nose builds up on the proximal side of a laterally extensive bedrock step. (b) and (e) Streamlined tail builds up on the distal side of a bedrock knob, owing to enhanced flow around the flanks. The drumlin forms a standing wave when the subglacial sediment is thick, although sediment does move through the drumlin. When subglacial sediment (drift) patches move over the rock substrate, the drumlin can move past the retarding knob. (c) Drumlin noses build up on the proximal side of a step of uniform height. The locations of noses reflect subglacial sediment 'streams'. Where gaps occur in the scarp, sediment streams over and flows into low-pressure points in the lee of the inter-gap knobs. (f) Subglacial sediment patches moving over bedrock are retarded by crags to produce crag and tails and detached tails. (g) and (h) Drumlins initiated by stiff sediment obstacles that form deforming cores and which may be fixed or mobile. (i) Drumlins initiated by stiff sediment obstacles which form rigid, fixed cores. Weaker sediment to either side is eroded by subsole deformation. (j) An example of how drumlin distribution patterns can reflect original sedimentary inhomogeneities (e.g. coarse gravel at ice-contact outwash fan apices and gravel bars). Such drumlins can become de-rooted and mobile (After Boulton, 1987).

thickness increases in areas where glacier ice overrode pre-existing till outliers isolated by glacifluvial erosion prior to an advance of the glacier (fig. 11.22). This nicely illustrates the importance of substrate drainage characteristics to the initiation of drumlinized terrain and complements the hypothetical scenario depicted in figures 11.20 and 11.21.

Boulton (1987) used the drumlins of northern Saskatchewan to illustrate his theory (fig. 11.23). Several of these drumlins appear to be connected and to form

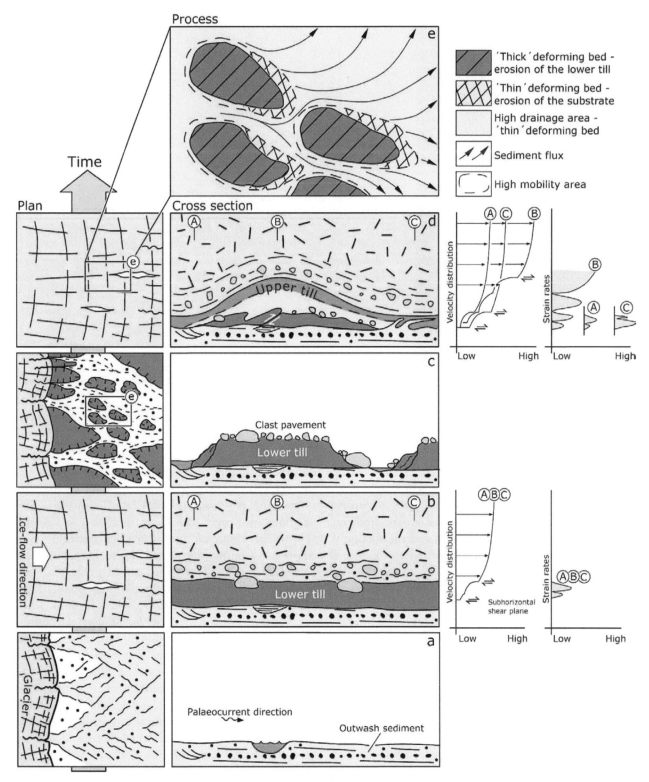

Figure 11.22 Reconstruction of the relationships between substrate materials and deforming till thickness based on the foreland of Sléttjökull, Iceland (Kjær et al., 2003b).

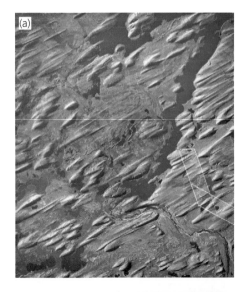

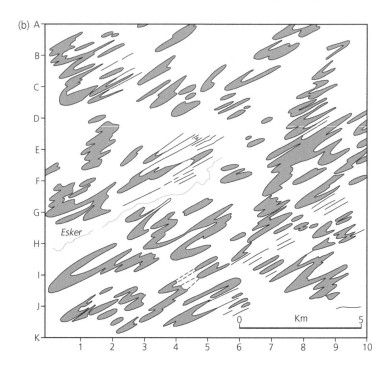

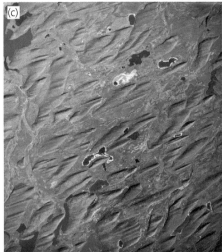

Figure 11.23 The Livingstone Lake drumlins, Saskatchewan. (a) Aerial photograph of parabolic, transverse asymmetrical and spindle drumlins (ice flow was from the north-east). (b) Sketch of the drumlins and eskers from the aerial photograph. (Boulton, 1987) The shapes are thought to represent fold traces and the major drumlins represent stable sediment masses around which other material has deformed to produce attenuated fold limbs. A transverse lineation occurs between I6 and B9, documenting a possible earlier ice flow directional indicator (fluting) which has been partially remoulded. (c) Aerial photograph of transverse asymmetrical drumlins (Aerial photographs from the National Aerial Photograph Library, Ottawa).

traces of shear folds on a horizontal plane. Boulton suggested that the barchanoid bodies represent competent, stiff sediment cores and the limbs represent attenuated deformable sediment. The broken lines on figure 11.23b outline lineations which underlie the drumlins and which obviously have persisted as rigid core areas. Interestingly, Shaw (1989) cited this same drumlin field as evidence for a completely different theory of drumlin formation (section 11.2.8).

Although the examples in figures 11.19–11.21 represent extreme heterogeneity of the glacier substrate, Boulton's (1987) theory appears to explain the formation of drumlins with cores of rock, till and stratified sediment. Cores of rock and consolidated till can form stiff, resistant regions around which sheaths of deforming sediment can flow. The operation of a subglacial mosaic of deforming and sliding bed processes (Piotrowski et al., 2004) could conceivably result in the switching on and off of drumlin seeding points. Although well-drained stratified sediment will also form resistant cores (e.g. Menzies and Brand, 2007), it may reveal very few signs of

disturbance, despite having been overridden by glacier ice (fig. 11.21i). Modern Icelandic examples of drumlins formed from pre-existing glacifluvial sediments have been described by Krüger and Thomsen (1984) and Krüger (1994). The drumlins formed from an inter-stream area of coarse bar deposits which acted as a subglacial obstacle. Signs of disturbance in stratified cores are commonly best developed in the tail of the drumlin (e.g. Stanford and Mickelson, 1985; Stephan, 1987), where Boulton (1987) suggested strong convergence should dominate during drumlin formation. That stratified sediments can resist very high stresses during glacial transport and can retain much of their original structure is illustrated by the presence of undeformed boudins in glacitectonites (Hart and Roberts, 1994; Hart, 1995a; Benn and Evans, 1996). General descriptions of drumlin sediment, as reviewed by Karrow (1981), suggest that drumlins tend to form in sandy rather than clay-rich tills. Boulton (1987) considered this to be compatible with his theory because the coarser-grained sediments are better drained and more likely to form resistant cores. Resistant cores composed of a wide range of sediment types as well as coarse-grained materials were reported from the Peterborough drumlin field, Canada, by Boyce and Eyles (1991), who suggested that hollows between the cores were eroded by excavational deformation. It should also be remembered that landforms created during one glaciation might be reactivated during the next, and therefore may persist through numerous glacial cycles (e.g. Hättestrand et al., 2004).

Geophysical imaging of the morphology and evolution of landforms at the base of modern Antarctic ice streams (section 6.3.4; King et al., 2007; A. M. Smith et al., 2007) provides powerful evidence that drumlins and flutings can indeed form by the localized mobilization of subglacial deforming sediment. Modelling work suggests that till bedforms can form and grow even in the absence of initial variations in sediment type. Hindmarsh (1998a, b) has shown that regions of thicker basal till have higher effective pressures and, for certain till rheologies, lower deformation rates. This means that initial random variations in till thickness will become amplified through time, leading to the growth of bumps on the bed. This idea has been developed further by Fowler (2000) and Schoof (2007c), who showed that a till instability mechanism can lead to the formation two-dimensional waves at glacier beds. When generalized to include three-dimensional beds and the effects of water, this work could form the basis of a complete, quantitative and fully testable model of drumlin formation.

The occurrence of stratified sediments in some drumlins has prompted some researchers to erect genetic models that incorporate subglacial fluvial processes. For example, stratified sediments in numerous drumlins in Ireland have been interpreted as subglacial glacifluvial deposits, rather than pre-existing sediment cores (e.g. McCabe and Dardis, 1989; Dardis and Hanvey, 1994). These authors argue that downglacier-dipping, sorted sediments exposed near the lee side of drumlins are contemporaneous with drumlin streamlining, and were deposited as *lee-side stratification sequences* in water-filled cavities. Thus, according to this model, the drumlins behaved rather like roches moutonnées with lee-side cavities, in which water-sorted sediments were deposited (section 11.2.7). Subglacial shearing is thought to modify drumlin form during and following lee-side deposition, deforming the upper stratified beds, forming a till carapace and then producing a streamlined form. This model of drumlin modification by subglacial meltwater has some similarities with controversial ideas developed by John Shaw and co-workers in Canada, who attribute drumlins, mega-flutings (and many other types of glacial landforms) to catastrophic subglacial mega-floods (section 11.2.8).

The distribution and characteristics of MSGL indicate that they form in the fast-flowing trunk zones of palaeo-ice streams (e.g. Stokes and Clark, 1999, 2001, 2003a, b;

Canals et al., 2000; Clark and Stokes, 2001, 2003; Wellner et al., 2001, 2006; Ó Cofaigh et al., 2002a; Evans et al., 2004; Evans et al., 2008). Late Quaternary chronologies for the James Bay lowlands, which contain MSGL produced by the former Laurentide Ice Sheet, imply ice velocities in the region of 400–1600 m yr^{-1} (Clark, 1993).

Tulaczyk et al. (2001) and Clark et al. (2003) proposed that MSGL may be produced by the ploughing of soft subglacial sediment by keels in the base of the overriding ice. This is based on the idea that the ice–bed interface will develop transverse irregularities where it traverses a rough bedrock surface, and that these will be amplified where the ice converges and accelerates. These longitudinally oriented 'keels' then carve elongate grooves and deform/displace material sideways and upwards into intervening ridges. This model is consistent with known examples of MSGL in North America, Antarctica and Europe, which begin downflow of large bedrock outcrops. The groove-ploughing theory for fluting and MSGL production, however, does not explain all MSGL characteristics. For example, Ó Cofaigh et al. (2005) described MSGL from the Marguerite Trough, the drainage pathway for a palaeo-ice stream on the Antarctic continental shelf, and identified some characteristics that are inconsistent with formation by groove-ploughing. First, the MSGL do not always start at rough bedrock outcrops, the necessary prerequisite for ice keel formation. Second, instead of being related to the spatial frequency of the bedrock roughness, groove spacing changes downflow and ridges often bifurcate. This test of the groove-ploughing hypothesis indicates that not all MSGL originate by ice keel grooving, but rather than completely invalidating the concept, it merely shows that subglacial bedform production could be accomplished in a variety of ways. Ó Cofaigh et al. (2005) proposed that the Marguerite Trough MSGL were formed by both subglacial sediment deformation from point sources, such as areas of stiff till or sticky spots, and groove-ploughing; the predominance of one process over another is related to localized variability in properties like till extent, depth and strength, and the presence of bedrock outcrops.

Repeat seismic transects were taken across Rutford Ice Stream in 1991, 1997 and 2004. Between 1991 and 1997, a part of the bed underwent remarkably high erosion rates of 1 m a^{-1}, orders of magnitude higher than previously reported subglacial erosion rates of 0.1–100 mm a^{-1} (Hallet et al., 1996; Alley et al., 2003b). Between 1997 and 2004, the erosion stopped and, instead, a 10 m high mound formed on the glacier bed. A drop in the acoustic impedance around the area of the drumlin indicated that it was composed of deforming sediment. Also significant was the occurrence of an adjacent area of increased impedance due to sediment dewatering and compaction, indicating that localized changes in bed porosity were

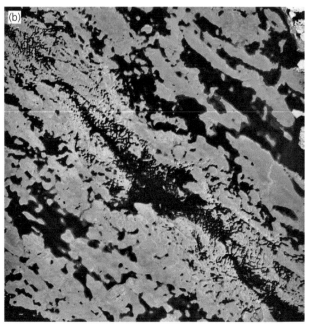

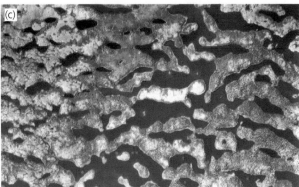

Figure 11.24 Ribbed terrain (Rogen moraine). (a) The 'type area' around Lake Rogen, Sweden. (J. Lundqvist): (b) Ribbed terrain near Rankin Inlet, Nunavut, Canada, giving the land surface a fish-scale appearance. (c) Details of ribbed terrain in Nunavut, Canada (National Aerial Photograph Library, Ottawa)

driving sediment mobilization. Smith et al. (2007) speculated that a subglacial bedform was evolving in one of two ways, both of which we have reviewed above: first, the positive relief feature could be a groove in the ice base, formed by flow over a bedrock protuberance further up-ice and filled by sediment mobilized from other parts of the bed. This is consistent with the theories of Boulton (1976) and Benn (1994a; see section 11.2.3); second, mounds are produced by rheological instabilities at the bed, a concept that has been modelled by Hindmarsh (1998a, b) and implies that the ice–bed interface may change over short distances from sliding to deforming modes.

11.2.5 RIBBED (ROGEN) TERRAIN

The term *Rogen moraine* was introduced to describe the distinctive morainic landscape around Lake Rogen in Sweden (fig. 11.24; Lundqvist, 1989a). Traditionally, the term was used to refer to fields of coalescent crescentic ridges up to 30 m high and up to 100 m wide, lying transverse to former ice flow. The arcuate forms are aligned with their outer limbs bent downglacier. In North

America, these features are usually referred to as *ribbed moraine* (Aylsworth and Shilts, 1989) or *ribbed terrain* (cf. Burgess et al., 2003). Individual ribs have asymmetric profiles with shallow up-ice and steep down-ice flanks, giving the impression of a fish-scale texture in aerial photographs (Aylsworth and Shilts, 1989). Local case studies have identified certain relationships between bedrock topography and ribbed terrain, but associations tend to be inconsistent. For example, fields of ribbed terrain in Scandinavia and in Labrador/Quebec and Newfoundland, Canada, lie in elongate depressions; but similar linear fields of ribbed terrain in Keewatin, Canada, appear to be independent of topography (Aylsworth and Shilts, 1989). The close association between ribbed terrain and drumlins, together with zones of transition between the two forms (figs 11.10 and 11.13), suggests that they may be deposited contemporaneously by similar subglacial processes.

An exhaustive inventory of ribbed terrain in Canada, Sweden and Ireland by Dunlop and Clark (2006a), involving the mapping of 33,000 individual forms, has produced the most comprehensive assessment of ridge morphology and distribution patterns to date. This study is significant because it has highlighted that our previous definitions of ribbed terrain are too restrictive. Specifically, they do not always have horns pointing down-ice or always have steeper distal slopes; their ridge crest heights are not accordant; ridge sizes are highly variable within fields and their occurrence is not related to topography. Dunlop and Clark (2006a) recognized 16 different types of ribbed terrain (fig. 11.25), including features such as *mega-scale ribbed terrain* (fig. 11.26) that are up to an order of magnitude longer than the features traditionally regarded as typical ribs. Additionally,

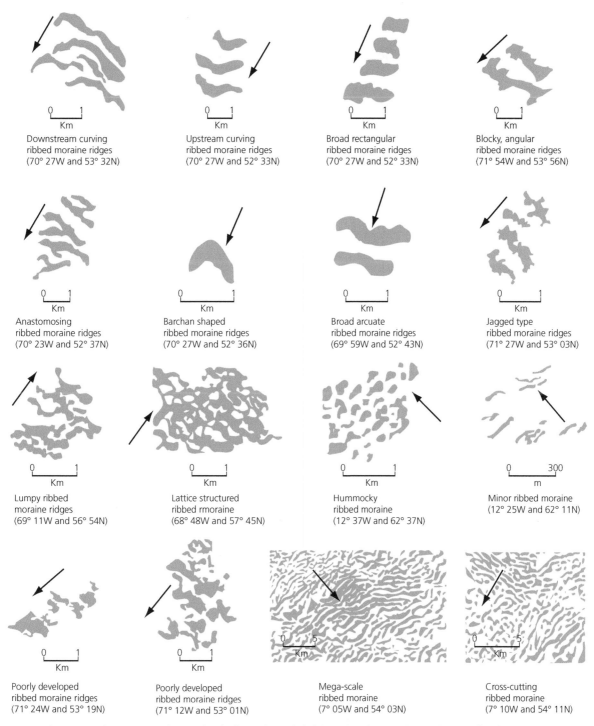

Figure 11.25 The 16 types of Rogen or ribbed terrain identified by Dunlop and Clark (2006a), with arrows showing direction of ice flow.

Dunlop and Clark (2006a) identified a wide variety of spatial relationships with other subglacial bedforms (fig. 11.27), involving not just transitions but also superimpositions, individual landform transitions (i.e. partially 'ribbed' lineations) and juxtapositions (e.g. *ladder-type associations*). The range of ridge morphologies may also be large over very small areas, with abrupt changes in ridge type over short distances. Spatial patterns and regional distributions have also been redefined by this type of work (cf. Knight and McCabe, 1997; Clark and Meehan, 2001; Dunlop and Clark, 2006b).

Ribbed terrain is composed of a variety of sediment facies, including massive and laminated diamicton, gravels, sands and silts, which commonly show evidence of deformation (Hättestrand, 1997; Lindén et al., 2008). In Sweden, ribbed terrain contains characteristic associations of laminated tills with attenuated inter-beds of sorted sediments known as *Kalix till* and *Sveg till*

Figure 11.26 A DEM of mega-scale ribbed terrain in County Monaghan, Ireland. Features in this image (30 km across) are up to 16 km long, making them the longest ribbed features ever recorded (Clark and Meehan, 2001).

(Shaw, 1979; Dreimanis, 1989). Laminae are wrapped around large clasts and augen-like inclusions, and the whole mass is deformed by folds and thrust planes that record downglacier overturning and transport. Similar facies associations in ribbed terrain in Newfoundland have been described by Fisher and Shaw (1992), although other associations also occur in this area, including disturbed masses of crudely bedded gravels and stratified stony diamicton. Möller (2006) reported ribbed terrain from Sweden that comprises stacked sequences of crudely bedded to massive diamictons, with minor inter-beds of contorted stratified sands and gravels, interpreted as successions of sediment gravity-flow deposits and minor fluvial deposits. These sediments are extensively folded

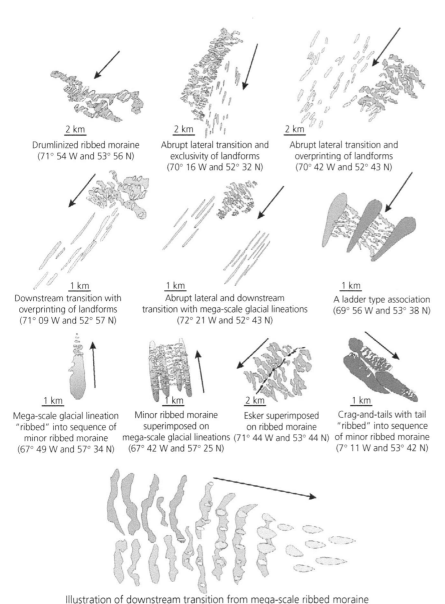

2 km
Drumlinized ribbed moraine
(71° 54 W and 53° 56 N)

2 km
Abrupt lateral transition and
exclusivity of landforms
(70° 16 W and 52° 32 N)

2 km
Abrupt lateral transition and
overprinting of landforms
(70° 42 W and 52° 43 N)

1 km
Downstream transition with
overprinting of landforms
(71° 09 W and 52° 57 N)

1 km
Abrupt lateral and downstream
transition with mega-scale glacial lineations
(72° 21 W and 52° 43 N)

1 km
A ladder type association
(69° 56 W and 53° 38 N)

1 km
Mega-scale glacial lineation
"ribbed" into sequence of
minor ribbed moraine
(67° 49 W and 57° 34 N)

1 km
Minor ribbed moraine
superimposed on
mega-scale glacial lineations
(67° 42 W and 57° 25 N)

2 km
Esker superimposed
on ribbed moraine
(71° 44 W and 53° 44 N)

1 km
Crag-and-tails with tail
"ribbed" into sequence
of minor ribbed moraine
(7° 11 W and 53° 42 N)

Illustration of downstream transition from mega-scale ribbed moraine
to drumlinized ribbed moraine to classic type drumlins
(6° 58 W and 54° 04 N)

Figure 11.27 Examples of the variety of spatial relationships between Rogen/ribbed terrain and other subglacially streamlined bedforms (Dunlop and Clark, 2006a).

and thrust in the outer horns of the ribs, suggesting that subglacial deformation has taken place preferentially in those areas of the landform. Lindén et al. (2008) have identified three component facies in northern Swedish ribbed terrain, comprising glacitectonized proximal ridge facies, a distal ridge facies of stratified clinoforms, and a carapace or 'draping facies' of subglacial till. This facies architecture was used by Lindén et al. (2008) to propose a formation sequence comprising:

(1) glacitectonic disturbance of pre-existing sediments to produce a proximal ridge;
(2) lee-side cavity deposition on the distal side of the ridge;
(3) stacking of proximal ridge facies to feed continued cavity filling;
(4) subglacial till deposition.

Various theories of ribbed terrain formation have been proposed and have been summarized by Lundqvist (1989b) and Hättestrand and Kleman (1999). A popular theory of ribbed terrain formation is the *debris-rich ice model*, which proposes that the ridges originate as thrust slices of basal ice, stacked up by compressive flow beneath the glacier either in topographic hollows or at the junction between warm-based and cold-based ice. In an analysis of the morphology of ribbed terrain, Hättestrand and Kleman (1999) argued for the fracturing of frozen till sheets at the boundary between frozen and thawed parts of the glacier bed. They identified ribbed terrain in which individual ridges appear to fit together like a jigsaw puzzle when the intervening depressions are removed and the ridges are closed up. However, Dunlop and Clark's

(2006) exhaustive mapping of ribbed terrain shows that jigsaw puzzle matching is rare.

Ribbed terrain is superimposed on flutings in some palaeo-ice streams, and it has been suggested that rib formation occurred during ice stream shutdown and bed freezing (Stokes et al., 2006a; cf. section 6.3.6). The close association of ribbed terrain with drumlins and flutings suggest that they can pre-date, post-date and be contemporaneous with each other, reflecting changing thermal conditions at both spatial and temporal scales at the ice–bed interface.

A *bed deformation model* was proposed by Boulton (1987), who interpreted ribbed terrain as part of a continuum of streamlined forms resulting from subsole sediment deformation. According to Boulton, ribbed terrain represents early stages of the drumlinization of transverse ridges of sediment on the glacier bed, and that they can develop from drumlins following a change in glacier flow direction (fig. 11.28). The deformation of weak bed materials around transverse ridges results in the preferential downglacier transport of the extremities of the features, producing the characteristic concave-downglacier planform. Different rates of sediment transport within the ridge causes fragmentation of the original ridge, creating numerous short crescentic ridges. The shape of ribbed terrain ridges is thus regarded as analogous to barchan dunes. Boulton argued that continued attenuation of ridges will result in barchanoid drumlins, then ellipsoidal drumlins, then flutings, as the bed adjusts to the new ice flow conditions. Thus, ribbed terrain and flutings are regarded as members of a continuum of bedforms, representing an evolutionary time series. Spatial transitions

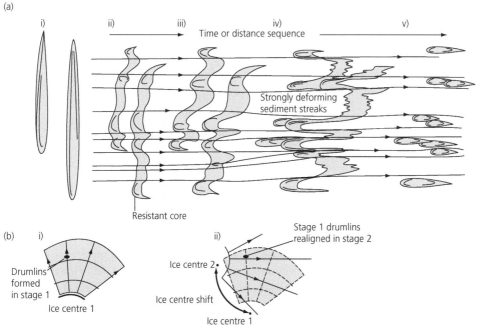

Figure 11.28 A schematic hypothetical reconstruction of the progressive transformation of flutings to ribbed terrain and drumlins by subglacial deformation. (a) (i) Original flutings produced by earlier ice flow direction. (ii–iii) Ribbed moraine stage. (iv–v) Drumlin stage. This can be a time sequence or a distance sequence. (b) Explanation of the change in ice flow direction in (a) by a shift in the ice dispersal centre (Modified from Boulton, 1987).

between moraine types can therefore be explained in terms of variations in the cumulative strain experienced by the bed materials, reflecting variations in basal ice velocity and bed strength.

Boulton (1987) did not consider the internal composition and structure of ribbed terrain, but the published evidence does appear to be consistent with his model. The Sveg tills appear to be very similar to glacitectonites formed by the shear and attenuation of pre-existing sediments (section 10.3.3; Hart and Roberts, 1994; Benn and Evans, 1996; Evans et al., 2006a), and so may record strain within the bed rather than within debris-rich ice. Similarly, Möller's (2006) sedimentological observations on some Swedish ribbed terrain appear to show a two-stage evolution involving the deformation of pre-existing debris flow deposits, interpreted by him as the product of localized subglacial reshaping of linear hummocky terrain originally deposited during an earlier glaciation. Lindén et al. (2008) also identified deformation of pre-existing sediments as essential to the initiation of individual ridges in ribbed terrain, implying greater affinities with cupola hills or overridden moraines (see sections 11.2.7 and 11.3.2) than subglacial bedforms.

The till instability model of drumlin formation (see page 463) has also been invoked to explain the origin of ribbed terrain. The bed ribbing instability explanation (BRIE) assumes that ice deforms internally and slides across the till surface, while at the same time the till deforms internally and slides on the underlying bedrock (Dunlop et al., 2008). Small perturbations in the till surface grow to become waves that migrate downstream. These ribs spontaneously grow under certain conditions and display wavelengths that are compatible with real world examples. Although this work suggests that both drumlins and ribbed terrain may be explained within a single framework, the very large variation in types of ribbed terrain indicates that it may actually be polygenetic (Dunlop and Clark, 2006a; Möller, 2006).

11.2.6 ICE STREAM SHEAR MARGIN MORAINES

Systematic mapping of the subglacial bedforms of former ice sheets has revealed long depositional ridges that demarcate the margins of former ice streams (fig. 11.29; Dyke and Morris, 1988; Dyke et al., 1992; Stokes and Clark, 2002b). The ridges are composed of drumlinized till and form discontinuous chains of segments up to 70 km long, up to 500 m wide and tens of metres high. The segments are sometimes offset slightly rather than running continuously end-on-end (Stokes and Clark, 2002b), but they clearly mark the outer boundaries of the flow sets of drumlins and mega-lineations that were produced by fast ice flow across formerly glacierized terrain. The drumlins and mega-lineations of the ice stream trunk

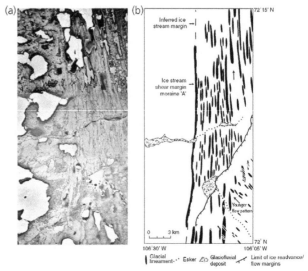

Figure 11.29 Aerial photograph and map of ice stream shear margin moraine from the former bed of the McClintock Channel ice stream, Arctic Canada (Stokes and Clark, 2002b; aerial photographs A16173-20-23, Energy, Mines and Resources, Canada).

usually run parallel to the ridges, although Kleman and Borgstrom (1994) have reported cases where flutings run obliquely up to the ridges. The term 'lateral shear moraine' was introduced by Dyke and Morris (1988) to describe these long ridge complexes at the margins of fast ice flow tracks, which were thought to be produced at the shearing zone between the warm-based ice of the fast flow track and the cold-based, sluggish ice beyond. Clark and Stokes (2001) and Stokes and Clark (2002b) have since refined the process–form relationships of these moraines, preferring instead to use the term *ice stream shear margin moraine*, in recognition of the fact that they appear to be associated exclusively with ice stream beds. Modelling of shear margin moraine formation by Hindmarsh and Stokes (2008) indicates that net sediment accumulation can occur in the transition zone where fast sliding ice meets cold-based ice. The conditions for moraine formation may be rare, however. In addition, given the highly dynamic nature of ice stream shear margins, it is likely that many shear margin moraines are destroyed or over-printed after formation.

11.2.7 SUBGLACIAL HUMMOCKY TERRAIN AND OVERRIDDEN MORAINES

During the 1950s and 1960s, several researchers argued that areas of hummocky terrain could be formed by the squeezing of fine-grained till into subglacial cavities. This model has been resurrected by Eyles et al. (1999), who proposed a two-stage process of hummocky terrain genesis by subglacial deformation (fig. 11.30). First, subglacial materials are fluted and drumlinized; then till is squeezed between foundering ice blocks, giving rise to till-cored hummocks, which eventually become juxtaposed with glacifluvial features such as kames and ice-walled lake

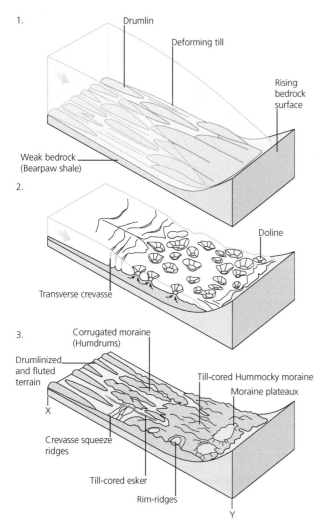

Figure 11.30 Conceptual model of subglacially moulded terrain with till-cored hummocky moraine and corrugated moraine on margins of topographic highs and drumlinized topography in lows (After Eyles et al., 1999).

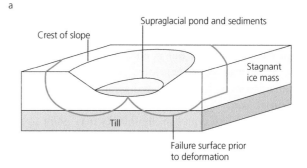

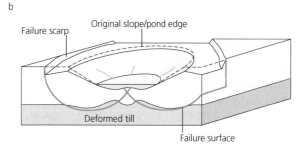

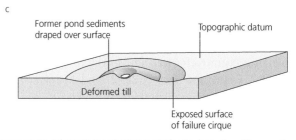

Figure 11.31 Schematic hypothetical model for the formation of hummocky moraine by subglacial deformation of fine-grained till below stagnant ice. (a) Initial conditions during deglaciation as stagnant ice with surface ponds rests on fine-grained deformable till bed. (b) Failure of ice resulting in deformation and mobilization of underlying water-saturated till. (c) Final landscape of circular 'doughnut' or 'rim ridge' which may, or may not be, draped by deformed pond sediments (Boone and Eyles, 2001, after Stalker, 1960).

plains (moraine plateaux). This conceptual model predicts that the former margins of ice streams in western Canada should be marked by a continuum of landforms that include, from the outer margin to the fast flow zone:

- moraine plateaux, till-cored hummocky moraine and rim ridges, ring forms or 'doughnuts';
- till-cored corrugated moraine or hummocky terrain with a clear ice-flow parallel grain, named 'humdrums';
- drumlinized and fluted terrain.

Boone and Eyles (2001) provided geotechnical constraints on the formation of the subglacial hummocky terrain, supporting the notion that subglacial till is pressed into shapes that reflect the dimensions of the cavities produced in the stagnating ice mass above. Specifically, humdrums reflect the late-stage modification of drumlins and rim ridges/ring forms, and more chaotic hummocky terrain reflect the perforations or 'dolines' that develop around the glacier margin and over topographic high points

during the earliest stages of deglaciation (fig. 11.31). Alternative interpretations of ring forms are that they record supraglacial sediment sloughing into sinkholes (see section 11.4.2).

The 'ice pressing' model of hummocky terrain development proposed by Eyles et al. (1999) and Boone and Eyles (2001) is consistent with several characteristics of the landforms in question. First, they are predominantly composed of diamictons which are identical in texture to the local subglacial tills and do not display the heterogeneity and contorted bedding typical of supraglacial landforms (see section 11.4.2). Second, they often contain deformation structures typical of loading by an overburden into a viscous fluid. Third, the shapes, sizes and till cores of rim ridges or doughnuts are entirely consistent with diapiric intrusion through perforations in stagnating ice. Fourth, the presence of a deforming till layer below modern glacier margins does result in the marginal thickening of diamicton and the squeezing of that diamicton into crevasse traces (Price, 1970; Evans and Hiemstra, 2005).

Despite the attractiveness of the pressing model, we must be cautious in our application of it to hummocky terrain genesis more generally. Not only is the western Canadian case study potentially unique to the regional bedrock and topography, but also the occurrence of deformation structures in hummocks may point to other subglacial origins. For example, some tracts of hummocky terrain in western Canada appear to be overridden glacitectonized moraine fields, and therefore relate to active glaciation rather than stagnation (Evans et al., 2008; see also section 11.3.2). This interpretation is especially attractive wherever hummocks are composed of folded and thrust faulted bedrock (Tsui et al., 1989). Examples of subglacially streamlined hummocks include Smith and Clark's (2005) 'ovoid forms' and Piotrowski and Vahldiek's (1991) 'elongated hills', features that should not surprise us, now that we have fully acknowledged the concept that glaciated landscapes are in essence palimpsests composed of overprinted bedform signatures.

Mapping of recently deglaciated forelands has recently drawn attention to the fact that many elongate/arcuate, transverse ridges are overridden push moraines. This has been demonstrated for Icelandic forelands by Evans et al. (1999a), Evans and Twigg (2002) and Evans (2003b), where overridden push moraines are manifest as wide and arcuate, low-amplitude ridges that are draped by sharp-crested push moraines. Complete overriding by glacier ice is also indicated by the presence of flutings that continue uninterrupted from the up-ice to down-ice slopes. These overridden ridges likely record the locations of push moraine construction during an earlier phase of ice expansion. Some 'hummocky moraine' in Scotland has also been interpreted as the product of partial to well-developed glacier streamlining of ice-contact fans (Lukas, 2005; see section 11.3.5).

11.2.8 MEGA-FLOOD EXPLANATION FOR SUBGLACIAL BEDFORM GENESIS

For over 25 years, John Shaw and co-workers have repeatedly argued that the full range of subglacial bedforms and some hummocky terrain were formed by subglacial mega-floods of gigantic proportions. Initially, Shaw (1983b) and Shaw and Kvill (1984) proposed that drumlins and ribbed terrain represent the infillings of giant scours cut upwards into basal ice by subglacial sheet floods (fig. 11.32). This interpretation is based on the premise that the similarity of the form of drumlins and scour marks made at the base of turbulent underflows is due to a common origin (fig. 11.33). According to this model, the scour marks are infilled with sediment during the waning stage of the flood event, providing an explanation for the occurrence in some drumlins of undisturbed

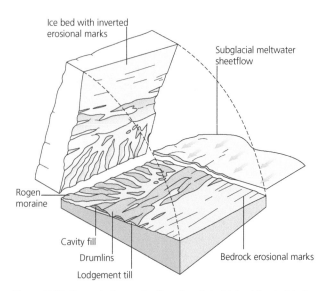

Figure 11.32 Conceptual reconstruction of a subglacial sheet flood origin for glacial landforms (Fisher and Shaw, 1992).

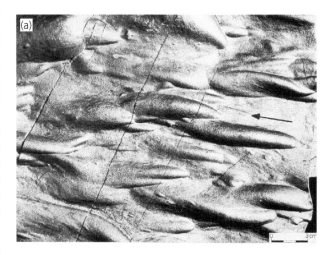

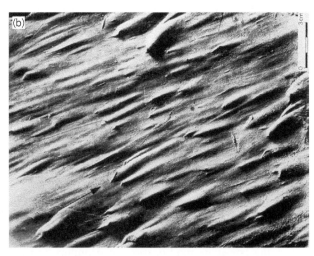

Figure 11.33 Streamlined erosional marks associated with turbidites. (a) Narrow, parabolic and spindle flute casts. (b) Longitudinal obstacle scour moulds cut behind small tool marks. The arrow indicates flow direction (Dzulinski and Walton, 1965).

stratified sediments whose bedding conforms to the surface shape of the landform. The integrity of the stratified sediments and the lack of deeply penetrating glacitectonic structures are interpreted as evidence that the material has not been subject to high stresses in the subglacial environment. The mega-flood explanation was later expanded by Shaw et al. (1989). To explain till-cored drumlins and flutings, which are inconsistent with the original hypothesis, they proposed that the subglacial meltwater can excavate the substrate *between* the bedforms, so that the drumlins or flutings are erosional remnants of pre-existing sediments. In this case, a form analogy is made between drumlin morphology and the lemniscate loops and streamlined forms of the fluvially eroded loessic hills of the Missoula scablands (see section 8.4.4). This erosional option within the mega-flood explanation facilitated its application to a wider range of subglacial bedforms and has latterly become the preferred option, covering streamlined bedforms (Shaw et al., 2000; Munro-Stasiuk and Shaw, 2002), ribbed terrain (Fisher and Shaw, 1992), fluted transverse ridges/overridden moraines (Beaney and Shaw, 2000) and hummocky terrain (Munro and Shaw, 1997; Munro-Stasiuk and Sjogren, 2006).

In each of these case studies, genetic interpretations ultimately rest on form analogy, because although the internal sedimentology and structures are described, they are then essentially discarded in the final explanation. So, despite the importance of stratified cores in the initial theory espoused by Shaw (1983b), they are explained away in the modified version of the theory, violating the principles of hypothesis testing in a Popperian sense (Benn and Evans, 2006). Hence, overridden moraines, hummocky terrain and push moraines become erosional ripple marks (Munro and Shaw, 1997) and flutings become spindle flutes carved by subglacial sheet floods (Munro-Stasiuk and Shaw, 2002). Where sedimentary exposures are used to validate the erosional explanation, there are significant inconsistencies in the purported process–form relationships. For example, some exposures said to be through the summits of hummocky terrain are not actually in the summits (Evans et al., 2006b). If we consider the form analogy argument, we can appreciate the superficial resemblance between drumlins and erosional sole marks (fig. 11.33), but streamlined and rippled forms are produced in a wide range of environments wherever two media shear past one another. They can be observed in cloud formations, riverbeds, windblown sand dunes and snow, at the base of turbidity currents and other mass movements, and glacier beds.

We have so far in this book reviewed a large body of knowledge of the dynamics of shear in the subglacial environment, and have demonstrated that the coupled shearing of basal ice and glacier substrata, as well as slip at the ice–bed interface, can produce a variety of streamlined,

transverse and non-aligned forms. The simplest explanation of streamlined forms is that obstacles below flowing media exhibit shadow effects, such that their presence influences patterns of erosion far downflow. This effect applies not only to turbulent flows, but also to non-turbulent flowing media, such as ice. Ancient drumlins and mega-flutings show a much stronger resemblance to streamlined subglacial landforms exposed by recent glacier retreat (in form and in scale) than they do to scours formed by turbulent media (Benn and Evans, 2006; Evans et al., 2006b). Moreover, historically produced fluting fields, such as those in front of Breiðamerkurjökull in Iceland, allow us to:

(1) relate sediment and landform characteristics to genetic processes with a high degree of confidence;
(2) appreciate that ice-margin parallel bands (flow sets) of flutings, separated by moraines, record known former ice flow directions and are slightly offset from one another due to their alignment at right angles to their contemporaneous moraine (Evans and Twigg, 2002);
(3) document that tills in the flutings commonly have erosional lower contacts with glacitectonized or undisturbed outwash.

Such clear modern analogues for ice sheet beds on modern glacier forelands seriously weaken the case for a mega-flood erosional origin, because they demonstrate that neither turbulent nor sheet water flows are necessary to explain streamlined subglacial landforms.

Each of these landform–sediment relationships can be explained satisfactorily by Boulton's (1987) deformation model of drumlin formation or Tulaczyk et al.'s (2001) ploughing mechanism of substrate fluting (see also Clark et al., 2003). Such processes can explain all drumlins and flutings in terms of the single process of streamlining of pre-existing bed materials, whether they are composed of till, rock or stratified sands and gravels. They also predict that subglacial bedforms should be mantled by glacitectonite or till, a prediction that is borne out in our experience. Where they exist beneath the glacitectonite/till carapace, stratified sediments act as well-drained cores that will not have undergone significant glacitectonic deformation if they were strong enough to withstand the stresses applied by overriding ice. An excellent example of the nature of sedimentary structures predicted by the Boulton (1987) model is the exposure through the Athabasca giant flutings in Alberta, Canada, reported and used as supporting evidence for the mega-flood erosional hypothesis by Shaw et al. (2000). The inter-bedded and disturbed tills and stratified sediments at the site have all of the characteristics of a vertical continuum, from undisturbed glacifluvial sediment at the base to a glacitectonite carapace, and therefore can be explained by subsole deformation.

Another problem with the mega-flood explanation for subglacial bedforms is the vast amount of water required to produce the extensive fields of glacially streamlined landforms. Shoemaker (e.g. 1992) proposed a possible meltwater reservoir at the centre of the Laurentide Ice Sheet, but this has raised several problems, and the idea is not taken seriously by most researchers (e.g. Clarke et al., 2005; Benn and Evans, 2006), although Evatt et al. (2006) have used numerical modelling to verify that large subglacial lakes, akin to those beneath the modern Antarctica Ice Sheet, may have existed under the Laurentide Ice Sheet. Widespread stratified diamictons in western Canada have been interpreted by Munro-Stasiuk (1999, 2000, 2003) and Shaw (2006) as melt-out tills, and thereby used to propose locations for subglacial lakes beneath the Laurentide Ice Sheet. As we have seen in sections 10.3.2 and 10.3.4, the very nature of melt-out till production (i.e. passive and slow release of englacial debris) precludes its association with the production of catastrophic subglacial meltwater floods (Benn and Evans, 2006).

Recent discoveries of large, episodically draining subglacial lakes beneath the Antarctic Ice Sheet (section 6.3.5) serve to remind us that large subglacial floods can be significant landscape modifiers. The catastrophic drainage of such reservoirs has been acknowledged in interpretations of large glacifluvial erosional features such as the Labyrinth in the Antarctic Dry Valleys and the many networks of tunnel channels from former ice sheet beds (section 8.4.4). Additionally, highly erosive subglacial drainage of large ice-marginal, supraglacial and subglacial lakes is well known from modern environments. Moreover, there is strong evidence that in some cases water is carried in distributed systems rather than conduits, and some Pleistocene flood events were of a much higher magnitude than any observed in historical times. It is, however, overly simplistic to regard the juxtaposition of these impressive glacifluvial landforms and subglacial bedforms such as drumlins, flutings, ribbed moraine and hummocky terrain as indicative that all were formed simultaneously by the same mechanisms. In essence, the mega-flood explanation of subglacial bedform genesis remains incompatible with the vast glacial research literature reviewed throughout this book.

11.2.9 CREVASSE-SQUEEZE RIDGES

Crevasse-squeeze ridges form where basal till infills fractures extending upwards from glacier beds. The process is apparently uniquely associated with glacier surges, where high stress gradients coincide with elevated basal water pressures (section 5.7; Sharp, 1985a). Basal crevasse-squeeze ridges have been reported from a number of recently deglaciated forelands, but particularly good examples occur at the margins of the Icelandic surging glaciers Brúarjökull and Eyjabakkajökull (Sharp, 1985a;

Evans and Rea, 1999, 2003; Evans et al., 2007b; fig. 11.34). The ridges are 1–2 m high, overlie fluted basal till and can be traced into the wasting glacier snout, where they occur as debris dykes continuing into crevasse traces. The ridges are arranged in a cross-cutting pattern that mimics the radial and transverse crevasse patterns on the glacier snout. The ridges have three different types of junction with flutings on the till surface:

(1) flutings exhibit strike-slip displacement along the line of ridges;
(2) flutings pass undisturbed through ridges;
(3) flute crest lines rise to intersect ridge crests.

These relationships indicate that the ridges were formed after or simultaneously with the flutings. In the first case, flutings appear to have been displaced along the line of crevasses while still enveloped by basal ice; in the second, the ridge appears to have been superimposed on the fluting; and in the third, both fluting and ridge were apparently squeezed up into a crevasse or other basal fracture.

Discontinuous and cross-cutting ridges have been described from a variety of Pleistocene glacial settings. Gravenor and Kupsch (1959) described *linear disintegration ridges*, consisting of linear intersecting ridges oriented parallel, transverse and oblique to former ice flow, from parts of western Canada (fig. 11.34). Although much larger than the modern Icelandic examples, the western Canadian ridges are similar in every respect to the crevasse-squeeze ridges at Brúarjökull and, consequently, Evans et al. (1999b) and Evans and Rea (1999, 2003) have proposed a similar origin. Basal crevasse-squeeze ridges should only be exposed where the former supraglacial debris cover was very thin, otherwise the low-amplitude ridges would be buried during ice wastage. At some locations, possible crevasse-squeeze ridges do disappear beneath supraglacially derived landforms, especially where they lie adjacent to end moraines (Gravenor and Kupsch, 1959). Because they have been exclusively associated with glacier surging (Sharp, 1985a, b; Evans and Rea, 1999, 2003), large networks of crevasse-squeeze ridges have good preservation potential due to the tendency of surged snouts to downwaste in situ.

Kleman (1988) has described a series of till ridges from the southern Norwegian-Swedish mountains which appear to be well-preserved crevasse-squeeze ridges. The ridges trace out rectilinear zigzag patterns, but in places resemble subglacial drainage networks in planform, and Kleman concluded that they were formed by squeezing of basal till into subglacial crevasses and meltwater conduits. Eskers composed of diamicton/subglacial till can occasionally be produced in this way (e.g. Larsen et al., 2006b). Squeezing of saturated sediment into transverse basal crevasses has been proposed as a mode of origin for some 'DeGeer moraines' (e.g. Zilliacus, 1989) (a concept that is discussed in section 11.6.2).

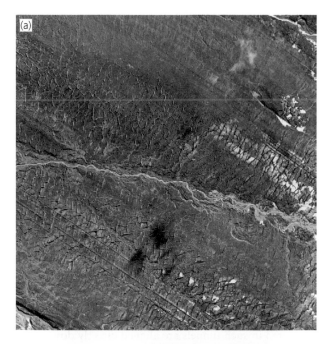

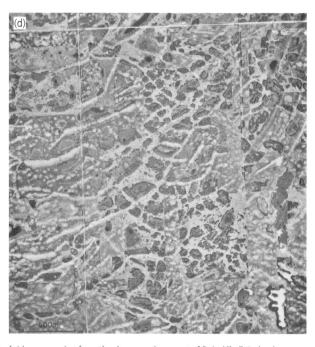

Figure 11.34 Examples of crevasse-squeeze ridges. (a) Aerial photograph extract of ridges emerging from the downwasting snout of Brúarjökull, Iceland (Loftmyndir ehf, Iceland). (b) Ground photograph of ridges at the margin of Eyjabakkajökull, Iceland. (c) Crevasse-squeeze ridge in the snout of Osbornebreen, Svalbard. (D.J.A. Evans) (d) Aerial photograph extract of ridges in Alberta, Canada (Energy, Mines and Resources, Canada).

11.2.10 ESKERS

The term *esker* (derived from the Irish word *eiscir* meaning 'ridge') refers to elongate, sinuous ridges of glacifluvial sand and gravel (fig. 11.35; Warren and Ashley, 1994). Some of the longest unbroken esker ridges in Canada are several hundreds of kilometres long (Shilts et al., 1987), can stand more than 50 m above the surrounding terrain and are up to several kilometres wide. Eskers are the infillings of ice-walled river channels or tunnel fills, and may

record deposition in subglacial, englacial or supraglacial drainage networks. A general genetic classification scheme proposed by Warren and Ashley (1994) and adapted by Brennand (2000) recognizes this range of depositional settings in four basic esker types (fig. 11.36):

(a) *tunnel fills*, formed in englacial and subglacial conduits, and exposed by ice ablation;

(b) *ice-channel fills*, deposited in subaerial (supraglacial) ice-walled channels;

(c) *segmented tunnel fills*, formed during pulsed glacier retreat;

(d) *beaded eskers*, consisting of successive subaqueous fans deposited during pulsed retreat of a water-terminating glacier.

Figure 11.35 An esker system in Lake Rorstromssjon, Angermandland, Sweden, illustrating bifurcations in the middle distance and the association between the esker and an elongate bedrock depression (E. Lindstrom).

Meltwater drains through a glacier via a number of pathways (sections 3.3 and 3.4) and this can result in the deposition of stratified sediments in ice-walled tunnels/channels at various levels in the glacier. Over time, eskers developing beneath a thinning glacier may break through the ice surface and thereby change from being a subglacial to a supraglacial system. Therefore the temporal evolution of glacial drainage networks results in the construction of eskers via a process–form continuum often more complex than that depicted in figure 11.36.

Where subglacial tunnel fills or eskers are the infillings of former subglacial conduit channels, they record the routing of water at the glacier bed where water flow can occur in pressurized conduits. Therefore, eskers formed in these conditions can have up-and-down long profiles, with some sections climbing over topographic obstacles (Shreve, 1972, 1985a, b; Syverson et al., 1994). Where tunnels are at atmospheric pressure, water flow follows the local slope just like subaerial streams, so that eskers deposited in such tunnels are aligned directly downslope. These are typical of thin glacier margins where tunnel creep closure rates are low (Johansson, 1994; Syverson et al., 1994). They are usually shorter and straighter than

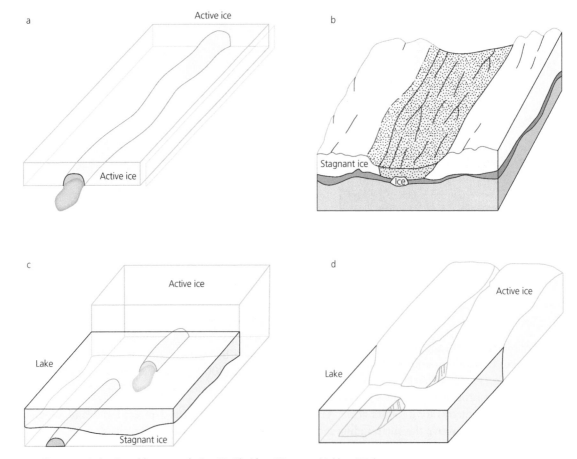

Figure 11.36 Diagrammatic sketches of four types of esker (Modified from Warren and Ashley, 1994).

normal eskers and are commonly associated with kame terraces or lateral drainage channels at their upper ends, and with kame and kettle topography at their lower ends (sections 11.4.3 and 11.4.5; Sugden and John, 1976).

Morphology and distribution

The planform of eskers is extremely variable, and can be represented by single continuous ridges of uniform cross-sectional profile, single ridges of variable height and width, single low ridges linking numerous mounds or beads, or complex braided systems of confluent and diffluent esker ridges (e.g. Warren and Ashley, 1994; Delaney, 2002; figs 11.37, 11.38). Dendritic networks of esker ridges may contain up to fourth order tributaries (Shilts et al., 1987). Most eskers are aligned sub-parallel to the direction of former glacier flow, reflecting former meltwater flow towards the ice margin. However, some eskers are aligned transverse to former glacier flow, a situation most common for former ice-walled channels guided in stagnant glacier snouts and/or ice-cored moraines in valley settings (Johansson, 1994). An interesting example deposited during the historical recession of Breiðamerkurjökull in Iceland allows us to demonstrate the importance of ice thickness changes in esker alignment. A multiple- to single-ridged, ice flow-parallel esker network terminates in an ice-marginal fan at the former western margin of the snout. However, a smaller esker ridge emerges from the margins of the larger esker system at the base of the ice-contact slope of the fan and runs parallel to the ice-contact face, oblique to the former glacier flow direction, for a distance of 2 km. This smaller esker relates to the development of a drainage network that was driven by the thinning of the glacier snout and the concomitant change in subglacial topography once the ice had thinned below the ice-contact face of the proglacial fan (Evans and Twigg, 2002).

Shreve (1985a) related the morphology of tunnel-fill eskers to hydraulic conditions beneath the glacier, particularly the rates of melting and freezing of the tunnel walls. He argued that *multiple-crested eskers*, consisting of several sub-parallel branches, form where the active channel has a tendency to migrate laterally rather than incise itself into the overlying ice. Such conditions are thought to be typical of gently ascending reaches, where tunnel melting rates are low due to the downstream increase in elevation head (section 3.2.2). Single thread, *broad-crested eskers* are thought to be associated with zones of net freezing of the walls in steeply ascending reaches, where tunnels are likely to be wide, low and stable. Finally, *sharp-crested eskers* were attributed to zones of net melting, where the tunnel can melt its way into the overlying ice as the floor fills with sediment. However, sharp-crested eskers may also form by the lowering of englacial tunnel fills onto the land surface (Price, 1973; see page 484 onwards).

The upstream or downstream ends of some subglacial tunnel fill eskers are connected to Nye channels eroded into bedrock or sediment, possibly recording synchronous patterns of deposition and erosion at the glacier bed (section 8.4.4). It is unlikely, however, that esker and channel systems extending over hundreds of kilometres were formed by a single tunnel system of that length. Rather,

Figure 11.37 The Carstairs esker system, Lanarkshire, Scotland, showing both single ridges and braiding pattern of esker development (Cambridge University Aerial Photograph Collection).

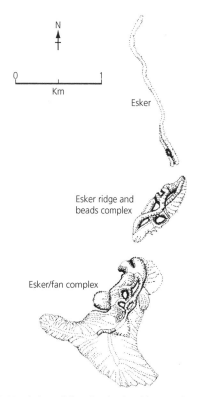

Figure 11.38 Morphology of the esker, bead and fan complex at Lanark, Ontario, Canada (Gorrell and Shaw, 1991, reproduced with permission of Elsevier).

they were probably formed in segments during ice sheet retreat, when the marginal zone of ablation and its associated channel system migrated towards the former ice sheet centre. Clear evidence for temporal changes is provided by eskers lying within Nye channels, indicating channel-bed erosion, followed by deposition and tunnel infilling, probably as a result of falling discharges (e.g. Booth and Hallet, 1993).

Exceptionally long esker ridges pose particular problems for glacial geomorphologists, specifically regarding the nature of tunnel occupation. The lack of any ice-marginal features over the length of very long eskers gives the impression that deposition took place synchronously along the entire length of the tunnel. As it is difficult to envisage such large drainage networks remaining stable in active ice, long dendritic eskers have been interpreted as indicative of regional ice stagnation by Brennand (2000). Where long ridges are punctuated by ice-marginal features such as subaqueous fans, outwash heads/ice-contact slopes or moraines, a case can be made for sequential construction in segments, the axis of tunnel sedimentation remaining stable from one melt season to the next. Boulton et al. (2007a, b) made a case for the maintenance of subglacial channel axes and deposition of long eskers (punctuated in some places by ice-marginal features) as segments beneath the outer zone of the margin of Breiðamerkurjökull, Iceland, over the last 100 years of active ice recession. This stability of subglacial drainage pathways was regarded by Boulton et al. (2007a, b) as a function of the strong coupling between drainage from groundwater catchments beneath the glacier, operating throughout the year, and summer melt. In this way, complex esker networks have been produced during active glacier recession and therefore do not necessarily reflect widespread glacier stagnation, as proposed by Brennand (2000) to explain the preservation of very long dendritic eskers in Canada. Nonetheless, low palaeoflow variability (based on sedimentary palaeocurrent data) and clear downflow cumulative trends in clast roundness over esker lengths of up to 70 km tend to support Brennand's proposition of synchronous channel infilling for some long eskers.

The morphology of apparently long eskers has been assessed by Hooke and Fastook (2007) using the southern Laurentide Ice Sheet landforms of Maine. They demonstrate that the eskers of the region are in relatively straight segments, each segment increasing in size down-ice to appear as a 'tadpole shape'. This is interpreted by Hooke and Fastook (2007) as the product of:

(1) the separate deposition of each segment in a sub-marginal tunnel;
(2) melt rates and concomitant sedimentation rates increasing in response to the increasing glacier surface slope near the margin;

(3) melt rates exceeding tunnel closure rates by an increasing amount towards the margin due to gradually thinning ice, which leads to a reduction in water flow velocity and the deposition of glacifluvial sediments.

Hooke and Fastook (2007) also provided some interesting insights into the changes in subglacial drainage characteristics during early stages of deglaciation and their implications for landform development. Specifically, they cited evidence for a change from the deposition of coalescent grounding line fans/moraines at the ice sheet margin (Kaplan, 1999) to a later style of deposition characterized by esker construction with no moraines. Additionally, this research concluded that esker production could not be explained by subglacial drainage alone, but needed significant input from supraglacial sources.

Tunnel-fill eskers in North Dakota have been associated with glacitectonically disturbed groundwater flow by Bluemle and Clayton (1984). Near the town of Anamoose, an esker starts at the margins of the depression in a hill-hole pair (see section 11.3.2). The depression has been excavated in a buried valley filled with sand and gravel, which acted as a closed aquifer prior to glacitectonism. The elevated groundwater pressures produced by glacier overriding of aquifers are critical to most cases of glacitectonic thrusting (see section 11.3.2). The formation of the esker in this situation was explained by Bluemle and Clayton (1984), using a vivid analogy of a popping cork from a champagne bottle. The pressure in the bottle (the aquifer) was released when the cork (the glacitectonized hill) was removed by thrusting. This allowed groundwater to escape from the aquifer, thus explaining the initiation of an esker at the margins of the excavated hole.

Segmented and beaded eskers are generally interpreted as time-transgressive landform assemblages deposited close to glacier margins during deglaciation. The distal ends of segments may mark the positions of ice-marginal portals during temporary glacier stillstands. Larger beads represent substantial depo-centres at the efflux points of subglacial tunnels, either in subaerial or subaqueous environments (e.g. Bannerjee and McDonald, 1975; Rust and Romanelli, 1975; Thomas, 1984; Warren and Ashley, 1994). A good example was presented by Hebrand and Åmark (1989), who described segmented eskers that comprise interlocking fans. Each esker segment consists of ridge, hummock and terrace components, composed of downstream-fining sequences of gravel and sand. Each sequence documents deposition in a restricted, debris-covered and largely stagnant outer margin of the ice sheet. The progressive recession of the active ice and the development of new stagnant ice led to the accretion of new fans in an up-ice direction. The downstream ridge–hummock–terrace transition in each segment records the deposition of an esker in an ice tunnel and then over the downwasting ice surface. Other examples of subaqueous

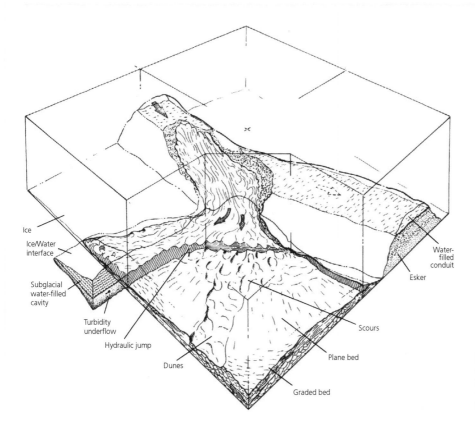

Figure 11.39 Reconstruction of the depositional environment in which lateral fans are deposited (Gorrell and Shaw, 1991, reproduced with permission of Elsevier).

beads have been described by Gorell and Shaw (1991) and Brennand (2000), who have described complex associations of esker ridges, beads and fans in Ontario, Canada, which they attributed to deposition in conduits, lateral water-filled cavities beneath the ice, and grounding-line fans beneath a floating glacier tongue (figs 11.38, 11.39). Wherever subglacial meltwater exits from the glacier into deep water, a subaqueous fan or ice-contact delta will be produced, depending on water depth (section 11.6.2). The likely occurrence of lateral, ephemerally water-filled cavities has been verified by modern glaciological observations (e.g. Gordon et al., 1998; Mair et al., 2002). These have been linked to the construction of anabranched reaches, subaqueous fans and hummocky zones along otherwise predominantly single-ridged eskers by Brennand (1994, 2000).

Eskers commonly develop in inter-lobate positions or ice lobe confluences on glaciers and ice sheets (e.g. Huddart et al., 1999). For example, at Breiðamerkurjökull, Iceland, the largest eskers were associated with portals emerging from beneath the medial moraines on the glacier surface (Price, 1973; Evans and Twigg, 2002). Similarly, the 1996 jökulhlaup at Skeiðarárjökull emerged in an inter-lobate position on the snout (Russell et al., 2001b, 2006). The tendency for glacial drainage systems to occupy such locations has been explained in terms of theoretical potential gradients derived from the hydraulic theory of Shreve (1972). If it is assumed that ice pressure exerts a fundamental control on hydraulic potential, then

we should expect that conduits will converge on inter-lobate positions where the ice surface descends towards elongate depressions in the glacier margin. The Shreve model, however, is based on several questionable assumptions (section 3.3.2), and the presence of conduits in inter-lobate positions probably reflects other processes. First, surface meltwater commonly drains to the bed at glacier confluences, providing major recharge points for subglacial drainage systems (Benn et al., 2009a). Second, surface streams running along low points between glacier lobes can incise to the bed, forming subglacial channels by a process of cut and closure (section 3.3.2; Gulley et al., 2009a).

Explanations of the development of ancient esker systems have often highlighted inter-lobate or ice confluence locations (e.g. Warren and Ashley, 1994; Punkari, 1997b; Thomas and Montague, 1997; Mäkinen, 2003), where early phases of esker sedimentation can develop into large inter-lobate 'moraine' construction. The glacifluvial nature of some of the largest inter-lobate 'moraines' in North America has been clearly demonstrated by Brennand and Shaw (1996) for the 'Harricana Moraine', and by Barnett et al. (1998), Pugin et al. (1999), Russell and Arnott (2003), Russell et al. (2003) and Sharp et al. (2007) for the 'Oak Ridges Moraine'. The Harricana Inter-lobate Moraine, formerly separating the Hudson Bay and Labrador sectors of the Laurentide Ice Sheet, stretches for 1000 km southwards from James Bay to Lake Simcoe, Ontario, and comprises a series of ridges up to 10 km wide and 100 m high. The sequential development of the Oak Ridges Moraine, Ontario, from esker to subaqueous fan to delta is a clear illustration of the evolution of inter-lobate landform complexes over the sites of esker construction (see section 12.4.5).

Composition

Eskers are composed of a wide variety of facies, ranging from sorted silts, sands, gravels and boulders to matrix-supported diamictons. As tunnels collapse or change shape, or streams change position or size, one depositional sequence may be truncated and partially infilled or overlaid by another (Terwindt and Augustinus, 1985). Consequently, cross sections through eskers reveal complex sequences of cross-bedded stratified sediments, often arranged with cut-and-fill geometries. Individual beds may be ripple cross-laminated, horizontally bedded, cross-bedded or massive, depending on flow conditions during deposition (section 10.4). Small-scale cross-bedded units in eskers represent the prograding avalanche faces of migrating bars (Bannerjee and McDonald, 1975), whereas larger cross-bedded units are likely to be deltaic foresets deposited in subglacial or proglacial pools or macroforms (fig. 11.40). Brennand and Shaw (1996) discussed how channel enlargement and infilling leads to the development of composite macroforms (fig. 11.40a) and oblique macroforms with characteristic oblique-accretion avalanche beds (fig. 11.40b).

Although the sedimentary structures observed in tunnel-fill eskers are similar to those found in open-channel fluvial deposits, there are some important distinctions relating to tunnel hydraulics. Bannerjee and McDonald (1975) pointed out that standing waves cannot form in full-pipe (conduit) flows, and that antidune bedforms cannot be deposited in such settings (section 10.4.5). Therefore, antidunes and backset bedding are usually taken to indicate flow in a free-surface stream rather than in water-filled tunnels. Brennand (1994), however, identified very large antidunes in the eskers of south-central Ontario, Canada, which she explained as closed conduit sediments because of the striking esker ridge continuity, low variability in palaeocurrent direction and up-and-down esker long profiles. Brennand explained these large bedforms as deposits from hyperconcentrated flood waters flowing from constricted to expanded reaches, where the flows were transformed from a supercritical to a subcritical state. Saunderson (1977) argued that matrix-supported gravels within eskers document sliding-bed transport of hyperconcentrated flows, and that such conditions may be unique to full pipe flows. More recent work has shown that matrix-supported gravels can be deposited in subaerial and subaqueous flows in a wide range of environments (e.g. Maizels, 1989a, b; Todd, 1989), and cannot therefore be used as diagnostic criteria for pipe flows.

The central core of eskers can display cyclic sequences of gravel and sand (Bannerjee and McDonald, 1975;

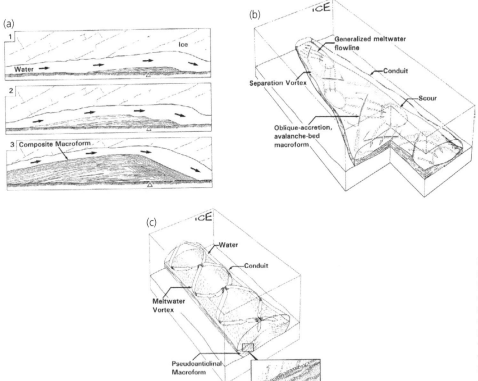

Figure 11.40 Macroform development in eskers deposited in R-channels. (a) Composite macroform deposition due to R-channel enlargement due to ice roof melting. (b) Oblique macroform deposition into an enlarged R-channel due to the combined effects of separation vortices and convergence scours. (c) Pseudo-anticlinal macroforms deposited by paired vortices in R-channel of uniform geometry (Brennand and Shaw, 1996).

Ringrose, 1982; Brennand, 1994; Brennand and Shaw, 1996), the base of each cyclic sequence often being marked by a scoured erosional surface. These sequences display fining upwards from cross-bedded and horizontally bedded, imbricated gravels to horizontally bedded, plane-bedded and trough cross-bedded sands. Such fining-upward sequences with erosive bases record fluctuations in discharge and sediment availability. The depositional cycles or rhythmicity in esker sediments have been interpreted as annual by Bannerjee and McDonald (1975) and as seasonal by Brennand (1994). Annual cyclic signals have been inferred for a Finnish esker by Mäkinen (2003), who described three vertically stacked lithofacies associations comprising summer deposits of massive to stratified coarse gravels, autumn to winter deposits of trough and ripple cross-stratified fine-grained sediments, and spring deposits of sandy stratified beds. Within the large cyclic sequences, cosets of climbing ripple drift or cross-lamination are believed to document anything between hours to tens of hours of deposition (Allen, 1971). This corresponds well to the typical discharge fluctuations of glacial meltwater streams.

Other depositional trends in esker sediments include coarsening-upward sequences and fining outwards from

the core (e.g. Saunderson, 1975). The fining outwards usually involves a gradation from cross-bedded sand and gravel near the core to horizontally bedded and cross-laminated sand with laminated silt inter-beds towards the margins (fig. 11.41). This is thought to represent side-wall melting and esker widening, an interpretation supported by the occurrence of faulting in the marginal zone. An esker may be 'widened' by sediment draping where water discharges from a tunnel portal into a subaqueous environment (fig. 11.41b). In all situations, some slumping will occur at the margins of an esker ridge once the ice walls are removed. The degree of internal disturbance by folding and faulting relates to the thickness of underlying ice or the magnitude of side slumping, and the most heavily disturbed esker sediments will be produced wherever the tunnel was located englacially or supraglacially (Price, 1973). The sediment sequences within eskers commonly form anticlinal structures or arched bedding when viewed in cross section. Anticlines have been interpreted as the product of slumping at the esker sides after removal of the supporting ice walls, and arched bedding as the result of simultaneous deposition on the crest ridge and flanks of the esker. An alternative interpretation of anticlinal structures is that they are in fact *pseudoanticlinal macroforms* deposited by paired vortices in an R-channel (fig. 11.40c; Brennand and Shaw, 1996; Brennand, 2000).

Subglacial meltwater exiting from a glacier into standing water often leads to the progradation of subaqueous fans or ice-contact deltas from esker ridges (section 11.6.2). In the absence of linking esker ridges, the fans or deltas are often described as beaded eskers (fig. 11.37). Sharpe (1988a) described glacimarine subaqueous fans on the floor of the former Champlain Sea Basin and on Victoria Island, Canada, which emanate from eskers and record the rapid deposition of sediment beyond the confines of an ice-walled conduit (fig. 11.41b). Exposures in these landforms show that esker cores continue beneath the fan forms and are overlain or draped by turbidites,

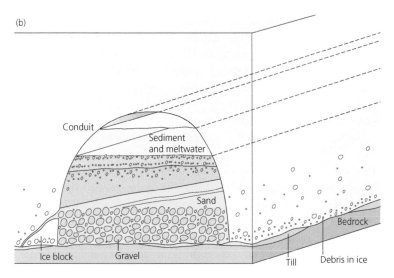

Figure 11.41 The evolution of eskers from tunnel fill ridges to subaqueous fans during ice recession in deep water. (a) The coarse-grained core of an esker fan, comprising horizontally bedded, polymodal cobble gravels and openwork cobble and pebble gravels which are inter-bedded with downwarped beds of laminated sands and clays, Rooskagh esker, Ireland. (D.J.A. Evans; see Delaney, 2001, 2002). (b) and (c) Conceptual diagrams to explain the deposition of an esker by tunnel fill and a subaqueous glacimarine fan in the late Quaternary Champlain Sea near Ottawa and Montreal, Canada (Modified from Sharpe, 1988).

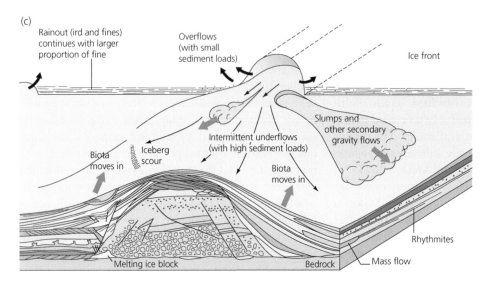

Figure 11.41 (*Continued*)

sediment flows and rhythmites (fig. 11.39). This partial burying of esker forms by subaqueous fan sediments probably attests to a combination of high sediment loads and slow glacier retreat rates. Detailed sedimentological descriptions of Pleistocene fan-deltas emanating from eskers have been given by Martini (1990).

Englacial and supraglacial esker construction

Sediments deposited in ice-walled channels have been described from many modern glacial environments (e.g. Price, 1973; Gustavson and Boothroyd, 1987; Kirkbride and Spedding, 1996; Huddart et al., 1999; Russell et al., 2001b). Such sediments are gradually lowered onto the substrate by melting of the underlying ice (Price, 1973). The draping of such glacifluvial landforms onto high-relief substrates provides an alternative explanation for eskers that climb over uphill sections of the former glacier bed. Supraglacially and englacially derived eskers will contain heavily disturbed if not structureless sediments after deposition, in contrast to the generally well-preserved stratified sediments of subglacial tunnel fills.

A number of studies have highlighted spatial complexity in esker morphology, necessitating interpretations that invoke supraglacial, englacial and subglacial origins. For example, Huddart et al. (1999) described an esker that has formed at the confluence zone of Vegbreen, Svalbard, which comprises an up-ice zone of multiple ridges grading down-ice to a single ridge, terminating in an ice-contact delta. The sedimentology, internal structures and morphology of this esker system indicate that it developed as a supraglacial trough fill (multiple-crested esker with flat tops) before narrowing down-ice into a channel that varied from supraglacial, englacial to subglacial along its length, thereby producing a single ridge esker before feeding into an ice-contact delta at the former glacier snout (fig. 11.42a). The Vegbreen esker provides us with an excellent modern analogue for ancient esker systems where changes from subglacial tunnels to supraglacial or subaerial ice-walled channels have taken place, often developing complex overprinted esker networks at one location or kame belts comprising numerous, discontinuous elongate ridges deposited predominantly in supraglacial troughs. Even though they are fragmented in the ancient landform record, the esker genesis of such features may be apparent in their linearity. An excellent example occurs on north-west Ellesmere Island, Canada (fig. 11.42b; Evans, 1990a), where sand and gravel ridges were deposited along the coalescence zone of two outlet glaciers during the last glaciation and now trend obliquely across the contours of a valley side.

Eskers often emerge from the chaotically downwasting surfaces of ice-cored outwash (Shilts et al., 1987; Evans and Twigg, 2002), and Moorman (2005) has shown that englacial meltwater conduits on Bylot Island, Canada, extend from a modern glacier snout into its ice-cored moraine, indicating that eskers may ultimately emerge from hummocky moraine after ice-core melt out. At Breiðamerkurjökull, Iceland, fan-shaped esker ridges emanate from a single large ridge (fig. 11.43), and Boulton et al. (2007a, b) proposed that they represent the locations of sediment-filled supraglacial channels on the outer toe of the glacier snout beyond the main stream portal. The portal therefore represents the break-out of a subglacial/englacial stream onto the glacier surface, resulting in the cutting of ice-walled channels that radiate out from the efflux point because the stream flow is suddenly at atmospheric pressure. Once the ice-walled channels fill up with sediment they are buried by an ice-contact fan. Once the buried snout starts to melt, the ice-contact fan collapses to reveal the thicker linear ridges of sediment at the former locations of the ice-walled channels.

In November 1996, glacial geomorphologists were treated to a spectacular demonstration of the evolution of an ice-walled esker system when a jökulhlaup erupted at the snout of Skeiðarárjökull, Iceland, and excavated a 500 m long and 100 m wide supraglacial channel that

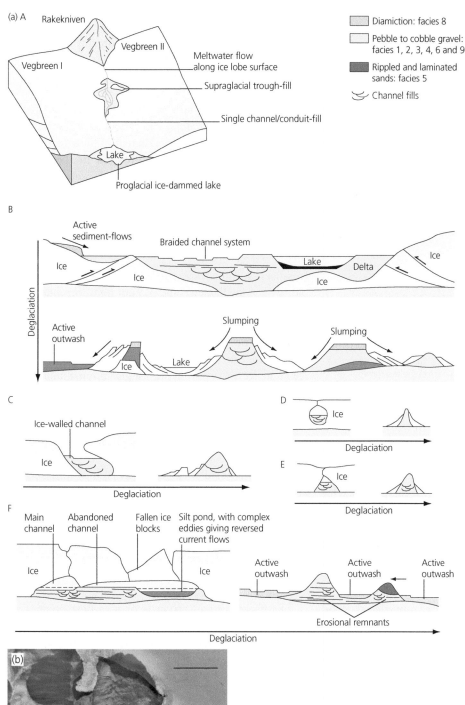

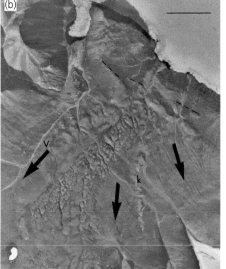

Figure 11.42 Examples of supraglacial eskers formed at the suture zones of High Arctic glaciers. (a) Diagrammatic representation of the formation of an esker complex at the suture zone of Vegbreen, Svalbard. (Huddart et al., 1999) Panel B shows the evolution of the deposits in a large supraglacial channel/trough, and panels C–F show the range of deposits formed by deposition in ice-walled channels and englacial and subglacial conduits down-ice of the trough. (b) A linear kame ridge (k) or supraglacial esker deposited at the confluence of two glacier snouts, Phillips Inlet, north-west Ellesmere Island, Arctic Canada. Note the inset lateral meltwater channels cut at the receding glacier margins (direction of ice recession shown by arrows; V = main valley). Scale bar at top right represents 1 km and broken lines highlight the former grounding line ridge where the glacier formerly floated in a glacioisostatically higher sea level (Aerial photograph reproduced with permission of the National Air Photograph Library, Ottawa).

Figure 11.43 Aerial photograph extract of an 'esker fan' on the foreland of Breiðamerkurjökull, Iceland. Note the branching of the small esker ridges from their apex at the lower end of a single large esker ridge, and the occurrence of hummocky topography between esker ridges caused by the recent collapse of the overlying outwash fan surface. The present glacier margin is off the top right of the image and the distance across the area shown is approximately 1.25 km (University of Glasgow and Landmælingar Islands, 1998).

filled up with 8 m of gravel macroforms (Russell et al., 2001b; Burke et al., 2008). Since the jökulhlaup, the morphology and sedimentology of the esker network has been documented and it has become clear that a single large esker ridge feeds into the ice-walled channel fill. The post-depositional melting of underlying ice has resulted in the formation of a relatively wide and flat-topped or stepped esker cross profile, and surface pitting is occurring wherever ice blocks were incorporated in the flood deposits (fig. 11.44). A similar jökulhlaup origin for a flat-topped esker has been proposed by Fard (2002).

Esker forms particularly diagnostic of supraglacial deposition are *zigzag eskers* (originally termed 'concertina eskers' by Knudsen, 1995). The ridges of zigzag eskers lack the sinuous or meandering pattern of regular eskers and are characterized by short straight segments that switch direction abruptly at 'elbows' (fig. 11.45). They are spectacularly developed on the foreland of the Icelandic surging glacier Brúarjökull (Knudsen, 1995; Evans et al., 2007b), where they have been associated with other landforms diagnostic of surging activity (Evans and Rea, 1999, 2003). Evans and Rea (2003) showed that these zigzag eskers were deposited in crevasse systems inherited from the surge. Bennett et al. (2000b) described another example from Skeiðarárjökull, Iceland, where ice-cored ridges comprising stratified sediments mark the locations of crevasses infilled by glacifluvial sediment during a surge in 1991. Interestingly, some of the crevasse fills at Skeiðarárjökull intersect frozen-in conduit fills which mark the location of englacial tunnels formed at the same time as the crevasse fills.

11.2.11 SUBGLACIAL VOLCANIC FORMS

The most instantly recognizable landforms produced by volcanic eruptions beneath glacier ice are isolated, flat-topped mountains known as *tuyas* (Björnsson, 1975; Smellie, 2000, 2007, in press; Smellie and Chapman, 2002). Other subglacial volcanic features include tephra mounds or ridges (*tindars*) and pillow mounds, ridges or sheets (fig. 11.46). The lateral extent of these features is a function of the type of lava involved in the eruption (Smellie, 2007, in press). Tuyas and other subglacial forms may be produced entirely during a single eruption or may build up during numerous eruptions beneath an ice mass that was capable of re-forming over the volcanic vent between eruptions. An excellent example of such a polygenetic tuya is Mount Haddington on James Ross Island, Antarctica, a stratovolcano 6.5 million years old, 60 km in diameter and 1.5 km high (Smellie, 2007; Smellie et al., 2008). Mount Haddington has developed to such a large size because of the long-term stability of the Antarctic Ice Sheet, whereas the prominent tuyas of Iceland were produced beneath the more extensive Icelandic Ice Sheet during the last glaciation, and in places where they extend above the modern-day glaciation level they are capped by plateau ice fields (Evans et al., 2006b; Smellie, 2007). The characteristic landforms and volcanic deposits of tuyas are indicative of lava eruption directly into a subglacial water reservoir. As the reservoir expands, it is filled with pillow lavas, hyalotuffs, hyaloclastites, conglomerates and compound lavas to form a steep-sided and flat-topped rock mass whose size and shape reflect the size of the reservoir. Eruption may cease before the tuya and associated cavity reach the surface of the glacier, or the tuya may develop into a Surtseyan or subaqueous-type volcano occupying a glacial lake (Smellie and Skilling, 1994).

The processes and products of tuya development are depicted in figure 11.47. During mafic tuya production, hyaloclastic, lava-fed deltas are deposited over an initial pillow mound. Such deltas generally do not develop in felsic tuyas, which are instead dominated by the accumulation of tephra or lava (Tuffen et al., 2002). Tephra-dominated felsic tuyas are characterized by thick cores of fine-grained tephra (hyalotuff) over which lies a flat cap of silicic lava, emplaced subaerially once the ice roof collapses. Lava flow-dominated felsic tuyas tend to be flat-topped columns or blade-like ridges composed of sub-horizontal lavas. Distinctions can also be made between tuyas erupted beneath temperate versus polar glaciers, as depicted in figure 11.47 (Smellie, 2007). Most significantly, the lack of subglacial meltwater evacuation routes in polar ice tends to result in the confinement of meltwater in a vault directly over the volcanic vent. The colder ice in polar systems also results in slower ice deformation rates and slower vault wall melting rates.

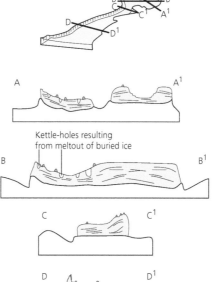

Peak flood
water surface

Water surface 1000h
November 6th

Buried ice blocks

Kettle-holes resulting
from meltout of buried ice

Rip-up clasts

Figure 11.44 The ice-walled supraglacial channel fill/esker deposited during the 1996 jökulhlaup at Skeiðarárjökull, Iceland. (a) Aerial view looking up ice of the ice-walled re-entrant and the flood waters during the jökulhlaup. A helicopter is visible for scale above the main channel. (O. Sigurðsson) (b) Cross sections of the ice-walled channel and associated deposits (left) during the flood event and (right) after three years. (Russell et al., 2001b). (c) Aerial view looking down ice of the sediment-filled embayments and their feeding esker ridge in 2003. Note also the pitted ice-contact fan located beyond the embayment fills – this fan is covered in icebergs in (a) (Modified from A. Gregory).

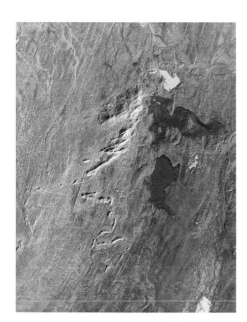

Figure 11.45 Aerial photograph extract of the foreland of Brúarjökull, Iceland, showing a zigzag esker (Loftmyndir ehf). Note the adjacent crevasse-squeeze ridges and long flutings. All of these landforms are diagnostic of the surging glacier landsystem (see section 12.3.8).

These factors in combination result in much slower enlargement rates for subglacial lakes and vaults in polar ice, forcing both water and ultimately lava onto the ice surface and resulting in taller tuyas than in temperate systems. More comprehensive descriptions of all subglacial volcanic forms and sediments have been provided by Smellie and Chapman (2002) and Smellie (in press, 2007).

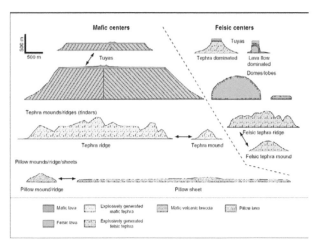

Figure 11.46 The cross-sectional morphologies and principal materials of different types of subglacial volcanic landforms (Smellie, 2007).

11.3 ICE-MARGINAL MORAINES

The present and former margins of glaciers and ice sheets are demarcated by a wide variety of depositional features produced by the complex interactions of numerous processes. In this section, we describe ice-marginal processes and moraines under four headings:

(1) proglacial glacitectonic landforms;
(2) push and squeeze moraines;
(3) dump moraines and ice-marginal aprons;
(4) latero-frontal fans and ramps.

These moraine types are formed by the deposition or deformation of sediment around the edge of active glacier snouts. It must be stressed that the four moraine types discussed in this section are not distinct, exclusive categories, and that in nature composite moraines are not uncommon. Moraines that originate supraglacially or englacially and are lowered onto the ground surface by glacier ablation are described in section 11.4, and ice-marginal moraines deposited in subaqueous settings are discussed in section 11.6.

Some general terms are applied to many moraine types, regardless of genesis. The outermost moraine ridge formed at the limit of a glacier advance is known as a *terminal moraine*. Younger moraines nested within a terminal moraine are termed *recessional moraines*, because they are

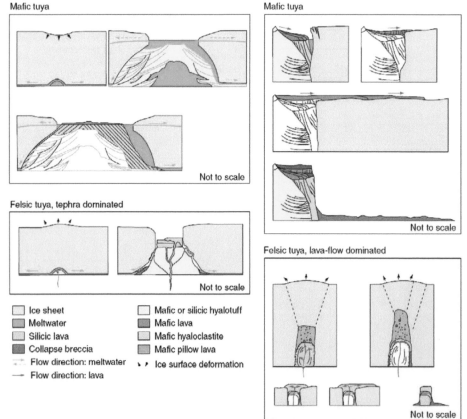

Figure 11.47 Idealized sketches of the evolutionary stages of mafic and felsic tuya production beneath (a) temperate glacier ice and (b) polar glacier ice. Note that the bedding structures in the mafic hyaloclastite represent the foresets of a lava-fed delta (Smellie, 2006, 2007).

(a)   (b)

formed during overall glacier recession, even though they may have been deposited or pushed up during minor readvances or stillstands. Terminal and recessional moraines may be subdivided into *frontal* and *lateral* components, or described as *latero-frontal moraines*. Where it can be demonstrated that recessional moraines formed on a yearly basis, they are known as *annual moraines*. In sections 11.3.2–11.3.5, we review the various landform–sediment assemblages produced in ice-marginal settings, but first we provide a synopsis of the processes and patterns of deposition common to such environments.

11.3.1 PROCESSES AND PATTERNS OF ICE-MARGINAL DEPOSITION

The most widespread processes of debris reworking and deposition in terrestrial ice-marginal environments are those driven by gravity, such as debris flow, falling, rolling and sliding, and deposition from flowing water. The relative importance of gravitational and glacifluvial processes varies from glacier to glacier, according to the availability of debris and meltwater.

At stationary or slowly changing ice margins, gravitational processes deposit debris as ice-contact aprons of scree or debris flows. Where debris supply is low, such aprons form small ramparts along the ice edge, whereas larger amounts of debris build up substantial fans and cones (fig. 11.48a). Debris supply is commonly variable along a glacier margin, and large debris cones may be deposited below re-entrants or crevasses in the ice margin, which funnel debris or debris concentrations on the surface (Eyles, 1979). Rates of debris supply are highest around the margins of debris-mantled glaciers, where huge aprons of scree, or *lateral moraines*, can be built up around the entire glacier tongue (fig. 11.48b; section 11.3.4; Boulton and Eyles, 1979; Small, 1983; Benn et al., 2003). Such large lateral moraines are common in high mountain environments such as the Himalaya, High Andes and the New Zealand Alps, where they can become sufficiently massive to dam back the glacier ice, impeding further advance. Therefore, periods of positive mass balance will cause the glacier to thicken, and if the rate of debris build-up at the margin keeps pace with the rate of ice thickening, the glacier can become perched high above the valley floor, hemmed in by its own moraines (fig. 11.49a). Successive episodes of glacier expansion will tend to terminate at the same moraine, so that it may be built up on several occasions, punctuated by non-active periods when vegetation, including trees, can colonize the outer slope (fig. 11.48b). The damming effects of large lateral moraines are strikingly demonstrated when the barrier is breached for any reason (fig. 11.49b). Lliboutry (1977) has argued that the crooked paths followed by some debris-mantled Peruvian glaciers reflect the breaching of the moraine barrier by lake outburst floods at times

Figure 11.48 Lateral debris accumulations. (a) Small rampart of ice-contact scree at the margin of the Nordenskjold Glacier, South Georgia. (D.J.A. Evans) (b) Large lateral moraine of the Ngozumpa Glacier, Nepal. The glacier surface formerly stood high above the side valley on the right, which is now occupied by a moraine-dammed lake. The downwasting glacier surface can be seen behind the moraine. The settlement of Gokyo stands on a former ice-contact fan (D.I. Benn).

of glacier recession. When the glacier expands once more, it is able to advance through the breach, but is restrained by the remaining moraine along the rest of its margin (fig. 11.50).

Substantial amounts of debris are reworked and deposited by *glacifluvial processes* at some ice margins, particularly the flanks of debris-mantled glaciers with well-developed supraglacial and englacial drainage networks (Gustavson and Boothroyd, 1987). Braided or meandering meltstreams can deposit sediment in troughs in the ice surface, burying the underlying ice and retarding ablation (Lawson, 1995; Huddart, 1999; Evans and Twigg, 2002). Topographic inversion and differential ablation can then result in the isolation of stagnant ice blocks which may become completely buried by outwash deposits accumulating around the ice margin. Eventual melting of such buried ice blocks leaves water-filled *kettle holes* in the outwash surface (fig. 11.51), whereas the melt-out of larger masses of ice buried below outwash produces a more irregular topography in which the areas between kettle holes are reduced to remnant upstanding ridges and mounds, which may follow the traces of

Figure 11.49 Moraines of debris mantled glaciers. (a) Debris-mantled glacier hemmed in by large lateral moraines, Cho La Glacier, Nepal. (b) Sequence of breach-lobe moraines marking the advance of a lobe of ice through a gap in a large lateral moraine, Ghiacciaio del Miage, Italy (D.I. Benn).

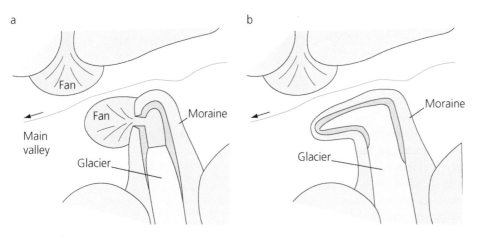

Figure 11.50 Lliboutry's (1977) conceptual model of the development of a crooked debris-mantled glacier. (a) A large lateral moraine is breached by the outburst of a moraine-dammed lake at a time of glacier thinning. (b) a subsequent glacier advance is channelled through the breach but constrained elsewhere.

former channels (*kame and kettle topography*; section 11.4.3). Topographic development during the ablation of some ice margins is complicated by the emergence of sediment-choked subglacial and englacial conduits, forming upstanding ridges of sand and gravel or eskers (section 11.2.10). The relocation of such glacifluvial materials by later glacier advances, either seasonally or during more prolonged periods of ice expansion, often results in the construction of morainic landforms composed of what we might regard as atypical sediments; for example, some push moraines comprise relatively well-sorted gravels and sands, highly contorted by glacier pushing (see sections 11.3.2 and 11.3.3), rather than till.

Characteristic aprons and ramps of debris form around the margins of some cold, polar and subpolar glaciers, where they are often referred to as *ice-contact screes* (Evans, 1989b; Fitzsimons, 1990, 2003; Ó Cofaigh et al., 2003b). When advancing, such glaciers typically terminate in steep ice cliffs, from which ice blocks and debris topple and accumulate around the margin (fig. 11.52). Snout advance over the aprons and ramps leads to the production of debris-rich basal ice sequences

Figure 11.51 Aerial photograph of the southern margin of Sandfellsjökull in 2007, showing the development of kettle holes in the marginal sandur surface (NERC UK).

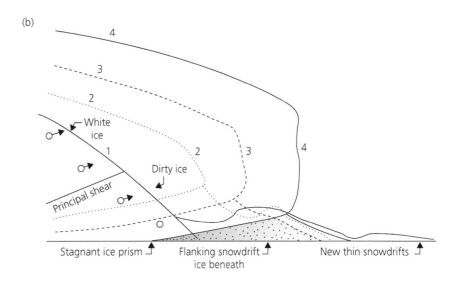

(b)

4

3

2

White ice

1   2   3   4

Dirty ice

Principal shear

Stagnant ice prism ⌐   Flanking snowdrift ⌐   New thin snowdrifts ⌐
                        ice beneath

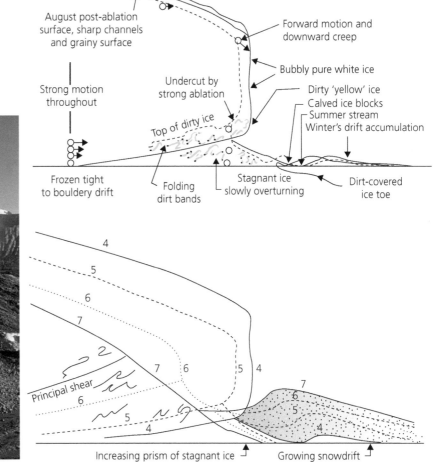

Forward and upward motion

June pre-ablation surface, smooth ridges, snow fill and glossy sublimation surface

August post-ablation surface, sharp channels and grainy surface

Forward motion and downward creep

Bubbly pure white ice

Strong motion throughout

Undercut by strong ablation

Dirty 'yellow' ice
Calved ice blocks
Summer stream
Winter's drift accumulation

Top of dirty ice

Frozen tight to bouldery drift

Folding dirt bands

Stagnant ice slowly overturning

Dirt-covered ice toe

4

5

6

7

7   6   5   4

Principal shear

6   7   6

5   5

4   4

Increasing prism of stagnant ice ⌐   Growing snowdrift ⌐

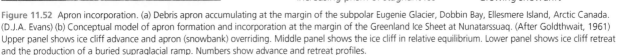

Figure 11.52 Apron incorporation. (a) Debris apron accumulating at the margin of the subpolar Eugenie Glacier, Dobbin Bay, Ellesmere Island, Arctic Canada. (D.J.A. Evans) (b) Conceptual model of apron formation and incorporation at the margin of the Greenland Ice Sheet at Nunatarssuaq. (After Goldthwait, 1961) Upper panel shows ice cliff advance and apron (snowbank) overriding. Middle panel shows the ice cliff in relative equilibrium. Lower panel shows ice cliff retreat and the production of a buried supraglacial ramp. Numbers show advance and retreat profiles.

(see section 9.3.2). During the brief summer ablation season, glacifluvial processes can also be important in such settings. Meltstreams deposit large amounts of sediment around the margins and at the foot of waterfalls that pour over the ice cliff.

11.3.2 PROGLACIAL GLACITECTONIC LANDFORMS

In this section we discuss ice-marginal (proglacial and sub-marginal) glacitectonic mechanisms, processes, structures and landforms. We classify glacitectonic sediment–

landform associations using the fourfold scheme of Aber et al. (1989):

(1) hill-hole pairs;
(2) composite ridges and thrust-block moraines;
(3) cupola hills;
(4) mega-blocks and rafts (fig. 11.53).

Glacitectonic moraine ridges are defined as those in which glacitectonized pre-tectonic and syn-tectonic sediments or bedrock constitute >25 per cent of the unit area of the moraine. Other types of moraine, such as push moraines, may contain small amounts of glacitectonized material

Landform	Height (m)	Area (km²)	Primary material	Primary morphology
Large composite ridge	100 to 200	20 to >100	Bedrock	Subparallel ridge and valley system, arcuate in plan
Hill–hole pair	20 to 200	<1 to >100	Variable	Ridged hill associated with source depression
Small composite ridge	20 to <100	1 to >100	Quaternary strata	Subparallel ridge and valley system, arcuate in plan
Cupola hill	20 to >100	1 to 100	Variable	Smoothed dome to elongate drumlin with till cover
Megablock/raft	0 to <30	<1 to 1000	Bedrock	Often concealed, flat buttes or irregular hills

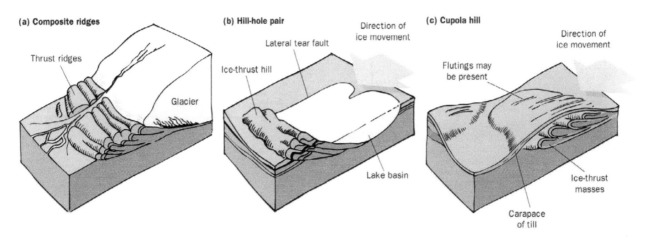

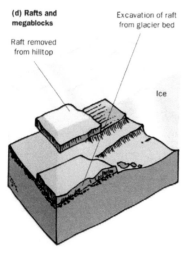

Figure 11.53 Characteristics of the main glacitectonic landforms (Upper table from Aber et al., 1989; lower sketches of the four main features from Evans and Benn, 2001).

(see section 11.3.3). Glacitectonic landforms may be composed of pre-Quaternary bedrock, pre-existing Quaternary sediments or contemporaneous sediment, and where they have been overridden by glacier ice after construction they may be covered by a carapace of glacitectonite or till (section 10.3.3). More prolonged and active glacier overriding will result in the streamlining and moulding of glacitectonic features to produce subglacial bedforms (section 11.2.2).

The mechanics of glacitectonism

Proglacial glacitectonics refers to the large-scale displacement of proglacial and sub-marginal materials due to stresses imposed by glacier ice, and involves ductile or brittle deformation or a combination of the two (fig. 11.53; Aber et al., 1989; Hart and Boulton, 1991; van der Wateren, 1995; Williams et al., 2001). Ductile deformation involves the production of large open folds in the sediments or rocks in front of an advancing glacier, which may develop into overfolds or begin to undergo internal thrusting due to continued ice advance (e.g. Kuriger et al., 2006). In contrast, brittle deformation involves the thrusting of semi-coherent blocks along discrete planes of failure. The lower limit of deformation is usually marked by a basal failure plane or *plane of décollement*, which often coincides with a sedimentary discontinuity or bedding plane or the base of the permafrost (Boulton et al., 1999). Although all materials are capable of undergoing both ductile and brittle failure, according to variations in the applied stress, temperature, strain rate and pore-water pressure, glacitectonic disruption of frozen sediments appears to be dominated by brittle failure (e.g. Evans and England, 1991). The compression of proglacial materials by glacitectonic deformation is very similar to the processes which occur during continental collision and mountain building, and proglacial glacitectonic landforms

can be regarded as scale models of mountain chains from which we can learn the well-founded principles of structural geology (see Twiss and Moores, 1992; Williams et al., 2001).

Proglacial glacitectonic deformation can be responsible for the formation of large thrust moraine complexes standing many tens of metres above the surrounding terrain. The elevation of such large thrust masses depends on a number of factors, the most important of which are low-strength proglacial sediments and high glacially imposed stresses. Sediment strength is dependent on grain-size and sorting, the existence of potential planes of failure, and pore-water pressures. High pore-water pressures can be encouraged by the existence of proglacial permafrost, which acts to confine water in underlying unfrozen aquifers (Mathews and Mackay, 1960). The presence of weakened sediments, however, is not a sufficient condition for proglacial glacitectonic deformation, which also requires stresses large enough to elevate large masses of sediment above the glacier margin. It is clear that in most situations the shear stresses beneath glacier margins are too low to produce the observed deformation. The solution to this problem was developed by Rotnicki (1976), van der Wateren (1985), Aber et al. (1989) and Williams et al. (2001), and is known as the *gravity spreading model*. According to this model, proglacial sediment failure results from the gradient of the weight of the glacier, which pushes sediment wedges away from the load and then upwards from their original position. The stress field producing a thrust block is therefore a function of the slope of the glacier front. (See Dahlen et al., 1984 for developments of critical wedge theory in structural geology.)

The development of lateral stresses near a sloping ice margin is illustrated in figure 11.54. At any point below the glacier, the downward-oriented stress produced by the

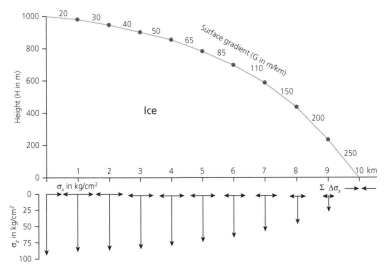

Figure 11.54 The gradient in normal stress (σ_z) at a glacier margin, leading to a net horizontal compressive stress ($\Sigma\Delta\sigma_x$) below the ice edge (Aber et al., 1989).

static weight of the ice column (the *normal stress or glacio-static stress*) is:

$$\sigma_z = \rho i \, g \, H \qquad (11.2)$$

where ρi is the density of the ice, g is gravitational acceleration ($9.81 \, \mathrm{m \, s^{-2}}$) and H is the ice thickness (m).
Since ρi is around $900 \, \mathrm{kg \, m^{-3}}$, equation 11.2 reduces to:

$$\sigma_z = 8829H \qquad (11.3)$$

For $H = 100 \, \mathrm{m}$, the normal stress would be approximately $882,900 \, \mathrm{Pa}$ or $882.9 \, \mathrm{kPa}$.

Part of the vertically oriented normal stress is transferred to a horizontal normal stress due to the tendency of subglacial materials to bulge outwards and press laterally against neighbouring particles in response to the load imposed by the overlying ice and sediment. The magnitude of the horizontal stress (σ_x) for any given glacio-static stress (σ_z) depends on the material properties, and is calculated using *Poisson's ratio, v^**:

$$\sigma_x = (v^* \sigma_z)/ 1 - v \qquad (11.4a)$$

Since Poisson's ratio is close to 0.2 for many unconsolidated sediments, the horizontal component of the glacio-static stress can be approximated as:

$$\sigma_x = 2200H \qquad (11.4b)$$

Because the glacio-static pressure is dictated by ice thickness, the horizontal component (σ_x) decreases in magnitude from the centre of the ice mass towards the ice margin, producing a lateral pressure gradient in the substratum (Rotnicki, 1976; van der Wateren, 1985; Aber et al., 1989). This lateral pressure gradient between two points is given by:

$$\sigma_x 1 - \sigma_x 2 = 2200(H_1 - H_2) \qquad (11.5)$$

The horizontal stress differences are cumulative – that is, the stress difference over a given interval is passed on and added to the stress difference of the next, resulting in a maximum horizontal compressive stress beneath the margin (fig. 11.54; Aber et al., 1989). To this stress must be added the basal shear stress or *glacio-dynamic stress*. Together, the glacio-static and glacio-dynamic stresses constitute the most important components of the total glacitectonic stress.

Failure will take place along a potential failure plane when the applied stress – in this case the total glacitectonic stress – equals or exceeds the *shearing resistance* (Aber et al., 1989). Failure conditions can be expressed using the familiar Coulomb equation (equation 4.5), and failure will occur when:

$$\sigma_{gt} \geq c + (P_i - P_w) \tan \Phi \qquad (11.6)$$

where σ_{gt} is the total glacitectonic stress; P_i is the ice overburden pressure; P_w is the pore-water pressure; c is cohesion; and Φ is the angle of internal friction.

It can easily be seen that, where the cohesion is small, failure is most likely where P_w is large: as the pore-water pressure approaches the overburden pressure, the stress required for the initiation of thrusting approaches zero. According to Aber et al. (1989), this situation arises either through the compaction of impermeable strata and the restriction of water escape routes, or the transmission of water under pressure into a confined aquifer. In an homogeneous material, thrust faults normally develop at an angle of 30° to the horizontal, because their internal angle of friction lies around that value. However, low-angle or sub-horizontal thrust faults will exploit pre-existing weaknesses in the substratum such as bedding planes, lithological boundaries, incompetent strata and permafrost boundaries. Steeper thrust fault angles (>30°) occurring in coarse-grained materials such as gravels attest to the existence of permafrost, because greater substrate rigidity is required to enable the development of such steep thrust fault angles. Movement of material along thrust planes due to gravity spreading results in the elevation of thrust blocks in front of the glacier margin (fig. 11.55).

In addition to gravity spreading, two other mechanisms may contribute to the development of thrust systems (Aber et al., 1989; Price and Cosgrove, 1990). First, *gravity gliding* or *gravity sliding* involves the deformation of sediment blocks as they move downslope under their own weight. This requires the uplifting of the ice-proximal zone by gravity spreading to produce a slope on which blocks can slide. Second, *compression* or *push-from-behind* may occur by the direct shoving of sediment wedges by the forward movement of the glacier. In reality, gravity spreading, gravity gliding and compression can act in unison in proglacial settings.

In summary, the likelihood of glacitectonic disturbance is dictated by various factors, the most important being:

(1) the slope of the proglacial area, particularly the presence of reverse slopes at the glacier margin;
(2) the presence of weak layers in the substratum, which form potential décollement planes;
(3) the nature of the ice–sediment contact in the proglacial area, specifically where glaciers are partially buried by the proglacial sediments so that snout advance drives a wedge of ice into the strata resulting in the production of thrust blocks, with sedimentary beds dipping away from rather than back towards the glacier snout (e.g. Evans and England, 1991);
(4) the subglacial and proglacial drainage.

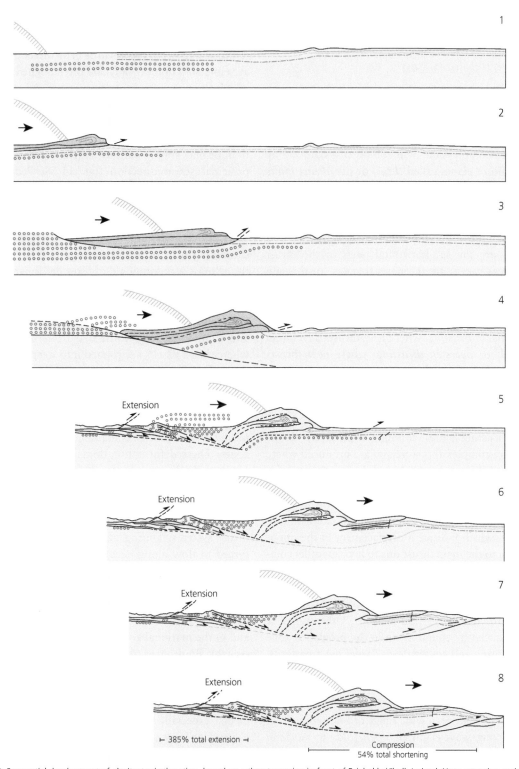

Figure 11.55 Sequential development of glacitectonic thrusting, based on a thrust moraine in front of Eyjabakkajökull, Iceland. Note extension and excavation beneath the margin, and compression and ridge construction on the foreland (Redrawn from Croot, 1988b).

As noted above, failure is encouraged by high pore-water pressures in proglacial and sub-marginal sediments and rocks. High water pressures can arise if impermeable sediments or permafrost occur at the glacier margin, both of which impede drainage and increase pore-water pressures in underlying aquifers (Mathews and Mackay, 1960; Mackay and Mathews, 1964). Even patchy permafrost or seasonally frozen ground can impact on the architecture of glacitectonic failure (e.g. Bennett et al., 2005; Waller and Tuckwell, 2005). High proglacial and sub-marginal pore-water pressures are also generated during glacier surges (section 5.7.5), so surging glaciers are very commonly associated with glacitectonic landforms (Evans and Rea 1999, 2003; Evans et al., 2007b).

Glacitectonic deformation

The uplift and shortening of proglacial sediments and rocks during glacitectonic deformation results in complex geological structures, analogous in many ways to tectonic thrust fold belts. Sub-horizontal substrata affected by glacitectonic compression will often dislocate along low-angle failure planes called *thrust* or *overthrust faults* (low-angle reverse fault), which separate the overriding *hanging wall* and the overridden *footwall*. Such faults normally produce a 'staircase' pattern whereby the sole thrust climbs up through the strata, producing *flats* and *ramps*; flats are the sub-horizontal parts of the thrust plane, whereas ramps are the parts that cut up through the strata at angles of around 30° (fig. 11.56). Continued compression by glacier advance will result in the development of multiple thrust sequences. These may involve either *piggyback thrusting*, where new thrusts develop in the footwall, or *overstep thrusting*, where new thrusts develop in the hanging wall (i.e. they develop in sequence either forwards or backwards from the initial thrust; figs 11.56c, d). Studies of glacitectonic structures by Rotnicki (1976) and van der Wateren (1995) suggest that piggyback thrusting is most common in proglacial deformation. More complex thrust masses are produced where *thrust slices* or *sheets* (horses) are imbricately stacked as a *duplex*. Duplexes therefore consist of a stack of horses bounded by a roof thrust and a floor thrust (fig. 11.56e). Other common structures in glacitectonized sediments are *back thrusts*, which indicate a displacement in the opposite direction to the main thrust due to layer-parallel compression (fig. 11.56f). Back thrusts will combine with frontal ramps to produce an uplifted hanging wall block called a *pop-up*, and may also truncate older thrust faults to produce *triangle zones* (fig. 11.56g). Other types of faulting produced by compression in the proglacial zone are *anastomosing* and *conjugate* or *Riedel shear* patterns. These failures are characterized by two sets of similar faults which cross-cut each other at a consistent angle.

Ductile deformation during glacitectonic compression and shortening produces a wide range of fold types. They range from long wavelength or *open folds* to *isoclinal* or *recumbent* (nappe) folds. Other fold types include chevron, sheath, kink-band, box and polyclinal folds (fig. 11.57). Complexities may arise where different sedimentary layers respond differently to stress and produce disharmonic folds, in which the fold wavelengths of some layers are smaller than those of others. Studies of the three-dimensional form of glacitectonic folds reveal that fold profiles typically vary along the fold axis (i.e. they are *non-cylindroidal*). Common examples of non-cylindroidal folds are *periclines*, *domes* and *basins* (fig. 11.57). Domes are simply anticlinal structures which plunge at similar angles in all directions, whereas structural basins are produced where strata dip inwards in all directions. Periclines are elongated domes which are often aligned in an *en echelon* fashion as a result of compression. Where strata have undergone complex deformation histories, some folds may be superimposed on pre-existing folds. This may involve either two or more separate phases of deformation, or may simply indicate shortening in more than one principal strain direction (fig. 11.58; Pedersen, 2000; Phillips et al., 2002).

Structures resulting from compressional deformation have been produced experimentally by Mulugeta and Koyi (1987), who subjected sand layers to 40 per cent shortening in a squeeze box (fig. 11.59). The resulting thrust mass was composed of three deformation domains. First to form is the *distal domain*, consisting of low-angle thrust faults demarcating thrust blocks with overturned drag folds, sheath folds, thrust faults, extension fractures and slumps. Next to form is the *intermediate domain*, where thrust blocks are rotated into steeper positions and thrust faults develop into concave-upward listric forms. Back kinking also develops in the upper portions of the thrust blocks. Finally, the *proximal domain* forms and is characterized by vertically orientated and laterally compacted thrust blocks containing back thrusts and kink zones. These deformation domains can be recognized in proglacially thrust sediments (e.g. Croot, 1988b; van der Wateren, 1995; Boulton et al., 1999; fig. 11.55).

The expulsion of groundwater during glacitectonism is thought to be important in the development of some thrust systems. During glacitectonic thrusting, water is forced to flow along décollement planes, where it facilitates further displacement, and into aquifers. This water may be expelled at the surface in springs or *blowouts* if aquifers are pressurized (Bluemle, 1993; Boulton and Caban, 1995). The patterns of groundwater flow beneath and in the marginal zones of ice sheets have been considered by Boulton et al. (1995) and Boulton and Caban (1995). Where confined aquifers become overpressurized due to the presence of glacier ice, groundwater and liquefied sediment may burst through proglacial sediments, especially in areas of discontinuous permafrost, to produce large-scale dewatering structures or *extrusion moraines* (section 12.5.6). Similar features, referred to as *blow-out structures* by Kjær et al. (2006), were created during the AD 1890 surge of Brúarjőkull in Iceland. Very little systematic work on groundwater patterns in front of glaciers has been undertaken, but Robinson et al. (2008) demonstrated that water tables are close to the surface on Skeiðarársandur, Iceland, and they are perched in areas of buried ice. This has significant implications for the responses of proglacial materials to glacier advance.

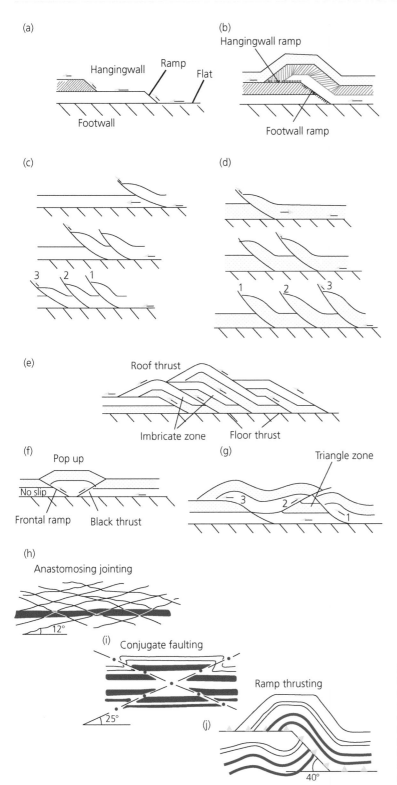

Figure 11.56 Large-scale structures typically found in thrust zones. (a) Shape of thrust surface with ramps and flats. (b) Hanging-wall geometry. (c) Piggyback thrust sequence. (d) Overthrust step sequence. (e) Duplex-imbricate thrust slices contained between a floor thrust and a roof thrust. (f) Pop-up structure formed by back thrusting. (g) Triangle zone formed by back thrusting. (h) Anastomosing jointing. (i) Conjugate faulting. (j) Ramp thrusting with distorted bedding in the footwall ramp (Modified from Park, 1983 and Pedersen, 1993).

Hill-hole pairs

A hill-hole pair is 'a discrete hill of ice-thrust material, often slightly crumpled, situated a short distance downglacier from a depression of similar size and shape' (Bluemle and Clayton, 1984; figs 11.53, 11.60). Ice-thrust hills may be found at distances of up to 5 km from source depressions, but source depressions can become infilled with younger sediment, and are not always visible as surface features. Conversely, glacitectonic depressions are sometimes found without associated hills (Ruszczynska-Szenajch, 1978), perhaps because positive relief features have been destroyed or heavily modified by subglacial

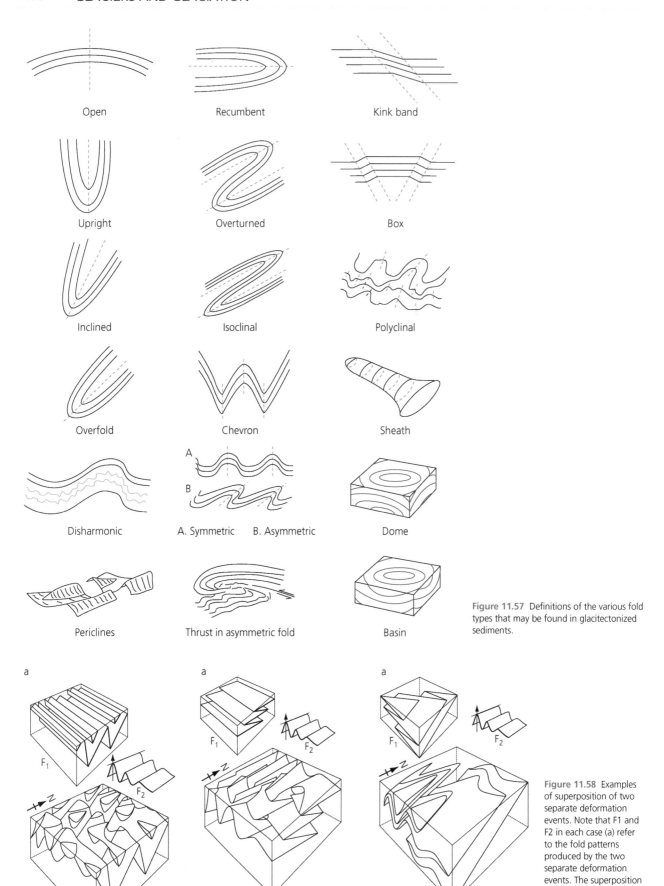

Figure 11.57 Definitions of the various fold types that may be found in glacitectonized sediments.

Figure 11.58 Examples of superposition of two separate deformation events. Note that F1 and F2 in each case (a) refer to the fold patterns produced by the two separate deformation events. The superposition produces the complex patterns in each (b) (Redrawn from Price and Cosgrove, 1990).

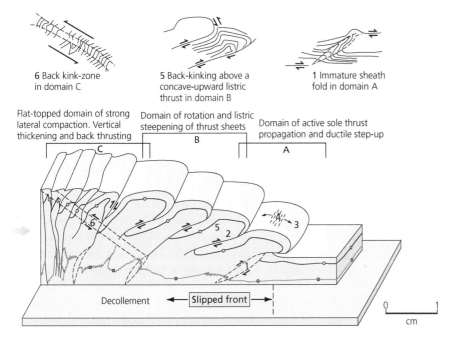

6 Back kink-zone
in domain C

5 Back-kinking above a
concave-upward listric
thrust in domain B

1 Immature sheath
fold in domain A

Flat-topped domain of strong
lateral compaction. Vertical
thickening and back thrusting

Domain of rotation and listric
steepening of thrust sheets

Domain of active sole thrust
propagation and ductile step-up

C

B

A

Decollement   Slipped front

cm

Figure 11.59 An example of piggyback-style thrusting, showing deformation of stratified sand after it was subjected to 40 per cent shortening in a squeeze box. (1) Initiation of thrust. (2) Thrust fault. (3) Slump zone. (4) Extension fractures. (5) Back kink fold. (6) Backthrust zone (Mulugeta and Koyi, 1987, reproduced with permission of the Geological Society of America).

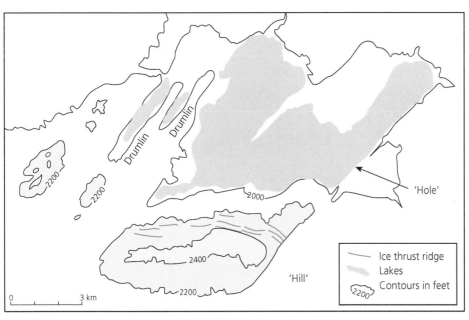

Drumlin

Drumlin

2200

2200

2000

'Hole'

2400

'Hill'

2200

— Ice thrust ridge
▒ Lakes
⟨2200⟩ Contours in feet

0   3 km

Figure 11.60 Simplified topographic map of the Wolf Lake hill-hole pair and associated drumlins, Alberta, Canada, with elevations in feet (After Aber et al., 1989).

erosion. The typical morphological features of the hill in a hill-hole pair comprise:

(1) an arcuate or crescentic planform, which is concave upglacier;

(2) a surface traversed by a series of transverse, sub-parallel ridges and depressions;

(3) an asymmetric cross profile, with the highest point and steeper slopes on the convex or downglacier side (Aber et al., 1989; Evans and Wilson, 2006).

The hole or topographic depression is approximately the same shape and area as the hill and is located on the concave or upglacier side. Parallel-sided margins of holes (fig. 11.60) are interpreted as strike-slip tear faults, marking the lateral boundaries of the thrust mass.

Herschel Island in Mackenzie Bay, Canada, is one of the largest hill-hole pairs so far recognized on the Earth's surface, with an area of >100 km^2 (Rampton, 1982), and is composed of preglacial Pleistocene sediments thrust up from a now submerged source depression. Herschel Island is arcuate in shape, with an upglacier concave face and an asymmetrical profile. Trellised stream patterns on the island record postglacial fluvial erosion along fracture zones (Aber et al., 1989). Sea cliffs reveal internal structures such as low-angle thrust faults, synclines and anticlines, tilted beds which occasionally approach sub-vertical angles, overturned folds, repeated strata, slickensides and segregated ice lenses, which originally formed horizontally but now parallel the deformed enclosing strata. The enclosed ice lenses indicate that the

Herschel Island sediments were permafrozen before being glacitectonized.

Composite ridges (thrust-block moraines)

The most common and distinctive type of glacitectonic landforms are *composite ridges*. These forms have been termed *push moraines* by some researchers (e.g. van der Wateren, 1995; Etzelmüller et al., 1996; Boulton et al., 1999; Bennett, 2001), but we reserve this term for smaller

moraines formed by bulldozing, or push-from-behind, rather than large-scale glacitectonic processes, which excavate and elevate proglacial materials (see section 11.3.3). Composite ridges are composed of multiple nappes or imbricated slices of up-thrust and contorted bedrock and/or unconsolidated sediments, which are commonly inter-layered with and overlain by glacigenic and glacifluvial sediment (figs 11.53, 11.61; Aber et al., 1989; van der Wateren, 1995, 2003). They have been subdivided by Aber et al. (1989) into large (>100 m of relief) and small (<100 m of relief) composite ridges. This size differentiation is essentially arbitrary, and the processes involved in the production of both large and small composite ridges are the same. However, Aber et al. (1989) suggested that large ridges usually contain a considerable volume of pre-Quaternary bedrock, whereas small ridges are composed mostly of unconsolidated Quaternary strata. The term *thrust-block moraine* was introduced by Evans and England (1991) to describe composite ridges comprising proglacially thrust masses of permafrozen blocks of Quaternary sediments at the margins of Arctic subpolar glaciers. Composite ridges are associated with

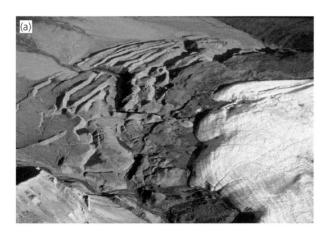

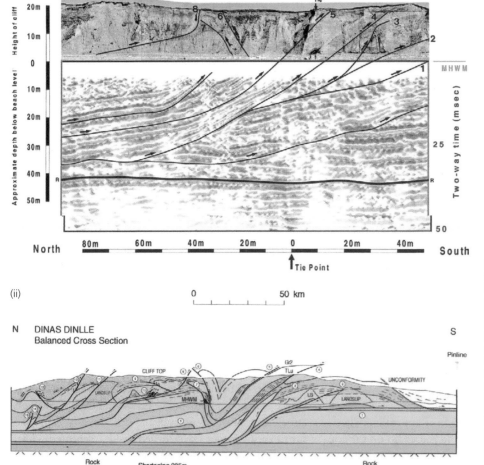

Figure 11.61 (a) Active proglacial thrusting of proglacial outwash in front of an Axel Heiberg Island piedmont lobe, Arctic Canada. (J. England) (b) Proglacially thrust blocks of stratified sediments at Dinas Dinlle, North Wales, with seismic reflection data below mean high water mark for the structures in the central part of a proglacial thrust moraine (i) and balanced cross section of the whole landform (ii), illustrating the bed parallel shearing in the sediment stack. The sediments are: lower till (blue), upper till (green), gravel 1 (orange), gravel 2 (brown), sand (yellow) and debris flow deposits (light blue) (Harris et al., 1997).

proglacial or sub-marginal glacitectonics and they mark the positions of glacier stillstands or readvances.

In planform, composite ridges comprise arcuate suites of sub-parallel ridges and intervening depressions, arranged in a concave upglacier pattern and conforming to the general shape of the glacier margin that produced them (figs 11.53, 11.61a). Individual ridge crests correspond to the crests of internal folds, the upturned ends or edges of thrust blocks/sheets or resistant upturned beds. The depth of disturbance associated with composite ridges is generally a few tens of metres, and may be up to 200 m for the very largest features (Aber et al., 1989; van der Wateren, 2003). The internal structures of large composite ridges are very similar to the products of thin-skinned tectonics in thrust and folded mountain belts like the Canadian Rockies, even though composite ridges are at least one order of magnitude smaller (Aber et al., 1989). The material in composite ridges commonly forms imbricately stacked sheets or nappes, with individual thrust slices showing varying degrees of internal deformation. Probably the most spectacular exposures through a large composite ridge are located at Møns Klint in Denmark, where several ice advances during the last glaciation have piled up numerous chalk scales with intervening Quaternary sediment, to produce a composite ridge with >150 m of structural relief. The structures within the cliffs at Møns Klint include imbricately thrust anticlines and individually folded and stacked chalk floes, which can be traced as a series of ridges on the rugged highland of Høje Møn. Composite ridges are increasingly being identified in seismic data (e.g. Huuse and Lykke-Andersen, 2000a; Andriashek and Atkinson, 2007), making it possible to map glacitectonic disturbance where surface topography is obscured or difficult to access. Additionally, seismic data can facilitate the complete assessment of structures beneath the exposed surfaces of composite ridges (Harris et al., 1997; Williams et al., 2001), an excellent example being that of Dinas Dinlle in North Wales (fig. 11.61b). In such cases, not only can we identify décollement surfaces and entire thrust and fold structures, but we can also 'restore' sections and quantify the cumulative strain (fig. 11.55).

Very extensive composite ridges occur in the area of the Missouri Coteau, a hilly upland belt stretching from North Dakota, USA, to southern Alberta and Saskatchewan, Canada. The boundary of the Missouri Coteau in Canada is demarcated in places by a prominent north-east-facing escarpment, marking the boundary between the Saskatchewan Plain and the Alberta Plain, which is underlain by more resistant rocks. During the overall retreat of the Laurentide Ice Sheet, readvances by ice lobes penetrated embayments in the escarpment, leading to the folding and thrusting of bedrock scales in an imbricate pattern to produce large composite ridges such as the Dirt Hills and Cactus Hills (fig. 11.62). Other large composite ridges and hill-hole pairs developed in the soft bedrock of western Canada are associated with the former margins of surge lobes in the receding Laurentide Ice Sheet (Evans et al., 2008). Other examples of large composite ridges include the Cromer Ridge in Norfolk, England (Hart, 1990), the Weichselian Main Stationary Line in Denmark (Pedersen et al., 1988) and the Dammer Berge in north Germany (van der Wateren, 2003).

There are numerous examples of small composite ridges, both in front of modern glacier snouts and in the ancient landform record. Spectacular composite ridges composed of outwash and glacimarine sediments form continuous moraines around the snouts of many Spitsbergen glaciers (fig. 11.63; van der Meer, 2004). In Spitsbergen and Iceland, composite ridges appear to be exclusively associated with surging glaciers, suggesting that conditions during surges are particularly conducive to proglacial tectonics (Sharp, 1985b; Croot, 1988a; Evans and Rea, 1999, 2003; Bennett et al., 2004; Evans et al., 2007b). The most important factors underlying this association are (1) rapid advance of the snout; (2) extreme compressive deformation at the margin; and (3) elevated pore-water pressures associated with the release of meltwater. The role of pore-water pressures is highlighted by the general absence of composite ridges where proglacial areas slope steeply away from the glacier, and water is able to drain away freely (Croot, 1988a). The internal

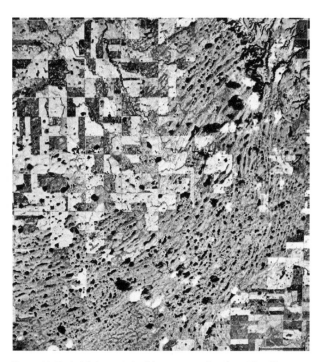

Figure 11.62 Aerial photograph of the eastern and northern Dirt Hills, Saskatchewan, Canada, showing the arcuate sub-parallel ridges and depressions arranged in a concave upglacier pattern. The scale bar at bottom right represents 2 km (Reproduced with permission of the National Air Photograph Library, Ottawa).

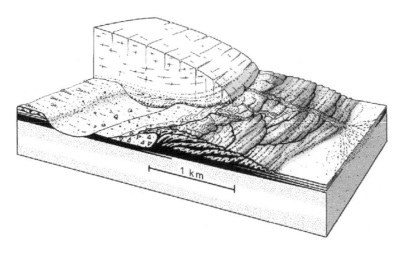

Figure 11.63 Diagrammatic representation of the composite ridge developed in front of Holmstrombreen, Svalbard, during a surge. The vertical scale is exaggerated, but shows a series of nappes developed in proglacial materials (van der Wateren, 1995; Boulton et al., 1999).

structures of recent composite ridges formed by Icelandic surging glaciers have been studied by Croot (1988b) and Benediktsson et al. (2009) (fig. 11.55).

Thrust-block moraines produced by proglacial thrusting of blocks of frozen outwash and raised marine sediments have been documented in the Canadian Arctic (e.g. Evans, 1989a; Evans and England, 1991; fig. 11.61a). In most examples, the bedding within individual blocks and block surfaces dips gently towards the glacier, suggesting that the moraines are composed of either imbricately stacked scales or partially rotated deep-seated blocks. In some cases, however, bedding may dip away from the glacier snout, suggesting deep-seated wedging of the advancing snout beneath proglacial sediments (Evans and England, 1991). The Canadian Arctic thrust-block moraines occur in areas where permafrost is often in excess of 700 m thick, but they are located in valley bottoms either below the marine limit or below former proglacial lake shorelines relating to the last glaciation. This suggests that permafrost may have been partially degraded below deep-water bodies prior to the Holocene advances responsible for thrusting. Therefore, block failure could have been initiated along an unfrozen *talik* between newly aggraded postglacial surface permafrost and the degraded top of the full glacial permafrost. Etzelmüller et al. (1996) have noted that thrust-block moraines in Svalbard, which is within the region of continuous permafrost, also occur only below the marine limit. They argued that the presence of salt plays an important role in maintaining liquid water in permafrozen glacimarine sediments, weakening them sufficiently to allow large-scale thrusting. Thrust-block moraines occur at the margins of cold-based glaciers in the Dry Valleys area, Antarctica, where they tend to be composed of lacustrine sediments. Fitzsimons (1996a) has argued that thrusting takes place when cold glaciers override weak, unfrozen marine or lake-floor sediments. Unfrozen areas occur below either saline lakes or wet-based lakes which have a permanent ice cover but do not undergo full-depth freezing. As the glacier enters the lake,

sediment blocks become frozen onto the base of the glacier, but can be thrust forward because their deeper layers remain unfrozen. In summary, therefore, the formation of thrust-block moraines in permafrost regions is thought to occur only within areas where unfrozen or only partially frozen sediments occur (Etzelmüller and Hagen, 2005).

Ancient examples of small composite ridges include the Brandon Hills in Manitoba, Canada (Aber et al., 1989), the Utrecht Ridge and associated hills and basins in the central Netherlands (van der Wateren, 1985), and Dinas Dinlle in North Wales (Harris et al., 1997). All are composed predominantly of glacifluvial sands and gravels. Based on structural evidence from the Utrecht Ridge, van der Wateren (1985) estimated that the imbricately stacked and gently folded thrust blocks have been pushed up at least 100 m from the basal décollement. The thrust blocks strike parallel to the ridge crest and dip at 35–40° towards the former ice margin. The composite ridges of the central Netherlands have probably been excavated from the glacial basins that are now buried beneath the post-Saalian sediment cover.

The structural architecture of composite ridge development has been demonstrated by van der Wateren (1995) and Boulton et al. (1999), using the surge moraine at the snout of Holmstrombreen, Svalbard (fig. 11.63) and the Dammer Berge moraine in Germany. The style of glacitectonic deformation varies with distance from the former glacier snout (fig 11.64). *Style A* represents the undeformed foreland, composed of outwash relating to the glacial advance. *Styles B, C* and *D* represent varying degrees of proglacial glacitectonic deformation, and *Style E* records subglacial deformation. Style E is further subdivided into compressive, or constructional, deformation (Ec), and extensional, excavational deformation (Ee). Constructional deformation is recorded by accretionary deformation tills and glacitectonites, and excavational deformation by areas of overdeepening, erosion and substrate streamlining. Oscillations of the ice margin will produce sediment successions in which different structural styles are superimposed. A single advance of the ice margin

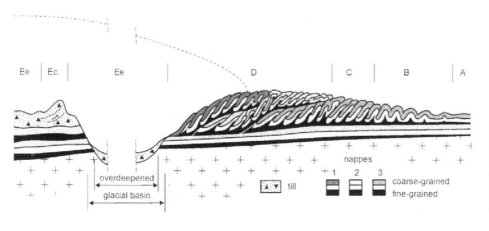

Figure 11.64 Glacitectonic structural styles A–E: A = undeformed foreland; B = Jura-style proglacial folding; C = low-angle thrust structures; D = nappe zone; E = subglacial zone, comprising compressional (Ec) and erosional (Ee) subzones (van der Wateren, 1995).

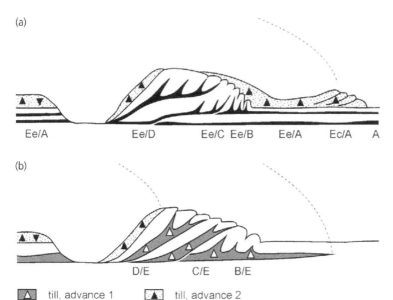

Figure 11.65 Overprinting of structural styles. (a) Advance sequence, where style E overprints styles A–D and the triangle pattern depicts till relating to the ice advance. (b) Readvance sequence, where styles B–D overprint style E and the black triangle pattern depicts till relating to older advance (van der Wateren, 1995).

results in the overprinting of styles A to D by style E, whereas ice-margin retreat then readvance will result in the overprinting of style E by styles B, C and D (fig. 11.65; see also section 12.2.6 on kineto- and tectono-stratigraphy).

Cupola hills

Cupola hill is the English translation of the Danish term *kuppelbakke*, which describes irregular hills with the general characteristics of glacitectonic landforms, but lacking hill-hole pair relationship and/or the transverse ridge morphology of composite ridges (Bluemle and Clayton, 1984). The characteristics of cupola hills are:

(1) a dome-like morphology varying from near circular to oval shapes, with lengths of 1–15 km and heights of 20–100 m;
(2) internal composition of detached and deformed floes of Quaternary sediments, older strata or bedrock;
(3) a carapace of till which truncates the underlying structures (figs 11.53, 11.66).

These characteristics indicate that cupola hills are subglacially overridden hill-hole pairs or composite ridges,

and represent early stages in the development of drumlins (section 11.2.4). In cases where the subglacial modification has been slight, transverse ridge morphology may be partially preserved and the cupola form will tend to lie transverse to former ice flow (cf. Pedersen et al., 1988; Rattas and Kalm, 2001). With increasing subglacial modification, ridge forms tend to be replaced by more smoothed, elongate forms.

Examples of cupola hills composed entirely of Quaternary sediments include the island of Ven in Sweden and Ristinge Klint, Denmark (Aber et al., 1989). The island of Ven is composed of gentle synclines disturbed by numerous thrust faults and overturned folds. The glacitectonic disturbance responsible for these structures has produced two major transverse ridges, the surface of which is muted by a cover of discordant till. In contrast, the topography of Ristinge Klint is subdued and drumlinized similar to cupola hills on the nearby island of Aero. Cliff sections reveal that the area of Ristinge is underlain by more than 30 imbricately stacked scales of Quaternary sediments, each scale being up to 20 m thick and containing the same stratigraphic

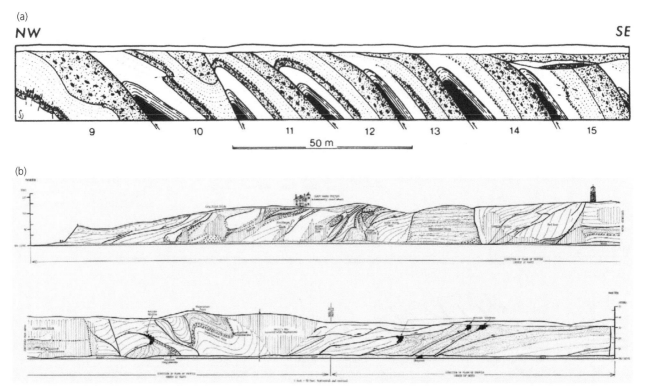

Figure 11.66 (a) The internal structures of a cupola hill, Ristinge Klint, Denmark, showing stacked scales of Quaternary strata (9–15) each separated by a thrust fault and associated drag folds. (Sjorring et al., 1982, reproduced with permission of the Danish Geological Survey) (b) Glacitectonic structures at Gay Head Cliff, Martha's Vineyard, as it appeared in 1959, extracted from the original USGS Open File report. The large-scale folds and faults are developed in Cretaceous, Tertiary and Quaternary deposits (Kaye, 1980, reproduced with permission of the USGS).

sequence of Saalian, Eemian and Weichselian strata (fig. 11.66a). The internal stratification of each scale is little disturbed, except for prominent drag folds associated with the thrust faults between scales. Thrusting of the scales is thought to have occurred at the ice margin during glacier advance and deposition of the upper, discordant tills; surface streamlining took place during subsequent ice overriding.

Excellent examples of cupola hills occur in the late Wisconsinan end moraine systems that make up the offshore islands of New England, USA. The islands consist of glacitectonically disturbed floes or thrust scales of Cretaceous, Tertiary and Quaternary strata (Oldale and O'Hara, 1984). Sections at Gay Head Cliff, Martha's Vineyard, expose a series of imbricated scales, each 20–30 m thick and dipping upglacier, interrupted by overturned folds, low-angle thrust faults and occasional down-dropped blocks or grabens (fig. 11.66b). In places, the strata have been uplifted by over 160 m. The imbricate structures are truncated by a thin and discordant till cover which was deposited when the Martha's Vineyard composite ridge was partially overrun by glacier ice. This overriding produced the cupola hill of Gay Head, which lies transverse to the former ice flow direction and possesses gently sloping sides and numerous irregular surface knobs and depressions.

Cupola hills may appear aligned in arcuate or transverse belts, representing the positions of former large composite ridges that have been modified by subglacial moulding. Numerous examples of this occur in Alberta, Canada, where the advancing Laurentide Ice Sheet margin underwent compression due to flow against local areas of higher topography. This resulted in the dislocation of the soft Jurassic, Cretaceous and Tertiary bedrock and the construction of various types of glacitectonic landform that were then overridden and subglacially moulded (Aber et al., 1989; Tsui et al., 1989; Evans, 2000b; Evans et al., 2008). Short periods of overriding and/or thin glacier ice coverage tends to result in better preservation of original glacitectonic forms, as illustrated by Evans and Twigg (2002) on the foreland of Breiðamerkurjökull, Iceland, where superficially fluted hill-hole pairs composed of fluvial, estuarine and organic materials are clearly visible just inside the Little Ice Age limit. At a larger scale, strongly fluted composite ridges occur at the southern margin of the Laurentide Ice Sheet in Alberta and can be compared to more strongly subglacially streamlined cupola hills at a location situated 250 km to the north on the bed of the same palaeo-ice stream (fig. 11.67).

Mega-blocks and rafts

Mega-blocks and rafts (fig. 11.53) are dislocated slabs of rock and unconsolidated strata which have been transported from their original position by glacier action (Ruszczynska-Szenajch, 1987; Bennett et al., 1999b). Traditionally, such rafts were assumed to have been

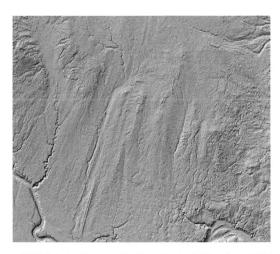

Figure 11.67 Subglacially streamlined thrust bedrock ridges (cupola hills) formed beneath a palaeo-ice stream in the south-west sector of the Laurentide Ice Sheet in Alberta, Canada. Sampled from SRTM imagery. Width of image is approximately 80 km.

Figure 11.68 Cretaceous bedrock rafts in till at Lethbridge, Alberta, Canada. Rod Smith is standing on the light-coloured shale raft, which has been incorporated in the upper layers of a grey-brown till. The section is capped by light brown glacilacustrine rhythmites (D.J.A. Evans).

frozen on to the base of cold-based ice sheets, although it is now recognized that failure may occur along a basal décollement in a subglacial deforming layer. Mega-blocks may originate from uprooted cupola hills, although it is often difficult, if not impossible, to demonstrate detachment from the parent substrate unless there is exceptionally good exposure. Where they are sub-horizontal, mega-blocks and rafts may form flat-topped buttes or plateaux and can be mistaken for bedrock outliers. Similarly, buried mega-blocks can be mistaken for bedrock if borehole logs or exposures are not available.

Although mega-blocks and rafts contain coherent masses of bedrock or sediment, they are also commonly traversed by shear zones, faults and brecciated zones, and may be folded (Aber et al., 1989). A limestone mega-block near Topeka, Kansas, is quite heavily fractured and contains some rotated blocks. The block measures 50 m × 150 m and is only 1–2 m thick. It is separated from underlying striated limestone by a 30 cm zone of brecciated shale (the shale is normally 12 m thick) and glacial sediment. This particular block has only been transported by horizontal sliding by less than 1 km, but some far-travelled mega-blocks have been documented. Rafts of Jurassic clay >20 m thick at Lukow, Poland, are derived from Lithuania, more than 300 km to the north-east, and are thought to have been transported in a frozen state (Ruszczynska-Szenajch, 1976). A similar transport distance has been calculated for the Cooking Lake mega-block near Edmonton, Alberta, Canada, which is 10 m thick and covers an area of approximately 10 km². A large number of smaller mega-blocks, composed of Cretaceous strata and sandwiched between Quaternary sediments, have been described from southern Alberta, Canada, by Stalker (1976; fig. 11.68).

Steeply dipping rafts are more likely to have some surface expression, although they may be buried and difficult to identify without extensive borehole data. A series of

detached chalk rafts occur in till overlying limestone in quarry exposures at Kvarnby in Sweden, suggesting that they were removed from the Baltic Sea bed and transported more than 25 km before being imbricately stacked at angles of up to 80° (Ringberg et al., 1984). These sub-surface rafts lie beneath a structurally controlled, low-amplitude hummocky topography, which would be difficult to differentiate from a cupola hill without exposures. Similarly, chalk rafts in the Anglian till of Hertfordshire, England, have been identified in borehole records, and as pale-toned patches in darker, freshly seeded fields on aerial photographs (Hopson, 1995). Unlike the Swedish examples, the Hertfordshire chalk rafts have been displaced only a short distance (mostly <1 km) from their origin by the southerly flowing Anglian Ice Sheet.

11.3.3 PUSH AND SQUEEZE MORAINES

A range of moraine-building processes are known to be operative in marginal to sub-marginal positions at glacier snouts, including bulldozing/pushing, squeezing, freeze-on, melt-out and mass flowage of sediments, the dominance of particular processes changing spatially and temporally, often on a seasonal basis.

Ice pushing is simply the bulldozing of ice-marginal sediment by an advancing glacier snout. During major glacier advances, sediment is usually overridden and becomes part of the subglacial environment, so that ice-marginal pushing is generally most important during winter readvances by glaciers undergoing overall retreat. In such situations, a series of annual *push moraines* is produced. Hewitt (1967) described the situation in the Karakoram, where summer/ablation season dumping of debris from supraglacial positions into aprons and fans along the margin is followed by winter/accumulation season glacier readvances, resulting in pushing and partial overriding of aprons. Similar demonstrations of the annual nature of push moraine formation have been reported by Evans and Twigg (2002) from the foreland of Breiðamerkurjökull, Iceland, where historical documentation and aerial photographs clearly identify the construction of one moraine every year since 1965. In some cases, bulldozing of proglacial sediments occurs in association with snowdrifts built up in the concavity in front of the ice margin (Birnie, 1977). The stresses imposed by the motion of the glacier can be transmitted through the snowbank, affecting proglacial sediments up to several metres in front of the glacier. Any moraines created by this process will not mark the position of the ice margin proper, but the former limit of the snowbank. Alternatively, the junction between the ice edge and the snowbank may buckle upwards, causing sediment to be pushed up between the two.

Ice-marginal squeezing takes place during the ablation season, when ice-marginal environments are often very wet, and meltwater streams and standing water are well developed (fig. 11.69a). As a result, ice-marginal sediments commonly have very high water contents, and in many cases are easily liquefied and displaced by the processes of squeezing and pushing. Pushing, as described above, involves a forward movement by the glacier shoving the material from behind, whereas squeezing only requires static loading of water-saturated material by the ice mass. Squeezing occurs in response to the pressure gradient that exists between sediment overlain by glacier ice and unconfined sediment on the glacier foreland or below low-pressure cavities under the glacier. Sediment is therefore squeezed out from beneath the ice margin or infills cavities and basal crevasses (fig. 11.69b). Extrusion of sediment from below glacier margins creates moraine ridges which mimic all the indentations of the ice margin and has been observed in front of several glaciers (e.g. Price, 1970, 1973). They are particularly common along the margins of the large southern outlet glaciers of Vatnajökull, Iceland, such as Fjallsjökull, Flåajökull, Skalafellsjökull, Breiðamerkurjökull and Heinabergsjökull, many parts of which are underlain by poorly drained, fine-grained till. Newly formed squeeze moraines have steep or vertical sides and are rarely more than 1 m high. Squeeze moraines tend to degrade rapidly, and are commonly subject to reworking by ice push, so that unmodified squeeze moraines are probably rare in the geomorphological record.

Squeezing often does not occur in isolation, and debris is commonly subject to relocation and reworking by several processes. A combination of squeezing and pushing has often been observed, because retreating glaciers will maintain some forward motion during the early part of the ablation period when subglacial sediments are becoming saturated by meltwater (Evans and Hiemstra, 2005).

Push moraines are broadly arcuate in planform, but in detail are often irregular and winding, reflecting the

Figure 11.69 (a) Standing water among freshly deposited sawtooth push moraines and discontinuous crevasse-squeeze ridges on the foreland of Svinafellsjökull, Iceland, in 2007. (b) Crevasse-squeeze ridge emerging on the surface of Flåajökull, Iceland (D.J.A. Evans).

morphology of the glacier snout. Where the glacier snout is indented by radial crevasses, push moraines have striking *saw-tooth* planforms, with downvalley-pointing teeth and upvalley-pointing notches (fig. 11.70; Price, 1970; Matthews et al., 1979; Sharp, 1984; Evans and Twigg, 2002). At Bødalsbreen, Norway, Little Ice Age sawtooth moraines have undulating crest lines, with high points at the notches and low points at the teeth. The teeth have asymmetric cross profiles with steeper distal slopes, but the notches tend to be more symmetrical. The greater heights of the notches was explained by Matthews et al. (1979) as the product of accumulation of bulldozed debris in the recesses formed by radial crevasses in the glacier margin, whereas the lower heights of the teeth reflect the spreading of debris around advancing projections of

Figure 11.70 Examples of push moraines. (a) Aerial view of the historical push moraines of Svinafellsjökull, Iceland, showing the unvegetated readvance moraines of the 1990s. Note the crevassed glacier snout which is responsible for the pecten and thus the sawtooth pattern of moraine ridge development. (A. Gregory) (b) The 'cross-valley' moraines of the Isortoq river, Baffin Island. (J.D. Ives) (c) Composite push moraine, comprising several stacked till layers, being constructed in 1993 at the margin of Flåajökull, Iceland, in response to a prolonged period of positive mass balance. (D.J.A. Evans) (d) Push moraine ridge being formed by a combination of pushing and squeezing at the snout of Fjallsjökull in 1965 (R.J. Price).

ice between crevasses. Micro-scale sawtooth forms also occur on the shallow proximal slopes of some sawtooth moraines, recording minor oscillations of the glacier snout. Due to the debris flow activity that characterizes freshly deposited moraines, such delicate forms do not have a good preservation potential. In New Zealand, sawtooth moraines with notch crests up to 30 m high occur within latero-frontal moraine complexes in the vicinity of Whataroa. It is unlikely, however, that such large features originated solely by pushing. Smaller features of a similar age are well preserved among the push moraines of the southern end of Lake Pukaki, on the eastern flank of the New Zealand Alps.

The planform of push moraines can also be dictated by glacier activity. It is common to find younger moraines overlying or completely consuming parts of older moraines wherever summer retreat is small and/or some winter readvances are more substantial than usual (fig. 11.70; Sharp, 1984; Evans and Twigg, 2002). Where glacier snouts are quasi-stable or have reached a position where they are in equilibrium with their environment, large terminal moraines may accumulate by annual accretion (Krüger,

1994; Evans and Hiemstra, 2005). Similarly, the absence of moraines relating to individual years within a field of annual moraines can be explained by unusually large winter advances or longer-term readvances. These can result in ice advance that exceeds several years of annual retreat, so winter push moraines can be destroyed or incorporated into a larger moraine ridge.

Push moraines vary widely in composition, and can consist of subglacial till, mass-movement deposits, water-sorted sediments, or large boulders, depending on the nature of the sediment on the glacier foreland. Where glacier margins are debris-covered, push moraines can consist mainly of supraglacial material dumped onto the forefield during the summer and pushed up during winter (Hewitt, 1967; Boulton and Eyles, 1979). Push moraines commonly have asymmetric cross profiles, with gentle proximal and steep distal slopes. The ice-proximal slopes of fresh push moraines are often covered with small flutings, recording glacier overriding and deformation (Evans and Twigg, 2002; Evans and Hiemstra, 2005). As the glacier retreats during the summer, sediment melting out from the ice base and material at the crests of the

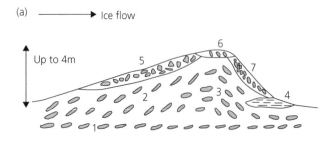

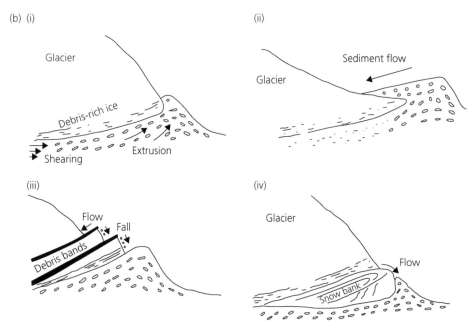

Figure 11.71 Conceptual models for annual push moraine formation based on examples at Skalafellsjökull, Iceland. (Sharp, 1984, reproduced with permission of the International Glaciological Society). (a) Idealized reconstruction of the internal clast fabrics of an annual push moraine ridge. (b) Process-form models for four types of push moraine, including (i) simple ridges comprising deformed subglacial till; (ii) ridges with an ice core incorporated by the flow of ridge top sediments back over the glacier margin; (iii) ridge with an ice core isolated beneath thick englacial debris bands which are exposed by backwasting of overlying ice; (iv) ridge formed at the distal end of a marginal snowbank which has been pushed forward by the glacier. The flow of debris from the glacier has incorporated the snowbank into the ridge.

moraines tend to feed subaerial debris flows, so delicate surface forms like the flutings get buried or destroyed.

Patterns of clast a-axis fabrics in push moraines at Skalafellsjökull, Iceland, have been used by Sharp (1984) to propose a general model of moraine construction (fig. 11.71a). Beneath the ridge, undisturbed lodgement till (or deformation till) has strong preferred clast orientations and low upglacier dips (1), contrasting with weaker fabrics and steeper upglacier dips of the proximal disturbed till (2) and the downglacier dips of the distal disturbed till (3). Surface sediments are characterized by weak fabrics of subaerial sediment gravity flows (5), weak ridge-parallel orientations of ice-slope colluvium dropped onto the ridge crest from the glacier (6) and size-sorting of distal face avalanche debris (7). These fabric patterns help to illustrate the internal structure of push moraines, which consists of an asymmetric fold with an axial plane dipping upglacier. This structure is a reflection of the overall cross-sectional morphology of the moraines. Sharp (1984) further identifies four types of push moraine ridge (fig. 11.71b). *Type A moraines* are the most common, and possess fluted proximal slopes of deformation till recording glacier overriding. *Type B moraines* contain a Type A moraine core, but also have an ice core and a surface

veneer of re-sedimented debris on the proximal slope, due to the burying of thin ice margins by sediment gravity flows. *Type C moraines* are also superimposed on Type A cores and are formed wherever debris bands crop out on the ice surface. As clean ice melts back, the debris bands insulate underlying ice to produce an ice core on the proximal slope of the moraine. Debris flows are then common as the ice slowly melts out. *Type D moraines* are produced wherever marginal snowbanks are pushed forward and overridden by the glacier (cf. Birnie, 1977).

Based on observations at the margin of Sléttjökull, Iceland, Krüger (1994) proposed that the penetration of a winter 'cold wave' from the glacier surface causes a slab of basal till to be frozen onto the glacier sole. This slab is carried forward by the winter readvance of the glacier, whereupon it shears over the proglacial sediments, forming a small ridge. The glacier melts back during the summer and the slab melts out. The return of cold winter temperatures initiates another episode of freezing-on and thrusting. If the glacier margin reaches approximately the same position in successive years, a moraine ridge will be constructed, consisting of imbricate slabs of melt-out till inter-bedded with debris-flow deposits. A variation on this process, the *double-layer annual-meltout model*, has

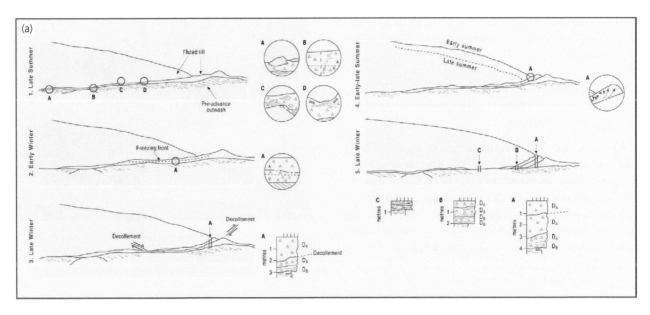

Figure 11.72 Push moraine formation by the incremental stacking of till slabs during an annual cycle of subglacial deformation, freeze-on and melt-out. (a) Conceptual model from Evans and Hiemstra (2005). In (1) late summer, subglacial processes include lodgement and sliding (A), bedrock and sediment plucking (B), subglacial deformation (C), and ice-keel ploughing (D), but note that locations at the bed are schematic and that these processes operate in temporally and spatially evolving mosaic. In (2) early winter, till slab freeze-on occurs in the more porous A horizon (A) which is followed in (B) by the detachment of a subglacial till slab along the decollement plane. In (4) early-late summer, the melt-out of the till slab (A) initiates pore-water migration, water escape, and sediment flow (small arrows) and sediment extrusion due to glacio-static and glacio-dynamic stresses. In (5), after a further seasonal cycle, winter freeze-on and marginal stacking of subglacial till produces vertical sequences of subglacially deformed and partially eroded A and B horizons (D_A and D_B in the vertical profile logs). Note the production of clast pavements in profile C, which occur in association with lodged clasts and flutings on the up-ice or thin end of the sub-marginal till wedges/slabs. (b) Winter freeze-on of sub-marginal sediment at the snout of Skeiðarárjökull, Iceland (R.I. Waller).

been proposed by Matthews et al. (1995) for the margin of a cirque glacier in Norway, whereby two stratigraphic layers are added to embryonic moraines at the end of each seasonal cycle, rather than the single till unit in Krüger's model. One layer results from the melt-out of frozen-on till at the ice base, and the other from the deposition of glacifluvial sands and gravels as a supraglacial unit. More recently, Evans and Hiemstra (2005) have proposed direct linkages between subglacial deformation processes,

ice-marginal squeezing and the freeze-on and melt-out of till slabs in the construction of push moraines at Icelandic glacier snouts. Deposition was inferred to occur in five stages (fig. 11.72):

(1) In late summer, the subglacial environment is characterized by a range of processes, including lodgement and sliding, bedrock and sediment plucking, subglacial deformation, and ice-keel ploughing in a temporally and spatially evolving process mosaic.

(2) During early winter, the thin part of the glacier snout freezes on to a slab of subglacial till.

(3) A later winter readvance initiates failure along a décollement plane, resulting in the transport of a slab of till onto the proximal side of the previous year's push moraine.

(4) During the summer, the till slab melts out and initiates pore-water migration, water escape, and sediment flow and sediment extrusion.

(5) By the following late winter period, winter freeze-on and marginal stacking of subglacial till occurs through the reworking of existing subglacial sediments and fresh materials advected to sub-marginal locations from up-ice. Repeated reworking of the thin end of sub-marginal till wedges produces overprinted strain signatures and clast pavements.

Freeze-on and melt-out release of sub-marginal sediment slabs appears to be a very efficient process in moraine construction in Antarctica, where Fitzsimons (1997b, 2003) and Fitzsimons et al. (2001) have described landforms that are often difficult to differentiate from glacitectonic thrust ridges. Indeed, as the above models have demonstrated, we should not be surprised that ice-marginal moraines lie on a process–form continuum (e.g. Bennett, 2001); variability in the final forms, even across a single glacier foreland, is dictated by the relative importance of a range of glacier sub-marginal processes, seasonal temperature ranges and local drainage conditions.

11.3.4 DUMP MORAINES/ICE-MARGINAL APRONS AND LATERO-FRONTAL MORAINES

Material accumulating at the surface of a glacier through the melt-out of debris-rich folia is ultimately subject to remobilization by mass flowage, fall or fluvial transport. Where such material exists at the glacier margin, its remobilization may result in it being dumped onto the adjacent terrain during ice recession. Depending on the rate of debris accumulation and glacier activity, a variety of sediment–landform associations will be created. For retreating glaciers, dumping of supraglacial material onto the former subglacial surface slowly emerging from beneath the ice produces a thin veneer of coarse-grained diamicton, or perhaps only sporadic collections of boulders (Eyles, 1979; Evans and Twigg, 2002). *Dump moraines* will form where the ice margin remains stationary during debris accumulation, although they will be bulldozed into push moraines if the glacier undergoes a subsequent re-advance, and it is common to see evidence of at least some glacier pushing of dumped material (Boulton and Eyles, 1979; Lukas, 2005). Dump moraine size is related to supraglacial debris volume and the length of the stillstand period. Small dump moraines will form where glaciers remain stationary during the winter but retreat during the summer, each moraine marking one winter's increment of debris accumulation. The largest and most spectacular dump moraines, however, form at the margins of debris-mantled valley glaciers which occupy similar positions for considerable periods.

The dumping of large quantities of debris around glacier margins builds up *latero-frontal dump moraines* (fig. 11.73; Boulton and Eyles, 1979; Small, 1983; Benn and Owen, 2002). Supraglacial debris slides, rolls, flows and falls from the glacier margins and is deposited as ice-contact scree (fig. 11.74; Owen and Derbyshire, 1989; Owen, 1994). In mountain environments, the debris may be predominantly passively transported rockfall material, consisting of coarse, angular clasts with little matrix. However, actively transported debris can also crop out near valley glacier margins due to the elevation of basal debris septa, introducing matrix-rich debris to some dump moraines (section 9.4.2; Small, 1983). On some latero-frontal moraines, there is a tendency for clast roundness to increase down-moraine, towards the glacier snout position (Benn and Ballantyne, 1994; Evans, 1999; Spedding and Evans, 2002; fig. 11.75). This may be due to:

- an increasing proportion of actively transported debris in the moraines closer to the former glacier centreline;

Figure 11.73 Latero-frontal moraines of the Hooker Glacier, New Zealand, showing the burial of older moraine ridges by paraglacial fans fed by avalanches and debris flows (D.J.A. Evans).

- preferential glacial entrainment of pre-existing glacifluvial sediment lying in valley bottoms; and/or
- the provision of meltwater-abraded debris by englacial conduits (Kirkbride and Spedding, 1996; Krüger and Aber, 1999; Spedding, 2000; Spedding and Evans, 2002; Swift et al., 2006).

In contrast, in lateral positions, debris in the moraine is dominated by passively transported rockfall material.

Breach-lobe moraines form where lobes of ice advance through breaches in lateral moraines (Benn et al., 2003; figs 11.49b, 11.50). Deline (1999a, b) described excellent examples of multiple breach-lobe moraines at the margin of the Miage Glacier, Italy, and other examples from debris-mantled glaciers have been reported (e.g. Kirkbride, 2000; Shroder et al., 2000). In such settings, the breach-lobe moraines form inset looped ridges in the gaps cut by glacier ice and/or moraine-dammed outburst floods through major latero-frontal moraines.

The accumulation of dumped material at the margin of a thickening glacier results in a wedge-shaped moraine, with crude internal bedding and clast fabrics dipping away from the glacier at angles between 10° and 40° (fig. 11.74a; Boulton and Eyles, 1979; Small, 1983). Inter-bedded sequences of diamictons and stratified sediments reflect intermittent debris flow and glacifluvial processes, and coarse, bouldery layers within the moraine may be derived from glacially transported rock avalanche material. The deposition of lateral dump moraines against valley sides creates swales between the moraine and the valley wall, previously termed *ablation valleys*, but now more appropriately called *valley-side depressions* (Hewitt, 1993) or *lateral morainic troughs* (Benn et al., 2003). During glacier thinning, dump moraines are abandoned and their inner faces are subject to collapse and reworking (Ballantyne and Benn, 1994b; Ballantyne, 2002a, 2003). A new inset dump moraine may be formed within the older one

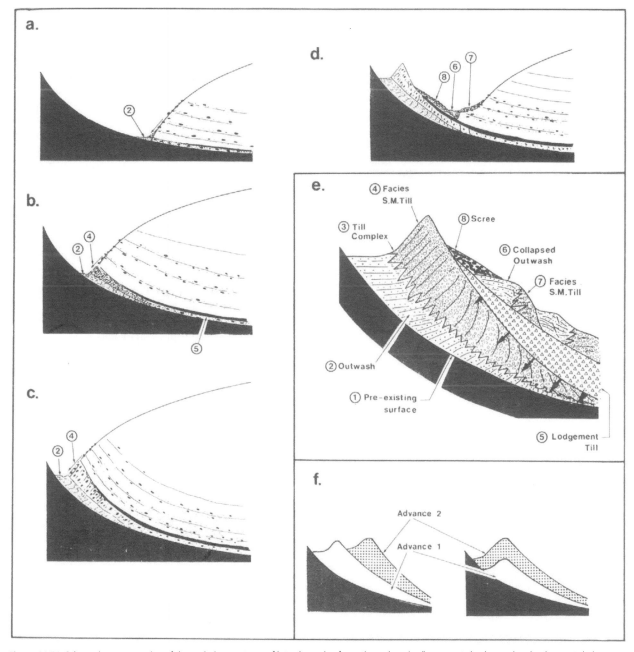

Figure 11.74 Schematic reconstruction of the evolutionary stages of lateral moraine formation, where (a–d) represent simple moraine development during a single glacier advance and retreat cycle. The numbers refer to the features labelled in (e), which depicts the sedimentology of the moraine produced in (d). The arrows in the SM till (supraglacial morainic till) indicate a zone of glacitectonic disturbance produced during ice overriding. Note that lake sediments may also be deposited between the moraine and the ice margin, even though they are not included in this model. In (f), the relationships between two separate landform-sediment suites are shown, where (left) advance 1 is more extensive than advance 2, and (right) advance 1 is less extensive than advance 2 (Boulton and Eyles, 1979, reproduced with permission of Balkema).

if the glacier restabilizes at a new position, but such moraines will be unstable if they are deposited on top of dead ice masses (fig. 11.74a). Alternatively, renewed thickening of the glacier may dump material on top of the old dump moraine, which may become completely buried. In this case, periods of non-deposition may be recorded by erosion surfaces, palaeosols or even buried

trees. Such buried organic material provides a valuable source of palaeoclimatic data in alpine environments (Grove, 1988).

Where dump moraines are deposited on steep valley sides, debris derived directly from the glacier is inter-digitated with sediment from the valley walls (fig. 11.76). For example, lateral moraines recently abandoned by the

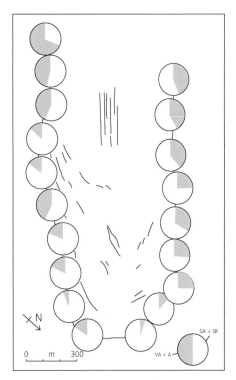

Figure 11.75 Storbreen clast roundness data. VA and A denote very angular and angular (passively transported debris), and SA and SR denote sub-angular and sub-rounded clasts (actively transported debris) (Redrawn from Benn and Ballantyne, 1994).

receding Tasman Glacier, New Zealand, contain inter-digitated beds of:

(1) stratified debris flows and fall-sorted debris avalanche material, with beds dipping away from the glacier margin;

(2) sub-horizontally bedded alluvium deposited as kame terraces along the ice margin;

(3) debris flows and fall-sorted screes from valley-side paraglacial fans. Removal of lateral support by glacier recession results in rotational failure of the inner slopes of the moraines.

In many cases, such failures clearly extend up into the screes which were deposited against the lateral moraines (fig. 11.76b). Where lateral morainic troughs exist, marginal streams act to rework debris delivered from hillslopes on one side and from the distal slopes of the lateral moraine on the other. This results in the inter-fingering of stratified and diamictic sediments. A trough will essentially act as a gutter between lateral moraine and hillslope, acting as a small drainage basin in its own right, collecting and partially to totally reworking the products of mass flows, snow and rock avalanches, nival melt and mountain streams (Hewitt, 1993; Benn et al., 2003). In situations where lateral moraines are later removed by paraglacial reworking, the lateral morainic trough fills may be mistaken for kame terraces (fig. 11.76a).

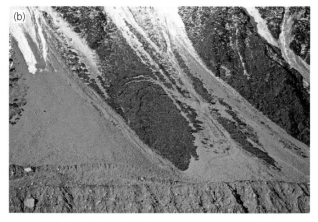

Figure 11.76 Lateral moraines, paraglacial alluvium and slope deposits of glacier margins in a tectonically active mountain environment. (a) Sub-horizontally bedded alluvium exposed by the rotational failure of lateral moraines at the receding margin of the Tasman Glacier, New Zealand. The more recent lateral moraines in the foreground and to the left of this view comprise inter-bedded supraglacial sediments dipping towards the valley side and reworked alluvium dipping towards the glacier. Older lateral moraines can be seen in the middle distance, where they have remained in contact with the alluvium. Because the alluvium was deposited between the lateral moraine and the valley wall, it is strictly a lateral morainic trough fill rather than a kame terrace. (b) The steep ice-contact slope of the left lateral moraine of the Tasman Glacier and rotational failure scars in partially vegetated screes. The rotational failure is a response to the removal of lateral support by the Tasman Glacier during its recent recession. (c) Section through a lateral moraine and debris flow fan at the margin of the Batal Glacier, Lahul Himalaya. Note the stratification and change in bedding angle in the vicinity of the debris flow fan, attesting to the interdigitation of glacial and non-glacial sediments in lateral moraines (D.J.A. Evans).

Figure 11.78 Trimline moraines comprising a veneer of bouldery rubble at the margins of subpolar glaciers, Ellesmere Island. Trimlines are clearly demarcated around the glacier snout in the main view and the inset photograph illustrates the sparsity of morainic debris (D.J.A. Evans).

Figure 11.77 The large latero-frontal moraine complex of Kvíárjökull, Iceland. The moraine exceeds 100 m in height in places and was produced as a result of large englacial and supraglacial debris concentrations within the glacier (See Spedding and Evans, 2002; aerial photograph provided by Landmælingar Islands, 1980).

Lateral dump moraines on steep valley sides have poor preservation potential due to rapid paraglacial reworking of their ice-proximal faces once the support of glacier ice is removed (Ballantyne and Benn, 1994b; Ballantyne, 2002, 2003). On the other hand, dump moraines deposited beyond the confines of cirques or steep valleys are more likely to be preserved because paraglacial activity is generally less significant after glacier recession. Such moraines can act as major barriers to subsequent glacier advances, and may be constructed over several glacial cycles, only being breached locally to produce breach-lobe moraines. For example, the 100 m high lateral moraines of Kvíárjökull, Iceland, are the product of incremental accumulation of Little Ice Age morainic debris over the top of an older, Holocene latero-frontal moraine (fig. 11.77; Gudmundsson, 1997; Spedding and Evans, 2002). The older moraine was responsible for restricting the flow of Kvíárjökull, forcing it to remain as a linear outlet glacier, unlike other outlet glaciers of the Vatnajökull ice cap, which descended onto coastal lowlands as piedmont lobes during the Little Ice Age.

Distinctive dump moraines known as *frontal-* or *ice-marginal aprons* form around the margins of some Arctic and polar glaciers (Evans, 1989b; Fitzsimons, 1997b; Ó Cofaigh et al., 2003b). These moraines form where debris and ice blocks accumulate at the base of steep, terminal ice cliffs by a combination of dry calving and limited melting. When forming, the aprons have steep ice-contact slopes and gentler ice-distal slopes, determined by the properties of the debris. However, they will be subject to considerable modification following deglaciation, due to the ablation of incorporated ice blocks, resulting in a drape or veneer of rubble on deglaciated land surfaces that appears as a faint trimline when viewed from a distance – hence termed a *trimline moraine* by Ó Cofaigh et al. (2003b; fig. 11.78). Care should be taken when differentiating such former marginal aprons from the remnants of englacial debris bands (controlled moraine) that are often concentrated in the marginal ice of subpolar glaciers and which can occur in discrete concentric bands after the melt-out of formerly stable snouts (see section 11.4.2).

Another important characteristic of lateral moraines is their tendency to be asymmetrically developed within individual valleys (fig. 11.79). Four causes of *within-valley asymmetry* have been identified (Matthews and Petch, 1982; Benn, 1989b; Bennett and Boulton, 1993; Evans, 1999):

(1) lateral moraines may be best developed on valley sides where there are extensive rockwalls, which supply larger amounts of debris to that side of the glacier;

(2) lateral moraine volume may reflect differences in the thickness and type of sediment on the foreland;

(3) cross-valley differences in lithology or structure can influence rates of debris supply, either to the surface or bed of the glacier;

(4) asymmetry may occur due to differences in shading or glacier dynamics on either side of a valley. For example, if one side of a glacier maintains a stable position for several seasons, a large moraine can be deposited, but if the other margin is in overall retreat, the glacier may leave a series of smaller moraines. In this case, the amount of debris deposited on each valley side may be the same, but an impression of asymmetry arises because deposition on one side is focused on a smaller area.

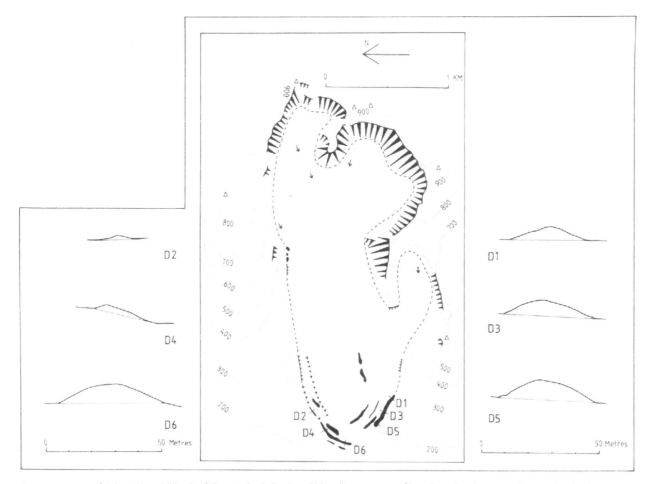

Figure 11.79 Map of Coire a'Ghreadaidh, Isle of Skye, Scotland, showing within-valley asymmetry of lateral moraines (Benn, 1989b, reproduced with permission of Wiley).

11.3.5 LATERO-FRONTAL FANS AND RAMPS

Deposition by debris flows and glacifluvial processes around stationary glacier margins can result in *latero-frontal fans and ramps*, consisting of coalescent debris fans which descend from the glacier snout (fig. 11.80; Owen and Derbyshire, 1989). The outer slopes of such forms have much shallower gradients than latero-frontal dump moraines built up by rockfall and related processes. The inner slopes are steep, consisting of the former ice-contact face, so that the whole landform has a pronounced asymmetric cross profile. In some situations where moraines are absent, fans and ramps provide the only evidence of former ice-marginal positions. Fans deposited predominantly by glacifluvial processes grade distally from *outwash heads*, or ice-contact faces, into *sandar* (Krzyszkowski, 2002; see also section 11.5.1).

Large debris-flow dominated ice-contact fans occur at the margins of many debris-mantled Karakoram valley glaciers (Owen, 1991; Owen and Derbyshire, 1993). Fan morphology develops over a considerable period, sometimes involving several generations of deposition, with intervening periods of abandonment. During fan accumulation, active depo-centres shift position, reflecting

Figure 11.80 Large debris flow-dominated ice-marginal fans ramping upwards to an ice-contact slope, Batal Glacier, Lahul Himalaya (D.J.A. Evans).

changes in the morphology of the ice margin and fan surface. As the fan progrades, surface angles tend to become progressively reduced. Meltwater exiting from englacial positions or draining the glacier surface locally excavates the ice-contact fan/ramp faces to produce inset meltwater fans. The Karakoram ice-contact fans can reach heights of hundreds of metres, attesting to very high rates of debris transfer in this high-altitude, tectonically active region.

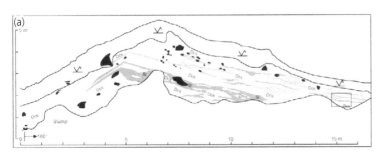

(c)

i Temporarily stationary ice margin

Glacier
Supraglacial debris flows
Stacked debris flows
Glaciofluvial 'wash' horizons

ii Retreating ice margin

Proximal rectilinear slope with material at angle of repose
Distal rectilinear slope (fan surface)

iii Readvancing ice margin

Possible addition of new material to fan
Proglacial deformation of units

iv Retreating ice margin

Proximal rectilinear slope with material at angle of repose
Distal rectilinear slope (fan surface)

v Readvancing ice margin and partial overriding

Subglacial deformation
Proglacial deformation

vi Readvancing ice margin and complete overriding

Subglacial deformation
Smoothing of moraine profile

Figure 11.81 Ice-contact fans in the north-west Highlands of Scotland. (a) Section sketch of an exposure through a non-glacitectonized ice-contact fan. Former ice flow was from left to right against the steep proximal slope. (b) Photograph of inter-bedded stratified diamictons and horizontally bedded sands, located by box in the section sketch. (c) Conceptual model of ice-contact fan/ramp development at actively receding glacier margins during the Scottish Younger Dryas. (i) and (ii) Fan/ramp development during glacier stillstand. (iii) and (iv) Glacitectonic disturbance of fan/ramp where glacier snout readvances. (v) and (vi) Glacitectonic and subglacial deformation in situations where the glacier snout partially and totally overrides the fan/ramp (Lukas, 2005).

Because debris-flow fans and ramps are deposited partly over glacier ice, they may be characterized by pitted or degraded surfaces after ice melt-out. When ice-contact fans are abandoned by glacier retreat, the inner faces are susceptible to collapse and may provide material for lateral moraine development during ice recession. Like lateral dump moraines, large fans and ramps may prove to be insurmountable obstacles for later glacier advances, and can act to constrain or channel the ice.

The sedimentology of the latero-frontal fans and ramps is dominated by massive, often bouldery diamictons, with beds up to tens of metres thick and dipping away from the glacier snout. Good examples of the internal structure of dump moraines have been provided by Johnson and Gillam (1995), who conducted a detailed study of late Pleistocene moraines at Durango, Colorado. Sections in the moraines expose inter-bedded debris-flow diamictons and water-sorted sediments dipping away from the former glacier margin, comprising a series of ice-marginal fans. Limited exposures of melt-out till in low-level, ice-proximal positions show that, in places,

sediment was deposited on top of debris-rich ice. Lukas (2003, 2005) has described Younger Dryas recessional moraines in Scotland, which consist of ice-contact fans that have been subjected to varying degrees of ice push. The sediments comprise matrix to clast-supported, stratified diamictons, tapering out laterally/distally and interfingering with horizontally bedded and laminated sands. The bedding of these sediments is sub-parallel with the ice-distal slopes of the moraine surface slopes, whereas the ice-proximal sides have been partially reworked following ice withdrawal (figs 11.81a, b). Some fans have not experienced any glacitectonic deformation, whereas others have been pervasively sheared and faulted during glacier overriding (fig. 11.81c).

Well-exposed examples of coalescent ice-contact fans up to several hundred metres high occur along the west coast of South Island, New Zealand (fig. 11.82). Coastal sections through the fans expose crude to well-developed inter-beds of clast-supported and matrix-supported diamictons, poorly sorted gravels and boulder- to cobble-sized rubble units, recording deposition from debris flows and other mass movements. Occasional pockets of glaci-lacustrine sediment, which have in some instances been glacitectonized, attest to the occurrence of short-lived proglacial lakes ponded between the ice margin and the

Figure 11.82 Stratified diamictons and crudely bedded gravels with lenses of fine-grained rhythmites deposited by avalanches, debris flows and slopewash in an ice-contact lateral debris fan, Gillespie's Beach, South Island, New Zealand (D.J.A. Evans).

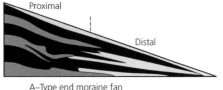

A–Type end moraine fan

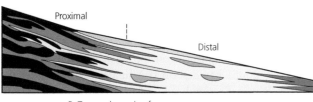

B–Type end moraine fan

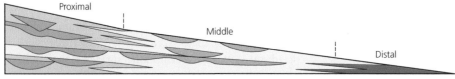

C–Type end moraine fan

■ Fine-grained diamicton
▨ Coarse-grained diamicton
▢ Massive to crudely bedded gravel
▨ Massive to low-angle bedded gravel and boulders
▢ Massive to crudely bedded sand or pebble sand in sheets
▨ Diamictic (pebble silty) sand
▨ Cross-stratified sand
■ Massive silt
▨ Laminated sand and silt

fans during ice retreat phases (section 11.4.4). The dip directions of the beds within the fans vary from one exposure to another, indicating deposition from a series of depo-centres located at the ice margin and partially reworked during glacial advance and retreat cycles. Similarly large ice-contact fans/ramps have been described from the Polish landscape by Zielinski and van Loon (1999a, b, 2000), Krzyszkowski (2002) and Krzyszkowski and Zielinski (2002), who call them 'end moraine fans', and differentiate them from ice-contact glacifluvial fans using the widespread occurrence of diamictons and diamictic gravels (cf. Pisarska-Jamrozy, 2006), fed directly from the supraglacial to the ice-marginal environment, and the asymmetric and steeper slope profiles of fans and ramps. Using modern alluvial fan analogues, Krzyszkowski and Zielinski (2002) proposed a threefold classification that reflects the dominance of mass flowage processes and high-energy sheet flows, with Type A fans containing exclusively diamictic deposits, Type B containing inter-fingered diamictons and stratified outwash, and Type C containing no diamictons but fining distally from massive and crudely bedded gravels to laminated sands and silts (fig. 11.83). The studies of Lukas (2003, 2005) and Krzyszkowski and Zielinski (2002) emphasize the deposition of fans and ramps at receding and/or oscillating glacier snouts, but it is evident that substantial accumulations of supraglacially derived material can be associated with advancing snouts where they have traditionally been called *hochsandur fans* (Krüger, 1997; Kjær et al., 2004); Hochsandur fans typically do not display the proximal-to-distal grain size transitions that characterize other outwash forms, and are deposited between moraine ridges as short, relatively steep conical

Figure 11.83 Classification scheme for ice-contact fans/ramps ('end-moraine fans') based on the predominance of supraglacially fed mass flows (Redrawn from Krzyszkowski and Zielinski, 2002).

fans. Clearly, it is often difficult to differentiate between a Type C fan/ramp and an ice-contact glacifluvial outwash fan (see section 11.5) in an ancient depositional setting.

11.4 SUPRAGLACIAL ASSOCIATIONS

This section describes sediment–landform associations that originate englacially or supraglacially, then evolve as they are progressively lowered onto the substratum by glacier ablation. This section begins with medial moraines, then goes on to discuss supraglacial hummocky moraine, controlled moraine, kame and kettle topography, ice-walled lake plains, kame terraces and pitted sandar. Many of these associations result from topographic inversion and/or glacier karst development on debris-mantled glaciers, and are commonly referred to as *ice-stagnation topography*. However, ice stagnation may occur only during a late stage in the development of such forms, and then only in a relatively narrow marginal zone, so that the presence of widespread supraglacial associations need not imply widespread glacier stagnation. Supraglacial hummocky moraine, kames and ice-walled lake plains are related features, which reflect deposition predominantly by mass movement, glacifluvial and glacilacustrine processes respectively. In supraglacial environments, however, these processes commonly operate in close proximity, and hybrid sediment–landform associations may be frequently observed.

11.4.1 MEDIAL MORAINES

Medial moraines are prominent features on the surfaces of many glaciers (see section 9.4.2), although they do not have good preservation potential, because they contain relatively little debris and are subject to intense reworking during glacier melt. They can sometimes be traced from modern glacier surfaces onto proglacial areas, where they may be recognizable as diffuse spreads of debris with distinctive lithological composition or particle morphologies. For example, the prominent medial moraines on Breiðamerkurjökull, Iceland, can be traced as linear boulder spreads on the proglacial foreland (Evans and Twigg, 2002); and linear debris stripes on the forelands of Svalbard glaciers represent the melted-out medial moraines formed by englacial folding of debris-rich folia (Glasser et al., 1999; Glasser and Hambrey, 2003). However, such moraines commonly have no marked surface expression, and may be very difficult to recognize once the glacier foreland becomes vegetated. Recognition of ancient medial moraines is easier where they consist of large boulders, which may be traced back to their source outcrops (fig. 11.84a). An unusually clear example of an ancient medial moraine emanating from mountainous

(a)

Figure 11.84 Examples of medial moraines. (a) Bouldery medial moraine near Glen Torridon, Scottish Highlands. The moraine is composed entirely of passively transported angular quartzite blocks. (D.I. Benn) (b) The Witch's Slide, a fossil medial moraine on the Isle of Jura, Scotland. (1) Bouldery ridges. (2) Quartzite bedrock escarpments. (3) Talus-covered slopes of Beinn an Oir. (4) Till. (5) Late Glacial marine limit. (6) Raised marine cliffs. (7) Raised marine beach ridges. The land surface slopes gradually from around 400 m in the south-east to sea level in the north-west. (Modified from Dawson, 1979) (c) Google Earth image of large lateral and medial moraines on the Payachatas massif, on the Chile/Bolivia border. The moraines demarcate a series of ice tongues that fringed the northern volcano Pomerape during the Late Pleistocene. The southern volcano Parinacota was created by more recent volcanic activity, and hence has no moraines of that age. (d) Digital elevation model of part of the west coast of South Island, New Zealand, showing the impressive latero-frontal and ice stream interaction medial moraines marked by black lines. Large quantities of supraglacial debris accumulated at the lateral margins of outlet glaciers debouching from the Southern Alps mountain ice field during the last glaciation. (e) Reconstruction of the sequential development of a medial moraine in the centre of the Athabasca Valley near Jasper, Alberta, Canada. (Upper) A well-developed medial moraine forms a substantial debris cover between the Athabasca and Miette Valley glaciers. (Middle) Bedrock benches on the valley sides become free of ice first, leaving a well-preserved fluted topography with little supraglacially derived material. Debris derived from the medial moraine and by melt-out of the medial debris septum accumulates on the ice surface and eventually develops into a linear, ice-cored moraine. Ice-marginal debris flow and fluvial deposits develop into kame terraces along the sides of the decaying, debris-covered ice mass. (Lower) During the final melt-out phase, a chain of kettle lakes forms in the area of the thickest supraglacial debris cover, and large gullies, faults and other collapse structures develop in the kame terrace deposits, owing to ice block decay. Morainal debris is also reworked by braided streams to produce a dissected and subdued series of kettle holes (Levson and Rutter, 1989b, reproduced with permission of *Canadian Journal of Earth Sciences*).

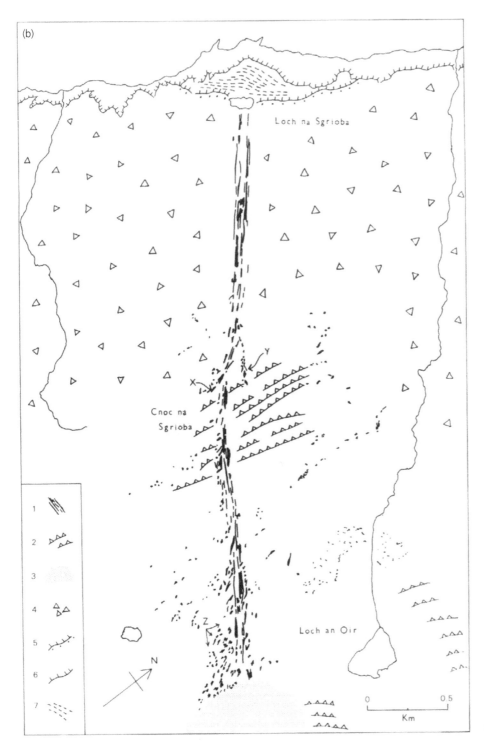

(b)

Loch na Sgrioba

Cnoc na
Sgrioba

Loch an Oir

1

2

3

4

5

6

7

N

0   0.5

Km

Figure 11.84 (Continued)

topography occurs on the Isle of Jura, Scotland (Dawson, 1979). The moraine is known locally as Scriob na Caillich, or the Witch's Slide, and consists of angular, quartzite blocks arranged in a series of parallel ridges which extend 3.5 km north-westward from the slopes of Beinn an Oir (fig. 11.84b). The moraine was produced by the last ice sheet, when Beinn an Oir stood above the ice surface as a nunatak.

In certain situations, specifically where glaciers coalesce after debouching from their individual feeder valleys in mountainous terrain, *ice stream interaction* type medial moraines have a high preservation potential (see also section 9.4.2). Very large ice stream interaction medial moraines, more than 100 m high in some areas, have been produced by the coalescence of ancient piedmont glaciers with well-developed lateral moraines, on the volcanic massifs of South America and on the west coast of New Zealand (figs 11.84c, d). Because these features accumulated essentially as lateral moraines, they contain sediments characteristic of ice-marginal

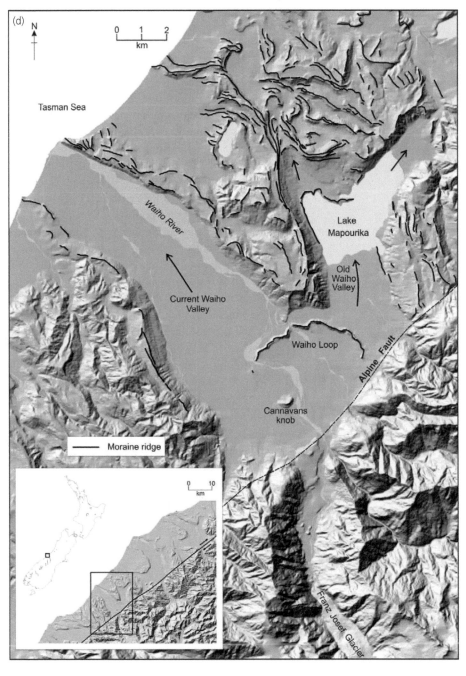

Figure 11.84 (Continued)

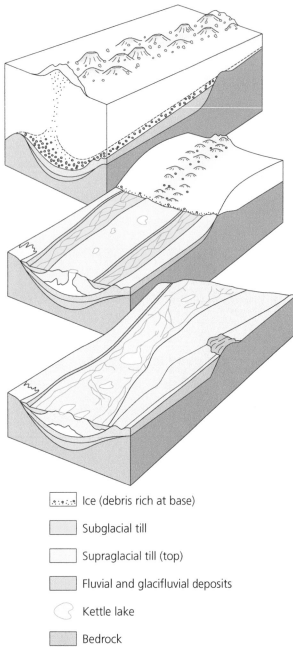

Ice (debris rich at base)

Subglacial till

Supraglacial till (top)

Fluvial and glacifluvial deposits

Kettle lake

Bedrock

Figure 11.84 (Continued)

Figure 11.85 Chaotic hummocky moraine marking the former surge limit of Tungnaárjökull, Iceland (D.J.A. Evans).

11.4.2 SUPRAGLACIAL HUMMOCKY MORAINE AND CONTROLLED MORAINE

The term *hummocky moraine* has been employed in a wide range of senses. As a purely descriptive term, it has been applied to moundy, irregular morainic topography 2–70 m in relief and exhibiting varying degrees of order, ranging from entirely chaotic assemblages of mounds, 15–400 m in diameter, to suites of nested transverse ridges (fig. 11.85; Sissons, 1967; Benn, 1992; Attig and Clayton, 1993; Bennett and Boulton, 1993; Johnson and Clayton, 2003). This broad definition encompasses landforms with a variety of origins, including deposition in association with active and stagnant ice, and glacitectonic deformation. Many authors, however, have employed the term 'hummocky moraine' in a more restrictive sense, to refer to moraines deposited during the melt-out of debris-mantled glaciers (e.g. Sharp, 1985b), and it is in this sense that we use the term. This genetic definition should only be applied to suites of moraines whose origin has been established by detailed mapping and sedimentological analyses, because generalized descriptions of surface morphology alone are insufficient to determine moraine genesis.

Hummocky moraine deposited from debris-mantled ice can appear chaotic or include linear elements, depending on the distribution of debris in the parent glacier and the patterns of debris redistribution and reworking during ice wastage. Occurrences with pronounced transverse linear elements are termed *controlled moraines*, because their morphology was controlled by the former pattern of debris concentrations in the parent ice. Controlled moraines are best developed in recently deglaciated terrains, where significant volumes of the parent ice constitute most of the landform relief. Early usage of the term *controlled moraine*, most influentially by Gravenor and Kupsch (1959), encompassed a range of landforms that we now classify separately and more securely with genetic

deposition (section 11.3.4). However, the medial moraine cores commonly display glacitectonic structures, and, in places, carapaces of subglacial till record glacier coalescence and overriding. The interesting implication is that such overridden inter-lobate features may well lie at the core of some mega-flutings. As noted in section 11.2.10, some medial moraines are associated with glacifluvial sediments. A low-relief kame-and-kettle topography centred on the bottom of the Athabasca Valley near Jasper in Alberta, Canada (fig. 11.84e), was interpreted as a former supraglacial medial moraine by Levson and Rutter (1989b), and a similar origin has been proposed by Carlson et al. (2005) for large elongate assemblages of kame mounds in Wisconsin.

terminology. For example, Gravenor and Kupsch (1959) included *crevasse-squeeze ridges* (section 11.2.9) in their family of controlled moraines.

The morphology and distribution pattern of *controlled moraines* are inherited from debris concentrations or septa within the parent glacier (section 9.4.2). A range of entrainment processes can produce linear debris concentrations on glacier surfaces, including basal freeze-on, hydrofracture filling, apron entrainment, folding, and shearing and stacking of marginal ice (Evans 1989b; Hambrey et al., 1997, 1999; Bennett et al., 1998a; Ó Cofaigh et al., 1999, 2003b; Glasser and Hambrey, 2003; Lacelle et al., 2007). Such transverse septa control the pattern of differential ablation on the glacier surface, producing linear ice-cored ridges which eventually become separated from the snout to form *ice-cored moraines* (Østrem, 1964; Fitzsimons, 2003; Ó Cofaigh et al., 2003b). The morphological expression of englacial structures is always most striking while the moraines retain ice cores. Fitzsimons (1990, 2003) described controlled moraines from Antarctic glacier margins where supraglacial deposition is clearly related to the concentration of debris bands in downwasting ice. In this arid polar setting, it is common to find the real outer ice edge separated from an 'apparent' cliffed ice edge by a shallow ramp comprising ice-cored moraine. Similar features are ubiquitous in the Canadian High Arctic (Ó Cofaigh et al., 2003b). Uneven sediment redistribution during ice-core wastage means that the final deposits tend to consist of discontinuous transverse ridges with intervening hummocks, preserving only a weak impression of the former englacial structure (Kjær and Krüger, 2001; fig. 11.86a). Reworking by supraglacial meltwater systems also acts to destroy large tracts of controlled moraine before they can be deposited on the ground surface (fig. 11.86b; Boulton, 1972). Some linearity may be imparted by localized glacier pushing of hummocky moraine during the later stages of melt-out, as has been observed at Kvíárjökull,

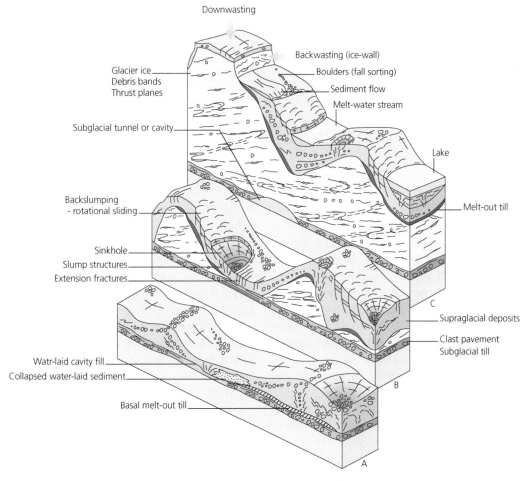

Figure 11.86 Models of controlled/hummocky moraine development. (a) Sequential development of supraglacial debris concentrations based on surveys of the downwasting snout of Kotlujökull, Iceland. (Kjær and Kruger, 2001) (b) Conceptual model of controlled moraine development from transverse debris concentrations in a glacier snout. (1) Debris released at glacier surface. (2) Foliation. (3) Debris band. (4) Shear fault. (5) Outwash collecting between ice ridges. (6) Supraglacial debris accumulating from the melt-out of debris bands. (7) Relatively clean ice surface resulting from mass flowage of melt-out debris. (8) Small mass flows on the surfaces of stagnant ice hummocks. (9) Subglacial till. (10) Supraglacial channels reworking mass flows from stagnant ice. (11) Large supraglacial mass flow. (12) Delta accumulation in supraglacial ponds. (13) Elongate ice ridge protected from melting by the accumulation of supraglacial sediment as debris bands melt out. (14) Absence of collapse structures in sediment deposited directly on to bedrock. (15) Collapse structures. (16) Interdigitating mass flow diamictons and glacifluvial sands and gravels. (17) Subglacial melt-out tills (Modified from Boulton, 1972).

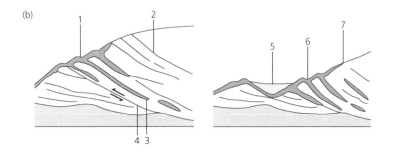

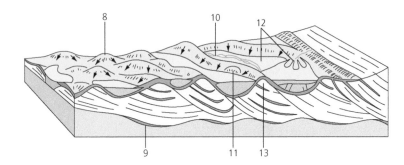

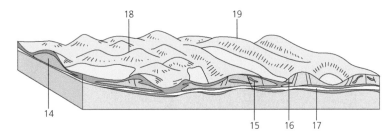

Figure 11.86 (*Continued*)

Iceland, by Spedding and Evans (2002), a process most likely to be common only in such temperate glaciers with debris covers restricted to their outer margins. Patterns of debris reworking and deposition can also be controlled by crevasses or other lines of weakness within the ice, which can guide the final stages of glacier disintegration. For example, Rains and Shaw (1981) suggested that the uneven distribution of debris on Antarctic glacier surfaces is determined by ice topography, rather than englacial debris concentrations, and debris consequently accumulates in *cusps* melted onto the glacier surface and between thrust blocks of ice near the margin.

Recent recession and thinning of Svalbard glaciers has provided us with an excellent field laboratory in which to investigate the evolution of controlled moraine in a mountain permafrost terrain (Sletten et al., 2001; Lyså and Lønne, 2001; Lukas et al., 2005). Many Svalbard glaciers are characterized by very striking linear debris concentrations around their margins (fig. 11.87a; Glasser and Hambrey, 2003), the origins of which have been explained by a variety of processes not necessarily operating in isolation, and some replacing others over time due

to climate change. These include freeze-on processes in polythermal snouts, passive transfer of rockfall material along flow lines, filling of englacial meltwater networks, crevasse infill during surging, thrusting of debris along shear planes, and englacial stacking of debris-rich ice facies through folding and thrusting. To these can be added the process of apron overriding, particularly the incorporation of buried glacier ice/ice-cored moraines and inter-bedded sequences of outwash and aufeis (river icings), common features of modern Svalbard glacier forelands. Specific historical glacier advances, relating to either climatic or surge triggers, are clearly recorded by moraines, but as these moraines are ice-cored and controlled, they are presently undergoing localized degradation due to melt-out (fig. 11.87b). This degradation is effectively destroying linearity through processes such as thermo-erosion, retrogressive debris flow and fluvial reworking, which are all at their most active in mountain terrains (Etzelmüller, 2000; Etzelmüller et al., 2000; Lukas et al., 2005). As a result, it is easy to get a false impression of moraine morphology and volume using the historical 'moraines' of Svalbard, which (excluding

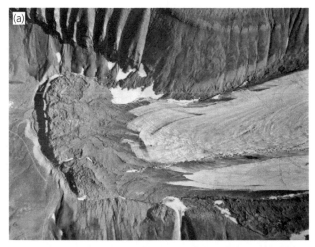

Figure 11.87 Controlled moraine on Svalbard. (a) Aerial photograph of controlled moraine on the snout of Longyearbreen, showing the gradation from debris-rich foliae on the ice surface to ice-cored (controlled) hummocky moraine marking the extent of historical (Little Ice Age) readvance. (Norsk Polarinstitutt) (b) Ground view of the surface of the ice-cored moraine of Longyearbreen, showing a retrogressive flow slide of supraglacial debris over buried glacier ice. Such flow slides are gradually fragmenting the linearity of the controlled ridges in the ice-cored moraine (D.J.A. Evans).

proglacial thrust moraines) are essentially zones of buried glacier ice adorned with controlled ridges. Therefore, the use of Svalbard glaciers as modern analogues for ancient glaciated landscapes, particularly in diagnosing ancient moraines as the pure products of englacial debris thrusting (cf. Hambrey et al., 1997, 1999; Bennett et al., 1998a; Graham et al., 2007), may not be entirely appropriate (Lukas 2005, 2007; Benn and Lukas, 2006). Schomacker and Kjær (2008) have calculated that retrogressive flows on stagnant glacier snouts are maintaining surface lowering rates of nearly 1 m/yr.

The ice-cored (controlled) morainic topography in the continuous permafrost zone of northern Canada constitutes an ideal modern analogue for the development of large tracts of ancient hummocky moraine in lowland terrains (Dyke and Savelle, 2000; Dyke and Evans, 2003), and, moreover, allows us to recognize that elongate, arcuate belts of ancient chaotic hummocks probably represent controlled moraines developed in the frozen outer margins of former glacier snouts (e.g. Attig et al., 1989; Clayton et al., 2001). The most important implication of this analogue is the realization that chaotic hummocky moraine does not necessarily record widespread ice stagnation, but rather the melt-out of controlled ridges in ice-cored moraine belts, produced incrementally in an *en echelon* pattern by actively receding frozen toe zones (Johnson and Clayton, 2003). In such settings, the profusion of debris bands in the outer snout is related to the range of debris entrainment processes that are inextricably linked with the switch from extensive, warm subglacial conditions to the narrow frozen bed zone of thin glacier margins receding in contact with permafrost (Clayton et al., 2001). Therefore, in non-permafrost terrains, the linearity of controlled moraines after the loss of ice cores is manifest in crudely aligned hummocks and ice-walled lake plains (section 11.4.4), visible only on aerial

images covering large areas. In permafrost zones or in recently deglaciated landscapes, controlled moraines are still ice-cored and, as a result, linearity is striking, and incipient melt-out of enclosing glacier ice is evident in the occurrence of kettle holes that later develop into ice-walled lake plains (e.g. Schomacker and Kjær, 2008).

Hummocky moraine is the end product of *topographic inversion* cycles during the ablation of debris-mantled ice. During ice ablation, debris is transferred away from topographic highs on the glacier surface by mass movements and meltwater, exposing ice cores to renewed melting and creating new depressions in former high points. Further debris reworking and topographic development is achieved by meltwater streams meandering between dirt cones, the collapse of englacial tunnels, and the enlargement of supraglacial lake basins. Several cycles of debris reworking and topographic inversion may occur before the debris is finally deposited in a series of irregular mounds and ridges (fig. 11.87; Evans and Rea, 2003; Evans et al., 2006c, 2007b).

An enigmatic type of supraglacial debris concentration is the circular rimmed mound. In hummocky moraine tracts these are referred to as *rim ridges*, *doughnuts* or *ring forms* (fig. 11.88a) and have been interpreted as either subglacially pressed features (see section 11.2.7) or the supraglacial infills of sinkholes/moulins (Clayton, 1967; Mollard, 2000; Johnson and Clayton, 2003; fig. 11.88b). *Circular moraine features* (CMF) are similar forms that have been reported from the upland plateaux of northern Norway and Greenland. These are round to oval-shaped debris accumulations between 20 and 170 m in diameter and up to 10 m high, resting directly on periglacial block-fields, from which they can be differentiated by their colour, grain size and erratic content. CMFs can take the form of rings, sheets and mounds, and were interpreted by Ebert and Kleman (2004) as former englacial debris concentrations, entrained at the transition zone between cold- and warm-based ice. This entrainment was most effective wherever thick valley ice was forced to flow over upland plateaux where cold-based thermal conditions prevailed. The englacial debris concentrations that lay directly above the plateaux at the onset of deglaciation were then let down onto the undisturbed blockfield, producing either mounds, sheets or rings, depending on the distance from the ice–bed interface (fig. 11.88c). CMFs essentially form one end of the hummocky moraine

continuum, being examples of patchy supraglacial debris spreads. At the other end of the continuum lies the high-relief, thick hummocky moraine produced by debris-covered mountain glaciers (Benn et al., 2003).

The sedimentology of hummocky moraine is complex, reflecting multiple cycles of redeposition during its formation. Typical facies associations consist of inter-bedded debris flows and other mass-movement deposits, laminated lacustrine sediments, and glacifluvial sands and gravels, in varying degrees of disturbance (Eyles, 1979; Andersson, 1998). The constituent debris may be actively or passively transported, depending on the position of the moraine relative to debris transport paths in the parent glacier. In some cases, debris from different sources and transport paths may remain segregated within the moraine, giving rise to compositional stratification (Benn, 1992). An exposure through a conical hummock in a field of chaotic hummocky moraine on the Isle of Skye, Scotland (fig. 11.89), reveals diamicton lenses inter-bedded with water-sorted gravels and sands, all displaying faults, slumps and other collapse structures, and is interpreted as a series of debris flow deposits and glacifluvial sediments which have undergone syn-depositional deformation and reworking relating to the melt-out of buried ice. Other well-exposed examples of hummocky moraine sediments are on the Lleyn Peninsula in North Wales (McCarroll and Harris, 1992), where inter-bedded

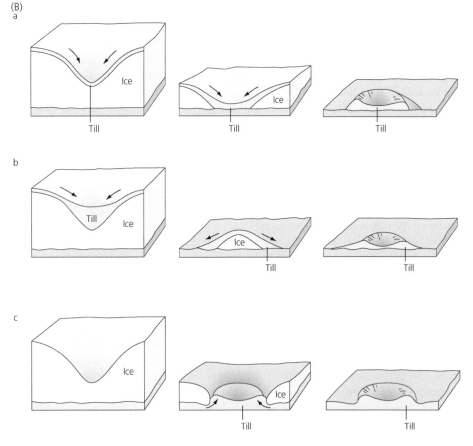

Figure 11.88 (A) Aerial photograph of rim ridges or ring forms in Saskatchewan, Canada, highlighted by the burn track of a grass fire. (Aerial photograph A6729-13, Energy, Mines and Resources; view is approximately 1 mile across) (B) Interpretations of rim ridges or ring forms due to sinkhole/moulin infill from above (a and b) and subglacial pressing into a perforation from below (c). (Johnson and Clayton, 2003) (c) Genetic interpretation of CMFs (After Ebert and Kleman, 2004).

debris-flow diamictons, sands and gravels overlie sub-glacial sediment associations. Such sediment successions are common to all degrading supraglacial debris accumulations on modern glaciers (cf. Kjær and Krüger, 2001; Lukas et al., 2005).

In certain situations, debris-rich basal ice may melt out contemporaneously with supraglacial sedimentation, resulting in a two-layer stratigraphy consisting of a lower assemblage of melt-out tills and an upper assemblage of debris flows, glacifluvial sands and gravels, and

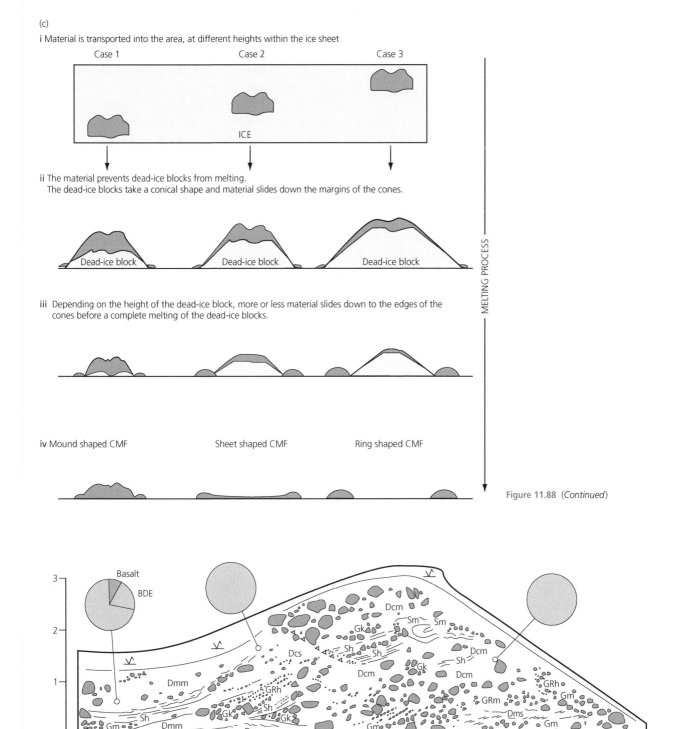

Figure 11.88 (*Continued*)

Figure 11.89 An exposure through hummocky moraine on the Isle of Skye, Scotland. Pie diagrams show proportions of basalt, Beinn Dearg (BDE) and Glamaig (GE) granites, and suggest compositional stratification (Modified from Benn, 1992).

glacilacustrine silts. An excellent example of this stratigraphic succession was described by Johnson et al. (1995) from the hummocky moraine of western Wisconsin, where supraglacial sediment is thought to have been lowered vertically onto a basal melt-out sequence. Variations in the thickness of basal debris concentrations, and thus the melt-out till, is considered to be important to the production of the final hummocky forms in western Wisconsin, rather than the effects of supraglacial topographic reversal alone.

11.4.3 KAME AND KETTLE TOPOGRAPHY

Kames, derived from the Scottish words *cam* or *kaim* (meaning crooked and winding or steep-sided mound), are variously shaped mounds, composed chiefly of sand and gravel, and formed by supraglacial or ice-contact glacifluvial deposition. The term originally encompassed eskers, and the close genetic associations between eskers and kames often make it difficult to differentiate them, and many transitional forms exist. As a result, the term 'kame' has been used in several and sometimes confusing

senses in the literature. It is commonly used as a noun, but has been used as an adjective, as in *kame plateau*, *kame terrace*, *kame delta* and *kame moraine*. Many of these terms are of limited usefulness, and duplicate more accurate terminology. Kame deltas, for example, are simply ice-contact deltas (see section 11.6.5). In this section, we discuss *kame and kettle topography*, which comprises assemblages of mounds and hollows (fig. 11.90). Kame terraces, or ice-contact valley-side terraces, are described in Section 11.4.5.

Kame and kettle topography forms tracts of mounds and ridges (kames) and intervening hollows (kettles or kettle holes). The kettle holes may be filled with water, and represent areas of subsidence caused by the melt of buried ice, whereas the kames are upstanding masses of glacifluvial deposits (fig. 11.91). Kettle holes also occur on otherwise continuous outwash surfaces, known as *pitted outwash* (section 11.4.6). The difference between kame and kettle topography and pitted outwash reflects the relative importance of glacifluvial deposition and the melt-out of buried ice in the form of the final landform assemblage. Kame and kettle topography may appear

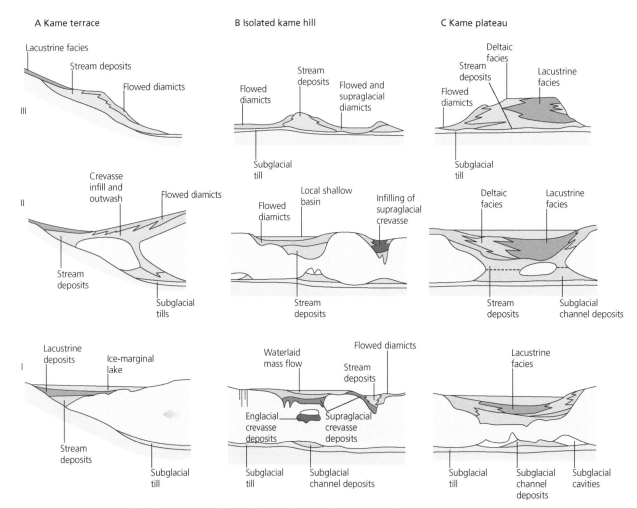

Figure 11.90 Schematic models for the various types of kame (After Brozikowski and van Loon, 1991).

completely chaotic, but some degree of lineation or pattern is not uncommon. Linear patterns provide evidence for *controlled deposition*, reflecting former englacial and supraglacial debris concentrations, crevasse patterns, and/or drainage systems (fig. 11.92; Thomas et al., 1985). Viewed on the ground from a distance, the accordant surfaces of the flat-topped residual kame mounds of a largely collapsed outwash surface may appear to form continuous terraces even when the whole assemblage is in fact chaotic in plan view. Many kame terraces only possess terrace forms during their early stages of existence, before ice cores melt out.

Kame and kettle topography evolves where large quantities of debris are reworked by supraglacial and englacial drainage systems during the final stages of glacier wastage (fig. 11.90; Clayton, 1964). Kames are commonly found in association with hummocky moraine (section 11.4.2), from which they can be differentiated by their morphology and composition. The presence of flat or gently sloping plateau surfaces, representing remnant glacifluvial terraces, is particularly diagnostic. Additionally, kames tend to have smoother slopes than hummocky moraine, due to less variable internal composition. Exposures in kames reveal bedded sand and gravel facies with variable internal geometry, depending on conditions during deposition. Inter-beds of lacustrine and mass-flow deposits may also be present (Price, 1973). The presence of glacifluvial sediments can often be inferred even where good exposures are lacking. One useful criterion is the presence of numerous rabbit burrows in mounds, indicating the occurrence of easily excavated sand. The morphology of surface clasts is also a useful indicator of

Figure 11.91 Kame and kettle topography on the foreland of Sandfellsjökull, Iceland (D.J.A. Evans).

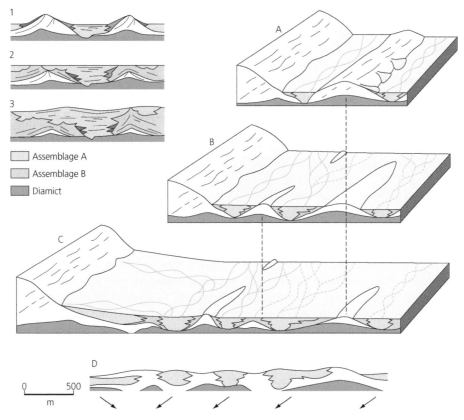

Figure 11.92 Depositional sequence produced in an ice-marginal supraglacial sandur based on interpretations of the Orrisdale Outwash Member on the Isle of Man. Fluvial sedimentation is guided by the changing supraglacial topography of melting ice. A–C show stages of sequential evolution of the sandur system to produce the final facies assemblage in D. The inset 1–3 shows the stages in the development of a marginal sandur and traces the progressive melt-out of stagnant ice ridges and the widening of the sedimentary system. Assemblage A comprises typical gravel and sand glacifluvial outwash, and assemblage B comprises predominantly overbank fines (Redrawn from Thomas et al., 1985).

water-worn material, because kames represent proximal glacifluvial deposits and therefore clasts tend to be less well rounded than in more distal outwash deposits, but better rounded than in subglacial facies. It should be noted, however, that clasts may have been subject to complex transport and depositional histories, and care must be taken not to attach too much significance to roundness statistics, unless large samples are taken (Benn, 2004a). There is usually some degree of internal folding and faulting of kame sediments due to the removal of supporting ice, particularly towards kame margins. The greatest amounts of disturbance will be associated with supraglacial or englacial deposits let down onto the substratum. The most common structures are steep normal faults bounding down-faulted masses (McDonald and Shilts, 1975).

11.4.4 ICE-WALLED LAKE PLAINS AND SUPRAGLACIAL LAKE DEPOSITS

During glacier wastage, large volumes of meltwater can become dammed between the ice margin and local topography, either in the form of valley walls or substantial accumulations of proglacial sediments such as latero-frontal moraines, fans and ramps (section 3.7). This results in the development of supraglacial lakes into which stratified sediments and mass flow diamictons accumulate, often very rapidly in debris-charged systems (Eyles et al., 1987; Schomacker and Kjær, 2008; fig. 11.93a). Ablation of the adjacent ice results in various degrees of disturbance of the sediment pile either post- or syn-depositionally. In the latter case, sediment-laden underflows can maintain continuous melting of stagnant ice on the lake bottom (Chikita et al., 2001).

Examples of ancient supraglacial lakes have been reported widely from the North American Cordillera (Eyles et al., 1987; Clague and Evans, 1994; Ward and Rutter, 2000) and the Southern Alps of New Zealand (Mager and Fitzsimons, 2007). The sedimentary sequences and structures observed in ancient supraglacial lake settings comprise thick units of inter-stratified sorted sediments and diamictons. Because the dominant grain size varies with distance from sediment influx point, upward-fining sequences from stratified diamictons to rhythmites reflect increasingly distal sedimentation as ice recedes from the basin. Cut-and-fill sequences often record subaqueous fan development and migration, localized mass flow incisions and delta progradation. These processes are common to all proximal glacilacustrine depositional environments, but some features are particularly diagnostic of supraglacial lake sedimentation. Sedimentologically, supraglacial lake infills are dominated by intra-formational soft-sediment deformation and water escape structures, caused by the continuous melt-out of underlying ice and ice walls (figs 11.93b, c, d). Geomorphologically they can be recognized by their undulatory to kettled surface topography on valley bottoms (e.g. Eyles et al., 2003) and may form arcuate belts where they accumulated along receding glacier margins. An excellent example of the latter is the 'end moraine' complex around Lake Pukaki in New Zealand, where supraglacial lakes developed between the receding valley glacier margin and its latero-frontal fan/ramp (Mager and Fitzsimons, 2007).

Ice-walled lake plains are irregularly shaped, flat-topped mounds containing rhythmically bedded fine-grained sediments and ice-contact deltas, which represent the infillings of former ice-walled lakes (Clayton et al., 2008). They are left as positive relief features when the surrounding ice melts, and are found in close association with other supraglacial landforms, such as hummocky moraine and ring forms (fig. 11.94; Johnson et al., 1995; Johnson and Clayton, 2003). A comparison of the ice-walled lake plains in North America, Denmark and Sweden by Johnson and Clayton (2003) revealed that they range in area from 0.1 to 30 km^2, can have a relief of between 2 m and 60 m, and contain lake sediments up to 50 m thick. Many ice-walled lake plains have raised rims standing 3–10 m above the central plateau. These are typically composed of glacifluvial or mass-movement deposits that inter-digitate with the finer-grained sediments of the lake plains and therefore record flowage from the ice slopes that surrounded the lakes. Stream- and wave-sorted sand and gravel also commonly occurs round the margins. The rims record late-stage deposition from the ice walls into the infilled lake basin, or subsidence of the central area due to melting of a central ice core. In many cases, however, the central parts of the mounds show little evidence of subsidence or collapse, indicating that deposition, at least in its final stages, took place on firm ground. The model of lake plain formation proposed by Clayton and Cherry (1967; fig. 11.94b) accounts for the morphology of the flat-topped mounds, in addition to the surrounding supraglacial landforms. Unstable lake plains are produced where supraglacial debris cover is relatively thin, resulting in rimmed mounds and low- to medium-relief hummocky moraine. Stable lake plains are produced in settings where supraglacial debris cover is thick and glacier melting is slower. This results in the deposition of thicker lake infills, a lack of mound rims and high-relief hummocky moraine.

11.4.5 KAME TERRACES

Kame terraces are gently sloping depositional terraces perched on valley sides, and are deposited by meltwater streams flowing between glacier margins and the adjacent valley wall (figs 11.90a, 11.95a). Flights or inset

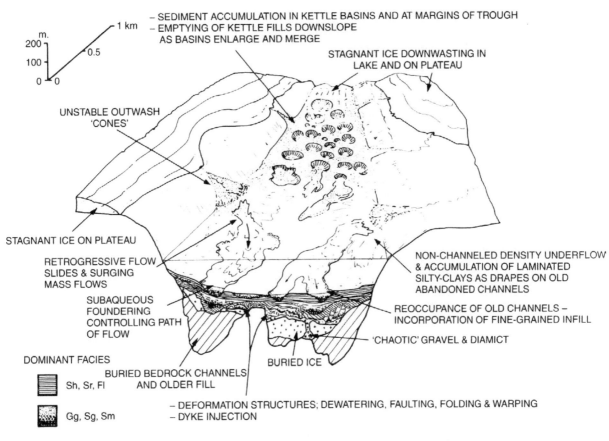

- SEDIMENT ACCUMULATION IN KETTLE BASINS AND AT MARGINS OF TROUGH
- EMPTYING OF KETTLE FILLS DOWNSLOPE
  AS BASINS ENLARGE AND MERGE

STAGNANT ICE DOWNWASTING IN LAKE AND ON PLATEAU

UNSTABLE OUTWASH 'CONES'

STAGNANT ICE ON PLATEAU

RETROGRESSIVE FLOW SLIDES & SURGING MASS FLOWS

SUBAQUEOUS FOUNDERING CONTROLLING PATH OF FLOW

NON-CHANNELED DENSITY UNDERFLOW & ACCUMULATION OF LAMINATED SILTY-CLAYS AS DRAPES ON OLD ABANDONED CHANNELS

REOCCUPANCE OF OLD CHANNELS – INCORPORATION OF FINE-GRAINED INFILL

'CHAOTIC' GRAVEL & DIAMICT

BURIED ICE

DOMINANT FACIES

Sh, Sr, Fl

Gg, Sg, Sm

BURIED BEDROCK CHANNELS AND OLDER FILL

- DEFORMATION STRUCTURES; DEWATERING, FAULTING, FOLDING & WARPING
- DYKE INJECTION

Figure 11.93 Examples of supraglacial lake deposits. (a) Conceptual reconstruction of sedimentation in a supraglacial lake environment in an area of moderate to high relief, based on the stratigraphy of the Fraser river valley, British Columbia, Canada. (Eyles et al., 1987, reproduced with permission of Blackwell): (b–d) Supraglacial lake deposits at Lake Pukaki, South Island, New Zealand, showing the range of sediment textures, from fine-grained rhythmites, through well-sorted gravelly foresets to coarse boulder mounds, and the localized heavy contortion produced by syn-depositional ice melt-out (D.J.A. Evans).

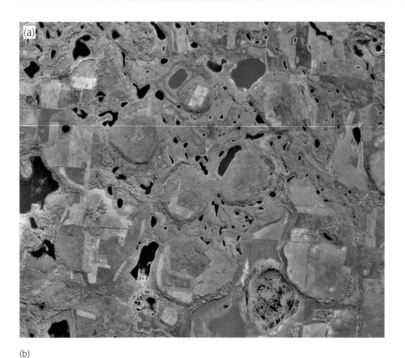

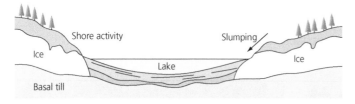

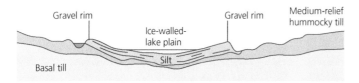

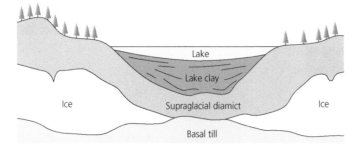

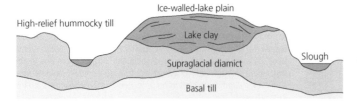

Figure 11.94 Ice-walled lake plains. (a) Aerial photograph of ice-walled lake plains near Witchekan Lake, Saskatchewan, Canada, surrounded by more chaotic ridges and kettle holes. (Aerial photograph A15882-138, National Aerial Photograph Library, Ottawa) (b) Conceptual models of ice-walled lake plain development in unstable and stable settings (After Clayton and Cherry, 1967).

Figure 11.95 (a) Eastward view across the outwash plain of the Tasman Glacier, north of Lake Pukaki, New Zealand, showing a staircase of kame terraces and lateral moraines dating back over the last glacial cycle. (b) Pitted surfaces on kame terraces deposited along the margin of Sandfellsjökull, Iceland (D.J.A. Evans).

sequences of kame terraces may occur on valley walls, documenting periodic reduction of the glacier surface during terrace formation. During deglaciation, the ice-contact faces of kame terraces are very unstable and are thus prone to collapse through debris flows and land-slides. Additionally, melt-out of buried ice within kame terraces often produces a kettled or pitted surface (fig. 11.95b).

Although composed predominantly of fluvial sands and gravels, kame terraces may also contain lacustrine sediments which collect in ephemeral ponds, and debris flow diamictons derived directly from the glacier surface (e.g. Levson and Rutter, 1989b). Kame terraces can some-times be confused with formerly continuous outwash surfaces that have been reduced to valley side terraces by postglacial incision (see section 11.5.1). However, out-wash terrace remnants on opposite valley sides should have similar gradients and altitudes, whereas paired kame terraces may have different gradients, reflecting differ-ences in ice-margin morphology (Gray, 1975).

A number of features identified as ancient kame ter-races may in fact be slope deposits which accumulated between lateral moraines and the valley side, but can eas-ily be differentiated from true kame terraces if exposures are available. An example of a terrace deposited on top of

a lateral moraine at the margin of the Tasman Glacier, New Zealand, is shown in figure 11.76a.

11.4.6 PITTED SANDAR

Pitted sandar, also referred to as *kettled sandar* or *kettled outwash plains*, are sandar which are cratered by hollows left by the melt-out of isolated buried blocks of glacier ice (Maizels, 1977). The ice blocks may originate in two main ways:

(1) as remnants of a glacier snout, detached from the rest of the glacier by differential ablation;
(2) as icebergs transported onto the sandur surface by flood waters, particularly jökulhlaups (Gustavson and Boothroyd, 1987; Russell, 1993; Fay, 2002a, b).

In either case, deposition of glacifluvial sediments may partially or completely bury the ice blocks, leaving hollows when the ice melts. Although case (2) refers to proglacial environments, pitted sandar formed by the melt-out of transported ice blocks are included in this section due to their close genetic similarity to supraglacial pitted sandar.

Where parts of ice-marginal sandar accumulate supraglacially, they will develop pitted zones after ice melt, and such zones may occur at various points along a deglaciated valley where overlapping fans were prograded sequentially. Hochsandur fans are very likely to develop pitted zones, because they are fed supraglacially (Kjær et al., 2004). Patterns of deposition on supraglacial sandar are strongly controlled by patterns of ablation of the underlying ice. For example, the Late Pleistocene Orrisdale Outwash Member of the Isle of Man has been interpreted as a supraglacial sandur in which patterns of sedimentation were controlled by transverse ridges of dead ice, perhaps reflecting englacial debris concentra-tions (fig. 11.92; Thomas et al., 1985). A series of mar-ginal sandar were built up between the dead ice ridges until melt-out and fluvial aggradation caused outwash widening and sandur coalescence. A particular type of pitted outwash sometimes occurs in mountainous terrain where large segments of wasting glacier lobes become isolated in valley bottoms and then buried by outwash

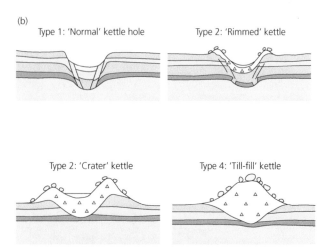

Figure 11.96 Ice block structures on sandur surfaces. (a) Ring structures indicative of iceberg melt-out after a jökulhlaup on a distal residual bar surface on Myrdalssandur, Iceland. Ring diameters are 20–25 m and the view is looking upstream. (J. Maizels) (b) Schematic reconstructions of the four types of ring structure produced by the in situ melting of ice blocks on sandur surfaces. Types 1–4 represent a progressive increase in sediment concentration of the parent ice block. (Modified from Maizels, 1992) (c) The morphology and general composition of a range of ice-block obstacle marks. Type 1: kettle-scours with proximal and lateral scour and tail formed by partial or total burial and subsequent partial exhumation. Type 2: semicircular obstacle marks formed by partial exhumation of partially buried ice blocks on the margins of the main channels. Types 3 and 4: obstacle marks characterized by a proximal and lateral scour crescent, a ridge stoss side of the scour crescent and a large aggradational tail whose structure is anticlinal. Truncation of anticlinal-shaped bedding produces the flat-topped morphology and armoured nature of Type 4 obstacle marks. Type 5: obstacle marks with fine-grained gravel in the immediate lee of the scour hollow and coarse gravel lag characterizing the distal part of the tail, formed by waning stage deposition followed by late waning stage erosion. Type 6: entirely erosional obstacle marks, formed by total exhumation of buried ice blocks or by scour around ice blocks grounded on the late waning stage. (d) Schematic diagram of antidune stoss-side strata forming around an ice block. (i) Flow is supercritical downflow of the ice block and generates an upstream-migrating standing wave. (ii) Plan view of the upstream-migrating standing wave. (iii) Antidune stoss beds deposited from upstream-migrating surface wave. (iv) 'Washout' situation with increasing bedform height. Flow on the upstream side slows and deepens, increasing the rate of sedimentation until antidune collapses. (v) Section after flood has receded. (e) Upstream-dipping gravel beds deposited on the side of a large deeply embedded ice block. View is from the ice block outwards into the flow. Upstream dipping beds represent antidune stoss-side sedimentation within a localized zone of supercritical flow around the ice block (Russell et al., 2006, reproduced with permission of Elsevier).

(Fleisher, 1986). Ice melt-out produces large depressions termed *dead-ice sinks*. Numerous closely spaced sinks can coalesce to form a single elongate depression with a complex bottom topography, referred to as a *dead-ice moat*. Similarly extensive, but more commonly shallower melt-out hollows in outwash can be produced where *naled* or river icing has been buried by proglacial outwash. Naled is very common at the margins of Svalbard glaciers, where chilled meltwaters exit from snouts and freeze along the banks of gravel channels (e.g. Bennett et al., 1998b).

Features resulting from the melt-out of ice blocks transported by jökulhlaups on Myrdalssandur, Iceland, have been studied in detail by Maizels (1992). Ice melt produces *boulder ring structures*, between 3 m and 40 m in diameter, consisting of near-circular, boulder-rich rims surrounding a central depression (fig. 11.96a). The rims are highest on the downstream flanks of the structures, and gaps may occur in the rim on the upstream flanks. Four principal types of ring structure were observed, which reflect the amount of sediment held within the ice block (fig. 11.96b).

- *Type 1* is a *normal kettle hole* produced by the collapse of outwash materials surrounding a clean ice block.
- *Type 2* is a *rimmed kettle* which possesses a diamicton drape in the central depression and low-amplitude bouldery diamicton rims.

- *Type 3* is a *crater kettle* which possesses a thick diamicton lining and thick, boulder-covered diamicton rims.
- *Type 4* is a *till-fill kettle*, produced by the melt-out of a heavily sediment-laden ice block and therefore constituting a mound of diamicton centred over a depression in the outwash.

The morphology and sedimentology of the boulder ring structures were linked to the sediment concentration of ice blocks with the aid of laboratory experiments, and Maizels concluded that the formation of Types 2, 3 and 4 ring structures require sediment concentrations of 5–20 per cent, 20–60 per cent and >60 per cent respectively.

Kettle holes may become completely infilled with sediments if sandur aggradation continues during the melt of buried ice blocks. In this case, no surface expression remains, but buried kettle holes are preserved in the sedimentary record as downfaulted blocks bounded by steep normal faults (fig. 11.96b: Type 1). The amount of displacement of such faults will decrease upwards if downfaulting occurred during sediment aggradation. This is because the deeper layers that were deposited immediately above the ice block will experience the greatest amount of subsidence, whereas shallower layers, deposited after some of the block has already melted, will experience less. The uppermost layers, deposited after complete melt-out of the block, will be undeformed.

(c)

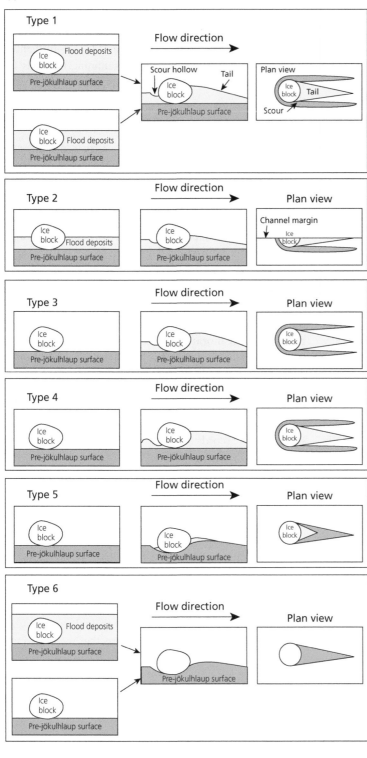

(d)

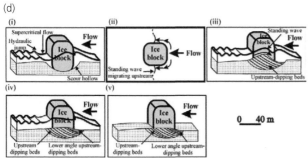

(e)

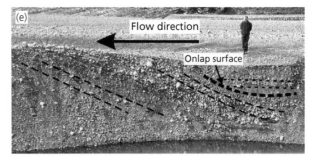

Figure 11.96 (Continued)

While ice blocks remain on the surface of a sandur they will interrupt water flow and initiate the production of *ice-block obstacle marks* (Russell, 1993; Fay, 2002a; Russell et al., 2005). Using the well-founded hydrological principles of fluid flow around obstacles like bridge piers, Fay (2002a) has demonstrated that horseshoe-shaped vortices and proximal scours will operate around ice blocks embedded on outwash surfaces, leading to the production of *kettle scours* and downstream tails of deposition termed *obstacle shadow ridges* (fig. 11.96c–e).

11.5 PROGLACIAL ASSOCIATIONS

This section discusses sediment–landform associations deposited in terrestrial proglacial environments by glacifluvial and slope processes, encompassing the products of direct meltwater deposition beyond ice margins in addition to the range of deposits conditioned by glacierization of drainage basins. The latter are traditionally grouped under the umbrella term 'paraglacial'.

11.5.1 SANDAR AND VALLEY TRAINS

Most proglacial rivers carry large amounts of suspended sediment and bedload, and this is characteristically deposited in extensive, gently sloping outwash plains known by the Icelandic term *sandar* (singular: *sandur*; fig. 11.97a). Narrow tracts of outwash hemmed in by valley sides in mountainous terrain are termed *valley trains* or *valley sandar* (fig. 11.97b). The streams responsible for sandar and valley trains are typically braided due to a combination of steep gradients, abundant bedload, cohesionless bank and bed material, and fluctuating discharges (Miall, 1992). The migration of braided channels and intervening bars during outwash accumulation results in very distinctive facies associations, careful study of which allows the former hydrological environment to be reconstructed in some detail. Some sandar are periodically inundated by GLOFs (section 3.7), and exhibit facies associations which differ in many respects from those deposited by 'normal' meltstreams fed by glacier ablation cycles (e.g. Sambrook-Smith, 2000). The morphological and sedimentological implications of these different discharge regimes has been summarized by Marren (2005). In essence, normal melt cycles impart low-magnitude/high-frequency signals in the glacifluvial outwash record, in contrast to GLOF-fed systems, where high-magnitude/low-frequency signals dominate the sedimentary record. In polar regions, smaller meltwater discharges and negligible precipitation result in very slow sandur development, with some stretches being characterized by coarse lag deposits, and continuous and thick sandar and valley trains are very rare (cf. Shaw and Healy, 1980; Mosley, 1988). This section examines the morphology of sandar and valley trains, goes on to describe the sedimentology of 'normal' or *ablation-controlled* and GLOF-influenced sandar, and concludes by briefly examining methods for reconstructing palaeodischarges from ancient glacifluvial deposits.

Morphology

The proximal ends of sandar and valley trains may be located at:

(1) former glacier terminus positions;
(2) the outlets of periodically draining ice-marginal lakes;
(3) some distance below former ice margins where stream velocities fall, such as where narrow, confined gorges open out into broad valley floors.

Where sandar commence at a former ice margin, feeder channels may occupy gaps in moraines; inset sequences of moraines on glacier forelands may also force proglacial outwash tracts to narrow where streams converge on breaches in the moraine ridges, or to change direction and run parallel to the moraines for short distances before exploiting a breach (Evans and Twigg, 2002; Marren, 2002). Alternatively, moraines may be absent, and the upper edge of the sandur may simply be marked by an abrupt upglacier-facing scarp known as an *ice-contact slope*, which records the location of the ice when the sandur was being deposited. Such ice-contact slopes are also known as *outwash heads* (Kirkbride, 2000), and may be the only evidence for former glacier limits in cases where ice-marginal deposition was dominated by glacifluvial processes.

Outwash heads sometimes form *ice-contact fans*, consisting of cone-shaped sediment accumulations radiating out from former feeder points and grading down onto a lower sandur surface (see section 11.3.5). The geometry of such fans reflects the routing of meltwater, which can be strongly constrained by the ice-marginal topography. Several authors have made strong pleas that we should not confuse ice-contact fans with proximal outwash fans/outwash heads (e.g. Zielinski and van Loon, 1999a, b, 2000; Krzyszkowski, 2002; Krzyszkowski and Zielinski, 2002; Pisarska-Jamrozy, 2006), although clearly there is a continuum between the two. Additionally, some confusion has arisen where the classic sandur models, based on outwash accumulating in unconfined settings (Boothroyd and Ashley, 1975), have been generally classified as analogous to alluvial fans, when in reality it is merely their morphology that has prompted the descriptive term 'sandur fan'. Zielinski and van Loon (2002, 2003) have emphasized that fan morphologies are characteristic of special modern circumstances, such as the short distances between glacier snouts and the sea, in settings like Alaska and southern Iceland, where classic models have been derived, and in reality it is more elongate braid-plains or valley trains that are most common in ancient depositional settings.

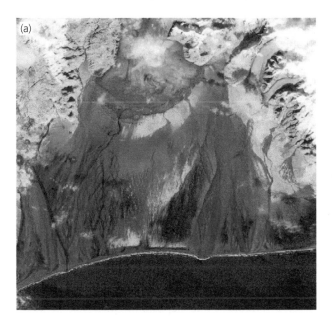

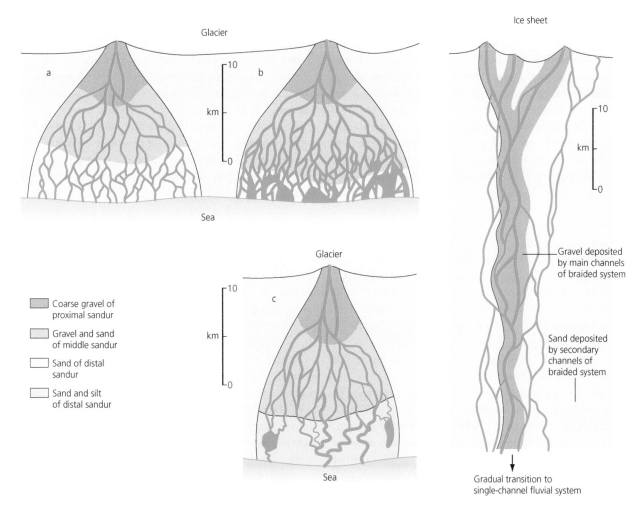

Figure 11.97 The morphology of proglacial sandar. (a) Landsat false colour image of Skeiðarársandur, Iceland. Darker colours on the fan represent areas with higher moisture content, and arcuate Little Ice Age moraines are visible a short distance beyond the glacier snout of Skeiðarárjökull. (NASA) (b) Incised and terraced valley sandur, Phillips Inlet, north-west Ellesmere Island, Arctic Canada. (D.J.A. Evans) (c) Summary of sandar morphologies based on modern and ancient examples (After Zielinski and van Loon, 2002, 2003).

The criticisms of Zielinski and van Loon (2002, 2003) have been tempered by Marren (2004), who emphasized that classic models have acknowledged both valley and fan-shaped sandar, and that braid-plains rather than fans have been identified in modern and ancient proglacial outwash (e.g. Bluck, 1974; Aitken, 1998a). Nonetheless, Zielinski and van Loon's (2002, 2003) summary diagram of conceptual models of proglacial sandar (fig. 11.97c) encapsulates their variability in gross morphology and sedimentology. Broadly, sandar can be subdivided into three zones. In the *proximal zone*, meltwater is confined to a few deep and narrow sediment-floored channels. In contrast, the *intermediate (middle) zone* comprises a complex network of wide and shallow braided channels, which shift position frequently and some of which only contain meltwater during periods of high discharge. Some parts of the sandur may become abandoned due to incision or channel switching, leading to the production of inactive channels or *palaeochannels*. In the *distal zone*, channels are very shallow and ill-defined, often merging to produce sheet flow during periods of high discharge. The long profiles of sandar are generally concave, exhibiting decreasing slope angles in a downstream direction. However, complexities may be introduced where tributaries join the main reach, or valley constrictions or moraines interrupt sandur development. The downflow decrease in gradient is generally associated with a decrease in sediment grain size, with gravels dominating in the proximal zone and gradually being replaced by sand or silt in the distal zone (Boothroyd and Ashley, 1975). The pattern of downstream-fining can break down to some extent in valley trains or confined sandar (fig. 11.97c), because channel migration is restricted, maintaining coarse gravel sedimentation for greater distances beyond the ice margin (e.g. Nicholas and Sambrook-Smith, 1998). Moreover, at smaller scales there are only weak correlations between channel slope and clast size, due to the intermittent role of GLOFs in transporting larger clasts over longer distances than during periods of normal discharge and the impact of sediment recycling by bank collapse (Maizels, 1995, 1997). Inputs of tributary streams will also dilute the downstream-fining trend (Aitken, 1998a).

When viewed from above over even short periods of time, sandar surfaces comprise complex networks of shifting channels and bars arranged in a braided pattern, although meandering and anastomosing patterns occur in distal reaches where banks are more stable (Boothroyd and Ashley, 1975). Braided reaches are characterized by steep gradients (2–50 m km^{-1}) and highly unstable channels with high width:depth ratios and low sinuosity, typical of stream beds in non-cohesive, erodible materials, with a lack of stabilizing vegetation. The channels are separated by *bars*, large dynamic stores of sand and gravel which become active and subject to erosion only at peak flows. In response to highly variable sediment loads, channel and bar forms may change positions on diurnal timescales, and rapid sediment throughput ensures that steady state is never reached. Anyone who has tried to cross a braided outwash stream on a daily basis can attest to the frustrations of constantly finding the best route through the shifting channels! Only the largest bar forms produced by the highest flow events such as GLOFs may become stable over periods of years.

Bar growth occurs when initially small irregularities in a channel are amplified by the deposition of sand or gravel, creating conditions favourable to continued accumulation of sediment. Conversely, bar erosion will occur when the bar form is no longer in equilibrium with flow conditions and sediment is removed from the system. Erosion and deposition can occur simultaneously, with sediment being lost from some parts of the bar and gained in others. Erosion and deposition can result in *downstream bar migration*, where the bar travels in a downstream direction by erosion of the upstream face and deposition on the advancing avalanche face; or *lateral accretion*, where sediment deposition occurs on one side of the bar.

Bar formation can occur by deposition of material in midstream, or by erosional dissection of pre-existing topographic highs (Ashmore, 1991; Ferguson, 1993; Maizels, 1995). *Bar initiation* by deposition appears to be a natural consequence of flow within wide, shallow channels, where the shear stress and transport capacity of the flow are low. The stream will deposit sediment preferentially in areas where the shear stress is lowest, forming an obstruction in mid-channel or against one bank. The presence of the obstruction will divert and concentrate the flow around the incipient bar, thus enlarging the channels to either side and isolating and encouraging further growth of the bar. Scouring of the floors and margins of channels remobilizes sediment, providing more material for bar growth in regions of low shear. Two main mechanisms of bar initiation have been recognized (fig. 11.98; Ashmore, 1991; Ferguson, 1993). The first is *central bar deposition*, the classic model of bar formation, in which elongated, migrating sheets of bedload stall in the centre of wide, shallow channels. Although central bar deposition is often regarded as the main process of bar initiation, Ashmore (1991) found that it was uncommon in flume experiments, and occurred only when shear stresses were close to the threshold for bedload transport. The second mechanism is *transverse bar conversion*, where bar initiation occurs in areas of flow divergence. Bedload eroded from an upstream scour pool forms a large lobe with prograding downstream avalanche faces. Thin bedload sheets stall on top of the lobe, which eventually emerges, and flow is then deflected laterally off its edges to form a symmetrical central bar. This process was originally recognized from small-scale, ephemeral pulses of

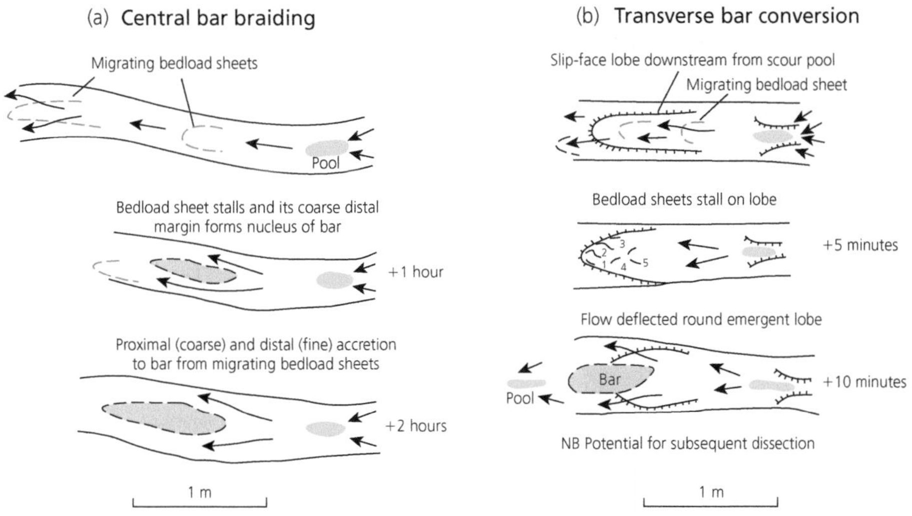

(a) **Central bar braiding**

Migrating bedload sheets

Pool

Bedload sheet stalls and its coarse distal margin forms nucleus of bar

+1 hour

Proximal (coarse) and distal (fine) accretion to bar from migrating bedload sheets

+2 hours

1 m

(b) **Transverse bar conversion**

Slip-face lobe downstream from scour pool

Migrating bedload sheet

Bedload sheets stall on lobe

+5 minutes

Flow deflected round emergent lobe

Bar

Pool

+10 minutes

NB Potential for subsequent dissection

1 m

Figure 11.98 Bar initiation by mid-channel deposition (a) Central deposition in which bedload stalls in areas where shear stress is low. Orange patches are pools and arrows show flow direction (b) Transverse bar conversion (After Ferguson, 1993).

sediment within channels (Southard et al., 1984), and is thought by many researchers (e.g. Ashmore, 1991; Bridge, 1993; Ferguson, 1993) to be the main process of bar accretion.

Erosional incision modifies existing bar forms, and is an important mechanism of bar evolution. At high stages, part of the flow may take a short cut across the bar surface, forming minor channels or *splays*. These splays have a steeper gradient than the main channels to either side, and tend to be preferentially enlarged to form *chutes* and can eventually form new major channels between bar remnants (fig. 11.99). Sediment eroded by splays and chutes can be redeposited in lobate forms up to several metres wide (lobes), which form the nucleus of new bars (Southard et al., 1984).

As river discharges fall from high to low flows, bars are draped by finer-grained sediments. Their largely sand-covered surfaces may then protrude from the stream flow. During the lower-stage flows, sand accumulation occurs on the downcurrent margins of gravel bars in the form of wedge-shaped cross-stratified deposits. Abandoned channels formed during late-stage incision of the bar may also be filled with sand and/or silt, and small-scale sand bed-forms often occur on the bar surface (McDonald and Bannerjee, 1971; Bluck, 1974; Gustavson, 1974).

The processes of sediment deposition, dissection and migration in dynamic channel systems give rise to a wide variety of bar forms, which is reflected in a broad range of alternative classification schemes (e.g. Ashley, 1990; Collinson, 1996). The most fundamental distinction is between *mid-channel bars*, which form within channels and cause bifurcation of flow, and *bank-attached bars*, which are deposited at channel margins and increase channel curvature. Such bars can be simple forms known as *unit bars*, with simple depositional histories controlled by local flow conditions, or larger, complex forms known as *compound bars* or *braid-bar complexes*, which have multi-stage histories of deposition and erosion.

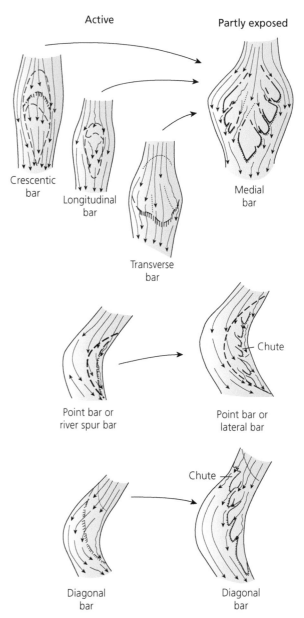

Active

Partly exposed

Crescentic bar

Longitudinal bar

Medial bar

Transverse bar

Point bar or river spur bar

Point bar or lateral bar

Chute

Diagonal bar

Diagonal bar

Chute

Figure 11.99 Bar evolution by the erosional dissection of mid-channel, lateral and diagonal bars (After Smith, 1985).

Figure 11.100 Aerial photograph of part of the Donjek river, Yukon Territory, Canada, showing the four topographic levels of the sandur surface identified by Williams and Rust (1969) (Reproduced with permission of the National Aerial Photograph Library, Ottawa).

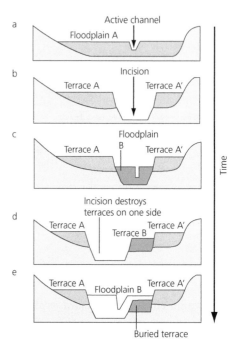

Figure 11.101 Glacifluvial terrace formation (After Lowe and Walker, 1984).

Mid-channel unit bars can be subdivided into *crescentic bars*, *longitudinal bars* and *transverse bars* (fig. 11.99). Crescentic bars are probably embryonic stages of longitudinal bars. These three bar types develop during high flows when the whole channel is submerged, and are dissected during lower flows to form compound *medial bars*. Bank-attached bars can take the form of *point bars*, which are produced on the insides of channel beds and are separated from the channel bank by a chute, and *diagonal bars* orientated oblique to river flow. During low flows, both types become dissected by chutes. Because proglacial outwash environments are very dynamic and individual stream channels are subject to constant changes in water discharges, most bars in sandar and valley trains are the products of multiple depositional and erosional events, and are therefore of a complex type (e.g. Bluck, 1974).

Successive stages of channel abandonment, due to lateral migration or vertical downcutting, are responsible for producing a series of topographic levels on sandur surfaces. Williams and Rust (1969) first identified such levels on the Donjek river, Yukon Territory, Canada (fig. 11.100). In their scheme:

- Level 1 is the main channel and principal sediment transport route, with little or no vegetation cover and bars exposed only at low flows.
- Level 2 is active only during flood stages, with few active channels at other times and a sparse vegetation cover.
- Level 3 takes only low-energy flow during flood stages and contains a moderate vegetation cover in humid climates.
- Level 4 comprises dry islands and interfluves with either a dense vegetation cover in humid climates or aeolian deflation features and migrating dunes.

A characteristic feature of many sandur surfaces is the presence of *terraces* at higher levels than that of the active channel. Indeed, in the case of ancient sandur plains, the entire surface may take the form of terraces perched high above the modern river channel. Flights of several terraces may be present (fig. 11.97b). Depositional terraces record episodes of sediment aggradation (sandur formation) followed by incision. Incision may be focused in the central part of the sandur, leaving *paired terraces* on both sides of the valley, but if incision is accompanied by migration of the active channel towards one valley side, a single *unpaired terrace* may be preserved on the opposite side (fig. 11.101). The switch from aggradation to incision may result from external forcing, such as changes in climate or base level, or from internal changes within the system, such as local sediment dynamics (Dawson and Gardiner, 1987). In glaciated catchments, the most common causes of terrace formation are fluctuations in sediment supply or fluvial discharge. For example, sandur aggradation will be encouraged during the paraglacial period in the early stages of deglaciation, when large quantities of unconsolidated sediment can enter the fluvial system (see section 11.5.2). When the supply of sediment begins to be exhausted, incision will occur if rivers maintain high capacity for sediment transport; this will be the case particularly if abundant meltwater is still available from receding glaciers. In Iceland, abandoned terraces perched high above contemporary sandar surfaces often relate to single, volcanically induced GLOF events, such as the 1918 *Katla-hlaup* deposits around south-east Myrdalsjőkull (Maizels 1989a, b, 1993).

Major terrace systems in many glaciated regions are known to have formed in full glacial conditions, when large amounts of sediment were introduced into lowland catchments. In Britain, the Main Terrace in the Severn basin can be traced southwards from ice-contact deposits at the limit of the last (Devensian) glaciation (Dawson and Gardiner, 1987), whereas some high fluvial terraces in the Thames basin and its northern precursor have been correlated with the earlier Anglian glaciation (Gibbard, 1988; Bridgland, 1994, 2006).

Incision in one part of a sandur will release sediment from storage, sending a pulse of sediment into lower parts of the system, which may initiate local aggradation. Thus terrace formation may occur at different times in proximal and distal parts of a catchment, depending on local sediment dynamics. The tendency for fluvial catchments to exhibit diachronous patterns of aggradation and incision has been termed *complex response* by Schumm (1977) and Patton and Schumm (1981), who warn against too simplistic an interpretation of terrace sequences in terms of climatic change. Further discussion of river terrace formation can be found in Dawson and Gardiner (1987) and Lowe and Walker (1997). Longer-term terrace development over several glacial cycles, over which uplift also plays a part, is encompassed in the widely cited model of Bridgland (2000, 2006; cf. Maddy et al., 2001).

Braided river facies associations

In braided river environments, the aggradation and migration of bedforms and larger-scale dunes, bars and channels produces a wide range of facies associations. Such associations can be usefully viewed as sediment packages at a range of scales (Allen, 1983; Miall, 1985, 1992; Bristow, 1996). At the smallest scale, sediment packages comprise individual *laminae* or *beds* laid down from single pulses of sediment (section 10.4). At larger scales, stratification is defined by *bounding surfaces*, which represent former surfaces of erosion or non-deposition lying between depositional units. Several orders of bounding surfaces can be recognized, with progressively wider lateral extent and environmental significance (Miall, 1985, 1992). *Sets* of strata deposited during the migration of single bedforms are bounded above and below by *first-order surfaces*; and *cosets*, or groups of sets deposited during migration of assemblages of bedforms are bounded by *second-order surfaces*. Successively higher-order surfaces (up to eighth order according to Miall, 1992) define progressively larger depositional units. According to Miall (1985, 1992), fourth and fifth order surfaces define *architectural elements*, representing large-scale components of the fluvial landscape, such as longitudinal bars and channel fills. Architectural elements combine to form depositional systems representative of different fluvial styles. Miall (1985, 1992) defined eight basic *architectural elements* characteristic of fluvial

deposits, which can then be used to reconstruct fluvial depositional styles from sediment exposures (fig. 11.102). The eight elements are:

(1) *Channels* (CH). This element represents the infills of channels. Typical facies include ripple cross-laminated sands and cross-bedded sands and gravels, recording the downstream migration of bedforms within the channel. With care, transverse sections can be used to reconstruct former channel dimensions, geometry and patterns of migration (Bristow, 1996).

(2) *Downstream accretion macroforms* (DA). This element comprises cosets of downstream-dipping cross beds resting on a flat or channelized base, representing the downstream migration of bar fronts. DA elements are particularly characteristic of braided streams. They commonly have a complex internal geometry, consisting of multiple cosets bounded by a hierarchy of erosion surfaces, which represent the migration of superimposed bedforms (e.g. ripples superimposed on dunes superimposed on bars).

(3) *Lateral accretion macroforms* (LA). These forms consist of sets of cross beds that dip transversely or obliquely to the main channel trend, recording the lateral migration of bank-attached or mid-channel bars. LA macroforms also develop on the insides of channel bends where deposition occurs, while erosion takes place on the opposite bank. Facies may be predominantly gravelly or sandy. The vertical thickness of LA and DA macroforms reflects the original bar top to channel height, minus any subsequent erosion, and is typically less than a few metres.

(4) *Gravel bars and bedforms* (GB). Low-relief gravel surfaces form a major element of many sandar, and are represented in sections as extensive tabular sheets of massive gravel and sets of trough and planar cross-bedded gravel (sections 10.4.4, 10.4.7). This element commonly occurs on top of LA or DA macroforms, recording the migration of bar tops over former channel margin positions.

(5) *Sediment gravity flow* (SG). This element consists of channelized or tabular lenses of poorly sorted gravel or diamicton, deposited by debris flows or related processes such as hyperconcentrated flows (Section 10.4.9). Such elements are common on the proximal reaches of some sandar and along the margins of some valley trains, where they represent mass flows from glacier margins or steep valley sides.

(6) *Sandy bedforms* (SB). This element includes lenses and sheets of horizontally bedded, ripple cross-laminated and cross-bedded sand, and records deposition on bar surfaces and in chutes and minor channels. This element commonly occurs on top of gravel bars and bedforms, where fining upward sequences may provide evidence for deposition during waning flow.

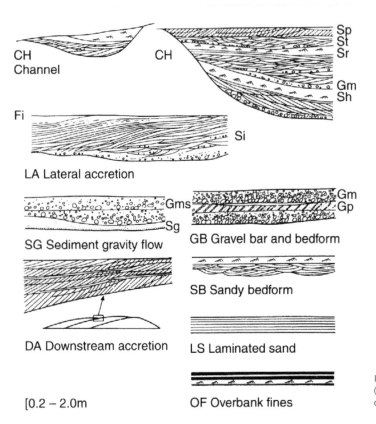

CH
Channel

Fi

Si

LA Lateral accretion

SG Sediment gravity flow

GB Gravel bar and bedform

DA Downstream accretion

SB Sandy bedform

LS Laminated sand

[0.2 – 2.0m

OF Overbank fines

Figure 11.102 Eight architectural elements of fluvial deposition (Miall 1992; reproduced with permission of Geological Association of Canada)

(7) *Laminated sand sheets* (LS). Laminated sand sheets are extensive tabular units of horizontally bedded sand, deposited by ephemeral sheet floods. They are uncommon in proximal sandar, but occur in some distal successions.

(8) *Overbank fines* (OF). This element consists of blanket-like drapes of massive and laminated silt, and forms a major component of the flood-plain deposits of meandering non-glacial rivers. However, OF elements occur locally within sandur deposits, particularly in distal environments where they infill abandoned channels.

Architectural elements can be recognized in many sandur and valley train deposits (e.g. Dawson and Gardiner, 1987), although not all researchers employ the terminology proposed by Miall (1985, 1992). For example, Siegenthaler and Huggenberger (1993) recognized three architectural elements in a study of last glacial (Würm) outwash in northern Switzerland. First, extensive tabular sheets of poorly sorted gravel with a matrix of silt and sand occur at high levels in the deposit as single beds or sets of beds up to several metres in thickness, and were interpreted as former 'traction carpets' deposited from hyperconcentrated floods (section 10.4.9). This is probably equivalent to Miall's sediment-gravity flow (SG) element. Second, horizontally bedded, well-sorted gravels form sheets a few metres to tens of metres wide and 1–2 dm thick. The sheets may grade laterally into cross-bedded gravels or abut inclined surfaces of older gravel.

This association is similar to Miall's GB element, and was interpreted as the deposits of migrating gravel sheets or incipient mid-channel bars within wide, shallow channel reaches. Third, large-scale troughs are filled with cross beds of sand and gravel, and range from a few metres to more than 100 m wide and 0.5–6 m deep. Cross-bedding is generally sub-parallel to the trough margins, giving the fills an onion-like appearance. This association is interpreted as the infills of deep, migrating scour pools, and is a special case of Miall's CH element. The pool and mid-channel gravel sheets were interpreted as the products of moderate-magnitude floods, and the poorly sorted gravel sheets as the deposits of high-magnitude floods. Facies and structures of low-magnitude flows were not recorded, indicating that differential sediment preservation can play an important role in the final character of sandur deposits.

Rigorous three- or two-dimensional analysis of glacifluvial sediments is very time consuming and requires extensive exposures. As a result, many studies have utilized one-dimensional *vertical profile analysis* to record and interpret glacifluvial successions, using Walther's law to infer spatial patterns of deposition from conformable facies associations (section 9.2.2). Repetitive vertical transitions between facies are regularly used to classify depositional styles (e.g. Zielinski and van Loon, 2003) and can be identified quantitatively by *Markov chain analysis*, which determines the probability that particular sequences will occur (e.g. Dawson and Bryant, 1987).

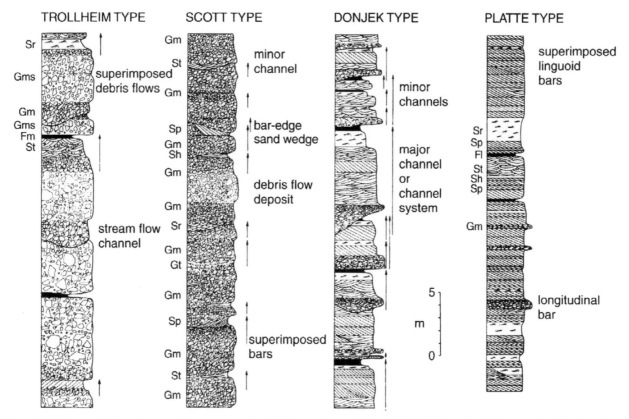

Figure 11.103 Four vertical facies type associations representative of different sandur environments (Miall, 1978, reproduced with permission of the Canadian Society of Petroleum Geologists).

Sequences with high probability of occurrence are interpreted as recurrent associations representing common depositional cycles. This approach has its limitations, as it ignores the type of bounding surface between units and therefore tends to group together facies that are not genetically related. Furthermore, it obscures the true geometrical relations between facies by a statistical averaging procedure. Nevertheless, it allows the identification of repetitive facies sequences which record common depositional events such as major floods, the migration of bars, and the infilling and abandonment of channels.

The range of variation in braided river deposits was illustrated by Miall (1977, 1978), who defined several type examples of braided river successions, based on published descriptions of well-known fluvial sequences (e.g. Rust, 1978; Boothroyd and Ashley, 1975). The type examples, or facies models, which are most representative of glaciofluvial sequences are named the Trollheim type, Scott type, Donjek type and Platte type assemblages, after the North American rivers where they were first recognized. Miall (1977, 1978) represented the models using vertical profiles, which can be equated with three-dimensional architectural element diagrams presented by Miall (1985), as shown in figures 11.103 and 11.104. The models do not represent distinct and separate styles of

deposition, but fixed points on a continuum of variation (cf. Zielinski and van Loon, 2002, 2003). Nevertheless, they are useful points of reference which exemplify contrasting styles of sedimentation in a range of sandur settings.

The *Trollheim type* vertical facies succession is dominated by massive, clast-supported gravels and matrix-supported gravels, representing braid bars and debris flow deposits, respectively. This succession is equivalent to stacked architectural elements GB and SG, and is typical of proximal settings where debris supply is large relative to water discharge.

The *Scott type* succession is named after the Scott outwash fan in Alaska studied by Boothroyd and Ashley (1975), and is typical of fluvially dominated proximal sandar. The dominant sediments are stacked massive to cross-bedded gravel units (GB), which record the aggradation and migration of longitudinal bars. Channel scours underlie gravel units but are often difficult to identify due to the similarity of grain sizes between units. Thin lenses of sand, representing deposition in abandoned channels or bar-edge sand wedges, are inter-bedded with the gravel units. Fining-upward gravel to sand units can also be observed and document deposition through the waning stages of high flow.

The *Donjek type* is typical of intermediate reaches of sandar, where average particle sizes are smaller than in proximal environments. Intermediate reaches also tend to exhibit several topographic levels, with a marked differentiation into active channels and interfluves which are only inundated during exceptional flood events. The succession consists of:

(1) massive gravels (GB);
(2) cross-bedded, horizontally bedded, and ripple cross-laminated sands (SB);
(3) drapes of fine-grained sediments (OF) in repeated fining-upward sequences.

These sequences document the aggradation and then abandonment of channels (deposition of gravel followed by sands and then fines), reflecting frequent channel and bar migration. An important component of this vertical sequence is the occurrence of diamicton clasts at the base of the massive gravel facies, interpreted as ice-proximal bar deposits. In contrast to the Scott type which contains more than 90 per cent gravel, the Donjek type contains anywhere between 10 and 90 per cent gravel, and therefore contains a wider range of recognizable bedforms.

The *Platte type* vertical facies succession is regarded as typical of distal reaches of sandar, although it is named after the non-glacial Platte river of Colorado and Nebraska. Runoff is spread between numerous, shallow distributaries, and predominantly sandy bedload is transported in migrating linguoid bars, dunes and other bedforms. The resulting facies association is dominated by sandy bedforms with abundant cross bedding (SB). Minor gravel lenses and overbank fines may also be present.

A fifth sandur model is applicable to the extreme distal reaches of glacially influenced rivers where the sediment source is predominantly wind-blown material (loess) and the resultant deposits are dominated by silts. The model was proposed by Rust (1978), based on studies of the Slims river in the Yukon, which is characterized by low-relief channels and bars. The resulting *Slims type* association consists of massive, laminated and ripple cross-laminated sandy silts deposited in drapes and migrating bedforms.

The *lithology*, *morphology* and *grain-size distribution* of particles in glacifluvial sediments provide important sources of information on sediment provenance and transport. *Clast lithological analysis*, or statistical analysis of the rock types of river gravels or finer particles, has been employed extensively to correlate widely separated terrace fragments or sediment exposures. For example, Rose et al. (1999) have used particle lithology to correlate scattered exposures of Middle Pleistocene gravels and sands in the English Midlands and East Anglia, to reconstruct the former existence of a major west–east flowing drainage system and its modification by subsequent glacial outwash. The appearance of extra-basinal, erratic material in fluvial sediments can be used to determine when glaciers first appeared in a catchment, providing a powerful method of palaeogeographic reconstruction.

On many sandar, particle *morphology* and *grain size* show systematic downstream variations as a result of

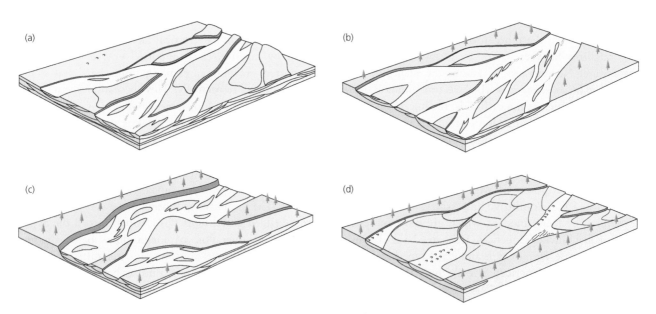

Figure 11.104 Block diagrams showing the braiding patterns associated with the type of fluvial facies successions shown in figure 11.103. (a) Proximal sandur with braid bars and debris flow deposits, equivalent to the Trollheim type vertical facies succession. (b) Fluvially dominated proximal sandur, equivalent to the Scott type succession. (c) Multi-level intermediate sandur, equivalent to the Donjek type succession. (d) Distal sandur, equivalent to the Platte type succession (After Miall, 1985).

selective transport of particles and progressive abrasive wear. For example, Whiteman (1986) demonstrated that glacifluvial gravels in front of two Norwegian glaciers display a downstream reduction in maximum grain size and an increase in particle roundness. Such trends may be blurred by several factors, including local variations in depositional environment, and the introduction of older sediment by bank erosion or tributary streams. Variability in local lithological factors may also influence general downstream changes in clast morphology, as has been demonstrated by Huddart (1994) for some Icelandic proglacial systems.

GLOF-dominated facies associations

Sandar that are periodically inundated by GLOFs or jökulhlaups (section 3.7) exhibit very distinctive facies successions, which can be used to reconstruct the type of flood event and flow characteristics. Maizels (1989a, b, 1993, 1997) compared the sedimentology and morphology of jökulhlaup-dominated and 'normal' sandar in southern Iceland, and further important work has been conducted by Russell and Marren (1999), Russell and Knudsen (2002), Marren (2005) and Russell et al. (2005).

Maizels recognized three types of facies succession on Icelandic sandar, controlled by the runoff and sediment supply regimes of each glacifluvial system. *Type I* or *non-jökulhlaup sandar* occur where runoff is ablation-related and seasonal, and have the characteristics described in the previous section. *Type II* or *limno-glacial jökulhlaup*

sandar develop where catastrophic floods are triggered by ice-dammed lake drainage, such as drainage of the lake Graenalon at the margin of Skeiðarárjökull (cf. Tweed and Russell, 1999). *Type III* or *volcano-glacial jökulhlaup sandar* occur where catastrophic runoff is triggered by subglacial volcanic eruptions, such as the 1918 jökulhlaup on Myrdalssandur following an eruption of Katla below Myrdalsjökull (Tómasson, 1996). Maizels identified several vertical facies associations, which she was able to relate to different flow regimes and sediment supplies. The associations formed by jökulhlaups were labelled Types A–C, with subtypes, and are illustrated in figure 11.105.

Limno-glacial jökulhlaup sandar are dominated, at least in their proximal zones, by coarsening-upward gravel units interpreted as flood surge deposits (Profile C5, fig. 11.105). The gravels are clast-supported and dominantly sub-rounded. These deposits are overlain by a fining-upward sequence of cobble and pebble gravels, then horizontally laminated pumice silts, sands and fine gravel, interpreted as a post-surge waning flow sequence. The whole inverse-to-normally graded succession records rising then falling discharges over a single jökulhlaup event. Rounded till balls have been observed in such deposits, indicating that local tills had been eroded by the flood, but deposition rates had been too rapid for complete disaggregation of the eroded blocks.

Volcano-glacial jökulhlaup sandar are dominated by lithofacies profile types A and B, but profile types C and

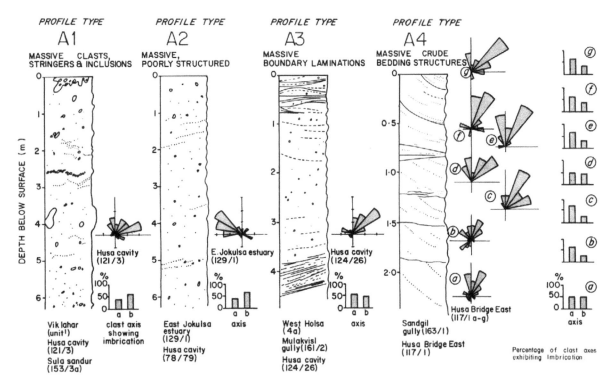

Figure 11.105 Vertical lithofacies profiles (Types A, B and C) from the jökulhlaup sandar of Iceland. Facies codes explained in figure 10.1 (Maizels, 1993, reproduced with permission of Elsevier).

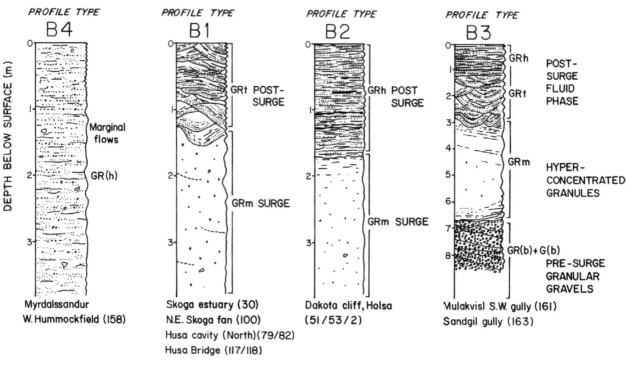

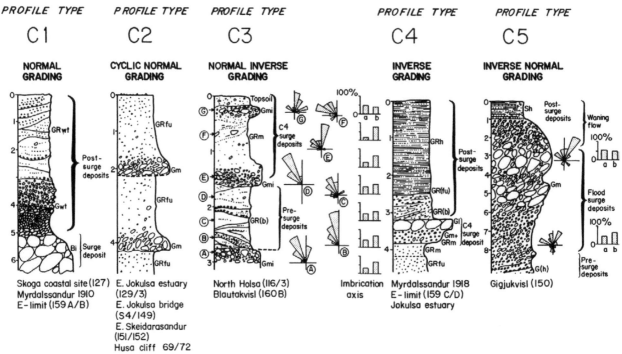

Figure 11.105 (Continued)

other facies associations may occur as minor components on a localized scale (fig. 11.105). *Type A* profiles consist of massive to crudely bedded granules of tephra and pumice, deposited by hyperconcentrated flows. The deposits may include clasts, stringers or inclusions of other materials that were easily incorporated into the flow. *Type B* profiles, especially B3, are the most widespread facies association on volcano-glacial jökulhlaup

sandar. They comprise massive granule gravels, capped and sometimes underlain by horizontally bedded or trough cross-bedded granule gravels. Profile Type B3 is characterized by four distinct units. Unit 1 consists of coarse-grained, crudely bedded, clast-supported gravels or pumice granules, interpreted as pre-surge sediments derived from pre-existing sandur deposits. Unit 2 is composed of massive to crudely bedded pumice granules,

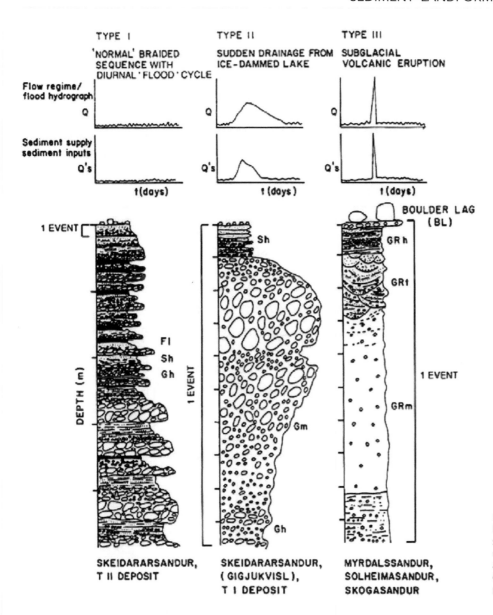

Figure 11.106 Representative vertical lithofacies profiles for Icelandic sandar, subject to different runoff regimes, with associated hydrographs and sediment supply characteristics (Maizels, 1993, reproduced with permission of Elsevier).

interpreted as the products of hyperconcentrated flows associated with the main flood surge. Unit 3 comprises trough cross-bedded pumice granules which are separated from unit 2 by an erosional contact, and unit 4 comprises horizontally bedded pumice granules and sands. Units 3 and 4 are interpreted as the products of post-surge conditions during which deposition was by increasingly shallow fluid flows.

The variability in vertical facies sequences from non-jökulhlaup and jökulhlaup sandar are summarized in figure 11.106, which relates the different runoff and sediment supply regimes of the three sandur types (I–III) to characteristic vertical profile models. The large difference in thickness of sedimentary units related to one discharge event between jökulhlaup and non-jökulhlaup sandar is very striking. The distinctive inverse-to-normal grading of the limno-glacial jökulhlaup sandur profile reflects the rising and falling flood stage. In contrast, the volcano-glacial jökulhlaup sandur profile has relatively constant

grain size distribution throughout, reflecting the predominance of easily disaggregated volcanic eruption products. As a result, rising and falling flood stages and changing sediment transport mechanisms are represented by differences in sediment structure.

Figure 11.107 shows how facies associations deposited by jökulhlaups vary systematically with flow type and grain size distribution of the available sediment. A continuum of flow types is shown along the horizontal axis, ranging from high-concentration, cohesive debris flows to low-concentration, turbulent fluid flows. Sediment grain size distribution is represented on the vertical axis, which shows the relative abundance of a coarse gravel mode (c) and a finer-grained pumice fragment mode (f). This continuum diagram simply demonstrates that deposits will increase in coarseness as the source materials get coarser, so that the granular sediments of volcano-glacial sandar plot towards the top of the diagram, and the gravelly sediments of the limno-glacial and non-jökulhlaup sandar

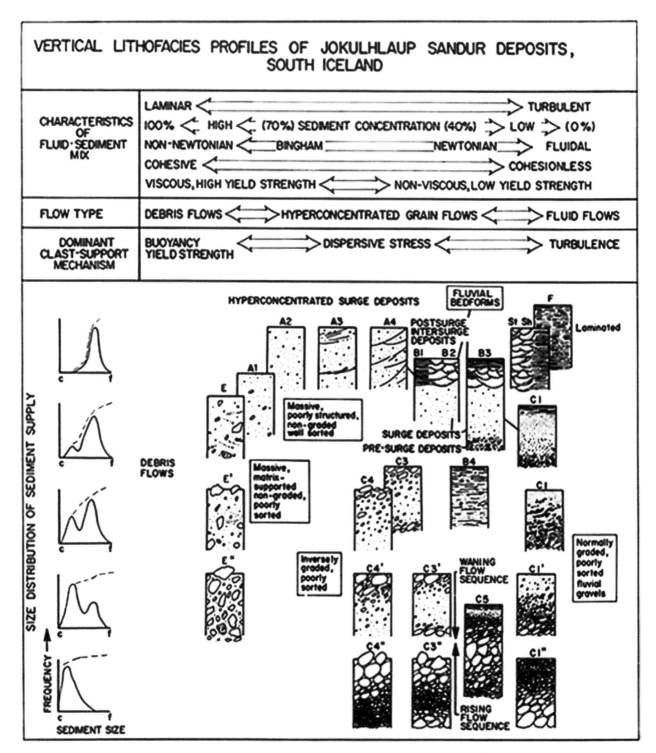

Figure 11.107 Typical vertical lithofacies profiles of Icelandic sandar deposits classified according to the characteristics of the fluid-sediment mix and relative availability of sediment (horizontal axis) and size distribution of the source sediments (vertical axis). On the sediment size curves, c = coarse-grained and f = fine-grained materials (Maizels, 1993, reproduced with permission of Elsevier).

plot towards the base. Furthermore, the profiles display increasing organization towards the right of the diagram, with decreasing sediment concentration. At lower sediment concentrations, particles are more able to move relative to one another, allowing the development of inverse or normal grading and other sedimentary structures.

Sediments deposited in association with jökulhlaup spillways on the North American Great Plains have been studied by Lord and Kehew (1987; see also section 8.4.4).

The outburst floods evacuated through the spillways were highly erosive, so deposits are rarely preserved. However, sediment eroded during spillway incision was deposited in wider reaches of channels in the form of *gravel bars*. Unlike bars in normal proglacial outwash, the spillway bars are composed of homogeneous masses of massive, matrix-supported and very poorly sorted pebbly cobble gravels. Boulders up to 3 m in diameter may also occur in the bar sediments. Lord and Kehew estimated that the outburst flows were *hyperconcentrated*, with sediment-water concentrations being as high as 40 per cent by weight. The conclusion that the flows were hyperconcentrated is similar to the findings of Maizels (1989a, b, 1993) in her studies of the deposits of the 1918 jökulhlaup from Katla. However, recent reconstructions of the Katla flood indicate that overall sediment concentrations were low, and that flow was fluidal rather than hyperconcentrated (Tómasson, 1996). One possible reason for this anomaly is that the Katla flood may have comprised an upper fluidal flow and a lower hyperconcentrated traction carpet (cf. Todd, 1989), and that the deposits only represent the lower flow component.

Palaeodischarges and changing discharge regimes

Several researchers have attempted to reconstruct palaeodischarges and velocities from sediment sequences (e.g. Church, 1978; Maizels, 1983; Dawson and Gardiner, 1987; Zielinski and van Loon, 2003). Such reconstructions are important, because it is of great interest to understand how long-term hydrological changes in a catchment influence landscape development and sediment yields.

Palaeohydrological methods employ empirical relationships between flow velocities and channel-form variables derived for modern rivers. By measuring variables that can be observed in ancient deposits, such as channel morphology and slope, sediment grain size and bed roughness, the methods then work backwards to derive likely values of velocities when the palaeochannel was bankfull. Discharges are then calculated by multiplying velocity by channel cross-section area. There are several potential sources of error in this approach, including the following (Dawson and Gardiner, 1987):

(1) Palaeochannel width and depth may be difficult to reconstruct from exposures, for several reasons. First, the upper part of the channel may have been truncated by subsequent erosion. Second, the available exposure may not provide a true transverse section, distorting the cross profile. Third, it may be difficult to identify contemporaneous surfaces on opposite channel banks due to overprinting during channel migration and infilling.

(2) Channel slope can be hard to measure or estimate.

(3) Bed friction is a very important factor, but is difficult to parameterize from sedimentary data.

(4) Suspended sediment concentration exerts a strong influence on flow viscosity and velocity, but is unknown for ancient rivers.

(5) For braided river systems, total discharge is distributed through several channels, but it may be impossible to determine from sedimentological evidence just how many channels were occupied simultaneously. One way of avoiding this difficulty is to identify the main channel and assume that it carried a fixed proportion of the total discharge. Mosley (1983) found that in the braided Rakaia river, New Zealand, the main channel carried between 47 and 93 per cent of the flow, with an average of 71 per cent. The very large range of these figures underlines the uncertainty of the method.

According to Dawson and Gardiner (1987), the errors introduced by these factors can be considerable – up to three orders of magnitude, with the major contribution being provided by uncertainties about channel form. Nevertheless, such methods do enable researchers to place bounds on the likely behaviour of former rivers during periods of sandur aggradation.

In order to put historical sandur development into the context of changing discharge regimes, Marren (2005) presented a summary of lithofacies sequencing and terrace aggradation/incision in relation to flood magnitude and frequency. This conceptual model relating sandur incision and aggradation to glacier activity (fig. 11.108) can be compared to Bridgland's (2000) model of terrace development on glacial/interglacial timescales. The influence of magnitude and frequency on sediment–landform associations is depicted for the three different scenarios of aggradation, incision and equilibrium, each scenario dictated by glacier advance, recession and stillstand respectively. Clearly, fluctuating glaciological conditions and magnitude–frequency regimes would result in more complex stratigraphies. During glacier advance, aggradation will take place in one of four styles according to the dominant magnitude–frequency regime: (A) where low-magnitude/high-frequency sedimentation produces thin, stacked bar core units that coarsen upwards; (B) where high-magnitude/high-frequency sedimentation in frequent floods partly reworks existing sediment, and the large-scale architecture reflects flood-scale bars and related features; (C) where rare cases of low-magnitude/low-frequency deposition may occur – for example, in surge fans; (D) where high-magnitude/low-frequency sedimentation produces jökulhlaup-type flood deposits.

During glacier stillstands, equilibrium conditions will prevail and result in three depositional styles: (E) where low-magnitude/high-frequency sedimentation results in a greater amount of sediment reworking, and channel and confluence sedimentation is more significant; (F) where high-magnitude/high-frequency sedimentation in repeated

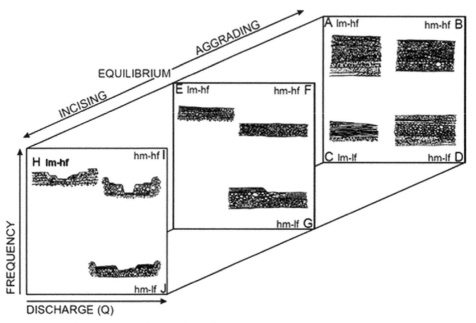

Figure 11.108 Schematic diagram to illustrate the role of magnitude and frequency in the development of sandur architecture. The magnitude and frequency regimes are illustrated for three depositional scenarios of aggradation, incision and equilibrium, which are in turn controlled by glacier advance, retreat and stillstand respectively. A–D = likely scenarios under advancing conditions; E–G = likely scenarios under equilibrium conditions; H–J = likely scenarios under retreating conditions (Marren, 2005).

floods introduces new sediment, but aggradation is more limited. Bar accretion is a more important process, and reactivation surfaces in gravel sediments reflect the multiple flood occupation of bar forms; (G) where high-magnitude/low-frequency sedimentation produces distinct topographic distinctions between flood sediments and intervening low-magnitude sediments, so that low-magnitude deposits are inset within flood terraces.

During glacier recession, incision will take place in one of three depositional styles: (H) where low-magnitude/high-frequency sedimentation results in channel and confluence deposits; bar core sediments occur as inset surfaces, fining inwards towards the most recently active; (I) where high-magnitude/high-frequency floods tend to erode rather than deposit and any depositional units consist of high-magnitude flood sediments, and there is little inter-flood reworking; (J) where high-magnitude/low-frequency flood deposits dominate terrace sequences, but low-magnitude deposits are better developed than in style (I), and there is a greater potential for inter-flood reworking.

11.5.2 PARAGLACIAL ASSOCIATIONS

Paraglacial activity, or the rapid readjustment of glaciated landscapes to non-glacial conditions following deglaciation, results in distinctive sediment–landform associations on slopes and valley floors. This subject has been comprehensively reviewed by Ballantyne (2002a, b, 2003), who considered a wide range of geomorphic systems, ranging from mountain slopes to barrier coasts. Here we consider paraglacial signatures only in directly

glaciated parts of catchments, rather than in lacustrine and marine environments.

The paraglacial period

As glaciers retreat from an area, newly deglaciated terrain is commonly subject to rapid change, as fluvial, slope and aeolian systems relax towards non-glacial states. The term *paraglacial* was defined by Church and Ryder (1972) to encompass 'nonglacial processes that are directly conditioned by glaciation', characteristic of recently deglaciated environments. These processes, however, are not unique to such environments, and it is perhaps more useful to use the term 'paraglacial' to refer to the *period* of rapid environmental readjustment following glacier retreat (Church and Ryder, 1972; Eyles and Kocsis, 1988; Ballantyne, 2002b). Paraglacial activity is distinct from *periglacial* processes, which are characteristic of all cold, non-glacial environments, regardless of whether glacier ice is or was present in the catchment (Ballantyne and Harris, 1994).

The *paraglacial period* is characterized by high rates of sediment delivery from slopes and into fluvial and aeolian systems. This period of rapid response is triggered by the instability of unconsolidated glacigenic sediments (e.g. in lateral moraines and kame terraces) and oversteepened rockslopes once their support of glacier ice is removed. Sediment yields and denudation rates are highest immediately following deglaciation, then decline through time as sediment supply becomes exhausted and slopes relax towards more stable profiles (Ballantyne and Benn, 1994b, 1996). The paraglacial period theoretically ends once sediment yields drop to rates typical of unglaciated

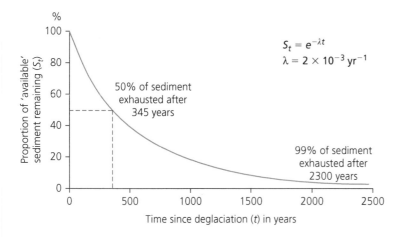

Figure 11.109 The paraglacial exhaustion model of Ballantyne (2002a). The rate of decline in sediment release is related to the proportion of available glacigenic sediment in a deglaciated catchment. The example chosen shows that 50 per cent of the initial available sediment is exhausted after 345 years and 99 per cent has been removed after 2300 years, thereby defining the paraglacial period for the basin.

catchments, although whether or not a landscape really fully adjusts following a glacial phase is difficult to ascertain. Certainly, delayed slope responses can occur many thousands of years after deglaciation (Ballantyne and Benn, 1996). Because glacigenic sediment supplies are not replenished during the paraglacial period, the rate of sediment reworking can be expected to diminish with time. Ballantyne (2002b) argued that sediment yields can be approximated using an exhaustion model, in which the amount of sediment remaining in the catchment follows a negative exponential function (fig. 11.109):

$$S_t = S_0 \, e^{-\lambda t}$$

where t is time elapsed since deglaciation, S_t is the proportion of available sediment remaining for reworking at time t, S_0 is the total available sediment at $t = 0$, and λ is the rate of change in the loss of available sediment by release and/or stabilization. Clearly, this function can only give a general picture of sediment yields through time, as it does not account for many important processes, such as climate change and colonization by vegetation.

The concept of the paraglacial period was developed from sediment yields reconstructed from valley-floor fans in North America, many of which were formed in the period following deglaciation in the Late Pleistocene or early Holocene (e.g. Church and Ryder, 1972; Beaudoin and King, 1994). In upland catchments close to sediment sources, sediment yields can be expected to peak as deglaciation commences, but in more distal reaches, sediment throughputs should peak later, reflecting a time lag as the slug of sediment travels through the system. Thus in downstream reaches of large catchments, sediment loads will not begin to increase until after peak yields in proximal reaches (Church and Slaymaker, 1989; Harbor and Warburton, 1993; Ballantyne, 2002b; fig. 11.110). Superimposed on the gradual declining curves of sediment yield are periods of either *rejuvenated* or *renewed paraglacial response* (Ballantyne, 2003; fig. 11.111). These can be triggered by climate change, climate extremes, neotectonics or human interference (e.g. Curry, 2000a, b).

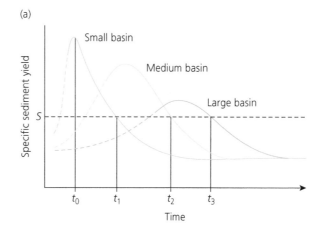

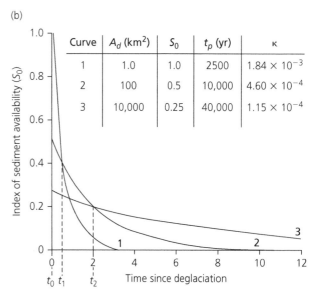

Figure 11.110 The (a) Harbor and Warburton (1993) and (b) Ballantyne (2002a) models for sediment yields in glaciated terrains. The Harbor and Warburton model shows that sediment yield at particular times (t_1, t_2, t_3) since deglaciation (t_0) is strongly conditioned by basin size, but basins of different size can have similar sediment yields (S) at different times after deglaciation. Ballantyne's model shows the impact of sediment exhaustion on basins of different size, where A_d = drainage area, S_0 = index of sediment availability at time of deglaciation, t_p = duration of paraglacial period, which terminates when 99 per cent of available sediment is exhausted, and κ = the rate of change of sediment reworking.

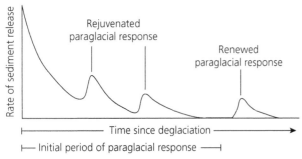

Figure 11.111 Diagram to convey the concepts of rejuvenated and renewed paraglacial responses on overall decline in sediment yield during the paraglacial period (Ballantyne, 2003).

Dating control on paraglacial sediments is often poor, and rates of sedimentation and sediment yield are therefore subject to large errors. Ballantyne and Benn (1994b, 1996), Ballantyne (1995) and Curry (1999, 2000a, b) have studied landform development and sediment yields in Norwegian valleys which have been deglaciated within the last two centuries, and demonstrated that very high rates of readjustment occurred immediately following the withdrawal of glacier ice. By comparing modern slope profiles with photographs taken earlier this century, they calculated minimum average rates of surface lowering of some debris slopes of 50–100 mm yr^{-1}, with rates as high as 200 mm yr^{-1} in some locations. Concomitant with slope erosion was the build-up of debris cones on valley floors. By bracketing the period of fan growth by the age of trees growing on their surfaces and the age of moraines that they overlie, Ballantyne (1995) derived aggradation rates of 8–29 mm yr^{-1}, equivalent to average erosion rates of 37–93 mm yr^{-1} in their catchment areas. In contrast with these high rates of sediment transfer, the cones studied by Ballantyne are now largely inactive, only 100–200 years after the withdrawal of glacier ice. In the Brecon Beacons of South Wales, Curry and Morris (2004) calculated that around 7 m (84 per cent) of overall rockwall retreat took place during the Lateglacial, immediately after deglaciation, in contrast with only 1.4 m (16 per cent) during the Holocene. This converts to mean retreat rates of 1.23 m ka^{-1} and 0.12 m ka^{-1} for the Lateglacial and Holocene periods respectively, and indicates that approximately half the talus had accumulated within 1 ka of deglaciation due to paraglacial rock mass instability.

Meigs et al. (2006) have reported rapid paraglacial responses by the tributary fluvial systems of Taan Fjord, Alaska, due to glacier recession up the fjord. In the last two decades, glacier recession has produced a base level fall of around 400 m, to which the tributary streams have evacuated valley fills and incised slot gorges into bedrock. This has resulted in the progradation of new fan deltas in the fjord; approximately 0.08 km^3 of the total 0.59 km^3 of stored alluvial and colluvial sediment in one of the tributary valleys has already been transferred to the fjord. Direct measurements of declining sediment yields on

Figure 11.112 A paraglacial fan derived from glacial sediments in the Lahul Himalaya (D.J.A. Evans).

recently deglaciated surfaces have been reported by Orwin and Smart (2004) from Small River Glacier, Canada. Suspended sediment data from surfaces of different ages revealed various responses to rainfall events of different magnitudes. Surfaces exposed since 1977 were increasingly vulnerable to mobilization during all rainfall events, but older surfaces exposed since 1910 yielded sediment for only those events with 5- and 30-year return periods. This indicates that intensity and duration of rainfall events has to increase in magnitude to mobilize sediment on surfaces of increasing maturity.

These data are important because they demonstrate that, in these locations, the most active period of paraglacial resedimentation occurs within decades of deglaciation, and that the transition to lower sediment yields occurs within one or two centuries. It has yet to be established whether these timescales also apply to other recently deglaciated environments. In high mountain environments such as the Himalaya, where catchments are larger and greater quantities of unconsolidated sediments are available (fig. 11.112), the period of rapid paraglacial readjustment appears to be as much as several centuries or more, but finer temporal resolution is gradually being achieved by systematic dating programmes on paraglacial processes and forms (e.g. Owen and Sharma, 1998; Watanabe et al., 1998; Barnard et al., 2004a, b, 2006a, b).

Slope deposits, fans and terraces

Catastrophic rockslope failures (RSFs) are common in recently deglaciated terrain, reflecting glacial over-steepening and unloading (e.g. Evans and Clague, 1994;

Fort, 2000; McSaveney, 2002; Holm et al., 2004; Geertsema et al., 2006; Hewitt, 2006; Cossart et al., 2008; Hewitt et al., 2008). RSFs play an important role in the long-term erosional evolution of glaciated mountain landscapes (Shroder and Bishop, 1998; Jarman, 2002, 2003, 2006; Hewitt, 2006; Turnbull and Davies, 2006; Korup et al., 2007). The collapse of mountain walls onto receding glaciers also inputs large volumes of supraglacial debris (section 9.3.1). Numerous RSFs have been reported from ancient deglaciated terrains. For example, Ballantyne (1997a), Shakesby and Matthews (1996) and Jarman (2003, 2006, 2007) have mapped numerous RSFs in the relatively stable mountainous terrains of the UK. Where these RSFs have been *cataclasmic* or broken up during transport, they are commonly represented by extensive boulder fields lying in valley bottoms and at the base of bedrock cliffs. Such deposits may have well-defined lobate fronts and arcuate transverse ridges on their surfaces, reflecting flow conditions just prior to final deposition. This appearance has led to the misidentification of many RSFs as potential rock glaciers (cf. Ballantyne and Stone, 2004) and even moraine sequences (e.g. Yi et al., 2006). The volume of some RSFs, especially where they have not undergone significant downslope disaggregation (*sub-cataclasmic*), renders them uncannily similar to moraines (e.g. Hewitt, 1999). Where RSF has not led to significant downslope transport, they are referred to as *arrested* and are recognizable as large dislocated blocks of bedrock lying short distances downslope from their failure headscarp. Slower responses by rockslopes to glacial unloading result in rockslope deformation or rock mass creep, a process that has been documented on deglaciated slopes above Affliction Glacier in British Columbia by Bovis (1990) and Bovis and Stewart (1998). Typical features include tension cracks, grabens or collapse pits and antiscarps at the tops of the failing slopes (Jarman and Ballantyne, 2002). Lower down, parts of the slopes can also appear to bulge in response to the gradual creep of the rock mass.

Talus deposits can accumulate rapidly in recently deglaciated terrain by rockfall, snow avalanching and debris flow (e.g. Augustinus, 1995; Andre, 1997; Hinchcliffe and Ballantyne, 1999; Cossart et al., 2008). Internally, talus deposits may consist of openwork gravels or diamictons, and commonly exhibit crude slope-parallel bedding. In permafrost environments, the bases of talus slopes may develop into protalus rock glaciers or lobes (Ballantyne, 2002a; Harrison et al., 2008).

Glacigenic sediments exposed on steep slopes by glacier retreat are subject to particularly rapid paraglacial reworking (fig. 11.113). The resulting *paraglacial slope deposits* form wedge- or cone-shaped slope-foot accumulations, the detailed morphology and sedimentology of which reflect the character of the source sediment and the processes of reworking and deposition (fig. 11.114; Curry

Figure 11.113 Rapidly degrading inner slopes of lateral moraines in the deglaciated valley of the Robertson Glacier, Kananaskis, Canadian Rockies (D.J.A. Evans).

and Ballantyne, 1999; Curry 1999, 2000a, b). Such accumulations can be distinguished from non-glacial slope deposits by the presence of erratics, and from in situ tills by slope-parallel bedding and downslope-oriented clast fabrics. Examples from the Karakoram Mountains have been described by Owen (1991) and from Fåbergstolsdalen, Norway, by Ballantyne and Benn (1994b). In higher-relief terrains, paraglacial fans are themselves subject to rapid reworking (Barnard et al., 2004a, b, 2006a, b).

Once glaciers have retreated from a drainage basin in part or whole, unstable sediments are made available to rivers, resulting in the rapid aggradation of thick valley fills and alluvial fans at the mouths of tributary valleys or gullies (fig. 11.115; Ballantyne, 2002a, 2003). Detailed descriptions of early Holocene fans and terraced valley fills in the mountainous terrain of British Columbia have been provided by Church and Ryder (1972), Eyles and Kocsis (1988) and others. The valley fills are deeply incised and modern river channels may lie over 200 m below the fan surfaces, reflecting reduced sediment supplies at the end of the paraglacial period. The occurrence of Mazama Ash, dating to 6.6 ka BP, at shallow depths on fans throughout south-central British Columbia indicates that most fan deposition took place during the paraglacial period of 10–6 ka BP. Rivers now largely transport only the material that is being made available by current denudation processes and so are in equilibrium with their normal weathering environment.

Aeolian reworking

The lack of stabilizing vegetation, the action of persistent glacier winds and large quantities of easily erodible sediment in proglacial areas make them prime sites for aeolian erosion, transportation and deposition (Seppälä, 2004). In addition, the fluctuating discharges of proglacial streams lead to intermittent drying and wind deflation of finer-grained alluvium. Measurements of

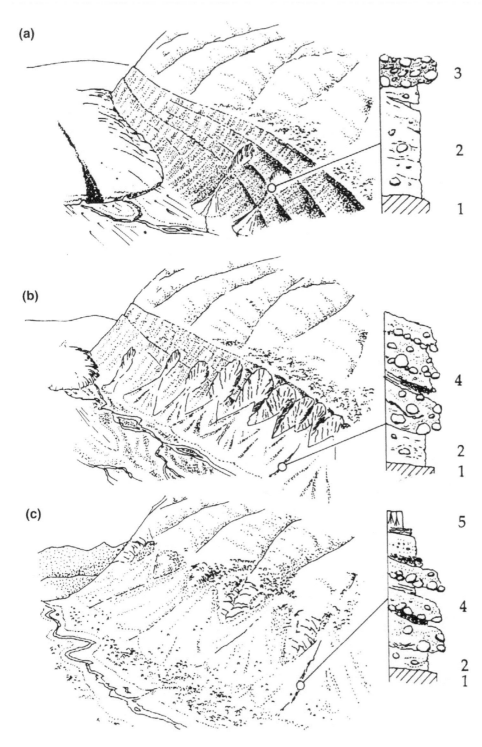

(a)

(b)

(c)

Figure 11.114 Schematic block diagrams of the landforms and sediments associated with three stages in the paraglacial reworking of steep drift slopes in glaciated valleys. (a) Initial slopes exposed by glacier recession, showing lateral moraines and the onset of gully incision. (b) Advanced gully development and deposition of coalescing debris fans downslope. (c) Exposed bedrock and stabilized, vegetated gullies and largely relict debris fans. By this stage, paraglacial reworking and slope adjustment have effectively ceased, owing to diminution of debris supply. Facies key: 1 = bedrock; 2 = subaerial sediments relating to an earlier episode of paraglacial sedimentation; 3 = ice-marginal deposits; 4 = paraglacially reworked sediment (debris flows and intercalated slopewash deposits); 5 = soil horizons (After Ballantyne and Benn, 1996).

summer katabatic winds at the margin of the Greenland Ice Sheet by van den Broeke et al. (1994) persistently exceed the threshold velocity for dry sediment, and average wind speeds for the marginal areas of the Antarctic Ice Sheet are more than double the threshold velocity for moist sediment (Parish and Bromwich, 1989); it is therefore important not to underestimate the impact of aeolian processes in ice marginal environments, with respect to both science and comfort!

In many environments, vegetation colonization tends to progressively stabilize deglaciated terrain, and the zone of maximum aeolian activity is generally located very close to the ice margin. In an investigation into the textural changes of Icelandic till through time, Boulton and Dent (1974) demonstrated that silts and clays in freshly exposed subglacial tills are commonly transported by wind onto surfaces deglaciated more than 20 years previously. This removal of fine-grained materials from the

Figure 11.115 Paraglacial incision of glacial deposits draping the lower slopes of the Fraser Valley, British Columbia, Canada. The Fraser river incised the valley fill during the Early Holocene, after the end of the last glaciation (Aerial photograph BC 1087-46, Province of British Columbia).

tills leaves behind stone pavements, which may contain wind-faceted clasts (*ventifacts*), due to the sand-blasting effect of the strong localized winds (Knight, 2008). These features are common in Antarctica (e.g. French and Guglielmin, 1999; Sugden et al., 1999) and are similar to the *desert pavements* produced in warm arid environments which protect underlying fine-grained materials from continued deflation.

Ice-proximal to distal zonation of aeolian processes is reflected in the wind-blown sediments and forms of proglacial areas. The waning strength of glacier winds with distance from the ice margin is reflected in the grain size distributions of the aeolian deposits, which grade from medium to fine sands (*coversand*) into silts and clays (*loess*) with distance (fig. 11.116). On a regional scale, the extensive and often thick accumulations of loess and coversand are invaluable archives of climatic and environmental changes during the Quaternary period. Such material in mid-latitude North America and Europe travelled large distances from freshly deglaciated and periglacial terrain during phases of glaciation.

Depositional patterns in such catchments provide important information on glacio-aeolian processes. For example, Roberts and Cunningham (1992) documented wedges of loess draping the lower slopes of the South Thompson river valley in British Columbia, Canada, which they interpreted as the product of two components of wind flow (fig. 11.117). The dominant downvalley (mountain and katabatic) winds and tangential secondary updraughts (slope winds) combine to carry silt from the thick glacilacustrine sediments deposited during deglaciation up the valley wall. The resulting wedge of loess feathers out upslope. The interaction of aeolian processes and

landforms with glacial and proglacial processes is summarized in figure 11.116, which illustrates the numerous complex combinations of transportational and depositional histories that are involved in the glacio-aeolian system.

11.6 GLACILACUSTRINE AND GLACIMARINE ASSOCIATIONS

Glacilacustrine and glacimarine depositional environments are very diverse, ranging from the margins of small ice-marginal lakes in glaciated valleys to the floors of the deep oceans. Sediment facies deposited in such subaqueous settings have been reviewed in chapter 10. In this section, we describe the landform–sediment associations typically found in glacigenic subaqueous depositional environments, including: (a) grounding-line fans; (b) subaqueous moraines, including morainal banks, coalescent subaqueous fans and De Geer moraines; (c) grounding-line wedges; (d) ice shelf moraines; (e) deltas; (f) submarine channel fills; and (g) lake floors. Larger-scale patterns of subaqueous sedimentation are discussed in section 12.5.2.

11.6.1 GROUNDING-LINE FANS

Grounding-line fans (also known as *subaqueous outwash fans*) are fan-shaped deposits that build up at efflux points where subglacial meltwater exits from ice-walled tunnels directly into deep water, and sediment load is deposited rapidly as a result of the sudden drop in stream velocity (Rust and Romanelli, 1975; Cheel and Rust, 1982; Powell, 1990; Powell and Domack, 1995; Powell and Alley, 1997; fig. 11.118). Grounding-line fans differ from deltas (section 11.6.5) in that the coarse sediment enters at the base of the water column instead of at the top, although grounding-line fans can build up to the water surface to form deltas if a glacier remains quasi-stationary for a period of time and sedimentation rates are high enough (Powell, 1990; Nemec et al., 1999; Thomas and Chiverrell, 2006). Because of their association with subglacial tunnels, grounding-line fans are often found at points along esker systems where they mark former glacier margin positions; in such cases, they may be referred to as *esker beads* (Bannerjee and McDonald, 1975). Subaqueous fans formed in glacimarine and glacilacustrine environments may be very similar, and may be indistinguishable from internal characteristics alone. However, in glacimarine environments, the buoyancy of inflowing waters tends to produce plumes which carry fine-grained sediments away from the fan, unless sediment concentrations are very high (Powell, 1990). The removal of fine-grained sediments in plumes means

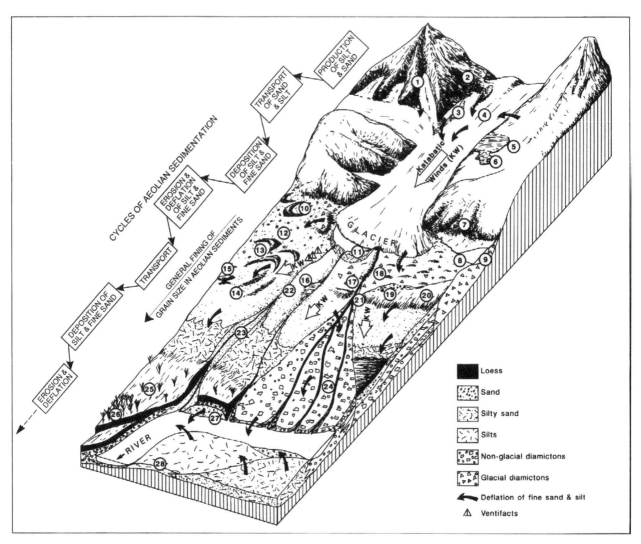

Figure 11.116 The process and sediment-landform associations in glacio-aeolian environments. (1) Silt and sand formation by weathering on high, steep valley slopes. (2) Silts and sands produced and supplied to the glacier by rockfall and other mass movement processes. (3) Silts deposited in a proglacial lake. (4) Silts, sands and other rock debris falling into crevasses. (5) Silts and sands washed into a small ice-marginal lake. (6) Terraces comprising lacustrine silts and sands produced when lake dries up or drains. (7) Fine sediments produced by overland flow and deposited at the base of the slope. (8) Meltwater stream feeding a proglacial lake. (9) Proglacial lake into which lacustrine silts and sands are deposited. (10) Parabolic dune. (11) Ice-contact lake. (12) and (13) Barchan dunes. (14) Longitudinal dunes. (15) Deflation hollow. (16) Deflation of lacustrine silts and sands in a dried-up proglacial lake. (17) Rock-strewn surface. (18) Hummocky moraine. (19) Aeolian sand infilling depressions within till ridges. (20) End moraine. (21) Meltwater stream dissecting end moraine. (22) Coversands. (23) Flood-plain sands and gravels. (24) Alluvial fan. (25) River terraces capped by loess. (26) Vegetated surface with formation of palaeosols. (27) Fines being deflated from flood-plain sediments. (28) Loess hills (Derbyshire and Owen, 1996, reproduced with permission of Butterworth-Heinemann).

that glacimarine fans tend to be predominantly coarse-grained, with a general lack of underflow deposits (turbidites).

The positioning, general shapes and sedimentary characteristics of grounding-line fans vary according to changes in tunnel jet discharges and plume production. Coarse bedload is deposited close to the tunnel mouth as the efflux jet decelerates, building up steep, unstable slopes at the glacier margin. Slope failures and renewed sediment discharge from the tunnel mouth feed mass flows that transport sediment radially away from the margin. Flows are likely to be cohesionless, due to the removal of fine-grained matrix material in suspension. As flows travel over the fan surface, turbulent mixing with

the overlying water results in their progressive dilution, and they may evolve into turbidity currents.

Powell (1990, 2003) has considered the effects of varying water and sediment discharge on fan evolution in glacimarine environments (fig. 11.119). For low discharges, efflux jets have low momentum and so rise immediately to the surface. Deposition on the fan is dominated by mass flows fed by bedload dumped at the conduit mouth, supplemented by rain-out from the plume. At moderate water discharges and high sediment discharges, efflux jets can travel across the fan surface before rising, resulting in a zone of traction deposition at the fan apex, flanked by a zone of mass flow activity. At high discharges, jet momentum is high enough to

Figure 11.117 A reconstruction of the environmental setting for loess deposition in the Early Holocene (late paraglacial reworking of valley fills), based on the South Thompson river valley, British Columbia, Canada. (1) Upper limit of loess deposition. (2) Mountain and katabatic winds. (3) Alluvial fans. (4) Loess cap. (5) Glacilacustrine silts. (6) Debris flow deposits. (7) Slope winds. (8) Floodplain (Roberts and Cunningham, 1992, reproduced with permission of Wiley).

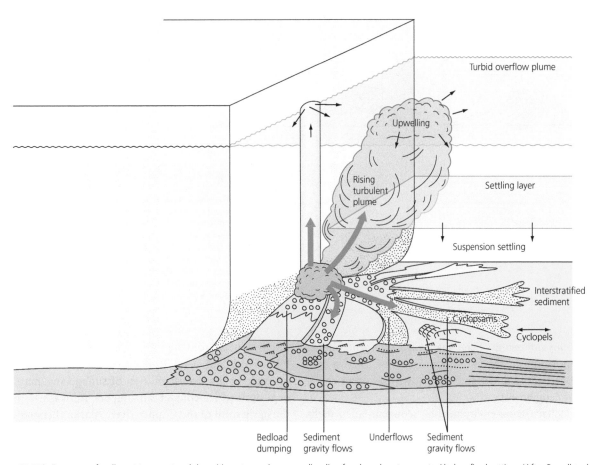

Figure 11.118 Processes of sediment transport and deposition at a marine grounding-line fan, based on temperate Alaskan fjord settings (After Powell and Domack, 1995; Powell, 2003).

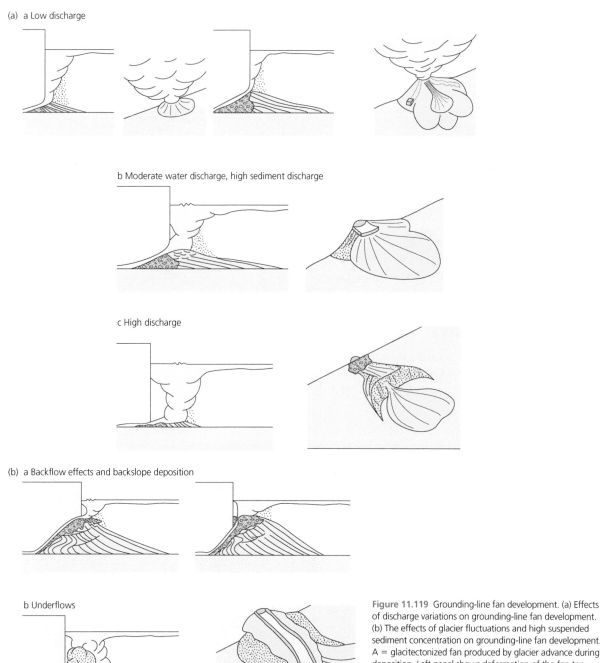

Figure 11.119 Grounding-line fan development. (a) Effects of discharge variations on grounding-line fan development. (b) The effects of glacier fluctuations and high suspended sediment concentration on grounding-line fan development. A = glacitectonized fan produced by glacier advance during deposition. Left panel shows deformation of the fan top, with horizontal jet flow causing a backflow near the fan apex. Right panel shows glacier recession and jet discharge onto the fan backslope, which fills with coarse sediment. B = underflows produced when high suspended sediment load falls back onto the fan surface (Modified from Powell, 1990).

delay buoyant rising of the suspended sediment plume, and deposition on the fan is dominated by tractive currents. Barchanoid dunes and mass flows may occur at the point of detachment.

In glacilacustrine environments, sediment-laden meltwater is generally denser than the surrounding lake water, and will tend to hug the lake floor as underflows (section 10.6.1). Deposition on grounding-line fans is therefore likely to be dominated by mass flows, with comparatively

minor inputs from high-level suspended sediment (e.g. Plink-Björkland and Ronnert, 1999). A Quaternary example of the inter-digitation of subaqueous mass flow deposits and ice-contact subaqueous outwash is that of the basin infill of the Copper river, Alaska (Bennett et al., 2002). The complex architecture of the sedimentary sequences (fig. 11.120) was explained by Bennett et al. (2002) as the product of basinward transport of subaqueous fans and inter-fan diamictic sediments (debris flows),

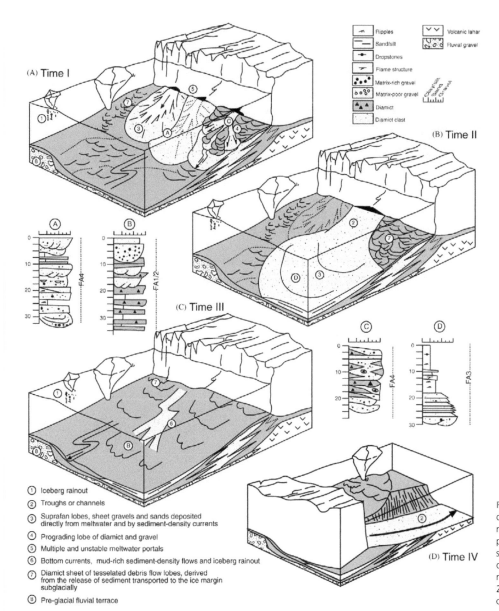

(A) Time I

(B) Time II

(C) Time III

(D) Time IV

① Iceberg rainout
② Troughs or channels
③ Suprafan lobes, sheet gravels and sands deposited directly from meltwater and by sediment-density currents
④ Prograding lobe of diamict and gravel
⑤ Multiple and unstable meltwater portals
⑥ Bottom currents, mud-rich sediment-density flows and iceberg rainout
⑦ Diamict sheet of tesselated debris flow lobes, derived from the release of sediment transported to the ice margin subglacially
⑧ Pre-glacial fluvial terrace

Ripples
Sand/silt
Dropstones
Flame structure
Matrix-rich gravel
Matrix-poor gravel
Diamict
Diamict clast
Volcanic lahar
Fluvial gravel

Figure 11.120 Reconstructed depositional environment and resultant facies sequences A–F, produced by the deposition of subaqueous fans and debris flows during deglaciation of the Copper river basin, Alaska (Bennett et al., 2002, reproduced with permission of Elsevier).

whose injection points changed position in relation to glacier marginal oscillations. Cycles of basin filling by these mechanisms are separated by erosional unconformities manifest in buried river terraces and channels.

Internally, grounding-line fans exhibit crude to well-developed *foreset bedding* dipping down from the fan apex parallel to the fan surface, generally at angles of 10–30° (fig. 11.121). Near the fan apex, beds commonly consist of coarse-grained, poorly sorted, massive to normally graded gravels, deposited from high-density, cohesionless debris flows and gravelly traction carpets. Units may have channelized, erosive bases, recording scour and fill on the fan surface. Due to the dynamic nature of grounding-line fans and the oversteepening of fan surfaces by rapid sedimentation rates, sediment remobilization by gravity flows is common, producing stratified diamicton units occupying scoured channels (Cheel and

Rust, 1982; Thomas and Chiverrell, 2006). Coarse sediments at the fan apex grade distally into better sorted, finer-grained facies, resulting from episodic turbidity currents and cohesionless debris flows, triggered by influxes of sediment-laden meltwater or slope failures on the fan surface (Postma et al., 1983). Various types of cross bedding and cross lamination may be present, recording the downflow migration of dunes and ripples at the base of turbid underflows (Cheel and Rust, 1982). Proximal to distal fining reflects the drop-off of flow velocities with distance from the tunnel mouth, whereas multiple vertical sequences of upward-fining units record waning discharges in individual flow events. Towards the fan margin, high-suspension sedimentation rates produce climbing ripple drift and massive sands. Ball and pillow and dish structures are common due to dewatering, as sediment is compacted by fan aggradation. Ice-rafted

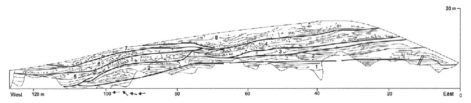

Figure 11.121 Section through Tullywee Bridge quarry, County Galway, Ireland, showing the architecture of the sediments in a subaqueous, grounding-line fan. The ice proximal slope is to the right; the numbers 1–8 on the section sketch refer to separate sedimentary packages of tunnel mouth fan deposits, comprising pebble and cobble gravels in sheets, lenses and channels (Thomas and Chiverrell, 2006, reproduced with permission of Elsevier).

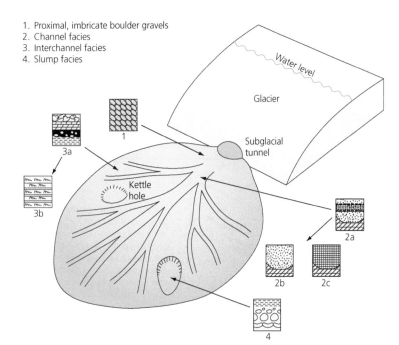

1. Proximal, imbricate boulder gravels
2. Channel facies
3. Interchannel facies
4. Slump facies

Figure 11.122 A depositional model for a single subaqueous outwash fan fed directly by a subglacial meltwater tunnel. (1) Proximal gravel facies. (2) Channel facies. (3) Inter-channel facies. (4) Slump facies. The palaeocurrent distribution from all facies of the fan is shown at the fan front (After Cheel and Rust, 1982).

debris up to boulder size may also occur in fan deposits, as may till clasts, dropped from subglacial till layers adhering to the base of icebergs released from floating glacier tongues (Powell, 2003).

A schematic model of facies distribution on glacilacustrine grounding-line fans was developed by Cheel and Rust (1982), based on Pleistocene examples from Canada (fig. 11.122). At the fan apex are *proximal gravels* (1), with upflow imbrication recording deposition from jets. Clast-supported diamictons may also be deposited from hyperconcentrated flows in such locations. Further down-fan is a system of steep-sided, bifurcating channels filled with *channel facies* (2). Three types of channel facies were recognized by Cheel and Rust (1982). *Type A channels* are filled with horizontally stratified sand, with discontinuous beds of imbricate gravel along channel

bases and between sand units. *Type B channels* are filled with massive sands; and *Type C channels* are filled with sub-horizontally laminated sands. The deeper Type A channels are the more proximal and contain sediments documenting erosive currents carrying gravel bedload. This gravel was excavated from the sediment surrounding the channels, explaining its discontinuous nature. Some inversely graded sands occur in all channel types, recording deposition by grain flows, derived either from the channel walls or directly from the tunnel mouth. Between channels, two types of *interchannel facies* (3) are found. *Type A* consists of proximal cross-stratified and massive to normally graded sands and fine gravels, whereas *Type B* consists of distal climbing ripple drift with occasional silt drapes. The massive to normally graded sands in the proximal facies were deposited by sediment gravity flows during periods of high sediment discharge, whereas the silt drapes of the distal facies record periods of suspension sedimentation when discharges were low.

A good Quaternary example of downflow facies changes associated with distance from jet efflux points was described by Hornung et al. (2007). The proximal facies are coarse, clast-supported gravelly deposits of hyperconcentrated flows and cohesionless debris flows, and are associated with the conduit or immediately proximal jet outflow zone of flow establishment (ZFE; cf. Powell, 1990). The intermediate facies are dominated by normally graded and cross-stratified gravels, with basal scour structures arranged in bar-like forms, and grading downcurrent into planar parallel-stratified and planar and trough cross-stratified sands and pebbly sands (including antidunes), with abundant scour structures and intercalated layers of fine sand/silt and silty mud; these intermediate deposits are associated with the zone of flow transition (ZFT). The most distal facies include trough cross-stratified sands and pebbly sands and are interpreted as deposits of the zone of established flow (ZEF). These examples are typical of high-discharge grounding-line fans where the depo-centre is located a short distance from the efflux point and sediment was deposited by subcritical to super-critical flow with a hydraulic jump (cf. Russell and Arnott, 2003).

A conceptual model for the construction of grounding-line fans at one location has been presented by Lønne (1995, 2001) and Lønne et al. (2001), based on their stratigraphic architecture. The model identifies stacked *allostratigraphic units* separated by erosional discontinuities, with each unit comprising ice-proximal and ice-distal slope deposits, together with glacitectonic structures associated with ice advance into the fan (fig. 11.123). Oscillations of the glacier margin may introduce further complications to the development of grounding-line fans (fig. 11.119). Glacier advance may produce glacitectonic thrusts and folds in the outwash and may even lead to till deposition, thereby illustrating that grounding-line fans and subaqueous moraines (see section 11.6.2) occupy points on the same process–form continuum (Bennett et al., 2000a; Cummings and Occhietti, 2001; Menzies, 2001; Russell et al., 2003). Nemec et al. (1999) described an interesting example of the evolution of a subaqueous fan sequence that has been progressively glacitectonized as a result of glacier readvance, culminating in the progradation of an ice-contact delta.

Deformation structures are common in grounding-line fan sediments (e.g. Powell, 1990, 2003; Thomas and Chiverrell, 2006) and may be so pervasive that primary structures can be difficult to distinguish; and in some cases proximal fan deposits have a completely churned diamictic appearance, with only isolated, deformed remnants of sorted facies. The melt-out of buried ice blocks in a grounding-line fan can result in faulting and fluidization

(McDonald and Shilts, 1975; Cheel and Rust, 1982). The grounding of icebergs can also result in sediment disturbance on fan surfaces (see section 10.6.6; fig. 11.123).

11.6.2 SUBAQUEOUS MORAINES

Subaqueous moraines are transverse moraines deposited at or close to the grounding lines of water-terminating glaciers. Alternative terms for such moraines include *morainal banks*, *sub-lacustrine moraines*, *cross-valley moraines* and *DeGeer moraines*. These terms have been used in overlapping senses, and there is no universally agreed set of definitions. In this book, the terms 'morainal bank' and 'composite grounding-line fans' are used for large subaqueous moraine ridges formed mainly by deposition at glacier grounding lines, and the term 'De Geer moraine' for smaller, narrow, sharp-crested ridges which commonly occur in fields of closely spaced sub-parallel moraines.

In planform, individual subaqueous moraines may be linear or have sinuous crest lines (fig. 11.124). In some cases, they are concave upglacier, reflecting the form of calving bays in the ice front (e.g. Benn, 1996; Ottesen et al., 2008a). Some degree of glacier pushing is a characteristic of most subaqueous moraines, and prominent glacitectonic structures and forms can be produced during glacier advances (e.g. Bennett et al., 2000a). Complete overriding of subaqueous moraines will produce smoothed or fluted transverse forms. The morphology and sedimentology of subaqueous moraines are highly variable, and a large number of depositional models have been proposed, the more important of which are discussed in the following sections.

Morainal banks and coalescent subaqueous fans

Wherever glacier grounding lines remain quasi-stable for long periods of time and/or sediment fluxes to the glacier margin are high, subaqueous moraines will form by the accumulation of sediment along the glacier margin. *Morainal banks* are elongate masses of sediment formed along glacier grounding lines by a variety of processes. Primary sediment delivery to morainal banks can be from several sources (fig. 11.125; Powell and Domack, 1995; Powell, 2003). First, supraglacial debris can slump or fall into the water down the terminal cliff. This process is particularly effective during calving events, when it is known as *calve dumping* (Powell, 1990). Second, englacial and subglacial debris bands melt-out below the water line, releasing debris directly into the water. Third, debris can be dispersed over a wider area by iceberg melt-out and overturning. Fourth, unfrozen sediment can emerge from beneath the glacier margin, either as the result of squeezing or as the output from subglacial deforming layers. At oscillating glacier margins, morainal

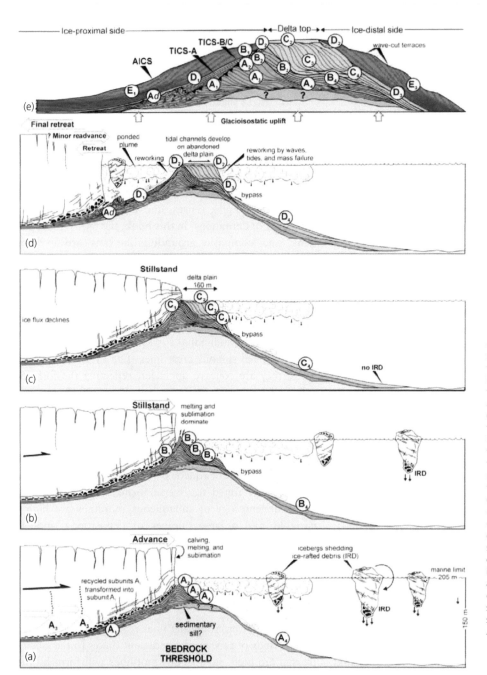

Figure 11.123 Reconstruction of the depositional history of the Mona moraine, southern Norway, based on the conceptual model of grounding-line development of Lønne (1995), from which the letter-coded allostratigraphic units are derived. (a) A submarine ice-contact fan is formed on the bedrock threshold at Mona during the advance of a temperate glacier with a calving tidewater front. (b) As the glacier margin stabilizes, the submarine fan aggrades to the sea surface. (c) An ice-contact, Gilbert-type delta develops. (d) The delta is abandoned due to a rapid ice front retreat by calving and reworking by waves and slope failures begins. (e) The moraine emerges due to regional glacio-isostatic uplift and its slopes accumulate regressive foreshore deposits. AICS = apparent ice-contact surface. TICS = true ice-contact surface (indicated for units A–C). Ad = deformed unit A.

Figure 11.124 Sinuous, low-amplitude morainal bank deposited at the grounding line of a subpolar glacier at Cape Armstrong, north-west Ellesmere Island, Arctic Canada. The small volume of debris in the ridge attests to the small debris turnover in a subpolar glacier advancing over bedrock (D.J.A. Evans).

banks can be built up or modified by ice-push or thrusting (Boulton, 1986b). Finally, as for grounding-line fans, considerable quantities of sediment can be delivered by melt-streams. Indeed, deposition on some morainal banks is dominated by subaqueous meltwater discharge, either because subglacial drainage is by sheet flow and emerges simultaneously along large parts of the margin, or conduit mouths migrate rapidly back and forth along the grounding line (Powell, 1990). Given the large variety of depositional and deformational processes active at glacier grounding lines, morainal banks are best regarded not as distinct depositional systems, but as part of a continuum ranging from isolated grounding-line fans and meltwater-dominated morainal banks to frontal dump, squeeze and push moraines (Powell and Domack, 1995; Powell, 2003).

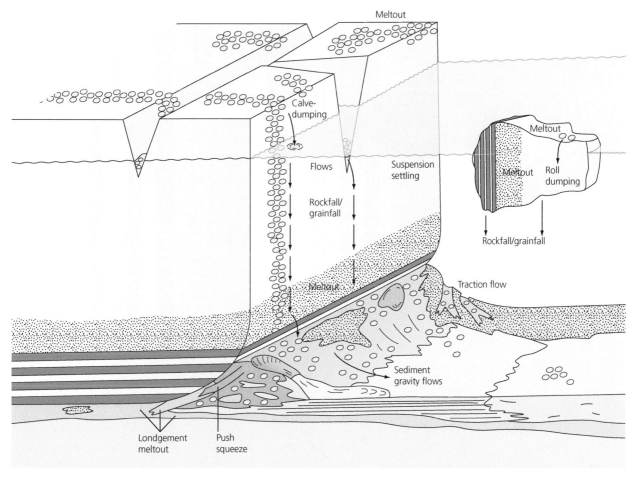

Figure 11.125 Processes of sediment transport and deposition at a morainal bank, based on temperate Alaskan fjord settings (After Powell and Domack, 1995; Powell 2003).

Some subaqueous moraines mainly consist of *coalescent grounding-line fans*, deposited contemporaneously at adjacent meltwater effluxes. The Salpausselka Moraines of Finland are spectacular examples, consisting of transverse chains of ice-contact fans and deltas deposited during stillstands of the southern margins of the Scandinavian Ice Sheet where it contacted deep water (Fyfe, 1990; section 12.5.2). Large-scale moraines formed of partially glacitectonized grounding-line fans occur in eastern Maine, USA (Ashley et al., 1991). The fans are arranged laterally along former grounding-line positions and are associated with tunnel mouth and esker deposits. The whole depositional complex covers an area of 250 km^2 over a 40 km wide zone and contains a volume of approximately 5 billion m^3 of sediment. The complex marks a series of stillstand positions of an actively retreating inter-lobate ice margin, deposited over the 1,000-year period from 13.5 ka to 12.5 ka BP. Parts of the Oak Ridges Moraine in Ontario, Canada, comprise subaqueous fans built at the former positions of meltwater portals in the receding ice front (Russell and Arnott, 2003; Sharpe et al., 2007; fig. 11.126).

Some subaqueous moraines are formed mainly by *dumping* and other *gravitational processes* at the foot of terminal ice cliffs. This is especially the case where polythermal glaciers with thick basal debris sequences contact the sea, releasing large amounts of debris in environments where there is only limited subglacial meltwater (Powell, 1984). At many glacier margins, however, gravitational deposits are inter-bedded with subaqueous outwash, particularly at the margins of temperate glaciers where meltwater is abundant. Exposures in a subaqueous moraine deposited near the margin of a Lateglacial ice-dammed lake at Achnasheen, Scotland, were described by Benn (1996). The ice-proximal portion of the moraine consists of deformation tills and glacitectonites derived from pre-existing glacilacustrine deposits, whereas the ice-distal portion is underlain by subaqueous outwash inter-bedded with cohesive to cohesionless debris flow deposits (fig. 11.127). The debris flows grade laterally into the deformation till, indicating that sediment was delivered from the glacier to the lake from the subglacial deforming layer, as well as from meltwater systems.

The size of moraines built up by outwash and gravitational processes is ultimately limited by debris supply from the glacier. Former limits of fjord glaciers on northwest Ellesmere Island, Canada, are marked by subaqueous moraines less than 1 m high (Evans, 1990a, b). These occur where former glaciers advanced from plateaux onto

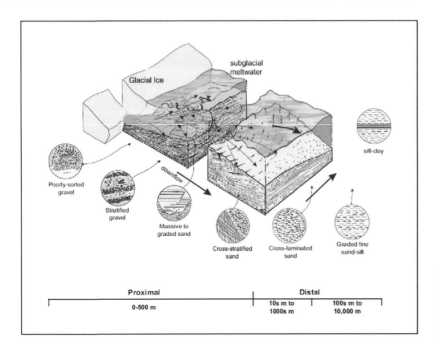

Figure 11.126 A conceptual model of subaqueous fan deposition during the production of the Oak Ridges Moraine, Ontario, Canada. This illustrates the rapid downflow grain size changes, in addition to changes perpendicular to flow (After Russell and Arnott, 2003; Sharpe et al., 2007).

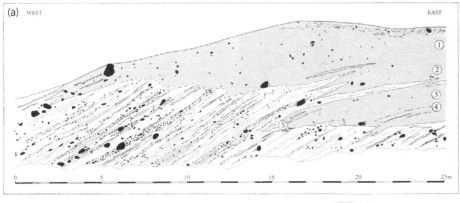

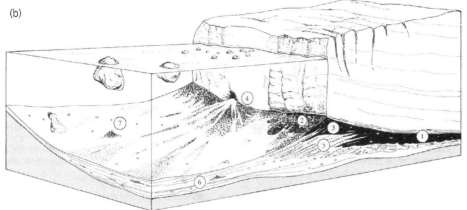

Figure 11.127 Morainal bank deposited in an ice-dammed lake at Achnasheen, Scotland, during the Lateglacial. (a) Exposure through the morainal bank (numbers refer to clast fabric samples). (b) Reconstruction of the depositional environment, where 1 = subglacial till; 2 = mass flows; 3 = mass flow diamictons; 4 = subglacial meltwater portal; 5 = subaqueous outwash; 6 = laminated lake sediments; 7 = iceberg dump mound (Benn, 1996, reproduced with permission of Taylor and Francis).

lowlands covered by very patchy sediment veneers. Supraglacial and subglacial debris loads were thus so insignificant that only low-amplitude diamicton ridges were deposited over bedrock at the grounding line (fig. 11.124). In contrast, large, extensive subaqueous moraines can be produced over short periods of time where debris is abundant, provided glacier stability is encouraged by glacio-dynamic or topographic factors. The development of subaqueous moraines is also influenced by bathymetry. In deep water, where calving rates are relatively high, rapid ice retreat commonly only allows time for the deposition of small ridges at the ice margin. In contrast, in shallow water where calving rates are lower, more substantial subaqueous fans can develop.

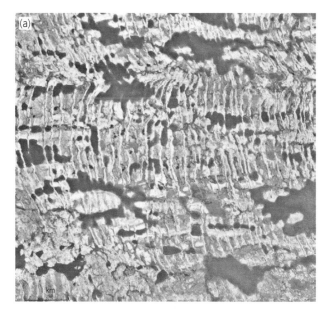

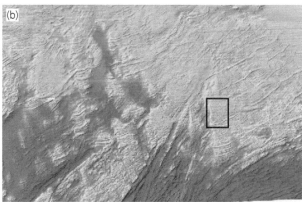

Figure 11.128 (a) De Geer moraines on the lowlands surrounding Hudson Bay, Canada. (National Aerial Photograph Library, Ottawa) (b) Multi-beam topographic image of sequences of De Geer moraines on the German Bank of the Scotian Shelf off the coast of Atlantic Canada. Sets of minor ridges are separated by larger single moraines indicative of readvances and/or stillstands. Enlarged area shows the relationship between De Geer moraines and bathymetry, with an area of cross-cutting ridges (crevasse-squeeze ridges) occurring on a topographic high point at the top of the image (Todd et al., 2007).

If such fans aggrade to sea level, ice-contact Gilbert-type deltas will form (section 11.6.5; Powell, 1990; Nemec et al., 1999).

De Geer moraines

De Geer moraines are named after the pioneering Swedish geologist Baron Gerard Jacob De Geer (1858–1943). They commonly occur in fields of closely spaced ridges in association with subaqueous sediments, and mark the intermittent retreat of water-terminating glaciers (fig. 11.128; e.g. Boulton, 1986b; Larsen et al., 1991; Aartolahti et al., 1995; Ottesen and Dowdeswell, 2006). For some modern examples, it is possible to show that subaqueous moraines were deposited annually by matching individual moraines to known ice-margin positions (Ottesen and Dowdeswell, 2006).

Most researchers regard De Geer moraines as former ice-marginal features, where sediment is deposited or pushed up during brief stillstands or minor readvances,

although other mechanisms have been proposed (e.g. Zilliacus, 1989; Beaudry and Prichonnet, 1991; King et al., 1991; Lundqvist, 2000). Boulton (1986b) described sequences of closely spaced subaqueous moraines on the seabed in front of Alpha Glacier, Cambridge Fiord, east Baffin Island, which appear to document annual winter pushing. These moraines are characterized by steeper distal than proximal slopes, lobate forms with acute inter-lobe angles pointing up glacier, and truncated sections due to partial overriding by later moraines. All of these characteristics are similar to those of terrestrial push moraines (as described in section 11.3.3). A wide range of sediment types has been reported from exposures through De Geer moraines, from tills to stratified sediments and various combinations of the two. Blake (2000) and Lindén and Möller (2005) showed that De Geer moraines formed by a combination of sediment accumulation and ice push. On the ice-distal side of the moraines, sediment gravity flow deposits inter-digitate with glacilacustrine sediments, whereas the ice-proximal sides are characterized by subglacial traction tills and associated canal fills that have been locally glacitectonized (fig. 11.129).

Fields of De Geer moraines formed during the retreat of surging glaciers can occur in juxtaposition with cross-cutting crevasse-squeeze ridges (Ottesen and Dowdeswell, 2006; Ottesen et al., 2008a; see section 11.2.9). Indeed, in some cases it can be difficult to distinguish De Geer moraines from crevasse-squeeze ridges on morphology alone, because the calving fronts of surging glaciers are commonly controlled by crevasse patterns and can have very complex planforms (Ottesen et al., 2008a). Todd et al. (2007) provided stunning multi-beam topographic images of sequences of De Geer moraines on the German Bank of the Scotian Shelf (fig. 11.128b), in which the moraines occur beneath deeper water and cross cutting (crevasse-squeeze ridges) occurs on topographic high points. This association between topographic high points and crevasse-squeeze ridges has been identified on the forelands of terrestrial surging glaciers, and is explained by intense deformation and crevassing of glacier ice as it flows over large perturbations or bumps in its bed (Evans and Rea, 2003; Evans et al., 2007b).

11.6.3 GROUNDING-LINE WEDGES: TILL DELTAS, TILL TONGUES AND TROUGH-MOUTH FANS

The term *grounding-line (zone) wedge* was introduced by Powell and Domack (1995) to refer to dipping diamicton beds overlain by horizontal sheets of diamicton, interpreted as subglacial till. The term was intended to replace the earlier term *till delta* (Alley et al., 1989), to avoid defining the sediment as till, and to avoid any association with sea level that could be inferred from the word 'delta'. Grounding-line wedges are thought to be characteristic of settings where large amounts of debris are available, but meltwater discharge is negligible, such as the grounding lines of high-latitude ice streams (Powell and Domack, 1995). Where meltwater is abundant, such as at the tidewater margins of temperate glaciers, deposition is more likely to produce morainal banks or grounding-line fans (see section 11.6.1).

Grounding-line wedges form near the grounding lines of glaciers where glacigenic sediment is redistributed by subaqueous debris flows, producing diamicton beds that

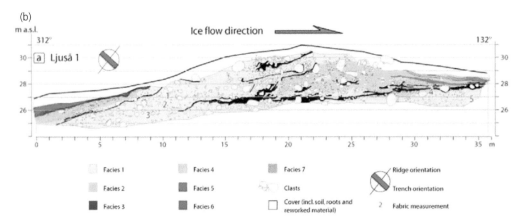

Figure 11.129 Characteristics of De Geer moraines in northern Sweden. (a) Conceptual model of the formation of De Geer moraines, showing the influence of calving and the following development of a grounding-line ridge fed by deforming bed sediments. F1–F5 are major facies recognized in moraine stratigraphic sections. (b) Typical cross section through a De Geer moraine showing major facies. Facies 1 = subglacial traction till; Facies 2 = subaqueously reworked till; Facies 3 = subglacial canal fills; Facies 4 = glacifluvial cobble gravels; Facies 5 = fine-grained glacilacustrine deposits (Lindén and Möller, 2005).

dip away from the margin (fig. 11.130; Anderson, 1999; Domack et al., 1999; Licht et al., 1999; Shipp et al., 2002; Ó Cofaigh et al., 2005). The debris flows produced in such environments, especially on the surfaces of trough-mouth fans (see below), have been termed *glacigenic debris flows*, or GDFs by King et al. (1998). Successive debris flows build out a prograding wedge of sediment beyond the grounding line, and subsequent grounding-line advance over these deposits causes them to be disconformably overlain by subglacial till. Using seismic data, Alley et al. (1989) identified examples of grounding-line wedges near the junction between Ice Stream B and the Ross Ice Shelf, Antarctica. In this case, the grounding-line wedge is thought to be a sediment sink for deforming till emerging from beneath the fast-flowing ice stream at the point where the ice begins to float. Some grounding-line wedges have distinctive surface morphologies resembling De Geer moraines (referred to as 'corrugation moraines' by Shipp et al., 2002). Sea floor images show widely spaced transverse ridges separated by fluted surfaces (fig. 11.131a). In cross profile, the ridges are asymmetrical with shallow proximal and steep distal faces (fig. 11.131b). The internal structure of grounding-line wedges will reflect patterns of grounding-line advance and retreat. Glacier advance will result in progressive onlap of subglacial facies over the top of old mass flow deposits (fig. 11.130), whereas retreat will result in the deposition of undermelt diamictons and mass flows on top of till sheets. Oscillation of the grounding line will therefore be recorded by inter-digitation of subglacial and subaqueous deposits (Powell et al., 2000; Powell, 2003). Furthermore, distinctive stratigraphies should

result from episodes of sea-level rise and fall (Boulton, 1990; section 12.5.1).

Powell and Domack (1995) argued that the *till tongues* reported from the Scotian and Norwegian continental shelves by King et al. (1991) also represent ancient grounding-line wedges. These 'till tongues' were identified from high-resolution seismic records, and consist of tongues of massive diamicton inter-digitating with stratified glacimarine deposits, including rhythmites and dropstone diamictons. Although King et al. (1991) interpreted 'till tongues' in a slightly different way to grounding-line wedges, the processes are much the same as those outlined above.

Major wedges of sediment consisting of outward-dipping diamicton beds overlain by subglacial till occur near the mouths of major troughs. Boulton (1990) referred to such accumulations as *trough-mouth fans*, and suggested that they may be characteristic depositional systems at the tidewater or floating termini of fast-flowing ice streams and outlet glaciers (fig. 11.132). Recently, grounding-line wedges have been identified by Vanneste et al. (2007) in amphitheatre-like embayments that lie between major troughs to the west of Svalbard. These features were previously regarded as large submarine slope failures, but seismic data have revealed that the embayments comprise prograding glacigenic sequences being input to the marine environment by slower-moving inter-stream ice.

Significant trough-mouth fans (TMFs) have since been identified at the edges of glaciated continental shelves and are now regarded as the major depo-centres for marine-terminating ice streams (Hiscott and Aksu, 1994; Stoker,

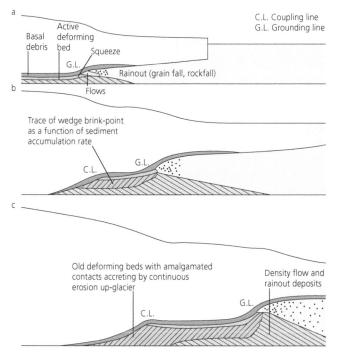

Figure 11.130 Depositional model for the formation of grounding-line wedges at the margin of a polar ice stream. (a) Till emerging from a subglacial deforming layer feeds prograding subaqueous debris flows. (b) Deforming till accumulates behind the grounding line, overlapping older debris flow deposits. The inner limit of subsole deformation is known as the coupling line. (c) Till erosion above the coupling line and migration of the grounding line produce the overall geometry of the grounding-line wedge (After Powell and Domack, 1995).

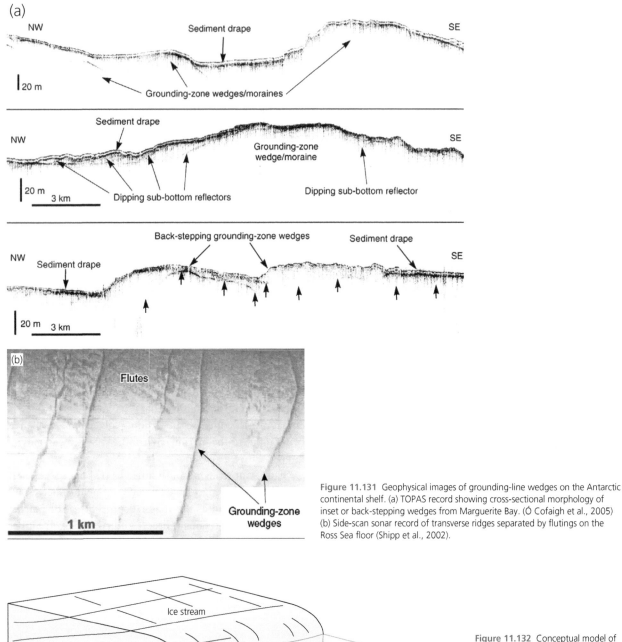

(a)

Figure 11.131 Geophysical images of grounding-line wedges on the Antarctic continental shelf. (a) TOPAS record showing cross-sectional morphology of inset or back-stepping wedges from Marguerite Bay. (Ó Cofaigh et al., 2005) (b) Side-scan sonar record of transverse ridges separated by flutings on the Ross Sea floor (Shipp et al., 2002).

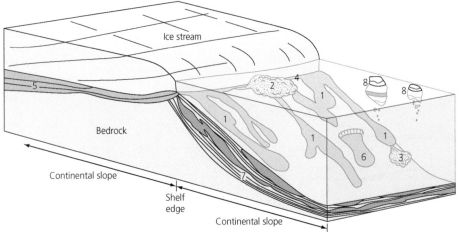

Figure 11.132 Conceptual model of trough-mouth fan development in front of an ice stream at the continental shelf edge. (1) Debris flows sourced from grounding line. (2) Buoyant turbid meltwater plume. (3) Turbidity current at downslope edge of debris flow. (4) Subglacial till extruded at grounding line. (5) Subglacial till. (6) Debris flows generated by slump on fan front. (7) Stratified sediments relating to suspension settling, contourites and turbidites. (8) Iceberg rafting (After Ó Cofaigh, 2007).

1995; Vorren and Laberg, 1997; Dowdeswell et al., 1998, 2008b; Vorren et al., 1998; Taylor et al., 2002a; Vorren, 2003; Ottesen et al., 2008a). TMFs cover very large areas of formerly glaciated continental shelf margins, the largest on the north-west European margin being the Bear Island TMF between Svalbard and mainland Norway, at 215,000 km². Although TMFs have accumulated over numerous glaciations, they did not begin to develop their present morphology until the mid-Pleistocene (Laberg and Vorren, 1996; Vorren et al., 1998). Systematic study of the stratigraphic architecture of TMFs has revealed that they are built up sequentially by the deposits of glacigenic debris flows (GDFs, which during the last glaciation accounted for up to 20 per cent of the sediment delivered to individual offshore basins (fig. 11.132; Laberg and Vorren, 1995, 2000; King et al., 1998; Taylor et al., 2002b; Bowles et al., 2003). GDFs differ from normal subaqueous debris flows in that they are characterized by extremely long run-out distances of up to 200 km on low slope gradients often <1°, and maintain a uniform thickness along their length. The GDFs are fed by instabilities on grounding-line wedges. Once deposited, GDFs resemble massive homogeneous diamictons (*debrites*) separated by variable thicknesses of laminated to massive suspension sediments (*plumites*), *turbidites* and *contourites* (e.g. Laberg and Vorren, 1995; Hesse et al., 1997, 1999; Laberg et al., 1999; Davison and Stoker, 2002; J. Evans et al., 2002; Ó Cofaigh et al., 2002b, 2004). Estimates of GDF frequency range from one every 35–75 years (Laberg and Vorren, 1995, 2000) to one every 2000 years (Taylor et al., 2002a, b). Additionally, very large submarine mass-wasting events can affect glacimarine depo-centres after ice recession; for example, the well-documented Storegga Slide (Haflidason et al., 2004), which involved the failure of ~3000 km^3 of sediment on the Norwegian continental margin in the early Holocene, and was responsible for tsunami waves around the north-west Atlantic shorelines (e.g. Smith et al., 2004).

11.6.4 ICE SHELF GROUNDING-LINE DEPOSITS AND ICE SHELF MORAINES

Ice shelves only form in high polar settings (section 5.6), so meltwater-related processes tend to be of much less importance at ice shelf grounding lines than at temperate grounding-line cliffs. Instead, sediment transport and deposition is dominated by a combination of gravitational processes and ice-marginal tectonics (Powell and Domack, 1995). Not surprisingly, modern ice shelf grounding lines are difficult and expensive to reach, and few direct observations have been made. Powell et al. (1996) conducted a pioneering study beneath the floating tongue of Mackay Glacier, Antarctica, using a robotic submersible equipped with a camera. Their observations showed that the most widespread depositional process beneath the floating margin is rain-out of debris released by melting of basal debris-rich ice (fig. 11.133). This produces a thin and patchy drape of sediment which overlies fluted subglacial till deposited when grounded ice extended further seawards. At the grounding line itself, ridges of rubble up to several metres high are produced by dumping and ice-push, and soft till emerging from beneath the glacier is redeposited in low-angle debris

flows. Apart from these observations, understanding of sedimentary processes beneath ice shelves is based on hypothetical modelling and interpretation of seismic profiles and ancient sedimentary successions.

Wherever floating glacier tongues meet localized shoals, they can ground and thicken to produce ice rises (section 1.2.3). Very little work has been undertaken on the nature of this interaction between ice shelves and offshore shoals, but the implications are nonetheless significant. For example, recent sea floor imaging supported by icebreakers and nuclear submarines in the Arctic Ocean have revealed remarkably well-preserved glacial features on the Chukchi Borderland (Northwind Ridge) and Lomonosov Ridge, reinvigorating a long established and unresolved debate about the extent of former glaciations in the region. In addition to iceberg scour marks in water depths of less than 400 m, the features include subglacial bedforms such as drumlins, flutings and MSGL, morainic ridges and grounding-line wedges to depths of 1 km (Jakobsson, 1999; Polyak et al., 2001; Jakobsson et al., 2005, 2008; Engels et al., 2008). In some places, the glacial lineations are overprinted in much the same style as their terrestrial counterparts. The occurrence of these features on the high points of the Arctic Ocean sea floor was interpreted by Jakobsson et al. (2008) as the geomorphic imprint of ice rise development, where a pan-arctic ice shelf, fed by ice sheets surrounding the Arctic Ocean and buttressed by thickening sea ice (cf. Grosswald and Hughes, 1999), locally grounded and thickened.

Ice shelf moraines are formed where debris is deposited or relocated around the margins of ice shelves or floating glacier tongues. The crest lines of such moraines tend to be horizontal or have very low gradients, reflecting the low gradient of ice shelves which have zero basal drag (section 5.6.2). The gradients of ancient ice-shelf moraines, however, may be steeper due to post-glacial isostatic tilting, so their gradients should be corrected with reference to the local sea-level record.

Modern examples of ice shelf moraines have been described from Antarctica. Sugden and Clapperton (1981) described an example that formed at the margin of the George VI Sound ice shelf, where it flowed from the Antarctic Peninsula across George VI Sound to Alexander Island. The horizontal moraine ridge was formed over a distance of 120 km along the promontories of Alexander Island, where basal debris was brought to the glacier surface by ablation and thrusting. Sediments within the moraine included:

(1) subglacial debris entrained on Alexander Island and the Antarctic Peninsula;
(2) glacifluvial and lacustrine sediments from ice marginal streams and ponds;
(3) marine sediments and fauna derived from freeze-on at the grounding line on Alexander Island.

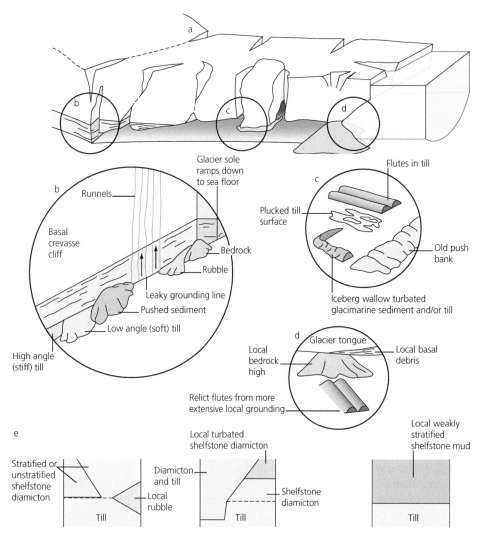

Figure 11.133 Conceptual model of important sedimentary processes beneath a glacier ice shelf, based on observations made by a robotic submersible beneath the Mackay Glacier Tongue, Antarctica. (a) General overview. (b) Grounding line. (c) Sea floor beyond grounding line. (d) Pinning point. (e) Idealized facies sequences. Not to scale (Modified from Powell et al., 1996).

More recently, Glasser et al. (2006) have made a thorough investigation of the McMurdo Ice Shelf, Antarctica, identifying similar complex sources of debris entrainment, as well as the onshore development of ice shelf moraines on the lower slopes of nearby Bratina Island (fig. 11.134). The moraines form an inset sequence of single- or multiple-crested ridges containing a variety of material, and in some cases probable ice cores. They record the sequential thinning of the ice shelf during the later part of the Holocene. An important aspect of Glasser et al.'s (2006) model of ice shelf moraine formation is the entrainment of debris by anchor ice formation over offshore shoals. This material is mixed with medial moraine debris on the ice shelf surface through the development of ice-cored ablation ridges and cones which the ice shelf contemporaneously pushes onshore.

England et al. (1978) and England (1999) described ancient ice-shelf moraines from north-east Ellesmere Island, Arctic Canada (fig. 11.135a) The horizontal moraines mark the former margins of floating glacier tongues, and contain marine shell fragments thought to have been entrained at the grounding line or transferred to the surface of the ice shelf by bottom freezing of water. Fjord-based ice shelf moraines have been described from Arctic Norway by D. J. A. Evans et al. (2002). Striking ancient examples of ice shelf moraines occur at the former northern margins of the Laurentide Ice Sheet, particularly in Viscount Melville Sound (Hodgson and Vincent, 1984; Dyke et al., 1992). The moraines consist of extensive linear belts of till between 60 and 150 m above present sea level, along the shores of Byam Martin, Melville, Banks and Victoria Islands and extending eastwards to north Baffin Island (fig. 11.135b).

11.6.5 DELTAS

Deltas are masses of sediment built out into standing water by subaerial streams, by a combination of fluvial processes above the water line and some combination of gravitational mass movement and suspension settling below water level. They are major features on sea coasts and lake shorelines in all climatic zones, although here we concentrate specifically on those deltas produced in

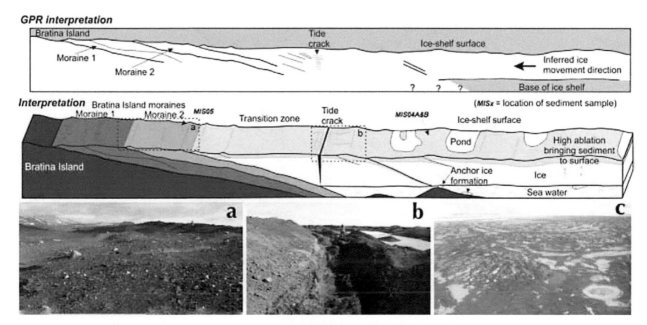

Figure 11.134 Ice shelf moraine development based on the McMurdo Ice Shelf/Bratina Island case study. The upper GPR profile reveals the raw data from which the interpretation below is compiled. Bottom diagram shows a conceptual model of moraine development. Photographs are: (a) Bratina Island ice shelf moraines; (b) tide crack between the ice shelf and Bratina Island; (c) aerial view of ice-cored ridges and cones on the surface of the ice shelf (Glasser et al., 2006, reproduced with permission of the International Glaciological Society).

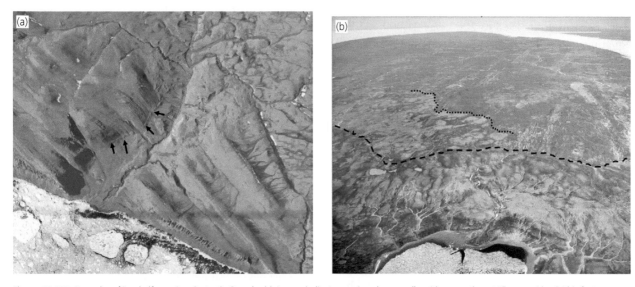

Figure 11.135 Examples of ice shelf moraines in Arctic Canada. (a) Arrows indicate moraine along a valley side on north-east Ellesmere Island. This feature was interpreted initially (by England et al., 1978) as a marine ice shelf, with Ellesmere Island glacier ice floating in a higher sea level than that at the present day (sea ice visible to bottom of image). It has since been reassessed by England (1999) as a glacial lake feature formed in a valley dammed by Ellesmere Island ice retreating towards the top of the image, and the Greenland Ice Sheet margin retreating off the coast. (b) The Viscount Melville Sound ice shelf moraine (Winter Harbour till, dashed line) and the limits of the Bolduc till (dotted line), perhaps from an earlier ice shelf, at Winter Harbour, Melville Island (Aerial photographs provided by the National Aerial Photograph Library, Ottawa).

glacierized basins (fig. 11.136). Deltas form by a combination of fluvial *aggradation* above water level, and *progradation* on the delta front where bedload is rapidly deposited when the incoming stream decelerates on contact with standing water. Fine-grained suspended sediment is carried beyond the delta by underflows, interflows and overflows (depending on plume buoyancy), to be deposited in the deeper parts of the lake or marine basin. Delta-front profiles usually display steep upper slopes, giving way to easier-angled slopes towards the base of the delta, reflecting proximal-to-distal decreases in sedimentation rates and grain size.

Deltas are important depositaries for sediment discharged from glaciers. Therefore, quantification of delta volume can provide valuable estimates on erosion rates in glaciated catchments. Østrem et al. (2005) measured the growth of a delta and distal varves in a proglacial lake in front of Nigardsbreen, Norway, since its initiation in

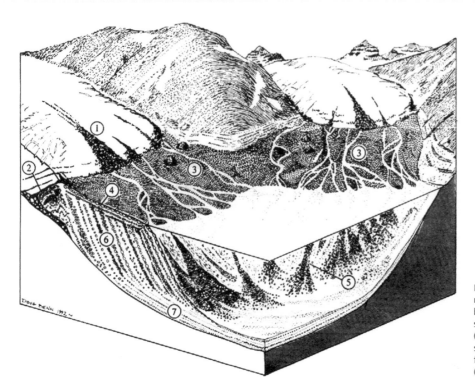

Figure 11.136 Block diagram of reconstructed Gilbert-type deltas based on Scottish Lateglacial case studies. (1) Supraglacial debris. (2) Subglacial debris. (3) Braided sandur surface. (4) Topsets. (5) Delta front. (6) Foresets. (7) Bottomsets (Benn, 1992).

1968. The annual deposition of suspended sediment and delta deposits from 1968 to 2003 was $21{,}000 \times 10^3$ kg, equating to a volume of approximately $10{,}500$ m³. At this annual rate of deposition, the lake will fill up with sediment in 500 years. The average annual growth of the delta over 36 years was calculated at $11{,}800 \times 10^3$ kg, giving a glacial erosion rate for the catchment of Nigardsbreen of approximately 0.3 mm/year.

Deltas may form in association with glaciers in two main settings:

(1) *glacier-fed deltas* receive sediment from proglacial meltwater streams following transport across an intervening land surface;

(2) *ice-contact deltas* are built out directly from glacier margins, after evolving from grounding-line fans or other subaqueous depositional systems.

The emergence of a delta at the margin of Hubbard Glacier, Alaska, is shown in figure 11.137. The sedimentology of deltas in these two settings is reviewed in the following sections. *Non-glacial deltas* receive no input from glacier melt, but may still have high sediment input from paraglacial reworking of relict glacigenic deposits.

Glacier-fed deltas

The morphology and sedimentology of deltas is strongly influenced by the gradient of the feeder river and water depth, leading to a range of delta types (fig. 11.138; Postma, 1990). Additionally, temporal changes in glacier coverage in feeder catchments can be reflected in sedimentary sequences (e.g. Dirszowsky and Desloges, 2004), and

in many situations such changes are coupled with the impacts of sea-level oscillations (Lønne and Nemec, 2004). The most common deltas in glacierized catchments are shallow-water *Hjulström-type deltas* with gently sloping fronts, and intermediate- to deep-water *Gilbert-type deltas* and *subaqueous debris cones* with steep fronts. Deltas intermediate in character between Hjulström-type and Gilbert-type deltas have been termed *Salisbury-type deltas* (Church and Gilbert, 1975). In glacierized landscapes, depositional environments do not remain stable for long periods of time, so deltas may evolve from one type to another in response to changing sea levels, proximity of glacier ice, sediment availability and river discharges.

Hjulstrom-type deltas are deposited in shallow waters at the distal end of sandur plains. They are characterized by three physiographic zones, consisting of:

(1) a subaerial delta plain;

(2) a gently sloping delta front dominated by coarse-load deposition and commonly influenced by waves;

(3) a very gently sloping pro-delta dominated by hemipelagic sedimentation.

Gilbert-type deltas are named after the pioneering American geologist Grove Karl Gilbert (1843–1918), and are the most common in glaciated basins. They occur as both glacier-fed and ice-contact types, and therefore the details reviewed here are relevant also to page 574. They have three main components:

(1) *topsets*, consisting of fluvial sediments deposited on the subaerial delta surface;

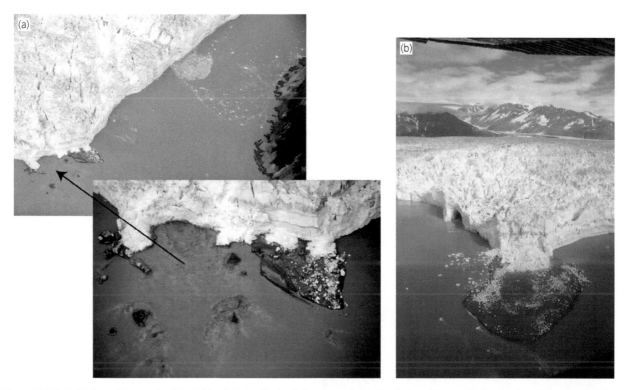

Figure 11.137 Aerial views of the margin of the Hubbard Glacier, Alaska, during the most recent readvance of the margin, showing the growth of a grounding-line fan to a delta. (a) 11 August 2007, when sediments start to appear at the efflux point of a subglacial stream. (b) 26 August 2007, when a delta had aggraded from a grounding-line fan and prograded so that its surface lay above high tide. A minimum estimate of the average sediment flux over a 3-week period of sedimentation is $0.16 \times 10^6 \, m^3 \, d^{-1}$, equating to a volume of $3.3 \times 10^6 \, m^3$ of sediment deposited by the subglacial stream (Motyka et al., 2008).

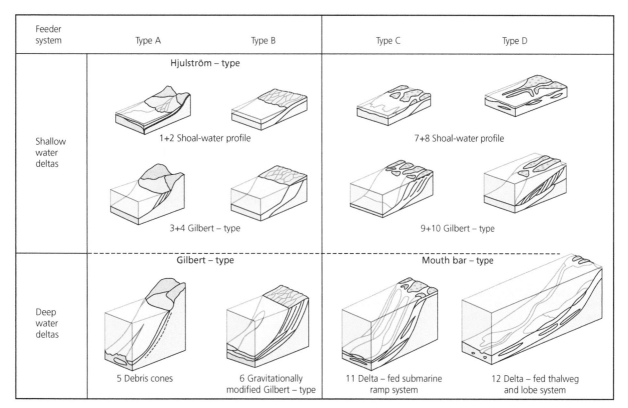

Figure 11.138 Classification scheme for deltas, showing delta morphology as a function of water depth and the gradient of the feeder river system. Feeder system types are: (A) very steep; (B) steep; (C) moderate; (D) low-gradient (After Postma, 1990).

(a)

(b)

(c)

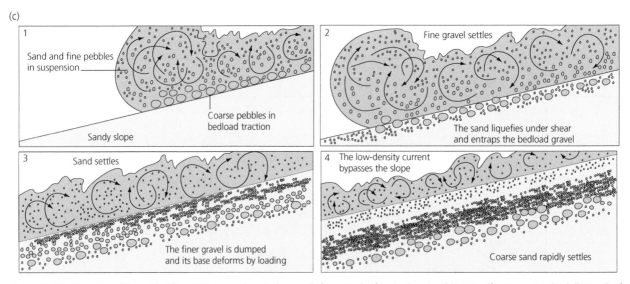

Figure 11.139 The nature of foreset bedding. (a) Exposure through the gravelly foreset beds of a raised marine delta in Hvalfjörður, west Iceland. (b) Details of sand and gravel foreset beds in a glacilacustrine delta, Drymen, Scotland. Note the grading within individual gravel beds and the soft sediment clast above the hammer. (D.J.A. Evans) (c) Explanation of the origin of normally graded foreset beds deposited by cohesionless debris flows or high-density turbidity currents, in this case resulting in couplets of a graded fine pebble to granule gravel overlain by a homogeneous sand with floating large pebbles. (1) Gravel-laden turbidity current arrives on a sandy delta slope. (2) Basal shear stress liquefies the sandy substratum, which then entraps the coarse bedload gravel. (3) Decrease of energy by bottom shear causes rapid settling of fine gravel from suspension. (4) Deceleration and rapid dumping of coarser sand and removal of finer load by remaining low-density current to the delta toe (After Nemec et al., 1999).

(2) *foresets*, deposited by gravitational processes on the delta foreslope;

(3) *bottomsets*, deposited by underflows on gentle slopes at the foot of the delta front (Postma, 1995).

Topsets are essentially sandur deposits, and incorporate the range of facies described in sections 10.4 and 11.5.1.

Delta foresets may be composed of a variety of gravel and sand facies, depending on sediment supply and depositional mechanism (figs 11.139a, b). In general, sorting is better than in grounding-line fan foresets, due to the effects of subaerial fluvial transport on grain size distribution. Remobilization of debris on the delta foreslope can,

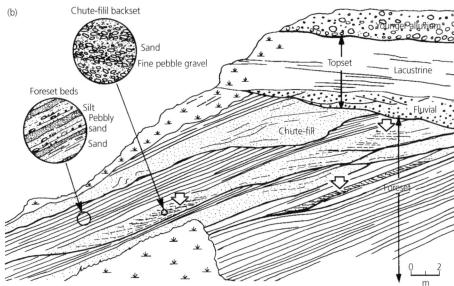

Figure 11.140 Examples of backset bedding in the foresets of Gilbert-type deltas. (a) Unit of pebbly sand (arrowed) with upslope dipping cross-strata (backsets) on an outwash delta foreslope in the Netherlands. The delta prograded towards the left. (b) Backsets (open arrows) in a Gilbert-type delta in Calabria, Italy. The delta prograded to the left (Nemec, 1990).

however, result in remixing of sediment, producing poorly sorted beds. Common facies include massive to inversely graded openwork gravels, sandy gravels, gravelly sands, and sands deposited by cohesionless debris flows; fall-sorted gravels deposited by debris falls; and normally graded gravels and sands deposited by high-density turbulent underflows (Martini, 1990; Nemec, 1990; Nemec et al., 1999). Due to the high-energy conditions on upper delta foreslopes, fine-grained sands and silts are rare and clays are absent. Delta foresets commonly form repeated fining-upward units, often separated by *welded contacts* (figs 11.139b, c), documenting fluctuations in sediment delivery on a variety of timescales. Such fluctuations may result from diurnal or weather-related changes in fluvial discharge, or switches in sediment influx positions on the delta surface (e.g. Gustavson et al., 1975; Mastalerz, 1990). Cross bedding, recording the downflow migration of dunes, is rare, although examples of *backsets* have been reported (Martini, 1990; Nemec et al., 1999; fig. 11.140).

Erosive events on upper delta foreslopes result in downslope-oriented *chutes* and *channels*, cut by turbidity currents and erosive debris flows originating at bowl-shaped slump scars or stream channel mouths. These chutes and channels become infilled with turbidites,

debris flows and suspension deposits, which may all contain intra-clasts of strata originating as blocks that slump from the channel walls and become buried by the infill (Nemec, 1990; Lønne, 1997; Nemec et al., 1999). In section, former chutes and channels are recognizable as cut-and-fill sequences within the foreset bedding, or by units with *welded contacts* at their bases recording partial incorporation of the underlying sediment.

The foresets of Gilbert-type deltas become easier-angled and finer-grained downslope. On lower delta foreslopes, typical facies are sands and silts deposited by turbid underflows, triggered by the collapse of upper delta foresets or fresh sediment influx events (Gustavson et al., 1975; Postma, 1995). Multiple fining-upward sequences, a few centimetres to a few tens of centimetres thick, are common, typically consisting of massive or planar laminated sands, grading upwards into ripple cross-laminated sands and silt drapes, recording progressively waning flow during individual underflows (Ashley et al., 1982). Partial or repeated sequences can result from fluctuations and surges within flows. In three dimensions, such sequences form overlapping lobes of sediment emanating from the shifting influx points of distributary delta surface channels (Smith and Ashley, 1985). The *bottomsets* of Gilbert-type deltas consist of distal turbidites and fine-grained silts and clays deposited from high-level suspension (sections 10.5.5, 10.6.1).

The fronts of deep-water Gilbert-type deltas may be subject to frequent slope failures and gravitational reworking. In such *gravitationally modified Gilbert-type deltas* (fig. 11.138), coarse, poorly sorted material is transferred to the delta foot by slumps and cohesionless debris flows. Typical structures include compressional slump folds; low-angle, rising and conjugate shear faults; sediment-filled tension cracks; shear zones beneath sediment slides; and distorted beds disturbed by water escaping from underlying strata (Nemec, 1990). The downslope fluidization of subaqueous mass flows will lead to incisions on the delta front and their infilling by turbidites (e.g. Lønne, 1997).

Subaqueous debris cones or *underwater conical deltas* lack a subaerial delta plain component (figs 11.138, 11.141; Nemec, 1990; Postma, 1990). Continued sediment delivery to an underwater conical delta often leads eventually to the shallowing of water depths until foresets are able to prograde at water level, allowing the development of a delta plain. Thus, conical underwater deltas may act as precursors to Gilbert-type deltas. The development of underwater conical and Gilbert-type deltas under non-steady state conditions in fjords in British Columbia, Canada, has been studied by Prior and Bornhold (1990), using side-scan sonar imagery. Delta morphology changes through time in response to changes in sediment supply, related to annual snowmelt and outburst floods, and subaqueous relief and fan gradients. In British Columbia, sediment supply has declined during the Holocene due to glacier recession and the stabilization of deglaciated slopes in the drainage basins, which essentially starves the offshore delta slopes. Second, fan gradients have gradually been reduced during postglacial time, as the deltas have aggraded and prograded into formerly deep water, reducing the gravitational stresses on the delta surfaces considerably. Taken together, reduced sediment supplies and gravitational stresses have conspired to replace high-energy sediment dispersal with low-energy processes. According to this model, changing processes of sediment dispersal result in a fourfold facies succession, consisting of:

(1) coarse-grained avalanche deposits with slopes in excess of 20°;
(2) cohesionless debris flows (inertia flows) on slopes averaging 12°;
(3) turbidites;
(4) reworked sediments resulting from gravitational instability on the delta front (fig. 11.142).

During this succession, the delta front is increasingly starved of coarser material, and slope instability increases in importance as the foreslope experiences loading by distributary mouth-bar growth and the passage of turbid underflows (fig. 11.143).

Ice-contact deltas

Ice-contact deltas (sometimes referred to as *kame deltas*; section 11.4.3) may be deposited in ice-marginal or supraglacial lakes, or in the sea. They form by similar processes to glacier-fed deltas and exhibit similar facies successions, although there are some important differences in morphology and facies architecture. Following deglaciation, ice-contact deltas form isolated plateaux terminating in steep bluffs or *ice-contact slopes* at their ice-proximal ends. They may be perched on topographic highs, where the only plausible source of sediment and water is a former glacier margin (Clayton et al., 2008). Pitting or kettle holes are common on ice-proximal surfaces, recording the melt-out of buried glacier ice (fig. 11.144). In glacimarine settings, ice-contact deltas are particularly useful in reconstructing glacial and relative sea-level histories (e.g. Ó Cofaigh, 1998; England et al., 2000, 2004; Ó Cofaigh et al., 2003b; Lønne and Nemec, 2004; Thomas and Chiverrell, 2006).

The sedimentology and internal structure of ice-contact deltas reflect the close proximity of glacier ice. Debris flow diamictons may be widespread both on the delta surface and in ice-proximal foresets, reflecting direct debris input from the glacier surface (Benn and Evans, 1993; Lønne, 1993; Nemec et al., 1999). The majority of ice-contact deltas originate as subaqueous grounding-line fans, which then aggrade to water level (Powell, 1990), so that their ice-proximal cores will consist of the sedimentary sequences described in section 11.6.1 (e.g. Thomas, 1984; Lønne, 1993; Nemec et al., 1999; Thomas and Chiverrell, 2006). The ice-distal topsets, foresets and bottomsets of ice-contact deltas, however, may be indistinguishable from those of glacier-fed deltas, although palaeocurrent indicators may be more varied than in other delta sequences, due to changing influx points at the glacier margin. Depending on glacier margin recession rates, ice-contact deltas may not always evolve from subaqueous fans, as demonstrated by Winsemann et al. (2007), using an example of a subaqueous fan isolated in front of an ice-contact delta in north-west Germany. Figure 11.145 shows a longitudinal section through an ice-contact delta deposited at the margin of a piedmont glacier that occupied Loch Lomond, Scotland, during the Younger Dryas. At the central and ice-proximal (right-hand) parts of the section, the deposits consist of cross-cutting sandy and gravelly subaqueous flows, interpreted as subaqueous fans. These deposits pass distally into gravel and sand foreset beds, which are overlain by horizontal topsets. At the ice-proximal end, the sequence has undergone extensive subglacial deformation, whereby the fan deposits have been remoulded into glacitectonite and subglacial traction till (sections 10.3.3 and 10.3.4), whereas the distal parts of the delta are traversed by low-angled thrust faults. The section records the evolution of subaqueous fans into an ice-contact delta, which was

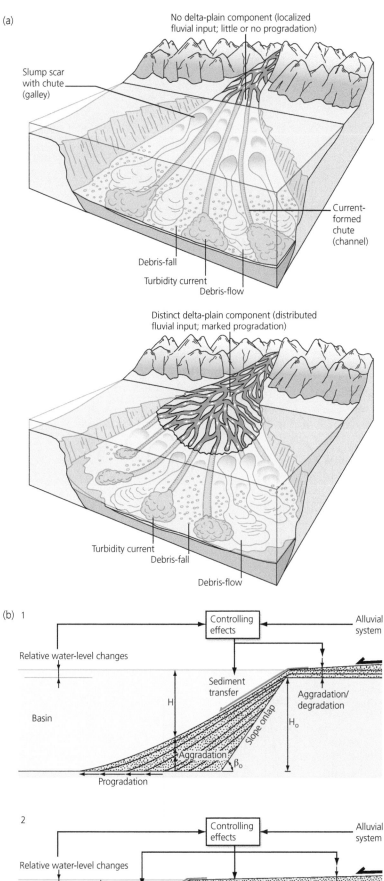

(a)

No delta-plain component (localized fluvial input; little or no progradation)

Slump scar with chute (galley)

Current-formed chute (channel)

Debris-fall

Turbidity current

Debris-flow

Distinct delta-plain component (distributed fluvial input; marked progradation)

Turbidity current

Debris-fall

Debris-flow

(b) 1

Controlling effects

Alluvial system

Relative water-level changes

Sediment transfer

Aggradation/ degradation

H

H_o

Basin

Slope onlap

Aggradation

β_o

Progradation

2

Controlling effects

Alluvial system

Relative water-level changes

Sediment transfer

Progradation

Aggradation/ degradation

H

H_o

Basin

Progradation

Aggradation

Slope onlap

β_o

Figure 11.141 (a) Three-dimensional sketches and (b) profiles through (1) an underwater conical delta and (2) a Gilbert-type delta (Modified from Nemec, 1990).

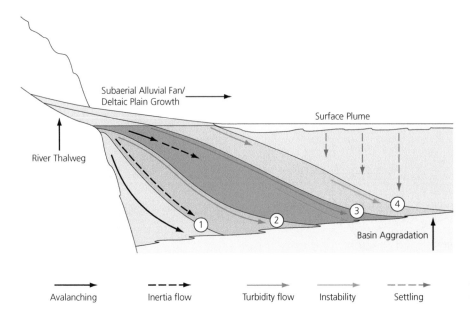

Figure 11.142 Sequential evolution of glacier-fed deltas, based on examples in British Columbian fjords. The sequence begins with debris avalanching and then progresses to inertia flows, to turbidity flows and finally to slope instability (Redrawn from Prior and Bornhold, 1990).

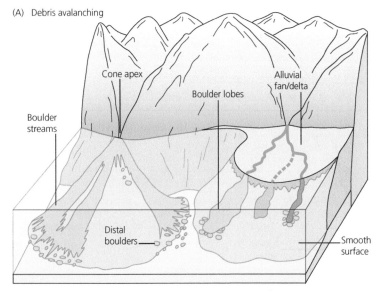

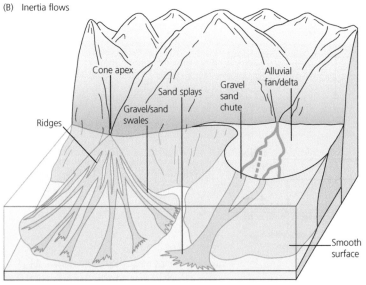

Figure 11.143 Schematic morphology of some subaqueous mass movements. (A) Debris avalanching (coarse-grained cohesionless debris flow and fall). (B) Inertia flows (sandy cohesionless debris flow). (C) Turbidity flows (underflows). (D) Slope instability (slumps, slides and cohesive debris flows) (Redrawn from Prior and Bornhold, 1990).

(C) Turbidity flows

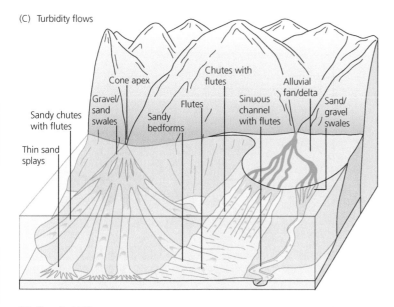

(D) Slope instability

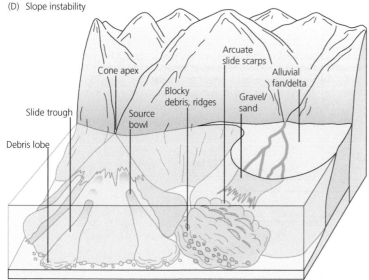

Figure 11.143 (*Continued*)

Figure 11.144 Ice-contact delta on the Knud Peninsula, Ellesmere Island. The delta (bottom right with surface pond) is connected to a contemporaneous sub-horizontal lateral moraine which extends into the distance along the fjord wall (identified by arrows further up fjord). Both features were deposited at the margins of a receding fjord glacier at the end of the last glaciation (D.J.A. Evans).

subsequently partially overridden and deformed by a minor readvance of the Loch Lomond glacier (Phillips et al., 2002; Benn et al., 2004).

The proximal parts of ice-contact deltas undergo slumping due to the melt-out of buried glacier ice or the removal of support from the ice-contact face. Typical structures include slump folds and normal faults, and it can sometimes be demonstrated that slumping occurred syn-depositionally – that is, delta deposition took place while underlying and adjacent ice was wasting. For supraglacial deltas, disturbance may be so extensive that the original morphology of the delta is almost totally destroyed.

Relationships between developmental sequences in ice-contact deltas and ice retreat rates have been considered by Shaw (1977). At relatively stable glacier termini, normal Gilbert-type deltas can develop because of the low ice retreat to sedimentation ratio (fig. 11.146A). If the terminus then retreats a short distance to another stable position, as often happens in glacilacustrine and

Figure 11.145 Ice-contact delta sediments from a glacilacustrine sedimentary sequence at Drumbeg Quarry, near Drymen, Scotland. A full section sketch is presented in figure 10.9a. (a) Subaqueous gravelly mass flow units cross-cutting subaqueous cross-bedded sands at the centre of the section (LFA 1a in fig. 10.9a). (b) Delta foreset beds overlying the cross-bedded sands (LFA 1b in fig. 10.9a). Glacitectonic disturbance is evident as low-angle shear planes (D.J.A. Evans).

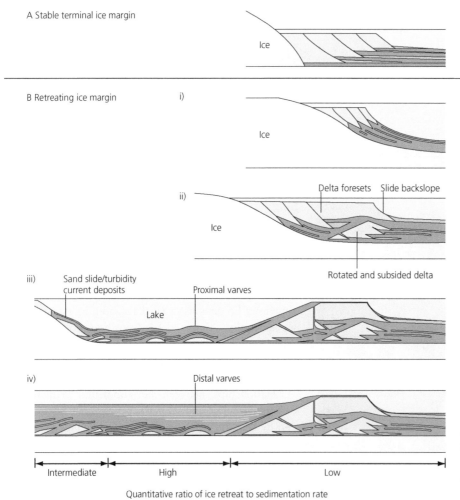

Figure 11.146 Reconstructions of depositional sequences in alpine glacial lakes during (A) glacier occupancy and ice-contact delta production; and (B) glacier retreat and ice-contact/supraglacial deposition of deltaic and distal sediments. Note delta collapse and the qualitative ratio of ice retreat to sedimentation rate (Redrawn from Shaw, 1977).

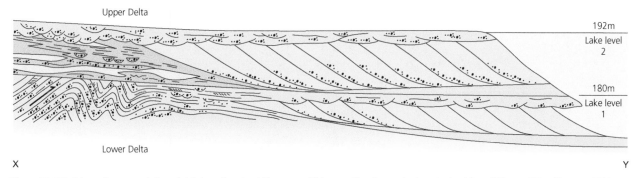

Figure 11.147 Schematic representation of deltaic sediments at Rhosesmor, Wales, resulting from a rise in water level from 180 m to 192 m (Thomas, 1984, reproduced with permission of Wiley).

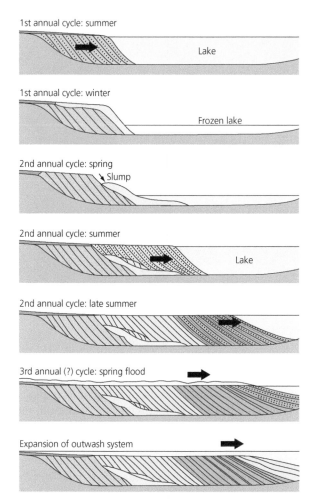

Figure 11.148 Model of seasonally controlled delta development based on sediments exposed near Neple, Poland (Mastalerz, 1990, reproduced with permission of the International Association of Sedimentologists).

glacimarine settings, the first delta will subside and be buried by a second deltaic sequence (fig. 11.146B). Subsidence in each delta occurs only on the ice-contact face. If the glacier margin retreats rapidly, the increased ice retreat to sedimentation ratio means that delta formation is replaced by deposition of grounding-line fans and distal rhythmites, including varves (fig. 11.146B: iii and iv).

Changing water depth during delta formation can also be recorded in internal structures (Helle, 2004; Russell, 2007). Falling water level results in delta incision and the formation of inset deltas graded to the new base level, or the production of channel-mouth bars, whereas rising water level results in 'stacked' delta sequences one above the other (fig. 11.147), or the draping of coarse-grained depo-centres by silts and muds. Mastalerz (1990) has interpreted Pleistocene deltaic sequences near Neple, eastern Poland, in terms of an annual cycle of water-level changes. According to his interpretation, the cycle consists of the following stages (fig. 11.148):

(1) *Summer*: water levels are high because glacier ablation rates are at a maximum, and deposition is by normal foreset progradation.

(2) *Winter*: ablation ceases, water levels fall and the ground surface freezes. Deltaic deposition ceases.

(3) *Spring*: the ground thaws, but water levels remain low, and large-scale slumping occurs on the exposed foreslope.

(4) *Summer*: water levels rise once more, and foreset progradation is resumed.

Similarly, periodic changes in ice-dammed lake water levels due to GLOFs from the margin of the Greenland Ice Sheet have been linked by Russell (2007) to changes in delta form and sediments.

CHAPTER 12
LANDSYSTEMS AND PALAEOGLACIOLOGY

Hummocky moraine & prairie sky, southern Alberta, Canada (D.J.A. Evans)

12.1 INTRODUCTION

The study of large-scale patterns of glacial sediments and landforms and their significance is one of the most fascinating and controversial branches of glacial geology. When viewed at the scale of an entire glaciated basin, the character and distribution of glacial sediments and landforms can reveal shifting patterns and processes of glaciation over complete glacial cycles, and provide a valuable complement to short-term studies of modern glacial environments. This information can then be used to inform the types of numerical ice sheet reconstructions reviewed in section 5.3 (e.g. Marshall et al., 2000, 2002; Siegert and Dowdeswell, 2004; Hubbard et al., 2005, 2006; Boulton and Hagdorn, 2006). However, reconstructions of former glacial environments are only as good as the observations and assumptions that

go into them, and it is at the largest scale that errors in interpretation become most magnified. For example, if a certain diamicton unit is interpreted as a basal till, there are very different implications for former glacier extent and dynamics than if it is interpreted as a distal glacimarine deposit. Furthermore, the interpretation of ancient landforms relies on analogues with modern examples, and an inappropriate choice of analogue can lead to potentially misleading conclusions. This problem has become less serious due to our widening range and deeper understanding of modern glacial landsystems, but there is always room for improvements.

Despite the difficulties raised by the interpretation of ancient landscapes, painstaking work over the past few decades has revealed a very detailed picture of ancient ice masses and their evolution. This is particularly the case for the great Pleistocene ice sheets of North America and

Europe, where the geological record documents complex but coherent patterns of environmental change over much of the last glacial cycle. Furthermore, the ice sheet record is being increasingly closely linked with regional and global climatic and oceanographic changes, and as a result we have a clearer impression of both long and short timescale oscillations of ice masses and their sediment–landform signatures.

In this chapter we review basin-scale distributions of glacial landforms and sediment packages, focusing on what such features reveal about patterns of environmental change. This provides an opportunity to synthesize the material presented in previous chapters, and to consider the evidence for former climates, hydrology, glacier dynamics, patterns of erosion and deposition, and sea-level change over complete glacial cycles in a variety of settings. Although we focus on the glacial geology of the Pleistocene, the period for which the evidence is most complete and accessible, it should not be forgotten that the Earth's glacial record extends into the much more distant past, and that an understanding of glacial systems can reveal a great deal about the geography of ancient continents and seas. In the following sections, we first introduce the concepts of palaeoglaciology, landsystems and sequence stratigraphy, and then synthesize the evidence used in reconstructing former glacier beds and margins. We then review glacial landsystems under headings defined by a combination of glaciation style, glacier dynamics and depositional setting. This is followed by a section on reconstructing glaciers and ice sheets. We conclude by reviewing evidence for past glaciations of Mars, where glacial geomorphological studies meet their most challenging frontier.

12.2 APPROACHES TO LARGE-SCALE INVESTIGATIONS

Investigations of glacial geology at regional and continental scales have a long history. Such studies now routinely emphasize the links between sediment–landform associations and genetic processes, which, in turn, are informed by research on modern glacial processes. This provides us with a framework for using geological evidence to reconstruct ice sheets and glaciers of the past (palaeoglaciology) and thereby contribute to a growing body of knowledge of long-term environmental change. Two main approaches are discussed here: (1) the study of glacial landsystems and process–form models; and (2) sequence stratigraphy. To put these in spatial and temporal context, we first review the more general themes of long-term tectonic settings for glaciation and the basic ingredients used in palaeoglaciological reconstruction.

12.2.1 TECTONIC SETTINGS FOR GLACIATION

The geological history of a region, including its ongoing tectonic evolution, combines with climatic cycles to dictate the style of glaciation (e.g. mountain ice fields, ice sheets) and the erosional and depositional impact of glaciation through time. This is best illustrated for the Earth's long timescale glacial record in figure 12.1

Age	Trench	Forearc		Backarc	Foreland	Intracratonic/Aulacogenic	Passive margin
L. Cenozoic <36 Ma	Gulf of Alaska		Bransfield Strait			Kleszczow Basin Poland North Sea Rift Ross Sea/Weddell Sea Rift Alaskan Interior	Eastern Canadian Continental margin N.W. European Continental margin
L. Palaeozoic 350–250 Ma	Palaeo-Pacific margin of Gondwana: Eastern Australia Antarctica			Karoo Basin S. Africa		Kalahari Basin Arabian Peninsula Parana Basins, Brazil, Indian Basins Australian Interior Basins	
Ordovician c.400 Ma				West Africa? Central Saudi Arabia			
L. Proterozoic 800–550 Ma	Damara mobile belt Arabian shield Paraguay-Araguaia fold belt Tiddiline basin, North Africa		Gaskiers F'M. NFLD Boston Bay Group	Bakoye GP Jbeliat GP West Africa			Palaeo-Atlantic margin of Laurentia Palaeo-Pacific margin of Laurentia
E. Proterozoic 2100–1800 Ma							Huronian supergroup: Gowganda FM
Archean >2500 Ma			Witswatersrand Basin S. Africa				

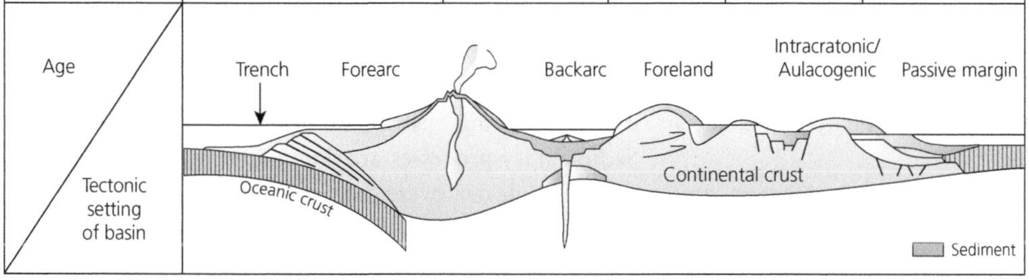

Figure 12.1 The geological/tectonic settings of the Earth's long timescale glacial record, demonstrating relationships between depositional environment and basin characteristics (modified from Eyles, 1993). Each of the geographical case studies is reviewed in detail in N. Eyles and Young (1994) and Eyles (1993), and summarized in Benn and Evans (1998); here we review the general characteristics of sedimentary sequences in each of the tectonic settings, based on such case studies.

(a) (i)

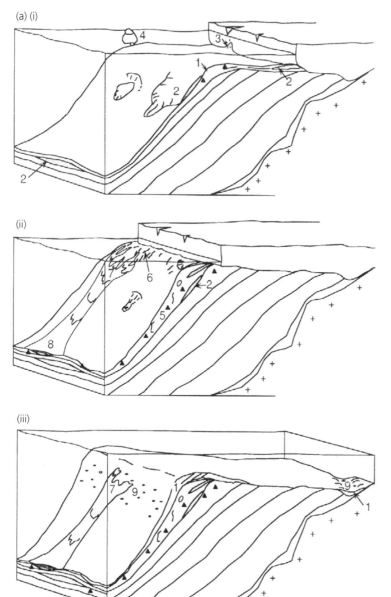

(ii)

(iii)

Figure 12.2 Models of glacial sedimentation (a) on active forearc and (b) passive continental margins: (1) proglacial muds; (2) massive, lenticular debris flows; (3) subglacial meltwater stream outlet; (4) icebergs and ice-rafted debris; (5) small debris flows; (6) cut-and-fill channels; (7) large submarine channels; (8) submarine channels on the middle slope, losing channel geometry; (9) gas-escape structures (pockmarks); (10) ice-grounding on the outer shelf in forearc basin only (Gipp, 1994; reproduced with permission of Cambridge University Press).

(Eyles, 1993). This idealized cross section conveys contrasting glacial depositional environments in a range of geologic/tectonic settings through geological time. The occurrence of large-scale geologic structures such as mountain chains and cratons provides the inception centres for ice masses and the bordering depositional basins for glacial sedimentation. We can view the long-term depositional history of the Earth's surface through the principles of basin analysis (Miall, 2000), and glaciated basins are no exception, other than that they are often more complex than other sedimentary systems. This complexity is as much a function of glaciation style as basin characteristics, but ultimately it can be regarded as a product of geological setting.

Contrasting glacial depositional styles at active and passive plate margins are illustrated in figure 12.2 (Gipp, 1994). An example of the former is the Yakataga Formation in the Gulf of Alaska, where glacimarine sedimentation has continued in a subsiding forearc basin from the late Miocene to the present (e.g. Eyles and Lagoe, 1990). The absence of a broad continental shelf prevents glacier ice from advancing far offshore, so a thick sediment pile accumulates around the ice margins (fig. 12.2a). The filling of a forearc basin results in an aggradational shelf extension, but ongoing subsidence results in the preservation of long sedimentary sequences. In contrast, at passive continental margins, glacier ice can advance across wider continental shelves, prograding sediments in seaward-dipping beds and truncating their tops, especially due to lowering eustatic sea levels during glacial stages. Sedimentary processes are dominated by debris flows initiated by glacier overriding and deformation of the prograding sediment pile (fig. 12.2b). The depositional record at both active and passive plate margins can be contrasted,

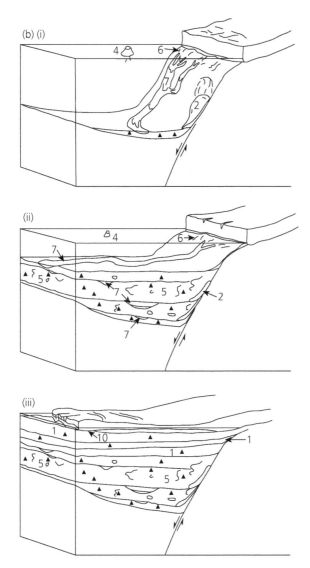

Figure 12.2 (*Continued*)

in turn, with the glacigenic sedimentary record of stable cratons at mid to high latitudes, where repeated occupancy by ice sheets during glacial stages results in the net removal of sediment and the exposure of areally scoured bedrock surfaces over vast regions. More detailed examples of the glacigenic sedimentation in different tectonic settings are given in section 12.5.2.

12.2.2 INGREDIENTS FOR PALAEOGLACIOLOGICAL RECONSTRUCTION

The reconstruction of former ice sheets and glaciers requires a wide range of complimentary information, including the extent, thickness and flow patterns of the ice, and their changes through time. It is therefore both a spatial and a temporal problem that integrates glacial geomorphology, glaciology and geochronology (Kleman et al., 1997, 2006; C. D. Clark et al., 2000, 2006; Napieralski et al., 2007). For such reconstructions to be

reliable, inferences about the process–form relationships of landform–sediment assemblages must be firmly grounded in our knowledge of modern glacier systems. Andrews (1982) and, later, Clark (1997) and Kleman et al. (2006) identified the main ingredients required for effective palaeoglaciological reconstruction, highlighting in particular the need to focus on whole systems rather than research areas of map-sheet scale. The main ingredients are as follows:

(1) the ice sheet/glacier footprint (subglacial bedforms, drainage patterns, till characteristics and distribution, erosional forms);
(2) the ice sheet/glacier margins (moraines, meltwater networks, glacifluvial forms);
(3) the ice sheet/glacier thickness (upper limit of till/glacial landforms, palaeonunataks);
(4) the chronology of events (dating ice margins, relative dating of flow events).

The task of using the ingredients to reconstruct an ice body is made difficult by the time-transgressive nature of ice sheet growth and decay, and the superimposition and partial-to-total reworking of the evidence of older events. Glacial landscapes commonly consist of superimposed depositional systems of different ages, resting on older erosional surfaces. Such landscapes have been compared to a *palimpsest*, a parchment that was reused by early monastic scholars when writing materials were in short supply (Kleman, 1992, 1994). Older text was erased to make way for other writing, but the erasing was not totally effective, and the older inscriptions tend to be partially visible still. Thus a palimpsest provides a vivid metaphor for landscapes in which the products of ancient processes are partially visible below more recent forms. Each new generation of the landscape inherits the topography, substrate type, drainage characteristics and other factors established by the previous generation, so that older landscape components act as important constraints on subsequent landscape evolution. In glaciated terrains, large-scale erosional forms are generally the cumulative effect of multiple glacial cycles, whereas the onshore sedimentary record typically records only the most recent glaciation and fragments of older events. This is particularly true of continental interiors, where widespread erosion and sediment reworking during glacier advances ensure that older glacigenic sequences rarely survive within younger glacial limits. Glacimarine sequences have greater preservation potential, particularly in subsiding basins such as the central graben of the North Sea, where >1 km of Quaternary sediments are preserved. On tectonically stable shelves, glacimarine sequences are more likely to be at least partially removed by glacial or proglacial erosion (sections 12.2.1, 12.5; Eyles and Eyles, 1992; Eyles, 1993; Gipp, 1994). Nonetheless, significant advances have been made in reconstructing temporally

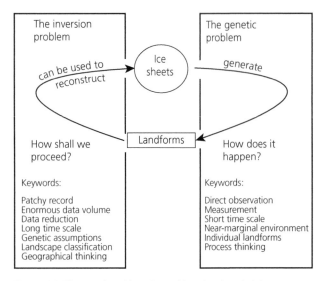

Figure 12.3 The genetic and inversion problems in palaeoglaciology as depicted in a closed flow diagram by Kleman and Borgström (1996) and Kleman et al. (2006). In chapter 11 we concentrated on the genetic problem; in this chapter we develop this by addressing the inversion problem.

and spatially complex ice sheet systems, particularly where systematic and consistent mapping programmes have covered large glaciated regions and the emerging patterns in glacigenic landform–sediment assemblages have been reconciled with modern glacial systems.

A set of protocols has been set up by palaeoglaciologists in order to facilitate a consistent approach to ice sheet reconstruction. These protocols stem from the 'inversion' concept, which is the application of genetic models to glacial landform–sediment assemblages so as to determine the form and dynamics of the ice sheet responsible for their production (fig. 12.3; Kleman and Borgström, 1996; Kleman et al., 2006). In chapters 8–11, we dealt predominantly with genetic issues, whereby we assessed the process–form relationships of glacial landforms and sediments; these are depicted on the right-hand side of the flow diagram in figure 12.3. In this chapter we will elaborate more on the inversion concept, the left-hand side of figure 12.3.

Although its primitive roots can be traced back to early ice sheet reconstructions in the late nineteenth century, the development and application of the inversion concept has resulted in increasingly sophisticated ice sheet reconstructions over the last two decades (fig. 12.4). Procedures vary slightly between researchers, but certain assumptions are widely adopted, resting on the glaciological principles set out in part I of this book and the process–form relationships reviewed in chapters 8–11. The following assumptions are acknowledged by Kleman and Borgström (1996), Kleman et al. (1997, 2006) and Clark (1999):

(1) the location of the ice/water phase boundary at the ice base is the primary control on landform creation, preservation and destruction;

(2) basal sliding requires a thawed bed;

(3) subglacial lineations only form when basal sliding occurs;

(4) subglacial lineations (erosional and depositional) are aligned with local ice flow and perpendicular to ice surface contours at the time of their formation;

(5) frozen bed conditions and/or areas of low basal ice velocity (e.g. ice divides) will inhibit the rearrangement of the subglacial landscape;

(6) regional deglaciation is accompanied by the production of spatially coherent but diachronous meltwater features, such as channels, eskers, lake shorelines and spillways;

(7) eskers form time-transgressively behind a receding ice front;

(8) lateral meltwater channels form the major landform record of deglaciation during frozen bed conditions.

In order to convert this knowledge – which is by necessity most often based on reductionist-scale investigations – into palaeoglaciological reconstructions, we need to develop holistic conceptual models for the genesis of landform–sediment assemblages. Hence we now review the landsystems approach to investigating glaciated terrain.

12.2.3 LANDSYSTEMS AND PROCESS–FORM MODELS

Landsystems are strictly defined as areas of common terrain attributes, different from those of adjacent areas, in which recurring patterns of topography, soils and vegetation reflect the underlying geology, past erosional and depositional processes, and climate. The identification of landsystems therefore constitutes a holistic approach to terrain evaluation, wherein the geomorphology and subsurface materials that characterize a landscape are genetically related to the processes involved in their development. The *landsystem* is the upper level in a hierarchy of terrain classification. At the lowest level are *land elements*, or individual landforms; at the intermediate level are *land facets* or groups of land elements; and *landsystems* are composites of linked land facets. Glacial examples of land elements, facets and landsystems could be, respectively, drumlins, drumlin fields and the whole assemblage of forms representing the former glacier bed. Terrain evaluation is usually restricted to the uppermost few metres of the ground, but Eyles (1983a) stressed that glacial landsystems must also include all genetically related subsurface sediments. The regional development of landsystem models allows subsurface conditions to be predicted from surface morphology, an ability that has profound implications for resource assessment and civil engineering works (Eyles, 1983a). Since the introduction of *glacial landsystems* as glacigenic landform–sediment assemblages based on engineering properties, they have been expanded, refined and redefined, first in the 1970s to 1980s, based on styles of glaciation (i.e. *subglacial,*

Figure 12.4 Four examples of ice sheet reconstruction using the inversion principle. (a) Boulton et al. (1985). (b) A single snapshot from Dyke and Prest's (1987) sequence of maps. (c) Boulton and Clark (1990a, b). (d) Kleman et al. (1997). In (c) and (d), numerous 'flow sets' of subglacial bedforms are overprinted. A comparison of the blended subglacial signature in (a) and the multi-phase depiction of bedforms in (c) clearly highlights our progress in understanding the mobility of ice sheet dispersal centres (Kleman et al., 2006; reproduced with permission of Blackwell).

supraglacial, glaciated valley landsystem; e.g. Eyles, 1983a), and then in the book *Glacial Landsystems* (Evans, 2003a), based on the recognition of a continuum of glaciation styles and dynamics.

No actual landscape can be expected to possess all the characteristics of a particular landsystem model, and will almost certainly display unique features associated with local conditions. Additionally, landsystem development will change through time depending on changes to thermal regimes and glacier morphology. Rather than constitute a weakness in the landsystem approach, this demonstrates that we can identify complex behaviour through landform–sediment signatures; this is crucial to our long-term goal of recognizing process–form continuums and building spatial and temporal variability into our palaeoglaciological reconstructions.

The landsystem approach can be placed in a dynamic context in *process–form models*. Such models were first applied to glacial landscapes by Clayton and Moran (1974), who used spatial patterns of glacial landforms

and sediments in North Dakota to infer changes in glacial dynamics during ice margin advance and retreat. This formed a natural development of the practice of mapping and classifying glacial landforms according to common origin and age (e.g. *landform associations*). Process–form models emphasize the interrelationships of specific landform–sediment associations at both local and regional scales, and their genetic significance in terms of migrating zones of deposition and erosion. Process–form models can be applied at a wide range of scales, from the marginal zone of a fluctuating valley glacier to palaeoglaciological reconstructions of whole ice sheets (e.g. Sugden, 1977; Dyke et al., 1982; Sharp, 1985b; Boulton and Clark, 1990a, b; Mooers, 1990b; Kleman et al., 2006). An expanding database on the evolution of sediment–landform assemblages in recently deglaciated terrain affords us greater confidence in applying the principle of actualism (uniformitarianism) in palaeoglaciology, especially where highly detailed and high-resolution surveys facilitate a more quantitative approach (e.g. Kjær and

Krüger, 2001; Schiefer and Gilbert, 2007; Schomacker and Kjær, 2008). In fact, confidence in these procedures is so buoyant that the actualist methodology is now routinely applied to the surface of Mars (section 12.6)!

12.2.4 SEQUENCE STRATIGRAPHY

A powerful approach to the study of glacial depositional systems has arisen from the concepts of *sequence stratigraphy*, which deals with large-scale depositional cycles in the rock record (Emery and Myers, 1996; Catuneanu, 2006). Originally developed by petroleum geologists from the Exxon Production Research Company in Houston, USA, as a tool for locating potential hydrocarbon reservoirs, sequence stratigraphy seeks to identify large-scale *sequences* of conformable, genetically related strata bounded by unconformities or other major discontinuities (van Wagoner et al., 1988; Walker, 1992). Within such sequences, strata may be grouped into *systems tracts*, defined as linkages of contemporaneous depositional systems (fig. 12.5). The original emphasis was on shallow marine environments, and systems tracts were originally defined in terms of major erosional and depositional episodes associated with global cycles of sea-level change. Three types of systems tract were recognized, formed during periods of marine transgression, highstand and regression.

Some principles of sequence stratigraphy have been applied to glacial depositional systems by Boulton (1990),

Eyles and Eyles (1992) and Martini and Brookfield (1995). It has been found necessary, however, to modify the approach before it can be applied to the glacial case. First, patterns of local sea-level change in glaciated areas are complicated by the isostatic depression of the crust by ice loading, and variation in crustal loading across a glaciated basin can cause one part to experience a fall in relative sea level while sea levels are rising in another. Hence the impact of a glacio-isostatically induced sea-level trend will vary across glaciated basins, and the *forced regressions* of glacio-isostatic rebound can be interrupted by short-lived transgressions due to local ice oscillations; for example, Helle (2004) described a complex sequence stratigraphy in an ice-contact depo-centre in Norway, reflecting crustal reloading during the Younger Dryas. The grouping of contemporaneous depositional systems into systems tracts on the basis of regional marine erosion surfaces is thus impractical (Eyles and Eyles, 1992). Second, glacial depositional systems located on land, in inland lakes or below grounded glacier ice, are generally controlled by factors other than global sea-level change.

Brookfield and Martini (1999) have summarized this problem with respect to ice contact deep-water successions, by comparing the main controls on sequence stratigraphy in non-glacial and glacial settings. Those controls are: (1) the water level relative to the depositional surface, and hence the *accommodation space* for sediment accumulation, and (2) the point of sediment injection. In non-glacial settings, these controls will vary

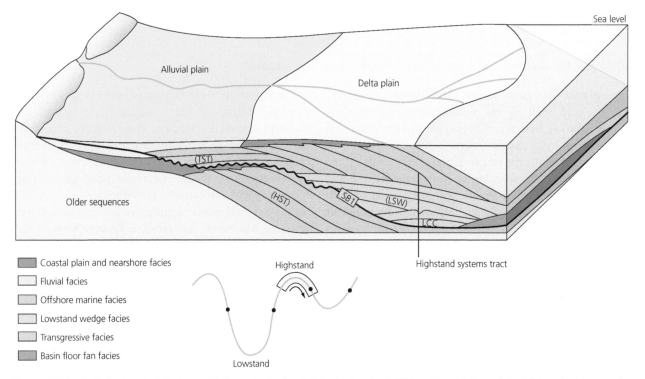

Figure 12.5 Hypothetical systems tract development during sea-level cycles. Highstand systems tracts (HST) are formed during periods of high sea level; transgressive systems tracts (TST) are formed during rising sea level; and lowstand wedges (LSW) are formed at times of low sea level. Sequence boundary 1 (SB1) is an erosion surface separating two *sequences*. LCC = leveed channel complex (Redrawn from Walker, 1992).

in tandem, allowing systems tracts to be defined by water-level rise and fall (fig. 12.5), whereas in glaciated basins, the accommodation space and the injection points are controlled by water level and the position of the glacier terminus (fig. 12.6a; Winsemann et al., 2004). Further complications arise due to the common situation where the injection of sediment into the water column from a glacier tends to be below the water level rather than at its surface (fig. 12.6a), rendering the sequence stratigraphic terms 'highstand' and 'lowstand' meaningless in ice contact settings. In order to avoid the unnecessarily complicated terminology that would arise from the recognition of both water level and glacier input controls on glacial sequence stratigraphy, Brookfield and Martini (1999) proposed the use of *allostratigraphy*, or descriptive and non-genetic sequence stratigraphy, wherein mappable sedimentary bodies are separated by discontinuities (fig. 12.6). In glacial systems, such discontinuities or *sequence boundaries* can be erosional breaks caused by

glacier readvance (fig. 12.6b) or the emptying of glacial lakes (e.g. Bennett et al., 2002).

The logical basis for defining *glacial systems tracts* is the *glacier advance-and-retreat cycle*. Within the maximum glacier limits, depositional systems deposited during advance and retreat will be separated by a time-transgressive subglacial erosion surface or a wedge of subglacially deposited or deformed sediment (Berthelsen, 1978; Boulton, 1996a, b). Beyond the ice limit, complete depositional sequences may be preserved in the form of distal-proximal-distal successions, which, if they are directly glacier-fed, will record the migration of depositional systems in response to glacier advance and retreat (Boulton, 1990; Eyles and Eyles, 1992). In glacilacustrine settings, unconformities can arise through lake drainage, which may record short-lived events or more lengthy (i.e. interstadial, interglacial) episodes. The application of sequence stratigraphy/allostratigraphy to a complex glacigenic depositional setting is well illustrated by the

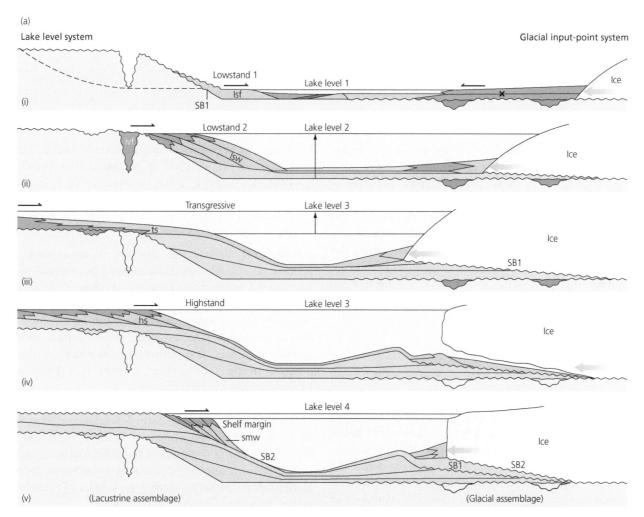

Figure 12.6 Idealized sequence stratigraphy to explain the correlation and complexity of facies assemblages due to changing lake levels and glacier input points due to calving. (a) Localized control of successive units due to lake level change and ice front calving (lsf = lowstand fan; SB = sequence boundary; lsw = lowstand wedge; ts = transgressive surface; hs = highstand; smw = shelf margin wedge of shelf margin system tract; ivf = incised valley fill). (b) Correlation of facies assemblages in lake level system and glacial input system showing systems tracts and bounding surfaces in space (A) and time (B) (WL = water level; hst = highstand systems tract; tst = transgressive systems tract; mfs = maximum flooding surface) (After Brookfield and Martini (1999)).

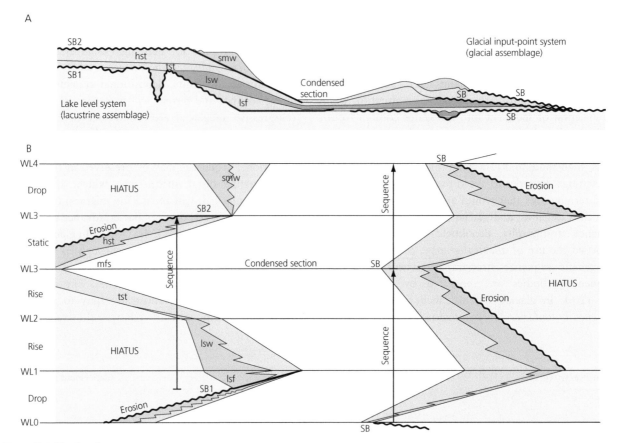

Figure 12.6 (*Continued*)

work of Thomas et al. (2004), based on the Quaternary of the Isle of Man. For example, in the local Jurby Formation they identified four offlapping packages of lithofacies assemblages separated from one another by unconformities or sequence boundaries (fig. 12.7). These packages record four readvances during the overall recession of the British-Irish Ice Sheet (BIIS). Throughout this chapter, we emphasize the role of the glacial cycle in controlling the large-scale sedimentary architecture of glaciated basins, but recognize also that shorter timescale events, such as glacier readvances, also introduce stratigraphic signatures.

12.2.5 STRATIGRAPHIC ARCHITECTURE

The large-scale form or *stratigraphic architecture* of glacigenic depositional systems is essential to our understanding of environmental change in glacially influenced basins (e.g. Boulton, 1990; Eyles and Eyles, 1992; Hambrey, 1994; Dowdeswell et al., 2002). Although the lateral and vertical relationships between glacial sediment–landform associations are very varied, they reflect the interaction of relatively few groups of variables. The most important are:

(1) the topography and tectonic setting of the basin;

(2) the extent, configuration and dynamics of glacier and floating ice and the position of associated depo-centres;

(3) changes in water depth due to eustatic and isostatic cycles and the life cycle of proglacial lakes (Powell, 1984, 1990, 1991, 2003; Boulton, 1990; Eyles and Eyles, 1992; Martini and Brookfield, 1995; Powell and Domack, 1995; Brookfield and Martini, 1999; Powell and Cooper, 2002; Teller, 2003; Vorren, 2003).

Powell (1984) defined four depositional zones, each of which is associated with distinctive sets of processes and stratigraphic architecture. These are: (1) subglacial; (2) ice-proximal; (3) ice-shelf; and (4) ice-distal. The zones were originally defined for glacimarine environments, but are also applicable to large proglacial lakes. Typical stratigraphic sequences that reflect the interplay of these processes in space are idealized in figure 12.8 (Eyles and Lazorek, 2007).

The *subglacial zone* is reviewed in detail in section 12.4. Facies deposited in this zone form downglacier-thickening sheets of basal till and associated lenses of water-sorted material. Sediment dynamics in the subglacial zone exert an important control on the delivery of sediment to the glacier margin, either by meltwater, subglacial till deformation or within sequences of basal ice.

The *ice-proximal zone* encompasses terrestrial and subaqueous environments adjacent to the glacier grounding line. The thickness of sediment packages deposited in the ice-proximal zone is related to rates of sediment

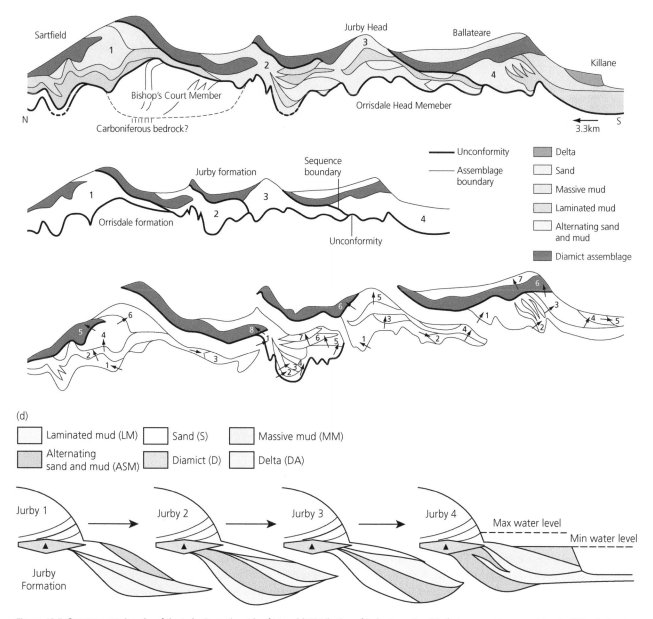

Figure 12.7 Sequence stratigraphy of the Jurby Formation, Isle of Man. (a) Distribution of Jurby Formation lithofacies assemblages overlying the Bishop's Court and Orrisdale Head Members of the Orrisdale Formation. (b) Summary of the four lithofacies assemblage packages. (c) Exploded diagram of each package showing the order of deposition. Note that the diamict assemblage is interpreted as a subglacial till. (d) Simplified reconstruction of the marginal readvances recorded by the four packages (Redrawn from Thomas et al., 2004).

delivery and removal, and rates of advance or retreat of the glacier front. At calving margins, for example, high rates of sediment delivery and slow retreat or quasi-stable conditions will encourage the accumulation of large volumes of sediment, whereas during rapid calving retreat a large proportion of debris will be rafted away, restricting the amount of ice-proximal deposition. At the tidewater margins of temperate glaciers, large amounts of coarse debris are deposited in grounding-line fans, and fine-grained sediments are carried away in turbid plumes to form cyclopsams and cyclopels. In contrast, grounding lines below ice shelves in polar environments are more commonly associated with grounding-line wedges composed of mass-flow deposits due to the

limited availability of subglacial meltwater (sections 10.5, 10.6). Sediment gravity flows are common in the ice-proximal zone, due to high sedimentation rates, ice push, iceberg calving, storm waves, tidally induced pressures below ice shelves and morainal bank collapse (Powell, 2003). Subaqueous outwash and sediment flows both commonly form prograding beds dipping away from the glacier margin, known as *clinoforms*. The upper parts of such clinoforms may be erosionally truncated during glacier advance, forming a sub-horizontal erosion surface overlain by basal till (cf. Lønne, 1995, 2001; Lønne et al., 2001).

The *ice shelf zone* occurs in polar climates where ice below the pressure-melting point floats and produces

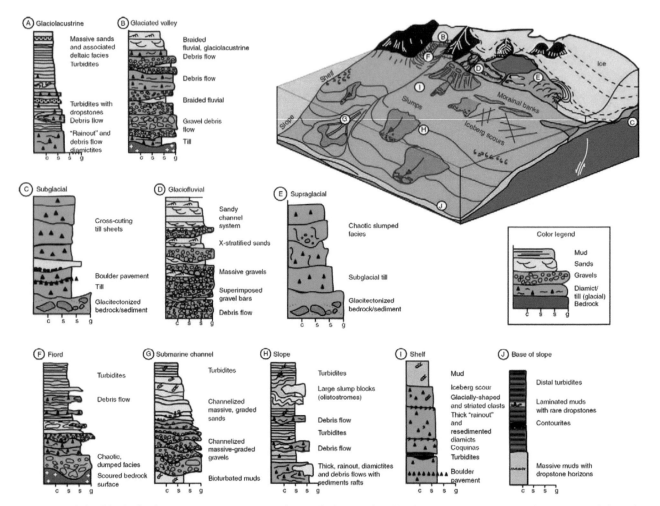

Figure 12.8 Idealized family of sedimentary sequences representing the interplay between depositional processes in an ice-contact subaquatic setting (Eyles and Lazorek, 2007; reproduced with permission of Elsevier).

glacier and sea-ice ice shelves. Deposition in this zone is dominated by undermelt and rain-out, forming drapes of dropstone diamicton and mud (section 10.6). Patterns of deposition vary with the debris content of the ice and the location of basal melting and freezing zones beneath the ice shelf, but in general deposits are thickest near the grounding line and thin distally. Below some ice shelves, basal freeze-on of seawater allows significant amounts of basal debris to be transported to the ice shelf edge, forming dropstone deposits in more distal locations. The geometry of undermelt deposits is to a large extent determined by the local topography. In areas with significant relief, reworking by mass movements relocates sediment and forms thick infills in topographic basins.

The *subaqueous ice-distal zone* is dominated by ice-rafted debris and muds formed by the settling of suspended mineral grains and microfaunal remains (section 10.6). Eyles and Eyles (1983) and Eyles et al. (1985) argued that sedimentary sequences in distal glacilacustrine and glacimarine environments are determined by the relative importance of rain-out, current reworking and gravitational resedimentation (fig. 12.9). In glacimarine

environments, deposition and reworking by these three processes can produce sequences of massive and stratified sediment tens to hundreds of metres thick and spanning several glacial cycles. The long-term preservation potential of such sequences is high, and they are well represented in the ancient geological record (Eyles, 1993). Iceberg scouring is common in high-latitude distal glacimarine environments, churning the bottom sediments and producing massive, structureless ice-keel turbates (section 10.6.6).

Depositional systems deposited in each of these zones can be superimposed, recording glacial advance-and-retreat cycles. This is illustrated in figures 12.10 and 12.11, which show changing patterns of sedimentation associated with the growth and decay of a water-terminating ice sheet (Boulton, 1990; Eyles and Eyles, 1992). During ice sheet advance, the ice-proximal zone migrates over the former ice-distal zone, depositing ice contact facies associations on top of distal sediments. Such associations tend to be eroded, transported or redeposited in the subglacial zone once glacier ice advances over the site. The ice sheet build-up phase is therefore typically

(a)

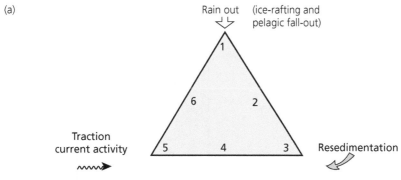

Dominant processes	Diamict lithofacies	Associated lithofacies
1 Rain out	Predominantly massive (Dmm) Planar geometries	Massive muds (Fm)
2 Rain out + resedimentation	Massive and stratified diamicts with flow structures, abundant silt and clay clasts. rafts (Dmm, Dms)	Massive muds (Fm), laminated silts and clays (turbidites) (Fl, Fld)
3 Resedimentation	Stratified and massive diamicts with flow structures abundant silt and clay clasts, rafts, variable grading characteristics (Dmm, Dms, Dmg). Fills and flattens irregular topography	Laminated silts and clays (turbidites) (Fl, Fld), graded and massive sands (Sg, Sm).
4 Resedimentation + traction current activity	Predominantly stratified diamicts with evidence of resedimentation and traction current activity (winnowed units, silt and sand stringers, rippled sands, flow structures, variable grading) (Dms, Dcs)	Traction bedded sands (Sr, St, Sp) graded sands (Sg) and deformed units of silty sand (Sd)
5 Traction current activity	Winnowed diamicts predominantly stratified (Dcs, Dms) Channelized geometries	Traction bedded gravels and sands (Gm, Sr, St, Sp)
6 Traction current activity + rain out	Massive and stratified diamicts with silt and sand stringers, rippled sands, some winnowed units (Dmm, Dms, Dcs)	Traction bedded sands and gravels (Sr, St, Sp, Gm). Some mud drapes (Fm)

Figure 12.9 Conceptual model of distal subaqueous sedimentation as a continuum between three end-member processes: rain-out (from suspended and ice-rafted sediment), traction current activity and gravitational resedimentation (Modified from Eyles et al., 1985).

represented by a sheet of subglacial till, and underlying glacimarine deposits are only likely to survive close to or beyond the maximum position of the margin. As the ice sheet retreats, the ice-proximal and distal zones migrate, in turn, over the former subglacial zone, commonly blanketing subglacial forms with a fining-upward drape. More complex successions result from ice-marginal oscillations and where deposition is also influenced by sediment from non-glacierized catchments or glacial meltstreams.

Water depth exerts a strong control on the distribution and character of subaqueous depositional systems, due to its influence on ice margin stability, the accommodation space available for sedimentation, and base levels for subaerial erosion (Eyles and Eyles, 1992; Lønne, 1995, 2001; Martini and Brookfield, 1995; Brookfield and Martini, 1999; Lønne et al., 2001). In glacimarine environments, patterns of sedimentation are influenced by isostatic depression and rebound during glacial cycles, which may be in phase or out of phase with global eustatic sea-level changes. The influence of sea-level change and sedimentation rates on depositional architecture in glacimarine environments has been considered by Bednarski (1988), using concepts similar to those of sequence stratigraphy. Figure 12.12 shows a hypothetical glacimarine depositional sequence produced during a sea-level transgression–regression cycle. The cycle begins during deglaciation, when receding ice allows marine waters to flood the basin, forming an erosional transgression surface (VII). A sequence of marine deposits is then laid down on top of the erosion surface, inter-fingering with terrestrial glacigenic sediments at the landward end (VI and V). The marine limit

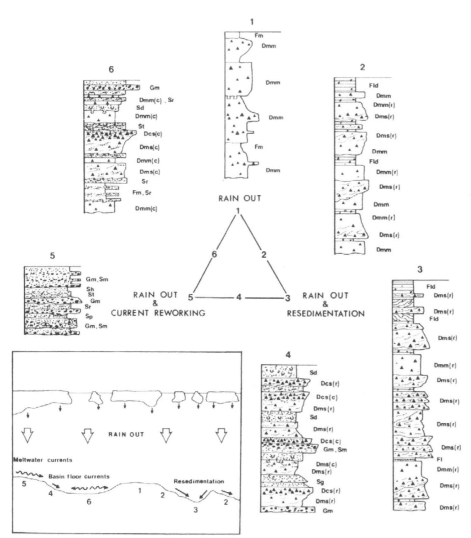

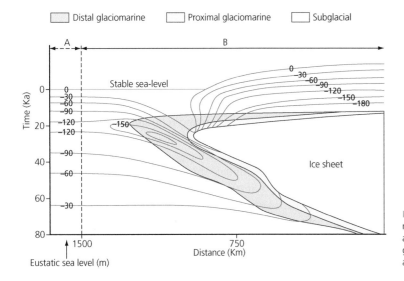

Figure 12.9 (*Continued*)

Figure 12.10 Time–distance diagram showing the migration of subglacial and glacimarine environments during a glacial cycle. The isolines show changes in sea level due to glacio-eustasy and glacio-isostasy (Eyles and Eyles, 1992, after Boulton, 1990).

marks the highest level reached by the sea. Subsequent regression can result in erosion, which truncates glacimarine sediments deposited during transgression and highstand (I). Alternatively, glacimarine sediment can be reworked into littoral features such as beaches, storm ridges or tidal flats (II, III and IV). Such shoreface reworking is very common around isostatically uplifted coasts, limiting the amount of glacimarine sediment

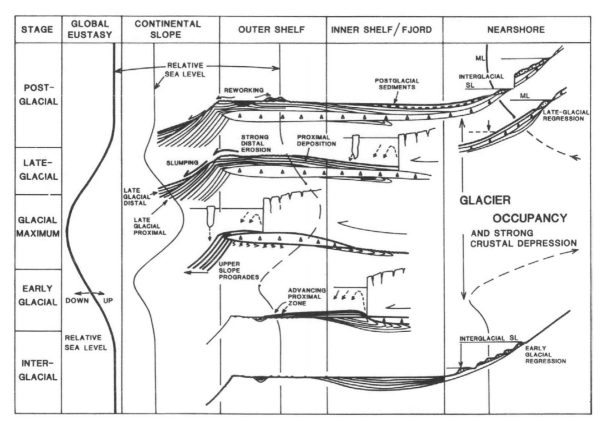

Figure 12.11 Conceptual model of glacimarine architecture in space and time, showing sediments produced on different parts of a continental margin throughout a glacial cycle. Relative sea levels are also depicted for each zone (Boulton, 1990; reproduced with permission of the Geological Society of London).

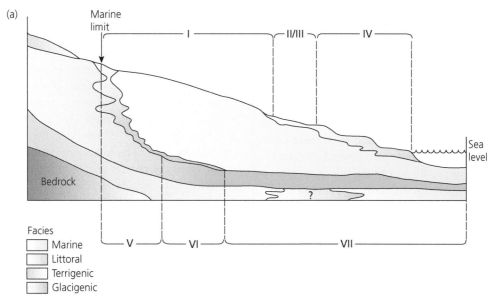

Figure 12.12 The stratigraphy of glacimarine deposition on a deglaciated coastline. (a) Idealized depositional sequences I–VII produced on a deglaciated coast during a glacial–deglacial cycle. (b) Classification of transgressions and regressions based on the work of Curray (1964), with sedimentary sequences I–VII identified (Bednarski, 1988; reproduced with permission of Les Presses de l'Université de Montreal).

exposed at the surface. In glacilacustrine environments, the opening and closing of outlets by ice or sediment can cause multiple rapid changes in water level, resulting in complex depositional systems separated by erosion surfaces.

Water-level changes have the largest effect on patterns of sedimentation in the shallower parts of a basin. In contrast, changing water levels may have little effect in deeper areas, where facies variations depend more simply on the proximity of glacier ice.

(b)

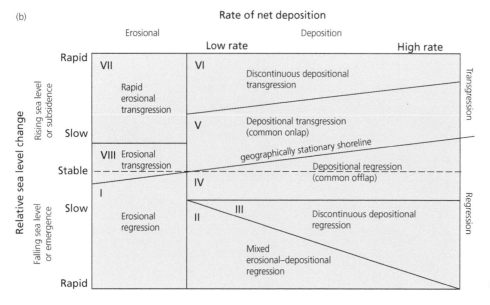

Figure 12.12 *(Continued)*

12.2.6 KINETO-STRATIGRAPHY (TECTONO-STRATIGRAPHY)

Evidence of large-scale patterns of ice flow and subglacial conditions may be preserved in the geologic record in the form of distinct till units, glacitectonic deformations, subglacial bedforms and surfaces of erosion or non-deposition. Assemblages of subglacial deposits and landforms created during the advance and retreat of an ice sheet are typically exceedingly complex, and often defy analysis by the traditional methods of sedimentary stratigraphy. In areas with soft substrata, major glacitectonic dislocations, responsible for features such as mega-blocks and rafts (see section 11.3.2), can provide considerable problems for stratigraphers and mineral prospectors (Aber et al., 1989). Furthermore, the products of several glacial advances may be superimposed, and glacitectonic dislocation and subglacial deformation often rework all or part of any pre-existing glacial and non-glacial strata. Due to partial erosion or patchy deposition, tills relating to any one glaciation can be discontinuous, so that individual events are only recognizable through the type, style and orientation of glacitectonic structures. As a result, new methods of analysis have been developed by glacial geologists to help them unravel the ice sheet record.

Before discussing such methods, it is useful to consider theoretical patterns of subglacial deposition and erosion created during an ice sheet advance-and-retreat cycle. A helpful overview has been provided by Boulton (1996a), who considered patterns of erosion and deposition along an ice sheet flow line by assuming that debris transport occurs entirely by subsole deformation at rates scaled to the balance velocity. This results in an outer zone of deposition near the ice sheet margin (where balance velocity decreases downflow and debris inputs exceed outputs) and an inner zone of erosion (where

balance velocity increases downflow and inputs are less than outputs). In addition, there is no erosion below the ice divide where the balance velocity is zero. This is clearly highly simplified, and does not deal explicitly with other important subglacial processes, such as the erosion of hard substrata, the entrainment of debris into basal ice layers, and deposition by lodgement and melt-out. However, Boulton's work does provide interesting insights into large-scale time-transgressive patterns of erosion and deposition, which can act as a basis for examining more complex cases.

Hypothetical patterns of erosion and deposition are shown in a time–space diagram in figure 12.13. The ice sheet is initiated at a nucleation point at time $(t) = 0$, and expands to the north and south, reaching its maximum 12,000 years after inception. It then contracts into the area where it first developed, finally vanishing at $t = 24,000$ years. The history of deposition and erosion at any given site reflects the migration of the depositional and erosional zones during ice sheet advance and retreat. At each site, deposition will occur during ice sheet advance. The resulting till will be removed from most sites by the subsequent erosion phase, although it will be preserved near the ice sheet limit. As the ice sheet retreats, the zone of deposition passes over each site in turn, depositing a retreat-phase till. After final deglaciation, the overall pattern of deposition and erosion comprises four major zones:

- *Zone 1.* The ice-divide zone, with slight erosion due to ice-divide migration and a thin till deposited during the retreat phase (sites E and F).
- *Zone 2.* A zone of strong erosion, in which the advance-phase till and part of the preglacial substratum are removed. The erosion surface is capped by the retreat-phase till (sites B, C, D, G and H).

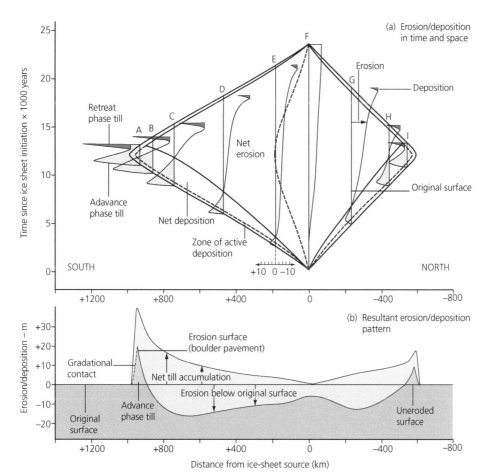

Figure 12.13 (a) Time–distance diagram showing conceptually modelled shifting patterns of subglacial erosion and deposition associated with the growth and decay of an ice sheet. Graphs A–I represent deposition and erosion at selected locations. (b) Resultant pattern of deposition and erosion at the end of the glacial cycle (After Boulton, 1996a).

- *Zone 3*. A zone where both advance- and retreat-phase tills are preserved, but with an intervening erosion surface. Some of the advance-phase till has been removed, but erosion was insufficient to bite down into the preglacial surface (sites A and I).
- *Zone 4*. A zone of continuous till deposition and no erosion. This zone only occurs very close to the ice sheet limit.

More complex stratigraphies will result from oscillations of the ice margin and the migration of different basal thermal regimes (Boulton, 1996a, b). Below real ice sheets, patterns of erosion and deposition are also strongly influenced by lateral variations in flow conditions, such as the presence of ice streams or variations in substrate geology and drainage. Nevertheless, figure 12.13 does highlight the fact that subglacial till units are time-transgressive deposits, and do not represent a 'snapshot' of conditions at an instant in time.

Berthelsen (1978) developed a powerful geological approach to the study of time-transgressive subglacial sequences. The approach is known as *kineto-stratigraphy*, or *tectono-stratigraphy*, and focuses on the *structural* record of subglacial environments in much the same way as event stratigraphy is used by structural geologists to study the evolution of orogenic belts. This approach subdivides subglacial sequences into *kineto-stratigraphic*

units, defined as sedimentary units 'deposited by an ice sheet or stream possessing a characteristic pattern and direction of movement'. Individual units are therefore grouped according to their directional elements, such as fabrics, folds and faults (fig. 12.14). The lower limit of a kineto-stratigraphic unit is defined as the lower boundary of the sediments deposited by the ice flow event, or the lower limit of associated penetrative and intense subglacial deformation. Thus kineto-stratigraphic units may consist of basal tills and underlying penetrative glacitectonite (Banham, 1977; Berthelsen, 1978; Pedersen, 1989; Benn and Evans, 1996). Ice flow events may also cause non-penetrative deformation of older strata; such disturbance is termed *extra-domainal deformation*, and does not form part of the overlying kineto-stratigraphic unit. Deformation of sediment within kineto-stratigraphic units is referred to as *domainal deformation*.

Multiple kineto-stratigraphic units, defined on the basis of their directional elements, could be formed within a single glacier advance–retreat cycle. In Boulton's (1996a, b) model, this may be represented by two units formed by the advance and retreat of an ice sheet with an intervening period of erosion (fig. 12.13). Ice-marginal oscillations may also impart deformation events on ice contact sediments. For example, Phillips et al. (2002) identified four separate tectono-stratigraphic units within

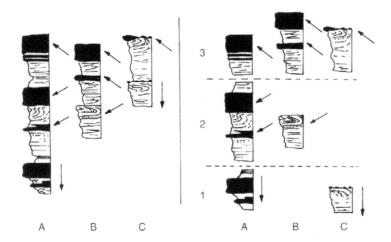

Figure 12.14 Hypothetical kineto-stratigraphic units at three sites A, B and C. At site A, three units are preserved, recording ice flow from the north (unit 1), north-east (unit 2) and south-east (unit 3). At site B, unit 1 is missing and unit 2 has been erosionally truncated and is represented by penetrative deformation structures which are unrelated to the overlying unit 3. At site C, unit 2 is missing and there is a major erosional hiatus at the top of unit 1. In some cases it is possible to infer 'missing' kineto-stratigraphic units from surviving extra-domainal deformation structures (Modified from Berthelsen, 1978).

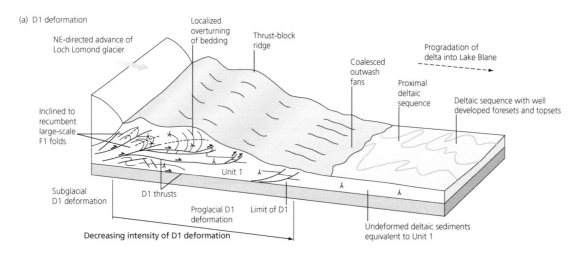

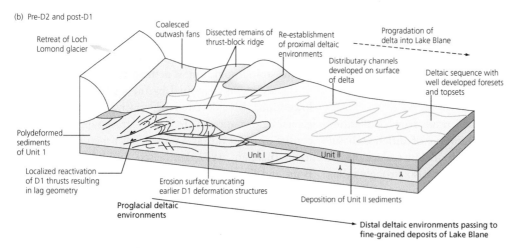

Figure 12.15 Reconstruction of the depositional environment at the oscillating margin of the Lomond Glacier, Scotland, during the Younger Dryas, based on the interpretation of tectono-stratigraphic units. (a) Ice-contact lake sediments (Unit I) are compressed (D1) by glacier advance to form a thrust block moraine. (b) Glacier recession allows moraine dissection and deposition of proximal deltaic sediment (Unit II). (c) Glacitectonic disturbance of Unit II by glacier overriding (D2) and the deposition of subglacial glacitectonite/till (Unit III). (d) Final recession of ice margin and deposition of outwash/deltaic sediments (Unit IV) (Phillips et al., 2002).

a sequence of ice contact lake sediments and overlying glacitectonite and till near Loch Lomond, Scotland. Each unit records a separate phase of deposition and/or deformation, the deformation phases being numbered D1 and D2 (fig. 12.15). Alternatively, multiple units could be formed due to changes in ice flow caused by ice-divide migration or internal reorganization of flow. Kineto-stratigraphy and similar approaches and methods have been adopted by many researchers (e.g. Aber, 1979; Eyles et al., 1982; Albino and Dreimanis, 1988;

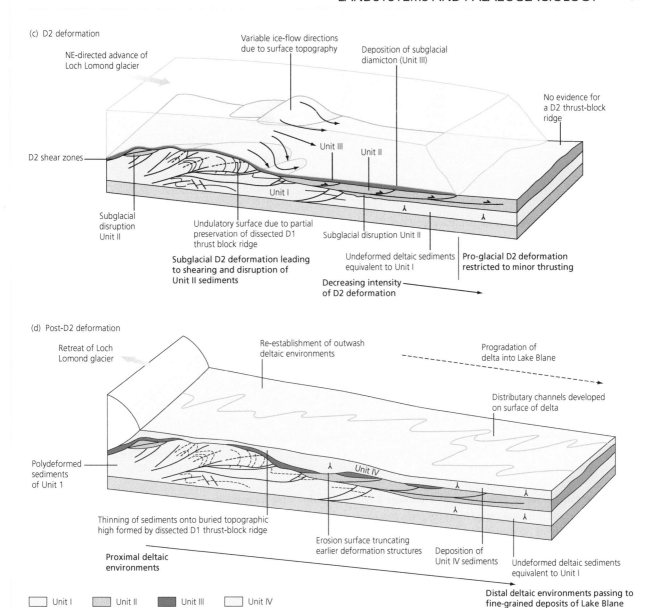

(c) D2 deformation

NE-directed advance of
Loch Lomond glacier

Variable ice-flow directions
due to surface topography

Deposition of subglacial
diamicton (Unit III)

No evidence for
a D2 thrust-block
ridge

D2 shear zones

Unit III   Unit II

Unit I

Subglacial
disruption
Unit II

Undulatory surface due to partial
preservation of dissected D1
thrust block ridge

Subglacial disruption Unit II

**Subglacial D2 deformation leading
to shearing and disruption of
Unit II sediments**

Undeformed deltaic sediments
equivalent to Unit I

**Pro-glacial D2 deformation
restricted to minor thrusting**

**Decreasing intensity
of D2 deformation**

(d) Post-D2 deformation

Retreat of Loch
Lomond glacier

Re-establishment of outwash
deltaic environments

Progradation of
delta into Lake Blane

Distributary channels developed
on surface of delta

Polydeformed
sediments
of Unit 1

Unit IV

Thinning of sediments onto buried topographic
high formed by dissected D1 thrust-block ridge

Erosion surface truncating
earlier deformation structures

Deposition of
Unit IV sediments

Undeformed deltaic sediments
equivalent to Unit I

**Proximal deltaic
environments**

**Distal deltaic environments passing to
fine-grained deposits of Lake Blane**

☐ Unit I   ☐ Unit II   ■ Unit III   ☐ Unit IV

Figure 12.15 (*Continued*)

Houmark-Nielsen, 1988; D. J. A. Evans, 1994; Hicock and Fuller, 1995). When combined with the study of contemporaneous erosional surfaces, bedforms and extra-domainal deformation, kineto-stratigraphy and related methods provide very powerful tools for unravelling the dynamic history of ice sheets.

12.3 GLACIER, ICE CAP AND ICE FIELD LANDSYSTEMS

The depositional record of glaciers, ice caps and ice fields is highly variable, reflecting the diversity of glacier types, thermal regime, climate and topography. Knowledge of the relationships between process and form can, in some cases, allow details of former climate and glacier dynamics to be reconstructed from the sediment and landform

record. In the following sections, we summarize the landsystems associated with valley glaciers, ice caps and ice fields, focusing first on the influence of glacier thermal regime, then on the role of topography.

12.3.1 THERMAL REGIME PROCESS–FORM CONTINUUM

The role of basal thermal regime in moraine formation can be summarized using the process–form continuum identified by Evans (2009; fig. 12.16). At one end of the continuum are warm-based glacier termini, found in mid-latitude settings such as Iceland and southern Norway, where recessional dump and push moraines record margin oscillations on annual or longer timescales. At the other end of the spectrum are the termini of predominantly cold polythermal glaciers, where deformation and thickening

a) Polythermal glacier

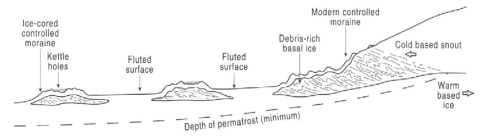

b) Polythermal glacier with seasonal-decadal signal

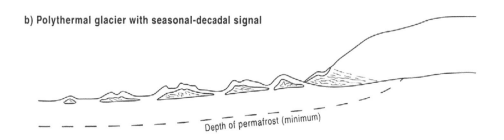

c) Temperate glacier with seasonal signal

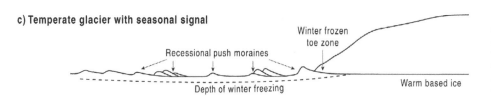

Figure 12.16 Process–form continuum for the production of end moraines constrained by basal thermal regime. Scenario a) is based on Svalbard glaciers and scenario c) on Icelandic piedmont lobes (Evans, 2009).

of basal ice take place in the marginal cold-based zone, allowing accretion of thick sequences of debris-rich ice and the development of controlled moraine. Response to climate change by such glacier systems is sluggish, and therefore end moraine belts record significant periods of sustained climatic cooling rather than shorter-term events. In the middle of the process–form continuum lie moraines constructed in areas of discontinuous or sporadic permafrost, by glacier margins that can respond to decadal or seasonal climate forcings (fig. 12.16b). In such settings, moraines can be produced on an annual basis, or certainly more regularly than in colder environments, but they generally contain larger proportions of glacier ice than in temperate settings because of the greater importance of sub-marginal freezing and debris transfer.

At the warm-based end of the continuum, dump and push moraines retain most of their linearity and planforms, such as sawtooth patterns, after deglaciation. In contrast, at the cold-based end of the continuum, the striking linearity of controlled moraine ridges is drastically altered after complete de-icing. This results in a landform signature comprising discontinuous transverse ridges with intervening hummocks, preserving only a weak impression of the former englacial structure (Kjær and Krüger, 2001; Evans, 2009). Supraglacial meltwater systems also destroy large tracts of controlled moraine, producing glacifluvial features such as ice-walled lake

plains and supraglacial trough fills. In essence, we can expect an improvement in the preservation of moraine linearity from panel A to panel C in figure 12.16.

It is tempting to add glacitectonic landforms to the extreme cold-based/permafrost end of the continuum, because modern examples predominantly lie in high-latitude environments and some researchers have invoked permafrost conditions as a necessary requirement for proglacial thrusting. However, permafrost is not a sufficient condition for glacitectonic disturbance, although it can facilitate it in certain circumstances (section 11.3.2). Rather, high sub-marginal and proglacial pore-water pressures and stress gradients are more fundamental to proglacial glacitectonic deformation. The geographical distribution of modern glacitectonic landforms may also be misleading in that they occur not only in high latitudes, but also around surging glacier snouts, raising the possibility that they may be uniquely diagnostic of surging (see section 12.3.8). As a result, it is difficult to place glacitectonic landforms on a moraine process–form continuum dictated by basal thermal regime, as they may occur throughout the continuum.

12.3.2 ACTIVE TEMPERATE GLACIERS

Temperate glacier margins are defined as those which are mainly at pressure-melting point for at least part of the year, and generally occur in areas of discontinuous or no permafrost (Evans and Twigg, 2002). Narrow frozen zones develop below the margins of many temperate glaciers during the winter, due to the penetration of a *cold wave* from the atmosphere through the thin ice at the edge of the glacier (Krüger, 1994; Evans and Hiemstra, 2005; fig. 12.16c). Additionally, small and discontinuous areas of net freezing can exist below some temperate margins in areas of discontinuous permafrost (e.g. Lawson, 1979a). At present, broad temperate glacier lobes are found mainly in southern Alaska, Iceland and Patagonia. Mountain glaciers in many mid- and low-latitude areas also have temperate margins, but the landsystems they create are generally strongly influenced by topography and supraglacial debris supply (see section 12.3.5).

Below temperate glaciers, sequences of debris-rich basal ice are typically thin or absent (Hubbard and Sharp, 1989, 1993), with the exception of some localized concentrations of debris related to glacio-hydraulic supercooling (e.g. Alley et al., 1997; Lawson et al., 1998; section 9.3.2). Some englacial thrusting and folding can also thicken debris-rich ice sequences where seasonally frozen ice occurs at the glacier margin. Moraine formation at temperate glacier margins is typically by a combination of dumping and ice-push, with localized squeezing (sections 11.3.3, 11.3.4). Active temperate glaciers respond rapidly to seasonal temperature fluctuations, and commonly undergo winter advances even when in overall recession. Consequently, glacier margins tend to oscillate on an annual basis and produce suites of annual recessional moraines during deglaciation, except where glacier velocities are low and/or ablation continues throughout the year (Sharp, 1984; Boulton, 1986b; Krüger, 1994; Evans and Twigg, 2002). The areas between recessional moraines represent periods of rapid glacier retreat, and commonly exhibit well-preserved subglacial landforms (fig. 12.17a). Imbricate thrusting in moraines, more commonly associated with polythermal and surging glaciers, does occur at some temperate ice margins. Modern examples have been described from Iceland, where thrusting mainly affects ice contact fans built up against the ice front, but can also relocate slabs of subglacial till (Humlum, 1985b; Boulton 1986b; Krüger, 1994; Evans and Hiemstra, 2005). The sole thrusts of these imbricate moraines are at shallow depth (1–2 m), and may be associated with the base of a seasonally frozen layer. In contrast, the sole thrusts of tectonized ridges in front of surging glaciers may be several tens of metres below the surface (section 12.3.8).

Meltwater has a profound influence on the landsystems formed at temperate glacier margins. Where subglacial and englacial drainage systems are well developed, sands and gravels can accumulate within conduits, and meltstreams emerging at the glacier surface can deposit considerable quantities of glacifluvial sediment in supraglacial positions. Indeed, on many temperate glaciers, glacifluvial sediment constitutes the major part of the englacial and supraglacial sediment load (section 9.4.3; Gustavson and Boothroyd, 1987; Kirkbride and Spedding, 1996; Spedding and Evans, 2002). Sediment deposited within conduits and in supraglacial outwash fans and lakes produces a complex assemblage of landforms, including esker systems, kame ridges and plateaux, and pitted outwash (sections 11.2.10, 11.4). The surface topography of such assemblages is commonly highly irregular, consisting of discontinuous ridges, mounds and meandering erosional channels, with a relative relief up to several tens of metres; and landforms at different topographic levels and locations record changing depositional environments during glacier downwasting and recession (Gustavson and Boothroyd, 1987; Johnson and Menzies, 1996; Evans and Twigg, 2002; fig. 11.43). Distinctive spatial associations of glacifluvial landforms deposited during ice retreat have been termed *morphosequences*. In such sequences, proglacial outwash passes upvalley into kame and kettle topography, which is inset within ice-marginal kame terraces and moraines. Kame and kettle topography is locally refashioned into suites of river terraces by proglacial and postglacial streams. In this way, subglacial, ice-marginal and supraglacial landsystems are commonly substantially modified by proglacial glacifluvial deposition and erosion during and following deglaciation (fig. 12.17b). Landforms such as drumlins, eskers and moraines may be heavily reworked by meltwater and/or buried by outwash sediments so that they are left as remnants or inliers within sandur plains. Even large ice-marginal moraines can be dissected or partially destroyed by meltstreams directly after emplacement (Price, 1973; Krüger, 1994; Marren, 2002). In some cases, virtually all the depositional legacy of a temperate ice lobe can take the form of glacifluvial sediments.

Using Icelandic lowland glacier forelands as case studies, Evans and Twigg (2002) and Evans (2003b), recognized three depositional domains associated with active temperate glaciers (fig. 12.17c): (1) marginal morainic; (2) subglacial; and (3) glacifluvial and glacilacustrine. The *marginal morainic domain* includes probably the most instantly recognizable landform signature of active temperate glaciers, the inset sequences of minor push moraines, usually documenting annual advances (fig. 12.17). They are widely spaced where glacier recession rates are high, or partially overprinted where the glacier has readvanced. Extended stillstands will result in the stacking of push moraines to produce large, complex ridges, comprising vertically accreted till wedges (Krüger, 1993, 1994; Evans and Hiemstra, 2005). Depending on local conditions, processes of push moraine formation range from winter freeze-on and summer melt-out of till slabs, to bulldozing of proglacial materials, to squeezing of saturated sub-marginal till. Overridden push moraines also constitute very conspicuous features on the forelands of some active temperate glaciers. These appear as arcuate, low-amplitude ridges adorned with flutings, and arranged

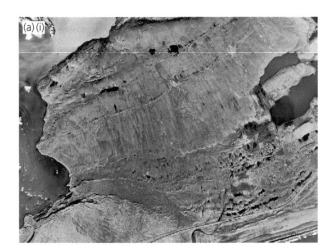

sub-parallel or parallel to the more recent superimposed recessional push moraines (Evans and Twigg, 2000, 2002; Evans, 2003b; section 11.2.7).

The *subglacial domain* is characterized by densely spaced flutings and drumlins that in places contain 'windows' through to underlying heavily abraded and quarried bedrock (fig. 12.17). Flutings are commonly arranged in arcuate strips aligned at right angles to push moraines at their downglacier end. This suggests that each individual arcuate strip of flutings and associated push moraine records individual flow events, possibly of seasonal status. Internally, flutings and drumlins consist of subglacial traction tills, commonly overlying glacitectonite and older undeformed sediments (Boulton, 1987; Evans, 2000a; Evans and Twigg, 2002). Vertical dykes, injection features and hydrofracture fills are also common in the subglacial sediments due to locally high pore-water

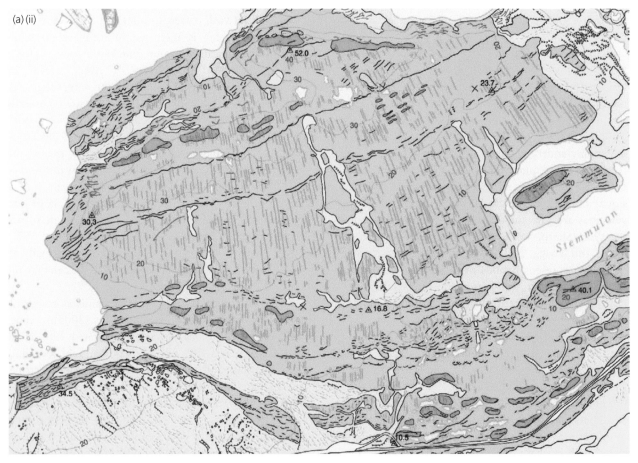

Figure 12.17 The active temperate glacial landsystem. (a) Aerial photograph extract (University of Glasgow/Landmælingar Islands, 1998) and surficial geology/geomorphology map extract (Evans and Twigg, 2000) from the recently deglaciated foreland of Breiðamerkurjökull, Iceland, showing inset sequences of recessional push moraines and flutings draped over drumlins and overridden moraine arcs. These landforms have been locally dissected by proglacial streams, which have also deposited terraced outwash and sandur fans. (b) Aerial photograph extract (Landmælingar Islands, 1989) of part of the deglaciated foreland of Heinabergsjökull, Iceland, showing the extensive destruction of a large part of the marginal morainic and subglacial domains by proglacial meltwater as it migrated from a proglacial to an ice-marginal drainage direction (cf. Evans et al., 1999a). The undisturbed foreland displays recessional push moraines and flutings draping overridden moraine arcs. The large area of flooded, interlinked kettle holes on either side of the present-day river reveals the location of a significant body of buried glacier ice. The linear chains of elongate hummocks along the shore of the proglacial lake form a network of melt-out hollows and push moraines produced where the snout oscillated in contact with an ice-contact linear sandur. (c) Conceptual landsystems model for active temperate glacier margins (Evans and Twigg, 2002). Landforms are numbered according to their domain (1 = morainic domain; 2 = glacifluvial domain; 3 = subglacial domain): (1a) small, often annual, push moraines; (1b) superimposed push moraines; (1c) hummocky moraine; (2a) ice-contact sandur fans; (2b) spillway-fed sandur fan; (2c) ice margin-parallel outwash tract/kame terrace; (2d) pitted sandur; (2e) eskers; (2f) entrenched ice-contact outwash fans; (3a) overridden (fluted) push

pressures associated with ice overriding and, in some cases, localized frozen ground (van der Meer et al., 1999).

The *glacifluvial and glacilacustrine domain* is not unique to active temperate glaciers, as similar associations also occur in front of polythermal glaciers (fig. 12.17). Sandur and hochsandur fans emanate from breaches in push moraines and may contain extensive areas of pitting at their apices, due to their development over buried ice (sections 11.3.5, 11.5.1). If the direction of drainage remains stable for significant periods during overall recession, fans will become incised and extensively reworked (fig. 12.17b). Terraced and locally pitted outwash tracts develop parallel to former ice fronts, as a result of

proglacial drainage directed by large morainic and outwash fan accumulations. Kame terraces occur along the margins of active temperate glaciers, receding into moderate-relief terrain (figs 11.51, 11.95b). Esker networks demarcate the locations of subglacial and englacial drainage networks, and also occur as part of ice contact fan accumulations (fig. 11.43; Price, 1973, Boulton et al., 2007a). Finally, glacilacustrine deposits and landforms reflect proglacial and ice-dammed lakes. In the case of ice-dammed lakes, glacilacustrine sediment–landform associations are prone to overprinting by proglacial glacifluvial deposition and erosion. Substantial accumulations of glacilacustrine sediment can occur in overdeepenings

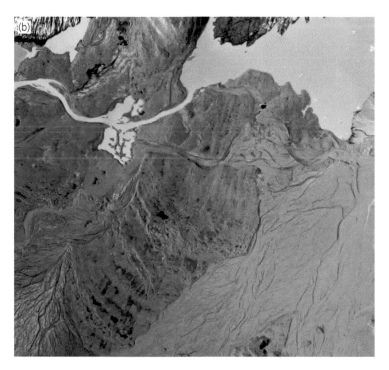

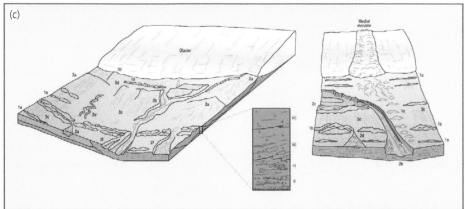

Figure 12.17 (*Continued*)

moraines; (3b) overridden, pre-advance ice-contact outwash fan; (3c) flutes; (3d) drumlins. The idealized stratigraphic section log shows a typical depositional sequence recording glacier advance over glacifluvial sediments, comprising: (i) undeformed outwash; (ii) glacitectonized outwash/glacitectonite; (iii) massive, sheared till with basal inclusions of pre-advance peat and glacifluvial sediment; (iv) massive sheared till with basal erosional contact; (d) conceptual models for styles of glacifluvial and glacilacustrine deposition based on the Malaspina Glacier, Alaska (after Gustavson and Boothroyd, 1987). Panel A: sandur fan fed by subglacial meltwater fountain. Panel B: sandur fan fed directly by subglacial meltwater. Panel C: subaqueous fan deposition by subglacial tunnel exiting into a proglacial lake. Panel D: subaqueous fan fed by englacial tunnel following earlier englacial–supraglacial drainage pathway and esker formation. Panel E: englacial tunnel feeding supraglacial meltwater stream that then deposits an ice-contact delta. In each case, eskers can develop in the feeder tunnels, and in the englacial drainage scenarios eskers can become entombed within the ice buried by fan/delta growth.

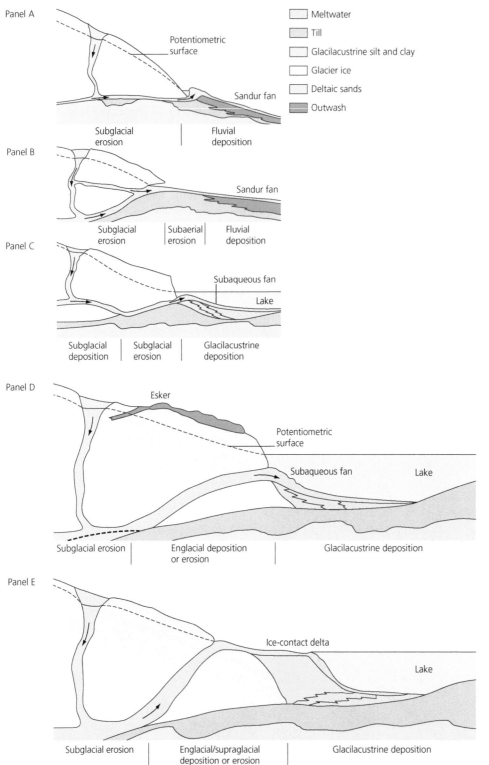

Panel A

Potentiometric surface

Sandur fan

Subglacial erosion | Fluvial deposition

Meltwater
Till
Glacilacustrine silt and clay
Glacier ice
Deltaic sands
Outwash

Panel B

Sandur fan

Subglacial erosion | Subaerial erosion | Fluvial deposition

Panel C

Subaqueous fan

Lake

Subglacial deposition | Subglacial erosion | Glacilacustrine deposition

Panel D

Esker

Potentiometric surface

Subaqueous fan | Lake

Subglacial erosion | Englacial deposition or erosion | Glacilacustrine deposition

Panel E

Ice-contact delta

Lake

Subglacial erosion | Englacial/supraglacial deposition or erosion | Glacilacustrine deposition

Figure 12.17 (*Continued*)

exposed by glacier recession (e.g. Boulton et al., 1982) and in side valleys in moderate- to high-relief terrain (e.g. Bennett et al., 2000c; section 10.6). The processes and forms that characterize the glacifluvial and glacilacustrine domain have been summarized by Gustavson and Boothroyd (1987; fig. 12.17d).

12.3.3 POLYTHERMAL GLACIERS

Polythermal glaciers are very diverse, and the extent of frozen and temperate ice at the bed varies according to environmental conditions and glacier thickness (section 2.3.3; Blatter and Hutter, 1991). Ice caps growing on thick permafrost must initially be entirely cold-based, but wet-based zones can develop in response to subglacial heating from geothermal heat and strain heating produced by glacier flow (section 2.3.2). Thawing of the bed is therefore most likely where ice is thick and glacier velocities are highest – for example, in the vicinity of the glacier equilibrium line and in areas of flow convergence (Dyke, 1993). Thus, wet-based ice will be most extensive below glaciers with relatively high mass throughput and/or strong converging flow. In the upper parts of glacier accumulation areas and beneath ice divides, cold-based zones are likely to persist due to low ice velocities and downward advection of cold firn. Cold-based ice occurs also beneath the glacier margins where ice is thin, velocities are low and proglacial permafrost extends below the glacier. Thus, in permafrost regions, small, low-activity glaciers are likely to be cold-based throughout (e.g. Etzelmüller et al., 2000).

The presence of wet-based ice upglacier from a frozen margin has a profound impact on processes of debris entrainment, transport and deposition. First, erosion below wet-based, sliding ice provides a source of debris that can be transported towards the margin in basal ice or a subglacial deforming layer. Erosion rates may be small where sliding velocities are low, but can be significant below outlet glaciers (Dyke, 1993; Hallet et al., 1996). As this debris is transported into the frozen zone at the glacier margin, sequences of debris-rich basal ice develop as a result of basal freeze-on (section 9.3.2; Alley et al., 1997). These debris sequences are commonly thickened and elevated towards the glacier surface by strong compressive deformation resulting from ice deceleration at the frozen margin, and thick basal debris sequences also form

where glaciers override buried ice dating from earlier glacial events. The long survival of buried ice in permafrost environments means that debris-rich ice may be recycled many times over several glacial cycles as glaciers advance and retreat.

Landsystem signatures of polythermal glaciers vary according to their position on the thermal regime continuum. Here, we focus on two examples:

(1) outlet glaciers and piedmont lobes of the Canadian and Greenland High Arctic, representative of the cold and arid end of the continuum;

(2) the polythermal snouts of Svalbard, representing the milder and less arid parts of the subpolar regions. In both settings, but particularly in Svalbard, we need to be aware that surging activity often complicates the polythermal glacier landsystem signature (see section 12.3.8).

High Arctic outlet and piedmont glaciers

On outlets of mountain and plateau ice fields in the Canadian and Greenland High Arctic, extraglacial debris sources are limited, and even in settings where steep rock-walls do exist above glaciers, the aridity of the climate generally restricts periglacial weathering and the accumulation of supraglacial debris (fig. 12.18). Supraglacial debris covers are typically produced around glacier margins by melt-out of debris-rich basal ice (sections 11.4.2, 12.5.1). This debris cover commonly protects the terminus from complete de-icing, because it eventually exceeds the active layer thickness, forming impressive ice-cored controlled moraines (section 11.4.2). The long-term preservation potential of the controlled moraine is low,

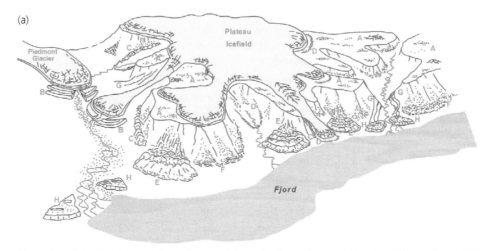

Figure 12.18 The arid subpolar outlet and piedmont glacier landsystem. (a) Conceptual landsystems model based on plateau and mountain ice field examples in the Canadian and Greenland High Arctic. (A) Blockfield/residuum. (B) Thrust block moraine. (C) Ice-cored lateral moraine. (D) Trimline moraine. (E) Glacier ice-cored protalus rock glacier. (F) Periglacial protalus rock glacier. (G) Lateral meltwater channels. (H) Raised, ice-contact deltas. (Ó Cofaigh et al., 2003b) (b) Part of aerial photograph of Blind Fjord, south-central Ellesmere Island, showing inset lateral meltwater channels and associated marine limit deltas declining in altitude up-fjord. (National Air Photograph Library, Ottawa; after Ó Cofaigh, 1998) (c) Oblique aerial photograph of Augusta Bay, central Ellesmere Island, showing the 'drift belt' of ice-contact subaqueous depo-centres and moraines and lateral meltwater channels (National Air Photograph Library, Ottawa; Hodgson, 1985).

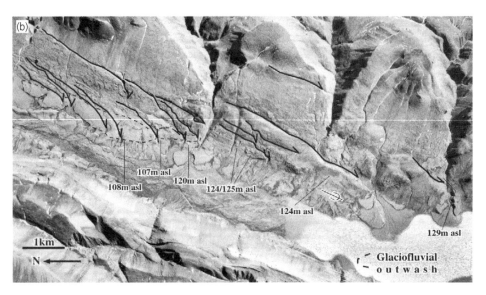

Figure 12.18 (Continued)

however, and complete de-icing typically produces only a thin veneer of debris (Evans, 1989b; fig. 11.78). As predominantly cold polythermal glaciers respond slowly to climate forcing, they do not produce complex sequences of recessional moraines. Although basal tills are produced beneath larger trunk glaciers, basal debris production is usually limited and till veneers are generally <50 cm thick (Evans, 1990a).

In areas of thick sediment such as fjord heads and valley bottoms, compressive stresses at the glacier margin can produce composite ridges and thrust-block moraines (section 11.3.2). Thrusting is most effective where fine-grained glacimarine and lacustrine sediments crop out, although thrusting may also affect frozen gravels (Evans and England, 1991). It was shown in section 11.3.2 that the conditions for the production of glacitectonic landforms are most readily satisfied in surging glacier systems. The distribution of surging glaciers in the High Arctic is only now being realized (Copland et al., 2003b), and it is possible that glacitectonic landforms in such settings

reflect surging behaviour rather than other factors (section 12.3.8). Proglacial thrusting is commonly accompanied by englacial thrusting and the elevation of basal debris to the ice surface to produce controlled moraine. The combination of proglacial and englacial thrusting and other debris entrainment mechanisms can thereby create a landsystem in which outer thrust-block moraines pass upvalley into ice-cored chaotic hummocky moraine, controlled moraine and/or kames (Evans, 2009). The generally low discharges of proglacial meltwater streams in Arctic environments mean that such landsystems are only locally reworked into outwash deposits, and tend to be well preserved.

Former polythermal glacier margins are commonly marked by *lateral meltwater channels* (section 8.4.4). Dyke (1993) described spectacular examples of lateral channels from the central Canadian Arctic, where they commonly form nested sets that clearly delineate successive positions of retreating ice lobes (figs 8.37 and 8.38). Indeed, inset sequences of lateral meltwater channels constitute the most extensive evidence of subpolar outlet glacier recession in the Canadian and Greenland High Arctic. Large areas of formerly cold-based ice and associated patterns of recession have been identified in the Canadian High Arctic, based on dense flights of inset lateral meltwater channels (Dyke et al., 1992; Dyke, 1993, 1999). On Devon Island, for example, the marginal origin of the channels is evident in their association with end moraines and ice contact deltas developed at the interface between receding snouts and marine limits. Because the deglacial chronology is well constrained in the region, Dyke (1993, 1999) has proposed that each meltwater channel in every flight represents one year's melting.

The gradients of the long profiles of these channels are related to the dynamics of the former glacier margins. Low-gradient channels are produced along the margins of glaciers occupying fjord heads where ice is drawn down

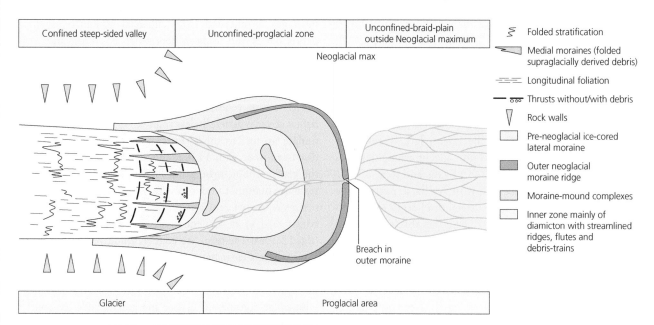

(b)

Figure 12.19 The Svalbard subpolar outlet and piedmont glacier landsystem. (a) Landsystem model for typical terrestrial Svalbard polythermal glacier margin. (After Glasser and Hambrey, 2003) (b) Aerial view of Midre Lovenbreen, showing the outer moraine arc of ice-cored (controlled) ridges with kettle holes indicative of ongoing de-icing and inner zone of linear debris stripes produced by a combination of subglacial streamlining under former warm-based ice and the draping of the landscape by supraglacial, flow-parallel linear debris bands. A large streamlined mound with large kettle depressions in the middle of the streamlined foreland likely represents overridden ice-cored moraine, attesting to an older ice-marginal position (D.I. Benn).

into the marine environment, and steeper gradients record recession of terrestrially based tongues. The scarcity of eskers in the Canadian Arctic reflects the predominantly marginal rather than subglacial drainage pathways, although supraglacial eskers do form at the junctions of coalescent lobes (fig. 11.42b). Kame terraces and ice contact deltas are commonly associated with the downstream ends of meltwater channels, and the decline in altitude of marine limits associated with the bases of meltwater channels clearly demonstrates sequential ice-marginal recession (figs 11.144, 12.18b and c).

Svalbard cirque and valley glaciers

The polythermal glacier margins of Svalbard are characterized by well-developed controlled moraines, forming a wide latero-frontal band of debris-covered ice (section 11.4.2; fig. 11.87; Lyså and Lønne, 2001; Sletten et al., 2001; Glasser and Hambrey, 2003; Lukas et al., 2005). On some glacier forelands, areas of fluted till and overridden moraines are found between the frontal loop of controlled moraines and the present-day terminus, attesting to generally warmer basal conditions than those found in the Canadian High Arctic (fig. 12.19). Svalbard controlled moraines have been termed 'moraine-mound complexes', and attributed to englacial glacitectonic processes at the junction between warm-based and cold-based ice (see section 9.4.2; Hambrey et al., 1997; Glasser and Hambrey, 2003). According to this model, debris is elevated to the glacier surface by englacial folding and thrusting in a zone of compressive flow at the thermal boundary. It should be noted, however, that similar belts of controlled moraine occur in front of glaciers that are known to have surged (e.g. Sletten et al., 2001). According to some estimates, up to 90 per cent of Svalbard glaciers are surge-type (Lefauconnier and Hagen, 1991; Sund et al., 2009), and it is possible that much of the controlled moraine in the archipelago reflects former surges rather than thermal boundaries in non-surging glaciers (section 12.3.8). Indeed, Midre Lovenbreen, where one of the type examples of Svalbard moraine-mound complexes is located, may have surged early in the twentieth century (Hansen, 2003), although this is considered unlikely by Hambrey et al. (2005).

Although linearity in the moraine-mound complexes is presently very striking in many cases, ongoing de-icing is effectively destroying linearity through processes such as thermo-erosion, retrogressive debris flow and fluvial reworking (Bennett et al., 2000d; Etzelmüller, 2000; Etzelmüller et al., 2000; Lukas et al., 2005; Schomacker and Kjær, 2007). The preservation potential of such moraines is unlikely to be high. For this reason, morainic evidence of late Pleistocene–Holocene glacier recession in Svalbard is very sparse (Lønne, 2005), and reconstructing past glacier fluctuations in the region largely relies on raised shorelines and offshore glacimarine deposits (cf. Andersson et al., 1999, 2000; Evans and Rea, 2005).

Although precipitation is low in Svalbard, particularly in the summer months, high fluvial discharges are maintained by snow and ice melt during the ablation season. Where glacier velocities are low, as is the case on the majority of cirque glaciers and quiescent surge-type glaciers, much more sediment is transported through glacier systems by glacifluvial processes than by glacier flow. Additionally, fluvial reworking has a large impact on landform development on glacier forelands, and widespread glacifluvial fans and sandar attest to efficient sediment transport away from ice margins. As is the case in the Canadian High Arctic, meltwater channels commonly incise into bedrock at or near the margins of Svalbard glaciers, and flights of parallel channels are commonly formed during glacier recession.

12.3.4 POLAR-CONTINENTAL GLACIERS

Distinctive glacial landsystems are formed around the margins of glaciers in the polar deserts fringing parts of the Antarctic Ice Sheet, such as the Vestfold Hills and Bunger Hills in east Antarctica, and the Dry Valleys in the Transantarctic Mountains. These areas are cold, windy and have a moisture deficit. For example, in the Wright Valley in the Dry Valleys region, MAT is −19.8 °C, and precipitation is only 10 mm yr^{-1} (Denton et al., 1993). Because of the low temperature and relative humidity, 40–80 per cent of glacier ablation is by sublimation (Lewis et al., 1998). Rates of ablation are extremely low (Fountain et al., 2006). Small local glaciers are entirely cold-based, and are nearly free of debris, although some of the larger terrestrial outlet tongues of the Antarctic Ice Sheet are partly wet-based with cold-based margins, and have stacked basal debris sequences similar to those of Arctic-type glaciers (Denton et al., 1993; Fitzsimons et al., 1999; Lorrain et al., 1999; Fitzsimons, 2003).

The extreme aridity and limited melting in this environment has important implications for glacial deposition and landsystem development. Where thick basal ice sequences crop out, marginal aprons accumulate by gravitational processes, but with very limited reworking by meltwater. Within the marginal aprons, belts of hummocky and controlled moraines can develop from stacked and thrust englacial debris sequences. The topography of controlled moraines is inherited from englacial structures, such as debris bands and folds in debris-rich basal ice, and commonly forms ridges transverse to glacier flow (Rains and Shaw, 1981; Fitzsimons, 2003). Features that appear to be terminal and hummocky moraines in fact commonly constitute a thin veneer of debris overlying buried ice. The debris cover is generally less than 0.5 m thick, but can accumulate up to 2 m thick (Fitzsimons, 1990, 2003). As we discussed in section 11.4.2, this debris cover often masks the glacier snout, prompting Fitzsimons (2003) to use the term 'ice margin' to refer to wide zones – up to several kilometres wide – of debris-covered ice where numerous debris bands feed inset sequences of controlled moraine ridges (fig. 12.20a). The term 'ice edge' is employed only in situations where a clear marginal cliff exists.

In addition to these controlled moraines (termed 'inner moraines' by Fitzsimons, 2003), other features associated with polar continental glacier margins include ice contact fans and screes (Shaw, 1977; Fitzsimons, 1997a) and thrust-block moraines (Fitzsimons, 1996a, 1997b). Entirely cold-based glaciers without stacked basal debris sequences leave very little imprint on the landscape, and end moraines commonly consist of little more than low ridges, although they can persist in the landscape for millions of years (fig. 12.20b; Sugden et al., 1995; Marchant et al., 2002). Ice contact fans and screes accumulate below ice cliffs, where they are fed by debris released from debris bands emerging at the glacier surface. On ice recession they are left as ice-cored, arcuate ridges, composed of poor to well-sorted, stratified to weakly stratified sediments and diamictons, indicative of alluvial and colluvial processes (Fitzsimons, 2003). Thrust-block moraines are up to a few tens of metres high, and consist of upglacier-dipping blocks of gravels, sands and occasional organic muds. The formation of these moraines appears to require special conditions, because most proglacial sediment is deeply frozen and ice-cemented and has a high shear strength. The thrust-block moraines studied by Fitzsimons are adjacent to ice-marginal lakes or marine inlets, and he argued that they formed by the proglacial thrusting of unfrozen deeper water sediments. Lake floors in the Dry Valleys can remain unfrozen below surface ice cover, and Fitzsimons (1996a) argued that when glaciers advanced into wet-based lakes, ice and debris were frozen onto the glacier sole and thrust upwards at the margin (see section 11.3.1). The presence of the lakes, therefore, appears to have created rare conditions conducive to thrusting in this extremely cold, arid environment. Some thrust blocks appear to have been detached in sub-marginal positions (Fitzsimons et al., 1999), indicating that there may be a unique thrust-block forming mechanism at work beneath these glaciers.

Like the cold margins of polythermal glaciers in the northern hemisphere, polar continental ice margins will develop ice shelf moraines where they float in large lakes

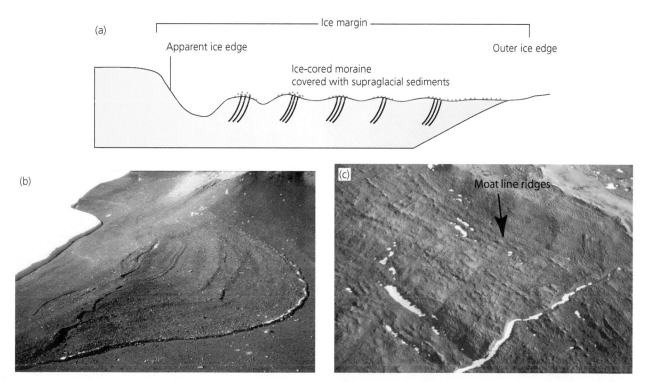

Figure 12.20 Features of the polar continental glacial landsystem. (a) Typical cross section through a polar continental glacier margin where numerous englacial debris concentrations rise to the snout surface to feed the accumulation of controlled moraine. The term 'ice margin' in this context encompasses the wide belt of debris-covered ice that lies outside the apparent ice edge. (Fitzsimons, 2003) (b) Moraine sequence formed at the margin of the Taylor Glacier, Dry Valleys, Antarctica. The ages of the moraines are between 100 ka and 5 Ma years BP. (D.E. Sugden) (c) Lake ice conveyor landforms or 'moat-line ridges' deposited in moats at the margins of lake ice-covered proglacial lakes in the Antarctic Dry Valleys (Hall et al., 2006).

or the sea (Sugden and Clapperton, 1981; Glasser et al., 2006; section 11.6.4; fig. 11.134). Sediment transport by lake ice in Antarctica gives rise to some curious sediment–landform assemblages called *lake-ice conveyor deposits*, which are linear dump mounds formed where surface debris falls through 'moats' (section 10.6.6; fig. 10.82). The resulting suite of landforms is very similar in appearance to features normally associated with oscillating glacier tongues (fig. 12.20c). Indeed, some radical reinterpretations of local glacial history have been necessitated by the recent recognition of lake-ice conveyor deposits (cf. Kelly et al., 2002; Hall et al., 2006). Lake-ice conveyors are capable of transporting glacial debris tens of kilometres across proglacial lakes. This landform assemblage, now recognized as diagnostic of proglacial lake ice development in very cold, arid environments, includes mounds that are either clustered or solitary and resemble terrestrial ice stagnation topography, sinuous longitudinal ridges, moat-line ridges that superficially resemble shorelines, and cross-valley ridges that can resemble moraines. Future research in Antarctic glacial geomorphology will very likely re-evaluate previous interpretations of process–form relationships in light of this new knowledge of the arid polar lake ice landform signature.

Although meltwater discharges are small in Antarctica, proglacial streams are capable of undercutting ice-cored moraines and accelerating the ablation process. Where

such streams dissect sublimating ice-cored ridges, they deposit small quantities of stratified sands and gravels, which inter-digitate with mass flow diamictons liberated from the buried ice. Unlike Arctic environments, marginal meltwater channels are rare, and extensive, thick proglacial outwash features are absent. The extreme aridity of polar desert environments, combined with very slow melt rates, means that glacial landforms undergo very little modification after deglaciation. As is the case for northern hemisphere permafrost regions, it looks likely that full de-icing of moraines will only occur under extreme climate warming scenarios. In that event, the final depositional products will be similar to those envisaged for Arctic-controlled moraine, albeit more subtle due to the lower debris concentrations in Antarctica.

Strong katabatic winds in polar continental environments winnow out finer material and shape surface clasts into faceted forms or *ventifacts*, but otherwise there is little paraglacial activity (Fitzsimons, 1996b). Denton et al. (1993) have presented convincing evidence that the Dry Valleys region is essentially a fossil landscape in which Pliocene and even Miocene landforms are preserved virtually intact. Hambrey et al. (2007) reached the same conclusion for the Radok Lake area of East Antarctica, concluding that pre-Pleistocene glaciations were wet-based and therefore created more glacial signatures than the cold-based Pleistocene glaciations. For example,

(a)

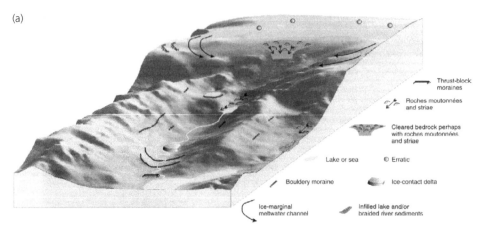

(b)

Figure 12.21 The plateau ice field landsystem. (a) Schematic diagram of landforms produced by a plateau ice field. In this case, the outlet valley descends below the marine limit, so an ice-contact delta has been formed in a glacio-isotatically higher sea level once the outlet glacier had receded upvalley. (Rea and Evans, 2003) (b) View up one of the outlet valleys draining the Øksfjordjøkelen plateau, north Norway, showing the bouldery latero-frontal moraines produced by the expanded Younger Dryas plateau ice field (Evans et al., 2002; Rea and Evans, 2007; photo D.J.A. Evans).

sequences of small recessional moraines near the margin of the Taylor Glacier record minor glacier fluctuations over the last 2.2 million years, and some surface till units in the area predate volcanic ashes deposited 4.4 million years ago (Marchant et al., 1993). The preservation of such forms shows that these polar continental settings have experienced a cold, arid climate since at least the late Miocene (Denton et al., 1993).

12.3.5 MOUNTAIN GLACIER LANDSYSTEMS

Plateau ice fields

Glaciated and glacierized plateaux are best developed in passive margin settings such as the fringes of the North Atlantic (e.g. Norway, the British Isles, eastern Canada, Iceland and Greenland) and in the Canadian Arctic Archipelago. Plateaux are ideal collection areas for permanent snowfields at the onset of glacial conditions. In areas where the plateau margins have precipitous drops into fjords or troughs, 'fall glaciers' or 'reconstituted glaciers' (Benn and Lehmkuhl, 2000) are nourished below the regional ELA by dry calving at the plateau cliff edge (Gellatly et al., 1986). During the more advanced stages of local glaciation, glaciers develop in the valley bottoms between plateaux, and it is at this stage that the most significant landforms are likely to be produced. Because they host glacier ice masses that are often largely cold-based and lack extraglacial debris sources, plateau ice fields tend to leave very subtle geomorphic imprints on deglaciation. This has resulted in misinterpretations of the geometry of former ice coverage in palaeoglaciological reconstructions.

A plateau ice field landsystem model has been compiled by Rea and Evans (2003), based on research on modern plateau ice fields in Norway, Arctic Canada and Iceland (fig. 12.21a; Evans, 1990a; Rea et al., 1998; Evans et al., 2002, 2006b; Rea and Evans, 2007). The model encompasses some of the features recognized in the polythermal glacier landsystem, with which it overlaps. Localized patches of warm-based ice, particularly around valley heads that drain plateau summits, will result in erosion and therefore debris transport and moraine construction on plateau margins (Rea et al., 1998). This contrasts with the preservation of blockfield and Tertiary weathering residues beneath cold-based summit ice (Rea et al., 1996a, b; Whalley et al., 1997). Erratics can be emplaced during periods of regional ice flow over the plateaux when the summits remain frozen. Recession of cold-based ice on the plateau summits is often only recorded by lateral meltwater channels (see section 8.4.4; Benn and Ballantyne, 2005). These channels may ultimately feed into proglacial channels that connect to steep alluvial fans at the base of plateau edge gullies (Evans et al., 2006b).

The largest accumulations of glacigenic material occur in lateral and latero-frontal moraines in the valleys draining plateaux (Evans et al., 2002), or as hummocky dump and push moraines at the valley heads (Rea et al., 1998; McDougall, 2001; Evans et al., 2006b). Valley moraines tend to be dominated by large, cobble- to boulder-size, angular material, even where there is evidence of basal sliding, indicating that the dominant debris source is rockfall (fig. 12.21b). Active free faces can provide sufficient debris to produce supraglacial lateral moraines, which may develop into rock-glacierized ice-cored moraines after deglaciation. Advance of outlet lobes into sediment-filled lowlands can result in the construction of large end moraine sequences, particularly where proglacial thrusting takes place.

The þorisjökull plateau ice field in Iceland contains a range of glacier–landform associations, because its present-day and Little Ice Age maximum margins display an asymmetric distribution dictated by topo-climatic controls (Rea and Evans, 2003; Evans et al., 2006b; fig. 12.22a). Its most restricted margin faces the south-west and lies on the plateau summit, where it has developed a closely spaced set of bouldery ice-cored moraines, lying directly downflow of rubbly flutings (fig. 12.22b). This landform assemblage indicates that the ice field has a polythermal regime. At its northern and eastern margins, the ice field feeds outlet glaciers that have flowed into the surrounding lowlands. The outlet lobes are fringed by asymmetrically developed supraglacial lateral moraines and rubbly

latero-frontal moraines below extensive cliffs (fig. 12.22c). The west-facing slopes of the plateau contain ice bodies and landforms that lie in the mid range of the spectrum of plateau glacier–landform associations, consisting of niche glaciers fronted by significant rock glacierized latero-frontal moraines (fig. 12.22d). At times of greater ice extent, summit ice linked with these niche glaciers to produce outlet lobes, which over time would excavate a valley head into the plateau edge.

Recognition of plateau ice field landsystems has important implications for reconstruction of Pleistocene glaciers. For example, Rea et al. (1998) and McDougall (2001) reinterpreted the style of Younger Dryas glaciation proposed for the English Lake District by Sissons (1980). Rather than the small trough head and cirque glaciers envisaged by Sissons, based on the distribution of morainic spreads in lower valleys, the new reconstructions extend the ice cover onto the surrounding plateaux, based on the extension of glacial landforms onto the plateau margins (fig. 12.23). This also makes the ice cover more glaciologically plausible, because the plateaux lie above the palaeo-ELA implied by the alpine style reconstruction.

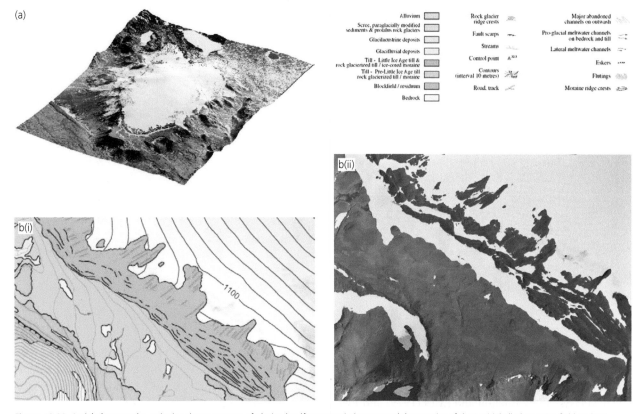

Figure 12.22 Aerial photographs and related map extracts of glacier–landform associations around the margins of the þorisjökull plateau ice field, Iceland. (a) Aerial photograph rendered DEM of the ice field viewed from the south-west, showing the asymmetrical development of the margins at the present day. (b) Aerial photograph (ii) and map extract (i) of plateau summit margin, showing Little Ice Age ice-cored moraines, flutings and meltwater channels. (c) Aerial photograph (ii) and map extract (i) of northern outlet glaciers, showing remnant Little Ice Age ice-cored moraines, fluted till surface and asymmetrically developed lateral and latero-frontal moraines with localized rock glacierization. (d) Aerial photograph (ii) and map extract (i) of niche glacier and its rock-glacierized Little Ice Age latero-frontal moraine loop (Evans et al., 2006b).

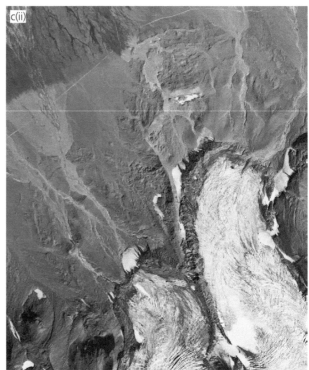

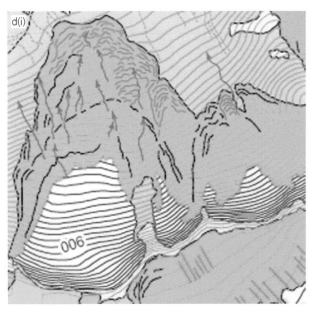

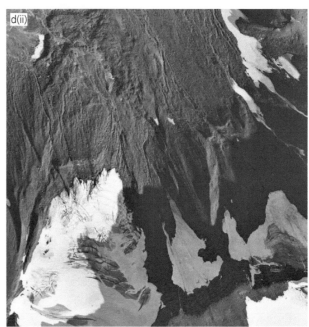

Figure 12.22 (*Continued*)

Glaciated valley landsystems

Valley glaciers produce very distinctive assemblages of sediments and landforms, referred to as the *glaciated-valley landsystem* by Boulton and Eyles (1979) and Eyles (1983b). The glaciated valley landsystem may contain ice-marginal, supraglacial, subglacial, proglacial and subaquatic landform–sediment assemblages in close proximity or superimposed on one another, recording the migration of very different depositional environments.

Landform assemblages formed in glaciated valleys, therefore, can share characteristics with other landsystems, with appropriate modifications to account for differences in scale and topographic setting. Glaciated valley settings, however, are unique, due to the importance of valley sides as debris sources and topographic confinement of depositional basins. The influence of these factors varies from valley to valley, and we can make generalizations about glaciated valley landsystems in low-relief or high-relief

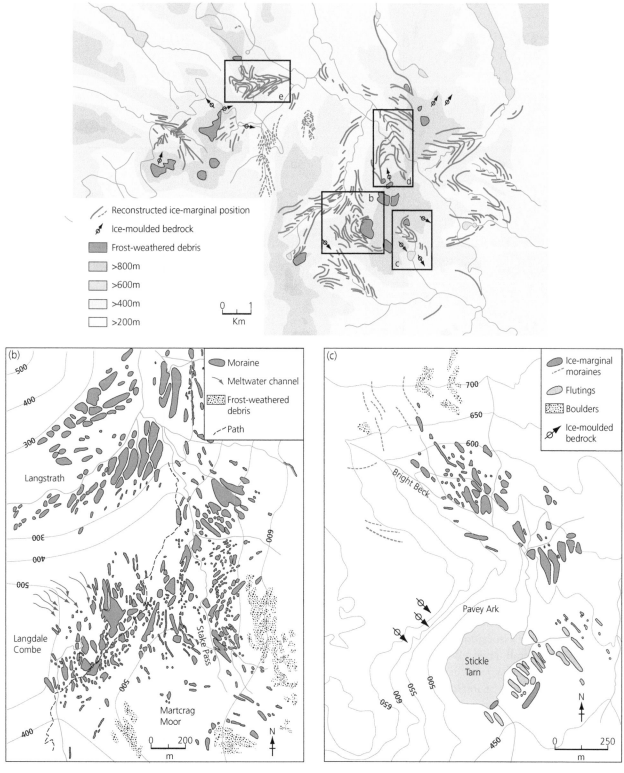

Figure 12.23 Younger Dryas plateau ice field palaeoglaciology based on glacial geomorphology and the glaciological controls specified by Manley (1955, 1959). (a) Summary map of reconstructed glacier margins in the central fells of the English Lake District, based on detailed glacial geomorphology maps. Example areas of glacial geomorphology identified by boxes are: (b) Stake Pass and Langdale Combe; (c) Pavey Ark/Stickle Tarn; (d) Greenup Gill; and (e) Honister Pass/Little Gatesgarthdale (Redrawn from McDougall, 2001)

settings. Low-relief settings include valleys in ancient mountain belts where the vertical distance between valley floor and summit ridges is generally less than 1000 m (e.g. the glaciated valleys of Scotland, Norway and Labrador). High-relief settings are characterized by extensive steep valley sides, rising thousands of metres above the valley floor. Such environments are commonly associated with young or tectonically active fold mountains

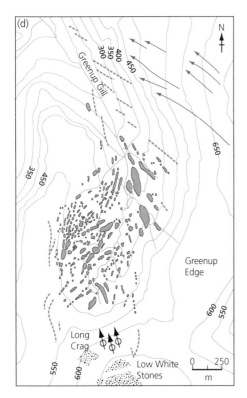

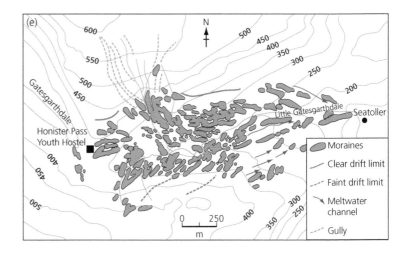

Figure 12.23 (Continued)

such as the European and New Zealand Alps, the High Andes and the Himalaya, where relative relief is commonly 3000 m or more. It should be emphasized, however, that the high- and low-relief classification is applied at the scale of individual valleys, and that both types may occur within a single mountain massif (e.g. Owen and Derbyshire, 1989, 1993). Moreover, even though higher-relief terrain will predominantly produce larger volumes of extraglacial debris, we should view the different signatures of valley glaciation as a continuum defined by the debris supply and the ice supply. This is demonstrated by figure 12.24 (Benn et al., 2003), which illustrates the importance of sediment transport in the construction of landform–sediment associations in mountain landscapes, regardless of overall relief. We can therefore classify glaciated valley landsystems according to their debris cover, whereby various supraglacial debris loads produce a range of depositional scenarios, from 'clean' to debris-covered glaciers and then to rock glaciers.

Although subglacial processes are just as active in valley glaciers as in other glacier systems, the imprints of such processes are only observed where supraglacial debris loads are relatively small and therefore the former glacier bed escapes burial during deglaciation (Eyles, 1979). On some lithologies, the beds of some former valley glaciers are so clearly preserved that they provide splendid opportunities to study subglacial systems at several scales, from that of the bed roughness up to the entire glacier system (e.g. Sharp et al., 1989a). Figure 12.25 shows the bed of a former cirque glacier on the Isle of

Skye, Scotland. Broadly, the bed can be divided into three zones which occur in sequence from the head of the glacier to the snout. First, the *erosional zone* on the upper part of the cirque floor is characterized by extensive areas of ice-moulded bedrock, recording net erosion of the bed. Striations, roches moutonnées, and an overdeepened rock basin document abrasion and quarrying of the bed by sliding, debris-charged basal ice. Second, the *intermediate zone* occurs downvalley and displays evidence for both erosion and deposition. Tills are thin and discontinuous, and are restricted to the swales between roches moutonnées. Lee-side cavity fills occur on the downglacier flanks of rock knobs (sections 10.3.4 and 11.2.7). Third, the *depositional zone* is in the lower part of the cirque, and is characterized by continuous till cover (see also fig. 11.6a). The till is overconsolidated and has a fissile structure, and is interpreted as a subglacial traction till (section 10.3.4). On aerial photographs, the till has a faintly lineated surface (fig. 12.25b), although flutings are not easily recognizable on the ground. These three zones bring to mind the patterns of subglacial erosion and deposition associated with continental ice sheets, albeit on a much smaller scale (section 12.4), and record downglacier transport of debris towards the margin.

According to Eyles (1979, 1983b), the majority of the debris transported by valley glaciers is derived from mass wasting of valley walls. However, in many low-relief glaciated valleys, subglacially entrained debris may be of equal or greater importance. For example, Benn (1992) used clast lithological analysis to show that Lateglacial

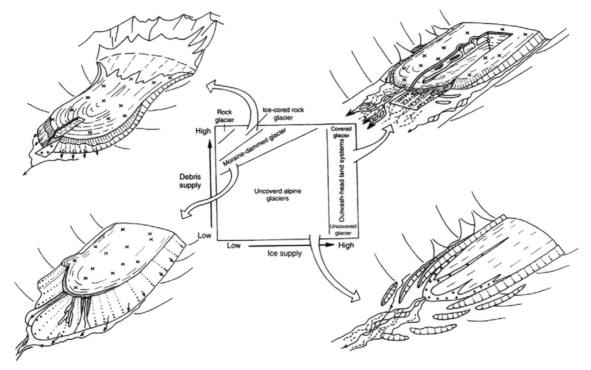

Figure 12.24 End members of the glaciated valley landsystem continuum defined by debris and ice supply (Benn et al., 2003).

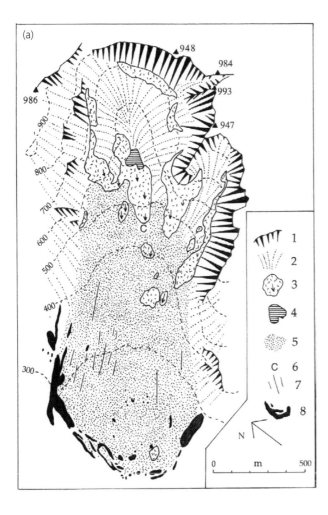

Figure 12.25 The subglacial signature of a cirque basin in a low-relief mountain setting with small amounts of supraglacial debris.
(a) Geomorphology map of Coire Lagan, Isle of Skye, Scotland, showing the distribution of ice-scoured bedrock and basal till formed during the Loch Lomond Readvance (Younger Dryas). In the upper part of the valley, large areas of bedrock are blanketed by postglacial scree. (1) Frost-shattered rockwalls. (2) Scree. (3) Ice-scoured rock with striae. (4) Rock basin lake. (5) Basal till. (6) Lee-side cavity fill. (7) Flutings. (8) Moraine ridges and mounds. Contours in metres. (b) Aerial photograph extract of the cirques below Beinn Bhan in the Applecross/Glenshieldaig area of north-west Scotland. (Copyright © RCAHMS, Edinburgh) The image shows strongly fluted cirque floors through which the geological bedding structures can still be seen orientated at right angles to fluting long axes.

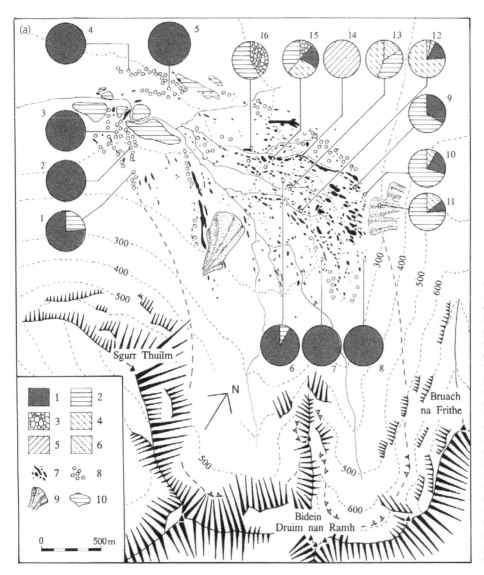

Figure 12.26 Evidence for subglacial derivation of debris during the Loch Lomond Stade, Coire na Creiche, Isle of Skye, Scotland. (a) Map of glacial landforms and debris lithology. The pie charts show the lithology of large surface boulders (samples 1–8) and clasts within moraine ridges (samples 9–16). (1) Gabbro. (2) Basalt. (3) Acid breccia. (4) Rhyolite. (5) Feldsparphyric dolerite. (6) Trachyte. (7) Moraine ridges and mounds. (8) Glacially transported boulders. (9) Debris fans. (10) Outwash terraces. The cirque backwalls are composed mainly of gabbro and basalt, and the other lithologies crop out on the cirque floor. (b) Reconstruction of Coire na Creiche glacier during deglaciation. (A) Glacier limit. (B) Gabbro boulders delivered from cirque backwalls. (C) Accumulation of scree at glacier margins. (D) Moraines composed of debris eroded from the bed (Benn, 1992).

moraines in many glaciated valleys on the Isle of Skye, Scotland, are composed predominantly of debris eroded from the bed (fig. 12.26). Clast form analyses on latero-frontal moraines in Norwegian mountain glaciers also show that large volumes of debris are transported sub-glacially (Matthews and Petch, 1982; Shakesby, 1989; Benn and Ballantyne, 1994; Evans, 1999). In these cases, subglacial debris entrainment appears to have been most effective below the glacier margins and in areas underlain by well-jointed rocks. In the case of Kvíárjökull, Iceland, Spedding and Evans (2002) demonstrated that the major-ity of debris in the large latero-frontal moraine loop (fig. 11.77) is derived from the bed and from englacial drainage networks.

The margins of valley glaciers are commonly delimited by lateral-frontal moraines and ice contact fans and ramps (sections 11.3.4, 11.3.5; fig 11.73). With respect to the volume of sediment and the nature of accumulation at the margins of valley glaciers, it is important to recog-nize the role of meltwater reworking and transport, in addition to the ice and debris supply depicted in figure 12.24. This was demonstrated by Shroder et al. (2000), who showed that the efficiency of fluvial processes at the ice margin dictates whether debris accumulates in latero-frontal moraines or outwash-head systems (see section 11.5.1). Moraine characteristics are strongly influenced by catchment lithology, and in valleys underlain by resist-ant, crystalline rocks, lateral moraines may be little more than ridges of boulders. Thrust moraines may occur where glaciers come into contact with thick, unconsoli-dated sediments such as glacimarine clays and silts (e.g. Benn and Evans, 1993; Evans and Wilson, 2006).

Many glaciated valleys in Scotland contain a very dis-tinctive assemblage of moraine ridges and mounds, known locally as 'hummocky moraine' (Sissons, 1967; Benn, 1992; Bennett and Boulton, 1993; Bennett, 1994; Benn and Lukas, 2006). The moraines commonly form converging cross-valley pairs of sharp-crested ridges and chains of hummocks up to 12 m high, but mostly only 2–3 m high. The ridges have an undulating long profile,

Figure 12.26 (*Continued*)

which is responsible for the hummocky appearance when viewed from the ground. The sedimentology, cross profiles and planforms of the moraines indicate that they are dump and push moraines or ice contact fans, displaying varying degrees of tectonic deformation (Lukas, 2003, 2005; section 11.3.5). The moraines were formed during oscillations of the ice margins during active retreat, rather than widespread stagnation as was previously believed. Pockets of chaotic hummocky moraine do occur within fields of recessional moraines, possibly due to the isolation and stagnation of sediment-covered ice remnants (Benn, 1992). For description and discussion of an alternative model of 'hummocky moraine' formation, see Bennett et al. (1998a), Graham et al. (2007) and Lukas (2007).

The occurrence of chaotic hummocky moraine is not necessarily indicative of widespread glacier stagnation, and many examples of 'hummocky moraine' have been observed at active ice margins (e.g. Eyles, 1979, 1983b; Evans and Twigg, 2002). The way in which *incremental marginal stagnation* can produce large areas of hummocky moraine has been reviewed in sections 11.4.2 and 12.3.1. Transverse lineations may be constructed within the resulting hummocky moraine either by crude preservation of controlled ridges or by thrusting of marginal debris-rich ice by more active ice upglacier (Evans, 2009). Staged stagnation has been proposed by Braun (2006) to explain the 'beaded valleys' of the Appalachian Plateau of north-east Pennsylvania, USA. The beaded appearance of valley bottoms is a function of 'till knobs' separated by

Figure 12.27 Latero-frontal moraines deposited by a piedmont glacier tongue in Green Creek, Sierra Nevada, USA. 1–3 indicate lateral moraines of different phases of glaciation; EM is the end moraine of lateral moraine 3 (R.P. Sharp, 1988; reproduced with permission of Cambridge University Press).

lake basins, indicative of successive ice stagnation zones up to 5 km wide.

Valley glaciers in high-relief settings typically have extensive covers of supraglacial debris, and sedimentation around their margins can form huge lateral-frontal dump moraines and ice contact fans and ramps (fig. 12.27; Boulton and Eyles, 1979; Small, 1983; Owen and

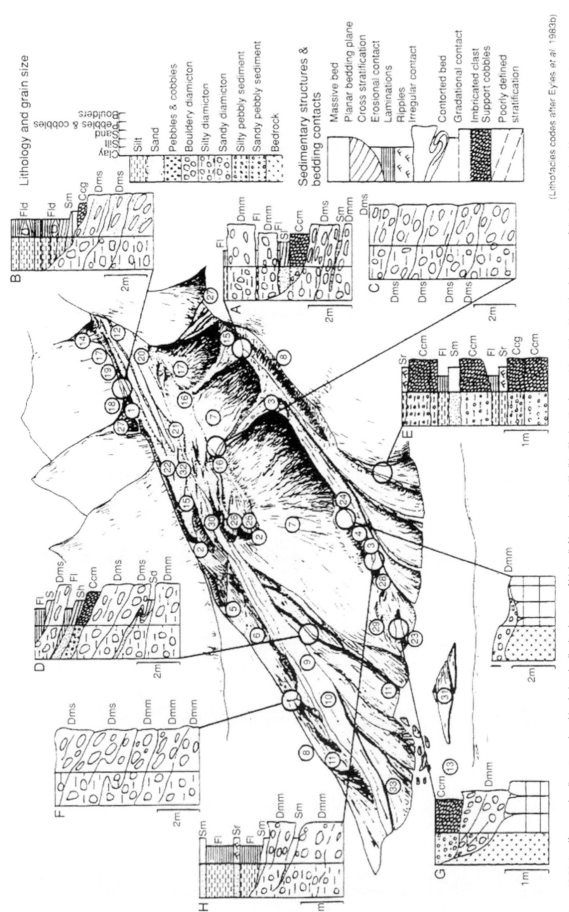

Figure 12.28 Landforms and sediments produced in glaciated valley landsystems of high relief based on the Ghulkin Glacier, Karakoram Mountains. Note the dominance of diamictons produced by mass movements: 1) truncated scree; 2) and 5) latero-frontal dump moraine; 3) laterally drained outwash channel; 4) glacifluvial outwash channel; 6) debris flow cones; 7) slide-modified lateral moraine; 8) abandoned lateral outwash fan; 9) meltwater channel; 10) meltwater fan; 11) abandoned meltwater fan; 12) bare ice; 13) trunk valley river; 14) debris flow; 15) flow slide; 16) gullied lateral moraine; 17) lateral moraine; 18) "ablation valley"; 19) "ablation valley"; 20) supraglacial lake; 21) supraglacial stream; 22) ice-contact terrace; 23) subglacial till exposure; 24) roche moutonnée; 25) flutings; 26) diffluence col; 27) high level till remnant; 28) diffluence col lake; 29) fines washed out from supraglacial debris; 30) ice cored moraines; 31) alluvium; 32) supraglacial debris; 33) dead ice (Owen and Derbyshire 1993; reproduced with permission of Routledge).

(Lithofacies codes after Eyles et al. 1983b)

Derbyshire 1989, 1993; Owen, 1994; Benn and Owen, 2002). Such landforms can constitute major barriers to glacier flow, so that repeated glacier advances may terminate at the same location and contribute to moraine building, resulting in landforms which can exceed 100 m in height and even the perching of glacier lobes high above trunk valley floors (fig. 11.49). Breach-lobe moraines are then created wherever the glacier locally exploits a gap in a major ridge (section 11.3.4). A depositional model for the margins of debris-covered glaciers has been presented by Owen and Derbyshire (1988, 1993), who termed it the *Ghulkin-type association*, after the glacier of that name in the Karakoram Mountains (fig. 12.28). The ice contact sediments consist of stacked debris flows dipping away from the glacier margin (sections 11.3.4, 11.3.5), and following glacier retreat these are dissected by meltstreams and overlain by glacifluvial fans and valley trains. Following ice retreat, the inner faces of lateral moraines are typically oversteepened and highly susceptible to paraglacial reworking. Indeed, where such moraines are overlooked by steep valley sides, they may not survive for very long after deglaciation, and the only record of glacier occupancy may be in the form of glacial erratics in debris fans (Owen, 1991). In low-relief mountains and/or areas of resistant bedrock, latero-frontal moraines typically range from 2 m to a few tens of metres high, depending on the availability of debris and the activity of the glacier terminus (Matthews and Petch, 1982; Benn, 1989b; Evans, 1999).

Debris-mantled glaciers can be expected to have relatively muted responses to climatic change, because large latero-frontal moraines limit the amount of glacier advance, and extensive supraglacial debris cover inhibits ablation of the underlying ice (Kirkbride and Warren, 1999; Nakawo et al., 1999; Kirkbride, 2000; Naito et al., 2000). Such glaciers are therefore very unlikely to oscillate in response to minor climatic fluctuations, and will tend to remain at the limits imposed by lateral-frontal moraines until advance or retreat is triggered by significant climate change. Thus the landform record of a retreating debris-mantled glacier might consist of major moraine complexes, representing periods of stability, separated by extensive tracts of hummocky moraine deposited during episodes of ice margin wastage and stagnation. In contrast, retreat of 'clean' valley glaciers may be punctuated by numerous oscillations of the ice margin, so that their recession is often recorded by push moraines superimposed on subglacial sediments and landforms. The size and spacing of such moraines reflects the availability of debris and the frequency and duration of stillstands or readvances of the margin (Eyles, 1979). It has been calculated that in the north-west Scottish Highlands, one moraine was produced every 3–23 years during recession from the Younger Dryas maximum, and readvances resulting in the reoccupation of

moraine positions by snout margins were very common (Benn and Lukas, 2006; Lukas and Benn, 2006). Rapid retreat and/or a sparse supraglacial debris cover will result in a thin, patchy veneer laid down on the subglacial surface. The exceptions to this are where debris cover thickens locally, such as beneath medial moraines, supraglacial lateral moraines or controlled moraines. The most prominent fossil medial moraines generally extend downvalley from spurs between confluent valleys (section 11.4.1).

At the highly debris-charged end of the glaciated valley landsystem continuum (fig. 12.24) lie rock glaciers. These are lobate masses of angular debris that resemble small glaciers, and move downslope as a consequence of the deformation of internal ice lenses or frozen sediments. They commonly have ridges, furrows and sometimes lobes on their surfaces, and have steep fronts down which debris slides and tumbles, to be overridden by the advancing mass (fig. 12.29). The origin of rock glaciers has inspired an astonishingly large and contentious literature (e.g. Giardino et al., 1987; Ballantyne and Harris, 1994; Humlum, 1998). The ice content in rock glaciers may originate in different ways, but it is clear that a major class of these landforms is closely genetically related to debris-covered mountain glaciers (Humlum, 1998, 2005). Humlum (2005) has shown how, in permafrost environments, snow avalanche tongues can become progressively debris-covered as they ablate during the summer months. By the end of the melt season, the snow content has been reduced to ice lenses buried beneath a mantle of coarse debris. Over time, annual increments of snow and rock build up ice-cored talus, which then deforms and flows downslope. This process is very similar to the accumulation of avalanche-fed glaciers. In both cases, accumulation of snow and debris occurs by avalanching from valley sides (sections 2.4.1, 9.3.1), and the end products differ only in the relative proportions of these inputs. Indeed, glaciers may evolve into rock glaciers, and vice versa, in response to changes in snow inputs (Humlum, 1998). Rock glaciers can also evolve from ice-cored moraines (Vere and Matthews, 1985; Evans, 1993; Ó Cofaigh et al., 2003b).

Temporary lakes are common features in glacierized mountain environments (section 3.7) and can have profound effects on the landscape. Lakes may be infilled rapidly by silts and other sediments, forming extensive terraces stretching upvalley from the dam. For example, during the winter of 1857/58, a landslide at Serat Pungurh in the Hunza Valley, Karakoram Mountains, dammed back a lake approximately 10 km long. The dam was breached about 6 months later, releasing a large flood down the Indus, causing much loss of life. During this 6-month period, up to 110 m of silt accumulated in the lake basin, and it now forms the lowest of three terraces in the valley (Owen and Derbyshire, 1993). These high

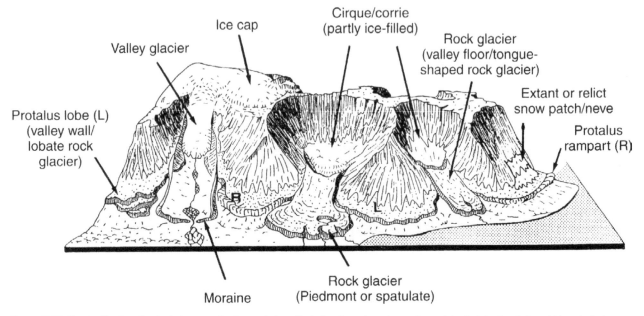

Figure 12.29 The classification of rock glaciers according to morphology, illustrating the various stages of mountain glacierization during which rock glaciers can form (Humlum, 1982; reproduced with permission of *Norsk Geografisk Tidsskrift*).

deposition rates are typical of the Karakoram region. Glacial lake outburst floods (GLOFs) commonly have high geomorphological impacts in mountain regions, eroding and reworking large volumes of sediment (see sections 3.7 and 12.3.6). As we saw in section 3.7, catastrophic floods following the failure of moraine dams constitute significant natural hazards in high-relief mountain regions (Reynolds, 2000; Richardson and Reynolds, 2000a, b). Flood tracks from ancient GLOFs may be preserved in the landscape in the form of channels and bouldery debris fans extending downvalley from the moraine breach (Clague and Evans, 1994; Coxon et al., 1996; Blair, 2002; Benn et al., 2006).

Valley fills including extensive supraglacial and proglacial lake deposits have been described from the mountains of British Columbia, Canada (Eyles et al., 1987; Ryder et al., 1991; Johnsen and Brennand, 2006), where the wastage of the Cordilleran Ice Sheet led to the isolation of stagnant glacier ice in deep valley bottoms. Clague and Evans (1994) have argued that environmental conditions were similar to those occurring today in the St Elias Mountains near the Canadian-Alaskan border, where major debris-mantled glacier tongues have thinned dramatically during the last few hundred years and stagnated in situ. Sediment sequences deposited in former supraglacial lakes in British Columbia record deposition in multiple basins, and are dominated by subaquatic mass flow deposits. The sediments are affected by numerous deformation structures caused by the melt-out of buried ice, and in some areas, valley-side delta and kame terraces are preserved, recording former lake levels or positions of the ice surface (fig. 11.93a).

Glacifluvial deposits are commonly well preserved in low- to moderate-relief glaciated valleys. The focusing of meltwater flow by valley sides results in the erosion of gorges or the deposition of ribbon-like valley trains along valley axes, where subglacial or ice-marginal landsystems tend to be buried or reworked. Maizels (1995) has mapped the valley train in front of the Glacier des Bossons in the French Alps, where glacifluvial activity has almost completely reworked any ice-marginal deposits. River terraces are striking components of many glaciated valley landscapes (see sections 11.5.1). Staircases of terraces occur along the floors of many glaciated valleys, and the highest members may show signs of ice-marginal deposition (*kame terraces*; section 11.4.5). Younger, lower terraces record deposition and erosion by proglacial meltwater and, in deglaciated basins, postglacial streams.

Following glacier retreat, ice-marginal and supraglacial landforms are subject to paraglacial reworking (section 11.5.2). As rivers adjust to new base levels and sediment loads, they incise existing sediment piles and redeposit debris in terraced valley trains (fig. 12.30). Paraglacial processes are particularly effective due to the abundance of steep slopes, the action of high-energy processes such as snow avalanching and ablation-triggered floods, and the widespread availability of unconsolidated sediment and unstable bedrock. Paraglacial reworking of glacial landforms and sediments is less effective where glaciers advanced out from high-relief mountainous regions to the foothills, and in such settings the preservation potential of the substantial ice-marginal landforms is greater. For example, the classic latero-frontal moraines of the Pinedale Glaciation in the USA were deposited in the foothills of the Rocky Mountains, where paraglacial processes were responsible for merely cosmetic changes (fig. 12.27). Similarly, some of the most impressive lateral-frontal moraines on Earth occur around the fringes of the New Zealand Alps, where glaciers were supplied by large quantities of debris by steep, tectonically active slopes. In the long valley systems emanating from the eastern side of the Southern Alps divide, numerous inset lateral moraines and kame terraces form continuous ice-marginal deposits which stretch over tens of kilometres. The valley glaciers that coalesced to form these landforms, Fox, Franz Josef, Tasman, Hooker and Mueller Glaciers, have now receded to their own steep-sided valleys, where paraglacial processes are heavily reworking any valley-side glacigenic accumulations. This change in

Figure 12.30 A valley train or proglacial outwash tract emanating from a breach in the Little Ice Age terminal moraine of the Tasman Glacier, New Zealand, and fed directly by proglacial/supraglacial lake waters. Note the incision and abandonment of older outwash surfaces (D.J.A. Evans).

landform–sediment signature is depicted in a conceptual time–space diagram by Benn et al. (2003), based on the Ben Ohau Range in New Zealand (fig. 12.31). The diagram reflects the gradual increase in the interruption of debris transfer through glacial systems from the LGM to the Holocene, as more glaciers have become restricted to steep-sided and confined valleys.

Trimlines and weathering zones

The upper limits of glacier occupancy in glaciated valleys can be preserved as erosional 'tidemarks', or *trimlines*, on valley sides (Ballantyne 1997b, 1998, 2007; Ballantyne et al. 1997). In recently deglaciated terrain, trimlines are often instantly recognizable from differences in vegetation cover either side of the former glacier limit. Below the limit, all pre-advance vegetation has been stripped off, leaving bare rock, glacigenic deposits and sparse pioneer vegetation; whereas above the limit, diverse plant communities or even forests record a much longer period of vegetational development. With the passing of centuries, such *vegetation trimlines* become harder to recognize, as plant communities within glacier limits approach the local climax vegetation. As a result, vegetation trimlines are not preserved in landscapes that have been deglaciated since the early to mid Holocene. In older glacial landscapes, *periglacial trimlines* may be preserved where frost-shattered terrain on upper slopes gives way abruptly to ice-scoured bedrock on lower ground (fig. 12.32). Periglacial trimlines owe their origin to the removal of pre-existing regolith by a glacier advance and/or periglacial weathering of the upper slopes during the period of glacier occupancy.

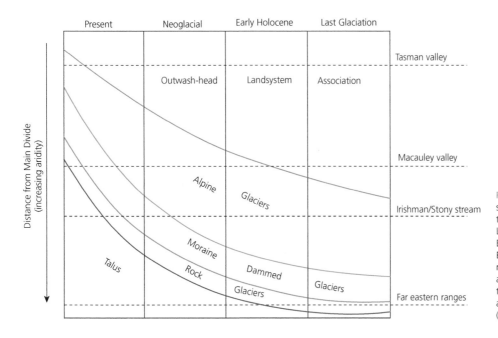

Figure 12.31 Distribution of different styles of glaciated valley landsystem in time and space, as related to the LGM–Holocene glacial phases in the Ben Ohau Range of New Zealand. ELAs have risen since the LGM, resulting in the uncoupling of glacial and fluvial transport networks and the increased dumping of debris around and on glacier snouts (Benn et al., 2003).

Figure 12.32 Periglacial trimline and granite palaeonunatak on Mount Hoffman in Yosemite National Park, California. A clear trimline separates a summit blockfield from ice-moulded rock on lower slopes (C.K. Ballantyne).

Figure 12.33 Aerial photograph stereopair of moraines and drift limit isolating a palaeonunatak in the Glenlyon area, Yukon Territory, Canada (Duk-Rodkin et al., 1986; reproduced with permission of the National Aerial Photograph Library, Ottawa).

Several criteria may be used to identify periglacial trimlines and/or to determine significant age differences between areas above and below them, including:

(1) the distribution of landforms such as roches moutonnées, and periglacial blockfields and tors;
(2) the depth of rock joints, which tend to be deeper above a trimline;
(3) an abrupt downslope limit to the distribution of long-term weathering products, such as the clay mineral gibbsite, or a greater variety of clay minerals above trimlines;
(4) small-scale weathering characteristics, such as rock texture and intact hardness and soil development;
(5) surface exposure dating from cosmogenic radionuclide concentrations (Ballantyne et al., 1997, 1998; Stone et al., 1998; Walden and Ballantyne, 2002; Rae et al., 2004; Golledge and Hubbard, 2005; Stone and Ballantyne, 2006).

The clarity of trimlines varies with rock type and structure, and the degree of postglacial weathering and slope activity. They are also best developed where glacier ice is fast flowing and topographically constrained, such as in maritime mountain ranges. The clearest trimlines are to be found on truncated spurs composed of resistant crystalline rocks (Thorp, 1981; Benn, 1989b). Palaeonunataks may also be delimited, at least in part, by ice-marginal moraines (e.g. Rae et al., 2004; fig. 12.33).

Abrupt changes in weathering intensity on mountain slopes can develop by other mechanisms, the most important of which is the selective preservation of relict periglacial features beneath thin, cold-based ice (section 8.6.3; Kleman and Borgström, 1994; Rea et al., 1996a; Briner et al., 2003; Marquette et al., 2004; Sugden et al., 2005; Phillips et al., 2006). The advent of cosmogenic nuclide dating has been most illuminating in this context, in that it has allowed the dating of erratics on tors and blockfields, often revealing that deeply weathered summits with indicators of prolonged periods of weathering

were covered during the most recent glaciation (e.g. Marquette et al., 2004). This shows that some transitions between periglacially weathered and ice-scoured terrain represent the thermal boundary between cold-based and wet-based ice below an ice sheet, rather than a former ice sheet surface. Such 'thermal trimlines' should be elevated on the upglacier sides of high ground and depressed on the downglacier sides, due to the effect of longitudinal stresses on the pressure-melting point of ice. In contrast, periglacial trimlines might be expected to have smooth regional gradients consistent with theoretical ice sheet profiles (Ballantyne and McCarroll, 1995). Two other explanations have been put forward for periglacial trimlines. The first relates to the impact of glacier readvance by warm-based ice after a period of ice sheet glaciation, so that relict periglacial features on many upland surfaces survived beneath cold-based ice only to be trimmed by the readvance. The second suggests that the lower limit of frost-weathered debris is climatically determined.

Taken together with lateral moraines and other ice-marginal landforms, periglacial trimlines allow the dimensions of former valley glaciers to be reconstructed with considerable accuracy, yielding important input for glaciological models or palaeoclimatic studies. Good examples of periglacial trimlines have been described by Ballantyne (1989), who used the distribution of periglacial and glacial landforms to determine the limits of Loch Lomond (Younger Dryas) Stadial glaciers on the Isle of Skye, Scotland. More recently, Ballantyne (1997b, 1998) and Ballantyne et al. (1998) have shown that there is an older, higher set of trimlines in Scotland, probably marking the upper limit of the last (Late Devensian) ice sheet. Thus, three weathering zones can be recognized. First, above the upper trimlines, blockfields and gibbsitic soils represent the weathered surfaces of nunataks that stood above the ice surface. Weathering has proceeded at least since the penultimate glaciation. Second, at lower altitudes, but outside the Loch Lomond Readvance limits, is a zone of 'frost-fretted' terrain, in which rock surfaces display the gross forms of glacial erosion but have been subject to minor joint widening and disaggregation by periglacial weathering since ice sheet deglaciation. Third, within the limits of the Loch Lomond Readvance, rock surfaces have only been subject to weathering in the Holocene, and glacial features are well preserved. Cosmogenic nuclide dating has provided some verification for this chronology. For example, Stone et al. (1998) and Stone and Ballantyne (2006) showed that rock exposure ages above the uppermost trimline in north-west Scotland are much older than the LGM, in contrast to those below the trimline which yield ages younger than the LGM; these data indicate that palaeonunataks probably existed above the northern margin of the BIIS. Nevertheless, cold-based ice, because of its non-erosive nature, can be invoked to explain such data, especially if uninterrupted exposure cannot be clearly demonstrated (Stone et al., 1998; Stone and Ballantyne, 2006). The most unequivocal evidence for all-pervasive ice coverage can be provided only by dating large numbers of erratics to find the youngest ages (e.g. Briner et al., 2003; Marquette et al., 2004; Phillips et al., 2006).

Mountain ice field landsystems

The identification of glaciated valley landsystems, palaeonunataks and plateau-centred ice masses allow us to make larger-scale reconstructions of mountain glaciations that involve the mapping of whole ice field systems. Examples include the Patagonian Ice Field in South America, which expanded far beyond its present limits at the LGM (e.g. Clapperton, 1993; Glasser et al., 2005, 2008; K. J. Turner et al., 2005; fig. 12.34), and the Younger Dryas ice fields in the Highlands and Islands of Scotland, where no glaciers exist today (Ballantyne, 1989; Benn et al., 1992; Bennett and Boulton, 1993; Golledge and Hubbard, 2005; Finlayson, 2006; Golledge, 2006, 2007; Lukas and Lukas, 2006; fig. 12.35). Expansion of outlet lobes from such ice centres involves the spatial and temporal development of glaciated valley landsystems depicted in figure 12.31, and at their maximum extent many such lobes will produce lowland landsystems signatures similar to those reviewed in sections 12.3.2 and 12.3.3. In other words, mountain ice field landsystems, because they cover large areas of diverse topography and can develop in a range of climatic settings, will encompass elements of other smaller-scale landsystems.

A landsystems approach has been employed by Golledge (2007) to compile a set of eight criteria with which ice caps can be differentiated from ice fields. The first two are glaciological constraints, whereby ice caps are >1000 km² in size and have domed accumulation areas

with steep outlet glaciers, in contrast to ice fields, which are >100 km² in size and have non-domed accumulation areas and low surface profiles around their margins. The remaining six geomorphological signatures are dictated by contrasting glacier-topographic relationships. They are:

(1) deglacial ice-marginal landforms and sediments, which are discordant with local topographic slope around ice caps but concordant around ice fields;
(2) subglacial sediments, which in ice cap systems are asymmetrically deposited on valley sides and thick on valley floors where pre-existing deposits are preserved; in ice field systems these deposits are symmetrically distributed on valley sides and pre-existing sediments are eroded from valley floors;
(3) modification of high-level cols and interfluves, which are streamlined beneath ice caps, but only streamlined in central areas and at low altitudes beneath ice fields;
(4) ice flow indicators, which are determined by ice surface slope beneath ice caps and by topography beneath ice fields;
(5) ice-marginal drainage features, which drain either with or against topography around ice caps, but generally follow topography around ice fields;
(6) ice flow patterns of the whole ice mass, which are predominantly radial from a local dome or ice shed beneath ice caps and are only radial beneath ice fields where flow is unconfined; otherwise flow beneath an ice field will be directed by topography.

The margins of mountain ice fields are commonly delineated by substantial accumulations of glacigenic material, due to the large debris fluxes in high-relief settings and/or the availability of thick Quaternary sediment sequences in fringing lowlands. Even in situations where outlet glaciers carry little debris from upland accumulation centres, large debris loads around terminal zones can result from reworking of overridden materials and the incorporation of subglacial debris at thermal boundaries (sections 11.4.2, 12.3.1), and the glacitectonic disturbance of valley fills (section 11.3.2). It is becoming increasingly apparent that overridden moraines (section 11.2.7) are common features in areas affected by mountain ice fields. For example, they have been widely reported from the beds of Younger Dryas ice fields in Scotland, where they appear as barchanoid ridges/ribbed moraine (Golledge, 2006, 2007) or fluted hummocks containing glacitectonites (Wilson and Evans, 2000; Benn and Lukas, 2006; Lukas et al., 2007). This emphasizes the point that readvancing glaciers invariably encounter and modify pre-existing deposits.

Quaternary expansion of the Patagonian Ice Field resulted in the advance of piedmont outlet glaciers, which were responsible for the scouring and streamlining of large tracts of bedrock, the excavation of overdeepenings,

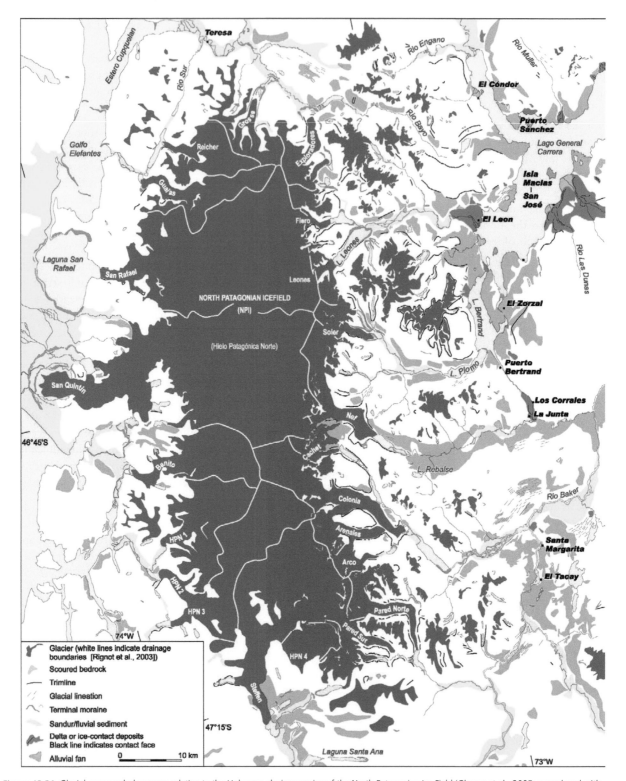

Figure 12.34 Glacial geomorphology map relating to the Holocene glacier margins of the North Patagonian Ice Field (Glasser et al., 2005; reproduced with permission of Elsevier).

and the construction of large arcuate moraine ridges (fig. 12.34; Glasser et al., 2005, 2008; Turner et al., 2005). Landform assemblages recording advance and retreat of the Magellan ice lobe are illustrated in figure 12.36 (Benn and Clapperton, 2000a, b). The ice terminated at a belt of complex moraine ridges, comprising major thrust ridges whose proximal slopes are adorned with minor ridges, kettle holes and irregular mounds. Inside this arc of moraines lie kame and kettle topography and glacifluvial terraces, which encircle two large, overdeepened basins now flooded by the sea. Immediately up-ice of the kame and kettle topography and overdeepened basins is a zone

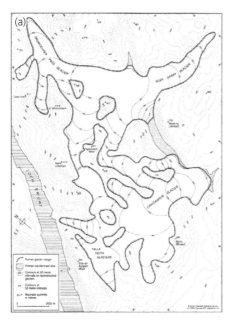
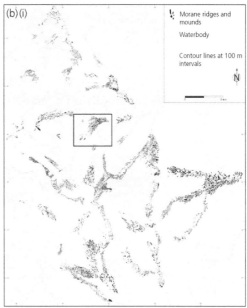

Figure 12.35 Examples of mountain ice field landsystems of the Scottish Younger Dryas. (a) The Younger Dryas Drumochter icefield in the Drumochter Hills of the Central Highlands of Scotland (Benn & Ballantyne 2005) (b) Glacial geomorphology and palaeoglaciological reconstructions from the north-west Highlands. (i) Map of moraine ridges and mounds in the valleys draining the Ben Hee and Foinaven uplands. (ii) Details of area boxed on main map, showing moraines (upper box) and palaeo-ice fronts (lower box) of the receding Easaidh Glacier. (c) The reconstructed mountain ice field, including the Easaidh Glacier (Lukas and Benn, 2006; Lukas and Lukas, 2006; reproduced with permission of Taylor & Francis).

of drumlins, composed of subglacially deformed materials and subglacial traction tills and draped by small moraine ridges. Prominent meltwater channels wind between upstanding sediment masses with streamlined, drumlinized surfaces and dissect the moraines. The juxtaposition of drumlins, excavational basins, thrust-block moraines and marginal meltwater channels suggests that the Magellan ice lobe may have been characterized by a cold-based margin and a wet-based interior. However, it is also possible that the landform assemblage reflects a surge (cf. section 12.3.8).

The importance of thick, pre-existing sediment sequences for moraine construction around the Patagonian Ice Field is well illustrated by the Llanquihue Drift in the southern Lake District of Chile (Andersen et al., 1999; Denton et al., 1999; Schluchter et al., 1999; Turbek and Lowell, 1999; fig. 12.37). This complex assemblage of moraine ridges, glacitectonic landforms, kame terraces, outwash plains and spillways documents the advance of piedmont lobes across large marine straits and embayments and onto offshore islands, and subsequent ice retreat. While in contact with deeper water, ice-proximal sedimentation was dominated by subaqueous processes. The dominance of glacifluvial sediments around the former terrestrial margins attests to glacier termini that were well coupled with their proglacial drainage networks (fig. 12.24). The occurrence of an isochronous relict land surface pre-dating the LGM allows us to appreciate the overprinting of landform–sediment assemblages relating to different glaciations.

Ice-dammed lakes are common features in glaciated mountains, being produced wherever glacier ice dams tributary or trunk valleys (section 3.7). They also occur in the overdeepenings created by outlet glaciers from mountain ice fields, and numerous excellent examples can be seen around the North Patagonian Ice Field (Glasser et al., 2005, 2008; Turner et al., 2005; fig. 12.34). During the last glaciation, the coalescence of the expanded North and South Patagonian Ice Fields resulted in the damming of large proglacial lakes on the eastern flanks of the Andean mountain chain. Most significantly, the greatly enlarged lakes of Lago Cochrane and Lago General Carrera in the Rio Baker drainage basin were drained by spillways that initiated a major river, more than 350 km long, draining to the South Atlantic. This drainage route was shut off after the two ice fields decoupled around 13 ka BP and the Rio Baker switched its drainage direction westwards to the Pacific (Turner et al., 2005). Sedimentation rates in such lakes can be increased significantly where mountain outlet glaciers contact their shorelines, as demonstrated by Gilbert et al. (2006) at Atlin Lake, a lake that has been uncovered by a retreating lobe of the Juneau Ice Field in northern British Columbia. Holocene sediments in the proximal basins of the lake are 120 m thick, translating into a mean sedimentation rate of >1 cm per year since deglaciation. Beyond the sill that separates the proximal basins from the main Atlin Lake, the sedimentation rates are only 1–4 mm per year.

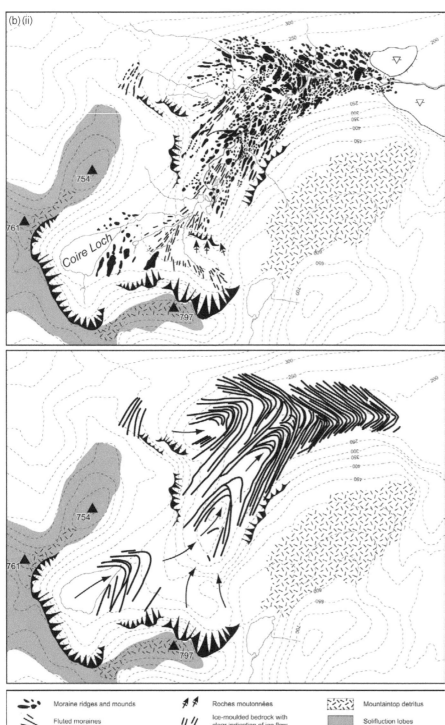

(b)(ii)

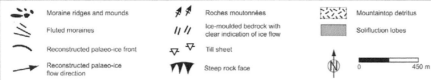

Moraine ridges and mounds		Roches moutonnées		Mountaintop detritus	
Fluted moraines		Ice-moulded bedrock with clear indication of ice flow		Solifluction lobes	
Reconstructed palaeo-ice front		Till sheet			
Reconstructed palaeo-ice flow direction		Steep rock face		N 0 450 m	

Figure 12.35 (*Continued*)

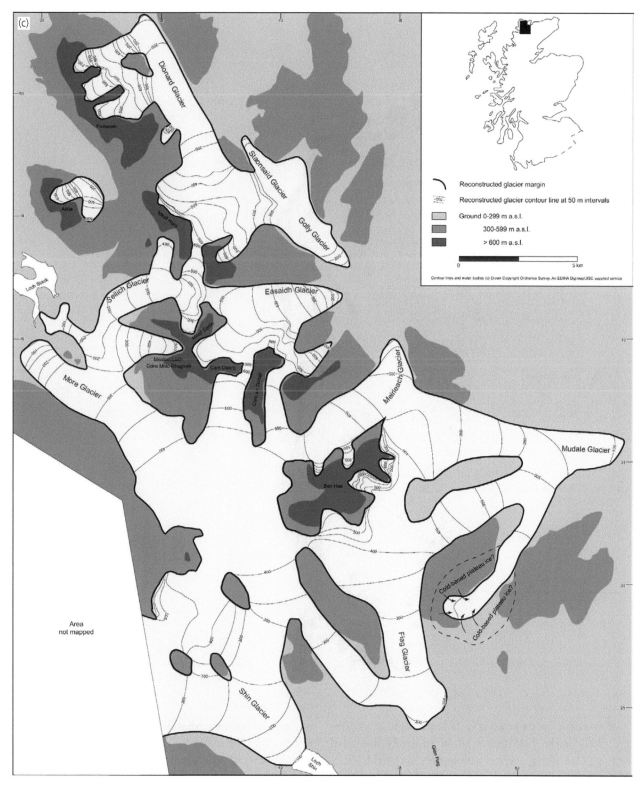

Figure 12.35 (*Continued*)

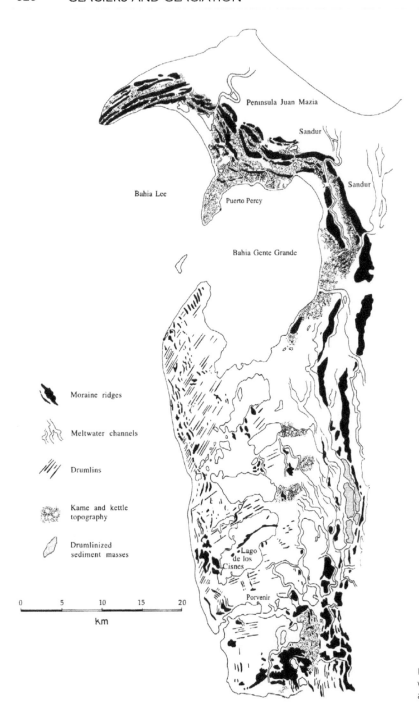

Peninsula Juan Mazia

Sandur

Sandur

Bahia Lee

Puerto Percy

Bahia Gente Grande

Moraine ridges

Meltwater channels

Drumlins

Kame and kettle topography

Drumlinized sediment masses

Lago de los Cisnes

Porvenir

0   5   10   15   20

km

Figure 12.36 Glacial landform assemblage on the western shoreline of northern Tierra del Fuego (Benn and Clapperton, 2000b).

Pleistocene ice-dammed lake sediments and landforms are likely to be best preserved in relatively low-relief settings bordering former mountain ice fields (e.g. McCulloch et al., 2005; Turner et al., 2005). Celebrated examples occur in the Scottish Highlands, in Glen Roy, at Achnasheen, and south of Loch Lomond (figs 11.136, 12.38; Benn, 1989a, 1996; Phillips et al., 2002; Palmer et al., 2008). In Glen Roy, water level was controlled by cols on the watershed, and the lake waters rose and fell as successive cols were blocked by glacier advance or exposed by retreat. The former water levels are recorded by very prominent shorelines known as the 'Parallel Roads', which remain strikingly clear after approximately 11,000 years of weathering and erosion. At Achnasheen, lake level was controlled by an ice dam, and was consequently less stable. This is reflected in the shorelines, which are much more numerous and less distinct that those at Glen Roy (Benn, 1989a). Around south Loch Lomond the Lomond glacier lobe expanded into a lowland drainage basin and prograded an ice contact delta landwards into Glacial Lake Blane (figs 10.48, 12.15). At these locations, assemblages of morainal banks, deltas,

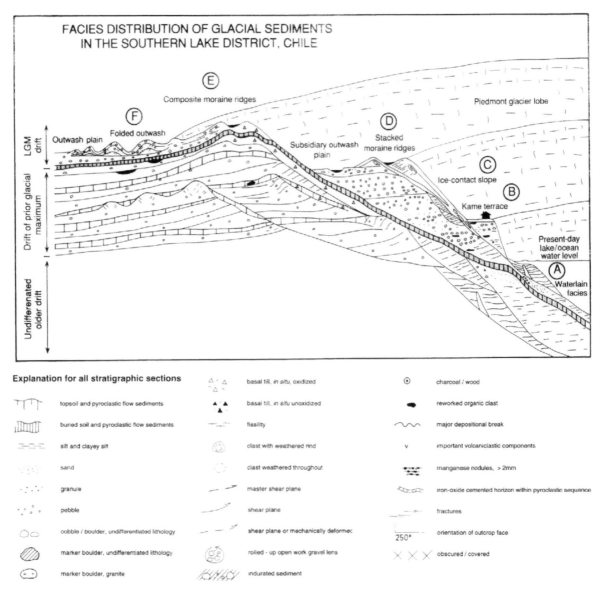

Figure 12.37 The Llanquihue Drift of the Chilean Lake District. Main sketch shows stratigraphic architecture of LGM deposits overlying older ice-contact sequence. Circled letters refer to each of the following panels, in which particular parts of the stratigraphy are related to specific depositional environments. (A) Locally glacitectonized waterlain facies from a proglacial lake. (B) Kame terrace deposition, with relic land surface showing more than one phase of sedimentation. (C) Ice-contact slope processes and sediments, including tills and glacitectonized stratified sediments. (D) Stacked moraine ridges, with tills on proximal slopes and mass flows on distal slopes. (E) Composite moraine ridges produced by the reoccupation of the same site by an ice margin. Erosion on the proximal slopes exposes older sediments at the land surface. (F) Glacitectonically folded outwash (Denton et al., reproduced with permission of Wiley and Blackwell).

glacitectonites and distal glacilacustrine sediments preserve a record of glacier fluctuations and shifting depo-centres during the lifespan of the lakes.

12.3.6 GLOF-DOMINATED GLACIAL LANDSYSTEM

In chapters 8 and 11 (sections 8.4.4, 11.4.6, 11.5.1), we reviewed the landform and sedimentary evidence for GLOFs in proglacial areas, and in section 11.2.10, we highlighted the rapidity of esker sedimentation during such floods. These landform and sediment imprints are

clearly recognizable in the landscape, allowing us to chart the impact of palaeo-GLOFs with increasing confidence (e.g. Longva and Thoresen, 1991; Coxon et al., 1996; A. J. Russell et al., 2003). At the very largest scale, we can include such features as the Channelled Scablands of Glacial Lake Missoula floods and the mega-scale depositional forms of the Altay region of Siberia, both of which are discussed in chapter 8. We turn here to the proglacial landform–sediment associations produced by GLOFs in modern glacierized catchments.

The best-known modern example of a GLOF-dominated landsystem is at Skeiðarárjökull, Iceland,

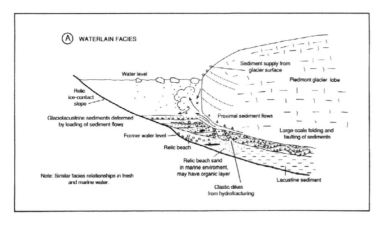

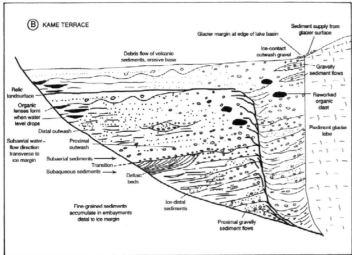

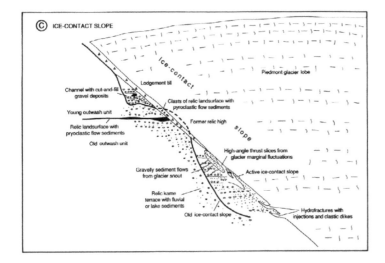

Figure 12.37 (Continued)

where the 1996 jökulhlaup provided a unique opportunity to assess the evolution of a coupled glacial and proglacial drainage system during a major flood event (Gomez et al., 2000, 2002). The sediment-filled englacial drainage network that evolved throughout the flood event is well recorded in the complex esker and supraglacial ice-walled canyon fill presently melting out from Skeiðarárjökull (fig. 11.44). Intensive ground-penetrating radar surveys by Burke et al. (2008) allowed the three-dimensional architecture of the sediment infill to be determined, and related to changing depositional conditions during the flood event (fig. 12.39). Successive depositional units record conduit unroofing and the development of a supraglacial ice-walled channel, a process that occurs in normal eskers but is associated with significantly more rapid deposition and ice burial during GLOFs.

Russell et al. (2006) proposed a general landsystem model for GLOF-dominated glacier margins, based on

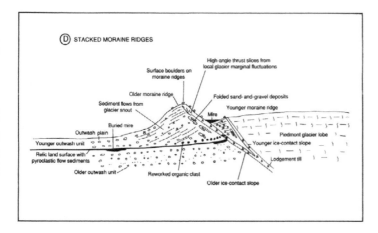

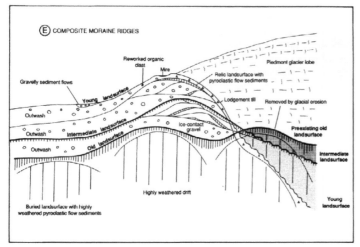

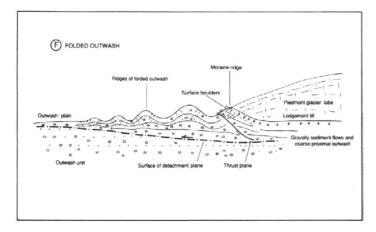

Figure 12.37 (*Continued*)

evidence from Skeiðarárjökull and Solheimajökull flood pathways. Russell et al. recognized both *unconfined* and *moraine-confined* variants of the landsystem (fig. 12.40). Unconfined systems are characterized by flooding directly from the glacier over most of the foreland, whereas in confined systems, floods are routed through ice-marginal lakes in the topographic lows behind prominent moraines. The latter results in greater spatial variability in sedimentary successions, determined by the occurrence of backwater conditions which may be either shallow and fast or deep and sluggish. Additionally, the flood waters

only coalesce to cover the entire sandur surface once they have broken through the moraine dam. Typical cross sections through the landform–sediment associations produced in both moraine-confined and unconfined subsystems are presented in figure 12.41. A prominent feature of both landform–sediment assemblages is the occurrence of densely spaced pits produced by the melting of icebergs (figs 11.96, 12.40, 12.42). These tend to be organized in linear clusters and, where densely spaced, can ultimately melt out to produce a chaotic or hummocky assemblage.

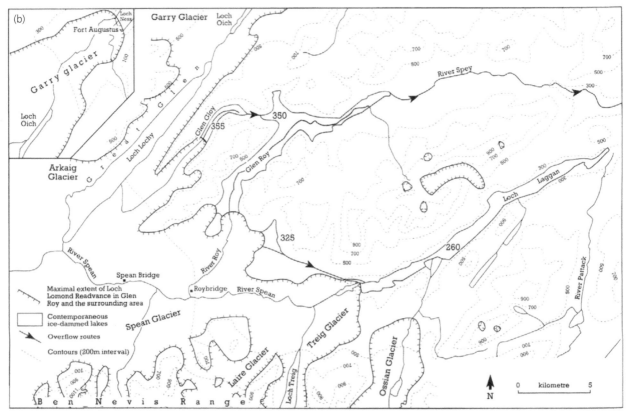

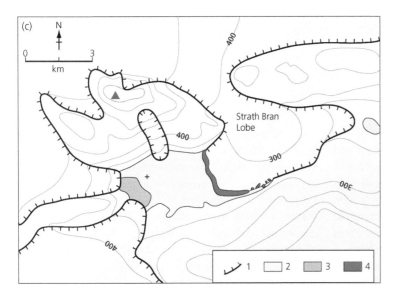

Figure 12.38 Loch Lomond (Younger Dryas) Stade ice-dammed lakes in the Scottish Highlands. (a) The 'Parallel Roads' or glacial lake shorelines of Glen Roy. (D.J.A. Evans) (b) Palaeoglaciological reconstruction of the glacier dams and lakes in the Glen Roy area. (Sissons, 1979) (c) Reconstruction of the Achnasheen ice-dammed lake, showing the locations of deltas and subaqueous moraines. (1) Glacier margin. (2) Lake. (3) Delta. (4) Moraine (Benn, 1996).

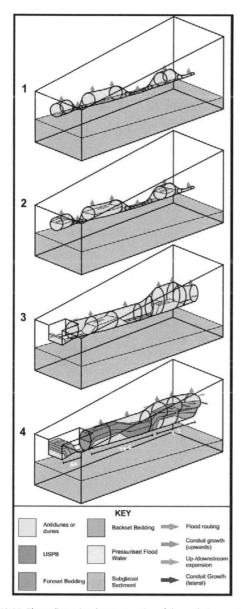

Figure 12.39 Three-dimensional reconstruction of the evolutionary stages of esker development during the 1996 jökulhlaup at Skeiðarárjökull based on ground-penetrating radar surveys. (1) Pressurized englacial flood discharge results in localized ice excavation and cavity generation. (2) Early englacial deposition in newly created accommodation space. (3) Expansion of upflow conduit and growth of downstream conduit upwards into the ice to produce unroofing and an ice-walled canyon at the glacier margin. (4) Conduit infilling and growth with continued conduit unroofing at the margin and headward growth of ice-walled canyon (Burke et al., 2008).

The relationships between depositional and erosional signatures of GLOFs have been identified by Carrivick (2007), using a downflow continuum constrained by flow depth, flow velocity, shear stress and stream power (fig. 12.43). The occurrence of all parts of the continuum in one landscape is likely to be fulfilled only in settings where bedrock topography confines the flow so that ice-marginal sandur fans do not absorb the flood energy.

Carrivick's (2007) example is from an Icelandic volcanically influenced setting where GLOFs result in spectacular landscape change. This is well illustrated by the Kota Fan in southern Iceland, where an eruption from Oræfi in AD 1727 resulted in the rapid accumulation of huge volumes of volcanic and glacial debris by lahars carrying large ice blocks.

In addition to the stratigraphic sequences outlined in chapter 11 (section 11.4.6), other sedimentary features that are common in GLOF-fed sandar are clastic dykes and hydrofracture fills (section 10.5.6), produced where groundwater pressures are elevated. Also common within GLOF deposits are diamictic (till) blocks, ripped up and carried along by the flow but not disaggregated before deposition in hyperconcentrated flows (section 10.4.9).

12.3.7 FJORDS

Many outlet glaciers from high-latitude ice caps and ice fields terminate in fjords, and the same was true of mid-latitude ice caps and ice fields during Quaternary glacial periods. Consequently, fjord-floor sediments are an important component of the depositional record of modern and ancient glaciers.

Stratigraphic architecture in fjords is strongly influenced by topography. Overdeepened basins act as sediment sinks, whereas bedrock sills provide pinning points where ice-proximal depositional systems can accumulate. Furthermore, the presence of steep flanking slopes encourages gravitational reworking of sediment and debris input from ice-free terrain (e.g. Powell, 2003; Stoker et al., 2006). Fjords can be subdivided into high- and low-relief types. High-relief fjords have lateral slopes which are too steep for the accumulation of sediment, and tend to be characterized by precipitous rockslopes flanking a sediment-floored trough. In contrast, low-relief fjords, such as the Scottish lochs (e.g. Dix and Duck, 2000; Stoker et al., 2006), contain more extensive low-gradient surfaces on which sediment can accumulate, encouraging the preservation of a wide variety of sediment–landform associations.

Fjord relief has a large influence on the geometry of sedimentary infills (e.g. Boulton, 1990; Sexton et al., 1992; Powell, 2003). In high-relief fjords, slumping from steep margins relocates sediment on the trough floor, producing flat-lying infill sequences such as that in Cambridge Fjord, Baffin Island (Boulton, 1990). In low-relief fjord environments, sediment cover is more widely distributed, forming an extensive *draped sequence* (e.g. Svalbard fjords and Scottish lochs). There may be still large local differences in sediment thickness, with the thickest sequences occurring close to sediment sources and in topographic lows.

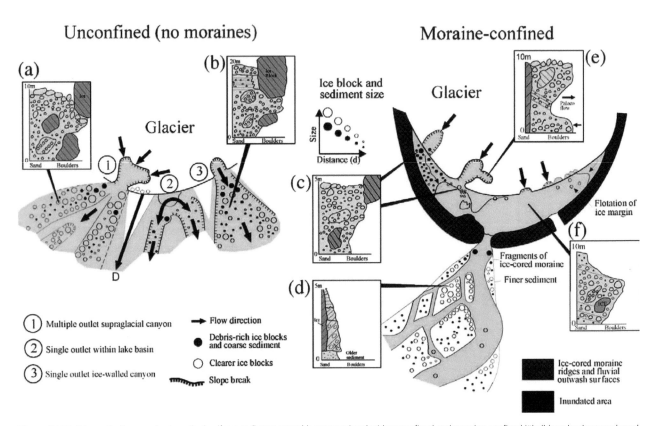

Figure 12.40 Schematic diagrams to show the landform-sediment assemblages associated with unconfined and moraine-confined jökulhlaup landsystems based on Icelandic piedmont lobes. Vertical sediment logs show increased spatial variability in moraine-confined settings and finer-grained deposits on proglacial fan surfaces than in unconfined settings (A.J. Russell et al., 2006; reproduced with permission of Elsevier).

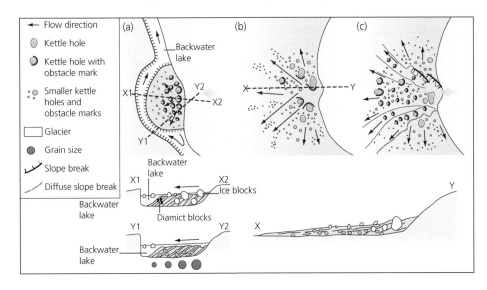

Figure 12.41 Schematic cross profiles through landforms produced in (a) moraine-confined, backwater settings and (b, c) unconfined settings. Rising stage deposition is depicted in (a) and (b), whereas (c) shows the dissection of the fan surface during falling flow stage (Modified from Russell et al., 2006).

Figure 12.42 Aerial view of the Gigjukvisl fan emanating from the supraglacial ice-walled channel produced by the 1996 jökulhlaup at Skeiðarárjökull. Note the linear chains of kettle holes and obstacle marks and patches of hummocky topography (H. Fay).

	Flow depth (m)	0	→	7	→	13	→	19	→	26
	Flow velocity (m s⁻¹)	0	→	5	→	11	→	15	→	22
	Shear stress (N m⁻²)	10²	→	10³	→		10⁴	→		10⁵
	Stream power (W m⁻²)	10³	→	10⁴	→		10⁵	→		10⁶

Figure 12.43 Downflow continuum of jökulhlaup-related landforms in topographically confined settings, based on Kverkfjallarani, Iceland (Carrivick, 2007; reproduced with permission of the International Glaciological Society).

The sedimentary infill of fjord basins can be usefully divided into three broad units, deposited during glacier advance, maximum and retreat:

(1) *Advance phase.* Within the glacier limit, the advance stage of a glacial cycle is typically represented by a basal till unit or an erosion surface. Till thickness generally increases down-fjord towards the glacier limit, reflecting erosion and downglacier transport of pre-existing sediment and bedrock fragments (Boulton, 1990). Beyond the glacier limit, extensive blankets of mud and diamicton may record distal glacimarine deposition during the advance phase, but are typically buried by younger glacimarine sediments.

(2) *Maximum phase.* Glacier limits within fjords are commonly located at pinning points where reduced calving rates encourage ice margin stability (section 5.5.4). Typical sediment–landform associations

include push or thrust moraines, grounding-line fans, morainal banks and deltas (Powell and Domack, 1995; section 11.6). In low- to intermediate-relief fjord basins, frontal depo-centres may continue into lateral moraines on the fjord walls, excellent examples being the moraine assemblages of the aptly named Moraine Fjord, South Georgia (Bentley et al., 2007; fig. 12.44). In contrast, in high-relief settings, lateral moraines are generally rapidly destroyed by paraglacial reworking. Lateral moraines with exceptionally low gradients indicate the former extent of floating ice shelves or glacier tongues close to flotation (figs 11.135, 11.144; section 11.6.4). Beyond the ice limit, blanket-like drapes of fine-grained sediments and diamictic muds record suspension sedimentation from turbid overflows and dumping from icebergs (Powell and Molnia, 1989). The relative amount of

Figure 12.44 Lateral moraines on the outer part of Moraine Fjord, South Georgia, deposited when the tidewater glaciers visible in the distance advanced down-fjord around 3 ka BP (D.J.A. Evans).

suspended sediment and ice-rafted debris varies with topographic setting, calving rates, and the availability of debris and meltwater. The highest rates of suspension sedimentation are associated with temperate fjord glaciers, where discharges of turbid meltwater are high during the ablation season, and thick sequences of cyclopels and cyclopsams can accumulate rapidly (Cowan and Powell, 1990).

(3) *Retreat phase.* During glacier retreat, depositional zones migrate up-fjord, and progressively more distal facies are laid down on top of older units. Pauses in glacier retreat may be marked by push or thrust moraines, grounding-line fans or morainal banks (Boulton, 1986; Dix and Duck, 2000; Ottesen and Dowdeswell, 2006). Substantial ice-marginal accumulations, such as large morainal banks and deltas, tend to be associated with topographic pinning points. At trough margins, glacimarine facies may be overlain by subaqueous fans fed by subaerial streams on the valley sides. Fan aggradation is generally most rapid in the paraglacial period immediately after deglaciation, when large amounts of readily entrained glacigenic sediment are available.

Depositional systems associated with the retreat phase can be illustrated by a set of models developed by Powell (1981) for Alaskan fjord glaciers (fig. 12.45):

- *Facies association I* results from rapid calving retreat in deep water. Large amounts of sediment are transported by icebergs, producing widespread dropstone muds and diamictons. Small morainal banks are formed during brief glacier stillstands or readvances.
- *Facies association II* is characteristic of slowly retreating or stationary glacier margins in shallow water. These conditions encourage the deposition of large grounding-line fans or morainal banks.
- *Facies association III* is formed by glaciers terminating in very shallow water. In such situations, the glacier margin melts more rapidly than it calves, resulting in a gently sloping front. Deposition from meltstreams or mass flows produces ice contact subaqueous fans which pass distally into turbidites.
- *Facies association IV* forms at fjord margins which receive sediment from a land-based glacier margin. Glacier-fed deltas are produced where glacial meltstreams enter the fjord, and deltaic bottomsets, foresets and topsets prograde over older facies.

These four associations can develop in sequence during glacier retreat into shallowing water, or alternate if the retreating margin migrates through a series of basins (fig. 12.46).

Complexity is introduced to the depositional zones discussed above wherever glacier margins oscillate, resulting in a stratigraphic architecture that is best analysed using a sequence stratigraphy approach (Powell et al., 2000; Powell and Cooper, 2002; Powell, 2003). This is illustrated by a hypothetical case in figure 12.47, based on Antarctic glacimarine sedimentary sequences, which can be applied to both fjord and continental shelf environments (section 12.5.2). To interpret such sequences, Powell and Cooper (2002) identified glacial systems tracts (GST) in relation to glacier advance and recession. At each stage reproduced in figure 12.47, a diagnostic sedimentary signature can be recognized and classified as glacial maximum (GMaST), glacial recession (GRST), glacial minimum (GMiST) and glacial advance (GAST) systems tracts. These are separated by bounding discontinuities, including the grounding-line retreat surface (GRS), the maximum glacial retreat surface (MRS) and the glacial advance surface (GAS), or an unconformity represented by the glacial erosion surface (GES).

Such complex stratigraphic signatures in the glacimarine deposits and landforms in fjords may be raised above sea level by isostatic uplift following deglaciation, thereby providing us with a readily accessible source of information on former ice margins in regions where terrestrial ice-marginal accumulations are commonly not well preserved. Emergent morainal banks and grounding-line fans have been the primary source of palaeoglaciological reconstructions in a range of high-latitude settings. In the Norwegian fjords, such depo-centres have been called *Tronder moraines*, and have facilitated an allostratigraphic approach towards reconstructing glacier advance and recession history by Lønne (1995, 2002) and Lønne et al. (2001; section 11.6; fig. 11.123). The morphology and distribution of such features are also significant. For example, Larsen et al. (1991) mapped the distribution of emergent De Geer moraines in the Møre area of western Norway, and showed that they yield a very detailed picture of deglaciation. Glacier retreat rates varied between fjords, as a consequence of differences in topography and ice discharge. Similarly, grounding-line fans, morainal banks and ice-contact deltas have been widely

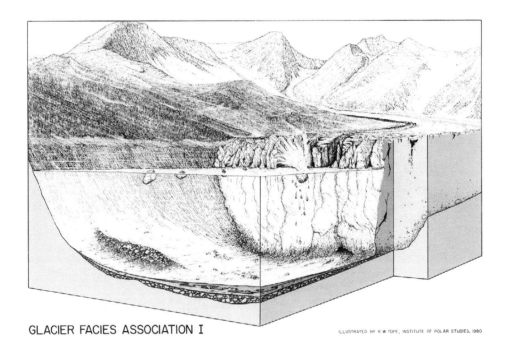

GLACIER FACIES ASSOCIATION I

ILLUSTRATED BY R W TOPE, INSTITUTE OF POLAR STUDIES, 1980

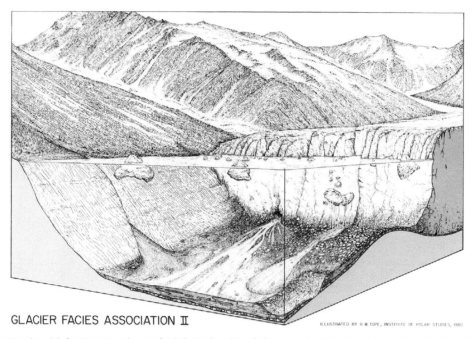

GLACIER FACIES ASSOCIATION II

ILLUSTRATED BY R W TOPE, INSTITUTE OF POLAR STUDIES, 1980

Figure 12.45 Depositional models for retreating tidewater fjord glaciers based on Alaskan examples. (I) Rapidly retreating glacier in deep water. (II) Slowly retreating glacier in shallow water. (III) Slowly retreating glacier in very shallow water. (IV) Terrestrial glacier margin supplying sediment to a delta (Drawings by R.W. Tope, from Powell, 1981; reproduced with permission of the International Glaciological Society).

employed to map and date the recession patterns of fjord glaciers in the Canadian Arctic, because larger amounts of sediment were deposited at subaqueous glacier margins than at terrestrial margins (e.g. Bednarski, 1988; Evans, 1990a; Ó Cofaigh, 1998; Ó Cofaigh et al., 1999; England et al., 2000).

Emergent glacimarine sediments and landforms are subject to modification by waves, currents and icebergs in the near-shore zone (fig. 12.12). Raised beaches and

deltas are formed where there is abundant sediment supply, and spectacular flights of terraces occur along the margins of some isostatically raised fjord coastlines. The altitude of raised shorelines provides important data on the history of sea-level change and glacier fluctuations. In fjords, the marine limit commonly declines towards the ice accumulation centre, recording the occupancy of inner basins by glacier ice during isostatic uplift of deglaciated outer coasts (section 7.3.2).

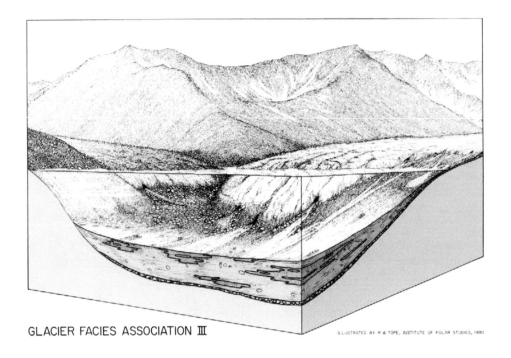

GLACIER FACIES ASSOCIATION Ⅲ

ILLUSTRATED BY R W TOPE, INSTITUTE OF POLAR STUDIES, 1980

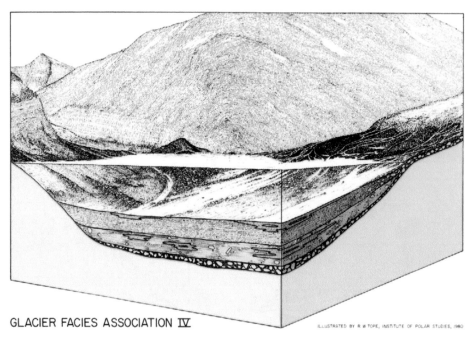

GLACIER FACIES ASSOCIATION Ⅳ

ILLUSTRATED BY R W TOPE, INSTITUTE OF POLAR STUDIES, 1980

Figure 12.45 *(Continued)*

12.3.8 SURGING GLACIERS

Glacier surges (section 5.7) produce very distinctive suites of landforms and sediments, due to a potent combination of erosional, deformational and depositional processes (Sharp 1985b, 1988b; Evans and Rea, 1999, 2003). The debris content of the margins of surging glaciers is typically very high, for a number of reasons. First, rapid sliding and the widespread development of cavities at the glacier bed are conducive to high rates of abrasion and quarrying (Clapperton, 1975; Humphrey and Raymond,

1994; Sharp et al., 1994; Hallet et al., 1996). Data from the 1982–83 surge of Variegated Glacier, Alaska, indicate that subglacial erosion rates during surges are among the highest in the world. In contrast, erosion rates during quiescent phases are likely to be very low, due to low sliding rates. In soft-bedded systems, high subglacial water pressures give rise to disruption of the substrate by hydrofracturing (Kjær et al., 2006), making it more susceptible to glacial erosion and directly increasing sediment loads of meltwater streams.

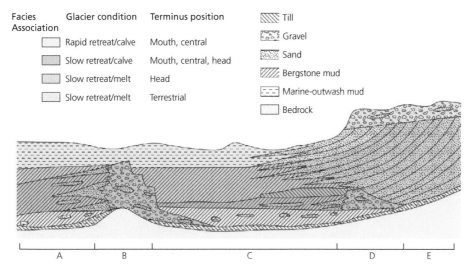

Figure 12.46 Hypothetical section through glacimarine sediment stratigraphy associated with a retreating glacier snout. The depositional environments of Facies Association I–IV are depicted in figure 12.45. Interval A: ice front terminated in deep water and retreated rapidly by calving. Interval B: calving continued but recession was slowed by a channel constriction. Interval C: ice once again calved rapidly in deep water. Interval D: ice had reached fjord head and recession slowed. Calving was replaced by surface melting. Interval E: ice front became terrestrial and an outwash delta prograded out over all previous facies (Powell, 1981; reproduced with permission of the International Glaciological Society).

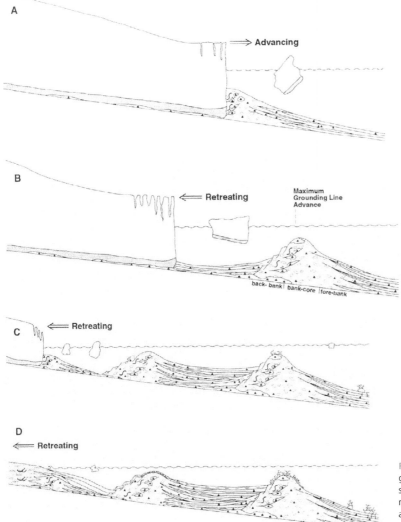

Figure 12.47 Idealized model of the production of a glacial advance, recession and readvance sedimentary sequence where facies are related to glacier activity and not sea level change. Vertical scale is exaggerated (Powell and Cooper, 2002, after Powell et al., 2000; reproduced with permission of the Geological Society, London).

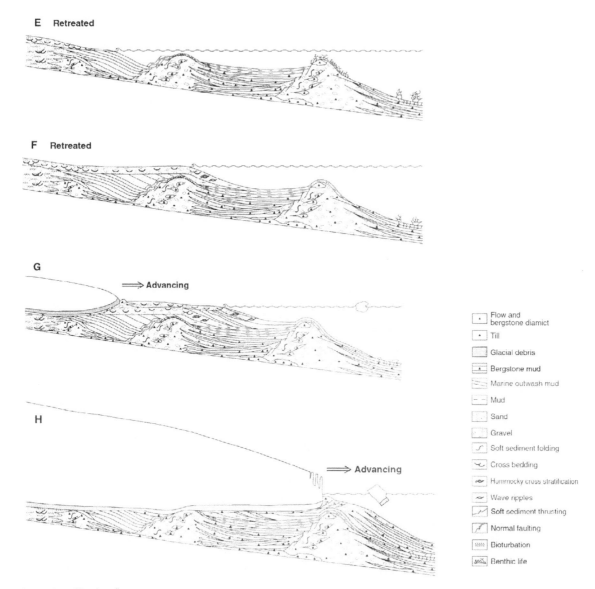

Figure 12.47 (Continued)

Second, severe compressional deformation at the advancing termini of surging glaciers results in extensive thrust faulting and folding, which thicken and elevate basal ice sequences (Clapperton, 1975; Sharp et al., 1988, 1994; Andrzejewski, 2002; Evans et al., 2006c). Near the ice front, basal debris is commonly elevated to the glacier surface, forming a zone of debris-mantled ice which expands upglacier when the glacier ablates (Bennett et al., 1996b; Hambrey et al., 1996; Murray et al., 1997; Glasser et al., 1998). Third, the very high water pressures associated with surging can result in spectacular ice fracturing (Roberts et al., 2000, 2001). These fractures, along with the extensive crevasse and fracture networks produced by surging, fill with sediment and then anneal to form englacial debris bands (Bennett et al., 2000b).

Fourth, the wastage of debris-mantled ice during the quiescent period of a surge cycle produces extensive areas of ice-cored moraines and glacifluvial outwash, which may be overridden and reincorporated into the glacier during the next surge (Raymond et al., 1987; Evans and Rea, 1999, 2003; Evans et al., 2006c, 2007b; Schomacker and Kjær, 2007). In this case, the supraglacial debris mantle produced by one surge forms part of the englacial debris load of the next. Evans and Rea (2003) and Schomacker et al. (2006) have reported the drumlinization of ice-cored features at Brúarjökull, Iceland.

Surging glaciers also tend to be associated with widespread subglacial and proglacial glacitectonic deformation. Where the glacier bed is composed of till or other unconsolidated sediment, high pore-water pressures during surges encourage subsole deformation, transporting weak, dilatant sediments and streamlining them into long flutings (Clarke et al., 1984; Clarke, 1987c; Christoffersen et al., 2005; Benediktsson et al., 2008). Fuller and Murray (2000), however, have presented evidence from Iceland that fast basal motion can occur mainly by bed decoupling and sliding, with little transport of subglacial till. Stresses transmitted to proglacial sediments commonly result in extensive tectonic thrusting, particularly if the substratum has been weakened by high pore-water pressures (Sharp, 1985b; Croot, 1988a; Evans and Rea, 1999, 2003; Russell et al., 2001a). In permafrost areas, significant bodies of ice can occur in proglacial areas, including segregated and intrusive ground ice, naled and buried glacier ice (e.g. Yde et al., 2005; Roberts et al., 2009). In sub-marginal and marginal locations, pressurized water can escape along hydrofractures, locally disrupting sedimentary structures and emerging from hydrodynamic blow-outs in proglacial deposits (Kjær et al., 2006; Benediktsson et al., 2008; section 10.5.6). In addition, large discharges of meltwater and sediment associated with glacier surges are responsible for major changes in deposition rates in proglacial lakes and sandar (Humphrey et al., 1986; Humphrey and Raymond, 1994; Fleischer et al., 2003). Stagnating surge lobes can also be extensively buried by proglacial outwash, giving rise to the development of pitted sandar (Evans and Rea, 1999, 2003; Evans et al., 2006c). If it is overridden by a further surge before buried ice blocks melt out, an embryonic pitted sandur can be sealed by a subglacial till layer. This has given rise to peculiar 'circular depressions' on the streamlined subglacial till surface at Skeiðarárjökull, Iceland, which were explained by Waller et al. (2008) as the products of continued melting of ice blocks in the underlying outwash.

The process-form relationships identified above give rise to a diagnostic suite of landforms and sediments that can be employed in the recognition of palaeo-surges. A terrestrial surge landsystem model was first developed by Sharp (1985b), and has since been refined by Evans and Rea (1999, 2003). Recently, Ottesen and Dowdeswell (2006) and Ottesen et al. (2008a) have developed a fjord-floor surge landsystem model.

Terrestrial surging glacier landsystems typically include eight characteristic landform types (fig. 12.48):

(1) glacitectonic composite ridges (see section 11.3.2);
(2) zigzag eskers (see section 11.2.10);
(3) crevasse-squeeze ridges (section 11.2.9);
(4) long flutings (section 11.2.3);
(5) supraglacial hummocky moraine (section 11.4.2);
(6) blow-out structures or extrusion moraines (see section 10.5.6);
(7) ice-cored sandur and glacilacustrine depo-centres (section 11.4.6);
(8) overridden composite ridges (see sections 11.2.7 and 11.3.2).

An interesting feature identified by Christoffersen et al. (2005) on the foreland of Elisebreen, a Svalbard surging glacier, is a meandering till ridge or till esker, thought to have been produced in much the same way as crevasse-squeeze ridges, when saturated till oozed into cavities at surge termination. In addition, complex till and glacitectonite stratigraphies are associated with the widespread disruption of pre-surge deposits (Boulton et al., 1996; Evans and Rea, 2003; Kjær et al., 2006; Benediktsson et al., 2008). Only zigzag eskers and crevasse-squeeze ridges appear to be uniquely diagnostic of surging, although there is a very high coincidence between glacitectonic composite ridges and surging glaciers. However, it is the *combination* of landforms and sediments – and their spatial organization – that characterizes the terrestrial surging glacier landsystem (fig. 12.49).

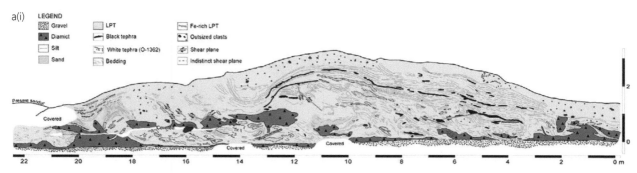

Figure 12.48 Landforms of the surging glacier landsystem. (a) Details of the composite ridges/large push moraines on the forelands of Iceland surging glaciers. (i) shows predominantly ductile sedimentary structures in the Brúarjökull 1890 surge moraine. (Benediktsson et al., 2008) (ii) shows contorted peat, loess and tephra in the same moraine. (D.J.A. Evans) (iii) shows overthrust folds in outwash sands and gravels in the 1890 surge moraine at Eyjabakkajökull. (D.J.A. Evans) (b) Long flutings on the foreland of Brúarjökull. (D.J.A. Evans) (c) Hummocky moraine or ice-cored landforms on the Brúarjökull foreland mapped on a 2003 DEM. Left image shows features prior to the 1964 surge, based on 1945 aerial photographs. Right image shows features after the 1964 surge. (After Schomacker and Kjær, 2007) (d) Blow-out features at the 1890 surge moraine of Brúarjökull. (Kjær et al., 2006) (e) Ice-cored sandur fans and associated features at the southern margin of Tungnaarjökull, Iceland. (i) Extract from an aerial orthophotograph; (ii) the same area mapped for glacial geomorphology and surficial geology. During a surge in 1945, an arc of thrust block moraines (A) was constructed in a collapsing sandur fan and now fronts a large expanse of ice-cored moraine. This suggests that the 1945 surge disturbed an older melting, outwash-covered ice mass dating to the 1880–90 surge. The limit of this older surge is marked by the pitted ice-contact face of a sandur fan that lies beyond the thrust moraine arc (B). Ice flow-parallel and sinuous controlled ridges behind the thrust block moraines mark the locations of eskers in the melting glacier ice (C). These eskers mark the locations of englacial drainage systems that, during the 1945 surge, fed sediment into the apices of sandur fans developing behind and between the incised remnants of the 1880–90 ice-contact fans (Evans et al., 2006c).

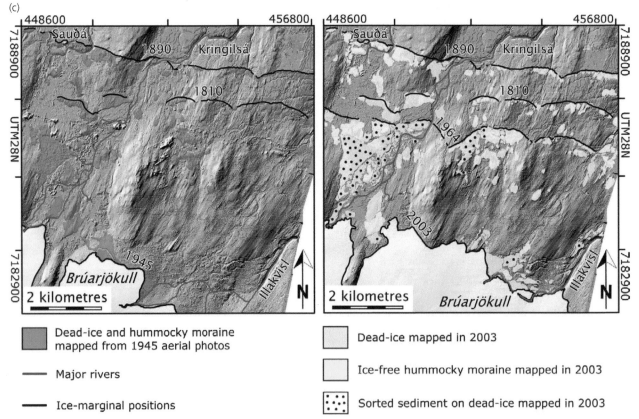

Dead-ice and hummocky moraine mapped from 1945 aerial photos

— Major rivers

— Ice-marginal positions

Dead-ice mapped in 2003

Ice-free hummocky moraine mapped in 2003

Sorted sediment on dead-ice mapped in 2003

Figure 12.48 (Continued)

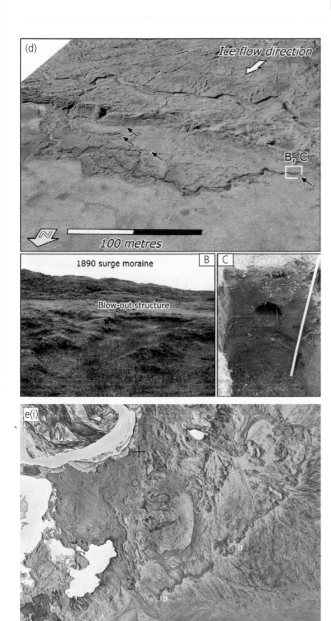

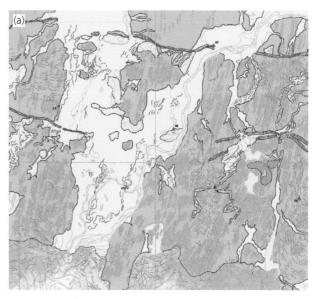

Figure 12.49 The glacial geomorphology and surficial geology of the foreland of Brúarjökull, an Icelandic surging glacier. (a) Extract from the map of Evans et al. (2007) (b) Extract from aerial orthophotograph of the same area. Note that the ice-cored terrain composed of zigzag eskers is classified as glacifluvial outwash (yellow). These are classified as 'dead-ice' by Schomacker and Kjær (2007) in figure 12.48c.

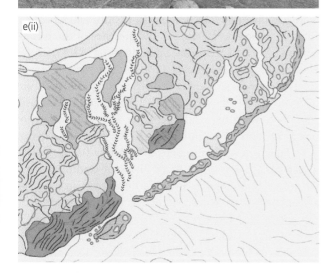

Figure 12.48 (Continued)

Evans and Rea (1999, 2003) recognized three overlapping zones organized in broad arcs around surging glacier margins (fig. 12.50). *The outermost zone* represents the limit of the surge and is composed of glacitectonically pushed and thrust pre-surge sediments. Elevated porewater pressures may give rise to water escape structures and hydrofracture fills in thrust-block moraines, and may also vent in the proglacial area to produce blow-out features. *The intermediate zone* consists of patchy hummocky moraine, often draped on the ice-proximal slopes of the thrust-block and push moraines, but also occurring wherever older, subglacially streamlined ice-cored debris is melting out. *The inner zone* is dominated by subglacial traction tills and long, low-amplitude flutings, produced

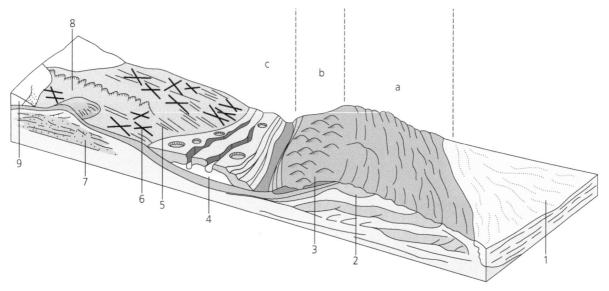

Figure 12.50 The surging glacier landsystem. (A) Outer zone of proglacially thrust pre-surge sediment which may grade into small push moraines in areas of thin sediment cover. (B) Zone of weakly developed chaotic hummocky moraine located on the down-ice sides of topographic depressions. (C) Zone of flutings, crevasse-squeeze ridges and concertina eskers. (1) Proglacial outwash fan. (2) Thrust-block moraine. (3) Hummocky moraine. (4) Stagnating surge snout covered by pitted and channelled outwash. (5) Flutings. (6) Crevasse-squeeze ridge. (7) Overridden and fluted thrust-block moraine. (8) Concertina esker. (9) Glacier with crevasse-squeeze ridges emerging at surface (Evans and Rea, 2003, after Evans et al., 1999b; Evans and Rea, 1999).

by subsole deformation during the surge, as well as crevasse-squeeze ridges and zigzag eskers. Although the preservation potential of zigzag eskers is likely to be poor, discontinuous gravel spreads and mounds in the ancient landform record may represent zigzag eskers. Some diagnostic forms of surging are intra-zonal, either because they are palimpsests of older surges (e.g. overridden moraines), because they relate to the location of proglacial outwash fans and streams (ice-cored, collapsed outwash), or because they occur in ponded topographic depressions on the foreland (collapsed lake plains). Additionally, normal meandering or anabranched eskers can develop within the stagnating ice during quiescence. This landsystems model has been applied to ancient terrestrial settings by Evans et al. (1999b), Evans and Rea (2003), Kovanen and Slaymaker (2004) and Kehew et al. (2005), where more than one of the landform–sediment assemblages has been recognized at ice sheet scale.

Fjord-floor surging glacier landsystems share many characteristics with their terrestrial counterparts, but there are also some major differences. Some submarine landform assemblages may also be diagnostic of surge activity. In Svalbard fjords, the imprints of former surges consist of large terminal moraines, long flutings, subglacial crevasse-squeeze ridges and overridden moraines (fig. 12.51; Ottesen and Dowdeswell, 2006; Ottesen et al., 2008b). The subglacial landform components are commonly much better preserved than on land, due to the lack of high-energy reworking processes. Beyond the outer moraine, iceberg furrows criss-cross the sea floor, recording the passage of icebergs calved from the surge front. Terminal moraines in Svalbard fjords can be over

1 km wide, and have asymmetric cross profiles with steeper proximal slopes and gentle distal slopes. Few data are available on their internal structure, but their morphology and lateral continuity with terrestrial thrust-block moraines strongly suggests that they formed by proglacial glacitectonic deformation of fjord-floor sediments (section 11.3.2). The moraines typically have large debris flows on their ice-distal sides. These can be several metres thick, and extend from the moraine crest into the deeper water beyond the moraine front. Kristensen et al. (2009) have argued that the debris flows record the failure of water-saturated moraine surfaces during and following proglacial glacitectonic deformation. In some cases, where surging glaciers advanced over fjords and then onto dry land, similar mud aprons occur in association with terrestrial thrust moraine complexes.

A striking difference between fjord floor and terrestrial surging glacier landsystems is the presence in the former of closely spaced recessional moraines (fig. 12.51; cf. De Geer moraines; section 11.6.2). In terrestrial settings, recessional moraines are not formed during the quiescent period, because glacier tongues stagnate and downwaste in situ following surge advances. In contrast, the fronts of quiescent fjord-terminating glaciers retreat by calving, allowing small push moraines to be formed at successive glacier margin positions. Assemblages of landforms similar to those observed in Svalbard fjords also occur in the ancient sea floor record, allowing palaeo-surges to be recognized (e.g. Lafferty et al., 2006).

Finally, it should be noted that the characteristic components of surging glacier landsystems (e.g. crevasse-squeeze ridges, thrust-block moraines) form where surging

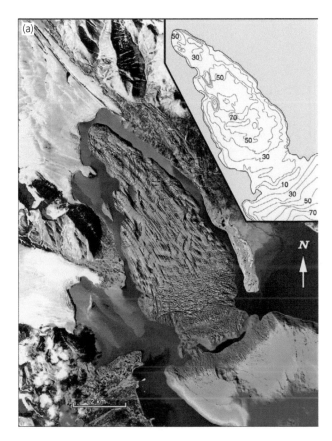

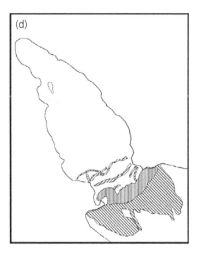

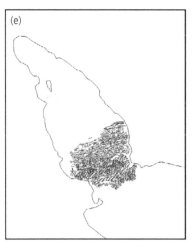

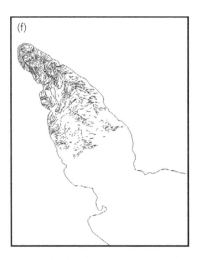

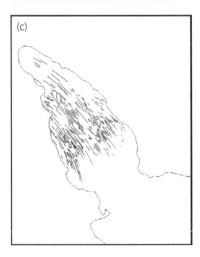

Figure 12.51 Submarine imprint of glacier surging in Borebukta, Svalbard. (a) Swath bathymetry for the Borebukta sea floor, showing shaded relief image and surge features and inset map of bathymetry with contours in metres. (b–f) Landform elements in the order of formation based on cross-cutting relationships, with (b) large transverse ridges/overridden moraines; (c) streamlined bedforms; (d) large terminal ridges, likely to be thrust moraines, with proglacial flow lobes; (e) 'rhombohedral ridges' interpreted as crevasse-squeeze ridges; (f) regularly spaced transverse ridges interpreted as recessional push moraines. The latter features, like De Geer moraines, are not observed in terrestrial surging glacier landsystems; (g) landsystem model for Svalbard tidewater surging glaciers, with order of formation of individual landform assemblages numbered in sequence (Ottesen and Dowdeswell, 2006).

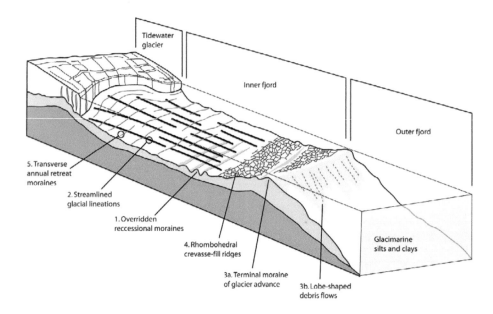

Figure 12.51 (*Continued*)

glaciers advance into areas underlain by thick accumulations of soft sediments. Where surging glaciers are underlain by thin, coarse-grained tills, these components may be absent, making the depositional record of glacier surges harder to recognize. The 2001 surge of the debris-covered Belvedere Glacier in the Italian Alps created sediments and landforms that are indistinguishable from those of many non-surging mountain glaciers. This supports the view that glacier surges may be under-recognized, both at the present time and in the geologic record.

12.3.9 CLIMATIC RECONSTRUCTION FROM PALAEOGLACIERS

Reconstructed ELAs of former glaciers have been widely used to infer palaeoclimatic conditions (Sissons and Sutherland, 1976; Porter and Orombelli, 1982; Benn and Lehmkuhl, 2000; Nesje and Dahl, 2000; Ballantyne, 2002c; Benn and Ballantyne, 2005; Benn et al., 2005). Such reconstructions depend on a number of assumptions, the most important of which are that the former glacier extent and steady-state ELA can be reconstructed accurately; and that the relationship between glacier ELA and climate is known and stable through time.

Former glaciers are easiest to reconstruct where their extent at a given point in time is clearly marked by moraines in the ablation zone and periglacial trimlines in the accumulation area. Hence regional glacier–climate reconstructions tend to concentrate on upland glaciation. However, even plateau ice fields can be reconstructed from the evidence of morainic landforms in surrounding valleys, providing modern systems are used to guide those parts of the reconstructions that lie on the plateau summits (see section 12.3.5; fig. 12.23). The reconstruction procedure involves delimiting the margins of former

ice masses based on: end and lateral moraines, drift limits and the outer limits of hummocky moraine spreads on lower ground; and periglacial trimlines, drift limits and meltwater channels on higher ground (e.g. Sissons, 1967; Benn and Ballantyne, 2005). Once glacier outlines have been defined, contours can be extrapolated from known ice surface elevations at the margins across the glacier surface (fig. 12.52). Contours should be drawn at right angles to ice flow indicators such as striae or flutings, although it must be demonstrated that such features are contemporary with the glacier limits. Contouring can also be constrained by midline ice surface elevations defined from a modelled surface profile (section 5.3.2). If neither midline elevations nor flow-line indicators are available, approximate contours can be drawn with convex forms on the lower glacier and increasingly concave towards the headwall, with straight contours occurring at the median altitude of the glacier (figs 12.23, 12.35).

Several methods for reconstructing former steady-state ELAs have been developed, most of which rely to a greater or lesser extent on the idea that the ELA can be derived from the geometry of former glaciers as defined by geomorphological evidence. Recent reviews of methods of ELA reconstruction have been provided by Nesje and Dahl (2000), Porter (2000) and Benn et al. (2005). The simplest use of geomorphological evidence in reconstructing a former ELA is identifying the *maximum elevation of lateral moraines* (MELM), based on the principle that lateral moraines are only deposited below the ELA, where debris is released by the melting of ice. Thus the highest altitude of an abandoned lateral moraine marks the palaeo-ELA. Problems arise, however, when glacier retreat is slow, and continuous provision of debris results in the incremental deposition of lateral moraines in an upvalley direction, although the concomitant thinning of

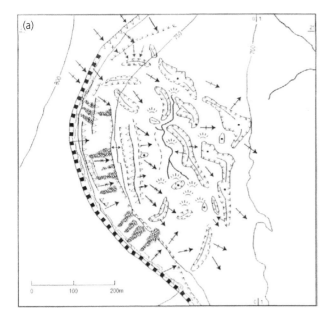

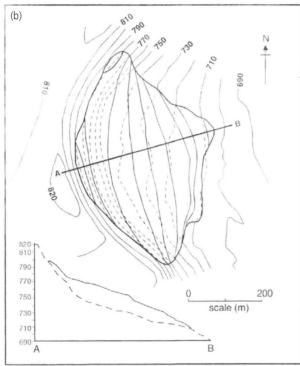

Figure 12.52 An example of palaeoglacier reconstruction based on a simple niche glacier in Cwm Gwaun Taf in the Brecon Beacons of South Wales. (a) Geomorphology map of Cwm Gwaun Taf, with inset depositional ridges representing the latero-frontal moraines of the former niche glacier. (b) Reconstructed glacier in Cwm Gwaun Taf, with cross section through line A-B (Carr, 2001b).

The most rigorous methods of palaeo-ELA reconstruction are based on the three-dimensional form of the glacier surface, combined with assumed mass balance–altitude relationships. The *accumulation area ratio method* is based on the assumption that, under steady-state conditions, the accumulation area of the glacier (i.e. the area above the ELA) occupies some fixed proportion of the total glacier area. Former steady-state ELAs, therefore, can be determined from contour maps of reconstructed glaciers using the accumulation area ratio (AAR). If the AAR is assumed to be 0.6, for example, the ELA is set as the altitude of the surface contour that lies below 0.6 (60 per cent) of the total area (fig 12.52). For modern mid- and high-latitude glaciers, steady-state AARs generally lie in the range 0.5–0.8, with typical values lying in the range 0.55–0.65 (Porter, 1975, 1977). The steep ablation gradients and shallow accumulation gradients of tropical glaciers means that they tend to have higher steady-state AARs (~0.8; Kaser and Osmaston, 2002). On the other hand, glaciers with debris-covered ablation areas have rather low AARs, due to the effect of debris on lowering ablation and increasing the relative size of the ablation area. Values for modern debris-covered glaciers in the Himalaya lie in the range 0.2–0.4 (Kulkarni, 1992; Benn and Lehmkuhl, 2000). Because of the wide range of possible values, it is clearly desirable to have some independent means of constraining the AARs of vanished glaciers. Kaser and Osmaston (2002) approached this problem by calculating ELAs for homogeneous groups of former glaciers in the Rwenzori Mountains, Africa, using a range of AARs. The AAR that resulted in the lowest variance of calculated ELAs was assumed to be the most appropriate value.

ELAs have been estimated by calculating the area-weighted mean altitude or median elevation of glaciers (MEG) (Sissons and Sutherland, 1976). Sutherland (1984) found that, for a sample of modern Norwegian glaciers, this method consistently overestimated ELAs. The method is useful for comparing results with those of older studies (e.g. Ballantyne, 2002c; Benn and Ballantyne, 2005), but is generally inferior to the AAR method.

One shortcoming of the AAR and MEG methods is that they take little account of variations in glacier shape, particularly the distribution of glacier area over its altitudinal range, or *hypsometry*. AARs on modern glaciers are influenced by glacier hypsometry, so that a glacier with a wide accumulation area and a narrow snout, for example, will have a different AAR than a glacier with a narrow accumulation basin and a broad snout, even if the ELAs are the same (Furbish and Andrews, 1984). Therefore, former glacier ELAs based on a uniform assumed AAR value may be subject to significant errors if there is a wide range of glacier types and shapes in the area under consideration. To overcome this difficulty, Furbish and Andrews (1984)

the glacier will eventually result in inset moraines. A further problem is that lateral moraines can degrade rapidly on deglaciation and disappear from steep slopes, thus giving a spuriously low ELA. However, the moraine method may be the best approach in areas where former glaciers were debris covered (Richards et al., 2000; Benn and Owen, 2002).

developed the *balance ratio method*, which takes account of both glacier hypsometry and the shape of the mass balance curve (section 2.5.4). This has more recently been termed the *area altitude balance ratio* (AABR) by Osmaston (2005). The method is based on the fact that, for equilibrium conditions, the total annual accumulation above the ELA must exactly balance the total annual ablation below the ELA. This can be expressed in terms of the areas above and below the ELA multiplied by the average accumulation and ablation, respectively:

$$\bar{b}_{nb} A_b = \bar{b}_{nc} A_c \qquad (12.1)$$

where $\bar{b}_{nb}$ and $\bar{b}_{nc}$ are the average net annual mass balance in the ablation area and accumulation area, respectively, and A_b and A_c are their respective areas.

If the ablation and accumulation gradients are assumed to be linear, $\bar{b}_{nb}$ and $\bar{b}_{nc}$ are equal to the ablation and accumulation at the area-weighted mean altitudes of the ablation area (z_b) and the accumulation area (z_c), respectively. By convention, z_b and z_c are measured positively from the ELA, and are shown as the heights associated with $1/2A_b$ and $1/2A_c$ in figure 12.53. For equilibrium conditions, the altitudes of z_b and z_c are determined by the balance ratio (equation 2.19). The steady-state ELA is defined as the altitude that satisfies equation 12.1 for the balance ratio representative of the area under study. The procedure can be accomplished quickly and easily using a Microsoft Excel® spreadsheet developed by Osmaston (2005). Representative AABRs for modern glaciers in a range of climate regimes have been calculated by Rea (2009), allowing the most appropriate values to be chosen, depending on the region in which palaeoglacier reconstruction is being undertaken.

If it is not possible to reconstruct former ice surfaces in detail, estimates of the steady-state ELA can be obtained using the toe-to-headwall altitude ratio (THAR). This method is based on the assumption that the ELA is at some fixed proportion of the distance between the toe of the glacier and the top of the valley headwall. The THAR method is very crude, as it takes no account of glacier hypsometry or climatic factors, although it does provide a quick method of estimating former ELAs in remote regions where topographic maps are unavailable or unreliable (Benn and Lehmkuhl, 2000).

Once palaeo-ELAs have been calculated, the next stage is to determine their climatic significance. Glacier ELAs can move up or down in response to a wide range of climatic factors, but to a good approximation, changes in ELAs can be regarded as the result of changes in summer temperature and snow accumulation. Warmer summers result in greater ablation totals and tend to increase the ELA, whereas greater snowfall tends to lower the ELA (section 2.5.5). Changes in glacier ELAs, therefore, can

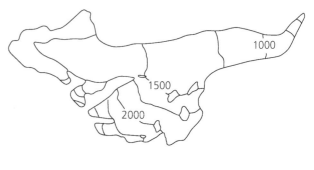

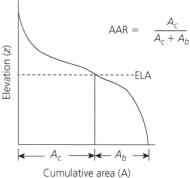

Figure 12.53 Contoured map and hypsometric curve for a valley glacier. These data can be used in conjunction with an assumed value of the accumulation area ratio to derive an estimate of the former steady-state equilibrium line altitude. For example, when the AAR is set at 0.6, the hypsometric curve is used to read off the altitude that lies below 60 per cent of the former glacier area (Modified from Furbish and Andrews, 1984).

result from changes in both of these variables, and there is no unique solution for any given shift in ELA. However, Ohmura et al. (1992) have shown that there is a good correlation between summer temperature and annual precipitation at glacier ELAs (fig. 2.31), providing a means of determining one of these variables if the other is already known. Palaeoprecipitation is very hard to quantify, but palaeotemperatures can be estimated from several types of proxy data, so temperature–precipitation relationships at ELAs provide very useful means of obtaining estimates of past precipitation (e.g. Nesje and Dahl, 2000). For example, Benn and Ballantyne (2005) used values of summer temperature derived from sub-fossil chironomids (midges) to determine precipitation at the ELAs of Younger Dryas ice caps in Scotland (cf. Brooks and Birks, 2001). They showed that the precipitation gradient across Scotland was steeper at that time than at present, possibly because of efficient scavenging of snowfall by the West Highland Ice Field.

It must be emphasized that temperature–precipitation relationships are based on the implicit assumption that precipitation is correlated with snow accumulation. In mountain areas, accumulation totals are very strongly influenced by topographic factors and wind redistribution, and in sheltered locations accumulation can be substantially greater than winter precipitation. As a result, palaeoprecipitation estimates derived from the

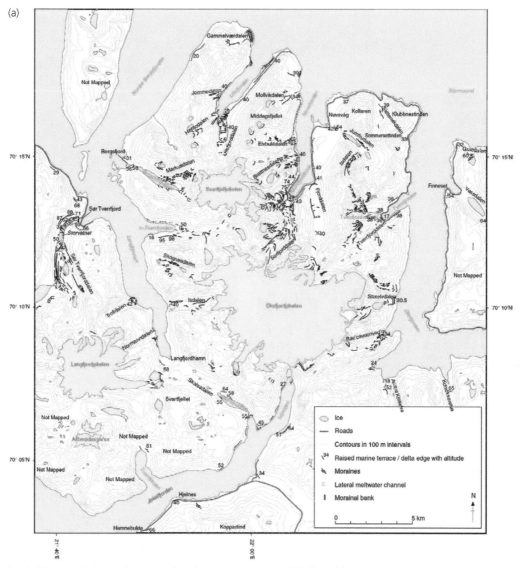

Figure 12.54 The glacial geomorphology and reconstruction of Younger Dryas extent of Øksfjordjøkelen, north Norway. (a) Glacial geomorphology map of the Øksfjordjøkelen plateau, showing the extent of the modern plateau ice field. (b) Modelled reconstruction of Øksfjordjøkelen during the Younger Dryas, showing the hypsometry of each outlet glacier. Only the extent of Sörfjorddalen has been dated directly through the radiocarbon dating of raised marine sediments associated with downvalley moraines (Evans et al., 2002; Rea and Evans, 2007).

ELAs of former cirque and valley glaciers are likely be in error. The ELAs of former ice caps can be expected to provide more reliable estimates, since topographic factors exert a smaller influence on average accumulation (Nesje and Dahl, 2000; Ballantyne and Benn, 2005).

Reconstructed glacier ELAs can also be used to estimate the velocities of former glaciers. As discussed in section 5.2.1, the balance velocity through any cross section is determined by the total mass that must be discharged. Balance velocities, therefore, can be calculated if the ELA, ice surface form and balance gradients are known or can be estimated (Sharp et al., 1989a; Carr and Coleman, 2007). An example of this approach has been provided by Rea and Evans (2007), who reconstructed the balance velocities of outlet glaciers of a Younger Dryas ice field in northern Norway (fig. 12.54).

An alternative approach to investigating the climatic implications of former glaciers, ice caps and ice fields is to use time-evolving numerical models to find which climate scenarios can produce glaciers of known dimensions (section 5.3.3). For example, Golledge et al. (2008b) compared the results of numerical modelling experiments with mapped and dated Younger Dryas glacier limits in Scotland. Such models have the great advantage of adding the time dimension to glacier reconstructions, in contrast with the 'snapshot' reconstructions discussed above, and can be used to estimate likely timescales for glacier advance-and-retreat cycles. Additionally, time-evolving models can also be used to develop testable hypotheses about former glacier behaviour, and highlight areas where further field data are needed.

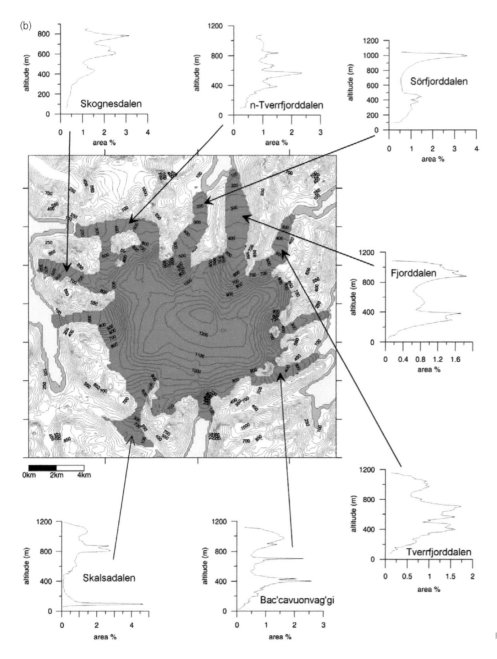

Figure 12.54 (Continued)

12.4 ICE SHEET BEDS

In this section, we discuss the characteristics of hard and soft glacier beds, examine the regional distributions of subglacial landforms, assess the evidence for cross-cutting lineations and their implications for palaeo-ice sheet dynamics, and then review concepts of sediment dispersal patterns and till lithology. In order to tie all these concepts together, we finish with an analysis of ice sheet beds as thermal regime palimpsests.

12.4.1 'HARD' AND 'SOFT' BEDS

Large parts of the areas occupied by the former Pleistocene ice sheets are, in effect, fossil glacier beds, providing a

window onto processes and environments that are only now being investigated with precision at the base of modern ice sheets. Large-scale palaeoglaciological studies and numerical modelling exercises have shown that the Laurentide and Scandinavian Ice Sheets were highly dynamic entities with complex histories of advance and retreat, ice-divide migration and internal dynamics, and that they shifted their configuration to adapt to climatic change, the position and depth of proglacial water bodies and basal conditions (e.g. Boulton and Clark, 1990a, b; Kleman and Borgström, 1996; Kleman et al., 1997, 2002, 2006; Clark et al., 2000, 2006; Boulton et al., 2001b; Boulton and Hagdorn, 2006; Kleman and Glasser, 2007).

At the scale of continental ice sheets, 'hard' and 'soft' bedded areas tend to be geographically distinct. At the

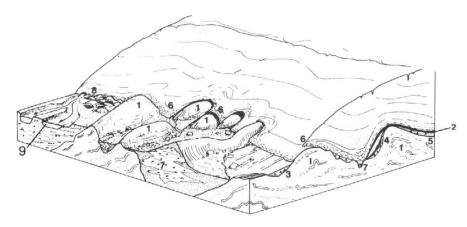

Figure 12.55 Subglacial deposits in an area of hard substrata. (1) Abraded and streamlined rock knobs. (2) Basal debris. (3) Subglacial till on low-relief rock surface. (4) Lee-side cavity fill. (5) Basal melt-out till. (6) Debris melting out at ice surface and dumped by gravity on the freshly exposed subglacial surface. (7) Subglacial esker with gravel core. (8) Hummocky or kettled outwash surface produced by the melt-out of ice buried by outwash fans. (9) Proglacial stream carrying subglacial abrasion products (Eyles, 1983a).

largest scale, former hard beds existed in *shield terrain*, or ancient continental crust, where the predominant rock types are igneous and metamorphic rocks, mostly of Precambrian age (Eyles et al., 1983b). At intermediate scales, ice sheet beds are mosaics of hard and soft components; the hard bed of the Antarctic Ice Sheet, for example, contains ribbons of soft subglacial till, above which ice streams flow. Formerly glaciated shield areas, such as the Canadian and Scandinavian Shields and the Lewisian basement of north-west Scotland, also contain ribbons of soft sediment and subglacially streamlined bedforms. At smaller scales, especially outside areas of former ice streaming, subglacial deposits over hard beds tend to be thin and patchy, contain numerous freshly plucked blocks, and overlie landscapes of areal scour or the floors of troughs (*bedrock-drift complexes*; Eyles, 1983a; fig. 12.55). As a result, the landscapes are dotted with lakes and ponds and deranged drainage networks, and bedrock and ancient meteorite impact structures are clearly visible (e.g. Veillette et al., 1999). Lee-side cavity fills can occur on the down-ice side of bed protrusions (section 11.2.3). Sediment cover is thickest between bedrock highs, and is commonly streamlined at a wide range of scales, forming flutings, mega-flutings, mega-scale lineations and rock-cored drumlins (e.g. Veillette et al., 1999; Jansson, 2005; Jansson and Glasser, 2005a; Hättestrand and Clark, 2006; De Angelis, 2007; Golledge et al., 2008a). Bedrock mega-grooves also occur (section 8.3.4) and bedrock structure can often be accentuated into strong flutings if lineament orientation coincides with predominant ice flow direction (Livingstone et al., 2008).

At a regional scale, patterns of glacial sediment thickness on hard bedrock terrains reflect the long-term evolution of ice sheet beds and the progressive transport and eventual removal of preglacial weathered regolith. For example, during the Pleistocene, easily erodable materials were cleared from the central zone of successive

Laurentide ice sheets, producing thickening wedges of subglacial tills and large trough-mouth fans around their outer zones. These zones of soft sediment would have been available for sub-marginal reworking by the ice sheet during every glaciation, and this is important when we consider the genesis of thick and/or multiple till sequences in the glacial depositional record. The removal of soft bed materials from the inception zone of the Laurentide Ice Sheet in Hudson Bay has been implicated as the driver of the Mid Pleistocene Transition (MPT; 1250–700 ka) when the Earth's 100 ka ice-volume cycle replaced the previous 41 ka cycle (section 1.4.2; Clark and Pollard, 1998; Clark et al., 1999; Roy et al., 2004). According to this view, early Pleistocene ice sheets were underlain by deformable sediments, so that basal shear stresses and surface gradients were generally low. In contrast, the central zones of later ice sheets were cleared of deformable sediments, and the resulting hard beds supported higher basal shear stresses. Maximum ice volumes, therefore, would have been greater in the late Pleistocene, even though ice sheet extents were similar. This shift may have brought about changes in ice sheet dynamics, and associated feedbacks between ice sheet growth–decay cycles and other components of the Earth system. In Scandinavia, Kleman et al. (2008) identified two zones of thick glacigenic sediments: an outer zone located near the long-term margins of the ice sheet, and an inner zone near the ice sheet centre (fig. 12.56). This inner zone was interpreted as the marginal deposits of mountain-centred ice fields that developed during the early and middle Quaternary, and later resisted the erosion by larger, more easterly centred Fennoscandian ice sheets, due to the fact that they were located beneath ice divides.

Soft substrata occur in areas underlain by weak sedimentary rocks, or where accumulations of Quaternary sediments mask hard lithologies. Extensive sedimentary bedrock lowlands exist beyond the margins of the

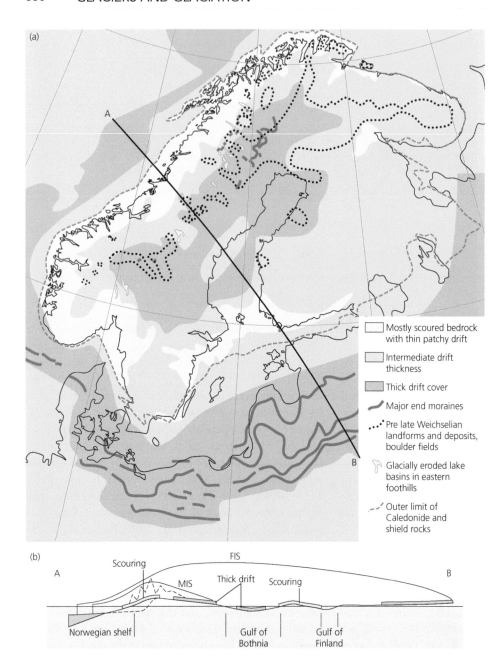

Figure 12.56 The relationship of drift thickness to ice sheet erosion history on a hard bed. (a) Map of drift cover thickness and distribution, different types and intensity of glacial erosion, and occurrences of relict landscapes in Fennoscandia. (b) Location of thick drift cover zones and scouring zones along the transect A-B in the map and their association with the morphology of mountain-style ice sheets (MIS) and larger Fennoscandian ice sheets (FIS) (Redrawn from Kleman et al., 2008).

North American and Scandinavian Shields, which provided an outer zone of potentially deformable sediment below the great Pleistocene ice sheets (Marshall et al., 1996). In soft bed areas, the subglacial landform signature typically consists of fluted and drumlinized terrain underlain by varying thicknesses of glacigenic sediments, and late-stage drainage events may be recorded by eskers and/or meltwater channels (figs 11.4, 12.57; Eyles and Eyles, 1992). Studies at modern glacier margins show that subglacial sediments commonly comprise thin till units perched above older sedimentary successions, as at Breiðamerkurjökull, Iceland, where approximately 2 m of fluted subglacial till overlies several metres of proglacial gravels and sands. In many areas occupied by Pleistocene ice sheets, however, the subglacial surface

may be underlain by tens of metres of vertically stacked subglacial tills and intervening stratified sediments. Till units may be laterally continuous for many hundreds of kilometres and may display remarkably uniform sedimentological characteristics and lithological composition (e.g. Kemmis, 1981). However, small-scale variability usually arises through the reworking of local materials, whose provenance indicates subtle changes in ice flow direction (e.g. Fish and Whiteman, 2001) or subglacial erosion of fresh strata exposed by glacial overdeepening (Boulton, 1996b). Widespread, largely uniform till units have been attributed to efficient debris mixing in the subglacial shear zone (section 10.3; Kemmis, 1981; Boulton, 1987; Alley, 1991; Hart and Roberts, 1994; Lian and Hicock, 2000; Evans et al., 2006a). At present, many researchers

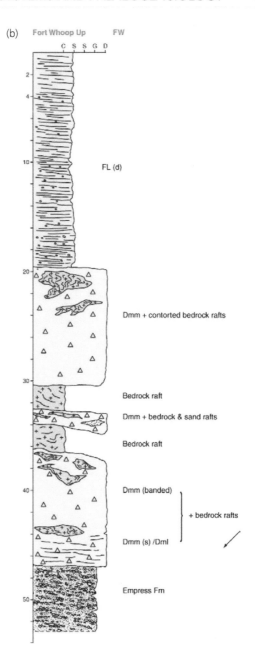

Figure 12.57 Stratigraphic sequence of typical soft ice sheet bed (b) is vertical profile log of photograph in (a) near the former margin of the Laurentide Ice Sheet in the Western Canada Sedimentary Basin, showing Cretaceous mega-rafts in contorted tills. Facies codes in log are explained in section 10.2.1 (photo D.J.A. Evans; after Evans et al., 2008).

favour the view that extensive, massive tills deposited in soft bed regions are subglacial traction tills, and that thickening towards former ice margins documents net long-term transfer of subglacial materials from erosional to depositional zones beneath the ice by a variety of processes (Alley et al., 1997; Evans et al., 2006a). This is encapsulated in Boulton's (1996a, b, 2006) concept of a 'till wave', which migrates and thickens margin-wards and eventually exhausts the supply of soft and/or erodible sediment up-ice, leading to an overdeepening. We can envisage this taking place during one glacier advance (e.g. Evans and Hiemstra, 2005) or over longer timescales – for example, during the numerous glaciations of the Early to Mid Pleistocene, when the net result was the evacuation of soft material from the Laurentian Shield to produce a hard bed over time. The occurrence of till wedges thickening towards ice margins and their relationships with moraine construction is discussed in section 12.2.6.

12.4.2 PALAEO-ICE STREAMS

In sections 6.2.4 and 6.3.4, we saw that most of the ice draining from the Greenland and Antarctic Ice Sheets flows in ice streams and outlet glaciers. In recent years, it has become clear that the same was true of the great Quaternary ice sheets, and that palaeo-ice streams can be identified from swarms of streamlined subglacial bedforms (figs 12.58, 12.59; Stokes and Clark, 1999, 2001; Clark and Stokes, 2003). Patterns of subglacial landforms show that the structure of palaeo-ice streams was similar to that of their modern counterparts, with onset zones, tributaries, trunk zones and lateral shear margins. Large numbers of palaeo-ice streams have been identified, both

on land and the continental shelves, revealing a fascinating picture of former ice sheet structure and its evolution through time.

Former ice stream onset zones are recorded in the landform record as shorter streamlined bedforms such as drumlins and flutings, arranged in a fan shape and converging on a narrower corridor of more elongate bedforms (e.g. De Angelis and Kleman, 2008). This has been likened to an hourglass shape, and is commonly mirrored by changes in till lithological properties (e.g. Dyke, 2008; section 12.4.6). In some cases, patterns of streamlined bedforms clearly record convergence of tributary flow units with a main trunk stream (Sejrup et al., 1998, 2003; Larsen et al., 2000; Ottesen et al., 2005a, b; Bradwell et al., 2007, 2008a; Dyke, 2008). The occurrence of a clear onset zone and areas of tributary convergence have

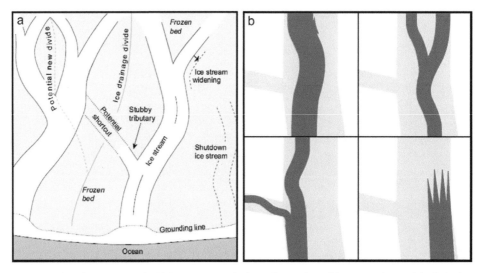

Figure 12.58 Changes in thermal conditions associated with ice streaming at the base of an ice sheet. (a) Conceptual map of the likely changes to ice streaming and intervening frozen bed zones beneath an ice sheet over short timescales of 10^1–10^3 ka, including the widening and shutdown of ice streams. (b) Conceptual diagram to show the various ways in which streaming ice may impact on an ice sheet bed. Ice stream corridors are depicted in grey and streaming activity at various times is depicted in red. This shows that sliding conditions may be evident in subglacial bedforms over whole corridors, but did not necessarily operate continuously (Kleman and Glasser, 2007; reproduced with permission of Elsevier).

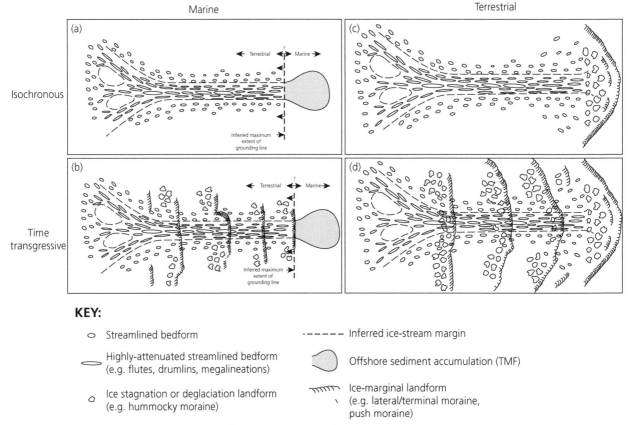

Figure 12.59 The palaeo-ice stream landsystem, compiled by Clark and Stokes (2003). Four end members are recognized: (a) marine isochronous; (b) marine time-transgressive; (c) terrestrial isochronous; and (d) terrestrial time-transgressive.

been used to identify a palaeo-ice stream draining the interior of the BIIS through the Irish Sea Basin, even though the trunk zone is submerged and presently unmapped at bedform scale (Stokes and Clark, 2001; Evans and Ó Cofaigh, 2003; Roberts et al., 2007).

As is the case for modern ice streams, former onset zones commonly coincide with the transition from hard, crystalline bedrock to thick infills of large sedimentary basins or marine shelves (Wellner et al., 2001; Canals et al., 2002; Ó Cofaigh et al., 2002a; Ottesen et al.,

2005a, b). Ottesen et al. (2008b) highlighted the importance of the relationship between substrate materials and ice stream driving stress and its manifestation in the landsystem record. Over the hard bedrock of onset zones, the driving stress is largely supported by basal drag, but where basal drag decreases over the softer substrate located downstream, the driving stress must be supported by lateral shear. Progressive changes in bedform morphology and elongation in a hard-bedded onset zone has been reported from The Minch in north-west Scotland by Bradwell et al. (2007, 2008a). In this area, ribbed terrain passes downflow into streamlined bedrock and mega-grooves, with elongation ratios that increase from 2:1 to 4:1, and ultimately to 7:1 and >10:1 in the ice stream trunk. This change is related to the increasing dominance of warm-based conditions downflow, with meltwater being implicated in the excavation of some rock basins. Similarly, Ó Cofaigh et al. (2002), Lowe and Anderson (2003) and Anderson and Fretwell (2008) reported well-developed meltwater channel systems in the onset zones of the Marguerite Bay and Pine Island Bay palaeo-ice streams off Antarctica. These meltwater features include anastomosing, radial and relatively straight channels, which are juxtaposed with glacially streamlined bedrock and become more narrowly focused in a downflow direction. This suggests that the subglacial drainage network was well organized. Palaeo-onset zones have also been inferred from variations in the preservation of pre-existing landforms, suggestive of thermal boundaries at the former ice sheet bed (De Angelis and Kleman, 2008).

There is commonly a strong association between grooved bedrock highs in onset zones and mega-scale glacial lineations in soft sediment downflow (e.g. Canals et al., 2000; Wellner et al., 2001, 2006; Amblas et al., 2006; Ottesen et al., 2008b). This association has been used by Tulaczyk et al. (2001) and Clark et al. (2003) to support the groove-ploughing mechanism for MSGL generation on ice stream beds, whereby keels are carved into the ice base as it passes over the bedrock high and these keels are then able to groove the softer substrate down-ice (see section 11.2.4). Patterns of bedform elongation in ice stream trunks vary both longitudinally and laterally, with elongation rising with distance downflow, all the way to the terminus in marine-terminating ice streams (fig. 11.18), but falling off again towards the lobate margins of terrestrial-terminating ice streams. Elongation also rises towards the centreline of the fast-flow trunk due to the maximum velocities at that location (Clark and Stokes, 2003; fig. 12.60). This pattern is easier to discern in isochronous than in time-transgressive palaeo-ice stream signatures (fig. 12.61).

The most characteristic feature of palaeo-ice stream trunk zones is MSGL (section 11.2.4), although other features include smoothed transverse ridges or overridden

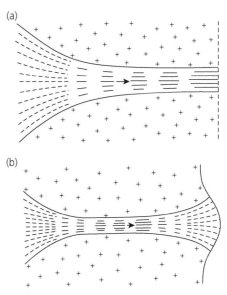

Figure 12.60 Idealized patterns of subglacial bedform elongation in (a) marine-terminating ice stream where velocity and bedform lineation increase towards the grounding line; and (b) terrestrially terminating ice stream where bedform lineation decreases in the lobate marginal zone due to falling velocities at the terminus. Across both types of ice stream, bedform lineation increases slightly towards to the centre of the fast flow trunk due to reduced friction (Clark and Stokes, 2003).

moraines (e.g. Evans et al., 2008; fig. 11.67) and ribbed terrain (e.g. Stokes et al., 2006a, 2008). In section 11.2.5, we presented the case for some ribbed terrain being the manifestation of ice stream shutdown, where the formerly thawed and fast-flowing ice changed its thermal properties and froze to its deforming bed (Stokes et al., 2006a, 2008). This gives rise to the superimposition of ribbed terrain over MSGL, as exemplified by the bed of the Dubawnt Lake palaeo-ice stream in northern Canada (fig. 12.62). Additionally, the consolidation of subglacial till has been identified as a sedimentological manifestation of ice stream bed freezing and shutdown by Christoffersen and Tulaczyk (2003a, b). Bedrock bumps will also form *sticky spots* on ice stream beds, leading to the development of crag–and-tail features such as those reported by Ottesen et al. (2008b) or larger bumps that deflect ice flow lineations, such as those proposed by Stokes et al. (2007). Figure 12.63 depicts different types of sticky spot identified on modern and ancient ice stream beds, each one related to a specific type of ice bed drainage condition, till rheology or thermal regime change.

Sharp margins to the trunk zones of palaeo-ice streams are delineated by various features. First, abrupt changes in the density of streamlined bedforms have been used to demarcate palaeo-ice streams (e.g. Dyke and Morris, 1988; Dyke et al., 1992; Hodgson, 1994; Clark and Stokes, 2001; Jørgensen and Piotrowski, 2003; Stokes et al., 2006b). Second, *shear margin moraines* record deposition at the boundary between fast-flowing and sluggish or stationary ice (section 11.2.6; fig. 11.29). Third, zones of

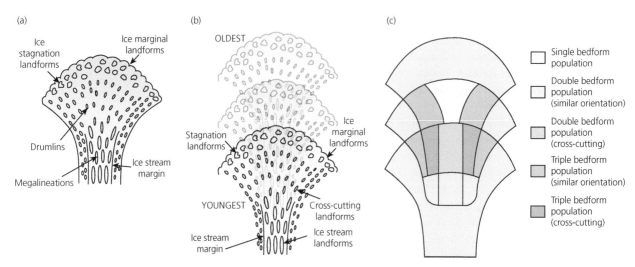

Figure 12.61 Idealized diagrams to convey the concept of isochronous and time-transgressive swarm/flow set production using an ice stream landsystem. (a) The rubber stamp imprint produced by isochronous flow set generation. (b) The smudged imprint resulting from time-transgressive recession. (c) The bedform palimpsest resulting from time-transgressive flow set production (Clark and Stokes, 2003).

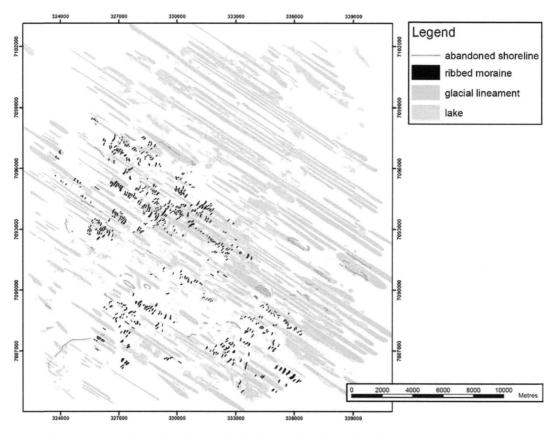

Figure 12.62 Map of ribbed terrain superimposed over the streamlined bedforms of the Dubawnt Lake palaeo-ice stream (Stokes et al., 2006a; reproduced with permission of the *Journal of Maps*).

greater than average subglacial erosion can be demarcated by rock corridors bordered by *linear till scarps* (Dyke, 2008) or *lee-side scarps* (Kleman et al., 1999), interpreted as lateral sliding boundaries. Fourth, measurable and often clearly visible changes in till lithology may reveal sharp palaeo-ice stream margins (e.g. Dyke and Morris, 1988; Dyke et al., 1992; Dredge, 2000; Dyke, 2008).

12.4.3 CROSS-CUTTING PALAEO-ICE FLOW INDICATORS AND PALAEO-ICE SHEET DYNAMICS

Regional mapping of subglacial landforms, in places complemented with studies of till lithological characteristics, can reveal large-scale patterns of ice-sheet flow

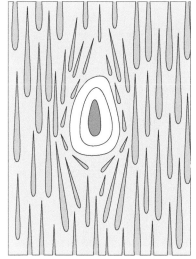

(a) (i) Bedrock bump sticky spot

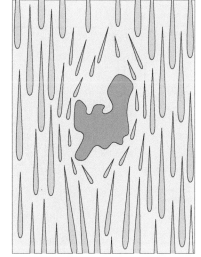

(a) (ii) Till free sticky spot

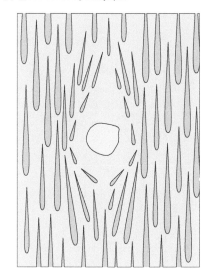

(a) (iii) Strong till sticky spot

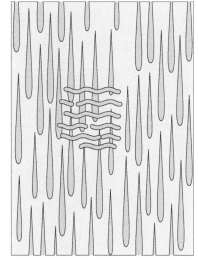

(a) (iv) Freeze-on sticky spot

Figure 12.63 Sticky spots on ice stream beds. (a) Conceptual models of sticky spots and their relationships to streamlined bedforms. Bedrock bumps, till-free areas or patches of relatively strong till produce shortening and alignment deviation of subglacial bedforms, whereas ribbed terrain are the product of freeze-on during ice stream shutdown. (b) Schematic flow diagram to show the evolution of sticky spots under various scenarios dictated by meltwater and sediment flux, and on hard versus soft substrates (After Stokes et al., 2007).

and their changes through time, yielding valuable information on ice sheet evolution and dynamics (fig. 12.64). Reconstructed flow patterns for ice sheet maxima demonstrate that ice sheets were not simple domes from which ice spread radially, but rather comprised multiple accumulation and dispersal centres (e.g. Dyke et al., 1982; Aylsworth and Shilts, 1989; Punkari, 1997a; Boulton et al., 2001b; Clark et al., 2006; De Angelis, 2007). Furthermore, it is possible to reconstruct the migration of ice divides and the switching of ice streams during a glacial cycle by mapping cross-cutting and superimposed subglacial bedforms, tracing indicator erratics, and undertaking kineto-stratigraphic analysis of exposures (sections 12.2.6, 12.4.6). In other words, the study of superimposed subglacial landsystems on a regional scale allows us to view the evolution of the last great ice sheets through time (fig. 12.4). Such studies have been greatly facilitated by the advent of satellite imagery and detailed sea floor bathymetry, but one should not overlook the

importance of many decades of painstaking observations made at the local scale by field geologists.

Superimposed drumlins and mega-flutings document changing ice flow directions through time (e.g. Rose, 1989; Mitchell, 1994; see section 11.2.4; fig. 11.28). This smearing of the glacier bed has been documented also in the spatial patterns of till geochemistry (Parent et al., 1996). Complex lineation patterns, relating to different flow directions at various stages in the life cycle of continental ice sheets, have been widely reported (e.g. Kleman et al., 1997, 2002, 2006; Veillette et al., 1999; Clark et al., 2000; Stokes et al., 2005, 2006b; Dyke, 2008). Additionally, it has been argued that transitions from one subglacial bedform to another record remoulding by changing ice flows in response to migrating ice divides (e.g. Boulton, 1987; Boulton and Clark, 1990a, b; Clark, 1993, 1994, 1997; Clark and Meehan, 2001; Jansson et al., 2002; Jansson, 2005). In hard bedrock regions, depositional landforms are often used in conjunction with

(b)

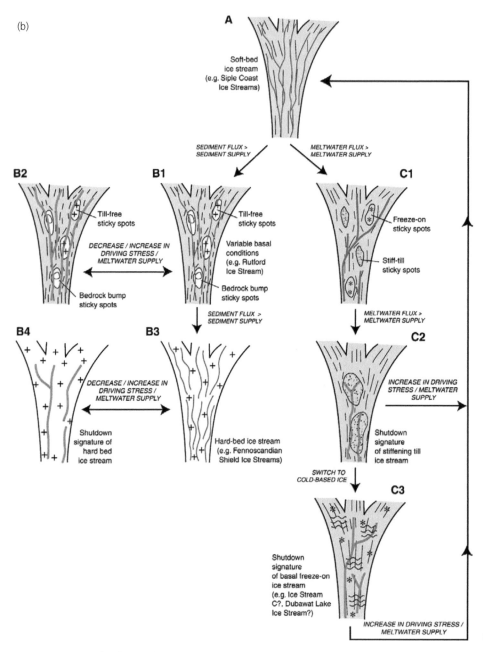

A

Soft-bed
ice stream
(e.g. Siple Coast
Ice Streams)

SEDIMENT FLUX >
SEDIMENT SUPPLY

MELTWATER FLUX >
MELTWATER SUPPLY

B2

Till-free
sticky spots

DECREASE / INCREASE IN
DRIVING STRESS /
MELTWATER SUPPLY

Bedrock bump
sticky spots

B1

Till-free
sticky spots

Variable basal
conditions
(e.g. Rutford
Ice Stream)

Bedrock bump
sticky spots

SEDIMENT FLUX >
SEDIMENT SUPPLY

C1

Freeze-on
sticky spots

Stiff-till
sticky spots

MELTWATER FLUX >
MELTWATER SUPPLY

B4

DECREASE / INCREASE IN
DRIVING STRESS /
MELTWATER SUPPLY

Shutdown
signature of
hard bed
ice stream

B3

Hard-bed ice stream
(e.g. Fennoscandian
Shield Ice Streams)

C2

INCREASE IN DRIVING
STRESS / MELTWATER
SUPPLY

Shutdown
signature
of stiffening till
ice stream

SWITCH TO
COLD-BASED ICE

C3

Shutdown
signature
of basal freeze-on
ice stream
(e.g. Ice Stream
C?, Dubawat Lake
Ice Stream?)

INCREASE IN DRIVING STRESS /
MELTWATER SUPPLY

Figure 12.63 (*Continued*)

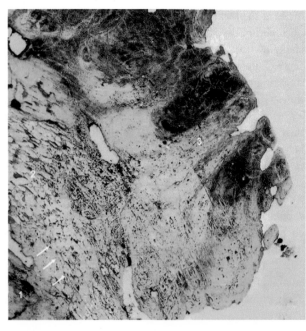

Figure 12.64 Landsat image of south-eastern Prince of Wales Island in the central Canadian Arctic, showing three superimposed drumlin fields (1–3 in order of decreasing age). A lateral shear moraine is marked by arrows (Image provided by the Geological Survey of Canada; geomorphology is from Dyke and Morris, 1988).

other ice flow directional information, such as glacial erosional features (striae, crag and tails, rock drumlins, roches moutonées) and till lithological characteristics (e.g. Dyke et al., 1992; Veillette et al., 1999; Jansson et al., 2002; Dyke, 2008; section 12.4.6). The preservation of subglacially streamlined features dating from older glaciations has also been reported from the beds of former ice sheets (e.g. Veillette et al., 1999; Graham et al., 2007).

The analysis of satellite images and aerial photographs of the areas occupied by the Pleistocene ice sheets has facilitated the reconstruction of changing ice sheet flow directions, using the inversion approach outlined in section 12.2.2. The major landform components of palaeo-ice sheet beds can be grouped into *swarms* or *flow sets*, which are simplified representations of large numbers of landforms (fig. 11.16; Stokes and Clark, 2003b; Kleman

et al., 2006). The assemblage of landforms in a swarm may include drumlins, mega-flutings and/or mega-scale lineations, and swarms may be superimposed on or cross-cut each other (figs 12.4, 12.65). Clark (1993, 1994) identified two relative age indicators which enable the sequence of ice flows to be reconstructed:

(1) *simple superimposition*, where one set of streamlined lineations is superimposed over another set with a different orientation;
(2) *pre-existing lineation deformation*, where deformation during a more recent ice flow phase alters the form or continuity of pre-existing lineations.

The degree of modification of earlier lineations forms a continuum, ranging from: (1) no modification of pre-existing lineation; to (2) superimposition of smaller forms on older lineations; to (3) substantial breaching

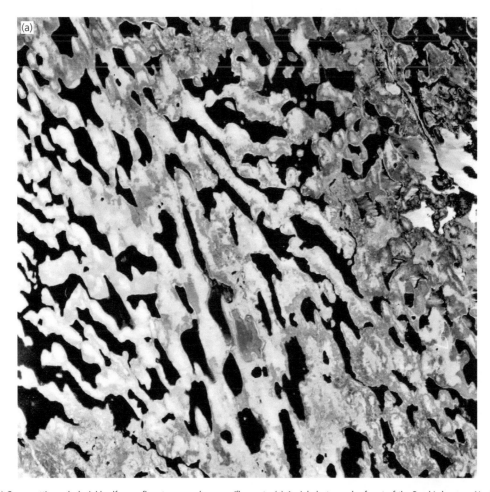

Figure 12.65 (a) Cross-cutting subglacial bedforms, flow traces and swarms/flow sets. (a) Aerial photograph of part of the Boyd Lake area, Northwest Territories, Canada, showing cross-cutting drift lineations. Drumlins, flutings and ribbed terrain can all be recognized in this view. (National Air Photo Library, Ottawa) (b) Diagrammatic representation of swarm identification from flow trace mapping. A swarm is identified using longitudinal continuity lines drawn along the flow traces, or coherently aligned bedforms and up-ice and down-ice transverse boundaries drawn oblique to the continuity lines. (Kleman et al., 2006) (c) Flow sets/swarms in the Canadian Arctic Archipelago. Each flow set represents a prominent ice flow direction at some time during the life cycle of the northern Laurentide Ice Sheet. Note that many flow sets are at least partially overprinted. (Stokes et al., 2005; reproduced with permission of Elsevier) (d) Landsat image showing details of overprinted subglacial bedforms and flow sets/swarms on Wollaston Peninsula, southern Victoria Island. (i) Three interpreted flow sets and their temporal sequencing. (ii) Detailed view of area outlined in (i), showing the 'herringbone' pattern that arises from the overprinting of three sets of drumlins. (iii) Higher-resolution extract from top of image in (ii), showing how the smaller bedforms of flow set fs-53 (yellow) overprint fs-55 (red), even though they run parallel with them in some areas. Not all bedforms have been marked in these images (Stokes et al., 2006b).

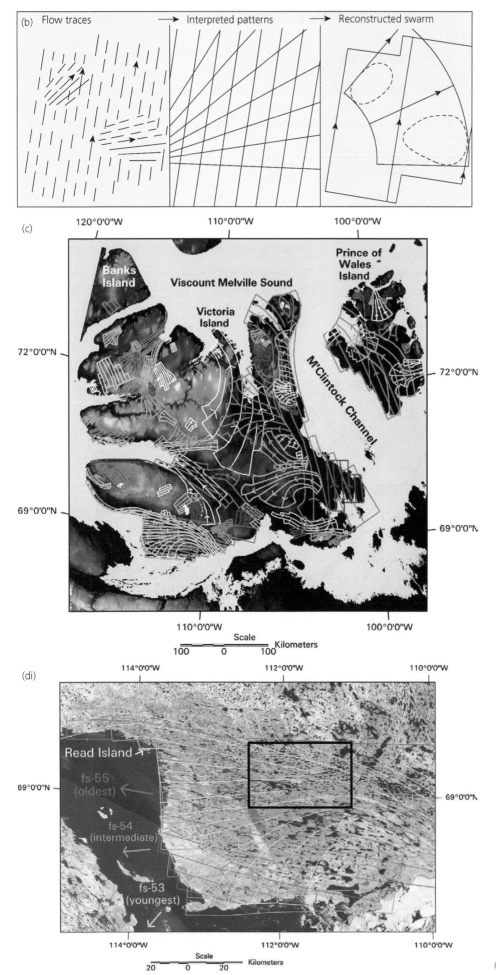

Figure 12.65 (*Continued*)

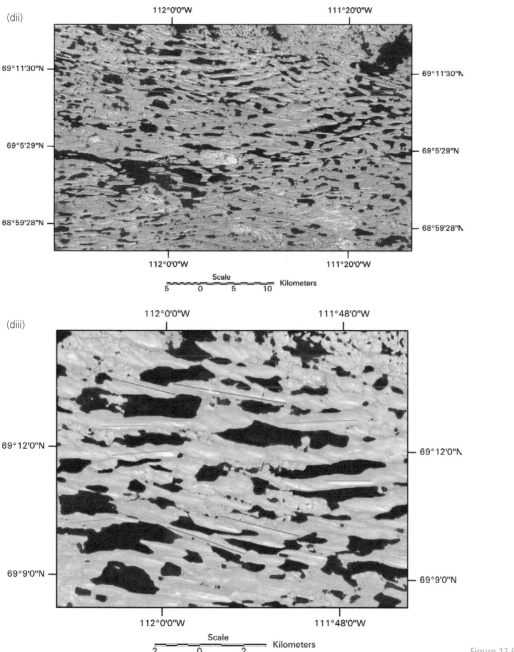

Figure 12.65 (*Continued*)

or deformation of pre-existing lineation; to (4) total reorganization of sediment into new orientation (fig. 12.66a).

A number of factors are thought to influence the degree of bedform modification consequent on changes in the ice flow direction, including ice flow velocity, sediment properties, ice thickness and the duration of individual flow phases. The influence of flow velocity is illustrated in figure 12.66b, which shows the theoretical velocity distribution along a flow line of a simple ice sheet (Clark, 1994). The balance velocity is zero below the ice divide and increases downflow to reach a maximum at the equilibrium line, from where it decreases outwards towards the margin. If ice flow velocity is the dominant factor controlling the degree of bedform modification when flow changes direction, the greatest amount of modification should be expected close to the margin. Thus bedforms in this area are likely to record only the most recent flow direction, whereas cross-cutting lineations tend to be expected to be best preserved further upglacier. Larger numbers of preserved lineations document less intense modification, and some of these are located in the former areas of ice divides, such as Keewatin, west of Hudson Bay. This model also explains why drumlins are commonly formed in belts upglacier from recessional moraines, and record the ice flow directions that prevailed during deglaciation (see section 11.2.4; Mooers, 1989b; Patterson and Hooke, 1996). Differences in the degree of bedform modification could also result from lateral variations in ice flow velocity, such as those near the margins of ice streams.

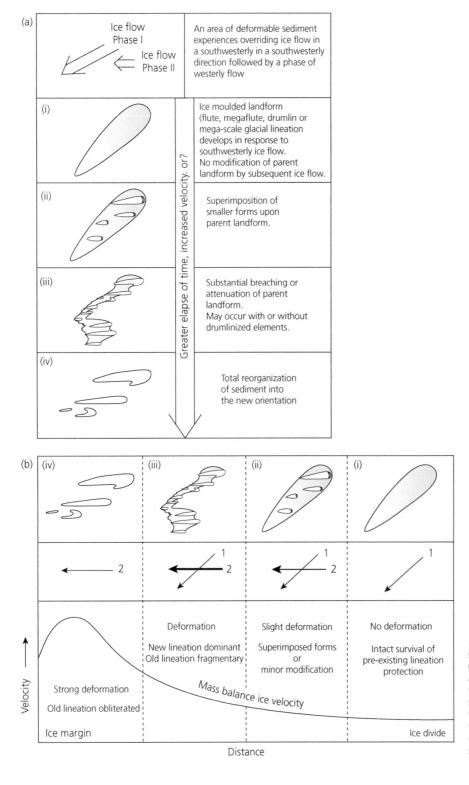

Figure 12.66 Theoretical development of subglacial bedforms due to remoulding. (a) Continuum of subglacial modification of a lineation when overridden by ice flowing from a different direction. (b) Continuum of subglacial modification transferred to a transect from ice divide to margin. Note that this assumes ice flow velocity is the main determinant in the degree of subglacial modification (After Clark, 1993).

Cross-cutting lineations also provide evidence for the nature of ice flow changes. If the changes in ice flow direction were gradual, and there was a continuous adjustment to changing flow, we should expect to see either a whole series of intermediate orientations or a single final orientation. However, only a small number of discrete ice flow directions are observed in the landform record in any one area, which led Clark (1993) to suggest two possible explanations:

(1) *punctuated ice-flow shifts*, in which periods of relative stability and lineation formation are separated by brief episodes during which ice flow direction changes rapidly;

(2) an *on/off mechanism* of lineation generation, which involves a steady change in the ice flow direction, but with an on/off lineation production mechanism that records snapshots of the changing flow regime.

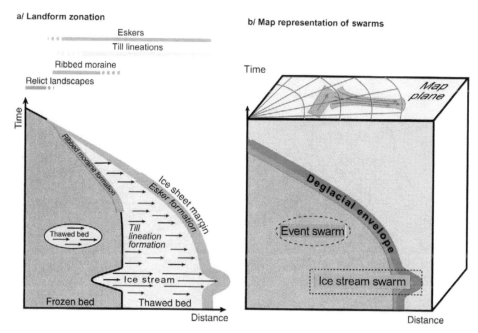

Figure 12.67 Time–distance diagram to convey the spatial and temporal aspects of subglacial landform production. (a) Locations of landform production related to basal thermal conditions over time. Note that eskers will form long, almost continuous ridges due to time-transgressive development in the deglacial envelope (Hooke and Fastook, 2007). Ribbed terrain can also form in the event and ice stream swarm due to late-stage ice stream shutdown. (Stokes et al., 2006a) (b) The relationship of landform production and map representation. The red lines on the map plane represent the deglacial envelope, which is superimposed on the event swarm. The ice stream swarm is the oldest recognizable flow set (Kleman et al., 2006).

Clark (1993) favoured the punctuated ice-flow model because it best explains continent-wide coherent sets of lineations. However, the 'on/off mechanism' may be applicable where basal conditions were near the threshold for bed erosion, deposition or deformation.

The spatial and temporal evolution of swarms and other landform suites associated with ice sheet occupancy has been depicted in a conceptual time–distance diagram by Kleman et al. (2006; fig. 12.67). This highlights *ice stream swarms*, that draw down large volumes of ice in a convergent head zone when ice sheets are at their maximum size, and *event swarms*, created in response to the development of thawed spots on the ice sheet bed. In addition, Kleman et al. (2006) identified the *deglacial envelope* as either a wet bed or frozen bed deglaciation swarm. Wet bed envelopes comprise inset sequences of sub-marginal and marginal landforms, such as flutings and drumlins and eskers, aligned oblique to contemporaneous marginal moraines and recording inward-transgressive flow sets (Kleman et al., 1997). Modern glacier forelands like that at Breiðamerkurjökull serve as excellent modern analogues for wet bed deglacial envelopes (cf. Evans and Twigg, 2002; section 12.3.2). Frozen bed envelopes comprise marginal meltwater channels superimposed on older surfaces that contain flow traces from older events. A modification of the time–distance diagram in figure 12.67 may be necessary with respect to the development of ribbed terrain. Kleman et al. (2006) proposed that ribbed moraine develops during periods when the ice sheet bed is changing from frozen

to thawed, which is potentially compatible with the BRIE model of Dunlop et al. (2008), but does not account for ribbing development in ice stream tracks or in an event swarm. Stokes et al. (2006a) suggested that ribbed moraine development in ice stream tracks relates to ice stream shut-down and therefore reversion to frozen bed conditions after a period of thawing.

Ice sheet scale maps of successive ice flow directions, based on the distribution of cross-cutting lineations, have been compiled for a number of regions, including northern Canada (e.g. Dyke et al., 1992; Veillette et al., 1999; Clark et al., 2000) and Fennoscandia (Kleman et al., 1997; Boulton et al., 2001b, fig. 12.68). The sequence of ice flow events depicted in such maps must be compatible with till fabric data and erratic dispersal (fig. 12.69; Clark, 1997). First, changes in ice flow through time are linked to changes in the positions of ice divides and therefore changes in the shape of the various sectors of the ice sheet. If shifts in ice dispersal are not taken into account when interpreting ice flow lineation maps, 'bogus' ice divides will exist in the ice sheet reconstruction (Boulton and Clark, 1990b). Second, the 'switching on' or activation of ice streams produces event swarms/flow sets that overprint convergence patterns on older lineations. Third, the recession of a lobate margin produces deglacial envelopes with misaligned lineations. The clarity of the imprint of a flow set and the complexity of the subglacial bedform palimpsest in an area is also conditioned by the temporal nature of the ice flow event; an isochronous event will produce a 'rubber stamp' imprint, whereas a

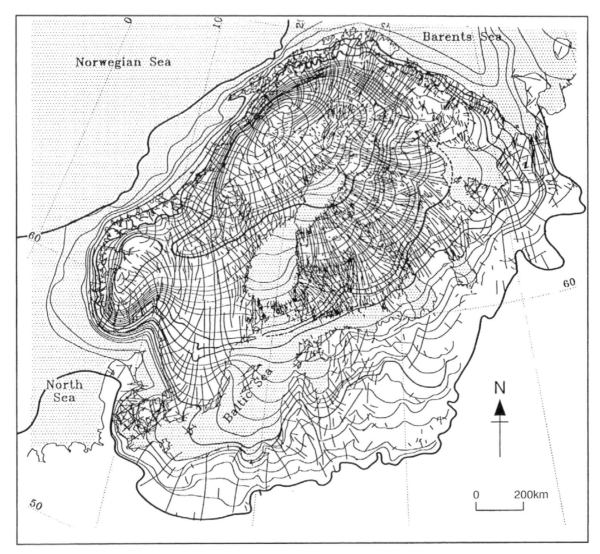

Figure 12.68 Inferred pattern of ice sheet recession over Fennoscandia, based on the dominant patterns of subglacial bedform lineation (with recession lines drawn normal to the lineations on the Shield) and moraines in the area outside the Shield (Boulton et al., 2001b).

time-transgressive set of events will result in a 'smudged imprint' (Clark, 1999; Clark and Stokes, 2003; fig. 12.61). Obviously, large numbers of early ice flow indicators are obliterated by later ice flows and this constitutes a major problem in producing a reconstruction of multiple flow events. If an ice divide does not shift significantly during a glacial cycle, older forms may be preserved beneath it, because the amount of erosion is limited by low ice velocities.

The resulting difficulties in deciphering ice sheet dynamics from the evidence for ice flow directions are well illustrated by the case of the Quebec-Labrador sector of the Laurentide Ice Sheet. In this region, the location of the 'Labrador ice divide' has long been recognized, based on its strong imprint as a horseshoe-shaped zone separating ice flow features that converge northwards towards Ungava Bay and radiate southwards and westwards (fig. 12.70). Early notions that this represented the

long-term ice divide in this sector of the ice sheet were dismissed, based on evidence of cross-cutting ice flow indicators in the region. An excellent example of the diversity of this complex database is represented by figure 12.71, which depicts the pattern of cross-cutting ice flows in the area of the eastern arm of the 'horseshoe', based on striae, subglacial bedforms and till geochemistry (Klassen and Thompson, 1993; Veillette et al., 1999). This type of data has been used by Veillette et al. (1999) to reconstruct an ice flow chronology that involves a significant migration of the ice divide through time, with the 'horseshoe' being a zone of intersection (ZI) that developed during the later stages of glaciation (fig. 12.72a). Although they agree that ice sheet divide migration was responsible for the complex overprinting of ice flow directional indicators in the region, Clark et al. (2000) presented a different palaeoglaciological reconstruction (fig. 12.72b). In contrast to both studies, Kleman et al. (1994) proposed that

Ice divide migration

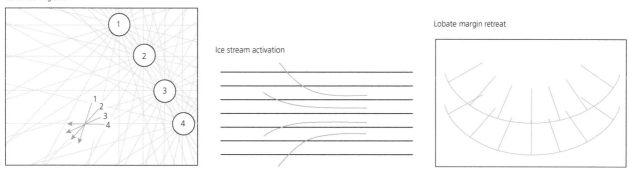

Figure 12.69 Three scenarios where cross-cutting bedforms may occur: (left) Ice divide migration from positions 1–4 and an example (cross-cutting arrows) of how each shift would be manifest in a bedform palimpsest at one location; (middle) How regular sheet flow can be overprinted by the onset zone of an ice stream when it becomes activated. Ribbed terrain can also be overprinted on both sets of lineations when the ice stream shuts down; (right) The time-transgressive sequence of deglacial envelopes produced along a lobate ice margin (Redrawn from Clark, 1997).

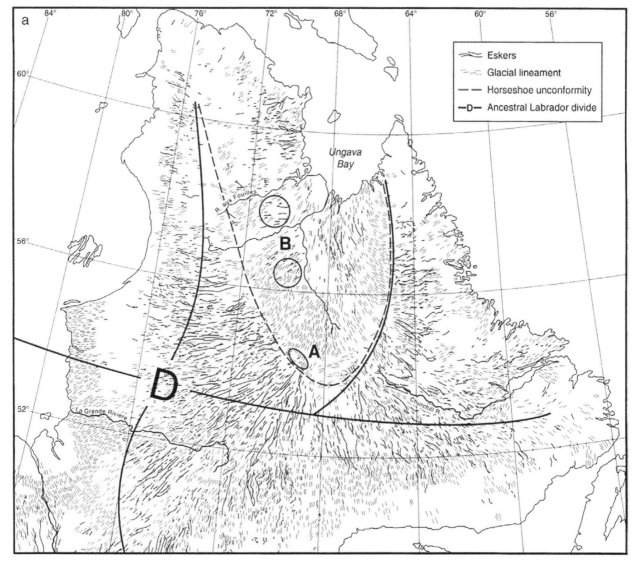

Figure 12.70 The main subglacial landforms of the Quebec-Labrador Sector of the Laurentide Ice Sheet, showing the 'horseshoe' zone or unconformity and ancestral Labrador ice divide. (Dyke and Prest, 1987) Circled area at (A) includes cross-cutting lineations and areas marked (B) include anomalously aligned eskers and lineations (Clark et al., 2000).

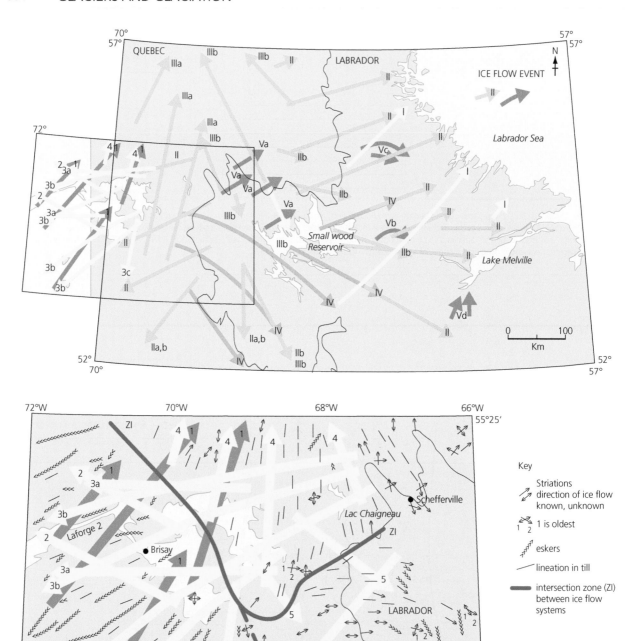

Figure 12.71 Maps of ice flow directional indicators, based on the work of Klassen and Thompson, 1993) (main lower map) and Veillette et al., 1999 (upper map, located by box on lower map). Five major flow events (I–V) identified by Klassen and Thompson are based on striae, subglacial bedforms and till composition. Five major ice flows (1–5) identified by Veillette et al. are based on striae, subglacial bedforms and eskers. ZI = zone of intersection. Flow 1 and Event I date from a pre-Wisconsinan glaciation and are thought to have continued through the ZI from the south (broken arrows). Flow 3 correlates with Event II and Flow 5 with Event IV. Flow 3 was captured by Flow 4 due to the southward migration of the Labrador horseshoe divide.

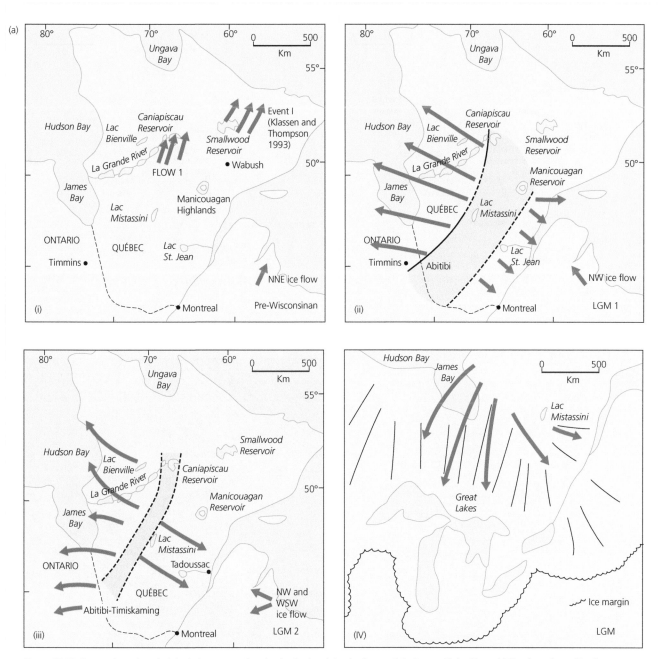

Figure 12.72 Contrasting palaeoglaciological reconstructions for the Quebec-Labrador Sector of the Laurentide Ice Sheet. (a) The chronology of ice flow proposed by Veillette et al. (1999). (A) Pre-Wisconsinan Flow 1/Event I. (B) Flow 2, with solid line representing the most easterly known limit of Flow 2 and short arrows marking the probable contemporaneous flow on the other side of the divide. (C) Mid-Wisconsinan position of the ice divide. (D) Ice margins and flow lines over James Bay and the Great Lakes, requiring an ice dispersal centre east of James Bay. (Dyke and Prest, 1987) (E) Location of Late Wisconsinan ice divide over the Labrador Trough producing early Flow 3 and Event II. (F) Capture of Flow 3 by Flow 4 around 10 ka BP, at which time the inter-lobate Harricana Moraine was being deposited. (b) Time–space diagram showing the margins, flow patterns and ice divides of Quebec-Labrador based on the data presented in Clark et al., 2000 (After Clark et al., 2006).

the subglacial landforms north of the 'horseshoe' were from a previous glaciation and preserved beneath cold-based ice during the last glaciation.

The large literature now available on research into ice sheet beds reveals that 'smudged' landform imprints are likely to be more common than 'rubber stamp' imprints. This is because ice flow events are time-transgressive and misaligned, due not only to the temporal migration of ice sheet dispersal centres, but also to the changing influence

of bed topography on flow patterns during ice recession. This is well illustrated in the offshore record, where seismic techniques and swathe bathymetry reveal overprinted and misaligned flow sets, associated in some places with ice-marginal landforms (e.g. Graham et al., 2007; Andreassen et al., 2008). The use of three-dimensional seismic surveys of offshore regions now facilitates the visualization of superimposed subglacial bedform layers, even if they lie below intensively iceberg-scoured sea

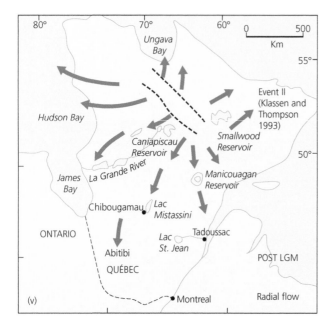

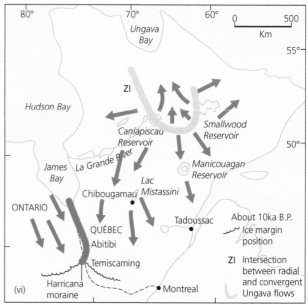

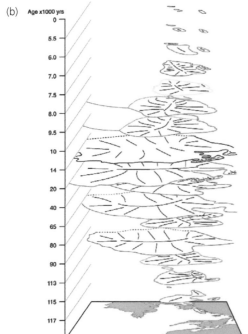

Figure 12.72 (*Continued*)

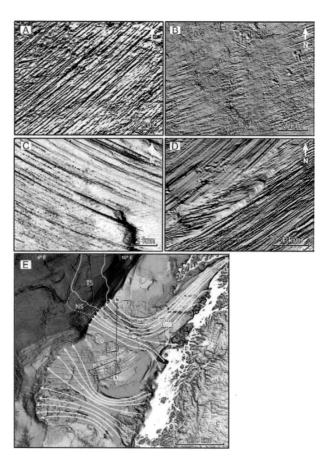

floors (e.g. Rafaelsen et al., 2002). An excellent example of overprinted palaeo-ice stream beds on the Norwegian continental shelf was provided by Dowdeswell et al. (2006; fig. 12.73). These flow sets date from different glaciations and it is thought that sediment accumulation and concomitant changes in accommodation space on the shelf were just as important as changes in ice sheet dynamics in forcing the switching of flow direction. The role of topography on ice flow directional changes in the offshore record was demonstrated by Ottesen et al. (2008b), who showed that cross-cutting bedforms occur on the shelf, where ice stream flow paths had the space to

Figure 12.73 Streamlined sedimentary bedforms produced beneath palaeo-ice streams on the mid-Norwegian continental shelf. (A) Buried surface ~100 m deep within upper Naust Formation. (B) ~200 m deep within upper Naust Formation. (C) Late Weichselian sediments at modern sea floor of Traenadjupet. (D) Sea floor of Vestfjorden (swath bathymetry). (E) Map of changing ice-stream flow directions inferred from orientation of streamlined bedforms. Red lines are Elsterian-Saalian and white lines are Weichselian ice stream flow directions. TS = Traenadjupet Slide; NS = Nyk Slide. Locations of surfaces in A–D are shown in E (labelled a–d). Seismic line is marked N-S (Dowdeswell et al., 2006; reproduced with permission of the Geological Society of America).

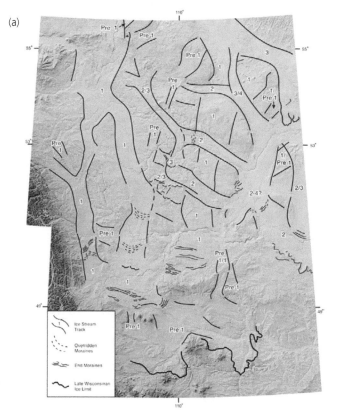

(a)

migrate, in contrast to the uniform alignment of bedforms produced by ice once it had receded into the topographic confinement of the fjords. Smudged imprints of flow sets have been reported also from terrestrial ice sheets. For example, Evans et al. (2008) and Ó Cofaigh et al. (in press) provided examples from western Canada of ice stream flow tracks and marginal landforms that were produced by temporal separation and overprinting, in places due to ice stream surging (fig. 12.74a).

Early interpretations of the subglacial bedforms produced by the BIIS were confounded by the apparent contradictory evidence of reversed flow directions. This is best illustrated by the cross-cutting bedforms and erratic transport pathways in north-west England and south-west Scotland. We now understand that the complexity of bedform overprinting in the region is related to the changing dominance of ice dispersal centres located over the English Pennines, the Lake District and the Southern Uplands, in addition to regional ice flow from the centre of the Scottish sector of the BIIS. Early southerly flows from Scotland were responsible for the transport of Scottish erratics through the Tyne Gap, up the Vale of Eden and over Stainmore, towards the eastern English coast. Later stages of ice flow were directed northwards down the Vale of Eden and into the Solway Lowlands.

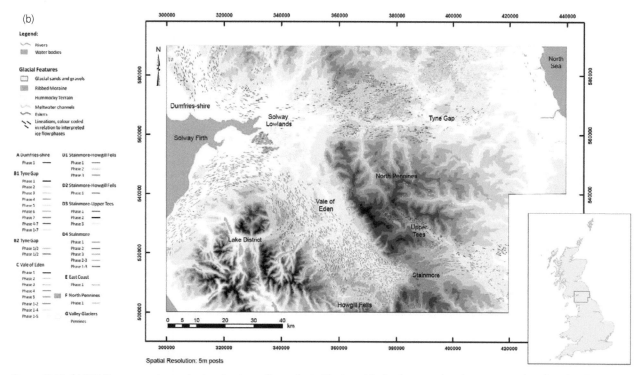

Figure 12.74 (a) SRTM imagery and map of palaeo-ice stream flow paths in Alberta and Saskatchewan and northern Montana, based on the relationships between cross-cutting flow sets. The relative flow sequencing is numbered from 1 to 4 (oldest to youngest), with some ice stream tracks having been occupied through more than one flow phase. Areas between ice stream tracks do contain streamlined bedforms which relate to non-streaming ice flow. Overridden moraines are thought to mark the margins of the Laurentide Ice Sheet during initial advance into western Canada. (Ó Cofaigh et al., in press) (b) Map of colour-coded flow sets relating to the dynamics of the central sector of the British-Irish Ice Sheet (Livingstone et al., 2008; reproduced with permission of the *Journal of Maps*).

Livingstone et al. (2008) and D. J. A. Evans et al. (2009) have identified seven to eight major flow phases relating to ice dispersal in the Lake District/Tyne Gap region (fig. 12.74b), which can be integrated into the eastern-most of six major topographically unconstrained flow phases identified by Salt and Evans (2004) for south-west Scotland and five flow phases draining through the Pennines via Stainmore. Evolving research on the BIIS is beginning to uncover more widespread evidence of numerous ice flow shifts, particularly in Ireland (Greenwood and Clark, 2008).

12.4.4 ICE SHEET BEDS AS THERMAL REGIME PALIMPSESTS

Patterns of deposition and erosion below ice sheets are dictated by the *subglacial thermal organization* through time. Thawed bed regions, particularly beneath ice streams, may widen or narrow through time or shut down altogether (section 6.3.6; fig. 12.58). Additionally, ice dispersal centres shift location through the life cycle of an ice sheet, giving rise to the temporal overprinting of warm and cold beds, and areas affected by largely stable dispersal centres can be manifest as regions largely devoid of subglacial bedforms (e.g. Aylsworth and Shilts, 1989; Hättestrand and Clark, 2006; De Angelis, 2007).

In essence, the preservation of unmodified older flow traces on ice sheet beds is dependent on the development of frozen bed conditions once glacier flow shuts down and a new flow set is initiated. The former lateral boundaries

of warm-based ice can then be identified using over-printed bedforms or the juxtaposition of bedforms and largely preserved preglacial land surfaces (Kleman and Hättestrand, 1999; Kleman et al., 1999). For example, figure 12.75a shows a former lateral boundary between cold- and warm-based ice, interpreted from the existence of cross-cutting streamlined bedforms. Both sets of lineations need warm-based sliding ice, but the preservation of the older flow set requires that the later phase of warm-based ice flow was restricted to the right of the dashed line. Similarly, figure 12.75b shows the margin of a warm-based sliding zone based on the preservation of an older till cover.

The importance of frozen bed zones in the evolution of ice sheet bed footprints has been demonstrated clearly in the Canadian Arctic Archipelago, where shifting ice flow directions have been reconstructed from drift lineations, regional till composition and erratic dispersal patterns (figs 12.64, 12.76). For example, the inferred ice flow history of Prince of Wales Island can be explained by the migration of the M'Clintock Ice Divide in the northern Laurentide Ice Sheet (Dyke et al., 1992; Hodgson, 1994). Overprinted flow sets demarcate the sequential changes in the locations of warm- and cold-based conditions through time. The edges of individual ice streams dating to different flow phases are very clear on satellite imagery where bedforms cross-cut each other (fig. 12.64), and the till plumes produced by each flow are lithologically distinct. The final stages of ice retreat on Prince of Wales Island were characterized by the stabilization of the ice divide

Sliding boundary, Great Bear Lake, Canada

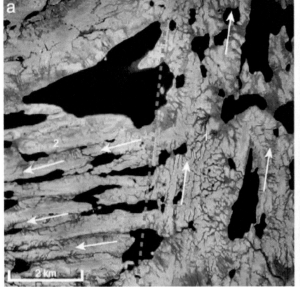

Contact between frozen-bed patch and sliding zone, Långfjället, Sweden

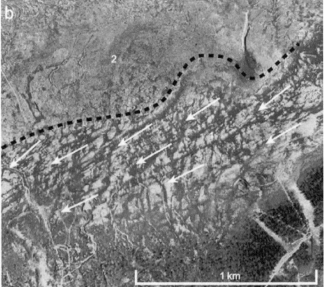

Figure 12.75 Images of former boundaries of warm- and cold-based conditions on ice sheet beds. (a) Aerial photograph of cross-cutting subglacial lineations near Great Bear Lake, northern Canada. The red line shows the boundary between two flow sets and is the location of a shear zone between the warm-based ice that produced flow set 1 and cold-based ice that protected flow set 2 after it was moulded by an earlier warm-based flow phase. (b) Satellite image of the former bed of the Scandinavian Ice Sheet in western Sweden, showing a sharp lateral boundary (broken black line) between the edge of sliding bed lineations of area 1 and an old till cover protected beneath former cold-based ice in area 2 (Kleman and Glasser, 2007).

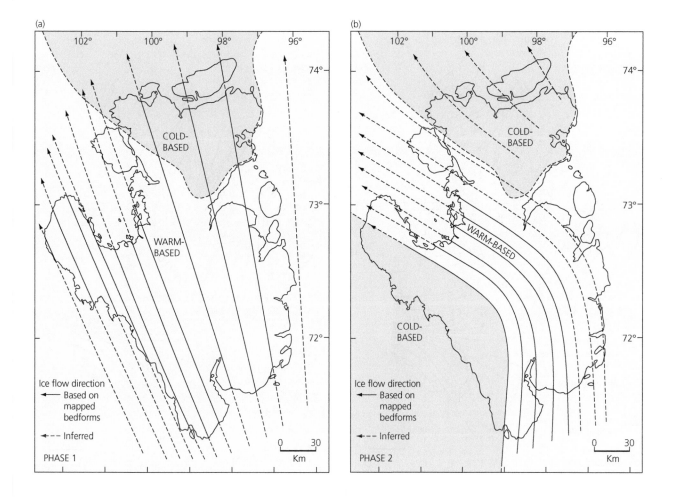

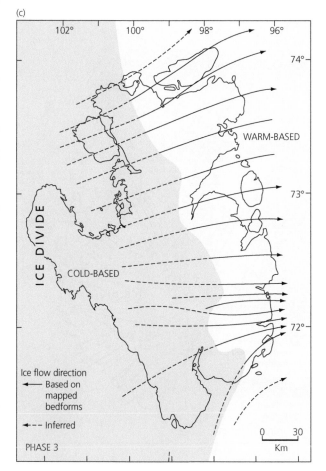

Figure 12.76 The relationships between successive ice flow events and changing basal thermal regimes beneath the northern Laurentide Ice Sheet, exemplified by the landform evidence on Prince of Wales Island, Canadian Arctic Archipelago. (a–c) Shifting flow conditions during the last glacial maximum. (d–g) Flow conditions and associated ice margins for 11, 10, 9.6 and 9.3 ka BP. The whole sequence shows the migration of ice divides and ice streams, the distribution of cold- and warm-based ice and the extent of the deglacial high sea level. The solid flow lines represent flow sets based on glacial bedforms that are visible even though later flows from different directions overrode them (Redrawn from Dyke et al., 1992).

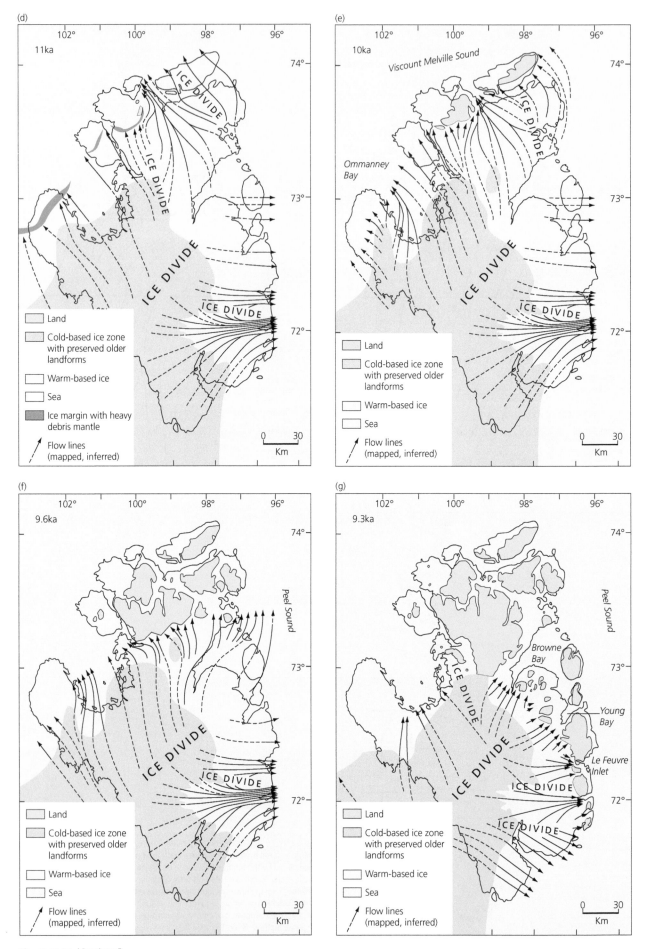

Figure 12.76 (*Continued*)

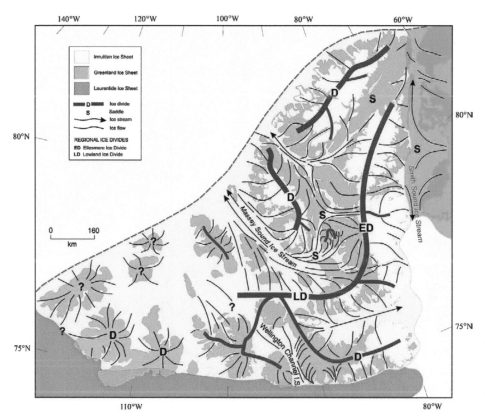

Figure 12.77 Reconstruction of the Innuitian Ice Sheet over the Canadian High Arctic during the LGM, showing ice divides, saddles and ice streams. Note that large areas of the ice sheet bed downflow of the main divides would have been warm-based at this time, but would have turned cold-based during ice sheet thinning, particularly over upland areas (England et al., 2006; reproduced with permission of Elsevier).

towards the south of the island, and the drawdown of ice into calving bays due to glacio-isostatically higher sea levels. This initiated the production of smaller fields of drumlins and flutings in topographic depressions. The palaeoglaciology of extensive areas of the northern Laurentide Ice Sheet has now been reconstructed, based on the same principles set out in the Prince of Wales Island case study (cf. Dredge, 2000; De Angelis and Kleman, 2005; Stokes et al., 2005, 2006b; Dyke, 2008; figs 12.65c, d).

In the Canadian High Arctic, the recognition of large areas of cold-based ice over the numerous islands of the archipelago, in contrast to the warm-based trunk ice of intervening inter-island channels, has been critical to the palaeoglaciological reconstruction of the Innuitian Ice Sheet (Dyke, 1999; England, 1999; England et al., 2006; fig. 12.77). The dispersal of granite erratics and subglacial lineations and ice-scoured bedrock indicate that large ice streams drained the Innuitian Ice Sheet at its maximum extent, the largest being the Massey Sound and Wellington Channel ice streams (Dyke, 1999; Ó Cofaigh et al., 2000; Atkinson, 2003). Early stages of ice sheet break-up in the west of the archipelago were driven by rapid calving in the inter-island channels and larger fjords, with significant areas of intervening high ground containing only meltwater channels indicative of recession of cold-based local ice caps (section 8.4.4; Dyke, 1993; Bednarski, 1998; Ó Cofaigh, 1998; Ó Cofaigh et al., 1999; England et al., 2004). Although some of these

upland ice caps only turned cold-based during later stages of deglaciation when they thinned significantly, many likely remained largely cold-based throughout the last glaciation, as evidenced by the preservation of deeply weathered terrains and even pre-Quaternary drainage networks.

12.4.5 ICE SHEET DRAINAGE

In chapter 6 we saw how recent studies of modern ice sheets are revealing evidence of varied and dynamic hydrological systems. Former ice sheet beds retain imprints of subglacial hydrologic processes, although this record is fragmentary, time-transgressive and sometimes difficult to interpret. Aspects of former ice sheet hydrology have been discussed in chapters 10 and 11; here, we briefly review large-scale patterns of ice sheet drainage, focusing on the influence of substrate lithology, thermal regime and water sources.

The nature of the substrate is an important first-order control on regional patterns of subglacial drainage. 'Hard' beds have low transmissivities and meltwater will mainly flow at the ice–bed interface, whereas substantial groundwater flow is possible through porous rocks and sediments (Boulton, 2006; Piotrowski, 2006; Person et al., 2007). Clark and Walder (1994) have noted that, in both North America and Eurasia, long esker systems are common in areas of hard substrata, and argued that this is because stable subglacial conduits are favoured by a

(a)

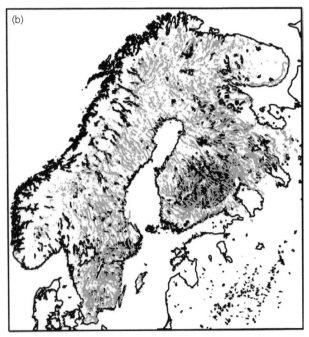

(b)

Figure 12.78 The regional distribution of eskers on the beds of the former Northern Hemisphere ice sheets. (a) North America, showing the limit of the Wisconsinan Laurentide Ice Sheet (heavy line) and the boundary between Canadian Shield rocks and younger sedimentary rocks (lighter line). Note the preferential occurrence of eskers on shield rocks. (Clark and Walder, 1994; reproduced with permission of the Geological Society of America) (b) Fennoscandia, showing the distribution of eskers over the terrestrial sector of ice cover and the increasing density of eskers at greater distances from the former ice divide (Boulton, 2006).

rigid bed. Eskers tend to be rarer in areas underlain by deformable substrata where, Clark and Walder argued, drainage can be expected to consist of a distributed canal system or pore-water flow (fig. 12.78). In detail, however, the picture is much more complex. Eskers do occur on soft substrata, in some cases superimposed on drumlins and other bedforms, and it is likely that on both hard and soft beds subglacial drainage switched between channelized and distributed systems as basal conditions changed over the course of glacial cycles. It should be noted that the 'hard-bedded' shield areas to the east and west of Hudson Bay were also the locations of the last, shrinking remnants of the Laurentide Ice Sheet, and that the esker distribution shown in figure 12.78 likely also reflects a temporal dimension to the hydrology of the ice sheet.

Significant subglacial groundwater flow can occur in areas underlain by permeable rocks and sediments. During Quaternary glaciations, aquifers within such regions were significantly recharged with groundwater and regional patterns of groundwater flow underwent major reorganizations (fig. 12.79; Boulton and Caban, 1995; Boulton et al., 1995; Grasby et al., 2000; Grasby and Chen, 2005; Piotrowski, 2006; Person et al., 2007; Lemieux et al., 2008). It is clear that near the southern margins of the great Quaternary ice sheets, groundwater flow alone was insufficient to discharge all of the water produced at or reaching the bed. Eskers and channels (including tunnel valleys)

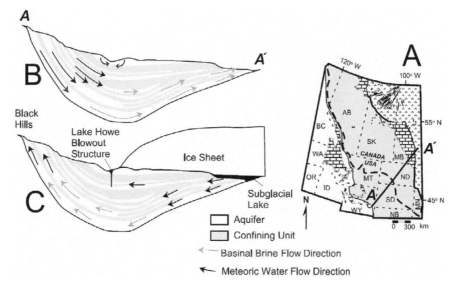

Figure 12.79 The relationship between bedrock permeability and subglacial/proglacial drainage pathways, illustrated by the case of the Western Canada Sedimentary Basin. (A) Map of bedrock types and distribution of eskers (sinuous black lines), which gives a clear impression of changing permeability in the ice sheet substrate across the boundary between the crystalline terrain of the Shield (stipple pattern) and the Cretaceous shales (shaded pattern) and Palaeozoic carbonates (brick pattern) of the prairies. (B) The modern and (C) the glacial hydrology of the Williston Basin along the cross section A-A′ in (A), showing groundwater recharge from the ice sheet and the concomitant reversal of brines in the aquifers. Over-pressurization of aquifers results in blow-out structures such as those at Lake Howe (After Grasby and Chen, 2005 and Grasby et al., 2000; reproduced with permission of the Geological Society of America).

show that subglacial conduits constituted important components of subglacial drainage systems below many 'soft-bedded' ice sheet areas (sections 8.4.4, 11.2.10).

Subglacial thermal regime appears to have influenced ice sheet drainage in two main ways. First, persistent cold-based regions in ice sheet interiors were largely isolated from both surface and subglacial sources of meltwater, and subglacial eskers and channels are absent (e.g. Aylsworth and Shilts, 1989; Jansson, 2005; De Angelis, 2007). Second, cold-based regions near ice sheet margins acted as barriers to water flow from warm-based regions located upglacier, elevating water pressures in confined subglacial aquifers (Boulton et al., 1995). As discussed in section 11.3.2, artesian water pressures associated with sub-marginal permafrost play an important role in large-scale glacitectonism (Oldale and O'Hara, 1984; Piotrowski, 2006). The break-out of pressurized water through permafrost barriers appears to be recorded by proglacial 'blow-out structures' (see section 11.3.2; Bluemle, 1993; Boulton and Caban, 1995) and some tunnel valleys (see section 8.4.4; Clayton et al., 1999; Cutler et al., 2002; Hooke and Jennings, 2006).

The character and spatial organization of subglacial drainage systems might be expected to reflect patterns of recharge. Study of the subglacial hydrology of modern ice sheets is still in its infancy, although there is strong circumstantial evidence for the seasonal development of efficient, surface-fed subglacial conduits below the Greenland Ice Sheet (section 6.2.3; Das et al., 2008), and longer-term cycles of storage and transfer in basally recharged drainage systems below Antarctica (section 6.3.5; Fricker et al., 2007). Evidence for surface- and

basally fed drainage systems has also been recognized on the beds of palaeo-ice sheets, although their large-scale distributions in space and time remain poorly known. Sedimentological evidence for rapidly fluctuating discharge in eskers is indicative of recharge from surface melting (e.g. Allen, 1971; Bannerjee and McDonald, 1975; Kleman et al., 1997; Arnold and Sharp, 2002), and indeed, the stable conduit systems recorded by many esker networks would have been encouraged by surface-to-bed recharge at discrete points (Brennand, 2000). However, channelized drainage systems can apparently also form in subglacially recharged aquifers (Boulton et al., 2007a, b), so do not necessarily imply that surface meltwater reached the ice sheet bed.

Several possible examples of subglacially recharged drainage systems have been described in the literature. For example, Anderson and Fretwell (2008) have described an extensive system of convergent channels in the onset zone of the former Marguerite Ice Stream, on the continental shelf off the Antarctic Peninsula. The system becomes increasingly organized in a downstream direction into fewer, straighter and broader channels, and may have served to discharge water from subglacial lakes similar to those below the modern ice sheet.

12.4.6 SEDIMENT DISPERSAL PATTERNS AND TILL LITHOLOGY

Rock fragments far removed from their bedrock source have long been used to reconstruct patterns of glacial transportation. For example, fragments of microgranite originating from the small island of Ailsa Craig off the

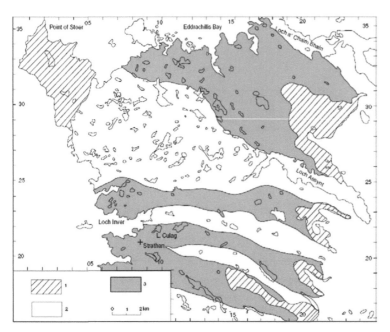

Figure 12.80 The boulder trains of Assynt, north-west Scotland, documenting glacier flow from east to west. Torridonian sandstone erratics are found in discrete plumes or trains emanating from the down-ice flanks of Torridonian sandstone rock outcrops. (1) Torridonian sandstone outcrops. (2) Areas devoid of Torridonian sandstone erratic boulders. (3) Areas where Torridonian sandstone erratic boulders have been found (Lawson, 1995; reproduced with permission of the Quaternary Research Association).

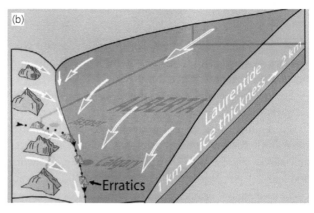

Figure 12.81 The Foothills Erratics Train, Alberta, Canada. (a) The Okotoks erratic near Calgary, southern Alberta. This huge quartzite boulder is part of the erratics train and attests to a transport distance of more than 375 km. (D.J.A. Evans) (b) Schematic reconstruction of the transport pathway of the erratics train at the suture zone of Laurentide and Cordilleran ice (Reproduced with permission of the Geological Survey of Canada).

west coast of Scotland occur in the tills surrounding the Irish Sea Basin, documenting glacier ice flow southwards from Scotland towards Ireland, Wales and west-central England (Sissons, 1967; Walden et al., 1992). Various exotic rock fragments originating from Scandinavia occur in the tills of eastern England (Ehlers and Gibbard, 1991), northern Germany, Denmark and the Netherlands (Overweel, 1977). Other erratic trains have an immediately obvious provenance where they emanate directly from their source outcrop, such as the Assynt sandstone boulder trains of north-west Scotland, which extend down-ice from mountains of Torridonian Sandstone (fig. 12.80; Lawson, 1990), or the 'omars' that were dispersed by the Laurentide Ice Sheet from the Belcher Islands in Hudson Bay (Prest et al., 2000). Probably the largest glacially transported blocks are the quartzites of the *Foothills erratics train*, which lie on the Cretaceous bedrock prairie of southern Alberta, more

than 375 km from their source in Jasper National Park (fig. 12.81). The largest block is estimated to weigh approximately 16,000 tons.

The distribution patterns of erratics also allow us to make reconstructions of former ice dispersal in complex ice sheets, often in the absence of other evidence. For example, the persistence of small upland ice dispersal centres in large ice sheets has been widely demonstrated. It has long been understood that the Cross Fell plateau in the English north Pennines hosted its own ice dispersal centre during the last glaciation, explaining the exclusion from the upper Wear and Tees valleys of far-travelled Scottish and Lake District erratics, which were otherwise emplaced to the north in Tyne Gap and to the south in Stainmore, by regional ice flowing across the Pennines and out to the eastern English coast (Mitchell, 2007). Similarly, in the Canadian High Arctic, Ó Cofaigh et al. (2000) have demonstrated that local plateau ice fields

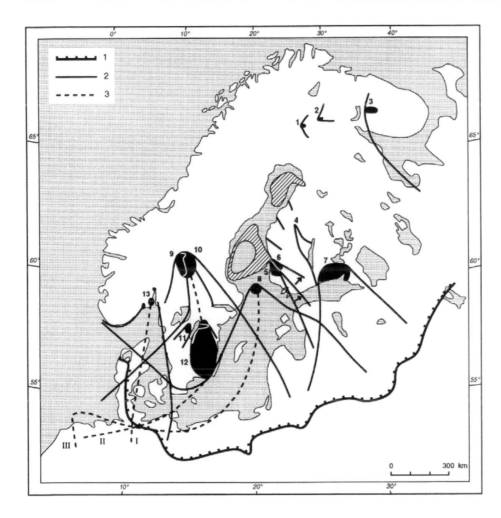

Figure 12.82 Glacial erratic indicator fans produced by the Scandinavian Ice Sheet based on various sources. (1) Outer limit of Weichselian glaciation. (2) Lateral limit of fan. (3) Path of dispersal. Source areas: (1) Jatulian sandstone and conglomerate; (2) Nattanen granite; (3) Umptek and Lujarv-Urt nepheline syenite; (4) Lappajarvi impactite; (5) Vehmaa and Laitila rapakivi granite; (6) Jotnian sandstone (Satakunta); (7) Viipuri rapakivi granite; (8) Aland rapakivi granite; (9) Jotnian (Dala) sandstone; (10) Dala porphyries; (11) Cambro-Silurian limestone; (12) Smaland granite; (13) Oslo rhomb porphyries and larvikite (Donner, 1995; reproduced with permission of Cambridge University Press).

persisted throughout regional ice sheet glaciation, because regionally transported granite erratics are found only in major valleys and fjords, and not on adjacent glaciated uplands.

Erratics and erratic trains, therefore, can be used to reconstruct former ice transportation pathways, although it should be remembered that debris transport histories may involve several glacial cycles during which the ice divides and ice flow vectors may shift dramatically. Nonetheless, the distribution of erratics and fine-grained components of till provide glacial geologists with a powerful tool for reconstructing the patterns and history of ice dispersal in studies of ice sheet dynamics (Boulton, 1996a; Punkari, 1997a; Kjær et al., 2003a). Additionally, the study of dispersal patterns has great practical applications for mineral prospecting in glaciated terrains (e.g. Shilts, 1982; DiLabio and Coker, 1989; Kujansuu and Saarnisto, 1990).

Indicator erratics are those for which a definite source area is known, and *indicator fans* are regional fields over which the erratics from particular source rocks have been dispersed by glacier flow (fig. 12.82). Indicator fans encompass the range of transport directions produced by shifting ice divides and dispersal centres, and can be

identified using a range of sediment sizes, from erratic blocks down to the fine-grained matrix of tills. The concentrations of indicator erratics vary systematically along ice flow lines. Within indicator outcrops, concentrations increase rapidly downglacier, reflecting the addition of new material from the glacier bed, but concentrations drop off rapidly down-ice of the outcrop margin. The transport distances of the majority of indicator erratics are relatively short. A good index of transport distance is the *half-distance value*, or the distance from the nearest possible source to the point at which the frequency of indicator erratics is half of its maximum. The half-distance value varies with the material resistance to abrasion and crushing and the depth to the source rock beneath superficial sediment (Bouchard and Salonen, 1990). Salonen (1986) found that the average half-distances for particles in tills in Finland are only 5 km. Very far-travelled erratics (up to 1200 km in northern Europe) may have been transported during several glacial cycles, and part of their journey may have been by iceberg rafting (e.g. Overweel, 1977).

This simple picture may be complicated by shifting patterns of erosion and sediment accumulation during glacier advance and retreat, and in some cases the peak in

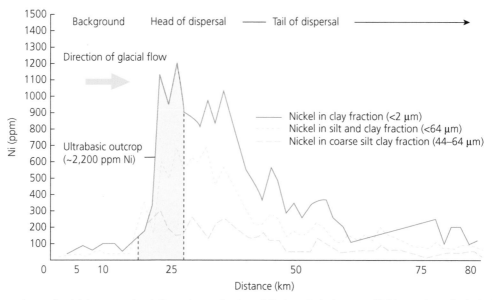

Figure 12.83 Dispersal curves for nickel concentrations in fine and coarse fractions of tills downglacier from an ophiolitic complex at Thetford Mines, Quebec, Canada. Nickel is particularly rich in the clay fraction because of the dominance of nickel-rich serpentine, which is preferentially reduced to clay sizes (Adapted from Shilts, 1993).

erratic concentration may be displaced down-ice from the boundary of the source outcrop (Boulton, 1996a). The up-ice and down-ice limits of an indicator plume are known as the *head* and *tail*, respectively. Debris dispersal by modern glaciers has been studied by DiLabio and Shilts (1979), who demonstrated that head and tail zones are identifiable in the lateral moraines of glaciers crossing from one bedrock type to another. The distance over which an indicator is identifiable, even in a till sheet of one glacial advance, will vary according to its comminution rate and its dilution rate down the flow line. This has been numerically modelled by Larson and Mooers (2004), who pointed out that the dilution rate is a function of erosivity of the glacier substrate over which the indicator material is being transported. Substrate erosivity impacts on indicator dilution, because it relates to the amount of material eroded along a glacier flow line, and, therefore, the volume of subglacial material added to the traction load downflow of the indicator source outcrop. Where glaciers advance into soft-bedded lowlands, they may incorporate, deform and stack thick sequences of pre-existing sediment, including older tills, thereby producing repetition in geochemical signatures in vertical sections (e.g. Boston et al., in prep.).

The characteristics of indicator plumes are demonstrated by figure 12.83, which shows variations in nickel concentrations in till in part of Quebec, Canada (Shilts, 1993). The dispersal curves show only background levels of nickel in the area located up-ice from the source ultrabasic bedrock outcrop, in contrast to the higher concentrations in the 'head' and 'tail' zones down-ice from the outcrop. The concentrations of nickel vary with grain size as a result of differential abrasion and crushing during

subglacial transport (section 9.5.1; Dreimanis and Vagners, 1971; Haldorsen, 1983). Each mineral has a different mode at which its chemical signature is strongest (*chemical partitioning*), and this must be taken into account when analysing till samples for their geochemical signatures. Anomalous geochemical signals can be detected in till samples due to weathering and pedogenesis, the depth of which is dictated by such factors as the water table and grain size. Therefore, till samples for geochemical analysis and indicator tracing must be taken from unweathered horizons in order to avoid post-depositional alterations, such as the destruction of labile ore minerals and their depletion in till samples from above the water table.

Vertical changes in till composition at a particular site may result from:

(1) the complete erosion or burial of some source outcrops;

(2) a shift in ice flow direction so that material is transported from different source areas;

(3) time-dependent patterns of erratic transport along a flow line.

Boulton (1996b) has shown that till composition can reflect the length of time available to transport debris from distant locations. Theory suggests that till composition should be dominated by local lithologies near the base and shift to predominantly far-travelled lithologies at the top, reflecting the passage of successive waves of debris from further and further upglacier. Boulton (1996b) applied his sediment transport model to till sequences in Illinois described by Johnson and Hansel (1990), and showed that the observed sequence could

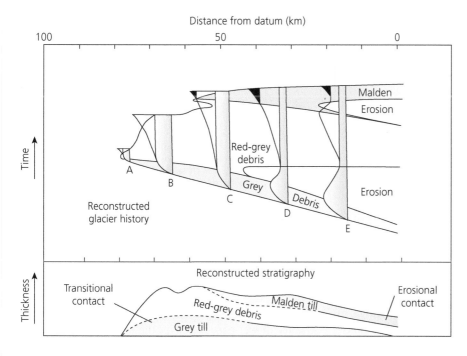

Figure 12.84 Schematic diagram showing how the tripartite till stratigraphy observed by Johnson and Hansel (1990) could have originated during a single glacial cycle (Boulton, 1996b; reproduced with permission of the International Glaciological Society).

have been created during a single glacial cycle, during which progressively more distant debris was delivered to the marginal zone (fig. 12.84). Dispersal patterns beneath former ice divides tend to be complex, and subglacial material can be transported in several different directions as the result of ice divide migration. The resulting dispersal pattern consists of multiple plumes emanating from a source outcrop, termed an *amoeboid pattern* by Shilts (1993).

Indicator fans or *dispersal fans* provide important clues to the location of mineral outcrops or ore bodies in glaciated terrain, where elongate plumes of mineral-enriched tills extend down-ice from source outcrops (Shilts, 1993). The mineral content of the plumes is known as a mineral *float*. Ore enrichment in tills can be assessed by pebble counts and/or geochemical and mineralogical analyses of the fine-grained matrix. Because ore bodies are often hidden by till, the geochemical or mineralogical characteristics of till samples, collected on a regular grid and then contoured, often reveal plume-shaped areas of metal enrichment or dispersal trains. The largest dispersal trains can be hundreds of kilometres long, but derived from very small ore bodies; they are hundreds to thousands of times larger than their source area, thus representing a much larger exploration target for prospectors than the ore body itself. Some dispersal trains can be identified visually on the basis of their colour, such as the carbonate till plumes that stretch across Prince of Wales Island, Somerset Island and Boothia Peninsula in the central Canadian Arctic archipelago (figure 12.64; Dyke et al., 1992) and on Baffin Island (Dredge, 2000; Dyke, 2008).

Dyke and Morris (1988) suggested a twofold classification scheme for erratic trains based on examples in Canada (fig. 12.85).

(1) *The Dubawnt type* is named after the Dubawnt sandstone in central Keewatin, which is a relatively restricted source outcrop from which material was dispersed as a plume over the surrounding bedrock by ice moving at a uniform rate over an entire region.

(2) *The Boothia type*, based on examples on the Boothia Peninsula, is characterized by debris plumes that extend from small parts of large source areas by zones of more rapid ice flow (ice streams) within the ice sheet. Complex ice flow histories can be manifest in palimpsest dispersal trains, whereby initial elongate trains are smeared in different directions by later changes in ice flow (Parent et al., 1996).

Changes in the lithological composition of glacifluvial materials have also been used to determine transport patterns in glacial depositional systems (e.g. Ryder, 1995). Studies on the lithology of materials in Scandinavian eskers have revealed that transport distances from source areas vary according to the comminution rate of the different rock types. In eskers, the increase in clasts of a particular rock type over its outcrop area begins further downflow from the up-ice margin of the outcrop than in tills. In addition, the concentration of rocks from such sources is far higher in esker sediments than in tills (Lillieskold, 1990). In glacifluvial systems, lithologies that disintegrate easily to form smaller grain sizes will travel further.

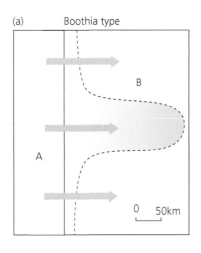

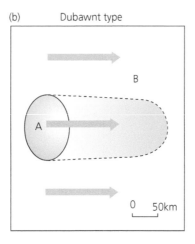

Figure 12.85 The twofold classification scheme for dispersal trains, based on the shape of the source outcrop and ice flow characteristics. The Boothia type is formed under ice streams and the Dubawnt type under sustained regional ice flow. Two different rock types are marked A and B; ice flow direction is indicated by arrows; and the dispersal of debris from rock type A is shaded orange (Adapted from Dyke and Morris, 1988).

12.5 ICE SHEET MARGINS

Ice-marginal environments migrate through time as ice sheets expand and contract. During ice sheet advance, ice-marginal sediments and landforms are overridden, and overprinted or destroyed by subglacial processes. Conversely, during ice sheet retreat, former subglacial environments are overprinted by ice-marginal processes. Ice-marginal sediments and landforms, therefore, represent the most recent and commonly best preserved component of glacial landscapes, and were collectively termed the *deglacial envelope* by Kleman et al. (2006). The deglacial envelope can consist of terrestrial, glacilacustrine or glacimarine associations, and can be a rich source of information on ice sheet response to environmental change.

In the following sections, we consider ice-marginal and sub-marginal sediments and landforms at the landscape scale, beginning with terrestrial ice margins, then turning to glacilacustrine and glacimarine margins.

12.5.1 TERRESTRIAL ICE SHEET MARGINS

Marginal lobes

In North America and Eurasia, marginal lobes of the great Quaternary ice sheets are demarcated by complex suites of sediments and landforms. Ice sheet advance over an area can be recorded by distinctive landform–sediment zones, particularly in regions underlain by weak rocks or unconsolidated sediments (fig. 12.86a). For example, in the northern Great Plains of North America, arcuate belts of hummocky ice-thrust terrain and excavational basins commonly lie at the distal edges of drumlin fields. Some of the thrust belts have a subdued, streamlined appearance, and contain cupola hills and drumlins with low length:width ratios (fig. 11.67). The excavational basins and thrust ridges document glacitectonic disturbance near the ice margin, probably encouraged by cold-based conditions below the snout or the impact of surging (Clayton

and Moran, 1974; Bluemle and Clayton, 1984; Bluemle et al., 1993; Evans et al., 1999b, 2008); and the streamlined terrain and drumlin belts record the subsequent advance of wet-based ice over the former marginal zone. In regions where subglacial deformation was intense and prolonged, glacitectonized thrust masses may lay buried, their upper surfaces intensely sheared and incorporated within subglacial sediments (see section 11.3.2). The amount of subglacial modification commonly decreases down-ice, and landform assemblages exhibit a continuum from drumlins with high length:width ratios, through drumlins with low length:width ratios and cupola hills, to largely unmodified thrust masses. Such landform associations apparently represent progressively less effective or short-lived subglacial moulding towards the ice limit (fig. 11.17; Boyce and Eyles, 1991).

The depositional and erosional record shows that ice-marginal lobes varied widely in their characteristics and behaviour, depending on climatic conditions, local topography and geology, basal thermal regime, ice dynamics and other factors (e.g. Mooers, 1990b; Colgan, 1999; Socha et al., 1999; Cutler et al., 2000; Colgan et al., 2003; Winguth et al., 2004). The influence of substrate type on glacitectonic landforms in the north European Plains has been considered by van der Wateren (2003; fig. 12.87). In areas underlain by weak Mesozoic-Cenozoic sedimentary strata, there are substantial thrust moraines, the size of which decreases towards the shallowest parts of the sedimentary basin. At the very edge of the basin in the adjacent Variscan highlands, thrust moraines are not found, and are replaced by dump moraines and ice contact fans and ramps. These patterns partly reflect systematic variations in subglacial groundwater characteristics on different substrata.

The thermal regime process–form continuum identified in section 12.3.1 can be applied to the terrestrial margins of former ice sheets, as well as those of smaller ice masses (Evans, 2009), so that the spatial distributions of push and controlled moraines can be used as indicators of

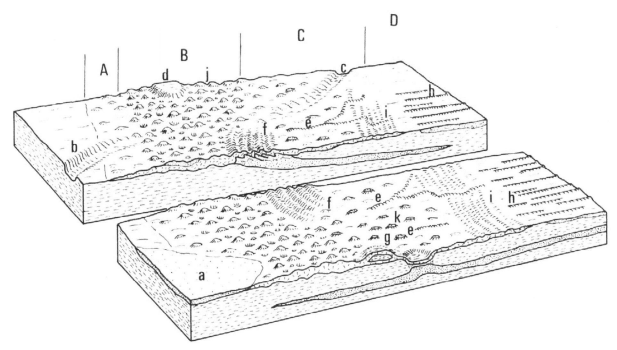

Figure 12.86 Zonation of glacial landforms associated with ice sheet occupancy. (a) Idealized zonation of glacial landforms and sediments on the prairies of North Dakota, USA. (A) Proglacial suite. (B) Supraglacial suite. (C) Transitional (sub-marginal suite). (D) Subglacial suite. (a) Proglacial lake. (b) Proglacial meltwater channel. (c) Subglacial meltwater channel or tunnel valley. (d) Ice-walled lake plain. (e) Esker. (f) Overridden transverse thrust moraines (cupola hills). (g) Prairie mound. (h) Flutings. (i) Recessional moraines. (j) Hummocky moraine. (k) Isolated kames (modified from Clayton and Moran, 1974). (b) Preferential substrate erosion by ice sheet margins impinging on soft substrates. (A) Schematic diagram to show the production of hummocky moraine after soft bedrock is detached and incorporated into basal ice at the junction of soft and hard substrates. (B) The puncturing of the Niagara Escarpment, Ontario, Canada by ice to produce through valleys or small troughs blanketed by moraine. (C) The southward displacement of the chalk escarpment in south-east England by glacial erosion (Eyles and Menzies, 1983).

temporal changes in sub-marginal thermal conditions. For example, broad belts of controlled moraine near the southern margin of the Wisconsinan Laurentide Ice Sheet have been interpreted as the product of compressive flow across a frozen toe zone (Clayton and Moran, 1974; Ham and Attig, 1996; Clayton et al., 2001). The controlled moraine belts commonly give way to push moraines in the direction of ice recession, suggestive of a shift to more temperate sub-marginal conditions through time.

Similarly, landform zonation in western Canada has been explained by D. J. A. Evans (2009) as a product of changing thermal regimes in the receding south-west margin of the Laurentide Ice Sheet. Figure 12.88 depicts part of the McGregor Lake moraine belt in Alberta. The moraine belt contains three partially overprinted suites of moraines and associated landforms, indicating that readvances took place during overall ice sheet recession. The innermost of these suites contains flutings (fig. 12.88a) and can be traced westwards to join the prominent Frank Lake push moraines described by Evans (2003b) and Evans et al. (1999b, 2008). The outermost suite is characterized by less continuous moraine ridges, numerous kettle holes and no flutings (fig. 12.88b). The terrain lying between the innermost and outermost suites contains some larger and more continuous moraine ridges, and clusters of closely spaced kettle holes. Based

on these characteristics, Evans (2009) interpreted the outermost suite as the remnants of controlled moraine, produced at a time when the ice sheet margin was cold-based. The innermost suite appears to consist of recessional push moraines constructed at a predominantly warm-based ice margin. It is possible that the intermediate zone of moraines relates to a period when ice-marginal conditions were changing from polythermal to temperate, and patchy permafrost led to the construction of larger push moraines in some areas.

It should be borne in mind, however, that some landforms can be produced by more than one process (equifinality), and that care must be exercised when considering the palaeo-environmental significance of the geologic record at the largest scales. For example, kettled, hummocky terrain underlain by stratified diamictons in the Valparaiso Moraine, Lake Michigan, has been interpreted as evidence of melt-out from thick sequences of debris-laden, stratified basal ice formed by glacio-hydraulic supercooling below the margin of the Laurentide Ice Sheet (Larson et al., 2006). This interpretation is supported by the fact that the local topographic conditions would have been suitable for supercooling (i.e. a steep adverse slope would have encouraged rapid depressurization of subglacial meltwater). Similar sediment–landform assemblages, however, can also form by

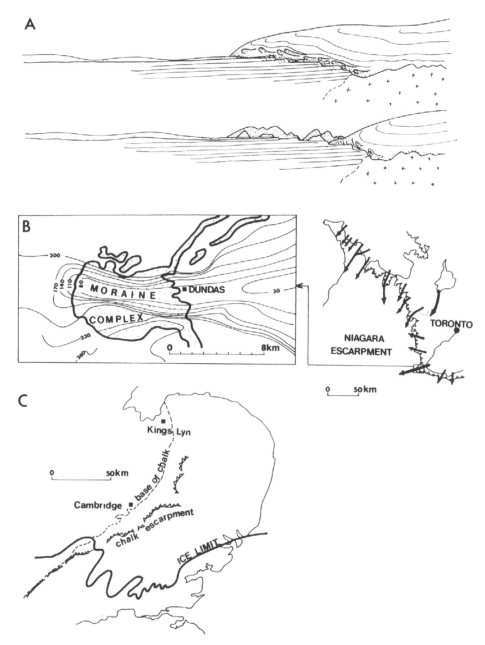

Figure 12.86 (*Continued*)

the elevation of basal debris at polythermal glacier margins, and thrusting of sediment and ice at the margins of surging glaciers, and it can be a challenging task to identify evidence uniquely diagnostic of any one process regime.

Some marginal lobes had low surface gradients, indicative of low basal drag; and subglacial sediment–landform associations located upglacier strongly suggest that they were the terrestrial termini of ice streams. All modern ice streams terminate in the ocean, so the geologic record of terrestrially terminating ice streams provides important information on the possible range of ice sheet behaviour (Patterson, 1997, 1998; Kjær et al., 2003a; Jennings, 2006; Evans et al., 2008; fig. 12.59). The role of a soft

bed in the dynamics of regional glacier flow has been demonstrated for the Michigan Lobe of the Laurentide Ice Sheet by Breemer et al. (2002). They argued that aquifers beneath the ice lobe were not capable of draining the likely volume of basal meltwater, and that the excess flowed either in the subglacial till or as a water film at the ice–bed interface. In both cases, this will have resulted in low basal drag and fast flow (sections 4.4, 4.5). In some cases, terrestrial palaeo-ice stream margins underwent rapid advances which appear to have been unrelated to regional climatic forcing. It is possible that some of these advances were surges, because elements of the surging glacier landsystem, such as crevasse-squeeze ridges and proglacial thrust masses, have been identified on the beds

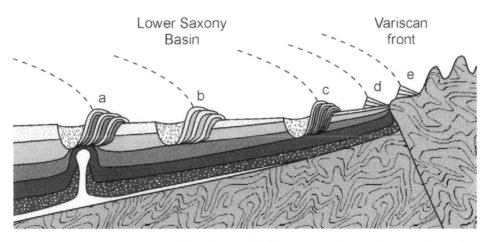

Figure 12.87 Schematic cross section with exaggerated scale through the bedrock and thrust moraines of the north European plains and Variscan highlands, showing the relationship between substrate and end moraine type/size produced at successive ice margins. Thrust moraines occur where there is a suitable shallow décollement in fine-grained clastic or lignite layers. (a) Salt diapir brings up layers that are too deep elsewhere to act as a décollement. (b) Shallowing of the basin provides a décollement in stratigraphically higher layers, including Pleistocene materials. (c) Near the basin margin, stratigraphically deeper layers are incorporated in thrust moraines. (d, e) Dump moraines and ice contact fans and ramps are produced where a suitable décollement is lacking. Note that thrust moraine size decreases towards the basin margins due to the shallowing of deformable substrate (van der Wateren, 2003)

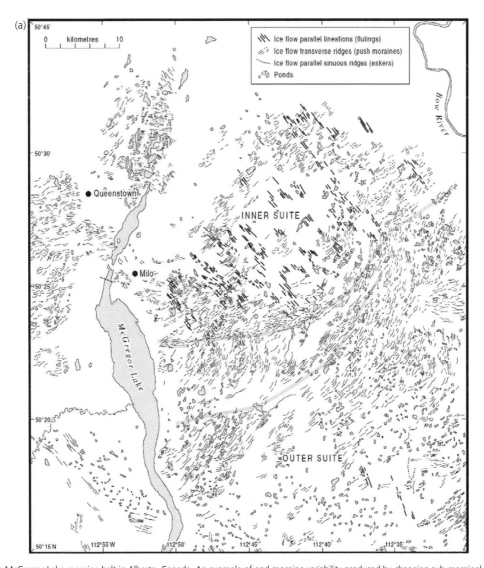

Figure 12.88 The McGregor Lake moraine belt in Alberta, Canada. An example of end moraine variability produced by changing sub-marginal thermal regimes during ice sheet recession. (a) Glacial geomorphological map of the area showing three partially overprinted suites of moraines and associated landforms (after Evans et al., 2006a). (b) Aerial photograph extract of the outermost moraine suite, showing less continuous moraine ridges, numerous kettle holes and no flutings in the lower right half of the image. The upper left half of the image shows part of the intermediate moraine suite and its larger and more continuous moraine ridges and lack of kettle holes (Evans, 2009)

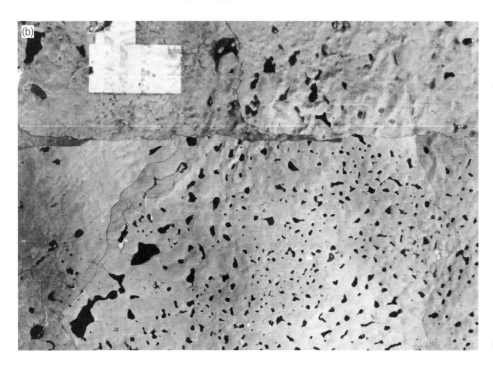

(b)

Figure 12.88 *(Continued)*

of terrestrially terminating palaeo-ice streams in western Canada (Evans et al., 1999b, 2008; Ó Cofaigh et al., in press).

Colgan et al. (2003) identified three characteristic landform–sediment assemblages formed by the lobes of the southern Laurentide Ice Sheet, interpreted as evidence of changing thermal regime and ice dynamics. *Landsystem A* comprises fluted till plains and low-relief push moraines (e.g. Ham and Attig, 2001) indicative of active ice recession. In *Landsystem B* (fig. 12.89a), drumlinized zones grade into moderate- to high-relief moraines and ice-walled lake plains, indicative of a polythermal ice sheet margin where sliding and deforming bed processes give way to a marginal frozen toe zone. Finally, *Landsystem C* includes many of the diagnostic landforms and sediments of surging activity (section 12.3.8). The case study of the Green Bay Lobe of the southern Laurentide Ice Sheet margin in Wisconsin (fig. 12.89b) shows how this glacial landform record can be combined with a chronology of ice-marginal oscillations and palaeo-permafrost evidence to numerically model the sequential changes in thermal conditions, balance velocities and ice surface profiles through time (Colgan et al., 2003; Winguth et al., 2004). The model shows that as marginal and sub-marginal permafrost degraded during the early stages of deglaciation, increasingly abundant subglacial meltwater resulted in progressively more significant amounts of basal sliding and substrate streamlining/drumlinization, eventually giving way to glacier surging.

Some marginal landform–sediment assemblages are associated with ice stream flow onshore from marine basins. For example, the Irish Sea Ice Stream of the BIIS

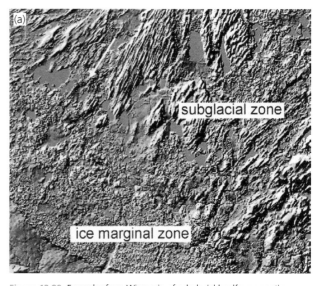

(a)

subglacial zone

ice marginal zone

Figure 12.89 Examples from Wisconsin of subglacial landform zonation beneath the southern lobes of the Laurentide Ice Sheet. (a) Shaded relief image of the sub-marginal imprint of polythermal conditions beneath the Langlade Lobe at the LGM limit, showing the large end moraine belt and associated ice-walled lake plains indicative of a frozen ice margin, and the drumlinized terrain of the warm-based subglacial zone. (b) Map of the glacial and permafrost landforms of the Green Bay Lobe, showing the progressive change from landforms indicative of a frozen margin in the south, through landforms of active temperate recession, to surge imprints in the north (Colgan et al., 2003). (c) Conceptual models of spatial and temporal variations in glacial landform development based on evidence from central Minnesota, USA, and documenting the impacts of the Rainy and Superior lobes of the southern Laurentide Ice Sheet. (A) The Rainy Lobe situation, where conditions varied little during ice recession. Zone 1 is an area of debris-covered ice due to proficient englacial thrusting and stacking of debris-rich ice sequences. Zone 2 is an area of basal freeze-on. Zone 3 is the thawed bed area. (B) The Superior Lobe situation, where suites of landforms indicate pronounced variations in landform development during ice recession. Zone 4 is an ice marginal area with sediment cover of variable thickness. A frozen toe was present at the maximum limit, but absent during recession. Zone 5 is an area of tunnel valley and esker formation, while drumlins were forming in Zone 6 (Mooers, 1990b; reproduced with permission of the Geological Society of America)

(b)

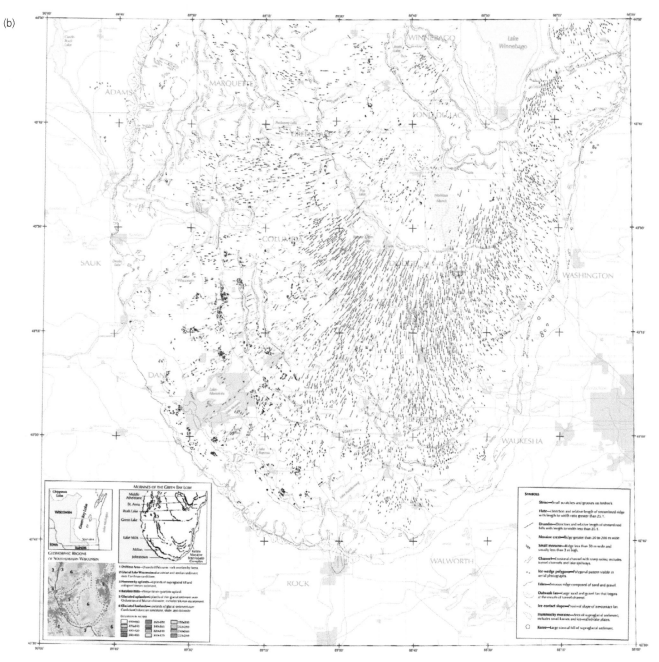

Figure 12.89 (*Continued*)

repeatedly emplaced reworked marine sediments around the basin margins (Evans and Ó Cofaigh, 2003; Thomas et al., 2004; Roberts et al., 2007), creating the misleading impression of formerly high sea levels (cf. Eyles and McCabe, 1989; McCarroll, 2001).

Oscillations of ice sheet margins are commonly recorded in complex glacifluvial depositional sequences. This is well illustrated by the deposits associated with the advance of the mid-Pleistocene Anglian Ice Sheet in north Norfolk, England, where a series of variably deformed basins of glacifluvial outwash (Runton Sand and Gravel Member) were laid down between proglacial thrust ridges that developed in front of the advancing ice margin

(fig. 12.90a; Phillips et al., 2008). The complexity of deformation structures within the outwash are explained as a product of the growth of 'till prisms', which were constructed by proglacial thrusting beneath a sandur plain and eventually punctured the sandur surface in some places, to produce sub-basins of ice margin parallel deposition (fig. 12.90b). The ice-marginal outwash therefore records pre-, syn- and post-tectonic sedimentation. The highly variable styles and intensities of proglacial and subglacial deformation in the exposures in Quaternary deposits along this coast are related to the juxtaposition of a wide range of sediment rheologies and their location relative to the ice margin.

(c)

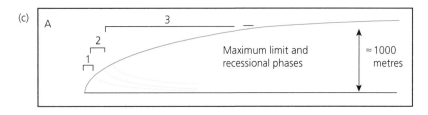

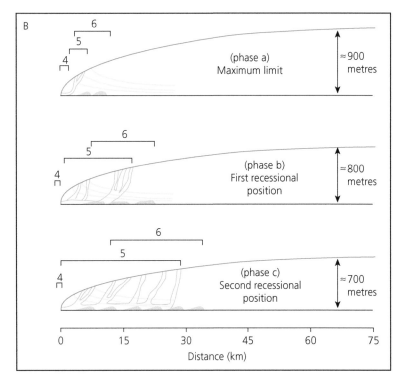

Figure 12.89 (*Continued*)

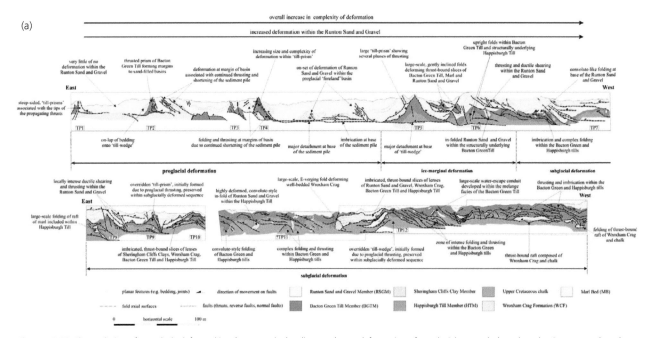

Figure 12.90 The evolution of complexly deformed ice sheet marginal sediments due to deformation of proglacial outwash, based on the Quaternary deposits of north Norfolk, England. (a) Section sketch of the coastal cliffs between West Runton and Sheringham, north Norfolk, showing the variation in style and intensity of deformation from east (ice-distal) to west (ice-proximal) in the tills and glacifluvial sediments, particularly the proglacial sandur deposits of the Runton Sand and Gravel Member. (b) Schematic illustration of the ice-marginal depositional environment in which the Runton Sand and Gravel Member was deposited in sub-basins between proglacial thrust moraine ridges (Redrawn from Phillips et al., 2008).

(b)

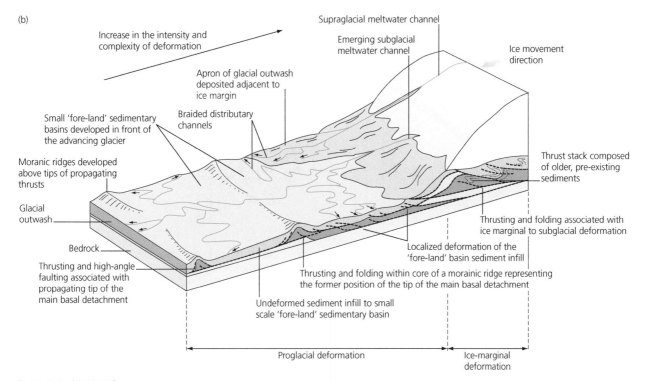

Figure 12.90 *(Continued)*

Inter-lobate landform complexes

Linear concentrations of large volumes of glacifluvial sediment have been used to demarcate the former suture zones between different flow units in ice sheets. It was pointed out in section 11.2.10 that eskers often coincide with medial moraines. Figure 12.91 illustrates how part of the Oak Ridges Moraine in Ontario, Canada, developed from a tunnel valley infill to a wide inter-lobate ridge due to deposition in a widening re-entrant in the ice sheet margin (Barnett et al., 1998; H. A. J. Russell et al., 2003). Such re-entrants probably originate as supraglacial troughs between confluent ice flow units (Huddart, 1999). The troughs then widen during ice recession and their bases can collapse into subglacial and englacial tunnels produced by the meltwater that has been concentrated at the suture zone. This can result in the burying of eskers by supraglacial trough fills, giving rise to vertical and lateral changes in sedimentary architecture from ice-walled tunnel fills (eskers) to sub-horizontally bedded, pitted outwash (section 11.4).

The glacifluvial nature of some of the largest 'inter-lobate moraines' in North America has been clearly demonstrated by Brennand and Shaw (1996) for the Harricana Moraine, and by Barnett et al. (1998), Pugin et al. (1999), Russell and Arnott (2003), Russell et al. (2003) and Sharp et al. (2007) for parts of the Oak Ridges Moraine. The Harricana Inter-lobate Moraine, formerly separating the Hudson Bay and Labrador sectors of the Laurentide Ice Sheet, stretches for 1000 km southwards from James Bay to Lake Simcoe, Ontario, and comprises a series of ridges up to 10 km wide and 100 m high. Despite the fact that the regional assemblages of glacial landforms all indicate an inter-lobate origin for the Harricana Moraine, Brennand and Shaw (1996) reject the inter-lobate hypothesis, based on their preferred interpretation of all subglacial landforms as mega-flood products (section 11.2.8). The Burntwood-Knife Inter-lobate Moraine, formerly separating the Hudson Bay and Keewatin sectors of the receding Laurentide Ice Sheet, covers a distance in excess of 500 km on the west side of Hudson Bay, and is composed of broad ridges of glacifluvial sediment and till. Individual segments are up to 12 km long, 4 km wide and 60 m high (fig. 12.92; Dyke and Dredge, 1989). The large quantities of sands and gravels comprising these moraines were deposited in a complex suite of landforms, including time-transgressive esker segments linked to proglacial lacustrine depocentres, located in a northward migrating ice-marginal re-entrant in the receding Laurentide Ice Sheet. A similar depositional environment was envisaged by Mäkinen (2003) for inter-lobate eskers in south-west Finland. The eskers comprise chains of ice contact subaqueous fans that were deposited at tunnel mouths exiting into a proglacial lake. The fans accumulated at the apices of up-ice migrating re-entrants located between coalescent ice lobes. Punkari (1997b) used these inter-lobate features to aid in the demarcation of flow sets on the bed of the Fennoscandian Ice Sheet.

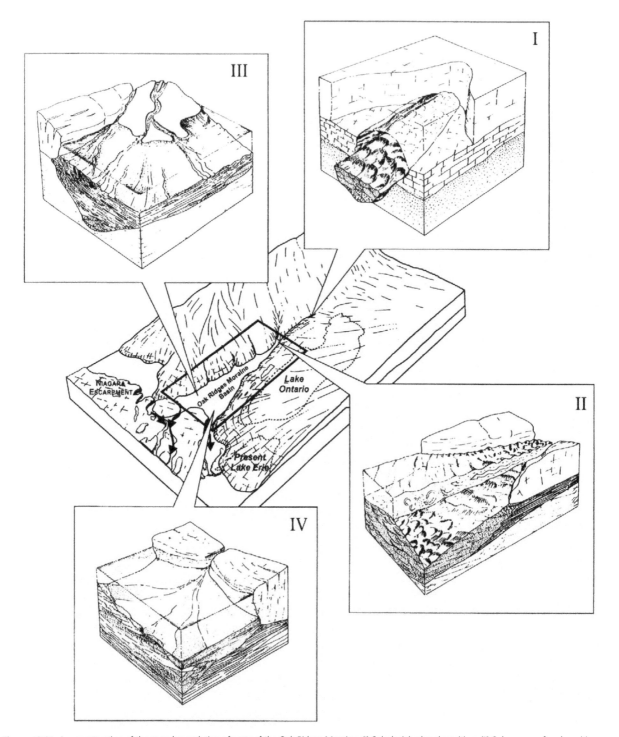

Figure 12.91 A reconstruction of the complex evolution of parts of the Oak Ridges Moraine. (I) Subglacial esker deposition. (II) Subaqueous fan deposition. (III) Fan to delta deposition. (IV) Ice-marginal fan sedimentation (Barnett et al., 1998).

The occurrence of anomalous subglacially streamlined landforms may also be related to former ice coalescence zones. For example, a mega-fluting complex at the base of the Central Alberta palaeo-ice stream in western Canada is unusual in that it is the only sharply defined fluting in a streamlined corridor that otherwise lacks sharp-crested lineations (fig. 12.93; Evans, 1996; Evans et al., 2008).

It also displays an unusual cross profile in that its summit comprises two parallel, flat-topped lineaments, separated by a central groove. The core of the feature is composed of contorted sands and gravels beneath a till carapace, which prompted Evans (1996) to propose that it is an esker that has been streamlined and grooved by fast-flowing ice. Credence to this theory and evidence that the

(a)

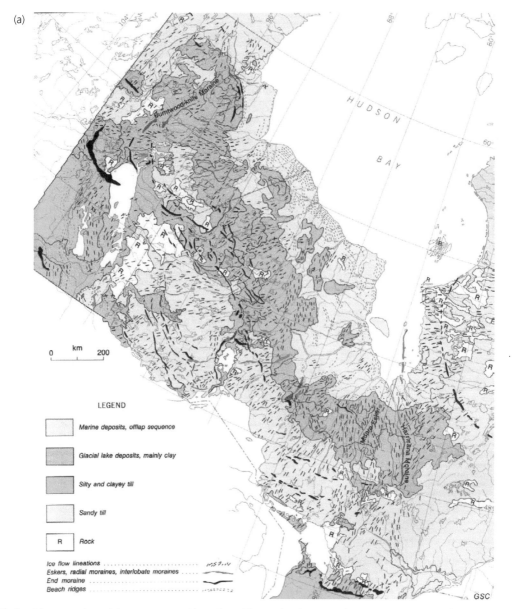

LEGEND

Marine deposits, offlap sequence

Glacial lake deposits, mainly clay

Silty and clayey till

Sandy till

R   Rock

Ice flow lineations ....
Eskers, radial moraines, interlobate moraines ......
End moraine ....
Beach ridges ....

Figure 12.92 The interlobate 'moraines' of the eastern Laurentide Ice Sheet. (a) Map of surface materials, moraines, eskers and ice flow indicators on the southwest Canadian Shield, showing the positions of the interlobate Burntwoodknife and Harricana moraines. (b) Aerial photograph stereopair of a segment of the Burntwoodknife interlobate 'moraine' near Settee Lake, northern Manitoba. The ridge is characterized by a number of attached deltas. The flow directions of the Hudson Bay and Keewatin ice bodies are indicated (Dredge and Cowan, 1989; reproduced with permission of the Geological Survey of Canada).

esker was formed as an inter-lobate feature prior to ice stream reactivation is provided by the fact that the mega-fluting changes down-ice into a complex esker network at a re-entrant in lobate morainic ridges (Evans, 2000b), and that the mega-fluting is located immediately up-ice from the ice-marginal re-entrant.

Subpolar ice sheet margins

In cold, high-latitude environments, ice sheets create similar suites of landforms to those described in section 12.3.4, but on a larger scale. Lateral meltwater channels, controlled moraine and glacitectonic landforms are common around the former margins of subpolar ice sheets,

but generally occur over larger areas and are more prominent in the landscape. The whole range of landforms has been represented in a landsystems model for northern Canada by Dyke and Evans (2003), who identified a regional zonation of landform–sediment assemblages dictated by thermal regime (fig. 12.94). The subglacial components of this landsystem are similar to those described in section 12.4, and we focus here on the marginal zones.

The outermost zone is dominated by inset sequences of lateral meltwater channels and local occurrences of ice-thrust/glacitectonic landforms. Otherwise, the landscape can bear little or no evidence of glacial modification, other

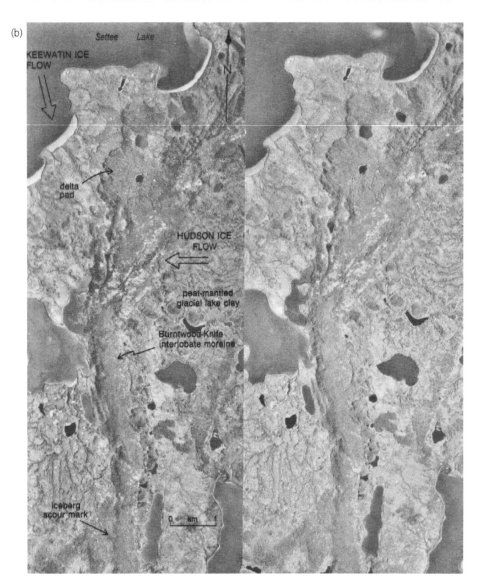

Figure 12.92 (*Continued*)

than a thin till veneer, scattered erratics or small, localized moraines. This paucity of glacial evidence in what is the most marginal glacierized terrain is as much a function of the duration of glaciation as the thermal regime. In other words it is not a prime location for the long-term modification under *average* glacial conditions (Porter, 1989). It is therefore unsurprising that palaeoglaciological reconstruction in these regions has been a controversial topic.

Inside this zone lie belts of ice-cored moraine entombed within the permafrost, but slowly undergoing the process of de-icing (Dyke and Savelle, 2000; Alexanderson et al., 2002; Henriksen et al., 2003). Close scrutiny of massive ice exposures in the permafrost has revealed that many ice bodies contain debris concentrations including large striated boulders, debris bands and folded folia, which in combination are diagnostic of glacier ice (e.g. Dyke et al., 1992; St Onge and McMartin, 1999; Worsley, 1999; Murton et al., 2004, 2005). Surface kettles on moraine surfaces have also been used to infer

glacier ice cores (Dyke and Savelle, 2000; Dyke and Hooper, 2001; Alexanderson et al., 2002; Dyke and Evans, 2003; Henriksen et al., 2003). Glacier ice becomes permafrost when surface melting generates enough supraglacial debris to equal the active layer thickness (~1 m thick). The largest ice-cored moraine belts are characterized by chaotic hummocks, but sub-parallel linear ridges impart a broad organization to the belts. This is interpreted as the manifestation of multiple ice fronts, or of large englacial structures or supraglacial debris-covered ridges parallel to the glacier margin. The supraglacial ridges are commonly arranged in nested suites lying transverse to former glacier flow. The ice-cored moraine belts are undergoing very slow melting from the surface downwards, locally enhanced by the development of flow slides (ground ice slumps) and active layer detachments that expose the ice cores. Old slump scars greatly outnumber fresh slumps, suggesting that ice core degradation was at a maximum during earliest

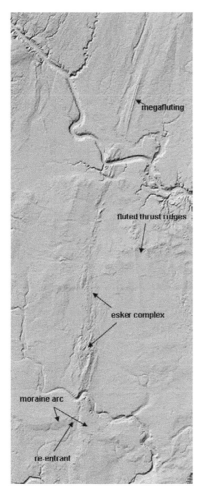

Figure 12.93 Mega-fluting complex on the bed of the Central Alberta palaeo-ice stream, western Canada. The mega-fluting terminates at a re-entrant in lobate morainic ridges immediately south of the Red Deer river, where it is replaced by an esker complex. Sampled from SRTM imagery. Width of view is approximately 45 km (Evans, 2000b; Evans et al., 2008).

post-glacial time, when moraine slopes were steepest and climate was generally warmer than at present. The survival of ice-cored terrain through interglacials can be expected to provide large, regional debris sources to advancing cold-based ice fringes during subsequent glaciations.

Similar to the landform zonation on Svalbard glacier forelands, the outer belt of ice-cored hummocks/controlled moraine lies in front of a landscape characterized by drumlinized/fluted till and esker networks. Inset sequences of ice-cored moraine belts have been interpreted as the products of incremental emplacement by actively receding frozen toe zones (Dyke and Savelle, 2000; Dyke and Evans, 2003). The abundant debris of the ice-marginal zone is likely to have been generated in the warm-based zone up-ice, so that ice-cored moraine belts are continuously draped over warm bed features as the cold-based ice migrates across the landscape during glacier recession. Variations arise where the ice margin locally recedes in contact with deep coastal waters, resulting

in the development of ice shelf moraines, grounding-line fans and subaqueous depo-centres. These marginal assemblages provide us with an invaluable chronological framework for ice recession through the radiocarbon dating of marine shells, allowing regional patterns of active ice recession to be determined (e.g. Dyke et al., 1992; Dyke and Savelle, 2000).

Ice-marginal channels can provide evidence for regional patterns of ice sheet flow and retreat (see section 8.4.4; Dyke et al., 1992; Dyke, 1999). In Britain, for example, marginal and sub-marginal channels have been employed by C. D. Clark et al. (2006), Greenwood et al. (2007) and Hughes (2008), in reconstructions of ice sheet recession (figs 12.95, 12.96). The clustering of marginal drainage features in some locations may reflect the changing nature of glacier hydrology through time (i.e. cold-based margins may have developed at different stages of recession).

The development of an ice sheet cover through a glacial cycle results in not only the migration of cold- and warm-based zones through time, but also the encroachment of warm-based regional ice sheets into areas covered by cold-based local ice caps (Dyke, 1999; England et al., 2006). This gives rise to the development of moraine systems at the junctions of the two ice masses. For example, Atkinson (2003) presented evidence for the encroachment of the warm-based Massey Sound Ice Stream on to the eastern coast of Amund Ringnes Island in the Canadian High Arctic. A coast-parallel strip of subglacial till (Ringnes Diamict), onlapping bedrock with sparse till veneer and lateral meltwater channels, demarcates the maximum incursion of warm-based ice over the area covered initially by a cold-based local ice cap. Similarly, an appreciation of the different signatures of juxtaposed warm- and cold-based ice masses has led to a new interpretation of the Keiva moraines on the Kola Peninsula, north-west Russia (fig. 12.97; Hättestrand et al., 2007). The moraines form part of the Keiva ice-marginal zone (KIZ), which also includes marginal meltwater features. On the Kola Peninsula side of the KIZ, the sparsity of subglacial landforms and localized development of lateral meltwater channels indicates the predominance of a cold-based ice cap (Hättestrand and Clark, 2006). This contrasts with the abundant subglacial bedforms and esker networks to the south of the KIZ, which document the former presence of warm-based ice over the White Sea. Drumlinization of large parts of the KIZ indicates that the moraine assemblage was formed prior to the region being overwhelmed by the northern margin of the Fennoscandian Ice Sheet.

Hättestrand and Johansen (2005) have described controlled moraines on cirque floors in Antarctica, which record ice sheet advance into mountainous areas with a sparse local ice cover. They argued that these landforms

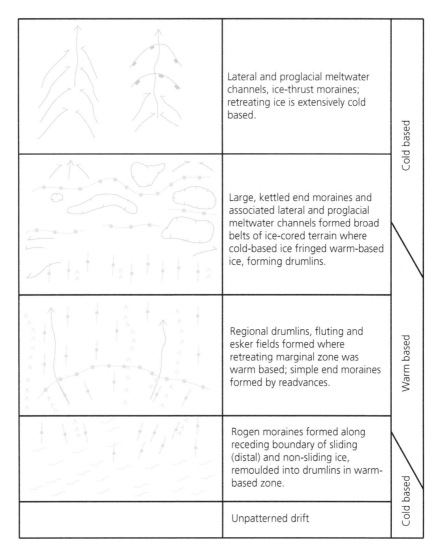

Figure 12.94 Conceptual model of the spatial organization of subpolar ice sheet marginal landforms. Meltwater channels are either proglacial (arrows) or lateral (single barbs). Glacitectonically thrust moraines are lines with boxes, and moraine ridges (controlled moraines) are lines with black dots. Enclosed depressions are large kettles forming in the controlled moraine zone. Sinuous lines of arrow heads are eskers; lines with single dots are drumlins and flutings; and the transverse short sinuous lines are ribbed terrain (Redrawn from Dyke and Evans, 2003).

provide modern analogues for Pleistocene end moraines in the Khibiny Mountains of the Kola Peninsula, which appear to have been produced by ice sheet flow against the topography, thereby backfilling the lower slopes of cirques. Such a style of regional glaciation relates to periods when mountain glaciers recede, and often even disappear altogether, due to their relatively short lag time in responding to climatic forcing. This allows ice sheet margins to occupy the lower slopes of mountain valleys and cirques, and deposit moraines that have traditionally been interpreted as the end moraines of cirque glaciers receding in the opposite direction. This has prompted Heyman and Hättestrand (2006) and Jansson and Glasser (2008) to radically reinterpret moraines traditionally regarded as the products of cirque and niche glaciers as the signatures of ice sheet advance into deglaciated mountains.

12.5.2 MARINE AND LACUSTRINE ICE SHEET MARGINS

Subaquatic depositional systems form a very important component of the record left by former ice sheets, for several reasons. First, large parts of the Pleistocene ice sheets terminated in the sea or large proglacial lakes, and large volumes of sediment were deposited in standing water at and beyond the ice margins (Powell, 2003; Teller, 2003; Vorren, 2003; fig. 12.98). Second, subaquatic sediments tend to have high preservation potential, and commonly preserve a more complete record of glaciation than terrestrial depositional systems. The continental shelves off northern Europe, North America and Antarctica are draped by thick sequences of glacimarine sediments, and much of the evidence for pre-Cenozoic glaciations consists of

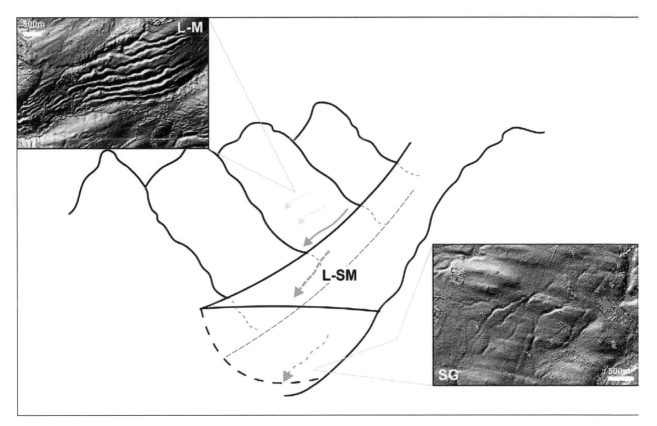

Figure 12.95 Schematic reconstruction of the occurrence of different meltwater channels in topographically constrained glacial settings, using illustrations from Britain. Lateral meltwater channels (L-M) flow subaerially along the ice margin, illustrated by examples near Blairgowrie, Scotland. Sub-marginal channels (L-SM) flow beneath the ice but are guided by the lateral margin. Subglacial channels (SG) flow along the bed and are best exemplified by hummocky long profiles, here illustrated by channels north of North Ugie Water, Scotland. Note that the DEM images have been transposed to fit the schematic diagram (Greenwood et al., 2007).

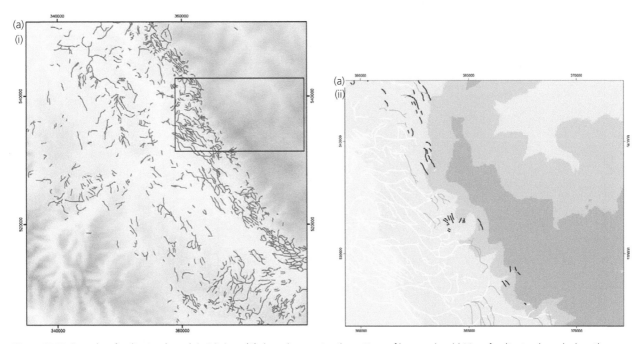

Figure 12.96 Examples of meltwater channels in Britain and their use in reconstructing patterns of ice recession. (a) Map of meltwater channels along the English Pennine Escarpment above the Vale of Eden (Hughes, 2008), with inset map of classifications from Greenwood et al. (2007). British National Grid coordinates; interval between tick marks is 20 km. In the inset map, the white channels are subglacial; grey channels are sub-marginal; and black channels are lateral. (b) Map of meltwater channels in Strathallan, near Dunblane, central Scotland, showing (i) zonation of lateral (light blue), sub-marginal (green) and subglacial (dark blue) channels. (Hughes, 2008). Topographic context is used to summarize flights of lateral channels and subglacial networks as arrows (ii) in the inferred direction of ice flow (after Hättestrand and Clark, 2006). Curvilinear chains of glacifluvial mounds extend from the downvalley ends of the sub-marginal channels. (c) Map of ice-marginal retreat pattern for the British Ice Sheet, based on lateral meltwater channels. Dotted lines connect margins inferred to be of same age based on their relationship with topography (Hughes, 2008)

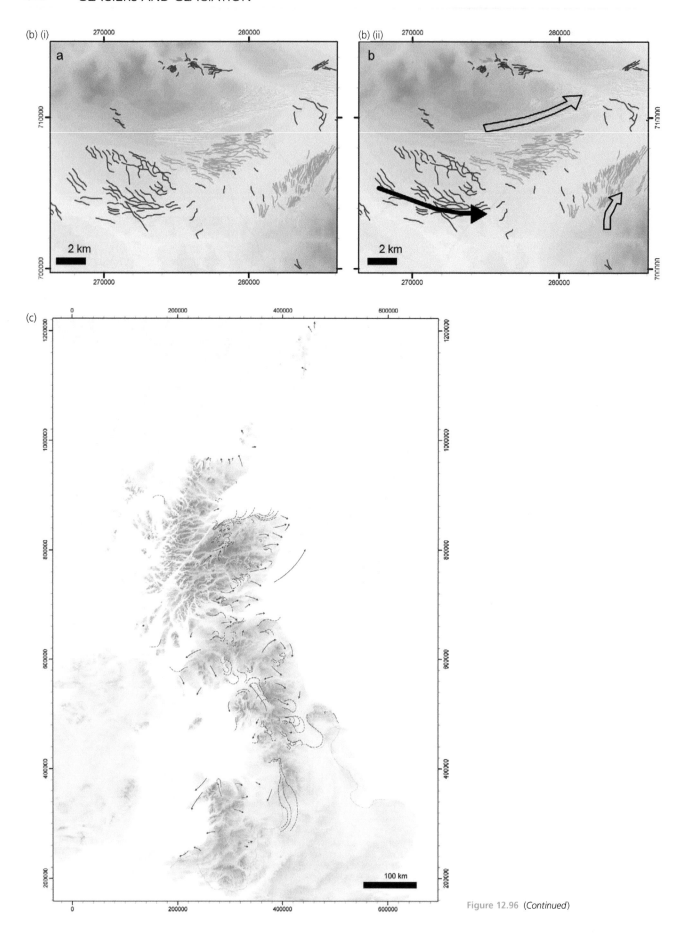

Figure 12.96 (*Continued*)

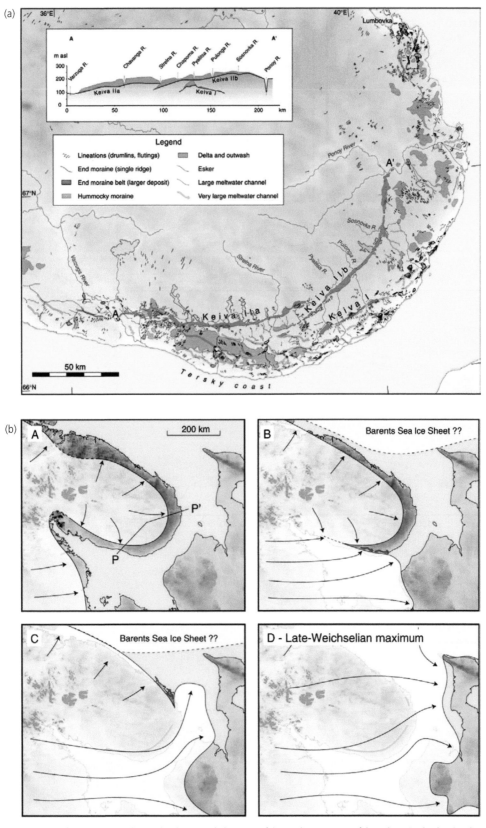

Figure 12.97 The Keiva moraines of the Kola Peninsula. (a) Glacial geomorphology map of the south-eastern part of the Kola Peninsula, showing the sparsity of subglacial landforms to the north of the Keiva ice marginal zone (KIZ) and the abundance of such landforms and meltwater features within the zone (Hättestrand and Clark, 2006). (b) Palaeoglaciology of the region, showing pre-Late Weichselian formation of the KIZ at the junction of the cold-based Kola Peninsula ice cap and the warm-based Fennoscandian ice, and later stages of overriding and patterns of recession. The red lines represent the position of the KIZ after its formation. (c) Cross profiles located at P-P¹ on (b), showing the evolution of ice cover in the region and the relationships between ice margins and KIZ production (Hättestrand et al., 2007b)

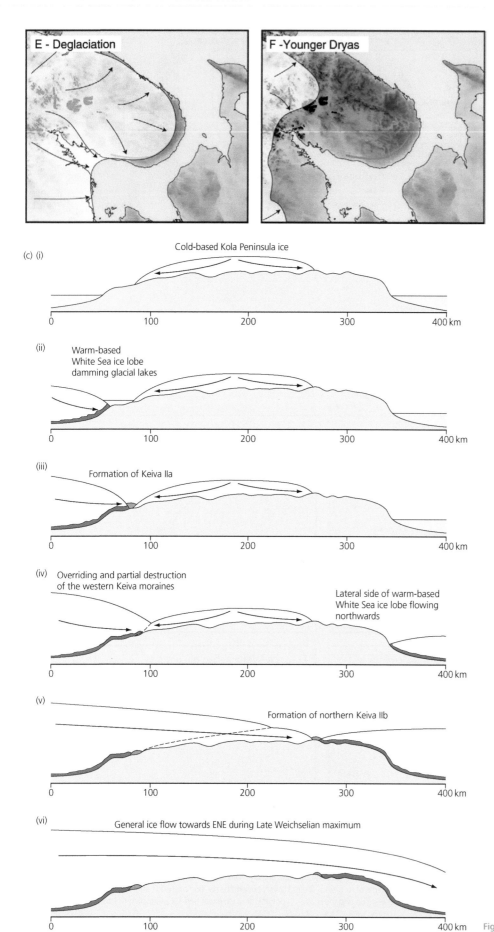

E - Deglaciation

F -Younger Dryas

(c) (i)

Cold-based Kola Peninsula ice

| 0 | 100 | 200 | 300 | 400 km |

(ii)

Warm-based
White Sea ice lobe
damming glacial lakes

| 0 | 100 | 200 | 300 | 400 km |

(iii)

Formation of Keiva IIa

| 0 | 100 | 200 | 300 | 400 km |

(iv)

Overriding and partial destruction
of the western Keiva moraines

Lateral side of warm-based
White Sea ice lobe flowing
northwards

| 0 | 100 | 200 | 300 | 400 km |

(v)

Formation of northern Keiva IIb

| 0 | 100 | 200 | 300 | 400 km |

(vi)

General ice flow towards ENE during Late Weichselian maximum

| 0 | 100 | 200 | 300 | 400 km |

Figure 12.97 (*Continued*)

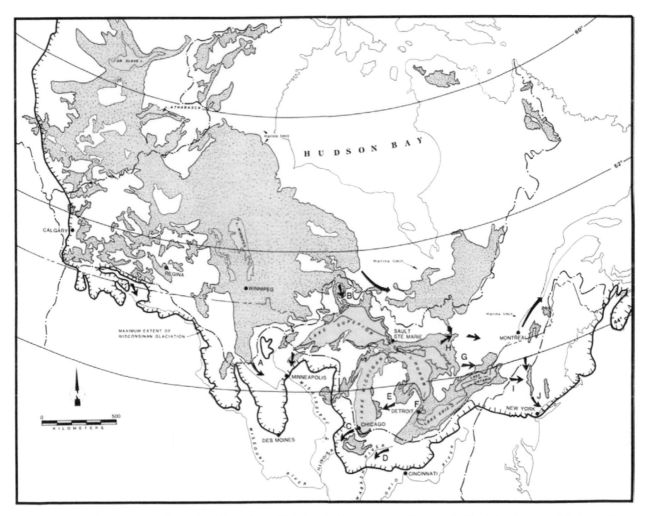

Figure 12.98 Map of the area covered by proglacial lake sediment (blue), deposited in time-transgressive proglacial lakes during the retreat of the Laurentide Ice Sheet, and the major drainage pathways (spillways) draining the lakes into the Mississippi river and Atlantic Ocean (arrows). The pathways are: (A) Minnesota river valley; (B) eastern Agassiz outlets; (C) Chicago outlet; (D) Wabash river valley; (E) Grand river valley; (F) Port Huron outlet; (G) Kirkfield (Fenelon Falls) outlet; (H) North Bay outlet; (I) Mohawk Valley; (J) Hudson Valley. Lake sediments in the St Lawrence valley and Hudson Bay are covered by later raised marine sediments and are therefore not marked (Teller, 1987; reproduced with permission of the Geological Society of America)

subaquatic facies (Eyles, 1993; Hambrey et al., 2002). Third, the study of subaquatic depositional systems yields important insights into the dynamics of former ice sheets, including the patterns of deglaciation, the location of the grounding lines of former ice streams, periods of increased meltwater runoff, and episodes of rapid calving retreat and ice-rafting (e.g. Zaragosi et al., 2001, 2006; Andrews and MacLean, 2003; Toucanne et al., 2008). In this section, we discuss the depositional record on continental shelves and in former large proglacial lakes and epicontinental seas.

Continental shelves and slopes

Patterns of sedimentation on continental shelves are dependent on a number of interrelated factors, including the extent, proximity and configuration of glacier ice, basin history and tectonic setting, bathymetry, oceanography, glacial thermal regime and sediment supply (fig. 12.8; Powell, 1984; Eyles and Lagoe, 1990; Eyles, 1993; Dowdeswell et al., 1996, 1998, 2002; O'Grady and

Syvitski, 2002). The impact of glaciation may be *direct* where the shelf has hosted a grounded glacier mass, or *indirect* where it merely received distal glacimarine sediments during a glacial cycle.

The direct impact of glaciation varies systematically with ice sheet configuration. Palaeo-ice streams leave very distinctive landform assemblages, consisting of streamlined bedrock onset zones, passing downflow into drumlins and then mega-scale glacial lineations (Wellner et al., 2001, 2006; Ottesen et al., 2005a, b, 2008b; Ottesen and Dowdeswell, 2009; section 12.4.2; fig. 12.99). Many marine-based palaeo-ice streams terminate in trough-mouth fans (section 11.6.3; fig. 11.132). Impressive examples occur at the former limits of the Barents Shelf and Fennoscandian ice sheets, including the Bear Island Fan and North Sea Fan, which record multiple episodes of sedimentation (Laberg and Vorren, 1996; Sejrup et al., 2003). The latter marks the terminus of the Norwegian Channel Ice Stream, which drained the southern interior of the Fennoscandian Ice Sheet along the south-west

(a) Inner shelf

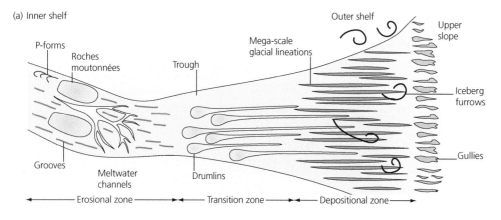

P-forms

Roches
moutonnées

Trough

Mega-scale
glacial lineations

Outer shelf

Upper
slope

Iceberg
furrows

Gullies

Grooves

Meltwater
channels

Drumlins

Erosional zone — Transition zone — Depositional zone

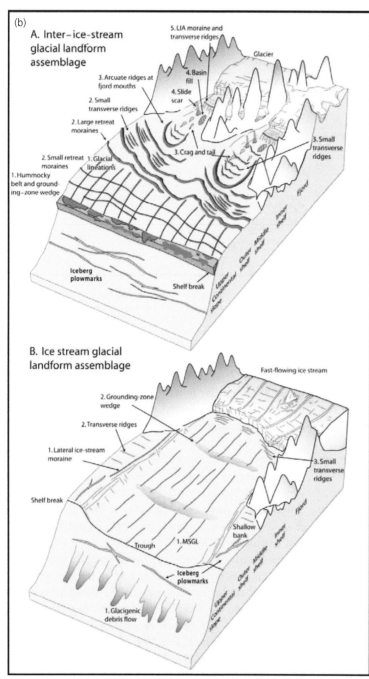

(b)

A. Inter–ice-stream
glacial landform
assemblage

5. LIA moraine and
transverse ridges

Glacier

3. Arcuate ridges at
fjord mouths

4. Basin
fill

4. Slide
scar

2. Small
transverse ridges

3. Small
transverse
ridges

2. Large retreat
moraines

3. Crag and tail

2. Small retreat
moraines

1. Glacial
lineations

1. Hummocky
belt and ground-
ing-zone wedge

Fjord

Outer Middle Inner
shelf shelf shelf

Iceberg
plowmarks

Shelf break

Upper
Continental shelf slope

B. Ice stream glacial
landform assemblage

Fast-flowing ice stream

2. Grounding-zone
wedge

2. Transverse ridges

1. Lateral ice-stream
moraine

3. Small
transverse
ridges

Shelf break

Fjord

Shallow
bank

Trough

1. MSGL

Outer Middle Inner
shelf shelf shelf

Iceberg
plowmarks

1. Glacigenic
debris flow

Upper
Continental shelf slope

Figure 12.99 (a) Palaeo-ice stream landsystem of continental
shelf settings (Redrawn from Wellner et al., 2006).
(b) Landsystems model of the offshore impacts of palaeo-ice
stream and inter-stream areas based on seabed features on
the Barents Shelf (Ottesen and Dowdeswell, 2009)

Norwegian coast during numerous glacial cycles. Subglacial till units commonly thicken towards former ice stream margins, forming grounding zone wedges (section 11.6.3; Shipp et al., 1999; Howat and Domack, 2003; Ottesen et al., 2005a, b, 2007, 2008b).

Inter-ice stream landform assemblages, formed by non-streaming parts of the marine-based Weichselian Ice Sheet off Svalbard, have been described by Ottesen and Dowdeswell (2009; fig. 12.99). Ice sheet grounding zones are marked by discontinuous belts of hummocky topography, and major fans are absent. Subtle subglacial lineations within the ice sheet limit are overprinted by numerous transverse moraines, recording slow ice retreat, punctuated by multiple stillstands. The moraines become arcuate in form where the ice retreated into individual fjords.

Glacigenic landform assemblages on Arctic and Antarctic continental shelves and fjords indicate that, at the end of the last glacial period, deglaciation of marine-based ice sheets occurred in three main ways:

(1) *rapidly*, by flotation and break-up;
(2) *episodically*, by stillstands and/or grounding events punctuating rapid retreat;
(3) *slow retreat* of grounded ice (Dowdeswell et al., 2008b; Ó Cofaigh et al., 2008; fig. 12.100).

Rapid retreat is implied by the presence of undisturbed subglacial landsystems, and an absence of recessional moraines. However, on many polar shelves, mega-scale glacial lineations are overprinted by moraines or grounding-zone wedges, showing that ice stream retreat was more often episodic than catastrophic.

The indirect impact of glaciation on continental shelves depends on the supply of suspended sediment and ice-rafted debris. In the waters around Antarctica, suspended sediment concentrations are low due to the limited discharge from glacial meltstreams from the predominantly cold-based ice sheet. In contrast, continental shelves adjacent to humid temperate glacierized regions, such as the northern Gulf of Alaska, receive substantial amounts of suspended sediment carried offshore by glacial meltwater (Carlson et al., 1990). The importance of ice-rafted debris also varies considerably, according to the number and debris content of icebergs, the subordinate nature of IRD relative to turbidites and suspension sediments in ice-proximal settings, and iceberg drift tracks – the latter being a function of regional currents (e.g. Ó Cofaigh et al., 2001; J. Evans et al., 2002). Around Antarctica, icebergs are steered by easterly winds on the inner shelf, then by westerlies as they drift northwards. Sedimentation rates, however, are low, due to generally low debris content. In the northern hemisphere, most icebergs are trapped in fjords and do not contribute substantially to sediment budgets on many

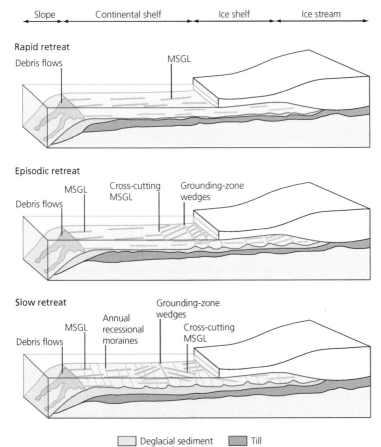

Figure 12.100 Family of landsystems models for ice stream recession from continental shelves, showing the contrasting landform assemblages relating to different recession rates (After Ó Cofaigh et al., 2008).

parts of the continental shelves (Hambrey, 1994). There are some exceptions to this, such as Baffin Bay, which receives sediment rafted by icebergs from outlet glaciers in east and west Greenland and small glaciers in the Canadian Arctic.

Further afield, iceberg rafting and the production of ice-rafted debris (IRD) in the deep oceans are now widely recognized as a reliable record of ice sheet activity in Quaternary climate reconstructions (e.g. Bond et al., 1992, 1993; Andrews, 1998, 2000; Diekmann et al., 2004; Hemming, 2007). For example, iceberg rafting was very important in the North Atlantic during several episodes during the last glaciation (*Heinrich Events*), when massive calving from expanded ice sheets transported carbonate debris to around 40° N (Bond et al., 1992; Rashid et al., 2003; Hemming, 2004). Analysis of ocean cores shows that sedimentation rates decreased towards the east and south, away from the main source of the icebergs in Hudson Strait, thereby relating Heinrich Events to oscillations of the Laurentide Ice Sheet (Andrews and Tedesco, 1992; Bond et al., 1992; MacAyeal, 1993; Andrews, 1998, 2000; Hulbe et al., 2004). The European Ice Sheets responded to Heinrich Events by producing a synchronized IRD signal, and also initiated their own IRD events (Dowdeswell et al., 1999a; Grousset et al., 2000; Scourse et al., 2000; Wilson and Austin, 2002; Lekens et al., 2005; Peck et al., 2007; Walden et al., 2007). Additionally, Zaragosi et al. (2006), Eynaud et al. (2007) and Toucanne et al. (2008) have documented turbidite pulses along the Celtic-Armorican margin of the North Atlantic that were triggered by BIIS recession. Similar pulses of sediments from north-west Laurentide Ice Sheet palaeo-ice stream oscillations have been proposed to explain IRD layers in the Arctic Ocean (Stokes et al., 2005). The role of ice streams in focusing the delivery of IRD on the Svalbard continental shelf edge has been elucidated by Dowdeswell and Elverhøi (2002).

Because turbidity currents can be initiated by meltwater draining from receding ice sheet margins, pulses of suspended sediment that they introduce to the ocean will likely be synchronized with Heinrich Events and other IRD events (Knutz et al., 2001; Hesse et al., 2004). Reworking by debris flows, turbidity currents and other gravitational processes is an important process on glacially influenced shelves, due to the low strength of rapidly deposited fine-grained sediment on the sea floor (Eyles and Eyles, 1992; Davison and Stoker, 2002; Elverhøi et al., 2002; Ó Cofaigh et al., 2002). Slope instability can be triggered by seismic shocks and stresses imposed by grounding icebergs. Sediment is also redistributed and sorted by deep ocean currents, and the resulting *contourites* cloak the bottoms of submarine channels and basins (section 11.6.3; Laberg et al., 1999; Knutz et al., 2002).

Vast areas of the high-latitude continental shelves are scoured by the keels of drifting icebergs (section 10.6.6). Iceberg scours are common to depths of around 450 m, but are rare at depths greater than 500–600 m (Dowdeswell and Bamber, 2007). At the extreme limit, scours have been observed in water depths of 850 m in the Arctic Ocean (Vogt et al., 1994). The depth of iceberg scouring clearly depends on iceberg size, which in turn is a function of ice thickness and other characteristics of the calving margin. Dowdeswell and Bamber (2007) found three peaks in the frequency distribution of the depth of iceberg scours off Antarctica. These are:

- 140–200 m, associated with small ice shelves fringing Antarctica;
- 250–300 m, from calving of tabular bergs from large ice shelves;
- 500–600 m, from calving of fast-flowing ice-sheet outlet glaciers.

Dowdeswell and Bamber concluded that ancient iceberg scours at water depths in excess of 500 m are probably indicative of calving from fast-flowing ice sheet outlet glaciers; topographically confined ice shelves fed from major interior basins; or a rapidly retreating or collapsing ice sheet.

Biological productivity is high in polar seas in areas where suspended sediment concentrations are low and sea ice breaks up during the summer, due to upwelling of nutrient-rich bottom water (Hambrey, 1994). In a zone several hundred kilometres wide around Antarctica, the principal source of sediment is the skeletons of *diatoms*, which rain to the sea floor, forming thick deposits of *siliceous ooze* (Hambrey, 1994; Abrantes, 2007). Areas of high diatom productivity tend to be less widespread in the Arctic, largely due to the extent of perennial sea ice. Other components of polar fauna, such as foraminifera, radiolaria and molluscs, also supply sediment to the sea floor, but rarely account for significant amounts of the total. Such fauna, however, are sensitive indicators of water salinity and temperature, and their fossils in sea floor sediment provide valuable information on past environmental change (Andrews et al., 1996; Lowe and Walker, 1997; Bradley, 1999; Anderson, 2007).

Tectonic setting exerts a strong influence on the style and preservation of glacigenic sediments on continental shelves (Eyles, 1993; Gipp, 1994; Hambrey, 1994; Powell et al., 2001; Hambrey et al., 2002). The tectonic setting of continental shelves can be subdivided into:

(1) passive margins;
(2) convergent margins, where oceanic crust is subducted below continental crust;
(3) rift basins, where the crust is under tension and the basin floor subsides (figs 12.1, 12.2).

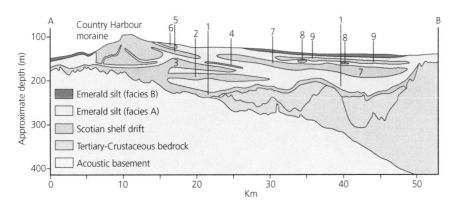

Figure 12.101 Interpretive diagram of seismic profiles from part of the Scotian Shelf, showing a major moraine complex and till tongues inter-bedded with glacimarine silts (Redrawn from Josenhans and Fader, 1989).

A well studied example of a glacially influenced *passive continental margin* is the eastern Canadian shelf, extending from the Scotian Shelf off Nova Scotia to Baffin Bay (figs 12.8, 12.101; Josenhans and Fader, 1989; Piper and Normark, 1989). The physiography of the shelf records repeated episodes of erosion under both glacial and non-glacial conditions, including Neogene fluvial incision, moderate glacial scouring and intense selective linear erosion. Ice-rafted debris first appears off eastern Canada in sediments dating to the late Pliocene (approximately 3 Ma), but despite this long record of glaciation, the quantity of Quaternary sediment is meagre, amounting to only a few tens of metres in many areas, and thickening to around 200 m in the offshore extensions of fjord basins. The limited amount of sediment reflects glacial, marine and subaerial erosion of the shelf, encouraged by low glacial sea levels, which have severely limited the survival of sediment over multiple glacial cycles. It is thought that large amounts of sediment have been evacuated from the shelf by grounded glaciers on several occasions, causing sediment to accumulate on the continental slope by a process of progradation (Eyles, 1993). According to Josenhans and Fader (1989), the uppermost glacial–postglacial sequence on the eastern Canadian Shelf thins towards the north. Off Nova Scotia and Newfoundland, former terminus positions of the Laurentide Ice Sheet are marked by large submerged moraines and thick diamicton sequences, which inter-finger with distal glacimarine deposits (fig. 12.101). Fields of De Geer moraines are common within basins inside the ice sheet limits (fig. 11.128; Todd et al., 2007). Further north, off the coast of Labrador, sediment cover is thinner and till tongues (grounding-line wedges) are not evident. This contrast is thought to result from differences in glacier thermal regime and ice dynamics; in the northern part of the shelf, deposition rates may have been limited by low sediment supply due to colder conditions, and rapid ice sheet retreat. Much of the continental shelf off eastern Canada is traversed by iceberg scour marks. Iceberg scours can be attributed to different glacial stages, providing an important source of environmental information.

For example, modern iceberg scours are abundant off Labrador, but only relict scours occur off Nova Scotia, recording a more southerly extension of iceberg drifting under full glacial conditions.

Another example of a glacially influenced passive continental margin is the Barents Shelf between Spitzbergen and mainland Scandinavia. Geophysical studies in this area have revealed striking evidence for a Weichselian marine-based ice sheet, similar in many ways to the modern West Antarctic Ice Sheet (Solheim et al., 1990; Vorren et al., 1990; Andreassen et al., 2008). The stratigraphic architecture of the glacigenic succession on the Barents Shelf is complex, reflecting shifting ice sheet configuration and activity (Vorren et al., 1990, 1998; Vorren, 2003; Andreassen et al., 2004). In places, stratified clinoforms are overlain by extensive horizontal diamicton sheets, possibly recording the progradation of grounding-line wedges and subsequent overriding by grounded ice (section 11.6.3). Elsewhere on the shelf, there is widespread evidence for subglacial erosion, including a vast areal scour zone through which glacier ice was funnelled to the western shelf edge. At its maximum, the ice sheet extended to the shelf break, where large amounts of proximal glacimarine sediment were deposited by sliding and other gravitational processes (e.g. Dowdeswell et al., 1996, 2002; Elverhøi et al., 1997; Dimakis et al., 2000; Laberg and Vorren, 2000; O'Grady and Syvitski, 2002). The thickest sediment sequences are in trough-mouth fans located on the shelf slope immediately below major troughs which acted as efficient pathways for the transport of debris via ice streams (e.g. Taylor et al., 2000, 2002b; Sejrup et al., 2003; section 11.6.3). Cross-cutting sets of mega-scale glacial lineations and moraine loops record several generations of ice streams (fig. 12.102). In many parts of the Barents Shelf, subglacial sediments and landforms have undergone very little modification, suggesting that final deglaciation was rapid, possibly by ice sheet destabilization and catastrophic calving retreat.

A 5 km thick sequence of glacimarine sediment deposited at a *convergent plate margin* is preserved in the Yakataga Formation in the Gulf of Alaska. The Pacific

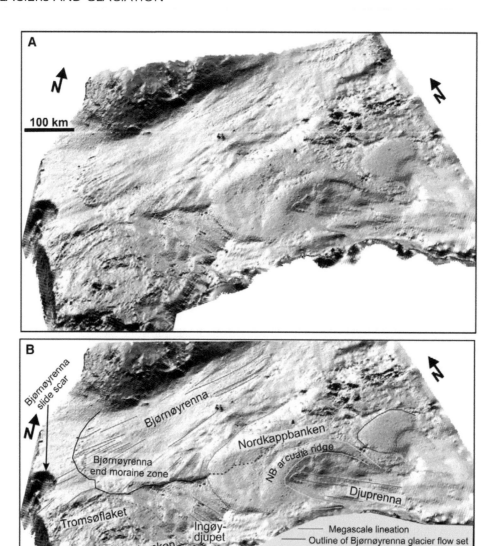

Figure 12.102 Seabed images of cross-cutting sets of mega-scale glacial lineations and associated moraine loops, relating to several generations of offshore palaeo-ice streams (Andreassen et al., 2008).

Plate is being subducted northwards below the North American continental plate, producing a structurally complex forearc basin (fig. 12.103a). Temperate glaciers flowing from the rapidly uplifting coastal mountains have delivered large volumes of fine-grained sediments to the shelf (Powell and Molnia, 1989; Eyles and Lagoe, 1990; Powell and Domack, 1995). The Yakataga Formation records the rapid infilling of the forearc basin on the landward side of an *accretionary wedge* (Eyles et al., 1991; Eyles, 1993). Basin infilling has resulted in a shallowing-upward succession from deep-water debris flows and turbidites to glacially influenced shallow marine facies deposited in a 'ponded' basin. Rapid uplift has raised parts of the Yakataga Formation above sea level, and impressive sections can be examined on the south coast

of Alaska and adjacent offshore islands, most notably Middleton Island, where exceptionally well-exposed successions have been dated to 1.8–0.7 Ma (Eyles and Lagoe, 1990). The succession on Middleton Island consists mainly of massive and stratified diamicton units, with associated boulder pavements and coquinas. Channel fills of gravelly sand are common at the base of the succession, and extensive blankets of mud occur at the top. Eyles and Lagoe (1990) argued that the sediments record changing depositional environments during full-glacial to interglacial conditions, and proposed a threefold interpretive sequence (fig. 12.103b):

(1) *Full-glacial stage.* Striated boulder pavements indicate that grounded or partially grounded ice extended to the shelf edge, although apparently no basal tills are

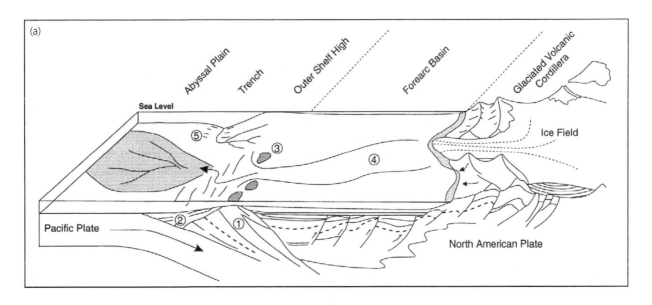

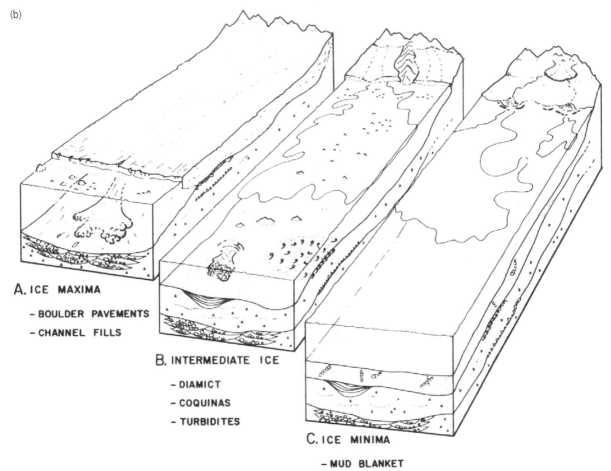

Figure 12.103 Sedimentation on the Gulf of Alaska active plate margin. (a) Landsystem model of the Gulf of Alaska basin. Note the tectonic disturbance of a thick Pleistocene stratigraphy, forming an accretionary wedge and an infilled forearc basin. (1) Accretionary wedge. (2) Slope basins in areas of compressional ridging. (3) Outer shelf high above accretionary wedge. (4) Shallow marine trough crossing the shelf. (5) Dendritic gully on outer slope (Modified from Eyles, 1993). (b) Idealized evolutionary sequence of glacimarine sedimentation based on the stratigraphy of Middleton Island. (A) Maximum ice cover. (B) Intermediate ice cover. (C) Minimal ice cover, similar to the present day (Eyles and Lagoe, 1990, reproduced with permission of the Geological Society of London)

exposed on Middleton Island. Beyond the ice front, channel-fill associations were deposited on the continental slope, possibly fed by meltwater debouching from the glacier.

(2) *Intermediate ice cover.* Extensive muddy diamicton units were deposited on the shelf, recording deposition from icebergs and the settling of suspended sediment. The abundance of ice-rafted debris suggests that many coastal glaciers reached tidewater, but did not extend far onto the shelf. At present, large quantities of suspended sediment are carried into the gulf by glacial meltstreams, and accumulation rates may have been even higher when ice cover was more extensive. Sedimentation rates were variable, however, and coquinas located on topographic highs indicate that such areas were starved of sediment.

(3) *Ice minima.* Extensive mud drapes with few dropstones record deposition during interglacial conditions, when sediment was delivered mainly by turbid rivers discharging into the gulf. The restricted amount of ice-rafted debris in these sediments indicates that few glaciers had tidewater margins. Such conditions are similar to those of today and may have been typical of interglacial periods in this area.

Subsiding rift basins contain the thickest sequences of glacimarine sediments. For example, the Ross and Weddell Sea basins off Antarctica form parts of a single rift system that contains a record of glaciation extending back to the Early Oligocene (approximately 36 Ma; Hamilton et al., 2001; Powell et al., 2001; Hambrey et al., 2002). Drilling programmes in this area have recovered up to 1500 m of glacially influenced strata containing a rich record of glacial fluctuations and environmental change. The lower part of the succession is dominated by turbidites and thin diamictites, and represents the infill of a deep basin adjacent to the uplifting Transantarctic Mountains, whereas the upper part is predominantly diamictic and records repeated advances and recessions of grounded ice across the continental shelf. Eyles (1993) has made the interesting suggestion that the presence of fast-flowing ice streams draining into the Ross Embayment (including the intensively studied Ice Stream B) reflects structural controls in this rifted basin. Another example of a glacially influenced subsiding basin is the North Sea, where the Pleistocene sequence reaches a maximum thickness of around 920 m in the central graben (Eyles, 1993; Hambrey, 1994).

Large proglacial lakes and epicontinental seas

During deglaciation of the Pleistocene ice sheets, ice margins were commonly in contact with large freshwater bodies. These were either proglacial lakes ponded between the ice and topographic barriers, like glacial lakes Agassiz and McConnell in North America (fig. 12.98; Lemmen

et al., 1994; Teller, 1995, 2003), or isolated arms of the sea, such as the Younger Dryas Baltic Ice Lake in northwest Europe (Jakobsson et al., 2007). Vast areas of northern mainland Europe and Siberia were covered by proglacial lakes dammed along the southern margin of the Barents-Kara Ice Sheet, resulting in continent-wide drainage reversals across the watershed of the Baltic Basin southwards into the Caspian Sea via the river Volga and into the Aral Sea (Mangerud et al., 2001a, 2004; Svendsen et al., 2004). The sequential development of vast lakes dammed by the receding margins of the Quebec-Labrador sector of the Laurentide Ice Sheet has been reconstructed using extensive shorelines (Jansson, 2003). Large elongate lake bodies are also produced where ice sheets flow onshore, damming the drainage of intermediate relief valleys – for example, in the case of glacial Lake Teifi in south-west Wales, which was dammed by the Irish Sea Ice Stream flowing parallel to the coast during the late Devensian glaciation (Fletcher and Siddle, 1998; Hambrey et al., 2001; Etienne et al., 2006). The extent and evolution of many of these water bodies has been reconstructed from the sedimentary record, in combination with shoreline evidence, pollen and microfaunal records, and dating programmes (e.g. Teller, 1995; Mangerud et al., 2001b; Andren et al., 2002). The presence of large proglacial water bodies is thought to have driven rapid ice margin retreat by calving, and may have triggered some episodes of ice advance or surges, by encouraging low effective pressures below the glacier margin (e.g. Fyfe, 1990; Stokes and Clark, 2003c). It is therefore important to understand the sedimentary record of proglacial lakes and epicontinental seas, in order to distinguish climatic and non-climatic controls on ice sheet behaviour.

Depositional systems in such settings are similar to those of continental shelves (Ashley, 1995), although there are some important differences:

(1) Fresh lake water and brackish shallow marine conditions reduce the importance of suspended sediment transport in overflows and interflows, so that underflow deposits are better represented in ice-proximal and distal systems.

(2) In lakes, the seasonal alternation between deposition by underflows in summer, and settling of clay particles in winter produces annual *varves* (section 10.6.3). Varves form extensive blankets in parts of North America, Europe and Scandinavia, and record annual cycles of sedimentation in proglacial lakes during deglaciation. When studied in combination with moraine sequences and biostratigraphic data, varves can yield very detailed chronologies, allowing deglaciation histories to be constructed at a resolution not often possible for terrestrial settings.

(3) Proglacial water bodies are prone to sudden changes in water level, due to the opening and closing of

outlets by glacier margin fluctuations and periodic outburst floods and recharge events. Cyclic water-level changes can be recorded by discrete sediment packages bounded above and below by erosion surfaces (fig. 12.6), and can add considerable complexity to stratigraphic sequences associated with glacier fluctuations (Martini and Brookfield, 1995).

(4) Proglacial lakes and epicontinental seas can receive substantial amounts of sediment from recently deglaciated landsurfaces surrounding the basin, as well as from the glacier margin (Martini and Brookfield, 1995; Johnsen and Brennand, 2006). In lakes and arms of the sea, even ice-distal parts of the basin can receive coarse river-borne sediment. Sediment sequences at basin margins can consist of inter-fingering assemblages derived directly from the ice and from the deglaciated area, and may contain a combined record of water-level changes and ice margin fluctuations.

Distal glacilacustrine environments tend to have low sedimentation rates, mainly from underflows, interflows and overflows, and, in glacier-contact lakes, ice rafting. Underflows are most active during the summer ablation season, when they are generated by flood events in glacial and non-glacial streams and slope failures in the lake basin. Cold, sediment-laden water sinks down the sides of the lake basin, and can travel considerable distances below the lake. Sediment dispersal is strongly guided by topography, which focuses underflows into low points on the lake floor (Ashley, 1995). Deposition from overflows and interflows tends to be strongly seasonal, governed by lake stratification and the presence of surface ice. Suspension settling is generally most rapid following the breakdown of stratification and lake overturn in the autumn. Settling of the clay fraction is also encouraged during winter, when the formation of surface ice suppresses wind shear and surface mixing. The dominance of underflows in the summer and suspension settling from overflows and interflows in the autumn and winter gives rise to a characteristic annual depositional cycle in many distal glacilacustrine environments. This cycle produces the classic *varve* couplets, consisting of variable but generally silty summer layers and blanket-like clay-rich winter layers.

In general, sedimentation rates and grain size both decline in a down-lake direction. This pattern, however, may be overprinted by other factors, such as patterns of circulation in the lake and subaqueous or subaerial slope failures. In glacier-contact lakes, sediment is transported into distal positions by drifting icebergs. Patterns of sediment accumulation depend on the rate of iceberg melting and their drifting velocity. In some lakes, accumulation of ice-rafted debris is greatest in the most distal part, near the lake outflow point, where bergs are carried by surface currents (McManus and Duck, 1988). Lake bathymetry is also important because bergs will become grounded in water shallower than the critical depth required for flotation, so that sills may act as barriers to the transfer of ice-rafted debris.

In southern Scandinavia and North America, some major moraine systems were deposited at the margins of proglacial lakes and epicontinental seas during pauses in the retreat or readvances of the last ice sheets (e.g. Hillaire-Marcel et al., 1981; Fyfe, 1990; Brandal and Heder, 1991). Some of the moraine systems record climatically controlled oscillations of the ice margin. For example, the Salpausselkä moraines in southern Finland delimit the margins of a large ice lobe that advanced into the deep waters of the Baltic Ice Lake during the Younger Dryas (fig. 12.104; Saarnisto and Saarinen, 2001; Rinterknecht et al., 2004). The Salpausselkä moraines are composed almost entirely of ice-contact glacifluvial and glacilacustrine sediments deposited in a variety of water depths. Fyfe (1990) showed that the depositional systems making up the outermost moraine (Salpausselkä I) show systematic relationships with water depth and the inferred hydrology of the ice sheet margin. Where the ice terminated in shallow water, the moraine consists of discrete grounding-line fans or deltas connected to large esker networks, recording sediment delivery via conduits. Sedimentation was focused around channel exits, allowing deposits to build up to water level in places. In contrast, where the ice terminated in deeper water, the moraine forms a narrow, nearly continuous ridge, and feeder esker systems are absent. Sediment was delivered to the ice margin at many points by a distributed subglacial drainage system, possibly a braided canal network. Fyfe (1990) argued that distributed drainage systems developed below the ice lobe where sub-marginal water pressures were increased by deep water at the terminus. Spatial differences in subglacial drainage may have influenced ice dynamics, and caused local variations in ice margin behaviour that were unrelated to climatic forcing.

Some moraine systems deposited during pauses in the retreat of the Pleistocene ice sheets appear to have been controlled by spatial and temporal changes in water level rather than climatic events. For example, moraines such as the Lac Daigle-Manitou-Matamek Moraine, Roulier Moraine, and the 600 km long St Narcisse Moraine in eastern Canada are located on topographic highs where the retreating Wisconsinan ice margin became temporarily anchored (Hillaire-Marcel and Occhietti, 1980). Other moraines apparently formed when ice margins stabilized following a drop in water level. For example, the 500 km long Sakami Moraine in Quebec documents subaqueous deposition in response to rapidly shallowing water depths along the western margin of the receding Labrador sector of the Laurentide Ice Sheet when Glacial Lake Ojibway

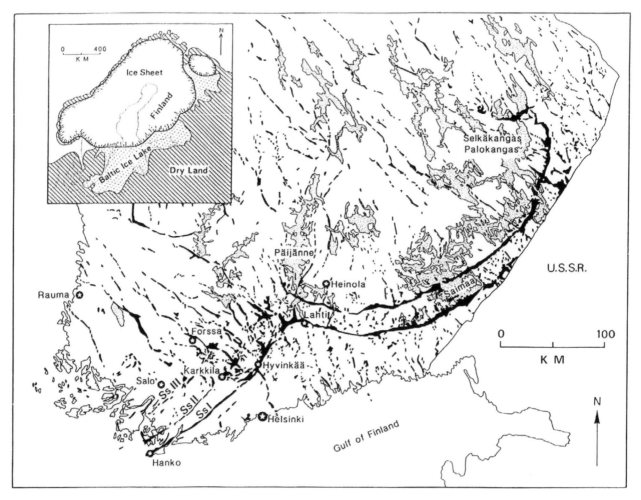

Figure 12.104 The Salpausselkä I, II and III moraines and associated eskers in southern Finland. Inset map shows reconstructed palaeogeography at the time of moraine deposition (Fyfe, 1990, after Eronen, 1983)

drained northwards into the Tyrrell Sea, the precursor to the present Hudson Bay (Hillaire-Marcel et al., 1981). The sudden drop in water level halted the retreat of the western margin of the Labrador sector, allowing the Sakami Moraine to build up while the ice sheet profile readjusted. Hillaire-Marcel et al. (1981) proposed the term *re-equilibration moraines* for moraines formed in such circumstances. Extensive, closely spaced, smaller moraines occur on both sides of the Sakami Moraine, documenting active retreat at rates of 173–239 m yr^{-1} in Lake Ojibway and at 200 m yr^{-1} in the Tyrrell Sea (Vincent, 1989).

The floors of some former proglacial lakes display striking iceberg scour marks, resembling those on high-latitude continental shelves (fig. 12.105; Woodworth-Lynas and Guigné, 1990; Teller, 2003). Grooves and their marginal ridges cross-cut each other and can bend or even follow circuitous pathways, reflecting palaeo-wind flow directions and keel shapes. Some grooves terminate at plough marks or *berms*, where the marginal ridges

converge to create a U-shaped ridge, attesting to the cessation of iceberg drift and/or flotation. In some instances, these berms contain substantial deformation structures that could be mistaken for glacitectonic disturbance (Eden and Eyles, 2001; Winsemann et al., 2003).

A variety of more enigmatic forms have been identified on the plains of former glacial lakes in North America by Mollard (1983, 2000; fig. 12.105). These take a variety of forms, including circular, reticulate, polygonal, cellular, orbicular, vermicular, doughnut-like and brain-like types. A variety of mechanisms have been proposed to account for such lake floor patterns, including: melt-out and buoyant eruption of ice blocks buried below accumulating lake sediments; settling of icebergs into soft lake muds following lake drainage; formation and melting of lenses and wedges of ground ice; differential shrinkage of montmorillonite clays; desiccation; artesian groundwater piping; ground shaking generated by earthquake shock waves; and scouring and pitting of bottom muds by drift ice.

Figure 12.105 Typical former lake floor patterns and floating ice marks on aerial photographs (provided by the National Aerial Photograph Library, Ottawa). (a) Iceberg scour marks on the former floor of Glacial Lake Agassiz. (b) Orbicular patterns near Maryfield, Saskatchewan. (Mollard, 1983). (c) Brain-like pattern on the floor of Glacial Lake Saskatchewan, near Zenon Park, Saskatchewan. (d) Raised rims around piping orifices, near Rosetown, Saskatchewan (average 100 m across). (e) Soft-sediment striations and micro-ridges from the St Lawrence river tidal mudflats (Dionne, 1974), used by Eyles et al. (2005) as modern analogues for lake ice imprints on glacial lake floors.

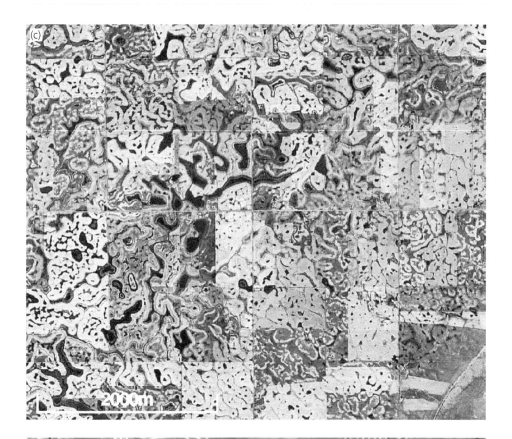

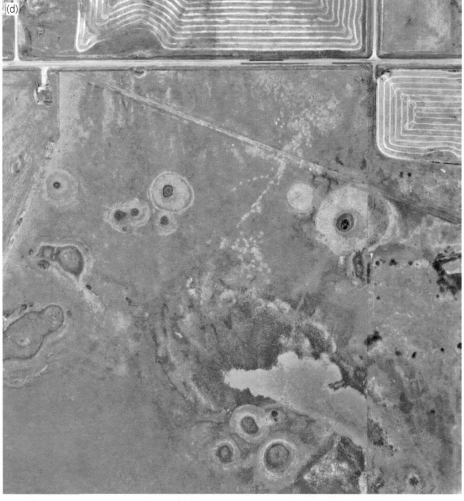

Figure 12.105 (*Continued*)

Figure 12.105 (*Continued*)

12.6 THE MARTIAN GLACIAL LANDSYSTEM: THE FINAL FRONTIER

It is appropriate to end this book by reviewing the newest and most exotic field of glacial study: the nature of glaciation on other planets. Ever-improving technological capabilities allow us to view extraterrestrial geomorphology at increasingly greater levels of detail, particularly on one of our nearest neighbours, Mars. The picture that is emerging from Mars suggests a long and complex geological history, possibly including multiple glaciations (e.g. Head et al., 2005; Dickson et al., 2008).

It has long been known that permanent ice caps exist in both polar regions of Mars, although they are quite unlike Earth's ice caps in appearance, composition and behaviour (e.g. Milkovitch and Head, 2005; P. C. Thomas et al., 2005). The south polar cap is rich in CO_2 ice, whereas the north polar cap is dominantly water ice. Over the winter pole, blizzards of frozen CO_2 deposit a few metres of dry ice on the ice cap at temperatures as low as $-123\,°C$, while at the opposite pole the ice cap gives up CO_2 by sublimation. Although it has been hypothesized that temperatures never rise high enough to melt water ice, surface water has been identified on the south polar cap by Bibring et al. (2004). The ice caps oscillate in size, even on an annual basis (e.g. Benson and James, 2005), but by different processes to those on Earth (P. C. Thomas et al., 2005). The extreme cold and aridity of the Martian polar environment means that the ice caps probably do not undergo widespread basal sliding, so are currently incapable of making a clear subglacial geomorphological impact on the planet's surface. Nevertheless, some remarkable landforms clearly show that glaciers were more extensive in the past (Kargel et al., 1995; Mangold and Allemand, 2001; Baker, 2003, 2005; Head and Marchant, 2003; Carr, 2006), and there is now

consensus in Martian geomorphology that environmental conditions were very different in the past due to the greater abundance of flowing water and ice. Furthermore, the areas immediately beyond the current ice caps have an uncharacteristically low density of meteorite impact craters, suggesting that these areas were protected or modified by expanded ice cover. This evidence raises the intriguing possibility that in the ancient past, Mars was affected by glacial cycles similar to those of the Earth.

Extensive Martian glaciations are now understood to have taken place between 2.1 and 0.4 Ma, and, like the Earth, the Martian surface is going through an interglacial at present (Head et al., 2003; Hauber et al., 2005). The initiation of glaciation on Mars, again like the Earth, is linked to its orbital parameters, but changes to the obliquity (axial tilt) are thought to be particularly significant in triggering Martian glaciation (Head et al., 2003, 2005; Greve and Mahajan, 2005; Forget et al., 2006). A Martian ice age hypothesis has been put forward by Head et al (2003). This hypothesis proposes that higher obliquity results in the warming of the Martian polar ice caps, and hence increased sublimation of water ice and the migration of water vapour to lower latitudes. Glaciation of lower latitudes then proceeds as water vapour nucleates on the abundant dust particles in the atmosphere or at the surface, depositing an ice- and dust-rich mantle. More recently, Forget et al. (2006) have produced a numerical climate model, which predicts glacier ice accumulation at lower latitudes through condensation and precipitation in the Martian atmosphere at times of high obliquity. During the present Martian interglacial, glacier ice is being gradually removed, predominantly by sublimation (e.g. Levrard et al., 2004), producing a glacial landscape similar to that of the arid polar setting of Antarctica (section 12.3.4; Head et al., 2005; Marchant and Head, 2007). An interesting non-climatic theory on the evolution of some glacier-like features has been proposed by Gillespie et al. (2005), in order to explain the accumulation and flowage of ice on relatively flat surfaces where there is no clear accumulation zone. This theory invokes aufeis formation, whereby groundwater erupts from beneath the Martian surface. This explains the unusual glacier morphology and the fact that such glaciers occur in isolation even though other similar sites are located nearby.

Early work on Martian surface features (e.g. Kargel and Strom, 1990, 1991, 1992; Ruff and Greeley, 1990; Kargel et al., 1991; Metzger, 1991) highlighted close similarities to glacial landforms found on Earth, implying that moraines, kame and kettle topography, sinuous esker ridges, tunnel valleys, sandar, glacilacustrine basins, streamlined grooves and mega-flutings, rock glaciers, cirques, arêtes and horns all unequivocally documented former glaciations. These forms are open to other interpretations; sinuous ridges, for example, have been variously interpreted as eskers, lava tubes, exhumed dikes, linear sand dunes or spits and bars (e.g. Head and Pratt, 2001; Tanaka and Kolb, 2001). More recent imagery, however, clearly depicts features that are identical to some Earth-based glacial landforms. Additionally, the occurrence of Martian landform associations which are assembled in a very similar fashion to some glacial landsystems on Earth circumvents the thorny issue of reconstructing glaciations based on the occurrence of isolated landforms. In other words, the similarity between some Martian landforms and glacial features on Earth applies at all observable scales. Early reports of Martian glacial landsystems, such as the mountain ice field and piedmont lobe landsystem recognized by Kargel and Strom (1992), are now being regularly replicated and convincingly compared with Earth-based landsystems (e.g. Head and Marchant, 2003; Arfstrom and Hartmann, 2005; Head et al., 2006a, b; Marchant and Head, 2007; fig. 12.106). Complex landform assemblages created by

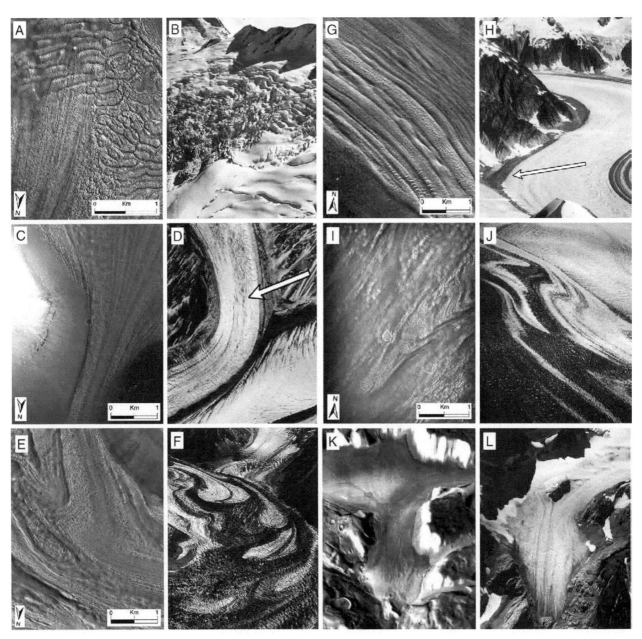

Figure 12.106 Comparison of images of glacial features from the Mars Orbiter Camera and from Post and Lachapelle's (2000) collection of Earth-based glacier systems (Head et al., 2006). (A, B) Upper accumulation zone, with A showing tabular blocks being deformed by downslope streaming, and B showing crevasses and seracs on Bishop Glacier, British Columbia. (C, D) Constriction and convergence of flow lines in C compared to constriction of glacier flow in D (arrow) due to valley narrowing on the Yentna Glacier, Alaska Range. (E, F) Convergence and folding, with E showing convergence of flow from two directions and folding due to differential flow velocities, and F showing looped and folded surge moraines on the Yanert Glacier, Alaska Range. (G, H) Medial and lateral moraines and stranded marginal deposits, with G showing linear ridges produced by differential sublimation and a marginal terrace (lower left), and H showing medial and lateral moraines and terraces (arrow) on Speel Glacier, south-east Alaska. (I, J) Convergence, folding and shear, with I showing very tight folds produced where flow is constricted in a narrowing valley, and J showing convergence and deformation on the Bering Glacier, Alaska. (K, L) Lobate-lineated valley fill and glacier snout, with L showing the Earth-based analogue of the retreating Honeycomb Glacier, North Cascade Range, with its lineated surface produced by differential ablation of flow-aligned septa and lobate margin.

the interaction between glacier and volcanic processes can also be deciphered using Earth analogues (e.g. Wilson and Head, 2002; Gregg et al., 2007; Kadish et al., 2008).

Active features strikingly similar to rock glaciers or debris-covered glaciers exist under the present interglacial conditions (Colaprete and Jakosky, 1998; Mangold et al.,

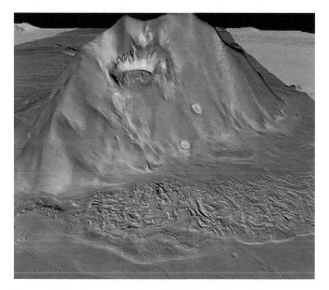

Figure 12.107 Mars Express High Resolution Stereo Camera image (30x vertical exaggeration) of a lobate debris apron (LDA) on the rim of the Hellas Basin of Mars. The massif feeding the flow is 3.75 km high. This is very similar to Earth-based forms produced by protalus rock glacierization or the rock glacierization of debris-covered snouts (Head et al., 2005).

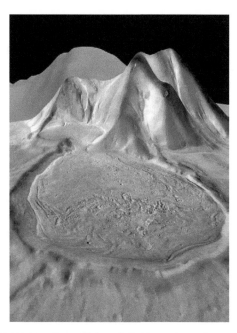

Figure 12.108 Mars Express High Resolution Stereo Camera image (30x vertical exaggeration) of an hourglass-shaped viscous flow feature composed of ice and debris on the margins of a 4 km high massif in the Hellas basin of Mars, with flow lines showing movement from one crater to another. The flow lines emanate from the steep backwall of the higher crater and constrict at a neck joining the two craters. They then form lobate structures on the surface of the lower snout area, resembling the characteristics of Earth-based piedmont glacier lobes. Note also the surface pits and depression on the snout relating to ice melt-out (Head et al., 2005).

2002; Whalley and Azizi, 2003; Gillespie et al., 2005; Head et al., 2005; Mahaney et al., 2007; Marchant and Head, 2007; fig. 12.107). Lineated, lobate forms give a clear impression of flow of debris-rich water ice from an accumulation zone to a snout zone (fig. 12.108; Kreslavsky and Head, 2002; Milliken et al., 2003). The lineations have been referred to as lineated valley fills (LVFs) (Head et al., 2006a, b; Levy et al., 2007; Dickson et al., 2008), and in planform they emerge from alcoves or re-entrants cut into uplands and then converge and coalesce, continuing as parallel ridges for up to 100 km before terminating in lobate forms (fig. 12.109). LVFs have been compared to glaciated valley landsystems (see section 12.3.5). Strong banding is produced where different ice flow units coalesce and extend into trunk valleys, giving rise to medial moraines (see section 11.4.1). In Martian geomorphic terms, the lineations are regarded as deformed flow ridges of debris. The lobations, called lobate debris aprons (LDAs) (Head et al., 2006a, b;

Figure 12.109 Mars Orbiter Camera image of a lineated valley fill (LVF) terminating as lobate patterns and thought to be a 4 km long, tongue-shaped glacier lobe. Note the chevron-shaped ridges pointing downglacier, indicative of higher velocities towards the central flow line, and the occurrence of patterned ground outside the lobe limits (Marchant and Head, 2007).

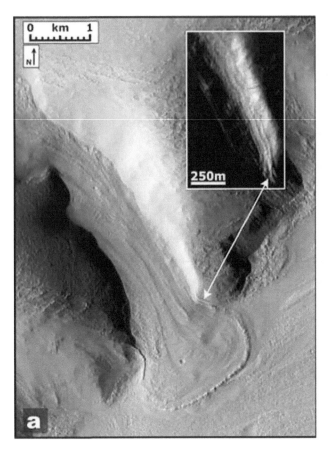
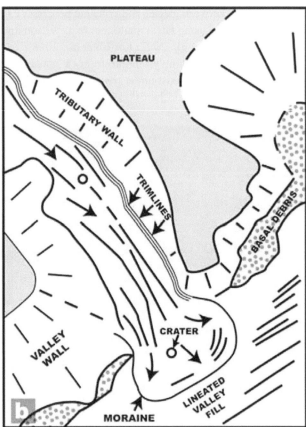

Figure 12.110 Mars Orbiter Camera image showing a debris-covered glacier lobe and its lateral moraines extending from a tributary basin and superimposing a lineated valley fill (LVF) in the main valley. The sketch map shows the geomorphic interpretations of the image. Arrows on the glacier lobe show flow directions (Dickson et al., 2008).

Levy et al., 2007; Dickson et al., 2008), are parallel concentric ridges arranged in lobate forms or tongue-shaped lobes extending from individual alcoves and onto adjacent lowlands (fig. 12.107). This morphology is similar in every respect to debris-covered piedmont glaciers and rock glaciers on Earth, which extend from cirque feeder basins (Head and Marchant, 2003; Head et al., 2005; Shean et al., 2005). Tightening and folding of parallel debris ridges occurs wherever the flow is restricted – for example, where lobes abut each other or topographic obstacles. Where LVF merges into LDA terrain, they show a remarkable similarity to the latero-frontal moraine loops of mountain glaciers on Earth (Arfstrom and Hartmann, 2005).

Evidence of continuing melt-out on these glacial landforms in the shape of circular to elongate pits, 'pit-and-butte' surfaces, elongate chains of knobs and hummocks ('knobby facies') and undeformed, inverted craters, indicates that they still contain large volumes of ice buried beneath a significant cover of surface debris (Mangold, 2003; Head et al., 2005; Aharonson and Schorghofer, 2006; Gregg et al., 2007; Levy et al., 2007; Kadish et al., 2008). The aridity of the Martian environment is therefore producing glacial landforms typical of

the purest form of melt-out, preserving englacial structures or *controlled moraines* (section 11.4.2; Marchant et al., 2002; Shean et al., 2005; Kadish et al., 2008). The hyper-arid polar continental landsystem of Antarctica is therefore our best analogue for glacial landform evolution on Mars (Head and Marchant, 2003; Marchant and Head, 2007). Nevertheless, the release of liquid water from these glaciers has been identified by Dickson and Head (2006) in the form of what appear to be kettle holes in ice-contact lake deposits.

Changing dynamics of Martian glaciers has been interpreted from the pattern of LVF and LDA landforms. Complex folds in LVF can occur where tributary flows join trunk systems. This is similar to the looped moraines we see produced on Earth where surging tributary glaciers are responsible for differential flow velocities in complex trunk glaciers (section 5.7.1). Levy et al. (2007) and Dickson et al. (2008) identified superimposed debris ridges, where tributary glaciers have advanced over older LVF, and inset lateral ridges at the margins of LVF, thought to be representative of lateral moraines (fig. 12.110) Another similarity between Martian and Earth-based glacial systems is the tendency for ice to change its flow pattern during deglaciation due to the

Figure 12.111 Mars Orbiter Camera image of a lobate flow of ice moving into a box canyon due to deglacial reorganization of regional ice flow. The lobate form is picked out by the looped controlled moraines on the glacier surface (Dickson et al., 2008).

increasing influence of topography (Hättestrand and Johansen, 2005; Dickson et al., 2008; fig. 12.111).

Many parts of Mars exhibit spectacular channels and canyons which must have been cut by surface stream systems (Baker, 2001; Carr, 2006; fig. 12.112). In addition, large *outflow channels* and *streamlined residuals* have been identified by Sharp and Malin (1975) and compared to the Channelled Scablands of Washington (e.g. Baker, 1979; Komatsu and Baker, 1996). They are significantly larger than our largest Earth-based flood channels (Baker, 2001), and peak discharges are estimated to have equalled those of Earth's Gulf Stream! Martian channels thought likely to be related to meltwater outflows have trough-shaped cross profiles, containing evidence of large-scale fluid flows and emanating from areas of 'chaotic terrain'. The channels are broadly anastomosing and split by residual 'island' remnants of the pre-flood surface. The channels have low sinuosity and high width:depth ratios, and record flow expansions and constrictions along their length. Also similar to Earth flood tracks are longitudinal grooves, inner channels, cataract complexes, scablands and bar complexes.

The close resemblance of the Martian channel systems to landscapes eroded by large catastrophic floods raises the interesting problem of where flood waters could have originated. The evidence for glaciers and ice-covered lakes (Kraal et al., 2008) on Mars inevitably leads to conjecture that some floods could have been GLOFs. Flood tracks emerging from the margin of the north polar cap have been compared to Icelandic glacio-volcanic jökulhlaup features by Hovius et al. (2008). Because the polar caps are thought to be too thin to induce basal melting (Clifford, 1987; Clifford and Parker, 2001), a combination of geothermal and glacier processes may be the most appropriate genetic interpretation for many Martian flood features. However, it has also been suggested that water could have burst out from highly pressurized subsurface aquifers, due

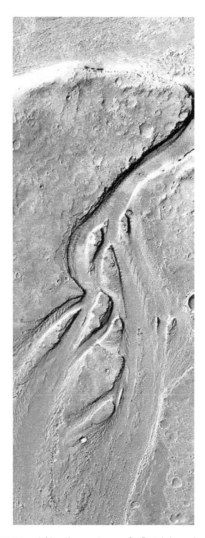

Figure 12.112 Mars Orbiter Camera image of a fluvial channel system up to 4 km wide, located south of Cerberus Rupes. Visible are anastomosing channels, streamlined uplands, terraces and transverse fluvial bedforms with a wavelength of 60 m, which are all typical of glacier flood pathways on Earth such as the Channelled Scablands (Baker, 2001).

to a range of possible triggers not all necessarily linked to glaciation history (Malin and Carr, 1999; Gulick, 2001; Max and Clifford, 2001; Fassett and Head, 2008).

Obviously, the very inaccessibility of the Martian surface forces us to make geomorphic interpretations based on form analogy. This entails a necessary and not altogether safe assumption, that the terrestrial forms and processes that we know so well are suitable analogues for those of other planets. Nonetheless, the case that has been made repeatedly for Martian glacial landsystems being similar to Earth's polar continental landsystem is a compelling one. This is due not only to landform similarity, but also to the climatic characteristics of extreme aridity, which result in the persistence of ice beneath sublimation debris for millennia (Marchant et al., 2002; Kowalewski et al., 2006). The inevitable corollary is that the glaciers of Mars remain entombed in permafrost, being disrupted apparently only occasionally during interglacials by meteorite impacts and subsurface water expulsion. A rapidly expanding literature, based on the exciting discoveries of our extraterrestrial explorations, is demonstrating that a full understanding of Martian cold-climate geomorphology relies on the achievement of synergy between Earth-based analogues and new process–form models for a planet–atmosphere system fundamentally different to our own. Thereby, the Martian surface continues to be the ultimate landsystems challenge for those who specialize in palaeoglaciology. Proposals to launch manned missions to Mars raise the possibility that glaciological fieldwork might one day be conducted there, perhaps by someone now reading this book.

APPENDIX

List of symbols used in equations

a	amplitude of subglacial bumps
A	prefactor in Glen's Flow Law
A_0	reference value for the flow law prefactor $(9.302 \times 10^{-2} \mathrm{Pa^{-3}yr^{-1}})$
A_B	bulk transfer coefficient
A_F	area of bed in contact with debris
A_G	glacier contributing area
A_{ocean}	area of global ocean
A_r	channel or glacier cross-section area
$\dot{A}$	abrasion rate
$\dot{b}$	balance rate
b_{nb}	mass balance in the ablation area
b_{nc}	mass balance in the accumulation area
B	total mass balance
BR	balance ratio
c	cohesion
c_i	specific heat capacity of ice ($2.05 \mathrm{J\,g^{-1}K^{-1}}$ at $-10\,°\mathrm{C}$)
c_P	specific heat capacity of air ($1.0035 \mathrm{J\,g^{-1}K^{-1}}$ for dry air at $0\,°\mathrm{C}$)
c_w	specific heat capacity of water ($4.2 \mathrm{J\,g^{-1}K^{-1}}$)
C	bed geometry parameter
C_d	concentration of particles in basal ice
C_P	climate sensitivity to precipitation changes
C_T	climate sensitivity to temperature changes
d	crevasse depth
d_w	depth of water in crevasses
d_{wc}	depth in the water column
$\dot{d}$	channel downcutting rate
D	particle diameter
D_V	valley depth
D_W	water depth
e	base of the natural logarithms (~ 2.71828)
e_A	vapour pressure in the air
e_S	vapour pressure at the surface
e^*	sediment voids ratio
E	drumlin elongation ratio
E_a	annual evaporation total
f	shape factor
F	valley form ratio
G	geothermal heat flux
g	gravitational acceleration ($9.81 \mathrm{m\,s^{-2}}$)
h_d	debris layer thickness
h_i	elevation of ice surface
h_w	elevation of water surface
H	ice thickness, high-angle shears
H_F	flotation thickness
I	radiation flux
k	thermal conductivity
K	hydraulic conductivity, constant in Clague–Mathews relationship
L	distance along flow path, glacier length, low-angle shears
l_b	bedform length
L_f	latent heat of fusion/melting ($330 \mathrm{J\,g^{-1}}$)
L_T	trough length
L_V	latent heat of vaporization/condensation ($2500 \mathrm{J\,g^{-1}}$)
LW	net longwave radiation
LW_{in}	incoming longwave radiation

LW_{out}	outgoing longwave radiation
M	energy used to melt ice or freeze water
$\dot{M}$	melt rate
n	exponent in Glen's Flow Law (~ 3)
n^*	sediment porosity
n^*_{rock}	rock porosity
$\tilde{n}$	Manning's roughness coefficient
N	effective pressure
p	precipitation
p_a	annual precipitation total
p_s	snowfall
p^*	channel 'wetted perimeter'
P_i	ice pressure
P_0	reference value of air pressure
P_s	separation pressure (for basal cavity formation)
P_W	water pressure
q	water flux ($\mathrm{m\,sec^{-1}}$)
Q	discharge
Q^*	activation energy for ice creep ($7.88 \times 10^4 \mathrm{J\,mol^{-1}}$)
QE	latent heat transfer
QH	sensible heat transfer
Q_{max}	peak discharge
QR	energy flux from rain
QT	energy used for temperature change in the ice
r^*	conduit radius
$\dot{r}$	rainfall rate (metres per second)
R	isostatic rebound
R^*	hydraulic radius
R_a	annual runoff total
R_g	gas constant ($8.314 \mathrm{J\,mol^{-1}K^{-1}}$)
R_n	Richardson number
S	channel slope
SW	net shortwave radiation
SW_{in}	incoming shortwave radiation
SW_{out}	outgoing shortwave radiation
t	time
t_R	response time
T	temperature
T_0	reference value for ice temperature in flow law prefactor equation
T_A	air temperature
T_i	ice temperature
T_r	rain temperature
T_s	surface temperature ($273 \mathrm{K}$ for a melting ice surface)
T_{sum}	mean summer air temperature
$\bar{T}_d$	mean daily temperature
$\dot{T}_m$	rate of change in ice melting temperature with pressure
u	ice velocity in x direction
u_W	wind speed
U_b	rate of basal motion
U_c	centreline velocity
U_C	calving rate
U_i	ice velocity due to creep
U_T	glacier velocity at the terminus
U_w	water velocity
v	ice velocity in y direction
v^*	Poisson's ratio

v_f	particle fall velocity	Λ	(lambda) ratio between basal sliding velocity and the effective pressure raised to the power of n
v_p	particle velocity		
V	sediment volume	λ	(lambda) wavelength of subglacial bumps
V_t	total water volume drained during an outburst flood	μ	(mu) viscosity
V_v	volume of voids	π	(pi) ratio between circumference and diameter of a circle ($\sim$3.14159)
w	glacier width		
w_b	maximum bedform width	ρ_0	(rho) reference value of air density
w_v	valley width	ρ_a	air density
W	glacier half-width	ρ_i	ice density
W_V	valley half-width	ρ_m	mantle density
z	elevation	ρ_p	particle density
Z_B	bed elevation	ρ_w	water density
α	(alpha) albedo	σ	(sigma) Stephan–Boltzmann constant (5.67×10^{-8})
ΔS_g	(delta S_g) change in glacial water/ice storage	σ_e	effective stress
ΔSL	change in mean global sea level	σ_{yield}	sediment yield strength
ΔS_o	change in non-glacial water/ice stores	$\sigma'_{xx}\sigma'_{yy}\sigma'_{zz}$	deviatoric stresses
ΔV_i	change in global ice volume	$\sigma_{xx}\sigma_{yy}\sigma_{zz}$	normal stress components
$\Delta\phi_f$	head loss due to friction between flowing water and channel walls	τ	(tau) shear stress
		τ_b	basal shear stress
ε	(epsilon) emissivity	τ_d	driving stress
$\dot{\varepsilon}$	shear strain rate	$\bar{\tau}_{xt}$	lateral shear stress
$\dot{\varepsilon}_{xx}\dot{\varepsilon}_{yy}\dot{\varepsilon}_{zz}$	strain rate components	Φ	(phi) angle of internal friction
θ	(theta) angle of high-angle shears	ϕ	(phi) hydraulic potential
Θ	(theta) vertical temperature gradient in ice	ψ	(psi) angle of low-angle shears
κ	(kappa) permeability		

REFERENCES

Aartolahti, T., Koivisto, M. and Nenonen, K. 1995. *De Geer Moraines in Finland*. Geological Survey of Finland, Special Paper 20, 67–74.

Abdalati, W., Krabill, W. B., Frederick, E. B., Manizade, S. S., Martin, C. F., Sonntag, J. G., Swift, R. N., Thomas, R. H., Wright, W. and Yungel, J. G. 2002. Aircraft laser altimetry mapping of the Greenland ice sheet: application to mass balance assessment. *Journal of Geodynamics* 34, 391–403.

Aber, J. S. 1979. Kineto-stratigraphy at Hvideklint, Mon, Denmark and its regional significance. *Bulletin of the Geological Society of Denmark* 28, 81–93.

Aber, J. S. and Ber, A. 2007. *Glaciotectonism*. Developments in Quaternary Science 6. Elsevier, Amsterdam.

Aber, J. S., Croot, D. G. and Fenton, M. M. 1989. *Glaciotectonic Landforms and Structures*. Kluwer, Dordrecht.

Abrantes, F. 2007. Marine diatoms. In Elias, S. A. (ed.), *Encyclopedia of Quaternary Science*. Elsevier, Amsterdam, 1668–78.

ACIA, 2005. Arctic Climate Impact Assessment: Scientific Report. Cambridge University Press.

Adalgeirsdottir, G., Smith, A. M., Murray, T., King, M. A., Makinson, K., Nicholls, K. W. and Behar, A. E. 2008. Tidal influence on Rutford Ice Stream, West Antarctica: observations of surface flow and basal processes from closely spaced GPS and passive seismic stations. *Journal of Glaciology* 54 (187), 715–24.

Adam, S., Pietroniro, A. and Brugman, M. M. 1997. Glacier snow line mapping using ERS-1 SAR imagery. *Remote Sensing of Environment* 61, 46–54.

Addison, K. 1981. The contribution of discontinuous rock-mass failure to glacier erosion. *Annals of Glaciology* 2, 3–10.

Ageta, Y. and Higuchi, K. 1984. Estimation of mass balance components of a summer-accumulation type glacier in the Nepal Himalaya. *Geografiska Annaler* 66A, 249–55.

Ageta, Y., Iwata, S., Yabuki, H., Naito, N., Sakai, A., Narama, C. and Karma 2000. Expansion of glacier lakes in recent decades in the Bhutan Himalayas. In Nakawo, M., Raymond, C. F. and Fountain, A. (eds), *Debris-Covered Glaciers*. IAHS Publication 264, 165–75.

Aharonson, O. and Schorghofer, N. 2006. Subsurface ice on Mars with rough topography. *Journal of Geophysical Research* 111, E11007, doi:10.1029/2005JE002636.

Aitken, J. F. 1998a. Sedimentology of Late Devensian glaciofluvial outwash in the Don Valley, Grampian Region. *Scottish Journal of Geology* 34, 97–117.

Aitken, M. J. 1998b. An Introduction to Optical Dating: The Dating of Quaternary Sediments by the Use of Photon-stimulated Luminescence, Oxford University Press.

Aizen, V. B., Aizen, E. M. and Melack, J. M. 1997. Snow distribution and melt in Central Tien Shan, Susamir Valley. *Arctic and Alpine Research* 29, 403–13.

Albert, M. A. and Perron, F. E. 2000. Ice layer and surface crust permeability in a seasonal snowpack. *Hydrological Processes* 14, 3207–14.

Albino, K. and Dreimanis, A. 1988. A time-transgressive kinetostratigraphic sequence spanning 180° in a single section at Bradtville, Ontario, Canada. In Croot, D. G. (ed.), *Glaciotectonics. Forms and Processes*. Balkema, Rotterdam, 11–20.

Alexander, J., Bridge, J. S., Cheel, R. J. and Leclair, S. F. 2001. Bed forms and associated sedimentary structures formed under supercritical water flows over aggrading sand beds. *Sedimentology* 48, 133–52.

Alexanderson, H., Adrielsson, L., Hjort, C., Möller, P., Antonov, O., Eriksson, S. and Pavlov, M. 2002. Depositional history of the North Taymyr ice-marginal zone, Siberia – a landsystem approach. *Journal of Quaternary Science* 17, 361–82.

Alho, P., Russell, A. J., Carrivick, J. L. and Käyhkö, J. 2005. Reconstruction of the largest Holocene jökulhlaup within Jökulsá á Fjöllum, NE Iceland. *Quaternary Science Reviews* 24, 2319–34.

Allen, J. R. L. 1971. A theoretical and experimental study of climbing-ripple cross-lamination, with a field application to the Uppsala esker. *Geografiska Annaler* 53A, 157–87.

Allen, J. R. L. 1982. *Sedimentary Structures: Their Character and Physical Basis*. Developments in Sedimentology 30. Elsevier, Amsterdam.

Allen, J. R. L. 1983. Studies in fluviatile sedimentation: bars, bar complexes and sandstone sheets (low-sinuosity braided streams) in the Brownstones (L. Devonian), Welsh Borders. *Sedimentary Geology* 33, 237–93.

Allen, J. R. L. 1985. *Principles of Physical Sedimentology*. Allen and Unwin, London.

Alley, R. B. 1991. Deforming bed origin for southern Laurentide till sheets? *Journal of Glaciology* 37, 67–76.

Alley, R. B. 1992a. Flow-law hypotheses for ice-sheet modeling. *Journal of Glaciology* 38, 245–56.

Alley, R. B. 1992b. How can low-pressure channels and deforming tills coexist subglacially? *Journal of Glaciology* 38, 200–7.

Alley, R. B. 1993. In search of ice stream sticky spots. *Journal of Glaciology* 39, 447–54.

Alley, R. B. 2000. Continuity comes first: recent progress in understanding subglacial deformation. In Maltman, A. J., Hubbard, B. and Hambrey, M. J. (eds), *Deformation of Glacial Materials*. Geological Society, London, Special Publication 176, 171–9.

Alley, R. B. and Bindschadler, R. A. 2001. The West Antarctic ice sheet and sea-level change. In Alley, R. B. and Bindschadler, R. A. (eds), *The West Antarctic Ice Sheet: Behavior and Environment*. American Geophysical Union, Antarctic Research Series 77, 1–11.

Alley, R. B., Blankenship, D. D., Bentley, C. R. and Rooney, S. T. 1986. Deformation of till beneath ice stream B, West Antarctica. *Nature* 322, 57–9.

Alley, R. B., Blankenship, D. D., Bentley, C. R. and Rooney, S. T. 1987. Till beneath Ice Stream B 3. Till deformation: evidence and implications. *Journal of Geophysical Research* 92, 8921–9.

Alley, R. B., Blankenship, D. D., Rooney, S. T. and Bentley, C. R. 1989. Sedimentation beneath ice shelves – the view from ice stream B. *Marine Geology* 85, 101–20.

Alley, R. B., Anandakrishnan, S., Bentley, C. R. and Lord, N. 1994. A water-piracy hypothesis for the stagnation of Ice Stream C, Antarctica. *Annals of Glaciology* 20, 187–94.

Alley, R. B., Cuffey, K. M., Evenson, E. B., Strasser, J. C., Lawson, D. E. and Larson, J. C. 1997. How glaciers entrain and transport basal sediment: physical constraints. *Quaternary Science Reviews* 16, 1017–38.

Alley, R. B., Lawson, D. E., Evenson, E. B., Strasser, J. C. and Larson, J. C. 1998. Glaciohydraulic supercooling: a freeze-on mechanism to create stratified, debris-rich basal ice. 2. Theory. *Journal of Glaciology* 44, 563–9.

Alley, R. B., Strasser, J. C., Lawson, D. E., Evenson, E. B. and Larson, G. J. 1999. Glaciological and geological implications of basal-ice accretion in overdeepenings. In Mickelson, D. M. and Attig, J. W. (eds), *Glacial Processes Past and Present*. Geological Society of America, Special Paper 337, 1–9.

Alley, R. B., Lawson, D. E., Evenson, E. B. and Larson, G. J. 2003a. Sediment, glaciohydraulic supercooling, and fast glacier flow. *Annals of Glaciology* 36, 135–41.

Alley, R. B., Lawson, D. E., Larson, G. J., Evenson, E. B. and Baker, G. S. 2003b. Stabilizing feedbacks in glacier-bed erosion. *Nature* 424, 758–60.

Alley, R. B., Anandakrishnan, S., Dupont, T. K. and Parizek, B. R. 2004. Ice streams – fast and faster? *Comptes Rendus Physique* 5, 723–34.

Alley, R. B., Clark, P. U., Huybrechts, P. and Joughin, I. 2005a. Ice-sheet and sea-level changes. *Science* 310, 456–60.

Alley, R. B., Dupont, T. K., Parizek, B. R. and Anandakrishnan, S. 2005b. Access of surface meltwater to beds of sub-freezing glaciers: preliminary insights. *Annals of Glaciology* 40 (7), 8–14.

Alt, D. 2001. *Glacial Lake Missoula and its Humongous Floods*. Mountain Press, Missoula.

Ambach, W. 1988. Interpretation of the positive-degree-days factor by heat balance characteristics – West Greenland. *Nordic Hydrology* 19, 217–24.

Amblas, D., Urgeles, R., Canals, M., Calafat, A. M., Rebesco, M., Camerlenghi, A., Estrada, F., De Batist, M. and Hughes-Clarke, J. E. 2006. Relationship between continental rise development and palaeo-ice sheet dynamics, northern Antarctic Peninsula Pacific margin. *Quaternary Science Reviews* 25, 933–44.

Amerson, B. E., Montgomery, D. R. and Meyer, G. 2008. Relative size of fluvial and glaciated valleys in central Idaho. *Geomorphology* 93, 537–47.

Anandakrishnan, S. 2003. Dilatant till layer near the onset of streaming flow of Ice Stream C, West Antarctica, determined by AVO (amplitude vs offset) analysis. *Annals of Glaciology* 36, 283–6.

Anandakrishnan, S. and Alley, R. B. 1994. Ice Stream C, Antarctica, sticky-spots detected by microearthquake monitoring. *Annals of Glaciology* 20, 183–6.

Anandakrishnan, S. and Alley, R. B. 1997. Stagnation of ice stream C, West Antarctica by water piracy. *Geophysical Research Letters* 24, 265–8.

Anandakrishnan, S., Blankenship, D. D., Alley, R. B. and Stoffa, P. L. 1998. Influence of subglacial geology on the position of a West Antarctic ice stream from seismic observations. *Nature* 394, 62–6.

Anandakrishnan, S., Alley, R. B., Jacobel, R. W. and Conway, H. 2001. The flow regime of Ice Stream C and hypotheses concerning its recent stagnation. In Alley, R. B. and Bindschadler, R. A. (eds), *The West Antarctic Ice Sheet: Behavior and Environment*. American Geophysical Union, Antarctic Research Series, vol. 77, 283–94.

Andersen, B. G. and Borns, H. W. 1994. *The Ice Age World*. Scandinavian University Press, Oslo.

Andersen, B. G., Denton, G. H. and Lowell, T. V. 1999. Glacial geomorphologic maps of Llanquihue Drift in the area of the southern Lake District, Chile. *Geografiska Annaler* 81A, 155–66.

Andersen, L. T., Hansen, D. L. and Huuse, M. 2005. Numerical modelling of thrust structures in unconsolidated sediments: implications for glaciotectonic deformation. *Journal of Structural Geology* 27, 587–96.

Anderson, C. H, Vining, M. R. and Nichols, C. M. 1994. Evolution of the Paradise/Stevens Glacier ice caves. *National Speleological Society Bulletin* 56, 70–81.

Anderson, D. M. 2007. Paleoceanography. In Elias, S. A. (ed.), *Encyclopedia of Quaternary Science*. Elsevier, Amsterdam, 1599–609.

Anderson, J. B. 1999. *Antarctic Marine Geology*. Cambridge University Press.

Anderson, J. B. and Fretwell, L. O. 2008. Geomorphology of the onset area of a palaeo-ice stream, Marguerite Bay, Antarctic Peninsula. *Earth Surface Processes and Landforms* 33, 503–12.

Anderson, J. B. and Shipp, S. S. 2001. Evolution of the West Antarctic Ice Sheet. In Alley, R. B. and Bindschadler, R. A. (eds), *The West Antarctic Ice Sheet: Behavior and Environment*. American Geophysical Union, Antarctic Research Series, vol. 77, 45–57.

Anderson, J. B., Kurtz, D. D., Domack, E. W. and Balshaw, K. M. 1980. Antarctic glacial marine sediments. *Journal of Geology* 88, 399–414.

Anderson, J. B., Shipp, S. S., Lowe, A. L., Smith Wellner, J. and Mosola, A. B. 2002. The Antarctic Ice Sheet during the Last Glacial Maximum and its subsequent retreat history: a review. *Quaternary Science Reviews* 21, 49–70.

Anderson, L. W. 1978. Cirque glacier erosion rates and characteristics of Neoglacial tills, Pangnirtung Fiord area, Baffin Island, NWT, Canada. *Arctic and Alpine Research* 10, 749–60.

Anderson, M. G. and MacDonnell, J. J. (eds), 2005. *Encyclopedia of Hydrological Sciences*. Wiley, London.

Anderson, R. S. 2000. A model of ablation-dominated medial moraines and the generation of debris-mantled glacier snouts. *Journal of Glaciology* 46 (154), 459–69.

Anderson, R. S., Hallet, B., Walder, J. and Aubry, B. F. 1982. Observations in a cavity beneath Grinell Glacier. *Earth Surface Processes and Landforms* 7, 63–70.

Anderson, R. S., Anderson, S. P., MacGregor, K. R., Waddington, E. D., O'Neel, S., Riihimaki, C. A. and Loso, M. G. 2004. Strong feedbacks between hydrology and sliding of a small alpine glacier. *Journal of Geophysical Research* 109, doi:10.1029/2004JF000120.

Anderson, R. S., Molnar, P. and Kessler, M. A. 2006. Features of glacial valley profiles simply explained. *Journal of Geophysical Research* 111, F01004, doi:10.1029/2005JF000344.

Anderson, S. P. 2005. Glaciers show direct linkage between erosion rate and chemical weathering fluxes. *Geomorphology* 67, 147–57.

Anderson, S. P., Drever, J. J. and Humphrey, N. F. 1997. Chemical weathering in glacial environments. *Geology* 25, 399–402.

Anderson, S. P., Drever, J. J., Frost, C. D. and Holden, P. 2000. Chemical weathering in the foreland of a retreating glacier. *Geochimica et Cosmochimica Acta* 64, 1173–89.

Anderson, S. P., Walder, J. S., Anderson, S., Kraal, E. R., Cunico, M., Fountain, A. G. and Trabant, D. 2003a. Integrated hydrologic and hydrochemical observations of Hidden Creek Lake jökulhlaups, Kennicott Glacier, Alaska. *Journal of Geophysical Research* 108, doi:10.1029/2002JF000004.

Anderson, S. P., Longacre, S. A. and Kraal, E. R. 2003b. Patterns of water chemistry and discharge in the glacier-fed Kennicott River, Alaska: evidence for subglacial water storage cycles. *Chemical Geology* 202, 297–312.

Andersson, G. 1998. Genesis of hummocky moraine in the Bolmen area, southwestern Sweden. *Boreas* 27, 55–67.

Andersson, T., Forman, S. L., Ingólfsson, Ó. and Manley, W. F. 1999. Late Quaternary environmental history of central Prins Karls Forland, western Svalbard. *Boreas* 28, 292–307.

Andersson, T., Forman, S. L., Ingólfsson, Ó. and Manley, W. F. 2000. Stratigraphic and morphologic constraints on the Weichselian glacial history of northern Prins Karls Forland, western Svalbard. *Geografiska Annaler* 82A, 455–70.

André, M.-F. 1997. Holocene rockwall retreat in Svalbard: a triplerate evolution. *Earth Surface Processes and Landforms* 22, 423–40.

André, M.-F. 2004. The geomorphic impact of glaciers as indicated by tors in North Sweden (Aurivaara, 68° N). *Geomorphology* 57, 403–21.

Andreassen, K., Nilssen, L. C., Rafaelsen, B. and Kuilman, L. 2004. Three-dimensional seismic data from the Barents Sea margin reveal evidence of past ice streams and their dynamics. *Geology* 32, 729–32.

Andreassen, K., Laberg, J. S. and Vorren, T. O. 2008. Seafloor geomorphology of the SW Barents Sea and its glaci-dynamic implications. *Geomorphology* 97, 157–77.

Andren, T., Bjorck, S. and Johnsen, S. 1999. Correlation of Swedish glacial varves with the Greenland (GRIP) oxygen isotope record. *Journal of Quaternary Science* 14, 361–71.

Andren, T., Lindeberg, G. and Andren, E. 2002. Evidence of the final drainage of the Baltic Ice Alek and the brackish phase of the Yoldia Sea in glacial varves from the Baltic Sea. *Boreas* 31, 226–38.

Andrews, J. T. 1970. A geomorphological study of Post-Glacial uplift with particular reference to Arctic Canada. *Institute of British Geographers*, Special Publication 2.

Andrews, J. T. 1972. Glacier power, mass balances, velocities and erosion potential. *Zeitschrift für Geomorphologie* Suppl. Bd 13, 1–17.

Andrews, J. T. 1982. On the reconstruction of Pleistocene ice sheets: a review. *Quaternary Science Reviews* 1, 1–30.

Andrews, J. T. 1998. Abrupt changes (Heinrich events) in late Quaternary North Atlantic marine environments: a history and review of data and concepts. *Journal of Quaternary Science* 13, 3–16.

Andrews, J. T. 2000. Icebergs and iceberg rafted detritus (IRD) in the North Atlantic: facts and assumptions. *Oceanography* 13, 100–8.

Andrews, J. T. 2006. Glaciers, oceans, atmosphere and climate. In Knight, P. G. (ed.), *Glacier Science and Environmental Change*. Blackwell, Oxford, 96–113.

Andrews, J. T. and Dugdale, R. E. 1971. Quaternary history of northern Cumberland Peninsula, Baffin Island, NWT. Part V: Factors affecting corrie glacierization in Okoa Bay. *Quaternary Research* 1, 532–51.

Andrews, J. T. and LeMasurier, W. E. 1973. Rates of Quaternary glacial erosion and corrie formation, Marie Byrd Land, Antarctica. *Geology* 1, 75–80.

Andrews, J. T. and MacLean, B. 2003. Hudson Strait ice streams: a review of stratigraphy, chronology, and links with North Atlantic Heinrich events. *Boreas* 32, 4–17.

Andrews, J. T. and Tedesco, K. 1992. Detrital carbonate-rich sediments, northwestern Labrador Sea: implications for ice-sheet dynamics and iceberg-rafting (Heinrich) events in the North Atlantic. *Geology* 20, 1087–90.

Andrews, J. T., Austin, W. E. N., Bergsten, H. and Jennings, A. E. (eds). 1996. *Late Quaternary Palaeoceanography of the North Atlantic Margins*. Geological Society, London, Special Publication 111.

Andriashek, L. D. and Atkinson, N. 2007. *Buried Channels and Glacial-drift Aquifers in the Fort McMurray Region, Northeast Alberta*. EUB/AGS Earth Sciences Report 2007-01, Alberta Energy and Utilities Board.

Andrzejewski, L. 2002. The impact of surges on the ice-marginal landsystem of Tungnaárjökull, Iceland. *Sedimentary Geology* 149, 59–72.

Aniya, M. and Naruse, R. 1985. Structure and morphology of Solar Glacier. In Nakajima, C. (ed.), *Glaciological Studies in Patagonia Northern Icefield, 1983–1984*. Data Center for Glacier Research, Japanese Society for Snow and Ice, 70–9.

Aniya, M. and Welch, R. 1981. Morphological analyses of glacial valleys and estimates of sediment thicknesses on the valley floor: Victoria Valley system, Antarctica. *Antarctic Record* 71, 76–95.

Anketell, J. M., Cegla, J. and Dzulynski, S. 1970. On the deformational structures in systems with reversed density gradients. *Géologique de Pologne, Annales Société* 40, 3–30.

Anundsen, K. 1990. Evidence of ice movement over southwest Norway indicating an ice dome over the coastal district of west Norway. *Quaternary Science Reviews* 9, 99–116.

Arendt, A. A. and Sharp, M. J. 1999. Energy balance measurements on a Canadian high Arctic glacier and their implication for mass balance modelling. *IAHS Publication* 256, 165–72.

Arendt, A. A., Echelmeyer, K. A., Harrison, W. D., Lingle, C. S. and Valentine, V. B. 2002. Rapid wastage of Alaska glaciers and their contribution to rising sea level. *Science* 297, 382–6.

Arfstrom, J. and Hartmann, W. K. 2005. Martian flow features, moraine-like ridges, and gullies: terrestrial analogs and interrelationships. *Icarus* 174, 321–35.

Arkhipov, S. A., Ehlers, J., Johnson, R. G. and Wright, H. E. 1995. Glacial drainage towards the Mediterranean during the Middle and Late Pleistocene. *Boreas* 24, 196–206.

Arnold, N. S. 2005. Investigating the Sensitivity of Glacier Mass-Balance/Elevation Profiles to Changing Meteorological Conditions: Model Experiments for Haut Glacier D'Arolla, Valais, Switzerland. *Arctic, Antarctic, and Alpine Research* 37, 139–45.

Arnold, N. S., Richards, K. S., Willis, C. and Sharp, M. J. 1998. Initial results from a physically based, distributed model of glacier hydrology. *Hydrological Processes* 12, 191–219.

Arnold, N. S. and Sharp, M. 2002. Flow variability in the Scandinavian Ice Sheet: modelling the coupling between ice sheet flow and hydrology. *Quaternary Science Reviews* 21, 485–502.

Arrigo, K. R. and van Dijken, G. L. 2003. Impact of iceberg C-19 on Ross Sea primary production. *Geophysical Research Letters* 30, doi:10.1029/2003GL017721.

Arsenault, A. M. and Meigs, A. J. 2005. Contribution of deep-seated bedrock landslides to erosion of a glaciated basin in southern Alaska. *Earth Surface Processes and Landforms* 30, 1111–25.

Arthern, R. A. and Wingham, D. J. 1998. The natural fluctuations of firn densification and their effect on the geodetic determination of ice sheet mass balance. *Climatic Change* 40, 605–24.

Arthurton, R. S. and Wadge, A. J. 1981. *Geology of the Country around Penrith*. Memoir of the British Geological Survey. HMSO, London.

Aschwanden, A. and Blatter, H. 2005. Meltwater production due to strain heating in Storglaciåren, Sweden. *Journal of Geophysical Research* 110, doi:10.1029/2005JF000328.

Ashley, G. M. 1975. Rhythmic sedimentation in glacial Lake Hitchcock, Massachusetts-Connecticut. In Jopling, A. V. and MacDonald, B. C. (eds), *Glaciofluvial and Glaciolacustrine Sedimentation*. SEPM Special Publication 23, 304–20.

Ashley, G. M. 1990. Classification of large-scale subaqueous bedforms: a new look at an old problem. *Journal of Sedimentary Petrology* 60, 160–72.

Ashley, G. M. 1995. Glaciolacustrine environments. In Menzies, J. (ed.), *Glacial Environments. Volume 1: Modern Glacial Environments: Processes, Dynamics and Sediments*. Butterworth-Heinemann, Oxford, 417–44.

Ashley, G. M. and Smith, N. D. 2000. Marine sedimentation at a calving glacier margin. *Geological Society of America Bulletin* 112, 657–67.

Ashley, G. M., Southard, J. B. and Boothroyd, J. C. 1982. Deposition of climbing-ripple beds: a flume simulation. *Sedimentology* 29, 67–79.

Ashley, G. M., Boothroyd, J. C. and Borns, H. W. 1991. Sedimentology of late Pleistocene (Laurentide) deglacial-phase deposits, eastern Maine: an example of a temperate marine grounded ice-sheet margin. In Anderson, J. B. and Ashley, G. M. (eds), *Glacial Marine Sedimentation: Paleoclimatic Significance*. Geological Society of America, Special Paper 261, 107–25.

Ashmore, P. E. 1991. How do gravel-bed rivers braid? *Canadian Journal of Earth Sciences* 28, 326–41.

Ashworth, P. J. and Ferguson, R. I. 1986. Interrelationships of channel processes, changes and sediments in a proglacial river. *Geografiska Annaler* 68A, 361–71.

Asku, A. E. and Hiscott, R. N. 1992. Shingled Quaternary debris flow lenses on the north east Newfoundland slope. *Sedimentology* 39, 193–206.

Atkins, C. B., Barrett, P. J. and Hicock, S. R. 2002. Cold glaciers erode and deposit: evidence from Allan Hills, Antarctica. *Geology* 30, 659–62.

Atkinson, B. K. 1984. Subcritical crack growth in geological materials. *Journal of Geophysical Research* 89B, 4077–114.

Atkinson, B. K. (ed.). 1987. *Fracture Mechanics of Rock*. Academic Press, London.

Atkinson, N. 2003. Late Wisconsinan glaciation of Amund and Ellef Ringnes islands, Nunavut: evidence for the configuration, dynamics and deglacial chronology of the northwest sector of the Innuitian Ice Sheet. *Canadian Journal of Earth Sciences* 40, 351–63.

Attig, J. W. and Clayton, L. 1993. Stratigraphy and origin of an area of hummocky glacial topography, northern Wisconsin, USA. *Quaternary International* 18, 61–7.

Attig, J. W., Mickelson, D. M. and Clayton, L. 1989. Late Wisconsin landform distribution and glacier bed conditions in Wisconsin. *Sedimentary Geology* 62, 399–405.

Augustinus, P. C. 1992a. The influence of rock mass strength on glacial valley cross profile morphometry: a case study from the Southern Alps, New Zealand. *Earth Surface Processes and Landforms* 17, 39–51.

Augustinus, P. C. 1992b. Outlet glacier trough size-drainage area relationships, Fiordland, New Zealand. *Geomorphology* 4, 347–61.

Augustinus, P. C. 1995. Rock mass strength and the stability of some glacial valley slopes. *Zeitschrift für Geomorphologie* 39, 55–68.

Aylsworth, J. M. and Shilts, W. W. 1989. Bedforms of the Keewatin Ice Sheet, Canada. *Sedimentary Geology* 62, 407–28.

Badino, G. 2002. The glacial karst. Proceedings of Vth International Symposium on Glacier Caves and Cryokarst. *Nimbus* 23, 82–93.

Badino, G., De Vivo, A. and Piccini, L. 2007. *Caves of Sky: A Journey in the Heart of Glaciers*. La Venta, Esplorazioni geografiche, Magliano Sabina.

Bagnold, R. A. 1956. The flow of cohesionless grains in fluids. *Royal Society of London Philosophical Transactions*, Series A 249, 235–97.

Bahr, D. B., Meier, M. F. and Peckham, S. D. 1997. The physical basis of glacier volume-area scaling. *Journal of Geophysical Research* 102 (B9), 20355–62.

Bahr, D. B., Pfeffer, W. T., Sassolas, C. and Meier, M. 1998. Response time of glaciers as a function of size and mass balance: 1. Theory. *Journal of Geophysical Research* 103, B5, 9777–82.

Baker, V. R. 1979. Erosional processes in channelized water flows on Mars. *Journal of Geophysical Research* 84, 7985–93.

Baker, V. R. 1981. *Catastrophic Flooding: The Origin of the Channeled Scabland*. Dowden, Hutchinson and Ross, Stroudsburg, PA.

Baker, V. R. 2001. Water and the Martian landscape. *Nature* 412, 228–36.

Baker, V. R. 2003. Icy Martian mysteries. *Nature* 426, 779–80.

Baker, V. R. 2005. Picturing a recently active Mars. *Nature* 434, 280–3.

Baker, V. R. 2002. High energy megafloods: planetary settings and sedimentary dynamics. In Martini, I. P., Baker, V. R. and Garzon, M. (eds), *Flood and Megaflood Deposits: Recent and Ancient Examples*. International Association of Sedimentologists Special Publication.

Baker, V. R., Greeley, R., Komar, P. D., Swanson, D. A. and Waitt, R. B. 1987. Columbia and Snake River Plains. In Graf, W. L. (ed.), *Geomorphic Systems of North America*. Geological Society of America, Centennial Special Volume 2, 403–68.

Baker, V. R., Benito, G. and Rudoy, A. N. 1993. Paleohydrology of Late Pleistocene superflooding, Altay Mountains, Siberia. *Science* 259, 348–50.

Bales, R. C., Davis, E. E. and Stanley, D. A. 1989. Ionic elution through shallow, homogeneous snow. *Water Resources Research* 25, 1869–77.

Bales, R. C., McConnell, J. R., Mosley-Thompson, E. and Csatho, B. 2001. Accumulation over the Greenland ice sheet from historical and recent records. *Journal of Geophysical Research* 106, D24, 33813–26.

Ballantyne, C. K. 1982. Aggregate clast form characteristics of deposits near the margins of four glaciers in the Jotumheimen Massif, Norway. *Norsk Geografisk Tidsskrift* 36, 103–13.

Ballantyne, C. K. 1987. Some observations of the morphology and sedimentology of two active protalus ramparts, Lyngen, northern Norway. *Arctic and Alpine Research* 19, 167–74.

Ballantyne, C. K. 1989. The Loch Lomond Readvance on the Isle of Skye, Scotland: glacier reconstruction and palaeoclimatic implications. *Journal of Quaternary Science* 4, 95–108.

Ballantyne, C. K. 1995. Paraglacial debris-cone formation on recently-deglaciated terrain. *The Holocene* 5, 25–33.

Ballantyne, C. K. 1997a. Holocene rock slope failures in the Scottish Highlands. *Palaoklimaforschung* 19, 197–205.

Ballantyne, C. K. 1997b. Periglacial trimlines in the Scottish Highlands. *Quaternary International* 38/39, 119–36.

Ballantyne, C. K. 1998. Age and significance of mountain-top detritus. *Permafrost and Periglacial Processes* 9, 327–45.

Ballantyne, C. K. 2002a. Paraglacial geomorphology. *Quaternary Science Reviews* 21, 1935–2017.

Ballantyne, C. K. 2002b. A general model of paraglacial landscape response. *The Holocene* 12, 371–6.

Ballantyne, C. K. 2002c. The Loch Lomond Readvance on the Isle of Mull, Scotland: glacier reconstruction and palaeoclimatic implications. *Journal of Quaternary Science* 17, 759–71.

Ballantyne, C. K. 2003. Paraglacial landsystems. In Evans, D. J. A. (ed.), *Glacial Landsystems*. Arnold, London, 432–61.

Ballantyne, C. K. 2006. Loch Lomond Stadial glaciers in the Uig Hills, western Lewis, Scotland. *Scottish Geographical Journal* 122, 256–73.

Ballantyne, C. K. 2007. Trimlines and palaeonunataks. In Elias, S. A. (ed.), *Encyclopedia of Quaternary Science*. Elsevier, Oxford, 892–903.

Ballantyne, C. K. and Benn, D. I. 1994a. Glaciological constraints on protalus rampart development. *Permafrost and Periglacial Processes* 5, 145–53.

Ballantyne, C. K. and Benn, D. I. 1994b. Paraglacial slope adjustment and resedimentation following recent glacier retreat, Fåbergstølsdalen, Norway. *Arctic and Alpine Research* 26, 255–69.

Ballantyne, C. K. and Benn, D. I. 1996. Paraglacial slope adjustment during recent deglaciation: implications for slope evolution in formerly glaciated terrain. In Brooks, S. and Anderson, M. G. (eds), *Advances in Hillslope Processes*, Wiley, Chichester, 1173–95.

Ballantyne, C. K. and Harris, C. 1994. *The Periglaciation of Great Britain*. Cambridge University Press.

Ballantyne, C. K. and McCarroll, D. 1995. The vertical dimensions of Late Devensian glaciation on the mountains of Harris and southeast Lewis, Outer Hebrides, Scotland. *Journal of Quaternary Science* 10, 211–23.

Ballantyne, C. K. and Stone, J. O. 2004. The Beinn Alligin rock avalanche, NW Scotland: cosmogenic [10]Be dating, interpretation and significance. *The Holocene* 14, 448–53.

Ballantyne, C. K., Black, N. M. and Finlay, D. P. 1989. Enhanced boulder weathering under late-lying snow patches. *Earth Surface Processes and Landforms* 14, 745–50.

Ballantyne, C. K., McCarroll, D., Nesje, A. and Dahl, S.-O. 1997. Periglacial trimlines, former nunataks and the dimensions of the last ice sheet in Wester Ross, northwest Scotland. *Journal of Quaternary Science* 12, 225–38.

Ballantyne, C. K., McCarroll, D., Nesje, A., Dahl, S.-O. and Stone, J. O. 1998. The last ice sheet in northwest Scotland: reconstruction and implications. *Quaternary Science Reviews* 17, 1149–84.

Ballantyne, C. K., Schnabel, C. and Xu, S. 2008. Exposure dating and reinterpretation of coarse debris accumulations ('rock glaciers') in the Cairngorm Mountains, Scotland. *Journal of Quaternary Science* 24, 19–31.

Bamber, J. L. and Kwok, R. 2004. Remote sensing techniques. In Bamber, J. L. and Payne, A. J. (eds), *Mass Balance of the Cryosphere: Observations and Modeling of Contemporary and Future Changes*. Cambridge University Press, 59–113.

Bamber, J. L. and Payne, A. J. (eds). 2004. *Mass Balance of the Cryosphere: Observations and Modeling of Contemporary and Future Changes*. Cambridge University Press.

Bamber, J. L. and Rivera, A. 2007. A review of remote sensing methods for glacier mass balance determination. *Global and Planetary Change* 59, 138–48.

Bamber, J. L., Alley, R. B. and Joughin, I. 2007. Rapid response of modern day ice sheets to external forcing. *Earth and Planetary Science Letters* 257, 1–13.

Bamber, J. L., Hardy, R. J. and Joughin, I. 2000. An analysis of balance velocities over the Greenland ice sheet and comparison with synthetic aperture radar interferometry. *Journal of Glaciology* 46 (152), 67–74.

Bamber, J. L., Krabill, W., Raper, V., Dowdeswell, J. A. and Oerlemans, J. 2005. Elevation changes measured on Svalbard glaciers and ice caps from airborne laser data. *Annals of Glaciology* 42, 202–8.

Banham, P. H. 1977. Glacitectonites in till stratigraphy. *Boreas* 6, 101–5.

Bannerjee, I. and McDonald, B. C. 1975. Nature of esker sedimentation. In Jopling, A. V. and McDonald, B. C. (eds), *Glaciofluvial and Glaciolacustrine Sedimentation*. SEPM Special Publication 23, 132–54.

Bard, E., Hamelin, B. and Fairbanks, G. 1990. U-Th ages obtained by mass spectrometry in corals from Barbados: sea level during the past 130,000 years. *Nature* 346, 456–8.

Bard, E., Hamelin, B., Arnold, M., Montaggioni, L., Cabioch, G., Faure, G. and Rougerie, F. 1996. Deglacial sea-level record from Tahiti corals and the timing of global meltwater discharge. *Nature* 382, 241–4.

Barker, P. F., Diekmann, B. and Escutia, C. 2007. Onset of Cenozoic Antarctic glaciation. *Deep Sea Research Part II: Topical Studies in Oceanography* 54, 2293–307.

Barnard, P. L., Owen, L. A. and Finkel, R. C. 2004a. Style and timing of glacial and paraglacial sedimentation in a monsoonal-influenced high Himalayan environment, the upper Bhagirathi Valley, Garwhal Himalaya. *Sedimentary Geology* 165, 199–221.

Barnard, P. L., Owen, L. A., Sharma, M. C. and Finkel, R. C. 2004b. Late Quaternary landscape evolution of a monsoon-influenced high Himalayan valley, Gori Ganga, Nanda Devi, NE Garwhal. *Geomorphology* 61, 91–110.

Barnard, P. L., Owen, L. A. and Finkel, R. C. 2006a. Quaternary fans and terraces in the Khumbu Himal south of Mount Everest: their characteristics, age and formation. *Journal of the Geological Society of London* 163, 383–99.

Barnard, P. L., Owen, L. A., Finkel, R. C. and Asahi, K. 2006b. Landscape response to deglaciation in a high relief, monsoon-influenced alpine environment, Langtang Himal, Nepal. *Quaternary Science Reviews* 25, 2162–76.

Barnett, P. J., Sharpe, D. R. and Russell, H. A. J. 1998. On the origin of the Oak Ridges Moraine. *Canadian Journal of Earth Sciences* 35, 1152–67.

Barnett, T. P., Adam, J. C. and Lettenmeier, D. P. 2005. Potential impacts of a warming climate on water availability in snow-dominated regions. *Nature* 438, 303–9.

Barrett, B. E., Murray, T., Clark, R. A. and Matsuoka, K. 2008. Distribution and character of water in a surge-type glacier revealed by multifrequency and multipolarization ground-penetrating radar. *Journal of Geophysical Research*, 113, F04011, 10.1029/2007JF000972.

Barrett, P. J. 1980. The shape of rock particles, a critical review. *Sedimentology* 27, 291–303.

Barrett, P. J. 1996. Antarctic palaeoenvironments through Cenozoic times: a review. *Terra Antarctica* 3, 103–19.

Barrett, P. J. 2003. Cooling a continent. *Nature* 421, 221–3.

Barry, R. G. 1992. *Mountain Weather and Climate*, 2nd edn. Routledge, London.

Barry, R. G. and Chorley, R. J. 2003. *Atmosphere, Weather and Climate*, 8th edn. Routledge, London.

Bartholomaus, T. C., Anderson, R. S. and Anderson, S. P. 2008. Response of glacier basal motion to transient water storage. *Nature Geoscience* 1, 33–7.

Bassis, J. N., Fricker, H. A., Coleman, R. and Minster, J.-B. 2008. An investigation into the forces that drive ice-shelf rift propagation on the Amery Ice Shelf, East Antarctica. *Journal of Glaciology* 54, 17–27.

Beaney, C. L. 2002. Tunnel channels in southeast Alberta, Canada: evidence for catastrophic channelized drainage. *Quaternary International* 90, 67–74.

Beaney, C. L. and Shaw, J. 2000. The subglacial geomorphology of southeast Alberta: evidence for subglacial meltwater erosion. *Canadian Journal of Earth Sciences* 37, 51–61.

Beaudoin, A. B. and King, R. H. 1994. Holocene palaeoenvironmental record preserved in a paraglacial alluvial fan, Sunwapta Pass, Jasper National Park, Alberta, Canada. *Catena* 22, 227–48.

Beaudry, L. M. and Prichonnet, G. 1991. Late glacial DeGeer moraines with glaciofluvial sediment in the Chapais area, Quebec, Canada. *Boreas* 20, 377–94.

BEDMAP Consortium 2002. http://www.antarctica.ac.uk//bas_research/data/access/bedmap/. BEDMAP homepage – A new ice thickness and subglacial topographic model of the Antarctic.

Bednarski, J. 1988. The geomorphology of glaciomarine sediments in a high arctic fiord. *Geographie Physique et Quaternaire* 42, 65–74.

Bednarski, J. 1998. Quaternary history of Axel Heiberg Island, bordering Nansen Sound, NWT, emphasizing the Last Glacial Maximum. *Canadian Journal of Earth Sciences* 35, 520–33.

Beer, J., Mende, W. and Stellmacher, R. 2000. The role of the sun in climate forcing. *Quaternary Science Reviews* 19, 403–15.

Bell, R. E. 2008. The role of subglacial water in ice-sheet mass balance. *Nature Geoscience* 1, 297–304.

Bell, R. E., Blankenship, D. D., Finn, C. A., Morse, D. L., Scambos, T. A., Brozena, J. M. and Hodge, S. M. 1998. Influence of subglacial geology on the onset of a West Antarctic ice stream from aerogeophysical observations. *Nature* 394, 58–62.

Bell, R. E., Studinger, M., Tikku, A. A., Clarke, G. K., Gutner, M. M. and Meertens. C. 2002. Origin and fate of Lake Vostok water frozen to the base of the East Antarctic ice sheet. *Nature* 416, 307–10.

Bell, R. E., Studlinger, M., Fahnestock, M. A. and Shuman, C. A. 2006. Tectonically controlled subglacial lakes on the flanks of the Gamburtsev Mountains, East Antarctica. *Geophysical Research Letters* 33, doi:10.1029/2005GL025207.

Bell, R. E., Studinger, M., Shuman, C. A., Fahnestock, M. A. and Joughin, I. 2007. Large subglacial lakes in East Antarctica at the onset of fast flowing ice streams. *Nature* 445, 904–7.

Benediktsson, Í. Ö., Ingólfsson, Ó., Schomacker, A. and Kjær, K. 2009. Formation of submarginal and proglacial end moraines: implications of ice-flow mechanism during the 1963–64 surge of Brúarjökull, Iceland. *Boreas* 38, 440–57.

Benediktsson, Í. Ö., Möller, P., Ingólfsson, Ó., van der Meer, J. J. M., Kjær, K. H. and Krüger, J. 2008. Instantaneous end moraine and sediment wedge formation during the 1890 glacier surge of Brúarjökull, Iceland. *Quaternary Science Reviews* 27, 209–34.

Benito, G. 1997. Energy expenditure and geomorphic work of the cataclysmic Missoula flooding in the Columbia River gorge, USA. *Earth Surface Processes and Landforms* 22, 457–72.

Benn, D. I. 1989a. Controls on sedimentation in a Late Devensian ice-dammed lake, Achnasheen, Scotland. *Boreas* 18, 31–42.

Benn, D. I. 1989b. Debris transport by Loch Lomond Readvance glaciers in northern Scotland, basin form and the within-valley asymmetry of lateral moraines. *Journal of Quaternary Science* 4, 243–54.

Benn, D. I. 1992. The genesis and significance of 'hummocky moraine': evidence from the Isle of Skye, Scotland. *Quaternary Science Reviews* 11, 781–99.

Benn, D. I. 1994a. Fabric shape and the interpretation of sedimentary fabric data. *Journal of Sedimentary Research* A64, 910–15.

Benn, D. I. 1994b. Fluted moraine formation and till genesis below a temperate glacier: Slettmarkbreen, Jotunheimen, Norway. *Sedimentology* 41, 279–92.

Benn, D. I. 1995. Fabric signature of subglacial till deformation, Breiðamerkurjökull, Iceland. *Sedimentology* 42, 735–47.

Benn, D. I. 1996. Subglacial and subaqueous processes near a glacier grounding line: sedimentological evidence from a former ice-dammed lake, Achnasheen, Scotland. *Boreas* 25, 23–36.

Benn, D. I. 2002. Clast fabric development in a shearing granular material: implications for subglacial till and fault gouge – discussion. *Geological Society of America Bulletin* 114, 382–3.

Benn, D. I. 2004a. Clast morphology. In Evans, D. J. A. and Benn, D. I. (eds), *A Practical Guide to the Study of Glacial Sediments*. Arnold, London, 78–92.

Benn, D. I. 2004b. Macrofabric. In Evans, D. J. A. and Benn, D. I. (eds), *A Practical Guide to the Study of Glacial Sediments*. Arnold, London, 93–114.

Benn, D. I. 2006. Interpreting glacial sediments. In Knight, P. (ed.), *Glacier Science and Environmental Change*. Blackwell, Oxford, 434–9.

Benn, D. I. and Ballantyne, C. K. 1993. The description and representation of particle shape. *Earth Surface Processes and Landforms* 18, 665–72.

Benn, D. I. and Ballantyne, C. K. 1994. Reconstructing the transport history of glacigenic sediments: a new approach based on the co-variance of clast shape indices. *Sedimentary Geology* 91, 215–27.

Benn, D. I. and Ballantyne, C. K. 2005. Palaeoclimatic reconstruction from Loch Lomond Readvance glaciers in the West Drumochter Hills, Scotland. *Journal of Quaternary Science* 20, 577–92.

Benn, D. I. and Clapperton, C. M. 2000a. Pleistocene glacitectonic landforms and sediments around central Magellan Strait, southernmost Chile: evidence for fast outlet glaciers with cold-based margins. *Quaternary Science Reviews* 19, 591–612.

Benn, D. I. and Clapperton, C. M. 2000b. Glacial sediment–landform associations and paleoclimate during the last glaciation, Strait of Magellan, Chile. *Quaternary Research* 54, 13–23.

Benn, D. I. and Evans, D. J. A. 1993. Glaciomarine deltaic deposition and ice-marginal tectonics: the 'Loch Don Sand Moraine', Isle of Mull, Scotland. *Journal of Quaternary Science* 8, 279–91.

Benn, D. I. and Evans, D. J. A. 1996. The interpretation and classification of subglacially-deformed materials. *Quaternary Science Reviews* 15, 23–52.

Benn, D. I. and Evans, D. J. A. 1998. *Glaciers and Glaciation*. Arnold, London.

Benn, D. I. and Evans, D. J. A. 2006. Subglacial megafloods: outrageous hypothesis or just outrageous? In Knight, P. G. (ed.), *Glacier Science and Environmental Change*. Blackwell, London, 42–6.

Benn, D. I. and Gemmell, A. M. D. 2002. 'Fractal dimensions' of diamictic particle-size distributions: simulations and evaluation. *Geological Society of America Bulletin* 114, 528–32.

Benn, D. I. and Gulley, J., Luckman, A., Adamek, A. and Glowacki, P. 2009. Englacial drainage systems formed by hydrologically driven crevasse propagation. *Journal of Glaciology* 55 (191), 513–523.

Benn, D. I. and Hulton, N. R. J. 2010. An Excel™ spreadsheet program for reconstructing the surface profile of former mountain glaciers and ice caps. *Computers and Geosciences*.

Benn, D. I. and Lehmkuhl, F. 2000. Mass balance and equilibrium-line altitudes of glaciers in high-mountain environments. *Quaternary International* 65/66, 15–29.

Benn, D. I. and Lukas, S. 2006. Younger Dryas glacial landsystems in North West Scotland: an assessment of modern analogues and palaeoclimatic implications. *Quaternary Science Reviews* 25, 2390–408.

Benn, D. I. and Owen, L. A. 2002. Himalayan glacial sedimentary environments: a framework for reconstructing and dating former glacial extents in high mountain regions. *Quaternary International* 97/98, 3–25.

Benn, D. I., Evans, D. J. A., Phillips, E. R., Hiemstra, J. F., Walden, J. and Hoey, T. B. 2004. The research project – a case study of Quaternary glacial sediments. In Evans, D. J. A. and Benn, D. I. (eds), *A Practical Guide to the Study of Glacial Sediments*. Arnold, London, 209–34.

Benn, D. I., Hulton, N. R. J. and Mottram, R. H. 2007a. 'Calving laws', 'sliding laws' and the stability of tidewater glaciers. *Annals of Glaciology* 46, 123–30.

Benn, D. I., Kirkbride, M. P., Owen, L. A. and Brazier. V. 2003. Glaciated valley landsystems. In Evans, D. J. A. (ed.), *Glacial Landsystems*. Arnold, London, 372–406.

Benn, D. I., Kristensen, L. and Gulley, J. 2009. Surge propagation constrained by a persistent subglacial conduit, Bakaninbreen-Paulabreen, Svalbard. *Annals of Glaciology* 50, 81–6.

Benn, D. I., Lowe, J. J. and Walker, M. J. C. 1992. Glacier response to climatic change during the Loch Lomond Stadial and early Flandrian: geomorphological and palynological evidence from the Isle of Skye, Scotland. *Journal of Quaternary Science* 7, 125–44.

Benn, D. I., Owen, L. A., Osmaston, H. A., Seltzer, G. O., Porter, S. C. and Mark, B. 2005. Reconstruction of equilibrium-line altitudes for tropical and sub-tropical glaciers. *Quaternary International* 138/139, 8–21.

Benn, D. I., Owen, L. A., Finkel, R. and Clemmens, S. 2006. Pleistocene lake outburst floods and fan formation along the eastern Sierra Nevada, California: implications for the interpretation of intermontane lacustrine records. *Quaternary Science Reviews* 25, 2729–48.

Benn, D. I., Warren, C. R. and Mottram, R. H. 2007b. Calving processes and the dynamics of calving glaciers. *Earth Science Reviews* 82, 143–79.

Benn, D. I., Wiseman, S. and Hands, K., 2001. Growth and drainage of supraglacial lakes on the debris-mantled Ngozumpa Glacier, Khumbu Himal. *Journal of Glaciology*, 47, 626–38.

Bennett, M. R. 1990. *The Cwms of Snowdonia: A Morphometric Analysis*. Queen Mary and Westfield College Research Papers in Geography 2.

Bennett, M. R. 1994. Morphological evidence as a guide to deglaciation following the Loch Lomond Readvance: a review of research approaches and models. *Scottish Geographical Magazine* 110, 24–32.

Bennett, M. R. 2001. The morphology, structural evolution and significance of push moraines. *Earth Science Reviews* 53, 197–236.

Bennett, M. R. 2003. Ice streams as the arteries of an ice-sheet: their mechanics, stability and significance. *Earth Science Reviews* 61, 309–39.

Bennett, M. R. and Boulton, G. S. 1993. A reinterpretation of Scottish 'hummocky moraine' and its significance for the deglaciation of the Scottish highlands during the Younger Dryas or Loch Lomond Stadial. *Geological Magazine* 130, 301–18.

Bennett, M. R. and Glasser, N. F. 1996. *Glacial Geology: Ice Sheets and Landforms*. Wiley, Chichester.

Bennett, M. R., Doyle, P. and Mather, A. E., 1996a. Dropstones: their origin and significance. *Palaeogeography, Palaeoclimatology, Palaeoecology* 121, 331–9.

Bennett, M. R., Hambrey, M. J., Huddart, D. and Ghienne, J. F. 1996b. The formation of a geometrical ridge network by the surge-type glacier Kongsvegen, Svalbard. *Journal of Quaternary Science* 11, 437–49.

Bennett, M. R., Hambrey, M. J., Huddart, D. and Glasser, N. F. 1998a. Glacial thrusting and moraine-mound formation in Svalbard and Britain: the example of Coire a' Cheud-chnoic (Valley of a Hundred Hills). *Quaternary Proceedings* 6, 17–34.

Bennett, M. R., Huddart, D., Hambrey, M. J. and Ghienne, J. F. 1998b. Modification of braided outwash surfaces by aufeis: an example from Pedersenbreen, Svalbard. *Zeitschrift für Geomorphologie* 42, 1–20.

Bennett, M. R., Waller, R. I., Glasser, N. F., Hambrey, M. J. and Huddart, D. 1999a. Glacigenic clast fabric: genetic fingerprint or wishful thinking? *Journal of Quaternary Science* 14, 125–35.

Bennett, M. R., Huddart, D. and Glasser, N. F. 1999b. Large-scale bedrock displacement by cirque glaciers. *Arctic, Antarctic and Alpine Research* 31, 99–107.

Bennett, M. R., Huddart, D. and McCormick, T. 2000a. An integrated approach to the study of glaciolacustrine landforms and sediments: a case study from Hagavatn, Iceland. *Quaternary Science Reviews* 19, 633–65.

Bennett, M. R., Huddart, D. and Waller, R. I. 2000b. Glaciofluvial crevasse and conduit fills as indicators of supraglacial dewatering during a surge, Skeiðarárjökull, Iceland. *Journal of Glaciology* 46, 25–34.

Bennett, M. R., Huddart, D. and McCormick, T. 2000c. The glacio-lacustrine landform-sediment assemblage at Heinabergsjökull, Iceland. *Geografiska Annaler* 82A, 1–16.

Bennett, M. R., Huddart, D., Glasser, N. F. and Hambrey, M. J. 2000d. Resedimentation of debris on an ice-cored lateral moraine in the high arctic (Kongsvegen, Svalbard). *Geomorphology* 35, 21–40.

Bennett, M. R., Huddart, D. and Thomas, G. S. P. 2002. Facies architecture within a regional glaciolacustrine basin: Copper river, Alaska. *Quaternary Science Reviews* 21, 2237–79.

Bennett, M. R., Waller, R. I., Midgley, N. G., Huddart, D., Gonzalez, S., Cook, S. J. and Tomio, A. 2003. Subglacial deformation at sub-freezing temperatures? Evidence from Hagafellsjökull-Eystri, Iceland. *Quaternary Science Reviews* 22, 915–23.

Bennett, M. R., Huddart, D., Waller, R. I., Cassidy, N., Tomio, A., Zukowskyj, P., Midgley, N. G., Cook, S. J., Gonzalez, S. and Glasser, N. F. 2004. Sedimentary and tectonic architecture of a large push moraine: a case study from Hagafellsjökull-Eystri, Iceland. *Sedimentary Geology* 172, 269–92.

Bennett, M. R., Huddart, D. and Waller, R. I. 2005. The interaction of a surging glacier with a seasonally frozen foreland: Hagafellsjökull-Eystri, Iceland. In Harris, C. and Murton, J. B. (eds), *Cryospheric Systems: Glaciers and Permafrost*. Geological Society, London, Special Publication 242, 51–62.

Bennett, M. R., Huddart, D. and Waller, R. I. 2006. Diamict fans in subglacial water-filled cavities – a new glacial environment. *Quaternary Science Reviews* 25, 3050–69.

Benson, J. L. and James, P. B. 2005. Yearly comparisons of the Martian polar caps: 1999–2003 Mars Orbiter Camera observations. *Icarus* 174, 513–23.

Bentley, C. R. 1987. Antarctic ice streams: a review. *Journal of Geophysical Research* 92, 8843–58.

Bentley, C. R. 2004. Mass balance of the Antarctic ice sheet: observational aspects. In Bamber, J. L. and Payne, A. J. (eds), *Mass Balance of the Cryosphere: Observations and Modeling of Contemporary and Future Changes*. Cambridge University Press, 459–89.

Bentley, M. J., Sugden, D. E., Hulton, N. R. J. and McCulloch, R. D. 2005. The landforms and pattern of deglaciation in the Strait of Magellan and Bahia Inutil, southernmost South America. *Geografiska Annaler* 87A, 313–33.

Bentley, M. J., Evans, D. J. A., Fogwill, C. J., Hansom, J. D., Sugden, D. E. and Kubik, P. W. 2007. Glacial geomorphology and chronology of deglaciation, South Georgia, sub-Antarctic. *Quaternary Science Reviews* 26, 644–77.

Benvenuti, M. and Martini, I. P. 2002. Analysis of terrestrial hyperconcentrated flows and their deposits. In Martini, I. P., Baker, V. R. and Garzon, G. (eds), *Flood and Megaflood Processes and Deposits: Recent and Ancient Examples*. International Association of Sedimentologists, Special Publication 32, Blackwell, Oxford, 167–93.

Berger, A. L., Gulick, S. P. S., Spotila, J. A., Upton, P., Jaeger, J. M., Chapman, J. B., Worthington, L. A., Pavlis, T. L., Ridgway, K. D., Willems, B. A. and McAleer, R. J. 2008. Quaternary tectonic response to intensified glacial erosion in an orogenic wedge. *Nature Geoscience* 1, 793–99.

Berthelsen, A. 1978. The methodology of kineto-stratigraphy as applied to glacial geology. *Bulletin of the Geological Society of Denmark* 27, 25–38.

Berthelsen, A. 1979. Recumbent folds and boudinage structures formed by subglacial shear: an example of gravity tectonics. *Geologie en Mijnbouw* 58, 253–60.

Bibring, J.-P., Langevin, Y., Poulet, F., Gendrin, A., Gondet, B., Berthé, M., Soufflot, A., Drossart, P., Combes, M., Bellucci, G., Moroz, V., Mangold, N., Schmitt, B. and the OMEGA team. 2004. Perennial water ice identified in the south polar cap of Mars. *Nature* 428, 627–30.

Bickel, L. 2000. *Mawson's Will*. Steerforth Press, South Royalton, Vermont.

Bik, M. J. J. 1969. The origin and age of prairie mounds of southern Alberta. *Biuletyn Peryglacjalny* 19, 85–130.

Billups, K. and Schrag, D. P. 2003. Application of benthic foraminiferal Mg/Ca ratios to questions of Cenozoic climate change. *Earth and Planetary Science Letters* 209, 181–95.

Bindoff, N. L. et al. 2007. Observations: oceanic climate change and sea level. In Solomon, S. et al. (eds), *Climate Change 2007: The Physical Science Basis*. Contribution of Working Group I to the Fourth Assessment of the Intergovernmental Panel on Climate Change. Cambridge University Press.

Bindschadler, R. A. 1982. A numerical model of temperate glacier flow applied to the quiescent phase of a surge-type glacier. *Journal of Glaciology* 28, 238–65.

Bindschadler, R. A. 1998. Monitoring ice sheet behavior from space. *Reviews of Geophysics* 36, 79–104.

Bindschadler, R. A. 2005. Changes in the ice plain of Whillans Ice Stream, West Antarctica. *Journal of Glaciology* 51 (175), 620–36.

Bindschadler, R. A. 2006. The environment and evolution of the West Antarctic ice sheet: setting the stage. *Philosophical Transactions of the Royal Society* A 364, 1583–605.

Bindschadler, R. A. and Choi, H. 2007. Increased water storage at ice-stream onsets: a critical mechanism? *Journal of Glaciology* 53 (181), 163–71.

Bindschadler, R. A., Harrison, W. D., Raymond, C. F. and Crosson, R. 1977. Geometry and dynamics of a surge-type glacier. *Journal of Glaciology* 18, 181–94.

Bindschadler, R. A., Jesek, K. C. and Crawford, J. 1987. Glaciological investigations using the synthetic aperture radar imaging system. *Annals of Glaciology* 9, 11–19.

Bindschadler, R. A., Dowdeswell, J., Hall, D. and Winther, J.-G. 2001. Glaciological applications with Landsat-7 imagery: early assessments. *Remote Sensing of Environment* 78, 163–79.

Bindschadler, R. A., King, M. A., Alley, R. B., Anandakrishnan, S. and Padman, L. 2003a. Tidally controlled stick-slip discharge of a West Antarctic ice stream. *Science* 301, 1087–93.

Bindschadler, R. A., Vornberger, P. L., King, M. A. and Padman, L. 2003b. Tidally driven stick-slip motion in the mouth of Whillans Ice Stream, Antarctica. *Annals of Glaciology* 36, 263–72.

Bindschadler, R. A., Vornberger, P. and Gray, L. 2005. Changes in the ice plain of Whillans Ice Stream, West Antarctica. *Journal of Glaciology* 51 (175), 620–36.

Bingham, R. G., Nienow, P. W., Sharp, M. J. and Boon, S. 2005. Subglacial drainage processes at a High Arctic polythermal glacier. *Journal of Glaciology* 51, 15–24.

Bintanja, R. 1998. The contribution of snowdrift sublimation to the surface mass balance of Antarctica. *Annals of Glaciology* 27, 251–9.

Bintanja, R. and van de Wal, R. S. W. 2008. North American ice-sheet dynamics and the onset of 100,000-year glacial cycles. *Nature* 454, 869–71.

Birnie, R. V. 1977. A snow-bank push mechanism for the formation of some 'annual' moraine ridges. *Journal of Glaciology* 18, 77–85.

Biscaye, P. E., Grousset, F. E., Revel, M., van der Gaast, S., Zielinski, G. A., Vaars, A. and Kukla, G. 1997. Asian provenance of glacial dust (stage 2) in the Greenland Ice Sheet Project 2 Ice Core, Summit, Greenland. *Journal of Geophysical Research* 102 (C12), 26765–81.

Bitz, C. M. and Battisti, D. S. 1999. Interannual to decadal variability in climate and the glacier mass balance in Washington, western Canada, and Alaska. *Journal of Climate* 12, 865–78.

Björnsson, H. 1974. Explanation of jökulhlaups from Grimsvötn, Vatnajökull, Iceland. *Jökull* 24, 1–26.

Björnsson, H. 1975. Subglacial water reservoirs, jökulhlaups and volcanic eruptions. *Jökull* 25, 1–12.

Björnsson, H. 1996. Scales and rates of glacial sediment removal: a 20 km long, 300 m deep trench created beneath Breiðamerkurjökull during the Little Ice Age. *Annals of Glaciology* 22, 141–6.

Björnsson, H. 1998. Hydrological characteristics of the drainage system beneath a surging glacier. *Nature* 395, 771–4.

Björnsson, H. 2002. Subglacial lakes and jökulhlaups in Iceland. *Global and Planetary Change* 35, 255–71.

Björnsson, H., Pálsson, F. and Guðmundsson, S. 2001. Jokulsárlon at Breiðamerkursandur, Vatnajökull, Iceland: 20th century changes and future outlook. *Jökull*, 50, 1–18.

Björnsson, H., Pálsson, F., Sigurdsson, O. and Flowers, G. E. 2003. Surges of glaciers in Iceland. *Annals of Glaciology* 36, 82–90.

Bjornstad, B. N., Fecht, K. R. and Pluhar, C. J. 2001. Long history of Pre-Wisconsin, ice age cataclysmic floods: evidence from southeastern Washington State. *Journal of Geology* 109, 695–713.

Blair, T. C. 2002. Alluvial-fan sedimentation from a glacial-outburst flood, Lone Pine, California, and contrasts with meteorological flood deposits. In Martini, I. P., Baker, V. R. and Garzon, G. (eds), *Flood and Megaflood Processes and Deposits: Recent and Ancient Examples*. Special Publications of Int. Ass. Sediment 32, 113–40.

Blair, T. C. and McPherson, J. G. 1994. Alluvial fan processes and forms. In Abrahams, A. D. and Parsons, A. J. (eds), *Geomorphology of Desert Environments*. Chapman and Hall, London, 354–402.

Blake, E. W., Clarke, G. K. C. and Gérin, M. C. 1992. Tools for examining subglacial bed deformation. *Journal of Glaciology* 38, 388–96.

Blake, E. W., Fischer, U. H. and Clarke, G. K. C. 1994. Direct measurement of sliding at the glacier bed. *Journal of Glaciology* 40, 595–9.

Blake, K. P. 2000. Common origin for De Geer moraines of variable composition in Raudvassdalen, northern Norway. *Journal of Quaternary Science* 15, 633–44.

Blankenship, D. D. and Bentley, C. R. 1987. The crystalline fabric of polar ice sheets inferred from seismic anisotropy. In Waddington, E. E. and Walder, J. S. (eds), *The Physical Basis of Ice Sheet Modelling*. IAHS Publication 170, 17–28.

Blankenship, D. D., Bentley, C. R., Rooney, S. T. and Alley, R. B. 1987. Till beneath Ice Stream B. I. Properties derived from seismic travel times. *Journal of Geophysical Research* 92, 8903–11.

Blatter, H. 1995. Velocity and stress fields in grounded glaciers: a simple algorithm for including longitudinal stress gradients. *Journal of Glaciology* 41, 333–43.

Blatter, H. and Hutter, K. 1991. Polythermal conditions in arctic glaciers. *Journal of Glaciology* 37, 261–9.

Bloom, A. L., Broecker, W. S., Chappell, J., Mathews, R. K. and Mesolella, K. J. 1974. Quaternary sea level fluctuations on a tectonic coast: new 230Th/234U dates from the Huon Peninsula, New Guinea. *Quaternary Research* 4, 185–205.

Bluck, B. J. 1974. Structural and directional properties of some valley sandur deposits in southern Iceland. *Sedimentology* 21, 533–54.

Bluck, B. J. 1999. Clast assembling, bed-forms and structure in gravel beaches. *Transactions of the Royal Society of Edinburgh* 89, 291–323.

Bluemle, J. P. 1993. Hydrodynamic blowouts in North Dakota. In Aber, J. S. (ed.), *Glaciotectonics and Mapping Glacial Deposits*. Canadian Plains Research Center, University of Regina, Saskatchewan, 259–66.

Bluemle, J. P. and Clayton, L. 1984. Large-scale glacial thrusting and related processes in North Dakota. *Boreas* 13, 279–99.

Bluemle, J. P., Lord, M. L. and Hunke, N. T. 1993. Exceptionally long, narrow drumlins formed in subglacial cavities, North Dakota. *Boreas* 22, 15–24.

Blunier, T., Schwander, J., Chappellaz, J., Parrenin, F. and Barnola, J. M. 2004. What was the surface temperature in central Antarctica during the last glacial maximum? *Earth and Planetary Science Letters* 218, 379–88.

Bogen, J. 1983. Morphology and sedimentology of deltas in fjord and fjord valley lakes. *Sedimentary Geology* 36, 245–67.

Bogen, J. 1996. Erosion rates and sediment yields of glaciers. *Annals of Glaciology* 22, 48–52.

Bøggild, C. E., Olesen, O. B., Ahlstrom, A. P. and Jørgensen, P. 2004. Automatic glacier ablation measurements using pressure transducers. *Journal of Glaciology* 50, 303–4.

Bøggild, C. E., Forsberg, R. and Reeh, N. 2005. Meltwater retention in a transect across the Greenland ice sheet. *Annals of Glaciology* 40, 169–73.

Bohren, C. F. 1983. Colors of snow, frozen waterfalls, and icebergs. *Journal of the Optical Society of America* 73, 1646–52.

Bolch, T., Buchroither, M. F., Peters, J., Baessler, M. and Bajracharya, S. 2008a. Identification of glacier motion and potentially dangerous lakes in the Mt. Everest region/Nepal using spaceborne imagery. *Natural Hazards and Earth System Sciences* 8, 1329–40.

Bolch, T., Buchroithner, M., Pieczonka, T. and Kunert, A. 2008b. Planimetric and volumetric glacier changes in the Khumbu Himal, Nepal, since 1962 using Corona, Landsat TM and ASTER data. *Journal of Glaciology* 54 (187), 592–600.

Bond, G. et al. 1992. Evidence for massive discharges of icebergs into the North Atlantic Ocean during the last glacial period. *Nature* 360, 245–9.

Bond, G., Broeker, W., Johnsen, S., McManus, J., Labeyrie, L., Jouzel, J. and Bonani, G. 1993. Correlations between climate records from North Atlantic sediments and Greenland ice. *Nature* 365, 143–7.

Bonow, J. M. 2005. Re-exposed basement landforms in the Disko region, West Greenland – disregarded data for estimation of glacial erosion and uplift modelling. *Geomorphology* 72, 106–27.

Boon, S. and Sharp, M. J. 2003. The role of hydrologically-driven ice fracture in drainage system evolution on an Arctic glacier. *Geophysical Research Letters* 30 (18), 1916 (10.1029/2003GL018034).

Boon, S., Sharp, M. and Nienow, P. 2003. Impact of an extreme melt event on the runoff and hydrology of a high arctic glacier. *Hydrological Processes* 17, 1051–72.

Boone, S. J. and Eyles, N. 2001. Geotechnical model for great plains hummocky moraine formed by till deformation below stagnant ice. *Geomorphology* 38, 109–24.

Booth, D. B. and Hallet, B. 1993. Channel networks carved by subglacial water: observations and reconstruction in the eastern Puget Lowland of Washington, *Geological Society of America Bulletin* 105, 671–82.

Booth, D. B., Goetz Troost, K., Clague, J. J. and Waitt, R. B. 2004. The Cordilleran Ice Sheet. In Gillespie, A. R., Porter, S. C. and Atwater, B. F. (eds), *The Quaternary Period in the United States*. Elsevier, Amsterdam, 17–44.

Boothroyd, J. C. and Ashley, G. M. 1975. Processes, bar morphology, and sedimentary structures on braided outwash fans, northeastern Gulf of Alaska. In Jopling, A. V. and McDonald, B. C. (eds), *Glaciofluvial and Glaciolacustrine Sedimentation*. SEPM Special Publication 23, 193–222.

Boston, C. M., Evans, D. J. A. and Ó Cofaigh, C. Submitted. Styles of till deposition at the margin of the LGM North Sea Lobe of the British-Irish Ice Sheet: an assessment based on geochemical properties of glacigenic deposits in eastern England. *Earth Science Reviews*.

Bouchard, M. A. and Salonen, V.-P. 1990. Boulder transport in shield areas. In Kujansuu, R. and Saarnisto, M. (eds), *Glacier Indicator Tracing*. Balkema, Rotterdam, 87–107.

Bougamont, M. and Tulaczyk, S. 2003. Glacial erosion beneath ice streams and ice stream tributaries: constraints on temporal and spatial distribution of erosion from numerical simulations of a West Antarctic ice stream. *Boreas* 32, 178–90.

Bougamont, M., Tulaczyk, S. and Joughin, I. 2003a. Response of subglacial sediments to basal freeze-on: 2. Application in numerical modelling of the recent stoppage of Ice Stream C, West Antarctica. *Journal of Geophysical Research* 108 (B4), 2223, doi:10.1029/2002JB001936.

Bougamont, M., Tulaczyk, S. and Joughin, I. 2003b. Numerical investigations of the slow-down of Whillans Ice Stream, West Antarctica: is it shutting down like Ice Stream C? *Annals of Glaciology* 37, 239–46.

Bougamont, M., Bamber, J. L. and Greuell, W. 2005. A surface mass balance model for the Greenland ice sheet. *Journal of Geophysical Research* 110, doi:10.1029/2005JF000348.

Bougamont, M., Bamber, J. L., Ridley, J. K., Gladstone, R. M., Greuell, W., Hanna, E., Payne, A. J. and Rutt, I. 2007. Impact of model physics on estimating the surface mass balance of the Greenland ice sheet. *Geophysical Research Letters* 34, L17501, doi:10/1029/2007GL030700.

Boulton, G. S. 1970a. On the origin and transport of englacial debris in Svalbard glaciers. *Journal of Glaciology* 9, 213–29.

Boulton, G. S. 1970b. On the deposition of subglacial and melt-out tills at the margins of certain Svalbard glaciers. *Journal of Glaciology* 9, 231–45.

Boulton, G. S. 1971. Till genesis and fabric in Svalbard, Spitsbergen. In Goldthwait, R. P. (ed.), *Till – A Symposium*. Ohio State University Press, 41–72.

Boulton, G. S. 1972. Modern arctic glaciers as depositional models for former ice sheets. *Journal of the Geological Society of London* 128, 361–93.

Boulton, G. S. 1974. Processes and patterns of subglacial erosion. In Coates, D. R. (ed.), *Glacial Geomorphology*. University of New York, Binghampton, 41–87.

Boulton, G. S. 1975. Processes and patterns of subglacial sedimentation: a theoretical approach. In Wright, A. E. and Moseley, F. (eds), *Ice Ages: Ancient and Modern*. Seel House, Liverpool, 7–42.

Boulton, G. S. 1976. The origin of glacially-fluted surfaces – observations and theory. *Journal of Glaciology*, 17, 287–309.

Boulton, G. S. 1978. Boulder shapes and grain-size distributions of debris as indicators of transport paths through a glacier and till genesis. *Sedimentology* 25, 773–99.

Boulton, G. S. 1979. Processes of glacier erosion on different substrata. *Journal of Glaciology*, 23, 15–38.

Boulton, G. S. 1982. Subglacial processes and the development of glacial bedforms. In Davidson-Arnott, R., Nickling, W. and Fahey, B. D. (eds), *Research in Glacial, Glacio-fluvial, and Glacio-lacustrine Systems*. Geobooks, Norwich, 1–31.

Boulton, G. S. 1986a. A paradigm shift in glaciology. *Nature* 322, 18.

Boulton, G. S. 1986b. Push moraines and glacier contact fans in marine and terrestrial environments. *Sedimentology* 33, 677–98.

Boulton, G. S. 1987. A theory of drumlin formation by subglacial sediment deformation. In Menzies, J. and Rose, J. (eds), *Drumlin Symposium*. Balkema, Rotterdam, 25–80.

Boulton, G. S. 1990. Sedimentary and sea level changes during glacial cycles and their control on glacimarine facies architecture. In Dowdeswell, J. A. and Scourse, J. D. (eds), *Glacimarine Environments: Processes and Sediments*. Geological Society, London, Special Publication 53, 15–52.

Boulton, G. S. 1996a. Theory of glacial erosion, transport and deposition as a consequence of subglacial sediment deformation. *Journal of Glaciology* 42, 43–62.

Boulton, G. S. 1996b. The origin of till sequences by subglacial sediment deformation beneath mid-latitude ice sheets. *Annals of Glaciology* 22, 75–84.

Boulton, G. S. 2006. Glaciers and their coupling with hydraulic and sedimentary processes. In Knight, P. (ed.), *Glacier Science and Environmental Change*. Blackwell, Oxford, 3–22.

Boulton, G. S. and Caban, P. E. 1995. Groundwater flow beneath ice sheets: Part II – Its impact on glacier tectonic structures and moraine formation. *Quaternary Science Reviews* 14, 563–87.

Boulton, G. S. and Clark, C. D. 1990a. A highly mobile Laurentide Ice Sheet revealed by satellite images of glacial lineations. *Nature* 346, 813–17.

Boulton, G. S. and Clark, C. D. 1990b. The Laurentide Ice sheet through the last glacial cycle: drift lineations as a key to the dynamic behaviour of former ice sheets. *Transactions of the Royal Society of Edinburgh, Earth Sciences* 81, 327–47.

Boulton, G. S. and Dent, D. L. 1974. The nature and rates of postdepositional changes in recently deposited till from south-east Iceland. *Geografiska Annaler* 56A, 121–34.

Boulton, G. S. and Dobbie, K. E. 1998. Slow flow of granular aggregates: the deformation of sediments beneath glaciers. *Philosophical Transactions of the Royal Society of London* A356, 2713–45.

Boulton, G. S. and Eyles, N. 1979. Sedimentation by valley glaciers: a model and genetic classification. In Schluchter, C. (ed.), *Moraines and Varves*. Balkema, Rotterdam, 11–23.

Boulton, G. S. and Hagdorn, M. 2006. Glaciology of the British Isles Ice Sheet during the last glacial cycle: form, flow, streams and lobes. *Quaternary Science Reviews* 25, 3359–90.

Boulton, G. S. and Hindmarsh, R. C. A. 1987. Sediment deformation beneath glaciers: rheology and sedimentological consequences. *Journal of Geophysical Research*, 92, B9, 9059–82.

Boulton, G. S. and Jones, A. S. 1979. Stability of temperate ice caps and ice sheets resting on beds of deformable sediment. *Journal of Glaciology* 24, 29–43.

Boulton, G. S. and Zatsepin, S. 2006. Hydraulic impacts of glacier advance over a sediment bed, *Journal of Glaciology* 52, 497–527.

Boulton, G. S., Caban, P. E. and van Gijssel, K. 1995. Groundwater flow beneath ice sheets. Part 1 – Large-scale patterns. *Quaternary Science Reviews* 14, 545–62.

Boulton, G. S., Dent, D. L. and Morris, E. M. 1974. Subglacial shearing and crushing, and the role of water pressures in tills from south-east Iceland. *Geografiska Annaler*, 56A, 135–45.

Boulton, G. S., Dobbie, K. E. and Zatsepin, S. 2001a. Sediment deformation beneath glaciers and its coupling to the subglacial hydraulic system. *Quaternary International* 86, 3–28.

Boulton, G. S., Dongelmans, P., Punkari, M. and Broadgate, M. 2001b. Palaeoglaciology of an ice sheet through a glacial cycle: the European ice sheet through the Weichselian. *Quaternary Science Reviews* 20, 591–625.

Boulton, G. S., Harris, P. W. V. and Jarvis, J. 1982. Stratigraphy and structure of a coastal sediment wedge of glacial origin inferred from sparker measurements in glacial Lake Jökulsárlón in south-eastern Iceland. *Jökull* 32, 37–47.

Boulton, G. S., Lunn, R., Vidstrand, P. and Zatsepin, S. 2007a. Subglacial drainage by groundwater-channel coupling, and the origin of esker systems. Part I – Glaciological observations. *Quaternary Science Reviews* 26, 1067–90.

Boulton, G. S., Lunn, R., Vidstrand, P. and Zatsepin, S. 2007b. Subglacial drainage by groundwater-channel coupling, and the origin of esker systems. Part 2 – Theory and simulation of a modern system. *Quaternary Science Reviews* 26, 1091–105.

Boulton G. S., van der Meer, J. J. M., Hart, J. K., Beets, D., Ruegg, G. H. J., van der Wateren, F. M. and Jarvis, J. 1996. Till and moraine emplacement in a deforming bed surge: an example from a marine environment. *Quaternary Science Reviews* 15, 961–87.

Boulton, G. S., van der Meer, J. J. M., Beets, D. J., Hart, J. K. and Ruegg, G. H. J. 1999. The sedimentary and structural evolution of a recent push moraine complex: Holmstrombreen, Spitsbergen. *Quaternary Science Reviews* 18, 339–71.

Bouma, A. H. 1962. *Sedimentology of some flysch deposits*. Elsevier, Amsterdam.

Bourgeois, O., Dauteuil, O. and van Vliet-Lanoe, B. 2000. Geothermal control on flow patterns in the Last Glacial Maximum ice sheet of Iceland. *Earth Surface Processes and Landforms* 25, 59–76.

Bovis, M. J. 1990. Rockslope deformation at Affliction Creek, southern Coast Mountains, British Columbia. *Canadian Journal of Earth Sciences* 27, 243–54.

Bovis, M. J. and Stewart, T. W. 1998. Long-term deformation of a glacially undercut rock slope, southwest British Columbia. In Moore, D. and Hungr, O. (eds), *Proceedings, 8th International Congress, International Association of Engineering Geology and the Environment*. Balkema, Rotterdam, 1267–76.

Bowles, F. A., Faas, R. W., Vogt, P. R., Sawyer, W. B. and Stephens, K. 2003. Sediment properties, flow characteristics, and depositional environment of submarine mud flows, Bear Island Fan. *Marine Geology* 197, 63–74.

Box, J. E. 2002. Survey of Greenland instrumental temperature records: 1873–2001. *International Journal of Climatology* 22, 1829–47.

Box, J. E. and Ski, K. 2007. Remote sounding of Greenland supraglacial melt lakes: implications for subglacial hydraulics. *Journal of Glaciology* 53, 257–65.

Box, J. E., Bromwich, D. H., Veenhuis, B. A., Bai, L.-S., Stroeve, J. C., Rogers, J. C., Steffen, K., Haran, T. and Sheng-Hung Wang 2006. Greenland Ice Sheet surface mass balance variability (1988–2004) from calibrated Polar MM5 output. *Journal of Climate* 19, 2783–800.

Boyce, E. S., Motyka, R. J. and Truffer, M. 2007. Flotation and retreat of a lake-calving terminus, Mendenhall Glacier, southeast Alaska, USA. *Journal of Glaciology* 53, 211–24.

Boyce, J. I. and Eyles, N. 1991. Drumlins carved by deforming till streams below the Laurentide ice sheet. *Geology* 19, 787–90.

Boyce, J. I. and Eyles, N. 2000. Architectural element analysis applied to glacial deposits: internal geometry of a late Pleistocene till sheet, Ontario, Canada. *Geological Society of America Bulletin* 112, 98–118.

Bradley, R. S. 1999. *Paleoclimatology: Reconstructing Climates of the Quaternary*. Academic Press, San Diego, CA.

Bradwell, T. 2005. Bedrock megagrooves in Assynt, NW Scotland. *Geomorphology* 65, 195–204.

Bradwell, T. 2006. The Loch Lomond Stadial glaciation in Assynt: a reappraisal. *Scottish Geographical Journal* 122 (4), 274–92.

Bradwell, T., Stoker, M. and Krabbendam, M. 2008a. Megagrooves and streamlined bedrock in NW Scotland: the role of ice streams in landscape evolution. *Geomorphology* 97, 135–56.

Bradwell, T., Stoker, M. and Larter, R. 2007. Geomorphological signature and flow dynamics of The Minch palaeo-ice stream, northwest Scotland. *Journal of Quaternary Science* 22, 609–17.

Bradwell, T., Stoker, M. S., Golledge, N. R., Wilson, C. K., Merritt, J. W., Long, D., Everest, J. D., Hestvik, O. B., Stevenson, A. G., Hubbard, A. L., Finlayson, A. G. and Mathers, H. E. 2008b. The northern sector of the last British Ice Sheet: maximum extent and demise. *Earth Science Reviews* 88, 207–26.

Braithwaite, R. J. 1995. Positive degree-day factors for ablation on the Greenland ice sheet studied by energy-balance modelling. *Journal of Glaciology* 41 (137), 153–60.

Braithwaite, R. J. 2002. Glacier mass balance: the first 50 years of international monitoring. *Progress in Physical Geography* 26, 76–95.

Braithwaite, R. J. 2008. Temperature and precipitation climate at the equilibrium-line altitude of glaciers expressed by the degree-day factor for melting snow. *Journal of Glaciology* 53 (186), 437–44.

Braithwaite, R. J. and Olesen, O. B. 1989. Calculation of glacier ablation from air temperature, West Greenland. In Oerlemans, J. (ed.), *Glacier Fluctuations and Climatic Change*, Kluwer, Dordrecht, 219–33.

Braithwaite, R. J. and Olesen, O. B. 1990. Response of the energy balance on the margin of the Greenland ice sheet to temperature changes. *Journal of Glaciology* 36, 217–21.

Braithwaite, R. J. and Zhang, Y. 2000. Sensitivity of mass balance of five Swiss glaciers to temperature changes assessed by tuning a degree-day model. *Journal of Glaciology* 46 (152), 7–14.

Braithwaite, R. J. and Zhang, Y. 2003. Modelling changes in glacier mass balance that may occur as a result of climate changes. *Geografiska Annaler* 81A, 489–96.

Braithwaite, R. J., Konzelmann, T., Marty, C. and Olesen, O. B. 1998. Reconnaissance study of glacier energy balance in North Greenland, 1993–94. *Journal of Glaciology* 44, 239–47.

Brandal, M. K. and Heder, E. 1991. Stratigraphy and sedimentation of a terminal moraine deposited in a marine environment – two examples from the Ra-ridge in Østfold, southeast Norway. *Norsk Geologisk Tidsskrift* 71, 3–14.

Braun, D. D. 2006. Deglaciation of the Appalachian Plateau, northeastern Pennsylvania – till shadows, till knobs forming 'beaded valleys': revisiting systematic stagnation-zone retreat. *Geomorphology* 75, 248–65.

Braun, J., Zwartz, D. and Tomkin, J. 1999. A new surface-processes model combining glacial and fluvial erosion. *Annals of Glaciology* 28, 282–90.

Braun, L. N., Weber, M. and Schulz, M. 2000. Consequences of climate change for runoff from Alpine regions. *Annals of Glaciology* 31, 19–25.

Brayshaw, A. C. 1984. Characteristics and origin of cluster bedforms in coarse-grained alluvial channels. In Koster, E. H. and Steel, R. J. (eds), *Sedimentology of Gravels and Conglomerates*. Canadian Society of Petroleum Geologists Memoir 10, 77–85.

Brazier, V., Kirkbride, M. P. and Gordon, J. E. 1998a. Active ice sheet deglaciation and ice-dammed lakes in the northern Cairngorm Mountains, Scotland. *Boreas* 27, 297–310.

Brazier, V., Kirkbride, M. P. and Owens, I. F. 1998b. The relationship between climate and rock glacier distribution in the Ben Ohau Range, New Zealand. *Geografiska Annaler* 80A, 193–207.

Breemer, C. W., Clark, P. U. and Haggerty, R. 2002. Modeling the subglacial hydrology of the late Pleistocene Lake Michigan Lobe, Laurentide Ice Sheet. *Geological Society of America Bulletin* 114, 665–74.

Brennand, T. A. 1994. Macroforms, large bedforms and rhythmic sedimentary sequences in subglacial eskers, south-central

Ontario: implications for esker genesis and meltwater regime. *Sedimentary Geology* 91, 9–55.

Brennand, T. A. 2000. Deglacial meltwater drainage and glaciodynamics: inferences from Laurentide eskers, Canada. *Geomorphology* 32, 263–93.

Brennand, T. A. and Shaw, J. 1994. Tunnel channels and associated landforms, south-central Ontario: their implications for ice-sheet hydrology. *Canadian Journal of Earth Sciences* 31, 505–22.

Brennand, T. A. and Shaw, J. 1996. The Harricana glaciofluvial complex, Abitibi region, Quebec: its genesis and implications for meltwater regime and ice sheet dynamics. *Sedimentary Geology* 102, 221–62.

Bretz, J. H. 1923. The Channeled Scabland of the Columbia Plateau. *Journal of Geology* 31, 617–49.

Bretz, J. H., Smith, H. T. U. and Neff, G. E. 1956. Channeled Scabland of Washington: new data and interpretations. *Geological Society of America Bulletin* 67, 957–1049.

Bridge, J. S. 1993. The interaction between channel geometry, water flow, sediment transport and deposition in braided rivers. In Best, J. L. and Bristow, C. S. (eds), *Braided Rivers*. Geological Society, London, Special Publication 75, 13–71.

Bridge, J. S. 2003. *Rivers and Floodplains: Forms, Processes and Sedimentary Record*. Blackwell, Oxford.

Bridge, J. S. and Demicco, R. V. 2008. *Earth Surface Processes, Landforms and Sediment Deposits*. Cambridge University Press.

Bridgland, D. R. 1994. *Quaternary of the Thames*, Geological Conservation Review Series, 7, Chapman and Hall, London.

Bridgland, D. R. 2000. River terrace systems in north-west Europe: an archive of environmental change, uplift and early human occupation. *Quaternary Science Reviews* 19, 1293–303.

Bridgland, D. R. 2006. The Middle and Upper Pleistocene sequence in the Lower Thames: a record of Milankovitch climatic fluctuation and early human occupation of southern Britain. *Proceedings of the Geologists' Association* 117, 281–305.

Briner, J. P. 2007. Supporting evidence from the New York drumlin field that elongate subglacial bedforms indicate fast ice flow. *Boreas* 36, 143–7.

Briner, J. P. and Kaufman, D. S. 2008. Late Pleistocene mountain glaciation in Alaska: key chronologies. *Journal of Quaternary Science* 23, 659–70.

Briner, J. P., Miller, G. H., Davis, P. T., Bierman, P. R. and Caffee, M. 2003. Last Glacial Maximum ice sheet dynamics in Arctic Canada inferred from young erratics perched on ancient tors. *Quaternary Science Reviews* 22, 437–44.

Bristow, C. 1996. Reconstructing fluvial channel morphology from sedimentary sequences. In Carling, P. A. and Dawson, M. R. (eds), *Advances in Fluvial Dynamics and Stratigraphy*. Wiley, Chichester, 351–71.

Broccoli, A. J. and Manabe, S. 1987. The contributions of continental ice, atmospheric CO_2, and land albedo to the climate of the last glacial maximum. *Climate Dynamics* 1, 87–99.

Brock, B. W. 2004. An analysis of short-term albedo variations at Haut Glacier d'Arolla, Switzerland. *Geografiska Annaler* 86A, 53–66.

Brock, B. W., Willis, I. C. and Sharp, M. J. 2000. Measurement and parameterisation of albedo variations at Haut Glacier d'Arolla, Switzerland. *Journal of Glaciology*, 46, 675–88.

Brocklehurst, S. H. and Whipple, K. X. 2002. Glacial erosion and relief production in the Eastern Sierra Nevada, California. *Geomorphology* 42, 1–24.

Brocklehurst, S. H. and Whipple, K. X. 2004. Hypsometry of glaciated landscapes. *Earth Surface Processes and Landforms* 29, 907–26.

Brodzikowski, K. and van Loon, A. J. 1991. *Glacigenic Sediments*. Elsevier, Amsterdam.

Brodzikowski, K., Haluszczak, A., Krzyszkowski, D. and van Loon, A. J. 1987. Genesis and diagnostic value of large-scale gravity-induced penecontemporaneous deformation horizons in

Quaternary sediments of the Kleszcozw Graben (central Poland). In Jones, M. E. and Preston, R. M. F. (eds), *Deformation of Sediments and Sedimentary Rocks*. Geological Society, London, Special Publication 29, 287–98.

Broecker, W. S. and Denton, G. H. 1989. The role of ocean-atmosphere reorganizations in glacial cycles. *Geochimica and Cosmochimica Acta* 53, 2465–501.

Broecker, W. S. Andree, M., Wolfi, W., Oescher, H., Bonani, G., Kennet, J. and Peteet, D. 1988. The chronology of the last deglaciation: implications to the cause of the Younger Dryas event. *Palaeoceanography* 3, 1–19.

Bromley, R. G. 1996. *Trace Fossils: Biology, Taphonomy and Applications*, 2nd edn. Routledge, London.

Bromwich, D. H., Guo, Z., Bai, L. and Chen, Q.-S. 2004a. Modeled Antarctic precipitation. Part I: Spatial and temporal variability. *Journal of Climate* 17, 427–47.

Bromwich, D. H., Toracinta, E. R., Oglesby, R. J., Fastook, J. L. and Hughes, T. J. 2005. LGM summer climate on the southern margin of the Laurentide Ice Sheet: wet or dry? *Journal of Climate* 18, 3317–38.

Bromwich, D. H., Toracinta, E. R., Wei, H., Oglesby, R. J., Fastook, J. L. and Hughes, T. J. 2004b. Polar MM5 simulations of the winter climate of the Laurentide Ice Sheet at the LGM. *Journal of Climate* 17, 3415–33.

Brook, M. S., Brock, B. W. and Kirkbride, M. P. 2003. Glacial outlet valley size-ice drainage area relationships: some considerations. *Earth Surface Processses and Landforms* 28, 645–53.

Brook, M. S., Kirkbride, M. P. and Brock, B. W. 2004. Rock strength and development of glacial valley morphology in the Scottish Highlands and northwest Iceland. *Geografiska Annaler* 86A, 225–34.

Brookfield, M. E. and Martini, I. P. 1999. Facies architecture and sequence stratigraphy in glacially influenced basins: basic problems and water-level/glacier input-point controls (with an example from the Quaternary of Ontario, Canada). *Sedimentary Geology* 123, 183–97.

Brooks, G. R. 1994. The fluvial reworking of Late Pleistocene drift, Squamish River drainage basin, southwest British Columbia. *Geographie Physique et Quaternaire* 48, 51–68.

Brooks, S. J. and Birks, H. J. B. 2001. Chironomid-inferred air temperatures from late-glacial and Holocene sites in north-west Europe: progress and problems. *Quaternary Science Reviews* 20, 1723–41.

Broster, B. E. and Hicock, S. R. 1985. Multiple flow and support mechanisms and the development of inverse grading in a subaquatic glacigenic debris flow. *Sedimentology* 32, 645–57.

Brown, C. S., Meier, M. F. and Post, A. 1982. Calving speed of Alaska tidewater glaciers with applications to the Columbia Glacier, Alaska. *U.S. Geological Survey Professional Paper* 1258-C.

Brown, G. H. 2002. Glacier meltwater hydrochemistry. *Applied Geochemistry* 17, 855–83.

Brown, G. H., Tranter, M. and Sharp, M. J. 1996. Experimental investigations of the weathering of suspended sediment by Alpine glacial meltwater. *Hydrological Processes* 10, 579–97.

Brown, N. E., Hallet, B. and Booth, D. B. 1987. Rapid soft-bed sliding of the Puget glacial lobe. *Journal of Geophysical Research* 92, 8985–97.

Brubaker, K., Rango, A. and Kustas, W. 1996. Incorporating radiation inputs into the snowmelt runoff model. *Hydrological Processes* 10, 1329–43.

Brugman, M. M. and Post, A. 1981. Effects of volcanism on the glaciers of Mount St Helens. *USGS Circular* 850-D.

Bryn, P., Berg, K., Forsberg, K. F., Solheim, A. and Kvalstad, T. J. 2005. Explaining the Storegga Slide. *Marine and Petroleum Geology* 22, 11–19.

Budd, W. F., Keage, P. L. and Blundy, N. A. 1979. Empirical studies of ice sliding. *Journal of Glaciology* 23, 157–70.

Budd, W. F., Corry, M. J. and Jacka, T. H. 1982. Results of the Amery Ice Shelf Project. *Annals of Glaciology* 3, 36–41.

Buoncristiani, J.-F. and Campy, M. 2001. Late Pleistocene Detrital Sediment Yield of the Jura Glacier, France. *Quaternary Research* 56, 51–61.

Burgess, D. O., Shaw, J. and Eyton, J. R. 2003. Morphometric comparisons between Rogen terrain and hummocky terrain. *Physical Geography* 24, 319–36.

Burke, M. J., Woodward, J., Russell, A. J., Fleisher, P. J. and Bailey, P. K. 2008. Controls on the sedimentary architecture of a single event englacial esker: Skeiðarárjökull, Iceland. *Quaternary Science Reviews* 27, 1829–47.

Busschers, F. S., van Balen, R. T., Cohen, K. M., Kasse, C., Weerts, H. J. T., Wallinga, J. and Bunnik, F. P. M. 2008. Response of the Rhine-Meuse fluvial system to Saalian ice sheet dynamics. *Boreas* 37, 377–98.

Cai, J., Powell, R. D., Cowan, E. A. and Carlson, P. 1997. Lithofacies and seismic-reflection interpretation of temperate glacimarine sedimentation from Tarr Inlet, Glacier Bay, Alaska. *Marine Geology* 143, 5–37.

Calkin, P. E., Wiles, G. C. and Barclay, D. J. 2001. Holocene coastal glaciation of Alaska. *Quaternary Science Reviews* 20, 449–61.

Campbell, I., Jacobel, R., Welch, B. and Pettersson, R. 2008. The evolution of surface flow stripes and stratigraphic folds within Kamb Ice Stream: why don't they match? *Journal of Glaciology* 54, 421–7.

Canals, M., Casamor, J. L., Urgeles, R., Calafat, A. M., Domack, E. W., Baraza, J., Farran, M. and De Batist, M. 2002. Seafloor evidence of subglacial sedimentary system off the northern Antarctic Peninsula. *Geology* 30, 603–6.

Canals, M., Urgeles, R. and Calafat, A. M. 2000. Deep sea floor evidence of past ice streams off the Antarctic Peninsula. *Geology* 28, 31–4.

Carey, M. 2007. The history of ice: how glaciers became an endangered species. *Environmental History* 12, 497–527.

Carey, M. 2008. Disasters, development, and glacial lake control in twentieth-century Peru. In Wiegandt, E. (ed.), *Mountains: Sources of Water, Sources of Knowledge*. Advances in Global Change Research, vol. 31. Springer, Netherlands, 181–96.

Carling, P. A. 1996. Morphology, sedimentology and palaeohydraulic significance of large gravel dunes: Altai Mountains, Siberia. *Sedimentology* 43, 647–64.

Carling, P. A., Kirkbride, A. D., Parnachov, S., Borodavko, P. S. and Berger, G. W. 2002. Late Quaternary catastrophic flooding in the Altai Mountains of south-central Siberia: a synoptic overview and an introduction to flood deposit sedimentology. In Martini, I. P., Baker, V. R. and Garzon, G. (eds), *Flood and Megaflood Processes and Deposits: Recent and Ancient Examples*. International Association of Sedimentologists, Special Publication 32. Blackwell, Oxford, 17–35.

Carlson, A. E. 2004. Genesis of dewatering structures and its implications for melt-out till identification. *Journal of Glaciology* 50, 17–24.

Carlson, A. E., Mickelson, D. M., Principato, S. M. and Chapel, D. M. 2005. The genesis of the northern Kettle Moraine, Wisconsin. *Geomorphology* 67, 365–74.

Carlson, P. R., Bruns, T. R. and Fisher, M. A. 1990. Development of slope valleys in the glacimarine environment of a complex subduction zone, northern Gulf of Alaska. In Dowdeswell, J. A. and Scourse, J. D. (eds), *Glacimarine Environments: Processes and Sediments*. Geological Society, London, Special Publication 53, 139–53.

Carlson, P. R., Powell, R. D. and Phillips, A. C. 1992. Submarine sedimentary features on a fjord delta front, Queen Inlet, Glacier Bay, Alaska. *Canadian Journal of Earth Sciences* 29, 565–73.

Carr, M. H. 2006. *The Surface of Mars*. Cambridge University Press.

Carr, S. J. 2001a. A glaciological approach for the discrimination of Loch Lomond Stadial glacial landforms in the Brecon Beacons, South Wales. *Proceedings of the Geologists' Association* 112, 253–62.

Carr, S. J. 2001b. Micromorphological criteria for discriminating subglacial and glacimarine sediments: evidence from a contemporary tidewater glacier, Spitsbergen. *Quaternary International* 86, 71–9.

Carr, S. J. 2004. Micro-scale features and structures. In Evans, D. J. A. and Benn, D. I. (eds), *A Practical Guide to the Study of Glacial Sediments*. Arnold, London, 115–44.

Carr, S. J. and Coleman, C. 2007. An improved technique for the reconstruction of former glacier mass-balance and dynamics. *Geomorphology* 92, 76–90.

Carr, S. J. and Goddard, M. A. 2007. Role of particle size in till fabric characteristics: systematic variation in till fabric from Vestari-Hagafellsjökull, Iceland. *Boreas* 36, 371–85.

Carr, S. J. and Rose, J. 2003. Till fabric patterns and significance: particle response to subglacial stress. *Quaternary Science Reviews* 22, 1415–26.

Carrasco, J. F., Casassa, G. and Rivera, A. 2002. Meteorological and climatological aspects of the Southern Patagonian Icefield. In Casassa, G., Sepúlveda, F. V. and Sinclair, R. M. (eds), *The Patagonian Icefields*. New York, Kluwer, 29–41.

Carrivick, J. L. 2007. Modelling coupled hydraulics and sediment transport of a high-magnitude flood and associated landscape change. *Annals of Glaciology* 45, 143–54.

Carrivick, J. L., Russell, A. J. and Tweed, F. S. 2004. Geomorphological evidence for jökulhlaups from Kverkfjöll volcano, Iceland. *Geomorphology* 63, 81–102.

Carsey, F., Behar, A., Lane, A. L., Realmuto, V. and Engelhardt, H. 2002. A borehole camera system for imaging the deep interior of ice sheets. *Journal of Glaciology* 48 (163), 622–8.

Carter, C. L., Dethier, D. P. and Newton, R. L. 2003. Subglacial environment inferred from bedrock-coating siltskins, Mendenhall Glacier, Alaska, USA. *Journal of Glaciology* 49, 568–76.

Carter, L., Mitchell, J. S. and Day, N. J. 1981. Suspended sediment beneath permanent and seasonal ice, Ross Ice Shelf, Antarctica. *New Zealand Journal of Geology and Geophysics* 24, 249–62.

Carter, S. P., Blankenship, D. D., Peters, M. E., Young, D. A., Holt, J. W. and Morse, D. L. 2007. Radar-based subglacial lake classification in Antarctica. *Geochemistry, Geophysics, Geosystems* 8, doi:10.1029/2006GC001408.

Casassa, G., Lopez, P., Pouyaud, B. and Escobar, F. 2009. Detection of changes in glacial runoff in alpine basins: examples from North America, the Alps, central Asia and the Andes. *Hydrological Processes* 23, 31–41.

Catania, G. A., Conway, H., Raymond, C. F. and Scambos, T. A. 2005. Surface morphology and internal layer stratigraphy in the downstream end of Kamb Ice Stream, West Antarctica. *Journal of Glaciology* 175, 423–31.

Catania, G. A., Neumann, T. A. and Price, S. F. 2008. Characterizing englacial drainage in the ablation zone of the Greenland ice sheet. *Journal of Glaciology* 54 (187), 567–78.

Catania, G. A., Scambos, T. A., Conway, H. and Raymond, C. F. 2006. Sequential stagnation of Kamb Ice Stream, West Antarctica. *Geophysical Research Letters* 33, doi:10.1029/2006GL026430.

Catuneanu, O., 2006. *Principles of Sequence Stratigraphy*. New York, Elsevier.

Cazenave, A. and Nerem, R. S. 2004. Present-day sea level change: observations and causes. *Reviews of Geophysics* 42, RG3001, doi:10.1029/2003RG000139.

Cenderelli, D. A. and Wohl, E. E. 2001. Peak discharge estimates of glacial-lake outburst floods and 'normal' climatic floods in the Mount Everest region, Nepal. *Geomorphology* 40, 57–90.

Cenderelli, D. A. and Wohl, E. E. 2003. Flow hydraulics and geomorphic effects of glacial lake outburst floods in the Mount Everest region, Nepal. *Earth Surface Processes and Landforms* 28, 385–407.

Chandler, D. M., Waller, R. I. and Adam, W. G. 2005. Basal ice motion and deformation at the ice-sheet margin, West Greenland. *Annals of Glaciology* 42, 67–70.

Chappell, J. 2002. Sea-level changes forced ice breakouts in the last glacial cycle: new results from coral terraces. *Quaternary Science Reviews* 21, 1229–40.

Chappell, J. and Shackleton, N. J. 1986. Oxygen isotopes and sea level. *Nature* 324, 137–40.

Chappell, J., Ohmura, A., Esat, T., McCulloch, M., Pandolfi, J., Ota, Y. and Pillans, B. 1996. Reconciliation of late Quaternary sea levels derived from coral terraces at Huon Peninsula with deep sea oxygen isotope records. *Earth Planetary Science Letters*, 141, 227–36.

Charbit, S., Paillard, D. and Ramstein, G. 2008. Amount of CO_2 emissions irreversibly leading to the total melting of Greenland. *Geophysical Research Letters* 35, L12503, doi:10.1029/GL033472.

Cheel, R. J. and Rust, B. R. 1982. Coarse grained facies of glacio-marine deposits near Ottawa, Canada. In Davidson-Arnott, R., Nickling, W. and Fahey, B. D. (eds), *Research in Glaciofluvial and Glaciolacustrine Systems*. Geobooks, Norwich, 279–95.

Chen, J. L., Wilson, C. R., Tapley, B. D., Famiiglietti, J. S. and Rodell, M. 2005. Seasonal global mean sea level change from satellite altimeter, GRACE, and geophysical models. *Journal of Geodesy* 79 (9), 532–9.

Chikita, K., Jha, J. and Yamada, T. 2001. Sedimentary effects on the expansion of a Himalayan supraglacial lake. *Global and Planetary Change* 28, 23–34.

Chinn, T. J. 1980. Glacier balances in the Dry Valleys area, Victoria Land, Antarctica. *IAHS Publication* 125, 237–47.

Chinn, T. J. 1988. The Dry Valleys of Victoria Land. In *Satellite Image Atlas of Glaciers of the World: Antarctica*. USGS Prof. Pap. 1386-B.

Chinn, T. J., Winkler, S., Salinger, M. J. and Haakensen, N. 2005. Recent glacier advances in Norway and New Zealand: a comparison of their glaciological and meteorological causes. *Geografiska Annaler* 87A, 141–57.

Chorley, R. J. 1973. *The History of the Study of Landforms or the Development of Geomorphology*. Methuen, London.

Chorley, R. J., Schumm, S. A. and Sugden, D. E. 1984. *Geomorphology*. Methuen, London.

Christoffersen, P. and Tulaczyk, S. 2003a. Thermodynamics of basal freeze-on: predicting basal and subglacial signatures beneath stopped ice streams and interstream ridges. *Annals of Glaciology* 36, 233–43.

Christoffersen, P. and Tulaczyk, S. 2003b. Signature of palaeo-ice stream stagnation: till consolidation induced by basal freeze-on. *Boreas* 32, 114–29.

Christoffersen, P., Piotrowski, J. A. and Larsen, N. K. 2005. Basal processes beneath an arctic glacier and their geomorphic imprint after a surge, Elisebreen, Svalbard. *Quaternary Research* 64, 125–37.

Christoffersen, P., Tulaczyk, S., Carsey, F. D. and Behar, A. E. 2006. A quantitative framework for interpretation of basal ice facies formed by ice accretion over subglacial sediment. *Journal of Geophysical Research* 111, F01017, doi:10.1029/2005JF000363.

Church, J. A. and Gregory, J. 2001. Changes in sea level. In Houghton, J. T. et al. (eds), *Climate Change 2001. The Scientific Basis*. Contribution of Working Group I to the Third Assessment Report of the Intergovernmental Panel on Climate Change. Cambridge University Press, 641–93.

Church, J. A. and White, N. J. 2006. A 20th century acceleration in global sea-level rise. *Geophysical Research Letters* 33, doi:10.1029/2005GL024826.

Church, J. A., White, N. J., Aarup, T., Wilson, W. S., Woodworth, P. L., Domingues, C. M., Hunter, J. R. and Lambeck, K. 2008. Understanding global sea levels: past, present and future. *Sustainability Science* 3, 9–22.

Church, M. 1978. Palaeohydrological reconstructions from a Holocene valley fill. In Miall, A. D. (ed.), *Fluvial Sedimentology*. Canadian Society of Petroleum Geologists Memoir 5, 743–72.

Church, M. and Gilbert, R. 1975. Proglacial fluvial and lacustrine sediments. In Jopling, A. V. and McDonald, B. C. (eds), *Glaciofluvial and Glaciolacustrine Sedimentation*. SEPM, Special Publication 23, 22–100.

Church, M. and Ryder, J. M. 1972. Paraglacial sedimentation: a consideration of fluvial processes conditioned by glaciation. *Geological Society of America Bulletin* 83, 3059–72.

Church, M. and Slaymaker, O. 1989. Disequilibrium of Holocene sediment yield in glaciated British Columbia. *Nature* 337, 452–4.

Churski, Z. 1973. Hydrographic features of the proglacial area of Skeiðararjökull. *Geographia Polonica* 26, 209–54.

Chutko, K. J. and Lamoureux, S. F. 2008. Identification of coherent links between interannual sedimentary structures and daily meteorological observations in Arctic proglacial lacustrine varves: potentials and limitations. *Canadian Journal of Earth Sciences* 45, 1–13.

Clague, J. J. and Evans, S. G. 1994. Historic retreat of Grand Pacific and Melbern Glaciers, Saint Elias Mountains, Canada: an analogue for decay of the Cordilleran ice sheet at the end of the Pleistocene? *Journal of Glaciology* 39, 619–24.

Clague, J. J. and Evans, S. G. 2000. A review of catastrophic drainage of moraine-dammed lakes in British Columbia. *Quaternary Science Reviews* 19, 1763–83.

Clague, J. J. and Mathews, W. H. 1973. The magnitude of jokulhlaups. *Journal of Glaciology* 12, 501–4.

Clapperton, C. M. 1975. The debris content of surging glaciers in Svalbard and Iceland. *Journal of Glaciology* 14, 395–406.

Clapperton, C. M. 1989. Asymmetrical drumlins in Patagonia, Chile. *Sedimentary Geology* 62, 387–98.

Clapperton, C. M. 1993. *Quaternary Geology and Geomorphology of South America*. Elsevier, Amsterdam.

Clarhäll, A. and Jansson, K. N. 2003. Time perspectives on glacial landscape formation – glacial flow chronology at Lac aux Goélands, northeastern Quebec, Canada. *Journal of Quaternary Science* 18, 441–52.

Clarhäll, A. and Kleman, J. 1999. Distribution and glaciological implications of relict surfaces on the Ultevis plateau, northwestern Sweden. *Annals of Glaciology* 28, 202–8.

Clark, C. D. 1993. Mega-scale lineations and cross-cutting ice-flow landforms. *Earth Surface Processes and Landforms* 18, 1–29.

Clark, C. D. 1994. Large-scale ice-moulding: a discussion of genesis and glaciological significance. *Sedimentary Geology* 91, 253–68.

Clark, C. D. 1997. Reconstructing the evolutionary dynamics of former ice sheets using multi-temporal evidence, remote sensing and GIS. *Quaternary Science Reviews* 16, 1067–92.

Clark, C. D. 1999. Glaciodynamic context of subglacial bedform generation and preservation. *Annals of Glaciology* 28, 23–32.

Clark, C. D. and Meehan, R. T. 2001. Subglacial bedform geomorphology of the Irish Ice Sheet reveals major configuration changes during growth and decay. *Journal of Quaternary Science* 16, 483–96.

Clark, C. D. and Stokes, C. R. 2001. Extent and basal characteristics of the M'Clintock Channel ice stream. *Quaternary International* 86, 81–101.

Clark, C. D. and Stokes, C. R. 2003. Palaeo-ice stream landsystem. In Evans, D. J. A. (ed.), *Glacial Landsystems*. Arnold, London, 204–27.

Clark, C. D., Greenwood, S. L. and Evans, D. J. A. 2006. Palaeo-glaciology of the last British-Irish Ice Sheet: challenges and some recent developments. In Knight, P (ed.), *Glacier Science and Environmental Change*. Blackwell, Oxford, 248–64.

Clark, C. D., Hughes, A. L. C., Greenwood, S. L., Spagnolo, M. and Ng, F. S. L. 2009. Size and shape characteristics of drumlins, derived from a large sample, and associated scaling laws. *Quaternary Science Reviews* 28, 677–92.

Clark, C. D., Knight, J. K. and Gray, J. T. 2000. Geomorphological reconstruction of the Labrador sector of the Laurentide Ice Sheet. *Quaternary Science Reviews* 19, 1343–66.

Clark, C. D., Tulaczyk, S. M., Stokes, C. R. and Canals, M. 2003. A groove-ploughing theory for the production of mega-scale lineations, and implications for ice stream mechanics. *Journal of Glaciology* 49, 240–56.

Clark, P. U. 1991. Striated clast pavements: products of deforming subglacial sediment? *Geology* 19, 530–3.

Clark, P. U. and Hansel, A. K. 1989. Clast ploughing, lodgement and glacier sliding over a soft glacier bed. *Boreas* 18, 201–7.

Clark, P. U. and Pollard, D. 1998. Origin of the middle Pleistocene transition by ice sheet erosion of regolith. *Palaeoceanography* 13, 1–9.

Clark, P. U. and Walder, J. S 1994. Subglacial drainage, eskers, and deforming beds beneath the Laurentide and Eurasian ice sheets. *Geological Society of America Bulletin* 106, 304–14.

Clark, P. U., Alley, R. B. and Pollard, D. 1999. Northern Hemisphere ice-sheet influences on global climate change. *Science* 286, 1104–11.

Clark, P. U., Archer, D., Pollard, D., Blum, J. D., Rial, J. A., Brovkin, V., Mix, A. C., Pisias, N. G. and Roy, M. 2006. The middle Pleistocene transition: characteristics, mechanisms, and implications for long-term changes in atmospheric pCO$_2$. *Quaternary Science Reviews* 25, 3150–84.

Clark, P. U., McCabe, A. M., Mix, A. C. and Weaver, A. J. 2004. Rapid rise of sea level 19,000 years ago and its global implications. *Science* 304, 1141–4.

Clark, P. U., Mitrovica, J. X., Milne, G. A. and Tamisiea, M. E. 2002. Sea-level fingerprinting as a direct test for the source of Global Meltwater Pulse IA. *Science* 295, 2438–41.

Clarke, G. K. C. 1982. Glacier outburst flood from 'Hazard lake', Yukon Territory, and the problem of flood magnitude prediction. *Journal of Glaciology* 28, 3–21.

Clarke, G. K. C. 1986. Professor Mathews, outburst floods, and other glaciological disasters. *Canadian Journal of Earth Sciences* 23, 859–68.

Clarke, G. K. C. 1987a. Subglacial till: a physical framework for its properties and processes. *Journal of Geophysical Research* 92, 9023–36.

Clarke, G. K. C. 1987b. A short history of scientific investigations on glaciers. *Journal of Glaciology*, Special Issue, 4–24.

Clarke, G. K. C. 1987c. Fast glacier flow: ice streams, surging and tidewater glaciers. *Journal of Geophysical Research* 92, 8835–41.

Clarke, G. K. C. 1991. Length, width and slope influences on glacier surging. *Journal of Glaciology* 37, 236–46.

Clarke, G. K. C. 1996. Lumped-element analysis of subglacial hydraulic circuits. *Journal of Geophysical Research* 101, B8, 17547–59.

Clarke, G. K. C. 2003. Hydraulics of subglacial outburst floods: new insights from the Spring-Hutter formulation. *Journal of Glaciology* 49 (165), 299–313.

Clarke, G. K. C. 2005. Subglacial processes. *Annual Review of Earth and Planetary Sciences* 33, 247–76.

Clarke, G. K. C. and Blake, E. W. 1991. Geometric and thermal evolution of a surge-type glacier in its quiescent state: Trapridge Glacier, Yukon Territory, Canada, 1969–89. *Journal of Glaciology* 37, 158–69.

Clarke, G. K. C., Collins, S. G. and Thompson, D. E. 1984. Flow, thermal structure and subglacial conditions of a surge-type glacier. *Canadian Journal of Earth Sciences* 21, 232–40.

Clarke, G. K. C., Leverington, D. W., Teller, J. T. and Dyke, A. S. 2004. Paleohydraulics of the last outburst flood from glacial Lake Agassiz and the 8200 BP cold event. *Quaternary Science Reviews* 23, 389–407.

Clarke, G. K. C., Leverington, D. W., Teller, J. T., Dyke, A. S. and Marshall, S. J. 2005. Fresh arguments against the Shaw megaflood hypothesis. A reply to comments by David Sharpe on 'Paleohydraulics of the last outburst flood from glacial Lake Agassiz and the 8200 BP cold event'. *Quaternary Science Reviews* 24, 1533–41.

Clarke, G. K. C., Marshall, S. J., Hillaire-Marcel, C., Bilodeau, G. and Veiga-Pires, C. 1999. A glaciological perspective on Heinrich events. In Clark, P. U., Webb, R. S. and Keigwin, L. D. (eds), *Mechanisms of Climate Change at Millennial Timescales*. American Geophysical Union, Geophysical Monograph 112, 243–62.

Clayton, J. A. and Knox, J. C. 2008. Catastrophic flooding from Glacial Lake Wisconsin. *Geomorphology* 93, 384–97.

Clayton, K. M. 2000. Glacial erosion of the Wash and Fen basin and the deposition of the chalky till of eastern England. *Quaternary Science Reviews* 19, 811–22.

Clayton, L. 1964. Karst topography on stagnant glaciers. *Journal of Glaciology* 5, 107–12.

Clayton, L. 1967. Stagnant glacier features of the Missouri Coteau in North Dakota. *North Dakota Geological Survey*, Miscellaneous Series 30, 25–46.

Clayton, L. and Cherry, J. A. 1967. Pleistocene superglacial and ice-walled lakes of west-central North America. *North Dakota Geological Survey*, Miscellaneous Series 30, 47–52.

Clayton, L. and Moran, S. R. 1974. A glacial process-form model. In Coates, D. R. (ed.), *Glacial Geomorphology*. State University of New York, Binghampton, 89–119.

Clayton, L., Attig, J. W. and Mickelson, D. M. 1999. Tunnel channels formed in Wisconsin during the last glaciation. In Mickelson, D. M. and Attig, J. W. (eds), *Glacial Processes Past and Present*. Geological Society of America, Special Paper 337, 69–82.

Clayton, L., Attig, J. W. and Mickelson, D. M. 2001. Effects of Late Pleistocene permafrost on the landscape of Wisconsin, USA. *Boreas* 30, 173–88.

Clayton, L., Attig, J. W., Ham, N. R., Johnson, M. D., Jennings, C. E. and Syverson, K. M. 2008. Ice-walled-lake plains: implications for the origin of hummocky glacial topography in middle North America. *Geomorphology* 97, 237–48.

Clayton, L., Mickelson, D. M. and Attig, J. W. 1989. Evidence against pervasively deformed bed material beneath rapidly moving lobes of the southern Laurentide Ice Sheet. *Sedimentary Geology* 62, 203–8.

Clifford, S. M. 1987. Polar basal melting on Mars. *Journal of Geophysical Research* 92, 9135–52.

Clifford, S. M. and Parker, T. 2001. The evolution of the Martian hydrosphere: implications for the fate of a potential ocean and the current state of the northern plains. *Icarus* 154, 40–79.

Cogley, J. G. 1999. Effective sample size for glacier mass balance. *Geografiska Annaler* 81A, 497–507.

Cogley, J. G. 2004. Greenland accumulation: an error model. *Journal of Geophysical Research* 109, doi:10.1029/JD004449.

Cogley, J. G. and Adams, W. P. 1998. Mass balance of glaciers other than the ice sheets. *Journal of Glaciology* 44, 315–25.

Cohen, D. 2000. Rheology of ice at the bed of Engabreen, Norway. *Journal of Glaciology* 46, 611–21.

Cohen, D., Hooyer, T. S., Iverson, N. R., Thomason, J. F. and Jackson, M. 2006. Role of transient water pressure in quarrying: a subglacial experiment using acoustic emissions. *Journal of Geophysical Research* 111, doi:10.1029/2005JF 000439.

Cohen, D., Iverson, N. R., Hooyer, T. S., Fischer, U. H., Jackson, M. and Moore, P. L. 2005. Debris-bed friction of hard-bedded glaciers. *Journal of Geophysical Research* 110, doi:10.1029/ 2004JF000228.

Colaprete, A. and Jakosky, B. M. 1998. Ice flow and rock glaciers on Mars. *Journal of Geophysical Research* 103, 5897–909.

Cole-Dai, J., Mosley-Thompson, E., Wight, S. P. and Thompson, L. G. 2000. A 4100-year record of explosive volcanism from an East Antarctica ice core. *Journal of Geophysical Research* 105, I24431-24442.

Colgan, P. M. 1999. Reconstruction of the Green Bay Lobe, Wisconsin, United States, from 26,000 to 13,000 radiocarbon years BP. In Mickelson, D. M. and Attig, J. W. (eds), *Glacial Processes: Past and Present*. Geological Society of America, Special Paper 337, 137–50.

Colgan, P. M., Mickelson, D. M. and Cutler, P. M. 2003. Ice-marginal terrestrial landsystems: southern Laurentide Ice Sheet margin. In Evans, D. J. A. (ed.), *Glacial Landsystems*. Arnold, London, 111–42.

Colhoun, E. and Shulmeister, J. 2007. Glaciations: Late Pleistocene of the SW Pacific Region. *Encyclopedia of Quaternary Science*, Elsevier, Amsterdam, 1066–76.

Collins, D. N. 1979. Hydrochemistry of meltwaters draining from an Alpine glacier. *Arctic and Alpine Research* 11, 307–24.

Collins, D. N. 1984. Water and mass balance measurements in glacierized drainage basins. *Geografiska Annaler* 66A, 197–214.

Collins, D. N. 2008. Climatic warming, glacier recession and runoff from Alpine basins after the Little Ice Age maximum. *Annals of Glaciology* 48, 119–24.

Collinson, J. D. 1996. Alluvial sediments. In Reading, H. G. (ed.), *Sedimentary Environments and Facies*, 3rd edn. Oxford, Blackwell, 37–82.

Comiso, J. C. and Nishio, F. 2008. Trends in sea ice cover using enhanced and compatible AMSR-E, SSM/I and SMMR data. *Journal of Geophysical Research* 113, doi:10.1029/2007JC004257.

Comiso, J. C., Yang, J., Honjo, S. and Krishfield, R. A. 2003. Detection of change in the Arctic using satellite and in situ data. *Journal of Geophysical Research* 108, doi:10.1029/2002JC001347.

Conway, H. and Rasmussen, L. 2000. Summer temperature profiles within supraglacial debris on Khumbu Glacier, Nepal. *Debris-Covered Glaciers*, IASH Publ. 264, 89–98.

Conway, H., Catania, G., Raymond, C. F., Gades, A. M., Scambos, T. A. and Engelhardt, H. 2002. Switch of flow direction in an Antarctic ice stream. *Nature* 419, 465–7.

Cook, A. J., Fox, A. J., Vaughan, D. G. and Ferrigno, J. G. 2005. Retreating glacier fronts on the Antarctic Peninsula over the past half-century. *Science* 308, 541–4.

Cook, S. J., Knight, P. G., Waller, R. I., Robinson, Z. P. and Adam, W. G. 2007. The geography of basal ice and its relationship to glaciohydraulic supercooling: Svinafellsjökull, southeast Iceland. *Quaternary Science Reviews* 26, 2309–15.

Cook, S. J., Waller, R. I. and Knight, P. G. 2006. Glaciohydraulic supercooling: the process and its significance. *Progress in Physical Geography* 30 (5), 577–88.

Copland, L. and Sharp, M. J. 2001. Mapping thermal and hydrological conditions beneath a polythermal glacier with radio-echo sounding. *Journal of Glaciology* 47, 232–42.

Copland, L., Harbor, J., Gordon, S. and Sharp, M. J. 1997. The use of borehole video in investigating the hydrology of a temperate glacier. *Hydrological Processes* 11, 211–24.

Copland, L., Sharp, M. J. and Nienow, P. W. 2003a. Links between short-term velocity variations and the subglacial hydrology of a predominantly cold polythermal glacier. *Journal of Glaciology* 49, 337–48.

Copland, L., Sharp, M. J. and Dowdeswell, J. A. 2003b. The distribution and flow characteristics of surge-type glaciers in the Canadian High Arctic. *Annals of Glaciology* 36, 73–81.

Corripio, J. G. and Purves, R. S. 2005. Surface energy balance of high altitude glaciers in the central Andes: the effect of snow penitentes. In de Jong, C., Collins, D. and Ranzi, R. (eds),

Climate and Hydrology in Mountain Areas. Wiley and Sons, London.

Cossart, E., Braucher, R., Fort, M., Bourlès, D. L. and Carcaillet, J. 2008. Slope instability in relation to glacial debuttressing in alpine areas (Upper Durance catchment, southeastern France): evidence from field data and [10]Be cosmic ray exposure ages *Geomorphology* 95, 3–26.

Coude, A. 1989. Comparative study of three drumlin fields in western Ireland: geomorphological data and genetic implications. *Sedimentary Geology* 62, 321–35.

Cowan, E. A. and Powell, R. D. 1990. Suspended sediment transport and deposition of cyclically interlaminated sediment in a temperate glacial fiord, Alaska, USA. In Dowdeswell, J. A. and Scourse, J. D. (eds), *Glacimarine Environments: Processes and Sediments*. Special Publications of the Geologists' Association 53, 75–89.

Cowan, E. A. and Powell, R. D. 1991. Suspended sediment transport and deposition of cyclically interlaminated sediment in a temperate glacial fiord, Alaska, USA. In Dowdeswell, J. A. and Scourse, J. D. (eds), *Glaciomarine Environments: Processes and Sediments*. Geological Society, London, Special Publication 53, 75–89.

Cowan, E. A., Cai, J., Powell, R. D., Clark, J. D. and Pitcher, J. N. 1997. Temperate glacimarine varves: an example from Disenchantment Bay, southern Alaska. *Journal of Sedimentary Research* 67, 536–49.

Cowan, E. A., Powell, R. D. and Smith, N. D. 1988. Marine event sedimentation from a rainstorm at tidewater front of a temperate glacier. *Geology* 16, 409–12.

Cowan, E. A., Seramur, K. C., Cai, J. and Powell, R. D. 1999. Cyclic sedimentation produced by fluctuations in meltwater discharge, tides and marine productivity in an Alaskan fjord. *Sedimentology* 46, 1109–26.

Coxon, P. 2005. The late Tertiary landscapes of western Ireland. *Irish Geography* 38, 111–27.

Coxon, P., Owen, L. A. and Mitchell, W. A. 1996. A late Quaternary catastrophic flood in the Lahul Himalayas. *Journal of Quaternary Science* 11, 495–510.

Croot, D. G. 1988a. Glaciotectonics and surging glaciers: a correlation based on Vestspitsbergen, Svalbard, Norway. In Croot, D. G. (ed.), *Glaciotectonics: Forms and Processes*. Balkema, Rotterdam, 49–61.

Croot, D. G. 1988b. Morphological, structural and mechanical analysis of neoglacial ice-pushed ridges in Iceland. In Croot, D. G. (ed.), *Glaciotectonics: Forms and Processes*. Balkema, Rotterdam, 33–47.

Crowell, J. C., 1999. *Pre-Mesozoic Ice Ages: Their Bearing on Understanding the Climate System*. Geological Society of America, Memoir 192.

Csatho, B., Schenk, T., van der Veen, C. J. and Krabill, W. B. 2008. Intermittent thinning of Jakobshavn Isbrae, west Greenland, since the Little Ice Age. *Journal of Glaciology* 54 (184), 131–44.

Cuffey, K. M. and Marshall, S. J. 2000. Substantial contribution to sea-level rise during the last interglacial from the Greenland ice sheet. *Nature* 404, 591–4.

Cuffey, K. M., Conway, H., Hallet, B., Gades, A. M. and Raymond, C. F. 1999. Interfacial water in polar glaciers, and sliding at −17°C. *Geophysical Research Letters* 26, 751–4.

Cullen, N. J., Mölg, T., Hardy, D. R., Steffen, K. and Kaser, G. 2007. Energy balance model validation on the top of Kilimanjaro using eddy correlation data. *Annals of Glaciology* 46, 227–33.

Cummings, D. and Occhietti, S. 2001. Late Wisconsinan sedimentation in the Quebec City region: evidence for energetic subaqueous fan deposition during initial deglaciation. *Geographie Physique et Quaternaire* 55, 257–73.

Cunningham, F. F. 1990. *James David Forbes: Pioneer Scottish Glaciologist*. Scottish Academic Press, Edinburgh.

Curran, K. J., Hill, P. S., Milligan, T. G., Cowan, E. A., Syvitski, J. P. M. and Konings, S. M. 2004. Fine-grained sediment flocculation below the Hubbard Glacier meltwater plume, Disenchantment Bay, Alaska. *Marine Geology* 203, 83–94.

Curray, J. R. 1964. Transgressions and regressions. In Miller, R. L. (ed.), *Papers in Marine Geology*. MacMillan, New York, 175–203.

Curry, A. M. 1999. Paraglacial modification of slope form. *Earth Surface Processes and Landforms* 24, 1213–28.

Curry, A. M. 2000a. Observations on the distribution of paraglacial reworking of glacigenic drift in western Norway. *Norsk Geografisk Tidsskrift* 54, 139–47.

Curry, A. M. 2000b. Holocene reworking of drift-mantled hillslopes in the Scottish Highlands. *Journal of Quaternary Science* 15, 529–41.

Curry, A. M. and Ballantyne, C. K. 1999. Paraglacial modification of glacigenic sediments. *Geografiska Annaler* 81A, 409–19.

Curry, A. M. and Morris, C. J. 2004. Lateglacial and Holocene talus slope development and rockwall retreat on Mynydd Du, UK. *Geomorphology* 58, 85–106.

Cutler, K. B., Edwards, R. L., Taylor, F. W., Cheng, H., Adkins, J., Gallup, C. D., Cutler, P. M., Burr, G. S. and Bloom, A. L. 2003. Rapid sea-level fall and deep-ocean temperature change since the last interglacial period. *Earth and Planetary Science Letters* 206, 253–71.

Cutler, P. M. and Munro, D. S. 1996. Visible and near-infrared reflectivity during the ablation period on Peyto Glacier, Alberta, Canada. *Journal of Glaciology* 42 (141), 333–40.

Cutler, P. M., Colgan, P. M. and Mickelson, D. M. 2002. Sedimentologic evidence for outburst floods from the Laurentide Ice Sheet margin in Wisconsin, USA: implications for tunnel-channel formation. *Quaternary International* 90, 23–40.

Cutler, P. M., MacAyeal, D. R., Mickelson, D. M., Parizek, B. R. and Colgan, P. M. 2000. A numerical investigation of ice lobe–permafrost interaction around the southern Laurentide ice sheet. *Journal of Glaciology* 46, 311–25.

D'Agata, C. and Zanutta, A. 2007. Reconstruction of the recent changes of a debris-covered glacier (Brenva Glacier, Mont Blanc Massif, Italy) using indirect sources: methods, results and validation. *Global and Planetary Change* 56, 57–68.

Dahl, R. 1965. Plastically sculptured detail forms on rock surfaces in northern Nordland, Norway. *Geografiska Annaler* 47, 83–140.

Dahl-Jensen, D. and Johnsen, S. J. 1986. Palaeotemperatures still exist in the Greenland ice sheet. *Nature* 320, 250–2.

Dahlen, F. A., Suppe, J. and Davis, D. M. 1984. Mechanics of fold and thrust belts and accretionary wedges: cohesive Coulomb theory. *Journal of Geophysical Research* 89, 10087–101.

Damon, P. E. and Peristykh, A. N. 2005. Solar forcing of global temperature change since AD 1400. *Climatic Change* 68, 101–11.

Dansgaard, W. 1964. Stable isotopes in precipitation. *Tellus* 16, 436–68.

Dansgaard, W., Clausen, H. B., Gundestrup, N., Hammer, C. U., Johnsen, S. J., Kristinsdottir, P. and Reeh, N. 1973. Stable isotope glaciology. *Medd. Grønland* 2, 1–53.

Dansgaard, W., Johnsen, S. J., Clausen, H. B., Dahl-Jensen, D., Gundestrup, N. S., Hammer, C. U., Hvidberg, C. S., Steffensen, J. P., Sveinbjörnsdottir, A. E., Jouzel, J. and Bond, G. 1993. Evidence for general instability of past climate from a 250-kyr ice-core record. *Nature* 364, 218–20.

Dardis, G. F. 1985. Till facies associations in drumlins and some implications for their mode of formation. *Geografiska Annaler* 67A, 13–22.

Dardis, G. F. 1987. Sedimentology of late Pleistocene drumlins in south-central Ulster, Northern Ireland. In Menzies J. and Rose, J. (eds), *Drumlin Symposium*. Balkema, Rotterdam, 215–24.

Dardis, G. F. and Hanvey, P. M. 1994. Sedimentation in a drumlin lee-side subglacial wave cavity, northwest Ireland. *Sedimentary Geology* 91, 97–114.

Dardis, G. F. and McCabe, A. M. 1983. Facies of subglacial channel sedimentation in late Pleistocene drumlins, Northern Ireland. *Boreas* 12, 263–78.

Dardis, G. F. and McCabe, A. M. 1987. Subglacial sheetwash and debris flow deposits in late Pleistocene drumlins, Northern Ireland. In Menzies, J. and Rose, J. (eds), *Drumlin Symposium*. Balkema, Rotterdam, 225–40.

Dardis, G. F., McCabe, A. M. and Mitchell, W. I. 1984. Characteristics and origins of lee-side stratification sequences in late Pleistocene drumlins, Northern Ireland. *Earth Surface Processes and Landforms* 9, 409–24.

Das, S. B. and Alley, R. B. 2005. Characterization and formation of melt layers in polar snow: observations and experiments from West Antarctica. *Journal of Glaciology* 51, 307–12.

Das, S. B., Joughin, I., Behn, M. D., Howat, I. M., King, M. A., Lizarralde, D. and Bhatia, M. P. 2008. Fracture propagation to the base of the Greenland ice sheet during supraglacial lake drainage. *Science* 320, 778–81.

Davis, D. M., Suppe, J. and Dahlen, F. A. 1983. Mechanics of fold and thrust belts and accretionary wedges. *Journal of Geophysical Research* 88, 1153–72.

Davis, P. T., Briner, J. P., Coulthard, R. D., Finkel, R. W. and Miller, G. H. 2006. Preservation of Arctic landscapes overridden by cold-based ice sheets. *Quaternary Research* 65, 156–63.

Davison, S. and Stoker, M. S. 2002. Late Pleistocene glacially-influenced deep-marine sedimentation off NW Britain: implications for the rock record. In Dowdeswell, J. A. and Ó Cofaigh, C. (eds), *Glacier-influenced Sedimentation on High Latitude Continental Margins*. Geological Society Special Publication 203, 129–47.

Dawson, A. G. 1979. A Devensian medial moraine in Jura. *Scottish Journal of Geology* 15, 43–8.

Dawson, A. G. 1992. *Ice Age Earth: Late Quaternary Geology and Climate*. Routledge, London.

Dawson, A. G., Matthews, J. A. and Shakesby, R. A. 1987. Rock platform erosion on periglacial shores: a modern analogue for Pleistocene rock platforms in Britain. In Boardman, J. (ed.), *Periglacial Processes and Landforms in Britain and Ireland*. Cambridge University Press, 173–82.

Dawson, M. R. and Bryant, I. D. 1987. Three-dimensional facies geometry in Pleistocene outwash sediments, Worcestershire, UK. In Ethridge, F. G., Flores, R. M. and Harvey, M. D. (eds), *Recent Developments in Fluvial Sedimentology*. SEPM Special Publication 39. Tulsa, OK, 191–6.

Dawson, M. R. and Gardiner, V. 1987. River terraces: the general model and a palaeohydrological and sedimentological interpretation of the terraces of the River Severn. In Gregory, K. J., Lewin, J. and Thornes, J. B. (eds), *Palaeohydrology in Practice*. Wiley, Chichester, 269–305.

De Angelis, H. 2007. Glacial geomorphology of the east-central Canadian Arctic. *Journal of Maps* 2007, 323–41.

De Angelis, H. and Kleman, J. 2005. Palaeo-ice streams in the northern Keewatin sector of the Laurentide Ice Sheet. *Annals of Glaciology* 42, 135–44.

De Angelis, H. and Kleman, J. 2007. Palaeo-ice streams in the Foxe/Baffin sector of the Laurentide Ice Sheet. *Quaternary Science Reviews* 26, 1313–31.

De Angelis, H. and Kleman, J. 2008. Palaeo-ice stream onsets: examples from the north-eastern Laurentide Ice Sheet. *Earth Surface Processes and Landforms* 33 (4), 560–72.

De Angelis, H. and Skvarca, P. 2003. Glacier surge after ice shelf collapse. *Science* 299, 1560–2.

DeConto, R. M. and Pollard, D. 2003. Rapid Cenozoic glaciation of Antarctica induced by declining atmospheric CO_2. *Nature* 421, 245–9.

Delaney, C. 2001. Morphology and sedimentology of the Rooskagh esker, Co. Roscommon. *Irish Journal of Earth Sciences* 19, 5–22.

Delaney, C. 2002. Sedimentology of a glaciofluvial landsystem, Lough Ree area, central Ireland: implications for ice margin characteristics during Devensian glaciation. *Sedimentary Geology* 149, 111–26.

Deline, P. 1999a. La mise en place de l'amphitéâtre morainique du Miage (Val Veny, Val d'Aoste). *Geomorphologie: Relief, Processus, Environnement* 1, 59–72.

Deline, P. 1999b. Les variations Holocènes récentes du Glacier du Miage (Val Veny, Val d'Aoste). *Quaternaire* 10, 5–13.

Delmonte, B., Petit, J. R. and Maggi, V. 2002. Glacial to Holocene implications of the new 27000-year dust record from the EPICA Dome C (East Antarctica) ice core. *Climate Dynamics* 18, 647–60.

Denman, K. L. et al. 2007. Couplings between changes in the climate system and biogeochemistry. In Solomon, S. et al. (eds), *Climate Change 2007: The Physical Science Basis.* Contribution of Working Group I to the Fourth Assessment of the Intergovernmental Panel on Climate Change. Cambridge University Press.

Denton, G. H. and Sugden, D. E. 2005. Meltwater features that suggest Miocene ice sheet overriding of the Transantarctic Mountains in Victoria Land, Antarctica. *Geografiska Annaler* 87A, 67–85.

Denton, G. H., Lowell, T. V., Heusser, C. J., Schluchter, C., Andersen, B. G., Heusser, L. E., Moreno, P. I. and Marchant, D. R. 1999. Geomorphology, stratigraphy and radiocarbon chronology of Llanquihue Drift in the area of the southern Lake District, Seno Reloncaví, and Isla Grande de Chiloé, Chile. *Geografiska Annaler* 81A, 167–229.

Denton, G. H., Sugden, D. E., Marchant, D. R., Hall, B. L. and Wich, T. I. 1993. East Antarctic ice sheet sensitivity to Pliocene climatic change from a Dry Valleys perspective. *Geografiska Annaler* 75A, 155–204.

De Woul, M. and Hock, R. 2005. Static mass-balance sensitivity of Arctic glaciers and ice caps using a degree-day approach. *Annals of Glaciology* 42, 217–24.

Deynoux, M., Miller, J. M. G., Domack, E. W., Eyles, N., Fairchild, I. and Young, G. M. (eds). 2004. *Earth's Glacial Record.* Cambridge University Press.

Dickson, J. L. and Head, J. W. 2006. Evidence for an Hesperian-aged south circum-polar lake margin environment on Mars. *Planetary and Space Science* 54, 251–72.

Dickson, J. L., Head, J. W. and Marchant, D. R. 2008. Late Amazonian glaciation at the dichotomy boundary on Mars: evidence for glacial thickness maxima and multiple glacial phases. *Geology* 36, 411–14.

Diekmann, B., Futterer, D. K. and Grobe, H. 2004. Terrigenous sediment supply in the polar to temperate South Atlantic: land-ocean links of environmental changes during the late Quaternary. In Wefer, G., Mulitza, S. and Ratmeyer, V. (eds), *The South Atlantic in the Late Quaternary: Reconstruction of Material Budgets and Current Systems.* Springer-Verlag, Berlin, 375–99.

Dietrich, R., Maas, H.-G., Baessler, M., Rülke, A., Richter, A., Schwalbe, E. and Westfield, P. 2007. Jakobshavn Isbrae, West Greenland: Flow velocities and tidal interaction of the front area from 2004 field observations. *Journal of Geophysical Research* 112, doi:10.1029/2006JF000601.

Dikau, R., Brunsden, D., Schrott, L. and Ibsen, M.-L. (eds). 1996. *Landslide Recognition: Identification, Movement and Courses.* Wiley, Chichester.

DiLabio, R. N. W. and Coker, W. B. (eds). 1989. *Drift Prospecting.* Geological Survey of Canada, Paper 89-20.

DiLabio, R. N. W. and Shilts, W. W. 1979. Composition and dispersal of debris by modern glaciers, Bylot Island, Canada. In Schluchter, C. (ed.), *Moraines and Varves.* Balkema, Rotterdam, 145–55.

Diolaiuti, G., Smiraglia, C., Vassena, G. and Motta, M. 2004. Dry calving processes at the ice cliff of Strandline Glacier, northern Victoria Land, Antarctica. *Annals of Glaciology* 39, 201–8.

Dimakis, P., Elverhøi, A., Høeg, K., Solheim, A., Harbitz, C., Laberg, J. S., Vorren, T. O. and Marr, J. 2000. Submarine slope stability on high latitude glaciated Svalbard-Barents Sea margin. *Marine Geology* 162, 303–16.

Dionne, J.-C. 1974. Polished and striated mud surfaces in the St Lawrence tidal flats, Quebec. *Canadian Journal of Earth Sciences* 11, 860–6.

Dionne, J.-C. 1985. Drift-ice abrasion marks along rocky shores. *Journal of Glaciology* 31, 237–41.

Dionne, J.-C. 1987. Tadpole rock (rocdrumlin): a glacial streamline moulded form. In Menzies, J. and Rose, J. (eds), *Drumlin Symposium.* Balkema, Rotterdam, 149–59.

Dirszowsky, R. W. and Desloges, J. R. 2004. Evolution of the Moose Lake Delta, British Columbia: implications for Holocene environmental change in the Canadian Rocky Mountains. *Geomorphology* 57, 75–93.

Dix, J. K. and Duck, R. W. 2000. A high-resolution seismic stratigraphy from a Scottish sea loch and its implications for Loch Lomond Stadial deglaciation. *Journal of Quaternary Science* 15, 645–56.

Doake, C. S. M., Corr, H. F. J., Jenkins, A., Makinson, K., Nicholls, K. W., Nath, C., Smith, A. M. and Vaughan, D. G. 2001. Rutford Ice Stream, Antarctica. In Alley, R. B. and Bindschadler, R. A. (eds), *The West Antarctic Ice Sheet: Behavior and Environment.* American Geophysical Union, Antarctic Research Series, vol. 77, 221–35.

Dolgushin, L. D. and Osipova, G. B. 1975. Glacier surges and the problem of their forecast. *International Association of Hydrological Sciences Publication* 104, 292–304.

Domack, E. W. and Lawson, D. E. 1985. Pebble fabric in an ice-rafted diamicton. *Journal of Geology* 93, 577–92.

Domack, E. W., Duran, D., Leventer, A., Ishman, S., Doanel, S., McCallum, S., Amblas, D., Ring, J., Gilbert, R. and Prentice, M. 2005. Stability of the Larsen B ice shelf on the Antarctic Peninsula during the Holocene epoch. *Nature* 436, 681–5.

Domack, E. W., Jacobson, E. K., Shipp, S. and Anderson, J. B. 1999. Sedimentological and stratigraphic signature of the Late Pleistocene/Holocene fluctuation of the West Antarctic Ice Sheet in the Ross Sea: a new perspective. Part 2. *Geological Society of America Bulletin* 111, 1486–516.

Donnadieu, Y., Goddéris, Y., Ramstein, G., Nédélec, A. and Meert, J. 2004. A 'snowball Earth' climate triggered by continental break-up through changes in runoff. *Nature* 428, 303–6.

Donner, J. J. 1995. *The Quaternary History of Scandinavia.* Cambridge University Press.

Douglas, B. C. 2001. Sea level change in the era of the recording tide gauge. In Douglas, B. C., Kearney, M. S. and Leatherman, S. P. (eds), *Sea Level Rise, History and Consequences.* Academic Press, San Diego, CA, 37–64.

Dowdeswell, J. A. 1987. Processes of glacimarine sedimentation. *Progress in Physical Geography* 11, 52–90.

Dowdeswell, J. A. 1989. On the nature of Svalbard icebergs. *Journal of Glaciology* 35, 224–34.

Dowdeswell, J. A. and Bamber, J. L. 2007. Keel depths of modern Antarctic icebergs and implications for sea-floor scouring in the geological record. *Marine Geology* 243, 120–31.

Dowdeswell, J. A. and Benham, T. J. 2003. A surge of Perseibreen, Svalbard, examined using aerial photography and ASTER high resolution satellite imagery. *Polar Research* 22, 373–83.

Dowdeswell, J. A. and Dowdeswell, E. K. 1989. Debris in icebergs and rates of glacimarine sedimentation: observations from Spitsbergen and a simple model. *Journal of Geology* 97, 221–31.

Dowdeswell, J. A. and Elverhøi, A. 2002. The timing of initiation of fast-flowing ice streams during a glacial cycle inferred from glacimarine sedimentation. *Marine Geology* 188, 3–14.

Dowdeswell, J. A. and Murray, T. 1990. Modelling rates of sedimentation from icebergs. In Dowdeswell, J. A. and Scourse, J. D. (eds), *Glacimarine Environments: Processes and Sediments.* Special Publication of the Geologists' Association 53, 121–37.

Dowdeswell, J. A. and Ó Cofaigh, C. (eds). 2002. *Glacier-Influenced Sedimentation on High Latitude Continental Margins*. Geological Society, London, Special Publication 203.

Dowdeswell, J. A. and Siegert, M. J. 2002. The physiography of modern Antarctic subglacial lakes. *Global and Planetary Change* 35, 221–36.

Dowdeswell, J. A., Benham, T. J., Strozzi, T. and Hagen, J. O. 2008a. Iceberg calving flux and mass balance of the Austfonna ice cap on Nordaustlandet, Svalbard. *Journal of Geophysical Research* 113, doi:10.1029/2007JF000905.

Dowdeswell, J. A., Drewry, D. J., Liestol, O. and Orheim, O. 1984. Radio echo sounding of Spitsbergen glaciers: problems in the interpretation of layer and bottom returns. *Journal of Glaciology* 30, 16–21.

Dowdeswell, J. A., Elverhøi, A. and Andrews, J. T. 1999a. Asynchronous deposition of ice-rafted layers in the Nordic seas and the North Atlantic Ocean. *Nature* 400, 348–51.

Dowdeswell, J. A., Elverhøi, A. and Spielhagen, R. 1998. Glacimarine sedimentary processes and facies on the Polar North Atlantic margins. *Quaternary Science Reviews* 17, 243–72.

Dowdeswell, J. A., Gorman, M. R., Glazovski, A. F. and Macheret, Y. Y. 1994a. Evidence for floating ice shelves in Franz Josef land, Russian High Arctic. *Arctic and Alpine Research* 26, 86–92.

Dowdeswell, J. A., Hambrey, M. J. and Wu, R. 1985. A comparison of clast fabric and shape in Late Precambrian and modern glacigenic sediments. *Journal of Sedimentary Petrology*, 55, 691–704.

Dowdeswell, J. A., Hamilton, G. S. and Hagen, J. O. 1991. The duration of the active phase of surge-type glaciers: contrasts between Svalbard and other regions. *Journal of Glaciology* 37, 388–400.

Dowdeswell, J. A., Hodgkins, R., Nuttall, A.-M., Hagen, J. O. and Hamilton, G. S. 1995. Mass balance change as a control on the frequency and occurrence of glacier surges in Svalbard, Norwegian High Arctic. *Geophysical Research Letters* 22, 2909–12.

Dowdeswell, J. A., Kenyon, N. H., Elverhøi, A., Laberg, J. S., Hollender, F. J., Mienert, J. and Siegert, M. J. 1996. Large scale sedimentation on the glacier-influenced polar North Atlantic margins: long range side-scan sonar evidence. *Geophysical Research Letters* 23, 3535–8.

Dowdeswell, J. A., Ó Cofaigh, C., Noormets, R., Larter, R. D., Hillenbrand, C.-D., Benetti, S., Evans, J. and Pudsey, C. J. 2008c. A major trough-mouth fan on the continental margin of the Bellingshausen Sea, West Antarctica: The Belgica Fan. *Marine Geology* 252, 129–40.

Dowdeswell, J. A., Ó Cofaigh, C., Taylor, J., Kenyon, N. H., Mienert, J. and Wilken, M. 2002. On the architecture of high-latitude continental margins: the influence of ice sheet and sea ice processes in the Polar North Atlantic. In Dowdeswell, J. A. and Ó Cofaigh, C. (eds), *Glacier-Influenced Sedimentation on High Latitude Continental Margins*. Geological Society, London, Special Publication 203, 105–28.

Dowdeswell, J. A., Ottesen, D., Evans, J., Ó Cofaigh, C. and Anderson, J. B. 2008b. Submarine glacial landforms and rates of ice-stream collapse. *Geology* 36, 819–22.

Dowdeswell, J. A., Ottesen, D. and Rise, L. 2006. Flow switching and large-scale deposition by ice streams draining former ice sheets. *Geology* 34, 313–16.

Dowdeswell, J. A., Unwin, B., Nuttall, A.-M. and Wingham, D. J. 1999b. Velocity structure, flow instability and mass flux on a large Arctic ice cap from satellite radar interferometry. *Earth and Planetary Science Letters* 167, 131–40.

Dowdeswell, J. A., Villinger, H., Whittington, R. J. and Marienfeld, P. 1993. Iceberg scouring in Scoresby Sund and on the East Greenland continental shelf. *Marine Geology* 111, 37–53.

Dowdeswell, J. A., Whittington, R. J., Jennings, A. E., Andrews, J. T., Mackensen, A. and Marienfeld, P. 2000. An origin for laminated glacimarine sediments through sea-ice build-up and suppressed iceberg rafting. *Sedimentology* 47, 557–76.

Dowdeswell, J. A., Whittington, R. J. and Marienfeld, P. 1994b. The origin of massive diamicton facies by iceberg rafting and scouring, Scoresby Sund, East Greenland. *Sedimentology* 41, 21–35.

Dredge, L. A. 2000. Carbonate dispersal trains, secondary till plumes, and ice streams in the west Foxe Sector, Laurentide Ice Sheet. *Boreas* 29, 144–56.

Dredge, L. A. and Cowan, W. R. 1989. Quaternary geology of the southwestern Canadian shield. In Fulton, R. J. (ed.), *Quaternary Geology of Canada and Greenland*. Geological Survey of Canada, Geology of Canada no. 1, 214–35.

Dreimanis, A. 1989. Tills: their genetic terminology and classification. In Goldthwait, R. P. and Matsch, C. L. (eds), *Genetic Classification of Glacigenic Deposits*. Balkema, Rotterdam, 17–84.

Dreimanis, A. and Gibbard, P. 2005. Stratigraphy and sedimentation of the stratotype sections of the Catfish Creek Drift Formation between Bradtville and Plum Point, north shore, Lake Erie, southwestern Ontario, Canada. *Boreas* 34, 101–22.

Dreimanis, A. and Rappol, M. 1997. Late Wisconsinan subglacial clastic intrusive sheets along Lake Erie Bluffs at Bradtville, Ontario, Canada. *Sedimentary Geology* 11, 225–48.

Dreimanis, A. and Vagners, U. J. 1971. Bimodal distribution of rock and mineral fragments in basal tills. In Goldthwait, R. P. (ed.), *Till: A Symposium*. Ohio State University Press, Columbus, 237–50.

Dreimanis, A., Hamilton, J. P. and Kelly, P. E. 1986. Complex subglacial sedimentation of Catfish Creek till at Bradtville, Ontario, Canada. In van der Meer, J. J. M. (ed.), *Tills and Glaciotectonics*. Balkema, Rotterdam, 73–87.

Drewry, D. J. 1975. Initiation and growth of the East Antarctic Ice Sheet. *Journal of the Geological Society of London* 131, 255–73.

Drewry, D. J. 1983. *Antarctica: Glaciological and Geophysical Folio*. University of Cambridge.

Drewry, D. J. 1986. *Glacial Geologic Processes*. Edward Arnold, London.

Drewry, D. J. and Cooper, A. P. R. 1981. Processes and models of Antarctic glaciomarine sedimentation. *Annals of Glaciology* 2, 117–22.

Duk-Rodkin, A. and Hughes, O. L. 1994. Tertiary-Quaternary drainage of the pre-glacial Mackenzie basin. *Quaternary International* 22/23, 221–41.

Duk-Rodkin, A., Jackson, L. E. and Rodkin, O. 1986. *A Composite Profile of the Cordilleran Ice Sheet During McConnell Glaciation, Glenlyon and Tay River Map Areas, Yukon Territory*. Geological Survey of Canada, Paper 86-1B, 257–62.

Dunlop, P. and Clark, C. D. 2006a. The morphological characteristics of ribbed moraine. *Quaternary Science Reviews* 25, 1668–91.

Dunlop, P. and Clark, C. D. 2006b. Distribution of ribbed moraine in the Lac Naococane region, central Quebec, Canada. *Journal of Maps* 2006, 59–70.

Dunlop, P., Clark, C. D. and Hindmarsh, R. C. A. 2008. The Bed Ribbing Instability Explanation (BRIE): testing a numerical model of ribbed moraine formation arising from coupled flow of ice and subglacial sediment. *Journal of Geophysical Research*, doi:10.1029/2007JF000954.

Duplessy, J.-C. 1978. Isotope studies. In Gribbin, J. (ed.), *Climatic Change*. Cambridge University Press, 46–67.

DuPont, T. K. and Alley, R. B. 2005. Assessment of the importance of ice-shelf buttressing to ice-sheet flow. *Geophysical Research Letters* 32, L04503, doi:10.1029/2004GL022024.

Duval, P. 1981. Grain growth and the mechanical behaviour of polar ice. *Annals of Glaciology* 6, 79–82.

Dyke, A. S. 1993. Landscapes of cold-centred Late Wisconsinan ice caps, Arctic Canada. *Progress in Physical Geography* 17, 223–47.

Dyke, A. S. 1999. The last glacial maximum and the deglaciation of Devon Island: support for an Innuitian Ice Sheet. *Quaternary Science Reviews* 18, 393–420.

Dyke, A. S. 2004. An outline of North American deglaciation with emphasis on central and northern Canada. In Ehlers, J. and Gibbard, P. L. (eds), *Quaternary Glaciations – Extent and Chronology. Part II: North America.* Elsevier, Amsterdam, 373–424.

Dyke, A. S. 2008. The Steensby Inlet Ice Stream in the context of the deglaciation on Northern Baffin Island, Eastern Arctic Canada. *Earth Surface Processes and Landforms* 33, 573–92.

Dyke, A. S. and Dredge, L. A. 1989. Quaternary geology of the northwestern Canadian shield. In Fulton, R. J. (ed.), *Quaternary Geology of Canada and Greenland.* Geological Survey of Canada, Geology of Canada, no. 1, 178–214.

Dyke, A. S. and Evans, D. J. A. 2003. Ice-marginal terrestrial landsystems: northern Laurentide and Innuitian ice sheet margins. In Evans, D. J. A. (ed.), *Glacial Landsystems.* Arnold, London, 143–65.

Dyke, A. S. and Hooper, M. J. G. 2001. *Deglaciation of Northwest Baffin Island, Nunavut.* Geological Survey of Canada, 1:500,000 scale map 1999A.

Dyke, A. S. and Morris, T. F. 1988. Drumlin fields, dispersal trains and ice streams in arctic Canada. *Canadian Geographer* 32, 86–90.

Dyke, A. S. and Peltier, W. R. 2000. Forms, response times and variability of relative sea-level curves, glaciated North America. *Geomorphology* 32, 315–33.

Dyke, A. S. and Prest, V. K. 1987. Late Wisconsinan and Holocene history of the Laurentide Ice Sheet. *Geographie Physique et Quaternaire* XLI, 237–63.

Dyke, A. S. and Savelle, J. M. 2000. Major end moraines of Younger Dryas age on Wollaston Peninsula, Victoria Island, Canadian arctic: implications for palaeoclimate and for formation of hummocky moraine. *Canadian Journal of Earth Sciences* 37, 601–19.

Dyke, A. S., Andrews, J. T., Clark, P. U., England, J., Miller, G. H., Shaw, J. and Veillette, J. J. 2002. The Laurentide and Innuitian ice sheets during the Last Glacial Maximum. *Quaternary Science Reviews* 21, 9–31.

Dyke, A. S., Dredge, L. A. and Vincent, J.-S. 1982. Configuration and dynamics of the Laurentide Ice Sheet during the Late Wisconsin maximum. *Geographie Physique et Quaternaire* 36, 5–14.

Dyke, A. S., Morris, T. F. and Green, D. E. C. 1991. *Postglacial Tectonic and Sea Level History of the Central Canadian arctic.* Geological Survey of Canada, Bulletin 397.

Dyke, A. S., Morris, T. F., Green, D. E. C. and England, J. 1992. *Quaternary Geology of Prince of Wales Island, Arctic Canada.* Geological Survey of Canada, Memoir 433.

Dyurgerov, M. 2002. *Glacier mass balance and regime: data of measurements and analysis.* Institute of Arctic and Alpine Research, University of Colorado, Occasional Paper no. 55.

Dyurgerov, M. 2003. Mountain and subpolar glaciers show an increase in sensitivity to climate warming and intensification of the water cycle. *Journal of Hydrology* 282, 164–76.

Dyurgerov, M. 2005. *Supplement to Occasional Paper No. 55.* Institute of Arctic and Alpine Research, University of Colorado.

Dyurgerov, M. and Meier, M. F. 2004. Glaciers and the study of climate and sea-level change. In Bamber, J. L. and Payne, A. J. (eds), *Mass Balance of the Cryosphere: Observations and Modeling of Contemporary and Future Changes.* Cambridge University Press, 579–621.

Dyurgerov, M. and Meier, M. F. 2005. *Glaciers and the Changing Earth System: A 2004 Snapshot.* Institute of Arctic and Alpine Research, Boulder, Colorado, Occasional Paper no. 58.

Dzulynski, S. and Walton, E. K. 1965. *Sedimentary Features of Flysch and Greywackes.* Elsevier, Amsterdam.

Ebert, K. and Kleman, J. 2004. Circular moraine features on the Varanger Peninsula, northern Norway, and their possible relation to polythermal ice sheet coverage. *Geomorphology* 62, 159–68.

Echelmeyer, K. and Harrison, W. D. 1990. Jakobshavns Isbrae, West Greenland: seasonal variations in velocity – or lack thereof. *Journal of Glaciology* 36, 82–8.

Echelmeyer, K. and Zhongxiang, W. 1987. Direct observation of basal sliding and deformation of basal drift at sub-freezing temperatures. *Journal of Glaciology* 33, 83–98.

Echelmeyer, K., Clarke, T. S. and Harrison, W. D. 1991. Surficial glaciology of Jakobshavns Isbrae, West Greenland. Part I: Surface morphology. *Journal of Glaciology* 37 (127), 368–82.

Eden, D. J. and Eyles, N. 2001. Description and numerical model of Pleistocene iceberg scours and ice-keel turbated facies at Toronto, Canada. *Sedimentology* 48, 1079–102.

Ehlers, J. and Gibbard, P. L. 1991. Anglian glacial deposits in Britain and the adjoining offshore regions. In Ehlers, J., Gibbard, P. L. and Rose, J. (eds), *Glacial Deposits in Great Britain and Ireland.* Balkema, Rotterdam, 17–24.

Ehlers, J. and Gibbard, P. L. (eds). 2004a. *Quaternary Glaciations: Extent and Chronology. Part 1: Europe.* Developments in Quaternary Science 2. Elsevier, Amsterdam.

Ehlers, J. and Gibbard, P. L. (eds). 2004b. *Quaternary Glaciations: Extent and Chronology. Part II: North America.* Developments in Quaternary Science 2. Elsevier, Amsterdam.

Ehlers, J. and Gibbard, P. L. (eds). 2004c. *Quaternary Glaciations: Extent and Chronology. Part III: South America, Asia, Africa, Australia, Antarctica.* Developments in Quaternary Science 2. Elsevier, Amsterdam.

Ehlers, J. and Linke, G. 1989. The origin of deep buried channels of Elsterian age in Northwest Germany. *Journal of Quaternary Science* 4, 255–65.

Ehlers, J. and Stephan, H.-J. 1979. Forms at the base of till strata as indicators of ice movement. *Journal of Glaciology* 22, 345–56.

Ehlers, J. and Wingfield, R. 1991. The extension of the Late Weichselian/Late Devensian ice sheets in the North Sea Basin. *Journal of Quaternary Science* 6, 313–26.

Ehlers, J., Meyer, K. D. and Stephan, H. J. 1984. Pre-Weichselian glaciation of North-West Europe. *Quaternary Science Reviews* 3, 1–40.

Eichler, A., Schwikowski, M. and Gäggerler, H. W. 2001. Meltwater-induced relocation of chemical species in Alpine firn. *Tellus* 53B, 192–203.

Eisen, O., Harrison, W. D. and Raymond, C. F. 2001. The surges of Variegated Glacier, Alaska, U.S.A., and their connection to climate and mass balance. *Journal of Glaciology* 47, 351–8.

Eisen, O., Harrison, W. D., Raymond, C. F., Echelmeyer, K. A., Bender, G. A. and Gorda, J. L. D. 2005. Variegated Glacier, Alaska, USA: a century of surges. *Journal of Glaciology* 51, 399–406.

Eklund, A. and Hart, J. K. 1996. Glaciotectonic deformation within a flute from the Isfallsglaciären, Sweden. *Journal of Quaternary Science* 11, 299–310.

Ekstrom, G., Nettles, M. and Tsai, V. C. 2006. Seasonality and increasing frequency of Greenland glacial earthquakes. *Science* 311, 1756–8.

Elderfield, H. and Ganssen, G. 2000. Past temperature and $\Delta^{18}O$ of surface ocean waters inferred from foraminiferal Mg/Ca ratios. *Nature* 405, 442–5.

Elfström, Å. and Rossbacher, L. 1985. Erosional remnants in the Báldakatj area, Lapland, Northern Sweden. *Geografiska Annaler* 67A, 167–76.

Elson, J. A. 1989. Comment on glacitectonite, deformation till, and comminution till. In Goldthwait, R. P. and Matsch, C. L. (eds), *Genetic Classification of Glacigenic Deposits.* Balkema, Rotterdam, 85–8.

Elverhøi, A., De Blasio, F. V., Butt, F. A., Issler, D., Harbitz, C., Engvik, L., Solheim, A. and Marr, J. 2002. Submarine mass-wasting on glacially influenced continental slopes: processes and dynamics. In Dowdeswell, J. A. and Ó Cofaigh, C. (eds), *Glacier-influenced Sedimentation on High Latitude Continental Margins.* Geological Society, London, Special Publication 203, 73–87.

Elverhøi, A., Fjeldskaar, W., Solheim, A., Nyland-Berg, M. and Russwurm, L. 1993. The Barents Sea Ice Sheet: a model of its growth and decay during the last ice maximum. *Quaternary Science Reviews* 12, 863–73.

Elverhøi, A., Harbitz, C. B., Dimakis, P., Mohrig, D. and Marr, J. 2000. On the dynamics of subaqueous debris flows. *Oceanography* 13, 109–17.

Elverhøi, A., Lønne, O. and Sealand, R. 1983. Glaciomarine sedimentation in a modern fjord environment, Spitsbergen. *Polar Research* 1, 127–49.

Elverhøi, A., Norem, H., Andersen, E. S., Dowdeswell, J. A., Fossen, I., Haflidason, H., Kenyon, N. H., Laberg, J. S., King, E. L., Sejrup, H. P., Solheim, A. and Vorren, T. O. 1997. On the origin and flow behaviour of submarine slides on deep-sea fans along the Norwegian-Barents Sea continental margin. *Geo-Marine Letters* 17, 119–25.

Embleton, C. and King, C. A. M. 1975. *Glacial Geomorphology.* Edward Arnold, London.

Emery, D. and Myers, K. J. (eds). 1996. *Sequence Stratigraphy.* Blackwell, Oxford.

Engelhardt, H. 2004. Thermal regime and dynamics of the West Antarctic Ice Sheet. *Annals of Glaciology* 39, 85–92.

Engelhardt, H. and Kamb, B. 1993. Vertical temperature profile of Ice Steam B. *Antarctic Journal US 28, Review 1993,* 63–6.

Engelhardt, H. and Kamb, B. 1998. Basal sliding of Ice Stream B, West Antarctica. *Journal of Glaciology* 44 (147), 223–30.

Engelhardt, H., Humphrey, N., Kamb, B. and Fahnestock, M. 1990. Physical conditions at the base of a fast-moving Antarctic ice stream. *Science* 248, 57–9.

Engels, J. L., Edwards, M. H., Polyak, L. and Johnson, P. D. 2008. Seafloor evidence for ice shelf flow across the Alaska/Beaufort margin of the Arctic Ocean. *Earth Surface Processes and Landforms* 33 (7), 1047–63.

Engeset, R. V., Kohler, J., Melvold, K. and Lunden, B. 2002. Change detection and monitoring of glacier mass balance and facies using ERS SAR winter images over Svalbard. *International Journal of Remote Sensing* 23, 2023–50.

England, J. 1987. Glaciation and the evolution of the Canadian high arctic landscape. *Geology* 15, 419–24.

England, J. 1990. The late Quaternary history of Greely Fiord and its tributaries, west-central Ellesmere Island. *Canadian Journal of Earth Sciences* 27, 255–70.

England, J. 1999. Coalescent Greenland and Innuitian ice during the last glacial maximum: revising the Quaternary of the Canadian high arctic. *Quaternary Science Reviews* 18, 421–56.

England, J., Atkinson, N., Dyke, A. S., Evans, D. J. A. and Zreda, M. 2004. Late Wisconsinan buildup and wastage of the Innuitian Ice Sheet across southern Ellesmere Island, Nunavut. *Canadian Journal of Earth Sciences* 41, 39–61.

England, J., Atkinson, N., Bednarski, J., Dyke, A. S., Hodgson, D. A. and Ó Cofaigh, C. 2006. The Innuitian Ice Sheet: configuration, dynamics and chronology. *Quaternary Science Reviews* 25, 689–703.

England, J., Bradley, R. S. and Miller, G. H. 1978. Former ice shelves in the Canadian high Arctic. *Journal of Glaciology* 20, 393–404.

England, J., Smith, I. R. and Evans, D. J. A. 2000. The last glaciation of east-central Ellesmere Island, Nunavut: ice dynamics, deglacial chronology and sea level change. *Canadian Journal of Earth Sciences* 37, 1355–71.

EPICA community members, 2004. Eight glacial cycles from an Antarctic ice core. *Nature* 429, 623–8.

Eronen, M. 1983. Late Weichselian and Holocene shore displacement in Finland. In Smith, D. E. and Dawson, A. G. (eds), *Shorelines and Isostasy.* Academic Press, London, 183–208.

Escutia, A., De Santis, L., Donda, F., Dunbar, R. B., Cooper, A. K., Brancolini, G. and Eittreim, S. L. 2005. Cenozoic ice sheet history from East Antarctic Wilkes Land continental margin sediments. *Global and Planetary Change* 45, 51–81.

Etienne, J. L., Jansson, K. N., Glasser, N. F., Hambrey, M. J., Davies, J. R., Waters, R. A., Maltman, A. J. and Wilby, P. R. 2006. Palaeoenvironmental interpretation of an ice-contact glacial lake succession: an example from the late Devensian of southwest Wales, UK. *Quaternary Science Reviews* 25, 739–62.

Etzelmüller, B. 2000. Quantification of thermo-erosion in proglacial areas – examples from Svalbard. *Zeitschrift für Geomorphologie* 44, 343–61.

Etzelmüller, B. and Hagen, J. O. 2005. Glacier-permafrost interaction in arctic and alpine mountain environments, with examples from southern Norway and Svalbard. In Harris, C. and Murton, J. B. (eds), *Cryospheric Systems: Glaciers and Permafrost.* Geological Society, London, Special Publication 242, 11–27.

Etzelmüller, B., Hagen, J. O., Vatne, G., Mysterud, R. S., Tonning, T. and Sollid, J. L. 2000. Glacial characteristics and sediment transfer system of Longyearbreen and Larsbreen, western Spitsbergen. *Norsk Geografisk Tidsskrift* 54, 157–68.

Etzelmüller, B., Hagen, J. O., Vatne, G., Ødegård, R. S. and Sollid, J. L. 1996. Glacier debris accumulation and sediment deformation influenced by permafrost: examples from Svalbard. *Annals of Glaciology* 22, 53–62.

Evans, D. J. A. 1989a. The nature of glacitectonic structures and sediments at subpolar glacier margins, northwest Ellesmere Island, Canada. *Geografiska Annaler* 71A, 113–23.

Evans, D. J. A. 1989b. Apron entrainment at the margins of subpolar glaciers, northwest Ellesmere Island, Canadian high arctic. *Journal of Glaciology* 35, 317–24.

Evans, D. J. A. 1990a. The effect of glacier morphology on surficial geology and glacial stratigraphy in a high arctic mountainous terrain. *Zeitschrift für Geomorphologie* 34, 481–503.

Evans, D. J. A. 1990b. The last glaciation and relative sea level history of northwest Ellesmere Island, Canadian high arctic. *Journal of Quaternary Science* 5, 67–82.

Evans, D. J. A. 1993. High latitude rock glaciers: a case study of forms and processes in the Canadian arctic. *Permafrost and Periglacial Processes* 4, 17–35.

Evans, D. J. A. 1994. The stratigraphy and sedimentary structures associated with complex subglacial thermal regimes at the southwestern margin of the Laurentide Ice Sheet, southern Alberta, Canada. In Warren, W. P. and Croot, D. G. (eds), *Formation and Deformation of Glacial Deposits.* Balkema, Rotterdam, 203–20.

Evans, D. J. A. 1996. A possible origin for a mega-fluting complex on the southern Alberta prairies, Canada. *Zeitschrift für Geomorphologie,* Supplement Band 106, 125–48.

Evans, D. J. A. 1999. Glacial debris transport and moraine deposition: a case study of the Jardalen cirque complex, Sogn-og-Fjordane, western Norway. *Zeitschrift für Geomorphologie* 43, 203–34.

Evans, D. J. A. 2000a. A gravel outwash/deformation till continuum, Skalafellsjokull, Iceland. *Geografiska Annaler* 82A, 499–512.

Evans, D. J. A. 2000b. Quaternary geology and geomorphology of the Dinosaur Provincial Park area and surrounding plains, Alberta, Canada: the identification of former glacial lobes, drainage diversions and meltwater flood tracks. *Quaternary Science Reviews* 19, 931–58.

Evans, D. J. A. (ed.) 2003a. *Glacial Landsystems.* Arnold, London.

Evans, D. J. A. 2003b. Ice-marginal terrestrial landsystems: active temperate glacier margins. In Evans, D. J. A. (ed.), *Glacial Landsystems.* Arnold, London, 12–43.

Evans, D. J. A. (ed.). 2004. *Geomorphology: Critical Concepts in Geography. Volume IV: Glacial Geomorphology.* Routledge, London.

Evans, D. J. A. 2007. Glacitectonic structures and landforms. In Elias, S. A. (ed.), *Encyclopedia of Quaternary Science*. Elsevier, Rotterdam, 831–8.

Evans, D. J. A. 2009. Controlled moraines: origins, characteristics and palaeoglaciological implications. *Quaternary Science Reviews* 28, 183–208.

Evans, D. J. A. and Benn, D. I. 2001. Earth's giant bulldozers. *Geography Review* 14, 29–33.

Evans, D. J. A. and Benn, D. I. (eds). 2004a. *A Practical Guide to the Study of Glacial Sediments*. Arnold, London.

Evans, D. J. A. and Benn, D. I. 2004b. Facies description and the logging of sedimentary exposures. In Evans, D. J. A. and Benn, D. I. 2004 (eds), *A Practical Guide to the Study of Glacial Sediments*. Arnold, London, 11–51.

Evans, D. J. A. and Campbell, I. A. 1995. Quaternary stratigraphy of the buried valleys of the lower Red Deer River, Alberta, Canada. *Journal of Quaternary Science* 10, 123–48.

Evans, D. J. A. and England, J. 1991. Canadian Landform Examples 19: high arctic thrust block moraines. *Canadian Geographer* 35, 93–7.

Evans, D. J. A. and Hansom, J. D. 1996. Scottish Landform Example 15: The Edinburgh Castle crag-and-tail. *Scottish Geographical Magazine* 112, 129–31.

Evans, D. J. A. and Hiemstra, J. F. 2005. Till deposition by glacier submarginal, incremental thickening. *Earth Surface Processes and Landforms* 30, 1633–62.

Evans, D. J. A. and Ó Cofaigh, C. 2003. Depositional evidence for marginal oscillations of the Irish Sea Ice Stream in southern Ireland during the last glaciation. *Boreas* 32, 76–101.

Evans, D. J. A. and Rea, B. R. 1999. Geomorphology and sedimentology of surging glaciers: a landsystems approach. *Annals of Glaciology* 28, 75–82.

Evans, D. J. A. and Rea, B. R. 2003. Surging glacier landsystem. In Evans, D. J. A. (ed.), *Glacial Landsystems*. Arnold, London, 259–88.

Evans, D. J. A. and Rea, B. R. 2005. Late Weichselian deglaciation and sea level history of St Jonsfjorden, Spitsbergen: a contribution to ice sheet reconstruction. *Scottish Geographical Journal* 121, 175–201.

Evans, D. J. A. and Twigg, D. R. 2000. *Breidamerkurjokull 1998*. 1:30,000 scale map. University of Glasgow and Loughborough University.

Evans, D. J. A. and Twigg, D. R. 2002. The active temperate glacial landsystem: a model based on Breiðamerkurjökull and Fjallsjökull, Iceland. *Quaternary Science Reviews* 21, 2143–77.

Evans, D. J. A. and Wilson, S. B. 2006. Scottish Landform Example 39: The Lake of Menteith hill-hole pair. *Scottish Geographical Journal* 122, 352–64.

Evans, D. J. A., Archer, S. and Wilson, D. J. H. 1999a. A comparison of the lichenometric and Schmidt hammer dating techniques based on data from the proglacial areas of some Icelandic glaciers. *Quaternary Science Reviews* 18, 13–41.

Evans, D. J. A., Clark, C. D. and Rea, B. R. 2008. Landform and sediment imprints of fast glacier flow in the southwest Laurentide Ice Sheet. *Journal of Quaternary Science* 23, 249–72.

Evans, D. J. A., Hiemstra, J. F. and Ó Cofaigh, C. 2007a. An assessment of clast macrofabrics in glacigenic sediments based on A/B plane data. *Geografiska Annaler* 89A, 103–20.

Evans, D. J. A., Lemmen, D. S. and Rea, B. R. 1999b. Glacial landsystems of the southwest Laurentide Ice Sheet: modern Icelandic analogues. *Journal of Quaternary Science* 14, 673–9.

Evans, D. J. A., Livingstone, S. J., Vieli, A. and Ó Cofaigh, C. 2009. The palaeoglaciology of the central sector of the British and Irish Ice Sheet: reconciling glacial geomorphology and preliminary ice sheet modeling. *Quaternary Science Reviews* 28, 740–58.

Evans, D. J. A., Owen, L. A. and Roberts, D. 1995. Stratigraphy and sedimentology of Devensian (Dimlington Stadial) glacial deposits, east Yorkshire, England. *Journal of Quaternary Science* 10, 241–65.

Evans, D. J. A., Phillips, E. R., Hiemstra, J. F. and Auton, C. A. 2006a. Subglacial till: formation, sedimentary characteristics and classification. *Earth Science Reviews* 78, 115–76.

Evans, D. J. A., Rea, B. R. and Benn, D. I. 1998. Subglacial deformation and bedrock plucking in areas of hard bedrock. *Glacial Geology and Geomorphology* rp04/1998 (http://ggg.qub.ac.uk/papers/frame.htm).

Evans, D. J. A., Rea, B. R., Hansom, J. D. and Whalley, W. B. 2002. Geomorphology and style of plateau icefield deglaciation in fjord terrains: the example of Troms-Finnmark, north Norway. *Journal of Quaternary Science* 17, 221–39.

Evans, D. J. A., Rea, B. R., Hiemstra, J. F. and Ó Cofaigh, C. 2006b. A critical assessment of subglacial mega-floods: a case study of glacial sediments and landforms in south-central Alberta, Canada. *Quaternary Science Reviews* 25, 1638–67.

Evans, D. J. A., Salt, K. and Allen, C. S. 1999c. Glacitectonized lake sediments, Barrier Lake, Kananaskis Country, Canadian Rocky Mountains. *Canadian Journal of Earth Sciences* 36, 395–407.

Evans, D. J. A., Twigg, D. R. and Rea, B. R. 2006c. *Surging Glacier Landsystem of Tungnaárjökull, Iceland*. 1:25,000 scale map poster. Durham University.

Evans, D. J. A., Twigg, D. R. and Shand, M. 2006d. Surficial geology and geomorphology of the þorisjökull plateau icefield, west central Iceland. *Journal of Maps* 2006, 17–29.

Evans, D. J. A., Twigg, D. R., Rea, B. R. and Shand, M. 2007b. Surficial geology and geomorphology of the Bruarjökull surging glacier landsystem. *Journal of Maps* 2007, 349–67.

Evans, I. S. 1977. World-wide variations in the direction and concentration of cirque and glacier aspects. *Geografiska Annaler* 59A, 151–75.

Evans, I. S. 1987. A new approach to drumlin morphometry. In Menzies, J. and Rose, J. (eds), *Drumlin Symposium*. Balkema, Rotterdam, 119–30.

Evans, I. S. 1990. Climatic effects on glacier distribution across the southern Coast Mountains, B.C., Canada. *Annals of Glaciology* 14, 58–64.

Evans, I. S. 1994. Lithological and structural effects on forms of glacial erosion: cirques and lake basins. In Robinson, D. A. and Williams, R. B. G. (eds), *Rock Weathering and Landform Evolution*. Wiley, Chichester, 455–72.

Evans, I. S. 1996. Abraded rock landforms (whalebacks) developed under ice streams in mountain areas. *Annals of Glaciology* 22, 9–16.

Evans, I. S. 1999. Was the cirque glaciation of Wales time-transgressive, or not? *Annals of Glaciology* 28, 33–9.

Evans, I. S. 2006. Allometric development of glacial cirque form: geological, relief and regional effects on the cirques of Wales. *Geomorphology* 80, 245–66.

Evans, I. S. 2007. Glacial landforms, erosional features: major scale forms. In Elias, S. A. (ed.), *Encyclopedia of Quaternary Science*. Elsevier, Amsterdam, 838–52.

Evans, I. S. and Cox, N. J. 1974. Geomorphometry and the operational definition of cirques. *Area* 6, 150–3.

Evans, I. S. and Cox, N. J. 1995. The form of glacial cirques in the English Lake District, Cumbria. *Zeitschrift für Geomorphologie* 39, 175–202.

Evans, J., Dowdeswell, J. A. and Ó Cofaigh, C. 2004. Late Quaternary submarine bedforms and ice-sheet flow in Gerlache Strait and on the adjacent continental shelf, Antarctic Peninsula. *Journal of Quaternary Science* 19, 397–407.

Evans, J., Dowdeswell, J. A., Grobe, H., Niessen, F., Stein, R., Hubberten, H. W. and Whittington, R. J. 2002. Late Quaternary sedimentation in Kejser Franz Joseph Fjord and the continental margin of East Greenland. In Dowdeswell, J. A. and Ó Cofaigh, C. (eds), *Glacier-influenced Sedimentation on High Latitude Continental Margins*. Geological Society, London, Special Publication 203, 149–79.

Evans, J., Ó Cofaigh, C., Dowdeswell, J. A. and Wadhams, P. 2009. Marine geophysical evidence for former expansion and flow of the Greenland Ice Sheet across the north-east Greenland continental shelf. *Journal of Quaternary Science* 24, 279–93.

Evans, S. G. and Clague, J. J. 1994. Recent climatic change and catastrophic geomorphic processes in mountain environments. *Geomorphology* 10, 107–28.

Evatt, G. W., Fowler, A. C., Clark, C. D. and Hulton, N. R. J. 2006. Subglacial floods beneath ice sheets. *Philosophical Transactions of the Royal Society* A364, 1769–98.

Evenson, E. B. and Clinch, J. M. 1987. Debris transport mechanisms at active glacier margins: Alaskan case studies. In Kujansuu, R. and Saarnisto, M. (eds), INQUA Till Symposium, Finland, 1985. *Geological Survey of Finland Special Paper* 3, 111–36.

Evenson, E. B., Dreimanis, A. and Newsome, J. W. 1977. Subaquatic flow tills: a new interpretation for the genesis of some laminated till deposits. *Boreas* 6, 115–33.

Evenson, E. B., Lawson, D. E., Strasser, J. C., Larson, G. J., Alley, R. B., Ensminger, S. L. and Stevenson, W. E. 1999. Field evidence for the recognition of glaciohydrologic supercooling. In Mickelson, D. M. and Attig, J. W. (eds), *Glacial Processes Past and Present*. Geological Society of America, Special Paper 337, 23–35.

Everest, J. and Bradwell, T. 2003. Buried glacier ice in southern Iceland and its wider significance. *Geomorphology* 52, 347–58.

Ewing, K., Loomis, S. and Lougeay, R. 1967. Water-spout on Kaskawulsh Glacier, Yukon Territory. *Journal of Glaciology* 6, 956–9.

Eyles, C. H. 1988. A model for striated boulder pavement formation on glaciated shallow marine shelves, an example from the Yakataga Formation, Alaska. *Journal of Sedimentary Petrology* 58, 62–71.

Eyles, C. H. and Eyles, N. 1983. Glaciomarine model for upper Precambrian diamictites of the Port Askaig Formation, Scotland. *Geology* 11, 692–6.

Eyles, C. H. and Eyles, N. 2000. Subaqueous mass flow origin for Lower Permian diamictites and associated facies of the Grant Group, Barbwire Terrace, Canning Basin, Western Australia. *Sedimentology* 47, 343–56.

Eyles, C. H. and Lagoe, M. B. 1990. Sedimentation patterns and facies geometries on a temperate glacially-influenced continental shelf; the Yakataga Formation, Middleton Island, Alaska. In Dowdeswell, J. A. and Scourse, J. D. (eds), *Glacimarine Environments: Processes and Sediments*. Geological Society, London, Special Publication 53, 363–86.

Eyles, C. H., Eyles, N. and Lagoe, M. B. 1991. The Yakataga Formation: a late Miocene to Pleistocene record of temperate glacial marine sedimentation in the Gulf of Alaska. In Anderson, J. B. and Ashley, G. M. (eds), *Glacial Marine Sedimentation: Palaeoclimatic Significance*. Geological Society of America, Special Paper 261, 159–80.

Eyles, C. H., Eyles, N. and Miall, A. D. 1985. Models of glaciomarine sedimentation and their application to the interpretation of ancient glacial sequences. *Palaeogeography, Palaeoclimatology, Palaeoecology* 51, 15–84.

Eyles, N. 1978. Scanning electron microscopy and particle size analysis of debris from a British Columbian glacier: a comparative report. In Whalley, W. B. (ed.), *Scanning Electron Microscopy in the Study of Sediments*. Geo Abstracts, Norwich, 227–42.

Eyles, N. 1979. Facies of supraglacial sedimentation on Icelandic and alpine temperate glaciers. *Canadian Journal of Earth Sciences* 16, 1341–61.

Eyles, N. 1983a. Glacial geology: a landsystems approach. In Eyles, N. (ed.), *Glacial Geology*. Pergamon, Oxford, 1–18.

Eyles, N. 1983b. The glaciated valley landsystem. In Eyles, N. (ed.), *Glacial Geology*. Pergamon, Oxford, 91–110.

Eyles, N. 1993. Earth's glacial record and its tectonic setting. *Earth Science Reviews* 35, 1–248.

Eyles, N. and Boyce, J. I. 1998. Kinematic indicators in fault gouge: tectonic analog for soft-bedded ice sheets. *Sedimentary Geology* 116, 1–12.

Eyles, N. and Eyles, C. H. 1992. Glacial depositional systems. In Walker, R. G. and James, N. P. (eds), *Facies Models: Response to Sea-level Change*, 73–100. Geological Association of Canada, Toronto.

Eyles, N. and Kocsis, S. 1988. Sedimentology and clast fabric of subaerial debris flow facies in a glacially-influenced alluvial fan. *Sedimentary Geology* 59, 15–28.

Eyles, N. and Lazorek, M. 2007. Glacigenic lithofacies. In Elias, S. A. (ed.), *Encyclopedia of Quaternary Science*. Elsevier, Amsterdam, 920–32.

Eyles, N. and McCabe, A. M. 1989. The Late Devensian <22,000 BP Irish Sea Basin: the sedimentary record of a collapsed ice sheet margin. *Quaternary Science Reviews* 8, 307–51.

Eyles, N. and Menzies, J. 1983. The subglacial landsystem. In Eyles, N. (ed.), *Glacial Geology*. Pergamon, Oxford, 19–70.

Eyles, N. and Miall, A. D. 1984. Glacial Facies. In Walker, R. G. (ed.), *Facies Models*. Geoscience Canada Reprint Series 1, 15–38.

Eyles, N. and Rogerson, R. J. 1978. A framework for the investigation of medial moraine formation: Austerdalsbreen, Norway, and Berendon Glacier, British Columbia, Canada. *Journal of Glaciology* 20, 99–113.

Eyles, N. and Young, G. M. 1994. Geodynamic controls on glaciation in Earth history. In Deynoux, M., Miller, J. M. G., Domack, E. W., Eyles, N., Fairchild, I. J. and Young, G. M. (eds), *Earth's Glacial Record*. Cambridge University Press, 1–28.

Eyles, N., Boyce, J. I. and Barendregt, R. W. 1999. Hummocky moraine: sedimentary record of stagnant Laurentide Ice Sheet lobes resting on soft beds. *Sedimentary Geology* 123, 163–74.

Eyles, N., Clark, B. M. and Clague, J. J. 1987. Coarse-grained sediment-gravity flow facies in a large supraglacial lake. *Sedimentology* 34, 193–216.

Eyles, N., Dearman, W. R. and Douglas, T. D. 1983b. The distribution of glacial landsystems in Britain and North America. In Eyles, N. (ed.), *Glacial Geology*. Pergamon, Oxford, 213–28.

Eyles, N., Doughty, M., Boyce, J. I., Mullins, H. T., Halfman, J. D. and Koseoglu, B. 2003. Acoustic architecture of glaciolacustrine sediments deformed during zonal stagnation of the Laurentide Ice Sheet: Mazinaw Lake, Ontario, Canada. *Sedimentary Geology* 157, 133–51.

Eyles, N., Eyles, C. H. and Miall, A. D. 1983a. Lithofacies types and vertical profile models: an alternative approach to the description and environmental interpretation of glacial diamict and diamictite sequences. *Sedimentology* 30, 393–410.

Eyles, N., Eyles, C. H., Woodworth-Lynas, C. M. T. and Randall, T. A. 2005. The sedimentary record of drifting ice (early Wisconsin Sunnybrook deposit) in an ancestral ice-dammed Lake Ontario, Canada. *Quaternary Research* 63, 171–81.

Eyles, N., Sladen, J. A. and Gilroy, S. 1982. A depositional model for stratigraphic complexes and facies superimposition in lodgement tills. *Boreas* 11, 317–33.

Eynaud, F., Zaragosi, S., Scourse, J. D., Mojtahid, M., Bourillet, J. F., Hall, I. R., Penaud, A., Locascio, M. and Reijonen, A. 2007. Deglacial laminated facies on the NW European continental margin: the hydrographic significance of British Ice Sheet deglaciation and Fleuve Manche palaeoriver discharges. *Geochemistry, Geophysics, Geosystems* 8, doi:10.1029/2006GC001496.

Eythorsson, J. 1951. Jökla-mys. *Journal of Glaciology* 1, 503.

Fabel, D., Harbor, J., Dahms, D., James, A., Elmore, D., Horn, L., Daley, K. and Steel, C. 2004. Spatial patterns of glacial erosion at a valley scale derived from terrestrial cosmogenic [10]Be and [26]Al concentrations in rock. *Annals of the Association of American Geographers* 94, 241–55.

Fabel, D., Stroeven, A. P., Harbor, J., Kleman, J., Elmore, D. and Fink, D. 2002. Landscape preservation under Fennoscandian ice sheets determined from in situ produced [10]Be and [26]Al. *Earth and Planetary Science Letters* 201, 397–406.

Fahnestock, M. and Bamber, J. 2001. Morphology and surface characteristics of the West Antarctic ice sheet. In Alley, R. B. and Bindschadler, R. A. (eds), *The West Antarctic Ice Sheet: Behavior and Environment*. American Geophysical Union, Antarctic Research Series, vol. 77, 13–27.

Fahnestock, M., Abdalati, W., Joughin, I., Brozena, J. and Gogineni, P. 2001. High geothermal heat flow, basal melt, and the origin of rapid ice flow in central Greenland. *Science* 294, 2338–42.

Fahnestock, M., Bindschadler, R., Kwok, R. and Jezek, K. 1993. Greenland ice-sheet surface properties and ice dynamics from ERS-1 SAR imagery. *Science* 262, 1530–4.

Faillettaz, J., Pralong, A., Funk, M. and Deichmann, N. 2008. Evidence of log-periodic oscillations and increasing icequake activity during the breaking-off of large ice masses. *Journal of Glaciology* 54 (187), 725–37.

Fairbanks, R. G. 1989. A 17,000 year glacio-eustatic sea level record: influence of glacial melting rates on the Younger Dryas event and deep ocean circulation. *Nature* 342, 637–43.

Fard, A. M. 2002. Large dead-ice depressions in flat-topped eskers: evidence of a Preboreal jökulhlaup in the Stockholm area, Sweden. *Global and Planetary Change* 35, 273–95.

Fassett, C. I. and Head, J. W. 2008. The timing of Martian valley network activity: constraints from buffered crater counting. *Icarus* 195, 61–89.

Fastook, J. L. and Schmidt, W. F. 1982. Finite element analysis of calving from ice fronts. *Annals of Glaciology* 3, 103–6.

Fatland, D. R. and Lingle, C. S. 2002. InSAR observations of the 1993–95 Bering Glacier (Alaska, USA) surge and a surge hypothesis. *Journal of Glaciology* 48 (162), 439–51.

Fay, H. 2002a. Formation of ice-block obstacle marks during the November 1996 glacier outburst flood (jökulhlaup), Skeiðarársandur, southern Iceland. In Martini, I. P., Baker, V. R. and Garzon, G. (eds), *Flood and Megaflood Processes and Deposits: Recent and Ancient Examples*. International Association of Sedimentologists, Special Publication 32, 85–97.

Fay, H. 2002b. Formation of kettle holes following a glacial outburst flood (jökulhlaup), Skeiðarársandur, southern Iceland. In Snorasson, A., Finsdottir, H. P. and Moss, M. (eds), *The Extremes of the Extremes: Extraordinary Floods*. IAHS Special Publication 271, 205–10.

Ferguson, R. I. 1993. Understanding braiding processes in gravel-bed rivers: progress and unsolved problems. In Best, J. L. and Bristow, C. S. (eds), *Braided Rivers*. Geological Society, London, Special Publication 75, 73–87.

Fichefet, T., Poncin, C., Goosse, H., Huybrechts, P., Janssens, I. and Treut, H. L. 2003. Implications of changes in freshwater flux from the Greenland Ice Sheet for the climate of the 21st century. *Geophysical Research Letters* 30, 1911.

Ficker, E., Sonntag, G. and Weber, E. 1980. Ansätze zur mechanischen Deutung der Rissentstehung bei Parabelrissen und Sichelbrüchen auf glazialgeformten Felsoberflächen. *Zeitschrift für Gletscherkunde und Glazialgeologie* 16, 25–43.

Figge, K. 1983. Morainic deposits in the German Bight area of the North Sea. In Ehlers, J. (ed.), *Glacial Deposits in North-West Europe*. Balkema, Rotterdam, 299–304.

Finlayson, A. G. 2006. Glacial geomorphology of the Creag Meagaidh massif, western Grampian Highlands: implications for local glaciation and palaeoclimate during the Loch Lomond Stadial. *Scottish Geographical Journal* 122, 293–307.

Fischer, U. H. and Clarke, G. K. C. 1994. Ploughing of subglacial sediment. *Journal of Glaciology* 40, 97–106.

Fischer, U. H. and Clarke, G. K. C. 1997. Stick-slip sliding behaviour at the base of a glacier. *Annals of Glaciology* 24, 390–6.

Fischer, U. H. and Clarke, G. K. C. 2001. Review of subglacial hydro-mechanical coupling: Trapridge Glacier, Yukon Territory, Canada. *Quaternary International* 86, 29–43.

Fischer, U. H. and Hubbard, B. 2006. Borehole-based subglacial instrumentation. In Knight, P. (ed.), *Glacier Science and Environmental Change*. Blackwell, Oxford, 387–94.

Fischer, U. H., Clarke, G. K. C. and Blatter, H. 1999. Evidence for temporally varying 'sticky spots' at the base of Trapridge Glacier, Yukon Territory, Canada. *Journal of Glaciology* 45, 352–60.

Fischer, U. H., Porter, P. R., Schuler, T., Evans, A. J. and Gudmundsson, G. H. 2001. Hydraulic and mechanical properties of glacial sediments beneath Unteraargletscher, Switzerland: implications for glacier basal motion. *Hydrological Processes* 15, 3525–40.

Fish, P. R. and Whiteman, C. A. 2001. Chalk micropalaeontology and the provenancing of Middle Pleistocene Lowestoft Formation till in eastern England. *Earth Surface Processes and Landforms* 26, 953–70.

Fisher, B. S. et al. 2007. Issues related to mitigation in the long term context. In Metz, B. et al. (eds), *Climate Change 2007: Mitigation*. Contribution of Working Group III to the Fourth Assessment Report of the Intergovernmental Panel on Climate Change. Cambridge University Press.

Fisher, D. A. and Koerner, R. M. 2003. Holocene ice core history – a multi-variable approach. In Mackay, A., Battarbee, R., Birks, J. and Oldfield, F. (eds), *Global Change in the Holocene*. Arnold, London, 281–93.

Fisher, T. G. and Shaw, J. 1992. A depositional model for Rogen moraine, with examples from the Avalon Peninsula, Newfoundland. *Canadian Journal of Earth Sciences* 29, 669–86.

Fisher, T. G., Jol, H. M. and Boudreau, A. M. 2005. Saginaw Lobe tunnel channels (Laurentide Ice Sheet) and their significance in south-central Michigan, USA. *Quaternary Science Reviews* 24, 2375–91.

Fitzsimons, S. J. 1990. Ice-marginal depositional processes in a polar maritime environment, *Journal of Glaciology* 36 (122), 279–86.

Fitzsimons, S. J. 1991. Supraglacial eskers in Antarctica. *Geomorphology* 4, 293–9.

Fitzsimons, S. J. 1996a. Formation of thrust-block moraines at the margins of dry-based glaciers, south Victoria Land, Antarctica. *Annals of Glaciology* 22, 68–74.

Fitzsimons, S. J. 1996b. Paraglacial redistribution of glacial sediments in the Vestfold Hills, East Antarctica. *Geomorphology* 15, 93–108.

Fitzsimons S. J. 1997a. Depositional models for moraines in east Antarctic coastal oases. *Journal of Glaciology* 43, 256–64.

Fitzsimons, S. J. 1997b. Glaciotectonic deformation of glaciomarine sediments and formation of thrust block moraines at the margin of an outlet glacier, Vestfold Hills, East Antarctica. *Earth Surface Processes and Landforms* 22, 175–87.

Fitzsimons, S. J. 2003. Ice-marginal terrestrial landsystems: polar continental glacier margins. In Evans, D. J. A. (ed.), *Glacial Landsystems*. Arnold, London, 89–110.

Fitzsimons, S. J., Lorrain, R. D. and Vandergoes, M. J. 2000. Behaviour of subglacial sediment and basal ice in a cold glacier. In Maltman, A. J., Hubbard, B., Hambrey, M. J. (eds), *Deformation of Glacial Materials*. Geological Society, London, Special Publications 176, 181–90.

Fitzsimons, S. J., McManus, K. J. and Lorrain, R. 1999. Structure and strength of basal ice and substrate of a dry based glacier: evidence for substrate deformation at sub-freezing temperatures. *Annals of Glaciology* 28, 236–40.

Fitzsimons, S. J., McManus, K. J., Sirota, P. and Lorrain, R. D. 2001. Direct shear tests of materials from a cold glacier: implications for landform development. *Quaternary International* 86, 129–37.

Fjeldskaar, W., Lindholm, C., Dehls, J. F. and Fjeldskaar, I. 2000. Postglacial uplift, neotectonics and seismicity in Fennoscandia. *Quaternary Science Reviews* 19, 1413–22.

Fleisher, P. J. 1986. Dead-ice sinks and moats: environments of stagnant ice deposition. *Geology* 14, 39–42.

Fleisher, P. J., Bailey, P. K. and Cadwell, D. H. 2003. A decade of sedimentation in ice-contact, proglacial lakes, Bering Glacier, Alaska. *Sedimentary Geology* 160, 309–24.

Fleisher, P. J., Cadwell, D. H. and Muller, E. H. 1998. Tsivat Basin Conduit System persists through two surges, Bering Piedmont Glacier, Alaska. *Geological Society of America Bulletin* 110, 877–87.

Fleisher, P. J., Lachniet, M. S., Muller, E. H. and Bailey, P. K. 2006. Subglacial deformation of trees within overridden foreland strata, Bering Glacier, Alaska. *Geomorphology* 75, 201–11.

Fleming, K., Johnston, P., Zwartz, D., Yokoyama, Y., Lambeck, K. and Chappell, J. 1998a. Refining the eustatic sea-level curve since the Last Glacial Maximum using far- and intermediate-field sites. *Earth and Planetary Science Letters* 163, 327–42.

Fleming, K. M., Dowdeswell, J. A. and Oerlemans, J. 1998b. Modelling the mass balance of northwest Spitsbergen glaciers and responses to climate change. *Annals of Glaciology* 24, 203–10.

Fletcher, B. J., Brentnall, S. J., Anderson, C. W., Berner, R. A. and Beerling, D. J. 2008. Atmospheric carbon dioxide linked with Mesozoic and early Cenozoic climate change. *Nature Geoscience* 1, 43–8.

Fletcher, C. J. N. and Siddle, H. J. 1998. Development of glacial Llyn Teifi, west Wales: evidence for lake level fluctuations at the margins of the Irish Sea ice sheet. *Journal of the Geological Society of London* 155, 389–99.

Flint, R. F. 1971. *Glacial and Quaternary Geology*. John Wiley and Sons, New York.

Flint, R. F., Sanders, J. E. and Rodgers, J. 1960. Diamictite: a substitute term for symmictite. *Geological Society of America Bulletin* 71, 1809.

Flowers, G. E. and Clarke, G. K. C. 2000. An integrated modelling approach to understanding subglacial hydraulic release events. *Annals of Glaciology* 31, 222–8.

Flowers, G. E. and Clarke, G. K. C. 2002a. A multicomponent model of glacier hydrology 1. Theory and synthetic examples. *Journal of Geophysical Research* 107, doi:10.1029/2001JB001122.

Flowers, G. E. and Clarke, G. K. C. 2002b. A multicomponent model of glacier hydrology 2. Application to Trapridge Glacier, Yukon, Canada. *Journal of Geophysical Research* 107, doi:10.1029/2001JB001124.

Flowers, G. E., Björnsson, H. and Palsson, F. 2003. New insights into the subglacial and periglacial hydrology of Vatnajökull, Iceland, from a distributed physical model. *Journal of Glaciology* 49, 257–70.

Flowers, G. E., Marshall, S. J., Björnsson, H. and Clarke, G. K. C. 2005. Sensitivity of Vatnajökull ice cap hydrology and dynamics to climate warming over the next two centuries. *Journal of Geophysical Research* 110, doi:10.1029/2004JF000200.

Fogg, G. E. 2005. *A History of Antarctic Science*. Cambridge University Press.

Foght, J. M., Aislabie, J., Turner, S., Brown, C. E., Ryburn, J., Saul, D. J. and Lawson, W. 2004. Culturable bacteria in subglacial sediments and ice from two southern hemisphere glaciers. *Microbial Ecology* 47, 329–40.

Forbes, A. C. and Lamoureux, S. F. 2005. Climatic controls on streamflow and suspended sediment transport in three large middle arctic catchments, Boothia Peninsula, Nunavut, Canada. *Arctic, Antarctic and Alpine Research* 37, 305–15.

Ford, D. C., Lauritzen, S. E. and Worthington, S. R. H. 2000. Speleogenesis of Castleguard Cave, Rocky Mountains, Alberta, Canada. In Klimchouk, A. B., Ford, D. C., Palmer, A. N. and Dreybrodt, W. (eds), *Speleogenesis: Evolution of Karst Aquifers*. National Speleological Society, Huntsville, AL, 332–7.

Forget, F. R. M., Haberle, F., Montmessin, B., Levrard, B. F. and Head, J. W. 2006. Formation of glaciers on Mars by atmospheric precipitation at high obliquity. *Science* 311, 368–71.

Forman, S. L. 1989. Late Weichselian glaciation and deglaciation of Forlandsundet area, western Spitsbergen, Svalbard. *Boreas* 8, 51–60.

Forman, S. L. 1990. Post-glacial relative sea level history of northwestern Spitsbergen, Svalbard. *Geological Society of America Bulletin* 102, 1580–90.

Fort, M. 2000. Glaciers and mass wasting processes: their influence on the shaping of the Kali Gandaki valley, Nepal. *Quaternary International* 65/66, 101–19.

Fountain, A. G. 1993. Geometry and flow conditions of subglacial water at South Cascade Glacier, Washington State, U.S.A., an analysis of tracer injections. *Journal of Glaciology* 39, 143–56.

Fountain, A. G. 1996. Effect of snow and firn hydrology on the physical and chemical characteristics of glacial runoff. *Hydrological Processes* 10, 509–21.

Fountain, A. G. and Vecchia, A. 1999. How many stakes are required to measure the mass balance of a glacier? *Geografiska Annaler* 81A, 563–74.

Fountain, A. G. and Walder, J. 1998. Water flow through temperate glaciers. *Reviews of Geophysics* 36, 299–328.

Fountain, A. G., Jacobel, R. W., Schlichting, R. and Jansson, P. 2005. Fractures as the main pathways of water flow in temperate glaciers. *Nature* 433 (7026), 618–21.

Fountain, A. G., Nylen, T. H., MacClune, K. L. and Dana, G. L. 2006. Glacier mass balances (1993–2001), Taylor Valley, McMurdo Dry Valleys, Antarctica. *Journal of Glaciology* 52, 451–62.

Fountain, A. G., Tranter, M., Nylen, T., Booth, D. and Lewis, K. 2004. Cryoconite holes on polar glaciers and their importance for meltwater runoff. *Journal of Glaciology* 50, 25–45.

Fowler, A. C. 1984. On the transport of moisture in polythermal glaciers. *Geophysical and Astrophysical Fluid Dynamics* 28, 99–140.

Fowler, A. C. 1986a. Sub-temperate basal sliding. *Journal of Glaciology* 32, 3–5.

Fowler, A. C. 1986b. A sliding law for glaciers of constant viscosity in the presence of subglacial cavitation. *Proceedings of the Royal Society*, Series A 407, 147–70.

Fowler, A. C. 1987. Sliding with cavity formation. *Journal of Glaciology* 33, 255–67.

Fowler, A. C. 1989. A mathematical analysis of glacier surges. *SIAM Journal of Applied Mathematics* 49, 246–63.

Fowler, A. C. 2000. An instability mechanism for drumlin formation. In Maltman, A., Hambrey, M. J. and Hubbard, B. (eds), *Deformation of Glacial Materials*. Geological Society, London, Special Publication 176, 307–19.

Fowler, A. C. 2003. Rheology of subglacial till [letter]. *Journal of Glaciology* 48 (163), 631–2.

Fowler, A. C., Murray, T. and Ng, F. S. L. 2001. Thermally-controlled glacier surging. *Journal of Glaciology* 47, 527–38.

Frakes, L. A., Francis, J. E. and Sykes, J. I. 1992. *Climate Modes of the Phanerozoic*. Cambridge University Press.

Francou, B., Ribstein, P., Saravia, R. and Tiriau, E. 1995. Monthly balance and water discharge on an intertropical glacier: the Zongo Glacier, Cordillera Real, Bolivia, 16° S. *Journal of Glaciology* 41, 61–7.

Frappé, T.-P. and Clarke, G. K. C. 2007. Slow surge of Trapridge Glacier, Yukon Territory, Canada. *Journal of Geophysical Research* 112, F03S32, doi:1029/2006JF000607.

Fredin, O. and Hättestrand, C. 2002. Relict lateral moraines in northern Sweden: evidence for an early mountain centred ice sheet. *Sedimentary Geology* 149, 145–56.

Free, M. and Robock, A. 1999. Global warming in the context of the Little Ice Age. *Journal of Geophysical Research* 104 (D16), 19057–70.

Freeze, R. A. and Cherry, J. A. 1979. *Groundwater*. Prentice Hall, Englewood Cliffs, NJ.

French, H. M. and Guglielmin, M. 1999. Observations on the ice marginal, periglacial geomorphology of Terra Nova Bay, northern Victoria Land, Antarctica. *Permafrost and Periglacial Processes* 10, 331–47.

Fricker, H. A., Popov, S., Allison, I. and Yound, N. 2001. Distribution of marine ice beneath the Amery Ice Shelf. *Geophysical Research Letters* 28, 2241–4.

Fricker, H. A., Scambos, T., Bindschadler, R. and Padman, L. 2007. An active subglacial water system in West Antarctica mapped from space. *Science* 315, 1544–8.

Fricker, H. A., Warner, R. C. and Allison, I. 2000. Mass balance of the Lambert Glacier-Amery Ice Shelf system, East Antarctica: a comparison of computed balance fluxes and measured fluxes. *Journal of Glaciology* 46 (155), 561–70.

Fricker, H. A., Young, N. W., Allison, I. and Coleman, R. 2002. Iceberg calving from the Amery Ice Shelf, East Antarctica. *Annals of Glaciology* 34, 241–6.

Fritz, W. J. and Moore, J. N. 1988. Basics of Physical Stratigraphy and Sedimentology. Wiley, New York.

Fuller, S. and Murray, T. 2000. Evidence against pervasive bed deformation during the surge of an Icelandic glacier. In Maltman, A. J., Hubbard, B. and Hambrey, M. J. (eds), *Deformation of Glacial Materials*. Geological Society, London, Special Publication 176, 203–16.

Funk, M., Echelmeyer, K. and Iken, A. 1994. Mechanisms of fast flow in Jakobshavns Isbrae, West Greenland. Part II: Modelling of englacial temperatures. *Journal of Glaciology* 40, 569–85.

Furbish, D. J. and Andrews, J. T. 1984. The use of hypsometry to indicate long-term stability and response of valley glaciers to changes in mass transfer. *Journal of Glaciology* 30, 199–211.

Fushimi, H., Yoshida, M., Watanabe, O. and Upadhyay, B. P. 1980. Distributions and grain sizes of supraglacial debris in the Khumbu Glacier, Khumbu Region, East Nepal. *Seppyo* 42, 18–25.

Fyfe, G. J. 1990. The effect of water depth on ice-proximal glaciolacustrine sedimentation: Salpausselka I, southern Finland. *Boreas* 19, 147–64.

Gagliardini, O., Cohen, D., Råback, P. and Zwinger, T. 2007. Finite-element modelling of subglacial cavities and related friction law. *Journal of Geophysical Research* 112, doi:10.1029/2006JF000576.

Gaidos, E., Lanoil, B., Thorsteinsson, T., Graham, A., Skidmore, M., Han, S. K., Rust, T. and Popp, B. 2004. A viable microbial community in a subglacial volcanic crater lake, Iceland. *Astrobiology* 4, 327–44.

Gardner, J. S. 1987. Evidence for headwall weathering zones, Boundary Glacier, Canadian Rocky Mountains. *Journal of Glaciology* 33, 60–7.

Gardner, J. S. and Hewitt, K. 1990. A surge of Bualtar Glacier, Karakoram Range, Pakistan: a possible landslide trigger. *Journal of Glaciology* 36, 159–62.

Gaur, U. N., Raturi, G. P. and Bhatt, A. B. 2003. Quantitative response of vegetation in glacial moraine of Central Himalaya. *The Environmentalist* 23, 1573–2991.

Geertsema, M., Clague, J. J., Schwab, J. W. and Evans, S. G. 2006. An overview of recent large catastrophic landslides in northern British Columbia, Canada. *Engineering Geology* 83, 120–43.

Geirsdóttir, A., Miller, G. H. and Andrews, J. T. 2007. Glaciation, erosion, and landscape evolution of Iceland. *Journal of Geodynamics* 43, 170–86.

Gellatly, A. F., Gordon, J. E., Whalley, W. B. and Hansom, J. D. 1988. Thermal regime and geomorphology of plateau ice caps in northern Norway: observations and implications. *Geology* 16, 983–6.

Gellatly, A. F., Whalley, W. B. and Gordon, J. E. 1986. Topographic control over recent changes in southern Lyngen Peninsula, north Norway. *Norsk Geografisk Tidsskrift* 40, 211–18.

Ghibaudo, G. 1992. Subaqueous sediment gravity flow deposits: practical criteria for their field description and classification. *Sedimentology* 39, 423–54.

Ghienne, J.-F. 2003. Late Ordovician sedimentary environments, glacial cycles, and post-glacial transgression in the Taoudeni Basin, West Africa. *Palaeogeography, Palaeoclimatology, Palaeoecology* 189, 117–45.

Giardino, J. R., Shroder, J. F. and Vitek, J. D. (eds). 1987. *Rock Glaciers*. Allen and Unwin, London.

Gibbard, P. L. 1980. The origin of stratified Catfish Creek till by basal melting. *Boreas* 9, 71–85.

Gibbard, P. L., Rose, J. and Bridgland, D. R. 1988. The history of the great northwest European rivers during the past three million years. *Philosophical Transactions of the Royal Society of London* B318, 559–602.

Gibbs, M. T. and Kump, L. R. 1994. Global chemical erosion during the last glacial maximum and the present: sensitivity to changes in lithology and hydrology. *Palaeoceanography* 9, 529–43.

Gilbert, R. 1975. Sedimentation in Lillooet Lake, British Columbia. *Canadian Journal of Earth Sciences* 12, 1697–711.

Gilbert, R. 1982. Contemporary sedimentary environments on Baffin Island, NWT, Canada: glaciomarine processes in fiords of eastern Baffin Island. *Arctic and Alpine Research* 14, 1–12.

Gilbert, R. 1990. Rafting in glaciomarine environments. In Dowdeswell, J. A. and Scourse, J. D. (eds), *Glacimarine Environments: Processes and Sediments*. Special Publications of the Geologists' Association 53, 105–20.

Gilbert, R. and Church, M. 1983. Contemporary sedimentary environments on Baffin Island, NWT, Canada: reconnaissance of lakes on Cumberland Peninsula. *Arctic and Alpine Research* 15, 321–32.

Gilbert, R. and Shaw, J. 1981. Sedimentation in proglacial Sunwapta Lake, Alberta. *Canadian Journal of Earth Sciences* 18, 81–93.

Gilbert, R., Desloges, J. R., Lamoureux, S. F., Serink, A. and Hodder, K. R. 2006. The sedimentary environment of Atlin Lake, northern British Columbia and the geomorphic and paleoenvironmental value of sedimentary records from large Cordilleran lakes. *Geomorphology* 79, 130–42.

Gillespie, A. R., Montgomery, D. R. and Mushkin, A. 2005. Are there active glaciers on Mars? *Nature* 438, doi:10.1038/nature04357.

Gipp, M. R. 1994. Architectural styles of glacially influenced marine deposits on tectonically active and passive continental margins. In Deynoux, M., Miller, J. M. G., Domack, E. W., Eyles, N., Fairchild, I. J. and Young, G. M. (eds), *Earth's Glacial Record*. Cambridge University Press, 109–20.

Gipp, M. R. 2000. Lift-off moraines: markers of last ice flow directions on the Scotian Shelf. *Canadian Journal of Earth Sciences* 37, 1723–34.

Gjessing, J. 1965. On 'plastic scouring' and 'subglacial erosion'. *Norsk Geografisk Tidsskrift* 20, 1–37.

Gjessing, J. 1967. Potholes in connection with plastic scouring forms. *Geografiska Annaler* 49A, 178–87.

GLACKIPR 2003. *Proceedings of VI Symposium of the International Commission Glacier Caves and Karst in Polar and High Mountain Regions of the International Union of Speleology*. Ny-Ålesund, Svalbard.

Glasser, N. F. 1995. Modelling the effect of topography on ice sheet erosion, Scotland. *Geografiska Annaler* 77A, 67–82.

Glasser, N. F. 2002a. Scottish Landform Example 28: The large roches moutonnées of upper Deeside. *Scottish Geographical Journal* 118, 129–38.

Glasser, N. F. 2002b. Newtondale (SE820930) and Hole of Horcum (SE845935). In Huddart, D. and Glasser, N. F. (eds), *Quaternary of Northern England*. GCR Series 25, Joint Nature Conservation Committee, Peterborough, 171–6.

Glasser, N. F. and Ghiglione, M. C. 2009. Structural, tectonic and glaciological controls on the evolution of fjord landscapes. *Geomorphology* 105, 291–302.

Glasser, N. F. and Hambrey, M. J. 2001. Styles of sedimentation beneath Svalbard valley glaciers under changing dynamic and thermal regimes. *Journal of the Geological Society of London* 158, 697–707.

Glasser, N. F. and Hambrey, M. J. 2002. Sedimentary facies and landform genesis at a temperate outlet glacier: Soler Glacier, North Patagonian Icefield. *Sedimentology* 49, 43–64.

Glasser, N. F. and Hambrey, M. J. 2003. Ice-marginal terrestrial landsystems: Svalbard polythermal glaciers. In Evans, D. J. A. (ed.), *Glacial Landsystems*. Arnold, London, 65–88.

Glasser, N. F. and Jansson, K. N. 2005. Fast-flowing outlet glaciers of the Last Glacial Maximum Patagonian Icefield. *Quaternary Research* 63, 206–11.

Glasser, N. F. and Scambos, T. A. 2008. A structural glaciological analysis of the 2002 Larsen B ice-shelf collapse. *Journal of Glaciology* 54, 3–16.

Glasser, N. F. and Warren, C. R. 1990. Medium scale landforms of glacial erosion in south Greenland: process and form. *Geografiska Annaler* 72A, 211–15.

Glasser, N. F., Bennett, M. R., Hambrey, M. J. and Huddart, D. 1999. The morphology, composition and origin of medial moraines on polythermal glaciers in Svalbard. *Glacial Geology and Geomorphology* (http://ggg.qub.ac.uk/papers/frame.htm).

Glasser, N. F., Etienne, J. L., Hambrey, M. J., Davies, J. R., Waters, R. A. and Wilby, P. R. 2004. Glacial meltwater erosion and sedimentation as evidence for multiple glaciations in west Wales. *Boreas* 33, 224–37.

Glasser, N. F., Goodsell, B., Copland, L. and Lawson, W. 2006. Debris characteristics and ice shelf dynamics in the ablation region of the McMurdo Ice Shelf, Antarctica. *Journal of Glaciology* 52, 223–34.

Glasser, N. F., Hambrey, M. J., Bennett, M. R. and Huddart, D. 2003. Comment: Formation and reorientation of structure in the surge-type glacier Kongsvegen. *Journal of Quaternary Science* 18, 95–7.

Glasser, N. F., Hambrey, M. J., Crawford, K. R., Bennett, M. R. and Huddart, D. 1998. The structural glaciology of Kongsvegen, Svalbard and its role in landform genesis. *Journal of Glaciology* 44, 136–48.

Glasser, N. F., Jansson, K. N., Harrison, S. and Kleman, J. 2008. The glacial geomorphology and Pleistocene history of South America between 38° S and 56° S. *Quaternary Science Reviews* 27, 365–90.

Glasser, N. F., Jansson, K. N., Harrison, S. and Rivera, A. 2005. Geomorphological evidence for variations of the North Patagonian Icefield during the Holocene. *Geomorphology* 71, 263–77.

Glen, J. W. 1955. The creep of polycrystalline ice. *Proceedings of the Royal Society*, Series A, 228, 519–38.

Gluckert, G. 1973. Two large drumlin fields in central Finland. *Fennia* 120, 37.

Gobbi, M., De Bernardi, F., Pelfini, M., Rossaro, B. and Brandmayr, P. 2006. Epigean Arthropod Succession along a 154-year Glacier Foreland Chronosequence in the Forni Valley (Central Italian Alps). *Arctic, Antarctic and Alpine Research* 38, 357–62.

Goldthwait, R. P. 1961. Regimen of an ice cliff on land in Northwest Greenland. *Folia Geographica Danica* 9, 107–15.

Goldthwait, R. P. 1979. Giant grooves made by concentrated basal ice streams. *Journal of Glaciology* 23, 297–307.

Golledge, N. R. 2006. The Loch Lomond Stadial glaciation south of Rannoch Moor: new evidence and palaeoglaciological insights. *Scottish Geographical Journal* 122, 326–43.

Golledge, N. R. 2007. An ice cap landsystem for palaeoglaciological reconstructions: characterizing the Younger Dryas in western Scotland. *Quaternary Science Reviews* 26, 213–29.

Golledge, N. R. and Hubbard, A. 2005. Evaluating Younger Dryas glacier reconstructions in part of the western Scottish Highlands: a combined empirical and theoretical approach. *Boreas* 34, 274–86.

Golledge, N. R. and Stoker, M. S. 2006. A palaeo-ice stream of the British Ice Sheet in eastern Scotland. *Boreas* 35, 231–43.

Golledge, N. R., Finlayson, A., Bradwell, T. and Everest, J. D. 2008a. The last glaciation of Shetland, North Atlantic. *Geografiska Annaler* 90A, 37–53.

Golledge, N. R., Hubbard, A. and Sugden, D. E. 2008b. High-resolution numerical simulation of Younger Dryas glaciation in Scotland. *Quaternary Science Reviews*, 27, 888–904.

Gomez, B. 1987. Bedload. In Gurnell, A. M. and Clark, M. J. (eds), *Glacio-Fluvial Sediment Transfer: An Alpine Perspective*. Wiley, Chichester, 355–76.

Gomez, B. and Small, R. J. 1985. Medial moraines of the Haut Glacier d'Arolla, Valais, Switzerland: debris supply and implications for moraine formation. *Journal of Glaciology* 31, 303–7.

Gomez, B., Dowdeswell, J. A. and Sharp, M. 1988. Microstructural control of quartz sand grain shape and texture: implications for the discrimination of debris transport pathways through glaciers. *Sedimentary Geology* 57, 119–29.

Gomez, B., Russell, A. J., Smith, L. C. and Knudsen, Ó. 2002. Erosion and deposition in the proglacial zone: the 1996 jokulhlaup on Skeiðarársandur, southeast Iceland. In Snorasson, A., Finnsdottir, H. and Moss, P. (eds), *The Extremes of the Extremes: Extraordinary Floods*. IAHS Special Publication 271, 217–21.

Gomez, B., Smith, L. C., Magilligan, F. J., Mertes, F. A. K. and Smith, N. D. 2000. Glacier outburst floods and outwash plain development: Skeiðarársandur, Iceland. *Terra Nova* 12, 126–31.

Gonzales, J. and Aydin, A. 2008. The origin of oriented lakes in the Andean foreland, Parque Nacional Torres del Paine (Chilean Patagonia). *Geomorphology* 97, 502–15.

Goodfellow, B. W. 2007. Relict non-glacial surfaces in formerly glaciated landscapes. *Earth Science Reviews* 80, 47–73.

Goodfellow, B. W., Stroeven, A. P., Hättestrand, C., Kleman, J. and Jansson, K. N. 2008. Deciphering a non-glacial/glacial landscape mosaic in the northern Swedish mountains. *Geomorphology* 93, 213–32.

Goodsell, B., Hambrey, M. J. and Glasser, N. F. 2002. Formation of band ogives and associated structures at Bas Glacier d'Arolla, Valais, Switzerland. *Journal of Glaciology* 48, 287–300.

Goodsell, B., Hambrey, M. J. and Glasser, N. F. 2005a. Debris transport in a temperate valley glacier: Haut Glacier d'Arolla, Valais, Switzerland. *Journal of Glaciology* 51 (172), 139–46.

Goodsell, B., Hambrey, M. J., Glasser, N. F., Nienow, P. and Mair, D. 2005b. The structural glaciology of a temperate valley glacier: Haut Glacier d'Arolla, Valais, Switzerland. *Arctic, Antarctic, and Alpine Research* 37, 218–32.

Gordon, J. 1976. *The New Science of Strong Materials, or Why You Don't Fall Through the Floor*. London, Penguin.

Gordon, J. 1978. *Structures, or Why Things Don't Fall Down*. London, Penguin.

Gordon, J. E. 1977. Morphometry of cirques in the Kintail-Affric-Cannich area of northwest Scotland. *Geografiska Annaler* 59A, 177–94.

Gordon, J. E. 1981. Ice-scoured topography and its relationships to bedrock structure and ice movement in parts of northern Scotland and west Greenland. *Geografiska Annaler* 63A, 55–65.

Gordon, J. E. 1993. The Cairngorms. In Gordon, J. E. and Sutherland, D. G. *Quaternary of Scotland*. London, Chapman and Hall, 259–76.

Gordon, J. E. and Birnie, R. V. 1986. Production and transfer of sub-aerially generated rock debris and resulting landforms on South Georgia: an introductory perspective. *British Antarctic Survey Bulletin* 72, 25–46.

Gordon, J. E., Birnie, R. V. and Timmis, R. 1978. A major rockfall and debris slide on the Lyell Glacier, South Georgia. *Arctic and Alpine Research* 10, 49–60.

Gordon, J. E., Whalley, W. B., Gellatly, A. F. and Vere, D. M. 1992. The formation of glacial flutes: assessment of models with evidence from Lyngsdalen, north Norway. *Quaternary Science Reviews* 11, 709–31.

Gordon, S., Sharp, M., Hubbard, B., Smart, C., Ketterling, B. and Willis, I. 1998. Seasonal reorganisation of subglacial drainage inferred from measurements in boreholes. *Hydrological Processes* 12, 105–33.

Gorrell, G. and Shaw, J. 1991. Deposition in an esker, bead and fan complex, Lanark, Ontario, Canada. *Sedimentary Geology* 72, 285–314.

Gosse, J. C. and Phillips, F. M. 2001. Terrestrial in situ cosmogenic nuclides: theory and application. *Quaternary Science Reviews* 20, 1475–560.

Gough, D. O. 1981. Solar interior structure and luminosity variations. *Solar Physics* 74, 21–34.

Gould, S. J. 1980. *The Panda's Thumb*. Norton, New York.

Gow, A. J. and Epstein, S. 1972. On the use of stable isotopes to trace the origins of ice in a floating ice tongue. *Journal of Geophysical Research* 77, 6552–7.

Graf, W., Reinwarth, O., Oerter, H., Mayer, C. and Lambrecht, A. 1999. Surface accumulation on Foundation Ice Stream, Antarctica. *Annals of Glaciology* 29, 23–8.

Graf, W. L. 1970. The geomorphology of the glacial valley cross-section. *Arctic and Alpine Research* 2, 303–12.

Graham, A. G. C., Lonergan, L. and Stoker, M. S. 2007. Evidence for Late Pleistocene ice stream activity in the Witch Ground Basin, central North Sea, from 3D seismic reflection data. *Quaternary Science Reviews* 26, 627–43.

Graham, D. J., Bennett, M. R., Glasser, N. F., Hambrey, M. J., Huddart, D. and Midgley, N. G. 2007. 'A test of the englacial thrusting hypothesis of "hummocky" moraine formation: case studies from the northwest Highlands, Scotland': Comments. *Boreas* 36, 103–7.

Grasby, S. E. and Chen, Z. 2005. Subglacial recharge into the Western Canada Sedimentary Basin: impact of Pleistocene glaciation on basin hydrodynamics. *Geological Society of America Bulletin* 117, 500–14.

Grasby, S. E., Osadetz, K., Betcher, R. and Render, F. 2000. Reversal of the regional scale flow system of the Williston Basin in response to Pleistocene glaciation. *Geology* 28, 635–8.

Grattan, J., Durand, M. and Taylor, S. 2003. Illness and elevated human mortality in Europe coincident with the Laki fissure eruption. In Oppenheimer, C., Pyle, D. M. and Barclay, J. (eds), *Volcanic Degassing*. Geological Society, London, Special Publication 213, 401–14.

Gravenor, C. P. and Kupsch, W. O. 1959. Ice disintegration features in western Canada. *Journal of Geology* 67, 48–64.

Gravenor, C. P., von Brunn, V. and Dreimanis, A. 1984. Nature and classification of waterlain glaciogenic sediments, exemplified by Pleistocene, Late Palaeozoic and Late Precambrian deposits. *Earth Science Reviews* 20, 105–66.

Gray, J. M. 1975. The Loch Lomond Readvance and contemporaneous sea-levels in Loch Etive and neighbouring areas of western Scotland. *Proceedings of the Geologists' Association* 86, 227–38.

Gray, J. M. 1981. P-forms from the Isle of Mull. *Scottish Journal of Geology* 17, 39–47.

Gray, J. M. 1992. Scarisdale: P-forms. In Walker, M. J. C., Gray, J. M. and Lowe, J. J. (eds), *The South-West Scottish Highlands. Field Guide*. Quaternary Research Association, Cambridge, 85–8.

Gray, L., Joughin, I., Tulaczyk, S., Spikes, V. B., Bindschadler, R. and Jezek, K. 2005. Evidence for subglacial water transport in the West Antarctic Ice Sheet through three-dimensional satellite radar interferometry. *Geophysical Research Letters* 32, doi:10.1029/2004GL021387.

Greenwood, S. L. and Clark, C. D. 2008. Subglacial bedforms of the Irish Ice Sheet. *Journal of Maps* 2008, 332–57.

Greenwood, S. L., Clark, C. D. and Hughes, A. L. C. 2007. Formalising an inversion methodology for reconstructing ice-sheet retreat patterns from meltwater channels: application to the British Ice Sheet. *Journal of Quaternary Science* 22, 637–45.

Gregg, T. K. P., Briner, J. P. and Paris, K. N. 2007. Ice-rich terrain in Gusev Crater, Mars? *Icarus* 192, 348–60.

Gregory, J. M. and Huybrechts, P. 2006. Ice-sheet contributions to future sea-level change. *Philosophical Transactions of the Royal Society* A, 364, 1709–31.

Gregory, J. M., Huybrechts, P. and Raper, S. C. B. 2004. Threatened loss of the Greenland ice-sheet. *Nature* 428, 616.

Greuell, W. and Genthon, C. 2004. Modelling land-ice surface mass balance. In Bamber, J. L. and Payne, A. J. 2004. (eds), *Mass Balance of the Cryosphere: Observations and Modeling of Contemporary and Future Changes*. Cambridge University Press, 117–68.

Greuell, W. and Konzelmann, T. 1994. Numerical modelling of the energy balance and the englacial temperature of the Greenland Ice Sheet. Calculations for the ETH-Camp location (West Greenland, 1155 m a.s.l.). *Global and Planetary Change* 9, 91–114.

Greuell, W., Knap, W. and Smeets, P. 1997. Elevational changes in meteorological variables along a midlatitude glacier during summer. *Journal of Geophysical Research* 102, 25941–54.

Greve, R. 1997. A continuum-mechanical formulation for shallow polythermal ice sheets. *Philosophical Transactions of the Royal Society London* A 355 (1726), 921–74.

Greve, R. and Mahajan, R. A. 2005. Influence of ice rheology and dust content on the dynamics of the north-polar cap of Mars. *Icarus* 174, 475–85.

Grootes, P. M., Stuiver, M., Thompson, L. G. and Mosely-Thompson, E. 1989. Oxygen isotope changes in tropical ice, Quelccaya, Peru. *Journal of Geophysical Research* 94 (D1), 1187–94.

Grosfield, K. and Sandhäger, H. 2004. The evolution of a coupled ice shelf–ocean system under different climate states. *Global and Planetary Change* 42, 107–32.

Grosswald, M. G. and Hughes, T. J. 1999. The case for an ice shelf in the Pleistocene Arctic Ocean. *Polar Geography* 23, 23–54.

Grousset, F. E., Pujol, C., Labeyrie, L., Auffret, G. and Boelaert, A. 2000. Were the North Atlantic Heinrich events triggered by the behaviour of the European ice sheets? *Geology* 28, 123–6.

Grove, J. M. 1988. *The Little Ice Age*. Routledge, London.

Grove, J. M. 2001. The Initiation of the 'Little Ice Age' in regions round the North Atlantic. *Climatic Change* 48, 53–82.

Grube, F. 1983. Tunnel valleys. In Ehlers, J. (ed.), *Glacial Deposits in North-West Europe*. Balkema, Rotterdam, 257–8.

Gruber, S., Hoelzle, M. and Haeberli, W. 2004. Permafrost thaw and destabilization of Alpine rock walls in the hot summer of 2003. *Geophysical Research Letters* 31/L13504.

Gude, M. and Barsch, D. 2005. Assessment of geomorphic hazards in connection with permafrost occurrence in the Zugspitze area (Bavarian Alps, Germany). *Geomorphology* 66, 85–93.

Gudmundsson, G. H. 1999. A three-dimensional numerical model of the confluence area of Unteraargletscher, Bernese Alps, Switzerland. *Journal of Glaciology* 45, 219–30.

Gudmundsson, G. H. 2003. Transmission of basal variability to a glacier surface. *Journal of Geophysical Research* 108 (B5), 2253, doi:10.1029/2002JB002107.

Gudmundsson, G. H. 2006. Fortnightly variations in the flow velocity of Rutford Ice Stream. *Nature* 444, 1063–4.

Gudmundsson, G. H., Raymond, C. F. and Bindschadler, R. 1998. The origin and longevity of flow stripes on Antarctic ice streams. *Annals of Glaciology* 27, 145–52.

Gudmundsson, H. J. 1997. A review of the Holocene environmental history of Iceland. *Quaternary Science Reviews* 16, 81–92.

Guilcher, A., Bodére, J.-C., Coudé, A., Hansom, J. D., Moign, A. and Peulvast, J.-P. 1986. Le problème des strandflats en cinq pays de hautes latitudes. *Revue de Geologie Dynamique et de Geographie Physique* 27, 47–79. English translation in Evans, D. J. A. (ed.), *Cold Climate Landforms*. Wiley, Chichester, 351–93.

Gulick, V. C. 2001. Origin of the valley networks on Mars: a hydrological perspective. *Geomorphology* 37, 241–68.

Gulley, J. 2009. Structural control of englacial conduits in the temperate Matanuska Glacier, Alaska, USA. *Journal of Glaciology* 55(192), 681–90.

Gulley, J. and Benn, D. I. 2007. Structural control of englacial drainage systems in Himalayan debris-covered glaciers. *Journal of Glaciology* 53, 399–412.

Gulley, J., Benn, D. I., Luckman, A. and Müller, D. 2009a. A cut-and-closure origin for englacial conduits on uncrevassed parts of polythermal glaciers. *Journal of Glaciology* 55 (89), 66–80.

Gulley, J., Benn, D. I., Screaton, L. and Martin, J. 2009b. Mechanisms of englacial conduit formation and implications for subglacial recharge. *Quaternary Science Reviews* 28, 1984–99.

Gupta, S., Collier, J. S., Palmer-Felgate, A. and Potter, G. 2007. Catastrophic flooding origin of shelf valley systems in the English Channel. *Nature* 448, 342–6.

Gurnell, A. M. 1982. The dynamics of suspended sediment concentration in an Alpine pro-glacial stream network. *Hydrological Aspects of Alpine and High Mountain Areas*. IASH Publication 138, 319–30.

Gurnell, A. M. 1987. Fluvial sediment yield from Alpine, glacierized catchments. In Gurnell, A. M. and Clark, M. J. (eds), *Glacio-Fluvial Sediment Transfer: An Alpine Perspective*. Wiley, Chichester, 415–20.

Gurnell, A. M. 1995. Sediment yield from Alpine glacier basins. In Foster, I. D. L., Gurnell, A. M. and Webb, B. W. (eds), *Sediment and Water Quality in River Catchments*. Wiley, Chichester, 407–35.

Gurnell, A. M. and Fenn, C. R. 1985. Spatial and temporal variations in electrical conductivity in a proglacial stream. *Journal of Glaciology* 31, 108–14.

Gurnell, A. M., Clark, M. J. and Hill, C. T. 1992. Analysis and interpretation of patterns within and between hydroclimatological time series in an Alpine glacier basin. *Earth Surface Processes and Landforms* 17, 821–39.

Gustavson, T. C. 1974. Sedimentation on gravel outwash fans, Malaspina Glacier foreland, Alaska. *Journal of Sedimentary Petrology* 44, 374–89.

Gustavson, T. C. 1975. Sedimentation and physical limnology in proglacial Malaspina Lake, southeastern Alaska. In Jopling, A. V. and McDonald, B. C. (eds), *Glaciofluvial and Glaciolacustrine Sedimentation*. SEPM Special Publication 23, 249–63.

Gustavson, T. C. and Boothroyd, J. C. 1987. A depositional model for outwash, sediment sources, and hydrologic characteristics, Malaspina Glacier, Alaska: a modern analog of the southeastern margin of the Laurentide Ice Sheet. *Geological Society of America Bulletin* 99, 187–200.

Gustavson, T. C., Ashley, G. M. and Boothroyd, J. C. 1975. Depositional sequences in glaciolacustrine deltas. In Jopling, A. V. and McDonald, B. C. (eds), *Glaciofluvial and Glaciolacustrine Sedimentation*. SEPM Special Publication 23, 264–80.

Haeberli, W. 1976. Isen temperaturen in den Alpen. *Zeitschrift für Gletscherkunde und Glazialgeologie* 11, 203–28.

Haeberli, W., Hoelzle, M. and Suter, S. (eds). 1998. *Into the Second Century of World Glacier Monitoring – Prospects and Strategies*. UNESCO Publishing.

Haeberli, W., Noetzli, J., Zemp, M., Baumann, S., Fraulenfelder, R. and Hoelzle, M. (eds). 2005. *Glacier Mass Balance Bulletin No. 8 (2002–2003)*. IUGG (CCS) – UNEP – UNESCO – WMO, Zurich.

Haflidason, H., Sejrup, H. P., Nygård, A., Mienert, J., Bryn, P., Lien, R., Forsberg, C. F., Berg, K. and Masson, D. 2004. The Storegga Slide: architecture, geometry and slide development. *Marine Geology* 213, 201–34.

Hagen, J. O., Liestøl, O., Roland, E. and Jørgensen, T. 1993. *Glacier Atlas of Svalbard and Jan Mayen*. Norsk Polar Institutt, Oslo.

Hagg, W., Braun, L. N., Kuhn, M. and Nesgaard, T. I. 2007. Modelling of hydrological response to climate change in glacierized Central Asian catchments. *Journal of Hydrology* 332, 40–53.

Haldorsen, S. 1981. Grain-size distribution of subglacial till and its relation to subglacial crushing and abrasion. *Boreas* 10, 91–105.

Haldorsen, S. 1983. Mineralogy and geochemistry of basal till and its relationship to till-forming processes. *Norsk Geologisk Tidsskrift* 63, 15–25.

Haldorsen, S. and Shaw, J. 1982. The problem of recognizing melt-out till. *Boreas* 11, 261–77.

Hall, A. M. and Glasser, N. F. 2003. Reconstructing the basal thermal regime of an ice stream in a landscape of selective linear erosion: Glen Avon, Cairngorm Mountains, Scotland. *Boreas* 32, 191–208.

Hall, A. M. and Phillips, W. M. 2006. Glacial modification of tors in the Cairngorm Mountains, Scotland. *Journal of Quaternary Science* 21, 811–30.

Hall, A. M. and Sugden, D. E. 1987. Limited modification of mid-latitude landscapes by ice sheets: the case of north-east Scotland. *Earth Surface Processes and Landforms* 12, 531–42.

Hall, A. M. and Sugden, D. E. 2007. The significance of tors in glaciated lands: a view from the British Isles. In Andre, M.-F. (ed.), *Du continent au bassin versant: theories et practiques en géographie physique (Hommage au Professeur Alain Godard)*. Presses Universitaires Blaise-Pascal, Clermont-Ferrand, 301–11.

Hall, B. L., Hendy, C. H. and Denton, G. H. 2006. Lake-ice conveyor deposits: geomorphology, sedimentology, and importance in reconstructing the glacial history of the Dry Valleys. *Geomorphology* 75, 143–56.

Hall, D. K., Williams, R. S., Luthcke, S. B. and Digirolamo, N. E. 2008. Greenland ice sheet surface temperature, melt and mass loss: 2000–06. *Journal of Glaciology* 54, 81–93.

Hallet, B. 1976a. Deposits formed by subglacial precipitation of $CaCO_3$. *Geological Society of America Bulletin* 87, 1003–15.

Hallet, B. 1976b. The effect of subglacial chemical processes on glacier sliding. *Journal of Glaciology* 17, 209–21.

Hallet, B. 1979a. Subglacial regelation water film. *Journal of Glaciology* 23, 321–34.

Hallet, B. 1979b. A theoretical model of glacial abrasion. *Journal of Glaciology* 23, 39–50.

Hallet, B. 1981. Glacial abrasion and sliding: their dependence on the debris concentration in basal ice. *Annals of Glaciology* 2, 23–8.

Hallet, B. 1996. Glacial quarrying: a simple theoretical model. *Annals of Glaciology* 22, 1–8.

Hallet, B. and Anderson, R. S. 1980. Detailed glacial geomorphology of a proglacial bedrock area at Castleguard Glacier, Alberta, Canada. *Zeitschrift für Gletscherkunde und Glazialgeologie* 16, 171–84.

Hallet, B., Hunter, L. and Bogen, J. 1996. Rates of erosion and sediment evacuation by glaciers: a review of field data and their implications. *Global and Planetary Change* 12, 213–35.

Hallet, B., Lorrain, R. and Souchez, R. 1978. The composition of basal ice from a glacier sliding over limestones. *Geological Society of America Bulletin* 89, 314–20.

Hallet, B., Walder, J. S. and Stubbs, C. W. 1991. Weathering by segregation ice growth in microcracks at subfreezing temperatures: verification from an experimental study using acoustic emissions. *Permafrost and Periglacial Processes* 2, 283–300.

Ham, N. R. and Attig, J. W. 1996. Ice wastage and landscape evolution along the southern margin of the Laurentide Ice Sheet, north-central Wisconsin. *Boreas* 25, 171–86.

Ham, N. R. and Attig, J. W. 2001. Minor end moraines of the Wisconsin Valley Lobe, north-central Wisconsin, USA. *Boreas* 30, 31–41.

Ham, N. R. and Mickelson, D. M. 1994. Basal till fabric and deposition at Burroughs Glacier, Glacier Bay, Alaska. *Geological Society of America Bulletin* 106, 1552–9.

Hambley, G. W. and Lamoureux, S. F. 2006. Recent summer climate recorded in complex varved sediments, Nicolay Lake, Cornwall Island, Nunavut. *Journal of Paleolimnology* 35, 629–40.

Hamblin, P. F. and Carmack, E. C. 1978. River-induced currents in a fjord lake. *Journal of Geophysical Research* 83, 885–99.

Hambrey, M. J. 1975. The origin of foliation in glaciers: evidence from some Norwegian examples. *Journal of Glaciology* 14, 181–5.

Hambrey, M. J. 1976. Structure of the glacier Charles Rabots Bre, Norway. *Geological Society of America Bulletin* 87, 1629–37.

Hambrey, M. J. 1994. *Glacial Environments*. UCL Press, London.

Hambrey, M. J. and Ehrmann, W. 2004. Modification of sediment characteristics during glacial transport in high alpine catchments: Mount Cook area, New Zealand. *Boreas* 33, 300–18.

Hambrey, M. J. and Glasser, N. F. 2003. The role of folding and foliation development in the genesis of medial moraines: examples from Svalbard glaciers. *Journal of Geology* 111, 471–85.

Hambrey, M. J. and Lawson, W. J. 2000. Structural styles and deformation fields in glaciers: a review. In Maltman, A. J., Hubbard, B. and Hambrey, M. J. (eds), *Deformation of Glacial Materials*. Geological Society, London, Special Publication 176, 59–83.

Hambrey, M. J. and McKelvey, B. 2000. Major Neogene fluctuations of the East Antarctic ice sheet: stratigraphic evidence from the Lambert Glacier region. *Geology* 28, 887–90.

Hambrey, M. J. and Milnes, A. G. 1975. Boudinage in glacier ice – some examples. *Journal of Glaciology* 14, 383–93.

Hambrey, M. J. and Müller, F. 1978. Structures and ice deformation in White Glacier, Axel Heiberg Island, NWT, Canada. *Journal of Glaciology* 20, 41–67.

Hambrey, M. J., Barrett, P. J. and Powell, R. D. 2002. Late Oligocene and early Miocene glacimarine sedimentation in the SW Ross Sea, Antarctica: the record from offshore drilling. In Dowdeswell, J. A. and Ó Cofaigh, C. (eds), *Glacier-Influenced Sedimentation on High Latitude Continental Margins*. Geological Society, London, Special Publication 203, 105–28.

Hambrey, M. J., Bennett, M. R., Dowdeswell, J. A., Glasser, N. F. and Huddart, D. 1999. Debris entrainment and transfer in polythermal glaciers. *Journal of Glaciology* 45 (149), 69–86.

Hambrey, M. J., Davies, J. R., Glasser, N. F., Waters, R. A., Dowdeswell, J. A., Wilby, P., Wilson, D. and Etienne, J. L. 2001. Devensian glacigenic sedimentation and landscape evolution in the Cardigan area of southwest Wales. *Journal of Quaternary Science* 16, 455–82.

Hambrey, M. J., Dowdeswell, J. A., Murray, T. and Porter, P. R. 1996. Thrusting and debris entrainment in a surging glacier: Bakaninbreen, Svalbard. *Annals of Glaciology* 22, 241–8.

Hambrey, M. J., Glasser, N. F., McKelvey, B. C., Sugden, D. E. and Fink, D. 2007. Cenozoic landscape evolution of an East Antarctic oasis (Radok Lake area, northern Prince Charles Mountains), and its implications for the glacial and climatic history of Antarctica. *Quaternary Science Reviews* 26, 598–626.

Hambrey, M. J., Huddart, D., Bennett, M. R. and Glasser, N. F. 1997. Genesis of 'hummocky moraine' by thrusting in glacier ice: evidence from Svalbard and Britain. *Journal of the Geological Society of London* 154, 623–32.

Hambrey, M. J., Murray, T., Glasser, N. F., Hubbard, A., Hubbard, B., Stuart, G., Hansen, S. and Kohler, J. 2005. Structure and changing dynamics of a polythermal valley glacier on a centennial timescale: Midre Lowénbreen, Svalbard. *Journal of Geophysical Research* 110, doi:10.1029/2004JF000128.

Hambrey, M. J., Quincey, D. J., Glasser, N. F., Reynolds, J. M., Richardson, S. J. and Clemmens, S. 2008. Sedimentological, geomorphological and dynamic context of debris-mantled glaciers, Mount Everest (Sagarmatha) region, Nepal. *Quaternary Science Reviews* 27, 2361–89.

Hamilton, G. S. and Dowdeswell, J. A. 1996. Controls on glacier surging in Svalbard. *Journal of Glaciology* 42, 157–68.

Hamilton, W. C. and Ibers, J. A. 1968 *Hydrogen Bonding in Solids: Methods of Molecular Structure Determination*. W. A. Benjamin, New York.

Hamilton, R. J., Luyendyk, B. P. and Sorlein, C. C. 2001. Cenozoic tectonics of the Cape Roberts rift basin and Transantarctic Mountains front, southwest Ross Sea. *Tectonic* 20, 325–42.

Hammer, C. U., Clausen, H. B. and Dansgaard, W. 1980. Greenland ice sheet evidence of post-glacial volcanism and its climatic impact. *Nature* 288, 230–5.

Hanebuth, T., Stattegger, K. and Grootes, P. M. 2000. Rapid flooding of the Sunda Shelf: a late glacial sea level record. *Science* 288, 1033–5.

Hanna, E., Huybrechts, P., Janssens, I., Cappelen, J., Steffen, K. and Stephens, A. 2005. Runoff and mass balance of the Greenland ice sheet: 1958–2003. *Journal of Geophysical Research* 110, doi:10.1029/2004JD005641.

Hanna, E., Huybrechts, P., Steffen, K., Cappelen, J., Huff, R., Shuman, C., Irvine-Fynn, T., Wise, S. and Griffiths, M. 2008. Increased runoff from the Greenland ice sheet: a response to global warming. *Journal of Climate* 21, 331–41.

Hanna, E., McConnel, J., Das, S. B., Cappelen, J. and Stephens, A. 2006. Observed and modelled Greenland ice sheet snow accumulation, 1958–2003, and links with regional climate forcing. *Journal of Climate* 19, 344–58.

Hansen, S. 2003. From surge-type to non-surge-type glacier behaviour: midre Lovénbreen, Svalbard. *Annals of Glaciology* 36, 97–102.

Hansom, J. D. 1983a. Shore platform development in the South Shetland Islands, Antarctica. *Marine Geology* 52, 211–29.

Hansom, J. D. 1983b. Ice-formed intertidal boulder pavements in the sub-Antarctic. *Journal of Sedimentary Petrology* 53, 135–45.

Hanson, B. and Hooke, R. Le B. 2000. Glacier calving: a numerical model of forces in the calving speed-water depth relation. *Journal of Glaciology* 46, 188–96.

Hanson, B. H. and Hooke, R. Le B. 2003. Buckling rate and overhang development at a calving face. *Journal of Glaciology* 49, 578–86.

Hanvey, P. M. 1987. Sedimentology of lee-side stratification sequences in late Pleistocene drumlins, north-west Ireland. In Menzies J. and Rose, J. (eds), *Drumlin Symposium*. Balkema, Rotterdam, 241–53.

Hanvey, P. M. 1989. Stratified flow deposits in a late Pleistocene drumlin in northwest Ireland. *Sedimentary Geology* 62, 211–21.

Haran, T., Bohlander, J., Scambos, T., Painter, T. and Fahnestock, M. (compilers). 2005, updated 2006. *MODIS Mosaic of Antarctica (MOA) Image Map*. National Snow and Ice Data Center, Boulder, CO.

Harbor, J. M. 1990. A discussion of Hirano and Aniya's (1988, 1989) explanation of glacial valley cross profile development. *Earth Surface Processes and Landforms* 15, 369–77.

Harbor, J. M. 1992. Numerical modelling of the development of U-shaped valleys by glacial erosion. *Geological Society of America Bulletin* 104, 1364–75.

Harbor, J. M. and Warburton, J. 1993. Relative rates of glacial and nonglacial erosion in alpine environments. *Arctic and Alpine Research* 25, 1–7.

Harbor, J. M., Hallet, B. and Raymond, C. F. 1988. A numerical model of landform development by glacial erosion. *Nature* 333, 347–9.

Harbor, J. M., Stroeven, A. P., Fabel, D., Clarhäll, A., Kleman, J., Li, Y., Elmore, D. and Fink, D. 2006. Cosmogenic nuclide evidence for minimal erosion across two subglacial sliding boundaries of the late glacial Fennoscandian ice sheet. *Geomorphology* 75, 90–9.

Haresign, E. C. 2004. *Glacio-limnological Interactions at Lake-calving Glaciers.* Unpublished PhD thesis, University of St Andrews.

Haresign, E. C. and Warren, C. R. 2005. Melt rates at calving termini: a study at Glaciar León, Chilean Patagonia. In Harris, C. and Murton, J. B. (eds), *Cryospheric Systems: Glaciers and Permafrost.* Geological Society, London, Special Publication 242, 99–110.

Harper, J. T. and Humphrey, N. F. 1995. Borehole video analysis of a temperate glacier's englacial and subglacial structure: implications for glacier flow models. *Geology* 23, 901–4.

Harper, J. T., Humphrey, N. F., Pfeffer, W. T. and Lazar, B. 2007. Two modes of accelerated glacier sliding related to water. *Geophysical Research Letters* 34, doi:10.1029/2007GL030233.

Harris, C. and Bothamley, K. 1984. Englacial deltaic sediments as evidence for basal freezing and marginal shearing, Leirbreen, Norway. *Journal of Glaciology* 30, 30–4.

Harris, C., Mühil, D. V., Isaksen, K., Haeberli, W., Sollid, J. L., King, L., Holmlund, P., Dramis, F., Gugliemin, M. and Palacios, D. 2003. Warming permafrost in European mountains. *Global and Planetary Change* 39, 215–25.

Harris, C., Williams, G., Brabham, P., Eaton, G. and McCarroll, D. 1997. Glaciotectonized Quaternary sediments at Dinas Dinlle, Gwynedd, North Wales, and their bearing on the style of deglaciation in the eastern Irish Sea. *Quaternary Science Reviews* 16, 109–27.

Harris, P. W. V. 1976. The seasonal temperature-salinity structure of a glacial lake: Jokulsarlon, southeast Iceland. *Geografiska Annaler* 58A, 329–36.

Harrison, S., Whalley, W. B. and Anderson, E. 2008. Relict rock glaciers and protalus lobes in the British Isles: implications for Late Pleistocene mountain geomorphology and palaeoclimate. *Journal of Quaternary Science* 23, 287–304.

Harrison, W. D. and Post, A. S. 2003. How much do we really know about glacier surges? *Annals of Glaciology* 36, 1–6.

Harrison, W. D., Echelmeyer, K. A., Chacho, E. F., Raymond, C. F. and Benedict, R. J. 1994. The 1897–88 surge of West Fork Glacier, Susitna Basin, Alaska, USA. *Journal of Glaciology* 40, 241–54.

Hart, J. K. 1990. Proglacial glaciotectonic deformation and the origin of the Cromer Ridge push moraine complex, North Norfolk, England. *Boreas* 19, 165–80.

Hart, J. K. 1994. Till fabric associated with deformable beds. *Earth Surface Processes and Landforms* 19, 15–32.

Hart, J. K. 1995a. Subglacial erosion, deposition and deformation associated with deformable beds. *Progress in Physical Geography* 19, 173–91.

Hart, J. K. 1995b. Recent drumlins, flutes and lineations at Vestari-Hagafellsjökull, Iceland. *Journal of Glaciology* 41, 596–606.

Hart, J. K. 1999. Identifying fast ice flow from landform assemblages in the geological record: a discussion. *Annals of Glaciology* 28, 59–66.

Hart, J. K. 2006. An investigation of subglacial processes at the microscale from Briksdalsbreen, Norway. *Sedimentology* 53, 125–46.

Hart, J. K. and Boulton, G. S. 1991. The inter-relation of glaciotectonic and glaciodepositional processes within the glacial environment. *Quaternary Science Reviews*, 10, 335–50.

Hart, J. K. and Roberts, D. H. 1994. Criteria to distinguish between subglacial glaciotectonic and glaciomarine sedimentation, I. Deformation styles and sedimentology. *Sedimentary Geology* 91, 191–213.

Hart, J. K. and Rose, J. 2001. Approaches to the study of glacier bed deformation. *Quaternary International* 86, 45–58.

Hart, J. K., Hindmarsh, R. C. A. and Boulton, G. S. 1990. Styles of subglacial glaciotectonic deformation within the context of the Anglian Ice Sheet. *Earth Surface Processes and Landforms* 15, 227–41.

Hart, J. K., Martinez, K., Ong, R., Riddoch, A., Rose, K. C. and Padhy, P. 2006. An autonomous multi-sensor subglacial probe: design and preliminary results from Briksdalsbreen, Norway. *Journal of Glaciology* 51, 389–97.

Hart, J. K., Rose, K. C., Martinez, K. and Ong, R. 2009. Subglacial clast behaviour and its implication for till fabric development: new results derived from wireless subglacial probe experiments. *Quaternary Science Reviews* 28, 597–607.

Hartzell, P. L., Nghiem, J. V., Richio, K. J. and Shain, D. H. 2005. Distribution and phylogeny of glacier ice worms (Mesenchytraeus solifugus and Mesenchytraeus solifugus rainierensis). *Canadian Journal of Zoology* 83, 1206–13.

Harvey, R. 2003. The origin and significance of Antarctic meteorites. *Chemie der Erde* 63, 93–147.

Hasnain, S. I. 1999. Runoff characteristics of a glacierized catchment, Garhwal Himalaya, India. *Hydrological Sciences* 44, 847–54.

Hasnain, S. I. and Thayyen, R. J. 1999a. Controls on the major ion chemistry of the Dokriani glacier meltwaters, Ganga basin, Garhwal Himalaya, India. *Journal of Glaciology* 45 (149), 87–92.

Hasnain, S. I. and Thayyen, R. J. 1999b. Discharge and suspended-sediment concentration of meltwaters draining from the Dokriani glacier, Garhwal Himalaya, India. *Journal of Hydrology* 218, 191–8.

Hastenrath, S. 1984. *The Glaciers of Equatorial East Africa.* Reidel, Dordrecht.

Hastenrath, S. and Kruss, P. D. 1992. The dramatic retreat of Mount Kenya's glaciers between 1963 and 1987: greenhouse forcing. *Annals of Glaciology* 16, 127–33.

Hattersley-Smith, G., Keys, J. E., Serson, H. and Mielke, J. E. 1970. Density stratified lakes in northern Ellesmere Island. *Nature* 225, 55–6.

Hättestrand, C. 1997. Ribbed moraines in Sweden – distribution pattern and palaeoglaciological implications. *Sedimentary Geology* 111, 41–56.

Hättestrand, C. and Clark, C. D. 2006. The glacial geomorphology of Kola Peninsula and adjacent areas in the Murmansk Region, Russia. *Journal of Maps* 2006, 30–42.

Hättestrand, C. and Johansen, N. 2005. Supraglacial moraines in Scharffenbergbotnen, Heimefrontfjella, Dronning Maud Land, Antarctica – significance for reconstructing former blue-ice areas. *Antarctic Science* 17, 225–36.

Hättestrand, C. and Kleman, J. 1999. Ribbed moraine formation. *Quaternary Science Reviews* 18, 43–61.

Hättestrand, C., Goodwillie, D. and Kleman, J. 1999. Size distribution of two cross-cutting drumlin systems in northern Sweden: a measure of selective erosion and formation time length. *Annals of Glaciology* 28, 146–52.

Hättestrand, C., Gotz, S., Naslund, J.-O., Fabel, D. and Stroeven, A. P. 2004. Drumlin formation time: evidence from northern and central Sweden. *Geografiska Annaler* 86A, 155–67.

Hättestrand, C., Kolka, V. and Johansen, N. 2007a. Cirque infills in the Khibiny Mountains, Kola Peninsula, Russia – palaeoglaciological interpretations and modern analogues in East Antarctica. *Journal of Quaternary Science* 23, 165–74.

Hättestrand, C., Kolka, V. and Stroeven, A. P. 2007b. The Keiva ice marginal zone on the Kola Peninsula, northwest Russia: a key component for reconstructing the palaeoglaciology of the northeastern Fennoscandian Ice Sheet. *Boreas* 36, 352–70.

Hauber, E., van Gasselt, S., Ivanov, B., Werner, S., Head, J. W., Neukum, G., Jaumann, R., Greeley, R., Mitchell, K. L., Muller, P. and The HRSC Co-investigator Team 2005. Discovery of a flank caldera and very young glacial activity at Hercates Tholus, Mars. *Nature* 434, 356–61.

Haynes, V. M. 1968. The influence of glacial erosion and rock structure on corries in Scotland. *Geografiska Annaler* 50A, 221–34.

Haynes, V. M. 1972. The relationship between the drainage areas and sizes of outlet troughs of Sukkertoppen Ice Cap, West Greenland. *Geografiska Annaler* 54A, 66–75.

Haynes, V. M. 1977. The modification of valley patterns by ice-sheet activity. *Geografiska Annaler* 59A, 195–207.

Haynes, V. M. 1983. Scotland's landforms. In Clapperton, C. M. (ed.), *Scotland: A New Study*. David and Charles, Newton Abbot, 28–63.

Haynes, V. M. 1995. Alpine valley heads on the Antarctic Peninsula. *Boreas* 24, 81–94.

Head, J. W. and Marchant, D. R. 2003. Cold-based mountain glaciers on Mars: Western Arsia Mons. *Geology* 31, 641–4.

Head, J. W. and Pratt, S. 2001. Extensive Hesperian-aged south polar ice sheet on Mars: evidence for massive melting and retreat, and lateral flow and ponding of meltwater. *Journal of Geophysical Research* 106, E6, 12275–99.

Head, J. W., Marchant, D. R., Agnew, M. C., Fassett, C. I. and Kreslavsky, M. A. 2006a. Extensive valley glacier deposits in the northern mid-latitudes of Mars: evidence for late Amazonian obliquity-driven climate change. *Earth and Planetary Science Letters* 241, 663–71.

Head, J. W., Mustard, J. F., Kreslavsky, M. A., Milliken, R. E. and Marchant, D. R. 2003. Recent ice ages on Mars. *Nature* 426, 797–802.

Head, J. W., Nahm, A. L., Marchant, D. R. and Neukum, G. 2006b. Modification of the dichotomy boundary on Mars by Amazonian midlatitude regional glaciation. *Geophysical Research Letters* 33, L08S03, doi:10.1029/2005GL024360.

Head, J. W., Neukum, G., Jaumann, R., Hiesinger, H., Hauber, E., Carr, M., Masson, P., Foing, B., Hoffmann, H., Kreslavsky, M., Werner, S., Milkovich, S., van Gasselt, S. and The HRSC Co-Investigator Team 2005. Tropical to mid-latitude snow and ice accumulation, flow and glaciation on Mars. *Nature* 434, 346–51.

Hebrand, M. and Åmark, M. 1989. Esker formation and glacier dynamics in eastern Skane and adjacent areas, southern Sweden. *Boreas* 18, 67–81.

Helle, S. K. 2004. Sequence stratigraphy in a marine moraine at the head of Hardangerfjorden, western Norway: evidence for a high-frequency relative sea-level cycle. *Sedimentary Geology* 164, 251–81.

Hellmer, H. H., Jacobs, S. S. and Jenkins, A. 1998. Ocean erosion of a floating Antarctic Glacier in the Amundsen Sea. *Antarctic Research Series* 75, 83–100.

Hemming, S. R. 2004. Heinrich events: massive late Pleistocene detritus layers of the North Atlantic and their global climate imprint. *Reviews of Geophysics* 42, RG1005.

Hemming, S. R. 2007. Terrigenous sediments. In Elias, S. A. (ed.), *Encyclopedia of Quaternary Science*. Elsevier, Amsterdam, 1776–85.

Henriksen, M., Mangerud, J., Matiouchkov, A., Paus, A. and Svendsen, J. I. 2003. Lake stratigraphy implies an 80,000 yr delayed melting of buried dead ice in northern Russia. *Journal of Quaternary Science* 18, 663–79.

Herman, F. and Braun, J. 2006. Fluvial response to horizontal shortening and glaciations: a study in the Southern Alps of New Zealand. *Journal of Geophysical Research* 111, F01008, doi:10.1029/2004JF000248.

Herman, F. and Braun, J. 2008. Evolution of the glacial landscape of the Southern Alps of New Zealand: insights from a glacial erosion model. *Journal of Geophysical Research* 113, F02009, doi:10.1029/2007JF000807.

Hertzfeld, U. C., Clarke, G. K. C., Mayer, H. and Greve, R. 2004. Derivation of deformation characteristics in fast-moving glaciers. *Computers and Geosciences* 30, 291–302.

Hesse, R., Khodabakhsh, S., Klaucke, I. and Ryan, W. B. F. 1997. Asymmetrical turbid surface-plume deposition near ice-outlets of the Pleistocene Laurentide Ice Sheet in the Labrador Sea. *Geo-Marine Letters* 17, 179–87.

Hesse, R., Klauck, I., Khodabakhsh, S. and Piper, D. 1999. Continental slope sedimentation adjacent to an ice margin. III. The upper Labrador Slope. *Marine Geology* 155, 249–76.

Hesse, R., Rashid, H. and Khodabakhsh, S. 2004. Fine grained sediment lofting from meltwater-generated turbidity currents during Heinrich events. *Geology* 32, 449–52.

Hewitt, K. 1967. Ice-front deposition and the seasonal effect: a Himalayan example. *Transactions of the Institute of British Geographers* 42, 93–106.

Hewitt, K. 1993. Altitudinal organisation of Karakoram geomorphic processes and depositional environments. In Shroder, J. F. (ed.) *Himalaya to the Sea*. Routledge, London, 159–83.

Hewitt, K. 1999. Quaternary moraines vs catastrophic rock avalanches in the Karakoram Himalaya, northern Pakistan. *Quaternary Research* 51, 220–37.

Hewitt, K. 2006. Disturbance regime landscapes: mountain drainage systems interrupted by large rockslides. *Progress in Physical Geography* 30, 365–93.

Hewitt, K. 2007. Tributary glacier surges: an exceptional concentration at Panmah Glacier, Karakoram Himalaya. *Journal of Glaciology* 53 (181), 181–8.

Hewitt, K., Clague, J. J. and Orwin, J. F. 2008. Legacies of catastrophic rock slope failures in mountain landscapes. *Earth Science Reviews* 87, 1–38.

Heyman, J. and Hättestrand, C. 2006. Morphology, distribution and formation of relict marginal moraines in the Swedish mountains. *Geografiska Annaler* 88A, 253–65.

Hicock, S. R. 1990. Genetic till prism. *Geology* 18, 517–19.

Hicock, S. R. 1991. On subglacial stone pavements in till. *Journal of Geology* 99, 607–19.

Hicock, S. R. 1992. Lobal interactions and rheologic superposition in subglacial till near Bradtville, Ontario, Canada. *Boreas* 21, 73–88.

Hicock, S. R. 1993. Glacial octahedron. *Geografiska Annaler* 75A, 35–9.

Hicock, S. R. and Dreimanis, A. 1992. Deformation till in the Great Lakes region: implications for rapid flow along the south-central margin of the Laurentide Ice Sheet. *Canadian Journal of Earth Sciences* 29, 1565–79.

Hicock, S. R. and Fuller, E. A. 1995. Lobal interactions, rheologic superposition, and implications for a Pleistocene ice stream on the continental shelf of British Columbia. *Geomorphology* 14, 167–84.

Hicock, S. R. and Lian, O. B. 1999. Cordilleran Ice Sheet lobal interactions and glaciotectonic superposition through stadial maxima along a mountain front in southwestern British Columbia, Canada. *Boreas* 28, 531–42.

Hicock, S. R., Goff, J. R., Lian, O. B. and Little, E. C. 1996. On the interpretation of subglacial till fabric. *Journal of Sedimentary Research* 66, 928–34.

Hiemstra, J. F. and Rijsdijk, K. F. 2003. Observing artificially induced strain: implications for subglacial deformation. *Journal of Quaternary Science* 18, 373–83.

Hiemstra, J. F. and van der Meer, J. J. M. 1997. Porewater controlled grain fracturing as an indicator for subglacial shearing in tills. *Journal of Glaciology* 43, 446–54.

Hiemstra, J. F., Evans, D. J. A. and Ó Cofaigh, C. 2007. The role of glacitectonic rafting and comminution in the production of subglacial tills: examples from southwest Ireland and Antarctica. *Boreas* 36, 386–99.

Hiemstra, J. F., Rijsdijk, K. F., Evans, D. J. A. and van der Meer, J. J. M. 2005. Integrated micro- and macro-scale analyses of Last Glacial Maximum Irish Sea Diamicts from Abermawr and Traeth y Mwnt, Wales, UK. *Boreas* 34, 61–74.

Hiemstra, J. F., Zaniewski, K., Powell, R. D. and Cowan, E. A. 2004. Strain signatures of fjord sediment sliding: micro-scale examples from Yakutat Bay and Glacier Bay, Alaska, USA. *Journal of Sedimentary Research* 74, 760–9.

Hildes, D. H. D., Clarke, G. K. C., Flowers, G. E. and Marshall, S. J. 2004. Subglacial erosion and englacial sediment transport modelled for North American ice sheets. *Quaternary Science Reviews* 23, 409–30.

Hillaire-Marcel, C. and Occhietti, S. 1980. Chronology, palaeogeography, and palaeoclimatic significance of the late and postglacial events in eastern Canada. *Zeitschrift für Geomorphologie* 24, 373–92.

Hillaire-Marcel, C., Occhietti, S. and Vincent, J.-S. 1981. Sakami moraine, Quebec: a 500 km long moraine without climatic control. *Geology* 9, 210–14.

Hinchcliffe, S. and Ballantyne, C. K. 1999. Talus accumulation and rockwall retreat, Trotternish, Isle of Skye, Scotland. *Scottish Geographical Journal* 115, 53–70.

Hindmarsh, R. C. A. 1996. Sliding of till over bedrock: scratching, polishing, comminution and kinematic wave theory. *Annals of Glaciology* 22, 41–8.

Hindmarsh, R. C. A. 1997. Deforming beds: viscous and plastic scales of deformation. *Quaternary Science Reviews* 16, 1039–56.

Hindmarsh, R. C. A. 1998a. The stability of a viscous till sheet coupled with ice flow, considered at wavelengths less than the ice thickness. *Journal of Glaciology* 44, 285–92.

Hindmarsh, R. C. A. 1998b. Drumlinization and drumlin-forming instabilities: viscous till mechanisms. *Journal of Glaciology* 44, 293–314.

Hindmarsh, R. C. A. 2004. A numerical comparison of approximations to the Stokes equations used in ice sheet and glacier modeling. *Journal of Geophysical Research* 109, F01012, doi:10.102912003JF000065.

Hindmarsh, R. C. A. 2006. The role of membrane-like stresses in determining the stability and sensitivity of the Antarctic ice sheets: back pressure and grounding line motion. *Philosophical Transactions of the Royal Society of London, A,* 364, 1733–67.

Hindmarsh, R. C. A. and LeMeur, E. 2001. Dynamical processes involved in the retreat of marine ice sheets. *Journal of Glaciology* 47 (157), 271–82.

Hindmarsh, R. C. A. and Rijsdijk, K. F. 2000. Use of a viscous model of till rheology to describe gravitational loading instabilities in glacial sediments. In Maltman, A. J., Hubbard, B. and Hambrey, M. J. (eds), *Deformation of Glacial Materials.* Geological Society, London, Special Publication 176, 191–201.

Hindmarsh, R. C. A. and Stokes, C. R. 2008. Formation mechanisms for ice-stream lateral shear margin moraines. *Earth Surface Processes and Landforms* 33, 610–26.

Hindmarsh, R. C. A., van der Wateren, F. M. and Verbers, A. L. L. M. 1998. Sublimation of ice through sediment in Beacon Valley, Antarctica. *Geografiska Annaler* 80A, 209–19.

Hirano, M. and Aniya, M. 1988. A rational explanation of cross-profile morphology for glacial valleys and of glacial valley development. *Earth Surface Processes and Landforms* 13, 707–16.

Hirano, M. and Aniya, M. 1989. A rational explanation of cross-profile morphology for glacial valleys and of glacial valley development: a further note. *Earth Surface Processes and Landforms* 14, 173–4.

Hirano, M. and Aniya, M. 1990. A reply to 'A discussion of Hirano and Aniya's (1988, 1989) explanation of glacial valley cross

profile development' by Jonathan M. Harbor. *Earth Surface Processes and Landforms* 15, 379–81.

Hirano, M. and Aniya, M. 2005. Response to Morgan's comment. *Earth Surface Processes and Landforms* 30, 514–15.

Hiscott, R. N. and Aksu, A. E. 1994. Submarine debris flows and continental slope evolution in front of Quaternary ice sheets, Baffin Bay, Canadian arctic. *Bulletin of the American Association of Petroleum Geology* 78, 445–60.

Hjort, C., Bentley, M. J. and Ingólfsson, Ó. 2001. Holocene and pre-Holocene temporary disappearance of the George VI Ice Shelf, Antarctic Peninsula. *Antarctic Science* 13, 296–301.

Hjulstrom, F. 1935. Studies of the morphological activity of rivers as illustrated by the River Fyris. *Bulletin of the Geological Institute, University of Uppsala* 25, 221–527.

Hock, R. 1999. A distributed temperature-index ice- and snowmelt model including potential direct solar radiation. *Journal of Glaciology* 45, 101–11.

Hock, R. 2003. Temperature index melt modeling in mountain areas. *Journal of Hydrology* 282, 104–15.

Hock, R. 2005. Glacier melt: a review of processes and their modelling. *Progress in Physical Geography* 29, 362–91.

Hock, R. and Hooke, R. le B. 1993. Evolution of the internal drainage system in the lower part of the ablation area of Storglaciären, Sweden. *Geological Society of America Bulletin* 105, 537–46.

Hodgkins, R. 1997. Glacier hydrology in Svalbard, Norwegian High Arctic. *Quaternary Science Reviews* 16, 957–73.

Hodgkins, R. and Dowdeswell, J. A. 1994. Tectonic processes in Svalbard tide-water glacier surges: evidence from structural geology. *Journal of Glaciology* 40, 553–60.

Hodgkins, R., Tranter, M. and Dowdeswell, J. A. 1998. The hydrochemical behaviour of runoff from a 'cold-based' glacier in the high Arctic (Scott Turnerbreen, Svalbard). *Hydrological Processes* 12, 87–104.

Hodgkins, R., Tranter, M. and Dowdeswell, J. A. 2004. The characteristics and formation of a high-Arctic proglacial icing. *Geografiska Annaler* 86A, 265–75.

Hodgson, D. A. 1985. The last glaciation of west-central Ellesmere Island, arctic archipelago, Canada. *Canadian Journal of Earth Sciences* 22, 347–68.

Hodgson, D. A. 1994. Episodic ice streams and ice shelves during retreat of the northwesternmost sector of the late Wisconsinan Laurentide Ice Sheet over the central Canadian Arctic Archipelago. *Boreas* 23, 14–28.

Hodgson, D. A. and Vincent, J.-S. 1984. A 10,000 yr BP extensive ice shelf over Viscount Melville Sound, Arctic Canada. *Quaternary Research* 22, 18–30.

Hodgson, G. J., Lever, J. H., Woodworth-Lynas, C. M. T. and Lewis, C. F. M. (eds). 1988. The dynamics of iceberg grounding and scouring (DIGS) experiment and repetitive mapping of the eastern Canadian continental shelf. *Environmental Studies Research Funds Report* 94.

Hodson, A. J., Anesio, A. M., Ng, F., Watson, R., Quirk, J., Irvine-Fynn, T., Dye, A., Clark, C., McCloy, P., Kohler, J. and Sattler, B. 2007. A glacier respires: quantifying the distribution and respiration CO_2 flux of cryoconite across an entire Arctic supraglacial ecosystem. *Journal of Geophysical Research* 112, G04S36, doi:10.1029/2007JG000452.

Hodson, A. J., Anesio, A. M., Tranter, M., Fountain, A., Osborn, M., Priscu, J., Laybourn-Parry, J. and Sattler, B. 2008. Glacial ecosystems. *Ecological Monographs* 78, 41–67.

Hodson, A. J., Tranter, M., Dowdeswell, J. A., Gurnell, A. M. and Hagen, J. O. 1997. Glacier thermal regime and suspended sediment yield: a comparison of two high arctic glaciers. *Annals of Glaciology* 24, 32–7.

Hodson, A. J., Tranter, M., Gurnell, A., Clark, M. and Hagen, J. O. 2002. The hydrochemistry of Bayelva, a high Arctic proglacial stream in Svalbard. *Journal of Hydrology* 257, 91–114.

Hodson, A. J., Tranter, M. and Vatne, G. 2000. Contemporary rates of chemical denudation and atmospheric CO_2 sequestration in glacier basins: an Arctic perspective. *Earth Surface Processes and Landforms* 25, 1447–71.

Hoey, T. B. 2004. The size of sedimentary particles. In Evans, D. J. A. and Benn, D. I. (eds), *A Practical Guide to the Study of Glacial Sediments*. Arnold, London, 52–77.

Hoffman, P. F. and Schrag, D. P. 2002. The snowball Earth hypothesis: testing the limits of global change. *Terra Nova* 14, 129–55.

Hoinkes, H. C. 1969. Surges of the Vernagtferner in the Ötztal Alps since 1599. *Canadian Journal of Earth Sciences* 6, 853–61.

Holland, D. M. 2002. Computing marine-ice thickness at an ice-shelf base. *Journal of Glaciology* 48, 9–19.

Holland, D. M., Thomas, R. H., de Young, B., Ribergaard, M. H. and Lyberth, B. 2008. Acceleration of Jakobshavn Isbrae triggered by warm subsurface ocean waters. *Nature Geoscience* 1, 659–64.

Holm, K., Bovis, M. and Jakob, M. 2004. The landslide response of alpine basins to post-Little Ice Age glacial thinning and retreat in southwestern British Columbia. *Geomorphology* 57, 201–16.

Holmes, J. A. 1993. Present and past patterns of glaciation in the northwest Himalaya: climatic, tectonic and topographic controls. In Shroder, J. F. (ed.), *Himalaya to the Sea*. Routledge, London, 72–90.

Holmlund, P. 1988. Internal geometry and evolution of moulins, Storglaciären, Sweden. *Journal of Glaciology* 34, 242–8.

Holmlund, P. and Hooke, R. Le B. 1983. High water-pressure events in moulins, Storglaciären, Sweden. *Geografiska Annaler* 65A, 19–25.

Holmlund, P. and Naslund, J. O. 1994. The glacially sculptured landscape in Dronning Maud Land, Antarctica, formed by wet-based mountain glaciation and not by the present ice sheet. *Boreas* 23, 139–48.

Holzhauser, H., Magny, M. J. and Zumbühl, H. J. 2005. Glacier and lake-level variations in west-central Europe over the last 3500 years. *The Holocene* 15, 789–801.

Hooke, R. Le B. 1973. Flow near the margin of the Barnes Ice Cap, and the development of ice-cored moraines. *Geological Society of America Bulletin* 84, 3929–48.

Hooke, R. Le B. 1981. Flow law for polycrystalline ice in glaciers: comparison of theoretical predictions, laboratory data, and field measurements. *Reviews of Geophysics* 19, 664–72.

Hooke, R. Le B. 1991. Positive feedbacks associated with erosion of glacial cirques and overdeepenings. *Geological Society of America Bulletin* 103, 1104–8.

Hooke, R. Le B. 2005. *Principles of Glacier Mechanics*. Cambridge University Press.

Hooke, R. Le B. and Elverhøi, E. 1996. Sediment flux from a fjord during glacial periods, Isfjorden, Spitsbergen. *Global and Planetary Change* 12, 237–49.

Hooke, R. Le B. and Fastook, J. 2007. Thermal conditions at the bed of the Laurentide Ice Sheet in Maine during deglaciation: implications for esker formation. *Journal of Glaciology* 53, 646–58.

Hooke, R. Le B. and Hudleston, P. J. 1978. Origin of foliation in glaciers. *Journal of Glaciology* 20, 285–99.

Hooke, R. Le B. and Iverson, N. R. 1995. Grain-size distribution in deforming subglacial tills: role of grain fracture: *Geology* 23, 57–60.

Hooke, R. Le B. and Jennings, C. E. 2006. On the formation of the tunnel valleys of the southern Laurentide ice sheet. *Quaternary Science Reviews* 25, 1364–72.

Hooke, R. Le B., Hanson, B., Iverson, N. R., Jansson, P. and Fischer, U. H. 1997. Rheology of till beneath Storglaciären, Sweden. *Journal of Glaciology* 43, 172–9.

Hooke, R. Le B., Holmlund, P. and Iverson, N. R. 1987. Extrusion flow demonstrated by borehole deformation measurements over a riegel, Storglaciären, Sweden. *Journal of Glaciology* 33, 72–8.

Hooke, R. Le B., Laumann, T. and Kohler, J. 1990. Subglacial water pressures and the shape of subglacial conduits. *Journal of Glaciology* 36, 67–71.

Hooyer, T. S. and Iverson, N. R. 2000. Clast fabric development in a shearing granular material: implications for subglacial till and fault gouge. *Geological Society of America Bulletin* 112, 683–92.

Hooyer, T. S., Iverson, N. R., Lagroix, F. and Thomason, J. F. 2008. Magnetic fabric of sheared till: a strain indicator for evaluating the bed deformation model of glacier flow. *Journal of Geophysical Research* 113, F02002, doi:10.1029/2007JF000757.

Hope, G. S., Peterson, J. A., Radok, U. and Allison, I. (eds). 1976. *The Equatorial Glaciers of New Guinea*. Balkema, Rotterdam.

Hoppe, G. and Schytt, V. 1953. Some observations on fluted moraine surfaces. *Geografiska Annaler* 35, 105–15.

Hopson, P. M. 1995. Chalk rafts in Anglian till in north Hertfordshire. *Proceedings of the Geologists Association* 106, 151–8.

Hornung, J. J., Asprion, U. and Winsemann, J. 2007. Jet-efflux deposits of a subaqueous ice-contact fan, glacial Lake Rinteln, northwestern Germany. *Sedimentary Geology* 193, 167–92.

Horton, P., Schaefli, B., Mezghani, A., Hingray, B. and Musy, A. 2006. Assessment of climate-change impacts on alpine discharge regimes with climate model uncertainty. *Hydrological Processes* 20, 2091–109.

Höskuldsson, A. and Sparks, R. S. J. 1997. Thermodynamics and fluid dynamics of effusive subglacial eruptions. *Bulletin of Vulcanology* 59, 219–30.

Houmark-Nielsen, M. 1988. Glaciotectonic unconformities in Pleistocene stratigraphy as evidence for the behaviour of former Scandinavian ice sheets. In Croot, D. G. (ed.), *Glaciotectonics: Forms and Processes*. Balkema, Rotterdam, 91–9.

Hovius, N., Lea-Cox, A. and Turowski, J. M. 2008. Recent volcano–ice interaction and outburst flooding in a Mars polar cap re-entrant. *Icarus* 197, 24–38.

Howat, I. M. and Domack, E. W. 2003. Reconstructions of western Ross Sea palaeo-ice stream grounding lines from high resolution acoustic stratigraphy. *Boreas* 32, 56–75.

Howat, I. M., Joughin, I., Tulaczyk, S. and Gogineni, S. 2005. Rapid retreat and acceleration of Helheim Glacier, east Greenland. *Geophysical Research Letters* 32, doi:10.1029/2005GL024737.

Howat, I. M., Joughin, I. and Scambos, T. A. 2007. Rapid changes in ice discharge from Greenland outlet glaciers. *Science* 315, 1559–61.

Howat, I. M., Joughin, I., Fahnestock, M., Smith, B. E. and Scambos, T. A. 2008. Synchronous retreat and acceleration of southeast Greenland outlet glaciers 2000–06: ice dynamics and coupling to climate. *Journal of Glaciology* 54 (187), 646–60.

Hubbard, A. 2000. The verification and significance of three approaches to longitudinal stresses in high-resolution models of glacier flow. *Geografiska Annaler* 82A, 471–87.

Hubbard, A. 2006. The validation and sensitivity of a model of the Icelandic ice sheet. *Quaternary Science Reviews* 25, 2297–313.

Hubbard, A., Blatter, H., Nienow, P., Mair, D. and Hubbard, B. 1998. Comparison of a three-dimensional model for glacier flow with field data from Haut Glacier d'Arolla, Switzerland. *Journal of Glaciology* 44, 368–78.

Hubbard, A., Hein, A. S., Kaplan, M. R., Hulton, N. R. J. and Glasser, N. F. 2005. A modelling reconstruction of the late glacial maximum ice sheet and its deglaciation in the vicinity of the Northern Patagonian Icefield, South America. *Geografiska Annaler* 87A, 375–91.

Hubbard, A., Sugden, D. E., Dugmore, A., Norddahl, H. and Pétursson, H. G. 2006. A modelling insight into the Icelandic Last Glacial Maximum ice sheet. *Quaternary Science Reviews* 25, 2283–96.

Hubbard, B. 1991. Freezing-rate effects on the physical characteristics of basal ice formed by net adfreezing. *Journal of Glaciology* 37, 339–47.

Hubbard, B. 2006. On the relationships between field data and numerical models of ice-mass motion. In Knight, P. (ed.), *Glacier Science and Environmental Change*. Blackwell, Oxford, 338–45.

Hubbard, B. and Glasser, N. 2005. *Field Techniques in Glaciology and Glacial Geomorphology*. Wiley, Chichester.

Hubbard, B. and Hubbard, A. 1998. Bedrock surface roughness and the distribution of subglacially precipitated carbonate deposits: implications for formation at Glacier de Tsanfleuron, Switzerland. *Earth Surface Processes and Landforms* 23, 261–70.

Hubbard, B. and Maltman, A. J. 2000. Laboratory investigations of the strength, static hydraulic conductivity and dynamic hydraulic conductivity of glacial sediments. In Maltman, A. J., Hambrey, M. J. and Hubbard, B. (eds), *Deformation of Glacial Materials*. Geological Society, London, Special Publication 176, 231–42.

Hubbard, B. and Nienow, P. 1997. Alpine subglacial hydrology. *Quaternary Science Reviews* 16, 939–55.

Hubbard, B. and Sharp, M. J. 1989. Basal ice formation and deformation: a review. *Progress in Physical Geography*, 13, 529–58.

Hubbard, B. and Sharp, M. J. 1993. Weertman regelation, multiple refreezing effects and the isotopic evolution of the basal ice layer. *Journal of Glaciology* 39, 275–91.

Hubbard, B. and Sharp, M. J. 1995. Basal ice facies and their formation in the Western Alps. *Arctic and Alpine Research* 27, 301–10.

Hubbard, B., Cook, S. and Coulson, H. 2009. Basal ice facies: a review and unifying approach. *Quaternary Science Reviews* 28, 1956–69.

Hubbard, B., Hubbard, A., Mader, H. M., Ison, J.-L, Grust, K. and Nienow, P. 2003. Spatial variability in the water content and rheology of temperate glaciers: Glacier de Tsanfleuron, Switzerland. *Annals of Glaciology* 37, 1–6.

Hubbard, B., Sharp, M. J., Willis, I. C., Neilsen, M. K. and Smart, C. C. 1995. Borehole water-level variations and the structure of the subglacial hydrological system of Haut Glacier d'Arolla, Valais, Switzerland. *Journal of Glaciology* 41 (139), 572–83.

Hubbard, T. D. and Reid, J. R. 2006. Analysis of flute forming conditions using ice sheet reconstructions and field techniques. *Geomorphology* 74, 137–51.

Huber, C., Leuenberger, M., Spahni, R., Flückigera, J., Schwander, J., Stocker, T. F., Johnsen, S., Landais, A. and Jouzel, J. 2006. Isotope calibrated Greenland temperature record over Marine Isotope Stage 3 and its relation to CH_4. *Earth and Planetary Science Letters* 243, 504–19.

Huddart, D. 1981. Fluvioglacial systems in Edenside (middle Eden Valley and Brampton kame belt). In Boardman, J. (ed.), *Eastern Cumbria – Field Guide*. Quaternary Research Association, London, 81–103.

Huddart, D. 1994. Rock type controls on downstream changes in clast parameters in sandur systems in southeast Iceland. *Journal of Sedimentary Research* A64, 215–25.

Huddart, D. 1999. Supraglacial trough fills, southern Scotland: origins and implications for deglacial processes. *Glacial Geology and Geomorphology* (http://ggg.qub.ac.uk/ggg/papers/full/1999/rp041999/rp04.html).

Huddart, D. and Bennett, M. R. 1997. The Carstairs Kames (Lanarkshire, Scotland): morphology, sedimentology and formation. *Journal of Quaternary Science* 12, 467–84.

Huddart, D., Bennett, M. R. and Glasser, N. F. 1999. Morphology and sedimentology of a high-arctic esker system: Vegbreen, Svalbard. *Boreas* 28, 253–73.

Hudleston, P. J. 1976. Recumbent folding in the base of the Barnes Ice Cap, Baffin Island, Northwest Territories, Canada. *Geological Society of America Bulletin* 87, 1684–92.

Hughes, A. L. C. 2008. *The Last British Ice Sheet: A Reconstruction Based on Glacial Landforms*. Unpublished PhD thesis, University of Sheffield.

Hughes, P. D. and Woodward, J. C. 2008. Timing of glaciation in the Mediterranean mountains during the last cold stage. *Journal of Quaternary Science* 23, 575–88.

Hughes, T. J. 1986. The marine ice transgression hypothesis. *Geografiska Annaler* 69A, 237–50.

Hughes, T. J. 1998. *Ice Sheets*. Oxford University Press.

Hughes, T. J. 2002. Calving bays. *Quaternary Science Reviews* 21, 267–82.

Hulbe, C. L. and Fahnestock, M. 2004. West Antarctic ice-stream discharge variability: mechanism, controls and pattern of grounding-line retreat. *Journal of Glaciology* 50 (171), 471–84.

Hulbe, C. L. and Fahnestock, M. 2007. Century-scale discharge stagnation and reactivation of the Ross ice streams, Antarctica. *Journal of Geophysical Research* 112, doi:10.1029/2006JF000603.

Hulbe, C. L., MacAyeal, D. R., Denton, G. H., Kleman, J. and Lowell, T. V. 2004. Catastrophic ice shelf breakup as the source of Heinrich event icebergs. *Palaeoceanography* 19, doi:10.1029/2003PA000890.

Hulbe, C. L., Wang, W., Joughin, I. R. and Siegert, M. J. 2003. The role of lateral and vertical shear in tributary flow toward a West Antarctic ice stream. *Annals of Glaciology* 36, 244–50.

Hulton, N. R. J. and Mineter, M. J. 2000. Modelling self-organization on ice streams, *Annals of Glaciology* 30, 127–36.

Hulton, N. R. J., Purves, R. S., McCulloch, R. D., Sugden, D. E. and Bentley, M. J. 2002. The last glacial maximum and deglaciation in southern South America. *Quaternary Science Reviews* 21, 233–41.

Humlum, O. 1982. Rock glacier types on Disko, central west Greenland. *Norsk Geografisk Tidsskrift* 82, 59–66.

Humlum, O. 1985a. Changes in texture and fabric of particles in glacial traction with distance from source, Myrdalsjökull, Iceland. *Journal of Glaciology* 31, 150–6.

Humlum, O. 1985b. Genesis of an imbricate push moraine, Höfdabrekkujökull, Iceland. *Journal of Geology* 93, 185–95.

Humlum, O. 1998. The climatic significance of rock glaciers. *Permafrost and Periglacial Processes* 9, 375–95.

Humlum, O. 2005. Holocene permafrost aggradation in Svalbard. In Harris, C. and Murton, J. B. (eds), *Cryospheric Systems: Glaciers and Permafrost*. Geological Society, London, Special Publication 242, 119–30.

Humlum, O., Elberling, B., Hormes, A., Fjordheim, K., Hansen, O. H. and Heinemeier, J. 2005. Late-Holocene glacier growth in Svalbard, documented by subglacial relict vegetation and living soil microbes. *The Holocene* 15, 396–407.

Humphrey, N. F. and Raymond, C. F. 1994. Hydrology, erosion and sediment production in a surging glacier: Variegated Glacier, Alaska, 1982–83. *Journal of Glaciology* 40, 539–52.

Humphrey, N. F., Kamb, B., Fahnestock, M. and Engelhardt, H. 1993. Characteristics of the bed of the lower Columbia Glacier, Alaska. *Journal of Geophysical Research* 98, 837–46.

Humphrey, N. F., Raymond, C. F. and Harrison, N. 1986. Discharges of turbid water during mini-surges of Variegated Glacier, Alaska, USA. *Journal of Glaciology* 32, 195–207.

Hunter, L. E. and Powell, R. D. 1998. Ice foot development at temperate tidewater margins in Alaska. *Geophysical Research Letters* 25, 1923–6.

Hunter, L. E., Powell, R. D. and Lawson, D. E. 1996. Morainal-bank sediment budgets and their influence on the stability of tidewater termini of valley glaciers entering Glacier Bay, Alaska, USA. *Annals of Glaciology* 22, 211–16.

Huss, M., Bauder, A., Werder, M., Funk, M. and Hock, R. 2007. Glacier-dammed lake outburst events of Gornersee, Switzerland. *Journal of Glaciology* 53, 189–200.

Hutter, K. 1983. *Theoretical Glaciology: Material Science of Ice and the Mechanics of Glaciers and Ice Sheets*. Reidel, Tokyo.

Huuse, M. and Lykke-Andersen, H. 2000a. Large-scale glaciotectonic thrust structures in the eastern Danish North Sea. In Maltman, A. J., Hubbard, B. and Hambrey, M. J. (eds), *Deformation of Glacial Materials*. Geological Society, London, Special Publication 176, 293–305.

Huuse, M. and Lykke-Andersen, H. 2000b. Overdeepened Quaternary valleys in the eastern Danish North Sea: morphology and origin. *Quaternary Science Reviews* 19, 1233–53.

Huybrechts, P. 1990. A 3-D model for the Antarctic Ice Sheet: a sensitivity study on the glacial–interglacial contrast. *Climate Dynamics* 5, 79–92.

Huybrechts, P. and de Wolde, J. 1999. The dynamic response of the Greenland and Antarctic ice sheets to multiple-century climatic warming. *Journal of Climate* 12, 2169–88.

Huybrechts, P., Payne, A. J. and EISMINT Intercomparison Group. 1996. The EISMINT benchmarks for testing ice-sheet models. *Annals of Glaciology* 23, 1–12.

Huybrechts, P., Gregory, J., Janssens, I. and Wild, M. 2004. Modelling Antarctic and Greenland volume changes during the 20th and 21st centuries by GCM time slice integrations. *Global and Planetary Change* 42, 83–105.

Iken, A. 1977. Movement of a large ice mass before breaking off. *Journal of Glaciology* 19, 595–605.

Iken, A. and Bindschadler, R. A. 1986. Combined measurements of subglacial water pressure and surface velocity of Findelengletscher, Switzerland: conclusions about drainage system and sliding mechanism. *Journal of Glaciology* 32, 101–19.

Iken, A. and Truffer, M. 1997. The relationship between subglacial water pressure and surface velocity of Findelengletscher, Switzerland, during its advance and retreat. *Journal of Glaciology* 43 (144), 328–38.

Iken, A., Röthlisberger, H., Flotron, A. and Haeberli, W. 1983. The uplift of Unteraargletscher at the beginning of the melt season – a consequence of water storage at the bed? *Journal of Glaciology* 23, 28–47.

Ildefonse, B. and Mancktelow, N. S. 1993. Deformation around rigid particles: the influence of slip at the particle/matrix interface. *Tectonophysics* 221, 345–59.

Ildefonse, B., Launeau, P., Bouchez, J. L. and Fernandez, A. 1992. Effects of mechanical interactions on the development of preferred orientations: a two-dimensional experimental approach. *Journal of Structural Geology* 14, 73–83.

Imbrie, J., Berger, A., Boyle, E. A., Clemens, S. C., Duffy, A., Howard, W. R., Kukla, G., Kutzbach, J., Martinson, D. G., McIntyre, A., Mix, A. C., Molfino, B., Morley, J. J., Peterson, L. C., Pisias, N. G., Prell, W. L., Raymo, M. E., Shackleton, N. J. and Toggweiler, J. R. 1993. On the structure and origin of major glaciation cycles 2. The 100,000 year cycle. *Palaeoceanography* 8, 699–735.

Imbrie, J., Boyle, E. A., Clemens, S. C., Duffy, A., Howard, W. R., Kukla, G., Kutzbach, J., Martinson, D. G., McIntyre, A., Mix, A. C., Molfino, B., Morley, J. J., Peterson, L. C., Pisias, N. G., Prell, W. L., Raymo, M. E., Shackleton, N. J. and Toggweiler, J. R. 1992. On the structure and origin of major glaciation cycles 1. Linear responses to Milankovitch forcing. *Palaeoceanography* 7, 701–38.

Ingólfsson, Ó., Norðdahl, H. and Schomacker, A. 2009. Deglaciation and Holocene glacial history of Iceland. In Schomacker, A., Krüger, J. and Kjaer, K. (eds), *The Mýrdalsjökull Ice Cap, Iceland: Glacial Processes, Sediments and Landforms on an Active Volcano*. Elsevier, Amsterdam and Oxford, 51–68.

Inoue, J. and Yoshida, M. 1980. Ablation and heat exchange over the Khumbu Glacier. *Seppyo* 42, 26–33.

IPCC. 2007. *Climate Change 2007: The Physical Science Basis*. Working Group I Contribution to the Fourth Assessment Report of the Intergovernmental Panel on Climate Change. Cambridge University Press.

Iqbal, M. 1983. *An Introduction to Solar Radiation*. Academic Press, Orlando, FL.

Irvine-Fynn, T. D. L., Moorman, B. J., Willis, I. C., Sjogren, D. B., Hodson, A. J., Mumford, P. N., Walter, F. S. A. and Williams, J. L. M. 2005. Geocryological processes linked to High Arctic proglacial stream suspended sediment dynamics: examples

from Bylot Island, Nunavut, and Spitsbergen, Svalbard. *Hydrological Processes* 19, 115–35.

Isenko, E., Naruse, R. and Mavlyudov, B. 2005. Water temperature in englacial and supraglacial channels: change along the flow and contribution to ice melting on the channel wall. *Cold Regions Science and Technology* 42, 53–62.

Iturrizaga, L. 2005. New observations on present and prehistoric glacier-dammed lakes in the Shimshal valley (Karakoram Mountains). *Journal of Asian Earth Sciences* 25, 545–55.

Iturrizaga, L. 2008. Paraglacial landform assemblages in the Hindukush and Karakoram Mountains. *Geomorphology* 95, 27–47.

Iverson, N. R. 1990. Laboratory simulations of glacial abrasion: comparison with theory. *Journal of Glaciology* 36, 304–14.

Iverson, N. R. 1991a. Potential effects of subglacial water pressure fluctuations on quarrying. *Journal of Glaciology* 37, 27–36.

Iverson, N. R. 1991b. Morphology of glacial striae: implications for abrasion of glacier beds and fault surfaces. *Geological Society of America Bulletin* 103, 1308–16.

Iverson, N. R. 1993. Regelation of ice through debris at glacier beds: implications for sediment transport. *Geology* 21, 559–62.

Iverson, N. R. 1995. Processes of erosion. In Menzies, J. (ed.), *Modern Glacial Environments: Processes, Dynamics and Sediments*. Butterworth-Heinemann, Oxford, 241–60.

Iverson, N. R. 1999. Coupling between a glacier and a soft bed. II: Model results. *Journal of Glaciology* 45, 41–53.

Iverson, N. R. 2000. Regelation of ice through debris at glacier beds: implications for sediment transport. *Geology* 21, 559–62.

Iverson, N. R. and Hooyer, T. S. 2002. Clast fabric development in a shearing granular material: implications for subglacial till and fault gouge – reply. *Geological Society of America Bulletin* 114, 383–4.

Iverson, N. R. and Iverson, R. M. 2001. Distributed shear of subglacial till due to Coulomb slip. *Journal of Glaciology* 47, 481–8.

Iverson, N. R. and Semmens, D. 1995. Intrusion of ice into porous media by regelation: a mechanism of sediment entrainment by glaciers. *Journal of Geophysical Research* 100, 10219–30.

Iverson, N. R., Baker, R. W. and Hooyer, T. S. 1998. Ring-shear studies of till deformation: Coulomb-plastic behavior and distributed strain in glacier beds. *Journal of Glaciology* 44, 634–42.

Iverson, N. R., Cohen, D., Hooyer, T. S., Fischer, U. H., Jackson, M., Moore, P. L., Lappegard, G. and Kohler, J. 2003. Effects of basal debris on glacier flow. *Science* 301, 81–4.

Iverson, N. R., Hanson, B., Hooke, R. Le B. and Jansson, P. 1995. Flow mechanism of glaciers on soft beds. *Science* 267, 80–1.

Iverson, N. R., Hooyer, T. S. and Hooke, R. Le B. 1996. A laboratory study of sediment deformation: stress heterogeneity and grain-size evolution. *Annals of Glaciology* 22, 167–75.

Iverson, N. R., Hooyer, T. S., Fischer, U. H., Cohen, D., Moore, P. L., Jackson, M., Lappegard, G. and Kohler, J. 2007. Soft-bed experiments beneath Engabreen, Norway: regelation infiltration, basal slip and bed deformation. *Journal of Glaciology* 53, 323–40.

Iverson, N. R., Hooyer, T. S., Thomason, J. F., Graesch, M. and Shumway, J. R. 2008. The experimental basis for interpreting particle and magnetic fabrics of sheared till. *Earth Surface Processes and Landforms* 33, 627–45.

Iverson, N. R., Jansson, P. and Hooke, R. Le B. 1994. In-situ measurement of the strength of deforming subglacial till. *Journal of Glaciology* 40, 497–503.

Ivins, E. R. and Wolf, D. 2008. Glacial isostatic adjustment: new developments from advanced observing systems and modeling. *Journal of Geodynamics* 46, 69–77.

Ivy-Ochs, S., Kerschner, H., Reuther, A., Maisch, M., Sailer, R., Schaefer, J., Kubik, P. W., Synal, H.-A. and Schlüchter, C. 2006. The timing of glacier advances in the northern European Alps based on surface exposure dating with cosmogenic [10]Be, [26]Al, [36]Cl and [21]Ne. In Siame, L. L., Bourles, D. L. and Brown, E. T. (eds), *In Situ-Produced Cosmogenic Nuclides and Quantification of Geological Processes*. Geological Society of America, Special Paper 415, 43–60.

Ivy-Ochs, S., Kerschner, H., Reuther, A., Preusser, F., Heine, K., Maisch, M., Kubik, P. W. and Schlüchter, C. 2008. Chronology of the last glacial cycle in the European Alps. *Journal of Quaternary Science* 23, 559–73.

Iwata, S., Aoki, T., Kadota, T., Seko, K. and Yamaguchi, S. 2000. Morphological evolution of the debris cover on Khumbu Glacier, Nepal, between 1978 and 1995. In Nakawo, M., Raymond, C. F. and Fontain, A. (eds), *Debris Covered Glaciers*. IAHS Publication 264, 3–11.

Jacobel, R. W., Welch, B. C., Steig, E. J. and Schneider, D. P. 2005. Glaciological and climatic significance of Hercules Dome, Antarctica: an optimal site for deep ice core drilling. *Journal of Geophysical Research* 110, doi:10.1029/2004JF000188.

Jacobson, A. D., Blum, J. D., Chamberlain, C. P., Poage, M. A. and Sloan, V. F. 2002. Ca/Sr and Sr isotope systematics of a Himalayan glacial chronosequence: carbonate versus silicate weathering rates as a function of landscape surface age. *Geochimica et Cosmochimica Acta* 66, 13–27.

Jaedicke, C. and Sandvik, A. D. 2002. High resolution snow distribution data from complex Arctic terrain: a tool for model validation. *Natural Hazards and Earth System Sciences* 2, 147–55.

Jaeger, J. M. and Nittrouer, C. A. 1999. Sediment deposition in an Alaskan fjord: controls on the formation and preservation of sedimentary structures in Icy Bay. *Journal of Sedimentary Research* 69, 1011–26.

Jahns, R. H. 1943. Sheet structure in granites: its origin and use as a measure of glacial erosion in New England. *Journal of Geology* 51, 71–98.

Jakobsson, M. 1999. First high-resolution chirp sonar profiles from the central Arctic Ocean reveal erosion of Lomonosov Ridge sediments. *Marine Geology* 158, 11–123.

Jakobsson, M., Björck, S., Alm, G., Andren, T., Lindeberg, G. and Svensson, N. O. 2007. Reconstructing the Younger Dryas ice dammed lake in the Baltic Basin: bathymetry, area and volume. *Global and Planetary Change* 57, 355–70.

Jakobsson, M., Gardner, J. V., Vogy, P., Mayer, L. A., Armstrong, A., Backman, J., Brennan, R., Calder, B., Hall, J. K. and Kraft, B. 2005. Multibeam bathymetric and sediment profiler evidence for ice grounding on the Chukchi Borderland, Arctic Ocean. *Quaternary Research* 63, 150–60.

Jakobsson, M., Polyak, L., Edwards, M., Kleman, J. and Coakley, B. 2008. Glacial geomorphology of the central Arctic Ocean: the Chukchi Borderland and the Lomonosov Ridge. *Earth Surface Processes and Landforms* 33, 526–45.

James, L. A. 1996. Polynomial and power functions for glacial valley cross-section morphology. *Earth Surface Processes and Landforms* 21, 413–32.

Jamieson, S. S. R., Hulton, N. R. J. and Hagdorn, M. 2008. Modelling landscape evolution under ice sheets. *Geomorphology* 97, 91–108.

Jamieson, S. S. R., Hulton, N. R. J., Sugden, D. E., Payne, A. J. and Taylor, J. 2005. Cenozoic landscape evolution of the Lambert basin, East Antarctica: the relative role of rivers and ice sheets. *Global and Planetary Change* 45, 35–49.

Jania, J. 1996. *Zrozumiec lodowce*. PWN, Warsaw.

Janssen, M., Zuidema, J. and Wanhill, R. J. H. 2004. *Fracture Mechanics*, 2nd edn. Taylor and Francis, London.

Jansson, K. N. 2003. Early Holocene glacial lakes and ice marginal retreat pattern in Labrador/Ungava, Canada. *Palaeogeography, Palaeoclimatology, Palaeoecology* 193, 473–501.

Jansson, K. N. 2005. Map of the glacial geomorphology of north-central Quebec-Labrador, Canada. *Journal of Maps* 2005, 46–55.

Jansson, K. N. and Glasser, N. F. 2005a. Palaeoglaciology of the Welsh sector of the British-Irish Ice Sheet. *Journal of the Geological Society of London* 162, 25–37.

Jansson, K. N. and Glasser, N. F. 2005b. Using Landsat 7 ETM+ imagery and digital terrain models for mapping glacial lineaments on former ice sheet beds. *International Journal of Remote Sensing* 26, 3931–41.

Jansson, K. N. and Glasser, N. F. 2008. Modification of peripheral mountain ranges by former ice sheets: the Brecon Beacons, southern UK. *Geomorphology* 97, 178–89.

Jansson, K. N. and Kleman, J. 1999. The horned crag-and-tails of the Ungava Bay landform swarm, Quebec-Labrador, Canada. *Annals of Glaciology* 28, 168–74.

Jansson, K. N., Kleman, J. and Marchant, D. R. 2002. The succession of ice-flow patterns in north-central Quebec-Labrador, Canada. *Quaternary Science Reviews* 21 (2002), 503–23.

Jansson, K. N., Stroeven, A. P. and Kleman, J. 2003. Configuration and timing of Ungava Bay ice streams, Labrador-Ungava, Canada. *Boreas* 32, 256–62.

Jansson, P., Hock, R. and Schneider, T. 2003. The concept of glacier storage: a review. *Journal of Hydrology* 282, 116–29.

Jansson, P., Naslund, J.-O., Pettersson, R., Richardson-Naslund, C. and Holmlund, P. 2000. Debris entrainment and polythermal structure in the terminus of Storglaciaren. In Nakawo, M., Raymond, C. F. and Fountain, A. (eds), *Debris-Covered Glaciers*. IAHS Publication 264, 143–51.

Jansson, P., Richardson, C. and Jonsson, S. 1999. Assessment of requirements for cirque formation in northern Sweden. *Annals of Glaciology* 28, 16–22.

Jarman, D. 2002. Rock slope failure and landscape evolution in the Caledonian Mountains, as exemplified in the Abisko area, northern Sweden. *Geografiska Annaler* 84A, 213–24.

Jarman, D. 2003. Paraglacial landscape evolution: the significance of rock slope failure. In Evans, D. J. A. (ed.), *The Quaternary of the Western Highland Boundary – Field Guide*. QRA, London, 50–68.

Jarman, D. 2006. Large rock slope failures in the Highlands of Scotland: characterisation, causes and spatial distribution. *Engineering Geology* 83, 161–82.

Jarman, D. 2007. Introduction to the mass movements in the older mountain areas of Great Britain. In Cooper, R. G. (ed.), *Mass Movements in Great Britain*. Geological Conservation Review Series 33, Peterborough, 33–56.

Jarman, D. and Ballantyne, C. K. 2002. Beinn Fhada, Kintail: an example of large scale paraglacial rock slope deformation. *Scottish Geographical Journal* 118, 59–68.

Jeffery, G. B. 1922. The motion of ellipsoidal particles immersed in a viscous fluid. *Proceedings of the Royal Society of London* 102, 169–79.

Jeffries, M. O. 2002. Ellesmere Island ice shelves and ice islands. In Williams, R. S. and Ferrigno, J. G. (eds), *Satellite Image Atlas of Glaciers of the World: North America*. USGS Professional Paper 1386-J, 147–62.

Jenkins, A., Vaughan, D. G., Jacobs, S. S., Hellmer, H. H. and Keys, J. R. 1997. Glaciological and oceanographic evidence of high melt rates beneath Pine Island Glacier, West Antarctica. *Journal of Glaciology* 43, 114–21.

Jennings, C. E. 2006. Terrestrial ice streams: a view from the lobe. *Geomorphology* 75, 100–24.

Jezek, K. C. and Liu, H. X. 2005. Structure of southeastern Antarctic Peninsula ice shelves and ice tongues from synthetic aperture radar imagery. *Journal of Glaciology* 51 (174), 373–6.

Jibson, R. W., Harp, E. L., Schulz, W. and Keefer, D. K. 2006. Large rock avalanches triggered by the M7.9 Denali Fault, Alaska, earthquake of 3 November 2002. *Engineering Geology* 83, 144–60.

Jiskoot, H., Murray, T. and Boyle, P. 2000. Controls on the distribution of surge-type glaciers in Svalbard. *Journal of Glaciology* 46 (154), 412–22.

Jiskoot, H., Murray, T. and Luckman, A. 2003. Surge potential and drainage basin characteristics in East Greenland. *Annals of Glaciology* 36, 142–8.

Jobard, S. and Dzikowski, M. 2006. Evolution of glacial flow and drainage during the ablation season. *Journal of Hydrology* 330, 663–71.

Jóhannesson, T., Raymond, C. and Waddington, E. 1989a. Timescale for adjustment of glaciers to changes in mass balance. *Journal of Glaciology* 35, 355–69.

Jóhannesson, T., Raymond, C. and Waddington, E. 1989b. A simple method for determining the response time of glaciers. In Oerlemans, J. (ed.), *Glacier Fluctuations and Climatic Change*. Kluwer, Dordrecht, 343–52.

Johansson, M., Migon, P. and Olvmo, M. 2001a. Development of joint-controlled rock basins in Bohus granite, SW Sweden. *Geomorphology* 40, 145–61.

Johansson, M., Olvmo, M. and Lidmar-Bergström, K. 2001b. Inherited landforms and glacial impact of different palaeosurfaces in southwest Sweden. *Geografiska Annaler* 83A, 67–89.

Johansson, P. 1994. The subglacially engorged eskers in the Lutto river basin, northeastern Finnish Lapland. In Warren, W. P. and Croot, D. G. (eds), *Formation and Deformation of Glacial Deposits*. Balkema, Rotterdam, 89–94.

Johnsen, S. J., Dahl-Jensen, D., Gundestrup, N., Steffensen, J. P., Clausen, H. B., Miller, H., Masson-Delmotte, V., Sveinbjörnsdottir, A. E. and White, J. 2001. Oxygen isotope and palaeotemperature records from six Greenland ice-core stations: Camp Century, Dye-3, GRIP, GISP2, Renland and NorthGRIP. *Journal of Quaternary Science* 16, 299–307.

Johnsen, T. F. and Brennand, T. A. 2006. The environment in and around ice-dammed lakes in the moderately high relief setting of the southern Canadian Cordillera. *Boreas* 35, 106–25.

Johnson, A. M. and Rodine, J. R. 1984. Debris flow. In Brunsden, D. and Prior, D. B. (eds), *Slope Instability*. Wiley, Chichester, 257–361.

Johnson, J. V., Prescott, P. R. and Hughes, T. J. 2004. Ice dynamics preceding catastrophic disintegration of the floating part of Jakobshavn Isbrae, Greenland. *Journal of Glaciology* 50 (171), 492–504.

Johnson, M. D. and Clayton, L. 2003. Supraglacial landsystems in lowland terrain. In Evans, D. J. A. (ed.), *Glacial Landsystems*. Arnold, London, 228–58.

Johnson, M. D. and Gillam, M. L. 1995. Composition and construction of late Pleistocene end moraines, Durango, Colorado. *Geological Society of America Bulletin* 107, 1241–53.

Johnson, M. D., Mickelson, D. M., Clayton, L. and Attig, J. W. 1995. Composition and genesis of glacial hummocks, western Wisconsin, USA. *Boreas* 24, 97–116.

Johnson, P. G. 1997. Spatial and temporal variability of ice-dammed lake sediments in alpine environments. *Quaternary Science Reviews* 16, 635–47.

Johnson, W. H. and Hansel, A. K. 1990. Multiple Wisconsinan glacigenic sequences at Wedron, Illinois. *Journal of Sedimentary Petrology* 60, 26–41.

Johnson, W. H. and Menzies, J. 1996. Pleistocene supraglacial and ice-marginal deposits and landforms. In Menzies, J. (ed.), *Past Glacial Environments. Sediments, Forms and Techniques*. Butterworth-Heinemann, Oxford, 137–60.

Jones, H. G. 1999. The ecology of snow-covered ecosystems: a brief overview of nutrient cycling and life in the cold. *Hydrological Processes* 13, 2135–47.

Jones, I. W., Munhoven, G., Tranter, M., Huybrechts, P. and Sharp, M. J. 2002. Modelled glacial and non-glacial HCO_3-, Si and Ge fluxes since the LGM: little potential for impact on atmospheric CO_2 concentrations and a potential proxy of continental chemical erosion, the marine Ge/Si ratio. *Global and Planetary Change* 33, 139–53.

Jopling, A. V. and Walker, R. G. 1968. Morphology and origin of ripple-drift cross-lamination, with examples from the Pleistocene of Massachusetts. *Journal of Sedimentary Petrology* 38, 971–84.

Jordan, R. E. and Stark, J. A. 2001. *Capillary Tension in Rotting Ice Laters*. Technical Report ERDC/CRREL TR-01-13. US Army Corps of Engineers, Cold Regions Research and Engineering Laboratory.

Jørgensen, F. and Piotrowski, J. A. 2003. Signature of the Baltic Ice Stream on Funen Island, Denmark during the Weichselian glaciation. *Boreas* 32, 242–55.

Jørgensen, F. and Sandersen, P. B. E. 2006. Buried and open tunnel valleys in Denmark – erosion beneath multiple ice sheets. *Quaternary Science Reviews* 25, 1339–63.

Josenhans, H. W. and Fader, G. B. J. 1989. A comparison of models of glacial sedimentation along the eastern Canadian margin. *Marine Geology* 85, 273–300.

Josenhans, H. W. and Zevenhuizen, J. 1990. Dynamics of the Laurentide Ice Sheet in Hudson Bay, Canada. *Marine Geology* 92, 1–26.

Joughin, I. and MacAyeal, D. R. 2005. Calving of large tabular icebergs from ice shelf rift systems. *Geophysical Research Letters* 32, L02501, doi:1029/2004GL020978.

Joughin, I. and Tulaczyk, S. 2002. Positive mass balance of the Ross Ice Streams, West Antarctica. *Science* 295, 476–9.

Joughin, I., Abdalati, W. and Fahnestock, M. 2004a. Large fluctuations in speed on Greenland's Jakobshavn Isbrae Glacier. *Nature* 432, 608–10.

Joughin, I., Bindschadler, R. A., King, M. A., Voight, D., Alley, R. B., Anandakrishnan, S., Horgan, H., Peters, L., Winberry, P., Das, S. B. and Catania, G. 2005. Continued deceleration of Whillans Ice Stream, West Antarctica. *Geophysical Research Letters* 32, doi:10.1029/2005GL024319.

Joughin, I., Das, S. B., King, M. A., Smith, B. E., Howat, I. M. and Moon, T. 2008a. Seasonal speedup along the western flank of the Greenland ice sheet. *Science* 320, 781–3.

Joughin, I., Fahnestock, M., MacAyeal, D., Bamber, J. and Gogineni, P. 2001. Observation and analysis of ice flow in the largest Greenland ice stream. *Journal of Geophysical Research* 106 (D24), 34021–34.

Joughin, I., Howat, I. M., Fahnestock, M., Smith, B., Krabill, W., Alley, R. B., Stern, H. and Truffer, M. 2008b. Continued evolution of Jakobshavn Isbrae following its rapid speedup. *Journal of Geophysical Research* 113, doi:10.1029/2008JF001023.

Joughin, I., Howat, I., Alley, R. B., Ekstrom, G., Fahnestock, M., Moon, T., Nettles, M., Truffer, M. and Tsai, V. C. 2008c. Ice-front variation and tidewater behavior on Helheim and Kangerdlugssuaq Glaciers, Greenland. *Journal of Geophysical Research* 113, F01004, doi:10.1029/2007JF000837.

Joughin, I., MacAyeal, D. R. and Tulaczyk, S. 2004b. Basal shear stress of the Ross ice streams from control method inversions. *Journal of Geophysical Research* 109, doi:10.1029/2003JB002960.

Joughin, I., Rignot, E., Rosanova, C. E., Lucchitta, B. K. and Bohlander, J. 2003. Timing of recent accelerations of Pine Island Glacier, Antarctica. *Geophysical Research Letters* 30, doi:10.1029/2003GL017609.

Joughin, I., Tulaczyk, S., Bindschadler, R. and Price, S. F. 2002. Changes in west Antarctic ice stream velocities: observation and analysis. *Journal of Geophysical Research* 107, doi:10.1029/2001JB001029.

Joughin, I., Tulaczyk, S., MacAyeal, D. R. and Engelhardt, H. 2004c. Melting and freezing beneath the Ross ice streams, Antarctica. *Journal of Glaciology* 50 (168), 96–108.

Jouzel, J., Alley, R. B., Cuffey, K. M., Dansgaard, W., Grootes, P., Hoffman, G., Johnsen, S. J., Koster, R. D., Peel, D., Shuman, C. A., Stievenard, M., Stuiver, M. and White, J. 1997. Validity of the temperature reconstruction from water isotopes in ice cores. *Journal of Geophysical Research* 102, 26471–87.

Jouzel, J., Hoffmann, G., Koster, R. D. and Masson, V. 2000. Water isotopes in precipitation: data/model comparison for present-day and past climates. *Quaternary Science Reviews* 19, 363–79.

Jouzel, J. et al. 2007. Orbital and millennial Antarctic climate variability over the past 800,000 years. *Science* 317, 793–6.

Juen, I., Kaser, G. and Georges, C. 2007. Modelling observed and future runoff from a glacierized tropical catchment (Cordillera Blanca, Peru). *Global and Planetary Change* 50, 37–48.

Kääb, A. 2005. Combination of SRTM3 and repeat ASTER data for deriving alpine glacier flow velocities in the Bhutan Himalaya. *Remote Sensing of Environment* 94, 463–74.

Kääb, A., Lefauconnier, B. and Melvold, K. 2006. Flow field of Kronebreen, Svalbard, using repeated Landsat 7 and ASTER data. *Annals of Glaciology* 42, 7–13.

Kadish, S. J., Head, J. W., Parsons, R. L. and Marchant, D. R. 2008. The Ascraeus Mons fan-shaped deposit: volcano–ice interactions and the climatic implications of cold-based tropical mountain glaciation. *Icarus* 197, 84–109.

Kadota, T., Fujita, K., Seko, K., Kayastha, R. B. and Ageta, Y. 1997. Monitoring and prediction of shrinkage of a small glacier in the Nepal Himalaya. *Annals of Glaciology* 24, 90–4.

Kamb, B. 1970. Sliding motion of glaciers: theory and observation. *Reviews of Geophysics and Space Physics* 8, 673–728.

Kamb, B. 1987. Glacier surge mechanism based on linked cavity configuration of the basal water conduit system. *Journal of Geophysical Research* 92, 9083–100.

Kamb, B. 1991. Rheological nonlinearity and flow instability in the deforming bed mechanism of Ice Stream motion. *Journal of Geophysical Research* 96, 585–95.

Kamb, B. 2001. Basal zone of the West Antarctic Ice Streams and its role in lubrication of their rapid motion. In Alley, R. B. and Bindschadler, R. A. (eds), *The West Antarctic Ice Sheet: Behavior and Environment*. American Geophysical Union, Antarctic Research Series, vol. 77, 157–99.

Kamb, B. and Engelhardt, H. 1987. Waves of accelerated motion in a glacier approaching surge: the mini surges of Variegated Glacier, Alaska, USA. *Journal of Glaciology* 33, 27–46.

Kamb, B. and LaChapelle, E. 1964. Direct observation of the mechanism of glacier sliding over bedrock. *Journal of Glaciology* 5, 159–72.

Kamb, B., Engelhardt, H., Fahnestock, M. A., Humphrey, N., Meier, M. and Stone, D. 1994. Mechanical and hydrologic basis for the rapid motion of a large tidewater glacier 2. Interpretation. *Journal of Geophysical Research* 99, 15231–44.

Kamb, B., Raymond, C. F., Harrison, W. D., Engelhardt, H., Echelmeyer, K. A., Humphrey, N., Brugman, M. M. and Pfeffer, T. 1985. Glacier surge mechanism: 1982–1983 surge of Variegated Glacier, Alaska. *Science* 227, 469–79.

Kaplan, M. R. 1999. Retreat of a tidewater margin of the Laurentide Ice Sheet in eastern coastal Maine between ca. 14,000 and 13,000 14C yr BP. *Geological Society of America Bulletin* 111, 620–32.

Kaplan, M. R., Hein, A. S., Hubbard, A. and Lax, S. M. 2009. Can glacial erosion limit the extent of glaciation? *Geomorphology* 103, 172–9.

Kargel, J. S. and Strom, R. G. 1990. Ancient glaciation on Mars. *Lunar and Planetary Science* 21, 598–9.

Kargel, J. S. and Strom, R. G. 1991. Terrestrial glacial eskers: analogs for Martian sinuous ridges. *Lunar and Planetary Science* 21, 598–9.

Kargel, J. S. and Strom. R. G. 1992. Ancient glaciation on Mars. *Geology* 20, 3–7.

Kargel, J. S., Baker, V. R., Begét, J. E., Lockwood, J. F., Péwé, T. L., Shaw, J. S. and Strom, R. G. 1995. Evidence of ancient continental glaciation in the Martian northern plains. *Journal of Geophysical Research* 100 (E3), 5351–68.

Kargel, J. S., Strom, R. G. and Johnson, N. 1991. Glacial geology of the Hellas region on Mars. *Lunar and Planetary Science* 22, 687–8.

Karrow, P. F. 1981. Till texture in drumlins. *Journal of Glaciology* 27, 497–502.

Kaser, G. and Osmaston, H. 2002. *Tropical Glaciers*. Cambridge University Press.

Kaser, G., Fountain, A. and Jansson, P. 2003a. *A Manual for Monitoring the Mass Balance of Mountain Glaciers*. International Hydrological Program – Technical Developments in Hydrology 59. UNESCO, Paris.

Kaser, G., Juen, I., Georgesa, C., Gómezb, J. and Tamayob, W. 2003b. The impact of glaciers on the runoff and the reconstruction of mass balance history from hydrological data in the tropical Cordillera Blanca, Perú. *Journal of Hydrology* 282, 130–44.

Kattelmann, R. 2003. Glacial lake outburst floods in the Nepal Himalaya: a manageable hazard? *Natural Hazards* 28, 145–54.

Kaufman, D. S., Porter, S. C. and Gillespie, A. R. 2004. Quaternary alpine glaciation in Alaska, the Pacific Northwest, Sierra Nevada and Hawaii. In Gillespie, A. R., Porter, S. C. and Atwater, B. F. (eds), *The Quaternary Period in the United States*. Elsevier, Amsterdam, 77–104.

Kavanaugh, J. L. and Clarke, G. K. C. 2001. Abrupt glacier motion and reorganization of basal shear stress following the establishment of a connected drainage system. *Journal of Glaciology* 47, 472–80.

Kavanaugh, J. L. and Clarke, G. K. C. 2006. Discrimination of the flow law for subglacial sediment using in situ measurements and an interpretation model. *Journal of Geophysical Research* 111, doi:10.1029/2005JF000346.

Kayastha, R. B., Ageta, Y. and Nakawo, M. 2000a. Positive degree-day factors for ablation on glaciers in the Nepalese Himalayas: case study on Glacier AX010 in Shorong Himal, Nepal. *Bulletin of Glaciological Research* 17, 1–10.

Kayastha, R. B., Ohata, T. and Ageta, Y. 1999. Application of a mass-balance model to a Himalayan glacier. *Journal of Glaciology* 45, 559–67.

Kayastha, R. B., Takeuchi, Y., Nakawo, M. and Ageta, Y. 2000b. Practical prediction of ice melting beneath various thickness of debris cover on Khumbu Glacier, Nepal, using a positive degree-day factor. In Nakawo, M., Raymond, C. F. and Fountain, A. (eds), *Debris-Covered Glaciers*. International Association of Hydrological Sciences Publication 264, 71–81.

Kaye, C. A. 1980. Geologic profile of Gay Head Cliff, Martha's Vineyard, Massachusetts. US Geological Survey, Open File Report, 80–148.

Kehew, A. E. and Kozlowski, A. L. 2007. Tunnel channels of the Saginaw Lobe, Michigan, USA. In Johansson, P. and Sarala, P. (eds), *Applied Quaternary Research in the Central Part of Glaciated Terrain*. Geological Survey of Finland, Special Paper 46, 69–78.

Kehew, A. E. and Lord, M. L. 1986. Origin and large-scale erosional features of glacial-lake spillways in the northern Great Plains. *Geological Society of America Bulletin* 97, 162–77.

Kehew, A. E. and Lord, M. L. 1987. Glacial lake outbursts along the mid-continent margins of the Laurentide ice-sheet. In Mayer, L. and Nash, D. (eds), *Catastrophic Flooding*. Allen and Unwin, Boston, 95–120.

Kehew, A. E., Beukema, S. P., Bird, B. C. and Kozlowski, A. L. 2005. Fast flow of the Lake Michigan Lobe: evidence from sediment-landform assemblages in southwestern Michigan, USA. *Quaternary Science Reviews* 24, 2335–53.

Kehew, A. E., Nicks, L. P. and Straw, W. T. 1999. Palimpsest tunnel valleys: evidence for relative timing of advances in an interlobate area of the Laurentide Ice Sheet. *Annals of Glaciology* 28, 47–52.

Kekonen, T., Moore, J. C., Mulvaney, R., Isaksson, E., Pohjola, V. and van De Wal, R. S. W. 2002. A 800 year record of nitrate from the Lomonosovfonna ice core, Svalbard. *Annals of Glaciology* 35, 261–5.

Kellerer-Pirklbauer, A. 2008. The supraglacial debris system at Pasterze Glacier, Austria: spatial distribution, characteristics and transport of debris. *Zeitschrift für Geomorphologie*, Suppl. Bd 52, 3–25.

Kelly, M. A., Denton, G. H. and Hall, B. L. 2002. Late Cenozoic paleoenvironment in southern Victoria Land, Antarctica, based on a polar glaciolacustrine deposit in western Victoria Valley. *Geological Society of America Bulletin* 114, 605–18.

Kemmis, T. J. 1981. Importance of the regelation process to certain properties of basal tills deposited by the Laurentide ice sheet in Iowa and Illinois, USA. *Annals of Glaciology* 2, 147–52.

Kemmis, T. J. and Hallberg, T. J. 1984. Lithofacies types and vertical profile models: an alternative approach to the description and environmental interpretation of glacial diamict and diamictite sequences. Discussion. *Sedimentology* 31, 886–90.

Kenneally, J. P. and Hughes, T. J. 2006. Calving giant icebergs: old principles, new applications. *Antarctic Science* 18, 409–19.

Kennedy, M. G. 2003. *Advance Rates of Columbia Glacier During the Last 1000 years, Prince William Sound, Alaska.* Unpublished MS thesis, College of Wooster, Ohio.

Kennett, J. P. 1977. Cenozoic evolution of Antarctic glaciation, the circum-Antarctic ocean, and their impact on global paleoceanography. *Journal of Geophysical Research* 82, 3843–59.

Kershaw, J. A., Clague, J. J. and Evans, S. G. 2005. Geomorphic and sedimentological signature of a two-phase outburst flood from moraine-dammed Queen Bess Lake, British Columbia, Canada. *Earth Surface Processes and Landforms* 30, 1–25.

Kessler, M. A., Anderson, R. S. and Briner, J. P. 2008. Fjord insertion into continental margins driven by topographic steering of ice. *Nature Geoscience* 1, 365–9.

Keys, J. E. 1978. *Water Regime of Disraeli Fiord, Ellesmere Island, Canada.* Department of National Defence, Defence Research Establishment, Report 792.

Khatwa, A. and Tulaczyk, S. 2001. Microstructural interpretations of modern and Pleistocene subglacially deformed sediments: the relative role of parent material and subglacial process. *Journal of Quaternary Science* 16, 507–17.

Khatwa, A., Hart, J. K. and Payne, A. J. 1999. Grain textural analyses across a range of glacial facies. *Annals of Glaciology* 28, 111–17.

King, E. C., Hindmarsh, R. C. A. and Stokes, C. R. 2009. Formation of mega-scale glacial lineations observed between a West Antarctic ice stream. *Nature Geoscience* 2, 585–96.

King, E. C., Woodward, J. and Smith, A. M. 2004. Seismic evidence for a water-filled canal in deforming till beneath Rutford Ice Stream, West Antarctica. *Geophysical Research Letters* 31, L20401, doi:10.1029/2004GL020379.

King, E. C., Woodward, J. and Smith, A. M. 2007. Seismic and radar observations of subglacial bed forms beneath the onset zone of Rutford Ice Stream, Antarctica. *Journal of Glaciology* 53 (183), 665–72.

King, E. L., Haflidason, H., Sejrup, H. P. and Løvlie, R. 1998. Glacigenic debris flows on the North Sea trough-mouth fan during ice stream maxima. *Marine Geology* 152, 217–46.

King, L. H., Rokoengen, K., Fader, G. B. J. and Gunleiksrud, T. 1991. Till tongue stratigraphy. *Geological Society of America Bulletin* 103, 637–59.

Kirkbride, M. P. 1993. The temporal significance of transitions from melting to calving termini at glaciers in the central Southern Alps of New Zealand. *The Holocene* 3, 232–40.

Kirkbride, M. P. 1995a. Processes of glacial transportation. In Menzies, J. (ed.), *Modern and past glacial environments.* Butterworth-Heinemann, Oxford, 147–69.

Kirkbride, M. P. 1995b. Ice flow vectors on the debris-mantled Tasman Glacier, 1957–1986. *Geografiska Annaler* 77A, 147–57.

Kirkbride, M. P. 2000. Ice marginal geomorphology and Holocene expansion of debris-covered Tasman Glacier, New Zealand. In Nakawo, M., Raymond, C. and Fountain, A. (eds), *Debris-Covered Glaciers.* IAHS Publication 264, 211–17.

Kirkbride, M. P. and Matthews, D. 1997. The role of fluvial and glacial erosion in landscape evolution: the Ben Ohau Range, New Zealand. *Earth Surface Processes and Landforms* 22, 317–27.

Kirkbride, M. P. and Spedding, N. F. 1996. The influence of englacial drainage on sediment-transport pathways and till texture in temperate valley glaciers. *Annals of Glaciology* 22, 160–6.

Kirkbride, M. P. and Sugden, D. E. 1992. New Zealand loses its top. *Geographical Magazine* July, 30–4.

Kirkbride, M. P. and Warren, C. R. 1997. Calving processes at a grounded ice cliff. *Annals of Glaciology* 24, 116–21.

Kirkbride, M. P. and Warren, C. R. 1999. Tasman Glacier, New Zealand: 20th century thinning and predicted calving retreat. *Global and Planetary Change* 22, 11–28.

Kjær, K. H. and Krüger, J. 1998. Does clast size influence fabric strength? *Journal of Sedimentary Research* 68, 746–9.

Kjær, K. H. and Krüger, J. 2001. The final phase of dead ice moraine development: processes and sediment architecture, Kotlujökull, Iceland. *Sedimentology* 48, 935–52.

Kjær, K. H., Houmark-Nielsen, M. and Richardt, N. 2003a. Ice flow patterns and dispersal of erratics at the southwestern margin of the last Scandinavian Ice Sheet: signature of palaeo-ice streams. *Boreas* 32, 130–48.

Kjær, K. H., Krüger, J. and van der Meer, J. J. M. 2003b. What causes till thickness to change over distance? Answers from Myrdalsjökull, Iceland. *Quaternary Science Reviews* 22, 1687–700.

Kjær, K. H., Larsen, E., van der Meer, J. J. M., Ingólfsson, Ó., Krüger, J., Benediktsson, Í. Ö., Knudsen, C. G. and Schomacker, A. 2006. Subglacial decoupling at the sediment/bedrock interface: a new mechanism for rapid flowing ice. *Quaternary Science Reviews* 25, 2704–12.

Kjær, K. H., Sultan, L., Krüger, J. and Schomacker, A. 2004. Architecture and sedimentation of outwash fans in front of the Myrdalsjökull ice cap, Iceland. *Sedimentary Geology* 172, 139–63.

Klassen, R. A. and Thompson, F. J. 1993. *Glacial History, Drift Composition, and Mineral Exploration, Central Labrador.* Geological Survey of Canada, Bulletin 435.

Klein, A. G. and Isacks, B. L. 1999. Spectral mixture analysis of Landsat Thematic Mapper images applied to the detection of the transient snowline on tropical Andean glaciers. *Global and Planetary Change* 22, 139–54.

Klein, A. G., Selzer, G. O. and Isacks, B. L. 1999. Modern and last local glacial maximum snowlines in the Central Andes of Peru, Bolivia and Northern Chile. *Quaternary Science Reviews* 18, 63–84.

Kleman, J. 1988. Linear till ridges in the southern Norwegian-Swedish mountains: evidence for a subglacial origin. *Geografiska Annaler* 70A, 35–45.

Kleman, J. 1990. On the use of glacial striae for reconstruction of paleo-ice sheet flow patterns, with application to the Scandinavian ice sheet. *Geografiska Annaler* 72A, 217–36.

Kleman, J. 1992. The palimpsest glacial landscape in northwestern Sweden: Late Weichselian deglaciation landforms and traces of older west-centred ice sheets. *Geografiska Annaler* 74A, 305–25.

Kleman, J. 1994. Preservation of landforms under ice sheets and ice caps. *Geomorphology* 9, 19–32.

Kleman, J. and Borgström, I. 1994. Glacial landforms indicative of a partly frozen bed. *Journal of Glaciology* 40, 255–64.

Kleman, J. and Borgström, I. 1996. Reconstruction of palaeo-ice sheets: the use of geomorphological data. *Earth Surface Processes and Landforms* 21, 893–900.

Kleman, J. and Glasser, N. F. 2007. The subglacial thermal organization (STO) of ice sheets. *Quaternary Science Reviews* 26, 585–97.

Kleman, J. and Hättestrand, C. 1999. Frozen-bed Fennoscandian and Laurentide ice sheets during the last glacial maximum. *Nature* 402, 63–6.

Kleman, J., Borgström, I. and Hättestrand, C. 1994. Evidence for a relict glacial landscape in Quebec-Labrador. *Palaeogeography, Palaeoclimatology, Palaeoecology* 111, 217–28.

Kleman, J., Fastook, J. and Stroeven, A. P. 2002. Geologically and geomorphologically constrained numerical model of Laurentide Ice Sheet inception and build up. *Quaternary International* 95/96, 87–98.

Kleman, J., Hättestrand, C., Borgström, I. and Stroeven, A. P. 1997. Fennoscandian palaeoglaciology reconstructed using a glacial geological inversion model. *Journal of Glaciology* 43, 283–99.

Kleman, J., Hättestrand, C. and Clarhäll, A. 1999. Zooming in on frozen-bed patches: scale-dependent controls on Fennoscandian ice sheet basal thermal zonation. *Annals of Glaciology* 28, 189–94.

Kleman, J., Hättestrand, C., Stroeven, A. P., Jansson, K. N., De Angelis, H. and Borgström, I. 2006. Reconstruction of paleo-ice sheets – inversion of their glacial geomorphological record. In Knight, P. G. (ed.), *Glacier Science and Environmental Change*. Blackwell, Oxford, 192–8.

Kleman, J., Marchant, D. and Borgström, I. 2001. Geomorphic evidence for Late Glacial ice dynamics on southern Baffin Island and in outer Hudson Strait, Nunavut, Canada. *Arctic, Antarctic and Alpine Research* 33, 249–57.

Kleman, J., Stroeven, A. P. and Lundqvist, J. 2008. Patterns of Quaternary ice sheet erosion and deposition in Fennoscandia and a theoretical framework for explanation. *Geomorphology* 97, 73–90.

Klemann, V. and Wolf, D. 2005. The eustatic reduction of shoreline diagrams: implications for the inference of relaxation-rate spectra and the viscosity stratification below Fennoscandia. *Geophysical Journal International* 162, 249–56.

Kneller, B. and Buckee, C. 2000. The structure and fluid mechanics of turbidity currents: a review of some recent studies and their geological implications. *Sedimentology* 47, Suppl. 1, 62–94.

Knight, J. 2008. The environmental significance of ventifacts: a critical review. *Earth-Science Reviews* 86, 89–105.

Knight, J. and McCabe, A. M. 1997. Identification and significance of ice flow-transverse subglacial ridges (Rogen moraines) in north central Ireland. *Journal of Quaternary Science* 12, 519–24.

Knight, J., McCarron, S. G. and McCabe, A. M. 1999. Landform modification by palaeo-ice streams in east central Ireland. *Annals of Glaciology* 28, 161–7.

Knight, P. G. 1989. Stacking of basal debris layers without bulk freezing on: isotopic evidence from West Greenland. *Journal of Glaciology* 35, 214–16.

Knight, P. G. 1994. Two-facies interpretation of the basal ice layer of the Greenland ice sheet contributes to a unified model of basal ice formation. *Geology* 22, 971–4.

Knight, P. G. 1997. The basal ice layer of glaciers and ice sheets. *Quaternary Science Reviews* 16, 975–93.

Knighton, A. D. 1972. Meandering habit of supraglacial streams. *Geological Society of America Bulletin* 83, 201–4.

Knighton, A. D. 1998. *Fluvial Forms and Processes: A New Perspective*. Arnold, London.

Knudsen, Ó. 1995. Concertina eskers, Bruarjökull, Iceland: an indicator of surge-type glacier behaviour. *Quaternary Science Reviews* 14, 487–93.

Knutz, P. C., Austin, W. E. N. and Jones, E. J. W. 2001. Millennial scale depositional cycles related to British ice sheet variability and North Atlantic palaeocirculation since 45 kyr BP, Barra Fan, UK margin. *Paleoceanography* 16, 53–64.

Knutz, P. C., Jones, E. J. W., Austin, W. E. N. and van Weering, T. C. E. 2002. Glacimarine slope sedimentation, contourite drifts and bottom current pathways on the Barra Fan, UK North Atlantic margin. *Marine Geology* 188, 129–46.

Koerner, R. M. 1970. The mass balance of the Devon Island ice cap, NWT, Canada, 1961–66. *Journal of Glaciology* 9, 325–36.

Koerner, R. M. 1977. Ice thickness measurements and their implications with respect to past and present ice volumes in the Canadian high arctic ice caps. *Canadian Journal of Earth Sciences* 14, 2697–705.

Koerner, R. M. and Taniguchi, H. 1976. Artificial radioactivity layers in the Devon Island Ice Cap, NWT. *Canadian Journal of Earth Sciences* 13, 1251–5.

Kohler, J., James, T. D., Murray, T., Nuth, C., Brandt, O., Barrand, N. E., Aas, H. F. and Luckman, A. 2007. Acceleration in thinning rate on western Svalbard glaciers. *Geophysical Research Letters* 34, L18502, doi:10.1029/2007GL030681.

Köhler, P. and Fischer, H. 2006. Simulating low frequency changes in atmospheric CO_2 during the last 740 000 years. *Climate of the Past* 2, 57–78.

Komatsu, G. and Baker, V. R. 1996. Channels in the Solar System. *Planetary and Space Science* 44, 801–15.

Konya, K., Matsumoto, T. and Naruse, R. 2004. Surface heat balance and spatially distributed ablation modelling at Koryto Glacier, Kamchatka Peninsula, Russia. *Geografiska Annaler* 86A, 337–48.

Konzelmann, T., van de Wal, R. S. W., Greuell, W., Bintanja, R., Henneken, E. A. C. and Abe-Ouchi, A. 1994. Parameterization of global and longwave incoming radiation for the Greenland Ice Sheet. *Global and Planetary Change* 9, 143–64.

Kopp, R. E., Kirschvink, J. L., Hilburn, I. A. and Nash, C. Z. 2005. The Paleoproterozoic snowball Earth: a climate disaster triggered by the evolution of oxygenic photosynthesis. *Proceedings of the National Academy of Sciences* 102, 11131–6.

Kor, P. S. G., Shaw, J. and Sharpe, D. R. 1991. Erosion of bedrock by subglacial meltwater, Georgian Bay, Ontario: a regional view. *Canadian Journal of Earth Sciences* 28, 623–42.

Korup, O., Clague, J. J., Hermanns, R. L., Hewitt, K., Strom, A. L. and Weidinger, J. T. 2007. Giant landslides, topography, and erosion. *Earth and Planetary Science Letters* 261, 578–89.

Kotlyakov, V. M., Arkhipov, S. M., Henderson, K. A. and Nagornov, O. V. 2004. Deep drilling of glaciers in Eurasian Arctic as a source of palaeoclimatic records. *Quaternary Science Reviews* 23, 1371–90.

Kovanen, D. J. and Slaymaker, O. 2004. Glacial imprints of the Okanagon Lobe, southern margin of the Cordilleran Ice Sheet. *Journal of Quaternary Science* 19, 547–65.

Kowalewski, D. E., Marchant, D. R., Levy, J. S. and Head, J. W. 2006. Quantifying low rates of summertime sublimation for buried glacier ice in Beacon Valley, Antarctica. *Antarctic Science* 18, 421–8.

Kozlowski, A. L., Kehew, A. E. and Bird, B. C. 2005. Outburst flood origin of the Central Kalamazoo River Valley, Michigan, USA. *Quaternary Science Reviews* 24, 2354–74.

Kraal, E. R., van Dijk, M., Postma, G. and Kleinhans, M. G. 2008. Martian stepped-delta formation by rapid water release. *Nature* 451, 973–6.

Krabill, W. B., Abdalati, W., Frederick, E. B., Manizade, S. S., Martin, C. F., Sonntag, J. C., Swift, R. N., Thomas, R. H., Wright, W. and Yungel, J. K. 2000. Greenland Ice Sheet: high elevation balance and peripheral thinning. *Science* 289, 428–30.

Krabill, W. B., Abdalati, W., Frederick, E. B., Manizade, S. S., Martin, C. F., Sonntagg, J. C., Swift, R. N., Thomas, R. H., Yungel, J. K., 2002. Aircraft laser altimetry measurements of changes of the Greenland ice sheet: technique and accuracy assessment. *Journal of Geodynamics* 34, 357–76.

Krabill, W. B., Frederick, E. B., Manizade, S. S., Martin, C. F., Sonntag, J. C., Swift, R. N., Thomas, R. H., Wright, W. and Yungel, J. K. 1999. Rapid thinning of parts of the southern Greenland Ice Sheet. *Science* 283, 1522–4.

Krabill, W. B. et al. 2004. Greenland ice sheet: increased coastal thinning. *Geophysical Research Letters* 31, doi:10.1029/2004GL021592.

Kreslavsky, M. A. and Head, J. W. 2002. Mars: nature and evolution of young latitude-dependent water-rich-ice mantle. *Geophysical Research Letters* 29, doi:10.1029/2002GL015392.

Krimmel, R. M. 1999. Analysis of difference between direct and geodetic mass balance measurements at South Cascade Glacier, Washington. *Geografiska Annaler* 81A, 653–8.

Krimmel, R. M. 2001. *Photogrammetric Data Set, 1957–2000, and Bathymetric Measurements for Columbia Glacier, Alaska.* USGS Water-Resources Investigations Report 01-4089. U.S. Geological Survey, Washington, DC.

Krimmel, R. M. and Trabant, D. C. 1992. The terminus of Hubbard Glacier, Alaska. *Annals of Glaciology* 16, 151–7.

Krimmel, R. M. and Vaughn, B. H. 1987. Columbia Glacier, Alaska: changes in velocity 1977–1986. *Journal of Geophysical Research* 92 (B9), 8961–8.

Krinsley, D. H. and Doornkamp, J. C. 1973. *Atlas of Quartz Sand Surface Textures.* Cambridge University Press.

Kristensen, L., Benn, D. I., Hormes, A. and Ottesen, D. 2009. Mud aprons in front of Svalbard surge moraines: evidence of subglacial deforming layers or proglacial tectonics? *Geomorphology* 111, 206–21.

Kristensen, T. B., Huuse, M., Piotrowski, J. A. and Clausen, O. R. 2007. A morphometric analysis of tunnel valleys in the eastern North Sea based on 3D seismic data. *Journal of Quaternary Science* 22, 801–15.

Kristensen, T. B., Piotrowski, J. A., Huuse, M., Clausen, O. R. and Hamberg, L. 2008. Time-transgressive tunnel valley formation indicated by infill sediment structure, North Sea – the role of glaciohydraulic supercooling. *Earth Surface Processes and Landforms* 33, 546–59.

Krüger, J. 1984. Clasts with stoss-lee form in lodgement tills: a discussion. *Journal of Glaciology* 30, 241–3.

Krüger, J. 1993. Moraine-ridge formation along a stationary ice front in Iceland. *Boreas* 22, 101–9.

Krüger, J. 1994. Glacial processes, sediments, landforms, and stratigraphy in the terminus region of Myrdalsjökull, Iceland. *Folia Geographica Danica* 21, 1–233.

Krüger, J. 1997. Development of minor outwash fans at Kötlujökull, Iceland. *Quaternary Science Reviews* 16, 649–59.

Krüger, J. and Aber, J. S. 1999. Formation of supraglacial sediment accumulations on Kötlujökull, Iceland. *Journal of Glaciology* 45, 400–2.

Krüger, J. and Kjær, K. H. 1999. A data chart for field description and genetic interpretation of glacial diamicts and associated sediments – with examples from Greenland, Iceland and Denmark. *Boreas* 28, 386–402.

Krüger, J. and Kjær, K. H. 2000. De-icing progression of ice-cored moraines in a humid, subpolar climate, Kötlujökull, Iceland. *The Holocene* 10, 737–47.

Krüger, J. and Thomsen, H. H. 1984. Morphology, stratigraphy and genesis of small drumlins in front of the glacier Myrdalsjökull, south Iceland. *Journal of Glaciology* 30, 94–105.

Krzyszkowski, D. 2002. Sedimentary successions in ice-marginal fans of the Late Saalian glaciation, southwestern Poland. *Sedimentary Geology* 149, 93–109.

Krzyszkowski, D. and Zielinski, T. 2002. The Pleistocene end moraine fans: controls on their sedimentation and location. *Sedimentary Geology* 149, 73–92.

Kuhle, M., 2007. The past valley glacier network in the Himalayas and the Tibetan ice sheet during the last glacial period and its glacial-isostatic, eustatic and climatic consequences. *Tectonophysics* 445, 116–44.

Kuhn, M. 1984. Mass budget imbalances as criterion for a climatic classification of glaciers. *Geografiska Annaler* 66A, 229–38.

Kuhn, M. 2001. The nutrient cycle through snow and ice, a review. *Aquatic Sciences* 63, 150–67.

Kuhn, M. 2003. Redistribution of snow and glacier mass balance from a hydrometeorological model. *Journal of Hydrology* 282, 95–103.

Kulessa, B., Hubbard, B., Williamson, M. and Brown, G. H. 2005. Hydrogeological analysis of slug tests in boreholes. *Journal of Glaciology* 51, 269–80.

Kulkarni, A. V. 1992. Mass balance of Himalayan glaciers using AAR and ELA methods. *Journal of Glaciology* 38, 101–4.

Kujansuu, R. and Saarnisto, M. 1990. (eds) *Handbook of Glacial Indicator Tracing.* Balkema, Rotterdam.

Kuriger, E. M., Truffer, M., Motyka, R. J. and Bucki, A. K. 2006. Episodic reactivation of large-scale push moraine in front of the advancing Taku Glacier, Alaska. *Journal of Geophysical Research* 111, F01009, doi:10.1029/2005JF000385.

Laberg, J. S. and Vorren, T. O. 1995. Late Weichselian submarine debris flow deposits on the Bear Island trough-mouth fan. *Marine Geology* 127, 45–72.

Laberg, J. S. and Vorren, T. O. 1996. The middle and late Pleistocene evolution of the Bear Island trough-mouth fan. *Global and Planetary Change* 12, 309–30.

Laberg, J. S. and Vorren, T. O. 2000. Flow behaviour of the submarine glacigenic debris flows on the Bear Island trough-mouth fan, western Barents Sea. *Sedimentology* 47, 1105–17.

Laberg, J. S., Vorren, T. O. and Knutsen, S. M. 1999. The Lofoten contourite drift off Norway. *Marine Geology* 159, 1–6.

Labeyrie, L. D., Duplessy, J. C. and Blanc, P. L. 1987. Variations in mode of formation and temperature of oceanic deep waters over the past 125,000 years. *Nature* 327, 477–82.

Lacelle, D., Lauriol, B., Clark, I. D., Cardyn, R. and Zdanowicz, C. 2007. Nature and origin of a Pleistocene-age massive ground-ice body exposed in the Chapman Lake moraine complex, central Yukon Territory, Canada. *Quaternary Research* 68, 249–60.

Lachniet, M. S. and Vazquez-Selem, L. 2005. Last glacial maximum equilibrium line altitudes in the circum-caribbean (Mexico, Guatemala, Costa Rica, Colombia, and Venezuela). *Quaternary International* 138, 129–44.

Lachniet, M. S., Larson, G. J., Lawson, D. E., Evenson, E. B. and Alley, R. B. 2001. Microstructures of sediment flow deposits and subglacial sediments. *Boreas* 30, 254–62.

Lafferty, B., Quinn, R. and Breen, C. 2006. Subglacial imprints associated with the isolation and decay of an ice mass in the lower Lough Erne basin, Co. Fermanagh, NW Ireland. *Journal of the Geological Society of London* 163, 421–30.

Lafrenière, M. and Sharp, M. 2003. Wavelet analysis of inter-annual variability in the runoff regimes of glacial and nival stream catchments, Bow Lake, Alberta. *Hydrological Processes* 17, 1093–118.

Lambeck, K. 1990. Glacial rebound, sea level change and mantle viscosity. *Quarterly Journal of the Royal Astronomical Society* 31, 1–30.

Lambeck, K. 1993a. Glacial rebound and sea level change: an example of a relationship between mantle and surface processes. *Tectonophysics* 223, 15–37.

Lambeck, K. 1993b. Glacial rebound of the British Isles. I: Preliminary model results. *Geophysical Journal International* 115, 941–59.

Lambeck, K. 1993c. Glacial rebound of the British Isles. II: A high resolution, high precision model. *Geophysical Journal International* 115, 960–90.

Lambeck, K. 1995. Late Devensian and Holocene shorelines of the British Isles and North Sea from models of glacio-hydro-isostatic rebound. *Journal of the Geological Society of London* 152, 437–48.

Lambeck, K. 1996. Glaciation and sea level change for Ireland and the Irish Sea since late Devensian/Midlandian time. *Journal of the Geological Society of London* 153, 853–72.

Lambeck, K. 2004. Sea-level change through the last glacial cycle: geophysical, glaciological and palaeogeographic consequences. *Comptes Redus Geoscience* 336, 677–89.

Lambeck, K. and Chappell, J. 2001. Sea level change through the last glacial cycle. *Science* 292, 679–86.

Lambeck, K., Anzidei, M., Antonioli, F., Benini, A. and Esposito, A. 2004. Sea level in Roman time in the Central Mediterranean and implications for recent change. *Earth and Planetary Science Letters* 224, 563–75.

Lambeck, K., Purcell, A., Johnston, P., Nakada, M. and Yokoyama, Y. 2003. Water-load definition in the glacio-hydro-isostatic sea-level equation. *Quaternary Science Reviews* 22, 309–18.

Lambeck, K., Yokoyama, Y. and Purcell, T. 2002. Into and out of the Last Glacial Maximum: sea-level change during Oxygen Isotope Stages 3 and 2. *Quaternary Science Reviews* 21, 343–60.

Lamoureux, S. F. 2001. Varve chronology techniques. In Last, W. M. and Smol, J. P. (eds), *Developments in Paleoenvironmental Research. Volume 2: Tracking Environmental Change Using Lake Sediments: Physical and Chemical Techniques.* Kluwer, Dordrecht, 247–60.

Lamoureux, S. F. and Gilbert, R. 2004a. Long term variability in autumn snowfall and temperature, and winter storminess recorded in varved sediments, Bear Lake, Devon Island, Nunavut. *Quaternary Research* 61, 134–47.

Lamoureux, S. F. and Gilbert, R. 2004b. Physical and chemical properties and proxies of high latitude lake sediments. In Pienitz, R., Douglas, M. S. V. and Smol, J. P. (eds), *Developments in Paleoenvironmental Research. Volume 8: Long-term Environmental Change in Arctic and Antarctic Lakes.* Springer, Dordrecht, 53–87.

Lamoureux, S. F., Stewart, K. A., Forbes, A. C. and Fortin, D. 2006. Multidecadal variations and decline in spring discharge in the Canadian middle Arctic since 1550 AD. *Geophysical Research Letters* 33, L02403, doi:10.1029/2005GL024942.

Landvik, J. Y., Bondevik, S., Elverhøi, A., Fjeldskaar, W., Mangerud, J., Salvigsen, O., Siegert, M. J., Svendsen J.-I. and Vorren, T. O. 1998. The last glacial maximum of Svalbard and the Barents sea area: ice sheet extent and configuration. *Quaternary Science Reviews* 17, 43–75.

Landvik, J. Y., Ingólfsson, Ó., Mienert, J., Lehman, S. J., Solheim, A., Elverhøi, A. and Ottesen, D. 2005. Rethinking Late Weichselian ice-sheet dynamics in coastal NW Svalbard. *Boreas* 34, 7–24.

Lanoil, B., Skidmore, M., Priscu, J. C., Han, S., Foo, W., Vogel, S. W., Tulaczyk, S. and Engelhardt, H. 2009. Bacteria beneath the West Antarctic Ice Sheet. *Environmental Microbiology* 11 (3), 609–15.

Lappegard, G. and Kohler, J. 2005. Determination of basal hydraulic systems based on subglacial high-pressure pump experiments. *Annals of Glaciology* 40, 37–42.

Lappegard, G., Kohler, J., Jackson, M. and Hagen, J. O. 2006. Characteristics of subglacial drainage systems deduced from load-cell measurements. *Journal of Glaciology* 52, 137–48.

Larour, E., Rignot, E. and Aubry, D. 2004. Processes involved in the propagation of rifts near Hemmen Ice Rise, Ronne Ice Shelf, Antarctica. *Journal of Glaciology* 50, 329–41.

Larsen, C. F., Motyka, R. J., Freymueller, J. T., Echelmeyer, K. A. and Ivins, E. R. 2005. Rapid viscoelastic uplift in southeast Alaska caused by post-Little Ice Age glacial retreat. *Earth and Planetary Science Letters* 237, 548–60.

Larsen, E. and Mangerud, J. 1981. Erosion rate of a Younger Dryas cirque glacier at Krakanes, western Norway. *Annals of Glaciology* 2, 153–8.

Larsen, E., Longva, O. and Follestad, B. A. 1991. Formation of De Geer moraines and implications for deglaciation dynamics. *Journal of Quaternary Science* 6, 263–77.

Larsen, E., Sejrup, H. P., Janocko, J., Landvik, J. Y., Stalsberg, K. and Steinsund, P. I. 2000. Recurrent interaction between the Norwegian Channel Ice Stream and terrestrial-based ice across southwest Norway. *Boreas* 29, 185–203.

Larsen, N. K. and Piotrowski, J. A. 2003. Fabric pattern in a basal till succession and its significance for reconstructing subglacial processes. *Journal of Sedimentary Research* 73, 725–34.

Larsen, N. K., Piotrowski, J. A. and Christiansen, F. 2006a. Microstructures and microshears as proxy for strain in subglacial diamicts: implications for basal till formation. *Geology* 34, 889–92.

Larsen, N. K., Piotrowski, J. A. and Kronborg, C. 2004. A multi-proxy study of a basal till: a time-transgressive accretion and deformation hypothesis. *Journal of Quaternary Science* 19, 9–21.

Larsen, N. K., Piotrowski, J. A., Christoffersen, P. and Menzies, J. 2006b. Formation and deformation of basal till during a glacier surge, Elisebreen, Svalbard. *Geomorphology* 81, 217–34.

Larsen, S. H., Davies, T. R. H. and McSaveney, M. J. 2005. A possible coseismic landslide origin of late Holocene moraines of the Southern Alps, New Zealand. *New Zealand Journal of Geology and Geophysics* 48, 311–14.

Larson, G. J., Lawson, D. E., Evenson, E. B., Alley, R. B., Knudsen, Ó., Lachniet, M. S. and Goetz, S. L. 2006. Glaciohydraulic supercooling in former ice sheets? *Geomorphology* 75, 20–32.

Larson, P. C. and Mooers, H. D. 2004. Glacial indicator dispersal processes: a conceptual model. *Boreas* 33, 238–49.

Laumann, T. and Reeh, N. 1993. Sensitivity to climate change of the mass balance of glaciers in southern Norway. *Journal of Glaciology* 39, 656–65.

Laverdière, C., Guimont, P. and Dionne, J. C. 1985. Les formes et les marques de l'érosion glaciaire du plancher rocheux: signification, terminologie, illustration. *Palaeogeography, Palaeoclimatology, Palaeoecology* 51, 365–87.

Laverdière, C., Guimont, P. and Pharand, M. 1979. Marks and forms on glacier beds: formation and classification. *Journal of Glaciology* 23, 414–16.

Lawson, D. E. 1979a. *Sedimentological Analysis of the Western Terminus Region of the Matanuska Glacier, Alaska.* Cold Regions Research and Engineering Laboratory, Report 79-9, Hanover, NH.

Lawson, D. E. 1979b. A comparison of the pebble orientations in ice and deposits of the Matanuska Glacier, Alaska. *Journal of Geology* 87, 629–45.

Lawson, D. E. 1981a. Distinguishing characteristics of diamictons at the margin of the Matanuska Glacier, Alaska. *Annals of Glaciology* 2, 78–84.

Lawson, D. E. 1981b. *Sedimentological Characteristics and Classification of Depositional Processes and Deposits in the Glacial Environment.* Cold Regions Research and Engineering Laboratory, Report 81-27, Hanover, NH.

Lawson, D. E. 1982. Mobilisation, movement and deposition of sub-aerial sediment flows, Matanuska Glacier, Alaska. *Journal of Geology* 90, 279–300.

Lawson, D. E. 1989. Glacigenic resedimentation: classification concepts and application to mass-movement processes and deposits. In Goldthwait, R. P. and Matsch, C. L. (eds), *Genetic Classification of Glacigenic Deposits.* Balkema, Rotterdam, 147–69.

Lawson, D. E. 1995. Sedimentary and hydrologic processes within modern terrestrial valley glaciers. In Menzies, J. (ed.), *Modern Glacial Environments: Processes, Dynamics and Sediments.* Butterworth-Heinemann, Oxford, 337–63.

Lawson, D. E., Strasser, J. C., Evenson, E. B., Alley, R. B., Larson, J. G. and Arcone, S. A. 1998. Glaciohydraulic supercooling: a freeze-on mechanism to create stratified, debris-rich basal ice. 1: Field evidence. *Journal of Glaciology* 44, 547–62.

Lawson, T. J. 1990. Former ice movement in Assynt, Sutherland, as shown by the distribution of glacial erratics. *Scottish Journal of Geology* 26, 25–32.

Lawson, W. J. 1996. Structural evolution of Variegated Glacier, Alaska, USA, since 1948. *Journal of Glaciology* 42, 261–70.

Lawson, W. J., Sharp, M. and Hambrey, M. J. 1994. The structural geology of a surge-type glacier. *Journal of Structural Geology* 16, 1447–62.

Lawson, W. J., Sharp, M. J. and Hambrey, M. J. 2000. Deformation histories and structural assemblages of glacier ice in a non-steady flow regime. In Maltman, A. J., Hubbard, B. and Hambrey, M. J. (eds), *Deformation of Glacial Materials.* Geological Society, London, Special Publication 176, 55–96.

Lazzara, M. A., Jezek, K. C., Scambos, T. A., MacAyeal, D. R. and van der Veen, C. J. 1999. On the recent calving of icebergs from the Ross Ice Shelf. *Polar Geography* 23, 201–12.

Le Heron, D. P. and Etienne, J. L. 2005. A complex subglacial clastic dyke swarm, Solheimajökull, southern Iceland. *Sedimentary Geology* 181, 25–37.

Le Meur, E. and Huybrechts, P. 1996. A comparison of different ways of dealing with isostasy: examples from modelling the Antarctic ice sheet during the last glacial cycle, *Annals of Glaciology* 23, 309–17.

Le Roux, J. P. 2003. Can dispersive pressure cause inverse grading in grain flows?: DISCUSSION. *Journal of Sedimentary Research* 73, 333–4.

Lea, D. W., Martin, P. A., Pak, D. K. and Spero, H. J. 2002. Reconstructing a 350 ky history of sea level using planktonic Mg/Ca and oxygen isotope records from a Cocos Ridge core. *Quaternary Science Reviews* 21, 283–93.

Lear, C. H., Elderfield, H. and Wilson, P. A. 2000. Cenozoic deep-sea temperatures and global ice volumes from Mg/Ca in benthic foraminiferal calcite. *Science* 287, 269–72.

Lee, A. G. G. and Rutter, E. H. 2004. Experimental rock-on-rock frictional wear: application to subglacial abrasion. *Journal of Geophysical Research* 109, doi:10.1029/2004JB003059.

Lefauconnier, B. and Hagen, J. O. 1990. Glaciers and climate in Svalbard, statistical analysis and reconstruction of the Brogger Glacier mass balance for the last 77 years. *Annals of Glaciology* 14, 148–52.

Lefauconnier, B. and Hagen, J. O. 1991. *Surging and Calving Glaciers in Eastern Svalbard*. Meddelelser 116. Norsk Polarinstitutt, Oslo.

Lefauconnier, B., Hagen, J. O., Pinglot, J. F. and Pourchet, M. 1994. Mass-balance estimates on the glacier complex Kongsvegen and Sveabreen, Spitsbergen, Svalbard, using radioactive layers. *Journal of Glaciology* 40, 368–76.

Legros, F. 2002. Can dispersive pressure cause inverse grading in grain flows? *Journal of Sedimentary Research* 72, 166–70.

Lekens, W. A. H., Sejrup, H. P., Haflidason, H., Petersen, G. O., Hjelstuen, B. and Knorr, G. 2005. Laminated sediments preceding Heinrich event 1 in the northern North Sea and southern Norwegian Sea: origin, processes and regional linkage. *Marine Geology* 216, 27–50.

Lemieux, J.-M., Sudicky, E. A., Peltier, W. R. and Tarasov, L. 2008. Simulating the impact of glaciations on continental groundwater flow systems. 1: Relevant processes and model formulation, *Journal of Geophysical Research* 113, F03017, doi:10.1029/2007JF000928.

Lemmen, D. S., Duk-Rodkin, A. and Bednarski, J. 1994. Late glacial drainage systems along the northwestern margin of the Laurentide Ice Sheet. *Quaternary Science Reviews* 13, 805–28.

Leonard, E. M. 1997. The relationship between glacial activity and sediment production: evidence from a 4450 year varve record of neoglacial sedimentation in Hector Lake, Alberta, Canada. *Journal of Palaeolimnology* 17, 319–30.

Letreguilly, A. and Reynaud, L. 1989. Spatial patterns of mass balance fluctuations of North American glaciers. *Journal of Glaciology* 35, 163–8.

Levrard, B. F., Forget, F., Montmessin, F. and Laskar, J. 2004. Recent ice-rich deposits formed at high latitudes on Mars by sublimation of unstable equatorial ice during low obliquity. *Nature* 431, 1072–5.

Levson, V. M. and Rutter, N. W. 1989a. A lithofacies analysis and interpretation of depositional environments of montane glacial diamictons, Jasper, Alberta, Canada. In Goldthwait, R. P. and Matsch, C. L. (eds), *Genetic Classification of Glacigenic Deposits*. Balkema, Rotterdam, 117–40.

Levson, V. M. and Rutter, N. W. 1989b. Late Quaternary stratigraphy, sedimentology, and history of the Jasper townsite area, Alberta, Canada. *Canadian Journal of Earth Sciences* 26, 1325–42.

Levy, J. S., Head, J. W. and Marchant, D. R. 2007. Lineated valley fill and lobate debris apron stratigraphy in Nilosyrtis Mensae, Mars: evidence for phases of glacial modification of the dichotomy boundary. *Journal of Geophysical Research* 112, E08004, doi:10.1029/2006JE002852.

Lewis, A. R., Marchant, D. R., Kowalewski, D. E., Baldwin, S. L. and Webb, L. E. 2006. The age and origin of the Labyrinth, western Dry Valleys, Antarctica: evidence for extensive middle Miocene subglacial floods and freshwater discharge to the Southern Ocean. *Geology* 34, 513–16.

Lewis, K. J., Fountain, A. G. and Dana, G. L. 1998. Surface energy balance and meltwater production for a Dry Valley glacier, Taylor Valley, Antarctica. *Annals of Glaciology* 27, 603–9.

Leynaud, D., Mienert, J. and Vanneste, M. 2009. Submarine mass movements on glaciated and non-glaciated European continental margins: a review of triggering mechanisms and preconditions to failure. *Marine and Petroleum Geology* 26 (5), 618–32.

Leysinger Vieli, G. J.-M. C. and Gudmundsson, G. H. 2004. On estimating length fluctuations of glaciers caused by changes in climatic forcing. *Journal of Geophysical Research* 109, F01007, doi:10.1029/2003JF000027.

Li, D., Yi, C., Ma, B., Wang, P., Ma, C. and Cheng, G. (2006) Fabric analysis of till clasts in the upper Urumqi River, Tian Shan, China. *Quaternary International* 154/155, 19–25.

Li, J., Zwally, H. J., Cornejo, C. and Yi, D. H. 2003. Seasonal variation of snow-surface elevation in North Greenland as modeled and detected by satellite radar altimetry. *Annals of Glaciology* 37, 233–8.

Li, Y., Liu, G. and Cui, Z. 2001. Glacial valley cross profile morphology, Tian Shan Mountains, China. *Geomorphology* 38, 153–66.

Lian, O. and Hicock, S. R. 2000. Thermal conditions beneath parts of the last Cordilleran Ice Sheet near its centre as inferred from subglacial till, associated sediments and bedrock. *Quaternary International* 68/71, 147–62.

Licht, K. J., Dunbar, N. W., Andrews, J. T. and Jennings, A. E. 1999. Distinguishing subglacial till and glacial marine diamictons in the western Ross Sea, Antarctica: implications for last glacial maximum grounding line. *Geological Society of America Bulletin* 111, 91–103.

Lidmar-Bergström, K. 1988. Exhumed Cretaceous landforms in south Sweden. *Zeitschrift für Geomorphologie* 72, 21–40.

Lidmar-Bergström, K. 1997. A long-term perspective on glacial erosion. *Earth Surface Processes and Landforms* 22, 297–306.

Lile, R. C. 1978. The effect of anisotropy on the creep of polycrystalline ice. *Journal of Glaciology* 21, 475–83.

Lillieskold, M. 1990. *Lithology of some Swedish eskers*. University of Stockholm, Department of Quaternary Geology, Report 17.

Lindén, M. and Möller, P. 2005. Marginal formation of De Geer moraines and their implications to the dynamics of grounding line recession. *Journal of Quaternary Science* 20, 113–33.

Lindén, M., Möller, P. and Adrielsson, L. 2008. Ribbed moraine formed by subglacial folding, thrust stacking and lee-side cavity infill. *Boreas* 37, 102–31.

Lindstrom, E. 1988. Are roches moutonnées mainly preglacial forms? *Geografiska Annaler* 70A, 323–31.

Lingle, C. and Fatland, D. R. 2003. Does englacial water storage drive temperate glacier surges? *Annals of Glaciology* 36, 14–20.

Linsley, B. K., 1996. Oxygen-isotope record of sea level and climate variations in the Sulu Sea over the past 150,000 years. *Nature* 380, 234–7.

Linton, D. L. 1963. The forms of glacial erosion. *Transactions of the Institute of British Geographers* 33, 1–28.

Liston, G. E., Winther, J.-G., Bruland, O., Elvehøy, H. and Sand, K. 1999. Below-surface melt on the coastal Antarctic ice sheet. *Journal of Glaciology* 45, 273–85.

Liu, Z. and Bromwich, D. H. 1997. Dynamics of the katabatic wind confluence zone near Siple Coast, West Antarctica. *Journal of Applied Meteorology* 36, 97–118.

Livermore, R., Nankivell, A., Eagles, G. and Morris, P. 2005. Palaeogene opening of Drake Passage. *Earth and Planetary Science Letters* 236, 459–70.

Livingstone, S. J., Ó Cofaigh, C. and Evans, D. J. A. 2008. Glacial geomorphology of the central sector of the last British-Irish Ice Sheet. *Journal of Maps* 2008, 358–77.

Lliboutry, L. 1968. General theory of subglacial cavitation and sliding of temperate glaciers. *Journal of Glaciology* 7, 21–58.

Lliboutry, L. 1977. Glaciological problems set by the control of dangerous lakes in Cordillera Blanca, Peru. II: Movement of a covered glacier embedded within a rock glacier. *Journal of Glaciology* 18, 255–73.

Lliboutry, L. 1979. Local friction laws for glaciers: a critical review and new openings. *Journal of Glaciology* 23, 67–95.

Lliboutry, L. 1987. Realistic, yet simple bottom boundary conditions for glaciers and ice sheets. *Journal of Geophysical Research* 92, 9101–9.

Lliboutry, L. 1993. Internal melting and ice accretion at the bottom of temperate glaciers. *Journal of Glaciology* 39, 50–64.

Lliboutry, L. 1996. Temperate ice permeability, stability of water veins and percolation of internal meltwater. *Journal of Glaciology* 42, 201–11.

Lliboutry, L. and Duval, P. 1985. Various isotropic and anisotropic ices found in glaciers and polar ice caps and their corresponding rheologies. *Annales Geophysicae* 3, 207–24.

Lliboutry, L., Arnao, B. M., Pautre, A. and Schneider, B. 1977. Glaciological problems set by the control of dangerous lakes in Cordillera Blanca, Peru. II: Movement of a covered glacier embedded within a rock glacier. *Journal of Glaciology* 18, 255–73.

Llubes, M., Lanseau, C. and Rémy, F. 2006. Relations between basal condition, subglacial hydrological networks and geothermal flux in Antarctica. *Earth and Planetary Science Letters* 241, 655–62.

Løken, O. H. and Hodgson, D. A. 1971. On the submarine geomorphology along the east coast of Baffin Island. *Canadian Journal of Earth Sciences* 8, 185–95.

Long, A. J., Roberts, D. H. and Wright, M. R. 1999. Isolation basin stratigraphy and Holocene relative sea level change on Arveprinsen Ejland, Disko Bugt, west Greenland. *Journal of Quaternary Science* 14, 323–45.

Longva, O. and Thoresen, M. K. 1991. Iceberg scours, iceberg gravity craters and current erosion marks from a gigantic Preboreal flood in southeastern Norway. *Boreas* 20, 47–62.

Lønne, I. 1993. Physical signatures of ice advance in a Younger Dryas ice-contact delta, Troms, northern Norway: implications for glacier-terminus history. *Boreas* 22, 59–70.

Lønne, I. 1995. Sedimentary facies and depositional architecture of ice-contact glacimarine systems. *Sedimentary Geology* 98, 13–43.

Lønne, I. 1997. Facies characteristics of a proglacial turbiditic sand-lobe at Svalbard. *Sedimentary Geology* 109, 13–35.

Lønne, I. 2001. Dynamics of marine glacier termini read from moraine architecture. *Geology* 29, 199–202.

Lønne, I. 2005. Faint traces of high Arctic glaciations: an early Holocene ice-front fluctuation in Bolterdalen, Svalbard. *Boreas* 34, 308–23.

Lønne, I. and Lauritsen, T. 1996. The architecture of a modern push moraine at Svalbard as inferred from ground penetrating radar measurements. *Arctic and Alpine Research* 28, 488–95.

Lønne, I. and Nemec, W. 2004. High-arctic fan delta recording deglaciation and environment disequilibrium. *Sedimentology* 51, 553–89.

Lønne, I., Nemec, W., Blikra, L. H. and Lauritsen, T. 2001. Sedimentary architecture and dynamic stratigraphy of a marine ice-contact system. *Journal of Sedimentary Research* 71, 922–43.

Lord, M. L. and Kehew, A. E. 1987. Sedimentology and palaeohydrology of glacial-lake outbursts in southeastern Saskatchewan and northwestern North Dakota. *Geological Society of America Bulletin* 99, 663–73.

Lorius, C., Jouzel, J., Raynaud, D., Hansen, J. and Le Treut, H. 1990. The ice-core record: climate sensitivity and future greenhouse warming. *Nature* 347, 139–45.

Lorrain, R. D., Fitzsimons, S. J., Vandergoes, M. J. and Stievenard, M. 1999. Ice composition evidence for the formation of basal ice from lake water beneath a cold-based Antarctic glacier. *Annals of Glaciology* 28, 277–81.

Loulergue, L., Schilt, A., Spahni, R., Masson-Delmotte, V., Blunier, T., Lemieux, B., Barnola, J.-M., Raynaud, D., Stocker, T. F. and Chappellaz, J. 2008. Orbital and millennial-scale features of atmospheric CH_4 over the past 800,000 years. *Nature* 453, 383–6.

Lowe, A. L. and Anderson, J. B. 2003. Evidence for abundant subglacial meltwater beneath the paleo-ice sheet in Pine Island Bay, Antarctica. *Journal of Glaciology* 49, 125–38.

Lowe, D. R. 1975. Water escape structures in coarse-grained sediments. *Sedimentology* 22, 157–204.

Lowe, D. R. 1976a. Grain flow and grain flow deposits. *Journal of Sedimentary Petrology* 46, 188–99.

Lowe, D. R. 1976b. Subaqueous liquefied and fluidised sediment flows and their deposits. *Sedimentology* 23, 285–308.

Lowe, J. J. and Walker, M. J. C. 1984. *Reconstructing Quaternary Environments*. Longman, London.

Lowe, J. J. and Walker, M. J. C. 1997. *Reconstructing Quaternary Environments*, 2nd edn. Longman, London.

Lowell, T. et al. 2005. Testing the Lake Agassiz meltwater trigger for the Younger Dryas. *Eos* 86 (40), doi:10.1029/2005EO 400001.

Lucchitta, B. K. and Rosanova, C. E. 1998. Retreat of northern margins of George VI and Wilkins Ice Shelves, Antarctic Peninsula. *Annals of Glaciology* 27, 41–6.

Luckman, A. and Murray, T. 2005. Seasonal changes in velocity before retreat of Jakobshavn Isbrae, Greenland. *Geophysical Research Letters* 32, doi:10.1029/2005GL022519.

Luckman, A., Murray, T. and Strozzi, T. 2002. Surface flow evolution throughout a glacier surge measured by satellite radar interferometry. *Geophysical Research Letters* 29, doi:10.1029/2001GL014570.

Luckman, A., Murray, T., de Lange, R. and Hanna, E. 2006. Rapid and synchronous ice-dynamic changes in east Greenland. *Geophysical Research Letters* 33, doi:10.1029/2005GL024428.

Ludwig, W., Amiotte-Suchet, P. and Probst, J.-L. 1999. Enhanced chemical weathering of rocks during the last glacial maximum: a sink for atmospheric CO_2? *Chemical Geology* 159, 147–61.

Lukas, S. 2003. Scottish Landform Example 31: The moraines around the Pass of Drumochter. *Scottish Geographical Journal* 119, 383–93.

Lukas, S. 2005. A test of the englacial thrusting hypothesis of 'hummocky' moraine formation: case studies from the northwest Highlands, Scotland. *Boreas* 34, 287–307.

Lukas, S. 2007. 'A test of the englacial thrusting hypothesis of "hummocky" moraine formation: case studies from the northwest Highlands, Scotland': reply to comments. *Boreas* 36, 108–13.

Lukas, S. and Benn, D. I. 2006. Retreat dynamics of Younger Dryas glaciers in the far NW Scottish Highlands reconstructed from moraine sequences. *Scottish Geographical Journal* 122, 308–25.

Lukas, S. and Lukas, T. 2006. A glacial geological and geomorphological map of the far NW Highlands, Scotland. *Journal of Maps* 2006, 43–58.

Lukas, S., Nicholson, L. I., Ross, F. H. and Humlum, O. 2005. Formation, meltout processes and landscape alteration of High-Arctic ice-cored moraines: examples from Nordenskiöld Land, central Spitsbergen. *Polar Geography* 29, 157–87.

Lukas, S., Spencer, J. Q. G., Robinson, R. A. J. and Benn, D. I. 2007. Problems associated with luminescence dating of Late

Quaternary glacial sediments in the NW Scottish Highlands. *Quaternary Geochronology* 2, 243–8.

Lundqvist, J. 1989a. Till and glacial landforms in a dry, polar region. *Zeitschrift für Geomorphologie* 33, 27–41.

Lundqvist, J. 1989b. Rogen (ribbed) moraine: identification and possible origin. *Sedimentary Geology* 62, 281–92.

Lundqvist, J. 2000. Palaeoseismicity and De Geer moraines. *Quaternary International* 68/71, 175–86.

Luthcke, S. B., Arendt, A. A., Rowlands, D. D., McCarthy, J. J. and Larsen, C. F. 2008. Recent glacier mass changes in the Gulf of Alaska region from GRACE mascon solutions. *Journal of Glaciology* 54 (188), 767–77.

Luthcke, S. B., Zwally, H. J., Abdalati, W., Rowlands, D. D., Ray, R. D., Nerem, R. S., Lemoine, F. G., McCarthy, J. J. and Chinn, D. S. 2006. Recent Greenland ice mass loss by drainage system from satellite gravity observations. *Science* 314, 1286–9.

Lüthi, D., Le Floch, M., Bereiter, B., Blunier, T., Barnola, J.-M., Siegenthaler, U., Raynaud, D., Jouzel, J., Fischer, H., Kawamura, K. and Stocker, T. F. 2008. High-resolution carbon dioxide concentration record 650,000–800,000 years before present. *Nature* 453, 379–82.

Lüthje, M., Pedersen, L. T., Reeh, N. and Greuell, W. 2006. Modelling the evolution of supraglacial lakes on the West Greenland ice-sheet margin. *Journal of Glaciology* 52, 608–18.

Lykke-Andersen, H. 1986. On the buried Norrea Valley – a contribution to the geology of the Norrea Valley, Jylland – and other buried overdeepened valleys. *Geogskrifter* 24, 211–23.

Lyså, A. and Lønne, I. 2001. Moraine development at a small high arctic valley glacier: Rieperbreen, Svalbard. *Journal of Quaternary Science* 16, 519–29.

Lythe, M. B., Vaughan, D. G. and the BEDMAP Consortium. 2001. BEDMAP: a new ice thickness and subglacial topographic model of Antarctica. *Journal of Geophysical Research* 106, B6, 11335–51.

Maag, H. 1969. *Ice Dammed Lakes and Marginal Glacial Drainage on Axel Heiberg Island*. Axel Heiberg Island Research Report, McGill University, Montreal.

MacAyeal, D. R. 1993. Binge/purge oscillations of the Laurentide ice sheet as a cause of the North Atlantic's Heinrich Events. *Palaeoceanography* 8, 775–84.

MacAyeal, D. R., Rommelaere, V., Huybrechts, P., Hulbe, C. L., Determann, J. and Ritz, C. 1996. An ice-shelf model test based on the Ross Ice Shelf, Antarctica. *Annals of Glaciology* 23, 46–51.

MacAyeal, D. R., Scambos, T. A., Hulbe, C. L. and Fahnestock, M. A. 2003. Catastrophic ice shelf break-up by an ice-shelf-fragment-capsize mechanism. *Journal of Glaciology* 49, 22–36.

MacGregor, K. R., Anderson, R. S., Anderson, S. P. and Waddington, E. D. 2000. Numerical simulations of glacial-valley longitudinal profile evolution. *Geology* 28, 1031–4.

MacGregor, K. R., Riihimaki, C. A. and Anderson, R. S. 2005. Spatial and temporal evolution of rapid basal sliding on Bench Glacier, Alaska, USA. *Journal of Glaciology* 51, 49–63.

Mackay, J. R. 1998. Pingo growth and collapse, Tuktoyaktuk Peninsula area, western Arctic Coast, Canada. *Geographie Physique et Quaternaire* 52, 271–323.

Mackay, J. R. and Mathews, W. H. 1964. The role of permafrost in ice thrusting. *Journal of Geology* 72, 378–80.

Mackiewicz, N. E., Powell, R. D., Carlson, P. R. and Molnia, B. F. 1984. Interlaminated ice-proximal glacimarine sediments in Muir Inlet, Alaska. *Marine Geology* 57, 113–47.

Maddy, D., Bridgland, D. R. and Westaway, R. 2001. Uplift-driven valley incision and climate-controlled river terrace development in the Thames Valley, UK. *Quaternary International* 79, 23–36.

Mader, H. 1992. Observations of the water-vein system in polycrystalline ice. *Journal of Glaciology* 38, 333–47.

Mager, S. and Fitzsimons, S. 2007. Formation of glaciolacustrine Late Pleistocene end moraines in the Tasman Valley, New Zealand. *Quaternary Science Reviews* 26, 743–58.

Magnusson, E., Rott, H., Björnsson, H. and Palsson, F. 2007. The impact of jökulhlaups on basal sliding observed by SAR interferometry on Vatnajökull, Iceland. *Journal of Glaciology* 53 (181), 232–40.

Mahaney, W. C. and Kalm, V. 2000. Comparative scanning electron microscopy study of oriented till blocks, glacial grains and Devonian sands in Estonia and Latvia. *Boreas* 29, 35–51.

Mahaney, W. C., Miyamoto, H., Dohm, J. M., Baker, V. R., Cabrol, N. A., Grin, E. A. and Berman, D. C. 2007. Rock glaciers on Mars: Earth-based clues to Mars' recent paleoclimatic history. *Planetary and Space Science* 55, 181–92.

Mahaney, W. C., Stewart, A. and Kalm, V. 2001. Quantification of SEM microtextures useful in sedimentary environmental discrimination. *Boreas* 30, 165–71.

Mair, D., Nienow, P., Willis, I. and Sharp, M. 2001. Spatial patterns of glacier motion during a high-velocity event: Haut Glacier d'Arolla, Switzerland. *Journal of Glaciology* 47 (156), 9–20.

Mair, D., Sharp, M. J. and Willis, I. C. 2002. Evidence for basal cavity opening from analysis of surface uplift during a high-velocity event: Haut Glacier d'Arolla, Switzerland. *Journal of Glaciology* 48, 208–16.

Mair, D., Willis, I., Fischer, U. H., Hubbard, B., Nienow, P. and Hubbard, A. 2003. Hydrological controls on patterns of surface, internal and basal motion during three 'spring events': Haut Glacier d'Arolla, Switzerland. *Journal of Glaciology* 49, 555–67.

Mair, R. and Kuhn, M. 1994. Temperature and movement measurements at a bergschrund. *Journal of Glaciology* 40, 561–5.

Maizels, J. K. 1977. Experiments on the origin of kettle holes. *Journal of Glaciology* 18, 291–303.

Maizels, J. K. 1989a. Sedimentology, palaeoflow dynamics and flood history of jökulhlaup deposits: palaeohydrology of Holocene sediment sequences in southern Iceland sandur deposits. *Journal of Sedimentary Petrology* 59, 204–23.

Maizels, J. K. 1989b. Sedimentology and palaeohydrology of Holocene flood deposits in front of a jökulhlaup glacier, south Iceland. In Bevan, K. and Carling, P. J. (eds), *Floods: The Geomorphological, Hydrological and Sedimentological Consequences*. Wiley, New York, 239–51.

Maizels, J. K. 1992. Boulder ring structures produced during jökulhlaup flows: origin and hydraulic significance. *Geografiska Annaler* 74A, 21–33.

Maizels, J. K. 1993. Lithofacies variations within sandur deposits: the role of runoff regime, flow dynamics and sediment supply characteristics. *Sedimentary Geology* 85, 299–325.

Maizels, J. K. 1995. Sediments and landforms of modern proglacial terrestrial environments. In Menzies, J. (ed.), *Modern Glacial Environments*. Butterworth-Heinemann, Oxford, 365–416.

Maizels, J. K. 1997. Jökulhlaup deposits in proglacial areas. *Quaternary Science Reviews* 16, 793–819.

Mäkinen, J. 2003. Time transgressive deposits of repeated depositional sequences within interlobate glaciofluvial (esker) sediments in Köyliö, SW Finland. *Sedimentology* 50, 327–60.

Malin, M. C. and Carr, M. H. 1999. Groundwater formation of Martian valleys. *Nature* 397, 589–91.

Maltman, A. 1994. *The Geological Deformation of Sediments*. Chapman and Hall, Cambridge.

Mandl, G. and Harkness, R. M. 1987. Hydrocarbon migration by hydraulic fracturing. In Jones, M. E. and Preston, R. M. F. (eds), *Deformation of Sediments and Sedimentary Rocks*. Geological Society, London, Special Publication 29, 39–53.

Mangerud, J., Astakhov, V., and Svendsen, J.-I. 2002. The extent of the Barents–Kara ice sheet during the Last Glacial Maximum. *Quaternary Science Reviews* 21, 111–19.

Mangerud, J., Astakhov, V., Jakobsson, M. and Svendsen, J. I. 2001a. Huge Ice Age lakes in Russia. *Journal of Quaternary Science* 16, 773–7.

Mangerud, J., Astakhov, V., Murray, A. and Svendsen, J. I. 2001b. The chronology of a large ice-dammed lake and the Barents-Kara

Ice Sheet advances, Northern Russia. *Global and Planetary Change* 31, 319–34.

Mangerud, J., Jakobsson, M., Alexanderson, H., Astakhov, V., Clarke, G. K. C., Henriksen, M., Hjort, C., Krinner, G., Lunkka, J.-P., Möller, P., Murray, A., Nikolskaya, O., Saarnisto, M. and Svendsen, J. O. 2004. Ice-dammed lakes and rerouting of the drainage of northern Eurasia during the Last Glaciation. *Quaternary Science Reviews* 23, 1313–32.

Mangold, N. 2003. Geomorphic analysis of lobate debris aprons on Mars at Mars Orbiter Camera scale: evidence for ice sublimation initiated by fractures. *Journal of Geophysical Research* 108 (E4), 8021, doi:10.1029/2002JE001885.

Mangold, N. and Allemand, P. 2001. Topographic analysis of features related to ice on mars. *Geophysical Research Letters* 28, 407–10.

Mangold, N., Allemand, P., Duval, P., Geraud, Y. and Thomas, P. 2002. Experimental and theoretical deformation of ice-rock mixtures: implications on rheology and ice content of Martian permafrost. *Planetary and Space Science* 50, 385–401.

Manley, G. 1955. On the occurrence of ice domes and permanently snow-covered summits. *Journal of Glaciology* 2, 453–6.

Manley, G. 1959. The late-glacial climate of north-west England. *Liverpool and Manchester Geological Journal* 2, 188–215.

March, A. 1932. Mathematische Theorie der Regelung nach der Korngestalt bei affiner Deformation. *Zeitschrift für Kristallographie* 81, 285–97.

Marchant, D. R. and Head, J. W. 2007. Antarctic dry valleys: microclimate zonation, variable geomorphic processes, and implications for assessing climate change on Mars. *Icarus* 192, 187–222.

Marchant, D. R., Denton, G. H. and Swisher, C. C. 1993. Miocene Pliocene-Pleistocene glacial history of Arena Valley, Quartermain Mountains, Antarctica. *Geografiska Annaler* 75A, 269–302.

Marchant, D. R., Lewis, A. R., Phillips, W. M., Moore, E. J., Souchez, R. A. and Landis, G. P. 2002. Formation of patterned ground and sublimation till over Miocene glacier ice in Beacon Valley, southern Victoria Land, Antarctica. *Geological Society of America Bulletin* 114, 718–30.

Marienfeld, P. 1992. Recent sedimentary processes in Scoresby Sund, East Greenland. *Boreas* 21, 169–86.

Mark, B. G. and Osmaston, H. A. 2008. Quaternary glaciation in Africa: key chronologies and climatic implications. *Journal of Quaternary Science* 23, 589–608.

Mark, B. G., Harrison, S. P., Spessa, A., New, M., Evans, D. J. A. and Helmens, K. F. 2005. Tropical snowline changes at the last glacial maximum: a global assessment. *Quaternary International* 138/139, 168–201.

Marquette, G. C., Gray, J. T., Gosse, J. C., Courchesne, F., Stockli, L., Macpherson, G. and Finkel, R. 2004. Felsenmeer persistence under non-erosive ice in the Torngat and Kaumajet mountains, Quebec and Labrador, as determined by soil weathering and cosmogenic nuclide exposure dating. *Canadian Journal of Earth Sciences* 41, 19–38.

Marren, P. M. 2002. Glacier margin fluctuations, Skaftafellsjökull, Iceland: implications for sandur evolution. *Boreas* 31, 75–81.

Marren, P. M. 2004. Discussion: present-day sandurs are not representative of the geological record. Sedimentary Geology 152, 1–5 (2002). *Sedimentary Geology* 164, 335–40.

Marren, P. M. 2005. Magnitude and frequency in proglacial rivers: a geomorphological and sedimentological perspective. *Earth Science Reviews* 70, 203–51.

Marshall, S. J. and Clarke, G. K. C. 1997. A continuum mixture model of ice stream thermomechanics in the Laurentide Ice Sheet 1. Theory. *Journal of Geophysical Research* 102, 20599–614.

Marshall, S. J., and Clarke, G. K. C. 1999. Modeling North American freshwater runoff and proglacial lake history through the last glacial cycle. *Quaternary Research* 52 (3), 300–15.

Marshall, S. J., Clarke, G. K. C., Dyke, A. S. and Fisher, D. A. 1996. Geologic and topographic controls on fast flow in the Laurentide and Cordilleran Ice Sheets. *Journal of Geophysical Research* 101, B8, 17827–39.

Marshall, S. J., James, T. S. and Clarke, G. K. C. 2002. North American Ice Sheet reconstructions at the Last Glacial Maximum. *Quaternary Science Reviews* 21, 175–92.

Marshall, S. J., Pollard, D., Hostetler, S. and Clark, P. U. 2004. Coupling ice-sheet and climate models for simulation of former ice sheets. In Gillespie, A. R., Porter, S. C. and Atwater, B. F. (eds), *The Quaternary Period in the United States*. Elsevier, Amsterdam, 105–26.

Marshall, S. J., Tarasov, L., Clarke, G. K. C. and Peltier, W. R. 2000. Glaciological reconstruction of the Laurentide Ice Sheet: physical processes and modelling challenges. *Canadian Journal of Earth Sciences* 37, 769–93.

Martin, C., Hindmarsh, R. C. A. and Navarro, F. J. 2006. Dating ice flow change near the flow divide at Roosevelt Island, Antarctica, by using a thermomechanical model to predict radar stratigraphy. *Journal of Geophysical Research* 111, doi:10.1029/2005JF000326.

Martin, P. A., Lea, D. W., Rosenthal, Y., Shackleton, N. J., Sarnthein, M. and Paperfuss, T. 2002. Quaternary deep sea temperature histories derived from benthic foraminiferal Mg/Ca. *Earth and Planetary Science Letters* 198, 193–209.

Martinec, J. 1989. Hour-to-hour snowmelt rates and lysimeter outflow during an entire ablation period. In Colbeck, S. C. (ed.), *Glacier and Snow Cover Variations*. IAHS Publication 183, 19–28.

Martini, I. P. 1990. Pleistocene glacial fan deltas in southern Ontario, Canada. In Colella, A. and Prior, D. (eds), *Coarse Grained Deltas*. International Association of Sedimentologists, Special Publication 10, 281–95.

Martini, I. P. and Brookfield, M. E. 1995. Sequence analysis of upper Pleistocene (Wisconsinan) glaciolacustrine deposits of the north-shore bluffs of Lake Ontario, Canada. *Journal of Sedimentary Research*, B65, 388–400.

Maruyama, S. and Santosh, M. 2008. Models on Snowball Earth and Cambrian explosion: a synopsis. *Gondwana Research* 14, 22–32.

Maslin, M. A. and Ridgwell, A. 2005. Mid-Pleistocene revolution and the 'eccentricity myth'. In Head, M. J. and Gibbard, P. L. (eds), *Early-Middle Pleistocene Transitions: The Land-Ocean Evidence*. Geological Society, London, Special Publication 247, 19–34.

Maslin, M. A., Seidov, D. and Lowe, J. 2001. Synthesis of the nature and causes of sudden climate transitions during the Quaternary. In Seidov, D., Haupt, B. J. and Maslin, M. A. (eds), *The Oceans and Rapid Climate Change: Past, Present, and Future*. AGU Geophysical Monograph Series 126, 9–52.

Mastalerz, K. 1990. Diurnally and seasonally controlled sedimentation on a glaciolacustrine foreset slope: an example from the Pleistocene of eastern Poland. In Colella, A. and Prior, D. (eds), *Coarse Grained Deltas*. International Association of Sedimentologists, Special Publication 10, 297–309.

Mathews, W. H. 1973. Record of two jökulhlaups. In *Symposium at Cambridge 1969 – Hydrology of Glaciers*. International Association of Hydrological Sciences Publication 95, 99–110.

Mathews, W. W. and Mackay, J. R. 1960. Deformation of soils by glacier ice and the influence of pore pressures and permafrost. *Transactions of the Royal Society of Canada*, Section 3, 54, 27–36.

Matsuoka, N. 2001. Direct observation of frost wedging in alpine bedrock. *Earth Surface Processes and Landforms* 26, 601–14.

Matsuoka, N. and Murton, J. 2008. Frost weathering: recent advances and future directions. *Permafrost and Periglacial Processes* 19, 195–210.

Matthews, J. A. and Briffa, K. 2005. The 'Little Ice Age': re-evaluation of an evolving concept. *Geografiska Annaler* A, 87, 17–36.

Matthews, J. A. and Petch, J. R. 1982. Within-valley asymmetry and related problems of Neoglacial lateral moraine development at certain Jotunheimen glaciers, southern Norway. *Boreas* 11, 225–47.

Matthews, J. A., Cornish, R. and Shakesby, R. A. 1979. 'Saw-tooth' moraines in front of Bodalsbreen, southern Norway. *Journal of Glaciology* 22, 535–46.

Matthews, J. A., McCarroll, D. and Shakesby, R. A. 1995. Contemporary terminal-moraine ridge formation at a temperate glacier: Styggedalsbreen, Jotunheimen, southern Norway. *Boreas* 24, 129–39.

Matthews, J. B. and Quinlan, A. V. 1975. Seasonal characteristics of water masses in Muir Inlet, a fjord with tidewater glaciers. *Journal of the Fisheries Research Board Canada* 32, 1693–703.

Mattson, L. E., Gardner, J. S. and Young, G. J. 1993. Ablation on debris covered glaciers: an example from the Rakhiot Glacier, Panjab, Himalaya. *IAHS Publication* 218, 289–96.

Maule, C. F., Purucker, M. E., Olsen, N. and Mosegaard, K. 2005. Heat flux anomalies in Antarctica revealed by satellite magnetic data. *Science* 309, 464–7.

Mavlyudov, B. 2005. About new type of subglacial channels, Spitsbergen. In Mavlyudov, B. (ed.), *Glacier Caves and Glacial Karst in High Mountains and Polar Regions.* Proceedings of the 7th GLACIPR Symposium. Institute of Geography RAS, Moscow, 54–60.

Max, M. D. and Clifford, S. M. 2001. Initiation of Martian outflow channels: related to the dissociation of gas hydrate? *Geophysical Research Letters* 28, 1787–90.

Mayo, L. R. 1984. Glacier mass balance and runoff research in the USA. *Geografiska Annaler* 66A, 215–27.

Mayo, L. R., 1988. Advance of Hubbard Glacier and 1986 outburst of Russell Fiord, Alaska, U.S.A. *Annals of Glaciology* 13, 189–94.

McCabe, A. M. and Clark, P. U. 1998. Ice sheet variability around the North Atlantic Ocean during the last deglaciation. *Nature* 392, 373–7.

McCabe, A. M. and Dardis, G. F. 1989. Sedimentology and depositional setting of late Pleistocene drumlins, Galway Bay, western Ireland. *Journal of Sedimentary Petrology* 59, 944–59.

McCabe, A. M. and Haynes, J. R. 1996. A late Pleistocene intertidal boulder pavement from an isostatically emergent coast, Dundalk Bay, eastern Ireland. *Earth Surface Processes and Landforms* 21, 555–72.

McCabe, A. M., Clark, P. U. and Clark, J. 2005. AMS ^{14}C dating of deglacial events in the Irish Sea Basin and other sectors of the British-Irish Ice Sheet. *Quaternary Science Reviews* 24, 1673–90.

McCabe, A. M., Clark, P. U., Clark, J. and Dunlop, P. 2007. Radiocarbon constraints on readvances of the British-Irish Ice Sheet in the northern Irish Sea Basin during the last deglaciation. *Quaternary Science Reviews* 26, 1204–11.

McCabe, A. M., Knight, J. and McCarron, S. G. 1998. Evidence for Heinrich event 1 in the British Isles. *Journal of Quaternary Science* 13, 549–68.

McCabe, A. M., Knight, J. and McCarron, S. G. 1999. Ice flow stages and glacial bedforms in north central Ireland: a record of rapid environmental change during the last glacial termination. *Journal of the Geological Society of London* 156, 63–72.

McCabe, G. J. and Fountain, A. G. 1995. Relations between atmospheric circulation and mass balance of South Cascade Glacier, Washington, USA. *Arctic and Alpine Research* 27, 226–33.

McCarroll, D. 2001. Deglaciation of the Irish Sea Basin: a critique of the glaciomarine hypothesis. *Journal of Quaternary Science* 16, 393–404.

McCarroll, D. 2006. Average glacial conditions and the landscape of Snowdonia. In Knight, P. (ed.), *Glacier Science and Environmental Change.* Blackwell, Oxford.

McCarroll, D. and Harris, C. 1992. The glacigenic deposits of western Lleyn, north Wales: terrestrial or marine? *Journal of Quaternary Science* 7, 19–29.

McCarroll, D. and Rijsdijk, K. F. 2003. Deformation styles as a key for interpreting glacial depositional environments. *Journal of Quaternary Science* 18, 473–89.

McCarroll, D., Matthews, J. A. and Shakesby, R. A. 1989. 'Striations' produced by catastrophic subglacial drainage of a glacier-dammed lake, Mjolkedalsbreen, southern Norway. *Journal of Glaciology* 35, 193–6.

McCulloch, R. D., Bentley, M. J., Tipping, R. M. and Clapperton, C. M. 2005. Evidence for late glacial ice dammed lakes in the central Strait of Magellan and Bahia Inutil, southernmost South America. *Geografiska Annaler* 87A, 335–62.

McDonald, B. C. and Bannerjee, I. 1971. Sediments and bedforms on a braided outwash plain. *Canadian Journal of Earth Sciences* 8, 1282–301.

McDonald, B. C. and Shilts, W. W. 1975. Interpretation of faults in glaciofluvial sediments. In Jopling, A. V. and McDonald, B. C. (eds), *Glaciofluvial and Glaciolacustrine Sedimentation.* SEPM Special Publication 23, 123–31.

McDougall, D. A. 2001. The geomorphological impact of Loch Lomond (Younger Dryas) Stadial plateau icefields in the central Lake District, northwest England. *Journal of Quaternary Science* 16, 531–43.

McManus, J. and Duck, R. W. 1988. Localised enhanced sedimentation from icebergs in a proglacial lake in Briksdal, Norway. *Geografiska Annaler* 70A, 215–23.

McMillan, M., Nienow, P., Shepherd, A., Benham, T. and Sole, A. 2007. Seasonal evolution of supra-glacial lakes on the Greenland Ice Sheet. *Earth and Planetary Science Letters* 262, 484–92.

McSaveney, M. J. 2002. Recent rockfalls and rock avalanches in Mount Cook National Park, New Zealand. In Evans, S. G. and Degraff, I. V. (eds), *Catastrophic Landslides: Effects, Occurrence and Mechanisms.* Geological Society of America, Reviews in Engineering Geology 15, 35–70.

Meehl, G. A. et al. 2007. Global climate projections. In Solomon, S. et al. (eds), *Climate Change 2007: The Physical Science Basis.* Contribution of Working Group I to the Fourth Assessment of the Intergovernmental Panel on Climate Change. Cambridge University Press.

Meert, J. G. 2007. Testing the Neoproterozoic glacial models. *Gondwana Research* 11, 573–4.

Meier, M. F. 1997. The iceberg discharge process: observations and inferences drawn from the study of Columbia Glacier. In van der Veen, C. J. (ed.), *Calving Glaciers: Report of a Workshop, Feb. 28–March 2, 1997.* Ohio State University, Columbus. Byrd Polar Research Centre, report 15, 109–14.

Meier, M. F. and Post, A. S. 1969. What are glacier surges? *Canadian Journal of Earth Sciences* 6, 807–19.

Meier, M. F. and Post, A. S. 1987. Fast tidewater glaciers. *Journal of Geophysical Research* 92, 9051–8.

Meier, M. F., Lundstrom, S., Stone, D., Kamb, B., Engelhardt, H., Humphrey, N., Dunlap, W. W., Fahnestock, M. A., Krimmell, R. M. and Walters, R. 1994. Mechanical and hydrologic basis for the rapid motion of a large tidewater glacier 1. Observations. *Journal of Geophysical Research* 99, 15219–29.

Meigs, A., Krugh, W. C., Davis, K. and Bank, G. 2006. Ultra-rapid landscape response and sediment yield following glacier retreat, Icy Bay, southern Alaska. *Geomorphology* 78, 207–21.

Menzies, J. 1979. A review of the literature on the formation and location of drumlins. *Earth Science Reviews* 14, 315–59.

Menzies, J. 1984. *Drumlins: A Bibliography.* Geo Books, Norwich.

Menzies, J. 1989. Subglacial hydraulic conditions and their possible impact upon subglacial bed formation. *Sedimentary Geology* 62, 125–50.

Menzies, J. 2000. Micromorphological analyses of microfabrics and microstructures indicative of deformation processes in glacial sediments. In Maltman, A. J., Hubbard, B. and Hambrey, M. J. (eds), *Deformation of Glacial Materials.* Geological Society, London, Special Publication 176, 245–57.

Menzies, J. 2001. The Quaternary sedimentology and stratigraphy of small, ice-proximal, subaqueous grounding-line moraines in the central Niagara Peninsula, southern Ontario. *Geographie Physique et Quaternaire* 55, 75–86.

Menzies, J. and Brand, U. 2007. The internal sediment architecture of a drumlin, Port Byron, New York State, USA. *Quaternary Science Reviews* 26, 322–55.

Menzies, J. and Maltman, A. J. 1992. Microstructures in diamictons: evidence of subglacial bed conditions. *Geomorphology* 6, 27–40.

Menzies, J. and Rose, J. 1989. Subglacial bedforms: an introduction. *Sedimentary Geology* 62, 117–22.

Menzies, J. and Zaniewski, K. 2003. Microstructures within a modern debris flow deposit derived from Quaternary glacial diamicton: a comparative micromorphological study. *Sedimentary Geology* 157, 31–48.

Menzies, J., van der Meer, J. J. M. and Rose, J. 2006. Till as a glacial 'tectomict', its internal architecture, and the development of a 'typing' method for till differentiation. *Geomorphology* 75, 172–200.

Mercer, J. H. 1956. Geomorphology and glacial history of southernmost Baffin Island. *Geological Society of America Bulletin* 67, 553–70.

Mercer, J. H. 1978. West Antarctic Ice Sheet and CO_2 greenhouse effect: a threat of disaster? *Nature* 271, 321–5.

Metcalf, R. C. 1986. The cationic denudation rate of an alpine glacier catchment: Gornergletscher, Switzerland. *Zeitschrift für Gletscherkunde und Glazialgeologie* 22, 19–32.

Metzger, S. M. 1991. A survey of esker morphologies, the connection to New York State glaciation and criteria for subglacial meltwater channel deposits on the planet Mars. *Lunar and Planetary Science* 22, 891–2.

Miall, A. D. 1977. A review of the braided river depositional environment. *Earth Science Reviews* 13, 1–62.

Miall, A. D. 1978. Lithofacies types and vertical profile models in braided river deposits: a summary. In Miall, A. D. (ed.), *Fluvial Sedimentology*. Canadian Society of Petroleum Geologists, Memoir 5, 597–604.

Miall, A. D. 1985. Architectural-element analysis: a new method of facies analysis applied to fluvial deposits. *Earth Science Reviews* 22, 261–308.

Miall, A. D. 1992. Alluvial deposits. In Walker, R. G. and James, N. P. (eds), *Facies Models: Response to Sea-level Change*, 119–42. Geological Association of Canada, Toronto.

Miall, A. D. 2000. *Principles of Sedimentary Basin Analysis*, 3rd edn. Springer-Verlag, Berlin.

Mickelson, D. M. 1973. Nature and rate of basal till deposition in a stagnating ice mass, Burroughs Glacier, Alaska. *Arctic and Alpine Research* 5, 17–27.

Mickelson, D. M. 1986. Observed processes of glacial deposition in Glacier Bay. In Anderson, P. J., Goldthwait, R. P. and McKenzie, G. D. (eds), *Landform and till genesis in the eastern Burroughs Glacier-Plateau remnant area, Glacier Bay, Alaska.* Institute of Polar Studies, Ohio State University. Miscellaneous Publication 236, 47–61.

Mickelson, D. M. and Colgan, P. M. 2004. The southern Laurentide Ice Sheet. In Gillespie, A. R., Porter, S. C. and Atwater, B. F. (eds), *The Quaternary Period in the United States*. Elsevier, 1–16.

Mickelson, D. M., Ham, N. R. and Ronnert, L. 1992. Comment on 'Striated clast pavements: Products of deforming subglacial sediment?' *Geology* 20, 285.

Mikucki, J. A., Foreman, C. H., Sattler, B., Lyons, W. B. and Priscu, J. A. 2004. Geomicrobiology of Blood Falls: an iron-rich saline discharge at the terminus of Taylor Glacier, Antarctica. *Aquatic Chemistry* 10, 199–220.

Milkovitch, S. and Head, J. W. 2005. North polar cap of Mars: polar layered deposit characterization and identification of a fundamental climate signal. *Journal of Geophysical Research* 110, E01005, doi:10.1029/2004JE002349.

Miller, J. W. 1972. Variations in New York drumlins. *Annals of the American Association of Geographers* 62, 418–23.

Milliken, R. E., Mustard, J. F. and Goldsby, D. L. 2003. Viscous flow features on the surface of Mars: observations from high-resolution Mars Orbiter Camera (MOC) images. *Journal of Geophysical Research* 108(E6), 5057, doi:10.1029/2002JE002005.

Mills, H. H. 1991. *Three-dimensional Clast Orientation in Glacial and Mass-movement Sediments*. U.S. Geological Survey, Open File Report 90–128.

Milne, G. A., Mitrovica, J. X. and Schrag, D. P. 2002. Estimating past continental ice volume from sea-level data. *Quaternary Science Reviews* 21, 361–76.

Mitchell, A. C., Brown, G. H. and Fuge, R. 2006. Minor and trace elements as indicators of solute provenance and flow routing in a subglacial hydrological system. *Hydrological Processes* 20, 877–97.

Mitchell, S. G. and Montgomery, D. R. 2006. Influence of a glacial buzzsaw on the height and morphology of the Cascade Range in central Washington State, USA. *Quaternary Research* 65, 96–107.

Mitchell, W. A. 1994. Drumlins in ice sheet reconstructions, with reference to the western Pennines, northern England. *Sedimentary Geology* 91, 313–31.

Mitchell, W. A. 1996. Significance of snowblow in the generation of Loch Lomond Stadial (Younger Dryas) glaciers in the western Pennines, northern England. *Journal of Quaternary Science* 11, 233–48.

Mitchell, W. A. 2006. Drumlin map of the western Pennines and southern Vale of Eden, northern England. *Journal of Maps* 2006, 10–16.

Mitchell, W. A. 2007. Reconstructions of the Late Devensian (Dimlington Stadial) British-Irish Ice Sheet: the role of the upper Tees drumlin field, north Pennines, England. *Proceedings of the Yorkshire Geological Society* 56, 221–34.

Mitrovica, J. X., Tamisiea, M. E., Davis, J. L. and Milne, G. A. 2001. Recent mass balance of polar ice sheets inferred from patterns of global sea level change. *Nature* 409, 1026–9.

Miyamoto, H., Komatsu, G., Baker, V. R., Dohm, J. M., Ito, K. and Tosaka, H. 2007. Cataclysmic Scabland flooding: insights from a simple depth-averaged numerical model. *Environmental Modelling and Software* 22, 1400–8.

Mölg, T. and Hardy, D. R. 2004. Ablation and associated energy balance of a horizontal glacier surface on Kilimanjaro. *Journal of Geophysical Research* 109, D16104, doi:10.1029/2003JD004338.

Mölg, T., Cullen, N. J., Hardy, D. R., Kaser, G. and Klok, L. 2008. Mass balance of a slope glacier on Kilimanjaro and its sensitivity to climate. *International Journal of Climatology* 28, 881–92.

Mollard, J. D. 1983. The origin of reticulate and orbiculate patterns on the floor of the Lake Agassiz basin. In Teller, J. T. and Clayton, L. (eds), *Glacial Lake Agassiz*. Geological Association of Canada, Special Paper 26, 355–74.

Mollard, J. D. 2000. Ice-shaped ring forms in Western Canada: their airphoto expressions and manifold polygenetic origins. *Quaternary International* 68/71, 187–98.

Möller, P. 2006. Rogen moraine: an example of glacial reshaping of pre-existing landforms. *Quaternary Science Reviews* 25, 362–89.

Molnar, P. and England, P. 1990. Late Cenozoic uplift of mountain ranges and global climate change: chicken or egg? *Nature* 346, 29–34.

Molnia, B. F., 2008. Glaciers of North America: Glaciers of Alaska. In Williams, R. S., Jr and Ferrigno, J. G. (eds), *Satellite Image Atlas of Glaciers of the World*. U.S. Geological Survey Professional Paper 1386-K.

Montgomery, D. R., Balco, G. and Willett, S. D. 2001. Climate, tectonics and the morphology of the Andes. *Geology* 29, 579–82.

Mooers, H. D. 1989a. On the formation of the tunnel valleys of the Superior Lobe, Central Minnesota. *Quaternary Research* 32, 24–35.

Mooers, H. D. 1989b. Drumlin formation: a time transgressive model. *Boreas* 18, 99–107.

Mooers, H. D. 1990a. Ice-marginal thrusting of drift and bedrock: thermal regime, subglacial aquifers, and glacial surges. *Canadian Journal of Earth Sciences* 27, 849–62.

Mooers, H. D. 1990b. A glacial-process model: the role of spatial and temporal variations in glacier thermal regime. *Geological Society of America Bulletin* 102, 243–51.

Mook, W. G. (ed.). 2000. *Environmental Isotopes in the Hydrological Cycle: Principles and Applications*. UNESCO, Paris.

Moorman, B. J. 2003. Glacier-permafrost hydrology interactions, Bylot Island, Canada. In Philips, M., Springman, S. M. and Arenson, L. U. (eds), *Proceedings of the 8th International Conference on Permafrost*. Balkema, Rotterdam, 783–8.

Moorman, B. J. 2005. Glacier-permafrost hydrological connectivity: Stagnation Glacier, Bylot Island, Canada. In Harris, C. and Murton, J. B. (eds), *Cryospheric Systems: Glaciers and Permafrost*. Geological Society, London, Special Publication 242, 63–74.

Moorman, B. J. and Michel, F. A. 2000. The burial of ice in the proglacial environment on Bylot Island, Arctic Canada. *Permafrost and Periglacial Processes* 11, 161–75.

Morgan, F. 2005. A note on cross-profile morphology for glacial valleys. *Earth Surface Processes and Landforms* 30, 513–14.

Morland, L. W. and Boulton, G. S. 1975. Stress in an elastic hump: the effects of glacier flow over elastic bedrock. *Proceedings of the Royal Society of London* Series A, 344, 157–73.

Morris, E. M. and Morland, L. W. 1976. A theoretical analysis of the formation of glacial flutes. *Journal of Glaciology* 17, 311–24.

Morris, E. M. and Vaughan, D. G. 2003. Spatial and temporal variation of surface temperature on the Antarctic Peninsula and the limit of variability of ice shelves. In Domack, E., Burnett, A., Leventer, A., Conley, P., Kirby, M. and Bindschadler, R. (eds), *Antarctic Peninsula Climate Variability: A Historical and Paleoenvironmental Perspective*. Antarctic Research Series, vol. 79, 61–8.

Mosley, M. P. 1983. The response of braided rivers to changing discharge. *New Zealand. Journal of Hydrology* 22, 18–67.

Mosley, M. P. 1988. Bedload transport and sediment yield in the Onyx River, Antarctica. *Earth Surface Processes and Landforms* 13, 51–67.

Mottram, R. H. and Benn, D. I. 2009. Testing crevasse depth models: a field study at Breiðamerkurjökull, Iceland. *Journal of Glaciology* 55(192), 746–52.

Motyka, R. J. 1997. Deep-water calving at Le Conte Glacier, Southeast Alaska. In van der Veen, C. J. (ed.), *Calving Glaciers: Report of a Workshop, Feb. 28–March 2, 1997*. Ohio State University, Columbus. Byrd Polar Research Centre, report 15, 115–18.

Motyka, R. J. and Begét, J. E. 1996. Taku Glacier, southeast Alaska, USA: Late Holocene history of a tidewater glacier. *Arctic and Alpine Research* 28, 42–51.

Motyka, R. J., Hunter, L., Echelmeyer, K. and Connor, C. 2003. Submarine melting at the terminus of a temperate tidewater glacier, LeConte Glacier, Alaska, U.S.A. *Annals of Glaciology* 36, 57–65.

Motyka, R. J., Lawson, D., Finnegan, D., Kalli, G., Molnia, B. and Arendt, A. 2008. Hubbard Glacier update: another closure of Russell Fiord in the making? *Journal of Glaciology* 54, 562–4.

Mueller, D. R., Vincent, W. F. and Jeffries, M. O. 2003. Break-up of the largest Arctic ice shelf and associated loss of an epishelf lake. *Geophysical Research Letters* 30, doi:10.1029/2003GL017931.

Mulder, T. and Alexander, J. 2001. The physical character of subaqueous sedimentary density flows and their deposits. *Sedimentology* 48, 269–99.

Müller, F. 1962. Zonation in the accumulation area of the glaciers of Axel Heiberg Island, N.W.T., Canada. *Journal of Glaciology* 4, 302–13.

Müller, F. and Iken, A. 1973. Velocity fluctuations and water regime of Arctic valley glaciers. *International Association of Hydrological Sciences Publication* 95, 165–82.

Mulugeta, G. and Koyi, H. 1987. Three-dimensional geometry and kinematics of experimental piggyback thrusting. *Geology* 15, 1052–6.

Munro, D. S. 1990. Comparison of melt energy computations and ablatometer measurements on melting ice and snow. *Arctic and Alpine Research* 22, 153–62.

Munro, M. J. and Shaw, J. 1997. Erosional origin of hummocky terrain in south-central Alberta, Canada. *Geology* 25, 1027–30.

Munro-Stasiuk, M. J. 1999. Evidence for water storage and drainage at the base of the Laurentide Ice Sheet. *Annals of Glaciology* 28, 175–80.

Munro-Stasiuk, M. J. 2000. Rhythmic till sedimentation: evidence for repeated hydraulic lifting of a stagnant ice mass. *Journal of Sedimentary Research* 70, 94–106.

Munro-Stasiuk, M. J. 2003. Subglacial Lake McGregor, south-central Alberta, Canada. *Sedimentary Geology* 160, 325–50.

Munro-Stasiuk, M. J. and Shaw, J. 2002. The Blackspring Ridge flute field, south-central Alberta, Canada: evidence for subglacial sheetflow erosion. *Quaternary International* 90, 75–86.

Munro-Stasiuk, M. J. and Sjogren, D. 2006. The erosional origin of hummocky terrain, Alberta, Canada. In Knight, P. G. (ed.), *Glacier Science and Environmental Change*. Blackwell, London, 33–6.

Munro-Stasiuk, M. J., Fisher, T. G. and Nitzsche, C. R. 2005. The origin of the western Lake Erie grooves, Ohio: implications for reconstructing the subglacial hydrology of the Great Lakes sector of the Laurentide Ice Sheet. *Quaternary Science Reviews* 24, 2392–409.

Murray, T. 1997. Assessing the paradigm shift: deformable glacier beds. *Quaternary Science Reviews* 16, 995–1016.

Murray, T. and Porter, P. R. 2001. Basal conditions beneath a soft-bedded polythermal surge-type glacier. *Quaternary International* 86, 103–16.

Murray, T., Corr, H., Forieri, A. and Smith, A. M. 2008. Contrasts in hydrology between regions of basal deformation and sliding beneath Rutford Ice Stream, West Antarctica, mapped using radar and seismic data. *Geophysical Research Letters* 35, doi:10.1029/2008GL033681.

Murray, T., Dowdeswell, J. A., Drewry, D. J. and Frearson, I. 1998. Geometric evolution and ice dynamics during a surge of Bakaninbreen, Svalbard. *Journal of Glaciology* 44 (147), 263–72.

Murray, T., Gooch, D. L. and Stuart, G. W. 1997. Structures within the surge front at Bakaninbreen using ground penetrating radar. *Annals of Glaciology* 24, 122–9.

Murray, T., Smith, A. M., King, M. A. and Weedon, G. P. 2007. Ice flow modulated by tides at up to annual frequencies at Rutford Ice Stream. *Geophysical Research Letters* 34, doi:10.1029/2008GL031207.

Murray, T., Strozzi, T., Luckman, A., Jiskoot, H. and Christakos, P. 2003. Is there a single surge mechanism? Contrasts in dynamics between glacier surges in Svalbard and other regions. *Journal of Geophysical Research* 108, B5 2237, doi:10.1029/2002JB001906.

Murray, T., Stuart, G. W., Miller, P. J., Woodward, J., Smith, A. M., Porter, P. R. and Jiskoot, H. 2000. Glacier surge propagation by thermal evolution at the bed. *Journal of Geophysical Research* 105, B6, 13491–507.

Murton, J. B., Waller, R. I., Hart, J. K., Whiteman, C. A., Pollard, W. H. and Clark, I. D. 2004. Stratigraphy and glaciotectonic structures of permafrost deformed beneath the northwest margin of the Laurentide Ice Sheet, Tuktoyaktuk Coastlands, Canada. *Journal of Glaciology* 50, 399–412.

Murton, J. B., Whiteman, C. A., Waller, R. I., Pollard, W. H., Clark, I. D. and Dallimore, S. R. 2005. Basal ice facies and supraglacial melt-out till of the Laurentide Ice Sheet, Tuktoyaktuk Coastlands, western Arctic Canada. *Quaternary Science Reviews* 24, 681–708.

Naito, N., Nakawo, M., Kadota, T. and Raymond, C. 2000. Numerical simulation of recent shrinkage of Khumbu Glacier, Nepal Himalaya. In Nakawo, M., Fountain, A. and Raymond, C. (eds), *Debris-Covered Glaciers*. IAHS Publication 264, 245–54.

Nakamura, T. and Jones, S. J. 1973. Mechanical properties of impure ice crystals. In Whalley, E., Jones, S. J. and Gold, L. W. (eds), *Physics and Chemistry of Ice*. Ottawa, Royal Society of Canada, 365–9.

Nakawo, M. and Rana, B. 1999. Estimate of ablation rate of glacier ice under a supraglacial debris layer. *Geografiska Annaler* 81A, 695–701.

Nakawo, M. and Young, G. J. 1981. Field experiments to determine the effect of a debris layer on ablation of glacier ice. *Annals of Glaciology* 2, 85–91.

Nakawo, M. and Young, G. J. 1982. Estimate of glacier ablation under a debris layer from surface temperature and meteorological variables. *Journal of Glaciology* 28, 29–34.

Nakawo, M., Raymond, C. F. and Fountain, A. (eds). 2000. *Debris-Covered Glaciers*. IAHS Publication 264.

Nakawo, M., Yabuki, H. and Sakai, A. 1999. Characteristics of Khumbu glacier, Nepal Himalaya: recent change in the debris-covered area. *Annals of Glaciology* 28, 118–22.

Napieralski, J., Harbor, J. and Yingkui, L. 2007. Glacial geomorphology and geographic information systems. *Earth Science Reviews* 85, 1–22.

Naruse, R. and Leiva, J. C. 1997. Preliminary study on the shape of snow penitents at Piloto Glacier, the central Andes. *Bulletin of Glacier Research* 15, 99–104.

Naruse, R., Fukami, H. and Aniya, M. 1992. Short-term variations in flow velocity of Glaciar Soler, Patagonia, Chile. *Journal of Glaciology* 38, 152–6.

Nelson, A. E., Willis, I. C. and Ó Cofaigh, C. 2005. Till genesis and glacier motion inferred from sedimentological evidence associated with the surge-type glacier, Brúarjökull, Iceland. *Annals of Glaciology* 42, 14–22.

Nemec, W. 1990. Aspects of sediment movement on steep delta slopes. In Colella, A. and Prior, D. (eds), *Coarse Grained Deltas*. International Association of Sedimentologists, Special Publication 10, 29–73.

Nemec, W. 1995. The dynamics of deltaic suspension plumes. In Oti, M. N. and Postma, G. (eds), *Geology of Deltas*. Balkema, Rotterdam, 31–93.

Nemec, W. and Steel, R. J. 1984. Alluvial and coastal conglomerates: their significant features and some comments on gravelly mass-flow deposits. In Koster, E. H. and Steel, R. J. (eds), *Sedimentology of Gravels and Conglomerates*. Canadian Society of Petroleum Geologists, Memoir 10, 1–31.

Nemec, W., Lønne, I. and Blikra, L. 1999. The Kregnes moraine in Gauldalen, west-central Norway: anatomy of a Younger Dryas proglacial delta in a palaeofjord basin. *Boreas* 28, 454–76.

Nereson, N. A. 2000. Elevation of ice-stream margin scars after stagnation. *Journal of Glaciology* 46 (152), 111–18.

Nesje, A. 2005. Briksdalsbreen in western Norway: AD 1900–2004 frontal fluctuations as a combined effect of variations in winter precipitation and summer temperature. *The Holocene* 15, 1245–52.

Nesje, A. and Dahl, S. O. 2000. *Glaciers and Environmental Change*. Oxford University Press.

Nesje, A. and Dahl, S. O. 2003. Glaciers as indicators of Holocene climatic change. In Mackay, A., Battarbee, R., Birks, J. and Oldfield, F. *Global Change in the Holocene*. Arnold, London, 264–80.

Nesje, A. and Whillans, I. M. 1994. Erosion of Sognefjord, Norway. *Geomorphology* 9, 33–45.

Nesje, A., Dahl, S. O., Valen, V. and Ovstedal, J. 1992. Quaternary erosion in the Sognefjord drainage basin, western Norway. *Geomorphology* 5, 511–20.

Newman, W. A. and Mickelson, D. M. 1994. Genesis of the Boston Harbor drumlins, Massachusetts. *Sedimentary Geology* 91, 333–43.

Ng, F. 2000. Canals under sediment-based ice sheets. *Annals of Glaciology* 30, 146–52.

Ng, F. and Björnsson, H. 2003. On the Claque-Mathews relation for jökulhlaups. *Journal of Glaciology* 49, 161–72.

Ng, F. and Conway, H. 2004. Fast-flow signature in the stagnated Kamb Ice Stream, West Antarctica. *Geology* 32 (6), 481–4.

Ng, F. and Hallet, B. 2002. Patterning mechanisms in subglacial carbonate dissolution and deposition. *Journal of Glaciology* 48, 386–400.

Ng, F., Liu, S., Mavlyudov, B. and Wang, Y. 2007. Climatic control on the peak discharge of glacier outburst floods. *Geophysical Research Letters* 34, L21503, doi:10.1029/2007GL031426.

NGRIP (North Greenland Ice Core Project) members. 2004. High-resolution record of Northern hemisphere climate extending into the last interglacial period. *Nature* 431, 147–51.

Nicholas, A. P. and Sambrook-Smith, G. H. 1998. Relationships between flow hydraulics, sediment supply, bedload transport and channel stability in the proglacial Virkisa River, Iceland. *Geografiska Annaler* 80A, 111–22.

Nichols, R. J., Sparks, R. S. J. and Wilson, C. J. N. 1994. Experimental studies of the fluidization of layered sediments and the formation of fluid escape structures. *Sedimentology* 41, 233–53.

Nicholson, L. A. 2004. Modelling Melt Beneath Supraglacial Debris: Implications for the Climatic Response of Debris-covered Glaciers. Unpublished PhD thesis, University of St Andrews.

Nicholson, L. A. and Benn, D. I. 2006. Calculating ablation beneath a debris layer from meteorological data. *Journal of Glaciology* 52 (187), 463–70.

Nick, F. M., van der Veen, C. J. and Oerlemans, J. 2007a. Controls on advance of tidewater glaciers: results from numerical modeling applied to Columbia Glacier. *Journal of Geophysical Research* 112, doi:10.1029/2006JF000551.

Nick, F. M., van der Kwast, J. and Oerlemans. J. 2007b. Simulation of the evolution of Breiðamerkurjökull in the late Holocene, *Journal of Geophysical Research* 112, B01103, doi:10.1029/2006JB004358.

Nick, F. M., Vieli, A., Howat, I. M. and Joughin, I. 2009. Large-scale changes in Greenland outlet glacier dynamics triggered at the terminus. *Nature Geoscience* doi:10.1038/NG0394.

Nickling, W. G. and Bennett, L. 1984. The shear strength characteristics of frozen coarse granular debris. *Journal of Glaciology* 30, 348–57.

Nienow, P. W., Hubbard, A. L., Hubbard, B. P., Chandler, D. M., Mair, D. W. F., Sharp, M. J. and Willis, I. C. 2005. Hydrological controls on diurnal ice flow variability in valley glaciers. *Journal of Geophysical Research* 110, doi:10.1029/2003JF000112.

Nienow, P. W., Sharp, M. and Willis, I. 1998. Seasonal changes in the morphology of the subglacial drainage system, Haut Glacier d'Arolla, Switzerland. *Earth Surface Processes and Landforms* 23, 825–43.

Noerdlinger, P. D. and Brower, K. R. 2007. The melting of floating ice raises the ocean level. *Geophysical Journal International* 170, 145–50.

Noetzli, J., Hoelzle, M. and Haeberli, W. 2003. Mountain permafrost and recent Alpine rock-fall events: a GIS-based approach to determine critical factors. *ICOP 2003 Permafrost: Proceedings of the Eighth International Conference on Permafrost, 21–25 July 2003, Zurich, Switzerland*, Balkema, Rotterdam, vol. 2, 827–32.

Nolan, M. 2003. The 'Galloping Glacier' trots: decadal-scale speed oscillations within the quiescent phase. *Annals of Glaciology* 36, 7–13.

Nolin, A. W. and Payne, M. C. 2007. Classification of glacier zones in western Greenland using albedo and surface roughness from the Multi-angle Imaging SpectroRadiometer (MISR). *Remote Sensing of Environment* 107, 264–75.

Nussbaumer, S. U., Zumbühl, H. J. and Steiner, D. 2007. Fluctuations of the 'Mer de Glace' (Mont Blanc area, France) AD 1500–2050: an interdisciplinary approach using new historical data and neural network simulations. *Zeitschrift für Gletscherkunde und Glazialgeologie* 40, 5–175.

Nye, J. F. 1952. The mechanics of glacier flow. *Journal of Glaciology* 2, 82–93.

Nye, J. F. 1957. The distribution of stress and velocity in glaciers and ice sheets. *Proceedings of the Royal Society* A 239, 113–33.

Nye, J. F. 1963. On the theory of the advance and retreat of glaciers. *Geophysical Journal of the Royal Astronomical Society* 7, 431–56.

Nye, J. F. 1976. Water flow in glaciers; jökulhlaups, tunnels and veins. *Journal of Glaciology* 17, 181–207.

Nye, J. F. 1989. The geometry of water veins and nodes in polycrystalline ice. *Journal of Glaciology* 35, 17–22.

Nye, J. F. 1997. Comments on 'Temperate ice permeability, stability of water veins and percolation of internal meltwater' by L. Lliboutry. *Journal of Glaciology* 43, 372.

Ó Cofaigh, C. 1996. Tunnel valley genesis. *Progress in Physical Geography* 20, 1–19.

Ó Cofaigh, C. 1998. Geomorphic and sedimentary signatures of early Holocene deglaciation in high arctic fiords, Ellesmere Island, Canada: implications for deglacial ice dynamics and thermal regime. *Canadian Journal of Earth Sciences* 35, 437–52.

Ó Cofaigh, C. 2007. Glacimarine sediments and ice-rafted debris. In Elias, S. (ed.), *Encyclopedia of Quaternary Science*. Elsevier, Amsterdam, 932–45.

Ó Cofaigh, C. and Dowdeswell, J. A. 2001. Laminated sediments in glacimarine environments: diagnostic criteria for their interpretation. *Quaternary Science Reviews* 20, 1411–36.

Ó Cofaigh, C., Dowdeswell, J. A., Allen, C. S., Hiemstra, J. F., Pudsey, C. J., Evans, J. and Evans, D. J. A. 2005. Flow dynamics and till genesis associated with a marine-based Antarctic palaeo-ice stream. *Quaternary Science Reviews* 24, 709–40.

Ó Cofaigh, C., Dowdeswell, J. A., Evans, J., Kenyon, N. H., Taylor, J., Mienert, J. and Wilken, M. 2004. Timing and significance of glacially influenced mass wasting in the submarine channels of the Greenland Basin. *Marine Geology* 207, 39–54.

Ó Cofaigh, C., Dowdeswell, J. A., Evans, J. and Larter, R. D. 2008. Geological constraints on Antarctic palaeo-ice stream retreat. *Earth Surface Processes and Landforms* 33, 513–25.

Ó Cofaigh, C., Dowdeswell, J. A. and Grobe, H. 2001. Holocene glacimarine sedimentation, inner Scoresby Sund, East Greenland: the influence of fast-flowing ice sheet outlet glaciers. *Marine Geology* 175, 103–29.

Ó Cofaigh, C., England, J. and Zreda, M. 2000. Late Wisconsinan glaciation of southern Eureka Sound: evidence for extensive Innuitian ice in the Canadian High Arctic during the Last Glacial Maximum. *Quaternary Science Reviews* 19, 1319–41.

Ó Cofaigh, C., Evans, J., Dowdeswell, J. A. and Larter, R. D. 2007. Till characteristics, genesis and transport beneath Antarctic palaeo-ice streams. *Journal of Geophysical Research* 112, F03006, doi:10.1029/2006JF000606.

Ó Cofaigh, C., Evans, D. J. A. and England, J. 2003b. Ice-marginal terrestrial landsystems: sub-polar glacier margins of the Canadian and Greenland high arctic. In Evans, D. J. A. (ed.), *Glacial Landsystems*. Arnold, London, 44–64.

Ó Cofaigh, C., Evans, D. J. A. and Smith, I. R. In press. Large-scale reorganisation and sedimentation of terrestrial ice: streams during Late Wisconsinan Laurentide Ice Sheet deglaciation. *Geological Society of America Bulletin*.

Ó Cofaigh, C., Lemmen, D. S., Evans, D. J. A. and Bednarski, J. 1999. Glacial landform-sediment assemblages in the Canadian

High Arctic and their implications for late Quaternary glaciation. *Annals of Glaciology* 28, 195–201.

Ó Cofaigh, C., Pudsey, C. J., Dowdeswell, J. A. and Morris, P. 2002a. Evolution of subglacial bedforms along a paleo-ice stream, Antarctic Peninsula continental shelf. *Geophysical Research Letters* 29, doi:10.1029/2001GL014488.

Ó Cofaigh, C., Taylor, J., Dowdeswell, J. A. and Pudsey, C. J. 2003a. Palaeo-ice streams, trough mouth fans and high latitude continental slope sedimentation. *Boreas* 32, 37–55.

Ó Cofaigh, C., Taylor, J., Dowdeswell, J. A., Rosell-Melé, A., Kenyon, N. H., Evans, J. and Mienert, J. 2002b. Sediment reworking on high latitude continental margins and its implications for palaeoceanographic studies: insights from the Norwegian-Greenland Sea. In Dowdeswell, J. A. and Ó Cofaigh, C. (eds), *Glacier-influenced Sedimentation on High Latitude Continental Margins*. Geological Society, London, Special Publication 203, 325–48.

O'Connor, J. 1993. *Hydrology, Hydraulics and Geomorphology of the Bonneville Flood*. Geological Society of America, Special Paper 274.

O'Grady, D. B. and Syvitski, J. P. M. 2002. Large scale morphology of Arctic continental slopes: the influence of sediment delivery on slope form. In Dowdeswell, J. A. and Ó Cofaigh, C. (eds), *Glacier-Influenced Sedimentation on High Latitude Continental Margins*. Geological Society, London, Special Publication 203, 11–31.

O'Neel, S., Echelmeyer, K. A. and Motyka, R. J. 2001. Short-term flow dynamics of a retreating tidewater glacier: LeConte Glacier, Alaska, U.S.A. *Journal of Glaciology* 47, 567–78.

O'Neel, S., Echelmeyer, K. A. and Motyka, R. J. 2003. Short-term variations in calving of a tidewater glacier: LeConte Glacier, Alaska, U.S.A. *Journal of Glaciology* 49, 587–98.

O'Neel, S., Marshall, H. P., McNamara, D. E. and Pfeffer, W. T. 2007. Seismic detection and analysis of icequakes at Columbia Glacier, Alaska. *Journal of Geophysical Research* 112, doi:10.1029/2006JF000595.

O'Neel, S., Pfeffer, W. T., Krimmel, R. and Meier, M. 2005. Evolving force balance at Columbia Glacier, Alaska, during its rapid retreat. *Journal of Geophysical Research*, 110, F03012, doi:10.1029/2005JF000292.

Obleitner, F. 1994. Climatological features of glacier and valley winds at the Hintereisferner (Ötztal Alps, Austria). *Theoretical and Applied Climatology* 49, 225–39.

Oerlemans, J. 1999. Comments on 'Mass balance of glaciers other than the ice sheets' by Cogley and Adams. *Journal of Glaciology* 45, 397–8.

Oerlemans, J. 2001. *Glaciers and Climate Change*. Balkema, Lisse.

Oerlemans, J. 2005. Extracting a climate signal from 169 glacier records. *Science* 308, 675–7.

Oerlemans, J. 2008. *Minimal glacier models*. Utrecht Publishing and Archiving Services, Universiteitsbibliotheek Utrecht (available online at: http://www.phys.uu.nl/~oerlemns/).

Oerlemans, J. and Fortuin, J. P. F. 1992. Sensitivity of glaciers and small ice caps to greenhouse warming. *Science* 258, 115–17.

Oerlemans, J. and Grisogono, B. 2002. Glacier winds and parameterisation of the related surface heat fluxes. *Tellus* A 54, 440–52.

Oerlemans, J. and van der Veen, C. J. 1984. *Ice Sheets and Climate*. Reidel, Dordrecht.

Oerlemans, J., Anderson, B., Hubbard, A., Huybrechts, P., Jóhannesson, T., Knap, W. H., Schmeits, M., Stroeven, A. P., van de Wal, R. S. W., Wallinga, J. and Zuo, Z. 1998. Modelling the response of glaciers to climate warming. *Climate Dynamics* 14, 267–74.

Oerlemans, J., Dahl-Jensen, D. and Masson-Delmotte, V. 2006. Ice sheets and sea level. *Science* 313, 1043–5.

Oerter, H., Kipfstuhl, J., Determann, J., Miller, H., Wagenbach, D., Minikin, A. and Graft, W. 1992. Evidence for basal marine ice in the Filchner–Ronne ice shelf. *Nature* 358, 399–401.

Ohlendorf, C., Niessen, F. and Weissert, H. 1997. Glacial varve thickness and 127 years of instrumental climate data – a comparison. *Climate Change* 36, 391–411.

Ohmura, A. 2001. Physical basis for the temperature-based melt-index method. *Journal of Applied Meteorology* 40, 753–61.

Ohmura, A., Kasser, P. and Funk, M. 1992. Climate at the equilibrium line of glaciers. *Journal of Glaciology* 38, 397–411.

Oke, T. R. 1987. *Boundary Layer Climates*. Routledge, London.

Oldale, R. N. and O'Hara, C. J. 1984. Glaciotectonic origin of the Massachusetts coastal end moraines and a fluctuating late Wisconsinan ice margin. *Geological Society of America Bulletin* 95, 61–74.

Oldfield, F. 1991. Environmental magnetism: a personal perspective. *Quaternary Science Reviews* 10, 73–85.

Olvmo, M. and Johansson, M. 2002. The significance of rock structure, lithology and pre-glacial deep weathering for the shape of intermediate scale glacial erosional landforms. *Earth Surface Processes and Landforms* 27, 251–68.

Olvmo, M., Lidmar-Bergström, K. and Lindberg, G. 1999. The glacial impact on an exhumed sub-Mesozoic etch surface in southwestern Sweden. *Annals of Glaciology* 28, 153–60.

Olvmo, M., Lidmar-Bergström, K., Ericson, K. and Bonow, J. M. 2005. Saprolite remnants as indicators of pre-glacial landform genesis in southeast Sweden. *Geografiska Annaler* 87A, 447–60.

Oppenheimer, M. 1998. Global warming and the stability of the west Antarctic Ice Sheet. *Nature* 393, 325–32.

Orheim, O. 1980. Physical characteristics and life expectancy of tabular Antarctic icebergs. *Annals of Glaciology* 1, 11–18.

Orwin, J. F. and Smart, C. C. 2004. The evidence for paraglacial sedimentation and its temporal scale in the deglacierizing basin of Small River Glacier, Canada. *Geomorphology* 58, 175–202.

Oskin, M. and Burbank, D. W. 2005. Alpine landscape evolution dominated by cirque retreat. *Geology* 33, 933–6.

Osmaston, H., 2005. Estimates of glacier equilibrium line altitudes by the area-altitude, the area-altitude balance ratio and the area-altitude balance index methods and their validation. *Quaternary International* 138/139, 22–31.

Østrem, G. 1959. Ice melting under a thin layer of moraine and the existence of ice in moraine ridges. *Geografiska Annaler* 41, 228–30.

Østrem, G. 1964. Ice-cored moraines in Scandinavia. *Geografiska Annaler* 46, 282–337.

Østrem, G. 1966. The height of the glaciation limit in southern British Columbia and Alberta. *Geografiska Annaler* 48A, 126–38.

Østrem, G. 1975. Sediment transport in glacial meltwater streams. In Jopling, A. V. and MacDonald, B. C. (eds), *Glaciofluvial and Glaciolacustrine Sedimentation*. SEPM Special Publication 23, 101–22.

Østrem, G. and Brugman, M. 1991. *Mass Balance Measurements: A Manual for Field and Office Work*. National Hydrology Research Institute, Scientific Report 4, Environment Canada, Saskatoon and Norges Vassdrags og Elektrisitetsvesen, Oslo.

Østrem, G. and Haakensen, N. 1993. Glaciers of Europe – Glaciers of Norway. In Williams, R. S. and Ferrigno, J. G. (eds), *Satellite Image Atlas of Glaciers of the World*. USGS Professional Paper 1386-E, 63–109.

Østrem, G. and Haakensen, N. 1999. Map comparison or traditional mass-balance measurements: which method is better? *Geografiska Annaler* 81A, 703–12.

Østrem, G., Haakensen, N. and Olsen, H. C. 2005. Sediment transport, delta growth and sedimentation in Lake Nigardsvatn, Norway. *Geografiska Annaler* 87A, 243–58.

Ottesen, D. and Dowdeswell, J. A. 2006. Assemblages of submarine landforms produced by tidewater glaciers in Svalbard. *Journal of Geophysical Research* 111, doi:10.1029/2005JF000330.

Ottesen, D. and Dowdeswell, J. A. 2009. An inter-ice-stream glaciated margin: submarine landforms and a geomorphic model based on marine-geophysical data from Svalbard. *Geological Society of America Bulletin* 121, 1647–65.

Ottesen, D., Dowdeswell, J. A. and Rise, L. 2005a. Submarine landforms and the reconstruction of fast-flowing ice streams within a large Quaternary ice sheet: the 2500-km-long Norwegian-Svalbard margin (57°–80° N). *Geological Society of America Bulletin* 117, 1033–50.

Ottesen, D., Dowdeswell, J. A., Landvik, J. Y. and Mienert, J. 2007. Dynamics of the Late Weichselian ice sheet on Svalbard inferred from high-resolution sea-floor morphology. *Boreas* 36, 286–306.

Ottesen, D., Dowdeswell, J. A., Benn, D. I., Kristensen, L., Christiansen, H. H., Christensen, O., Hansen, L., Lebesbye, E., Forwick, M. and Vorren, T. O. 2008a. Submarine landforms characteristic of glacier surges in two Spitsbergen fjords. *Quaternary Science Reviews* 27, 1583–99.

Ottesen, D., Rise, L., Knies, J., Olsen, L. and Henriksen, S. 2005b. The Vestfjorden-Trænadjupet palaeo-ice stream drainage system, mid-Norwegian continental shelf. *Marine Geology* 218, 175–89.

Ottesen, D., Stokes, C. R., Rise, L. and Olsen, L. 2008b. Ice-sheet dynamics and ice streaming along the coastal parts of northern Norway. *Quaternary Science Reviews* 27, 922–40.

Otto-Bliesner, B. L., Marshall, S. J., Overpeck, J. T., Miller, G. H., Hu, A. and CAPE Last Interglacial Project Members. 2006. Simulating arctic climate warmth and icefield retreat in the last interglaciation. *Science* 311, 1751–3.

Otero, J., Navarro, F.J., Matrin, C., Cuadrado, M.L. and Corcuera, M.I. 2010. A three-dimensional calving model: numerical experiments on Johnsons Glacier, Livingston Island, Antarctica. *Journal of Glaciology* 56 (196), 200–214.

Otvos, E. G. 2000. Beach ridges – definitions and significance. *Geomorphology* 32, 83–108.

Overpeck, J. T., Otto-Bliesner, B. L., Miller, G. H., Muhs, D. R., Alley, R. B. and Kiehl, J. T. 2006. Palaeoclimatic evidence for future ice sheet instability and rapid sea-level rise. *Science* 311, 1747–50.

Overweel, C. J. 1977. *Distribution and Transport of Fennoscandian Indicators*. Scripta Geologica 43, Leiden.

Owen, L. A. 1991. Mass movement deposits in the Karakoram Mountains. *Zeitschrift für Geomorphologie* 35, 401–24.

Owen, L. A. 1994. Glacial and non-glacial diamictons in the Karakoram Mountains and Western Himalayas. In Warren, W. P. and Croot, D. G. (eds) *The Formation and Deformation of Glacial Deposits*, Balkema, Rotterdam, 9–28.

Owen, L. A. and Derbyshire, E. 1988. Glacially-deformed diamictons in the Karakoram Mountains, northern Pakistan. In Croot, D. G. (ed.), *Glacitectonics: Forms and Processes*. Balkema, Rotterdam, 149–76.

Owen, L. A. and Derbyshire, E. 1989. The Karakoram glacial depositional system. *Zeitschrift für Geomorphologie*, Suppl. Bd 76, 33–73.

Owen, L. A. and Derbyshire, E. 1993. Quaternary and Holocene intermontane basin sedimentation in the Karakoram Mountains. In Shroder, J. F. (ed.), *Himalaya to the Sea*. Routledge, London, 108–31.

Owen, L. A. and England, J. 1998. Observations on rock glaciers in the Himalayas and Karakoram mountains of northern Pakistan and India. *Geomorphology* 26, 199–214.

Owen, L. A. and Sharma, M. C. 1998. Rates of paraglacial fan formation in the Garwhal Himalaya: implications for landscape evolution. *Geomorphology* 26, 171–84.

Owen, L. A., Caffee, M., Finkel, R. C. and Seong, Y. B. 2008. Quaternary glaciation of the Himalayan-Tibetan orogen. *Journal of Quaternary Science* 23, 513–31.

Owen, L. A., Derbyshire, E. and Scott, C. H. 2002a. Contemporary sediment production and transfer in high-altitude glaciers. *Sedimentary Geology* 155, 13–36.

Owen, L. A., Finkel, R. C., Caffee, M. W. and Gualtieri, L. 2002b. Timing of multiple glaciations during the late Quaternary in the Hunza Valley, Karakoram Mountains, northern Pakistan: defined by cosmogenic radionuclide dating of moraines. *Geological Society of America Bulletin* 114, 593–604.

Owen, L. A., Gualtieri, L., Finkel, R. C., Caffee, M. W., Benn, D. I. and Sharma, M. C. 2001. Cosmogenic radionuclide dating of glacial landforms in the Lahul Himalaya, northern India: defining the timing of late Quaternary glaciation. *Journal of Quaternary Science* 16, 555–63.

Owen, L. A., Robinson, R., Benn, D. I., Finkel, R. C., Yi, C., Putkonen, J., Li, D., Murray, A. S., and Davis, N. K. 2009. Quaternary glaciation of Mount Everest. *Quaternary Science Reviews* 28, 1412–33.

Owens, I. F. 1992. A note on the Mount Cook rock avalanche of 14 December 1991. *New Zealand Geographer* 48, 74–8.

Pälli, A., Moore, J. C., Jania, J., Kolondra, L. and Glowacki, P. 2003. The drainage pattern of Hansbreen and Werenskioldbreen, two polythermal glaciers in Svalbard. *Polar Research* 22, 355–71.

Palmer, A. P., Lowe, J. J. and Rose, J. 2008. *The Quaternary of Glen Roy and Vicinity – Field Guide*. Quaternary Research Association, London.

Palmer, A. P., Rose, J., Lowe, J. J. and Walker, M. J. C. 2007. The pre-Holocene succession of glaciolacustrine sediments within Llangorse basin. In Carr, S. J., Coleman, S. G., Humpage, A. J. and Shakesby, R. A. (eds), *Quaternary of the Brecon Beacons – Field Guide*. Quaternary Research Association, London, 220–9.

Pálsson, F., Björnsson, H., Eydal, G. P. and Haraldsson, H. H. 2001. *Vatnajökull: Mass Balance, Meltwater Drainage and Surface Velocity of the Glacial Year 1999–2000*. University of Iceland, Science Institute, National Power Company of Iceland, Reykjavik, Report RH-01-2001.

Parent, M., Paradis, S. J. and Doiron, A. 1996. Palimpsest glacial dispersal trains and their significance for drift prospecting. *Journal of Geochemical Exploration* 56, 123–40.

Parish, T. R. and Bromwich, D. H. 1989. Instrumented aircraft observations of the katabatic wind regime near Terra Nova Bay. *Monthly Weather Review* 117, 1570–85.

Parish, T. R. and Bromwich, D. H. 2007. Reexamination of the near surface airflow over the Antarctic continent and implications on atmospheric circulations at high southern latitudes. *Monthly Weather Review* 135, 1961–73.

Parish, T. R. and Cassano, J. J. 2003. The role of katabatic winds on the Antarctic surface wind regime. *Monthly Weather Review* 131, 317–33.

Parizek, B. R. and Alley, R. B. 2004a. Implications of increased Greenland surface melt under global-warming scenarios: ice sheet simulations. *Quaternary Science Reviews* 23, 1013–27.

Parizek, B. R. and Alley, R. B. 2004b. Ice thickness and isostatic imbalances in the Ross Embayment, West Antarctica: model results. *Global and Planetary Change* 42, 265–78.

Parizek, B. R., Alley, R. B. and MacAyeal, D. R. 2005. The PSU/UofC finite-element thermomechanical flowline model of ice-sheet evolution. *Cold Regions Science and Technology* 42, 145–68.

Park, R. G. 1983. *Foundations of Structural Geology*. Blackie, London.

Passchier, C. W. and Trouw, R. A. J. 1996. *Microtectonics*. Springer, Berlin.

Paterson, W. S. B. 1994. *The Physics of Glaciers*, 3rd edn. Pergamon, Oxford.

Paterson, W. S. B. and Reeh, N. 2001. Thinning of the ice sheet in northwest Greenland over the past forty years. *Nature* 414, 60–2.

Patterson, C. J. 1994. Tunnel-valley fans of the St. Croix moraine, east-central Minnesota, USA. In Warren, W. P. and Croot, D. G. (eds), *Formation and Deformation of Glacial Deposits*. Balkema, Rotterdam, 69–87.

Patterson, C. J. 1997. Southern Laurentide ice lobes were created by ice streams: Des Moines Lobe in Minnesota, USA. *Sedimentary Geology* 111, 249–61.

Patterson, C. J. 1998. Laurentide glacial landscapes: the role of ice streams. *Geology* 26, 643–6.

Patterson, C. J. and Boerboom, T. J. 1999. The significance of pre-existing, deeply weathered crystalline rock in interpreting the effects of glaciation in the Minnesota River valley, USA. *Annals of Glaciology* 28, 53–8.

Patterson, C. J. and Hooke, R. Le B. 1996. Physical environment of drumlin formation. *Journal of Glaciology* 41, 30–8.

Patton, P. C. and Schumm, S. A. 1981. Ephemeral-stream processes: implications for studies of Quaternary valley fills. *Quaternary Research* 15, 24–43.

Pattyn, F. 2003. A new three-dimensional higher-order thermomechanical ice sheet model: basic sensitivity, ice stream development and ice flow across subglacial lakes. *Journal of Geophysical Research*, 108, doi:10.1029/2002JB002329.

Pattyn, F. 2006. GRANTISM: an Excel™ model for Greenland and Antarctic ice-sheet response to climate change. *Computers and Geosciences* 32, 316–25.

Pattyn, F. and Decleir, H. 1995. Subglacial topography in the central Sør Rondane Mountains, East Antarctica: configuration and morphometric analysis of cross profiles. *Antarctic Record* 39, 1–24.

Pattyn, F., Huyghe, A., de Brabander, S. and De Smedt, B. 2006. Role of transition zones in marine ice sheet dynamics. *Journal of Geophysical Research* 111, doi:10.1029/JF000394.

Pattyn, F., Perichon, L., Aschwanden, A., Breuer, B., de Smedt, B., Gagliardini, O., Gudmundsson, G. H., Hindmarsh, R., Hubbard, A., Johnson, J. V., Kleiner, T., Konovalov, Y., Martin, C., Payne, A. J., Pollard, D., Price, S., Rückamp, M., Saito, F., Souček, O., Sugiyama, S. and Zwinger, T. 2008. Benchmark experiments for higher-order and full Stokes ice sheet models (ISMIP-HOM). *The Cryosphere* 2, 95–108.

Paul, M. A. and Evans, H. 1974. Observations on the internal structure and origin of some flutes in glaciofluvial sediments, Blomstrandbreen, north-west Spitsbergen. *Journal of Glaciology* 13, 393–400.

Paul, M. A. and Eyles, N. 1990. Constraints on the preservation of diamict facies (melt-out tills) at the margins of stagnant glaciers. *Quaternary Science Reviews* 9, 51–69.

Payne, A. J. 1999. A thermomechanical model of ice flow in West Antarctica. *Climate Dynamics* 15, 115–25.

Payne, A. J. and Dongelmans, P. W. 1997. Self-organization in the thermo-mechanical flow of ice sheets, *Journal of Geophysical Research* 102 (B6), 12219–33.

Payne, A. J., Holland, P. R., Shepherd, A. P., Rutt, I. C., Jenkins, A. and Joughin, I. 2007. Numerical modeling of ocean-ice interactions under Pine Island Bay's ice shelf, *Journal of Geophysical Research* 112, C10019, doi:10.1029/2006JC003733.

Payne, A. J., Huybrechts, P., Abe-Ouchi, A., Calov, R., Fastook, J. L., Greve, R., Marshall, S. J., Marsiat, I., Ritz, C., Tarasov, L. and Thomassen, S. J. 2000. Results from the EISMINT model intercomparison: the effects of thermomechanical coupling. *Journal of Glaciology* 46 (153), 227–38.

Pearce, J. T., Pazzaglia, F. J., Evenson, E. B., Lawson, D. E., Alley, R. B., Germanoski, D. and Denner, J. D. 2003. Bedload component of glacially discharged sediment: Insights from the Matanuska Glacier, Alaska. *Geology* 31, 7–10.

Pearson, P. N. and Palmer, M. R. 2000. Atmospheric carbon dioxide concentrations over the past 60 million years. *Nature* 406, 695–9.

Peck, V. L., Hall, I. R., Zahn, R., Grousset, F., Hemming, S. R. and Scourse, J. D. 2007. The relationship of Heinrich events and their European precursors over the last 60 ka BP: a multi-proxy ice-rafted debris provenance study in the NE Atlantic. *Quaternary Science Reviews* 26, 862–75.

Pedersen, S. A. S. 1989. Glacitectonite: brecciated sediments and cataclastic sedimentary rocks formed subglacially. In Goldthwait, R. P. and Matsch, C. L. (eds), *Genetic Classification of Glacigenic Deposits*, Balkema, Rotterdam, 89–91.

Pedersen, S. A. S. 1993. The glaciodynamic event and glaciodynamic sequence. In Aber, J. S. (ed.), *Glaciotectonics and Mapping*

Glacial Deposits. Canadian Plains Research Center, University of Regina, Saskatchewan, 67–85.

Pedersen, S. A. S. 2000. Superimposed deformation in glaciotectonics. *Bulletin of the Geological Society of Denmark* 46, 125–44.

Pedersen, S. A. S., Petersen, K. S. and Rasmussen, L. A. 1988. Observations on glaciodynamic structures at the Main Stationary Line in western Jutland, Denmark. In Croot, D. G. (ed.), *Glaciotectonics: Forms and Processes.* Balkema, Rotterdam, 177–83.

Peltier, W. R. 1985. New constraints on transient lower mantle rheology and internal mantle buoyancy from glacial rebound data. *Nature* 318, 614–17.

Peltier, W. R. 1987. Glacial isostasy, mantle viscosity, and Pleistocene climatic change. In Ruddiman, W. F. and Wright, H. E. (eds), *North America and Adjacent Oceans During the Last Deglaciation.* Geological Society of America, The Geology of North America, vol. K-3, 155–82.

Peltier, W. R. 1996. Mantle viscosity and ice-age ice sheet topography. *Science* 273, 1359–64.

Peltier, W. R. 1999. Global sea level rise and glacial isostatic adjustment. *Global and Planetary Change* 20, 93–123.

Peltier, W. R. 2002a. On eustatic sea level history: Last Glacial Maximum to Holocene. *Quaternary Science Reviews* 21, 377–96.

Peltier, W. R. 2002b. Global glacial isostatic adjustment: palaeogeodetic and space-geodetic tests of the ICE-4G (VM2) model. *Journal of Quaternary Science* 17, 491–510.

Peltier, W. R. and Andrews, J. T. 1983. Glacial geology and glacial isostasy of the Hudson Bay region. In Smith, D. E. and Dawson, A. G. (eds), *Shorelines and Isostasy.* Academic Press, London, 285–320.

Peltier, W. R. and Drummond, R. 2002. A 'broad-shelf effect' upon postglacial relative sea level history. *Geophysical Research Letters* 29 (8), 1169, doi:10.1029/2001GL014273.

Person, M., McIntosh, J., Bense, V. and Remenda, V. H. 2007. Pleistocene hydrology of North America: the role of ice sheets in reorganizing groundwater flow systems. *Reviews of Geophysics* 45, RG3007, doi:10.1029/2006RG000206.

Peteet, D. 1995. Global Younger Dryas? *Quaternary International* 28, 93–104.

Peters, L. E., Anandakrishnan, S., Alley, R. B., Winberry, J. P. and Voigt, D. E. 2006. Subglacial sediments as a control on the onset and location of two Siple Coast ice streams, West Antarctica. *Journal of Geophysical Research* 111, B01302. doi:10.1029/2005JB003766.

Petit, J. R. et al. 1999. Climate and atmospheric history of the past 420,000 years from the Vostok ice core, Antarctica. *Nature* 399, 429–36.

Pettersson, R. 2004. *Dynamics of the Cold Surface Layer of Polythermal Storglaciären, Sweden.* PhD thesis, Department of Physical Geography and Quaternary Geology, Stockholm University.

Pettersson, R., Jansson, P. and Holmlund, P. 2003. Cold surface layer thinning on Storglaciären, Sweden, observed by repeated ground penetrating radar surveys. *Journal of Geophysical Research* 108 (F1), doi:10.1029/2003JF000024.

Pettit, E. C., Jacobson, H. P. and Waddington, E. D. 2003. Effects of basal sliding on isochrones and flow near an ice divide. *Annals of Glaciology* 37, 370–6.

Pfeffer, W. T. 1992. Stress-induced foliation in the terminus of Variegated Glacier, Alaska, USA, formed during the 1982–83 surge. *Journal of Glaciology* 38, 213–22.

Pfeffer, W. T. 2007a. *The Opening of a New Landscape: Columbia Glacier at Mid-retreat.* American Geophysical Union, Washington, DC.

Pfeffer, W. T. 2007b. A simple mechanism for irreversible tidewater glacier retreat. *Journal of Geophysical Research* 112, doi:10.1029/2006JF000590.

Pfeffer, W. T., Cohn, J., Meier, M. F. and Krimmel, R. M. 2000. Alaskan glacier beats a rapid retreat. *EOS* 81 (48), 28 November.

Pfeffer, W. T., Harper, J. T. and O'Neel, S. 2008. Kinematic constraints on glacier contributions to 21st century sea-level rise. *Science* 321, 1340–3.

Pfeffer, W. T., Sassolas, C., Bahr, D. B. and Meier, M. 1998. Response time of glaciers as a function of size and mass balance: 2. Numerical experiments. *Journal of Geophysical Research* 103, B5, 9783–9.

Phillips, A. C., Smith, N. D. and Powell, R. D. 1991. Laminated sediments in prodeltaic deposits, Glacier Bay, Alaska. In Anderson, J. B. and Ashley, G. M. (eds), *Glacial Marine Sedimentation: Paleoclimatic Significance.* Geological Society of America, Special Paper 261, 51–60.

Phillips, E. R. 2006. Micromorphology of a debris flow deposit: evidence of basal shearing, hydrofracturing, liquefaction and rotational deformation during emplacement. *Quaternary Science Reviews* 25, 720–38.

Phillips, E. R. and Auton, C. A. 2008. Microtextural analysis of a glacially 'deformed' bedrock: implications for inheritance of preferred clast orientations in diamictons. *Journal of Quaternary Science* 23, 229–40.

Phillips, E. R., Evans, D. J. A. and Auton, C. A. 2002. Polyphase deformation at an oscillating ice margin following the Loch Lomond Readvance, central Scotland, UK. S*edimentary Geology* 149, 157–82.

Phillips, E. R., Lee, J. R. and Burke, H. 2008. Progressive proglacial to subglacial deformation and syntectonic sedimentation at the margins of the mid-Pleistocene British Ice Sheet: evidence from north Norfolk, UK. *Quaternary Science Reviews* 27, 1848–71.

Phillips, E. R., Merritt, J., Auton, C. A. and Golledge, N. 2007. Microstructures in subglacial and proglacial sediments: understanding faults, folds and fabrics, and the influence of water on the style of deformation. *Quaternary Science Reviews* 26, 1499–528.

Phillips, W. M., Hall, A. M., Mottram, R., Fifield, L. K. and Sugden, D. E. 2006. Cosmogenic ^{10}Be and ^{26}Al exposure ages of tors and erratics, Cairngorm Mountains, Scotland: timescales for the development of a classic landscape of selective linear glacial erosion. *Geomorphology* 73, 222–45.

Pickrill, R. A. and Irwin, J. 1982. Predominant headwater inflow and its control of lake-river interactions in Lake Wakatipu. *New Zealand Journal of Freshwater Research* 16, 201–13.

Pickrill, R. A. and Irwin, J. 1983. Sedimentation in a deep glacier-fed lake: Lake Tekapo, New Zealand. *Sedimentology* 30, 63–75.

Pierce, K. L. 2004. Pleistocene glaciations of the Rocky Mountains. In Gillespie, A. R., Porter, S. C. and Atwater, B. F. (eds), *The Quaternary Period in the United States.* Elsevier, Amsterdam, 63–76.

Pierrehumbert, R. T. 1999. Huascaran δ^{18}O as an indicator of tropical climate during the Last Glacial Maximum. *Geophysical Research Letters* 26, 1345–8.

Pierson, T. C. and Costa, J. E. 1987. A rheologic classification of subaerial sediment-water flows. In Costa, J. E. and Wieczorek, G. F. (eds), *Debris Flows/Avalanches: Process, Recognition, and Mitigation.* Geological Society of America, Reviews in Engineering Geology 7, 1–12.

Pillans, B. 2007. Quaternary stratigraphy – overview. *Encyclopedia of Quaternary Science.* Elsevier, Amsterdam, 2785–802.

Piotrowski, J. A. 1994. Waterlain and lodgement till facies of the lower sedimentary complex from the Danischer-Wohld cliff, Schleswig-Holstein, North Germany. In Warren, W. P. and Croot, D. G. (eds), *Formation and Deformation of Glacial Deposits.* Balkema, Rotterdam, 3–8.

Piotrowski, J. A. 1997a. Subglacial groundwater flow during the last glaciation in northwestern Germany. *Sedimentary Geology* 111, 217–24.

Piotrowski, J. A. 1997b. Subglacial hydrology in northwestern Germany during the last glaciation: groundwater flow, tunnel valleys and hydrological cycles. *Quaternary Science Reviews* 16, 169–85.

Piotrowski, J. A. 2006. Groundwater under ice sheets and glaciers. In Knight, P. G. (ed.), *Glacier Science and Environmental Change*. Blackwell, Oxford, 50–60.

Piotrowski, J. A. and Kraus, A. M. 1997. Response of sediment to ice sheet loading in northwestern Germany: effective stresses and glacier-bed stability. *Journal of Glaciology* 43, 495–502.

Piotrowski, J. A. and Tulaczyk, S. 1999. Subglacial conditions under the last ice sheet in northwest Germany: ice-bed separation and enhanced basal sliding? *Quaternary Science Reviews* 18, 737–51.

Piotrowski, J. A. and Vahldiek, J. 1991. Elongated hills near Schönhorst, Schleswig-Holstein: drumlins or terminal push moraines? *Bulletin of the Geological Society of Denmark* 38, 231–42.

Piotrowski, J. A., Geletneky, J. and Vater, R. 1999. Soft-bedded subglacial meltwater channel from the Welzow-Süd open-cast lignite mine, Lower Lusatia, eastern Germany. *Boreas* 28, 363–74.

Piotrowski, J. A., Larsen, N. J. and Junge, F. W. 2004. Reflections on soft subglacial beds as a mosaic of deforming and stable spots. *Quaternary Science Reviews* 23, 993–1000.

Piotrowski, J. A., Larsen, N. K., Menzies, J. and Wysota, W. 2006. Formation of subglacial till under transient bed conditions: deposition, deformation and basal decoupling under a Weichselian ice sheet lobe, central Poland. *Sedimentology* 53, 83–106.

Piotrowski, J. A., Mickelson, D. M., Tulaczyk, S., Krzyszowski, D. and Junge, F. W. 2001. Were deforming beds beneath past ice sheets really widespread? *Quaternary International* 86, 139–50.

Piotrowski, J. A., Mickelson, D. M., Tulaczyk, S., Krzyszowski, D. and Junge, F. W. 2002. Reply to comments by G. S. Boulton, K. E. Dobbie and S. Zatsepin on: deforming soft beds under ice sheets: how extensive were they? *Quaternary International* 97/98, 173–7.

Piper, D. J. W. and Normark, W. R. 1989. Late Cenozoic sea-level changes and the onset of glaciation: impact on continental slope progradation off eastern Canada. *Marine Petroleum Geology* 6, 336–48.

Pisarska-Jamrozy, M. 2006. Transitional deposits between the end moraine and outwash plain in the Pomeranian glaciomarginal zone of NW Poland: a missing component of ice-contact sedimentary models. *Boreas* 35, 126–41.

Plewes, L. A. and Hubbard, B. 2001. A review of the use of radio-echo sounding in glaciology. *Progress in Physical Geography* 25, 203–36.

Plink-Björkland, P. and Ronnert, L. 1999. Depositional processes and internal architecture of Late Weichselian ice-marginal submarine fan and delta settings, Swedish west coast. *Sedimentology* 46, 215–34.

Pohjola, V. A. 1994. TV-video observation of englacial voids in Storglaciären, Sweden. *Journal of Glaciology* 40, 231–40.

Pollard, D. and DeConto, R. M. 2005. Hysteresis is Cenozoic Antarctic ice-sheet variations. *Global and Planetary Change* 45, 9–21.

Polyak, L., Edwards, M. H., Coakley, B. J. and Jakobsson, M. 2001. Ice shelves in the Pleistocene Arctic Ocean inferred from glaciogenic deep-sea bedforms. *Nature* 410, 453–7.

Pomeroy, J. W. and Jones, H. G. 1996. Wind-blown snow: sublimation, transport and changes to polar snow. In Wolff, E. W. and Bales, R. C. (eds), *Chemical Exchange Between the Atmosphere and Polar Snow*. Springer-Verlag, Berlin, 453–89.

Porter, P. R., Murray, T. and Dowdeswell, J. A. 1997. Sediment deformation and basal dynamics beneath a glacier surge front: Bakininbreen, Svalbard. *Annals of Glaciology* 24, 21–6.

Porter, S. C. 1975. Equilibrium line altitudes of late Quaternary glaciers in the Southern Alps, New Zealand. *Quaternary Research* 5, 27–47.

Porter, S. C. 1977. Present and past glaciation threshold in the Cascade range, Washington State, USA: topographic and climatic controls, and paleoclimatic implications. *Journal of Glaciology* 18, 101–16.

Porter, S. C. 1989. Some geological implications of average Quaternary glacial conditions. *Quaternary Research* 32, 245–61.

Porter, S. C. 2000. Snowline depression in the Tropics during the last glaciation. *Quaternary Science Reviews* 20, 1067–91.

Porter, S. C. and Orombelli, G. 1981. Alpine rockfall hazards. *American Scientist* 69, 67–75.

Porter, S. C. and Orombelli, G. 1982. Late glacial ice advances in the western Italian Alps. *Boreas* 11, 125–40.

Posamentier, H. W., Jervey, M. T. and Vail, P. R. 1988. Eustatic controls on clastic deposition I – conceptual framework. In Wilgus, C. K. et al. (eds), *Sea-level Changes: An Integrated Approach*. SEPM Special Publication 42, 109–24.

Post, A. S. 1969. Distribution of surging glaciers in western North America. *Journal of Glaciology* 8, 229–40.

Post, A. S. and Lachapelle, E. R. 2000. *Glacier Ice*. University of Washington Press, Seattle.

Post, A. S. and Motyka, R. J. 1995. Taku and Le Conte glaciers, Alaska: calving-speed control of late-Holocene asynchronous advances and retreats. *Physical Geography* 16, 59–82.

Postma, G. 1986. Classification for sediment-gravity flow deposits based on flow conditions during sedimentation. *Geology* 14, 291–4.

Postma, G. 1990. Depositional architecture and facies of river and fan deltas: a synthesis. In Colella, A. and Prior, D. B. (eds), *Coarse-Grained Deltas*. Blackwell, Oxford. International Association of Sedimentologists, Special Publication 10, 13–27.

Postma, G. 1995. Causes of architectural variation in deltas. In Oti, M. N. and Postma, G. (eds), *Geology of Deltas*. Balkema, Rotterdam, 3–16.

Postma, G., Nemec, W. and Kleinspehn, K. L. 1988. Large floating clasts in turbidites: a mechanism for their emplacement. *Sedimentary Geology* 58, 47–61.

Postma, G., Roep, T. B. and Ruegg, G. H. J. 1983. Sandy gravelly mass flow deposits in an ice-marginal lake (Saalian, Leuvenumsche Beek Valley, Veluwe, The Netherlands) with emphasis on plug flow deposits. *Sedimentary Geology* 34, 59–82.

Powell, D. M. 1998. Patterns and processes of sediment sorting in gravel-bed rivers. *Progress in Physical Geography* 22, 1–32.

Powell, R. D. 1981. A model for sedimentation by tidewater glaciers. *Annals of Glaciology* 2, 129–34.

Powell, R. D. 1983. Glacial marine sedimentation processes and lithofacies of temperate tidewater glaciers, Glacier Bay, Alaska. In Molnia, B. F. (ed.), *Glacial-Marine Sedimentation*. Plenum Press, New York, 195–232.

Powell, R. D. 1984. Glacimarine processes and inductive lithofacies modelling of ice shelf and tidewater glacier sediments based on Quaternary examples. *Marine Geology* 57, 1–52.

Powell, R. D. 1990. Glacimarine processes at grounding-line fans and their growth to ice-contact deltas. In Dowdeswell, J. A. and Scourse, J. D. (eds), *Glacimarine Environments: Processes and Sediments*. Geological Society, London, Special Publication 53, 53–73.

Powell, R. D. 1991. Grounding-line systems as second-order controls on fluctuations of tidewater termini of temperate glaciers. In Anderson, J. B. and Ashley, G. M. (eds), *Glacial Marine Sedimentation: Paleoclimatic Significance*. Geological Society of America, Special Paper 261, 75–93.

Powell, R. D. 2003. Subaquatic landsystems: fjords. In Evans, D. J. A. (ed.), *Glacial Landsystems*. Arnold, London, 313–47.

Powell, R. D. and Alley, R. B. 1997. Grounding line systems: processes, glaciological inferences and the stratigraphic record. In Barker, P. F. and Cooper, A. C. (eds), *Geology and Seismic Stratigraphy of the Antarctic Margin, 2*. American Geophysical Union, Antarctic Research Series 71, 169–87.

Powell, R. D. and Cooper, J. M. 2002. A glacial sequence strati-graphic model for temperate, glaciated continental shelves. In Dowdeswell, J. A. and Ó Cofaigh, C. (eds), *Glacier-Influenced Sedimentation on High Latitude Continental Margins*. Geological Society, London, Special Publication 203, 215–44.

Powell, R. D. and Domack, E. 1995. Modern glaciomarine environments. In Menzies, J. (ed.), *Glacial Environments. Volume 1: Modern Glacial Environments: Processes, Dynamics and Sediments*. Butterworth-Heinemann, Oxford, 445–86.

Powell, R. D. and Molnia, B. F. 1989. Glacimarine sedimentary processes, facies and morphology of the south-southeast Alaska shelf and fjords. *Marine Geology* 85, 359–90.

Powell, R. D., Dawber, M., McInnes, J. N. and Pyne, A. R. 1996. Observations of the grounding-line area at a floating glacier terminus. *Annals of Glaciology* 22, 217–23.

Powell, R. D., Krissek, L. A. and van der Meer, J. J. M. 2000. Preliminary depositional environmental analysis of Cape Roberts 2/2A, Victoria Land Basin, Antarctica: palaeoglaciological and palaeoclimatic inferences. *Terra Antarctica* 7, 313–22.

Powell, R. D., Laird, M. G., Naish, T. R., Fielding, C. R., Krissek, L. A. and van der Meer, J. J. M. 2001. Depositional environments for strata cored in CRP-3 (Cape Roberts Project), Victoria Land Basin, Antarctica: palaeoglaciological and palaeoclimatological inferences. *Terra Antarctica* 8, 207–16.

Praeg, D. 2003. Seismic imaging of mid-Pleistocene tunnel-valleys in the North Sea Basin: high resolution from low frequencies. *Journal of Applied Geophysics* 53, 273–98.

Pralong, A. and Funk, M. 2006. On the instability of avalanching glaciers. *Journal of Glaciology* 52 (176), 31–48.

Prest, V. K. 1983. *Canada's Heritage of Glacial Features*. Geological Survey of Canada, Miscellaneous Report 28.

Prest, V. K., Donaldson, J. A. and Mooers, H. D. 2000. The omar story: the role of omars in assessing glacial history of west-central North America. *Geographie Physique et Quaternaire* 54, 257–70.

Price, N. J. and Cosgrove, J. W. 1990. *Analysis of Geological Structures*. Cambridge University Press.

Price, R. J. 1970. Moraines at Fjallsjokull, Iceland. *Arctic and Alpine Research* 2, 27–42.

Price, R. J. 1973. *Glacial and Fluvioglacial Landforms*. Oliver and Boyd, Edinburgh.

Price, S. F., Bindschadler, R. A., Hulbe, C. L. and Blankenship, D. D. 2002. Force balance along an inland tributary and onset to Ice Stream D, West Antarctica. *Journal of Glaciology* 48 (160), 20–30.

Price, S. F., Conway, H. and Waddington, E. D. 2007. Evidence for late Pleistocene thinning of Siple Dome, West Antarctica. *Journal of Geophysical Research* 112, doi:10.1029/2006JF000725.

Price, S. F., Payne, A. J., Catania, G. A. and Neumann, T. A. 2008. Seasonal acceleration of inland ice via longitudinal coupling to marginal ice. *Journal of Glaciology* 54, 213–19.

Prior, D. B. and Bornhold, B. D. 1988. Submarine morphology and processes of fjord fan deltas and related high-gradient systems: modern examples from British Columbia. In Nemec, W. and Steel, R. J. (eds), *Fan Deltas: Sedimentology and Tectonic Settings*. Blackie and Son, London, 125–43.

Prior, D. B. and Bornhold, B. D. 1989. Submarine sedimentation on a developing Holocene fan delta. *Sedimentology* 36, 1053–76.

Prior, D. B. and Bornhold, B. D. 1990. The underwater development of Holocene fan deltas. In Colella, A. and Prior, D. B. (eds), *Coarse-Grained Deltas*. International Association of Sedimentologists, Special Publication 10, 75–90.

Prior, D. B., Bornhold, B. D. and Coleman, J. M. 1983. *Geomorphology of a Submarine Landslide, Kitimat Arm, British Columbia*. Geological Survey of Canada, Open File Report 961.

Pritchard, H. D. and Vaughan, D. G. 2007. Widespread acceleration of tidewater glaciers on the Antarctic Peninsula. *Journal of Geophysical Research* 112, doi:10.1029/2006JF000597.

Pritchard, H. D., Arthern, R. J., Vaughan, D. G. and Edwards, L. A. 2009. Extensive dynamic thinning on the margins of the Greenland and Antarctic ice sheets. *Nature*, doi:10.1038/nature08471.

Pritchard, H. D., Murray, T., Luckman, A., Strozzi, T. and Barr, S. 2005. Glacier surge dynamics of Sortebrae, East Greenland, from synthetic aperture radar feature tracking. *Journal of Geophysical Research* 110, F03005, doi:10.1029/2004JF000233.

Pritchard, H. D., Murray, T., Strozzi, T., Barr, S. and Luckman, A. 2003. Surge-related topographic change of the glacier Sortebrae, East Greenland, derived from synthetic aperture radar interferometry. *Journal of Glaciology* 49 (166), 381–90.

Pudsey, C. J. and Evans, J. 2001. First survey of Antarctic sub-ice shelf sediments reveals mid-Holocene ice shelf retreat. *Geology* 29, 787–90.

Pugin, A., Pullan, S. E. and Sharpe, D. R. 1999. Seismic facies and regional architecture of the Oak Ridges Moraine area, southern Ontario. *Canadian Journal of Earth Sciences* 36, 409–32.

Pulina, M. and Rehak, J. 1991. Glacier caves in Spitsbergen. In Eraso, A. (ed.), *1st International Symposium of Glacier Caves and Karst in Polar Regions*. Instituto Tecnológico Geominero de España, Madrid, 93–117.

Punkari, M. 1997a. Subglacial processes of the Scandinavian Ice Sheet in Fennoscandia inferred from flow-parallel features and lithostratigraphy. *Sedimentary Geology* 111, 263–83.

Punkari, M. 1997b. Glacial and glaciofluvial deposits in the interlobate areas of the Scandinavian ice sheet. *Quaternary Science Reviews* 16, 741–53.

Quincey, D. J., Lucas, R. M., Richardson, S. D., Glasser, N. F., Hambrey, M. J. and Reynolds, J. M. 2005. Optical remote-sensing techniques in high-mountain environments: application to glacial hazards. *Progress in Physical Geography* 29, 475–505.

Quincey, D. J., Luckman, A. and Benn, D. I. 2009. Quantification of Everest-region glacier velocities between 1992 and 2002 using satellite radar interferometry and feature tracking. *Journal of Glaciology* 55 (192), 596–606.

Quincey, D. J., Richardson, S. D., Luckman, A., Lucas, R. M., Reynolds, J. M., Hambrey, M. J. and Glasser, N. F. 2007. Early recognition of glacial lake hazards in the Himalaya using remote sensing datasets. *Global and Planetary Change* 56, 137–52.

Rabassa, J. 2008. Late Cenozoic Glaciations in Patagonia and Tierra del Fuego. *Developments in Quaternary Science* 11, 151–204.

Rabus, B. T. and Echelmeyer, K. A. 1997. The flow of a polythermal glacier: McCall Glacier, Alaska, USA. *Journal of Glaciology* 43, 522–36.

Rabus, B. T. and Echelmeyer, K. A. 2002. Increase of 10 m ice temperature: climate warming or glacier thinning? *Journal of Glaciology* 48, 279–86.

Rack, W. and Rott, H. 2004. Pattern of retreat and disintegration of the Larsen B ice shelf, Antarctic Peninsula. *Annals of Glaciology* 39, 505–10.

Radic, V. and Hock, R. 2006. Modelling future glacier mass balance and volume changes using ERA-40 reanalysis and climate models: a sensitivity study at Storglaciären, Sweden. *Journal of Geophysical Research* 111, doi:10.1029/2005JF000440.

Rae, A. C., Harrison, S., Mighall, T. and Dawson, A. G. 2004. Periglacial trimlines and nunataks of the Last Glacial Maximum: the Gap of Dunloe, southwest Ireland. *Journal of Quaternary Science* 19, 87–97.

Rafaelsen, B., Andreassen, K., Kuilman, L. W., Lebesbye, E., Hogstad, K. and Midtbø, M. 2002. Geomorphology of buried glacigenic horizons in the Barents Sea from three-dimensional seismic data. In Dowdeswell, J. A. and Ó Cofaigh, C. (eds), *Glacier-influenced Sedimentation on High Latitude Continental Margins*. Geological Society, London, Special Publication 203, 259–76.

Ragotzkie, R. A. 1978. Heat budgets of lakes. In Lerman, A. (ed.), *Lakes – Chemistry, Geology, Physics*. Springer-Verlag, New York, 1–19.

Rains, R. B. and Shaw, J. 1981. Some mechanisms of controlled moraine development, Antarctica. *Journal of Glaciology* 27, 113–28.

Rains, R. B., Shaw, J., Sjogren, D. B., Munro-Stasiuk, M. J., Skoye, K. R., Young, R. R. and Thompson, R. T. 2002. Subglacial tunnel channels, Porcupine Hills, southwest Alberta, Canada. *Quaternary International* 90, 57–65.

Ramillien, G., Lombard, A., Cazenave, A., Ivins, E. R., Llubes, M., Remy, F. and Biancale, R. 2006. Interannual variations of the mass balance of the Antarctica and Greenland ice sheets from GRACE. *Global and Planetary Change* 53, 198–208.

Rampton, V. N. 1982. *Quaternary Geology of the Yukon Coastal Plain*. Geological Survey of Canada, Bulletin 317.

Rampton, V. N. 2000. Large-scale effects of subglacial meltwater flow in the southern Slave Province, Northwest Territories, Canada. *Canadian Journal of Earth Sciences* 37, 81–93.

Rana, B., Shrestha, A. B., Reynolds, J. M., Aryal, R., Pokhrel, A. P. and Budhathoki, K. P. 2000. Hazard assessment of the Tsho Rolpa Glacier Lake and ongoing remediation measures. *Journal of the Nepal Geological Society*, 22, 563–70.

Randall, D. A. et al. 2007. Climate models and their evaluation. In Solomon, S. et al. (eds), *Climate Change 2007: The Physical Science Basis*. Contribution of Working Group I to the Fourth Assessment of the Intergovernmental Panel on Climate Change. Cambridge University Press.

Raper, S. C. B. and Braithwaite, R. J. 2005. The potential for sea level rise: new estimates from glacier and ice cap area and volume distribution. *Geophysical Research Letters* 32, doi:10.1029/2004GL021981.

Rashid, H., Hesse, R. and Piper, D. J. W. 2003. Origin of unusually thick Heinrich layers in ice-proximal regions of the northwest Labrador Sea. *Earth and Planetary Science Letters* 208, 319–36.

Rasmussen, L. A. and Krimmell, R. M. 1999. Using vertical aerial photography to estimate mass balance at a point. *Geografiska Annaler* 81A, 725–33.

Rastas, J. and Seppälä, M. 1981. Rock jointing and abrasion forms on roches moutonnées, SW Finland. *Annals of Glaciology* 2, 159–63.

Rattas, M. and Kalm, V. 2001. Glaciotectonic deformation pattern in the hummocky moraine in the distal part of the Saadjärve drumlin field, east-central Estonia. *Slovak Geological Magazine* 7, 243–6.

Raymo, M. E. and Nisancioglu, K. 2003. The 41 kyr world: Milankovitch's other unsolved mystery. *Palaeoceanography* 18, doi:10.1029/2002PA000791.

Raymo, M. E. and Ruddiman, W. F. 1992. Tectonic forcing of late Cenozoic climate. *Nature* 359, 117–22.

Raymond, C. F. 1983. Deformation in the vicinity of ice divides. *Journal of Glaciology* 29 (103), 357–73.

Raymond, C. F. 1987. How do glaciers surge? A review. *Journal of Geophysical Research* 92, 9121–34.

Raymond, C. F. 1996. Shear margins in glaciers and ice sheets. *Journal of Glaciology* 42, 90–102.

Raymond, C. F. 2000. Energy balance of ice streams. *Journal of Glaciology* 46 (155), 665–74.

Raymond, C. F. and Harrison, W. D. 1985. Some observations on the behaviour of the liquid and gas phases in temperate ice. *Journal of Glaciology* 14, 213–33.

Raymond, C. F. and Harrison, W. D. 1988. Evolution of Variegated Glacier, Alaska, USA, prior to its surge. *Journal of Glaciology* 34, 154–69.

Raymond, C. F., Jóhannesson, T., Pfeffer, T. and Sharp, M. 1987. Propagation of a glacier surge into stagnant ice. *Journal of Geophysical Research* 92, 9037–49.

Raymond, C. F. and Nolan, M. 2000. Drainage of a glacial lake through an ice spillway. In Nakawo, N., Fountain, A. and Raymond, C. (eds), *Debris-Covered Glaciers*. IAHS Publication 264, 199–207.

Raymond, C. F., Echelmeyer, K. A., Whillans, I. M. and Doake, C. S. M. 2001. Ice stream shear margins. In Alley, R. B. and Bindschadler, R. A. (eds), *The West Antarctic Ice Sheet: Behavior and Environment*. American Geophysical Union, Antarctic Research Series vol. 77, 137–55.

Rea, B. R. 1994. Joint control in the formation of rock steps in the subglacial environment. In Robinson, D. A. and Williams, R. B. G. (eds), *Rock Weathering and Landform Evolution*. Wiley, Chichester, 473–86.

Rea, B. R. 1996. A note on the experimental production of a mechanically polished surface within striations. *Glacial Geology and Geomorphology* (available at: http://ggg.qub.ac.uk/papers/full/1997/tn011997/tn01.pdf).

Rea, B. R. 2009. Defining modern day area-altitude balance ratios (AABRs) and their use in glacier-climate reconstructions. *Quaternary Science Reviews* 28, 237–48.

Rea, B. R. and Evans, D. J. A. 1996. Landscapes of areal scouring in NW Scotland. *Scottish Geographical Magazine* 112, 47–50.

Rea, B. R. and Evans, D. J. A. 2003. Plateau icefield landsystems. In Evans, D. J. A. (ed.), *Glacial Landsystems*. Arnold, London, 407–31.

Rea, B. R. and Evans, D. J. A. 2007. Quantifying climate and glacier mass balance in North Norway during the Younger Dryas. *Palaeogeography, Palaeoclimatology, Palaeoecology*, 246 (2–4), 307–30.

Rea, B. R. and Whalley, W. B. 1994. Subglacial observations from Øksfjordjøkelen, north Norway. *Earth Surface Processes and Landforms* 19, 659–73.

Rea, B. R., Evans, D. J. A., Dixon, T. S. and Whalley, W. B. 2000. Contemporaneous, localized, basal ice-flow variations: implications for bedrock erosion and the origin of p-forms. *Journal of Glaciology* 46, 470–6.

Rea, B. R., Whalley, W. B., Evans, D. J. A., Gordon, J. E. and McDougall, D. A. 1998. Plateau icefields: geomorphology and dynamics. *Quaternary Proceedings* 6, 35–54.

Rea, B. R., Whalley, W. B. and Porter, E. M. 1996b. Rock weathering and the formation of summit blockfield slopes in Norway: examples and implications. In Anderson, M. G. and Brooks, S. M. (eds), *Advances in Hillslope Processes*. Wiley, Chichester, 1257–75.

Rea, B. R., Whalley, W. B., Rainey, M. M. and Gordon, J. E. 1996a. Blockfields, old or new? Evidence and implications from plateaus in northern Norway. *Geomorphology* 15, 109–12.

Reading, H. G. (ed.). 1996. *Sedimentary Environments and Facies*, 3rd edn. Blackwell, Oxford.

Reading, H. G. and Collinson, J. D. 1996. Clastic coasts. In Reading, H. G. (ed.), *Sedimentary Environments and Facies*, 3rd edn. Blackwell, Oxford, 154–231.

Reading, H. G. and Levell, B. K. 1996. Controls on the sedimentary rock record. In Reading, H. G. (ed.), *Sedimentary Environments and Facies*, 3rd edn. Blackwell, Oxford, 5–36.

Reeh, N. 1968. On the calving of ice from floating glaciers and ice shelves. *Journal of Glaciology* 7, 215–32.

Reeh, N. 2006. Current status and recent changes of the Greenland Ice Sheet. In Knight, P. (ed.), *Glacier Science and Environmental Change*. Blackwell, Oxford, 224–7.

Reeh, N., Madsen, S. N. and Mohr, J. J. 1999. Combining SAR interferometry and the equation of continuity to estimate the three-dimensional glacier surface-velocity vector. *Journal of Glaciology* 45 (151), 533–8.

Rees, H. G. and Collins, D. N. 2006. Regional differences in response of flow in glacier-fed Himalayan rivers to climatic warming. *Hydrological Processes* 20, 2157–69.

Reijmer, C. H. and Hock, R. 2008. Internal accumulation on Storglaciären, Sweden, in a multi-layer snow model coupled to a distributed energy- and mass-balance model. *Journal of Glaciology* 54 (184), 61–72.

Reimnitz, E. and Kempema, E. W. 1988. Ice rafting: an indication of glaciation? *Journal of Glaciology* 34, 254–5.

Remenda, V. H., Cherry, J. A. and Edwards, T. W. D. 1994. Isotopic composition of old ground water from Lake Agassiz: implications for Late Pleistocene climate. *Science* 266, 1975–8.

Rémy, F., Shaeffer, P. and Legrésy, B. 1999. Ice flow physical processes derived from ERS-1 high-resolution map of the Antarctica and Greenland ice sheets. *Geophysical Journal International* 139, 645–56.

Reinick, H. E. and Singh, I. B. 1980. *Depositional Sedimentary Environments*. Springer Verlag, Berlin.

Renssen, H., van Geel, B., van der Plicht, J. and Magny, M. 2000. Reduced solar activity as a trigger for the start of the Younger Dryas? *Quaternary International* 68/71, 373–83.

Retelle, M. J. 1986. Stratigraphy and sedimentology of coastal lacustrine basins, northeastern Ellesmere Island, NWT. *Geographie Physique et Quaternaire* 40, 117–28.

Reynolds, J. M. 1999. Photographic feature: glacial hazard assessment at Tsho Rolpa, Rolwaling, Central Nepal. *Quarterly Journal of Engineering Geology* 32 (3), 209–14.

Reynolds, J. M. 2000. On the formation of supraglacial lakes on debris-covered glaciers. In Nakawo, M., Raymond, C. F. and Fountain, A. (eds), *Debris-Covered Glaciers*. IAHS Publication 264, 153–61.

Ribstein, P., Titiau, E., Francou, B. and Saravia, R. 1995. Tropical climate and glacier hydrology: a case study in Bolivia. *Journal of Hydrology* 165, 221–34.

Richards, B. W. M., Benn, D. I., Owen, L. A., Rhodes, E. J. and Spencer, J. Q. 2000. Timing of Late Quaternary glaciations south of Mount Everest in the Khumbu Himal, Nepal. *Geological Society of America Bulletin* 112, 1621–32.

Richards, K., Sharp, M., Arnold, N., Gurnell, A., Clark, M., Tranter, M., Nienow, P., Brown, G., Willis, I. and Lawson, W. 1996. An integrated approach to modelling hydrology and water quality in glacierized catchments. *Hydrological Processes* 10, 479–508.

Richardson, C. and Holmlund, P. 1996. Glacial cirque formation in northern Scandinavia. *Annals of Glaciology* 22, 102–6.

Richardson, S. D. and Reynolds, J. M. 2000a. An overview of glacial hazards in the Himalayas. *Quaternary International* 65/66, 31–47.

Richardson, S. D. and Reynolds, J. M. 2000b. Degradation of ice-cored moraine dams: implications for hazard development. In Nakawo, M., Raymond, C. F. and Fountain, A. (eds), *Debris-Covered Glaciers*. IAHS Publication 264, 187–97.

Ridley, J. K., Huybrechts, P., Gregory, J. M. and Lowe, J. A. 2005. Elimination of the Greenland Ice Sheet in a high CO_2 climate. *Journal of Climate* 18, 3409–27.

Riedel, J. L., Haugerud, R. A. and Clague, J. J. 2007. Geomorphology of a Cordilleran Ice Sheet drainage network through breached divides in the North Cascades Mountains of Washington and British Columbia. *Geomorphology* 91, 1–18.

Rignot, E. 1998. Fast recession of a West Antarctic glacier. *Science* 281, 549–51.

Rignot, E. 2006. Changes in ice dynamics and mass balance of the Antarctic ice sheet. *Philosophical Transactions of the Royal Society*, Series A, 364, 1637–55.

Rignot, E. 2008. Changes in West Antarctic ice stream dynamics observed with ALOS PALSAR data. *Geophysical Research Letters* 35, L12505, doi:10.1029/2008GL033365.

Rignot, E. and Jacobs, S. S. 2002. Rapid bottom melting widespread near Antarctic ice sheet grounding lines. *Science* 296, 2020–3.

Rignot, E. and Kanagaratnam, P. 2005. Changes in the velocity structure of the Greenland Ice Sheet. *Science* 311, 986–90.

Rignot, E. and Steffen, K. 2008. Channelized bottom melting and stability of floating ice shelves, *Geophysical Research Letters* 35, L02503, doi:10.1029/2007GL031765.

Rignot, E. and Thomas, R. H. 2002. Mass balance of polar ice sheets. *Science* 297, 1502–6.

Rignot, E., Forster, R. and Isacks, B. 1996. Mapping of glacial motion and surface topography of Hielo Patagonico Norte, Chile, using Satellite SAR L-band interferometry data. *Annals of Glaciology* 23, 209–16.

Rignot, E. J., Box, J. E., Burgess, E. and Hanna, E. 2008. Mass balance of the Greenland ice sheet from 1958 to 2007. *Geophysical Research Letters* 35, doi:10.1029/2008GL035417.

Rignot, E. J., Gogineni, S. P., Krabill, W. B. and Ekholm, S. 1997. North and Northeast Greenland ice discharge from satellite radar interferometry. *Science* 276, 934–7.

Rignot, E. J., Rivera, A. and Casassa, G. 2003. Contribution of the Patagonia Icefields of South America to sea level rise. *Science* 302, 434–7.

Rignot, E. J., Vaughan, D. G., Schmeltz, M., Dupont, T. and MacAyeal, D. 2002. Acceleration of Pine Island and Thwaites Glaciers, West Antarctica. *Annals of Glaciology* 34, 189–94.

Riihimaki, C. A., MacGregor, K. R., Anderson, R. S., Anderson, S. P. and Loso, M. G. 2005. Sediment evacuation and glacial erosion rates at a small alpine glacier. *Journal of Geophysical Research* 110, F03003, doi:10.1029/2004JF000189.

Rijsdijk, K. F. 2001. Density-driven deformation structures in glacigenic consolidated diamicts: examples from Traeth y Mwnt, Cardiganshire, Wales, UK. *Journal of Sedimentary Research* 71, 122–35.

Rijsdijk, K. F., Owen, G., Warren, W. P., McCarroll, D. and van der Meer, J. J. M. 1999. Clastic dykes in over-consolidated tills: evidence for subglacial hydrofracturing at Killiney Bay, eastern Ireland. *Sedimentary Geology* 129, 111–26.

Ringberg, B., Holland, B. and Miller, U. 1984. Till stratigraphy and provenance of the glacial chalk rafts at Kvarnby and Angdala, southern Sweden. *Striae* 20, 79–90.

Ringrose, S. 1982. Depositional processes in the development of eskers in Manitoba. In Davidson-Arnott, R., Nickling, W. and Fahey, B. D. (eds), *Research in Glacial, Glacio-fluvial and Glacio-lacustrine Systems*. Geo Books, Norwich, 117–38.

Rinterknecht, V. R., Clark, P. U., Raisbeck, G. M., Yiou, F., Brook, E. J., Tschudi, S. and Lunkka, J. P. 2004. Cosmogenic [10]Be dating of the Salpausselkä I Moraine in southwestern Finland. *Quaternary Science Reviews* 23, 2283–9.

Rippin, D., Willis, I. and Arnold, N. 2005. Seasonal patterns of velocity and strain across the tongue of the polythermal glacier Midre Lovénbreen, Svalbard. *Annals of Glaciology* 42, 445–54.

Roberts, D. H. and Hart, J. K. 2005. The deforming bed characteristics of a stratified till assemblage in north East Anglia, UK: investigating controls on sediment rheology and strain signatures. *Quaternary Science Reviews* 24, 123–40.

Roberts, D. H. and Long, A. J. 2005. Streamlined bedrock terrain and fast flow, Jakobshavns Isbrae, West Greenland: implications for ice stream and ice sheet dynamics. *Boreas* 34, 25–42.

Roberts, D. H., Dackombe, R. V. and Thomas, G. S. P. 2007. Palaeo-ice streaming in the central sector of the British-Irish Ice Sheet during the Last Glacial Maximum: evidence from the northern Irish Sea Basin. *Boreas* 36, 115–29.

Roberts, D. H., Yde, J. C., Knudsen, N. T., Long, A. J. and Lloyd, J. M. 2009 Ice-marginal dynamics during surge activity, Kuannersuit Glacier, Disko Island, West Greenland. *Quaternary Science Reviews* 28, 209–22.

Roberts, M. C. and Cunningham, F. F. 1992. Post-glacial loess deposition in a montane environment: South Thompson River valley, British Columbia, Canada. *Journal of Quaternary Science* 7, 291–301.

Roberts, M. C. and Rood, R. M. 1984. The role of ice contributing area in the morphology of transverse fiords, British Columbia. *Geografiska Annaler* 66A, 381–93.

Roberts, M. J. 2005. Jökulhlaups: a reassessment of floodwater flow through glaciers. *Reviews of Geophysics* 43, doi:10.1029/2003RG000147.

Roberts, M. J., Pálsson, F., Gudmundsson, M. T., Björnsson, H. and Tweed, F. S. 2005. Ice-water interactions during floods from Grænalón glacier-dammed lake, Iceland. *Annals of Glaciology* 40, 133–8.

Roberts, M. J., Russell, A. J., Tweed, F. S. and Knudsen, Ó. 2000. Ice fracturing during jökulhlaups: implications for englacial floodwater routing and outlet development. *Earth Surface Processes and Landforms* 25, 1429–49.

Roberts, M. J., Russell, A. J., Tweed, F. S. and Knudsen, Ó. 2001. Controls on englacial sediment deposition during the November 1996 jökulhlaup, Skeiðarárjökull, Iceland. *Earth Surface Processes and Landforms* 26, 935–52.

Roberts, M. J., Tweed, F. S., Russell, A. J., Knudsen, Ó. and Harris, T. D. 2003. Hydrologic and geomorphic effects of temporary ice-dammed lake formation during jökulhlaups. *Earth Surface Processes and Landforms* 28, 723–37.

Roberts. M. J., Tweed, F. S., Russell, A. J., Knudsen, Ó., Lawson, D. E., Larson, G. J., Evenson, E. B., Björnsson, H. 2002. Glaciohydraulic supercooling in Iceland. *Geology* 30, 439–42.

Robertson, S. 2008. Structural composition and sediment transfer in a composite cirque glacier: Glacier de St Sorlin, France. *Earth Surface Processes and Landforms* 33, 1931–47.

Robin, G. de Q. 1976. Is the basal ice of a temperate glacier at the pressure melting point? *Journal of Glaciology* 16, 183–96.

Robin, G. de Q. 1981. Polar ice sheets: developments since Wegener. *International Journal of Earth Sciences* 70, 648–63.

Robin, G. de Q. 1983. The δ value-temperature relationship. In Robin, G. de Q. (ed.), *The Climatic Record in Polar Ice Sheets*. Cambridge University Press, 180–4.

Robin, G. de Q., Swithinbank, C. W. M. and Smith, B. M. E. 1970. Radio echo exploration of the Antarctic ice sheet. *International Association of Hydrological Sciences Publication* 86, 97–115.

Robinson, Z. P., Fairchild, I. J. and Russell, A. J. 2008. Hydrogeological implications of glacial landscape evolution at Skeiðarársandur, SE Iceland. *Geomorphology* 97, 218–36.

Röhl, K. 2006. Thermo-erosional notch development at fresh-water-calving Tasman Glacier, New Zealand. *Journal of Glaciology* 52, 203–13.

Röhl, K. 2008. Characteristics and evolution of supraglacial ponds on debris-covered Tasman Glacier, New Zealand. *Journal of Glaciology* 54 (188), 867–80.

Rolstad, C., Amilien, J., Hagen, J. O. and Lundén, B. 1997. Visible and near-infrared digital images for determination of ice velocities and surface elevation during a surge of Osbornebreen, a tidewater glacier in Svalbard. *Annals of Glaciology* 24, 255–61.

Rolstad, C., Whillans, I., Hagen, J. O. and Isaksson, E. 2000. Large-scale force budget of an outlet glacier: Jutulstraumen, Dronning Maud Land, Antarctica. *Annals of Glaciology* 30, 35–41.

Ronnert, L. and Mickelson, D. M. 1992. High porosity of basal till at Burroughs Glacier, southeastern Alaska. *Geology* 20, 849–52.

Rosanski, K., Johnsen, S. J., Schotterer, U. and Thompson, L. G. 1997. Reconstruction of past climates from stable isotope records of palaeo-precipitation preserved in continental archives. *Hydrological Sciences – Journal des Sciences Hydrologiques* 42, 725–45.

Rose, J. 1987. Drumlins as part of a glacier bedform continuum. In Menzies, J. and Rose, J. (eds), *Drumlin Symposium*. Balkema, Rotterdam, 103–16.

Rose, J. 1989. Glacier stress patterns and sediment transfer associated with the formation of superimposed flutes. *Sedimentary Geology* 62, 151–76.

Rose, J. 1992. Boulder clusters in glacial flutes. *Geomorphology* 6, 51–8.

Rose, J. 1994. Major river systems of central and southern Britain during the Early and Middle Pleistocene. *Terra Nova* 6, 435–43.

Rose, J. and Letzer, J. M. 1977. Superimposed drumlins. *Journal of Glaciology* 18, 471–80.

Rose, J. and Smith, M. J. 2008. Glacial geomorphological maps of the Glasgow region, western central Scotland. *Journal of Maps* 2008, 399–416.

Rose, J., Lee, J. A., Candy, I. and Lewis, S. G. 1999. Early and Middle Pleistocene river systems in eastern England: evidence from Leet Hill, southern Norfolk, England. *Journal of Quaternary Science* 14, 347–60.

Rose, K. E. 1979. Characteristics of ice flow in Marie Byrd Land, Antarctica. *Journal of Glaciology* 24, 63–75.

Röthlisberger, H. 1972. Water pressure in intra- and subglacial channels. *Journal of Glaciology* 11, 177–203.

Röthlisberger, H. and Iken, A. 1981. Plucking as an effect of water-pressure variations at the glacier bed. *Annals of Glaciology* 2, 57–62.

Röthlisberger, H. and Lang, H. 1987. Glacial hydrology. In Gurnell, A. M. and Clark, M. J. (eds), *Glacio-fluvial Sediment Transfer*. Wiley, New York, 207–84.

Rotnicki, K. 1976. The theoretical basis for and a model of glacio-tectonic deformations. *Quaestiones Geographicae* 3, 103–39.

Rott, H., Rack, W., Skvarca, P. and De Angelis, H. 2002. Northern Larsen Ice Shelf, Antarctica: further retreat after collapse. *Annals of Glaciology* 34, 277–82.

Rott, H., Skvarca, P. and Nagler, T. 1996. Rapid collapse of Northern Larsen Ice Shelf, Antarctica. *Science* 271, 788–92.

Roush, J. J., Lingle, C. S., Guritz, R. M., Fatland, D. R. and Voronina, V. A. 2003. Surge-front propagation and velocities during the early 1993–95 surge of Bering Glacier, Alaska, USA, from sequential SAR imagery. *Annals of Glaciology* 36, 37–44.

Roy, M., Clark, P. U., Raisbeck, G. M. and Yiou, F. 2004. Geochemical constraints on the origin of the middle Pleistocene transition from the glacial sedimentary record of the north-central US. *Earth and Planetary Science Letters* 227, 281–6.

Royer, D. L. 2006. CO_2 forced climate thresholds during the Phanerozoic. *Geochimica et Cosmochimica Acta* 70, 5665–75.

Royer, D. L., Berner, R. A., Montañez, I. P., Tabor, N. J. and Beerling, D. J. 2004. CO_2 as a primary driver of Phanerozoic climate. *GSA Today* 14, 4–10.

Rozanski, K., Johnsen, S. J., Schotterer, U. and Thompson, L. G. 1997. Reconstruction of past climates from stable isotope records of palaeo-precipitation preserved in continental archives. *Hydrological Sciences – Journal des Sciences Hydrologiques* 42, 725–45.

Ruddiman, W. F. (ed.). 1997. *Tectonic Uplift and Climate Change*, Plenum Press, New York.

Ruddiman, W. F. 2000. *Earth's Climate: Past and Future*, W. H. Freeman, New York.

Rudoy, A. N. 2002. Glacier-dammed lakes and geological work of glacial superfloods in the Late Pleistocene, southern Siberia, Altai Mountains. *Quaternary International* 87, 119–40.

Rudoy, A. N. and Baker, V. R. 1993. Sedimentary effects of cataclysmic late Pleistocene glacial outburst flooding, Altay Mountains, Siberia. *Sedimentary Geology* 85, 53–62.

Rudoy, A. N., Galachov, V. P. and Danilin, A. L. 1989. Reconstruction of glacial discharge in the head of the Chuja River and alimentation of ice-dammed lakes in the late Pleistocene. *Izvestiya Vsesoyuznogo Geograficheskogo Obshchestva* 121, 236–44.

Ruff, S. W. and Greeley, R. 1990. Sinuous ridges of the south polar region, Mars: possible origins. *Lunar and Planetary Science* 21, 1047–8.

Russell, A. J. 1993. Obstacle marks produced by flows around stranded ice blocks during a jökulhlaup in west Greenland. *Sedimentology* 40, 1091–111.

Russell, A. J. 2007. Controls on the sedimentology of an ice-contact jökulhlaup-dominated delta, Kangerlussuaq, west Greenland. *Sedimentary Geology* 193, 131–48.

Russell, A. J. and Knudsen, Ó. 1999. An ice-contact rhythmite (turbidite) succession deposited during the November 1996 catastrophic outburst flood (jökulhlaup), Skeiðarárjökull, Iceland. *Sedimentary Geology* 127, 1–10.

Russell, A. J. and Knudsen, Ó. 2002. The effects of glacier-outburst flood flow dynamics on ice-contact deposits: November 1996 jökulhlaup, Skeiðarársandur, Iceland. In Martini, I. P., Baker, V. R. and Garzon, G. (eds), *Flood and Megaflood Processes and Deposits: Recent and Ancient Examples*. Blackwell, Oxford. International Association of Sedimentologists, Special Publication 32, 67–83.

Russell, A. J. and Marren, P. M. 1999. Proglacial fluvial sedimentary sequences in Greenland and Iceland: a case study from active proglacial environments subject to jökulhlaups. In Jones, A. P., Tucker, M. E. and Hart, J. K. (eds), *The Description and Analysis of Quaternary Stratigraphic Field Sections*. Quaternary Research Association Technical Guide 7, 171–208.

Russell, A. J., Fay, H., Marren, P. M., Tweed, F. S. and Knudsen, Ó. 2005. Icelandic jökulhlaup impacts. In Caseldine, C. J., Russell, A. J., Hardardottir, J. and Knudsen, Ó. (eds), *Iceland: Modern Processes and Past Environments*. Developments in Quaternary Science 5, Elsevier, Amsterdam, 153–203.

Russell, A. J., Gregory, A. R., Large, A. R. G., Fleisher, P. J. and Harris, T. D. 2007. Tunnel channel formation during the November 1996 jökulhlaup, Skeiðarárjökull, Iceland. *Annals of Glaciology* 45, 95–103.

Russell, A. J., Knight, P. G. and van Dijk, T. A. G. P. 2001a. Glacier surging as a control on the development of proglacial, fluvial landforms and deposits, Skeiðarársandur, Iceland. *Global and Planetary Change* 28, 163–74.

Russell, A. J., Knudsen, Ó., Fay, H., Marren, P. M., Heinz, J. and Tronicke, J. 2001b. Morphology and sedimentology of a giant supraglacial, ice-walled, jökulhlaup channel, Skeiðarárjökull, Iceland: implications for esker genesis. *Global and Planetary Change* 28, 193–216.

Russell, A. J., Roberts, M. J., Fay, H., Marren, P. M., Cassidy, N. J., Tweed, F. S. and Harris, T. 2006. Icelandic jökulhlaup impacts: implications for ice sheet hydrology, sediment transfer and geomorphology. *Geomorphology* 75, 33–64.

Russell, A. J., Tweed, F. S. and Harris, T. 2003. High energy sedimentation, Craig Aoil, Spean Bridge, Scotland: implications for meltwater movement and storage during Loch Lomond Stadial (Younger Dryas) ice retreat. *Journal of Quaternary Science* 18, 415–30.

Russell, H. A. J. and Arnott, R. W. C. 2003. Hydraulic-jump and hyperconcentrated-flow deposits of a glacigenic subaqueous fan: Oak Ridges Moraine, southern Ontario, Canada. *Journal of Sedimentary Research* 73, 887–905.

Russell, H. A. J., Arnott, R. W. C. and Sharpe, D. R. 2003. Evidence for rapid sedimentation in a tunnel channel, Oak Ridges Moraine, southern Ontario, Canada. *Sedimentary Geology* 160, 33–55.

Russell-Head, D. S. 1980. The melting of free-drifting icebergs. *Annals of Glaciology* 1, 119–22.

Rust, B. R. 1975. Fabric and structure in glaciofluvial gravels. In Jopling, A. V. and McDonald, B. C. (eds), *Glaciofluvial and Glaciolacustrine Sedimentation*. SEPM Special Publication 23, 238–48.

Rust, B. R. 1977. Mass flow deposits in a Quaternary succession near Ottawa, Canada: diagnostic criteria for subaqueous outwash. *Canadian Journal of Earth Sciences* 14, 175–84.

Rust, B. R. 1978. Depositional models for braided alluvium. In Miall, A. D. (ed.), *Fluvial Sedimentology*. Canadian Society of Petroleum Geologists, Memoir 5, 605–25.

Rust, B. R. and Gibling, M. R. 1990. Three-dimensional antidunes as HCS mimics and a fluvial sandstone: the Pennsylvanian South Bar Formation near Sydney, Nova Scotia. *Journal of Sedimentary Petrology* 60, 540–8.

Rust, B. R. and Romanelli, R. 1975. Late Quaternary subaqueous outwash deposits near Ottawa, Canada. In Jopling, A. V. and McDonald, B. C. (eds), *Glaciofluvial and Glaciolacustrine Sedimentation*. SEPM Special Publication 23, 177–92.

Ruszczynska-Szenajch, H. 1976. Glacitektoniczne depresje i kry lodowcowe na tle budowy geologicznej poludniowo-wschodniego Mazowsza i poludniowego Podlasia. *Studia Geologica Polonica* 50, 1–106.

Ruszczynska-Szenajch, H. 1978. Glacitectonic origin of some lake basins in areas of Pleistocene glaciations. *Polskie Archiwum Hydrobiologii* 25, 373–81.

Ruszczynska-Szenajch, H. 1987. The origin of glacial rafts: detachment, transport, deposition. *Boreas* 16, 101–12.

Ruszczynska-Szenajch, H. 2001. 'Lodgement till' and 'deformation till'. *Quaternary Science Reviews* 20, 579–81.

Ruszczynska-Szenajch, H., Trzcinski, J. and Jarosinska, U. 2003. Lodgement till deposition and deformation investigated by macroscopic observation, thin section analysis and electron microscope study at site Dçbe, central Poland. *Boreas* 32, 399–415.

Rutt, I. C., Hagdorn, M., Hulton, N. R. J. and Payne, A. J. 2009. The Glimmer community ice sheet model. *Journal of Geophysical Research* 114, F02004, doi:10.1029/2008JF001015.

Ryder, J. M. 1995. Recognition and interpretation of flow direction indicators for former glaciers and meltwater streams. In Bobrowsky, P. T., Sibbick, S. J., Newell, J. M. and Matysek, P. F. (eds), *Drift Exploration in the Canadian Cordillera, British Columbia*. Ministry of Energy, Mines and Petroleum Resources, Paper 1995-2, 1–22.

Ryder, J. M., Fulton, R. J. and Clague, J. J. 1991. The Cordilleran ice sheet and the glacial geomorphology of southern and central British Columbia. *Geographie Physique et Quaternaire* 45, 365–77.

Sættem, J. 1994. Glaciotectonic structures along the southern Barents shelf margin. In Warren, W. P. and Croot, D. G. (eds), *Formation and Deformation of Glacial Deposits*. Balkema, Rotterdam, 95–113.

Saarnisto, M. and Saarinen, T. 2001. Deglaciation chronology of the Scandinavian Ice Sheet from the Lake Onega Basin to the Salpausselkä end moraines. *Global and Planetary Change* 31, 387–405.

Sagan, C. and Chyba, C. 1997. The early faint sun paradox: organic shielding of ultraviolet-labile greenhouse gases. *Science* 276, 1217–21.

Sakai, A., Nakawo, M. and Fujita, K. 2002. Distribution characteristics and energy balance of ice cliffs on debris-covered glaciers, Nepal Himalaya. *Arctic, Antarctic and Alpine Research* 34, 12–19.

Sakai, A., Takeuchi, N., Fujita, K. and Nakawo, M. 2000. Role of supraglacial ponds in the ablation process of a debris-covered glacier in the Nepal Himalaya. In Nakawo, N., Fountain, A. and Raymond, C. (eds), *Debris-Covered Glaciers*. IAHS Publication 264, 53–61.

Salonen, V.-P. 1986. Glacial transport distance distributions of surface boulders in Finland. *Geological Survey of Finland Bulletin* 338.

Salt, K. E. and Evans, D. J. A. 2004. Scottish Landform Example 32: Superimposed subglacially streamlined landforms of southwest Scotland. *Scottish Geographical Journal* 120, 133–47.

Sambrook-Smith, G. H. 2000. Small-scale cyclicity in alpine proglacial fluvial sedimentation. *Sedimentary Geology* 132, 217–31.

Sammis, C., King, G. and Biegel, R. 1987. The kinematics of gouge deformation. *Pure and Applied Geophysics* 125, 777–812.

Sarala, P. 2006. Ribbed moraine stratigraphy and formation in southern Finnish Lapland. *Journal of Quaternary Science* 21, 387–98.

Saunderson, H. C. 1975. Sedimentology of the Brampton esker and its associated deposits: an empirical test of theory. In Jopling, A. V. and McDonald, B. C. (eds), *Glaciofluvial and Glaciolacustrine Sedimentation*. SEPM Special Publication 23, 155–76.

Saunderson, H. C. 1977. The sliding bed facies in esker sands and gravels: a criterion for full-pipe (tunnel) flow? *Sedimentology* 24, 623–38.

Scambos, T. A., Bohlander, J. A., Shuman, C. A. and Skvarca, P. 2004. Glacier acceleration and thinning after ice shelf collapse in the Larsen B embayment, Antarctica. *Geophysical Research Letters* 31, L18402, doi:10.1029/2004GL020670.

Scambos, T., Hulbe, C. and Fahnestock, M. 2003. Climate-induced ice shelf disintegration in the Antarctic Peninsula. In Domack, E. W., Burnett, A., Leventer, A., Conley, P., Kirby, M. and Bindschadler, R. (eds), *Antarctic Peninsula Climate Variability: A Historical and Palaeoenvironmental Perspective.* American Geophysical Union, Antarctic Research Series 79, 79–92.

Scambos, T. A., Hulbe, C., Fahnestock, M., and Bohlander, J. 2000. The link between climate warming and break-up of ice shelves in the Antarctic Peninsula. *Journal of Glaciology* 46, 516–30.

Scambos, T., Sergienko, O., Sargent, A., MacAyeal, D. and Fastook, J. 2005. ICESat profiles of tabular iceberg margins and iceberg breakup at low latitudes. *Geophysical Research Letters* 32, doi:10.1029/2005GL023802.

Schiefer, E. and Gilbert, R. 2007. Reconstructing morphometric change in a proglacial landscape using historical aerial photography and automated DEM generation. *Geomorphology* 88, 167–78.

Schilling, D. H. and Hollin, J. T. 1981. Numerical reconstructions of valley glaciers and small ice caps. In Denton, G. H. and Hughes, T. J. (eds), *The Last Great Ice Sheets.* New York, Wiley, 207–20.

Schilling, S. P., Carrara, P. E., Thompson, R. A. and Iwatsubo, E. Y. 2004. Post eruption glacier development within the crater of Mount St Helens, Washington, USA. *Quaternary Research* 61, 325–9.

Schluchter, C., Gander, P., Lowell, T. V. and Denton, G. H. 1999. Glacially folded outwash near Lago Llanquihue, southern Lake District, Chile. *Geografiska Annaler* 81A, 347–58.

Schneeberger, C., Blatter, H., Abe-Ouchi, A. and Wild, M. 2003. Modelling changes in the mass balance of glaciers of the northern hemisphere for a transient $2 \times CO_2$ scenario. *Journal of Hydrology* 282, 145–63.

Schneider, T. 2000. Hydrological processes in the wet-snow zone of a glacier: a review. *Zeitschrift für Gletscherkunde und Glazialgeologie* 36, 89–105.

Schomacker, A. and Kjær, K. H. 2007. Origin and de-icing of multiple generations of ice-cored moraines at Brúarjökull, Iceland. *Boreas* 36, 411–25.

Schomacker, A. and Kjær, K. H. 2008. Quantification of dead-ice melting in ice-cored moraines at the high arctic glacier Holmströmbreen, Svalbard. *Boreas* 37, 211–25.

Schomacker, A., Krüger, J. and Kjær, K. H. 2006. Ice-cored drumlins at the surge-type glacier Brúarjökull, Iceland: a transitional state landform. *Journal of Quaternary Science* 21, 85–93.

Schoof, C. 2005. The effect of cavitation on glacier sliding. *Proceedings of the Royal Society,* Series A, 461, 609–27.

Schoof, C. 2007a. Ice sheet grounding line dynamics: steady states, stability and hysteresis. *Journal of Geophysical Research* 112, doi:10.1029/2006JF000664.

Schoof, C. 2007b. Marine ice-sheet dynamics. Part 1: The case of rapid sliding. *Journal of Fluid Mechanics* 573, 27–55.

Schoof, C. 2007c. Pressure-dependent viscosity and interfacial instability in coupled ice-sediment flow. *Journal of Fluid Mechanics* 570, 227–52.

Schrag, D. P., Adkins, J. F., McIntyre, K., Alexander, J. L., Hodell, D. A., Charles, C. D. and McManus, J. F. 2002. The oxygen isotopic composition of seawater during the Last Glacial Maximum. *Quaternary Science Reviews* 21, 331–42.

Schroeder, J. 1995. Les moulins du glacier Hans de 1988 a 1992. In Griselin, M. (ed.), *Actes du 3e Symposium International, Cavités Glaciaires et Cryokarst en Régions Polaires et de Haute Montagne.* Les Belles Lettres, Paris. Annales Littéraires de Université de Besançon 561, Série Géographie 34, 31–9.

Schuler, T., Fischer, U. H. and Gudmundsson, G. H. 2004. Diurnal variability of subglacial drainage conditions as revealed by tracer experiments. *Journal of Geophysical Research* 109, doi:10.1029/2003JF000082.

Schuler, T., Fischer, U. H., Sterr, R., Hock, R. and Gudmundsson, G. H. 2002. Comparison of modelled water input and measured discharge prior to a release event: Unteraargletscher, Bernese Alps, Switzerland. *Nordic Hydrology* 33, 27–46.

Schuler, T., Hock, R., Jackson, M., Elvehøy, H., Braun, M., Brown, I. and Hagen, J. O. 2005. Distributed mass balance and climate sensitivity modelling of Engabreen, Norway. *Annals of Glaciology* 42, 395–401.

Schumm, S. A. 1977. *The Fluvial System.* Wiley, New York.

Schumm, S. A. and Lichty, R. W. 1965. Time, space and causality in geomorphology. *American Journal of Science* 263, 110–19.

Schwab, W. C., Lee, H. J. and Molnia, B. F. 1987. Causes of varied sediment gravity flow types on the Alsek prodelta, northeast Gulf of Alaska. *Marine Geotechnology* 7, 312–42.

Schweizer, J. and Iken, A. 1992. The role of bed separation and friction in sliding over an undeformable bed. *Journal of Glaciology* 38, 77–92.

Schwikowski, M., Brütsch, S., Casassa, G. and Rivera, A. 2006. A potential high-elevation ice-core site at Hielo Patagonico Sur. *Annals of Glaciology* 43, 8–13.

Scourse, J. D. and Furze, M. F. A. 2001. A critical review of the glaciomarine model for Irish Sea deglaciation: evidence from southern Britain, the Celtic Sea and adjacent continental slope. *Journal of Quaternary Science* 16, 419–34.

Scourse, J. D., Hall, I. R., McCave, I. N., Young, J. R. and Sugdon, C. 2000. The origin of Heinrich layers: evidence from H2 for European precursor events. *Earth and Planetary Science Letters* 182, 187–95.

Seaberg, S. Z., Seaberg, J. Z., Hooke, R. Le B. and Wiberg, D. W. 1988. Character of the englacial and subglacial drainage system in the lower part of the ablation area of Storglaciären, Sweden, as revealed by dye-trace studies. *Journal of Glaciology* 34, 217–27.

Sejrup, H. P., Landvik, J., Larsen, E., Eiriksson, J., Janocko, J. and King, E. L. 1998. The Jæren area: a border zone of the Norwegian Channel Ice Stream. *Quaternary Science Reviews* 17, 801–12.

Sejrup, H. P., Larsen, E., Haflidason, H., Berstad, I. M., Hjelstuen, B. O., Jonsdottir, H. E., King, E. L., Landvik, J., Longva, O., Nygard, A., Ottesen, D., Raunholm, S., Rise, L. and Stalsberg, K. 2003. Configuration, history and impact of the Norwegian Channel Ice Stream. *Boreas* 32, 18–36.

Sejrup, H. P., Larsen, E., Landvik, J., King, E. L., Haflidason, H. and Nesje, A. 2000. Quaternary glaciations in southern Fennoscandia: evidence from southwestern Norway and the northern North Sea region. *Quaternary Science Reviews* 19, 667–85.

Selby, M. J. 1983. *Hillslope Materials and Processes.* Oxford University Press.

Selby, M. J. 1993. *Hillslope Materials and Processes,* 2nd edn. Oxford University Press.

Sella, G. F., Stein, S., Dixon, T. H., Craymer, M., James, T. S., Mazzotti, S. and Dokka, R. K. 2007. Observation of glacial isostatic adjustment in 'stable' North America with GPS. *Geophysical Research Letters* 34, L02306, doi:10.1029/2006GL027081.

Seppälä, M. 2004. *Wind as a Geomorphic Agent in Cold Climates.* Cambridge University Press.

Sexton, D. J., Dowdeswell, J. A., Solheim, A. and Elverhøi, A. 1992. Seismic architecture and sedimentation in northwest Spitzbergen fjords. *Marine Geology* 103, 53–68.

Shabtaie, S. and Bentley, C. R. 1987. West Antarctic ice streams draining into the Ross Ice Shelf: configuration and mass balance. *Journal of Geophysical Research* 92, 1311–36.

Shackleton, N. J. 1967. Oxygen isotope analyses and Pleistocene temperatures reassessed. *Nature* 215, 15–17.

Shackleton, N. J. 1987. Oxygen isotopes, ice volume and sea level. *Quaternary Science Reviews* 6, 183–90.

Shackleton, N. J., Hall, M. A. and Vincent, E. 2000. Phase relationships between millennial-scale events 64,000–24,000 years ago. *Paleoceanography* 15, 565–9.

Shain, D. H., Mason, T. A., Farrell, A. H. and Michalewicz, L. A. 2001. Distribution and behavior of ice worms (Mesenchytraeus solifugus) in south-central Alaska. *Canadian Journal of Zoology* 79 (10), 1813–21.

Shakesby, R. A. 1989. Variability in Neoglacial moraine morphology and composition, Storbreen, Jotunheim, Norway: within-moraine patterns and their implications. *Geografiska Annaler* 71A, 17–29.

Shakesby, R. A. and Matthews, J. A. 1996. Glacial activity and paraglacial landsliding in the Devensian lateglacial: evidence from Craig Cerrig-gleisiad and Fan Dringarth, Forest Fawr (Brecon Beacons), South Wales. *Geological Journal* 31, 143–57.

Shakesby, R. A., Matthews, J. A., McEwan, L. and Berrisford, M. S. 1999. Snow-push processes in pronival (protalus) rampart formation: geomorphological evidence from southern Norway. *Geografiska Annaler* 81A, 31–45.

Shanmugam, G. 1997. The Bouma Sequence and the turbidite mind set. *Earth Science Reviews* 42, 201–29.

Sharp, M. J. 1984. Annual moraine ridges at Skalafellsjökull, south-east Iceland. *Journal of Glaciology* 30, 82–93.

Sharp, M. J. 1985a. 'Crevasse-fill' ridges: a landform type characteristic of surging glaciers? *Geografiska Annaler* 67A, 213–20.

Sharp, M. J. 1985b. Sedimentation and stratigraphy at Eyjabakkajökull: an Icelandic surging glacier. *Quaternary Research* 24, 268–84.

Sharp, M. J. 1988a. Surging glaciers: behaviour and mechanisms. *Progress in Physical Geography* 12, 349–70.

Sharp, M. J. 1988b. Surging glaciers: geomorphic effects. *Progress in Physical Geography* 12, 533–59.

Sharp, M. J., Brown, G. H., Tranter, M., Willis, I. C. and Hubbard, B. 1995b. Comments on the use of chemically based mixing models in glacier hydrology. *Journal of Glaciology* 41, 241–6.

Sharp, M. J., Dowdeswell, J. A. and Gemmell, J. C. 1989a. Reconstructing past glacier dynamics and erosion from glacial geomorphic evidence: Snowdon, North Wales. *Journal of Quaternary Science* 4, 115–30.

Sharp, M. J., Gemmell, J. C. and Tison, J.-L. 1989b. Structure and stability of the former subglacial drainage system of the Glacier de Tsanfleuron, Switzerland. *Earth Surface Processes and Landforms* 14, 119–34.

Sharp, M. J., Jouzel, J., Hubbard, B. and Lawson, W. 1994. The character, structure and origin of the basal ice layer of a surge-type glacier. *Journal of Glaciology* 40, 327–40.

Sharp, M. J., Lawson, W. and Anderson, R. S. 1988. Tectonic processes in a surge-type glacier. *Journal of Structural Geology* 10, 499–515.

Sharp, M. J., Parkes, J., Cragg, B., Fairchild, I. J., Lamb, H. and Tranter, M. 1999. Widespread bacterial populations at glacier beds and their relationship to rock weathering and carbon cycling. *Geology* 27, 107–10.

Sharp, M. J., Tison, J.-L. and Fierens, G. 1990. Geochemistry of subglacial calcites: implications for the hydrology of the basal water film. *Arctic and Alpine Research* 22, 141–52.

Sharp, M. J., Tranter, M., Brown, G. H. and Skidmore, M. 1995a. Rates of chemical denudation and CO_2 drawdown in a glacier-covered catchment. *Geology* 23, 61–4.

Sharp, R. P. 1988. *Living Ice: Understanding Glaciers and Glaciation*. Cambridge University Press.

Sharp, R. P. and Malin, M. C. 1975. Channels on Mars. *Geological Society of America Bulletin* 86, 593–609.

Sharpe, D. R. 1988. Glaciomarine fan deposits in the Champlain Sea. In Gadd, N. R. (ed.), *The Late Quaternary Development of the Champlain Sea Basin*. Geological Association of Canada, Special Paper 35, 63–82.

Sharpe, D. R. and Shaw, J. 1989. Erosion of bedrock by subglacial meltwater, Cantley, Quebec. *Geological Society of America Bulletin* 101, 1011–20.

Sharpe, D. R., Russell, H. A. J. and Logan, C. 2007. A 3-dimensional geological model of the Oak Ridges Moraine area, Ontario, Canada. *Journal of Maps* 2007, 239–53.

Shaw, J. 1977. Tills deposited in arid polar environments. *Canadian Journal of Earth Sciences* 14, 1239–45.

Shaw, J. 1979. Genesis of the Sveg tills and Rogen moraines of central Sweden: a model of basal melt-out. *Boreas* 8, 409–26.

Shaw, J. 1982. Melt-out till in the Edmonton area, Alberta, Canada. *Canadian Journal of Earth Sciences* 19, 1548–69.

Shaw, J. 1983a. Forms associated with boulders in melt-out till. In Evenson, E. B., Schluchter, C. and Rabassa, J. (eds), *Tills and Related Deposits*. Balkema, Rotterdam, 3–12.

Shaw, J. 1983b. Drumlin formation related to inverted meltwater erosional marks. *Journal of Glaciology* 29, 461–79.

Shaw, J. 1987. Glacial sedimentary processes and environmental reconstruction based on lithofacies. *Sedimentology* 34, 103–16.

Shaw, J. 1988a. Subglacial erosional marks, Wilton Creek, Ontario. *Canadian Journal of Earth Sciences* 25, 1256–67.

Shaw, J. 1988b. Sublimation till. In Goldthwait, R. P. and Matsch, C. L. (eds), *Genetic Classification of Glacigenic Deposits*. Balkema, Rotterdam, 141–2.

Shaw, J. 1989. Drumlins, subglacial meltwater floods, and ocean responses. *Geology* 17, 853–6.

Shaw, J. 1994. Hairpin erosional marks, horseshoe vortices and subglacial erosion. *Sedimentary Geology* 91, 269–83.

Shaw, J. 2006. A glimpse at meltwater effects associated with continental ice sheets. In Knight, P. G. (ed.), *Glacier Science and Environmental Change*. Blackwell, London, 25–32.

Shaw, J. and Healy, T. R. 1980. Morphology of the Onyx River system, McMurdo Sound region, Antarctica. *New Zealand Journal of Geology and Geophysics* 23, 223–38.

Shaw, J. and Kvill, D. 1984. A glaciofluvial origin for drumlins of the Livingstone Lake area, Saskatchewan. *Canadian Journal of Earth Sciences* 12, 1426–40.

Shaw, J. and Sharpe, D. R. 1987. Drumlin formation by subglacial meltwater erosion. *Canadian Journal of Earth Sciences* 24, 2316–22.

Shaw, J., Faragini, D. M., Kvill, D. R. and Rains, R. B. 2000. The Athabasca fluting field, Alberta, Canada: implications for the formation of large scale fluting (erosional lineations). *Quaternary Science Reviews* 19, 959–80.

Shaw, J., Kvill, D. and Rains, R. B. 1989. Drumlins and catastrophic subglacial floods. *Sedimentary Geology* 62, 177–202.

Shean, D. E., Head, J. W. and Marchant, D. R. 2005. Origin and evolution of a cold-based tropical mountain glacier on Mars: the Pavonis Mons fan-shaped deposit. *Journal of Geophysical Research* 110, E05001, doi:10.1029/2004JE002360.

Shennan, I., Lambeck, K., Horton, B. P., Innes, J. B., Lloyd, J. M., McArthur, J. J., Purcell, T. and Rutherford, M. M. 2000. Late Devensian and Holocene records of relative sea-level changes in northwest Scotland and their implications for glacio-hydro-isostatic modelling. *Quaternary Science Reviews* 19, 1103–36.

Shepherd, A. 2004. Warm ocean is eroding West Antarctic Ice Sheet. *Geophysical Research Letters* 31, L23402, doi:10.1029/2004GL021106.

Shepherd, A. and Wingham, D. 2007. Recent sea-level contributions of the Antarctic and Greenland ice sheets. *Science* 315, 1529–32.

Shepherd, A., Wingham, D. and Rignot, E. 2004. Warm ocean eroding West Antarctic Ice Sheet. *Geophysical Research Letters* 31, L23402, doi:10.1029/2004GL021106.

Shepherd, A., Wingham, D. J., Mansley, J. A. D. and Corr, H. F. J. 2001. Inland thinning of Pine Island Glacier, West Antarctica. *Science* 291, 862–4.

Shepherd, A., Wingham, D. J. and Mansley, J. A. D. 2002. Inland thinning of the Amundsen Sea sector, West Antarctica. *Geophysical Research Letters* 19, doi:10.1029/2001GL014183.

Shepherd, A., Wingham, D., Payne, T. and Skvarca, P. 2003. Larsen ice shelf has progressively thinned. *Science* 302, 856–9.

Sheridan, M. F., Wohletz, K. and Dehn, J. 1987. Discrimination of grain-size subpopulations in pyroclastic deposits. *Geology* 15, 367–70.

Shilts, W. W. 1982. Glacial dispersal: principles and practical applications. *Geoscience Canada* 9, 42–8.

Shilts, W. W. 1993. Geological survey of Canada's contributions to understanding the composition of glacial sediments. *Canadian Journal of Earth Sciences* 30, 333–53.

Shilts, W. W., Aylsworth, J. M., Kaszycki, C. A. and Klassen, R. A. 1987. Canadian Shield. In Graf, W. L. (ed.), *Geomorphic Systems of North America*. Geological Society of America, Centennial Special Volume 2, 119–61.

Shindell, D. T., Schmidt, G. A., Miller, R. L. and Mann, M. E. 2003. Volcanic and solar forcing of climate change during the Preindustrial Era. *Journal of Climate* 16, 4094–107.

Shipp, S. S., Anderson, J. B. and Domack, E. W. 1999. Late Pleistocene-Holocene retreat of the West Antarctic Ice Sheet system in the Ross Sea. Part 1: Geophysical results. *Geological Society of America Bulletin* 111, 1486–516.

Shipp, S. S., Wellner, J. S. and Anderson, J. B. 2002. Retreat signature of a polar ice stream: subglacial geomorphic features and sediments from the Ross Sea, Antarctica. In Dowdeswell, J. A. and Ó Cofaigh, C. (eds), *Glacier-influenced Sedimentation on High Latitude Continental Margins*. Geological Society, London, Special Publication 203, 277–304.

Shoemaker, E. M. 1986a. Subglacial hydrology for an ice sheet resting on a deformable aquifer. *Journal of Glaciology* 32, 20–30.

Shoemaker, E. M. 1986b. The formation of fjord thresholds. *Journal of Glaciology* 32, 65–71.

Shoemaker, E. M. 1992. Water sheet outburst floods from the Laurentide Ice Sheet. *Canadian Journal of Earth Sciences* 29, 1250–64.

Shreve, R. L. 1972. Movement of water in glaciers. *Journal of Glaciology* 11, 205–14.

Shreve, R. L. 1985a. Esker characteristics in terms of glacier physics, Katahdin esker system, Maine. *Geological Society of America Bulletin* 96, 639–46.

Shreve, R. L. 1985b. Late Wisconsin ice-surface profile calculated from esker paths and types, Katahdin esker system, Maine. *Quaternary Research* 23, 27–37.

Shroder, J. F. and Bishop, M. P. 1998. Mass movement in the Himalaya: new insights and research directions. *Geomorphology* 26, 13–35.

Shroder, J. F., Bishop, M. P., Copland, L. and Sloan, V. F. 2000. Debris-covered glaciers and rock glaciers in the Nanga Parbat Himalaya, Pakistan. *Geografiska Annaler* 82A, 17–31.

Shumway, J. R. and Iverson, N. R. 2009. Magnetic fabrics of the Douglas Till of the Superior Lobe: exploring bed-deformation kinematics. *Quaternary Science Reviews* 28, 107–19.

Siddall, M., Rohling, E. J., Almogi-Labin, A., Hemleben, Ch., Meischner, D., Schmelzer, L. and Smeed, D. A. 2003. Sea-level fluctuations during the last glacial cycle. *Nature* 423, 853–8.

Siegenthaler, C. and Huggenberger, P. 1993. Pleistocene Rhine gravel: deposits of a braided river system with dominant pool preservation. In Best, J. L. and Bristow, C. S. (eds), *Braided Rivers*. Geological Society, London, Special Publication 75, 147–62.

Siegert, M. J. 2000. Antarctic subglacial lakes. *Earth Science Reviews* 50, 29–50.

Siegert, M. J. 2008. Antarctic subglacial topography and ice-sheet evolution. *Earth Surface Processes and Landforms* 33, 646–60.

Siegert, M. J. and Dowdeswell, J. A. 1995. Numerical modelling of the Late Weichselian Svalbard-Barents Sea Ice Sheet. *Quaternary Research* 42, 1–13.

Siegert, M. J. and Dowdeswell, J. A. 2004. Numerical reconstructions of the Eurasian ice sheet, its climate and glacial products during the Late Weichselian. *Quaternary Science Reviews* 23, 1273–83.

Siegert, M. J., Carter, S., Tabacco, I., Popov, S. and Blankenship, D. D. 2005. A revised inventory of Antarctic subglacial lakes. *Antarctic Science* 17, 453–60.

Siegert, M. J., Ellis-Evans, J. C., Tranter, M., Mayer, C., Petit, J.-R., Salamatin, A. and Priscu, J. C. 2001. Physical, chemical and biological processes in Lake Vostok and other Antarctic subglacial lakes. *Nature* 414, 603–9.

Siegert, M. J., Hindmarsh, R., Corr, H., Smith, A., Woodward, J., King, E. C., Payne, A. J. and Joughin, I. 2004a. Subglacial Lake Ellsworth: a candidate for in situ exploration in West Antarctica. *Geophysical Research Letters* 31, doi:10.1020/2004GL021477.

Siegert, M. J., Tranter, M., Ellis-Evans, J. C., Priscu, J. C. and Lyons, W. B. 2003. The hydrochemistry of Lake Vostok and the potential for life in Antarctic subglacial lakes. *Hydrological Processes* 17, 795–814.

Siegert, M. J., Welch, B., Morse, D., Vieli, A., Blankenship, D. D., Joughin, I., King, E. C., Leysinger Vieli, G. J.-M. C., Payne, A. J. and Jacobel, R. 2004b. Ice flow direction change in interior West Antarctica. *Science* 305, 1948–51.

Sigurdsson, O., Johnsson, T. and Jóhannesson, T. 2007. Relation between glacier-termini variations and summer temperature in Iceland since 1930. *Annals of Glaciology* 46, 170–6.

Simons, M. and Hager, B. H. 1997. Localization of the gravity field and the signature of glacial rebound. *Nature* 390, 500–4.

Singh, P. and Singh, V. P. 2001. *Snow and Glacier Hydrology*. Kluwer, Dordrecht.

Singh, P., Arora, M. and Goel, N. K. 2006. Effect of climate change on runoff of a glacierized Himalayan basin. *Hydrological Processes* 20, 1979–92.

Sissons, J. B. 1963. The glacial drainage system around Carlops, Peebleshire. *Transactions of the Institute of British Geographers* 32, 95–111.

Sissons, J. B. 1967. *The Evolution of Scotland's Scenery*. Oliver and Boyd, Edinburgh.

Sissons, J. B. 1971. The geomorphology of central Edinburgh. *Scottish Geographical Magazine* 87, 185–96.

Sissons, J. B. 1979. The limit of the Loch Lomond Advance in Glen Roy and vicinity. *Scottish Journal of Geology* 15, 31–42.

Sissons, J. B. 1980. The Loch Lomond Advance in the Lake District, northern England. *Transactions of the Royal Society of Edinburgh, Earth Sciences* 71, 13–27.

Sissons, J. B. 1983. Shorelines and isostasy in Scotland. In Smith, D. E. and Dawson, A. G. (eds), *Shorelines and Isostasy*. Academic Press, London, 209–25.

Sissons, J. B. and Cornish, R. 1982. Differential glacio-isostatic uplift of crustal blocks at Glen Roy, Scotland. *Quaternary Research* 18, 268–88.

Sissons, J. B. and Sutherland, D. G. 1976. Climatic inferences from former glaciers in the south-east Grampian Highlands. *Journal of Glaciology* 17, 325–46.

Sjogren, D. B., Fisher, T. G., Taylor, L. D., Jol, H. M. and Munro-Stasiuk, M. J. 2002. Incipient tunnel channels. *Quaternary International* 90, 41–56.

Sjorring, S., Nielsen, P. E., Frederiksen, J. K., Hegner, J., Hyde, G., Jensen, J. B., Morgensen, A. and Vortisch, W. 1982. Observationer fra Ristinge Klint, felt- og laboratorie- undersogelser. *Dansk Geologisk Forening*, Arsskrift for 1981, 135–49.

Skidmore, M. L. and Sharp, M. J. 1999. Drainage system behaviour of a High Arctic polythermal glacier. *Annals of Glaciology* 28, 209–15.

Skidmore, M. L., Anderson, S. P., Sharp, M. J., Foght, J. and Lanoil, B. D. 2005. Comparison of microbial community compositions of two subglacial environments reveals a possible role for micro-organisms in chemical weathering processes, *Applied and Environmental Microbiology* 71, 6986–97.

Skidmore, M. L., Foght, J. M. and Sharp, M. J. 2000. Microbial life beneath a high Arctic glacier. *Applied and Environmental Microbiology* 66, 3214–20.

Skvarca, P., De Angelis, H., Naruse, R., Warren, C. R. and Anaiya, M. 2002. Calving rates in freshwater: new data from southern Patagonia. *Annals of Glaciology* 34, 379–84.

Skvarca, P., Rack, W., Rott, H. and Ibarzabal y Donangelo, T. 1999. Climatic trend and disintegration of ice shelves on the Antarctic Peninsula: an overview. *Polar Research* 18, 151–7.

Slatt, R. M. and Eyles, N. 1981. Petrology of glacial sand: implications for the origin and mechanical durability of lithic fragments. *Sedimentology* 28, 171–83.

Sletten, K., Lyså, A. and Lønne, I. 2001. Formation and disintegration of a high arctic ice-cored moraine complex, Scott Turnerbreen, Svalbard. *Boreas* 30, 272–84.

Small, R. J. 1983. Lateral moraines of Glacier De Tsidjiore Nouve: form, development and implications. *Journal of Glaciology* 29, 250–9.

Small, R. J. 1987. Englacial and supraglacial sediment: transport and deposition. In Gurnell, A. M. and Clark, M. J. (eds), *Glacio-fluvial Sediment Transfer: An Alpine Perspective*. Wiley, Chichester, 111–45.

Small, R. J. and Gomez, B. 1981. The nature and origin of debris layers within Glacier de Tsidjiore Nouve, Valais, Switzerland. *Annals of Glaciology* 2, 109–13.

Small, R. J., Clark, M. J. and Cawse, T. J. P. 1979. The formation of medial moraines on Alpine glaciers. *Journal of Glaciology* 22, 43–52.

Smalley, I. J. and Unwin, D. J. 1968. The formation and shape of drumlins and their distribution and orientation in drumlin fields. *Journal of Glaciology* 7, 377–90.

Smalley, I. J. and Warburton, J. 1994. The shape of drumlins, their distribution in drumlin fields, and the nature of the sub-ice shaping forces. *Sedimentary Geology* 91, 241–52.

Smellie, J. L. 2000. Subglacial eruptions. In Sigurdsson, H. (ed.), *Encyclopedia of Volcanoes*. Academic Press, San Diego, CA, 403–18.

Smellie, J. L. 2007. Quaternary vulcanism, subglacial landforms. In Elias, S. A. (ed.), *Encyclopedia of Quaternary Science*. Elsevier, Rotterdam, 784–98.

Smellie, J. L. 2009. Terrestrial sub-ice volcanism: landform morphology, sequence characteristics and environmental influences, and implications for candidate Mars examples. In Chapman, M. G. and Keszthely, L. (eds), *Preservation of Random Mega-Scale Events on Mars and Earth: Influence on Geologic History*. Geological Society of America, Special Papers 453, 55–76.

Smellie, J. L. and Chapman, M. G. (eds). 2002. *Volcano–Ice Interactions on Earth and Mars*. Geological Society, London, Special Publication 202.

Smellie, J. L. and Skilling, I. P. 1994. Products of subglacial volcanic eruptions under different ice thicknesses: two examples from Antarctica. *Sedimentary Geology* 91, 115–29.

Smellie, J. L., Johnson, J. S., McIntosh, W. C., Esser, R., Gudmundsson, M. T., Hambrey, M. J. and van Wyk de Vries, B. 2008. Six million years of glacial history recorded in volcanic lithofacies of the James Ross Island Volcanic Group, Antarctic Peninsula. *Palaeogeography, Palaeoclimatology, Palaeoecology* 260, 122–48.

Smiraglia, C. 1989. The medial moraines of Ghiacciaio dei Forni, Valtellina, Italy: morphology and sedimentology. *Journal of Glaciology* 35, 81–4.

Smith, A. M. 2006. Microearthquakes and subglacial conditions. *Geophysical Research Letters* 33, doi:10.1029/2006GL028207.

Smith, A. M., Murray, T., Davison, B. M., Clough, A. F., Woodward, J. and Jiskoot, H. 2002. Late surge glacial conditions on Bakaninbreen, Svalbard, and implications for surge termination. *Journal of Geophysical Research* 107 (B8), 2152, doi:10.1029/2002JB000457.

Smith, A. M., Murray, T., Nicholls, K. W., Makinson, K., Athalgeirsdottir, G., Behar, A. and Vaughan, D. G. 2007. Rapid erosion and drumlin formation observed beneath a fast-flowing Antarctic ice stream. *Geology* 35, 127–30.

Smith, D. E. and Dawson, A. G. (eds). 1983. *Shorelines and Isostasy*. Academic Press, London.

Smith, D. E., Cullingford, R. A. and Firth, C. R. 2000. Patterns of isostatic land uplift during the Holocene: evidence from mainland Scotland. *The Holocene* 10, 489–501.

Smith, D. E., Shi, S., Cullingford, R. E., Dawson, A. G., Dawson, S., Firth, C. R., Foster, I. D. L., Fretwell, P. T., Haggart, B. A., Holloway, L. K. and Long, D. 2004. The Holocene Storegga Slide tsunami in the United Kingdom. *Quaternary Science Reviews* 23, 2291–321.

Smith, G. A. 1993. Missoula flood dynamics and magnitudes inferred from sedimentology of slack-water deposits on the Columbia Plateau, Washington. *Geological Society of America Bulletin* 105, 77–100.

Smith, I. R. 1999. Late Quaternary glacial history of Lake Hazen Basin and eastern Hazen Plateau, northern Ellesmere Island, Nunavut, Canada. *Canadian Journal of Earth Sciences* 36, 1547–65.

Smith, I. R. 2000. Diamictic sediments within high arctic lake sediment cores: evidence for lake ice rafting along the lateral glacial margin. *Sedimentology* 47, 1157–79.

Smith, J. A., Bentley, M. J., Hodgson, D. A. and Cook, A. J. 2007. George VI Ice Shelf: past history, present behaviour and potential mechanisms for future collapse. *Antarctic Science* 19, 131–42.

Smith, J. A., Mark, B. and Rodbell, D. T. 2008. The timing and magnitude of mountain glaciation in the tropical Andes. *Journal of Quaternary Science* 23, 609–34.

Smith, L. M. and Andrews, J. T. 2000. Sediment characteristics in iceberg dominated fjords, Kangerlussuaq region. East Greenland. *Sedimentary Geology* 130 (2), 11–25.

Smith, M. J. and Clark, C. D. 2005. Methods for the visualization of digital elevation models for landform mapping. *Earth Surface Processes and Landforms* 30, 885–900.

Smith, N. D. 1985. Proglacial fluvial environment. In Ashley, G. M., Shaw, J. and Smith, N. D. (eds), *Glacial Sedimentary Environments*. SEPM Short Course 16, 85–136.

Smith, N. D. and Ashley, G. M. 1985. Proglacial lacustrine environments. In Ashley, G. M., Shaw, J. and Smith, N. D. (eds), *Glacial Sedimentary Environments*. SEPM Short Course 16, 135–215.

Smith, N. D. and Syvitski, J. P. M. 1982. Sedimentation in a glacier-fed lake: the role of pelletisation on deposition of fine-grained suspensates. *Journal of Sedimentary Petrology* 52, 503–13.

Smith, N. D., Vendl, M. A. and Kennedy, S. K. 1982. Comparison of sedimentation regimes in four glacier-fed lakes of western Alberta. In Davidson-Arnott, R., Nickling, W. and Fahey, B. D. (eds), *Research in Glacial, Glaciofluvial, and Glaciolacustrine Systems*. Geobooks, Norwich, 203–38.

Sneed, W. A. and Hamilton, G. S. 2007. Evolution of melt pond volume on the surface of the Greenland Ice Sheet. *Geophysical Research Letters* 34, doi:10.1029/2006GL028697.

Socha, B. J., Colgan, P. M. and Mickelson, D. M. 1999. Ice-surface profiles and bed conditions of the Green Bay Lobe from 13,000 to 11,000 ^{14}C years BP. In Mickelson, D. M. and Attig, J. W. (eds), *Glacial Processes: Past and Present*. Geological Society of America, Special Paper 337, 151–8.

Sohn, H.-G., Jezek, K. C. and van der Veen, C. J. 1998. Jakobshavn Glacier, West Greenland: 30 years of spaceborne observations. *Geophysical Research Letters* 25, 2699–702.

Sole, A., Payne, T., Bamber, J., Nienow, P. and Krabill, W. 2008. Testing hypotheses of the cause of peripheral thinning of the Greenland ice sheet: is land-terminating ice thinning at anomalously high rates? *The Cryosphere* 2, 205–18.

Solheim, A., Russwurm, L., Elverhøi, A. and Nyland Berg, M. 1990. Glacial geomorphic features in the northern Barents Sea: direct evidence for grounded ice and implications for the pattern of deglaciation and late glacial sedimentation. In Dowdeswell, J. A. and Scourse, J. D. (eds), *Glacimarine Environments: Processes and Sediments*. Geological Society, London, Special Publication 53, 253–68.

Solomina, O., Haeberli, W., Kull, C. and Wiles, G. 2008. Historical and Holocene glacier–climate variations: General concepts and overview. *Global and Planetary Change* 60, 1–9.

Souchez, R. A. and Jouzel, J. 1984. On the isotopic composition in ∂D and $\partial^{18}O$ of water and ice during freezing. *Journal of Glaciology* 30, 369–72.

Souchez, R. A. and Lorrain, R. D. 1978. Origin of the basal ice layer from Alpine glaciers indicated by its geochemistry. *Journal of Glaciology* 20, 319–28.

Souchez, R. A. and Lorrain, R. D. 1987. The subglacial sediment system. In Gurnell, A. M. and Clark, M. J. (eds), *Glaciofluvial Sediment Transfer: An Alpine Perspective*. Wiley, Chichester, 147–63.

Souchez, R. A. and Lorrain, R. D. 1991. *Ice Composition and Glacier Dynamics*. Springer Verlag, Berlin.

Southard, J. B., Smith, N. D. and Kuhnle, R. A. 1984. Chutes and lobes: newly identified elements of braiding in shallow gravelly streams. In Koster, E. H. and Steel, R. J. (eds), *Sedimentology of Gravels and Conglomerates*. Canadian Society of Petroleum Geologists, Memoir 10, 51–9.

Spedding, N. F. 2000. Hydrological controls on sediment transport pathways: implications for debris-covered glaciers. In Nakawo, N., Fountain, A. and Raymond, C. (eds), *Debris-covered Glaciers*. IAHS Publication 264, 133–42.

Spedding, N. F. and Evans, D. J. A. 2002. Sediments and landforms at Kvíárjökull, southeast Iceland: a reappraisal of the glaciated valley landsystem. *Sedimentary Geology* 149, 21–42.

Spotila, J. A., Buscher, J. T., Meigs, A. J. and Reiners, P. W. 2004. Long-term glacial erosion of active mountain belts: example from the Chugach-St Elias Range, Alaska. *Geology* 32, 501–4.

Spring, U. and Hutter, K. 1981. Numerical studies of jökulhlaups. *Cold Regions Science and Technology* 4, 221–44.

Spring, U. and Hutter, K. 1982. Conduit flow of a fluid through its solid phase and its application to intraglacial channel flow. *International Journal of Engineering Science* 20, 327–63.

St Onge, D. A. 1984. Surficial deposits of the Redrock Lake area, District of Mackenzie. *Current Research*, Part A, Geological Survey of Canada, Paper 84-1A, 271–8.

St Onge, D. A. and McMartin, I. 1995. *Quaternary Geology of the Inman River Area, Northwest Territories*. Geological Survey of Canada, Bulletin 446.

St Onge, D. A. and McMartin, I. 1999. La moraine du Lac Bluenose (Territoires du Nord-Ouest), une moraine a noyau de glace de glacier. *Geographie Physique et Quaternaire* 53, 287–95.

Staiger, J. K. W., Gosse, J. C., Johnson, J. V., Fastook, J., Gray, J. T., Stockli, D. F., Stockli, L. and Finkel, R. 2005. Quaternary relief generation by polythermal glacier ice. *Earth Surface Processes and Landforms* 30, 1145–59.

Staiger, J. K. W., Gosse, J., Little, E. C., Utting, D. J., Finkel, R., Johnson, J. V. and Fastook, J. 2006. *Quaternary Geochronology* 1, 29–42.

Stalker, A. MacS. 1960. *Ice-pressed Drift Forms and Associated Deposits in Alberta*. Geological Survey of Canada, Bulletin 57.

Stalker, A. MacS. 1976. *Megablocks, or the Enormous Erratics of the Albertan Prairies*. Geological Survey of Canada, Paper 76-1C, 185–8.

Stanford, S. D. and Mickelson, D. H. 1985. Till fabric and deformational structures in drumlins near Waukesha, Wisconsin, USA. *Journal of Glaciology* 31, 220–8.

Stea, R. R. 1994. Relict and palimpsest glacial landforms in Nova Scotia, Canada. In Warren, W. P. and Croot, D. G. (eds), *Formation and Deformation of Glacial Deposits*. Balkema, Rotterdam, 141–58.

Stea, R. R. and Brown, Y. 1989. Variation in drumlin orientation, form and stratigraphy relating to successive ice flows in southern and central Nova Scotia. *Sedimentary Geology* 62, 223–40.

Stearns, L. A. and Hamilton, G. S. 2007. Rapid volume loss from two East Greenland outlet glaciers quantified using repeat stereo satellite imagery. *Geophysical Research Letters* 34, doi:10.1029/2006GL028982.

Stearns, L. A., Jezek, K. C. and van der Veen, C. J. 2005. Decadal-scale variations in ice flow along Whillans Ice Stream and its tributaries. *Journal of Glaciology* 51 (172), 147–57.

Steffen, K. and Box, J. 2001. Surface climatology of the Greenland ice sheet: Greenland Climate Network 1995–1999. *Journal of Geophysical Research* 106, D24, 33951–64.

Steiner, D., Pauling, A., Nussbaumer, S. U., Nesje, A., Luterbacher, J., Wanner, H. and Zumbühl, H. J. 2008. Sensitivity of European glaciers to precipitation and temperature – two case studies. *Climatic Change* 90, 413–41.

Steiner, D., Walter, A. and Zumbühl, H. J. 2005. The application of a non-linear back-propagation neural network to study the mass balance of Grosse Aletschgletscher, Switzerland. *Journal of Glaciology* 51, 313–23.

Stenborg, T. 1969. Studies of the internal drainage of glaciers. *Geografiska Annaler* 51A, 13–41.

Stenborg, T. 1973. Some viewpoints on the internal drainage of glaciers. In *Hydrology of Glaciers*. IAHS Publication 95, 117–29.

Stephan, H.-J. 1987. Form, composition, and origin of drumlins in Schleswig-Holstein. In Menzies, J. and Rose, J. (eds), *Drumlin Symposium*. Balkema, Rotterdam, 335–45.

Stewart, A. D. 1991. Torridonian. In Craig, G. Y. (ed.), *Geology of Scotland*, 3rd edn. Geological Society, London, 65–85.

Stibal, M., Sabacká, M. and Kastovská, K. 2006. Microbial communities on glacier surfaces in Svalbard: impact of physical and chemical properties on abundance and structure of cyanobacteria and algae. *Microbial Ecology* 52, 644–54.

Stoker, M. S. 1995. The influence of glacigenic sedimentation on slope-apron development on the continental margin off northwest Britain. In Scrutton, R. A., Stoker, M. S., Shimmield, G. B. and Tudhope, A. W. (eds), *The Tectonics, Sedimentation and Palaeoceanography of the North Atlantic Region*. Geological Society, London, Special Publication 90, 159–77.

Stoker, M. S., Bradwell, T., Wilson, C., Harper, C., Smith, D. and Brett, C. 2006. Pristine fjord landsystem revealed on the sea bed in the Summer Isles region, NW Scotland. *Scottish Journal of Geology* 42, 89–99.

Stokes, C. R. 2002. Identification and mapping of palaeo-ice stream geomorphology from satellite imagery: implications for ice stream functioning and ice sheet dynamics. *International Journal of Remote Sensing* 23, 1557–63.

Stokes, C. R. and Clark, C. D. 1999. Geomorphological criteria for identifying Pleistocene ice streams. *Annals of Glaciology* 28, 67–74.

Stokes, C. R. and Clark, C. D. 2001. Palaeo-ice streams. *Quaternary Science Reviews* 20, 1437–57.

Stokes, C. R. and Clark, C. D. 2002a. Are long subglacial bedforms indicative of fast glacier flow? *Boreas* 31, 239–49.

Stokes, C. R. and Clark, C. D. 2002b. Ice stream shear margin moraines. *Earth Surface Processes and Landforms* 27, 547–58.

Stokes, C. R. and Clark, C. D. 2003a. Giant glacial grooves detected on Landsat ETM+ satellite imagery. *International Journal of Remote Sensing* 24, 905–10.

Stokes, C. R. and Clark, C. D. 2003b. Laurentide ice streams on the Canadian Shield: a conflict with the soft-bedded ice stream paradigm? *Geology* 31, 347–50.

Stokes, C. R. and Clark, C. D. 2003c. The Dubawnt Lake palaeo-ice stream: evidence for dynamic ice sheet behaviour on the Canadian Shield and insights regarding the controls on ice stream location and vigour. *Boreas* 32, 263–79.

Stokes, C. R., Clark, C. D., Darby, D. A. and Hodgson, D. A. 2005. Late Pleistocene ice export events into the Arctic Ocean from the M'Clure Strait Ice Stream, Canadian Arctic Archipelago. *Global and Planetary Change* 49, 139–62.

Stokes, C. R., Clark, C. D., Lian, O. B. and Tulaczyk, S. 2006a. Geomorphological map of ribbed moraines on the Dubawnt Lake palaeo-ice stream bed: a signature of ice stream shut-down? *Journal of Maps* 2006, 1–9.

Stokes, C. R., Clark, C. D. and Winsborrow, M. C. M. 2006b. Subglacial bedform evidence for a major palaeo-ice stream and its retreat phases in Amundsen Gulf, Canadian Arctic Archipelago. *Journal of Quaternary Science* 21, 399–412.

Stokes, C. R., Clark, C. D., Lian, O. B. and Tulaczyk, S. 2007. Ice stream sticky spots: a review of their identification and influence beneath contemporary and palaeo-ice streams. *Earth-Science Reviews* 81, 217–49.

Stokes, C. R., Lian, O. B., Tulaczyk, S. and Clark, C. D. 2008. Superimposition of ribbed moraines on a palaeo-ice stream bed: implications for ice stream dynamics and shutdown. *Earth Surface Processes and Landforms* 33, 593–609.

Stone, J. O. and Ballantyne, C. K. 2006. Dimensions and deglacial chronology of the Outer Hebrides Ice Cap, northwest Scotland: implications of cosmic ray exposure dating. *Journal of Quaternary Science* 21, 75–84.

Stone, J. O., Ballantyne, C. K. and Fifield, K. 1998. Exposure dating and validation of periglacial weathering limits, northwest Scotland. *Geology* 26, 587–90.

Stone, R. 2004. Iceland's Doomsday Scenario? *Science* 306, 1278–81.

Strasser, U., Corripio, J., Pellicciotti, F., Burlando, P., Brock, B. and Funk, M. 2004. Spatial and temporal variability of meteorological variables at Haut Glacier d'Arolla (Switzerland) during the ablation season 2001: measurements and simulations. *Journal of Geophysical Research* 109, D03203, doi:10.1029/2003JD003973.

Stroeven, A. P., Fabel, D., Harbor, J., Hättestrand, C. and Kleman, J. 2002b. Quantifying the erosional impact of the Fennoscandian ice sheet in the Tornetrask-Narvik corridor, northern Sweden, based on cosmogenic radionuclide data. *Geografiska Annaler* 84A, 275–87.

Stroeven, A. P., Fabel, D., Harbor, J., Hättestrand, C. and Kleman, J. 2002c. Reconstructing the erosion history of glaciated passive margins: applications of in situ produced cosmogenic nuclide techniques. In Doré, A. G., Cartwright, J. A., Stoker, M. S., Turner, J. P. and White, N. (eds), *Exhumation of the North Atlantic Margin: Timing, Mechanisms and Implications for Petroleum Exploration*. Geological Society, London, Special Publication 196, 153–68.

Stroeven, A. P., Fabel, D., Hättestrand, C. and Harbor, J. 2002a. A relict landscape in the centre of Fennoscandian glaciation: cosmogenic radionuclide evidence of tors preserved through multiple glacial cycles. *Geomorphology* 44, 145–54.

Stuart, G. W., Murray, Y., Gamble, N., Hayes, K. and Hodson, A. 2003. Characterization of englacial channels by ground-penetrating radar: an example from Brøggerbreen, Svalbard. *Journal of Geophysical Research* 108, B11, 2525, doi:10.1029JB002435.

Stuart, G. W., Murray, T., Brisbourne, A., Styles, P. and Toon, S. 2005. Acoustic emissions from a surging glacier: Bakaninbreen, Svalbard. *Annals of Glaciology* 42, 151–7.

Studinger, M., Bell, R. E., Karner, G. D., Tikku, A. A., Holt, J. W., Morse, D. L., Richter, T. G., Kempf, S. D., Peters, M. E., Blankenship, D. D., Sweeney, R. E. and Rystrom, V. L. 2003. Ice cover, landscape setting, and geological framework of Lake Vostok, East Antarctica. *Earth and Planetary Science Letters* 205, 195–210.

Sturm, M. and Benson, C. S. 1997. Vapor transport, grain growth and depth-hoar development in the subarctic snow. *Journal of Glaciology* 43, 42–59.

Sturm, M., Holmgren, J., König, M. and Morris, K. 1997. The thermal conductivity of seasonal snow. *Journal of Glaciology* 43, 26–41.

Sugden, D. E. 1974. *Landscapes of Glacial Erosion in Greenland and their Relationship to Ice, Topographic and Bedrock Conditions*. Institute of British Geographers, Special Publication 7, 177–95.

Sugden, D. E. 1977. Reconstruction of the morphology, dynamics and thermal characteristics of the Laurentide ice sheet at its maximum. *Arctic and Alpine Research* 9, 27–47.

Sugden, D. E. 1978. Glacial erosion by the Laurentide ice sheet. *Journal of Glaciology* 20, 367–91.

Sugden, D. E. and Clapperton, C. M. 1981. An ice-shelf moraine, George VI Sound, Antarctica. *Annals of Glaciology* 2, 135–41.

Sugden, D. E. and John, B. S. 1976. *Glaciers and Landscape*. Arnold, London.

Sugden, D. E., Balco, G., Cowdery, S. G., Stone, J. O. and Sass, L. C. III. 2005. Selective glacial erosion and weathering zones in the coastal mountains of Marie Byrd Land, Antarctica. *Geomorphology* 67, 317–34.

Sugden, D. E., Denton, G. H. and Marchant, D. R. 1991. Subglacial meltwater channel systems and ice sheet overriding, Asgard Range, Antarctica. *Geografiska Annaler* 73A, 109–21.

Sugden, D. E., Denton, G. H. and Marchant, D. R., 1993. The case for a stable East Antarctic ice sheet: the background. *Geografiska Annaler* 75A (4), 151–4.

Sugden, D. E., Glasser, N. F. and Clapperton, C. M. 1992. Evolution of large roches moutonnées. *Geografiska Annaler* 74A, 253–64.

Sugden, D. E., Marchant, D. R., Potter, N. Jr, Souchez, R. A., Denton, G. H., Swisher, C. C. III and Tison, J. L. 1995. Preservation of Miocene glacier ice in East Antarctica. *Nature* 376, 412–14.

Sugden, D. E., Summerfield, M. A., Denton, G. H., Wilch, T. I., McIntosh, W. C., Marchant, D. R. and Rutford, R. H. 1999. Landscape development in the Royal Society Range, southern Victoria Land, Antarctica: stability since the mid-Miocene. *Geomorphology* 28, 181–200.

Sugiyama, S. and Gudmundsson, G. H. 2004. Short-term variations in glacier flow controlled by subglacial water pressure at Lauteraargletscher, Bernese Alps, Switzerland. *Journal of Glaciology* 50, 353–62.

Sugiyama, S., Bauder, A., Zahno, C. and Funk, M. 2007. Evolution of Rhonegletscher, Switzerland, over the past 125 years and in the future: application of an improved flowline model. *Annals of Glaciology* 46, 268–74.

Sund, M., Eiken, T., Hagen, J. O. and Kääb, A. 2009. Svalbard surge dynamics derived from geometric changes. *Annals of Glaciology* 50 (52), 50–60.

Suter, S. and Hoelzle, M. 2002. Cold firn in the Mont Blanc and Monte Rosa areas, European Alps: spatial distribution and statistical models. *Annals of Glaciology* 35, 9–18.

Sutherland, D. G. 1984. Modern glacier characteristics as a basis for inferring former climates with particular reference to the Loch Lomond Stadial. *Quaternary Science Reviews* 3, 291–309.

Svendsen, J. I., Alexanderson, H., Astakhov, V. I. et al. 2004. Late Quaternary ice sheet history of Northern Eurasia. *Quaternary Science Reviews* 23, 1229–71.

Swift, D. A. 2006. Haut Glacier d'Arolla, Switzerland: hydrological controls on subglacial sediment evacuation and glacial erosional capacity. In Knight, P. G. (ed.), *Glacier Science and Environmental Change*. Blackwell, Oxford, 23–5.

Swift, D. A., Evans, D. J. A. and Fallick, A. E. 2006. Transverse englacial debris-rich ice bands at Kviarjökull, southeast Iceland. *Quaternary Science Reviews* 25, 1708–18.

Swift, D. A., Nienow, P. W. and Hoey, T. B. 2005a. Basal sediment evacuation by subglacial meltwater: suspended sediment transport from Haut Glacier d'Arolla, Switzerland. *Earth Surface Processes and Landforms* 30, 867–83.

Swift, D. A., Nienow, P. W., Hoey, T. B. and Mair, D. W. F. 2005b. Seasonal evolution of runoff from Haut Glacier d'Arolla, Switzerland and implications for glacial geomorphic processes. *Journal of Hydrology* 309, 133–48.

Swift, D. A., Nienow, P. W., Spedding, N. and Hoey, T. B. 2002. Geomorphic implications of subglacial drainage configuration: rates of basal sediment evacuation controlled by seasonal drainage system evolution. *Sedimentary Geology* 149, 5–19.

Swift, D. A., Persano, C., Stuart, F. M., Gallagher, K. and Witham, A. 2008. A reassessment of the role of ice sheet glaciation in the long-term evolution of the East Greenland fjord region. *Geomorphology* 97, 109–25.

Swingedouw, D., Fichefet, T., Huybrechts, P., Goosse, H., Driesschaert, E. and Loutre, M.-F. 2008. Antarctic ice-sheet melting provides negative feedbacks on future climate warming. *Geophysical Research Letters* 35, L17705, doi:10.1029/2008GL034410.

Swithinbank, C. W. M. 1988. Antarctica. In Williams, R. S. and Ferrigno, J. G. (eds), *Satellite Image Atlas of Glaciers of the World*. USGS Professional Paper 1386-B.

Syverson, K. M. and Mickelson, D. M. 2009. Origin and significance of lateral meltwater channels formed along a temperate glacier margin, Glacier Bay, Alaska. *Boreas* 38, 132–45.

Syverson, K. M., Gaffield, S. J. and Mickelson, D. M. 1994. Comparison of esker morphology and sedimentology with former ice-surface topography, Burroughs Glacier, Alaska. *Geological Society of America Bulletin* 106, 1130–42.

Syvitski, J. P. M. and Farrow, G. G. 1989. Fjord sedimentation as an analogue for small hydrocarbon-bearing fan deltas. In Whateley, M. K. G. and Pickering, K. T. (eds), *Deltas: Sites and Traps for Fossil Fuels*. Geological Society, London, Special Publication 41, 21–43.

Syvitski, J. P. M. and Murray, J. W. 1981. Particle interaction in fjord suspended sediment. *Marine Geology* 39, 215–42.

Syvitski, J. P. M., Burrell, D. C. and Skei, J. M. 1987. *Fjords: Processes and Products*. Springer, New York.

Syvitski, J. P. M., Stein, A. B., Andrews, J. T. and Milliman, J. D. 2001. Icebergs and the sea floor of the East Greenland (Kangerlussuaq) continental margin. *Arctic, Antarctic and Alpine Research* 33, 52–61.

Tabacco, I. E., Cianfarra, P., Forieri, A., Salvini, F. and Zirizotti, A. 2006. Physiography and tectonic setting of the subglacial lake district between Vostok and Belgica subglacial highlands (Antarctica). *Geophysical Journal International* 165, 1029–40.

Takeuchi, N. and Koshima, S. 2000. Effect of debris cover on species composition of living organisms in supraglacial lakes on a Himalayan glacier. In *Debris Covered Glaciers*. IAHS Publication 264, 267–75.

Takeuchi, N., Kohshima, S. and Fujita, K. 1998. Snow algae community on a Himalayan glacier, Glacier AX010, East Nepal: relationship with summer mass balance. *Bulletin of Glacier Research* 16, 43–50.

Takeuchi, N., Kohshima, S. and Seko, K. 2001. Structure, formation and darkening process of albedo-reducing material (cryoconite) on a Himalayan glacier: a granular algal mat growing on the glacier. *Arctic, Antarctic and Alpine Research* 33, 115–22.

Talling, P. J., Peakall, J., Sparks, R. S. J., Ó Cofaigh, C., Dowdeswell, J. A., Felix, M., Wynn, R. B., Baas, J. H., Hogg, A. J., Masson, D. G., Taylor, J. and Weaver, P. P. E. 2002. Experimental constraints on shear mixing rates and processes: implications for the dilution of submarine debris flows. In Dowdeswell, J. A. and Ó Cofaigh, C. (eds), *Glacier-influenced Sedimentation on High Latitude Continental Margins*. Geological Society, London, Special Publication 203, 89–103.

Tamisiea, M. E., Mitrovica, J. X., Milne, G. A. and Davis, J. L. 2001. Global geoid and sea level changes due to present-day ice mass fluctuations. *Journal of Geophysical Research* 106 (B12), 30849–63.

Tanaka, K. L. and Kolb, E. J. 2001. Geologic history of the polar regions of Mars based on Mars Global Surveyor data. I: Noachian and Hesperian Periods. *Icarus* 154, 3–21.

Tangborn, W. V. 1997. Using low-altitude meteorological observations to calculate the mass balance of Alaska's Columbia Glacier and relate it to calving and speed. In van der Veen, C. J. (ed.), *Calving Glaciers: Report of a Workshop, Feb. 28–March 2, 1997*. Ohio State University, Columbus. Byrd Polar Research Centre, report 15, 141–61.

Tarasov, L. and Peltier, W. R. 2003. Greenland glacial history, borehole constraints, and Eemian extent. *Journal of Geophysical Research* 108, B3, doi:10.1029/2001JB001731.

Taylor, J., Dowdeswell, J. A., Kenyon, N. H., Whittington, R. J. and van Weering, T. C. E. 2000. Morphology and late Quaternary sedimentation on the North Faeroes slope and abyssal plain, North Atlantic. *Marine Geology* 168, 1–24.

Taylor, J., Dowdeswell, J. A., Kenyon, N. H. and Ó Cofaigh, C. 2002b. Late Quaternary architecture of trough-mouth fans: debris flows and suspended sediments on the Norwegian margin. In Dowdeswell, J. A. and Ó Cofaigh, C. (eds), *Glacier-influenced Sedimentation on High Latitude Continental Margins*. Geological Society, London, Special Publication 203, 55–71.

Taylor, J., Dowdeswell, J. A. and Siegert, M. J. 2002a. Late Weichselian depositional processes, fluxes and sediment volumes on the margins of the Norwegian Sea (62–75° N). *Marine Geology* 188, 61–78.

Teller, J. T. 1987. Proglacial lakes and the southern margin of the Laurentide Ice Sheet. In Ruddiman, W. F. and Wright, H. E. (eds), *North America and Adjacent Oceans During the Last Deglaciation*. Geological Society of America, Geology of North America, vol. K-3, 39–69.

Teller, J. T. 1995. History and drainage of large ice-dammed lakes along the Laurentide Ice Sheet. *Quaternary International* 28, 83–92.

Teller, J. T. 2003. Subaquatic landsystems: large proglacial lakes. In Evans, D. J. A. (ed.), *Glacial Landsystems*. Arnold, London, 348–71.

Teller, J. T., Leverington, D. W. and Mann, J. D. 2002. Freshwater outbursts to the oceans from glacial Lake Agassiz and their role in climate change during the last deglaciation. *Quaternary Science Reviews* 21, 879–87.

ter Wee, M. W. 1983. The Saalian glaciation in the northern Netherlands. In Ehlers, J. (ed.), *Glacial Deposits in North-West Europe*. Balkema, Rotterdam, 405–12.

Terwindt, J. H. J. and Augustinus, P. G. E. F. 1985. Lateral and longitudinal successions in sedimentary structures in the Middle Mause esker, Scotland. *Sedimentary Geology* 45, 161–88.

Thomas, G. S. P. 1984. A late Devensian glaciolacustrine fan-delta at Rhosesmor, Clwyd, North Wales. *Geological Journal* 19, 125–41.

Thomas, G. S. P. and Chiverrell, R. C. 2006. A model of subaqueous sedimentation at the margin of the Late Midlandian Irish Ice Sheet, Connemara, Ireland, and its implications for regionally high isostatic sea levels. *Quaternary Science Reviews* 25, 2868–93.

Thomas, G. S. P. and Chiverrell, R. C. 2007. Structural and depositional evidence for repeated ice-marginal oscillation along the eastern margin of the Late Devensian Irish Sea Ice Stream. *Quaternary Science Reviews* 26, 2375–405.

Thomas, G. S. P. and Connell, R. J. 1985. Iceberg drop, dump and grounding structures from Pleistocene glaciolacustrine sediments, Scotland. *Journal of Sedimentary Petrology* 55, 243–9.

Thomas, G. S. P. and Montague, E. 1997. The morphology, stratigraphy and sedimentology of the Carstairs esker, Scotland, UK. *Quaternary Science Reviews* 16, 661–74.

Thomas, G. S. P., Chester, D. K. and Crimes, P. 1998. The late Devensian glaciation of the eastern Lleyn Peninsula, North Wales: evidence for terrestrial depositional environments. *Journal of Quaternary Science* 13, 255–70.

Thomas, G. S. P., Chiverrell, R. C. and Huddart, D. 2004. Ice-marginal depositional responses to readvance episodes in the Late Devensian deglaciation of the Isle of Man. *Quaternary Science Reviews* 23, 85–106.

Thomas, G. S. P., Connaughton, M. and Dackombe, R. V. 1985. Facies variation in a late Pleistocene supraglacial outwash sandur from the Isle of Man. *Geological Journal* 20, 193–213.

Thomas, P. C., Malin, M. C., James, P. B., Cantor, B. A., Williams, R. M. E. and Gierasch, P. 2005. South polar residual cap of Mars: features, stratigraphy, and changes. *Icarus* 174, 535–59.

Thomas, R. H. 2004. Greenland: recent mass balance observations. In Bamber, J. L. and Payne, A. J. (eds), *Mass Balance of the Cryosphere: Observations and Modeling of Contemporary and Future Changes.* Cambridge University Press, 393–436.

Thomas, R. H. 2007. Tide-induced perturbations of glacier velocities. *Global and Planetary Change* 59, 217–24.

Thomas, R. H., Frederick, E., Krabill, W., Manizade, S., Martin, C. and Mason, A. 2005. Elevation changes on the Greenland ice sheet from comparison of aircraft and ICESat laser-altimeter data. *Annals of Glaciology* 42, 77–82.

Thomason, J. F. and Iverson, N. R. 2006. Microfabric and microshear evolution in deformed till. *Quaternary Science Reviews* 25, 1027–38.

Thomason, J. F. and Iverson, N. R. 2008. A laboratory study of particle ploughing and pore-pressure feedback: a velocity-weakening mechanism for soft glacier beds. *Journal of Glaciology* 54, 169–81.

Thompson, L. G. 2000. Ice core evidence for climate change in the Tropics: implications for our future. *Quaternary Science Reviews* 19, 19–35.

Thompson, L. G. 2001. Stable isotopes and their relationship to temperature as recorded in low-latitude ice cores. In Gerhard, L. C., Harrison, W. E. and Hanson, B. M. (eds), *Geological Perspectives of Global Change*, 99–119.

Thompson, L. G., Mosely-Thompson, E. and Arnao, B. M. 1984. Major El Niño/Southern Oscillation events recorded in stratigraphy of the tropical Quelccaya Ice Cap. *Science* 226, 50–2.

Thompson, L. G., Mosely-Thompson, E., Dansgaard, W. and Grootes, P. M. 1986. The Little Ice Age as recorded in the stratigraphy of the tropical Quelccaya Ice Cap. *Science* 234, 361–4.

Thompson, L. G., Mosley-Thompson, E., Davis, M. E., Henderson, K. A., Brecher, H. H., Zagorodnov, V. S., Mashiotta, T. A., Lin, P. N., Mikhalenko, V. N., Hardy, D. R. and Beer, J. 2002. Kilimanjaro ice core records: evidence of Holocene climate change in tropical Africa. *Science* 298, 589–93.

Thompson, L. G., Mosley-Thompson, E. and Henderson, K. A. 2000. Ice-core palaeoclimate records in tropical South America since the Last Glacial Maximum. *Journal of Quaternary Science* 15, 377–94.

Thompson, R. and Oldfield, F. 1986. *Environmental Magnetism.* Allen and Unwin, London.

Thorp, P. W. 1981. A trimline method for defining the upper limit of Loch Lomond Readvance glaciers: examples from the Loch Leven and Glencoe areas. *Scottish Journal of Geology* 17, 49–64.

Thorp, P. W. 1991. Surface profiles and basal shear stresses of outlet glaciers from a Lateglacial mountain icefield in western Scotland. *Journal of Glaciology* 37, 77–89.

Thwaites, F. T. 1946. *Outline of Glacial Geology.* Edwards Brothers, Ann Arbor, MI.

Thyssen, F., Bombosch, A. and Sandhager, H. 1993. Elevation, ice thickness and structure mark maps of the central part of the Fichner-Ronne Ice Shelf. *Polarforschung* 62, 17–26.

Tison, J.-L. and Hubbard, B. 2000. Ice crystallographic evolution at a temperate glacier: Glacier de tsanfleuron, Switzerland. In Maltman, A. J., Hubbard, B. and Hambrey, M. J. (eds), *Deformation of Glacial Materials.* Geological Society, London, Special Publication 176, 23–38.

Tison, J.-L., Petit, J.-R., Barnola, J.-M. and Mahaney, W. C. 1993. Debris entrainment at the ice–bedrock interface in sub-freezing temperatures, Terre Adélie, Antarctica. *Journal of Glaciology* 39, 303–15.

Titus, D. D., Larson, G. J., Strasser, J. C., Lawson, D. E., Evenson, E. B. and Alley, R. B. 1999. Isotopic composition of vent discharge from the Matanuska Glacier, Alaska: implications for the origin of basal ice. In Mickelson, D. M. and Attig, J. W. (eds), *Glacial Processes Past and Present.* Geological Society of America, Special Paper 337, 37–44.

Todd, B. J., Valentine, P. C., Longva, O. and Shaw, J. 2007. Glacial landforms on German Bank, Scotian Shelf: evidence for Late Wisconsinan ice-sheet dynamics and implications for the formation of De Geer moraines. *Boreas* 36, 148–69.

Todd, S. P. 1989. Stream-driven, high-density gravelly traction carpets: possible deposits in the Trabeg Conglomerate Formation, SW Ireland and some theoretical considerations of their origin. *Sedimentology* 36, 513–30.

Tómasson, H. 1996. The jökulhlaup from Katla in 1918. *Annals of Glaciology* 22, 249–54.

Tomkin, J. H. 2007. Coupling glacial erosion and tectonics at active orogens: a numerical modelling study. *Journal of Geophysical Research* 112, doi:10.1029/2005JF000332.

Tomkin, J. H. and Braun, J. 2002. The influence of alpine glaciation on the relief of tectonically active mountain belts. *American Journal of Science* 302, 169–90.

Tomkins, J. D. and Lamoureux, S. F. 2005. Multiple hydroclimatic controls over recent sedimentation in proglacial Mirror Lake, southern Selwyn Mountains, Northwest Territories, *Canadian Journal of Earth Sciences* 42, 1589–99.

Toucanne, S., Zaragosi, S., Bourillet, J. F., Naughton, F., Cremer, M., Eynaud, F. and Dennielou, B. 2008. Activity of the turbidite levees of the Celtic-Armorican margin (Bay of Biscay) during the last 30,000 years: imprints of the last European deglaciation and Heinrich events. *Marine Geology* 247, 84–103.

Trabant, D. C., Krimmel, R. M., Echelmeyer, K. A., Zirnheld, S. L. and Elsberg, D. H. 2003. The slow advance of a calving glacier: Hubbard Glacier, Alaska, U.S.A. *Annals of Glaciology* 36, 45–50.

Trabant, D. C., March, R. S. and Molnia, B. F. 2002. Growing and advancing calving glaciers in Alaska. *Eos* trans. AGU, 83 (47), Fall Meet. Suppl., Abstract C62A-0913 (available at: http://ak .water.usgs.gov/glaciology/hubbard/reports/index.htm).

Tranter, M. 2003. Geochemical weathering in glacial and proglacial environments. In Drever, J. I. (ed.), *Treatise on Geochemistry.* Elsevier, vol. 5, 189–205.

Tranter, M. and Jones, H. G. 2001. The chemistry of snow: processes and nutrient cycling. In Jones, H. G., Pomeroy, J. W., Walker, D. A. and Hoham, R. W. (eds), *Snow Ecology: An Interdisciplinary Examination of Snow-covered Ecosystems.* Cambridge University Press, 127–67.

Tranter, M. and Raiswell, R. 1991. The composition of the englacial and subglacial component in bulk meltwaters draining the Gornergletscher. *Journal of Glaciology* 37 (125), 59–66.

Tranter, M., Brown, G. H., Hodson, A. J. and Gurnell, A. 1996. Hydrochemistry as an indicator of subglacial drainage system structure: a comparison of alpine and sub-polar environments. *Hydrological Processes* 10, 541–56.

Tranter, M., Brown, G., Rainwell, R., Sharp, M. and Gurnell, A. 1993. A conceptual model of solute acquisition by Alpine glacial meltwaters. *Journal of Glaciology* 39, 573–81.

Tranter, M., Sharp, M. J., Lamb, H. R., Brown, G. H., Hubbard, B. P. and Willis, I. C. 2002. Geochemical weathering at the bed of Haut Glacier d'Arolla, Switzerland: a new model. *Hydrological Processes* 16, 959–93.

Tranter, M., Skidmore, M. and Wadham, J. 2005. Hydrological controls on microbial communities in subglacial environments. *Hydrological Processes* 19, 995–8.

Truffer, M. and Echelmeyer, K. A. 2003. Of isbrae and ice streams. *Annals of Glaciology* 36, 66–72.

Truffer, M. and Harrison, W. D. 2006. In situ measurements of till deformation and water pressure. *Journal of Glaciology* 52 (177), 175–82.

Truffer, M., Motyka, R. J., Harrison, W. D., Echelmeyer, K. A., Fisk, B. and Tulaczyk, S. 1999. Subglacial drilling at Black Rapids Glacier, Alaska, USA: drilling method and sample descriptions. *Journal of Glaciology* 45 (151), 495–505.

Tsui, P. C., Cruden, D. M. and Thomson, S. 1989. Ice-thrust terrains and glaciotectonic settings in central Alberta. *Canadian Journal of Earth Sciences* 26, 1308–18.

Tuffen, H., McGarvie, D. W., Gilbert, J. S. and Pinkerton, H. 2002. Physical volcanology of a subglacial-to-emergent rhyolite tuya at Raudufossafjöll, Torfajökull, Iceland. In Smellie, J. L. and Chapman, M. G. (eds), *Volcano–Ice Interactions on Earth and Mars*. Geological Society, London, Special Publication 202, 213–36.

Tulaczyk, S. M., Kamb, B. and Engelhardt, H. F. 2000a. Basal mechanics of Ice Stream B, West Antarctica. I: Till mechanics. *Journal of Geophysical Research* 105 (B1), 463–81.

Tulaczyk, S. M., Kamb, B. and Engelhardt, H. F. 2000b. Basal mechanics of Ice Stream B, West Antarctica. II: Undrained-plastic-bed model. *Journal of Geophysical Research* 105 (B1), 483–94.

Tulaczyk, S. M., Kamb, B., Scherer, R. P. and Engelhardt, H. F. 1998. Sedimentary processes at the base of a West Antarctic ice stream: constraints from textural and compositional properties of subglacial debris. *Journal of Sedimentary Research* 68, 487–96.

Tulaczyk, S. M., Scherer, R. P. and Clark, C. D. 2001. A ploughing model for the origin of weak tills beneath ice streams: a qualitative treatment. *Quaternary International* 86, 59–70.

Turbek, S. E. and Lowell, T. V. 1999. Glacial deposition along an ice-contact slope: an example from the southern Lake District, Chile. *Geografiska Annaler* 81A, 325–46.

Turnbull, J. M. and Davies, T. R. H. 2006. A mass movement origin for cirques. *Earth Surface Processes and Landforms* 31, 1129–48.

Turner, J., Colwell, S. R., Marshall, G. J., Lachlan-Cope, T. A., Carleton, A. M., Jones, P. D., Lagun, V., Reid, P. A. and Iagovkina, S. 2004. The SCAR READER Project: toward a high-quality database of mean Antarctic meteorological observations. *Journal of Climate* 17, 2890–8.

Turner, J., Colwell, S. R., Marshall, G. J., Lachlan-Cope, T. A., Carleton, A. M., Jones, P. D., Lagun, V., Reid, P. A. and Iagovkina, S. 2005. Antarctic climate change during the last 50 years. *International Journal of Climatology* 25, 279–94.

Turner, K. J., Fogwill, C. J., McCulloch, R. D. and Sugden, D. E. 2005. Deglaciation of the eastern flank of the North Patagonian Icefield and associated continental-scale lake diversions. *Geografiska Annaler* 87A, 363–74.

Tweed, F. S. and Russell, A. J. 1999. Controls on the formation and sudden drainage of glacier-impounded lakes: implications for jökulhlaup characteristics. *Progress in Physical Geography* 23, 79–110.

Tweed, F. S., Roberts, M. J. and Russell, A. J. 2005. Hydrologic monitoring of supercooled water from Icelandic glaciers. *Quaternary Science Reviews* 24, 2308–18.

Twiss, R. J. and Moores, E. M. 1992. *Structural Geology*. Freeman and Co., New York.

Uppala, S. M. et al. 2005. The ERA-40 re-analysis. *Quarterly Journal of the Royal Meteorological Society* 131, 2961–3012.

van de Wal, R. S. W. 2004. Greenland: modelling. In Bamber, J. L. and Payne, A. J. (eds), *Mass Balance of the Cryosphere*. Cambridge University Press, 437–57.

van de Wal, R. S. W. and Oerlemans, J. 1995. Response of valley glaciers to climate change and kinematic waves: a study with a numerical ice-flow model. *Journal of Glaciology* 41, 142–52.

van de Wal, R. S. W., Boot, W., van den Broeke, M. R., Smeets, C. J. P. P., Reijmer, C. H., Donker, J. J. A. and Oerlemans, J. 2008. Large and rapid melt-induced velocity changes in the ablation zone of the Greenland ice sheet. *Science* 321, 111–13.

van den Berg, W. J. R., van de Wal, R. S. W., Milne, G. A. and Oerlemans, J. 2008. The effect of isostasy on dynamical ice sheet modelling; a case study for Eurasia. *Journal of Geophysical Research*, doi:10.1029/2007JB004994.

van den Berg, W. J. R., van den Broeke, M. R., Reijmer, C. H. and van Meijgaard, E. 2006. Reassessment of the Antarctic surface mass balance using calibrated output of a regional atmospheric climate model. *Journal of Geophysical Research* 111, doi:10.1029/2005JD006495.

van den Broeke, M. R. 2005. Strong surface melting preceded collapse of Antarctic Peninsula ice shelf. *Geophysical Research Letters* 32, doi:10.1029/2005GL023247.

van den Broeke M. R., Duynkerke, P. G. and Oerlemans, J. 1994. The observed katabatic flow at the edge of the Greenland ice sheet during GIMEX-91. *Global and Planetary Change* 9, 3–15.

van den Broeke, M. R. 1997. Momentum, heat and moisture budgets of the katabatic wind layer over a midlatitude glacier in summer. *Journal of Applied Meteorology* 36, 763–74.

van der Meer, J. 1993. Microscopic evidence of subglacial deformation. *Quaternary Science Reviews* 12, 553–87.

van der Meer, J. J. M. 1997. Particle and aggregate mobility in till: microscopic evidence of subglacial processes. *Quaternary Science Reviews* 16, 827–31.

van der Meer, J. J. M. (ed.). 2004. *Spitsbergen Push Moraines*. Developments in Quaternary Science 4. Elsevier, Amsterdam.

van der Meer, J. J. M., Kjær, K. H. and Krüger, J. 1999. Subglacial water escape structures and till structures, Sléttjökull, Iceland. *Journal of Quaternary Science* 14, 191–205.

van der Meer, J. J. M., Menzies, J. and Rose, J. 2003. Subglacial till: the deforming glacier bed. *Quaternary Science Reviews* 22, 1659–85.

van der Veen, C. J. 1996. Tidewater calving. *Journal of Glaciology* 42, 375–85.

van der Veen, C. J. 1997a. Controls on the position of iceberg-calving fronts. In van der Veen, C. J. (ed.), *Calving Glaciers: Report of a Workshop, Feb. 28–March 2, 1997*. Ohio State University, Columbus. Byrd Polar Research Centre, report 15, 163–72.

van der Veen, C. J. 1997b. Backstress: what it is and how it affects glacier flow. In van der Veen, C. J. (ed.), *Calving Glaciers: Report of a Workshop, Feb. 28–March 2, 1997*. Ohio State University, Columbus. Byrd Polar Research Centre, report 15, 173–80.

van der Veen, C. J. 1998a. Fracture mechanics approach to penetration of surface crevasses on glaciers. *Cold Regions Science and Technology* 27, 31–47.

van der Veen, C. J. 1998b. Fracture mechanics approach to penetration of bottom crevasses on glaciers. *Cold Regions Science and Technology* 27, 213–23.

van der Veen, C. J. 1999a. *Fundamentals of Glacier Dynamics*. Balkema, Rotterdam.

van der Veen, C. J. 1999b. Crevasses on glaciers. *Polar Geography* 23, 213–45.

van der Veen, C. J. 2002a. Calving glaciers. *Progress in Physical Geography* 26, 96–122.

van der Veen, C. J. 2002b. Polar ice sheets and global sea level: how well can we predict the future? *Global and Planetary Change* 32, 165–94.

van der Veen, C. J. 2007. Fracture propagation as means of rapidly transferring surface meltwater to the base of glaciers. *Geophysical Research Letters* 34, L01501, doi:10.1029/2006GL028385.

van der Veen, C. J. and Payne, A. J. 2004. Modelling land-ice dynamics. In Bamber, J. L. and Payne, A. J. (eds), *Mass Balance of the Cryosphere*. Cambridge University Press, 169–225.

van der Veen, C. J. and Whillans, I. M. 1989a. Force budget I. Theory and numerical methods. *Journal of Glaciology* 35, 53–60.

van der Veen, C. J. and Whillans, I. M. 1989b. Force budget II. Application to two-dimensional flow along Byrd Station Strain Network, Antarctica. *Journal of Glaciology* 35, 61–7.

van der Veen, C. J. and Whillans, I. M. 1996. Model experiments on the evolution and stability of ice streams. *Annals of Glaciology* 23, 129–37.

van der Wal, W., Wu, P., Siderisa, M. G. and Shum, C. K. 2008. Use of GRACE determined secular gravity rates for glacial isostatic adjustment studies in North-America. *Journal of Geodynamics* 46, 144–54.

van der Wateren, F. M. 1985. A model of glacial tectonics, applied to the ice-pushed ridges in the central Netherlands. *Bulletin of the Geological Society Denmark* 34, 55–74.

van der Wateren, F. M. 1995. Structural geology and sedimentology of push moraines. *Mededelingen Rijks Geologische Dienst* 54.

van der Wateren, F. M. 2003. Ice-marginal terrestrial landsystems: southern Scandinavian ice sheet margin. In Evans, D. J. A. (ed.), *Glacial Landsystems*. Arnold, London, 166–203.

van der Wateren, F. M., Kluiving, S. J. and Bartek, L. R. 2000. Kinematic indicators of subglacial shearing. In Maltman, A., Hubbard, B. and Hambrey, M. J. (eds), *Deformation of Glacial Materials*. Geological Society, London, Special Publication 176, 259–78.

van Lipzig, N. P. M., van Meijgaard, E. and Oerlemans, J. 2002. The spatial and temporal variability of the surface mass balance in Antarctica: results from a regional climate model. *International Journal of Climatology* 22, 1197–217.

van Lipzig, N. P. M., Turner, J., Colwell, S. R. and van den Broeke, M. R. 2004. The near-surface wind field over the Antarctic continent. *International Journal of Climatology* 24, 1973–82.

van Loon, A. J. 2008. Could 'Snowball Earth' have left thick glaciomarine deposits? *Gondwana Research* 14, 73–81.

van Wagoner, J. C., Posamentier, H. W., Mitchum, R. M., Vail, P. R., Sarg, J. F., Loutit, T. S. and Hardenbol, J. 1988. An overview of the fundamentals of sequence stratigraphy and key definitions. In Wilgus, C. K. et al. (eds), *Sea-level Changes: An Integrated Approach*. SEPM Special Publication 42, 39–45.

van Weert, E. H. A., van Gijssel, K., Leijnse, A. and Boulton, G. S. 1997. The effects of Pleistocene glaciations on the geohydrological system of Northwest Europe. *Journal of Hydrology* 195, 137–59.

Vanneste, M., Berndt, C., Laberg, J. S. and Mienert, J. 2007. On the origin of large shelf embayments on glaciated margins: effects of lateral ice flux variations and glacio-dynamics west of Svalbard. *Quaternary Science Reviews* 26, 2406–19.

Vatne, G. 2001. Geometry of englacial water conduits, Austre Brøggerbreen, Svalbard. *Norsk Geografisk Tidsskrift* 55, 85–93.

Vaughan, D. G., 1993. Relating the occurrence of crevasses to surface strain rates. *Journal of Glaciology*, 39, 255–66.

Vaughan, D. G. 2008. West Antarctic Ice Sheet collapse: the fall and rise of a paradigm. *Climate Change* 91, 65–79.

Vaughan, D. G. and Doake, C. S. M. 1996. Recent atmospheric warming and retreat of ice shelves on the Antarctic Peninsula. *Nature* 379, 328–31.

Vaughan, D. G. and Spouge, J. R. 2002. Risk estimation of the collapse of the West Antarctic ice sheet. *Climate Change* 52, 65–91.

Vaughan, D. G., Bamber, J. L., Giovinetto, M. and Cooper, A. P. R. 1999. Reassessment of net surface mass balance in Antarctica. *Journal of Climate* 12, 933–46.

Vaughan, D. G., Smith, A. M., Nath, P. C. and Le Meur, E. 2003. Acoustic impedence and basal shear stress beneath four Antarctic ice streams. *Annals of Glaciology* 36, 225–32.

Vaughan, D. G. et al. 2006. New boundary conditions for the West Antarctic ice sheet: subglacial topography beneath Pine Island Glacier. *Geophysical Research Letters* 33, L09501, doi:10.1029/GL025588.

Vaughan, D. G., Rivera, A., Woodward, J., Corr, H. F. J., Wendt, J. and Zamora, R. 2007. Topographic and hydrologic controls on Subglacial Lake Ellsworth, West Antarctica. *Geophysical Research Letters* 34, doi:10.1029/2007GL030769.

Veblen, T. T., Ashton, D. H., Rubulis, S., Lorenz, D. C. and Cortes, M. 1989. Nothofagus stand development on in-transit moraines, Casa Pangue Glacier, Chile. *Arctic and Alpine Research* 21, 144–55.

Veillette, J., Mueller, D.R., Antoniades, D. and Vincent, W.F. 2008. Arctic epishelf lakes as sentinel ecosystems: Past, present and future. *Journal of Geophysical Research*, 113. Doi: 1029/2008JG000730.

Veillette, J. J., Dyke, A. S. and Roy, M. 1999. Ice-flow evolution of the Labrador sector of the Laurentide Ice Sheet: a review, with new evidence from northern Quebec. *Quaternary Science Reviews* 18, 993–1019.

Velicogna, I. and Wahr, J. 2005. Greenland mass balance from GRACE. *Geophysical Research Letters* 32, doi:10.1029/2005GL023955.

Velicogna, I. and Wahr, J. 2006. Acceleration of Greenland ice mass loss in spring 2004. *Nature* 443, 329–31.

Vere, D. M. and Benn, D. I. 1989. Structure and debris characteristics of medial moraines in Jotunheimen, Norway: implications for moraine classification. *Journal of Glaciology* 35, 276–80.

Vere, D. M. and Matthews, J. A. 1985. Rock glacier formation from a lateral moraine at Bukkeholsbreen, Jotunheimen, Norway: a sedimentological approach. *Zeitschrift für Geomorphologie* 29, 397–415.

Vieli, A. and Payne, A. J. 2005. Assessing the ability of numerical ice sheet models to simulate grounding-line migration. *Journal of Geophysical Research* 110, F1003. doi:10.1029/2004JF000202.

Vieli, A., Funk, M. and Blatter, H. 2000. Tidewater glaciers: frontal flow acceleration and basal sliding. *Annals of Glaciology* 31, 217–21.

Vieli, A., Funk, M. and Blatter, H. 2001. Flow dynamics of tidewater glaciers: a numerical modelling approach. *Journal of Glaciology* 47, 595–606.

Vieli, A., Jania, J. and Kolondra, L. 2002. The retreat of a tidewater glacier: observations and model calculations on Hansbreen, Spitsbergen. *Journal of Glaciology* 48, 592–600.

Vieli, A., Jania, J., Blatter, H. and Funk, M. 2004. Short-term velocity variations on Hansbreen, a tidewater glacier in Spitsbergen. *Journal of Glaciology* 50, 389–98.

Vieli, A., Payne, A. J., Du, Z. and Shepherd, A. 2006. Numerical modelling and data assimilation of the Larsen B ice shelf, Antarctic Peninsula. *Philosophical Transactions of the Royal Society*, Series A, 364, 1815–39.

Vieli, A., Payne, A. J., Shepherd, A. and Du, Z. 2007. Causes of precollapse changes of the Larsen B ice shelf: numerical modelling and assimilation of satellite observations. *Earth and Planetary Science Letters* 259, 297–306.

Vimeux, F., Cuffey, K. M. and Jouzel, J. 2002. New insights into southern hemisphere temperature changes from ice cores using deuterium excess correction. *Earth and Planetary Science Letters* 203, 829–43.

Vincent, J.-S. 1989. Quaternary geology of the southeastern Canadian shield. In Fulton, R. J. (ed.), *Quaternary Geology of Canada and Greenland*. Geological Survey of Canada, Geology of Canada, no. 1, 249–75.

Vincent, W. F., Gibson, J. A. E. and Jeffries, M. O. 2001. Ice shelf collapse, climate change and habitat loss in the Canadian High Arctic. *Polar Record* 37, 133–42.

Virkkunen, K., Moore, J. C., Isaksson, E., Pohjola, V., Perämäki, P., Grinsted, A. and Kekonen, T. 2007. Warm summers and ion concentrations in snow: comparison of present day with Medieval Warm Epoch from snow pits and an ice core from Lomonosovfonna, Svalbard. *Journal of Glaciology* 53 (183), 623–34.

Vivian, R. 2001. *Des Glaciers du Faucigny aux Glaciers de Mont Blanc*. La Fontaine de Siloé, Montmelian.

Vizcaino, M., Mikolajewicz, U., Gröger, M., Maier-Reimer, E., Schurgers, G. and Wingnuth, A. M. E. 2008. Long-term ice sheet–climate interactions under anthropogenic greenhouse forcing with a complex Earth System Model. *Climate Dynamics* 31, 665–90.

Vogel, S. W., Tulaczyk, S. and Joughin, I. R. 2003. Distribution of basal melting and freezing beneath tributaries of Ice Stream C: implication for the Holocene decay of the West Antarctic ice sheet. *Annals of Glaciology* 36, 273–82.

Vogel, S. W., Tulaczyk, S., Kamb, B., Engelhardt, H., Carsey, F. D., Behar, A. E., Lane, A. L. and Joughin, I. 2005. Subglacial conditions during and after stoppage of an Antarctic Ice Stream: is reactivation imminent? *Geophysical Research Letters* 32, doi:10.1029/2005GL022563.

Vogt, P. R., Crane, K. and Sundvor, E. 1994. Deep Pleistocene iceberg plowmarks on the Yermak Plateau: sidescan and 3.5 kHz evidence for thick calving ice fronts and a possible marine ice sheet in the Arctic Ocean. *Geology* 22, 403–6.

Vorren, T. O. 2003. Subaquatic landsystems: continental margins. In Evans, D. J. A. (ed.), *Glacial Landsystems*. Arnold, London, 289–312.

Vorren, T. O. and Kristoffersen, Y. 1986. Late Quaternary glaciation in the south-western Barents Sea. *Boreas* 15, 51–9.

Vorren, T. O. and Laberg, J. S. 1997. Trough-mouth fans: palaeoclimate and ice sheet monitors. *Quaternary Science Reviews* 16, 865–81.

Vorren, T. O., Hald, M., Edvardsen, M. and Lind-Hansen, O. W. 1983. Glacigenic sediments and sedimentary environments on continental shelves: general principles with a case study from the Norwegian shelf. In Ehlers, J. (ed.), *Glacial Deposits in North-west Europe*. Balkema, Rotterdam, 61–73.

Vorren, T. O., Laberg, J. S., Blaumme, F., Dowdeswell, J. A., Kenyon, N. H., Mienert, J., Rumohr, J. and Werner, F. 1998. The Norwegian-Greenland Sea continental margins: morphology and Late Quaternary sedimentary processes and environment. *Quaternary Science Reviews* 17, 273–302.

Vorren, T. O., Lebesbye, E., Andreassen, K. and Larsen, K. B. 1989. Glacigenic sediments on a passive continental margin as exemplified by the Barents Sea. *Marine Geology* 85, 251–72.

Vorren, T. O., Lebesbye, E. and Larsen, K. B. 1990. Geometry and genesis of the glacigenic sediments in the southern Barents Sea. In Dowdeswell, J. A. and Scourse, J. D. (eds), *Glacimarine Environments: Processes and Sediments*. Geological Society, London, Special Publication 53, 269–88.

Vuichard, D. and Zimmerman, M. 1987. The 1985 catastrophic drainage of a moraine-dammed lake, Khumbu Himal, Nepal: cause and consequences. *Mountain Research and Development* 7, 91–110.

Waddington, E. D. 1986. Wave ogives. *Journal of Glaciology* 32, 325–34.

Wadham, J. L. and Nuttall, A.-M. 2002. Multi-phase formation of superimposed ice during a mass-balance year at a maritime high-Arctic glacier. *Journal of Glaciology* 48, 545–51.

Wadham, J. L., Bottrell, S., Tranter, M. and Raiswell, R. 2004. Stable isotope evidence for microbial sulphate reduction at the bed of a polythermal High Arctic glacier. *Earth and Planetary Science Letters* 219, 341–55.

Wadham, J. L., Cooper, R. J., Tranter, M. and Hodgkins, R. 2001b. Enhancement of glacial solute fluxes in the proglacial zone of a polythermal glacier. *Journal of Glaciology* 47, 378–86.

Wadham, J. L., Hodgkins, R., Cooper, R. J. and Tranter, M. 2001a. Evidence for seasonal subglacial outburst events at a polythermal glacier, Finsterwalderbreen, Svalbard. *Hydrological Processes* 15, 2259–80.

Wadham, J. L., Hodson, A. J., Tranter, M. and Dowdeswell, J. A. 1998. Hydrochemistry of meltwaters draining a polythermal-based, high-Arctic glacier, south Svalbard. I: The ablation season. *Hydrological Processes* 12, 1825–48.

Wadham, J. L., Tranter, M. and Dowdeswell, J. A. 2000. Hydrochemistry of meltwaters draining a polythermal-based,

high-Arctic glacier, south Svalbard. II: Winter and early spring. *Hydrological Processes* 14, 1767–86.

Waelbroeck, C., Labeyrie, L., Michel, E., Duplessy, J. C., McManus, J. F., Lambeck, K., Balbon, E. and Labracherie, M. 2002. Sea-level and deep water temperature changes derived from benthic foraminifera isotopic records. *Quaternary Science Reviews* 21, 295–305.

Wagnon, P., Ribstein, P., Francou, B. and Sicart, J. E. 2001. Anomalous heat and mass budget of Glaciar Zongo, Bolivia, during the 1997/98 El Niño year. *Journal of Glaciology* 47, 21–8.

Wagnon, P., Ribstein, P., Kaser, G. and Berton, P. 1999. Energy balance and runoff seasonality of a Bolivian glacier. *Global and Planetary Change* 22, 49–58.

Wahr, J., Swenson, S. and Velicogna, I. 2006. Accuracy of GRACE mass estimates. *Geophysical Research Letters* 33, L06401, doi:10.1029/2005GL025305.

Waitt, R. B. 1980. About forty last-glacial Lake Missoula jökulhlaups through southern Washington. *Journal of Geology* 88, 653–79.

Waitt, R. B. 1985. Case for periodic, colossal jökulhlaups from glacial Lake Missoula. *Geological Society of America Bulletin* 95, 1271–86.

Wakahama, G., Kuroiwa, D., Hasemi, T. and Benson, C. S. 1976. Field observations and experimental and theoretical studies on the superimposed ice of McCall Glacier, Alaska. *Journal of Glaciology* 16(74), 135–49.

Walcott, R. I. 1970. Isostatic response to loading of the crust in Canada. *Canadian Journal of Earth Sciences* 7, 716–27.

Walden, J. 2004. Particle lithology (or mineral and geochemical analysis). In Evans, D. J. A. and Benn, D. I. (eds), *A Practical Guide to the Study of Glacial Sediments*. Arnold, London, 145–81.

Walden, J. and Ballantyne, C. K. 2002. Use of environmental magnetic measurements to validate the vertical extent of ice masses at the Last Glacial Maximum. *Journal of Quaternary Science* 17, 193–200.

Walden, J., Smith, J. P. and Dackombe, R. V. 1992. Mineral magnetic analyses as a means of lithostratigraphic correlation and provenance indication of glacial diamicts: intra- and inter-unit variation. *Journal of Quaternary Science* 7, 257–70.

Walden, J., Wadsworth, E., Austin, W. E. N., Peters, C., Scourse, J. D. and Hall, I. R. 2007. Compositional variability of ice-rafted debris in Heinrich layers 1 and 2 on the northwest European continental slope identified by environmental magnetic analyses. *Journal of Quaternary Science* 22, 163–72.

Walder, J. S. 1986. Hydraulics of subglacial cavities. *Journal of Glaciology* 32, 439–45.

Walder, J. S. and Costa, J. E. 1996. Outburst floods from glacier-dammed lakes: the effect of mode of lake drainage on flood magnitude. *Earth Surface Processes and Landforms* 21, 701–23.

Walder, J. S. and Fowler, A. 1994. Channelized subglacial drainage over a deformable bed. *Journal of Glaciology* 40, 3–15.

Walder, J. S. and Hallet, B. 1979. Geometry of former subglacial water channels and cavities. *Journal of Glaciology* 23, 335–46.

Walder, J. S. and Hallet, B. 1986. The physical basis for frost weathering: toward a more fundamental and unified perspective. *Arctic and Alpine Research* 18, 27–32.

Walker, R. G. 1990. Facies modeling and sequence stratigraphy. *Journal of Sedimentary Petrology* 60, 777–86.

Walker, R. G. 1992. Facies, facies models and modern stratigraphic concepts. In Walker, R. G. and James, N. P. (eds), *Facies Models: Response to Sea-level Change*. Geological Association of Canada, Toronto, 1–14.

Walker, R. T. and Holland, D. M. 2007. A two-dimensional coupled model for ice shelf–ocean interaction. *Ocean Modelling* 17, 123–39.

Waller, R. I. 2001. The influence of basal processes on the dynamic behaviour of cold-based glaciers. *Quaternary International* 86, 117–28.

Waller, R. I. and Tuckwell, G. W. 2005. Glacier-permafrost interactions and glaciotectonic landform generation at the margin of the Leverett Glacier, West Greenland. In Harris, C. and Murton, J. B. (eds), *Cryospheric Systems: Glaciers and Permafrost*. Geological Society, London, Special Publication 242, 39–50.

Waller, R. I., Hart, J. K. and Knight, P. G. 2000. The influence of tectonic deformation on facies variability in stratified debris-rich basal ice. *Quaternary Science Reviews* 19, 775–86.

Waller, R. I., van Dijk, T. A. G. P. and Knudsen, Ó. 2008. Subglacial bedforms and conditions associated with the 1991 surge of Skeiðarárjökull, Iceland. *Boreas* 37, 179–94.

Walters, R. A. and Dunlap, W. W. 1987. Analysis of time series of glacier speed: Columbia Glacier, Alaska. *Journal of Geophysical Research* 92, 8969–75.

Ward, B. and Rutter, N. 2000. Deglacial valley fill sedimentation, Pelly River, Yukon Territory, Canada. *Quaternary International* 68/71, 309–28.

Warneke, T., Croudace, I. W., Warwick, P. E. and Taylor, R. N. 2002. A new ground-level fallout record of uranium and plutonium isotopes for northern temperate latitudes. *Earth and Planetary Science Letters* 203, 1047–57.

Warren, C. R. 1992. Iceberg calving and the glacioclimatic record. *Progress in Physical Geography* 16, 253–82.

Warren, C. R. 1993. Rapid recent fluctuations of the calving San Rafael Glacier, Chilean Patagonia: climatic or non-climatic? *Geografiska Annaler* 75A, 111–25.

Warren, C. R. and Kirkbride, M. P. 2003. Calving speed and climatic sensitivity of New Zealand lake-calving glaciers. *Annals of Glaciology* 36, 173–8.

Warren, C. R. and Sugden, D. E. 1993. The Patagonian icefields: a glaciological review. *Arctic and Alpine Research* 25, 316–31.

Warren, C. R., Benn, D. I., Winchester, V. and Harrison, S. 2001. Buoyancy-driven lacustrine calving, Glaciar Nef, Chilean Patagonia. *Journal of Glaciology* 47, 135–46.

Warren, C. R., Glasser, N. F., Harrison, S., Winchester, V., Kerr, A. R. and Rivera, A. 1995. Characteristics of tide-water calving at Glaciar San Rafael, Chile. *Journal of Glaciology* 41, 273–89.

Warren, W. P. and Ashley, G. M. 1994. Origins of the ice-contact stratified ridges (eskers) of Ireland. *Journal of Sedimentary Research* A64, 433–49.

Watanabe, T., Dali, L. and Shiraiwa, T. 1998. Slope denudation and the supply of debris to cones in Langtang Himal, Central Nepal Himalaya. *Geomorphology* 26, 185–97.

Watanabe, T., Kameyama, S. and Sato, T. 1995. Imja glacier dead-ice melt rates and changes in a supraglacial lake, 1989–1994, Khumbu Himal, Nepal: danger of lake drainage. *Mountain Research and Development* 15, 293–300.

Watson, E., Luckman, B. H. and Bin Yu. 2006. Long-term relationships between reconstructed seasonal mass balance at Peyto Glacier, Canada, and Pacific sea surface temperatures. *The Holocene* 16, 783–90.

Weaver, A. J., Saenko, O. A., Clark, P. U. and Mitrovica, J. X. 2004. Meltwater Pulse 1A from Antarctica as a trigger of the Bølling-Allerød warm interval. *Science* 299, 1709–13.

Weber, J. E. 2000. Non-temperate glacier flow over wavy sloping ground. *Journal of Glaciology* 46, 453–8.

Weertman, J. 1964. The theory of glacier sliding. *Journal of Glaciology* 5, 287–303.

Weertman, J. 1974. Stability of the junction between an ice sheet and an ice shelf. *Journal of Glaciology* 13, 3–11.

Weertman, J. 1976. Glaciology's grand unsolved problem. *Nature* 260, 284–6.

Weertman, J. 1983. Creep deformation of ice. *Annual Review of Earth and Planetary Sciences* 11, 215–40.

Weirich, F. H. 1984. Turbidity currents: monitoring their occurrence and movement with a three-dimensional sensor network. *Science* 224, 384–7.

Weirich, F. H. 1986. The record of density-induced underflows in a glacial lake. *Sedimentology* 33, 261–77.

Welch, B. C., Pfeffer, W. T., Harper, J. T. and Humphrey, N. F. 1998. Mapping subglacial surfaces beneath temperate valley glaciers by 2-pass migration of radio echo sounding data. *Journal of Glaciology* 44, 164–70.

Wellner, J. S., Lowe, A. L., Shipp, S. S. and Anderson, J. B. 2001. Distribution of glacial geomorphic features on the Antarctic continental shelf and correlation with substrate: implications for ice behaviour. *Journal of Glaciology* 47, 397–411.

Wellner, J. S., Heroy, D. C. and Anderson, J. B. 2006. The death mask of the Antarctic Ice Sheet: comparison of glacial geomorphic features across the continental shelf. *Geomorphology* 75, 157–71.

Wen, J., Jezek, K. C., Csathó, B. M., Herzfeld, U. C., Farness, K. L. and Huybrechts, P. 2007. Mass budgets of the Lambert, Mellor and Fisher Glaciers and basal fluxes beneath their flowbands on Amery Ice Shelf. *Science in China Series D: Earth Sciences* 50, 1693–706.

Wendt, J., Dietrich, R., Fritsche, M., Wendt, A., Yuskevitch, A., Kokhanov, A., Lukin, V., Shibuya, K. and Doi, K. 2006. Geodetic observations of ice flow velocities over the southern part of subglacial Lake Vostok, Antarctica, and their glaciological implications. *Geophysical Journal International* 166, 991–8.

Werner, M., Mikolajewicz, U., Heimann, M. and Hoffmann, G. 2000. Borehole versus isotope temperatures on Greenland: seasonality does matter. *Geophysical Research Letters* 27, 723–6.

Werritty, A. 1992. Downstream fining in a gravel-bed river in southern Poland: lithologic controls and the role of abrasion. In Billi, P., Hey, R. D., Thorne, C. R. and Tacconi, P. (eds), *Dynamics of Gravel-bed Rivers*. Wiley, London, 333–50.

Whalley, W. B. 1996. Scanning electron microscopy. In Menzies, J. (ed.), *Past Glacial Environments. Sediments, Forms and Techniques*. Butterworth-Heinemann, Oxford, 357–75.

Whalley, W. B. and Azizi, F. 2003. Rock glaciers and protalus landforms: analogous forms and ice sources on Earth and Mars. *Journal of Geophysical Research* 108 (E4), 8032.

Whalley, W. B., Rea, B. R., Rainey, M. M. and McAlister, J. J. 1997. Rock weathering and blockfields: some preliminary data from mountain plateaus in north Norway. In Widdowson, M. (ed.), *Palaeosurfaces: Recognition, Reconstruction and Interpretation*. Geological Society, London, Special Publication 129, 133–45.

Whillans, I. M. and van der Veen, C. J. 1997. The role of lateral drag in the dynamics of Ice Stream B, Antarctica. *Journal of Glaciology* 43, 231–7.

Whillans, I. M., Bentley, C. R. and van der Veen, C. J. 2001. Ice Streams B and C. In Alley, R. B. and Bindschadler, R. A. (eds), *The West Antarctic Ice Sheet: Behavior and Environment*. American Geophysical Union, Antarctic Research Series, vol. 77, 257–81.

Whillans, I. M., Chen, Y. H., van der Veen, C. J. and Hughes, T. J. 1989. Force budget. III: Application to three-dimensional flow of Byrd Glacier, Antarctica. *Journal of Glaciology* 35, 68–80.

Whipple, K. X., Kirby, E. and Brocklehurst, S. H. 1999. Geomorphic limits to climate-induced increases in topographic relief. *Nature* 401, 39–43.

White, A. F., Blum, A. E., Bullen, T. D., Vivit, D. V., Schulz, M. and Fitzpatrick, J. 1999. The effect of temperature on experimental and natural chemical weathering rates of granitoid rocks. *Geochimica et Cosmochimica Acta* 63, 3277–91.

Whiteman, C. A. 1986. Variability of clast size and roundness in contemporary meltwater rivers at Okstindan, north Norway. In Bridgland, D. R. (ed.), *Clast Lithological Analysis*. Quaternary Research Association, Cambridge, 179–92.

Wiles, G. C., Calkin, P. E. and Post, A. 1995. Glacier fluctuations in the Kenai Fjords, Alaska, USA: an evaluation of controls on iceberg-calving glaciers. *Arctic and Alpine Research* 27, 234–45.

Wiles, G. C., D'Arrigo, R., Villalba, R., Calkin, P. and Barclay, D. J. 2004. Century-scale solar variability and Alaskan temperature change over the past millennium, *Geophysical Research Letters* 31 (L15203).

Williams, G. D., Brabham, P. J., Eaton, G. P. and Harris, C. 2001. Late Devensian glaciotectonic deformation at St Bees, Cumbria: a critical wedge model. *Journal of the Geological Society of London* 158, 125–35.

Williams, P. F. and Rust, B. R. 1969. The sedimentology of a braided river. *Journal of Sedimentary Petrology* 39, 649–79.

Williams, R. S. Jr and Ferrigno, J. G. (eds). 1999. *Satellite Image Atlas of Glaciers of the World: South America*. U.S. Geological Survey, Professional Paper 1386-I.

Williams, R. S. Jr and Ferrigno, J. G. 2009. State of the Earth's cryosphere at the beginning of the 21st century: glaciers, global snow cover, floating ice, and permafrost and periglacial environments. *Satellite Image Atlas of Glaciers of the World*. U.S. Geological Survey, Professional Paper 1386-A.

Williams, R. S., Hall, D. K. and Benson, C. S. 1991. Analysis of glacier facies using satellite techniques. *Journal of Glaciology* 37, 120–8.

Willis, I. C. 1995. Intra-annual variations in glacier motion: a review. *Progress in Physical Geography* 19, 61–106.

Willis, I. C., Sharp, M. J. and Richards, K. S. 1990. Configuration of the drainage system of Mitdalsbreen, Norway, as indicated by dye-tracing experiments. *Journal of Glaciology* 36, 89–101.

Wilson, L. and Head, J. W. 2002. Heat transfer and melting in subglacial basaltic volcanic eruptions: implications for volcanic deposit morphology and meltwater volumes. In Smellie, J. L. and Chapman, M. G. (eds), *Volcano–Ice Interaction on Earth and Mars*. Geological Society, London, Special Publication 202, 5–26.

Wilson, L. J. and Austin, W. E. N. 2002. Millennial and sub-millennial scale variability in sediment colour from the Barra Fan, NW Scotland: implications for British ice sheet dynamics. In Dowdeswell, J. A. and Ó Cofaigh, C. (eds), *Glacier-influenced Sedimentation on High Latitude Continental Margins*. Geological Society, London, Special Publication 203, 349–65.

Wilson, S. B. and Evans, D. J. A. 2001. Scottish Landforms Example 24: Coire a' Cheud-chnoic, the 'hummocky moraine' of Glen Torridon. *Scottish Geographical Journal* 116, 149–58.

Wingfield, R. T. R. 1990. The origin of major incisions within the Pleistocene deposits of the North Sea. *Marine Geology* 91, 31–52.

Wingham, D. J., Ridout, A. J., Scharroo, R., Arthern, R. J. and Shum, C. K. 1998. Antarctic elevation change from 1992 to 1996. *Science* 282, 456–8.

Wingham, D. J., Shepherd, A., Muir, A. and Marshall, G. J. 2006b. Mass balance of the Antarctic ice sheet. *Philosophical Transactions of the Royal Society*, Series A, 364, 1627–35.

Wingham, D. J., Siegert, M. L., Shepherd, A. and Muir, A. S. 2006a. Rapid discharge links Antarctic subglacial lakes. *Nature* 440, 1033–6.

Winguth, C., Mickelson, D. M., Colgan, P. M. and Laabs, B. J. C. 2004. Modeling the deglaciation of the Green Bay Lobe of the southern Laurentide Ice Sheet. *Boreas* 33, 34–47.

Winsemann, J., Asprion, U. and Meyer, T. 2004. Sequence analysis of early Saalian glacial lake deposits (NW Germany): evidence of local ice margin retreat and associated calving processes. *Sedimentary Geology* 165, 223–51.

Winsemann, J., Asprion, U., Meyer, T. and Schramm, C. 2007. Facies characteristics of Middle Pleistocene (Saalian) ice-margin subaqueous fan and delta deposits, glacial Lake Leine, NW Germany. *Sedimentary Geology* 193, 105–29.

Winsemann, J., Asprion, U., Meyer, T., Schultz, H. and Victor, P. 2003. Evidence of iceberg-ploughing in a subaqueous ice-contact fan, glacial Lake Rinteln, NW Germany. *Boreas* 32, 386–98.

Winther, J.-G., Gerland, S., Ørbaek, J. B., Ivanov, B., Blanco, A. and Boike, J. 1999. Spectral reflectance of melting snow in a high Arctic watershed on Svalbard: some implications for optical satellite remote sensing studies. *Hydrological Processes* 13, 2033–49.

Wohl, E. E. 1998. Bedrock channel morphology in relation to erosional processes. In Tinkler, K. J. and Wohl, E. E. (eds), *Rivers Over Rock: Fluvial Processes in Bedrock Channels*. American Geophysical Union, Washington, DC, 133–52.

Wolman, M. G. and Miller, J. P. 1960. Magnitude and frequency of forces in geomorphic processes. *Journal of Geology* 68, 54–74.

Woodward, J., Murray, T. and McCaig, A. 2002. Formation and reorientation of structure in the surge-type glacier Kongsvegen, Svalbard. *Journal of Quaternary Science* 17, 201–9.

Woodworth-Lynas, C. M. T. and Guigné, J. Y. 1990. Iceberg scours in the geological record: examples from glacial Lake Agassiz. In Dowdeswell, J. A. and Scourse, J. D. (eds), *Glacimarine Environments: Processes and Sediments*. Special Publications of the Geologists' Association 53, 217–23.

World Glacier Monitoring Service. 2005a. *Glacier Mass Balance Bulletin N-8 (2002–2005)*. IUGG (CCS) – UNEP – UNESCO – WMO, Zurich.

World Glacier Monitoring Service. 2005b. *Fluctuations of Glaciers 1995–2000 (vol. VIII)*. IUGG (CCS) – UNEP – UNESCO, Zurich.

Worsley, P. 1999. Context of relict Wisconsinan glacial ice at Angus Lake, SW Banks Island, western Canadian arctic and stratigraphic implications. *Boreas* 28, 543–50.

Wright, A. P., Siegert, M. J., Le Brocq, A. M. and Gore, D. B. 2008. High sensitivity of subglacial hydrological pathways in Antarctica to small ice-sheet changes. *Geophysical Research Letters* 35, doi:10.1029/2008GL034937.

Wu, P. and Peltier, W. R. 1983. Glacial isostatic adjustment and the free air gravity anomaly as a constraint upon deep mantle viscosity. *Geophysical Journal of the Royal Astronomical Society* 74, 377–449.

WWF. 2005. *An Overview of Glaciers, Glacier Retreat, and Subsequent Impacts in Nepal, India and China*. Worldwide Fund for Nature (available at: assets.panda.org/downloads/himalayaglaciersreport2005.pdf).

Wynn, P. M., Hodson, A. and Heaton, T. 2006. Chemical and isotopic switching within the subglacial environment of a High Arctic glacier. *Biogeochemistry* 78, 173–93.

Wysota, W. 1994. Morphology, internal composition and origin of drumlins in the southeastern part of the Chelmno-Dobrzyn Lakeland, North Poland. *Sedimentary Geology* 91, 345–64.

Yamada, T. 1998. *Glacier Lake and Its Outburst Flood in the Nepal Himalaya*. Data Centre for Glacier Research, Japanese Society of Snow and Ice, Tokyo Monograph 1.

Yao, T., Thompson, L. G., Mosley-Thompson, E., Zhihong, Y., Xingping, Z. and Lin, P.-N. 1996. Climatological significance of $\delta^{18}O$ in north Tibetan ice cores. *Journal of Geophysical Research* 101 (D23), 29531–7.

Yde, J. C., Knudsen, N. T., Larsen, N. K., Kronborg, C., Nielsen, O. B., Heinemeier, J. and Olsen, J. 2005. The presence of thrust-block naled after a major surge event: Kuannersuit Glacier, West Greenland. *Annals of Glaciology* 42, 145–50.

Yde, J. C., Riger-Kusk, M., Christiansen, H. H., Knudsen, N. T. and Humlum, O. 2008. Hydrochemical characteristics of bulk meltwater from an entire melt season, Longyearbreen, Svalbard. *Journal of Glaciology* 54 (185), 259–72.

Yi, C., Zhu, L., Seong, Y. B., Owen, L. A. and Finkel, R. C. 2006. A lateglacial rock avalanche event, Tianchi Lake, Tien Shan, Xinjiang. *Quaternary International* 154/155, 26–31.

Yokoyama, Y., Lambeck, K., De Dekker, P., Johnson, P. and Fifield, K. 2000. Timing for the maximum of the last glacial constrained by lowest sea-level observations. *Nature* 406, 713–16.

Young, G. M. 2004. Earth's two great Precambrian glaciations: aftermath of the 'Snowball Earth' hypothesis. In Eriksson, P. G., Altermann, W., Nelson, D. R., Mueller, W. U. and Catuneanu, O. (eds), *The Precambrian Earth: Tempos and Events*. Developments in Precambrian Geology. Elsevier, Amsterdam, vol. 12, 440–8.

Zachos, J., Pagani, M., Sloan, L., Thomas, E. and Billups, K. 2001. Trends, rhythms, and aberrations in global climate 65 Ma to present. *Science* 292, 686–93.

Zaragosi, S., Bourillet, J. F., Eynaud, F., Toucanne, S., Denhard, B., van Toer, A. and Lanfumey, V. 2006. The impact of the last European deglaciation on the deep-sea turbidite systems of the Celtic-Armorican margin (Bay of Biscay). *Geomarine Letters* 26, 317–29.

Zaragosi, S., Eynaud, F., Pujol, C., Auffret, G. A., Turon, J. L. and Garlan, T. 2001. Initiation of the European deglaciation as recorded in the northwestern Bay of Biscay slope environments (Meriadzek Terrace and Trevelyan Escarpment): a multi-proxy approach. *Earth and Planetary Science Letters* 188, 493–507.

Zhang, Y., Liu, S. and Ding, Y. 2006. Observed degree-day factors and their spatial variation on glaciers in western China. *Annals of Glaciology* 43, 301–6.

Zielinski, G. A. 1995. Stratospheric loading and optical depth estimates of explosive volcanism over the last 2100 years derived from the Greenland Ice Sheet Project 2 ice core. *Journal of Geophysical Research* 100 (D10), 20937–55.

Zielinski, G. A., Mayewski, P. A., Meeker, L. D., Whitlow, S., Twickler, M. S., Morrison, M., Meese, D. A., Gow, A. J. and Alley, R. B. 1994. Record of volcanism since 7000 B.C. from the GISP2 Greenland Ice Core and Implications for the Volcano-Climate System. *Science* 264, 948–52.

Zielinski, T. and van Loon, A. J. 1999a. Subaerial terminoglacial fans I: a semi-quantitative sedimentological analysis of the proximal environment. *Geologie en Mijnbouw* 77, 1–15.

Zielinski, T. and van Loon, A. J. 1999b. Subaerial terminoglacial fans II: a semi-quantitative sedimentological analysis of the middle and distal environments. *Geologie en Mijnbouw* 78, 73–85.

Zielinski, T. and van Loon, A. J. 2000. Subaerial terminoglacial fans III: overview of sedimentary characteristics and depositional model. *Geologie en Mijnbouw* 79, 93–107.

Zielinski, T. and van Loon, A. J. 2002. Present-day sandurs are not representative of the geological record. *Sedimentary Geology* 152, 1–5.

Zielinski, T. and van Loon, A. J. 2003. Pleistocene sandur deposits represent braidplains, not alluvial fans. *Boreas* 32, 590–611.

Zilliacus, H. 1989. Genesis of DeGeer moraines in Finland. *Sedimentary Geology* 62, 309–17.

Zotikov, I. A. 2006. *The Antarctic Subglacial Lake Vostok: Glaciology, Biology and Planetology*. Springer-Praxis, New York.

Zwally, H. J. and Li, J. 2002. Seasonal and interannual variations of firn densification and ice-sheet surface elevation at the Greenland summit. *Journal of Glaciology* 48, 199–207.

Zwally, H. J., Abdalati, W., Herring, T., Larson, K., Saba, J. and Steffen, K. 2002a. Surface melt-induced acceleration of Greenland Ice-sheet flow. *Science* 297, 218–22.

Zwally, H. J., Brenner, A. C., Major, J. A., Bindschadler, R. A. and Marsh, J. G. 1989. Growth of the Greenland ice sheet: measurement. *Science* 246, 1587–9.

Zwally, H. J., Schutz, B., Abdalati, W., Abshire, J., Bentley, C., Brenner, A., Bufton, J., Dezio, J., Hancock, D., Harding, D., Herring, T., Minster, B., Quinn, K., Palm, S., Spinhirne, J. and Thomas, R. 2002b. ICESat's laser measurements of polar ice, atmosphere, ocean, and land. *Journal of Geodynamics* 34, 405–45.

Zwally, H. J. et al. 2005. Mass changes of the Greenland and Antarctic ice sheets and shelves and contributions to sea-level rise: 1992–2002. *Journal of Glaciology* 51 (175), 509–27.

Zweck, C. and Huybrechts, P. 2003. Modeling the marine extent of Northern Hemisphere ice sheets during the last glacial cycle. *Annals of Glaciology* 37, 173–80.

INDEX

T - #0537 - 071024 - C816 - 276/210/38 - PB - 9780340905791 - Matt Lamination